Marko Kostić

Almost Periodic Type Solutions

De Gruyter Studies in Mathematics

Volume 101

Marko Kostić

Almost Periodic Type Solutions

to Integro-Differential-Difference Equations

DE GRUYTER

Mathematics Subject Classification 2020
Primary: 39A06, 42A75, 43A60; Secondary: 39A24, 35B15, 47D99

Author
Prof. Dr. Marko Kostić
University of Novi Sad
Faculty of Technical Sciences
Trg D. Obradovića 6
21125 Novi Sad
Serbia
marco.s@verat.net

ISBN 978-3-11-168728-5
e-ISBN (PDF) 978-3-11-168974-6
e-ISBN (EPUB) 978-3-11-169021-6
ISSN 0179-0986

Library of Congress Control Number: 2024952142

Bibliographic information published by the Deutsche Nationalbibliothek
The Deutsche Nationalbibliothek lists this publication in the Deutsche Nationalbibliografie; detailed bibliographic data are available on the Internet at http://dnb.dnb.de.

Typesetting: VTeX UAB, Lithuania

www.degruyter.com
Questions about General Product Safety Regulation:
productsafety@degruyterbrill.com

Preface

The theory of abstract Volterra integrodifferential equations and the theory of abstract Volterra difference equations are very attractive fields of research of many authors. The almost periodic features and the asymptotically almost periodic features of solutions to the abstract Volterra differential-difference equations in Banach spaces have been sought in many research articles published by now. The main aim of this monograph is to continue our previous work collected in the monographs [444], [447] and [448] by providing several new results about the existence and uniqueness of almost periodic type solutions to the abstract Volterra differential-difference equations, which could be solvable or unsolvable with respect to the highest derivative (order). We would like to particularly emphasize that this is probably the first research monograph devoted to the study of almost periodic type solutions to the abstract Volterra difference equations depending on several variables. We also consider here many new important spaces of (metrically) generalized almost periodic type spaces of sequences and functions, and their almost automorphic analogues. It is also worth noting that this is probably the first research monograph, which concerns the generalized almost periodic type sequences and their applications in a rather detailed manner; for the first time in the existing literature, we also present here some applications of results from the theory of C-regularized solution operator families to the abstract Volterra difference equations.

Fractional calculus and discrete fractional calculus are rapidly growing fields of theoretical and applied mathematics, which are incredibly important in modeling of various real phenomena appearing in different fields like aerodynamics, rheology, interval-valued systems, chaotic systems with short memory and image encryption and discrete-time recurrent neural networks. Many important research results regarding the abstract fractional differential equations and the abstract fractional difference equations in Banach spaces have recently been obtained by a great number of authors from the whole world. In this monograph, we also contribute to the theories of (discrete) fractional calculus, fractional differential-difference equations and multidimensional Laplace transform.

This monograph consists of the preliminary chapter and three individual parts, which are further divided into eight chapters. The chapters are broken down into a great number of sections and subsections. The numbering of definitions and statements is done by chapter and section, and the bibliography is by author in alphabetic order. The material presented in this book is, more or less, easily accessible to the authors and readers familiar with the essence of functional analysis and integration theory, the basic theory of abstract Volterra integrodifferential equations in Banach spaces, fractional calculus and the basic theory of almost periodic functions.

This research monograph may be interesting to PhD students in mathematics, the mathematicians working within the fields of abstract partial differential equations and the experts from all areas of functional analysis and fractional calculus. Although the monograph is far from being complete, we have decided to quote almost eight hundred

https://doi.org/10.1515/9783111689746-201

and thirty research articles, which could be of some importance to the interested readers for further developments of the theory established here. We have also tried the reference list to be kept away from any form of plagiarism.

The author would like to say a big thank to his family, friends, colleagues and advisor Professor S. Pilipović (Novi Sad, Serbia) for unstinting support of his work. My sincere appreciation also goes to V. Fedorov (Chelyabinsk, Russia), H. C. Koyuncuoğlu, T. Katican, Ö. Ö. Kaymak (Izmir, Turkey), B. Chaouchi (Khemis Miliana, Algeria), D. Velinov, P. Dimovski, B. Prangoski (Skopje, Macedonia), M. Žigić (Novi Sad, Serbia), F. Tomić (Novi Sad, Serbia), W.-S. Du (Kaohsiung, Taiwan), K. Khalil (La Havre, France), S. Abbas (Mandi, India), C. Chesneau (Caen, France), Y. Bagul (Manwath, India), V. Kumar (Waterloo, Canada), A. Chávez (Trujillo, Peru), P. T. Xuan (Hanoi, Vietnam), D. N. Cheban (Chisinau, Moldova), R. Ponce (Talca, Chile), C. Lizama (Santiago, Chile), P. J. Miana, L. Abadias (Zaragoza, Spain), M. Murillo-Arcila, J. A. Conejero, A. Peris, J. Bonet (Valencia, Spain), J. M. Sepulcre, T. Vidal (Alicante, Spain), E. M. A. El-Sayed (Alexandria, Egypt), M. S. Moslehian (Mashhad, Iran), M. Gümüs (Zongalduc, Turkey), V. Keyantuo (Rio Piedras Campus, Puerto Rico, USA), M. Hasler (Guadeloupe, France), T. Diagana (Huntsville, USA) and G. M. N'Guérékata (Baltimore, USA).

Loznica/Novi Sad
January 2025

Marko Kostić

Contents

Basic notation

$\mathbb{N}, \mathbb{Z}, \mathbb{Q}, \mathbb{R}, \mathbb{C}$: The natural numbers, integers, rationals, reals, complexes.
For any $s \in \mathbb{R}$, we denote $\lfloor s \rfloor = \sup\{l \in \mathbb{Z} : s \geqslant l\}$ and $\lceil s \rceil = \inf\{l \in \mathbb{Z} : s \leqslant l\}$.
$\operatorname{Re} z, \operatorname{Im} z$: The real and imaginary part of a complex number $z \in \mathbb{C}$; $|z|$: the module of z, $\arg(z)$: the argument of a complex number $z \in \mathbb{C} \smallsetminus \{0\}$.
$B(z_0, r) = \{z \in \mathbb{C} : |z - z_0| \leqslant r\}$ $(z_0 \in \mathbb{C}, r > 0)$.
$\Sigma_\alpha = \{z \in \mathbb{C} \smallsetminus \{0\} : |\arg(z)| < \alpha\}$, $\alpha \in (0, \pi]$.
$card(G)$: The cardinality of G.
$\mathbb{N}_0 = \mathbb{N} \cup \{0\}$.
$\mathbb{N}_n = \{1, \ldots, n\}$.
$\mathbb{N}_n^0 = \{0, 1, \ldots, n\}$.
$\mathbb{R}^n$: The real Euclidean space, $n \geqslant 2$.
$B(\mathbf{t}_0, l) \equiv \{\mathbf{t} \in \mathbb{R}^n : |\mathbf{t} - \mathbf{t}_0| \leqslant l\}$ $(\mathbf{t}_0 \in \mathbb{R}^n; l > 0)$.
If $\alpha = (\alpha_1, \ldots, \alpha_n) \in \mathbb{N}_0^n$ is a multiindex, then we denote $|\alpha| = \alpha_1 + \cdots + \alpha_n$.
$x^\alpha = x_1^{\alpha_1} \cdots x_n^{\alpha_n}$ for $x = (x_1, \ldots, x_n) \in \mathbb{R}^n$ and $\alpha = (\alpha_1, \ldots, \alpha_n) \in \mathbb{N}_0^n$.
$f^{(\alpha)} := \partial^{|\alpha|} f / \partial x_1^{\alpha_1} \cdots \partial x_n^{\alpha_n}$; $D^\alpha f := (-i)^{|\alpha|} f^{(\alpha)}$.
If (X, τ) is a topological space and $F \subseteq X$, then the interior, the closure, the boundary and the complement of F with respect to X are denoted by $\operatorname{int}(F)$ (or F°), $\overline{F}$, ∂F and F^c, respectively.
X: Complex Banach space, if not stated otherwise.
$L(E, X)$: The space of all continuous linear mappings between Banach spaces E and X, $L(X) = L(X, X)$.
X^*: The dual space of X.
A: A linear operator on X.
$\mathcal{A}$: A multivalued linear operator on X (MLO).
C: An injective continuous linear operator on X, if not stated otherwise.
If F is a subspace of X, then we denote by $\mathcal{A}_{|F}$ the part of $\mathcal{A}$ in F.
$D(\mathcal{A}), R(\mathcal{A}), \rho(\mathcal{A}), \sigma(\mathcal{A})$: The domain, range, resolvent set and spectrum of $\mathcal{A}$.
$N(\mathcal{A})$ or $\operatorname{Kern}(\mathcal{A})$: The null space of $\mathcal{A}$.
$\overline{\mathcal{A}}$: The closure of $\mathcal{A}$.
$\rho_C(\mathcal{A})$: The C-resolvent set of $\mathcal{A}$.
$\chi_\Omega(\cdot)$: The characteristic function, defined to be identically one on Ω and zero elsewhere.
$\Gamma(\cdot)$: The Gamma function.
If $\alpha > 0$, then $g_\alpha(t) = t^{\alpha-1}/\Gamma(\alpha)$, $t > 0$; $g_0(t) \equiv$ the Dirac delta distribution.
If $1 \leqslant p < \infty$, $(X, \|\cdot\|)$ is a complex Banach space, and $(\Omega, \mathcal{R}, \mu)$ is a measure space, then $L^p(\Omega, X, \mu)$ denotes the space which consists of those strongly μ-measurable functions $f : \Omega \to X$ such that $\|f\|_p := (\int_\Omega \|f(\cdot)\|^p d\mu)^{1/p}$ is finite; $L^p(\Omega, \mu) \equiv L^p(\Omega, \mathbb{C}, \mu)$.
$L^\infty(\Omega, X, \mu)$: The space which consists of all strongly μ-measurable, essentially bounded functions.
$\|f\|_\infty = ess \sup_{t \in \Omega} \|f(t)\|$, the norm of a function $f \in L^\infty(\Omega, X, \mu)$.
$L^p(\Omega : X) \equiv L^p(\Omega, X) \equiv L^p(\Omega, X, \mu)$, if $p \in [1, \infty]$ and $\mu = m$ is the Lebesgue measure; $L^p(\Omega) \equiv L^p(\Omega : \mathbb{C})$.
$L^p_{\mathrm{loc}}(\Omega : X)$: The space consisting of those Lebesgue measurable functions $u(\cdot)$ such that, for every bounded open subset Ω' of Ω, one has $u_{|\Omega'} \in L^p(\Omega' : X)$; $L^p_{\mathrm{loc}}(\Omega) \equiv L^p_{\mathrm{loc}}(\Omega : \mathbb{C})$ $(1 \leqslant p \leqslant \infty)$.
$C_0(\mathbb{R}^n)$: The space consisted of those functions $f \in C(\mathbb{R}^n)$ for which $\lim_{|x| \to \infty} |f(x)| = 0$, topologized by the norm $\|f\| := \sup_{x \in \mathbb{R}^n} |f(x)|$.
$W^{k,p}_{\mathrm{loc}}(\Omega : E)$: The space of those E-valued distributions $u \in \mathcal{D}'(\Omega : E)$ such that, for every bounded open subset Ω' of Ω, one has $u_{|\Omega'} \in W^{k,p}(\Omega' : E)$.
Assume that $I = \mathbb{R}$ or $I = [0, \infty)$. By $C_b(I : X)$ we denote the space consisting of bounded continuous functions from I into X; $C_0([0, \infty) : X)$ denotes the closed subspace of $C_b(I : X)$ consisting of functions vanishing at infinity. By $BUC(I : X)$, we denote the space consisted of all bounded uniformly continuous functions from I to X. The sup-norm turns these spaces into Banach's.

https://doi.org/10.1515/9783111689746-202

The abbreviation $AC_{loc}([0,\infty) : X)$ stands for the space of all X-valued functions that are absolutely continuous on closed subinterval of $[0,\infty)$.
$AC_{loc}([0,\infty)) \equiv AC_{loc}([0,\infty) : \mathbb{C})$.
$C^k(\Omega : X)$: The space of k-times continuously differentiable functions ($k \in \mathbb{N}_0$) from a nonempty subset $\Omega \subseteq \mathbb{C}$ into X; $C(\Omega : X) \equiv C^0(\Omega : X)$.
$Z\{f_k\}(z)$: Z-transform of $(f_k)_{k\in\mathbb{N}_0}$.
$P(u)$: Poisson transform of sequence u.
$\Delta^a u$: Riemman–Liouville fractional derivative of sequence u of order $a > 0$.
$\Delta^a_C u$: Caputo fractional derivative of sequence u of order $a > 0$.
$\Delta^a_{a,b} u$: Generalized Hilfer (a, b, a)-fractional derivative of sequence u of order $a > 0$.
$\Delta^a_W u$: Weyl fractional derivative of sequence u of order $a > 0$.
The Cesàro sequence $(k^a(v))_{v\in\mathbb{N}_0}$ is defined by

$$k^a(v) := \frac{\Gamma(v+a)}{\Gamma(a)v!}.$$

$\tilde{f}(\lambda) = \mathcal{L}f(\lambda)$: Laplace transform of function $f(\cdot)$.
$[\mathcal{L}^{-1}f(\lambda)](t)$: Inverse Laplace transform of function $f(\cdot)$.
$J^a_t u$: Riemann–Liouville fractional integral of order a.
$D^a_t u(t)$: Riemann–Liouville fractional derivative of order a.
$\mathbf{D}^a_t u(t)$: Caputo fractional derivative of order a.
$D^{a,\beta}_t u(t)$: Hilfer fractional derivative of order $a > 0$ and type $\beta \in [0,1]$.
$D^{\gamma}_{t,+} u(t)$: Weyl–Liouville fractional derivative $D^{\gamma}_{t,+} u(t)$ of order γ.
$E_{a,\beta}(z)$: Mittag-Leffler function.
$\Phi_\gamma(z)$: Wright function.

Introduction

The class of almost periodic functions was introduced by the Danish mathematician H. Bohr around 1924–1926 and later generalized by many other authors (see the research monographs [112, 246, 304, 348, 444, 447, 508, 633] and [805] for further information concerning almost periodic functions and their applications). Suppose that $(X, \|\cdot\|)$ is a complex Banach space and $F : \mathbb{R}^n \to X$ is a continuous function ($n \in \mathbb{N}$). Then we say that the function $F(\cdot)$ is almost periodic if for each $\varepsilon > 0$ there exists $l > 0$ such that for each $\mathbf{t}_0 \in \mathbb{R}^n$ there exists $\boldsymbol{\tau} \in B(\mathbf{t}_0, l) \equiv \{\mathbf{t} \in \mathbb{R}^n : |\mathbf{t} - \mathbf{t}_0| \leqslant l\}$ with

$$\|F(\mathbf{t} + \boldsymbol{\tau}) - F(\mathbf{t})\| \leqslant \varepsilon, \quad \mathbf{t} \in \mathbb{R}^n; \tag{1}$$

here, $|\cdot - \cdot|$ denotes the Euclidean distance in $\mathbb{R}^n$ and τ is usually called an ε-almost period of $F(\cdot)$. Any trigonometric polynomial in $\mathbb{R}^n$ is almost periodic and a continuous function $F(\cdot)$ is almost periodic if and only if there exists a sequence of trigonometric polynomials in $\mathbb{R}^n$, which converges uniformly to $F(\cdot)$.

An X-valued sequence $(x_k)_{k \in \mathbb{Z}^n}$ is called (Bohr) almost periodic if, for every $\varepsilon > 0$, there exists $l > 0$ such that for each $\mathbf{t}_0 \in \mathbb{Z}^n$ there exists $\boldsymbol{\tau} \in B(\mathbf{t}_0, l)$ with integer coordinates such that (1) holds for all $\mathbf{t} \in \mathbb{Z}^n$. Any almost periodic X-valued sequence is bounded and its range is relatively compact in X. The equivalent concept of Bochner almost periodicity of X-valued sequences can be introduced as well; see e. g., [683, Theorem 70, pp. 185–186] and [683, Theorems 71–73, pp. 186–188] for the one-dimensional setting. It is well known that a sequence $(x_k)_{k \in \mathbb{Z}^n}$ in X is almost periodic if and only if there exists an almost periodic function $F : \mathbb{R}^n \to X$ such that $x_k = F(k)$ for all $k \in \mathbb{Z}^n$; see, e. g., the proof of [252, Theorem 2] given in the one-dimensional case. It is not difficult to prove that, for every almost periodic sequence $(x_k)_{k \in \mathbb{N}^n}$ in X, there exists a unique almost periodic sequence $(\tilde{x_k})_{k \in \mathbb{Z}^n}$ in X such that $\tilde{x_k} = x_k$ for all $k \in \mathbb{N}^n$, so that a sequence $(x_k)_{k \in \mathbb{N}^n}$ in X is almost periodic if and only if there exists an almost periodic function $F : [0, \infty)^n \to X$ such that $x_k = F(k)$ for all $k \in \mathbb{N}^n$. For more details about infinite matrices, which sum every almost periodic sequence, we refer the reader to the old paper [693] by J. A. Siddiqi and references quoted therein.

Further on, the class of Stepanov-p-almost periodic functions, where $p \geqslant 1$, was introduced by W. Stepanoff in 1926 [724]. Suppose that the function $F : \mathbb{R}^n \to X$ is locally p-integrable, where $1 \leqslant p < \infty$; then we say that $F(\cdot)$ is Stepanov-p-almost periodic if for every $\varepsilon > 0$ there exists $l > 0$ such that for each $\mathbf{t}_0 \in \mathbb{R}^n$ there exists $\boldsymbol{\tau} \in B(\mathbf{t}_0, l) \cap \mathbb{R}^n$ with

$$\|F(\mathbf{t} + \boldsymbol{\tau} + \mathbf{u}) - F(\mathbf{t} + \mathbf{u})\|_{L^p([0,1]^n : X)} \leqslant \varepsilon, \quad \mathbf{t} \in \mathbb{R}^n.$$

The following known example of a Stepanov-2-almost periodic function has not been quoted in [444–448]: Suppose that there exists $c > 0$ such that the sequence $(\lambda_k)_{k \in \mathbb{Z}}$ of real numbers satisfies $\lambda_{k+1} - \lambda_k > c$ for all $k \in \mathbb{Z}$ and $(a_k)_{k \in \mathbb{Z}}$ is a sequence of real numbers such that $\sum_{k=-\infty}^{+\infty} |a_k|^2 < +\infty$. Then the function

https://doi.org/10.1515/9783111689746-001

$$f(t) = \sum_{k=-\infty}^{+\infty} a_k e^{i\lambda_k t}, \quad t \in \mathbb{R}$$

is Stepanov-2-almost periodic; see [508, Theorem 5.3.2, pp. 214–216]. It could be interesting to consider the general value of exponent $p \neq 2$ here.

The class of Stepanov almost periodic sequences, introduced by J. Andres and D. Pennequin in [56] for the case $n = 1$, reduces to the class of almost periodic sequences, which is not the case for the corresponding classes of functions; a similar statement holds in the higher-dimensional setting. This is no longer true for the class of equi-Weyl almost periodic sequences, which provides a proper extension of the class of almost periodic sequences; cf. A. Iwanik [393], who was the first author considering such sequences (in actual fact, A. Iwanik has considered the class of equi-Weyl-1-almost periodic sequences with values in compact metric spaces). If $1 \leqslant p < +\infty$ and $F : \mathbb{R}^n \to X$ is locally p-integrable, then we say that $F(\cdot)$ is:

(i) equi-Weyl-p-almost periodic if, for every $\varepsilon > 0$, there exist two finite real numbers $l > 0$ and $L > 0$ such that for each $\mathbf{t}_0 \in \mathbb{R}^n$ there exists $\tau \in B(\mathbf{t}_0, L) \cap \mathbb{R}^n$ with

$$\sup_{\mathbf{t}\in\mathbb{R}^n} \Big[l^{-\frac{n}{p}} \| F(\tau + \cdot) - F(\cdot) \|_{L^p(\mathbf{t}+l[0,1]^n:X)} \Big] < \varepsilon.$$

(ii) Weyl-p-almost periodic if, for every $\varepsilon > 0$, there exists a finite real number $L > 0$ such that for each $\mathbf{t}_0 \in \mathbb{R}^n$ there exists $\tau \in B(\mathbf{t}_0, L) \cap \mathbb{R}^n$ with

$$\limsup_{l\to+\infty} \sup_{\mathbf{t}\in\mathbb{R}^n} \Big[l^{-\frac{n}{p}} \| F(\tau + \cdot) - F(\cdot) \|_{L^p(\mathbf{t}+l[0,1]^n:X)} \Big] < \varepsilon.$$

Any Bohr almost periodic function is Stepanov-p-almost periodic and any Stepanov-p-almost periodic function is equi-Weyl-p-almost periodic; it is also well known that the class of Weyl-p-almost periodic functions contains the class of equi-Weyl-p-almost periodic functions.

The class of (equi-)Weyl-p-almost periodic sequences and the class of Doss-p-almost periodic sequences have recently been introduced and analyzed in our joint research article with W.-S. Du and D. Velinov [267]. A sequence $(x_k)_{k\in\mathbb{Z}^n}$ is said to be equi-Weyl-p-almost periodic, respectively, Weyl-p-almost periodic, if the following holds:

(e-M1) For every $\varepsilon > 0$, there exist $s \in \mathbb{N}$ and $L > 0$ such that, for every $\mathbf{t}_0 \in \mathbb{Z}^n$, the cube $I' \equiv \mathbf{t}_0 + [0, L]^n$ contains a point $\tau \in I' \cap \mathbb{Z}^n$, which satisfies

$$\sup_{\mathbf{t}_0\in\mathbb{Z}^n} s^{-n/p} \Big[\sum_{j\in(\mathbf{t}_0+[0,s]^n)\cap\mathbb{Z}^n} \| x_{j+\tau} - x_j \|^p \Big]^{1/p} < \varepsilon, \tag{2}$$

respectively,

(M1) For every $\varepsilon > 0$, there exists $L > 0$ such that, for every $\mathbf{t}_0 \in \mathbb{Z}^n$, the cube $I' \equiv \mathbf{t}_0 + [0, L]^n$ contains a point $\tau \in I' \cap \mathbb{Z}^n$, which satisfies that there exists an integer $s_\tau \in \mathbb{N}$ such that (2) holds for all integers $s \geqslant s_\tau$.

Furthermore, a sequence $(x_k)_{k\in\mathbb{Z}^n}$ is said to be Doss-p-almost periodic if, for every $\varepsilon > 0$, there exists $L > 0$ such that, for every $\mathbf{t}_0 \in \mathbb{Z}^n$, the cube $I' \equiv \mathbf{t}_0 + [0, L]^n$ contains a point $\tau \in I' \cap \mathbb{Z}^n$, which satisfies

$$\limsup_{s\to+\infty} s^{-n/p}\Bigg[\sum_{j\in[-s,s]^n\cap\mathbb{Z}^n} \|x_{j+\tau} - x_j\|^p\Bigg]^{1/p} < \varepsilon.$$

We introduce the class of Besicovitch-p-almost periodic functions $F : \mathbb{R}^n \to X$ in the following way: if $F \in L^p_{loc}(\mathbb{R}^n : X)$, then we first define

$$\|F\|_{\mathcal{M}^p} := \limsup_{t\to+\infty}\Bigg[\frac{1}{(2t)^n}\int_{[-t,t]^n} \|F(\mathbf{s})\|^p\, d\mathbf{s}\Bigg]^{1/p}.$$

Then we know that $\|\cdot\|_{\mathcal{M}^p}$ is a seminorm on the space $\mathcal{M}^p(\mathbb{R}^n : X)$ consisting of those $L^p_{loc}(\mathbb{R}^n : X)$-functions $F(\cdot)$ for which $\|F\|_{\mathcal{M}^p} < \infty$. Denote $K_p(\mathbb{R}^n : X) := \{f \in \mathcal{M}^p(\mathbb{R}^n : X); \|F\|_{\mathcal{M}^p} = 0\}$ and

$$M_p(\mathbb{R}^n : X) := \mathcal{M}^p(\mathbb{R}^n : X)/K_p(\mathbb{R}^n : X).$$

The seminorm $\|\cdot\|_{\mathcal{M}^p}$ on $\mathcal{M}^p(\mathbb{R}^n : X)$ induces the norm $\|\cdot\|_{M^p}$ on $M^p(\mathbb{R}^n : X)$ under which $M^p(\mathbb{R}^n : X)$ is complete; hence, $(M^p(\mathbb{R}^n : X), \|\cdot\|_{M^p})$ is a Banach space. We say that a function $F \in L^p_{loc}(\mathbb{R}^n : X)$ is Besicovitch-p-almost periodic if there exists a sequence of trigonometric polynomials (almost periodic functions, equivalently) converging to $F(\cdot)$ in $(M^p(\mathbb{R}^n : X), \|\cdot\|_{M^p})$. The vector space consisting of all Besicovitch-p-almost periodic functions, denoted by $B^p(\mathbb{R}^n : X)$, is closed in $M^p(\mathbb{R}^n : X)$ and, therefore, a Banach space itself. We know that any equi-Weyl-p-almost periodic function is Besicovitch-p-almost periodic as well as that there exists a Weyl-p-almost periodic function $f : \mathbb{R} \to \mathbb{R}$, which is not Besicovitch-p-almost periodic [447].

Concerning the existence and uniqueness of almost periodic type solutions for various classes of ordinary integrodifferential equations and partial integro-differential equations, we may refer, e. g., to [16, 17, 34, 40, 66–68, 102, 148, 169, 174, 183, 306, 318, 336, 358, 512, 513, 519, 551, 587, 655, 677, 731, 735, 745, 757, 772, 774, 775, 778, 785, 787, 788, 801, 813, 816] and the lists of references quoted in our previously published research monographs regarding the almost periodic functions and their applications. The abstract semilinear Cauchy problems have been analyzed in the recent research monograph [566] by P. Magal and S. Ruan; concerning the well-posedness and qualitative properties of solutions of differential-difference equations, the reader may consult the extensive research report [98] by R. Bellman and K. L. Cooke (see also the research article [210] by K. L. Cooke, K. R. Meyer and the references quoted there). The existence and uniqueness of almost periodic solutions for various classes of the abstract nonlinear integrodifferential equations have been analyzed by M. Hieber and his coauthors, among many others; for some results obtained in this direction, we may refer to [143, 144, 378–380]. It is also worthwhile to mention that H. S. Ding et al. have recently considered, in [262], the Kadets type and Loomis type theorems for asymptotically almost periodic functions; in particular, the authors have shown that a bounded primitive function of an asymptotically almost periodic function $f : \mathbb{R} \to X$ is remotely almost periodic if X does not contain an isomorphic copy of the space c_0, the Banach space of all numerical sequences vanishing at plus infinity, equipped with the sup-norm.

The existence and uniqueness of almost periodic type solutions for the so-called differential equations with "maxima" have been analyzed by many authors so far (see, e. g., [87, 88, 362, 644, 669, 751]), while the existence and uniqueness of almost periodic type solutions for certain kinds of differential equations with piecewise constant argument have been sought in [200, 298, 374, 802, 803]. In the work by M. Ayachi [77], a set of sufficient conditions is established to ensure the existence and global exponential stability of measure-pseudo almost periodic solutions within a specific class of bidirectional associative memory neural networks. Similarly, in the study conducted by T. Liang, Y. Q. Yang, Y. Liu and L. Li [522] specific sufficient conditions are provided for the existence and global exponential stability of almost periodic solutions in Cohen–Grossberg neural networks on time scales. These references, along with the cited references therein, underscore the broader significance of almost periodic solutions, transcending theoretical considerations and finding practical applications in diverse fields such as neuroscience, physics, biology and engineering; cf. also [77, 522] and [712]. As an open problem for our readers, stated at the very beginning of the monograph, we would like to ask whether the results of [213, Theorems 1, 2], established by C. Corduneanu, can be formulated and proved if the index $p \in [1, +\infty)$ in his analysis is not equal to two, and so on and so forth.

In the discrete setting, a sequence $(x_j)_{j\in\mathbb{Z}^n}$ is said to be Besicovitch-p-almost periodic if for every $\varepsilon > 0$ there exists a trigonometric polynomial $P(\cdot)$ such that

$$\limsup_{s\to+\infty} s^{-n/p}\Bigg[\sum_{j\in[-s,s]^n\cap\mathbb{Z}^n} \|x_j - P(j)\|^p\Bigg]^{1/p} < \varepsilon.$$

The class of one-dimensional Besicovitch almost periodic sequences has been introduced by A. Bellow, V. Losert [100] and further analyzed by V. Bergelson et al. in [106] and [267] (cf. also the research article [264] by T. Downarowicz and A. Iwanik, which concerns the notion of quasiuniform convergence in compact dynamical systems).

The notion of almost automorphy was introduced by S. Bochner in 1955, when he was studying problems related to differential geometry [121]. We know that the notion of almost automorphy is a generalization of the notion of almost periodicity: Suppose that $F : \mathbb{R}^n \to X$ is continuous, where $(X, \|\cdot\|)$ is a complex Banach space. Then it is said that $F(\cdot)$ is almost automorphic if for every sequence $(\mathbf{b}_k)$ in $\mathbb{R}^n$ there exist a subsequence $(\mathbf{a}_k)$ of $(\mathbf{b}_k)$ and a map $G : \mathbb{R}^n \to X$ such that

$$\lim_{k\to\infty} F(\mathbf{t} + \mathbf{a}_k) = G(\mathbf{t}) \quad \text{and} \quad \lim_{k\to\infty} G(\mathbf{t} - \mathbf{a}_k) = F(\mathbf{t}), \tag{3}$$

pointwisely for $\mathbf{t} \in \mathbb{R}^n$. In this case, the range of $F(\cdot)$ is relatively compact in X and the limit function $G(\cdot)$ is bounded on $\mathbb{R}^n$ but not necessarily continuous on $\mathbb{R}^n$. If the convergence of limits appearing in (3) is uniform on compact subsets of $\mathbb{R}^n$, then we say that $F(\cdot)$ is compactly almost automorphic. We know that an almost automorphic function $F(\cdot)$ is compactly almost automorphic if and only if $F(\cdot)$ is uniformly continuous [101]; furthermore, the Bochner criterion says that a continuous function $F : \mathbb{R}^n \to X$ is almost periodic if and only if the convergence of limits appearing in (3) is uniform on the whole Euclidean space $\mathbb{R}^n$. The first systematic study of almost automorphic functions on topological groups was conducted by W. A. Veech in [758, 759] (see also the reference list given in the Appendix section of [184]). In the last mentioned paper, we have recently introduced several new classes of multidimensional almost periodic type functions and emphasized that these classes can be considered on (semi)topological groups.

Let $1 \leqslant p < +\infty$. If $F : \mathbb{R}^n \to X$, then we introduce the multidimensional Bochner transform $\hat{F} : \mathbb{R}^n \to X^{\Omega}$ by

$$[\hat{F}_{\Omega}(\mathbf{t}; x)](u) := F(\mathbf{t} + \mathbf{u}; x), \quad \mathbf{t} \in \mathbb{R}^n, \ \mathbf{u} \in [0,1]^n, \ x \in X;$$

here, X^{Ω} denotes the set of all functions from Ω into X. A p-locally integrable function $F : \mathbb{R}^n \to X$ is said to be Stepanov-p-almost automorphic if the Bochner transform $\hat{F} : \mathbb{R}^n \to L^p([0,1]^n : X)$ is well-defined and almost automorphic. Various classes of multidimensional Weyl almost automorphic type functions, respectively, Besicovitch almost automorphic type functions, have recently been analyzed in [447], respectively [448].

In [483], we have analyzed the multidimensional almost automorphic sequences of the form $F : \mathbb{Z}^n \times X \to Y$, where $(Y, \|\cdot\|_Y)$ is likewise a complex Banach space. Specifically, a sequence $F : \mathbb{Z}^n \to Y$ is said to be almost automorphic if for every sequence $(\mathbf{b}_k)$ in $\mathbb{Z}^n$ there exist a subsequence $(\mathbf{a}_k)$ of $(\mathbf{b}_k)$ and a map $G : \mathbb{Z}^n \to X$ such that the limit equations in (3) hold pointwisely for $\mathbf{t} \in \mathbb{Z}^n$. The notion of Stepanov-p-almost automorphy for sequences is meaningful only in the metrical sense and this notion will be seriously considered in this monograph, where we will also provide the basic results about the metrically Weyl almost automorphic type sequences (functions) and the metrically Besicovitch almost automorphic type sequences (functions). Let us recall that S. Bochner has shown that the sequence $(\text{sign}(\cos(2\pi k\alpha)))_{k\in\mathbb{Z}}$ is almost automorphic but not almost periodic if $\alpha \notin \mathbb{Q}$; moreover, an easy construction of a class of almost automorphic functions was given by W. A. Veech in [758, Theorem 6.2.1], where the author proved that the function

$$F(z) = \left(\frac{\sin(\pi z)}{\pi}\right)^m \sum_{k\in\mathbb{Z}} \frac{f(k)}{(z-k)^m}, \quad z \in \mathbb{C} \setminus \mathbb{Z}$$

is almost automorphic in the horizontal strips $\operatorname{Im} z = c \neq 0$ if and only if $(f(k))_{k\in\mathbb{Z}}$ is an almost automorphic sequence ($m \geqslant 2$).

For the basic source of information about difference equations and their applications, the reader may consult the research monographs [19, 21, 25, 219, 272, 396, 424] and references cited therein. This is probably the first research monograph, which considers the almost periodic type solutions and almost automorphic type solutions of difference equations depending of several variables; it is our strong belief that the almost periodic type solutions of Volterra difference equations depending on several independent variables will receive considerable attention of the authors in the near future. Concerning this issue, we would like to emphasize that a large class of difference equations depending on two variables can be expressed in the form

$$F(a_{ij}, a_{i-1,j}, a_{i+1,j}, \ldots) = 0. \tag{4}$$

These equations are usually classified according to the number of terms in (4); for example, the discrete Laplace equation

$$a_{i+1,j} + a_{i,j+1} + a_{i-1,j} + a_{i,j-1} - 4a_{ij} = 0$$

and the discrete Poisson equation

$$a_{i+1,j} + a_{i,j+1} + a_{i-1,j} + a_{i,j-1} - 4a_{ij} = f_{ij}$$

are four-level equations, while the equation

$$a_{i+1,j} + a_{i+1,j} + a_{i,j+1} + a_{i-1,j} + a_{i,j-1} - 4a_{ij} = f_{ij}$$

is a five-level equation. We say that the difference equation (4) is linear if

$$\begin{aligned} &F(\alpha a_{ij} + \beta b_{ij}, \alpha a_{i-1,j} + \beta b_{i-1,j}, \alpha a_{i+1,j} + \beta b_{i+1,j}, \ldots) \\ &\quad = \alpha F(a_{ij}, a_{i-1,j}, a_{i+1,j}, \ldots) + \beta F(b_{ij}, b_{i-1,j}, b_{i+1,j}, \ldots) \end{aligned}$$

for all scalars $\alpha, \beta \in \mathbb{R}$. Given a one-dimensional sequence $(a_k)_{k\in\mathbb{Z}}$, the forward difference operator and the backward difference operator are defined by $\Delta a_k := a_{k+1} - a_k$ and $\nabla a_k := a_k - a_{k-1}$, respectively ($k \in \mathbb{Z}$). The discrete Laplacian is given by

$$\Delta^2 a_k := \Delta(\Delta a_k) := a_{k+2} - 2a_{k+1} + a_k, \quad k \in \mathbb{Z}.$$

Using the compositions of difference operators Δ and ∇, we can consider the difference operators of higher order; for example, we have

$$\Delta^2 \nabla a_k = a_{k+2} - 3a_{k+1} + 3a_k - a_{k-1}, \quad k \in \mathbb{Z}.$$

Also, we know that there are several important identities for the difference operators Δ and ∇; we will only recall here the Abel summation by parts formula

$$\sum_{k=r}^{s} b_{k+1} \Delta a_k = a_{s+1} b_{s+1} - a_r b_r - \sum_{k=r}^{s} a_k \Delta b_k.$$

The forward difference operators and the backward difference operators can be also introduced and analyzed for the multidimensional sequences; for example, in the two-dimensional setting, we can consider the forward difference operators

$$\Delta_1 a_{ij} := a_{i+1,j} - a_{ij} \quad \text{and} \quad \Delta_2 a_{ij} := a_{i,j+1} - a_{ij}.$$

Many important results of mathematical analysis, like Green's formula in the plane and Grönwall inequality, have analogues for the difference operators; see [197, pp. 23–25, 43–44] for more details in this direction. The finite convolution operator is defined by

$$(a *_0 b)(k) := \sum_{j=0}^{k} a_{k-j} b_j, \; k \in \mathbb{N}_0,$$

while the infinite convolution operator can be defined, under certain extra assumptions, by

$$(a * b)(k) := \sum_{j=-\infty}^{\infty} a_{k-j} b_j, \; k \in \mathbb{Z}.$$

Many results for the usual finite convolution product and the infinite convolution product of functions can be reformulated for the corresponding products of sequences. For example, if $a \in l^1(\mathbb{Z})$, i. e., $\sum_{k=-\infty}^{\infty} |a(k)| < +\infty$ and $(b_k)_{k\in\mathbb{Z}}$ is an almost periodic (automorphic) sequence, then $((a * b)(k))_{k\in\mathbb{Z}}$ is also an almost periodic (automorphic) sequence; see, e. g., the proof of [65, Theorem 2.13]. Concerning the numerical solutions of PDEs and finite difference methods, the monograph [707] by G. D. Smith et al. is worth of mentioning.

The main aim of invariance method, also known as the Lie analysis, is to find a group of transformations leaving a difference equation invariant. The Lie analysis enables one to lower the order of difference equation and to solve it analytically; sometimes the Lie analysis can be also helpful in the analysis of the existence and uniqueness of periodic solutions to difference equations. For more details on the subject, we refer the reader to the research monograph [388] by P. E. Hydon, the research article [307] by M. Folly-Gbetoula and references quoted therein.

Periodic solutions of nonlinear difference equations have been systematically analyzed in the research monograph [346] by E. A. Grove and G. Ladas (cf. also [104, 229, 241, 259, 286, 372, 426, 434, 516, 517, 533, 591, 659, 701, 704, 705, 708, 709, 746, 773, 782, 796, 818, 825], Springer Proceedings in Mathematics and Statistics (Vol. 180) [42], B. Ferguson and G. Lims' monograph [299] about discrete time dynamic economic models, and the references cited therein; concerning periodic solutions of so-called max-type difference equations, we also refer to the research articles [528–530] by A. Linero-Bas, V. Manosa, D. Nieves-Roldán and [163] by J. S. Cánovas, A. Linero Bas, G. Soler López). Periodic solutions of rational difference equations have been investigated in many research studies by now; see, e. g., the research articles by A. Anisimova, D. Batenkov, G. Binyamini and

I. Bula in [42], the research monographs [159] by E. Camouzis and G. Ladas, [503] by M. R. S. Kulenović and G. Ladas, the research articles [23, 52, 53, 160, 206, 228, 238, 265] as well as the references cited therein. Concerning the necessary and sufficient spectral criteria for the stability of the solutions for certain classes of systems of linear partial difference equations in Banach space, we refer the reader to [573].

For the sake of illustration, let us consider the following difference equation:

$$x(k+1) = -a(k)x(k) + b(k)\tanh(x(k)) + c(k), \quad k \in \mathbb{Z}, \tag{5}$$

which is the discrete analogue of the biological equation used for modeling a single artificial effective neuron by dissipation [334]; here, $a(k) > 0$ for all $k \in \mathbb{Z}$, $b(\cdot)$ and $c(\cdot)$ are bounded sequences, and the homogeneous part of (5) admits an exponential dichotomy. If $a(\cdot)$, $b(\cdot)$ and $c(\cdot)$ are antiperiodic sequences and there exists a common antiperiod $T \in \mathbb{N}$ such that $a(k+T) = -a(k)$ for all $k \in \mathbb{Z}$, $b(k+T) = -b(k)$ for all $k \in \mathbb{Z}$ and $c(k+T) = -c(k)$ for all $k \in \mathbb{Z}$, then the discrete dynamical system (5) has an antiperiodic solution of the same antiperiod T.

Consider now the system of difference equations

$$\Delta x_1(k) = -k_1x_1(k) + f_1(k) \quad \text{and} \quad \Delta x_2(k) = k_1x_1(k) - k_2x_2(k) + f_2(k), \quad k \in \mathbb{Z}, \tag{6}$$

where Δ designates the forward difference operator and $k_{1,2}$ are nonzero real constants. This is a linear perturbation of the problem

$$\Delta x_1(k) = -k_1x_1(k), \quad k \in \mathbb{Z} \quad \text{and} \quad \Delta x_2(k) = k_1x_1(k) - k_2x_2(k), \quad k \in \mathbb{Z},$$

which can be employed in modeling of the blood alcohol level on the discrete time by taking x_1 to be the concentration of alcohol in stomach and x_2 to be the concentration of alcohol in blood ([556]; cf. also [689] for slightly different applications of the discrete fractional calculus). If we assume that $f_1(k+T) = f_1(k)$ for all $k \in \mathbb{Z}$ and $f_2(k+T) = f_2(k)$ for all $k \in \mathbb{Z}$, with some $T \in \mathbb{N}$, then we can rewrite the system (6) in the matricial form and conclude that there exists a periodic solution of (6) with the period T.

A few interesting results about the existence and uniqueness of the (almost) periodic solutions to the systems of difference equations and the second-order nonautonomous difference equations can be found in the research articles [819, 820] by S. Zhang. In [819], the author has analyzed the existence of a unique (almost) periodic solution to the system of difference equations

$$x(k+1) = A(k)x(k) + b(k), \quad k \in \mathbb{Z}. \tag{7}$$

It has been shown that, if $(A(k))_{k\in\mathbb{Z}}$ is an almost periodic real matrix of format $s \times s$ and $(b(k))_{k\in\mathbb{Z}}$ is an almost periodic sequence in $\mathbb{R}^s$, then the system (7) has a unique almost periodic solution provided that

$$\overline{A} = \limsup_{k\to+\infty} \frac{1}{k}\sum_{j=1}^{k}\|A(j)\| < 1,$$

where $\|A(j)\|$ denotes the sum of all absolute values of the elements of $A(j)$; moreover, if $(A(k))_{k\in\mathbb{Z}}$ and $(b(k))_{k\in\mathbb{Z}}$ are of period $T \in \mathbb{N}$, then the solution $(x(k))_{k\in\mathbb{Z}}$ is also of period T. The method proposed in [819] is sometimes ideal for studying the existence and uniqueness of (almost) periodic solutions of the higher-order difference equations; for example, the second-order linear difference equation

$$x(k+1) = a(k)x(k) + b(k)x(k-1) + f(k), \quad k \in \mathbb{Z} \tag{8}$$

can be transformed into an equivalent system of the first-order difference equations and arguing so we can show that there exists a unique (almost) periodic solution of (8) provided that $(a(k))_{k\in\mathbb{Z}}$, $(b(k))_{k\in\mathbb{Z}}$ and $(f(k))_{k\in\mathbb{Z}}$ are (almost) periodic sequences and there exists a real number $\alpha > 0$ such that

$$\alpha + \limsup_{k\to+\infty} \frac{1}{k}\sum_{j=1}^{k}\|a(j)\| + \frac{\|b\|_\infty}{\alpha} < 1.$$

Let us also notice that C. González and A. Melado-Jiménez have investigated, in [325, 326], the asymptotic behavior of solutions of difference equations in Banach spaces (see also [433] and [590]). Chaotic solutions of delay difference equations have been studied by Z. Li and X. Zhu in [518]; cf. also [691, 692]. Concerning semilinear functional difference equations with infinite delay, we refer the reader to the excellent article [22] by R. P. Agarwal, C. Cuevas and M. V. S. Frasson; cf. also [308, 580] and the research monographs [510] and [642]. Also, for some applications of (fractional) nonlinear difference equations in mathematical biology, we refer the reader to [218, 730] and references quoted therein.

The abstract difference equations in normed spaces have been considered in the research monograph [320] by M. I. Gil, where it has been assumed that all operator coefficients are bounded linear operators. For example, if $A \in L(X)$, then a unique solution of the abstract difference equation

$$x(k+1) = Ax(k) + f(k), \quad k \in \mathbb{N}_0; \quad x(0) = x_0, \tag{9}$$

is given by

$$x(k) = A^k x_0 + \sum_{j=0}^{k-1} A^{k-1-j} f(j), \quad k \in \mathbb{N}.$$

If $T \in \mathbb{N}$, $f(\cdot)$ is of period T and $1 \notin \sigma(A^T)$, then the unique solution $(x(k))_{k\in\mathbb{N}_0}$ of (9) will be also of period T.

The abstract nondegenerate difference equations of higher order have been analyzed in [320, Chapter 13]. Consider, for example, the abstract higher-order difference equation

$$u(j+m) + \sum_{k=0}^{m-1} A_{m-k} u(j+k) = 0, \quad j \in \mathbb{N}_0; \quad u(k) = x_k, \quad 0 \leqslant k \leqslant m-1, \tag{10}$$

where A_{m-k} are bounded linear operators for $0 \leqslant k \leqslant m-1$. The corresponding polynomial operator pencil $K(\cdot)$ is defined by

$$K(z) := \sum_{k=0}^{m} A_k z^{m-k}, \quad z \in \mathbb{C},$$

where $A_0 = \mathrm{I}$. A point $z_0 \in \mathbb{C}$ is said to be a regular point of $K(\cdot)$ if $K(z_0)$ is boundedly invertible; otherwise, $z_0 \in \mathbb{C}$ is said to be a singular point of $K(\cdot)$. The union of all singular points of $K(\cdot)$ is called the spectrum of $K(\cdot)$. Finally, the polynomial operator pencil $K(\cdot)$ is said to be stable if its spectrum is inside the unit circle. If this is the case, then the difference equation (10) is exponentially stable, i. e., there exist a constant $M \geqslant 1$ and a constant $a \in (0,1)$ such that any solution $u(\cdot)$ of (10) satisfies $\|u(t)\| \leqslant Me^{a} \max_{0 \leqslant m \leqslant k-1} \|u(k)\|$ for all $t \in \mathbb{N}_0$.

An important tool in the analysis and the representation formula of solutions to the abstract higher-order difference equation (10) is the Z-transform of sequences. Suppose that a sequence $(f_k)_{k\in\mathbb{N}_0}$ in X satisfies $\limsup_{k\to+\infty} \|f_k\|^{1/k} < r < +\infty$. Then the function

$$F(z) := F\{f_k\}(z) := \sum_{k=0}^{\infty} \frac{f(k)}{z^k}, \quad |z| > r$$

is analytic and it is called Z-transform of $(f_k)_{k\in\mathbb{N}_0}$. If $(B_k)_{k\in\mathbb{N}_0}$ is a sequence in $L(X)$ and $\limsup_{k\to+\infty} \|B_k\|^{1/k} < r < +\infty$, then we can also consider the Z-transform of $(B_k)_{k\in\mathbb{N}_0}$, which is defined by

$$\Phi(z) := \sum_{k=0}^{\infty} \frac{B(k)}{z^k}, \quad |z| > r.$$

Then a simple computation yields that the sequence $((B *_0 f)(k))_{k\in\mathbb{N}_0}$ satisfies $\limsup_{k\to+\infty} \|(B *_0 f)(k)\|^{1/k} < r < +\infty$ and that its Z-transform is given by $F(\cdot)\Phi(\cdot)$.

Applying the Cauchy formula for the coefficients of Laurent series, we have

$$f(k) = \frac{1}{2\pi i} \oint_{|z|=r} z^{k-1} F(z)\, dz, \quad k \in \mathbb{N}_0.$$

Applying the Z-transform and the formula

$$Z\{f_{k+j}\}(z) = z^j \left[Z\{f_k\}(z) - \sum_{s=0}^{j-1} f_s z^{-s} \right], \quad |z| > r,$$

we can prove that the unique solution of the abstract higher-order difference equation

$$\sum_{k=0}^{m} A_{m-k} y(k+j) = f_j, \quad j \in \mathbb{N}_0; \quad y(k) = 0, \quad 0 \leqslant k \leqslant m-1$$

can be represented by $y(m) = (G *_0 f)(m)$, where

$$G(k) = \frac{1}{2\pi i} \oint_{|z|=r} z^{k-1} K^{-1}(z)\, dz, \tag{11}$$

provided that $\limsup_{k\to+\infty} \|f_k\|^{1/k} < r < +\infty$ and $K(\cdot)$ is boundedly invertible for $|z| \geqslant r$; see [320, Theorem 13.2.1].

We can also analyze the bilateral Z-transform of a sequence $(f_k)_{k\in\mathbb{Z}}$ in X by

$$F_b(z) := \sum_{k=-\infty}^{\infty} \frac{f(k)}{z^k}, \quad |z| > r.$$

The bilateral Z-transform is a linear transform, which is compatible with the infinite convolution of sequences and has certain time shifting properties. For more details about the Z-transform, we refer to the excellent monograph [413] by E. I. Jury (cf. also [412] and [752]), the research monograph [345] by A. C. Grove, Chapter 3 in the research monograph [650] by J. G. Proakis, D. G. Manolakis as well as the research articles [123] and [130].

There exists a vast amount of literature concerning the well-posedness and qualitative properties of solutions for various subclasses of the following general class of the abstract higher-order differential inclusions:

$$0 \in \mathcal{B}' u^{(m)}(t) + \sum_{i=0}^{m-1} \mathcal{A}_i' u^{(i)}(t) + F(t), \quad t \geqslant 0,$$

accompanied with certain initial conditions, where $m \in \mathbb{N}$, $\mathcal{B}'$ and $\mathcal{A}_i'$ $(0 \leqslant i \leqslant m-1)$ are multivalued linear operators on complex Banach space X and $F : [0,\infty) \to P(X)$ is a multivalued forcing term. On the other hand, there is only a few research articles concerning the well-posedness and qualitative properties of solutions of the difference equations with unbounded linear operators and multivalued linear operators (see [61–64] and [665] and references quoted therein for some results about the second-order difference inclusions with maximal monotone operators). For example, the qualitative features of solutions for the following abstract higher-order difference inclusion:

$$0 \in \mathcal{B}' \Delta^m u(k) + \sum_{j=0}^{m-1} \mathcal{A}_i' \Delta^i u(k) + F_\Delta(k), \quad k \in \mathbb{Z}, \tag{12}$$

which can be equivalently rewritten in the form

$$0 \in \mathcal{B} u(k+m) + \sum_{i=0}^{m-1} \mathcal{A}_i u(k+i) + F(k), \quad k \in \mathbb{Z}, \tag{13}$$

both equipped without initial conditions, where $\mathcal{B}$ and $\mathcal{A}_i$ $(0 \leqslant i \leqslant m-1)$ are multivalued linear operators on complex Banach space X and $F_\Delta : \mathbb{Z} \to P(X)$, respectively, $F : \mathbb{Z} \to P(X)$, is a multivalued forcing term, have not been well explored in the existing literature. It is clear that the difference inclusions (12)–(13) can be viewed as the discrete analogues of the differential inclusion

$$A_p u^{(p)}(t) + \sum_{i=0}^{p-1} A_i u^{(i)}(t) = f(t), \quad t \in \mathbb{R}$$

as well as that the difference inclusions (12)–(13) can be subjected with the initial conditions

$$u(0) = u_0, \ldots, u(m-1) = u_{m-1}; \tag{14}$$

furthermore, we can analyze the fractional analogues of (12)–(13). Without any doubt, the most important subcase of (12), respectively (13), is the following abstract higher-order difference equation:

$$B'\Delta^m u(k) + \sum_{i=0}^{m-1} A_i'\Delta^i u(k) + f_\Delta(k) = 0, \quad k \in \mathbb{Z},$$

respectively,

$$Bu(k+m) + \sum_{i=0}^{m-1} A_i u(k+i) + f(k) = 0, \quad k \in \mathbb{Z},$$

where B', B and A_i', A_i $(0 \leqslant i \leqslant m-1)$ are single-valued linear operators on X and $f_\Delta : \mathbb{Z} \to X$, respectively, $f : \mathbb{Z} \to X$, is a single-valued forcing term; these difference inclusions can be equipped with the initial conditions of type (14).

The analysis of the existence and uniqueness of asymptotically almost periodic (automorphic) solutions of the abstract difference equation with complex-valued coefficients

$$x_{n+k} + c_{k-1}x_{n+k-1} + \cdots + c_1 x_n + c_0 x_n = f_n; \quad x(i) = x_i,\ 0 \leqslant i \leqslant k-1, \tag{15}$$

where $(f_n)_{n\geqslant 0}$ is an asymptotically almost periodic (automorphic) sequence, is not a trivial problem; even if $k = 2$, the elementary theory shows that a solution of (15) can be asymptotically almost periodic (automorphic) but also unbounded. To illustrate this, let us consider the following second-order difference equation:

$$x_{n+2} - x_{n+1} + x_n = f_n; \quad x(0) = x_0,\ x(1) = x_1. \tag{16}$$

If $f_n \equiv \cos(na)$, where $a - (\pi/3) \notin 2\pi\mathbb{Z}$, then the unique solution of (16) has the form $x_n = ae^{in\pi/3} + be^{-in\pi/3} + c\cos(na) + d\sin(na)$, $n \in \mathbb{N}$ for certain complex numbers a, b,

c, d and, therefore, it is almost periodic; on the other hand, if $f_n \equiv \cos(n\pi/3)$, then the unique solution of (16) has the form $x_n = (a+cn)e^{in\pi/3} + (b+dn)e^{-in\pi/3}$, $n \in \mathbb{N}$ for certain complex numbers a, b, c, d and it is not bounded.

Further on, in many research papers published by now, A. G. Baskakov has investigated the spectral analysis of differential operators with unbounded operator-valued coefficients and qualitative properties of evolution systems by using the difference relations and semigroups of difference relations (cf. [93, 167] and references quoted therein; let us also note that, in a joint paper with A. Yu. Duplishcheva [94], the same author has investigated the abstract difference second-order equations with bounded linear coefficients in l^p spaces using the operator-valued matrices of the second order). In [116], M. S. Bichegkuev has investigated the well-posedness of the initial value problem

$$x(k+1) \in \mathcal{A}x(k) + f(k), \quad k \in \mathbb{N} \tag{17}$$

and proved the following results (the notion and notation will be explained in a part concerning the multivalued linear operators):

(i) The difference inclusion (17) has a unique solution for every sequence $(f_k)_{k\in\mathbb{N}} \in l^p(\mathbb{N} : X)$ if and only if $0 \in \rho(\mathcal{A})$ and $\sigma(\mathcal{A}) \subseteq \{\lambda \in \mathbb{C} : |\lambda| > 1\}$.
(ii) The difference inclusion (17) has a solution in $l^p(\mathbb{N} : X)$ for every sequence $(f_k)_{k\in\mathbb{N}} \in l^p(\mathbb{N} : X)$ if $\sigma(\mathcal{A}) \cap \{\lambda \in \mathbb{C} : |\lambda| = 1\} = \emptyset$.

In addition to the above, we will briefly describe the main results established in the research study [286] by L. Fang, N'gbo N'gbo and Y. Xia and the research study [781] by H. Wu et al. The authors have examined there a discrete nonautonomous Lotka–Volterra model and proved, under certain assumptions, the existence of a positive almost periodic solution of the considered system. The authors have also constructed the corresponding Lyapunov function and proved the exponential convergence of solutions. It is worth noting that many authors have analyzed a similar problematic before; for example, the existence of an attractive positive almost periodic solution for the discrete Lotka–Volterra system

$$x_{k+1} = \frac{x_k a_k}{1 + \beta_k x_k}, \quad k \in \mathbb{Z}$$

has been analyzed by K. Gopalsamy and S. Mohamad in [333] (see also the research monograph [332]). The relations between the global quasiuniform asymptotic stability of difference systems and the existence of almost periodic solutions have been investigated by Y. H. Xia and S. S. Chen in [784], where the authors have applied their results in the qualitative analysis of solutions to the following discrete Lotka–Volterra system:

$$x_i(k+1) = x_i(k)e^{r_i(k) - \sum_{j=1}^{m} a_{ij}(k)x_j(k)}, \quad 1 \leqslant i \leqslant m.$$

In [286], the authors have used the exponential dichotomy theory for difference equations and the Banach fixed-point theorem; in such a way, the authors have analyzed the existence and uniqueness of solutions for the discrete almost periodic Lotka–Volterra model

$$x_i(k+1) = x_i(k)e^{r_i(k)-\sum_{j=1}^{m} a_{ij}(k)x_j(k)-\sum_{j=1}^{m} b_{ij}(k)x_j(k-\tau(k))}, \quad 1 \leqslant i \leqslant m,$$

where $r_i(\cdot)$, $a_{ij}(\cdot)$, $b_{ij}(\cdot)$ and $\tau(\cdot)$ are almost periodic sequences for $1 \leqslant i,j \leqslant m$. See also [48–50, 588, 623, 798] and references quoted therein.

In [781], H. Wu et al. have investigated the existence of multiple periodic solutions to a class of fourth-order difference equations using the variational methods. The authors have been motivated by the results established in the well-known research article [353] by C. P. Gupta concerning the bending of an elastic beam with simply-supported ends under an external force; the discretization of the corresponding boundary value problem leads to $\Delta^4 x_{k-2} = e_k$. Speaking matter-of-factly, the authors have analyzed the existence of periodic solutions for the following nonlinear fourth-order difference equation:

$$\Delta^4 x_{k-2} + f(k, x_k) = 0, \quad k \in \mathbb{Z},$$

where the term $f(\cdot;\cdot)$ satisfies certain assumptions including the periodicity in the first variable. Concerning some other results about the existence and uniqueness of periodic solutions to the fourth-order difference equations, established by Chinese mathematicians, we can also recommend [117, 156, 158, 349–352, 791], the master thesis [733, 812, 824] and the references quoted therein.

The discrete generating series of one variable and its connection with the linear difference equations have recently been analyzed by V. S. Alekseev, S. S. Akhtamova and A. P. Lyapin in [39]. A discrete generating series for a series $F : \mathbb{Z}^n \to \mathbb{C}$ and its functional relations have been explored by S. S. Akhtamova, T. Cuchta and A. P. Lyapin in [37]. It is also worthwhile to mention that S. Chandragiri has investigated the difference equations and generating functions for some lattice path problems in [171, 172]; cf. also [60, 136, 138, 564, 565] for some other results obtained in this field. Also, we would like to mention the following:

(i) The characteristic polynomial of a difference equation

$$\sum_{\alpha \in A} c_\alpha F(x+\alpha) = 0, \quad x \in \mathbb{N}_0^n, \tag{18}$$

where A is a fixed subset of $\mathbb{N}_0^n$, c_α are some (constant) coefficients and $F : \mathbb{N}_0^n \to \mathbb{C}$ is an unknown sequence, is the polynomial

$$P(\lambda) := \sum_{\alpha \in A} c_\alpha \lambda^\alpha, \quad \lambda \in \mathbb{C}^n,$$

where $\lambda^\alpha \equiv \lambda_1^{\alpha_1} \cdot \lambda_2^{\alpha_2} \cdot \cdots \cdot \lambda_n^{\alpha_n}$. Using the notion of amoeba of the characteristic polynomial of (18) and the notion of a multiple Laurent series, E. K. Leinartas has established a description for the solution space of a multidimensional difference equation with constant coefficients; see [506, Theorems 1, 2].

(ii) The multidimensional versions of the famous Poincaré theorem for difference equations have been analyzed by E. K. Leinartas, M. Passare and A. K. Tsikh in [507].

(iii) Among many others, the Bessel difference equation

$$t(t-1)\Delta^2 y(t) + t\Delta y(t-1) + t(t-1)y(t) - n^2 y(t) = 0, \quad t \in \mathbb{Z}$$

has been studied by M. Bohner, T. Kuchta [127] and A. Slavík [697]. See also the research article [139] by R. H. Boyer, who analyzed the discrete Bessel functions, the research article [128] by M. Bohner and T. Kuchta, where the authors considered the generalized hypergeometric difference equation, and the recent research article [217] by T. Cuchta, D. Grow and N. Wintz, where the authors considered the discrete matrix hypergeometric functions.

The research article [95] is also worth of mentioning. In this paper, A. Bašić, L. Smajlović and Z. Šabanac have proved that for each $c \in \mathbb{C} \setminus \{i\alpha : |\alpha| \geqslant 1\}$ the function

$$J_{k,c} = \frac{(c/2)^k (t)_k}{k!} F_1\left(\frac{k+t}{2}, \frac{k+t+1}{2}; k+1; -c^2\right), \quad t \in \mathbb{Z},\ k \in \mathbb{N}_0$$

is a solution to the backward difference equation

$$t(t+1)[y_k(t+2) - 2y_k(t+1) + y_k(t)] + t[y_k(t+1) - y_k(t)] + c^2 t(t+1)y_k(t+2) = k^2 y_k(t),$$

as well as that for each $c \in \mathbb{C} \setminus \{\alpha : |\alpha| \geqslant 1\}$ the function

$$I_{k,c} = \frac{(c/2)^k (t)_k}{k!} F_1\left(\frac{k+t}{2}, \frac{k+t+1}{2}; k+1; c^2\right), \quad t \in \mathbb{Z},\ k \in \mathbb{N}_0$$

is a solution to the forward difference equation

$$t(t+1)[y_k(t+2) - 2y_k(t+1) + y_k(t)] + t[y_k(t+1) - y_k(t)] - c^2 t(t+1)y_k(t+2) = k^2 y_k(t),$$

where

$$F_1(\alpha, \beta; \gamma; z) = \sum_{k=0}^{+\infty} \frac{(\alpha)_k (\beta)_k}{(\gamma)_k k!} z^k$$

denotes the well-known Gaussian hypergeometric function; here, $(t)_k = \Gamma(t+k)/\Gamma(t)$ for $t \in \mathbb{R} \setminus -\mathbb{N}_0$. See also the research articles [157, 201], the doctoral dissertation [215] of T. Cuchta and [216].

The theory of linear nonautonomous second-order difference equations is still very unexplored and full of open problems.

Further on, the fractional equations on continuous and discrete time domains have been well studied. Fractional difference equations are discrete dynamic equations that describe how a quantity changes over a discrete time interval with a fractional order. In other words, the fractional difference equations focus on change between noninteger time steps rather than consecutive integer time steps. The fractional difference equations are commonly used for modeling discrete phenomena in different disciplines such as economics, physics, engineering and biology. In the qualitative theory of fractional difference equations, stability, boundedness, periodicity and attractors for the solutions are explored within the context of fractional-order differentiation. This theory is crucial for researchers since it provides valuable insights into the behavior for the solutions of nonlinear equations with complicated dynamics, which can be found in various scientific disciplines.

Neutral fractional difference equations incorporate both fractional differencing and neutral delay effects. They arise in systems where the evolution of a variable at a particular time depends not only on its current and past values but also on past values of another related variable with a time delay. This delay term introduces an additional layer of complexity, making the analysis and solution of these equations challenging. Such equations find applications in various fields, including population dynamics, ecology, control theory and neuroscience, where memory effects and delayed interactions play a crucial role in shaping system dynamics. The analysis of delayed neutral equations is significant since these equations provide a mathematical framework to models in nature and engineering, which involve time delays in the interactions between variables. In mathematical point of view, it is very important to analyze neutral equations with delay term since delay in the time may lead to stability issues for the solutions. In a joint research study with H. C. Koyuncuoğlu and J. M. Jonnalagadda [485], we turned the spotlight on the nonlinear delayed neutral fractional difference equations of the form

$$\nabla_0^{\alpha,\beta}[x(t) - g(t, x(t-\tau))] = f(t, x(t-\tau)), \quad t \in \mathbb{N}, \tag{19}$$

which involve the Hilfer nabla fractional difference $\nabla_0^{\alpha,\beta}$ and analyze their solutions qualitatively.

Freely speaking, the nabla fractional calculus is a branch of mathematics that extends traditional calculus to noninteger orders. Instead of derivatives and integrals, it deals with nabla operators, which are discrete counterparts of derivatives. This field has applications in various scientific domains, including physics, engineering and signal processing, offering insights into complex systems where the fractional behavior is prevalent. The fractional nabla calculus and fractional equations with nabla differences are well explored by many researchers. For the discrete nabla fractional calculus, we refer to readers the pioneering monograph [330] by C. Goodrich and A. C. Peterson and the important research articles [54] by G. A. Anastassiou, [74] and [75] by P. Eloe; cf. also [331, 356, 408, 409, 496] and [770] for an elaborative reading on the Hilfer nabla fractional difference operators and the Hilfer fractional nabla difference equations.

Set $\mathbb{N}_a := \{a, a+1, a+2, \dots\}$ for any $a \in \mathbb{R}$. We present the initial definitions as follows:

(i) Thebackward jump operator $\rho : \mathbb{N}_{a+1} \to \mathbb{N}_a$ is given by $\rho(t) := t - 1, t \in \mathbb{N}_{a+1}$.
(ii) The μth-order nabla fractional Taylor monomial is given by

$$H_\mu(t,a) := \frac{(t-a)^{\bar{\mu}}}{\Gamma(\mu+1)} := \frac{\Gamma(t-a+\mu)}{\Gamma(t-a)\Gamma(\mu+1)}, \quad \mu \in \mathbb{R}\backslash\{\dots,-2,-1\},$$

provided the right-hand side is well-defined.
(iii) Let $f : \mathbb{N}_a \to X$ and $N \in \mathbb{N}$. Then the first-order nabla difference of $f(\cdot)$ is given by $(\nabla f)(t) := f(t) - f(t-1)$ for $t \in \mathbb{N}_{a+1}$. Moreover, the higher-order nabla differences of $f(\cdot)$, $\nabla^N f$ are defined recursively.
(iv) Let $f : \mathbb{N}_a \to X$ and $\upsilon > 0$. Then υ^{th} nabla sum of $f(\cdot)$ based at a is introduced as

$$(\nabla_a^{-\upsilon} f)(t) := \sum_{s=a}^{t} H_{\upsilon-1}(t,\rho(s))f(s), \quad t \in \mathbb{N}_a.$$

In the next definition, we exhibit two well-known fractional difference operators.

Definition 0.0.1. Let $\upsilon > 0, N \in \mathbb{N}$ and $N - 1 < \upsilon \leqslant N$.
Riemann–Liouville fractional nabla difference Suppose that $f : \mathbb{N}_a \to X$. Then the R-L fractional difference of $f(\cdot)$ is given by

$$(\nabla_a^\upsilon f)(t) := (\nabla^N(\nabla_a^{-(N-\upsilon)} f))(t), \quad t \in \mathbb{N}_{a+N}. \tag{20}$$

Caputo nabla fractional difference Let $f : \mathbb{N}_{a-N} \to X$. Then the Caputo fractional difference of $f(\cdot)$ is given by

$$(\nabla_{*a}^\upsilon f)(t) := (\nabla_a^{-(N-\upsilon)}(\nabla^N f))(t), \quad t \in \mathbb{N}_a. \tag{21}$$

The basic structural properties of Riemann–Liouville fractional nabla differences and Caputo nabla fractional differences are given as follows: If $f : \mathbb{N}_a \to X$ and υ, $\mu > 0$, then we have:
(i) $(\nabla_a^{-\upsilon}\nabla_a^{-\mu} f)(t) = (\nabla_a^{-\upsilon-\mu} f)(t), t \in \mathbb{N}_a$.
(ii) $(\nabla_{a+1}^{-\upsilon}\nabla f)(t) = (\nabla\nabla_a^{-\upsilon} f)(t) - H_{\upsilon-1}(t,\rho(a))f(a), t \in \mathbb{N}_{a+1}$.
(iii) Let $\upsilon > 0$ and $\mu \in \mathbb{R}$. Then $\nabla_a^{-\upsilon} H_\mu(t,\rho(a)) = H_{\mu+\upsilon}(t,\rho(a)), t \in \mathbb{N}_a$.
(iv) Let υ, $\mu \in \mathbb{R}$ and $N \in \mathbb{N}$ so that $N - 1 < \upsilon \leqslant N$. Then we have

$$\nabla_a^\upsilon H_\mu(t,\rho(a)) = H_{\mu-\upsilon}(t,\rho(a)), \quad t \in \mathbb{N}_{a+N}.$$

(v) Let $X = \mathbb{R}$, $\upsilon > -1$ and $s \in \mathbb{N}_a$. Then the following holds:
 (v.1) If $t \in \mathbb{N}_{\rho(s)}$, then $H_\upsilon(t,\rho(s)) \geqslant 0$; moreover, if $t \in \mathbb{N}_s$, then we have $H_\upsilon(t,\rho(s)) > 0$.
 (v.2) If $t \in \mathbb{N}_{\rho(s)}$ and $\upsilon > 0$, then $H_\upsilon(t,\rho(s))$ is a decreasing function of s; if $t \in \mathbb{N}_s$ and $-1 < \upsilon < 0$, then $H_\upsilon(t,\rho(s))$ is an increasing function of s.

(v.3) If $t \in \mathbb{N}_{\rho(s)}$ and $\upsilon \geqslant 0$, then $H_\upsilon(t,\rho(s))$ is a nondecreasing function of t; if $\upsilon > 0$ and $t \in \mathbb{N}_s$, then $H_\upsilon(t,\rho(s))$ is an increasing function of t. Also, if $t \in \mathbb{N}_{s+1}$ and $-1 < \upsilon < 0$, then $H_\upsilon(t,\rho(s))$ is a decreasing function of t.

Let $f : \mathbb{N}_a \to X$, $0 \leqslant \beta \leqslant 1$ and $N \in \mathbb{N}$ so that $N-1 < \alpha \leqslant N$. Then α^{th}**-order,** β^{th}**-type Hilfer nabla fractional difference of** $f(\cdot)$ is defined by

$$(\nabla_a^{\alpha,\beta} f)(t) := (\nabla_{a+N}^{-\beta(N-\alpha)} \nabla^N \nabla_a^{-(1-\beta)(N-\alpha)} f)(t), \quad t \in \mathbb{N}_{a+N}. \tag{22}$$

In a similar fashion with the Hilfer fractional derivative on continuous time domains, the Hilfer nabla fractional difference operator defined in (22) is identical to the R-L nabla fractional difference in (20) when $\beta = 0$, and also, it turns into the Caputo nabla fractional difference represented in (21) when $\beta = 1$.

If $\mu \in \mathbb{R}$, $0 < \alpha \leqslant 1$ and $0 \leqslant \beta \leqslant 1$, then we have

$$\nabla_a^{\alpha,\beta} H_\mu(t,\rho(a)) = H_{\mu-\alpha}(t,\rho(a)) - H_{\beta(1-\alpha)-1}(t,\rho(a)), \quad t \in \mathbb{N}_{a+1},$$

provided that all terms are well-defined. Furthermore, if $f : \mathbb{N}_a \to X$, $0 < \alpha \leqslant 1$, $0 \leqslant \beta \leqslant 1$ and $\gamma = \alpha + \beta - \alpha\beta$, then we have:

(i) $(\nabla_a^{\alpha,\beta} f)(t) = (\nabla_{a+1}^{-\beta(1-\alpha)} \nabla_a^{\gamma} f)(t)$, $t \in \mathbb{N}_{a+1}$.
(ii) $(\nabla_{a+1}^{-\gamma} \nabla_a^{\gamma} f)(t) = (\nabla_{a+1}^{-\alpha} \nabla_a^{\alpha,\beta} f)(t)$, $t \in \mathbb{N}_{a+1}$.
(iii) $(\nabla_a^{\gamma} \nabla_a^{-\alpha} f)(t) = (\nabla_a^{\beta(1-\alpha)} f)(t)$, $t \in \mathbb{N}_{a+1}$.
(iv) $(\nabla_a^{\alpha,\beta} \nabla_a^{-\alpha} f)(t) = (\nabla_{a+1}^{-\beta(1-\alpha)} \nabla_a^{\beta(1-\alpha)} f)(t)$, $t \in \mathbb{N}_{a+1}$.

In our further work, we will primarily follow the notation used by C. Lizama and his coauthors, which slightly differs from the notation introduced above, and we will mainly work with the basis $a = 0$. If $\alpha > 0$ and $v \in \mathbb{N}_0$, then the Cesàro sequence $(k^\alpha(v))_{v\in\mathbb{N}_0}$ is defined by

$$k^\alpha(v) := \frac{\Gamma(v+\alpha)}{\Gamma(\alpha)v!}.$$

It is well known that, for every $\alpha > 0$ and $\beta > 0$, we have $k^\alpha *_0 k^\beta \equiv k^{\alpha+\beta}$ and $|k^\alpha(v) - g_\alpha(v)| = O(g_\alpha(v)|1/v|)$, $v \in \mathbb{N}$; in particular, $k^\alpha(v) \sim g_\alpha(v)$, $v \to +\infty$; if $0 < \alpha \leqslant 1$, then it is well known that

$$g_\alpha(v+1) < k^\alpha(v) < g_\alpha(v), \quad v \in \mathbb{N}.$$

Now, if $u : \mathbb{N}_0 \to X$ is a given sequence and $\alpha \in (0,\infty) \setminus \mathbb{N}_0$, then we define the fractional integral $\Delta^{-\alpha} u : \mathbb{N}_0 \to X$ by

$$\Delta^{-\alpha} u(v) := \sum_{j=0}^{v} k^\alpha(v-j) u(j), \quad v \in \mathbb{N}_0.$$

Set $m := \lceil \alpha \rceil$. Then the Riemann–Liouville fractional difference operator of order $\alpha > 0$, Δ^α shortly, is defined by

$$\Delta^\alpha u(v) := [\Delta^m(\Delta^{-(m-\alpha)}u)](v), \quad v \in \mathbb{N}_0,$$

while the Caputo fractional difference operator of order $\alpha > 0$, Δ_C^α for short, is defined by

$$\Delta_C^\alpha u(v) := [\Delta^{\alpha-m}(\Delta^m u)](v), \quad v \in \mathbb{N}_0.$$

Albeit seem different, these notions are the same as those introduced in Definition 0.0.1 with the basis $a = 0$.

Suppose now that $u : \mathbb{N}_0 \to X$, $\alpha > 0$, $m = \lceil \alpha \rceil$, $a : \mathbb{N}_0 \to X$ and $b : \mathbb{N}_0 \to X$. The following notion extends the notions of the Riemann–Liouville fractional difference operator of order $\alpha > 0$, the Caputo fractional difference operator of order $\alpha > 0$ and the Hilfer fractional difference operator of order $\alpha > 0$ and type $\beta \in (0, 1)$, with the basis $a = 0$.

Definition 0.0.2. The generalized Hilfer (a, b, α)-fractional derivative of sequence $u(\cdot)$, denoted shortly by $D_{a,b}^\alpha u$, is defined by

$$D_{a,b}^\alpha u(v) := (b *_0 \Delta^m(a *_0 u))(v), \quad v \in \mathbb{N}_0.$$

If $0 \leqslant \beta \leqslant 1$, then the usual Hilfer fractional derivative $D^{\alpha,\beta}u$ of order α and type β is defined as the generalized Hilfer (a, b, α)-fractional derivative of $u(\cdot)$, with $a(v) = k^{(1-\beta)(m-\alpha)}(v)$ and $b(v) = k^{\beta(1-\alpha)}(v)$. Define $D_{a,b}^0 u := a *_0 b *_0 u$.

In Subsection 7.4.2, we will enquire into the asymptotic behavior of solutions to the abstract nonlinear fractional difference equations with delay, involving the generalized Hilfer (a, b, α)-fractional derivatives; in such a way, we will extend the corresponding results about the asymptotic behavior of solutions to (19).

If $\alpha > 0$, $m = \lceil \alpha \rceil$ and $u : \mathbb{Z} \to X$ satisfies $\sum_{v=-\infty}^{\infty} \|u(v)\| \cdot (1 + |v|)^{m-\alpha-1} < +\infty$, then we define the Weyl fractional derivative

$$[\Delta_W^\alpha u](v) := [\Delta^m(\Delta_W^{-(m-\alpha)}u)](v), \quad v \in \mathbb{Z},$$

where

$$(\Delta_W^{-(m-\alpha)}u)(v) := \sum_{l=-\infty}^{v} k^{m-\alpha}(v-l)u(l), \quad v \in \mathbb{Z}.$$

In this monograph, we provide several new results concerning the existence and uniqueness of almost periodic type solutions for the following classes of abstract difference equations (inclusions):

(i)

$$u(k+1) \in \mathcal{A}u(k) + Cf(k), \quad k \in \mathbb{Z},$$

where $\mathcal{A}$ is a closed multivalued linear operator (MLO) in X and $C \in L(X)$ is an injective operator.

(ii)

$$\Delta^2 u(k) \in \mathcal{A}u(k) + Cf(k), \quad k \in \mathbb{Z},$$

where $\mathcal{A}$ is a closed MLO in X and $C \in L(X)$ is an injective operator.

(iii)

$$\Delta_W^{\alpha} u(k) \in \mathcal{A}u(k+1) + Cf(k), \quad k \in \mathbb{Z},$$

where $0 < \alpha \leqslant 1$, $\mathcal{A}$ is a closed MLO and $C \in L(X)$ is an injective operator.

(iv)

$$\Delta^{\alpha} u(v) \in \mathcal{A}u(v+2) + Cf(v); \quad u(0) = Cu_0,\ u(1) = Cu_1,$$

where $1 < \alpha \leqslant 2$, $\mathcal{A}$ is a subgenerator of a discrete (α, C)-resolvent family $(S_{\alpha}(v))_{v \in \mathbb{N}_0}$ and $C \in L(X)$ is an injective operator.

(v)

$$\Delta^{\alpha} u(v) \in \mathcal{A}u(v+1) + Cf(v); \quad u(0) = Cu_0,$$

where $0 < \alpha \leqslant 1$, $\mathcal{A}$ is a subgenerator of a discrete (α, C)-resolvent family $(S_{\alpha}(v))_{v \in \mathbb{N}_0}$ and $C \in L(X)$ is an injective operator.

(vi)

$$u(k_1+1, k_2+1, \dots, k_n+1) = \lambda_1 \lambda_2 \dots \lambda_n \cdot u(k_1, k_2, \dots, k_n) + F(k_1, k_2, \dots, k_n),$$

for all $(k_1, k_2, \dots, k_n) \in \mathbb{Z}^n$, and

$$u(k_1+1, \dots, k_n+1) = \lambda u(k_1, \dots, k_n) + f(k_1, \dots, k_n), \quad (k_1, \dots, k_n) \in \mathbb{Z}^n,$$

where the complex parameters λ, λ_j $(1 \leqslant j \leqslant n)$ possess certain values.

(vii) Nonconvolution type Volterra difference system with infinite delay given in the form

$$x(t+1) = A(t)x(t) + \sum_{j=-\infty}^{t} B(t,j)x(j) + f(t), \quad t \in \mathbb{Z},$$

where A and B are $n \times n$ matrix functions and $f(\cdot)$ is a vector function.

(vii) Abstract difference equation

$$u(k,m) = A(k,m)u(k-1,m-1) + f(k,m), \quad k,\ m \in \mathbb{N}$$

depending on two variables, where $A(k,m)$ are bounded linear operators on X.

(ix) Nonconvolution type Volterra difference equation

$$x(t+1) = A(t)x(t) + \sum_{s=-\infty}^{t-1} K(t,s)x(s), \quad t \in \mathbb{N}_0,$$

where $A(t) = [a_{i,j}(t)]$ is an invertible matrix function for all $t \in \mathbb{N}_0$ with bounded inverse and $K(t,s) = [k_{i,j}(t,s)]$;

(x)

$$Au_{k+1} + Bu_k = f_k, \quad k \in \mathbb{Z}$$

and

$$A_p u_{k+p} + A_{p-1}u_{k+p-1} + \cdots + A_0 u_k = f_k, \quad k \in \mathbb{Z},$$

where A, B and $A_p, \ldots, A_0$ are closed linear operators on X.

(xi)

$$u(v) \in f(v) + \mathcal{A}(a *_0 u)(v), \quad v \in \mathbb{N}_0,$$

where $\mathcal{A}$ is an MLO in X, $(a(v))_{v\in\mathbb{N}_0}$ and $(f(v))_{v\in\mathbb{N}_0}$ are given sequences in X.

(xii)

$$u(k+1) \in \mathcal{A} \sum_{j=-\infty}^{k} a(k-j)u(j+1) + \sum_{j=-\infty}^{k} b(k-j)f(k), \quad k \in \mathbb{Z},$$

where $\mathcal{A}$ is an MLO in X, $(a(v))_{v\in\mathbb{N}_0}$ and $(f(v))_{v\in\mathbb{N}_0}$ are given sequences in X.

(xiii) Various classes of the abstract multiterm fractional differential equations with Riemann–Liouville and Caputo derivatives.

(ixv) Various classes of the abstract multiterm fractional differential equations with Weyl derivatives on $\mathbb{Z}$, considered without initial conditions.

(xv) Various classes of the abstract multiterm discrete Volterra equations of nonscalar type.

(xvi) Various classes of the abstract multiterm discrete Volterra equations depending on several variables, etc.

Moreover, we investigate the semilinear and nonautonomous analogues of these equations.

For some recent results concerning the existence and uniqueness of solutions to the nonlinear fractional difference equations, we refer the reader to the research articles [51, 285, 563] and references quoted therein.

The first serious analysis of Poisson transform

$$v \mapsto \int_0^{+\infty} e^{-t} \frac{t^v}{v!} u(t)\, dt, \quad v \in \mathbb{N}_0,$$

has been conducted by C. Lizama in [535]. Let us observe that the above expression cannot be defined for the negative values of v; because of that we have suggested the use of the following Poisson-like transform:

$$v \mapsto y_{a,b,c,j,\omega}(v) := \int_0^{+\infty} e^{-b(ct^{-1}+at)^j} \frac{(\omega t)^{v-\frac{1}{2}}}{\Gamma(v+\frac{1}{2})} u(t)\, dt, \quad v \in \mathbb{Z},$$

where $a \in \mathbb{R}$, $b,\ c,\ \omega \in \mathbb{R} \setminus \{0\}$ and $j \in \mathbb{N}$. Since the Poisson transform cannot be defined for not-exponentially bounded functions $u(\cdot)$, we have also suggested here the use of the following transform:

$$v \mapsto \int_0^{+\infty} e^{-b(at)^j} \frac{(\omega t)^v}{v!} u(t)\, dt, \quad v \in \mathbb{N}_0.$$

It is well known that the Poisson-like transforms connect the solutions of the abstract (multiterm; fractional) differential equations and the abstract (multiterm; fractional) difference equations. More details about this important issue will be given in the second part of the monograph.

In this monograph, we also analyze the multidimensional vector-valued Laplace transform; this is the topic that has not received enough attention by now. If $f : [0,+\infty)^n \to X$ is a locally integrable function, then the multidimensional vector-valued Laplace transform of $f(\cdot)$, denoted by $F(\cdot) = \tilde{f} = \mathcal{L}f$, is defined by

$$\begin{aligned} F(\lambda_1,\ldots,\lambda_n) &:= \lim_{T\to+\infty} \int_{[0,T]^n} e^{-\lambda_1 t_1 - \cdots - \lambda_n t_n} f(t_1,\ldots,t_n)\, dt_1 \ldots dt_n \\ &:= \int_0^{+\infty} \cdots \int_0^{+\infty} e^{-\lambda_1 t_1 - \cdots - \lambda_n t_n} f(t_1,\ldots,t_n)\, dt_1 \ldots dt_n, \end{aligned}$$

if it is well-defined. We can simply prove that $F(\lambda_1,\ldots,\lambda_n)$ is well-defined for $\operatorname{Re}\lambda_1 > \omega_1,\ldots,\operatorname{Re}\lambda_n > \omega_n$, provided that there exist finite real constants $M \geqslant 1$ and $\omega_1 \in \mathbb{R},\ldots,\omega_n \in \mathbb{R}$ such that $\|f(t_1,\ldots,t_n)\| \leqslant M\exp(\omega_1 t_1 + \cdots + \omega_n t_n)$ for a. e. $t_1 \geqslant 0,\ldots,t_n \geqslant 0$. Consider now the following condition:

(GR) $f(\cdot)$ is Lebesgue measurable and there exist real constants $\omega_1 \in \mathbb{R}, \ldots, \omega_n \in \mathbb{R}$, $\eta_1 \in (-1, +\infty), \ldots, \eta_n \in (-1, +\infty)$ and $\zeta_1 \in (-1, +\infty), \ldots, \zeta_n \in (-1, +\infty)$ such that

$$\|f(t_1, \ldots, t_n)\| \leqslant M(t_1^{\eta_1} + t_1^{\zeta_1}) \cdot \cdots \cdot (t_n^{\eta_n} + t_n^{\zeta_n}) \exp(\omega_1 t_1 + \cdots + \omega_n t_n),$$
$$\text{for a. e. } t_1 \geqslant 0, \ldots, t_n \geqslant 0.$$

In this case, the Fubini theorem implies that the function $F(\lambda_1, \ldots, \lambda_n)$ is also well-defined for $\operatorname{Re} \lambda_1 > \omega_1, \ldots, \operatorname{Re} \lambda_n > \omega_n$ and the Lebesgue dominated convergence theorem implies that $F(\cdot)$ is analytic in this region of $\mathbb{C}^n$ (see L. Hörmander [385] for the basic introduction to the theory of analytic functions of several complex variables).

The uniqueness theorem for Laplace transform holds in the multidimensional framework. Concerning our original contributions regarding the multidimensional vector-valued Laplace transform, we will only note that we have proved here the complex inversion theorem for the multidimensional vector-valued Laplace transform and apply this result in the existence and uniqueness of solutions to the abstract partial fractional differential inclusions with Riemann–Liouville or Caputo derivatives. In our follow-up research studies, we will analyze the multidimensional vector-valued Laplace transform and its applications in a more detailed manner.

The structure of this monograph can be briefly outlined as follows. In the first chapter of monograph, we collect the basic definitions and results, which are needed for anything that follows. The first part of monograph, consisting of two chapters, is devoted to the study of various classes of almost periodic (automorphic) sequences and almost periodic (automorphic) solutions to the abstract difference equations, the abstract fractional difference equations and the abstract Volterra difference equation; we have already mentioned some classes of the abstract difference equations that we will consider here. The second part of monograph concerns several new classes of generalized metrically almost periodic (automorphic) functions and generalized metrically almost periodic (automorphic) solutions to the abstract (degenerate) Volterra integrodifferential equations. In this part, which consists of three chapters, we first clarify the metrical Bochner criterion and consider the notion of metrical Stepanov almost periodicity (cf. Section 4.1). We also analyze here the notion of metrical Stepanov almost automorphy and provide certain applications of this concept in Section 5.1 as well as the notion of metrical Weyl almost automorphy, the notion of metrical Besicovitch almost automorphy and provide certain applications of these concepts. Several new important results about Stepanov-p-almost periodic functions and Stepanov-p-almost automorphic functions with the general exponent $p > 0$ are presented in Section 5.4. In the remainder of the second part, we consider the following classes of generalized almost periodic functions and furnish some illustrative examples and applications:

(i) Metrically piecewise continuous ρ-almost periodic functions (cf. Section 4.2).
(ii) Almost periodic functions and almost automorphic functions in general measure (cf. Section 6.1 and Section 6.2).

(iii) (Equi-)Weyl-p-almost periodic functions [Doss-p-almost periodic functions] in general measure, with a general exponent $p > 0$ (cf. Section 6.3).

The third part of monograph is devoted to the study of abstract Volterra integrodifferential functional inclusions and their almost periodic type solutions. The main theoretical concept of (F, G, C)-resolvent operator families is developed in a joint research study with V. E. Fedorov; cf. Section 7.1 for more details in this direction. In the remainder of this part, we consider the following topics:

(i) Abstract fractional differential inclusions with Hilfer derivatives (cf. Section 7.2).
(ii) Abstract fractional differential inclusions with generalized Laplace derivatives (cf. Section 7.3).
(iii) Asymptotic behavior of solutions of abstract nonlinear fractional equations with delay and generalized Hilfer (a, b, α)-derivatives and asymptotically almost periodic type solutions of abstract nonlinear fractional differential equations with delay and generalized Hilfer (a, b, α)-derivatives (cf. Section 7.4 and Section 7.5).
(iv) Multidimensional fractional calculus and some classes of fractional partial differential inclusions (cf. Section 8.1).
(v) Multidimensional Poisson transform and its applications (cf. Section 8.2). Concerning this issue, we will only note for now that, if $u : [0, \infty)^n \to X$ is a given locally integrable function and the value of

$$[P(u)](v) := [P(u)](v_1, \dots, v_n) := \int\limits_{[0,\infty)^n} e^{-x_1-\cdots-x_n} \frac{x_1^{v_1}}{v_1!} \cdot \cdots \cdot \frac{x_n^{v_n}}{v_n!} u(x_1, \dots, x_n)\, dx_1\, dx_2 \dots dx_n$$

is well-defined for all $v_1 \in \mathbb{N}_0, \dots, v_n \in \mathbb{N}_0$, then the mapping $u \mapsto P(u)$ is called the multidimensional Poisson transform. It is clear that we have

$$[P(u)](v_1, \dots, v_n) = \left(\mathcal{L}\left[\frac{\cdot_1^{v_1}}{v_1!} \cdot \cdots \cdot \frac{\cdot_n^{v_n}}{v_n!} u(\cdot_1, \dots, \cdot_n)\right]\right)(1, \dots, 1), \quad (v_1, \dots, v_n) \in \mathbb{N}_0^n.$$

As any scientific research monograph, this monograph certainly contains some weak parts, typographical errors and deficiencies. For example, it may look like that the monograph is a little bit oversaturated with definitions and that some minor parts are written in a heuristical manner. The author would like to thank in advance for any in-depth comment, suggestion or objection of the readers.

We close the introductory part with the observation that the organization and main ideas of each section will be explained within itself (maybe with many technical details and a little bit overlong).

1 Preliminaries

1.1 Research methodology: basic concepts, tools and techniques

The main purpose of this section is to recall the basic definitions and results from functional analysis, fractional calculus, the theory of Laplace transform and the theory of Lebesgue spaces with variable exponents $L^{p(x)}(\Omega)$.

Notation and terminology

Suppose that $n \in \mathbb{N}$ as well as that X, Y, Z and T are given nonempty sets (here, n denotes the dimension of the underlying Euclidean space under our consideration but sometimes we also consider the sequences of the form (a_n) or (b_n); it is our strong belief that this will not cause any confusion henceforth). Let us recall that a binary relation between X into Y is any subset $\rho \subseteq X \times Y$. If $\rho \subseteq X \times Y$ and $\sigma \subseteq Z \times T$ with $Y \cap Z \neq \emptyset$, then we define $\rho^{-1} \subseteq Y \times X$ and $\sigma \cdot \rho = \sigma \circ \rho \subseteq X \times T$ by $\rho^{-1} := \{(y,x) \in Y \times X : (x,y) \in \rho\}$ and $\sigma \circ \rho := \{(x,t) \in X \times T : \exists y \in Y \cap Z \text{ such that } (x,y) \in \rho \text{ and } (y,t) \in \sigma\}$, respectively. As is well known, the domain and range of ρ are defined by $D(\rho) := \{x \in X : \exists y \in Y \text{ such that } (x,y) \in X \times Y\}$ and $R(\rho) := \{y \in Y : \exists x \in X \text{ such that } (x,y) \in X \times Y\}$, respectively; $\rho(x) := \{y \in Y : (x,y) \in \rho\}$ $(x \in X)$, $x \, \rho \, y \Leftrightarrow (x,y) \in \rho$. Set $\rho(X') := \{y : y \in \rho(x) \text{ for some } x \in X'\}$ $(X' \subseteq X)$, $\mathbb{N}_n := \{1,\ldots,n\}$ and $\mathbb{N}_n^0 := \{0,1,\ldots,n\}$. An unbounded subset $A \subseteq \mathbb{Z}$ is called syndetic if there exists a strictly increasing sequence (a_k) of integers such that $A = \{a_k : k \in \mathbb{Z}\}$ and $\sup_{k \in \mathbb{Z}}(a_{k+1} - a_k) < +\infty$. Set, for every $\mathbf{t}_0 \in \mathbb{R}^n$ and $l > 0$, $B(\mathbf{t}_0, l) := \{\mathbf{t} \in \mathbb{R}^n : |\mathbf{t} - \mathbf{t}_0| \leqslant l\}$, where $|\cdot - \cdot|$ denotes the Euclidean distance in $\mathbb{R}^n$. If $I \subseteq \mathbb{R}^n$ and $M > 0$, we set $I_M := \{\mathbf{t} \in I : |\mathbf{t}| \geqslant M\}$ and $I_M' := \{\mathbf{t} \in I : |\mathbf{t}| \leqslant M\}$. If $X_0 \subseteq X$, where $(X, \|\cdot\|)$ is a complex Banach space, then $CH(X_0)$ denotes the convex hull of X_0. If not explicitly stated otherwise, we will always assume henceforth that $(Y, \|\cdot\|_Y)$ and $(Z, \|\cdot\|_Z)$ are complex Banach spaces, as well. By I, we denote the identity operator on Y. If A is arbitrary set, then its power set is denoted by $P(A)$.

As is well known, the notion of a pseudometric space was introduced by Dj. Kurepa in [504] (1934). For the basic source of information about pseudometric spaces, we refer the reader to [435, 625, 694, 780] and the references quoted therein.

Vector-valued functions, closed operators

By $L(X, Y)$, we denote the space consisting of all continuous linear mappings from X into Y; $L(X) \equiv L(X,X)$. We topologize the spaces $L(X,Y)$ and X^*, the dual space of X, in the usual way.

A linear operator $A : D(A) \to X$ is said to be closed if the graph of the operator A, defined by $G_A := \{(x, Ax) : x \in D(A)\}$, is a closed subset of $X \times X$. The null space (kernel) and range of A are denoted by $N(A)$ and $R(A)$, respectively. Let us recall that a linear operator A is called closable if there exists a closed linear operator B such that $A \subseteq B$. If F is a linear submanifold of X, then we define the part of A in F by $D(A_{|F}) := \{x \in D(A) \cap F : Ax \in F\}$ and $A_{|F}x := Ax, x \in D(A_{|F})$.

https://doi.org/10.1515/9783111689746-002

The power A^n of A is defined inductively ($n \in \mathbb{N}_0$). If $\alpha \in \mathbb{C} \setminus \{0\}$, A and B are linear operators, then we define the operators αA, $A + B$ and AB in the usual way. The Gamma function is denoted by $\Gamma(\cdot)$ and the principal branch is always used to take the powers. Set, for every $\alpha > 0$,

$$g_\alpha(t) := t^{\alpha-1}/\Gamma(\alpha), \quad t > 0,$$

$g_0(t) \equiv$ the Dirac δ-distribution and $0^\zeta := 0$. Define $\Sigma_\alpha := \{z \in \mathbb{C} \setminus \{0\} : |\arg(z)| < \alpha\}$, $\alpha \in (0, \pi]$.

By $C(\Omega : X)$, we denote the space of all continuous functions $f : \Omega \to X$, where $\emptyset \neq \Omega \subseteq \mathbb{C}^n$; $C(\Omega) \equiv C(\Omega : \mathbb{C})$. If $s \in \mathbb{R}$, then we define $\lfloor s \rfloor := \sup\{l \in \mathbb{Z} : s \geqslant l\}$ and $\lceil s \rceil := \inf\{l \in \mathbb{Z} : s \leqslant l\}$. If X, $\mathrm{Y} \neq \emptyset$, then we set $\mathrm{Y}^{\mathrm{X}} := \{f \mid f : \mathrm{X} \to \mathrm{Y}\}$.

Suppose that $I = \mathbb{R}$ or $I = [0, \infty)$. By $C_b(I : X)$, we denote the vector space consisting of all bounded continuous functions from I into X; the abbreviation $C_0(I : X)$ denotes the vector subspace of $C_b(I : X)$ consisting of those functions $f : I \to X$ such that $\lim_{|t| \to \infty} \|f(t)\| = 0$. By $BUC(I : X)$, we denote the space of all bounded uniformly continuous functions from I to X; $C_b(I) \equiv C_b(I : \mathbb{C})$, $C_0(I) \equiv C_0(I : \mathbb{C})$ and $BUC(I) \equiv BUC(I : \mathbb{C})$. Equipped with the sup-norm, $C_b(I : X)$, $C_0(I : X)$ and $BUC(I : X)$ are Banach spaces.

Multivalued linear operators and solution operator families subgenerated by them

A multivalued map (multimap) $\mathcal{A} : X \to P(Y)$ is said to be a multivalued linear operator (MLO) if the following holds:

(i) $D(\mathcal{A}) := \{x \in X : \mathcal{A}x \neq \emptyset\}$ is a linear subspace of X.

(ii) $\mathcal{A}x + \mathcal{A}y \subseteq \mathcal{A}(x + y)$, x, $y \in D(\mathcal{A})$ and $\lambda \mathcal{A}x \subseteq \mathcal{A}(\lambda x)$, $\lambda \in \mathbb{C}$, $x \in D(\mathcal{A})$.

If $X = Y$, then we say that $\mathcal{A}$ is an MLO in X. Recall that, for every x, $y \in D(\mathcal{A})$ and λ, $\eta \in \mathbb{C}$ with $|\lambda| + |\eta| \neq 0$, we have $\lambda \mathcal{A}x + \eta \mathcal{A}y = \mathcal{A}(\lambda x + \eta y)$. Also, if $\mathcal{A}$ is an MLO, then $\mathcal{A}0$ is a linear submanifold of Y and $\mathcal{A}x = f + \mathcal{A}0$ for any $x \in D(\mathcal{A})$ and $f \in \mathcal{A}x$. Define $R(\mathcal{A}) := \{\mathcal{A}x : x \in D(\mathcal{A})\}$ and $N(\mathcal{A}) := \mathcal{A}^{-1}0 := \{x \in D(\mathcal{A}) : 0 \in \mathcal{A}x\}$ (we call that the range and kernel space of $\mathcal{A}$, respectively). The inverse $\mathcal{A}^{-1}$ of an MLO is defined by $D(\mathcal{A}^{-1}) := R(\mathcal{A})$ and $\mathcal{A}^{-1}y := \{x \in D(\mathcal{A}) : y \in \mathcal{A}x\}$. We know that $\mathcal{A}^{-1}$ is an MLO in X, as well as that $N(\mathcal{A}^{-1}) = \mathcal{A}0$ and $(\mathcal{A}^{-1})^{-1} = \mathcal{A}$. If $N(\mathcal{A}) = \{0\}$, i. e., if $\mathcal{A}^{-1}$ is single-valued, then $\mathcal{A}$ is said to be injective.

Assuming that $\mathcal{A}$, $\mathcal{B} : X \to P(Y)$ are two MLOs, we define its sum $\mathcal{A} + \mathcal{B}$ by $D(\mathcal{A} + \mathcal{B}) := D(\mathcal{A}) \cap D(\mathcal{B})$ and $(\mathcal{A} + \mathcal{B})x := \mathcal{A}x + \mathcal{B}x$, $x \in D(\mathcal{A} + \mathcal{B})$. Clearly, $\mathcal{A} + \mathcal{B}$ is likewise an MLO.

Suppose now that $\mathcal{A} : X \to P(Y)$ and $\mathcal{B} : Y \to P(Z)$ are two MLOs, where Z is a complex Banach space. The product of $\mathcal{A}$ and $\mathcal{B}$ is defined by $D(\mathcal{B}\mathcal{A}) := \{x \in D(\mathcal{A}) : D(\mathcal{B}) \cap \mathcal{A}x \neq \emptyset\}$ and $\mathcal{B}\mathcal{A}x := \mathcal{B}(D(\mathcal{B}) \cap \mathcal{A}x)$. We have that $\mathcal{B}\mathcal{A} : X \to P(Z)$ is an MLO and $(\mathcal{B}\mathcal{A})^{-1} = \mathcal{A}^{-1}\mathcal{B}^{-1}$. The scalar multiplication of an MLO $\mathcal{A} : X \to P(Y)$ with the number $z \in \mathbb{C}$, $z\mathcal{A}$ for short, is defined by $D(z\mathcal{A}) := D(\mathcal{A})$ and $(z\mathcal{A})(x) := z\mathcal{A}x$, $x \in D(\mathcal{A})$.

The integer powers of an MLO $\mathcal{A} : X \to P(X)$ are defined inductively as follows: $\mathcal{A}^0 =: I$; if $\mathcal{A}^{n-1}$ is defined, set

$$D(\mathcal{A}^n) := \{x \in D(\mathcal{A}^{n-1}) : D(\mathcal{A}) \cap \mathcal{A}^{n-1}x \neq \emptyset\},$$

and

$$\mathcal{A}^n x := (\mathcal{A}\mathcal{A}^{n-1})x = \bigcup_{y \in D(\mathcal{A}) \cap \mathcal{A}^{n-1}x} \mathcal{A}y, \quad x \in D(\mathcal{A}^n).$$

Suppose that $\mathcal{A} : X \to P(Y)$ and $\mathcal{B} : X \to P(Y)$ are two MLOs. Then the inclusion $\mathcal{A} \subseteq \mathcal{B}$ is equivalent to saying that $D(\mathcal{A}) \subseteq D(\mathcal{B})$ and $\mathcal{A}x \subseteq \mathcal{B}x$ for all $x \in D(\mathcal{A})$.

We say that an MLO operator $\mathcal{A} : X \to P(Y)$ is closed if for any sequences (x_k) in $D(\mathcal{A})$ and (y_k) in Y such that $y_k \in \mathcal{A}x_k$ for all $k \in \mathbb{N}$ we have that the suppositions $\lim_{k\to\infty} x_k = x$ and $\lim_{k\to\infty} y_k = y$ imply $x \in D(\mathcal{A})$ and $y \in \mathcal{A}x$. Any MLO has a closed linear extension, in contrast to the usually analyzed single-valued linear operators.

The following lemma will be used later on (see, e. g., [445, Theorem 1.2.3]).

Lemma 1.1.1. *Suppose that $\mathcal{A} : X \to P(Y)$ is a closed* MLO, *Ω is a locally compact, separable metric space and μ is a locally finite Borel measure defined on Ω. Let $f : \Omega \to X$ and $g : \Omega \to Y$ be μ-integrable, and let $g(x) \in \mathcal{A}f(x)$, $x \in \Omega$. Then $\int_\Omega f\, d\mu \in D(\mathcal{A})$ and $\int_\Omega g\, d\mu \in \mathcal{A}\int_\Omega f\, d\mu$.*

Suppose that $\mathcal{A}$ is an MLO in X and $C \in L(X)$. The C-resolvent set of $\mathcal{A}$, $\rho_C(\mathcal{A})$ for short, is defined as the union of those complex numbers $\lambda \in \mathbb{C}$ satisfying that:

(i) $R(C) \subseteq R(\lambda - \mathcal{A})$.
(ii) $(\lambda - \mathcal{A})^{-1}C$ is a single-valued linear continuous operator on X.

The operator $\lambda \mapsto (\lambda - \mathcal{A})^{-1}C$ is said to be the C-resolvent of $\mathcal{A}$. If $C = I$, then we say that $\rho(\mathcal{A}) \equiv \rho_C(\mathcal{A})$ is the resolvent set of $\mathcal{A}$ and the mapping $\lambda \mapsto R(\lambda : \mathcal{A}) \equiv (\lambda - \mathcal{A})^{-1}$ is called the resolvent of $\mathcal{A}$ ($\lambda \in \rho(\mathcal{A})$). For more details regarding the generalized resolvent equations, the analytical properties of C-resolvents of multivalued linear operators and fractional powers of multivalued linear operators, we refer the reader to [445].

We will use the following lemma [445].

Lemma 1.1.2. *We have*

$$(\lambda - \mathcal{A})^{-1}C\mathcal{A} \subseteq \lambda(\lambda - \mathcal{A})^{-1}C - C \subseteq \mathcal{A}(\lambda - \mathcal{A})^{-1}C, \quad \lambda \in \rho_C(\mathcal{A}).$$

The operator $(\lambda - \mathcal{A})^{-1}C\mathcal{A}$ is single-valued on $D(\mathcal{A})$ and $(\lambda - \mathcal{A})^{-1}C\mathcal{A}x = (\lambda - \mathcal{A})^{-1}Cy$, whenever $y \in \mathcal{A}x$ and $\lambda \in \rho_C(\mathcal{A})$.

We will use condition (C1) henceforward.

(C1) There exist finite constants c, $M > 0$ and $\beta \in (0,1]$ such that

$$\Psi := \Psi_c := \{\lambda \in \mathbb{C} : \operatorname{Re}\lambda \geqslant -c(|\operatorname{Im}\lambda| + 1)\} \subseteq \rho(\mathcal{A})$$

and

$$\|R(\lambda : \mathcal{A})\| \leqslant M(1 + |\lambda|)^{-\beta}, \quad \lambda \in \Psi.$$

If $A = \mathcal{A}$ is single-valued and satisfies condition (C1), then A is said to be almost sectorial; cf. [443, 444] and references cited therein for more details about this important class of operators.

Of concern is the following abstract degenerate Volterra inclusion:

$$\mathcal{B}u(t) \subseteq \mathcal{A}\int_0^t a(t-s)u(s)ds + \mathcal{F}(t), \quad t \in [0,\tau), \tag{1.1}$$

where $0 < \tau \leqslant \infty$, $a \in L^1_{\mathrm{loc}}([0,\tau))$, $a \neq 0$, $\mathcal{F}: [0,\tau) \to P(Y)$ and $\mathcal{A}: X \to P(Y)$, $\mathcal{B}: X \to P(Y)$ are two given mappings (possibly nonlinear). We need the following notion.

Definition 1.1.3 (cf. [445, Definition 3.1.1(i)]). (i) A function $u \in C([0,\tau) : X)$ is said to be a presolution of (1.1) if $(a * u)(t) \in D(\mathcal{A})$ and $u(t) \in D(\mathcal{B})$ for $t \in [0,\tau)$, as well as (1.1) holds.

(ii) A solution of (1.1) is any presolution $u(\cdot)$ of (1.1) satisfying additionally that there exist functions $u_{\mathcal{B}} \in C([0,\tau) : Y)$ and $u_{a,\mathcal{A}} \in C([0,\tau) : Y)$ such that $u_{\mathcal{B}}(t) \in \mathcal{B}u(t)$ and $u_{a,\mathcal{A}}(t) \in \mathcal{A}\int_0^t a(t-s)u(s)ds$ for $t \in [0,\tau)$, as well as

$$u_{\mathcal{B}}(t) \in u_{a,\mathcal{A}}(t) + \mathcal{F}(t), \quad t \in [0,\tau).$$

(iii) A strong solution of (1.1) is any function $u \in C([0,\tau) : X)$ satisfying that there exist two continuous functions $u_{\mathcal{B}} \in C([0,\tau) : Y)$ and $u_{\mathcal{A}} \in C([0,\tau) : Y)$ such that $u_{\mathcal{B}}(t) \in \mathcal{B}u(t)$, $u_{\mathcal{A}}(t) \in \mathcal{A}u(t)$ for all $t \in [0,\tau)$, and

$$u_{\mathcal{B}}(t) \in (a * u_{\mathcal{A}})(t) + \mathcal{F}(t), \quad t \in [0,\tau).$$

In the remainder of this subsection, we will analyze multivalued linear operators as subgenerators of (a,k)-regularized (C_1, C_2)-existence and uniqueness families and (a,k)-regularized C-resolvent families. Unless specified otherwise, we assume that $0 < \tau \leqslant \infty$, $k \in C([0,\tau))$, $k \neq 0$, $a \in L^1_{\mathrm{loc}}([0,\tau))$, $a \neq 0$, $\mathcal{A} : Y \to P(Y)$ is an MLO, $C_1 \in L(X,Y)$, $C_2 \in L(Y)$ is injective, $C \in L(Y)$ is injective and $C\mathcal{A} \subseteq \mathcal{A}C$.

We will use the following notion (see, e. g., [445, Definition 3.2.1, Definition 3.2.2]).

Definition 1.1.4. (i) It is said that $\mathcal{A}$ is a subgenerator of a (local, if $\tau < \infty$) mild (a,k)-regularized (C_1, C_2)-existence and uniqueness family $(R_1(t), R_2(t))_{t\in[0,\tau)} \subseteq L(X,Y) \times L(Y)$ if the mappings $t \mapsto R_1(t)y$, $t \geqslant 0$ and $t \mapsto R_2(t)x$, $t \in [0,\tau)$ are continuous for every fixed $x \in Y$ and $y \in X$, as well as the following conditions hold:

$$\left(\int_0^t a(t-s)R_1(s)y\,ds, R_1(t)y - k(t)C_1y\right) \in \mathcal{A}, \quad t \in [0,\tau),\ y \in X \quad \text{and} \tag{1.2}$$

$$\int_0^t a(t-s)R_2(s)y\,ds = R_2(t)x - k(t)C_2x, \quad \text{whenever } t \in [0,\tau) \text{ and } (x,y) \in \mathcal{A}. \tag{1.3}$$

(ii) Let $(R_1(t))_{t\in[0,\tau)} \subseteq L(X, Y)$ be strongly continuous. Then it is said that $\mathcal{A}$ is a subgenerator of a (local, if $\tau < \infty$) mild (a, k)-regularized C_1-existence family $(R_1(t))_{t\in[0,\tau)}$ if (1.2) holds.

(iii) Let $(R_2(t))_{t\in[0,\tau)} \subseteq L(Y)$ be strongly continuous. Then it is said that $\mathcal{A}$ is a subgenerator of a (local, if $\tau < \infty$) mild (a, k)-regularized C_2-uniqueness family $(R_2(t))_{t\in[0,\tau)}$ if (1.3) holds.

Definition 1.1.5. Suppose that $0 < \tau \leqslant \infty$, $k \in C([0, \tau))$, $k \neq 0$, $a \in L^1_{\mathrm{loc}}([0, \tau))$, $a \neq 0$, $\mathcal{A} : Y \to P(Y)$ is an MLO, $C \in L(Y)$ is injective and $C\mathcal{A} \subseteq \mathcal{A}C$. Then it is said that a strongly continuous operator family $(R(t))_{t\in[0,\tau)} \subseteq L(Y)$ is an (a, k)-regularized C-resolvent family with a subgenerator $\mathcal{A}$ if $(R(t))_{t\in[0,\tau)}$ is a mild (a, k)-regularized C-uniqueness family having $\mathcal{A}$ as subgenerator, $R(t)C = CR(t)$ and $R(t)\mathcal{A} \subseteq \mathcal{A}R(t)$ $(t \in [0, \tau))$.

If $\tau = \infty$, then $(R(t))_{t\geqslant 0}$ is said to be exponentially bounded (bounded) if there exists $\omega \in \mathbb{R}$ $(\omega = 0)$ such that the family $\{e^{-\omega t}R(t) : t \geqslant 0\}$ is bounded; the infimum of such numbers is said to be the exponential type of $(R(t))_{t\geqslant 0}$. The above notion can be simply understood for the classes of mild (a, k)-regularized C_1-existence families and mild (a, k)-regularized C_2-uniqueness families.

The integral generator of a mild (a, k)-regularized C_2-uniqueness family $(R_2(t))_{t\in[0,\tau)}$ (mild (a, k)-regularized (C_1, C_2)-existence and uniqueness family $(R_1(t), R_2(t))_{t\in[0,\tau)}$) is defined by

$$\mathcal{A}_{\mathrm{int}} := \left\{(x, y) \in X \times X : R_2(t)x - k(t)C_2x = \int_0^t a(t - s)R_2(s)y\, ds,\ t \in [0, \tau)\right\};$$

we define the integral generator of an (a, k)-regularized C-resolvent family $(R(t))_{t\in[0,\tau)}$ in the same way.

For simplicity, we will assume henceforth that any (a, k)-regularized C-resolvent family considered below is likewise a mild (a, k)-regularized C-existence family (subgenerated by $\mathcal{A}$). We refer the reader to [443–445] for several simple conditions ensuring this property. The basic facts about strongly continuous semigroups, integrated semigroups and C-regularized semigroups may be obtained by consulting the monographs [443, 445] and references cited therein.

Integration in Banach spaces

The following elementary definition can be found in many textbooks.

Definition 1.1.6. (i) We say that a function $f : I \to X$ is simple if there exist $k \in \mathbb{N}$, elements $z_i \in X$, $1 \leqslant i \leqslant k$ and Lebesgue measurable subsets Ω_k, $1 \leqslant i \leqslant k$ of I, such that $m(\Omega_i) < \infty$, $1 \leqslant i \leqslant k$ and

$$f(t) = \sum_{i=1}^{k} z_i \chi_{\Omega_i}(t), \quad t \in I. \tag{1.4}$$

(ii) We say that a function $f : I \to X$ is measurable if there exists a sequence (f_k) in X^I such that, for every $k \in \mathbb{N}$, $f_k(\cdot)$ is a simple function and $\lim_{k\to\infty} f_k(t) = f(t)$ for a. e. $t \in I$.

(iii) Let $-\infty < a < b < \infty$ and $a < \tau < \infty$. We say that a function $f : [a,b] \to X$ is absolutely continuous if for every $\varepsilon > 0$ there exists a number $\delta > 0$ such that for any finite collection of open subintervals (a_i, b_i), $1 \leqslant i \leqslant k$ of $[a,b]$ with $\sum_{i=1}^{k}(b_i - a_i) < \delta$, we have $\sum_{i=1}^{k} \|f(b_i) - f(a_i)\| < \varepsilon$; a function $f : [a,\tau) \to X$ is said to be absolutely continuous if for every $\tau_0 \in (a,\tau)$, the function $f_{|[a,\tau_0]} : [a,\tau_0] \to X$ is absolutely continuous. Furthermore, a function $f : [a,\infty) \to X$ is said to be locally absolutely continuous if for every $\tau_0 > a$, the function $f_{|[a,\tau_0]} : [a,\tau_0] \to X$ is absolutely continuous; by $AC_{\mathrm{loc}}([a,\infty))$, we denote the collection of all locally absolutely continuous functions $f : [a,\infty) \to X$.

If $f : I \to X$ and (f_k) is a sequence of measurable functions such that $\lim_{k\to\infty} f_k(t) = f(t)$ for a. e. $t \in I$, then the function $f(\cdot)$ is measurable as well. The Bochner integral of a simple function $f : I \to X$, $f(t) = \sum_{i=1}^{k} z_i \chi_{\Omega_i}(t)$, $t \in I$ is defined by

$$\int_I f(t)\,dt := \sum_{i=1}^{k} z_i m(\Omega_i).$$

The definition of Bochner integral does not depend on the representation (1.4), as easily shown.

We say that a measurable function $f : I \to X$ is Bochner integrable if there exists a sequence of simple functions (f_k) in X^I such that $\lim_{k\to\infty} f_k(t) = f(t)$ for a. e. $t \in I$ and

$$\lim_{k\to\infty} \int_I \|f_k(t) - f(t)\|\,dt = 0; \tag{1.5}$$

if this is the case, the Bochner integral of $f(\cdot)$ is defined by

$$\int_I f(t)\,dt := \lim_{k\to\infty} \int_I f_k(t)\,dt.$$

This definition does not depend on the choice of a sequence of simple functions (f_k) in X^I satisfying $\lim_{k\to\infty} f_k(t) = f(t)$ for a. e. $t \in I$ and (1.5). We know that $f : I \to X$ is Bochner integrable if and only if $f(\cdot)$ is measurable and the function $t \mapsto \|f(t)\|$, $t \in I$ is integrable. For any Bochner integrable function $f : [0,\infty) \to X$, we have $\int_0^\infty f(t)\,dt = \lim_{\tau\to+\infty} \int_0^\tau f_{|[0,\tau]}(t)\,dt$.

The space of all Bochner integrable functions from I into X is designated by $L^1(I : X)$; endowed with the norm $\|f\|_1 := \int_I \|f(t)\|\,dt$, $L^1(I : X)$ is a Banach space. It is said that a function $f : [0,\infty) \to X$ is locally (Bochner) integrable if $f(\cdot)_{|[0,\tau]}$ is Bochner integrable for every $\tau > 0$. The space of all locally integrable functions from $[0,\infty)$ into X is denoted by $L^1_{\mathrm{loc}}([0,\infty) : X)$. If $f : [a,b] \to X$ is Bochner integrable, where $-\infty < a < b < +\infty$, then the function $F(t) := \int_a^t f(s)\,ds$, $t \in [a,b]$ is absolutely continuous and $F'(t) = f(t)$ for a. e. $t \in [a,b]$.

We need the following fundamental results.

Theorem 1.1.7. (i) (The dominated convergence theorem)
Suppose that (f_k) is a sequence of Bochner integrable functions from X^I and that there exists an integrable function $g : I \to \mathbb{R}$ such that $\|f_k(t)\| \leqslant g(t)$ for a. e. $t \in I$ and $k \in \mathbb{N}$. If $f : I \to X$ and $\lim_{k\to\infty} f_k(t) = f(t)$ for a. e. $t \in I$, then $f(\cdot)$ is Bochner integrable, $\int_I f(t)\,dt = \lim_{k\to\infty} \int_I f_k(t)\,dt$ and $\lim_{k\to\infty} \int_I \|f_k(t) - f(t)\|\,dt = 0$.

(ii) (The Fubini theorem)
Let I_1 and I_2 be segments in $\mathbb{R}$ and let $I = I_1 \times I_2$. Suppose that $F : I \to X$ is measurable and $\int_{I_1} \int_{I_2} \|f(s,t)\|\,dt\,ds < \infty$. Then $f(\cdot,\cdot)$ is Bochner integrable, the repeated integrals $\int_{I_1} \int_{I_2} f(s,t)\,dt\,ds$ and $\int_{I_2} \int_{I_1} f(s,t)\,ds\,dt$ exist and equal to the integral $\int_I f(s,t)\,ds\,dt$.

Suppose now that $1 \leqslant p < \infty$ and $(\Omega, \mathcal{R}, \mu)$ is a measure space. By $L^p(\Omega : X)$, we denote the space of all strongly μ-measurable functions $f : \Omega \to X$ such that $\|f\|_p := (\int_\Omega \|f(\cdot)\|^p\,d\mu)^{1/p}$ is finite. The space $L^\infty(\Omega : X)$ consisting of all strongly μ-measurable, essentially bounded functions, is a Banach space equipped with the norm $\|f\|_\infty := ess\,\sup_{t\in\Omega} \|f(t)\|, f \in L^\infty(\Omega : X)$. The famous Riesz–Fischer theorem says that $(L^p(\Omega : X), \|\cdot\|_p)$ is a Banach space for all $p \in [1,\infty]$; furthermore, $(L^2(\Omega : X), \|\cdot\|_2)$ is a Hilbert space. Let us recall, if $\lim_{s\to\infty} f_s = f$ in $L^p(\Omega : X)$, then there exists a subsequence (f_{s_k}) of (f_s) such that $\lim_{k\to\infty} f_{s_k}(t) = f(t)$ μ-almost everywhere. If the Banach space X is reflexive, then $L^p(\Omega : X)$ is reflexive for all $p \in (1,\infty)$ and its dual is isometrically isomorphic to $L^{p/(p-1)}(\Omega : X)$.

Let $\emptyset \neq \Omega \subseteq \mathbb{R}^n$. The space $L^p_{loc}(\Omega : X)$ for $1 \leqslant p \leqslant \infty$ is defined in the usual way; $L^p_{loc}(\Omega) \equiv L^p_{loc}(\Omega : \mathbb{C})$. If Ω is open, then $C^k(\Omega : X)$ denotes the space of k-times continuously differentiable functions $f : \Omega \to X$.

Suppose now that $k \in \mathbb{N}$ and $p \in [1,\infty]$. Then the Sobolev space $W^{k,p}(\Omega : X)$ consists of those X-valued distributions $u \in \mathcal{D}'(\Omega : X)$ such that, for every multi-index $\alpha \in \mathbb{N}_0^n$ with $|\alpha| \leqslant k$, we have $D^\alpha u \in L^p(\Omega, X)$. Here, the derivative D^α is taken in the sense of distributions. By $W^{k,p}_{loc}(\Omega : X)$, we denote the space of those X-valued distributions $u \in \mathcal{D}'(\Omega : X)$ such that, for every bounded open subset Ω' of Ω, we have $u_{|\Omega'} \in W^{k,p}(\Omega' : X)$.

Fractional calculus and fractional differential equations

As already mentioned multiple times, the fractional calculus and fractional differential equations are extremely growing fields of research, which have invaluable importance in engineering, physics, chemistry, mechanics, electricity, control theory and many other branches of applied science. For more details about fractional calculus and fractional differential equations, we refer the reader to the monographs cited in [443, 445] and [447].

Suppose that $\alpha > 0$, $m = \lceil \alpha \rceil$ and $I = (0,T)$ for some $T \in (0,\infty]$. Then the Riemann–Liouville fractional integral J_t^α of order α is defined by

$$J_t^\alpha f(t) := (g_\alpha * f)(t), \quad f \in L^1(I : X),\ t \in I.$$

The Caputo fractional derivative $\mathbf{D}_t^\alpha u(t)$ is defined for those functions $u \in C^{m-1}([0,\infty) : X)$ for which $g_{m-\alpha} * (u - \sum_{k=0}^{m-1} u_k g_{k+1}) \in C^m([0,\infty) : X)$, by

$$\mathbf{D}_t^\alpha u(t) := \frac{d^m}{dt^m}\left[g_{m-\alpha} * \left(u - \sum_{k=0}^{m-1} u_k g_{k+1}\right)\right]. \tag{1.6}$$

It is worth noticing that the existence of Caputo fractional derivative $\mathbf{D}_t^\alpha u$ for $t \geqslant 0$ implies the existence of Caputo fractional derivative $\mathbf{D}_t^\zeta u$ for $t \geqslant 0$ and any $\zeta \in (0, \alpha)$.

Let us recall that the Weyl–Liouville fractional derivative $D_{t,+}^\gamma u(t)$ of order $\gamma \in (0,1)$ is defined for those continuous functions $u : \mathbb{R} \to X$ such that

$$t \mapsto \int_{-\infty}^{t} g_{1-\gamma}(t-s)u(s)\, ds, \quad t \in \mathbb{R}$$

is a well-defined continuously differentiable mapping, by

$$D_{t,+}^\gamma u(t) := \frac{d}{dt}\int_{-\infty}^{t} g_{1-\gamma}(t-s)u(s)\, ds, \quad t \in \mathbb{R}.$$

Define $D_{t,+}^1 u(t) := -(d/dt)u(t)$.

The Mittag-Leffler functions and the Wright functions are of crucial importance in fractional calculus. Let $\alpha > 0$ and $\beta \in \mathbb{R}$. Then the Mittag-Leffler function $E_{\alpha,\beta}(z)$ is defined by

$$E_{\alpha,\beta}(z) := \sum_{n=0}^{\infty} \frac{z^n}{\Gamma(\alpha n + \beta)}, \quad z \in \mathbb{C};$$

we define, for short, $E_\alpha(z) := E_{\alpha,1}(z)$, $z \in \mathbb{C}$. If $\gamma \in (0,1)$, then we define the Wright function $\Phi_\gamma(\cdot)$ by

$$\Phi_\gamma(z) := \sum_{n=0}^{\infty} \frac{(-z)^n}{n!\Gamma(1-\gamma-\gamma n)}, \quad z \in \mathbb{C}.$$

Let us recall that $\Phi_\gamma(\cdot)$ is an entire function as well as that:
(i) $\Phi_\gamma(t) \geqslant 0$, $t \geqslant 0$;
(ii) $\int_0^\infty e^{-\lambda t}\gamma s t^{-1-\gamma}\Phi_\gamma(t^{-\gamma}s)\, dt = e^{-\lambda^\gamma s}$, $\operatorname{Re}\lambda > 0$, $s > 0$; and
(iii) $\int_0^\infty t^r \Phi_\gamma(t)\, dt = \frac{\Gamma(1+r)}{\Gamma(1+\gamma r)}$, $r > -1$.

Fixed-point theorems

The fixed-point theory is a rapidly growing field of research. We will recall here the statements of the Banach contraction principle and the Schauder fixed-point theorem; for more details about the fixed-point theory, the reader may consult the monographs cited in [447].

Let (E,d) be a metric space. Then we say that $T : E \to E$ is a contraction mapping on E if there exists a constant $q \in [0,1)$ such that $d(T(x),T(y)) \leqslant qd(x,y)$ for all x, $y \in E$. For our further work, it will be necessary to recall the following well-known result.

Theorem 1.1.8 (The Banach contraction principle, 1922). *Let (E,d) be a complete metric space, and let $T : E \to E$ be a contraction mapping. Then T admits a unique fixed point x in X (i. e., $T(x) = x$).*

The first version of the following theorem was conjectured and proved on Banach spaces by J. Schauder in 1930; four years later, A. Tychonoff proved the theorem in the case that K is a compact convex subset of a locally convex space (see also Theorem 2.2.22 below).

Theorem 1.1.9 (The Schauder fixed-point theorem). *Suppose that K is a non-empty convex compact subset of a locally convex Hausdorff space V and T is a continuous mapping of K into itself. Then T has at least one fixed point.*

Laplace transform

Let $0 < \tau \leqslant \infty$ and $a \in L^1_{\mathrm{loc}}([0,\tau))$. Then we say that the function $a(t)$ is a kernel on $[0,\tau)$ if for each $f \in C([0,\tau))$ the assumption $\int_0^t a(t-s)f(s)\,ds = 0$, $t \in [0,\tau)$ implies $f(t) = 0$, $t \in [0,\tau)$. We need the following condition on the scalar-valued kernel $k(t)$:

(P1): $k(t)$ is Laplace transformable, i. e., it is locally integrable on $[0,\infty)$ and there exists $\beta \in \mathbb{R}$ such that $\tilde{k}(\lambda) := (\mathcal{L}k)(\lambda) := \lim_{b\to\infty}\int_0^b e^{-\lambda t}k(t)dt := \int_0^\infty e^{-\lambda t}k(t)dt$ exists for all $\lambda \in \mathbb{C}$ with $\operatorname{Re}\lambda > \beta$. Put $\operatorname{abs}(k) := \inf\{\operatorname{Re}\lambda : \tilde{k}(\lambda) \text{ exists}\}$.

Suppose now that E is a sequentially complete locally convex space over the field of complex numbers (we refer the reader to [445] for basic source of information concerning integration in locally convex spaces). Let us consider now the existence of Laplace integral of a function $f : [0,\infty) \to E$, that is,

$$(\mathcal{L}f)(\lambda) := \tilde{f}(\lambda) := \int_0^\infty e^{-\lambda t}f(t)\,dt := \lim_{\tau\to\infty}\int_0^\tau e^{-\lambda t}f(t)\,dt,$$

for $\lambda \in \mathbb{C}$. If $\tilde{f}(\lambda_0)$ exists for some $\lambda_0 \in \mathbb{C}$, then we define the abscissa of convergence of $\tilde{f}(\cdot)$ by $abs_X(f) := \inf\{\operatorname{Re}\lambda : \tilde{f}(\lambda) \text{ exists}\}$; otherwise, we set $abs_E(f) := +\infty$. We say that $f(\cdot)$ is Laplace transformable, or equivalently, that $f(\cdot)$ belongs to the class (P1)-E, if $abs_E(f) < \infty$; further on, we abbreviate $abs_E(f)$ to $abs(f)$, if there is no risk for confusion. The vector-valued Laplace transform has many interesting features; in the sequel, we will particularly use the following ones.

Lemma 1.1.10. *Let $f \in$ (P1)-E, $s < 0$ and $f_0 \in L^1[s,0]$.*

(i) *Put $f_s(t) := f(t-s)$, $t \geqslant 0$, $h_s(t) := f(t+s)$, $t \geqslant -s$ and $h_s(t) := f_0(t+s)$, $t \in [0,-s]$. Then $abs(f_s) = abs(h_s) = abs(f)$, $\tilde{f_s}(\lambda) = e^{-\lambda s}[\tilde{f}(\lambda) - \int_0^{-s} e^{-\lambda t}f(t)\,dt]$ and $\widetilde{h_s}(\lambda) = e^{\lambda s}[\tilde{f}(\lambda) - \int_0^s e^{-\lambda t}f_0(t)\,dt]$ ($\lambda \in \mathbb{C}$, $\operatorname{Re}\lambda > abs(f)$).*

(ii) *Let* $f \in$ (P1)-*E*, $h \in L^1_{\mathrm{loc}}([0,\infty))$ *and* $\mathrm{abs}(|h|) < \infty$. *Suppose, in addition, that* $f \in C([0,\infty) : E)$. *Put*

$$(h * f)(t) := \int_0^t h(t-s)f(s)\,ds, \quad t \geqslant 0.$$

Then the mapping $t \mapsto (h * f)(t)$, $t \geqslant 0$ *is continuous*, $h * f \in$ (P1)-*E*, *and*

$$\widetilde{h * f}(\lambda) = \tilde{h}(\lambda)\tilde{f}(\lambda), \quad \lambda \in \mathbb{C}, \quad \operatorname{Re}\lambda > \max(abs(|h|), abs(f)).$$

We say that a function $h(\cdot)$ belongs to the class $LT - E$ if there exist an exponentially bounded function $f \in C([0,\infty) : E)$ and a real number $a > 0$ such that $h(\lambda) = (\mathcal{L}f)(\lambda)$, $\lambda > a$. For more details about vector-valued Laplace transform, we refer the reader to [69, 445] and [792].

Lebesgue spaces with variable exponents $L^{p(x)}(\Omega)$

Concerning the Lebesgue spaces with variable exponents, the research monograph [256] by L. Diening et al. is of invaluable importance; cf. also [448], we will use the same notion and notation as there.

Let $\emptyset \neq \Omega \subseteq \mathbb{R}^n$ be a nonempty Lebesgue measurable set and let $M(\Omega : X)$ be the collection of all measurable functions $f : \Omega \to X$; $M(\Omega) := M(\Omega : \mathbb{R})$. Further on, let $\mathcal{P}(\Omega)$ be the vector space of all Lebesgue measurable functions $p : \Omega \to [1,\infty]$. For any $p \in \mathcal{P}(\Omega)$ and $f \in M(\Omega : X)$, we define

$$\varphi_{p(x)}(t) := \begin{cases} t^{p(x)}, & t \geqslant 0, \quad 1 \leqslant p(x) < \infty, \\ 0, & 0 \leqslant t \leqslant 1, \quad p(x) = \infty, \\ \infty, & t > 1, \quad p(x) = \infty \end{cases}$$

and

$$\rho(f) := \int_\Omega \varphi_{p(x)}(\|f(x)\|)\,dx.$$

We define the Lebesgue space $L^{p(x)}(\Omega : X)$ with variable exponent by

$$L^{p(x)}(\Omega : X) := \Big\{f \in M(\Omega : X) : \lim_{\lambda \to 0+} \rho(\lambda f) = 0\Big\}.$$

Equivalently,

$$L^{p(x)}(\Omega : X) = \{f \in M(\Omega : X) : \text{ there exists } \lambda > 0 \text{ such that } \rho(\lambda f) < \infty\};$$

see, e. g., [256, p. 73]. For every $u \in L^{p(x)}(\Omega : X)$, we introduce the Luxemburg norm of $u(\cdot)$ by

$$\|u\|_{p(x)} := \|u\|_{L^{p(x)}(\Omega:X)} := \inf\{\lambda > 0 : \rho(f/\lambda) \leqslant 1\}.$$

Equipped with the above norm, $L^{p(x)}(\Omega : X)$ is a Banach space (see, e. g., [256, Theorem 3.2.7] for the scalar-valued case), coinciding with the usual Lebesgue space $L^p(\Omega : X)$ in the case that $p(x) = p \geqslant 1$ is a constant function. Further on, for any $p \in M(\Omega)$, we define

$$p^- := \operatorname{essinf}_{x\in\Omega} p(x) \quad \text{and} \quad p^+ := \operatorname{esssup}_{x\in\Omega} p(x).$$

Define also

$$D_+(\Omega) := \{p \in M(\Omega) : 1 \leqslant p^- \leqslant p(x) \leqslant p^+ < \infty \text{ for a. e. } x \in \Omega\}.$$

For $p \in D_+([0,1])$, the space $L^{p(x)}(\Omega : X)$ behaves nicely; if this is the case, then we have

$$L^{p(x)}(\Omega : X) = \{f \in M(\Omega : X); \text{ for all } \lambda > 0 \text{ we have } \rho(\lambda f) < \infty\}.$$

Furthermore, we know the following:
(i) (The Hölder inequality)
Let $p,\ q,\ r \in \mathcal{P}(\Omega)$ such that

$$\frac{1}{q(x)} = \frac{1}{p(x)} + \frac{1}{r(x)}, \qquad x \in \Omega.$$

Then, for every $u \in L^{p(x)}(\Omega : X)$ and $v \in L^{r(x)}(\Omega)$, we have $uv \in L^{q(x)}(\Omega : X)$ and

$$\|uv\|_{q(x)} \leqslant 2\|u\|_{p(x)}\|v\|_{r(x)}.$$

(ii) Let Ω be of a finite Lebesgue's measure and let $p,\ q \in \mathcal{P}(\Omega)$ such $q \leqslant p$ a. e. on Ω. Then $L^{p(x)}(\Omega : X)$ is continuously embedded in $L^{q(x)}(\Omega : X)$, with the constant of embedding less or equal to $2(1 + m(\Omega))$.
(iii) Let $f \in L^{p(x)}(\Omega : X)$, $g \in M(\Omega : X)$ and $0 \leqslant \|g\| \leqslant \|f\|$ a. e. on Ω. Then $g \in L^{p(x)}(\Omega : X)$ and $\|g\|_{p(x)} \leqslant \|f\|_{p(x)}$.
(iv) Suppose that $f \in L^{p(x)}(\Omega : X)$ and $A \in L(X, Y)$. Then $Af \in L^{p(x)}(\Omega : Y)$ and $\|Af\|_{L^{p(x)}(\Omega:Y)} \leqslant \|A\| \cdot \|f\|_{L^{p(x)}(\Omega:X)}$.

On L^p-spaces ($p > 0$)

Let $0 < p < 1$ and let Ω' be any Lebesgue measurable subset of $\mathbb{R}^n$ with positive Lebesgue measure. Then the space $L^p(\Omega' : Y)$ consists of all Lebesgue measurable functions $f : \Omega' \to Y$ such that $\int_{\Omega'} \|f(\mathbf{u})\|^p \, d\mathbf{u} < +\infty$. The metric $d(\cdot;\cdot)$ on $L^p(\Omega' : Y)$ is given by

$d(f,g) := \int_{\Omega'} \|f(\mathbf{u}) - g(\mathbf{u})\|^p \, d\mathbf{u}$ for all $f,\ g \in L^p(\Omega' : Y)$; equipped with this metric, $L^p(\Omega' : Y)$ is a complete quasi-normed metric space. If $\nu : \Omega' \to (0,\infty)$ is a Lebesgue measurable function, then we define the pseudometric space $L^p_\nu(\Omega' : Y)$ as in the case that $p \geqslant 1$, when the basic properties of $L^p_\nu(\Omega' : Y)$ are well known.

Before proceeding any further, we would like to emphasize that the theory of Lebesgue spaces $L^{p(x)}$ with variable exponent $0 < p(x) < 1$ has not still been constituted; because of that, we will work with the constant coefficients $p \in (0,1)$ in the sequel (cf. [447] for many results concerning the generalized almost periodic type functions in the Lebesgue spaces with variable exponent $L^{p(x)}$, where $p(x) \geqslant 1$). We will use the following lemma, which might be known in the existing literature.

Lemma 1.1.11. *Suppose that* $\emptyset \neq \Omega \subseteq \mathbb{R}^n$ *is a Lebesgue measurable set,* $p > 0$ *and* $f \in L^p(\Omega)$. *Then we have*

$$\int_{\Omega} |f(x)|^p \, dx = p \int_0^\infty y^{p-1} m(\{x \in \Omega : |f(x)| > y\}) \, dy. \tag{1.7}$$

Proof. The formula (1.7) is well known in the case that $p \geqslant 1$; see, e. g., [80, pp. 7–8]. If $0 < p < 1$, then we can use (1.7) with $p = 1$ and the elementary change of variables $z = y^{1/p}$:

$$\begin{aligned}\int_{\Omega} |f(x)|^p \, dx &= \int_0^\infty m(\{x \in \Omega : |f(x)|^p > y\}) \, dy \\ &= \int_0^\infty m(\{x \in \Omega : |f(x)| > y^{1/p}\}) \, dy \\ &= p \int_0^\infty z^{p-1} m(\{x \in \Omega : |f(x)| > z\}) \, dz.\end{aligned}$$

□

Remark 1.1.12. The formula (1.7) will play an important role in our further work. We would like to notice that this formula does not have a satisfactory analogue in the theory of the Lebesgue spaces with variable exponent (see, e. g., the introductory part of [256]).

For more details concerning L^p-spaces for $0 < p < 1$, we refer the reader to the lectures of K. Conrad [209] and M. Rosenzweig [670]; cf. also [584, Chapter 13, Appendix A].

Part I: **Generalized almost periodic sequences and generalized almost automorphic sequences in $\mathbb{Z}^n$**

In this part, we analyze various classes of generalized multidimensional almost periodic type sequences and generalized multidimensional almost automorphic type sequences in complex Banach spaces. We also provide some applications to the abstract Volterra difference equations in Banach spaces depending on one (several) variables. As already mentioned in the preface, discrete fractional calculus is a very popular field of theoretical and applied mathematics, which is incredibly important in modeling various phenomena appearing in many fields. We consider here the abstract fractional difference equations in Banach spaces with Riemann–Liouville derivatives, Caputo derivatives and Weyl derivatives, as well.

We assume henceforth that $(X, \| \cdot \|)$, $(Y, \| \cdot \|_Y)$ and $(Z, \| \cdot \|_Z)$ are complex Banach spaces, $n \in \mathbb{N}$ and $\mathcal{B}$ is a certain collection of subsets of X, which satisfies that for each $x \in X$ there exists $B \in \mathcal{B}$ such that $x \in B$. In this part, we consider the sequences of the form $F : I \times X \to Y$, where $\emptyset \neq I \subseteq \mathbb{Z}^n$.

https://doi.org/10.1515/9783111689746-003

2 Multidimensional almost periodic type sequences, multidimensional almost automorphic type sequences and applications

In this chapter, we analyze various classes of multidimensional ρ-almost periodic type sequences, multidimensional almost automorphic type sequences and provide certain applications to the abstract difference equations with integer-order derivatives, the abstract Volterra difference equations and the abstract fractional difference equations.

2.1 Generalized ρ-almost periodic sequences and applications

In our joint study with W.-S. Du and D. Velinov [267], we have recently introduced and analyzed the classes of (equi-)Weyl-p-almost periodic sequences, Doss-p-almost periodic sequences and Besicovitch-p-almost periodic sequences with a general exponent $p \geqslant 1$, providing also certain applications to the abstract impulsive Volterra integrodifferential inclusions. Concerning some applications of generalized almost periodic type sequences in the qualitative analysis of solutions for various classes of impulsive Volterra integrodifferential equations, Volterra difference equations and ordinary differential equations, the reader may consult the research monographs [81] by D. Bainov, P. Simeonov, [683] by A. M. Samoilenko, N. A. Perestyuk, [723] by G. T. Stamov and the doctoral dissertation [760] by M. Veselý. The existence results for a class of impulsive semilinear functional differential inclusions have recently been analyzed by Y. Luo and W. Wang in [558].

The structural results concerning (multidimensional) c-almost periodic functions, where $c \in \mathbb{C}$ and $|c| = 1$, can be found in the research article [430] by M. T. Khalladi et al. and the research monograph [447]. The strong motivational factor for the genesis of paper [267], from which we have taken the material of this section, presents the fact that the class of c-almost periodic sequences has not been explored in the existing literature so far. Further on, in a joint research article [289] with M. Fečkan, M. T. Khalladi and A. Rahmani, the author has recently introduced and analyzed the class of multidimensional ρ-almost periodic type functions of the form $F : I \times X \to Y$, where $\emptyset \neq I \subseteq \mathbb{R}^n$, X and Y are complex Banach spaces and ρ is a general binary relation on Y. In that paper, we have assumed very mild conditions on the domain $I \times X$; for example, we have not assumed that the interior of I is nonempty or that the set I is unbounded in direction of some coordinate axes. Here, we specifically analyze the situation in which the following conditions hold true:

$$\emptyset \neq I' \subseteq \mathbb{Z}^n, \quad \emptyset \neq I \subseteq \mathbb{Z}^n \quad \text{and} \quad I + I' \subseteq I. \tag{30}$$

In particular, we introduce and analyze several new classes of Stepanov, Weyl, Besicovitch and Doss ρ-almost periodic type sequences. Following our research studies carried

https://doi.org/10.1515/9783111689746-004

out in [177, 448, 451, 454], we can further analyze many other classes of multidimensional ρ-almost periodic type sequences of the above form.

The organization of section can be briefly summarized as follows. Section 2.1.1 recalls the basic definitions and results about Weyl ρ-almost periodic type functions, Doss ρ-almost periodic type functions and Besicovitch almost periodic type functions in $\mathbb{R}^n$. In Section 2.1.2, we remind the readers of the already known notions of (metrical) ρ-almost periodicity for the sequences of the form $F : I \times X \to Y$; the term "sequence" used here is a little bit inappropriate in the case that X is not a trivial space. The first original contribution is Theorem 2.1.6, where we analyze the existence of a Bohr I'-almost periodic type function $\tilde{F} : \mathbb{R}^n \to Y$ such that $\tilde{F}(\mathbf{t}) = F(\mathbf{t})$ for all $\mathbf{t} \in I$, where $F : I \to Y$ is a given Bohr I'-almost periodic type sequence; cf. also Proposition 2.1.7 and Theorem 2.1.8. An analogue of Theorem 2.1.6 for T-almost periodic sequences, where $T \in L(Y)$ is a linear isomorphism, is clarified in Theorem 2.1.9; cf. also Corollary 2.1.10. The main structural results about the introduced classes of generalized ρ-almost periodic sequences are given in Proposition 2.1.12, Proposition 2.1.16, Theorem 2.1.18, Propositions 2.1.22–2.1.24 and Theorem 2.1.28; cf. also Corollary 2.1.20 and Corollary 2.1.29. Concerning the above mentioned results, we will only note here that it is very difficult to state any satisfactory result concerning the discretization of (equi)-Weyl-p-almost periodic type functions, Doss-p-almost periodic type functions and Besicovitch-p-almost periodic type functions. Several new applications to the abstract Volterra difference equations and the abstract impulsive Volterra integrodifferential equations are given here. As already mentioned, the theory of difference equations in several variables is still very unexplored (concerning this subject, we can also recommend the book chapter [736] by L. Székelyhidi and references cited therein). The research article [267] is probably the first research article, which investigates the almost periodic solutions of difference equations depending on several variables. In this section, we also propose many useful comments, illustrative examples and open problems about the notion under our consideration.

2.1.1 Weyl ρ-almost periodic type functions, Doss ρ-almost periodic type functions and Besicovitch almost periodic type functions in $\mathbb{R}^n$

In this subsection, we will always assume that $\rho \subseteq Y \times Y$ is a function. If $\emptyset \neq \Lambda \subseteq \mathbb{R}^n$, then $p(\Lambda)$ denotes the collection of all Lebesgue measurable functions from Λ into $[1, \infty]$. Let us assume that the following condition holds:

(WM1) Let $\emptyset \neq \Lambda \subseteq \mathbb{R}^n$ and $\emptyset \neq \Lambda' \subseteq \mathbb{R}^n$. Let $\emptyset \neq \Omega \subseteq \mathbb{R}^n$ be a Lebesgue measurable set such that $m(\Omega) > 0$, $p \in \mathcal{P}(\Lambda)$, $\Lambda' + \Lambda + l\Omega \subseteq \Lambda$, $\Lambda + l\Omega \subseteq \Lambda$ for all $l > 0$, $\phi : [0, \infty) \to [0, \infty)$ and $\mathbb{F} : (0, \infty) \times \Lambda \to (0, \infty)$.

We need the following notion [448].

Definition 2.1.1. (i) By $e - W^{(p(\mathbf{u}),\phi,\mathbb{F},\rho)}_{\Omega,\Lambda',\mathcal{B}}(\Lambda \times X : Y)$, we denote the set consisting of all functions $F : \Lambda \times X \to Y$ such that, for every $\varepsilon > 0$ and $B \in \mathcal{B}$, there exist two finite real numbers $l > 0$ and $L > 0$ such that for each $\mathbf{t}_0 \in \Lambda'$ there exists $\tau \in B(\mathbf{t}_0, L) \cap \Lambda'$ such that, for every $x \in B$, the mapping $\mathbf{u} \mapsto \rho(F(\mathbf{u}; x))$, $\mathbf{u} \in \mathbf{t} + l\Omega$ is well-defined and

$$\sup_{x\in B}\sup_{\mathbf{t}\in\Lambda} \mathbb{F}(l,\mathbf{t})\phi(\|F(\tau+\mathbf{u};x) - \rho(F(\mathbf{u};x))\|_Y)_{L^{p(\mathbf{u})}(\mathbf{t}+l\Omega)} < \varepsilon.$$

(ii) By $W^{(p(\mathbf{u}),\phi,\mathbb{F},\rho)}_{\Omega,\Lambda',\mathcal{B}}(\Lambda\times X : Y)$, we denote the set consisting of all functions $F : \Lambda\times X \to Y$ such that, for every $\varepsilon > 0$ and $B \in \mathcal{B}$, there exists a finite real number $L > 0$ such that for each $\mathbf{t}_0 \in \Lambda'$ there exists $\tau \in B(\mathbf{t}_0, L) \cap \Lambda'$ such that, for every $x \in B$, the mapping $\mathbf{u} \mapsto \rho(F(\mathbf{u}; x))$, $\mathbf{u} \in \mathbf{t} + l\Omega$ is well-defined, and

$$\limsup_{l\to+\infty}\sup_{x\in B}\sup_{\mathbf{t}\in\Lambda} \mathbb{F}(l,\mathbf{t})\phi(\|F(\tau+\mathbf{u};x) - \rho(F(\mathbf{u};x))\|_Y)_{L^{p(\mathbf{u})}(\mathbf{t}+l\Omega)} < \varepsilon.$$

Suppose now that Λ is a general nonempty subset of $\mathbb{R}^n$ as well as that $p \in \mathcal{P}(\Lambda)$ and the following condition holds:

$$\phi : [0,\infty) \to [0,\infty) \text{ is measurable,} \quad \mathrm{F} : (0,\infty) \to (0,\infty) \text{ and } p \in \mathcal{P}(\Lambda).$$

Set $\Lambda'' := \{\tau \in \mathbb{R}^n : \tau + \Lambda \subseteq \Lambda\}$ and assume $\emptyset \neq \Lambda' \subseteq \Lambda''$.

We also need the following notion [448].

Definition 2.1.2. Suppose that the function $F : \Lambda \times X \to Y$ satisfies that $\phi(\|F(\cdot + \tau; x) - \rho(F(\cdot; x))\|) \in L^{p(\cdot)}(\Lambda'_t)$ for all $t > 0$, $x \in X$ and $\tau \in \Lambda'$. Then we say that the function $F(\cdot;\cdot)$ is Doss-$(p, \phi, \mathrm{F}, \mathcal{B}, \Lambda', \rho)$-almost periodic if, for every $B \in \mathcal{B}$ and $\varepsilon > 0$, there exists $l > 0$ such that for each $\mathbf{t}_0 \in \Lambda'$ there exists a point $\tau \in B(\mathbf{t}_0, l) \cap \Lambda'$ such that, for every $t > 0$, $x \in B$ and $\cdot \in \Lambda_t$, we have

$$\limsup_{t\to+\infty} \mathrm{F}(t)\sup_{x\in B}[\phi(\|F(\cdot+\tau;x) - \rho(F(\cdot;x))\|_Y)]_{L^{p(\cdot)}(\Lambda'_t)} < \varepsilon.$$

Suppose, finally, that Λ is a general nonempty subset of $\mathbb{R}^n$ as well as that $p \in \mathcal{P}(\Lambda)$, the function $\phi : [0,\infty) \to [0,\infty)$ is Lebesgue measurable and $\mathrm{F} : (0,\infty) \to (0,\infty)$. Let $\emptyset \neq \Lambda' \subseteq \Lambda''$. Recall, a trigonometric polynomial $P : \Lambda\times X \to Y$ is any linear combination of functions like $(\mathbf{t}; x) \mapsto e^{i\langle\lambda,\mathbf{t}\rangle}c(x)$, where $c : X \to Y$ is a continuous function.

The following notion has recently been introduced in [455, Definition 2.1].

Definition 2.1.3. Suppose that $F : \Lambda\times X \to Y$, $\phi : [0,\infty)\to[0,\infty)$ and $\mathrm{F} : (0,\infty)\to(0,\infty)$. Then we say that the function $F(\cdot;\cdot)$ belongs to the class $e-(\mathcal{B},\phi,\mathrm{F})-B^{p(\cdot)}(\Lambda\times X : Y)$ if for each set $B \in \mathcal{B}$ there exists a sequence $(P_k(\cdot;\cdot))$ of trigonometric polynomials such that

$$\lim_{k\to+\infty}\limsup_{t\to+\infty}\mathrm{F}(t)\sup_{x\in B}[\phi(\|F(\mathbf{t};x) - P_k(\mathbf{t};x)\|_Y)]_{L^{p(\mathbf{t})}(\Lambda'_t)} = 0,$$

where we assume that the term in braces belongs to the space $L^{p(\mathbf{t})}(\Lambda'_t)$ for any compact set K.

2.1.2 Bohr $(\mathcal{B}, I', \rho)$-almost periodic type sequences

We start our work with the observation that we have recently introduced, in [289, Definitions 2.1, 2.22, 2.25], the notions of Bohr $(\mathcal{B}, I', \rho)$-almost periodicity, $(\mathcal{B}, I', \rho)$-uniform recurrence, $\mathbb{D}$-asymptotical Bohr $(\mathcal{B}, I', \rho)$-almost periodicity of type 1 and $\mathbb{D}$-asymptotical $(\mathcal{B}, I', \rho)$-uniform recurrence of type 1 for a function of the form $F : I \times X \to Y$. For the sake of completeness, we will only recall the following notion.

Definition 2.1.4. Suppose that $\emptyset \neq I' \subseteq \mathbb{R}^n$, $\emptyset \neq I \subseteq \mathbb{R}^n$, $F : I \times X \to Y$ is a continuous function, ρ is a binary relation on Y and $I + I' \subseteq I$. Then we say that:

(i) $F(\cdot;\cdot)$ is Bohr $(\mathcal{B}, I', \rho)$-almost periodic if for every $B \in \mathcal{B}$ and $\varepsilon > 0$ there exists $l > 0$ such that for each $\mathbf{t}_0 \in I'$ there exists $\tau \in B(\mathbf{t}_0, l) \cap I'$ such that, for every $\mathbf{t} \in I$ and $x \in B$, there exists an element $y_{\mathbf{t};x} \in \rho(F(\mathbf{t}; x))$ such that
$$\|F(\mathbf{t} + \tau; x) - y_{\mathbf{t};x}\|_Y \leqslant \varepsilon.$$

(ii) $F(\cdot;\cdot)$ is $(\mathcal{B}, I', \rho)$-uniformly recurrent if for every $B \in \mathcal{B}$ there exists a sequence (τ_k) in I' such that $\lim_{k\to+\infty} |\tau_k| = +\infty$ and that, for every $\mathbf{t} \in I$ and $x \in B$, there exists an element $y_{\mathbf{t};x} \in \rho(F(\mathbf{t}; x))$ such that
$$\lim_{k\to+\infty} \sup_{\mathbf{t}\in I; x\in B} \|F(\mathbf{t} + \tau_k; x) - y_{\mathbf{t};x}\|_Y = 0.$$

If (30) holds, then $F : I \times X \to Y$ is a continuous function if and only if for each $\mathbf{t} \in I$, $x \in B$ and $\varepsilon > 0$ there exists $\delta > 0$ such that, for every $y \in X$ with $\|x - y\| < \delta$, we have $\|F(\mathbf{t}; x) - F(\mathbf{t}; y)\|_Y < \varepsilon$; in particular, any function $F : I \to Y$ is already continuous. The notion introduced in [289, Definitions 3.1, 3.4], with $\omega \in \mathbb{Z}^n \setminus \{0\}$, $\omega_j \in \mathbb{Z} \setminus \{0\}$ for $1 \leqslant j \leqslant n$ and some extra assumptions being satisfied, can serve us to introduce the notion of (ω, ρ)-periodicity and the notion of $(\omega_j, \rho_j)_{j\in\mathbb{N}_n}$-periodicity of a sequence $F : I \to X$.

General abbreviation of notation

As in all recent research studies of multi-dimensional almost periodic (automorphic) type functions, our general agreements will be the following ones: In the analysis of various classes of generalized almost periodic (automorphic) functions sequences of the form $F : I \to Y$ ($F : \Lambda \to Y$), we omit the term "$\mathcal{B}$" from the notation. We omit the term "I'" from the notation if $I' = I''$ ("Λ'" from the notation if $\Lambda' = \Lambda''$) and the term "ρ" from the notation if $\rho = \mathrm{I}$; for example, a Bohr $\mathcal{B}$-almost periodic sequence is nothing else but a Bohr $(\mathcal{B}, I', \rho)$-almost periodic sequence with $I' = I$ and $\rho = \mathrm{I}$. We also write "c" in place of "$c\mathrm{I}$" if $c \in \mathbb{C}$.

Before proceeding any further, we would like to observe that almost all structural results from the first three sections of [289] hold in the discrete framework. All exceptions are listed below:

(A1) It is clear that the assertions of [289, Corollary 2.4, Theorems 2.14, 2.16, Propositions 3.7, 2.24] cannot be directly formulated in the discrete framework.
(A2) We should further examine the question whether the statements of [289, Propositions 2.18, 2.20] can be formulated with $I = \mathbb{Z}$ or $I = \mathbb{N}_0$ and $I' = \mathbb{N}$.
(A3) We should further examine the question whether the statements of [289, Theorem 2.28, Corollary 2.29] can be formulated with the condition (AP-E) replaced with the condition:
(AP-ED) For every $\mathbf{t}' \in \mathbb{Z}^n$, there exists a finite real number $M > 0$ such that $\mathbf{t}' + I_M \subseteq I$.

Remark 2.1.5. Before considering these questions, let us observe that the notion of strong $\mathcal{B}$-almost periodicity, introduced in [447, Definition 6.1.24], is meaningful in the discrete setting and that the statement of [447, Proposition 6.1.25] holds in the discrete framework. Concerning the notion of Bohr $(\mathcal{B}, c)$-almost periodicity and the notion of $(\mathcal{B}, c)$-uniform recurrence introduced in [447, Definition 7.1.6], we would like to note that the statements of [447, Proposition 7.1.9, Corollary 7.1.11, Propositions 7.1.13–7.1.16, Theorem 7.1.18] hold in the discrete framework. Keeping this in mind, we can simply prove that the statements of [430, Propositions 2.2, 2.6–2.9, 2.11, 2.17; Corollary 2.10; Theorem 2.13] continue to hold for c-almost periodic sequences (c-uniformly recurrent sequences); in particular, if a sequence $(x_k)_{k\in\mathbb{N}}$ is c-uniformly recurrent for some $c \in \mathbb{C}$, then we must have $|c| = 1$. The assertions of [447, Theorems 6.1.40, 7.1.25] can be directly formulated in the discrete framework, as well.

Concerning the question (A2), we would like to note that the statements of [289, Propositions 2.18, 2.20] continue to hold if $I = \mathbb{Z}$ or $I = \mathbb{N}_0$ and $I' = \mathbb{N}$. This follows from the same argumentation as in the continuous case. Concerning the question (A3), the situation is much more complicated. In connection with this problem, we will first state and prove the following analogue of [447, Theorem 6.1.37] in the discrete framework.

Theorem 2.1.6. *Suppose that $I' \subseteq I \subseteq \mathbb{Z}^n$, $I + I' \subseteq I$, the set I' is unbounded, $S \subseteq \mathbb{Z}^n$ is finite,* (AP-ED) *holds and $\Omega_S := [(I' \cup (-I')) + (I' \cup (-I'))] \cup S$. Then $F : I \to Y$ is a Bohr I'-almost periodic sequence, respectively, an I'-uniformly recurrent sequence if and only if there exists a Bohr I'-almost periodic, respectively, an I'-uniformly, recurrent function $\tilde{F} : \mathbb{R}^n \to Y$ such that $\tilde{F}(\mathbf{t}) = F(\mathbf{t})$ for all $\mathbf{t} \in I$. If this is the case, then $\tilde{F}(\cdot)$ is Bohr Ω_S-almost periodic, respectively, Ω_S-uniformly recurrent; furthermore, $R(\tilde{F}(\cdot)) \subseteq CH(\overline{R(F)})$ and the assumption that $F(\cdot)$ is bounded implies that $\tilde{F}(\cdot)$ is uniformly continuous.*

Proof. Suppose first that $F : I \to Y$ is a Bohr I'-almost periodic sequence, respectively, an I'-uniformly recurrent sequence. Repeating to the letter the argumentation given in the proof of the above mentioned result, we get that there exists a Bohr I'-almost periodic, respectively, an I'-uniformly recurrent, sequence $\tilde{F}_{\mathbb{Z}} : \mathbb{Z}^n \to Y$ such that $\tilde{F}_{\mathbb{Z}}(\mathbf{t}) = F(\mathbf{t})$ for all $\mathbf{t} \in I$. In order to extend the function $\tilde{F}_{\mathbb{Z}} : \mathbb{Z}^n \to Y$ to a Bohr I'-almost periodic function, respectively, an I'-uniformly recurrent function, $\tilde{F} : \mathbb{R}^n \to Y$

such that $\tilde{F}(\mathbf{t}) = \tilde{F}_{\mathbb{Z}}(\mathbf{t})$ for all $\mathbf{t} \in \mathbb{Z}^n$, we can argue as in the proof of [252, Theorem 2] with appropriate technical modifications. For the sake of convenience, we will present all relevant details in the case that $n = 2$, extending the proof of [252, Theorem 2] with $c = 1$ and $\delta = 1/2$ to the two-dimensional setting. If $t = (t_1, t_2) \in \mathbb{R}^2$ is given, then there exist the unique numbers $k \in \mathbb{Z}$ and $m \in \mathbb{Z}$ such that $t_1 \in [k, k+1)$ and $t_2 \in [m, m+1)$. We first define $\tilde{F}(t_1, m) := \tilde{F}_{\mathbb{Z}}(k, m)$ if $t_1 \in [k, k+(1/2))$ and $\tilde{F}(t_1, m) := 2(\tilde{F}_{\mathbb{Z}}(k+1, m) - \tilde{F}_{\mathbb{Z}}(k, m))(t_1 - k - (1/2)) + \tilde{F}_{\mathbb{Z}}(k, m)$ if $t_1 \in [k+(1/2), k+1)$; we similarly define $\tilde{F}(t_1, m+1) := \tilde{F}_{\mathbb{Z}}(k, m+1)$ if $t_1 \in [k, k+(1/2))$ and $\tilde{F}(t_1, m+1) := 2(\tilde{F}_{\mathbb{Z}}(k+1, m+1) - \tilde{F}_{\mathbb{Z}}(k, m+1))(t_1 - k - (1/2)) + \tilde{F}_{\mathbb{Z}}(k, m+1)$ if $t_1 \in [k + (1/2), k + 1)$. After that, we define $\tilde{F}(t_1, t_2) := \tilde{F}(t_1, m)$ if $t_2 \in [m, m + (1/2))$ and $\tilde{F}(t_1, t_2) := 2(\tilde{F}(t_1, m+1) - \tilde{F}(t_1, m))(t_2 - m - (1/2)) + \tilde{F}(t_1, m)$ if $t_2 \in [m + (1/2), m + 1)$. It can be simply verified that the function $\tilde{F}(\cdot)$ is continuous as well as that $R(\tilde{F}(\cdot)) \subseteq CH(\overline{R(F)})$ and the function $\tilde{F}(\cdot)$ is uniformly continuous if $F(\cdot)$ is bounded. Further on, let us assume that a point $\mathbf{t}_0 \in I'$ and a number $\varepsilon > 0$ are given; then there exist $l > 0$ and $\tau = (\tau_1, \tau_2) \in I' \cap B(\mathbf{t}_0, l)$ such that $\|\tilde{F}_{\mathbb{Z}}(\mathbf{s} + \tau) - \tilde{F}_{\mathbb{Z}}(\mathbf{s})\| < \varepsilon/9$, $\mathbf{s} \in \mathbb{Z}^2$. Now we will prove that $\|\tilde{F}(\mathbf{t} + \tau) - \tilde{F}(\mathbf{t})\| < \varepsilon$, $\mathbf{t} \in \mathbb{R}^2$. Suppose that $k \in \mathbb{Z}$, $m \in \mathbb{Z}$, $\mathbf{t} = (t_1, t_2)$, $t_1 \in [k, k + 1)$ and $t_2 \in [m, m + 1)$. There exist four possibilities:

(i) $t_1 \in [k, k + (1/2))$ and $t_2 \in [m, m + (1/2))$;
(ii) $t_1 \in [k, k + (1/2))$ and $t_2 \in [m + (1/2), m + 1)$;
(iii) $t_1 \in [k + (1/2), k + 1)$ and $t_2 \in [m, m + (1/2))$;
(iv) $t_1 \in [k + (1/2), k + 1)$ and $t_2 \in [m + (1/2), m + 1)$.

If (i) holds, then $t_1 + \tau_1 \in [k + \tau_1, k + \tau_1 + (1/2))$ and we have

$$\|\tilde{F}(\mathbf{t} + \tau) - \tilde{F}(\mathbf{t})\| = \|\tilde{F}_{\mathbb{Z}}(t_1 + \tau_1, m + \tau_2) - \tilde{F}_{\mathbb{Z}}(t_1, m)\| \leqslant \varepsilon/3,$$

where the last estimate follows from the estimate $\|\tilde{F}_{\mathbb{Z}}(\mathbf{s} + \tau) - \tilde{F}_{\mathbb{Z}}(\mathbf{s})\| < \varepsilon/9$, $\mathbf{s} \in \mathbb{Z}^2$ and the argumentation contained in the proof of [252, Theorem 2]. If (ii) holds, then we have $t_2 + \tau_2 \in [m + \tau_2 + (1/2), m + \tau_2 + 1)$ and, therefore,

$$\begin{aligned}
&\|\tilde{F}(\mathbf{t} + \tau) - \tilde{F}(\mathbf{t})\| \\
&\quad = \|2[\tilde{F}_{\mathbb{Z}}(t_1 + \tau_1, m + 1 + \tau_2) - \tilde{F}_{\mathbb{Z}}(t_1 + \tau_1, m + \tau_2)] \cdot (t_2 - m - (1/2)) + \tilde{F}_{\mathbb{Z}}(t_1 + \tau_1, m + \tau_2) \\
&\qquad - 2[\tilde{F}_{\mathbb{Z}}(t_1, m + 1) - \tilde{F}_{\mathbb{Z}}(t_1, m)] \cdot (t_2 - m - (1/2)) - \tilde{F}_{\mathbb{Z}}(t_1, m)\| \\
&\quad \leqslant \|\tilde{F}_{\mathbb{Z}}(t_1 + \tau_1, m + 1 + \tau_2) - \tilde{F}_{\mathbb{Z}}(t_1, m + 1)\| + \|\tilde{F}_{\mathbb{Z}}(t_1 + \tau_1, m + \tau_2) - \tilde{F}_{\mathbb{Z}}(t_1, m)\| \\
&\qquad + \|\tilde{F}_{\mathbb{Z}}(t_1 + \tau_1, m + \tau_2) - \tilde{F}_{\mathbb{Z}}(t_1, m)\| \leqslant 3 \cdot (\varepsilon/3) = \varepsilon.
\end{aligned}$$

The analysis of cases (iii) and (iv) is similar and, therefore, $\tilde{F}(\cdot)$ is Bohr I'-almost periodic, respectively, I'-uniformly recurrent; as in [447], this simply implies that $\tilde{F}(\cdot)$ is Bohr Ω_S-almost periodic, respectively, Ω_S-uniformly recurrent. Finally, it is clear that the existence of a Bohr I'-almost periodic, respectively, an I'-uniformly recurrent, function $\tilde{F} : \mathbb{R}^n \to Y$ such that $\tilde{F}(\mathbf{t}) = F(\mathbf{t})$ for all $\mathbf{t} \in I$ implies that $F(\cdot)$ is Bohr I'-almost periodic, respectively, I'-uniformly recurrent. □

There exist many other ways to extend the function $\tilde{F}_{\mathbb{Z}}(\cdot)$ to a function $\tilde{F}(\cdot)$ defined on the whole Euclidean plane, obeying all required properties from the formulation of Theorem 2.1.6 (we only need to change the values of parameters c and δ from the proof of [252, Theorem 2]). This readily implies that any nonempty subset I of $\mathbb{Z}^n$ cannot be admissible with respect to the almost periodic extensions (cf. [447, Definition 6.1.39] for the notion).

Now we will focus our attention to the case in which $I' = I = \mathbb{Z}^n$. We need the following result of independent interest (cf. also [112, pp. 54–59] for several related results given in the one-dimensional setting).

Proposition 2.1.7. *Suppose that $F : \mathbb{R}^n \times X \to Y$ is a $\mathcal{B}$-almost periodic function, where $\mathcal{B}$ is any collection of compact subsets of X. Then the function $F(\cdot;\cdot)$ is Bohr $(\mathcal{B}, \mathbb{Z}^n)$-almost periodic.*

Proof. The statement of proposition is trivial if $Y = \{0\}$; otherwise, there exists an element $y \in Y$ such that $\|y\|_Y = 1$. Let $\varepsilon > 0$ and $B \in \mathcal{B}$ be fixed. Then [447, Proposition 6.1.22] implies that there exists $\delta \in (0, \varepsilon)$ such that the assumption $|\mathbf{t} - \mathbf{t}'| + \|x - x'\| \leqslant \delta$ for some $\mathbf{t}$, $\mathbf{t}' \in \mathbb{R}^n$ and x, $x' \in X$ implies $\|F(\mathbf{t}; x) - F(\mathbf{t}', x')\|_Y \leqslant \varepsilon$. Furthermore, [447, Proposition 6.1.19] implies that there exists a relatively dense set of points $\tau = (\tau_1, \dots, \tau_n)$ in $\mathbb{R}^n$ such that $\|F(\mathbf{t} + \tau; x) - F(\mathbf{t}; x)\|_Y \leqslant \varepsilon$ for all $\mathbf{t} \in \mathbb{R}^n$ and $x \in B$, as well as that $\|G_j(\mathbf{t} + \tau; x) - G_j(\mathbf{t}; x)\|_Y \leqslant \varepsilon$ for all $\mathbf{t} \in \mathbb{R}^n$, $j \in \mathbb{N}_n$ and $x \in B$, where the Bohr $\mathcal{B}$-almost periodic function $G_j : \mathbb{R}^n \times X \to Y$ is defined as the usual periodic extension of the function by $G_{j;0}(\mathbf{t}; x) := (1 - |1 - t_j|)y$, $\mathbf{t} = (t_1, \dots, t_j, \dots, t_n) \in [0, 2]^n$, $x \in X$ to the space $\mathbb{R}^n \times X$. As in the one-dimensional setting, this yields that there exist two vectors $p \in \mathbb{Z}^n$ and $w = (w_1, \dots, w_n) \in B(0, \delta)$ such that $\tau = 2p + \omega$. Therefore, we have

$$\begin{aligned}
&\|F(\mathbf{t} + 2p; x) - F(\mathbf{t}; x)\|_Y \\
&\quad \leqslant \|F(\mathbf{t} + 2p; x) - F(\mathbf{t} + 2p + w; x)\|_Y + \|F(\mathbf{t} + 2p + w; x) - F(\mathbf{t}; x)\|_Y \\
&\quad \leqslant \varepsilon + \delta < 2\varepsilon, \quad \mathbf{t} \in \mathbb{R}^n,\ x \in B.
\end{aligned}$$

This completes the proof because the set consisting of all points $2p \in \mathbb{Z}^n$ with the above properties is relatively dense in $\mathbb{Z}^n$, which can be trivially shown. □

Keeping in mind Theorem 2.1.6 and Proposition 2.1.7, we can extend the statement of [252, Theorem 2] to the higher-dimensional setting.

Theorem 2.1.8. *Suppose that $F : \mathbb{Z}^n \to Y$. Then $F(\cdot)$ is a Bohr almost periodic sequence if and only if there exists a Bohr almost periodic function $\tilde{F} : \mathbb{R}^n \to Y$ such that $F(\mathbf{t}) = \tilde{F}(\mathbf{t})$ for all $\mathbf{t} \in \mathbb{Z}^n$.*

As a corollary of Theorem 2.1.8, we have that the set of all Bohr almost periodic sequences $F : \mathbb{Z}^n \to Y$ is a linear vector space with the usual operations.

Further on, if $S \subseteq \mathbb{Z}^n$ is finite, $c \in \mathbb{C} \setminus \{1\}$, $|c| = 1$ and $\arg(c)/\pi \in \mathbb{Q}$, then the set of all (Bohr) c-almost periodic sequences $F : \mathbb{Z}^n \to Y$ is not a linear vector space with the usual operations; we define the set Ω_S as it has been done on [447, p. 467]. Arguing as in the proof of Theorem 2.1.6, we can similarly deduce the following analogues of [289, Theorem 2.28] and [447, Theorem 7.1.26] in the discrete framework.

Theorem 2.1.9. *Suppose that $I' \subseteq I \subseteq \mathbb{Z}^n$, $I + I' \subseteq I$, the set I' is unbounded, $\rho = T \in L(Y)$ is a linear isomorphism, $S \subseteq \mathbb{Z}^n$ is finite and* (AP-ED) *holds. Then $F : I \to Y$ is a Bohr (I', T)-almost periodic function, respectively, an (I', T)-uniformly recurrent function if and only if there exists a Bohr (I', T)-almost periodic function, respectively, an (I', T)-uniformly recurrent function, $\tilde{F} : \mathbb{R}^n \to Y$ such that $\tilde{F}(\mathbf{t}) = F(\mathbf{t})$ for all $\mathbf{t} \in I$. Furthermore, $R(\tilde{F}(\cdot)) \subseteq CH(T^{-1}\overline{R(F)})$, the boundedness of $F(\cdot)$ implies that $\tilde{F}(\cdot)$ is uniformly continuous and the assumptions* $\arg(c)/\pi \in \mathbb{Q}$ *and* $\rho = c\mathrm{I}$ *imply that $\tilde{F}(\cdot)$ is Bohr (Ω_S, T)-almost periodic, respectively, (Ω_S, T)-uniformly recurrent.*

As an immediate consequence of Theorem 2.1.9, we have the following.

Corollary 2.1.10. *Suppose that $c \in \mathbb{C}$, $|c| = 1$ and $F : \mathbb{Z}^n \to Y$ is a c-almost periodic sequence. Then there exists a Bohr $(\mathbb{Z}^n, c)$-almost periodic function $\tilde{F} : \mathbb{R}^n \to Y$ such that $F(\mathbf{t}) = \tilde{F}(\mathbf{t})$ for all $\mathbf{t} \in \mathbb{Z}^n$.*

Further on, it is logical to ask the following questions with regard to Proposition 2.1.7 and Corollary 2.1.10.

PROBLEM. Let $c \in \mathbb{C} \setminus \{1\}$ and $|c| = 1$.

(QE1) Suppose that $F : \mathbb{R}^n \times X \to Y$ is a $(\mathcal{B}, c)$-almost periodic function, where $\mathcal{B}$ is any collection of compact subsets of X. Is it true that the function $F(\cdot;\cdot)$ is Bohr $(\mathcal{B}, \mathbb{Z}^n, c)$-almost periodic?

(QE2) Suppose that $F : \mathbb{R}^n \to Y$ is a c-almost periodic function. Is it true that $(F(\mathbf{t}))_{\mathbf{t}\in\mathbb{Z}^n}$ is a c-almost periodic sequence?

Suppose, finally, that $(\mathbf{v}_1, \ldots, \mathbf{v}_n)$ is a basis of $\mathbb{R}^n$,

$$I = \{\alpha_1\mathbf{v}_1 + \cdots + \alpha_n\mathbf{v}_n : \alpha_i \geqslant 0 \text{ for all } i \in \mathbb{N}_n\} \cap \mathbb{Z}^n$$

is a convex polyhedral in $\mathbb{R}^n \cap \mathbb{Z}^n$, and $I' \subseteq \mathbb{Z}^n$ is a proper convex subpolyhedral of I. We would like to stress that the set Ω_S from the formulation of Theorem 2.1.6 is relatively dense in $\mathbb{R}^n$, while the set Ω_S from the formulation of Theorem 2.1.9 is relatively dense in $\mathbb{R}^n$ provided that $\arg(c)/\pi \in \mathbb{Q}$. If this is the case, then the mean value $M(F)$, given by the expression (34) below, exists uniformly in $\mathbf{s} \in \mathbb{Z}^n$.

2.1.3 Generalized ρ-almost periodic type sequences

In this subsection, we analyze various classes of Stepanov, Weyl, Besicovitch and Doss ρ-almost periodic type sequences of the form $F : \Lambda \times X \to Y$, where $\emptyset \neq \Lambda \subseteq \mathbb{Z}^n$. We will always assume here that $\Lambda = \Lambda_1 \times \Lambda_2 \times \cdots \times \Lambda_n$, where for each $j \in \mathbb{N}_n$ there exists an integer $a \in \mathbb{Z}$ such that $\Lambda_j = \mathbb{Z}$, $\Lambda_j = \{\ldots, a-2, a-1, a\}$ or $\Lambda_j = \{a, a+1, a+2, \ldots\}$. Set $\Lambda'' := \{\mathbf{a} \in \mathbb{Z}^n : \mathbf{a} + \Lambda \subseteq \Lambda\}$. For every integer $l \in \mathbb{N}$, we introduce the set P_l consisting of all closed subrectangles of Λ, which contains exactly $(l+1)^n$ points with all integer coordinates. Suppose that a function $\mathbb{F}_l : \{l\} \times P_l \to [0, \infty)$ is given for each integer $l \in \mathbb{N}$.

The following notion generalizes the notion introduced by J. Andres and D. Pennequin in [56].

Definition 2.1.11. Suppose that $F : \Lambda \times X \to Y$ is a given sequence, $l \in \mathbb{N}$, $1 \leqslant p < +\infty$, $\Lambda' \subseteq \Lambda''$ and ρ is a binary relation on Y. Then we say that $F(\cdot;\cdot)$ is Stepanov-$(\mathcal{B}, \Lambda', \mathbb{F}_\cdot, p, \rho, l)$-almost periodic if, for every $\varepsilon > 0$ and $B \in \mathcal{B}$, there exists $L > 0$ such that, for every $\mathbf{t}_0 \in \Lambda'$, there exists a point $\tau \in \Lambda' \cap B(\mathbf{t}_0, L)$ which satisfies that, for every $J \in P_l$ and for every $j \in J$ and $x \in B$, there exists $z_{j,x} \in \rho(F(j;x))$ such that

$$\sup_{x \in B} \mathbb{F}_l(l, J) \left[\sum_{j \in J} \| F(j+\tau; x) - z_{j,x} \|^p \right]^{1/p} < \varepsilon. \tag{31}$$

In the classical concept, a sequence is almost periodic if and only if it is Stepanov almost periodic (see, e. g., [56, Consequence 3]). Furthermore, we can simply prove the following result.

Proposition 2.1.12. (i) *Suppose that $F : \Lambda \times X \to Y$ is a given sequence, $l \in \mathbb{N}$, $1 \leqslant p < +\infty$, $\Lambda' \subseteq \Lambda''$ and ρ is a binary relation on Y. If there exists a real number $c_l > 0$ such that $\mathbb{F}_l(l, J) \leqslant c_l l^{-n/p}$ for all $J \in P_l$ and $F(\cdot;\cdot)$ is Bohr $(\mathcal{B}, \Lambda', \rho)$-almost periodic, then $F(\cdot;\cdot)$ is Stepanov-$(\mathcal{B}, \Lambda', \mathbb{F}_\cdot, p, \rho, l)$-almost periodic.*

(ii) *Suppose that $F : \Lambda \times X \to Y$ is a given sequence, $l \in \mathbb{N}$, $1 \leqslant p < +\infty$, $\Lambda' \subseteq \Lambda''$ and ρ is a binary relation on Y. If there exists a real number $c_l > 0$ such that $\mathbb{F}_l(l, J) \geqslant c_l$ for all $J \in P_l$ and $F(\cdot;\cdot)$ is Stepanov-$(\mathcal{B}, \Lambda', \mathbb{F}_\cdot, p, \rho, l)$-almost periodic, then $F(\cdot;\cdot)$ is Bohr $(\mathcal{B}, \Lambda', \rho)$-almost periodic.*

Keeping in mind the above result, it becomes clear that the concept of Stepanov-$(\mathcal{B}, \Lambda', \mathbb{F}_\cdot, p, \rho, l)$-almost periodicity introduced above is not satisfactory enough; because of that, in the remainder of section, we will focus our attention mainly to the Weyl, Besicovitch and Doss classes of generalized ρ-almost periodic sequences.

The following notion generalizes the notion introduced in [106, 264, 267] and [393].

Definition 2.1.13. Suppose that $F : \Lambda \times X \to Y$ is a given sequence, $1 \leqslant p < +\infty$, $\Lambda' \subseteq \Lambda''$ and ρ is a binary relation on Y. Then we say that $F(\cdot;\cdot)$ is:

(i) equi-Weyl-$(\mathcal{B}, \Lambda', \mathbb{F}_\cdot, p, \rho)$-almost periodic if, for every $\varepsilon > 0$ and $B \in \mathcal{B}$, there exist $l \in \mathbb{N}$ and $L > 0$ such that, for every $\mathbf{t}_0 \in \Lambda'$, there exists a point $\tau \in \Lambda' \cap B(\mathbf{t}_0, L)$, which satisfies that, for every $J \in P_l$ and for every $j \in J$ and $x \in B$, there exists $z_{j,x} \in \rho(F(j;x))$ such that (31) holds.

(ii) Weyl-$(\mathcal{B}, \Lambda', \mathbb{F}_\cdot, p, \rho)$-almost periodic if, for every $\varepsilon > 0$ and $B \in \mathcal{B}$, there exists $L > 0$ such that, for every $\mathbf{t}_0 \in \Lambda'$, there exists a point $\tau \in \Lambda' \cap B(\mathbf{t}_0, L)$, which satisfies that there exists an integer $l_\tau \in \mathbb{N}$ such that, for every $l \geqslant l_\tau, J \in P_l, j \in J$ and $x \in B$, there exists $z_{j,x} \in \rho(F(j;x))$ such that (31) holds.

It is obvious that any equi-Weyl-$(\mathcal{B},\Lambda',\mathbb{F}_{\cdot},p,\rho)$-almost periodic sequence is Weyl-$(\mathcal{B},\Lambda',\mathbb{F}_{\cdot},p,\rho)$-almost periodic and any Weyl-$(\mathcal{B},\Lambda',\mathbb{F}_{\cdot},p,\rho)$-almost periodic sequence is Doss-$(\mathcal{B},\Lambda',\mathbb{F}_{\cdot},p,\rho)$-almost periodic, where the notion of Doss-$(\mathcal{B},\Lambda',\mathbb{F}_{\cdot},p,\rho)$-almost periodicity is introduced as follows.

Definition 2.1.14. Suppose that $F : \Lambda \times X \to Y$ is a given sequence, $1 \leqslant p < +\infty$, $\Lambda' \subseteq \Lambda''$ and ρ is a binary relation on Y. Then we say that $F(\cdot;\cdot)$ is Doss-$(\mathcal{B},\Lambda',\mathbb{F}_{\cdot},p,\rho)$-almost periodic if, for every $\varepsilon > 0$ and $B \in \mathcal{B}$, there exists $L > 0$ such that, for every $\mathbf{t}_0 \in \Lambda'$, there exists a point $\tau \in \Lambda' \cap B(\mathbf{t}_0, L)$, which satisfies that there exists an increasing sequence (l_k) of positive integers such that, for every $k \in \mathbb{N}, J \in P_{l_k}, j \in J$ and $x \in B$, there exists $z_{j,x} \in \rho(F(j;x))$ such that (31) holds with the number l replaced by the number l_k therein.

The situation in which the following condition holds:

(FV) There exists a function $\mathbb{F} : (0,\infty) \to (0,\infty)$ such that $\mathbb{F}(l,J) = \mathbb{F}(l)$ for all $l \in \mathbb{N}$ and $J \in P_l$

will be dominant in our analysis; in this case, an (equi-)Weyl-$(\mathcal{B},\Lambda',\mathbb{F}_{\cdot},p,\rho)$-almost periodic [Doss-$(\mathcal{B},\Lambda',\mathbb{F}_{\cdot},p,\rho)$-almost periodic] function is also called (equi-)Weyl-$(\mathcal{B},\Lambda',\mathbb{F},p,\rho)$-almost periodic [Doss-$(\mathcal{B},\Lambda',\mathbb{F},p,\rho)$-almost periodic]. The situation in which condition (FV) does not hold is far from being simple for consideration (cf. [447, Example 6.3.4 and pp. 425–428] for some applications made in the continuous framework).

Remark 2.1.15. We feel it is our duty to emphasize that the notion of a scalar-valued almost periodic sequence in the sense of Weyl approach, introduced by A. Bellow and V. Losert in [100], is completely misleading; in their approach, an almost periodic sequence $(x_k)_{k\in\mathbb{N}}$ in the sense of Weyl is nothing else but the usual asymptotically almost periodic sequence (by an asymptotically almost periodic sequence we mean a sum of an almost periodic sequence and a sequence vanishing at plus infinity; see [100, p. 316, Lemma 3.6]). It can be simply proved that any asymptotically almost periodic sequence $(x_k)_{k\in\mathbb{N}}$ is equi-Weyl-$(l^{-1/p},p)$-almost periodic, i. e., equi-Weyl-p-almost periodic in the usual sense ($p \geqslant 1$); on the other hand, the sequence $(x_k)_{k\in\mathbb{N}}$ given by $x_k := 1$ if there exists $l \in \mathbb{N}$ such that $k = l^3$, and $x_k := 0$, otherwise, is equi-Weyl-$(l^{-\sigma},p)$-almost periodic for any $\sigma > 1/2$ but not asymptotically almost periodic. Concerning asymptotically almost periodic sequences, we want also to note that N. V. Minh has analyzed, in [596], the asymptotic behavior of Volterra difference equations of the form

$$x(k+1) = Ax(k) + \sum_{l=0}^{k} B(k-l)x(l), \quad k \in \mathbb{N}_0$$

and

$$x(k+1) = Ax(k) + \sum_{l=0}^{k} B(k-l)x(l) + y(k), \quad k \in \mathbb{N}_0, \tag{32}$$

where A, B_k are linear continuous operators acting in X and $\sum_{k=0} \|B(k)\| < +\infty$. In [596, Theorem 4.14], the author has proved that the equation (32) has no asymptotically stable solution, respectively, asymptotically almost periodic solution, if $y \notin C_0(\mathbb{N}_0 : X)$, respectively, $y(\cdot)$ is not asymptotically almost periodic; cf. also [559] and [595].

We continue by stating the following result.

Proposition 2.1.16. *Suppose that $F : \Lambda \times X \to Y$ is a given sequence, $1 \leqslant p < +\infty$, $\Lambda' = \Lambda''$ and $\rho : Y \to Y$ is a continuous function. If* (FV) *holds and $F(\cdot;\cdot)$ is equi-Weyl-$(\mathcal{B}, \mathbb{F}, p, \rho)$-almost periodic, then for each bounded set $B \in \mathcal{B}$ the set $\{F(\mathbf{t}; x) : \mathbf{t} \in \Lambda; x \in B\}$ is bounded as well.*

Proof. Let $B \in \mathcal{B}$ be given and let $\varepsilon = 1$. Without loss of generality, we may assume that $\Lambda = \Lambda' = \mathbb{Z}^n$ or $\Lambda = \Lambda' = [0, \infty)^n$. Suppose first that $\Lambda = \Lambda' = \mathbb{Z}^n$. Then there exist $l \in \mathbb{N}$ and $L > 0$ such that, for every fixed $\mathbf{t} \in \mathbb{Z}^n$, there exists a point $\tau \in \mathbb{Z}^n \cap B(\mathbf{t}, L)$, which satisfies that, for every $J \in P_l$ and for every $j \in J$ and $x \in B$, (31) holds with $z_{j,x} = \rho(F(j; x))$. Then $\mathbf{t} - \tau \in B(0, L)$ and, by choosing an appropriate closed rectangle J in $\mathbb{R}^n$ with a vertex $\mathbf{t} - \tau$, we obtain that $\|F(\mathbf{t} - \tau + \tau; x) - \rho(F(\mathbf{t} - \tau; x))\|_Y \leqslant 1/\mathbb{F}(l)$ for all $x \in B$. This implies $F(\mathbf{t}; x) \in B(\rho(F(\mathbf{t} - \tau; x)), 1/\mathbb{F}(L))$, which gives the required conclusion since B is bounded and $\rho(\cdot)$ is continuous. Suppose now that $\Lambda = \Lambda' = [0, \infty)^n$. If $n = 1$, then the final conclusion follows similarly as in the proof of [267, Proposition 2], with the corresponding ε-period τ belonging to the segment $[t - 2L, t]$ for $t \geqslant 2L$. In the general case, any of the sequences $t \mapsto F(t, j_2, j_3, \ldots, j_n)$, $t \in \mathbb{N}_0$, $t \mapsto F(j_1, t, j_3, \ldots, j_n)$, $t \in \mathbb{N}_0$, $t \mapsto F(j_1, j_2, t, \ldots, j_n)$, $t \in \mathbb{N}_0, \ldots, t \mapsto F(j_1, j_2, j_3, \ldots, j_{n-1}, t)$, $t \in \mathbb{N}_0$ is equi-Weyl-$(\mathcal{B}, \mathbb{F}, p, \rho)$-almost periodic (the integers $j_1 \geqslant 0, \ldots, j_n \geqslant 0$ are fixed in advance). Taking into account the result established in the one-dimensional setting, it suffices to prove that the set $\{F(\mathbf{t}; x) : t_1 \geqslant 2L, \ldots, t_n \geqslant 2L; x \in B\}$ is bounded ($\mathbf{t} = (t_1, t_2, \ldots, t_n)$). This follows as in the case that $\Lambda = \Lambda' = \mathbb{Z}^n$, with the corresponding ε-period τ belonging to the cube $[t_1 - 2L, t_1] \times [t_2 - 2L, t_2] \times \cdots \times [t_n - 2L, t_n]$. □

In particular, any equi-Weyl-$(l^{-n/p}, p, \rho)$-almost periodic sequence $F : \mathbb{Z}^n \to Y$, where $\rho : Y \to Y$ is a continuous function and Y is a finite-dimensional space, has a relatively compact range. It seems very plausible that there exist an infinite-dimensional Banach space Y and an equi-Weyl-$(l^{-1/p}, p, \mathrm{I})$-almost periodic sequence $F : \mathbb{Z} \to Y$ whose range is not relatively compact in Y.

Example 2.1.17. (i) Let us observe that there exists a Weyl-$(l^{-1/p}, p, \mathrm{I})$-almost periodic real sequence $(y_k)_{k\in\mathbb{N}}$ [i. e., $(y_k)_{k\in\mathbb{N}}$ is Weyl-p-almost periodic in the usual sense], which is not (Besicovitch-p-)bounded, not equi-Weyl-$(l^{-1/p}, p, \mathrm{I})$-almost periodic and not Besicovitch-p-almost periodic in the sense of [267, Definition 9]; cf. [267, Example 4(ii)]. Concerning the sequence $(y_k)_{k\in\mathbb{N}}$ considered in [267, Example 4(i)], we would to note that $(y_k)_{k\in\mathbb{N}}$ is equi-Weyl-$(l^{-\sigma}, p, \mathrm{I})$-almost periodic for any $\sigma > 0$ and $p \geqslant 1$, as easily approved; let us also recall that for each $p \geqslant 1$ there exists a Besicovitch-p-almost periodic real sequence $(y_k)_{k\in\mathbb{N}}$, which is not Weyl-p-almost periodic (see [267, p. 23]).

(ii) Let $l \in \mathbb{N}$. Suppose that $(y_k)_{k\in\mathbb{N}}$ is a real sequence defined by $y_k := 0$ for $k = 1, 2, \ldots, l$; $y_{l+2k} := 1$ $(k \in \mathbb{N}_0)$ and $y_{l+2k+1} := -1$ $(k \in \mathbb{N}_0)$. Then $(y_k)_{k\in\mathbb{N}}$ is equi-Weyl-$(l^{-\sigma}, p, -\mathrm{I})$-almost periodic for any $\sigma > 0$ and $p \geqslant 1$, i. e., the sequence $(y_k)_{k\in\mathbb{N}}$ is equi-Weyl-p-almost antiperiodic.

(iii) Define the sequence $F : \mathbb{Z}^n \to \mathbb{R}$ by $F(k_1, \ldots, k_n) := 0$ if there exists an index $j \in \mathbb{N}_n$ such that $k_j < 0$ and $F(k_1, \ldots, k_n) := 1$, otherwise. Then it can be simply proved (cf. [447, Example 6.3.9] for the continuous case) that $F(\cdot)$ is Weyl-$(l^{-\sigma}, p, \mathrm{I})$-almost periodic for any $p \geqslant 1$ and $\sigma > (n-1)/p$.

We continue by raising the following issue.

Problem. In the continuous framework, we know that the space of all complex-valued equi-Weyl-$(l^{-n/p}, p, \mathrm{I})$-almost periodic functions $F : \mathbb{R} \to \mathbb{C}$ is not complete with respect to the Weyl-p-seminorm. If we denote by P the space consisting of all complex-valued equi-Weyl-$(l^{-n/p}, p, \mathrm{I})$-almost periodic sequences $F : \mathbb{Z} \to \mathbb{C}$, then it can be simply proved, as in the continuous framework, that the expression

$$d(G,H) := \lim_{l\to+\infty} l^{-n/p} \sup_{k\in\mathbb{Z}} \left[\sum_{j=k}^{k+l} \|G(j) - H(j)\|^p \right]^{1/p}, \quad G, H \in P$$

defines a pseudometric on P. Is (P, d) complete or not?

Further on, we set $\Lambda_j' := \mathbb{R}$ if $\Lambda_j = \mathbb{Z}$, $\Lambda_j' := (-\infty, a]$ if $\Lambda_j = \{\ldots, a-2, a-1, a\}$ for some $a \in \mathbb{Z}$ and $\Lambda_j' := [a, \infty)$ if $\Lambda_j = \{a, a+1, a+2, \ldots\}$ for some $a \in \mathbb{Z}$ $(1 \leqslant j \leqslant n)$. After that, we set $\Lambda_e := \Lambda_1' \times \Lambda_2' \times \cdots \times \Lambda_n'$. Now we are ready to state the following result concerning the extensions of (equi-)Weyl ρ-almost periodic type sequences and Doss ρ-almost periodic type sequences.

Theorem 2.1.18. *Suppose that $F : \Lambda \times X \to Y$ is a given sequence, $1 \leqslant p < +\infty$, $\Lambda' \subseteq \Lambda''$ and $\rho = T \in L(Y)$. If (FV) holds and $F(\cdot;\cdot)$ is (equi-)Weyl-$(\mathcal{B}, \Lambda', \mathbb{F}, p, \rho)$-almost periodic [Doss-$(\mathcal{B}, \Lambda', \mathbb{F}, p, \rho)$-almost periodic], where for each $j \in \mathbb{N}_n$ we have $\Lambda_j' := [a, \infty)$ for some $a \in \mathbb{Z}$ or $\Lambda_j' = \mathbb{R}$, then there exists a continuous function $\tilde{F} : \Lambda_e \times X \to Y$ such that $\tilde{F} \in (e-)W^{p,x,\mathbb{F}}_{[0,1]^n,\Lambda',\mathcal{B}}(\Lambda_e \times X : Y)$ [$\tilde{F}(\cdot;\cdot)$ is Doss-$(p, x, \mathbb{F}, \mathcal{B}, \Lambda', T)$-almost periodic] and $\tilde{F}(\mathbf{t}; x) = F(\mathbf{t}; x)$ for all $\mathbf{t} \in \Lambda$ and $x \in X$.*

Proof. We will present the proof only in the one-dimensional setting, for the class of equi-Weyl-$(\mathcal{B}, \Lambda', \mathbb{F}, p, \rho)$-almost periodic sequences $F : \Lambda \to Y$; the general result can be deduced similarly, following the argumentation contained in the proof of Theorem 2.1.6. Suppose first that $\Lambda_e = [a, \infty)$ for some $a \in \mathbb{Z}$. If $t \in [b, b+1)$ for some $b \in \mathbb{Z}$ with $b \geqslant a$, then we set $\tilde{F}(t) := F(b)$ for $t \in [b, b+(1/2))$ and $\tilde{F}(t) := 2(F(b+1)-F(b))(t-b-(1/2))+F(b)$ for $t \in [b+(1/2), b+1)$. Let $\varepsilon > 0$ be given. By our assumption, we can find an integer $l \in \mathbb{N}$

and a real number $L > 0$ such that, for every $t_0 \in \Lambda'$, there exists a point $\tau \in \Lambda' \cap B(t_0, L)$, which satisfies that, for every $j \in \mathbb{N} \cap [a, \infty)$, we have $\sum_{k=j}^{j+l} \|F(k+\tau) - TF(k)\|^p \leqslant \varepsilon^p [\mathbb{F}(l)]^{-1}$. We need to prove that, for every fixed real number $x \geqslant a$, we have

$$\int_x^{x+l} \|\tilde{F}(s+\tau) - T\tilde{F}(s)\|^p \, ds \leqslant \text{Const.} \cdot \varepsilon^p [\mathbb{F}(l)]^{-1}. \tag{33}$$

In order to show this, observe first that for each $t \in \mathbb{R}$ we have

$$\int_{t+(1/2)}^{t+1} (s - t - (1/2)) \, ds = 1/8;$$

keeping in mind the definition of $\tilde{F}(\cdot)$ and this equality, it follows that

$$\begin{aligned}
&\int_x^{x+l} \|\tilde{F}(s+\tau) - T\tilde{F}(s)\|^p \, ds \\
&\leqslant \int_{\lfloor x \rfloor}^{\lfloor x \rfloor+1} \|\tilde{F}(s+\tau) - T\tilde{F}(s)\|^p \, ds + \cdots + \int_{\lfloor x \rfloor+l}^{\lfloor x \rfloor+l+1} \|\tilde{F}(s+\tau) - T\tilde{F}(s)\|^p \, ds \\
&\leqslant \left[\int_{\lfloor x \rfloor}^{\lfloor x \rfloor+(1/2)} \|\tilde{F}(s+\tau) - T\tilde{F}(s)\|^p \, ds + \cdots + \int_{\lfloor x \rfloor+l}^{\lfloor x \rfloor+l+(1/2)} \|\tilde{F}(s+\tau) - T\tilde{F}(s)\|^p \, ds \right] \\
&\quad + \left[\int_{\lfloor x \rfloor+(1/2)}^{\lfloor x \rfloor+1} \|\tilde{F}(s+\tau) - T\tilde{F}(s)\|^p \, ds + \cdots + \int_{\lfloor x \rfloor+l+(1/2)}^{\lfloor x \rfloor+l+1} \|\tilde{F}(s+\tau) - T\tilde{F}(s)\|^p \, ds \right] \\
&\leqslant c_p \big[\|F(\lfloor x \rfloor + \tau) - TF(\lfloor x \rfloor)\|_Y^p + \cdots + \|F(\lfloor x \rfloor + l + \tau) - TF(\lfloor x \rfloor + l)\|_Y^p \big] \\
&\quad + c_p \cdot \big[\|F(\lfloor x \rfloor + \tau) - TF(\lfloor x \rfloor)\|_Y^p + \cdots + \|F(\lfloor x \rfloor + \tau + l + 1) - TF(\lfloor x \rfloor + l + 1)\|_Y^p \big] \\
&\leqslant 3c_p \sup_{j \geqslant a} \sum_{k=j}^{j+l} \|F(k+\tau) - TF(k)\|^p \leqslant 3c_p \varepsilon^p [\mathbb{F}(l)]^{-1},
\end{aligned}$$

where $c_p > 0$ is a finite real constant. Therefore, (33) holds true, which completes the proof in this case. The consideration is quite similar in the case that $\Lambda_e = \mathbb{R}$. □

Remark 2.1.19. It is also possible to assume that $\Lambda'_{j_0} := (-\infty, a]$ for some $a \in \mathbb{Z}$ and $j_0 \in \mathbb{N}$ but then we must replace the set $\Omega = [0,1]^n$ with the direct product of sets $\Omega_j = [0,1]$ or $\Omega_j = [-1,0]$ for $1 \leqslant j \leqslant n$, with the obvious choice $\Omega_{j_0} = [-1,0]$.

Making use of Proposition 2.1.16, Theorem 2.1.18, Remark 2.1.19 and the construction given in the proof of Theorem 2.1.6, we can formulate the following result.

Corollary 2.1.20. *Suppose that $F : \Lambda \to Y$ is a given sequence, $1 \leqslant p < +\infty$, $\Lambda' = \Lambda''$ and $\rho = \mathrm{I}$. Suppose further that for each $j \in \mathbb{N}_n$ we have $\Lambda_j' := [a,\infty)$ ($\Lambda_j' := (-\infty,a]$) for some $a \in \mathbb{Z}$ or $\Lambda_j' = \mathbb{R}$, and $F(l,J) \equiv l^{-n/p}$ for all $l \in \mathbb{N}$ and $J \in P_l$. Define $\Omega := \Omega_1 \times \cdots \times \Omega_n$, where $\Omega_j = [0,1]$ if $\Lambda_j' = [a,\infty)$ and $\Omega_j = [-1,0]$ if $\Lambda_j' = (-\infty,a]$ for some $a \in \mathbb{Z}$ ($1 \leqslant j \leqslant n$). If $F(\cdot)$ is equi-Weyl-($\mathcal{B},\Lambda',\mathbb{F},p,\rho$)-almost periodic, then the mean value*

$$M(F) := \lim_{T\to+\infty} \frac{1}{T^n} \sum_{\mathbf{t}\in(\mathbf{s}+T\Omega)\cap\mathbb{Z}^n} F(\mathbf{t}) \tag{34}$$

exists uniformly on $\mathbf{s} \in \Lambda$.

Proof. Without loss of generality, we may assume that $\Lambda_e = \mathbb{R}^n$. Let the function $\tilde{F}(\cdot;\cdot)$ be given by Theorem 2.1.18; then we know that the mean value

$$M(\tilde{F}) := \lim_{T\to+\infty} \frac{1}{T^n} \int_{\mathbf{s}+T\Omega} \tilde{F}(\mathbf{t})\,d\mathbf{t},$$

exists uniformly on $\mathbf{s} \in [0,\infty)^n$; cf. the proof of [447, Theorem 6.3.32] and [447, Remark 6.3.33]. Keeping in mind the way of construction of $\tilde{F}(\cdot)$, this implies the required conclusion after a simple computation involving the boundedness of sequence $F(\cdot)$. □

Remark 2.1.21. In contrast with the statements of Theorem 2.1.6 and Theorem 2.1.9, it is very difficult to state a satisfactory converse in Theorem 2.1.18 for the corresponding Weyl (Doss) class. In order to better explain this, let us notice that there exists an infinitely differentiable Stepanov-1-almost periodic function $f : \mathbb{R} \to \mathbb{R}$ such that the sequence $(f(k))_{k\in\mathbb{Z}}$ is unbounded and the sequence $(f(k+(1/2)))_{k\in\mathbb{Z}}$ is almost periodic (see [56, Example 4]). Due to Proposition 2.1.16, it follows that the sequence $(f(k))_{k\in\mathbb{Z}}$ cannot be equi-Weyl-almost periodic, i. e., equi-Weyl-($\mathbb{Z}, l^{-1}, 1, \mathrm{I}$)-almost periodic.

For the sequel, let us recall that A. Iwanik has investigated the equi-Weyl-1-almost periodic sequences with values in compact metric spaces [393]. We would like to point out that the assertions of [393, Lemma 1] holds for an arbitrary equi-Weyl-1-almost periodic sequence $g : \mathbb{Z} \to X$ such that $R(g)$ is contained in a compact convex subset of X as well as that the assumption that $R(g)$ is a relatively compact subset of X is slightly redundant in our framework. We will state and prove the following extension of [393, Lemma 1].

Proposition 2.1.22. *Suppose that $F : \Lambda \to Y$ is a given sequence such that $R(F) \subseteq K$ for some compact convex subset K of Y, $1 \leqslant p < +\infty$, $\Lambda' = \Lambda''$ and $\rho = \mathrm{I}$. Suppose further that $F(l,J) \equiv l^{-n/p}$ for all $l \in \mathbb{N}$ and $J \in P_l$. If $F(\cdot)$ is equi-Weyl-($\Lambda',\mathbb{F}_{\cdot},p,\rho$)-almost periodic, then for each $\varepsilon > 0$ there exist a Bohr almost periodic function $H : \Lambda \to Y$ with values in K and an integer $l \in \mathbb{N}$ such that, for every $J \in P_l$, we have*

$$l^{-n/p}\Big[\sum_{j\in J}\|F(j;x) - H(j;x)\|^p\Big]^{1/p} \leqslant \varepsilon. \tag{35}$$

Proof. We will outline the main details of proof only. If $\Lambda'_{j_0} := (-\infty, a]$ ($\Lambda'_{j_0} = [a, +\infty)$) for some $a \in \mathbb{Z}$, then we set $\Omega_j = [-1, 0]$ ($\Omega_j = [0, 1]$); if $\Lambda'_{j_0} := \mathbb{R}$, then we set $\Omega_j = [-1, 1]$; cf. also Remark 2.1.19. Set $\Omega := \Omega_1 \times \Omega_2 \times \cdots \times \Omega_n$ and assume $\varepsilon > 0$. Then we know that there exist $l \in \mathbb{N}$ and $L \in \mathbb{N}$ such that, for every $\mathbf{t}_0 \in \Lambda'$, there exists a point $\tau \in \Lambda' \cap B(\mathbf{t}_0, L)$, which satisfies that, for every $J \in P_l$ and for every $j \in J$, (31) holds with $z_{j,x} = F(j; x)$. We write the region Λ as a countable union of the closed rectangles $(\Lambda_j)_{j\in\mathbb{N}}$, which are translations of the cube $L\Omega$ in $\mathbb{R}^n$. Then for each $j \in \mathbb{N}$ there exists a point $\tau_j \in \Lambda_j$ such that, for every $J \in P_l$ and for every $j \in J$, (31) holds with $z_{j,x} = F(j; x)$ and $\tau = \tau_j$. It is clear that the set $J = \{\tau_j : j \in \mathbb{N}\}$ is syndetic in Λ, with the meaning clear. Define $J_k := J \cap k\Omega$ for all $k \in \mathbb{N}$. Let a point $j \in \Lambda$ be fixed. Then any member of the sequence $(|J_{kl}|^{-1} \sum_{\mathbf{t}\in J_{kl}} F(j + \mathbf{t}))_{k\in\mathbb{N}}$ belongs to K since $R(F) \subseteq K$ and K is convex. Since K is a compact subset of X, we obtain the existence of a strictly increasing sequence (k_m) of positive integers such that

$$\lim_{m\to+\infty} \frac{1}{|J_{k_m l}|} \sum_{\mathbf{t}\in J_{k_m l}} F(j + \mathbf{t}) =: H(j)$$

exists in K. Keeping in mind that

$$\liminf_{m\to+\infty}(a_m + b_m) \geqslant \liminf_{m\to+\infty} a_m + \liminf_{m\to+\infty} b_m$$

for any two sequences (a_m) and (b_m) of positive real numbers and the well-known inequality between the means

$$\left(\frac{a_1 + \cdots + a_m}{m}\right)^p \leqslant \frac{a_1^p + \cdots + a_m^p}{m}, \quad m \in \mathbb{N};\ a_j \geqslant 0,\ 1 \leqslant j \leqslant m,$$

we can argue in the same way as in [393] to conclude that the function $H : \Lambda \to Y$ is Bohr almost periodic and satisfies the required properties. □

Remark 2.1.23. The foregoing argumentation shows that, for every equi-Weyl-p-almost periodic sequence $g : \mathbb{Z} \to X$, there exists a uniformly continuous equi-Weyl-p-almost periodic function $\tilde{g} : \mathbb{R} \to X$ such that $\tilde{g}(t) = g(t)$ for all $t \in \mathbb{Z}$ as well as that $\tilde{g}(t) \in CH(R(g))$, $t \in \mathbb{R}$ ($1 \leqslant p < +\infty$). Then we can argue as in the proof of [447, Theorem 6.3.23] in order to see that for each $\varepsilon > 0$ there exists an almost periodic function $h : \mathbb{R} \to X$ such that $R(h) \subseteq \overline{CH(R(g))}$ and $D_W(\tilde{g}, h) < \varepsilon$, where $D_W(\cdot;\cdot)$ denotes the Weyl distance of functions. But, it is not clear how to prove that the last estimate implies that for each $\varepsilon > 0$ there exists $l > 0$ such that

$$D^p_{S_l}\big((\tilde{g}(k))_{k\in\mathbb{Z}}, (h(k))_{k\in\mathbb{Z}}\big) := \sup_{k\in\mathbb{Z}} \frac{1}{l} \sum_{j=k}^{k+l-1} \|\tilde{g}(j) - h(j)\|^p < \varepsilon.$$

Of course, if $g : \mathbb{Z} \to X$ is an almost periodic sequence, then we have $\|\tilde{g}(t) - h(t)\| < \varepsilon$ for all $t \in \mathbb{R}$; the same result can be clarified for the almost periodic sequences $g : \mathbb{Z}^n \to X$, providing thus an extension of [100, Fundamental Theorem II, p. 319] to the higher-dimensional setting.

The converse statement in Proposition 2.1.22 can be proved using a simple argumentation along with the decomposition

$$\begin{aligned}&\|F(\mathbf{t}+\tau)-F(\mathbf{t})\|_Y\\&\quad\leqslant \|F(\mathbf{t}+\tau)-H(\mathbf{t}+\tau)\|_Y+\|H(\mathbf{t}+\tau)-H(\mathbf{t})\|_Y+\|H(\mathbf{t})-F(\mathbf{t})\|_Y, \quad \mathbf{t}\in\Lambda,\ \tau\in\Lambda':\end{aligned}$$

Proposition 2.1.24. *Suppose that $F:\Lambda\to Y$ is a given sequence, $1\leqslant p<+\infty$, $\Lambda'=\Lambda''$ and $\rho=\mathrm{I}$. Suppose further that $F(l,J)\equiv l^{-n/p}$ for all $l\in\mathbb{N}$ and $J\in P_l$. If for each $\varepsilon>0$, there exist a Bohr almost periodic function $H:\Lambda\to Y$ and an integer $l\in\mathbb{N}$ such that, for every $J\in P_l$, we have* (35). *Then $F(\cdot)$ is equi-Weyl-$(\Lambda',\mathbb{F}.,p,\rho)$-almost periodic.*

Since the sum of two compact (convex) subsets of Y is likewise a compact (convex) subset of Y, combining Proposition 2.1.22 and Proposition 2.1.24, we get the following.

Proposition 2.1.25. *Denote by $e-W^{p,\Lambda'}_{ap;cc}(\Lambda:Y)$ the collection of all equi-Weyl-$(\Lambda',\mathbb{F}.,p,\rho)$-almost periodic sequences such that $F(l,J)\equiv l^{-n/p}$ for all $l\in\mathbb{N}$ and $J\in P_l$, $1\leqslant p<+\infty$, $\Lambda'=\Lambda''$, $\rho=\mathrm{I}$ and $R(F)$ is contained in a compact convex subset of Y. Then $e-W^{p,\Lambda'}_{ap;cc}(\Lambda:Y)$ is a vector space with the usual operations.*

Remark 2.1.26. Suppose that Y is a finite-dimensional space and the assumptions of Proposition 2.1.24 hold. Since the convex hull of a compact subset K of Y is compact, Proposition 2.1.16 implies that $F(\cdot)$ is equi-Weyl-$(\Lambda',\mathbb{F}.,p,\rho)$-almost periodic if and only if for each $\varepsilon>0$ there exist a Bohr almost periodic function $H:\Lambda\to Y$ and an integer $l\in\mathbb{N}$ such that, for every $J\in P_l$, we have (35). If this is the case, then $F(\cdot)$ is Besicovitch-$(l^{-n/p},p)$-almost periodic in the sense of Definition 2.1.27 below.

In connection with Proposition 2.1.25 and Remark 2.1.26, we would like to ask the following question (it could be also interesting to formulate an analogue of [393, Lemma 3] in our framework).

Problem. Denote by $e-W^{p,\Lambda'}_{ap}(\Lambda:Y)$ the set of all equi-Weyl-$(\Lambda',\mathbb{F}.,p,\rho)$-almost periodic sequences, where $F(l,J)\equiv l^{-n/p}$ for all $l\in\mathbb{N}$ and $J\in P_l$, $1\leqslant p<+\infty$, $\Lambda'=\Lambda''$ and $\rho=\mathrm{I}$. Is it true that $e-W^{p,\Lambda'}_{ap}(\Lambda:Y)$ is a vector space with the usual operations? Furthermore, is it true that the equivalence relation clarified in Remark 2.1.26 holds if the space Y is infinite-dimensional?

Now we will introduce the class of Besicovitch-$(\mathcal{B},\mathbb{F},p)$-almost periodic sequences (concerning some new references about Besicovitch-p-almost periodic functions, we will only quote here the research article [227] by L. I. Danilov, where the author has recently analyzed the Besicovitch almost periodic type selections of multivalued maps).

Definition 2.1.27. Suppose that $F:\Lambda\times X\to Y$ is a given sequence, $\mathbb{F}:(0,\infty)\to[0,\infty)$ and $1\leqslant p<+\infty$. Then we say that $F(\cdot;\cdot)$ is Besicovitch-$(\mathcal{B},\mathbb{F},p)$-almost periodic if, for every $\varepsilon>0$ and $B\in\mathcal{B}$, there exists a trigonometric polynomial $P(\cdot;\cdot)$ such that

$$\limsup_{l\to+\infty} \mathbb{F}(l) \sup_{x\in B}\left[\sum_{j\in[-l,l]^n\cap\Lambda} \|F(j;x)-P(j;x)\|^p\right]^{1/p} < \varepsilon.$$

If $\mathbb{F}(l) \equiv l^{-n/p}$, then we omit the term "$\mathbb{F}$" from the notation.

Since the operation $\limsup_{l\to+\infty}\cdot$ is subadditive, it follows that the set of all Besicovitch-$(\mathcal{B},\mathbb{F},p)$-almost periodic sequences is a vector space with the usual operations. The usual example of a Besicovitch-p-almost periodic sequence ($\Lambda = \mathbb{Z}^n$, $X = \{0\}$, $\mathbb{F}(l) \equiv l^{-n/p}$) is obtained in the one-dimensional framework by taking the Fourier coefficients of a complex Borel measure on the unit circle (see [100, p. 315]).

The following results can be established for the Besicovitch class (Corollary 2.1.29(ii) can be deduced using the argumentation contained in the proof of [100, Lemma 3.4(1)]).

Theorem 2.1.28. *Suppose that $F : \Lambda\times X \to Y$ is a given sequence, $\mathbb{F} : (0,\infty) \to [0,\infty)$, $1 \leqslant p < +\infty$ and $\Lambda' \subseteq \Lambda''$. If $F(\cdot;\cdot)$ is Besicovitch-$(\mathcal{B},\mathbb{F},p)$-almost periodic, where for each $j \in \mathbb{N}_n$ we have $\Lambda_j' := [a,\infty)$ for some $a \in \mathbb{Z}$ or $\Lambda_j' = \mathbb{R}$, then there exists a continuous function $\tilde{F} : \Lambda_e \times X \to Y$ such that $\tilde{F} \in e - (\mathcal{B},x,\mathbb{F}) - B^p(\Lambda_e\times X : Y)$ and $\tilde{F}(\mathbf{t};x) = F(\mathbf{t};x)$ for all $\mathbf{t} \in \Lambda$ and $x \in X$.*

Corollary 2.1.29. *Suppose that $F : \Lambda\times X \to Y$ is a given sequence, $1 \leqslant p < +\infty$ and $\Lambda' = \Lambda''$. Suppose further that for each $j \in \mathbb{N}_n$ we have $\Lambda_j' := [a,\infty)$ ($\Lambda_j' := (-\infty,a]$) for some $a \in \mathbb{Z}$ or $\Lambda_j' = \mathbb{R}$, and $\mathbb{F}(l) \equiv l^{-n/p}$ for all $l > 0$. Define $\Omega := \Omega_1 \times\cdots\times \Omega_n$, where $\Omega_j = [0,1]$ if $\Lambda_j' = [a,\infty)$ and $\Omega_j = [-1,0]$ if $\Lambda_j' = (-\infty,a]$ for some $a \in \mathbb{Z}$ ($1 \leqslant j \leqslant n$). If $F(\cdot)$ is Besicovitch-$(\mathcal{B},\mathbb{F},p)$-almost periodic, then the following holds:*

(i) *The set $\{F(\mathbf{t};x) : \mathbf{t} \in \Lambda,\ x \in B\}$ is Besicovitch-p-bounded for each bounded subset B of the collection $\mathcal{B}$, i. e.,*

$$\limsup_{l\to+\infty} \frac{1}{l^n} \sup_{x\in B} \sum_{\mathbf{t}\in[-l,l]^n\cap\Lambda} \|F(\mathbf{t};x)\|^p < +\infty.$$

(ii) *If $X = \{0\}$, then the mean value $M(F)$, given by* (34), *exists uniformly on $\mathbf{s} \in \Lambda$.*

It is worth noting that, besides the mean value $M(F)$, we can also define the Bohr-Fourier coefficients of $F(\cdot)$; cf. [100, Lemma 3.4(1)]. We ought to observe that the proofs of [100, Lemma 3.11: (1)(b); (2)] are not correct: Strictly speaking, the argumentation given in the cited monograph [112, pp. 107–109] of A. S. Besicovitch only shows that, for a given Besicovitch-p-almost periodic function $f : \mathbb{R} \to \mathbb{R}$ and a given number $\varepsilon > 0$, we have the existence of a sufficiently large positive real number $t_0(\varepsilon) > 0$ and a corresponding Bochner–Fejér trigonometric polynomial $\sigma_B^f(\cdot)$ such that

$$\int_{-t}^{t} \|f(s) - \sigma_B^f(s)\|^p\, ds \leqslant 2\varepsilon^p t, \quad t \geqslant t_0(\varepsilon).$$

But it is not clear why would the last inequality imply the existence of an integer $k_0(\varepsilon) \in \mathbb{N}$ such that

$$\sum_{j=-k}^{k} \|f(j) - \sigma_B^f(j)\|^p \leqslant 2\varepsilon^p k, \quad k \geqslant k_0(\varepsilon),$$

even if the all above terms are well-defined and $f(\cdot)$ is continuous. Therefore, it is clear that we must follow another approach in the discrete setting.

Remark 2.1.30. In the question [267, (Q4)], we have asked the following: Is it true that the sequence $(y_k)_{k\in\mathbb{Z}}$ $[(y_k)_{k\in\mathbb{N}}]$ is (equi-)Weyl-p-almost periodic [Doss-p-almost periodic/Besicovitch-p-almost periodic] ($1 \leqslant p < \infty$) if and only if there exists a continuous (equi-)Weyl-p-almost periodic [Doss-p-almost periodic/Besicovitch-p-almost periodic] function $f : \mathbb{R} \to X$ [$f : [0,\infty) \to X$] such that $y_k = f(k)$ for all $k \in \mathbb{Z}$ [$k \in \mathbb{N}$] (cf. the notion introduced above with $n = 1$, $\mathbb{F}_\cdot(l) \equiv l^{-1/p}$ and $\rho = \mathrm{I}$)?

In Theorem 2.1.18 and Theorem 2.1.28, we have proved the existence of a continuous (equi-)Weyl-p-almost periodic [Doss-p-almost periodic/Besicovitch-p-almost periodic] function $f(\cdot)$ obeying the required properties. On the other hand, in Remark 2.1.21, we have shown that the converse statement is not true for the class of equi-Weyl-p-almost periodic sequences; it seems very plausible that the same statement is not true for the classes of Doss-p-almost periodic sequences and Besicovitch-p-almost periodic sequences.

Concerning the completeness of the space of Besicovitch-$(\mathcal{B},p)$-almost periodic sequences, denoted here simply by P, we will only state the following direct consequence of [455, Theorem 2.3], which provides a discrete analogue of the famous result established by J. Marcinkiewicz in [571].

Theorem 2.1.31. *Suppose that* $1 \leqslant p < +\infty$, $\Lambda' = \Lambda''$ *and for each* $j \in \mathbb{N}_n$ *we have* $\Lambda_j' := [a,\infty)$ $(\Lambda_j' := (-\infty, a])$ *for some* $a \in \mathbb{Z}$ *or* $\Lambda_j' = \mathbb{R}$, *and* $\mathbb{F}(l) \equiv l^{-n/p}$ *for all* $l > 0$. *Define* $\Omega := \Omega_1 \times \cdots \times \Omega_n$, *where* $\Omega_j = [0,1]$ *if* $\Lambda_j' = [a,\infty)$ *and* $\Omega_j = [-1,0]$ *if* $\Lambda_j' = (-\infty, a]$ *for some* $a \in \mathbb{Z}$ $(1 \leqslant j \leqslant n)$. *Then, for every bounded set* B *of the collection* $\mathcal{B}$, *we have that* (P, d_B) *is a complete pseudometric space, where*

$$d_B(F,G) := \limsup_{l\to+\infty} l^{-n/p} \sup_{x\in B} \Big[\sum_{j\in[-l,l]^n\cap\Lambda} \|F(j) - G(j)\|^p\Big]^{1/p}, \quad F, G \in P.$$

Before proceeding to the next subsection, we will only note that the statements of [455, Propositions 1, 2] can be formulated in the discrete setting; cf. also [100, Lemma 3.1], which can be formulated if one of the corresponding sequences $\mathbf{b}(\cdot)$ or $\mathbf{c}(\cdot)$ is vector-valued. Details can be left to the enthusiastic readers.

2.1.4 Metrically generalized ρ-almost periodic sequences

In this subsection, we will continue the investigation raised in our previous subsection. We will reconsider and slightly generalize various classes of generalized ρ-almost periodic sequences examined so far by using the concept of metrical almost periodicity [466]. We analyze here the Stepanov, Weyl, Besicovitch and Doss classes of metrically generalized ρ-almost periodic sequences.

As before, we consider here the sequences of the form $F : \Lambda \times X \to Y$, where $\emptyset \neq \Lambda \subseteq \mathbb{Z}^n$. We assume that $\Lambda = \Lambda_1 \times \Lambda_2 \times \cdots \times \Lambda_n$, where for each $j \in \mathbb{N}_n$ there exists an integer $a \in \mathbb{Z}$ such that $\Lambda_j = \mathbb{Z}$, $\Lambda_j = \{\ldots, a-2, a-1, a\}$ or $\Lambda_j = \{a, a+1, a+2, \ldots\}$. Define $\Lambda'' := \{\mathbf{a} \in \mathbb{Z}^n : \mathbf{a} + \Lambda \subseteq \Lambda\}$. For every integer $l \in \mathbb{N}$, we define P_l to be the set consisting of all closed subrectangles of Λ, which contains exactly $(l+1)^n$ points with all integer coordinates. In the sequel, we will assume that condition (FV) automatically holds as well as that for each $l \in \mathbb{N}$ and $J \in P_l$ we have that $(P_{l,J}, d_{l,J})$ is a pseudometric space, where $P_{l,J} \subseteq Y^J$ is closed under the addition and subtraction of functions, and $0 \in P_{l,J}$. Define $\|f\|_{l,J} := d_{l,J}(f, 0)$ for all $f \in P_{l,J}$. We will cite here the easily accessible research article [476].

The following notion generalizes the notion introduced in [476, Definitions 3, 6, 7].

Definition 2.1.32. Suppose that $F : \Lambda \times X \to Y$ is a given sequence, $\mathbb{F} : \mathbb{N} \to [0, \infty)$, $\Lambda' \subseteq \Lambda''$ and ρ is a binary relation on Y. Then we say that $F(\cdot;\cdot)$ is:

(i) Stepanov-$(\mathcal{B}, \Lambda', \mathbb{F}, \mathcal{P}, \rho, l)$-almost periodic for some $l \in \mathbb{N}$ if, for every $\varepsilon > 0$ and $B \in \mathcal{B}$, there exists $L > 0$ such that, for every $\mathbf{t}_0 \in \Lambda'$, there exists a point $\tau \in \Lambda' \cap B(\mathbf{t}_0, L)$, which satisfies that, for every $J \in P_l$ and for every $j \in J$ and $x \in B$, there exists $z_{j,x} \in \rho(F(j;x))$ such that

$$\sup_{x \in B} \mathbb{F}(l) \|F(\cdot + \tau; x) - z_{\cdot,x}\|_{l,J} < \varepsilon. \tag{36}$$

(ii) equi-Weyl-$(\mathcal{B}, \Lambda', \mathbb{F}, \mathcal{P}, \rho)$-almost periodic if, for every $\varepsilon > 0$ and $B \in \mathcal{B}$, there exist $l \in \mathbb{N}$ and $L > 0$ such that, for every $\mathbf{t}_0 \in \Lambda'$, there exists a point $\tau \in \Lambda' \cap B(\mathbf{t}_0, L)$, which satisfies that, for every $J \in P_l$ and for every $j \in J$ and $x \in B$, there exists $z_{j,x} \in \rho(F(j;x))$ such that (36) holds.

(iii) Weyl-$(\mathcal{B}, \Lambda', \mathbb{F}, \mathcal{P}, \rho)$-almost periodic if, for every $\varepsilon > 0$ and $B \in \mathcal{B}$, there exists $L > 0$ such that, for every $\mathbf{t}_0 \in \Lambda'$, there exists a point $\tau \in \Lambda' \cap B(\mathbf{t}_0, L)$, which satisfies that there exists an integer $l_\tau \in \mathbb{N}$ such that, for every $l \geqslant l_\tau$, $J \in P_l$, $j \in J$ and $x \in B$, there exists $z_{j,x} \in \rho(F(j;x))$ such that (36) holds.

(iv) Doss-$(\mathcal{B}, \Lambda', \mathbb{F}, \mathcal{P}, \rho)$-almost periodic if, for every $\varepsilon > 0$ and $B \in \mathcal{B}$, there exists $L > 0$ such that, for every $\mathbf{t}_0 \in \Lambda'$, there exists a point $\tau \in \Lambda' \cap B(\mathbf{t}_0, L)$, which satisfies that there exists an increasing sequence (l_k) of positive integers such that, for every $k \in \mathbb{N}$, $J \in P_{l_k}$, $j \in J$ and $x \in B$, there exists $z_{j,x} \in \rho(F(j;x))$ such that (36) holds with the number l replaced by the number l_k therein.

We will not consider here the uniformly recurrent analogues of the notion introduced above; cf. also [476, Definitions 9, 10]. In Definition 2.1.32, the natural choice is

$$\|f\|_{l,J} \equiv \left[\sum_{j \in J} \|f(j)\|^p \nu^p(j)\right]^{1/p}, \quad f \in P_{l,J}, \tag{37}$$

for some $p \in [1, \infty)$ but we can also consider the notion with

$$\|f\|_{l,J} \equiv \sum_{j \in J} \|f(j)\|^p \nu^p(j), \quad f \in P_{l,J}, \tag{38}$$

where $p \in (0,1)$; here, $\nu : \mathbb{Z}^n \to [0, \infty)$ is an arbitrary weight sequence (in [476], we have always assumed that $\nu(\cdot) \equiv 1$ and $p \geqslant 1$). The similar pseudometrics will be used for the Besicovitch-$(\mathcal{B}, \mathbb{F}, \mathcal{P})$-almost periodic sequences introduced in Definition 2.1.38 below.

The uniform convergence (or the convergence in the pseudometric of space $P_{l,J}$) of metrically generalized ρ-almost periodic sequences can be analyzed. We continue by reexamining two examples from our previous work.

Example 2.1.33. (i) The sequence $(x_k)_{k \in \mathbb{N}}$, given by $x_k := 1$ if there exists $j \in \mathbb{N}$ such that $k = j^3$, and $x_k := 0$, otherwise, is equi-Weyl-$(l^{-\sigma}, p)$-almost periodic for any $\sigma > 1/2$ and $p > 0$; cf. (37)–(38) with $\nu(\cdot) \equiv 1$ and [476, Remark 2].

(ii) Let $l_0 \in \mathbb{N}$, let $(x_k)_{k \in \mathbb{N}}$ be a real sequence defined by $x_k := 0$ for $k = 1, 2, \ldots, l_0$; $x_{l_0+2k} := 1\,(k \in \mathbb{N}_0)$ and $x_{l_0+2k+1} := -1\,(k \in \mathbb{N}_0)$. Then $(x_k)_{k \in \mathbb{N}}$ is equi-Weyl-$(l^{-\sigma}, \mathcal{P}, -\mathrm{I})$-almost periodic for any $\sigma > 0$, where $P_{j,l}$ is given by (37)–(38) with $\nu(\cdot)$ being an arbitrary nonnegative function; cf. also [476, Example 1(ii)].

We have already explained that the notion of Stepanov-$(\mathcal{B}, \Lambda', \mathbb{F}, \mathcal{P}, \rho, l)$-almost periodicity is not satisfactory enough because it is in a close connection with the notion of metrical Bohr-$(\mathcal{B}, \Lambda', \rho)$-almost periodicity, where the pseudometric $\|\cdot\|_{l,J}$ is given by the formula (37). A similar statement holds for the corresponding classes of sequences with the exponents $p \in (0,1)$, when the pseudometric $\|\cdot\|_{l,J}$ is given by the formula (38). Furthermore, the statement of [476, Proposition 4] remains true with the general exponents $p > 0$ but an equi-Weyl-$(\mathbb{F}, \mathcal{P})$-almost periodic sequence $F : \mathbb{Z}^n \to Y$ need not be bounded if the pseudometric is given by the formula (37) or (38) and there is no constant $c > 0$ such that $\nu(\cdot) \geqslant c$, which can be approved by a great number of very simple counterexamples.

If $F : \Lambda \times X \to Y$ is Stepanov-$(\mathcal{B}, \Lambda', \mathbb{F}, \mathcal{P}, \rho, l)$-almost periodic for some $l \in \mathbb{N}$, then it is clear that $F(\cdot;\cdot)$ is equi-Weyl-$(\mathcal{B}, \Lambda', \mathbb{F}, \mathcal{P}, \rho)$-almost periodic. Furthermore, it is clear that every equi-Weyl-$(\mathcal{B}, \Lambda', \mathbb{F}, \mathcal{P}, \rho)$-almost periodic sequence is Weyl-$(\mathcal{B}, \Lambda', \mathbb{F}, \mathcal{P}, \rho)$-almost periodic as well as that every Weyl-$(\mathcal{B}, \Lambda', \mathbb{F}, \mathcal{P}, \rho)$-almost periodic sequence is Doss-$(\mathcal{B}, \Lambda', \mathbb{F}, \mathcal{P}, \rho)$-almost periodic. All these inclusions can be strict, as easily approved.

In [476, Theorems 4, 5], we have considered the extensions of (equi-)Weyl-p-almost periodic type sequences, Doss-p-almost periodic type sequences and Besicovitch-p-almost periodic type sequences, where $p \geqslant 1$; let us observe here that these results continue to hold for all exponents $p > 0$. Without going into full details, we will only mention that the argumentation contained in the proof of [476, Theorem 4] shows that the possible extensions can be considered even if the pseudometric on $P_{l,J}$ is given by (37) or (38); for example, we have the following result (see [448, Definitions 4.3.6, 6.2.11] for the corresponding notion).

Theorem 2.1.34. *Suppose that $F : \mathbb{Z} \times X \to Y$ is a given sequence, $p > 0$, $\Lambda' \subseteq \mathbb{Z}$ and $\rho = T \in L(Y)$. If $F(\cdot;\cdot)$ is (equi-)Weyl-$(\mathcal{B}, \Lambda', \mathbb{F}, \mathcal{P}, \rho)$-almost periodic [Doss-$(\mathcal{B}, \Lambda', \mathbb{F}, \mathcal{P}, \rho)$-almost periodic], where $P_{l,J}$ is given by* (37) *for $p \geqslant 1$ and* (38) *for $0 < p < 1$, with the function $\nu : \mathbb{Z} \to [0, \infty)$ such that there exists a finite real constant $c > 0$ with $\nu(k) \leqslant c\nu(k+1)$, $k \in \mathbb{Z}$. Define $\tilde{\nu}(t) := \nu(k)$, if $t \in [k, k+1)$ for some $k \in \mathbb{Z}$, $P := L^{\infty}(\mathbb{R} : \mathbb{C})$, $P_{t,l} := L^{p}_{\tilde{\nu}}([t, t+l] : Y)$ for all $t \in \mathbb{R}$, $l > 0$, and $P_t := L^{p}_{\tilde{\nu}}([-t, t] : Y)$ for all $t \in \mathbb{R}$. Then there exists a continuous function $\tilde{F} : \mathbb{R} \times X \to Y$ such that $\tilde{F} \in (e-)W^{(x,\mathbb{F},T,\mathcal{P}_{t,l},\mathcal{P})}_{[0,1],\Lambda',\mathcal{B}}(\mathbb{R} \times X : Y)$ [$\tilde{F}(\cdot;\cdot)$ is Doss-$(\mathcal{P}, x, \mathbb{F}, \mathcal{B}, \Lambda', T)$-almost periodic] and $\tilde{F}(t;x) = F(t;x)$ for all $t \in \mathbb{Z}$ and $x \in X$.*

Remark 2.1.35. We can also extend the weight sequence $\nu(\cdot)$ in the following way: $\tilde{\nu}(t) := \nu(k+1)$, if $t \in (k, k+1]$ for some $k \in \mathbb{Z}$, then a similar result holds true if we assume that there exists a finite real constant $c > 0$ such that $\nu(k) \geqslant c\nu(k+1)$, $k \in \mathbb{Z}$.

The statement of [476, Proposition 6] can be metrically generalized in the following way (the proof is almost the same and, therefore, omitted; observe only that the inequality between the means implies that the inequality stated on [476, p. 13, l. 13] holds in the reverse sense for $p \in (0,1)$ so that a similar extension cannot be formulated for these values of exponent p).

Proposition 2.1.36. *Suppose that $F : \Lambda \to Y$ satisfies that $R(F) \subseteq K$ for some compact convex subset K of Y, $1 \leqslant p < +\infty$, $\Lambda' = \Lambda''$ and $\rho = \mathrm{I}$. Suppose further that $\mathbb{F}(l) \equiv l^{-n/p}$ for all $l \in \mathbb{N}$. If $F(\cdot)$ is equi-Weyl-$(\Lambda', \mathbb{F}, \mathcal{P})$-almost periodic, where for each $l > 0$ and $J \in P_l$ we have that the pseudometric on $P_{l,J}$ is given by* (37) *and there exists a bounded sequence $\varphi : \Lambda'' \to [0, \infty)$ such that $\nu(x+y) \leqslant \nu(x)\varphi(y)$ for all $x \in \Lambda$ and $y \in \Lambda''$. Then for each $\varepsilon > 0$ there exist a Bohr ν-almost periodic function $H : \Lambda \to Y$ [i. e., for every $\varepsilon > 0$, there exists $L > 0$ such that, for every $\mathbf{t}_0 \in \Lambda''$, there exists $\tau \in B(\mathbf{t}_0, l) \cap \Lambda''$ such that $\|[H(\mathbf{t}+\tau) - H(\mathbf{t})] \cdot \nu(\mathbf{t})\|_Y \leqslant \varepsilon$ for all $\mathbf{t} \in \Lambda$] with values in K and an integer $l \in \mathbb{N}$ such that, for every $J \in P_l$, we have*

$$l^{-n/p}\Big[\sum_{j \in J} \|F(j) - H(j)\|^p \nu^p(j)\Big]^{1/p} \leqslant \varepsilon.$$

We will include the main details of the proof of the following extension of [476, Proposition 7].

Proposition 2.1.37. *Suppose that $F : \Lambda \to Y$ is a given sequence, $p > 0$, $\Lambda' = \Lambda''$, $\rho = T \in L(Y)$ and for each $\varepsilon > 0$ there exist a Bohr (T, ν)-almost periodic sequence $H : \Lambda \to Y$ [that is, for every $\varepsilon > 0$, there exists $L > 0$ such that, for every $\mathbf{t}_0 \in \Lambda''$, there exists $\tau \in B(\mathbf{t}_0, l) \cap \Lambda''$ such that $\|[H(\mathbf{t}+\tau) - TH(\mathbf{t})] \cdot \nu(\mathbf{t})\|_Y \leqslant \varepsilon$ for all $\mathbf{t} \in \Lambda$] and an integer $l \in \mathbb{N}$ such that, for every $J \in P_l$, we have*

$$\mathbb{F}(l)\Big[\sum_{j \in J} \|F(j;x) - H(j;x)\|^p \nu^p(j)\Big]^{1/p} \leqslant \varepsilon, \quad \text{if } p \geqslant 1, \quad \text{respectively,}$$

$$\mathbb{F}(l) \sum_{j \in J} \|F(j;x) - H(j;x)\|^p \nu^p(j) \leqslant \varepsilon, \quad \text{if } p \in (0,1).$$

Suppose further that there exists a bounded sequence $\varphi : \Lambda \to [0,\infty)$ *such that* $v(x+y) \leqslant v(x)\varphi(y)$ *for all* $x \in \Lambda$, $y \in \Lambda''$ *and*

$$\mathbb{F}(l)\Big[\sum_{j\in J} v^p(j)\Big]^{1/p} \leqslant \varepsilon, \quad \textit{if } p \geqslant 1, \quad \textit{respectively,} \tag{39}$$

$$\mathbb{F}(l)\sum_{j\in J} v^p(j) \leqslant \varepsilon, \quad \textit{if } p \in (0,1).$$

Then $F(\cdot)$ *is equi-Weyl-*$(\mathbb{F}, \mathcal{P}, T)$*-almost periodic, where the pseudometric on* $P_{l,J}$ *is given by* (37) *for* $p \geqslant 1$*, respectively,* (38) *for* $p \in (0,1)$.

Proof. Let $L > 0$ and $\tau \in B(\mathbf{t}_0, l) \cap \Lambda''$ be as in the formulation of proposition. Furthermore, let $\varepsilon > 0$ be given and let an integer $l > 0$ satisfy the prescribed assumption. Fix $J \in P_l$. Using the decomposition,

$$\begin{aligned}
&\|F(\mathbf{t}+\tau) - TF(\mathbf{t})\|_Y \\
&\quad\leqslant \|F(\mathbf{t}+\tau) - H(\mathbf{t}+\tau)\|_Y + \|H(\mathbf{t}+\tau) - TH(\mathbf{t})\|_Y \\
&\qquad + \|T\| \cdot \|H(\mathbf{t}) - F(\mathbf{t})\|_Y, \quad \mathbf{t} \in \Lambda,\ \tau \in \Lambda''
\end{aligned}$$

and our assumptions on the sequence $v(\cdot)$, we get the existence of a finite real constant $c_p > 0$ such that

$$\begin{aligned}
&\Big[\sum_{j\in J}\|F(j+\tau) - F(j)\|^p v^p(j)\Big]^{1/p} \leqslant c_p(1+\|\varphi\|_\infty) \\
&\times\Big\{\Big[\sum_{j\in J}\|F(j+\tau) - H(j+\tau)\|^p v^p(j+\tau)\Big]^{1/p} + \Big[\sum_{j\in J}\|H(j+\tau) - TH(j)\|^p v^p(j)\Big]^{1/p} \\
&+ \|T\| \cdot \Big[\sum_{j\in J}\|F(j) - H(j)\|^p v^p(j)\Big]^{1/p}\Big\}, \quad \text{if } p \geqslant 1.
\end{aligned}$$

Keeping in mind the estimate (39), this simply implies the required; similarly, we can consider the case $p \in (0,1)$. □

Suppose now that, for every $l \in \mathbb{N}$, (P_l, d_l) is a pseudometric space, where $P_l \subseteq Y^{[-l,l]^n \cap \Lambda}$ is closed under the addition and subtraction of functions, and $0 \in P_l$. Define $\|f\|_l := d_l(f, 0)$ for all $f \in P_l$. For our later purposes, we will introduce here the following metrical generalization of the notion considered in [476, Definition 8].

Definition 2.1.38. Suppose that $F : \Lambda \times X \to Y$ is a given sequence and $\mathbb{F} : (0,\infty) \to [0,\infty)$. Then we say that $F(\cdot;\cdot)$ is Besicovitch-$(\mathcal{B}, \mathbb{F}, \mathcal{P})$-almost periodic, if for every $\varepsilon > 0$ and $B \in \mathcal{B}$, there exists a trigonometric polynomial $P(\cdot;\cdot)$ such that

$$\sup_{x\in B}\limsup_{l\to+\infty} \mathbb{F}(l)\|F(\cdot;x) - P(\cdot;x)\|_l < \varepsilon.$$

We will not consider here the completeness of the space of Besicovitch-$(\mathcal{B}, \mathbb{F}, \mathcal{P})$-almost periodic sequences; cf. [476, Theorem 6] and [476, Corollary 3] for more details in this direction.

2.1.5 Some applications

In this subsection, we will provide certain applications of the introduced notion to the abstract Volterra difference equations. Before doing this, we would like to notice that we have recently analyzed the existence and uniqueness of Weyl, Besicovitch and Doss almost periodic type solutions to the abstract impulsive Volterra integrodifferential equations in [267]. Concerning the statement of [267, Theorem 8], we would like to make the following comment: Let us replace the condition (ew-M1), respectively, (w-M1), in the formulation of this result with the following condition:

(ew-M1-T) For every $\varepsilon > 0$, there exist $s \in \mathbb{N}$ and $L > 0$ such that every interval $I' \subseteq [0, \infty)$ of length L contains a point $\tau \in I'$, which satisfies that there exists an integer $q_\tau \in \mathbb{N}$ such that $|t_{i+q_\tau} - t_i - \tau| < \varepsilon$ for all $i \in \mathbb{N}$ and

$$\sup_{|J|=s} \left[\frac{1}{s} \sum_{j \in J} \| y_{j+q_\tau} - T y_j \|^p \right]^{1/p} < \varepsilon, \tag{40}$$

where the supremum is taken over all segments $J \subseteq \mathbb{N}$ of length s and $\rho = T \in L(X)$.

(w-M1-T) For every $\varepsilon > 0$, there exists $L > 0$ such that every interval $I' \subseteq [0, \infty)$ of length L contains a point $\tau \in I'$, which satisfies that there exist an integer $q_\tau \in \mathbb{N}$ and an integer $s_\tau \in \mathbb{N}$ such that $|t_{i+q_\tau} - t_i - \tau| < \varepsilon$ for all integers $i \in \mathbb{N}$ and (40) holds for all integers $s \geqslant s_\tau$, with $\rho = T \in L(X)$.

Then the function $G_2 : [0, \infty) \to X$ from the formulation of the above mentioned result will be (equi-)Weyl-(p, T)-almost periodic (see [447] for the notion); a similar comment can be made in the case of consideration of [267, Theorem 9]. Observe that it would be very difficult to say anything relevant if the term $1/s$ in (40) is replaced with the term $1/s^\sigma$, where $\sigma \in (0, 1)$. An attempt should be made to extend the result of [267, Theorem 8] for metrically (equi-)Weyl-p-almost periodic sequences.

Let us recall that the family of sequences $(t_k^j)_{k \in \mathbb{Z}}$ $[(t_k^j)_{k \in \mathbb{N}}]$, $j \in \mathbb{Z}$ $[j \in \mathbb{N}]$ is called equipotentially almost periodic, if for every $\varepsilon > 0$, there exists a relatively dense set Q_ε in $\mathbb{R}$ [in $[0, \infty)$] such that for each $\tau \in Q_\varepsilon$ there exists an integer $q \in \mathbb{Z}$ $[q \in \mathbb{N}]$ such that $|t_{i+q} - t_i - \tau| < \varepsilon$ for all $i \in \mathbb{Z}$ $[i \in \mathbb{N}]$. In connection with our results established in [267], we will also provide the following example.

Example 2.1.39. Suppose that the family of sequences $(t_k^j)_{k \in \mathbb{Z}}$, $j \in \mathbb{Z}$ is equipotentially almost periodic. Then we know that there exist $\zeta \in \mathbb{R} \setminus \{0\}$ and an almost periodic function $a : \mathbb{R} \to \mathbb{R}$ such that $t_k = \zeta k + a(k)$ for all $k \in \mathbb{Z}$. The function $g(x) := \zeta x + a(x)$,

$x \in \mathbb{R}$ has the property that the function $x \mapsto f(x) = g(x) - \zeta x$, $x \in \mathbb{R}$ is almost periodic; then we can apply the theorem of H. Bohr concerning the argument of a complex-valued almost periodic function in order to see that the function $x \mapsto \exp(ig(x))$, $x \in \mathbb{R}$ is almost periodic. This, in particular, implies that $(e^{it_k})_{k\in\mathbb{Z}}$ is an almost periodic sequence. Of course, the converse statement is not true since $(e^{i[2\pi k^2]})_{k\in\mathbb{Z}}$ is an almost periodic sequence but the sequence $(t_k \equiv 2\pi k^2)_{k\in\mathbb{Z}}$ does not satisfy that the family of sequences $(t_k^j)_{k\in\mathbb{Z}}$, $j \in \mathbb{Z}$ is equipotentially almost periodic.

It would be very difficult to summarize all relevant results concerning the almost periodic type solutions to the abstract Volterra difference equations. For more details on the subject, we refer the reader to the research monographs by R. P. Agarwal [19], R. P. Agarwal, C. Cuevas, C. Lizama [21], J. Banasiak [86], S. S. Cheng [197], S. Elaydi [272] as well as to the research articles [1–4, 24, 32, 33, 35, 56, 65, 82, 120, 186, 187, 212, 221, 261, 313, 327, 328, 333, 363, 364, 368–370, 375, 389, 425, 429, 436, 538, 541, 542, 637, 678, 696, 712–716] and [783, 784, 789, 819, 820, 822, 821]; cf. also the research monograph [20], the references quoted in [444, Section 3.11] and [447, p. 303] as well as the works of I. A. Trishina [674, 675] and [749, 750].

We will divide the remainder of this subsection into three individual parts:

1. On the abstract difference inclusion $u(k+1) \in \mathcal{A}u(k) + f(k)$

In [65, Section 3], D. Araya, R. Castro and C. Lizama have considered the almost automorphic solutions of the first-order linear difference equation

$$u(k+1) = Au(k) + f(k), \quad k \in \mathbb{Z}, \tag{41}$$

where $A \in L(X)$ and $(f_k \equiv f(k))_{k\in\mathbb{Z}}$ is an almost automorphic sequence (cf. also [115, 247, 581] for similar results obtained in the stochastic framework, with almost periodic forcing terms $(f_k)_{k\in\mathbb{Z}}$). In this part, we will expand the above mentioned research study by assuming that $(f_k)_{k\in\mathbb{Z}}$ is a generalized almost periodic sequence.

We will first assume that $A = \lambda I$, where $\lambda \in \mathbb{C}$ and $|\lambda| \neq 1$. Due to [65, Theorem 3.1], we know that the almost automorphy of sequence $(f_k)_{k\in\mathbb{Z}}$ implies the existence of an almost automorphic solution $u(\cdot)$ of (41), which is given by

$$u(k) = \sum_{m=-\infty}^{k} \lambda^{k-m} f(m-1), \quad k \in \mathbb{Z}, \tag{42}$$

if $|\lambda| < 1$, and

$$u(k) = -\sum_{m=k}^{\infty} \lambda^{k-m-1} f(m), \quad k \in \mathbb{Z}, \tag{43}$$

if $|\lambda| > 1$. Before proceeding any further, we would like to note that this is a unique almost automorphic solution of (41); in actual fact, any almost automorphic sequence is bounded and we only need to show the uniqueness of bounded solutions of (41). But this can be proved in an almost trivial way; furthermore, there exists a unique polynomially bounded solution of (42).

We will first prove the following result.

Theorem 2.1.40. *Suppose that* $\mathbb{F} : (0,\infty) \to (0,\infty)$, $1 \leqslant p < +\infty$, $\rho = T \in L(X)$, (FV) *holds and* $f(\cdot)$ *is equi-Weyl-*$(\mathbb{F},p,T)$*-almost periodic [polynomially bounded Weyl-*$(\mathbb{F},p,T)$*-almost periodic; polynomially bounded Doss-*$(\mathbb{F},p,T)$*-almost periodic]. Then a unique (equi-)Weyl-*$(\mathbb{F},p,T)$*-almost periodic [polynomially bounded Weyl-*$(\mathbb{F},p,T)$*-almost periodic; polynomially bounded Doss-*$(\mathbb{F},p,T)$*-almost periodic] solution of* (41) *is given by* (42) *if* $|\lambda| < 1$, *and* (43) *if* $|\lambda| > 1$.

Proof. We will consider the class of equi-Weyl-$(\mathbb{F},p,T)$-almost periodic sequences only. Due to Proposition 2.1.16, we know that $(f_k)_{k\in\mathbb{Z}}$ is a bounded sequence and it remains to be proved that the function $u(\cdot)$, given by (42) if $|\lambda| < 1$, and (43) if $|\lambda| > 1$, is an (equi-)Weyl-$(\mathbb{F}.,p,T)$-almost periodic solution of (41). Clearly, the function $u(\cdot)$ is well-defined and bounded; for simplicity, we will consider henceforth the case $|\lambda| < 1$. Let $\varepsilon > 0$ be given. Then we know that there exist $l \in \mathbb{N}$ and $L > 0$ such that, for every $t_0 \in \mathbb{Z}$, there exists $\tau \in B(t_0,l) \cap \mathbb{Z}$ such that

$$\sup_{k\in\mathbb{Z}} \mathbb{F}(l)\Biggl[\sum_{j=k}^{k+l}\|f(j+\tau) - Tf(j)\|^p\Biggr]^{1/p} \leqslant \varepsilon. \tag{44}$$

Let $k \in \mathbb{Z}$ be fixed. Then we have

$$\begin{aligned}
&\mathbb{F}(l)\Biggl[\sum_{j=k}^{k+l}\|u(j+\tau) - Tu(j)\|^p\Biggr]^{1/p} \\
&\quad = \mathbb{F}(l)\Biggl[\sum_{j=k}^{k+l}\Biggl\|\sum_{v=0}^{\infty}\lambda^v[f(j+\tau-v-1) - Tf(j-v-1)]\Biggr\|^p\Biggr]^{1/p} \\
&\quad \leqslant \mathbb{F}(l)\Biggl[\sum_{j=k}^{k+l}\Biggl|\sum_{v=0}^{\infty}|\lambda|^v\|f(j+\tau-v-1) - Tf(j-v-1)\|\Biggr|^p\Biggr]^{1/p}.
\end{aligned}$$

Suppose now that $\zeta > 1/p$. Using the Hölder inequality, we obtain that there exists a finite real constant $c_\lambda > 0$ such that

$$\Biggl|\sum_{v=0}^{\infty}|\lambda|^v\|f(j+\tau-v-1) - Tf(j-v-1)\|\Biggr|^p \leqslant c_\lambda^p\sum_{v=0}^{\infty}\frac{1}{(1+v^\zeta)^p}\|f(j+\tau-v-1) - Tf(j-v-1)\|^p,$$

so that the last estimate in the above calculation and (44) together imply that

$$
\begin{aligned}
&\mathbb{F}(l)\left[\sum_{j=k}^{k+l}\|u(j+\tau)-Tu(j)\|^p\right]^{1/p}\\
&\leqslant c_\lambda \mathbb{F}(l)\left[\sum_{j=k}^{k+l}\sum_{v=0}^{\infty}\frac{1}{(1+v^\zeta)^p}\|f(j+\tau-v-1)-Tf(j-v-1)\|^p\right]^{1/p}\\
&= c_\lambda \mathbb{F}(l)\left[\sum_{v=0}^{\infty}\frac{1}{(1+v^\zeta)^p}\sum_{j=k}^{k+l}\|f(j+\tau-v-1)-Tf(j-v-1)\|^p\right]^{1/p}\\
&\leqslant c_\lambda \mathbb{F}(l)\left[\sum_{v=0}^{\infty}\frac{1}{(1+v^\zeta)^p}(\varepsilon/\mathbb{F}(l))^p\right]^{1/p} = \varepsilon c_\lambda\left[\sum_{v=0}^{\infty}\frac{1}{(1+v^\zeta)^p}\right]^{1/p},
\end{aligned}
$$

completing the proof. □

We can similarly prove the following result.

Theorem 2.1.41. *Suppose that $\mathbb{F} : (0,\infty) \to (0,\infty)$, $1 \leqslant p < +\infty$, $\rho = T \in L(X)$ and $f(\cdot)$ is equi-Weyl-$(\mathbb{F},\mathcal{P},T)$-almost periodic [polynomially bounded Weyl-$(\mathbb{F},\mathcal{P},T)$-almost periodic; polynomially bounded Doss-$(\mathbb{F},\mathcal{P},T)$-almost periodic], where for each $l > 0$ and $J \in P_l$ the pseudometric on $P_{l,J}$ is given by* (37) *with the sequence $v : \mathbb{Z} \to [0,\infty)$ satisfying that there exist a sequence $\psi : \mathbb{Z} \to (0,\infty)$ and a number $\sigma > 0$ such that $v(x+y) \leqslant v(x)\psi(y)$ for all $x, y \in \mathbb{Z}$ and $\sum_{v=0}^{+\infty}[\psi(v+1)]^p/(1+v^\sigma)^p < +\infty$. Then a unique (equi-)Weyl-$(\mathbb{F},\mathcal{P},T)$-almost periodic [polynomially bounded Weyl-$(\mathbb{F},\mathcal{P},T)$-almost periodic; polynomially bounded Doss-$(\mathbb{F},\mathcal{P},T)$-almost periodic] solution of* (41) *is given by* (42) *if $|\lambda| < 1$, and* (43) *if $|\lambda| > 1$.*

In general case, we have the following result.

Theorem 2.1.42. *Suppose that $\mathbb{F} : (0,\infty) \to (0,\infty)$, $1 \leqslant p < +\infty$, $\rho = T \in L(X)$,* (FV) *holds and $f(\cdot)$ is equi-Weyl-$(\mathbb{F},p,T)$-almost periodic [polynomially bounded Weyl-$(\mathbb{F},p,T)$-almost periodic; polynomially bounded Doss-$(\mathbb{F},p,T)$-almost periodic]. Then there exists an (equi-)Weyl-$(\mathbb{F},p,T)$-almost periodic [polynomially bounded Weyl-$(\mathbb{F},p,T)$-almost periodic; polynomially bounded Doss-$(\mathbb{F},p,T)$-almost periodic] solution of* (41), *provided that $A \in L(X)$ and $\|A\| < 1$.*

Keeping in mind Theorem 2.1.40, the statement of [65, Theorem 3.2] can be simply reformulated for the generalized almost periodic solutions considered there. For simplicity, we will only state the following result.

Theorem 2.1.43. *Suppose that A is a complex matrix, which satisfies that its point spectrum is disjoint from the unit sphere $S_1 := \{\lambda \in \mathbb{C} : |\lambda| = 1\}$. Suppose further that $\mathbb{F} : (0,\infty) \to (0,\infty)$, $1 \leqslant p < +\infty$,* (FV) *holds and $f(\cdot)$ is equi-Weyl-$(\mathbb{F},p,\mathrm{I})$-almost periodic. Then there exists a unique equi-Weyl-$(\mathbb{F},p,\mathrm{I})$-almost periodic solution of* (41).

The interested reader is advised to extend the last two statements for metrically generalized ρ-almost periodic sequences.

We can similarly prove the analogues of Theorem 2.1.40, Theorem 2.1.42 and Theorem 2.1.43 for the generalized Besicovitch almost periodic solutions of (42). For the sake of brevity, we will only state and prove the following analogue of Theorem 2.1.42 here.

Theorem 2.1.44. *Suppose that* $\mathbb{F} : (0,\infty) \to (0,\infty)$, $1 \leqslant p < +\infty$, $\mathbb{F}_1 : (0,\infty) \to (0,\infty)$ *and* $f(\cdot)$ *is polynomially bounded Besicovitch-*$(\mathbb{F}, \mathcal{P})$*-almost periodic, where for each* $l > 0$ *the pseudometric on* P_l *is given by*

$$\|f\|_l \equiv \left[\sum_{j\in[-l,l]\cap\mathbb{Z}} \|f(j)\|^p v^p(j)\right]^{1/p}, \quad f \in P_l,$$

with the sequence $v : \mathbb{Z} \to [0,\infty)$ *satisfying that there exist a sequence* $\psi : \mathbb{Z} \to (0,\infty)$ *and a number* $\sigma > 0$ *such that* $v(x+y) \leqslant v(x)\psi(y)$ *for all* $x, y \in \mathbb{Z}$ *and* $\sum_{v=0}^{+\infty}[\psi(v+1)]^p/(1+v^\sigma)^p < +\infty$. *Suppose further that for each* $\varepsilon > 0$ *there exist* $l_0 > 0$, $c > 0$ *and* $k > 0$ *such that, for every* $l \geqslant l_0$ *and* $v \in \mathbb{N}_0$, *we have*

$$\frac{\mathbb{F}_1(l)}{\mathbb{F}(l+v)} \leqslant c(1+v)^k.$$

If

$$\limsup_{l\to+\infty} \mathbb{F}_1(l)\left[\sum_{j\in[-l,l]\cap\mathbb{Z}} v^p(j)\right]^{1/p} < +\infty, \tag{45}$$

then a unique polynomially bounded Besicovitch-$(\mathbb{F}_1, \mathcal{P})$*-almost periodic solution of* (41) *is given by* (42) *if* $|\lambda| < 1$, *and* (43) *if* $|\lambda| > 1$.

Proof. We will consider the case $|\lambda| < 1$ only. Let $\varepsilon > 0$ be fixed. Then there exists a trigonometric polynomial $p(\cdot)$ such that

$$\limsup_{l\to+\infty} \mathbb{F}(l)\left[\sum_{j\in[-l,l]\cap\mathbb{Z}} \|f(j)-p(j)\|^p\right]^{1/p} < \varepsilon.$$

The sequence $j \mapsto u_p(j) \equiv \sum_{v=0}^{+\infty} \lambda^v p(j-v-1)$, $j \in \mathbb{Z}$ is almost periodic, as simply approved. Keeping in mind the assumptions on the weight sequence $v(\cdot)$ and the function $\mathbb{F}_1(\cdot)$, we can repeat verbatim the argumentation contained in the proof of [476, Theorem 7] in order to see that

$$\limsup_{l\to+\infty} \mathbb{F}_1(l)\left[\sum_{j\in[-l,l]\cap\mathbb{Z}} \|u(j)-u_p(j)\|^p v^p(j)\right]^{1/p} < \varepsilon. \tag{46}$$

Now we can find a trigonometric polynomial $P(\cdot)$ such that $\|P_1(j) - u_p(j)\| \leqslant \varepsilon$ for all $j \in \mathbb{Z}$. Using this estimate, (45) and (46), we simply get

$$\limsup_{l\to+\infty} \mathbb{F}_1(l)\left[\sum_{j\in[-l,l]\cap\mathbb{Z}} \|u(j)-P(j)\|^p v^p(j)\right]^{1/p} < \text{Const.}\cdot\varepsilon.$$

□

We will not reconsider here the statements of [65, Theorems 3.6–3.9]. For the sequel, we need the following result.

Theorem 2.1.45. *Suppose that (f_k) is a (polynomially) bounded sequence, $\mathcal{A}$ is a closed* MLO *such that $\mathcal{A}^{-v}C \in L(X)$, $v \in \mathbb{N}$ and $\sum_{v=1}^{+\infty} \|\mathcal{A}^{-v}C\| < +\infty$. Then the abstract difference inclusion*

$$u(k+1) \in \mathcal{A}u(k) + Cf(k), \quad k \in \mathbb{Z}, \tag{47}$$

has a (polynomially) bounded solution $u(\cdot)$, given by

$$u(k) = -\sum_{m=k}^{\infty} \mathcal{A}^{k-m-1}Cf(m) = -\sum_{v=1}^{\infty} \mathcal{A}^{-v}Cf(k-1+v), \quad k \in \mathbb{Z}. \tag{48}$$

Moreover, if (f_k) is almost periodic (almost automorphic), then (u_k) is almost periodic (almost automorphic).

Proof. It is clear that $u(\cdot)$ is well-defined and (polynomially) bounded because (f_k) is a (polynomially) bounded sequence. Furthermore, $\mathcal{A}$ is closed and we have

$$\begin{aligned} u(k+1) &= -\sum_{m=k+1}^{\infty} \mathcal{A}^{k-m}Cf(m) = Cf(k) - \sum_{m=k}^{\infty} \mathcal{A}^{k-m}Cf(m) \\ &\in -\mathcal{A}\sum_{m=k}^{\infty} \mathcal{A}^{k-m-1}Cf(m) + Cf(k), \quad k \in \mathbb{Z}, \end{aligned}$$

which simply implies the required. If (f_k) is almost periodic, then we can simply prove that (u_k) is almost periodic, as well; the corresponding statement for the almost automorphic sequences follows from the dominated convergence theorem and the same argumentation as in the case that $\mathcal{A} = \lambda I$, where $|\lambda| < 1$. □

The assumptions of Theorem 2.1.45 are satisfied if $C = I$ and $\|\mathcal{A}^{-1}\| < 1$; in this case, we can simply show that there does not exist an eigenvalue of $\mathcal{A}$, which belongs to the unit sphere in $\mathbb{C}$. In particular, if $\mathcal{A} = A$ is a regular complex matrix such that $\|A^{-1}\| < 1$, then we can always employ [65, Theorem 3.2] in the study of almost periodic (almost automorphic) solutions of (47).

On the other hand, it is clear that Theorem 2.1.45 can be applied in the analysis of the existence and uniqueness of almost periodic and almost automorphic solutions for a large class of the abstract differential inclusions in Banach spaces, when the results established in [65] are inapplicable. For example, we can consider the following semidiscrete Poisson heat equation in the space $L^p(\Omega)$:

$$(ACP): \begin{cases} m(x)v(k+1,x) = \Delta v(k,x) - bv(k,x), & k \in \mathbb{Z},\ x \in \Omega; \\ v(k,x) = 0, & (k,x) \in \mathbb{Z} \times \partial\Omega, \end{cases}$$

where Ω is a bounded domain in $\mathbb{R}^n$, $b > 0$, $m(x) \geqslant 0$ a. e. $x \in \Omega$, $m \in L^\infty(\Omega)$ and $1 < p < \infty$. In actual fact, if B is the multiplication in $L^p(\Omega)$ with $m(x)$, and $A = \Delta - b$ acts with the Dirichlet boundary conditions, then the resolvent of multivalued linear operator $\mathcal{A} = AB^{-1}$ contains an obtuse angle with zero being part of its interior. After possible division of $\mathcal{A}$ with a certain positive constant $c > 0$, we obtain $\|(\mathcal{A}/c)^{-1}\| < 1$ and Theorem 2.1.45 can be applied, with $C = \mathrm{I}$, after making the substitution $u(k, x) \equiv m(x)v(k, x)$. If $C \neq \mathrm{I}$, then Theorem 2.1.45 can be applied to the abstract first-order semidiscrete Cauchy problems with the (noncoercive) differential operators with constant coefficients in $L^p(\mathbb{R}^n)$, $BUC(\mathbb{R}^n)$ or $C_b(\mathbb{R}^n)$; let us only recall that such operators can have the empty resolvent set; see [444, 445] for more details.

For more examples of multivalued linear operators $\mathcal{A}$ for which we have $0 \in \rho(\mathcal{A})$ and $\|\mathcal{A}^{-1}\| < 1$, we refer the reader to the monographs [287] and [445] (it is worth noting that, after a suitable translation of $\mathcal{A}$ and a division of $\mathcal{A}$ with a certain positive constant, the applications of Theorem 2.1.45 can be always make with any multivalued linear operator $\mathcal{A}$ such that $\rho(\mathcal{A}) \neq \emptyset$). It is worth noting that the requirements of Theorem 2.1.45 are satisfied provided that there exists a real number $R > 1$ such that the mapping $\lambda \mapsto (\lambda - \mathcal{A})^{-1}C \in L(X)$, $\lambda \in \{z \in \mathbb{C} : |z| < R\}$ is continuous; cf. the proof of Theorem 2.1.51(iii) below. Keeping in mind Theorem 2.1.45, the statements of Theorem 2.1.40, Theorem 2.1.41, Theorem 2.1.42 and Theorem 2.1.44 can be simply formulated for the abstract first-order difference inclusion (47), provided that $0 \in \rho(\mathcal{A})$ and $\|\mathcal{A}^{-1}\| < 1$; details can be left to the interested readers.

The following counterexample shows that we cannot so easily apply Theorem 2.1.45 in the analysis of the existence and uniqueness of generalized almost periodic (automorphic) solutions of the abstract second-order difference inclusions.

Example 2.1.46. Of concern is the following abstract second-order difference inclusion

$$\Delta^2 u(k) \in \mathcal{A}u(k) + f(k), \quad k \in \mathbb{Z}. \tag{49}$$

Plugging $v_k := \Delta u_k$, $k \in \mathbb{Z}$, this inclusion can be equivalently rewritten as the first-order difference inclusion

$$(u_{k+1}, v_{k+1})^T \in \mathcal{B}(u_k, v_k)^T + (0, f_k)^T, \quad k \in \mathbb{Z},$$

where the multivalued linear operator $\mathcal{B}$ acts on the Banach space X^2 and is defined through $D(\mathcal{B}) := D(\mathcal{A}) \times X$ and $\mathcal{B}(x, y) := \{(x + y, z + y) : z \in \mathcal{A}x\}$, $x \in D(\mathcal{A})$, $y \in X$. The proof of the following simple result can be left to the enthusiastic readers: Suppose that $1 \in \rho(\mathcal{A})$. Then $0 \in \rho(\mathcal{B})$ and we have

$$\mathcal{B}^{-1}(x, y)^T = ((\mathrm{I} - \mathcal{A})^{-1}x - (\mathrm{I} - \mathcal{A})^{-1}y,\ x - (\mathrm{I} - \mathcal{A})^{-1}x + (\mathrm{I} - \mathcal{A})^{-1}y)^T, \quad x, y \in X.$$

Keeping in mind this formula, it readily follows that

$$\|\mathcal{B}^{-1}\| \geqslant \sup_{\|x\| \leqslant 1} [\|(\mathrm{I} - \mathcal{A})^{-1}x\| + \|x - (\mathrm{I} - \mathcal{A})^{-1}x\|] \geqslant \sup_{\|x\| \leqslant 1} \|x\| = 1,$$

so that Theorem 2.1.45 cannot be applied in the analysis of the existence of almost periodic (almost automorphic) solutions of (49). A similar argumentation shows that we cannot apply Theorem 2.1.45 in the analysis of the existence of almost periodic (almost automorphic) solutions of the complete second-order difference inclusion

$$\Delta^2 u(k) \in B\Delta u(k) + \mathcal{A}u(k) + f(k), \quad k \in \mathbb{Z},$$

where $B \in L(X)$ and $1 \in \rho(\mathcal{A} + B)$.

For more details about the existence and uniqueness of almost automorphic solutions of second-order equations involving time scales with boundary conditions, we refer the reader to the recent paper [202] by M. F. Choquehuanca, J. G. Mesquita and A. Pereira; some interesting results about the existence and uniqueness of weighted pseudoasymptotic periodic solutions to the scalar second-order nonautonomous difference equations have been given by Z. Xia in [786] (cf. also the research articles [574] by D. Maroncelli and [575] by D. Maroncelli, J. Rodríguez).

Concerning the generalized almost periodic type solutions of the following semilinear analogue of (41):

$$u(k+1) = Au(k) + f(k; u(k)), \quad k \in \mathbb{Z},$$

where $A \in L(X)$, we will only note that the already established results can be used to deduce the variants of [65, Theorems 4.1, 4.3] for a class of bounded c-uniformly recurrent sequences (cf. [430, Theorems 2.28, 3.1]) and a class of bounded slowly oscillating sequences (cf. [451, Theorem 5] and the paragraph following it). We will consider the composition principles for the generalized Weyl, Besicovitch and Doss almost periodic sequences somewhere else (cf. [444–448] for continuous versions).

Before proceeding to the next part, we would like to recommend for the readers the doctoral dissertation of M. Veselý [760], where we have located a great number of results about the existence and uniqueness of almost periodic solutions of the abstract difference equation $u(k+1) = A_k u(k) + f(k)$, $k \in \mathbb{Z}$, where $(A_k)_{k\in\mathbb{Z}}$ is a sequence of closed linear operators satisfying certain properties.

2. On the abstract fractional difference inclusion $\Delta^\alpha u(k) \in \mathcal{A}u(k+1) + f(k)$
If $\alpha > 0$ and $v \in \mathbb{N}_0$, then the Cesàro sequence $(k^\alpha(v))_{v\in\mathbb{N}_0}$ is defined by

$$k^\alpha(v) := \frac{\Gamma(v+\alpha)}{\Gamma(\alpha)v!}.$$

Then we know that, for every $\alpha > 0$ and $\beta > 0$, we have $k^\alpha *_0 k^\beta \equiv k^{\alpha+\beta}$ and $|k^\alpha(v) - g_\alpha(v)| = O(g_\alpha(v)|1/v|)$, $v \in \mathbb{N}$; in particular, $k^\alpha(v) \sim g_\alpha(v)$, $v \to +\infty$; if $0 < \alpha \leqslant 1$, then we also know that

$$g_\alpha(v+1) < k^\alpha(v) < g_\alpha(v), \quad v \in \mathbb{N}.$$

Furthermore, if $0 < \alpha \leqslant 1$ and $\sum_{k\in\mathbb{Z}} \|u(k)\| \cdot g_\alpha(|k|) < +\infty$, then we define the Weyl fractional derivative $\Delta^\alpha u$ by

$$\Delta_W^\alpha u := \Delta\left(\sum_{j=-\infty}^{\cdot} k^\alpha(\cdot - v)u(j) \right).$$

In the recent research article [46], E. Alvarez, S. Díaz and C. Lizama have analyzed the existence and uniqueness of (N, λ)-periodic solutions for the abstract fractional difference equation

$$\Delta_W^\alpha u(k) = Au(k+1) + f(k), \quad k \in \mathbb{Z}, \tag{50}$$

where A is a closed linear operator on X, $0 < \alpha \leqslant 1$ and $\Delta^\alpha u(k)$ denotes the Weyl fractional difference operator of order α. The asymptotically almost periodic mild solutions to (50) have been also sought by J. Cao, B. Samet and Y. Zhou in [164]; cf. also [540], where C. Lizama, M. Murillo-Arcila and C. Leal have analyzed the Lebesgue regularity for differential difference equations with two different fractional damping terms, as well as [562], where W. Lv and J. Feng have analyzed some nonlinear discrete fractional mixed type sum-difference equation boundary value problems in Banach spaces with the Riemann–Liouville fractional derivatives. For the interested readers, we can also recommend the research articles [218, 526, 527, 560, 561, 647] and [823].

In this part, we will use the same notion and notation as in the above mentioned paper [46], providing a few important observations about [46, Theorem 4.2]. Suppose that A is a closed linear operator on X such that $1 \in \rho(A)$, where $\rho(A)$ denotes the resolvent set of A, and $\|(\mathrm{I} - A)^{-1}\| < 1$. Then [46, Theorem 3.4] implies that $\mathcal{A}$ generates a discrete (α, α)-resolvent sequence $\{S_{\alpha,\alpha}(v)\}_{v\in\mathbb{N}_0}$ such that $\sum_{v=0}^{+\infty} \|S_{\alpha,\alpha}(v)\| < +\infty$. If $(f_k)_{k\in\mathbb{Z}}$ is a bounded sequence, then we know that the function

$$u(k) = \sum_{l=-\infty}^{k-1} S_{\alpha,\alpha}(k-1-l)f(l), \quad k \in \mathbb{Z} \tag{51}$$

is a mild solution of (50). Since $\sum_{v=0}^{+\infty} \|S_{\alpha,\alpha}(v)\| < +\infty$, we can almost directly conclude that $u(t)$ will be T-almost periodic (T-uniformly recurrent), where $\rho = T \in L(X)$, provided that $(f_k)_{k\in\mathbb{Z}}$ is T-almost periodic (T-uniformly recurrent). Concerning the generalized almost periodic type solutions of (50), we are in a position to directly clarify certain results in the case that (FV) holds and the forcing term $f(\cdot)$ is equi-Weyl-$(\mathbb{F}, 1, T)$-almost periodic [bounded Weyl-$(\mathbb{F}, 1, T)$-almost periodic; bounded Doss-$(\mathbb{F}, 1, T)$-almost periodic; bounded Besicovitch-$(\mathbb{F}, 1)$-almost periodic]. In actual fact, if $f(\cdot)$ enjoys this feature, then a mild solution $u(\cdot)$ of (50), given by (51), enjoys the same feature as well. In order to see this, let us assume that $f(\cdot)$ is equi-Weyl-$(\mathbb{F}, 1, T)$-almost periodic, for example. Then $f(\cdot)$ is bounded and we have

$$\begin{aligned}
&\mathbb{F}(l)\sum_{j=k}^{k+l}\|u(j+\tau)-Tu(j)\| \\
&= \mathbb{F}(l)\sum_{j=k}^{k+l}\left\|\sum_{v=0}^{\infty} S_{a,a}(v)[f(j+\tau-1-v)-Tf(j-1-v)]\right\| \\
&\leqslant \mathbb{F}(l)\sum_{j=k}^{k+l}\sum_{v=0}^{\infty}\|S_{a,a}(v)\|\cdot\|f(j+\tau-1-v)-Tf(j-1-v)\| \\
&= \mathbb{F}(l)\sum_{v=0}^{\infty}\sum_{j=k}^{k+l}\|S_{a,a}(v)\|\cdot\|f(j+\tau-1-v)-Tf(j-1-v)\|,
\end{aligned}$$

which simply implies the required ($k, \tau \in \mathbb{Z}; l \in \mathbb{N}$). Unfortunately, the existence of an equi-Weyl-$(\mathbb{F}, p, T)$-almost periodic solution of (50), where $p > 1$, requires further investigations of the solution family $\{S_{a,a}(v)\}_{v\in\mathbb{N}_0}$; basically, the mild solution $u(\cdot)$ of (50), given by (51), enjoys the same feature as the forcing term $f(\cdot)$, provided that $\sum_{v=0}^{\infty}[\|S_{a,a}(v)\|^q \cdot (1+v^{\zeta})^q] < +\infty$ for some $\zeta > 1/p$, where $1/p+1/q = 1$ (this follows from our previous considerations from the first part of Section 2.1.5). We can similarly analyze the existence and uniqueness of equi-Weyl-$(\mathbb{F}, \mathcal{P}, T)$-almost periodic solutions of (50), where $T \in L(X)$ and for each $l > 0$ and $J \in P_l$ the pseudometric on $P_{l,J}$ is given by (37) with $p = 1$ and the sequence $\nu : \mathbb{Z} \to [0, \infty)$ satisfying that there exists a bounded sequence $\psi : \mathbb{Z} \to (0, \infty)$ such that $\nu(x + y) \leqslant \nu(x)\psi(y)$ for all $x, y \in \mathbb{Z}$.

Concerning the generalized almost periodic type solutions of the following semilinear analogue of (50):

$$\Delta_W^a u(k) = Au(k+1) + f(k; u(k)), \quad k \in \mathbb{Z},$$

we will only note that an analogue of [46, Theorem 4.5] can be formulated, e. g., for bounded c-uniformly recurrent sequences and bounded slowly oscillating sequences.

The notion of a discrete (a, C)-resolvent family introduced in the subsequent definition is an extension of the notion of a discrete (a, a)-resolvent family introduced by L. Abadias and C. Lizama in [3, Definition 3.1]; cf. also [46, Definition 2.5]:

Definition 2.1.47. Suppose that $\mathcal{A}$ is a closed multivalued linear operator and $C \in L(X)$ is an injective operator. Then we say that the operator $\mathcal{A}$ is a subgenerator of a discrete (a, C)-resolvent family $(S_a(v))_{v\in\mathbb{N}_0} \subseteq L(X)$ if $\mathcal{A}C \subseteq C\mathcal{A}$, $\mathcal{A}S_a(v) \subseteq S_a(v)\mathcal{A}$ and $CS_a(v) = S_a(v)C$ for all $v \in \mathbb{N}_0$, and the assumption $(x,y) \in \mathcal{A}$ implies

$$S_a(v)x = k^a(v)Cx + \sum_{j=0}^{v} k^a(v-j)S_a(j)y, \quad v \in \mathbb{N}_0. \tag{52}$$

The integral generator $\mathcal{A}_{\mathrm{int}}$ of $(S_a(v))_{v\in\mathbb{N}_0}$ is defined by

$$\mathcal{A}_{\mathrm{int}} := \{(x,y) \in X \times X : (52) \text{ holds}\}.$$

It is a closed MLO, which extends any other subgenerator of $(S_a(v))_{v\in\mathbb{N}_0}$ and satisfies $\mathcal{A}_{\text{int}} = C^{-1}\mathcal{A}_{\text{int}}C$; furthermore, it can be simply proved that for any subgenerator $\mathcal{A}$ of $(S_a(v))_{v\in\mathbb{N}_0}$ we have that $C^{-1}\mathcal{A}C$ is also subgenerator of $(S_a(v))_{v\in\mathbb{N}_0}$. Only after assuming certain extra conditions, we can prove that $\mathcal{A}_{\text{int}}$ is a subgenerator of $(S_a(v))_{v\in\mathbb{N}_0}$ as well as that for any subgenerator $\mathcal{A}$ of $(S_a(v))_{v\in\mathbb{N}_0}$ we have $C^{-1}\mathcal{A}C = \mathcal{A}_{\text{int}}$; cf. [445] for more details given in the continuous framework (it can be also very plausible that a discrete analogue of the functional equality [445, (274), p. 295] can be used to define the notion of a discrete (a, C)-resolvent family).

The following result will be useful henceforth.

Proposition 2.1.48. *Suppose that $\mathcal{A}$ is a subgenerator of a discrete (a, C)-resolvent family $(S_a(v))_{v\in\mathbb{N}_0}$. Then $S_a(v)x = 0$ for all $x \in \mathcal{A}0$ and $v \in \mathbb{N}_0$.*

Proof. Since $(0, x) \in \mathcal{A}$, (52) immediately implies $\sum_{j=0}^{v} k^a(v-j)S_a(j)x = 0$ for all $v \in \mathbb{N}_0$. Recursively, we obtain $S_a(v)x = 0$ for all $x \in \mathcal{A}0$ and $v \in \mathbb{N}_0$. □

Now we will state and prove the following result.

Proposition 2.1.49. *Suppose that $\mathcal{A}$ is a subgenerator of a discrete (a, C)-resolvent family $(S_a(v))_{v\in\mathbb{N}_0}$. If $\lambda \in \rho_C(\mathcal{A})$, then we have $C\sum_{j=0}^{v} k^a(v-j)S_a(j)x \in D(\mathcal{A})$ for all $x \in X$ and $v \in \mathbb{N}_0$, as well as*

$$S_a(v)Cx - k^a(v)C^2x \in C^{-1}\mathcal{A}0 + \mathcal{A}C\sum_{j=0}^{v} k^a(v-j)S_a(j)x, \quad v \in \mathbb{N}_0,\ x \in X. \tag{53}$$

In particular, if $\mathcal{A} = A$ is single-valued, then

$$S_a(v)x - k^a(v)Cx = C^{-1}AC\sum_{j=0}^{v} k^a(v-j)S_a(j)x, \quad v \in \mathbb{N}_0,\ x \in X. \tag{54}$$

Proof. Set $y := \lambda(\lambda-\mathcal{A})^{-1}Cx - Cx$. Then $y \in \mathcal{A}(\lambda-\mathcal{A})^{-1}Cx$ due to Lemma 1.1.2. Using the functional equality (52), we get that

$$S_a(v)(\lambda-\mathcal{A})^{-1}Cx = k^a(v)(\lambda-\mathcal{A})^{-1}C^2x + \sum_{j=0}^{v} k^a(v-j)S_a(j)[\lambda(\lambda-\mathcal{A})^{-1}C - C]x,$$

which simply implies along our commuting assumptions that

$$\begin{aligned}(\lambda-\mathcal{A})^{-1}C[S_a(v)x - k^a(v)Cx] &= \lambda(\lambda-\mathcal{A})^{-1}C\sum_{j=0}^{v} k^a(v-j)S_a(j)x \\ &\quad - C\sum_{j=0}^{v} k^a(v-j)S_a(j)x, \quad v \in \mathbb{N}_0,\ x \in X.\end{aligned}$$

Therefore, $C\sum_{j=0}^{v}k^{a}(v-j)S_{a}(j)x \in D(\mathcal{A})$ and $\sum_{j=0}^{v}k^{a}(v-j)S_{a}(j)x \in D((\lambda-\mathcal{A})(\lambda-\mathcal{A})^{-1}C\mathcal{A}C)$ for all $x \in X$ and $v \in \mathbb{N}_0$, as well as

$$S_{a}(v)C^{2}x \in k^{a}(v)C^{3}x + (\lambda-\mathcal{A})(\lambda-\mathcal{A})^{-1}C\mathcal{A}C\sum_{j=0}^{v}k^{a}(v-j)S_{a}(j)x, \quad x \in X,\ v \in \mathbb{N}_0.$$

This implies

$$(\lambda-\mathcal{A})^{-1}C[S_{a}(v)Cx - k^{a}(v)C^{2}x] \in (\lambda-\mathcal{A})^{-1}C\mathcal{A}C\sum_{j=0}^{v}k^{a}(v-j)S_{a}(j)x,$$

for any $x \in X$ and $v \in \mathbb{N}_0$. Since $(\lambda-\mathcal{A})^{-1}x = (\lambda-\mathcal{A})^{-1}y$ for some $x, y \in X$ if and only if $x-y \in \mathcal{A}0$, the above yields

$$C[S_{a}(v)Cx - k^{a}(v)C^{2}x] \in \mathcal{A}0 + C\mathcal{A}C\sum_{j=0}^{v}k^{a}(v-j)S_{a}(j)x, \quad x \in X,\ v \in \mathbb{N}_0$$

and (53). If $\mathcal{A} = A$ is single-valued, then (53) simply implies (54). □

Define now, for every $n \in \mathbb{N}_0$, the sequence $(\beta_{a,v}(j))_{1\leqslant j\leqslant v}$ in the same way as it has been done directly after [3, Definition 3.1]. The next result follows from a relatively simple argumentation involving the way of construction of sequence $(\beta_{a,v}(j))_{1\leqslant j\leqslant v}$, the validity of [3, Theorem 3.2] with $A = 0$, the argumentation contained in the proof of [46, Theorem 3.4] and Lemma 1.1.2; cf. also [4, Proposition 3.2].

Theorem 2.1.50. *Suppose that* $1 \in \rho(\mathcal{A})$. *Define* $S_{a}(0) := (1-\mathcal{A})^{-1}$ *and*

$$S_{a}(v) := \sum_{j=1}^{v}\beta_{a,v}(j)(1-\mathcal{A})^{-j-1}, \quad v \in \mathbb{N}.$$

Then $(S_{a}(v))_{v\in\mathbb{N}_0}$ *is a unique discrete* (a, I)*-resolvent family* $(S_{a}(v))_{v\in\mathbb{N}_0}$ *generated by* $\mathcal{A}$*; furthermore, the functional inclusion*

$$S_{a}(v)x \in k^{a}(v)Cx + \mathcal{A}\sum_{j=0}^{v}k^{a}(v-j)S_{a}(j)x, \quad x \in X,\ v \in \mathbb{N}_0 \tag{55}$$

holds with $C = \mathrm{I}$, $R(S_{a}(0)) \subseteq D(\mathcal{A})$ *and* $R(S_{a}(v)) \subseteq D(\mathcal{A}^{2})$ *for all* $v \in \mathbb{N}$. *Moreover, the assumptions* $\|(1-\mathcal{A})^{-1}\| < 1$ *and* $0 < a \leqslant 1$ *together imply* $\sum_{v=0}^{\infty}\|S_{a}(v)\| < +\infty$.

Suppose now $1 \in \rho_{C}(\mathcal{A})$. Define $S_{a}(0) := (\mathrm{I}-\mathcal{A})^{-1}C$ and

$$S_{a}(v) := \sum_{j=1}^{v}\beta_{a,v}(j)[(\mathrm{I}-\mathcal{A})^{-1}C]^{j+1}, \quad v \in \mathbb{N}.$$

Then the assumptions $\|(\mathrm{I}-\mathcal{A})^{-1}C\| < 1$ and $0 < \alpha \leqslant 1$ together imply $\sum_{v=0}^{\infty}\|S_\alpha(v)\| < +\infty$; see, e. g., the proof of [46, Theorem 3.4]. But, it is not generally true that $\mathcal{A}$ is a subgenerator of the discrete (α, C)-resolvent family $(S_\alpha(v))_{v\in\mathbb{N}_0}$ if $C \neq \mathrm{I}$. For example, we have $S_\alpha(1)x = \beta_{\alpha,1}[(\mathrm{I}-\mathcal{A})^{-1}C]^2x$, $x \in X$ but

$$\begin{aligned} &k^\alpha(1)Cx + k^\alpha(1)(\mathrm{I}-\mathcal{A})^{-1}Cy + k^\alpha(0)\beta_{\alpha,1}(1)[(\mathrm{I}-\mathcal{A})^{-1}C]^2y \\ &\quad = k^\alpha(1)(\mathrm{I}-\mathcal{A})^{-1}Cx + k^\alpha(0)\beta_{\alpha,1}(1)\{[(\mathrm{I}-\mathcal{A})^{-1}C]^2x - (\mathrm{I}-\mathcal{A})^{-1}C^2x\}, \quad x \in X. \end{aligned}$$

In connection with this observation, we will state and prove the following result.

Theorem 2.1.51. (i) *Suppose that $\mathcal{A}$ is a closed* MLO, *which generates a discrete (α, C)-resolvent family $(S_\alpha(v))_{v\in\mathbb{N}_0}$ satisfying* (55). *Then $C^{-j}[(1-\mathcal{A})^{-1}C]^{j+1} = (1-\mathcal{A})^{-j-1}C \in L(X)$ for all $j \in \mathbb{N}_0$, $S_\alpha(0) = (1-\mathcal{A})^{-1}C$ and $(S_\alpha(v))_{v\in\mathbb{N}_0}$ is given recursively by the equation* (56) *below. Furthermore, $(1-\mathcal{A})^{-m}S_\alpha(v) \in L(X)$ for all m, $v \in \mathbb{N}_0$ and the equation* (57) *below holds.*

(ii) *Suppose that $\mathcal{A}$ is a closed* MLO *and $(1-\mathcal{A})^{-n}C \in L(X)$ for all $n \in \mathbb{N}$. Define $S_\alpha(0) := (1-\mathcal{A})^{-1}C$ and $(S_\alpha(v))_{v\in\mathbb{N}_0}$ recursively by*

$$S_\alpha(v)x := k^\alpha(v)(1-\mathcal{A})^{-1}Cx + \sum_{j=0}^{v-1} k^\alpha(v-j)[-S_\alpha(j)x + (1-\mathcal{A})^{-1}S_\alpha(j)x], \tag{56}$$

for any $v \in \mathbb{N}$ and $x \in X$. Then the sequence $(S_\alpha(v))_{v\in\mathbb{N}_0} \subseteq L(X)$ is well-defined and satisfies that $(1-\mathcal{A})^{-m}S_\alpha(v) \in L(X)$ for all m, $v \in \mathbb{N}_0$; furthermore, $(S_\alpha(v))_{v\in\mathbb{N}_0}$ is a unique discrete (α, C)-resolvent family with subgenerator $\mathcal{A}$ and

$$S_\alpha(v) := \sum_{j=1}^{v} \beta_{\alpha,v}(j)C^{-j}[(1-\mathcal{A})^{-1}C]^{j+1}, \quad v \in \mathbb{N}. \tag{57}$$

(iii) *Suppose that $\mathcal{A}$ is a closed* MLO *and $0 < \alpha \leqslant 1$. If there exists $R > 1$ such that the mapping $\lambda \mapsto (\lambda-\mathcal{A})^{-1}C \in L(X)$, $\lambda \in \{z \in \mathbb{C} : |1-z| < R\}$ is continuous, then $(1-\mathcal{A})^{-n}C \in L(X)$ for all $n \in \mathbb{N}$ and the discrete (α, C)-resolvent family $(S_\alpha(v))_{v\in\mathbb{N}_0}$, defined in part* (ii), *satisfies $\sum_{v=0}^{+\infty}\|S_\alpha(v)\| < +\infty$.*

Proof. We will prove only (i) and (iii). In order to prove (i), let us first note that our functional equalities (54) and (55) with $v = 0$ simply imply that $S_\alpha(0) \subseteq (1-\mathcal{A})^{-1}C$ and $(1-\mathcal{A})^{-1}C$ is single-valued; hence, $1 \in \rho_C(\mathcal{A})$ and $S_\alpha(0) = (1-\mathcal{A})^{-1}C$. Further on, the functional equation (55) with $v = 1$ implies after a simple calculation that

$$S_\alpha(1)x \in k^\alpha(1)(1-\mathcal{A})^{-1}Cx + k^\alpha(1)(1-\mathcal{A})^{-1}\mathcal{A}(1-\mathcal{A})^{-1}Cx, \quad x \in X. \tag{58}$$

Multiplying this inclusion with C and using the first inclusion in Lemma 1.1.2, we infer that

$$CS_\alpha(1)x = k^\alpha(1)(1-\mathcal{A})^{-1}C^2x + k^\alpha(1)(1-\mathcal{A})^{-1}C(1-\mathcal{A})^{-1}Cx - C(1-\mathcal{A})^{-1}Cx,$$

for any $x \in X$. Hence, $(1-\mathcal{A})^{-1}C \cdot (1-\mathcal{A})^{-1}Cx \in R(C)$ and

$$C^{-1}(1-\mathcal{A})^{-1}C \cdot (1-\mathcal{A})^{-1}Cx = \frac{1}{k^{a}(1)}[S_a(1)x + (1-k^{a}(1))(1-\mathcal{A})^{-1}Cx],$$

for any $x \in X$. This implies $C^{-j}[(1-\mathcal{A})^{-1}C]^{j+1} \in L(X)$ for $j = 1$; on the other hand, (58) implies

$$y_x := \frac{1}{k^{a}(1)}[S_a(1)x - k^{a}(1)(1-\mathcal{A})^{-1}Cx] \in (1-\mathcal{A})^{-1}\mathcal{A}(1-\mathcal{A})^{-1}Cx, \quad x \in X.$$

As a consequence, we have

$$(1-\mathcal{A})^{-1}Cx \in (1-\mathcal{A})[(1-\mathcal{A})^{-1}Cx + y_x],$$

so that $(1-\mathcal{A})^{-1}Cx \in R(1-\mathcal{A}) = D((1-\mathcal{A})^{-1})$. Since $card((1-\mathcal{A})^{-j}Cx) \leqslant 1$ for all $j \in \mathbb{N}$ (see [445, p. 41]), we get that

$$(1-\mathcal{A})^{-2}Cx = (1-\mathcal{A})^{-1}Cx + y_x = C^{-1}(1-\mathcal{A})^{-1}C \cdot (1-\mathcal{A})^{-1}Cx, \quad x \in X.$$

This simply implies the validity of the equalities (56) and (57) with $v = 1$. Invoking this formulae and multiplying $S_a(2)$ with C^2, we easily get that $C^{-j}[(1-\mathcal{A})^{-1}C]^{j+1} \in L(X)$ for $j = 2$; starting from a modification of the equation (58) with $v = 2$, we can similarly prove that $(1-\mathcal{A})^{-3}C \in L(X)$ and, therefore, $(1-\mathcal{A})^{-3}C = C^{-2}[1-\mathcal{A})^{-1}C]^3$ since the left-hand side of this equality is clearly contained in the right-hand side of it. Repeating this procedure, we obtain the required statement; evidently, $(1-\mathcal{A})^{-m}S_a(v) \in L(X)$ for all $m, v \in \mathbb{N}_0$. In order to prove (iii), fix a number $r \in (1, R)$. Then an application of [445, Proposition 1.2.6(iii)] yields that the mapping $\lambda \to (\lambda-\mathcal{A})^{-1}C$, $|1-\lambda| < R$ is analytic in $L(X)$ as well as that $(1-\mathcal{A})^{-j}C \in L(X)$ for all $j \in \mathbb{N}$ and

$$\frac{d^j}{d\lambda^j}(\lambda-\mathcal{A})^{-1}C = (-1)^{j-1}(j-1)!(\lambda-\mathcal{A})^{-j}C, \quad j \in \mathbb{N},\ |1-\lambda| < R.$$

Applying this equality and the Cauchy integral formula for the derivatives, it follows that

$$\|(1-\mathcal{A})^{-j-1}Cx\| = \left\| \frac{1}{2\pi i} \oint_{|z-1|=r} \frac{(z-\mathcal{A})^{-j}Cx}{(z-1)^{j+1}}\,dz \right\| \leqslant \frac{M_r\|x\|}{r^{j+1}}, \quad j \in \mathbb{N},\ x \in X,$$

where $M_r := \sup_{|z-1|=r} \|(z-\mathcal{A})^{-1}C\|$. Keeping in mind the argumentation given in the proof of [46, Theorem 3.4] (let us only observe here that, in the proof of this result, we must use the Fubini theorem and the fact that $\sum_{n=0}^{\infty} \varphi_{a,0}(n,j) = 1$ for all $j \in \mathbb{N}_0$, with the same notation used), this simply implies the required. □

Remark 2.1.52. (i) Conditions $1 \in \rho(\mathcal{A})$ and $\|(1-\mathcal{A})^{-1}\| < 1$ imply that $\Omega := \{\lambda \in \mathbb{C} : |\lambda - 1| < 1/\|(1-\mathcal{A})^{-1}\|\} \subseteq \rho(\mathcal{A})$ and the mapping $\lambda \mapsto (\lambda - \mathcal{A})^{-1} \in L(X)$, $\lambda \in \Omega$ is analytic. Hence, the assumptions made in part (iii) hold with $C = \mathrm{I}$.

(ii) The interested reader may try to reconsider the statement of [4, Theorem 3.3] in our general framework.

In the remainder of this part, we analyze the well-posedness of the following abstract fractional difference inclusion

$$\Delta_W^{\alpha} u(k) \in \mathcal{A}u(k+1) + Cf(k), \quad k \in \mathbb{Z}, \tag{59}$$

where $0 < \alpha \leqslant 1$, $\mathcal{A}$ is a closed multivalued linear operator and $C \in L(X)$ is an injective operator. We will use the following notion (cf. also [3, Definitions 4.1, 4.4]).

Definition 2.1.53. (i) A sequence $(u(k))_{k\in\mathbb{Z}}$ is said to be a strong solution of (59) if $\sum_{k\in\mathbb{Z}} \|u(k)\| \cdot g_{\alpha}(|k|) < +\infty$, $u(k) \in D(\mathcal{A})$ for all $k \in \mathbb{Z}$ and (59) holds.

(ii) Suppose that $\mathcal{A}$ is the integral generator of a discrete (α, C)-resolvent family $(S_{\alpha}(v))_{v\in\mathbb{N}_0} \subseteq L(X)$. Then a sequence $(u(k))_{k\in\mathbb{Z}}$ is said to be a mild solution of (59) if

$$u(k) := \sum_{j=-\infty}^{k-1} S_{\alpha}(k-1-j)f(j), \quad k \in \mathbb{Z} \tag{60}$$

and the above series is absolutely convergent.

Let us consider now the condition (C1) and the following condition:

(C2) $\mathcal{A}$ is the integral generator of an exponentially decaying C-regularized semigroup $(T(t))_{t\geqslant 0}$.

Then we know the following (cf. [444, pp. 124–132] for more details):

(A1) If (C1) holds, then there exists a unique strongly continuous operator family $(R_{\alpha}(t))_{t>0} \subseteq L(X)$ such that

$$\int_0^{+\infty} e^{-\lambda t} R_{\alpha}(t)x\,dt = (\lambda^{\alpha} - \mathcal{A})^{-1}x, \quad \operatorname{Re}\lambda > 0,\ x \in X, \tag{61}$$

$\|R_{\alpha}(t)\| = O(t^{\alpha\beta-1})$, $t \in (0,1]$ and $\|R_{\alpha}(t)\| = O(t^{-1-\alpha})$, $t \geqslant 1$.

(A2) If (C2) holds, then there exists a unique strongly continuous operator family $(R_{\alpha}(t))_{t\geqslant 0} \subseteq L(X)$ such that (61) holds and $\|R_{\alpha}(t)\| = O(t^{-1-\alpha})$, $t \geqslant 1$; furthermore, $\mathcal{A}$ is the integral generator of an exponentially bounded (g_{α}, g_{α})-regularized C-resolvent family $(R_{\alpha}(t))_{t\geqslant 0}$.

If (C1) or (C2) holds, then we define (cf. [535] for more details)

$$S_{\alpha}(v)x := \int_0^{+\infty} e^{-t}\frac{t^{v}}{v!} R_{\alpha}(t)x\,dt, \quad v \in \mathbb{N}_0,\ x \in X. \tag{62}$$

The slight technical modifications of [3, Theorems 3.5, 4.2] show the following (concerning the part (i), we would like to emphasize that the existence and summability of the discrete (a, C)-resolvent family $(S_a(v))_{v\in\mathbb{N}_0} \subseteq L(X)$ is a simple consequence of Theorem 2.1.51(iv); here, we present a new representation of $(S_a(v))_{v\in\mathbb{N}_0}$, only; the representation formula established in [375, Theorem 3.1] also holds for the multivalued linear subgenerators of bounded C-regularized semigroups).

Theorem 2.1.54. (i) *Suppose that* (C1), *respectively,* (C2), *holds. Then* $\mathcal{A}$ *is the integral generator of a discrete* (a, I)*-resolvent family* $(S_a(v))_{v\in\mathbb{N}_0} \subseteq L(X)$, *respectively, a discrete* (a, C)*-resolvent family* $(S_a(v))_{v\in\mathbb{N}_0} \subseteq L(X)$, *given by* (62), *and* $\sum_{v=0}^{+\infty} \|S_a(v)\| < +\infty$.

(ii) *Suppose that* $\mathcal{A}$ *is the integral generator of a discrete* (a, C)*-resolvent family* $(S_a(v))_{v\in\mathbb{N}_0} \subseteq L(X)$ *and* $\sum_{v=0}^{+\infty} \|S_a(v)\| < +\infty$. *If* $f_i \in D(\mathcal{A})$ *for all* $i \in \mathbb{Z}$, $\sum_{i\in\mathbb{Z}} \|f_i\| < +\infty$ *and there exists a sequence* $(g_i)_{i\in\mathbb{Z}}$ *such that* $g_i \in \mathcal{A}f(i)$ *for all* $i \in \mathbb{Z}$ *and* $\sum_{i\in\mathbb{Z}} \|g_i\| < +\infty$, *then the function* $u : \mathbb{Z} \to X$, *given by* (60), *is a strong solution of* (59) *and* $\sum_{k\in\mathbb{Z}} \|u_k\| < +\infty$.

Under certain logical assumptions, the possible applications of Theorem 2.1.51(iv) and Theorem 2.1.54 can be given to the following fractional semidiscrete Poisson heat equation in the space $L^p(\Omega)$:

$$(ACP)_a : \begin{cases} m(x)\Delta_W^a v(k, x) = \Delta v(k+1, x) - bv(k+1, x), & k \in \mathbb{Z},\ x \in \Omega; \\ v(k, x) = 0, & (k, x) \in \mathbb{Z} \times \partial\Omega, \end{cases}$$

and a class of the abstract fractional semidiscrete Cauchy problems with the (non-coercive) differential operators with constant coefficients.

If a closed MLO $\mathcal{A}$ generates a summable (a, C)-resolvent family $(S_a(v))_{v\in\mathbb{N}_0}$, then we can clarify many structural results about the existence and uniqueness of almost periodic type solutions (almost automorphic type solutions) to (59). For example, we have the following (the notion used in (iii) will be explained later): Suppose that the function $u : \mathbb{Z} \to X$, given by (60), is a mild solution of (59) and $f(\cdot)$ is bounded. Then:

(i) $u(\cdot)$ is T-almost periodic (T-uniformly recurrent), where $T \in L(X)$, provided that $(f_k)_{k\in\mathbb{Z}}$ is T-almost periodic (T-uniformly recurrent).

(ii) $u(\cdot)$ is equi-Weyl-$(\mathbb{F}, 1, T)$-almost periodic [bounded Weyl-$(\mathbb{F}, 1, T)$-almost periodic; bounded Doss-$(\mathbb{F}, 1, T)$-almost periodic; bounded Besicovitch-$(\mathbb{F}, 1)$-almost periodic], provided that $f(\cdot)$ is.

(iii) $u(\cdot)$ is bounded and Levitan pre-(I', T)-almost periodic (Levitan T-almost periodic), provided that $f(\cdot)$ is; here, $T \in L(X)$.

We can similarly analyze the following class of semilinear Volterra difference equations with infinite delay:

$$u(k+1) = \alpha \sum_{l=-\infty}^{k} a(k-l)u(l) + f(k; u(k)), \quad k \in \mathbb{Z},\ \alpha \in \mathbb{C}; \tag{63}$$

cf. [45, Theorems 3.1, 3.3] for more details in this direction.

3. Abstract fractional difference inclusions with Riemann–Liouville fractional derivatives

Suppose that $1 < \alpha \leqslant 2$ and $(u_v)_{v\geqslant 0}$ is a given sequence. Then we define the Riemann–Liouville fractional derivative

$$\begin{aligned}\Delta^{\alpha}u(v) &:= \Delta^{2}(\Delta^{\alpha-2}u)(v)\\ &:= \Delta^{2}\sum_{j=0}^{v}k^{2-\alpha}(v-j)u(j)\\ &= \sum_{j=0}^{v+2}k^{2-\alpha}(v+2-j)u(j) - 2\sum_{j=0}^{v+1}k^{2-\alpha}(v+1-j)u(j) + \sum_{j=0}^{v}k^{2-\alpha}(v-j)u(j). \qquad (64)\end{aligned}$$

Even if $\mathcal{A} = A$ is a closed linear operator and $C = \mathrm{I}$, the following result provides a proper extension of [4, Theorem 3.5, Corollary 3.6] since we do not assume here that $u_0 \in D(\mathcal{A})$ or $u_1 \in D(\mathcal{A})$.

Theorem 2.1.55. *Suppose that $1 < \alpha \leqslant 2$ and $\mathcal{A}$ is a subgenerator of a discrete (α, C)-resolvent family $(S_\alpha(v))_{v\in\mathbb{N}_0}$. Then the abstract fractional inclusion*

$$\Delta^{\alpha}u(v) \in \mathcal{A}u(v+2) + Cf(v); \quad u(0) = Cu_0,\ u(1) = Cu_1 \qquad (65)$$

has a unique solution $(u_v)_{v\geqslant 0}$, which is given by

$$\begin{aligned}u(v) &:= \sum_{j=1}^{v}\beta_{\alpha,v}(j)(1-\mathcal{A})^{-j}Cu_0 - \alpha S_\alpha(v-1)u_0\\ &\quad + \sum_{j=1}^{v-1}\beta_{\alpha,v-1}(j)(1-\mathcal{A})^{-j}Cu_1 + (S_\alpha *_0 f)(v-2), \quad v\in\mathbb{N},\ v\geqslant 2. \qquad (66)\end{aligned}$$

Proof. Suppose that $(u_v)_{v\geqslant 0}$ is a solution of (65). Then (64) implies

$$\begin{aligned}&\sum_{j=0}^{v+2}k^{2-\alpha}(v+2-j)u(j) - 2\sum_{j=0}^{v+1}k^{2-\alpha}(v+1-j)u(j)\\ &\quad + \sum_{j=0}^{v}k^{2-\alpha}(v-j)u(j) \in \mathcal{A}u(v+2) + Cf(v), \quad v\in\mathbb{N},\ v\geqslant 2.\end{aligned}$$

Since $k^{2-\alpha}(0) = 1$ and $1 \in \rho_C(\mathcal{A})$ due to Theorem 2.1.51(i), this immediately implies that u_{v+2} is uniquely determined by $u_0, u_1, \ldots, u_{v+1}$ and $f(\cdot)$, as well as that

$$Cu_{v+2} = (1-\mathcal{A})^{-1}C\Bigg[-\sum_{j=0}^{v+1}k^{2-\alpha}(v+2-j)u(j) + 2\sum_{j=0}^{v+1}k^{2-\alpha}(v+1-j)u(j)$$

$$- \sum_{j=0}^{v} k^{2-a}(v-j)u(j) + Cf(v)\Bigg], \quad v \in \mathbb{N},\ v \geqslant 2.$$

It remains to be proved that the sequence $(u_v)_{v\geqslant 0}$, given by (66), is a solution of (65). We will provide the main details of this fact with $f \equiv 0$. In actual fact, the argumentation contained in the proof of [4, Theorem 3.5] essentially shows that we have $\Delta^a S_a(v) \subseteq \mathcal{A}S_a(v+2)$ for all $v \in \mathbb{N}_0$. Keeping this inclusion in mind and using Proposition 2.1.48, it is trivial to prove that the sequence

$$w(v) := S_a(v)[u_0 - x_0] - aS_a(v-1)u_0 + S_a(v-1)[u_1 - y_1], \quad v \in \mathbb{N},\ v \geqslant 2$$

is well-defined and does not depend on the values of $x_0 \in \mathcal{A}u_0$ and $y_1 \in \mathcal{A}u_1$ as well as that it solves the fractional inclusion (494) with $u(\cdot)$ replaced by $w(\cdot)$, provided that $u_0 \in D(\mathcal{A})$ and $u_1 \in D(\mathcal{A})$. If this is the case, then we can use Theorem 2.1.51(i), the equation (57) and Lemma 1.1.2 in order to see that the sequence $(u_v)_{v\geqslant 0}$, given by (66), is identically equal to the sequence $(w_v)_{v\geqslant 0}$ defined above. Performing a simple computation, this yields the final conclusion. □

Suppose now that $0 < a \leqslant 1$ and $(u_v)_{v\geqslant 0}$ is a given sequence. Then we define the Riemann–Liouville fractional derivative

$$\Delta^a u(v) := \Delta(\Delta^{a-1}u)(v) = \sum_{j=0}^{v+1} k^{1-a}(v+1-j)u(j) - \sum_{j=0}^{v} k^{1-a}(v-j)u(j), \quad v \geqslant 0.$$

Consider now the abstract fractional inclusion

$$\Delta^a u(v) \in \mathcal{A}u(v+1) + Cf(v); \quad u(0) = Cu_0. \tag{67}$$

Even if $\mathcal{A} = A$ is a closed linear operator and $C = \mathrm{I}$, it is not clear why one has to assume that $u_0 \in D(\mathcal{A})$ in the definition of a strong solution of (67), given in [375, Definition 4.1]. We say that a sequence $u : \mathbb{N}_0 \to X$ is a strong solution of (67) if $u(v) \in D(\mathcal{A})$ for all $v \in \mathbb{N}$ and (67) identically holds. The second main result of this subsection can be proved in a similar manner as Theorem 2.1.55 and it provides a proper extension of [375, Theorem 4.1].

Theorem 2.1.56. *Suppose that $0 < a \leqslant 1$ and $\mathcal{A}$ is a subgenerator of a discrete (a, C)-resolvent family $(S_a(v))_{v\in\mathbb{N}_0}$. Then the abstract fractional inclusion (67) has a unique strong solution $(u_v)_{v\geqslant 0}$, which is given by*

$$u(v) := \sum_{j=1}^{v} \beta_{a,v}(j)(1-\mathcal{A})^{-j}Cu_0 + (S_a *_0 f)(v-1), \quad v \in \mathbb{N}. \tag{68}$$

The formula (68) can be used in the analysis of the existence and uniqueness of asymptotically almost periodic (automorphic) type solutions of (67), provided that $(S_a(v))_{v\in\mathbb{N}_0}$ is summable.

4. Two multidimensional analogues of the abstract difference equation $u(k+1) = Au(k) + f(k)$

It is worth noticing that the statements of [65, Theorems 3.1, 3.2, 3.5] can be formulated in the multidimensional setting. Without going into full details, we will only explain here how one can formulate some multi-dimensional extensions of [65, Theorem 3.1(i)].

Suppose that $f : \mathbb{Z}^n \to X$, $\lambda_1, \lambda_2, \ldots, \lambda_n$ are given complex numbers and

$$\max(|\lambda_1|, |\lambda_2|, \ldots, |\lambda_n|) < 1.$$

Consider the function

$$\begin{aligned} u(k_1, k_2, \ldots, k_n) &:= \sum_{l_1 \leqslant k_1, l_2 \leqslant k_2, \ldots, l_n \leqslant k_n} \lambda_1^{k_1-l_1} \lambda_2^{k_2-l_2} \cdot \ldots \cdot \lambda_n^{k_n-l_n} f(l_1 - 1, l_2 - 1, \ldots, l_n - 1) \\ &= \sum_{v_1 \geqslant 0, v_2 \geqslant 0, \ldots, v_n \geqslant 0} \lambda_1^{v_1} \lambda_2^{v_2} \cdot \ldots \cdot \lambda_n^{v_n} f(k_1 - v_1 - 1, k_2 - v_2 - 1, \ldots, k_n - v_n - 1) \end{aligned}$$

defined for any $(k_1, k_2, \ldots, k_n) \in \mathbb{Z}^n$. Using the same argumentation as before, we can simply prove the following: If (FV) holds and the sequence $f(\cdot)$ is equi-Weyl-$(\Lambda', \mathbb{F}, p, T)$-almost periodic [polynomially bounded Weyl-$(\Lambda', \mathbb{F}, p, T)$-almost periodic; polynomially bounded Doss-$(\Lambda', \mathbb{F}, p, T)$-almost periodic], then the sequence $u(\cdot)$ enjoys the same property as $f(\cdot)$; a similar statement can be deduced for the class of generalized Besicovitch-p-almost periodic sequences ($\rho = T \in L(X)$).

On the other hand, it is very simple to find the form of function $F : \mathbb{Z}^n \to X$ such that

$$u(k_1 + 1, k_2 + 1, \ldots, k_n + 1) = \lambda_1 \lambda_2 \cdots \lambda_n \cdot u(k_1, k_2, \ldots, k_n) + F(k_1, k_2, \ldots, k_n),$$

for all $(k_1, k_2, \ldots, k_n) \in \mathbb{Z}^n$. More precisely, we have

$$\begin{aligned} & u(k_1 + 1, k_2 + 1, \ldots, k_n + 1) \\ &= \sum_{l_1 \leqslant k_1+1, l_2 \leqslant k_2+1, \ldots, l_n \leqslant k_n+1} \lambda_1^{k_1+1-l_1} \lambda_2^{k_2+1-l_2} \cdot \ldots \cdot \lambda_n^{k_n+1-l_n} f(l_1 - 1, l_2 - 1, \ldots, l_n - 1) \\ &= \sum_{l_2 \leqslant k_2+1, \ldots, l_n \leqslant k_n+1} \lambda_2^{k_2+1-l_2} \cdot \ldots \cdot \lambda_n^{k_n+1-l_n} f(k_1, l_2 - 1, \ldots, l_n - 1) \\ &\quad + \lambda_1 \sum_{l_1 \leqslant k_1, l_2 \leqslant k_2+1, \ldots, l_n \leqslant k_n+1} \lambda_1^{k_1+1-l_1} \lambda_2^{k_2+1-l_2} \cdot \ldots \cdot \lambda_n^{k_n+1-l_n} f(l_1 - 1, l_2 - 1, \ldots, l_n - 1) \\ &= \sum_{l_2 \leqslant k_2+1, \ldots, l_n \leqslant k_n+1} \lambda_2^{k_2+1-l_2} \cdot \ldots \cdot \lambda_n^{k_n+1-l_n} f(k_1, l_2 - 1, \ldots, l_n - 1) \\ &\quad + \lambda_1 \sum_{l_1 \leqslant k_1, l_3 \leqslant k_3+1, \ldots, l_n \leqslant k_n+1} \lambda_1^{k_1-l_1} \lambda_3^{k_3+1-l_3} \cdot \ldots \cdot \lambda_n^{k_n+1-l_n} f(l_1 - 1, k_2, \ldots, l_n - 1) \\ &\quad + \lambda_1 \lambda_2 \sum_{l_1 \leqslant k_1, l_2 \leqslant k_2, l_3 \leqslant k_3+1, \ldots, l_n \leqslant k_n+1} \lambda_1^{k_1-l_1} \lambda_2^{k_2-l_2} \lambda_3^{k_3+1-l_3} \cdot \ldots \cdot \lambda_n^{k_n+1-l_n} f(l_1 - 1, l_2 - 1, \ldots, l_n - 1) \\ &= \cdots . \end{aligned}$$

In the second approach, we consider the solution $u_j : \mathbb{Z} \to X$ of the equation $u_j(k+1) = \lambda u_j(k) + f_j(k)$, $k \in \mathbb{Z}$, where $f_j(\cdot)$ is a generalized almost periodic sequence ($1 \leqslant j \leqslant n$) and $\lambda \in \mathbb{C}$ satisfies $|\lambda| < 1$. Define $u(k_1,\ldots,k_n) = u_1(k_1) + u_2(k_2) + \cdots + u_n(k_n)$ and $f(k_1,\ldots,k_n) = f_1(k_1) + f_2(k_2) + \cdots + f_n(k_n)$ for all $k_j \in \mathbb{Z}$ ($1 \leqslant j \leqslant n$). Then we have

$$u(k_1+1,\ldots,k_n+1) = \lambda u(k_1,\ldots,k_n) + f(k_1,\ldots,k_n), \quad (k_1,\ldots,k_n) \in \mathbb{Z}^n;$$

moreover, the sequence $u(\cdot)$ has a similar almost periodic behavior as the forcing terms $f_j(\cdot)$. For example, if all sequences $f_j(\cdot)$ are equi-Weyl-p-almost periodic in the usual sense, then the sequence $u(\cdot)$ is likewise equi-Weyl-p-almost periodic in the usual sense ($1 \leqslant p < \infty$). We can similarly analyze the metrically generalized ρ-almost periodic solutions of the following abstract difference inclusion:

$$u(k_1+1,\ldots,k_n+1) \in \mathcal{A}u(k_1,\ldots,k_n) + f(k_1,\ldots,k_n), \quad (k_1,\ldots,k_n) \in \mathbb{Z}^n,$$

where $0 \in \rho(\mathcal{A})$ and $\|\mathcal{A}^{-1}\| < 1$.

In [542], C. Lizama and L. Roncal have investigated the almost periodicity for semidiscrete equations with the (fractional) Laplacian. It is worth noting that the argument contained in the proof of Theorem 2.1.40 can serve one to formulate an analogue of [542, Theorem 1.5(1)] for the forcing terms $g(t,\cdot)$ which are generalized Weyl almost periodic for each fixed number $t \geqslant 0$; an extension to the higher dimensions can be also formulated following the consideration given in [542, Remark 14]. Let us observe that we can also analyze the pointwise products of generalized almost periodic sequences and the invariance of generalized almost periodicity under the actions of the infinite convolution products (see, e. g., [65, Theorem 2.13]).

At the end of this section, we will mention some conclusions and final remarks about the introduced notion, some topics not considered here and some perspectives for the further expansion of the theory of generalized almost periodic sequences. Before proceeding any further, we would like to emphasize that many other classes of generalized ρ-almost periodic type sequences are introduced and analyzed in our recent research studies [177, 463, 448, 454]. For example, suppose that $F : I \times X \to Y$ and (30) holds. Then the notion introduced in [451, Definitions 1–6] can be used to provide the definitions of:

(1) $(S, \mathbb{D}, \mathcal{B})$-asymptotical (ω, ρ)-periodicity of $F(\cdot;\cdot)$;
(2) $(S, \mathcal{B})$-asymptotical $(\omega_j, \rho_j, \mathbb{D}_j)_{j\in\mathbb{N}_n}$-periodicity of $F(\cdot;\cdot)$;
(3) $\mathbb{D}$-quasiasymptotical $(\mathcal{B}, I', \rho)$-almost periodicity of $F(\cdot;\cdot)$;
(4) $\mathbb{D}$-quasiasymptotical $(\mathcal{B}, I', \rho)$-uniform recurrence of $F(\cdot;\cdot)$;
(5) $\mathbb{D}$-remotely $(\mathcal{B}, I', \rho)$-almost periodicity of $F(\cdot;\cdot)$;
(6) $\mathbb{D}$-remotely $(\mathcal{B}, I', \rho)$-uniform recurrence of $F(\cdot;\cdot)$;
(7) $(\mathbb{D}, \mathcal{B}, \rho)$-slowly oscillating property of $F(\cdot;\cdot)$, and
(8) $(\mathcal{B}, (\mathbb{D}_j, \rho_j)_{j\in\mathbb{N}_n})$-slowly oscillating property of $F(\cdot;\cdot)$.

Furthermore, the notion introduced in [177, Definition 2.1] can be used to provide the definition of a Levitan almost periodic sequence $F : I \times X \to Y$. The applications of Levitan almost periodic sequences (and the sequences introduced in [451, Definitions 1–6] mentioned above) to the abstract Volterra difference equations have not been considered in the existing literature by now; we will examine this problem somewhere else.

We would like to emphasize that we have not considered here the generalized ρ-uniformly recurrent type sequences. More precisely, we can introduce the following notion (compare to Definition 2.1.13 and Definition 2.1.14).

Definition 2.1.57. Suppose that $F : \Lambda \times X \to Y$ is a given sequence, $1 \leqslant p < +\infty$, ρ is a binary relation on Y and $(\tau_k)_{k\in\mathbb{N}}$ is a sequence in Λ'' such that $\lim_{k\to+\infty} |\tau_k| = +\infty$. Then we say that $F(\cdot;\cdot)$ is:

(i) equi-Weyl-$(\mathcal{B}, (\tau_k), \mathbb{F}., p, \rho)$-uniformly recurrent, if for every $\varepsilon > 0$ and $B \in \mathcal{B}$, there exist $l \in \mathbb{N}$ and $k_0 \in \mathbb{N}$ such that, for every $k \geqslant k_0$, $J \in P_l$ and for every $j \in J$ and $x \in B$, there exists $z_{j,x} \in \rho(F(j;x))$ such that (31) holds with the point τ replaced by the point τ_k therein.

(ii) Weyl-$(\mathcal{B}, (\tau_k), \mathbb{F}., p, \rho)$-uniformly recurrent, if for every $\varepsilon > 0$ and $B \in \mathcal{B}$, there exists $k_0 \in \mathbb{N}$ such that, for every $k \geqslant k_0$, there exists an integer $l_k \in \mathbb{N}$ such that, for every $l \geqslant l_k$, $J \in P_l$, $j \in J$ and $x \in B$, there exists $z_{j,x} \in \rho(F(j;x))$ such that (31) holds with the point τ replaced by the point τ_k therein.

Definition 2.1.58. Suppose that $F : \Lambda \times X \to Y$ is a given sequence, $1 \leqslant p < +\infty$, ρ is a binary relation on Y and $(\tau_k)_{k\in\mathbb{N}}$ is a sequence in Λ'' such that $\lim_{k\to+\infty} |\tau_k| = +\infty$. Then we say that $F(\cdot;\cdot)$ is Doss-$(\mathcal{B}, (\tau_k), \mathbb{F}., p, \rho)$-uniformly recurrent if, for every $\varepsilon > 0$ and $B \in \mathcal{B}$, there exists $k_0 \in \mathbb{N}$ such that, for every $k \geqslant k_0$, there exists an increasing sequence (l_m^k) of positive integers such that, for every $m \in \mathbb{N}$, $J \in P_{l_m^k}$, $j \in J$ and $x \in B$, there exists $z_{j,x} \in \rho(F(j;x))$ such that (31) holds with the point τ replaced by the point τ_k and the number l replaced by the number l_m^k therein.

It is very simple to rephrase Theorem 2.1.18 and the conclusion from Remark 2.1.19 to the Weyl and Doss generalized classes of ρ-uniformly recurrent type sequences; details can be left to the curious readers.

2.2 Generalized almost periodic solutions of Volterra difference equations

The main results of this section are taken from our joint work with H. C. Koyuncuoğlu [482]. We aim to continue here the research study [476] by investigating some classes of Levitan ρ-almost periodic type sequences and remotely ρ-almost periodic type sequences. We also provide certain applications of our results to the abstract Volterra difference equations.

The section is simply organized; after collecting the basic results about principal fundamental matrix solutions, Green functions and exponential dichotomies in Section 2.2.1, we analyze the Levitan ρ-almost periodic type sequences and the remotely ρ-almost periodic type sequences in Section 2.2.2 and Section 2.2.3, respectively. After that, we provide certain applications of the established results to the abstract Volterra difference equations as well as some conclusions and final remarks about the introduced notion.

Before proceeding further, we recall the following notion (cf. [448] for more details on the subject).

Definition 2.2.1. Suppose that $F : \mathbb{R}^n \to Y$ is a continuous function and $T \in L(Y)$. Then we say that the function $F(\cdot)$ is:

(i) Levitan T-pre-almost periodic if $F(\cdot)$ is for each $N > 0$ and $\varepsilon > 0$ there exists a finite real number $l > 0$ such that for each $\mathbf{t}_0 \in \mathbb{R}^n$ there exists $\boldsymbol{\tau} \in B(\mathbf{t}_0, l)$ such that

$$\|F(\mathbf{t} + \boldsymbol{\tau}) - TF(\mathbf{t})\| \leqslant \varepsilon \quad \text{for all } \mathbf{t} \in \mathbb{R}^n \text{ with } |\mathbf{t}| \leqslant N;$$

by $E(\varepsilon, T, N)$ we denote the set of all such points τ, which we also call (ε, N, T)-almost periods of $F(\cdot)$.

(ii) strongly Levitan T-almost periodic if $F(\cdot)$ is Levitan T-pre-almost periodic and, for every real numbers $N > 0$ and $\varepsilon > 0$, there exist a finite real number $\eta > 0$ and the relatively dense sets $E^j_{\eta;N}$ in $\mathbb{R}$ $(1 \leqslant j \leqslant n)$ such that the set $E_{\eta;N} \equiv \prod_{j=1}^n E^j_{\eta;N}$ consists solely of (η, N, T)-almost periods of $F(\cdot)$ and $E_{\eta;N} \pm E_{\eta;N} \subseteq E(\varepsilon, T, N)$.

2.2.1 Principal fundamental matrix solutions, Green functions and exponential dichotomies

In order to analyze the existence and uniqueness of solutions for a class of discrete dynamical systems, we shall first remind the readers of the notion of discrete exponential dichotomy, which plays an important role in the setup of the main results.

Definition 2.2.2 (cf. [496, Definition 5] and [497]). Let $X(t)$ be the principal fundamental matrix solution of the linear homogeneous system

$$x(t+1) = A(t)x(t), \quad t \in \mathbb{Z}; \quad x(t_0) = x_0 \in \mathbb{C}^n, \tag{69}$$

where $A(t)$ is a matrix function, which is invertible for all $t \in \mathbb{Z}$. Then we say that (69) admits an exponential dichotomy if there exist a projection P and positive constants α_1, α_2, β_1 and β_2 such that

$$\begin{aligned} \|X(t)PX^{-1}(s)\| &\leqslant \beta_1(1+\alpha_1)^{s-t}, \quad t \geqslant s, \\ \|X(t)(I-P)X^{-1}(s)\| &\leqslant \beta_2(1+\alpha_2)^{t-s}, \quad s \geqslant t. \end{aligned}$$

We define the Green function by

$$G(t,s) := \begin{cases} X(t)PX^{-1}(s) & \text{for } t \geqslant s, \\ X(t)(I-P)X^{-1}(s) & \text{for } s \geqslant t. \end{cases}$$

We will use the following result later on (cf. [496, Theorem 2]).

Theorem 2.2.3. *If the system* (69) *admits an exponential dichotomy and the function* $f(\cdot)$ *is bounded, then the nonhomogeneous system*

$$x(t+1) = A(t)x(t) + f(t), \quad t \in \mathbb{Z}; \quad x(t_0) = x_0 \tag{70}$$

has a bounded solution of the form

$$x(t) = \sum_{j=-\infty}^{\infty} G(t, j+1)f(j). \tag{71}$$

2.2.2 Levitan ρ-almost periodic type sequences

In a joint research article with B. Chaouchi and D. Velinov [177], the author has recently analyzed Levitan ρ-almost periodic type functions and uniformly Poisson stable functions. We will use the following notions (cf. also [177, Definition 2.1, Definition 2.13]).

Definition 2.2.4. Suppose that $\emptyset \neq I \subseteq \mathbb{Z}^n$, $\emptyset \neq I' \subseteq \mathbb{Z}^n$, $i + i' \in I$ for all $i \in I$, $i' \in I'$ and $F : I \times X \to Y$. Then we say that the sequence $F(\cdot;\cdot)$ is:

(i) Levitan-pre-$(\mathcal{B}, I', \rho)$-almost periodic if for every $\varepsilon > 0$, $B \in \mathcal{B}$ and $N > 0$, there exists $L > 0$ such that, for every $\mathbf{t}_0 \in I'$, there exists $\tau \in B(\mathbf{t}_0, l) \cap I'$ such that, for every $x \in B$ and $i \in I$ with $|i| \leqslant N$, there exists $y_{i;x} \in \rho(F(i;x))$ such that

$$\|F(i+\tau;x) - y_{i;x}\| \leqslant \varepsilon, \quad x \in B;$$

by $E_{\varepsilon;N;B}$ we denote the set consisting of all such numbers $\tau \in I'$.

(ii) Levitan $(\mathcal{B}, \rho)$-almost periodic if $F(\cdot;\cdot)$ is Levitan-pre-$(\mathcal{B}, I', \rho)$-almost periodic with $I' = I$, $\rho = \mathrm{I}$ and, for every $\varepsilon > 0$, $B \in \mathcal{B}$ and $N > 0$, there exist a number $\eta > 0$ and a relatively dense set $E_{\eta;N;B}$ in I (i. e., for every $\varepsilon > 0$ there exists $l > 0$ such that for each $\mathbf{t} \in I$ there exists $\tau \in B(\mathbf{t}, l) \cap E_{\eta;N;B}$) such that $E_{\eta;N;B} \subseteq I'$ and $E_{\eta;N;B} \pm E_{\eta;N;B} \subseteq E_{\varepsilon;N;B}$.

(iii) Strongly Levitan $(\mathcal{B}, \rho)$-almost periodic if $F(\cdot;\cdot)$ is Levitan $\mathcal{B}$-almost periodic and the set $E_{\eta;N;B}$ from the part (ii) can be written as $E_{\eta;N;B} = \prod_{j=1}^{n} E^j_{\eta;N;B}$, where the set $E^j_{\eta;N;B}$ is relatively dense in the j-th projection of the set I.

Using the same argumentation as in the proofs of [252, Theorem 2], [476, Theorem 2.3, Proposition 2.4, Theorem 2.6] and the fact that strongly Levitan N-almost periodic functions form the vector space with the usual operations, we may deduce the following important results (we will only outline the main details of proofs; by a strongly Levitan almost periodic sequence (function), we mean a strongly Levitan I-almost periodic sequence (function)).

Theorem 2.2.5. *Suppose that $\rho = T \in L(Y)$ and $F : \mathbb{Z}^n \to Y$. Then the following holds:*
(i) *If $F : \mathbb{Z}^n \to Y$ is a Levitan T-pre-almost periodic sequence, then there exists a continuous Levitan T-pre-almost periodic function $\tilde{F} : \mathbb{R}^n \to Y$ such that $R(\tilde{F}(\cdot)) \subseteq CH(\overline{R(F)})$ and $\tilde{F}(k) = F(k)$ for all $k \in \mathbb{Z}^n$. Furthermore, if $F(\cdot)$ is bounded, then $\tilde{F}(\cdot)$ is uniformly continuous.*
(ii) *If $F : \mathbb{Z}^n \to Y$ is a (strongly) Levitan T-almost periodic sequence, then there exists a continuous (strongly) Levitan T-almost periodic function $\tilde{F} : \mathbb{R}^n \to Y$ such that $R(\tilde{F}(\cdot)) \subseteq CH(\overline{R(F)})$ and $\tilde{F}(k) = F(k)$ for all $k \in \mathbb{Z}^n$. Furthermore, if $F(\cdot)$ is bounded, then $\tilde{F}(\cdot)$ is uniformly continuous.*

Proof. We will present all relevant details of the proof of (ii) in the two-dimensional setting; cf. also the proof of [252, Theorem 2] with $c = 1$ and $\delta = 1/2$. Consider first the statement (i). If $t = (t_1, t_2) \in \mathbb{R}^2$ is given, then there exist the uniquely determined numbers $k \in \mathbb{Z}$ and $m \in \mathbb{Z}$ such that $t_1 \in [k, k+1)$ and $t_2 \in [m, m+1)$. Define first $\tilde{F}(t_1, m) := \tilde{F}_{\mathbb{Z}}(k, m)$ if $t_1 \in [k, k+(1/2))$ and $\tilde{F}(t_1, m) := 2(\tilde{F}_{\mathbb{Z}}(k+1, m) - \tilde{F}_{\mathbb{Z}}(k, m))(t_1 - k - (1/2)) + \tilde{F}_{\mathbb{Z}}(k, m)$ if $t_1 \in [k+(1/2), k+1)$; we similarly define $\tilde{F}(t_1, m+1) := \tilde{F}_{\mathbb{Z}}(k, m+1)$ if $t_1 \in [k, k+(1/2))$ and $\tilde{F}(t_1, m+1) := 2(\tilde{F}_{\mathbb{Z}}(k+1, m+1) - \tilde{F}_{\mathbb{Z}}(k, m+1))(t_1 - k - (1/2)) + \tilde{F}_{\mathbb{Z}}(k, m+1)$ if $t_1 \in [k+(1/2), k+1)$. After that, we define $\tilde{F}(t_1, t_2) := \tilde{F}(t_1, m)$ if $t_2 \in [m, m+(1/2))$ and $\tilde{F}(t_1, t_2) := 2(\tilde{F}(t_1, m+1) - \tilde{F}(t_1, m))(t_2 - m - (1/2)) + \tilde{F}(t_1, m)$ if $t_2 \in [m+(1/2), m+1)$. Then the function $\tilde{F}(\cdot)$ is continuous, $R(\tilde{F}(\cdot)) \subseteq CH(\overline{R(F)})$, $\tilde{F}(k) = F(k)$ for all $k \in \mathbb{Z}^n$ and the function $\tilde{F}(\cdot)$ is uniformly continuous provided that $F(\cdot)$ is bounded. As in the proof of [476, Theorem 2.3], we may show that $\tilde{F}(\cdot)$ is a Levitan T-pre-almost periodic function provided that $F(\cdot)$ is a Levitan T-pre-almost periodic sequence. □

Theorem 2.2.6. *Suppose that $F : \mathbb{Z}^n \to Y$. If $F : \mathbb{R}^n \to Y$ is a strongly Levitan almost periodic function and $F(\cdot)$ is uniformly continuous, then $F_{|\mathbb{Z}^n} : \mathbb{Z}^n \to Y$ is a strongly Levitan almost periodic sequence.*

Proof. Let $\varepsilon > 0$ and $N > 0$ be given. The result is clearly true if $Y = 0$; we assume $Y \neq 0$ henceforth. Since $F(\cdot)$ is uniformly continuous, we can find a number $\delta \in (0, \varepsilon)$ such that the assumptions $x, y \in \mathbb{R}^n$ and $|x - y| \leqslant \delta$ imply $\|F(x) - F(y)\|_Y \leqslant \varepsilon$. Since the strongly Levitan almost periodic functions form a vector space with the usual operations, we know that there exists a number $\eta \in (0, \delta)$ and relatively dense sets $E^j_{\eta;N}$ in $\mathbb{R}$ such that the set $E_{\eta;N} \equiv \prod_{j=1}^n E^j_{\eta;N}$ consists solely of common (η, N)-almost periods of the function $F(\cdot)$ and the functions $G_j(\cdot)$ defined below ($1 \leqslant j \leqslant n$) as well as that $E_{\eta;N} \pm E_{\eta;N} \subseteq E(\varepsilon, N)(F, G_1, \ldots, G_n)$; here, we use the same notion and notation as in [177]. Therefore, if $\tau = (\tau_1, \ldots, \tau_n)$ in $E_{\eta;N}$, then we have $\|F(\mathbf{t} + \tau) - F(\mathbf{t})\|_Y \leqslant \eta$ for all $\mathbf{t} \in \mathbb{R}^n$ with $|\mathbf{t}| \leqslant N$, and $\|G_j(\mathbf{t} + \tau) - G_j(\mathbf{t})\|_Y \leqslant \eta$ for all $\mathbf{t} \in \mathbb{R}^n$ with $|\mathbf{t}| \leqslant N$ and $j \in \mathbb{N}_n$, where the Bohr

$\mathcal{B}$-almost periodic function $G_j : \mathbb{R}^n \to Y$ is defined as the usual periodic extension of the function $G_{j;0}(\mathbf{t}) := (1-|1-t_j|)y, \mathbf{t} = (t_1, \ldots, t_j, \ldots, t_n) \in [0,2]^n$ to the space $\mathbb{R}^n$ (the nonzero element $y \in Y$ is fixed in advance). As in the one-dimensional setting, this simply implies that there exist two vectors $p \in \mathbb{Z}^n$ and $w = (w_1, \ldots, w_n) \in B(0, \eta)$ such that $\tau = 2p + \omega$. Therefore, we have

$$\begin{aligned} &\|F(\mathbf{t} + 2p) - F(\mathbf{t})\|_Y \\ &\leqslant \|F(\mathbf{t} + 2p) - F(\mathbf{t} + 2p + w)\|_Y + \|F(\mathbf{t} + 2p + w) - F(\mathbf{t})\|_Y \\ &\leqslant \varepsilon + \eta < 2\varepsilon, \quad \mathbf{t} \in \mathbb{R}^n, \ |\mathbf{t}| \leqslant N. \end{aligned}$$

This simply yields that $F_{|\mathbb{Z}^n}(\cdot)$ is a Levitan almost periodic sequence and the second condition from the formulation of Definition 2.2.4(iii) holds, so that $F_{|\mathbb{Z}^n}(\cdot)$ is a strongly Levitan almost periodic sequence. □

Remark 2.2.7. It is very difficult to state a satisfactory analogue of Theorem 2.2.6 if the function $F(\cdot)$ is not uniformly continuous. In connection with this issue, we would like to mention that many intriguing examples of unbounded Levitan almost periodic functions $F : \mathbb{R} \to \mathbb{R}$, which are not uniformly continuous have recently been constructed by A. Nawrocki in [618]; the discretizations of such functions cannot be simply analyzed by means of Theorem 2.2.6; see also the recent research article [190] by D. N. Cheban.

We continue by providing the following illustrative example.

Example 2.2.8. Suppose that

$$F(t) := \frac{1}{2 + \cos t + \cos(\sqrt{2}t)}, \quad t \in \mathbb{R}. \tag{72}$$

Then we know that the function $F(\cdot)$ is Levitan almost periodic, unbounded and not uniformly continuous [508, 509]. Furthermore, the sequence $(F(k))_{k\in\mathbb{Z}}$ is unbounded, as easily approved, and Levitan almost periodic, which can be proved as follows (Theorem 2.2.6 is inapplicable here). The argumentation contained on [509, p. 59] shows that for each $\varepsilon > 0$ and $N > 0$ there exists a sufficiently small number $\delta > 0$ such that any integer, which is a δ-almost period of the function $2 + \cos\cdot + \cos(\sqrt{2}\cdot)$ is also a Levitan (ε, N)-almost period of the function $F(\cdot)$; it is well known that the set of all such integers, which are δ-almost periods is relatively dense in $\mathbb{R}$. If $\varepsilon > 0$ and $N > 0$ are given, then we can simply choose the number $\eta = \delta/2$ in Definition 2.2.4(ii) and the set $E_{\eta;N}$ consisting of all integer $(\delta/2)$-almost periods of the function $2+\cos\cdot+\cos(\sqrt{2}\cdot)$. Observe finally that, due to [619, Corollary 1], for each $\varepsilon > 0$ there exists $M_\varepsilon > 0$ such that $F(k) \leqslant M_\varepsilon |k|^{2+\varepsilon}$ for all $k \in \mathbb{Z}$.

The notion of a strongly Levitan almost periodic sequence and the notion of a Levitan almost periodic sequence coincide in the one-dimensional setting. Without going into any further details concerning the validity of the above result in the multidimensional setting, where the famous Bogolyubov theorem does not admit a satisfactory reformulation (cf. [177] for more details), we will only formulate here the following important consequence of Theorem 2.2.5 and Theorem 2.2.6.

Theorem 2.2.9. *Suppose that* $F : \mathbb{Z} \to Y$ *is bounded. Then* $(F(k))_{k\in\mathbb{Z}}$ *is a Levitan almost periodic sequence if and only if* $(F(k))_{k\in\mathbb{Z}}$ *is an almost automorphic sequence.*

Proof. If $(F(k))_{k\in\mathbb{Z}}$ is a Levitan almost periodic sequence, then Theorem 2.2.5(ii) implies that there exists a uniformly continuous, Levitan almost periodic function $\tilde{F} : \mathbb{R} \to Y$ such that $\tilde{F}(k) = F(k)$ for all $k \in \mathbb{Z}$. Due to [804, Theorem 3.1], we have that $\tilde{F} : \mathbb{R} \to Y$ is compactly almost automorphic so that $(F(k))_{k\in\mathbb{Z}}$ is an almost automorphic sequence. On the other hand, if $(F(k))_{k\in\mathbb{Z}}$ is an almost automorphic sequence, then there exists a compactly almost automorphic function $\tilde{F} : \mathbb{R} \to Y$ such that $\tilde{F}(k) = F(k)$ for all $k \in \mathbb{Z}$. Clearly, $\tilde{F}(\cdot)$ is uniformly continuous; applying again [804, Theorem 3.1], we get that $\tilde{F}(\cdot)$ is Levitan almost periodic. Therefore, the final conclusion simply follows from an application of Theorem 2.2.6. □

2.2.3 Remotely ρ-almost periodic type sequences

The following notion is a special case of the notion introduced in [454, Definition 4.1] (see also [493, Definitions 3.1, 3.2]).

Definition 2.2.10. Suppose that $\mathbb{D} \subseteq I \subseteq \mathbb{Z}^n$, $\emptyset \neq I' \subseteq \mathbb{Z}^n$, $\emptyset \neq I \subseteq \mathbb{Z}^n$, the sets $\mathbb{D}$ and I' are unbounded, $I + I' \subseteq I$ and $F : I \times X \to Y$ is a given function. Then we say that:

(i) $F(\cdot;\cdot)$ is $\mathbb{D}$-quasiasymptotically Bohr $(\mathcal{B}, I', \rho)$-almost periodic if for every $B \in \mathcal{B}$ and $\varepsilon > 0$ there exists a finite real number $l > 0$ such that for each $\mathbf{t}_0 \in I'$ there exists $\tau \in B(\mathbf{t}_0, l) \cap I'$ such that, for every $x \in B$, there exists a function $G_x \in Y^{\mathbb{D}}$, the set of all functions from $\mathbb{D}$ into Y, such that $G_x(\mathbf{t}) \in \rho(F(\mathbf{t}; x))$ for all $\mathbf{t} \in \mathbb{D}$, $x \in B$ and

$$\limsup_{|\mathbf{t}|\to+\infty;\mathbf{t}\in\mathbb{D}} \sup_{x\in B} \|F(\mathbf{t}+\tau; x) - G_x(\mathbf{t})\|_Y \leqslant \varepsilon.$$

(ii) $F(\cdot;\cdot)$ is $\mathbb{D}$-remotely $(\mathcal{B}, I', \rho)$-almost periodic if $F(\cdot;\cdot)$ is $\mathbb{D}$-quasiasymptotically Bohr $(\mathcal{B}, I', \rho)$-almost periodic and, for every $B \in \mathcal{B}$, the function $F(\cdot;\cdot)$ is bounded and uniformly continuous on $I \times B$.

Remark 2.2.11. If $X = \{0\}$ in (ii), then the boundedness and the uniform continuity on $I \times B$ is equivalent with the boundedness on I.

The usual notion is obtained by plugging $X = \{0\}$, $\mathbb{D} = I' = I$ and $\rho = \mathrm{I}$, when we also say that the function $F(\cdot)$ is quasiasymptotically almost periodic (remotely almost periodic). If $\mathbb{D}$, I', $I \subseteq \mathbb{R}^n$, then we accept the same terminology for the functions.

The following result, which establishes a bridge between remotely almost periodic functions on continuous and discrete time domains, can be deduced with the help of the argumentation contained in the proof of [817, Theorem 2.1].

Theorem 2.2.12. *A necessary and sufficient condition for a sequence $F : \mathbb{Z}^n \to Y$ to be remotely almost periodic is that there exists a remotely almost periodic function $H : \mathbb{R}^n \to Y$ so that $F(t) = H(t)$ for all $t \in \mathbb{Z}^n$.*

We perform the proof of the following composition principle by exactly pursuing the same direction of the proof of [811, Lemma 3.4]; the same proof works for the functions and can be adapted for the almost automorphic sequences (functions).

Theorem 2.2.13. *Suppose that $(Z, \|\cdot\|_Z)$ is a complex Banach space, $F : \mathbb{Z}^n \times X \to Y$ is $\mathcal{B}$-remotely almost periodic and $G : \mathbb{Z}^n \times Y \to Z$ is $\mathcal{B}'$-remotely almost periodic, where $\mathcal{B}$ denotes the family of all bounded subsets of X and $\mathcal{B}'$ denotes the family of all bounded subsets of Y. Suppose further that for each bounded subset B' of Y there exists a finite real constant $L_{B'} > 0$ such that*

$$\|G(\mathbf{t};y_1) - G(\mathbf{t};y_2)\|_Z \leqslant L_{B'}\|y_1 - y_2\|_Y, \quad \mathbf{t} \in \mathbb{Z}^n,\ y_1,\ y_2 \in B. \tag{73}$$

Then the sequence $H : \mathbb{Z}^n \times X \to Z$, defined by $H(\mathbf{t};x) := G(\mathbf{t};F(\mathbf{t};x))$, $\mathbf{t} \in \mathbb{Z}^n$, $x \in X$, is $\mathcal{B}$-remotely almost periodic.

Proof. Let $\varepsilon > 0$ and $B \in \mathcal{B}$ be given. Then the set $B' := \{F(\mathbf{t};x) : \mathbf{t} \in \mathbb{Z}^n,\ x \in B\}$ is bounded and there exists $L_{B'} > 0$ such that (73) holds. This yields

$$\begin{aligned}
&\|H(\mathbf{t}';x') - H(\mathbf{t};x)\|_Z \\
&\quad \leqslant \|G(\mathbf{t}';F(\mathbf{t}';x')) - G(\mathbf{t}';F(\mathbf{t};x))\|_Z + \|G(\mathbf{t}';F(\mathbf{t};x)) - G(\mathbf{t};F(\mathbf{t};x))\|_Z \\
&\quad \leqslant L_{B'}\|F(\mathbf{t}';x') - F(\mathbf{t};x)\|_Y + \sup_{y\in B'}\|G(\mathbf{t}';y) - G(\mathbf{t}';y)\|_Z, \quad \mathbf{t},\ \mathbf{t}' \in \mathbb{Z}^n,\ x,\ x' \in B,
\end{aligned}$$

which simply implies that the function $H(\cdot;\cdot)$ is bounded and uniformly continuous on $I \times B$. Further on, let us denote by $l_\infty(B : Y)$ the Banach space of all bounded functions from B into Y, equipped with the sup-norm. Then the function $F_B : \mathbb{Z}^n \to l_\infty(B : Y)$, given by $[F_B(\mathbf{t})](x) := F(\mathbf{t};x)$, $\mathbf{t} \in \mathbb{Z}^n$, $x \in B$ is remotely almost periodic and the function $G_{B'} : \mathbb{Z}^n \to l_\infty(B' : Y)$, given by $[G_{B'}(\mathbf{t})](y) := G(\mathbf{t};y)$, $\mathbf{t} \in \mathbb{Z}^n$, $y \in B'$ is remotely almost periodic. Consequently, the set $\tau(H,\varepsilon)$ consisting of all points $p \in \mathbb{Z}^n$ such that

$$\limsup_{|\mathbf{t}|\to\infty} \sup_{y\in B'}\|G(\mathbf{t}+p;y) - G(\mathbf{t};y)\|_Z + \limsup_{|\mathbf{t}|\to\infty} \sup_{x\in B}\|F(\mathbf{t}+p;x) - F(\mathbf{t};x)\|_Y < \varepsilon$$

is relatively dense in $\mathbb{Z}^n$. The final conclusion follows from the next computation ($\mathbf{t} \in \mathbb{Z}^n$, $x \in B$):

$$\|G(\mathbf{t}+p;F(\mathbf{t}+p;x)) - G(\mathbf{t};F(\mathbf{t};x))\|_Z \leqslant \|G(\mathbf{t}+p;F(\mathbf{t}+p;x)) - G(\mathbf{t};F(\mathbf{t}+p;x))\|_Z$$

$$
\begin{aligned}
&+ \|G(\mathbf{t}; F(\mathbf{t} + p; x)) - G(\mathbf{t}; F(\mathbf{t}; x))\|_Z \\
&\leqslant \|G(\mathbf{t} + p; F(\mathbf{t} + p; x)) - G(\mathbf{t}; F(\mathbf{t} + p; x))\|_Z + L_{B'} \|F(\mathbf{t} + p; x) - F(\mathbf{t}; x)\|_Y,
\end{aligned}
$$

and the subadditivity of operation $\limsup_{|\mathbf{t}|\to\infty} \cdot$. □

For more details about remotely almost periodic motions of dynamical systems, we refer the reader to the recent research article [189]. We end this subsection by noting that the space of remotely almost periodic sequences $RDAP(\mathbb{Z}^n : Y)$ is, in fact, a closed subspace of the Banach space of bounded sequences on $\mathbb{Z}^n$ so that $RDAP(\mathbb{Z}^n : Y)$ is a Banach space when endowed by the sup-norm.

In the following three parts, we will provide some applications of our results and introduced notion to the abstract Volterra difference equations.

1. On the abstract difference equation $u(k+1) = Au(k) + f(k)$, its fractional and multidimensional analogues

As already mentioned, D. Araya, R. Castro and C. Lizama have investigated, in [65, Section 3], the almost automorphic solutions of the first-order linear difference equation (41), where $A \in L(X)$ and $(f_k \equiv f(k))_{k\in\mathbb{Z}}$ is an almost automorphic sequence. We will reconsider here the obtained results by assuming that $(f_k)_{k\in\mathbb{Z}}$ is a Levitan almost periodic type sequence (cf. also [476]).

Suppose first that $A = \lambda I$, where $\lambda \in \mathbb{C}$ and $|\lambda| \neq 1$. We already know that the almost automorphy of sequence $(f_k)_{k\in\mathbb{Z}}$ implies the existence of a unique almost automorphic solution $u(\cdot)$ of (41), given by (42) if $|\lambda| < 1$, and (48) if $|\lambda| > 1$. We also have the following.

Proposition 2.2.14. *Suppose that $\rho = T \in L(X)$ and $f(\cdot)$ is a bounded Levitan pre-(I', T)-almost periodic sequence (Levitan T-almost periodic sequence). Then a unique bounded Levitan pre-(I', T)-almost periodic solution (Levitan T-almost periodic solution) of* (41) *is given by* (42) *if $|\lambda| < 1$, and* (48) *if $|\lambda| > 1$.*

Proof. We will only prove that the sequence $(u(k))_{k\in\mathbb{Z}}$ is bounded and Levitan pre-(I', T)-almost periodic provided that $|\lambda| < 1$ and $f(\cdot)$ is a bounded Levitan pre-(I', T)-almost periodic sequence. This is clear for the boundedness; suppose now that the numbers $\varepsilon > 0$ and $N > 0$ are fixed. Then there exists a natural number $v' \in \mathbb{N} \setminus \{1\}$ such that

$$
\sum_{v=v'}^{\infty} |\lambda|^v \|f(j+\tau-v-1) - Tf(j-v-1)\| \leqslant (1 + \|T\|)\|f\|_\infty \sum_{v=v'}^{\infty} |\lambda|^v < \varepsilon/2, \quad \tau \in \mathbb{Z}. \tag{74}
$$

Set $N' := N + 1 + v'$. Let $\tau \in I'$ be any $(\varepsilon(1 - |\lambda|)/2, N')$-almost period of the sequence $(f(k))_{k\in\mathbb{Z}}$, with the meaning clear. Then we have

$$
\|u(j+\tau) - Tu(j)\| \leqslant \sum_{v=0}^{\infty} |\lambda|^v \|f(j+\tau-v-1) - Tf(j-v-1)\|
$$

$$\leqslant \sum_{v=0}^{v'-1} |\lambda|^v \|f(j+\tau-v-1) - Tf(j-v-1)\|$$

$$+ \sum_{v=v'}^{\infty} |\lambda|^v \|f(j+\tau-v-1) - Tf(j-v-1)\|$$

$$\leqslant \sum_{v=0}^{v'-1} |\lambda|^v (\varepsilon(1-|\lambda|)/2) + (\varepsilon/2) \leqslant \varepsilon, \quad j \in \mathbb{Z},\ |j| \leqslant N.$$

This implies the required conclusion. □

Similarly, we can prove the following (the statement of [65, Theorem 3.2] can be simply rephrased for the bounded Levitan T-almost periodic type sequences, as well).

Proposition 2.2.15. *Suppose that $\rho = T \in L(X)$, $A \in L(X)$ and $f(\cdot)$ is a bounded Levitan pre-(I', T)-almost periodic sequence (Levitan T-almost periodic sequence) and $\|A\| < 1$. Then there exists a unique bounded Levitan pre-(I', T)-almost periodic solution (Levitan T-almost periodic solution) of* (41).

The analogues of Proposition 2.2.14 and Proposition 2.2.15 can be formulated for the abstract first-order difference inclusion (47); cf. Theorem 2.1.45 for more details. Before investigating some fractional difference equations below, we would like to make the following observations.

Remark 2.2.16. Suppose that there exist two finite real constants $M \geqslant 1$ and $k \in \mathbb{N}$ such that $\|f(j)\| \leqslant M(1+|j|)^k$ for all $j \in \mathbb{Z}$. Then the solution $u(\cdot)$ from Proposition 2.2.14 is still well-defined and we have $u(j) = \sum_{v=0}^{\infty} \lambda^v f(j-v-1)$ for all $j \in \mathbb{Z}$, so that

$$\|u(j)\| \leqslant M \sum_{v=0}^{\infty} |\lambda|^v (1+|j|+|v|)^k$$

$$\leqslant M3^{k-1} \sum_{v=0}^{\infty} |\lambda|^v (1+|j|)^k + M3^{k-1} \sum_{v=0}^{\infty} |\lambda|^v v^k \leqslant M'(1+|j|)^k, \quad j \in \mathbb{Z},$$

for some finite real constant $M' \geqslant 1$. But it is not clear how we can prove that $u(\cdot)$ is Levitan pre-(I', T)-almost periodic (Levitan T-almost periodic); in the newly arisen situation, the main problem is the existence of a sufficiently large natural number $v' \in \mathbb{N}$, depending only on $\varepsilon > 0$ and $N > 0$, such that (74) holds true. We have not been able to find a solution of this problem even for the Levitan almost periodic sequence $(F(k) \equiv 1/(2+\cos k + \cos(\sqrt{2}k)))_{k\in\mathbb{Z}}$ from Example 2.2.8, with $I' = \mathbb{Z}$ and $T = \mathrm{I}$.

2. Fractional analogues of $u(k+1) = Au(k) + f(k)$

We investigate here the abstract fractional difference equation (50), where A is a closed linear operator on X, $0 < \alpha < 1$ and $\Delta^\alpha u(k)$ denotes the Caputo fractional difference operator of order α. We will use the same notion and notation as in [46].

Let A be a closed linear operator on X such that $1 \in \rho(A)$, where $\rho(A)$ denotes the resolvent set of A, and let $\|(I - A)^{-1}\| < 1$. Due to [46, Theorem 3.4], we know that A generates a discrete (a, a)-resolvent sequence $\{S_{a,a}(v)\}_{v\in\mathbb{N}_0}$ such that $\sum_{v=0}^{+\infty} \|S_{a,a}(v)\| < +\infty$. Furthermore, if $(f_k)_{k\in\mathbb{Z}}$ is a bounded sequence, then we know that the function

$$u(k) = \sum_{l=-\infty}^{k-1} S_{a,a}(k-1-l)f(l), \quad k \in \mathbb{Z}$$

is a mild solution of (50). Since $\sum_{v=0}^{+\infty} \|S_{a,a}(v)\| < +\infty$, the argumentation contained in the proof of Proposition 2.2.14 enables one to deduce the following analogue of this result.

Proposition 2.2.17. *Suppose that $\rho = T \in L(X)$ and $f(\cdot)$ is a bounded Levitan pre-(I', T)-almost periodic sequence (Levitan T-almost periodic sequence). Then a mild solution of (50), given by (51), is bounded Levitan pre-(I', T)-almost periodic (Levitan T-almost periodic).*

Before proceeding further, we will only note that we can similarly analyze the existence and uniqueness of Levitan T-almost periodic type solutions of (63).

3. Multidimensional analogues of $u(k+1) = Au(k) + f(k)$

We have already analyzed some multidimensional analogues of the abstract difference equation $u(k+1) = Au(k) + f(k)$. In the first concept, we assume that $f : \mathbb{Z}^n \to X$, $\lambda_1, \lambda_2, \ldots, \lambda_n$ are given complex numbers and

$$\max(|\lambda_1|, |\lambda_2|, \ldots, |\lambda_n|) < 1.$$

Consider the function

$$\begin{aligned} u(k_1, k_2, \ldots, k_n) &:= \sum_{l_1\leqslant k_1, l_2\leqslant k_2, \ldots, l_n\leqslant k_n} \lambda_1^{k_1-l_1}\lambda_2^{k_2-l_2} \cdot\dots\cdot \lambda_n^{k_n-l_n} f(l_1-1, l_2-1, \ldots, l_n-1) \\ &= \sum_{v_1\geqslant 0, v_2\geqslant 0, \ldots, v_n\geqslant 0} \lambda_1^{v_1}\lambda_2^{v_2} \cdot\dots\cdot \lambda_n^{v_n} f(k_1-v_1-1, k_2-v_2-1, \ldots, k_n-v_n-1) \end{aligned} \tag{75}$$

defined for any $(k_1, k_2, \ldots, k_n) \in \mathbb{Z}^n$. Then we can simply find the form of function $F : \mathbb{Z}^n \to X$ such that

$$u(k_1+1, k_2+1, \ldots, k_n+1) = \lambda_1\lambda_2\cdots\lambda_n \cdot u(k_1, k_2, \ldots, k_n) + F(k_1, k_2, \ldots, k_n), \tag{76}$$

for all $(k_1, k_2, \ldots, k_n) \in \mathbb{Z}^n$. Arguing as in the proof of Proposition 2.2.14, we may conclude the following: If $\rho = T \in L(X)$ and $f(\cdot)$ is a bounded Levitan pre-(I', T)-almost periodic sequence (Levitan T-almost periodic sequence), then a mild solution of (76), given by (75), is bounded Levitan pre-(I', T)-almost periodic (Levitan T-almost periodic).

In the second concept, we consider the solution $u_j : \mathbb{Z} \to X$ of the equation $u_j(k+1) = \lambda u_j(k)+f_j(k)$, $k \in \mathbb{Z}$, where $f_j(\cdot)$ is a bounded Levitan pre-(I', T)-almost periodic sequence (Levitan T-almost periodic sequence) for $1 \leqslant j \leqslant n$ and $\lambda \in \mathbb{C}$ satisfies $|\lambda| < 1$. Define $u(k_1, \dots, k_n) := u_1(k_1) + u_2(k_2) + \dots + u_n(k_n)$ and $f(k_1, \dots, k_n) := f_1(k_1) + f_2(k_2) + \dots + f_n(k_n)$ for all $k_j \in \mathbb{Z}$ $(1 \leqslant j \leqslant n)$. Then we have

$$u(k_1 + 1, \dots, k_n + 1) = \lambda u(k_1, \dots, k_n) + f(k_1, \dots, k_n), \quad (k_1, \dots, k_n) \in \mathbb{Z}^n;$$

moreover, the sequence $u(\cdot)$ is likewise bounded Levitan pre-(I', T)-almost periodic sequence (Levitan T-almost periodic sequence); here, $\rho = T \in L(X)$. A similar conclusion can be given for the abstract first-order difference inclusions with multi-valued linear operators satisfying certain assumptions.

Before proceeding to the next subsection, we will only observe that all results established in this subsection can be formulated if the term "bounded Levitan pre-(I', T)-almost periodic" is replaced with the term "remotely (I', T)-almost periodic." Then the solution $u(\cdot)$ will be also remotely (I', T)-almost periodic; for example, in the case of consideration of Proposition 2.2.14, we can apply the following computation:

$$\begin{aligned}\limsup_{|j|\to+\infty}\|u(j+\tau) - Tu(j)\| &\leqslant \sum_{v=0}^{\infty} |\lambda|^v \limsup_{|j|\to+\infty}\|f(j+\tau-v-1) - Tf(j-v-1)\| \\ &= \sum_{v=0}^{\infty} |\lambda|^v \limsup_{|j|\to+\infty}\|f(j+\tau) - Tf(j)\| \leqslant \varepsilon \sum_{v=0}^{\infty} |\lambda|^v,\end{aligned}$$

where τ is a remotely ε-almost period of the forcing term $f(\cdot)$.

4. The existence and uniqueness of remotely ρ-almost periodic type solutions for the equation (70)

We start this subsection by stating the following result concerning the inhomogeneous discrete dynamical system (70).

Theorem 2.2.18. *Let $I' \subseteq \mathbb{Z}$, $\inf I' = -\infty$ and $\sup I' = +\infty$. Assume that $f : \mathbb{Z} \to \mathbb{R}^n$ is bounded and quasiasymptotically (I', T)-almost periodic, where $T \in L(\mathbb{C}^n)$, and the homogeneous part of* (70) *admits an exponential dichotomy. If for each $p \in I'$, we have*

$$\limsup_{|t|\to+\infty} \sum_{j\in\mathbb{Z}} \|G(t+p, j+p) - G(t,j)\| = 0, \tag{77}$$

then the bounded solution $x(t)$ of (70), *given by* (71), *is quasiasymptotically (I', T)-almost periodic.*

Proof. By Theorem 2.2.3, the bounded solution of (70) is given by

$$x(t) = \sum_{j=-\infty}^{\infty} G(t, j+1)f(j).$$

Further on, we have

$$\begin{aligned}
&\|x(t+p)-Tx(t)\| \\
&= \Big\|\sum_{j=-\infty}^{\infty} G(t+p,j+1)f(j) - T\sum_{j=-\infty}^{\infty} G(t,j+1)f(j)\Big\| \\
&= \Big\|\sum_{j=-\infty}^{\infty} G(t+p,j+p+1)f(j+p) - T\sum_{j=-\infty}^{\infty} G(t,j+1)f(j)\Big\| \\
&= \Big\|\sum_{j=-\infty}^{\infty} G(t+p,j+p+1)f(j+p) \pm \sum_{j=-\infty}^{\infty} G(t,j+1)f(j+p) - T\sum_{j=-\infty}^{\infty} G(t,j+1)f(j)\Big\| \\
&\leqslant \Big\|\sum_{j=-\infty}^{\infty} (G(t+p,j+p+1)-G(t,j+1))f(j+p)\Big\| + \Big\|\sum_{j=-\infty}^{\infty} G(t,j+1)(f(j+p)-Tf(j))\Big\| \\
&\leqslant \|f\|_{\infty}\sum_{j=-\infty}^{\infty}\|G(t+p,j+p+1)-G(t,j+1)\| + \sum_{j=-\infty}^{\infty}\beta(1+\alpha)^{-|t-j-1|}\|f(j+p)-Tf(j)\|.
\end{aligned}$$

Hence,

$$\begin{aligned}
\limsup_{t\to\pm\infty}\|x(t+p)-x(t)\| &\leqslant \|f\|_{\infty}\limsup_{t\to\pm\infty}\sum_{j=-\infty}^{\infty}\|G(t+p,j+p+1)-G(t,j+1)\| \\
&\quad + \limsup_{t\to\pm\infty}\sum_{j=-\infty}^{\infty}\beta(1+\alpha)^{-|t-j-1|}\|f(j+p)-f(j)\|.
\end{aligned}$$

Let $\varepsilon > 0$ be given. Then there exists $l > 0$ such that every interval I of length l contains a point p such that there exists an integer $M(\varepsilon,p) > 0$ such that

$$\|f(j+p)-Tf(j)\| \leqslant \varepsilon, \quad |j| \geqslant M(\varepsilon,p).$$

This implies

$$\begin{aligned}
&\limsup_{t\to\pm\infty}\|x(t+p)-x(t)\| \\
&\quad\leqslant \|f\|_{\infty}\limsup_{t\to\pm\infty}\sum_{j=-\infty}^{\infty}\|G(t+p,j+p+1)-G(t,j+1)\| \\
&\qquad + 2\|f\|_{\infty}\limsup_{t\to\pm\infty}\sum_{|j|<M(\varepsilon,p)}\beta(1+\alpha)^{-|t-j-1|} \\
&\qquad + \varepsilon\limsup_{t\to\pm\infty}\sum_{|j|\geqslant M(\varepsilon,p)}\beta(1+\alpha)^{-|t-j-1|} \\
&\quad= \|f\|_{\infty}\limsup_{t\to\pm\infty}\sum_{j=-\infty}^{\infty}\|G(t+p,j+p+1)-G(t,j+1)\|
\end{aligned}$$

$$+ \varepsilon \limsup_{t \to \pm\infty} \sum_{|j| \geqslant M(\varepsilon,p)} \beta(1+\alpha)^{-|t-j-1|}$$

$$\leqslant \|f\|_\infty \limsup_{t \to \pm\infty} \sum_{j=-\infty}^{\infty} \|G(t+p, j+p+1) - G(t, j+1)\| + 2\beta\varepsilon \sum_{j \in \mathbb{Z}} (1+\alpha)^{-|j|}.$$

An application of (77) completes the proof. □

Remark 2.2.19. The assumption that for each $p \in I'$ we have (77) is a little bit redundant. This assumption holds if the Green function $G(t, s)$ is bi-periodic in the usual sense, with appropriately chosen set I'; in particular, this situation occurs if the functions $A_\pm(\cdot)$ from the formulation of [579, Theorem 2] are p-periodic for some $p \in \mathbb{N}$ (see the equation [579, (21), Lemma 2]), when we can choose $I' := p\mathbb{N}$.

Consider now the situation in which the functions $A_\pm(\cdot)$ from the formulation of [579, Theorem 2] are remotely almost periodic and the sequence $f(\cdot)$ is remotely almost periodic ($I' = \mathbb{Z}, \rho = \mathrm{I}$). Then the remotely almost periodic extension $\tilde{f}(\cdot)$ of the sequence $f(\cdot)$ to the real line can share the same set of remotely ε-periods with the functions $A_\pm(\cdot)$. We can apply again the equation [579, (21), Lemma 2] and a simple calculation in order to see that the solution $x(\cdot)$ will be remotely almost periodic.

The proofs of [579, Theorems 3, 4] are not completely correct because the authors have not proved that, in general case, there exists a common set of remotely ε-bi-almost periods of $G(t, s)$ and remotely ε-almost periods of forcing term $f(\cdot)$.

We continue by stating the following result.

Theorem 2.2.20. *Consider the nonlinear discrete dynamical system*

$$x(t+1) = A(t)x(t) + g(t, x(t)), \quad x(t_0) = x_0, \tag{78}$$

where $g : \mathbb{Z} \times \mathbb{R}^n \to \mathbb{R}^n$ is $\mathcal{B}$-remotely almost periodic with $\mathcal{B}$ being the collection of all bounded subsets of $\mathbb{R}^n$, and the homogeneous part of (78) *admits an exponential dichotomy, which satisfies that for each $p \in \mathbb{Z}$ we have* (77)*. If the function $g(\cdot;\cdot)$ satisfies the Lipschitz condition*

$$\|g(t,x) - g(t,y)\| \leqslant L\|x - y\| \quad \text{for all } x \quad \text{and } y \in \mathbb{R}^n,$$

and

$$L\left(\frac{2\beta}{\alpha} + \beta\right) = \lambda < 1,$$

then the functional system (78) *has a unique remotely almost periodic solution.*

Proof. Suppose that $x(\cdot)$ is remotely almost periodic. Then Theorem 2.2.13 implies that the function $g(\cdot; x(\cdot))$ is remotely almost periodic. Let us introduce the mapping $H : RDAP(\mathbb{Z} : \mathbb{R}^n) \to RDAP(\mathbb{Z} : \mathbb{R}^n)$ by

$$[H(x(\cdot))](t) := \sum_{j=-\infty}^{\infty} G(t,j+1)g(j,x(j)), \quad t \in \mathbb{Z}.$$

Theorem 2.2.18 indicates that H maps $RDAP(\mathbb{Z} : \mathbb{R}^n)$ into itself. If $x, y \in RDAP(\mathbb{Z} : \mathbb{R}^n)$, then we have

$$\begin{aligned}
\|H(x) - H(y)\| &\leqslant \sum_{j=-\infty}^{\infty} \|G(t,j+1)\| \cdot \|g(j,x(j)) - g(j,y(j))\| \\
&= \sum_{j=-\infty}^{t-1} \|G(t,j+1)\| \cdot \|g(j,x(j)) - g(j,y(j))\| \\
&\quad + \sum_{j=t}^{\infty} \|G(t,j+1)\| \cdot \|g(j,x(j)) - g(j,y(j))\| \\
&\leqslant L\|x-y\| \left(\sum_{j=-\infty}^{t-1} \beta(1+\alpha)^{j+1-t} + \sum_{j=t}^{\infty} \beta(1+\alpha)^{t-j-1} \right) \\
&= L\|x-y\| \left(\frac{2\beta}{\alpha} + \beta \right) \\
&< \lambda \|x-y\|,
\end{aligned}$$

which shows that H is a contraction. The Banach fixed-point theorem implies that the nonlinear discrete dynamical system (78) has a unique remotely almost periodic solution. □

Now we turn our attention into a more specific discrete dynamical system, which is a nonconvolution type Volterra difference system with infinite delay given in the form

$$x(t+1) = A(t)x(t) + \sum_{j=-\infty}^{t} B(t,j)x(j) + f(t), \quad t \in \mathbb{Z}, \tag{79}$$

where A and B are $n \times n$ matrix functions and $f(\cdot)$ is a vector function. Indeed, almost periodic solutions of Volterra difference equations have taken prominent attention in the existing literature, and there is a vast literature based on the existence of discrete almost periodic solutions for numerous kinds of Volterra difference equations. In the pioneering paper of S. Elaydi (see [273]), the investigation of sufficient conditions for the existence of discrete almost periodic solutions was stated as an open problem, and [496] (2018) provided a solution to this open problem by using the discrete variant of exponential dichotomy and the fixed-point theory. It is clear that the space $RDAP(\mathbb{Z} : \mathbb{R}^n)$ is a much more larger space than the space of discrete almost periodic functions. In this part, we consider the remotely almost periodic solutions of (79).

By a remotely almost periodic solution of the Volterra system (79), we mean a vector-valued remotely almost periodic function $x^{\xi}(\cdot)$ on $\mathbb{Z}$, which satisfies (79) for all $t \in \mathbb{Z}_+$ and $x^{\xi}(t) = \xi(t)$ for all $t \in \mathbb{Z}_-$, where $\mathbb{Z}_-$ is the set of negative integers ($\mathbb{Z}_+$ is the set of nonnegative integers), and $\xi : \mathbb{Z}_- \to \mathbb{R}^n$ is the bounded initial vector function with $\sup_{t \in \mathbb{Z}_-} |\xi(t)| < U_{\xi} < \infty$.

Initially, we make the following assumption:

A1 The homogeneous part of the Volterra system (79) admits an exponential dichotomy.

As in [496], we define the following mapping:

$$(Tx^{\xi})(t) := \begin{cases} \xi(t), & t \in \mathbb{Z}_-, \\ \sum\limits_{j=-\infty}^{\infty} G(t,j+1)W(j,x(j)), & t \in \mathbb{Z}_+, \end{cases}$$

where

$$W(j,x(j)) := \sum_{k=-\infty}^{j} B(j,k)x(k) + f(j).$$

In the remainder, we assume the following conditions:

A2 The sequence $f(\cdot)$ is remotely almost periodic.

A3 For each $p \in \mathbb{Z}$, (77) holds with the function $G(\cdot;\cdot)$ replaced by the function $B(\cdot;\cdot)$. Also, we ask that there exists a positive constant $U_B > 0$ such that

$$0 < \sup_{t\in\mathbb{Z}_+} \sum_{k=-\infty}^{t} \|B(t,k)\| \leqslant U_B < \infty.$$

A4 For each $p \in \mathbb{Z}$, we have (77).

The following result follows from an application of Theorem 2.2.18.

Lemma 2.2.21. *If the function $x(\cdot)$ is remotely almost periodic, then $W(\cdot,x(\cdot))$ is remotely almost periodic, too.*

Theorem 2.2.22 (Schauder). *Let $\mathbb{B}$ be a Banach space. Assume that K is a closed, bounded and convex subset of $\mathbb{B}$. If $T : K \to K$ is a compact operator, then T has a fixed point in K.*

In order to establish the final outcome, we introduce the following set:

$$\Theta_U := \{x^{\xi} \in RDAP(\mathbb{Z} : \mathbb{R}^n) : \|x^{\xi}\| \leqslant U\}$$

for a fixed positive constant $U > 0$. Clearly, Θ_U is a bounded, closed and convex subset of $RDAP(\mathbb{Z} : \mathbb{R}^n)$. We have the following.

Theorem 2.2.23. *Assume that the conditions* (**A1–A4**) *are satisfied. Then the Volterra difference system* (79) *has a remotely almost periodic solution.*

Proof. As the initial task, we have to show that $T : \Theta_U \to \Theta_U$. Pick $x^{\xi} \in \Theta_U$. Then $W(\cdot,x(\cdot))$ is remotely almost periodic, and consequently, $T(x^{\xi})(\cdot)$ is remotely almost periodic. We skip the proof of this assertion since one may easily show this claim by exactly repeating the same steps of the proof of Theorem 2.2.18. Further on, we have

$$\begin{aligned}\|(Tx)(t)\| &\leqslant \sum_{j=-\infty}^{\infty} \|G(t,j+1)\| \cdot \|W(j,x(j))\| \\ &\leqslant U_W \sum_{j=-\infty}^{\infty} \|G(t,j+1)\| \\ &\leqslant U_W\left(\frac{2\beta}{\alpha}+\beta\right),\end{aligned}$$

where U_W stands for the upper bound of the remotely almost periodic function W. Set

$$U := \max\left\{U_\xi, U_W\left(\frac{2\beta}{\alpha}+\beta\right)\right\},$$

and observe that T maps Θ_U into itself. Suppose now that $\varphi_1, \varphi_2 \in \Theta_U$ and define $\delta = \delta(\varepsilon) > 0$ by

$$\delta := \frac{\varepsilon}{U_B(\frac{2\beta}{\alpha}+\beta)}.$$

Next, we pursue the proof by showing that T is continuous. If $\|\varphi_1 - \varphi_2\| < \delta$, then we have

$$\begin{aligned}\|(T\varphi_1)(t) - (T\varphi_2)(t)\| &\leqslant \sum_{j=-\infty}^{\infty} \|G(t,j+1)\| \cdot \|W(j,\varphi_1(j)) - W(j,\varphi_2(j))\| \\ &\leqslant U_B\|\varphi_1 - \varphi_2\| \sum_{j=-\infty}^{\infty} \|G(t,j+1)\| \\ &\leqslant U_B\|\varphi_1 - \varphi_2\|\left(\frac{2\beta}{\alpha}+\beta\right) \\ &< \varepsilon, \quad t \in \mathbb{Z},\end{aligned}$$

which implies the continuity of T.

As the final step of our proof, we aim to show that $T(\Theta_U)$ is precompact by using diagonalization. Suppose that the sequence $\{x_k\} \in \Theta_U$, and consequently, $\{x_k(t)\}$ is a bounded sequence for $t \in \mathbb{Z}$. Thus, it has a convergent subsequence $\{x_k(t_k)\}$. By repeating the diagonalization for each $k \in \mathbb{Z}_+$, we get a convergent subsequence $\{x_{k_l}\}$ of $\{x_k\}$ in Θ_U. Since T is continuous, $\{T(x_k)\}$ has a convergent subsequence in $T(\Theta_U)$; therefore, $T(\Theta_U)$ is precompact. The conclusion follows from Schauder's theorem, which shows that there exists a function $x \in \Theta_U$ so that $(Tx^\xi)(t) = x(t)$ for all $t \in \mathbb{Z}_+$. Equivalently, the nonconvolution type Volterra difference system has a remotely almost periodic solution. □

We can similarly analyze the existence of discrete almost automorphic solutions to (79). Let us finally mention a few topics not considered in our previous work and some perspectives for further investigations of the abstract Volterra difference equations:

1. Many recent papers analyze the class of almost periodic functions in view of the Lebesgue measure μ; cf. [618], the references cited therein and Chapter 6 below. Here, we will not consider the discretizations of the almost periodic functions in view of the Lebesgue measure μ; cf. also [618, Lemma 2.8].

2. Suppose that $\emptyset \neq I \subseteq \mathbb{Z}^n, \emptyset \neq I' \subseteq \mathbb{Z}^n, i+i' \in I$ for all $i \in I, i' \in I'$ and $F : I \times X \to Y$. The following notion is also meaningful: a sequence $F(\cdot;\cdot)$ is said to be Bebutov-$(\mathcal{B}, I', \rho)$-almost periodic, if for every $\varepsilon > 0, B \in \mathcal{B}$ and $N > 0$, there exist a sequence $(\tau_k)_{k\in\mathbb{N}}$ in I' such that $\lim_{k\to+\infty} |\tau_k| = +\infty$ and a positive integer $k_0 \in \mathbb{N}$ such that, for every $x \in B$ and $i \in I$ with $|i| \leqslant N$, there exists $y_{i;x} \in \rho(F(i;x))$ such that

$$\|F(i + \tau_k; x) - y_{i;x}\| \leqslant \varepsilon, \quad x \in B,\ k \geqslant k_0.$$

We will skip all details concerning the class of Bebutov-$(\mathcal{B}, I', \rho)$-almost periodic sequences.

3. It is worth noting that the notion of quasiasymptotically almost periodicity and the notion of remote almost periodicity have not been considered in the sense of Bochner's approach. We can also consider the following notion: Suppose that $\mathbb{D} \subseteq I \subseteq \mathbb{R}^n, \emptyset \neq I' \subseteq \mathbb{R}^n, \emptyset \neq I \subseteq \mathbb{R}^n$, the sets $\mathbb{D}$ and I' are unbounded, $I + I' \subseteq I$ and $F : I \times X \to Y$ is a given function. Then we say that:

(i) $F(\cdot;\cdot)$ is $\mathbb{D}$-quasiasymptotically Bochner $(\mathcal{B}, I', \rho)$-almost periodic, if for every $B \in \mathcal{B}$ and for every unbounded sequence $(\tau_k')_{k\in\mathbb{N}}$ in I', there exists a subsequence $(\tau_k)_{k\in\mathbb{N}}$ of $(\tau_k')_{k\in\mathbb{N}}$ such that, for every $x \in B$, there exists a function $G_x \in Y^{\mathbb{D}}$ such that $G_x(\mathbf{t}) \in \rho(F(\mathbf{t};x))$ for all $\mathbf{t} \in \mathbb{D}$, $x \in B$ and

$$\lim_{k\to+\infty} \limsup_{|\mathbf{t}|\to+\infty;\mathbf{t}\in\mathbb{D}} \sup_{x\in B} \|F(\mathbf{t} + \tau_k; x) - G_x(\mathbf{t})\|_Y = 0.$$

(ii) $F(\cdot;\cdot)$ is Bochner $\mathbb{D}$-remotely $(\mathcal{B}, I', \rho)$-almost periodic if $F(\cdot;\cdot)$ is $\mathbb{D}$-quasi-asymptotically Bochner $(\mathcal{B}, I', \rho)$-almost periodic and, for every $B \in \mathcal{B}$, the function $F(\cdot;\cdot)$ is uniformly continuous on $I \times B$.

We will consider this notion somewhere else.

2.3 Multidimensional almost automorphic type sequences and applications

Suppose that $F : \mathbb{R}^n \to X$ is continuous, where $(X, \|\cdot\|)$ is a complex Banach space. Let us recall that $F(\cdot)$ is said to be almost automorphic if for every sequence $(\mathbf{b}_k)$ in $\mathbb{R}^n$ there exist a subsequence $(\mathbf{a}_k)$ of $(\mathbf{b}_k)$ and a map $G : \mathbb{R}^n \to X$ such that

$$\lim_{k\to\infty} F(\mathbf{t} + \mathbf{a}_k) = G(\mathbf{t}) \quad \text{and} \quad \lim_{k\to\infty} G(\mathbf{t} - \mathbf{a}_k) = F(\mathbf{t}), \tag{80}$$

pointwisely for $\mathbf{t} \in \mathbb{R}^n$. In this case, the range of $F(\cdot)$ is relatively compact in X and the limit function $G(\cdot)$ is bounded on $\mathbb{R}^n$ but not necessarily continuous on $\mathbb{R}^n$. If the convergence of limits appearing in (80) is uniform on compact subsets of $\mathbb{R}^n$, then we say that $F(\cdot)$ is compactly almost automorphic. We know that an almost automorphic function $F(\cdot)$ is compactly almost automorphic if and only if it is uniformly continuous [101].

As already mentioned, the first systematic study of almost automorphic functions on topological groups was conducted by W. A. Veech in [758, 759] (see also the papers [594] by P. Milnes, [664] by A. Reich, [744] by R. Terras and the reference list given in the Appendix section of [184], where we have recently introduced several new classes of multidimensional almost periodic type functions, like $(\mathrm{R}, \mathcal{B})$-multialmost automorphic functions and $(\mathrm{R}_X, \mathcal{B})$-multialmost automorphic functions, and emphasized that the introduced notion can be considered on ((semi)topological groups). In a joint research article with H. C. Koyuncuoğlu [483], we have specifically analyzed the situation in which $G = \mathbb{Z}^n$, the discrete topological group endowed with the usual topology and addition: we consider here the multidimensional almost automorphic sequences of the form $F : \mathbb{Z}^n \times X \to Y$, where $(Y, \|\cdot\|_Y)$ is likewise a complex Banach space.

Concerning some applications of the one-dimensional almost automorphic sequences, weighted pseudo almost automorphic sequences and Stepanov pseudo almost automorphic sequences to the abstract Volterra difference equations, we refer the reader to [5, 9, 45, 65, 165]; cf. also the research article [173], which concerns the weighted pseudo asymptotically antiperiodic solutions to semilinear difference equations. The results established in the above mentioned research articles can be straightforwardly generalized to the class of bounded R-almost automorphic sequences, where R is any collection of sequences in $\mathbb{Z}$, which satisfies that for each sequence $(b_k) \in \mathrm{R}$ any its subsequence also belongs to R. Concerning the applications made in the one-dimensional setting by now, we have reached a conclusion that the class of asymptotically almost automorphic sequences has not been well explored in the existing literature. We provide here several new applications of asymptotically almost automorphic type sequences to the abstract Volterra difference equations in both, the one-dimensional setting and the multidimensional setting.

The organization of section can be simply described as follows. In Definition 2.3.1, we introduce the notions of an $(\mathrm{R}, \mathcal{B})$-multialmost automorphic sequence and an $(\mathrm{R}_X, \mathcal{B})$-multialmost automorphic sequence; Definition 2.3.2 introduces the notions of an $(\mathrm{R}, \mathcal{B}, W_{\mathcal{B},\mathrm{R}})$-multialmost automorphic sequence and an $(\mathrm{R}, \mathcal{B}, \mathrm{P}_{\mathcal{B},\mathrm{R}})$-multialmost automorphic sequence. Our first structural result is Theorem 2.3.3; after that, we provide some illustrations in Example 2.3.4(i)–(iii). The main result of Section 2.3.1 is Theorem 2.3.5, where we investigate the extensions of bounded R-multialmost automorphic sequences defined on $\mathbb{Z}^n$ to the bounded uniformly continuous, compactly R′-multialmost automorphic functions defined on the whole Euclidean space $\mathbb{R}^n$ (see also Corollary 2.3.6 and Theorem 2.3.7). In the continuation, we observe that many structural results established in [184] can be simply transferred to the discrete setting (cf. Theorem 2.3.8). We investigate certain applications of the introduced notion to the abstract Volterra difference equations in three separate subsections.

If R is a nonempty collection of sequences in $\mathbb{R}^n$ and R_X is a nonempty collection of sequences in $\mathbb{R}^n \times X$, then the notions of a (compactly) $(\mathrm{R}, \mathcal{B})$-multialmost automorphic function and a (compactly) $(\mathrm{R}_X, \mathcal{B})$-multialmost automorphic function have recently been introduced in [184, Definitions 2.1, 2.4]; we refer the reader to [184, Definition 2.2] for the notions of an $(\mathrm{R}, \mathcal{B}, W_{\mathcal{B},\mathrm{R}})$-multialmost automorphic function and an $(\mathrm{R}, \mathcal{B}, \mathrm{P}_{\mathcal{B},\mathrm{R}})$-multialmost automorphic function.

2.3.1 $(\mathrm{R}_X, \mathcal{B})$-multialmost automorphic type sequences

Unless stated otherwise, we will always assume henceforth that R is a nonempty collection of sequences in $\mathbb{Z}^n$ and R_X is a nonempty collection of sequences in $\mathbb{Z}^n \times X$. We start this section by introducing the following discrete analogues of [184, Definitions 2.1, 2.4].

Definition 2.3.1. (i) Suppose that $F : \mathbb{Z}^n \times X \to Y$ is continuous. Then we say that the sequence $F(\cdot;\cdot)$ is $(\mathrm{R}, \mathcal{B})$-multialmost automorphic if for every $B \in \mathcal{B}$ and for every sequence $(\mathbf{b}_k = (b_k^1, b_k^2, \ldots, b_k^n)) \in \mathrm{R}$ there exist a subsequence $(\mathbf{b}_{k_l} = (b_{k_l}^1, b_{k_l}^2, \ldots, b_{k_l}^n))$ of $(\mathbf{b}_k)$ and a sequence $F^* : \mathbb{Z}^n \times X \to Y$ such that

$$\lim_{l\to+\infty} F(\mathbf{t} + (b_{k_l}^1, \ldots, b_{k_l}^n); x) = F^*(\mathbf{t}; x) \tag{81}$$

and

$$\lim_{l\to+\infty} F^*(\mathbf{t} - (b_{k_l}^1, \ldots, b_{k_l}^n); x) = F(\mathbf{t}; x), \tag{82}$$

hold pointwisely for all $x \in B$ and $\mathbf{t} \in \mathbb{Z}^n$.

(ii) Suppose that $F : \mathbb{Z}^n \times X \to Y$ is continuous. Then we say that the sequence $F(\cdot;\cdot)$ is $(\mathrm{R}_X, \mathcal{B})$-multialmost automorphic if for every $B \in \mathcal{B}$ and for every sequence $((\mathbf{b}; \mathbf{x})_k = ((b_k^1, b_k^2, \ldots, b_k^n); x_k)) \in \mathrm{R}_X$ there exist a subsequence $((\mathbf{b}; \mathbf{x})_{k_l} = ((b_{k_l}^1, b_{k_l}^2, \ldots, b_{k_l}^n); x_{k_l}))$ of $((\mathbf{b}; \mathbf{x})_k)$ and a sequence $F^* : \mathbb{Z}^n \times X \to Y$ such that

$$\lim_{m\to+\infty} F(\mathbf{t} + (b_{k_m}^1, \ldots, b_{k_m}^n); x + x_{k_m}) = F^*(\mathbf{t}; x) \tag{83}$$

and

$$\lim_{l\to+\infty} F^*(\mathbf{t} - (b_{k_l}^1, \ldots, b_{k_l}^n); x - x_{k_l}) = F(\mathbf{t}; x), \tag{84}$$

hold pointwisely for all $x \in B$ and $\mathbf{t} \in \mathbb{Z}^n$. We say that the function $F(\cdot;\cdot)$ is compactly $(\mathrm{R}_X, \mathcal{B})$-multialmost automorphic if the convergence of limits in (83)–(84) is uniform on any compact subset K of $\mathbb{Z}^n \times X$, which belongs to $\mathbb{Z}^n \times B$.

In the case that $X = \{0\}$ and $\mathcal{B} = \{X\}$, i. e., if we consider the sequence $F : \mathbb{Z}^n \to Y$, then we also say that $F(\cdot)$ is R-multialmost automorphic (R_X-multi-almost automorphic). Further on, if R denotes the collection of all sequences in $\mathbb{Z}^n$, then we say that $F(\cdot)$ is almost automorphic (in the case that $n = 1$, we omit the term "multi" from the notation).

The following notion is the discrete analogue of [184, Definition 2.2].

Definition 2.3.2. Suppose that $F : \mathbb{Z}^n \times X \to Y$ is a continuous function as well as that for each $B \in \mathcal{B}$ and $(\mathbf{b}_k = (b_k^1, b_k^2, \ldots, b_k^n)) \in \mathrm{R}$ we have that $W_{B,(\mathbf{b}_k)} : B \to P(P(\mathbb{Z}^n))$ and $\mathrm{P}_{B,(\mathbf{b}_k)} \in P(P(\mathbb{Z}^n \times B))$. Then we say that $F(\cdot;\cdot)$ is:

(i) $(\mathrm{R}, \mathcal{B}, W_{\mathcal{B},\mathrm{R}})$-multialmost automorphic if for every $B \in \mathcal{B}$ and for every sequence $(\mathbf{b}_k = (b_k^1, b_k^2, \ldots, b_k^n)) \in \mathrm{R}$ there exist a subsequence $(\mathbf{b}_{k_l} = (b_{k_l}^1, b_{k_l}^2, \ldots, b_{k_l}^n))$ of $(\mathbf{b}_k)$ and a sequence $F^* : \mathbb{Z}^n \times X \to Y$ such that (81)–(82) hold pointwisely for all $x \in B$ and $\mathbf{t} \in \mathbb{Z}^n$ as well as that for each $x \in B$ the convergence in $\mathbf{t}$ is uniform for any element of the collection $W_{B,(\mathbf{b}_k)}(x)$.

(ii) $(\mathrm{R}, \mathcal{B}, \mathrm{P}_{\mathcal{B},\mathrm{R}})$-multialmost automorphic if for every $B \in \mathcal{B}$ and for every sequence $(\mathbf{b}_k = (b_k^1, b_k^2, \ldots, b_k^n)) \in \mathrm{R}$ there exist a subsequence $(\mathbf{b}_{k_l} = (b_{k_l}^1, b_{k_l}^2, \ldots, b_{k_l}^n))$ of $(\mathbf{b}_k)$ and a sequence $F^* : \mathbb{Z}^n \times X \to Y$ such that (81)–(82) hold pointwisely for all $x \in B$ and $\mathbf{t} \in \mathbb{Z}^n$ as well as that the convergence in (81)–(82) is uniform in $(\mathbf{t}; x)$ for any set of the collection $\mathrm{P}_{B,(\mathbf{b}_k)}$.

In the case that $X = \{0\}$ and $\mathcal{B} = \{X\}$, i. e., if we consider the sequence $F : \mathbb{Z}^n \to Y$, then we also say that $F(\cdot)$ is $(\mathrm{R}, W_{\mathrm{R}})$-multialmost automorphic ($(\mathrm{R}, \mathrm{P}_{\mathrm{R}})$-multialmost automorphic); in this case, we abbreviate $W_{B,(\mathbf{b}_k)}(x)$ to $W_{(\mathbf{b}_k)}$ and $\mathrm{P}_{B,(\mathbf{b}_k)}$ to $\mathrm{P}_{(\mathbf{b}_k)}$.

In the sequel, we will consider the following special cases:

C1d. Let $\mathrm{R} := \{b : \mathbb{N} \to \mathbb{Z}^n;$ for all $j \in \mathbb{N}$, we have $b_j \in \{(a, a, a, \ldots, a) \in \mathbb{Z}^n : a \in \mathbb{Z}\}\}$. If $n = 2$ and $\mathcal{B}$ is the collection of all bounded subsets of X, then we say that the sequence $F(\cdot;\cdot)$ is bi-almost automorphic.

C2d. R is a collection of all sequences $b(\cdot)$ in $\mathbb{Z}^n$. Then any $(\mathrm{R}, \mathcal{B})$-multialmost automorphic function is clearly $(\mathrm{R}_1, \mathcal{B})$-multialmost automorphic for any other collection R_1 of sequences $b(\cdot)$ in $\mathbb{Z}^n$.

The proof of following simple result is trivial and, therefore, omitted.

Theorem 2.3.3. (i) *Suppose that $F : \mathbb{R}^n \times X \to Y$ is $(\mathrm{R}, \mathcal{B})$-multialmost automorphic ($(\mathrm{R}_X, \mathcal{B})$-multialmost automorphic). Then the sequence $F_{|\mathbb{Z}^n} : \mathbb{Z}^n \times X \to Y$ is $(\mathrm{R}_{\mathbb{Z}^n}, \mathcal{B})$-multialmost automorphic ($(\mathrm{R}_{X,\mathbb{Z}^n}, \mathcal{B})$-multi-almost automorphic), where*

$$\mathrm{R}_{\mathbb{Z}^n} := \{(\mathbf{b}_k = (b_k^1, b_k^2, \ldots, b_k^n)) \in \mathrm{R} : b_k^j \in \mathbb{Z} \text{ for all } k \in \mathbb{N} \text{ and } j \in \mathbb{N}_n\} \tag{85}$$

and

$$\mathrm{R}_{X,\mathbb{Z}^n} := \{((\mathbf{b}; \mathbf{x})_k = ((b_k^1, b_k^2, \ldots, b_k^n); x_k)) \in \mathrm{R}_X : b_k^j \in \mathbb{Z} \text{ for all } k \in \mathbb{N} \text{ and } j \in \mathbb{N}_n\}.$$

(ii) *Suppose that* $F : \mathbb{R}^n \times X \to Y$ *is* $(\mathrm{R}, \mathcal{B}, W_{\mathcal{B},\mathrm{R}})$*-multialmost automorphic* ($(\mathrm{R}, \mathcal{B}, \mathrm{P}_{\mathcal{B},\mathrm{R}})$*-multialmost automorphic). Then the sequence* $F_{|\mathbb{Z}^n} : \mathbb{Z}^n \times X \to Y$ *is* $(\mathrm{R}_{\mathbb{Z}^n}, \mathcal{B}, W^{\mathbb{Z}^n}_{\mathcal{B},\mathrm{R}_{\mathbb{Z}^n}})$*-multialmost automorphic* ($(\mathrm{R}_{\mathbb{Z}^n}, \mathcal{B}, \mathrm{P}^{\mathbb{Z}^n}_{\mathcal{B},\mathrm{R}_{\mathbb{Z}^n}})$*-multialmost automorphic), where* $\mathrm{R}_{\mathbb{Z}^n}$ *is given by* (85) *and, for every* $B \in \mathcal{B}$ *and* $(\mathbf{b}_k = (b_k^1, b_k^2, \ldots, b_k^n)) \in \mathrm{R}_{\mathbb{Z}^n}$,

$$W^{\mathbb{Z}^n}_{\mathcal{B},(\mathbf{b}_k)}(x) := W_{\mathcal{B},(\mathbf{b}_k)}(x) \cap P(P(\mathbb{Z}^n)), \quad x \in B$$

and

$$\mathrm{P}^{\mathbb{Z}^n}_{\mathcal{B},(\mathbf{b}_k)} := \{D \in \mathrm{P}_{\mathcal{B},(\mathbf{b}_k)} : D \subseteq \mathbb{Z}^n \times B\}.$$

We continue by providing some illustrative examples.

Example 2.3.4. (i) In the multidimensional setting, many explanatory applications of Theorem 2.3.3 can be provided using the discretizations of functions considered in Example preceding [184, Remark 2.3] and the examples preceding [184, Definition 2.4].

(ii) Let us recall that R. Terras has constructed an almost automorphic sequence $f : \mathbb{Z} \to \mathbb{R}$ such that the limit

$$\lim_{N\to+\infty} \frac{1}{2N+1} \sum_{i=-N}^{N} f(i),$$

which is usually called the mean value of $f(\cdot)$, does not exist [744]. This example can be simply transferred to the multidimensional setting.

(iii) Define

$$F(t_1, t_2, \ldots, t_n) := \operatorname{sign}(\cos(2\pi t_1 \alpha_1)) \cdot \operatorname{sign}(\cos(2\pi t_2 \alpha_2)) \cdot \cdots \cdot \operatorname{sign}(\cos(2\pi t_n \alpha_n)),$$

for any $(t_1, t_2, \ldots, t_n) \in \mathbb{Z}^n$, where $\alpha_1, \alpha_2, \ldots, \alpha_n$ are irrational numbers. Then we can simply prove that $F(\cdot)$ is almost automorphic but not almost periodic [122].

We continue by stating the following result.

Theorem 2.3.5. *Suppose that* $F : \mathbb{Z}^n \to Y$ *is a bounded* R*-multialmost automorphic sequence and the assumption* $(b_l) \in \mathrm{R}$ *implies that any subsequence of* (b_l) *also belongs to* R*. Let* R' *be the collection of all sequences* (a_l) *in* $\mathbb{R}^n$ *satisfying that there exists a sequence* $(b_l) \in \mathrm{R}$ *such that* $\sup_{l\in\mathbb{N}} |a_l - b_l| < +\infty$*. Then there exists a bounded uniformly continuous, compactly* R'*-multialmost automorphic function* $\tilde{F} : \mathbb{R}^n \to Y$ *such that* $R(\tilde{F}(\cdot)) \subseteq CH(\overline{R(F)})$, $\|\tilde{F}\|_\infty = \|F\|_\infty$ *and* $\tilde{F}(k) = F(k)$ *for all* $k \in \mathbb{Z}^n$.

Proof. We will provide the main details of the proof in the two-dimensional setting. If $(t_1, t_2) \in \mathbb{R}^2$, then there exist integers $k, m \in \mathbb{Z}$ such that $t_1 \in [k, k+1)$ and $t_2 \in [m, m+1)$. We define first $\tilde{F}(t_1, m) := F(k, m) + (t_1 - k)[F(k+1, m) - F(k, m)]$, $\tilde{F}(t_1, m+1) := F(k, m+1) + (t_1 - k)[F(k+1, m+1) - F(k, m+1)]$ and after that $\tilde{F}(t_1, t_2) :=$

$\tilde{F}(t_1,m)+(t_2-m)[\tilde{F}(t_1,m+1)-\tilde{F}(t_1,m)]$. Since $F(\cdot)$ is bounded, we can simply prove that $\tilde{F}(\cdot)$ is bounded and uniformly continuous as well as that $R(\tilde{F}(\cdot)) \subseteq CH(\overline{R(F)})$, $\|\tilde{F}\|_\infty = \|F\|_\infty$ and $\tilde{F}(k) = F(k)$ for all $k \in \mathbb{Z}^n$. The only thing that remained to be proved is that $\tilde{F}(\cdot)$ is compactly R$'$-multialmost automorphic. So, let $K \subseteq \mathbb{R}^n$ be a compact set and let $(a_l') \in \mathrm{R}'$. Then there exists a sequence $(b_l') \in \mathrm{R}$ such that $\sup_{k\in\mathbb{N}} |a_l'-b_l'| < +\infty$. Due to our assumptions, we can find a subsequence $(b_l = (b_l^1, b_l^2)) \in \mathrm{R}$ of (b_l') and a function $G : \mathbb{Z}^2 \to Y$ such that $\lim_{l\to+\infty} F(k_1+b_l^1, k_2+b_l^2) = G(k_1,k_2)$ and $\lim_{l\to+\infty} G(k_1-b_l^1, k_2-b_l^2) = F(k_1,k_2)$ for all $k_1,\ k_2 \in \mathbb{Z}$. We define the limit function $\tilde{G} : \mathbb{R}^2 \to Y$ in the same way as the function $\tilde{F}(\cdot)$, with the function $F(\cdot)$ replaced therein with the function $G(\cdot)$. It is clear that the assumption $(a_k) \in \mathrm{R}'$ implies that any subsequence of (a_k) also belongs to R'; therefore, $(a_l = (a_l^1, a_l^2)) \in \mathrm{R}'$. Keeping in mind the Bolzano–Weierstrass theorem, we may assume without loss of generality that $\lim_{l\to+\infty}(a_l-b_l)$ exists and equals $c = (c_1, c_2)$. It suffices to show that

$$\lim_{l\to+\infty} \tilde{F}(t_1 + a_l^1, t_2 + a_l^2) = \tilde{G}(t_1 + c_1, t_2 + c_2) \tag{86}$$

and

$$\lim_{l\to+\infty} \tilde{G}(t_1 + c_1 - a_l, t_2 + c_2 - a_l^2) = \tilde{F}(t_1, t_2),$$

uniformly in $(t_1, t_2) \in K$. The both limit equalities can be proved in a similar manner and we will only show here (86). Since $\tilde{F}(\cdot)$ is uniformly continuous, it follows that there exists $l_0 \in \mathbb{N}$ such that, for every $l \geqslant l_0$ and $(t_1,t_2) \in \mathbb{R}^2$, we have $\|\tilde{F}(t_1 + a_l^1, t_2 + a_l^2) - \tilde{F}(t_1 + b_l^1 + c_1, t_2 + b_l^2 + c_2)\|_Y \leqslant \varepsilon/2$. Therefore, it suffices to show that there exists $l_1 \in \mathbb{N}$ such that, for every $l \geqslant l_1$ and $(t_1,t_2) \in K$, we have

$$\|\tilde{F}(t_1 + b_l^1 + c_1, t_2 + b_l^2 + c_2) - \tilde{G}(t_1 + c_1, t_2 + c_2)\|_Y \leqslant \varepsilon/2.$$

Clearly, there exists a compact set K' in $\mathbb{R}^2$ such that the assumptions $(t_1, t_2) \in K$, $t_1 + c_1 \in [A, A+1)$ and $t_2 + c_2 \in [B, B+1)$ for some integers $A,\ B \in \mathbb{Z}$ imply $(A,B) \in K'$. If this is the case, then we have $t_1+b_l^1+c_1 \in [A+b_l^1, A+b_l^1+1)$ and $t_2+b_l^2+c_2 \in [B+b_l^2, B+b_l^2+1)$ for all $l \in \mathbb{N}$; furthermore, we have

$$\begin{aligned}
&\tilde{F}(t_1 + b_l^1 + c_1, t_2 + b_l^2 + c_2)\\
&\quad = F(A + b_l^1, B + b_l^2) + (t_1 + c_1 - A)[F(A + 1 + b_l^1, B + b_l^2) - F(A + 1 + b_l^1, B + b_l^2)]\\
&\qquad + (t_2 + c_2 - B)[\{F(A + b_l^1, B + b_l^2 + 1) + (t_1 + c_1 - A)\\
&\qquad \times [F(A + b_l^1 + 1, B + b_l^2 + 1) - F(A + b_l^1, B + b_l^2 + 1)]\}\\
&\qquad - \{F(A + b_l^1, B + b_l^2) + (t_1 + c_1 - A)[F(A + b_l^1 + 1, B + b_l^2) - F(A + b_l^1, B + b_l^2)]\}]
\end{aligned}$$

for all $l \in \mathbb{N}$ and

$$\begin{aligned}
\tilde{G}(t_1 + c_1, t_2 + c_2) &= \tilde{G}(A, B) + (t_1 + c_1 - A)[\tilde{G}(A + 1, B) - \tilde{G}(A + 1, B)]\\
&\quad + (t_2 + c_2 - B)[\{\tilde{G}(A, B + 1) + (t_1 + c_1 - A)
\end{aligned}$$

$$\times \big[\tilde{G}(A+1,B+1) - \tilde{G}(A,B+1)\big]\big\}$$
$$- \big\{\tilde{G}(A,B) + (t_1 + c_1 - A)\big[F(A+1,B) - \tilde{G}(A,B)\big]\big\}\big],$$

which simply implies the required. □

Now we will state the following corollary of Theorem 2.3.5, which provides an extension of [817, Proposition 2.6] to the higher-dimensional setting.

Corollary 2.3.6. *Suppose that $F : \mathbb{Z}^n \to Y$ is an almost automorphic sequence. Then there exists a compactly almost automorphic function $\tilde{F} : \mathbb{R}^n \to Y$ such that $R(\tilde{F}(\cdot)) \subseteq CH(\overline{R(F)})$, $\|\tilde{F}\|_\infty = \|F\|_\infty$ and $\tilde{F}(k) = F(k)$ for all $k \in \mathbb{Z}^n$.*

The subsequent result follows from the argumentation contained in the proof of Theorem 2.3.5.

Theorem 2.3.7. *Suppose that $F : \mathbb{Z}^n \to Y$ is a bounded $(\mathrm{R}, W_{\mathrm{R}})$-multialmost automorphic sequence and the assumption $(b_l) \in \mathrm{R}$ implies that any subsequence of (b_l) also belongs to R. Let R' be the collection of all sequences (a_l) in $\mathbb{R}^n$ satisfying that there exists a sequence $(b_l) \in \mathrm{R}$ such that $\sup_{l\in\mathbb{N}} |a_l - b_l| < +\infty$. Then there exists a bounded uniformly continuous, $(\mathrm{R}', W_{\mathrm{R}'})$-multialmost automorphic function $\tilde{F} : \mathbb{R}^n \to Y$ such that $R(\tilde{F}(\cdot)) \subseteq CH(\overline{R(F)})$, $\|\tilde{F}\|_\infty = \|F\|_\infty$ and $\tilde{F}(k) = F(k)$ for all $k \in \mathbb{Z}^n$, provided that the following condition holds:*
(W) *For every sequence $(a_l) \in \mathrm{R}'$, for every element D_1 of the collection $W'_{(a_l)}$ and for every compact set $K \subseteq \mathbb{R}^n$, there exists a sequence $(b_l) \in \mathrm{R}$ such that $\sup_{l\in\mathbb{N}} |a_l - b_l| < +\infty$ and an element D of the collection $W_{(b_l)}$ such that $D_1 + (K \cap \mathbb{Z}^n) \subseteq D$.*

The following results can be proved as in the continuous setting (cf. [184, Propositions 2.5, 2.10, 2.11, Corollary 2.21]).

Theorem 2.3.8. (i) *Suppose that $F : \mathbb{Z}^n \times X \to Y$ is an $(\mathrm{R}, \mathcal{B})$-multialmost automorphic function, where R denotes the collection of all sequences in $\mathbb{Z}^n$ and $\mathcal{B}$ denotes any collection of compact subsets of X. If for every $B \in \mathcal{B}$, there exists a finite real constant $L_B > 0$ such that, for every $x, y \in B$ and $\mathbf{t} \in \mathbb{Z}^n$, we have*

$$\|F(\mathbf{t}; x) - F(\mathbf{t}; y)\|_Y \leqslant L_B\|x - y\|,$$

then, for every set $B \in \mathcal{B}$, the set $\{F(\mathbf{t}, x) : \mathbf{t} \in \mathbb{Z}^n,\ x \in B\}$ is relatively compact in Y.
(ii) *Suppose that $F : \mathbb{Z}^n \times X \to Y$ is an $(\mathrm{R}_X, \mathcal{B})$-multialmost automorphic function, where R_X denotes the collection of all sequences in $\mathbb{Z}^n \times X$ and $\mathcal{B}$ denotes any collection of compact subsets of X. Then, for every set $B \in \mathcal{B}$, the set $\{F(\mathbf{t}, x) : \mathbf{t} \in \mathbb{Z}^n,\ x \in B\}$ is relatively compact in Y.*
(iii) *(The supremum formula) Let $F : \mathbb{Z}^n \times X \to Y$ be $(\mathrm{R}, \mathcal{B})$-multialmost automorphic. Suppose that there exists a sequence $b(\cdot)$ in R whose any subsequence is unbounded. Then for any $a \geqslant 0$, we have*

$$\sup_{\mathbf{t}\in\mathbb{Z}^n, x\in X} \|F(\mathbf{t}; x)\|_Y = \sup_{\mathbf{t}\in\mathbb{Z}^n, |t|\geqslant a, x\in X} \|F(\mathbf{t}; x)\|_Y.$$

(iv) *Suppose that for each integer* $j \in \mathbb{N}$ *the sequence* $F_j(\cdot;\cdot)$ *is* $(\mathrm{R_X}, \mathcal{B})$*-multialmost automorphic and, for every sequence, which belongs to* $\mathrm{R_X}$*, any its subsequence also belongs to* $\mathrm{R_X}$*. If the sequence* $(F_j(\cdot;\cdot))$ *converges uniformly to a function* $F(\cdot;\cdot)$ *on* X*, then the function* $F(\cdot;\cdot)$ *is* $(\mathrm{R_X}, \mathcal{B})$*-multialmost automorphic. If, additionally, for each* $B \in \mathcal{B}$ *and* $(\mathbf{b};\mathbf{x}) \in \mathrm{R_X}$*, we have* $W_{B,(\mathbf{b};\mathbf{x})} : B \to P(P(\mathbb{Z}^n))$*,* $\mathrm{P}_{B,(\mathbf{b};\mathbf{x})} \in P(P(\mathbb{Z}^n \times B))$*,* $W_{B,(\mathbf{b};\mathbf{x})}(x) \subseteq W_{B,(\mathbf{b};\mathbf{x})'}(x)$ *and* $\mathrm{P}_{B,(\mathbf{b};\mathbf{x})} \subseteq \mathrm{P}_{B,(\mathbf{b};\mathbf{x})'}$ *for any* $x \in B$ *and any subsequence* $(\mathbf{b};\mathbf{x})'$ *of* $(\mathbf{b};\mathbf{x})$*, and the function* $F_j(\cdot;\cdot)$ *is* $(\mathrm{R}_X, \mathcal{B}, W_{\mathcal{B},\mathrm{R}_X})$*-multialmost automorphic, respectively,* $(\mathrm{R}_X, \mathcal{B}, \mathrm{P}_{\mathcal{B},\mathrm{R}_X})$*-multialmost automorphic, then the function* $F(\cdot;\cdot)$ *is likewise* $(\mathrm{R}_X, \mathcal{B}, W_{\mathcal{B},\mathrm{R}_X})$*-multialmost automorphic, respectively,* $(\mathrm{R}_X, \mathcal{B}, \mathrm{P}_{\mathcal{B},\mathrm{R}_X})$*-multialmost automorphic.*

(v) *Suppose that for each integer* $j \in \mathbb{N}$ *the sequence* $F_j(\cdot;\cdot)$ *is* $(\mathrm{R}, \mathcal{B})$*-multi-almost automorphic and, for every sequence which belongs to* R*, any its subsequence also belongs to* R*. If for each* $B \in \mathcal{B}$*, there exists* $\varepsilon_B > 0$ *such that the sequence* $(F_j(\cdot;\cdot))$ *converges uniformly to a function* $F(\cdot;\cdot)$ *on the set* $B^\circ \cup \bigcup_{x\in\partial B} B(x, \varepsilon_B)$*, then the function* $F(\cdot;\cdot)$ *is* $(\mathrm{R}, \mathcal{B})$*-multi-almost automorphic. If, additionally, for each* $B \in \mathcal{B}$ *and* $(\mathbf{b}_k) \in \mathrm{R}$ *we have* $W_{B,(\mathbf{b}_k)} : B \to P(P(\mathbb{Z}^n))$*,* $\mathrm{P}_{B,(\mathbf{b}_k)} \in P(P(\mathbb{Z}^n \times B))$*,* $W_{B,(\mathbf{b})}(x) \subseteq W_{B,(\mathbf{b})'}(x)$ *and* $\mathrm{P}_{B,(\mathbf{b})} \subseteq \mathrm{P}_{B,(\mathbf{b})'}$ *for any* $x \in B$ *and any subsequence* $(\mathbf{b})'$ *of* $(\mathbf{b})$*, and* $F_j(\cdot;\cdot)$ *is* $(\mathrm{R}, \mathcal{B}, W_{\mathcal{B},\mathrm{R}})$*-multialmost automorphic, respectively,* $(\mathrm{R}, \mathcal{B}, \mathrm{P}_{\mathcal{B},\mathrm{R}})$*-multialmost automorphic, then the function* $F(\cdot;\cdot)$ *is likewise* $(\mathrm{R}, \mathcal{B}, W_{\mathcal{B},\mathrm{R}})$*-multialmost automorphic, respectively,* $(\mathrm{R}, \mathcal{B}, \mathrm{P}_{\mathcal{B},\mathrm{R}})$*-multialmost automorphic.*

(vi) *Suppose that* $F : \mathbb{Z}^n \times X \to Y$ *is* $(\mathrm{R}, \mathcal{B})$*-multialmost automorphic and* $G : \mathbb{Z}^n \times Y \to Z$ *is* $(\mathrm{R}, \mathcal{B}, \mathrm{P}_{\mathrm{R},\mathcal{B}})$*-multialmost automorphic, where* R *is a collection of all sequences* $b : \mathbb{N} \to \mathbb{Z}^n$*,* $\mathcal{B}$ *is the collection of all compact subsets of* X*, as well as for every* $B \in \mathcal{B}$ *we have that* $\mathrm{P}_{\mathrm{R},\mathcal{B}}(B)$ *is the collection of all compact subsets of* $\mathbb{Z}^n \times X$*, and there exists a finite constant* $L > 0$ *such that*

$$\|G(\mathbf{t};x) - G(\mathbf{t};y)\|_Z \leqslant L\|x - y\|_Y, \quad \mathbf{t} \in \mathbb{Z}^n,\ x, y \in Y.$$

Then the sequence $W(\cdot;\cdot)$*, given by*

$$W(\mathbf{t};x) := G(\mathbf{t};F(\mathbf{t};x)), \quad \mathbf{t} \in \mathbb{Z}^n,\ x \in X,$$

is $(\mathrm{R}, \mathcal{B})$*-multialmost automorphic.*

The notion of a $\mathbb{D}$-asymptotically $(\mathrm{R}_X, \mathcal{B})$-multialmost automorphic sequence $F : \mathbb{Z}^n \times X \to Y$ can be introduced in the same way as in [184, Definition 2.13]. The statements of [184, Propositions 2.14, 2.15, Lemma 2.16, Proposition 2.17, Theorem 2.23] can be straightforwardly reformulated in the discrete setting. The weighted pseudo almost automorphic type sequences and their applications have been studied in [5, 9, 45]; the notion considered in these research articles can be further generalized using the various classes of discrete weighted ergodic components in general metric (cf. [453] and references cited therein for the continuous version).

Remark 2.3.9. Let $1 \leqslant p < +\infty$. In [9, Definition 2.4], S. Abbas, Y.-K. Chang and M. Hafayed have tried to introduce the notion of a Stepanov p-almost automorphic sequence. But, a very simple analysis shows that the S^p norm introduced on [9, l. 12–13, p. 101] is equivalent with the usual sup-norm as well as that a sequence $(f(k))_{k\in\mathbb{Z}}$ is Stepanov p-almost automorphic if and only if $(f(k))_{k\in\mathbb{Z}}$ is Bochner almost automorphic (this observation puts a constraint to the notion introduced in [9, Definition 2.6] and the results established in [9, Lemmas 3.1, 3.2, Section 4]). We will consider the Weyl-p-almost automorphic sequences and the Besicovitch-p-almost automorphic sequences somewhere else.

Concerning applications, we would like to recall that we have already analyzed the following problems and their semilinear analogues:

1. The first-order difference equation (41) and the first-order difference inclusion (47), where $(f_k \equiv f(k))_{k\in\mathbb{Z}}$ is an almost periodic type sequence or an almost automorphic type sequence.
2. The abstract fractional difference equation $\Delta^\alpha u(k) = Au(k+1)+f(k)$, $k \in \mathbb{Z}$, where A is a closed linear operator on X, $0 < \alpha < 1$ and $\Delta^\alpha u(k)$ denotes the Caputo fractional difference operator of order α; cf. [46, Definition 2.3] for the notion.
3. Some multidimensional analogues of (41) and (47) are also of concern.

In order to avoid any form of repeating or plagiarism, we want only to note that the similar results can be established in the case that the forcing term $f(k)$ or $F(k_1,\dots,k_n)$ is bounded and R-multialmost automorphic.

We will divide the remainder of this section into three individual subsections.

2.3.2 On the abstract difference equation $u(k+1) = A(k)u(k) + f(k)$, $k \geqslant 0$; $u(0) = u_0$

In this subsection, we will consider the abstract difference equation $u(k+1) = A(k)u(k)+f(k)$, $k \geqslant 0$; $u(0) = u_0$, its semilinear and multi-dimensional analogues. Here, we assume that $(A(k))_{k\geqslant 0}$ is a sequence of closed linear operators on X.

We will first consider the case in which $A(k) \equiv A \in L(X)$ and $\|A\| < 1$. Then we know (see, e. g., [272, Subsection 1.2.1] for the scalar-valued case) that a unique solution of the above equation is given by

$$u(k) = A^k u_0 + \sum_{l=0}^{k-1} A^{k-1-l} f(l), \quad k \geqslant 0. \tag{87}$$

We will assume here that a sequence $(f(k))_{k\geqslant 0}$ is bounded and asymptotically R-almost automorphic in the following sense: there exist an R-almost automorphic sequence $(g(k))_{k\in\mathbb{Z}}$ and a sequence $(q(k))_{k\geqslant 0}$ vanishing at plus infinity such that $f(k) = g(k)+q(k)$, $k \geqslant 0$. Then we have

$$u(k) = \sum_{l=0}^{+\infty} A^l g(k-1-l) + \left[A^k u_0 + \sum_{l=0}^{k-1} A^{k-1-l} q(l) - \sum_{l=k}^{+\infty} A^l g(k-1-l)\right]$$
$$:= G(k) + Q(k), \quad k \geqslant 0.$$

Similarly, as in the proof of [65, Theorem 2.13], we can show that $G(\cdot)$ is a bounded, R-almost automorphic sequence. As in the continuous setting (cf. [444, Proposition 2.6.13, Remark 2.6.14]), we can prove that the sequence $(Q(k))_{k\geqslant 0}$ vanishes at plus infinity, so that the solution $(u(k))_{k\geqslant 0}$ is bounded and asymptotically R-almost automorphic, as well.

Consider now the following semilinear difference equation:

$$u(k+1) = Au(k) + f(k; u(k)), \quad k \geqslant 0; \quad u(0) = u_0. \tag{88}$$

Then a simple computation shows that a unique solution of problem (88) has the form

$$u(k) = A^k u_0 + \sum_{l=0}^{k-1} A^{k-1-l} f(l; u(l)), \quad k \geqslant 0.$$

Our main result concerning the existence and uniqueness of bounded, asymptotically R-almost automorphic solutions of problem (88) is given as follows.

Theorem 2.3.10. *Suppose that $A \in L(X)$, $\mathcal{B}$ denotes the collection of all bounded subsets of X and* R *is any collection of sequences in $\mathbb{Z}$ such that the conditions* [D1]-[D2] *hold, where:*

D1. *There exists a sequence in* R *whose any subsequence is unbounded.*

D2. *For every sequence, which belongs to* R, *we have that any its subsequence belongs to* R.

Suppose further that the function $f_0 : \mathbb{Z} \times X \to X$ is (R, $\mathcal{B}$)-almost automorphic, $f_1 : \mathbb{N}_0 \times X \to X$ satisfies that for each bounded set B of X we have $\lim_{k\to+\infty} \sup_{x\in B} \|f_1(k; x)\| = 0$ and $f(k; x) = f_0(k; x) + f_1(k; x)$, $k \geqslant 0$, $x \in X$. Let the following conditions hold:

(i) *The set $\{f(k; x) : k \in \mathbb{N}_0,\ x \in B\}$ is bounded for any bounded subset B of X.*

(ii) *There exists a constant $L \in (0, 1)$ such that*

$$\max(\|f(k; x) - f(k; y)\|, \|f_0(k; x) - f_0(k; y)\|) \leqslant L\|x - y\|, \quad k \geqslant 0,\ x, y \in X.$$

(iii) $\|A\| < 1 - L$.

Then there exists a unique bounded, asymptotically R*-almost automorphic solution of* (88).

Proof. Denote by $AAA_{\mathrm{R},b}(\mathbb{N}_0 : X)$ the vector space of all bounded, asymptotically R-almost automorphic functions $h : \mathbb{N}_0 \to X$. Since we have assumed that [D1]-[D2] hold, an argumentation contained in the proof of [184, Proposition 2.15] shows that

$AAA_{\mathrm{R},b}(\mathbb{N}_0 : X)$ is a Banach space when equipped with the sup-norm. Due to (i), we have that the set $\{f_0(k;x) : k \in \mathbb{N}_0,\ x \in B\}$ is bounded for any bounded subset B of X. Keeping in mind this fact and the assumption that the function $f_0 : \mathbb{Z} \times X \to X$ is $(\mathrm{R}, \mathcal{B})$-almost automorphic, we can argue as in the proof of [184, Proposition 2.20] in order to see that, for every bounded R-almost automorphic sequence $g : \mathbb{Z} \to X$, the sequence $k \mapsto f_0(k; g(k))$, $k \in \mathbb{Z}$ is bounded and R-almost automorphic. Furthermore, if $h(k) = g(k) + q(k)$, $k \geqslant 0$ for some sequence $(q(k))$ vanishing at plus infinity, then we have

$$f(k;h(k)) = f_0(k;g(k)) + [f_0(k;h(k)) - f_0(k;g(k))] + f_1(k;h(k)), \quad k \geqslant 0.$$

Due to (ii), we have $\|f_0(k;h(k)) - f_0(k;g(k))\| \leqslant L\|q(k)\| \to 0$ as $k \to +\infty$ and the prescribed assumption on the function $f_1(\cdot;\cdot)$ shows that the sequence $k \mapsto f(k;h(k)) - f_0(k;g(k))$, $k \geqslant 0$ vanishes at plus infinity. Therefore, the sequence $(f(k;h(k)))$ is bounded and asymptotically R-almost automorphic. Since $\|A\| < 1$, our previous analysis of the linear case shows that the mapping $\mathcal{P} : AAA_{\mathrm{R},b}(\mathbb{N}_0 : X) \to AAA_{\mathrm{R},b}(\mathbb{N}_0 : X)$, given by

$$[\mathcal{P}(h)](k) := A^k u_0 + \sum_{l=0}^{k-1} A^{k-1-l} f(l;h(l)), \quad k \geqslant 0,$$

is well-defined. Since $\|A\| < 1 - L$, a simple computation shows that the mapping $\mathcal{P}$ is a contraction so that the final conclusion follows by applying the Banach contraction principle. □

Let us consider now the general case. Then the unique solution of the abstract difference equation $u(k+1) = A(k)u(k) + f(k)$, $k \geqslant 0$; $u(0) = u_0$ is given by

$$u(k) := B_{k-1,0}u_0 + B_{k-1,1}f(0) + B_{k-1,2}f(1) + \cdots + B_{k-1,k-1}f(k-2) + B_{k-1,k}f(k-1), \quad (89)$$

for any $k \geqslant 1$, where $B_{k-1,l} := A_{k-1} \cdot A_{k-2} \cdot \cdots \cdot A_l$ for $0 \leqslant l \leqslant k-1$ and $B_{k-1,k} := \mathrm{I}$ $(k \geqslant 1)$; cf. also the equation [272, (1.2.4)]. We will assume that the following conditions hold true:

(i) If $(b_k) \in \mathrm{R}$, then any subsequence of (b_k) also belongs to R.
(ii) There exist a bounded R-almost automorphic sequence $(g(k))_{k \in \mathbb{Z}}$ and a sequence $(q(k))_{k \geqslant 0}$ vanishing at plus infinity such that $f(k) = g(k) + q(k)$, $k \geqslant 0$.
(iii) For every $k \in \mathbb{Z}$ and $l \in \mathbb{N}_0$, there exists an operator $\mathbf{B}_{k-1,k-l} \in L(X)$ such that $\sup_{k \in \mathbb{Z}} \sum_{l=0}^{+\infty} \|\mathbf{B}_{k-1,k-l}\| < +\infty$, $\mathbf{B}_{k-1,l} = B_{k-1,l}$ for $k \geqslant 1$, $0 \leqslant l \leqslant k$ and (iv) holds, where:
(iv) For every sequence $(b'_m) \in \mathrm{R}$, there exists a subsequence (b_m) of (b'_m) such that, for every $k \in \mathbb{Z}$, we have

$$\lim_{m \to +\infty} \sum_{l=0}^{+\infty} \|\mathbf{B}_{k \pm b_m - 1, k \pm b_m - l} - \mathbf{B}_{k-1,k-l}\| = 0. \quad (90)$$

(v) $\lim_{k\to+\infty}\sum_{l=k}^{+\infty}\|\mathbf{B}_{k-1,k-l}\|=0$.
(vi) For every $\varepsilon>0$ and $M\in\mathbb{N}$, there exists $k_0\in\mathbb{N}$ such that, for every $k\geqslant\max(M,k_0)$, we have $\sum_{l=k-M}^{k-1}\|B_{k-1,k-l}\|\leqslant\varepsilon$.

Then we have the following.

Theorem 2.3.11. *There exist a bounded* R*-almost automorphic sequence* $(G(k))_{k\in\mathbb{Z}}$ *and a sequence* $(Q(k))_{k\geqslant 0}$ *vanishing at plus infinity such that* $u(k)=G(k)+Q(k)$, $k\geqslant 0$.

Proof. Using (89), we have

$$\begin{aligned} u(k) &:= \sum_{l=0}^{+\infty}\mathbf{B}_{k-1,k-l}g(k-1-l)\\ &\quad+\left[B_{k-1,0}u_0+\sum_{l=0}^{k-1}B_{k-1,k-l}q(k-1-l)-\sum_{l=k}^{+\infty}\mathbf{B}_{k-1,k-l}g(k-1-l)\right]\\ &:= G(k)+Q(k),\quad k\geqslant 0. \end{aligned}\tag{91}$$

Due to (v), we have $\lim_{k\to+\infty}B_{k-1,0}u_0 = 0$. Since $g(\cdot)$ is unbounded, we have $\|\sum_{l=k}^{+\infty}\mathbf{B}_{k-1,k-l}g(k-1-l)\|\leqslant\|g\|_\infty\sum_{l=k}^{+\infty}\|\mathbf{B}_{k-1,k-l}\|\to 0$ as $k\to+\infty$. Let $\varepsilon>0$ be given. Then there exists $M\in\mathbb{N}$ such that $\|q(k)\|\leqslant\varepsilon$ for all $k\geqslant M$. Using (vi) and the following calculation:

$$\begin{aligned} \left\|\sum_{l=0}^{k-1}B_{k-1,k-l}q(k-1-l)\right\| &\leqslant \varepsilon\sum_{l=0}^{k-1-M}\|B_{k-1,k-l}\|+\sum_{l=k-M}^{k-1}\|B_{k-1,k-l}\|\\ &\leqslant \varepsilon\sup_{k\in\mathbb{Z}}\sum_{l=0}^{+\infty}\|\mathbf{B}_{k-1,k-l}\|+\sum_{l=k-M}^{k-1}\|B_{k-1,k-l}\|, \end{aligned}$$

it follows that $\lim_{k\to+\infty}\sum_{l=0}^{k-1}B_{k-1,k-l}q(k-1-l)=0$ so that $Q(\cdot)$ vanishes at plus infinity. On the other hand, since $\sup_{k\in\mathbb{Z}}\sum_{l=0}^{+\infty}\|\mathbf{B}_{k-1,k-l}\|<+\infty$ and $g(\cdot)$ is bounded, we immediately obtain that $G(\cdot)$ is bounded. Due to (91), it remains to be proved that $G(\cdot)$ is R-almost automorphic. Toward this end, let a sequence $(b_m')\in\mathrm{R}$ be fixed. Taking into account (i) and (iv), we obtain the existence of a subsequence (b_m) of (b_m') and a mapping $g^*:\mathbb{Z}\to X$ such that $\lim_{m\to+\infty}g(k+b_m)=g^*(k)$ and $\lim_{m\to+\infty}g^*(k-b_m)=g(k)$ for all $k\in\mathbb{Z}$ as well as that, for every $k\in\mathbb{Z}$, we have (90). Set $u^*(k):=\sum_{l=0}^{+\infty}\mathbf{B}_{k-1,k-l}g^*(k-l-1)$, $k\in\mathbb{Z}$. The discrete version of the dominated convergence theorem, taken together with the decomposition

$$\begin{aligned} u(k+b_m)-u^*(k) &= \sum_{l=0}^{+\infty}[\mathbf{B}_{k+b_m-1,k+b_m-l}-\mathbf{B}_{k-1,k-l}]g(k+b_m-l-1)\\ &\quad+\sum_{l=0}^{+\infty}\mathbf{B}_{k-1,k-l}[g(k+b_m-l-1)-g^*(k-l-1)] \end{aligned}$$

and the assumption (iv), easily implies $\lim_{m\to+\infty} u(k+b_m) = u^*(k)$ for all $k \in \mathbb{Z}$. The second limit equality can be proved similarly. □

We will illustrate Theorem 2.3.11 by the following examples.

Example 2.3.12. Let $(a_k)_{k\in\mathbb{Z}}$ be any sequence of positive integers such that $a_{k+a} = a_k$, $k \in \mathbb{Z}$ for some $a \in \mathbb{N}$, let $A \in L(X)$ satisfy $\|A\| < 1$ and let $A(k) := A^{a_k}$, $k \in \mathbb{Z}$. Further on, let R be the collection of all sequences (b_k) of integer numbers such that there exists $k_0 \in \mathbb{N}$ such that all elements $b_{k_0}, b_{k_0+1}, \ldots$ are multiples of a. Then one can simply verify that the assumptions (i)–(vi) hold true.

Example 2.3.13. Let $A(k) \equiv A \in L(X)$, $\|A\| < 1$, let (i) hold and let there exist a bounded R-almost automorphic sequence $(g_j(k))_{k\in\mathbb{Z}}$ and a sequence $(q_j(k))_{k\geqslant 0}$ vanishing at plus infinity such that $f_j(k) = g_j(k) + q_j(k)$, $k \geqslant 0$ $(1 \leqslant j \leqslant n)$. Due to Theorem 2.3.11, we know that there exist a bounded R-almost automorphic sequence $(G_j(k))_{k\in\mathbb{Z}}$ and a sequence $(Q_j(k))_{k\geqslant 0}$ vanishing at plus infinity such that $u_j(k) = G_j(k) + Q_j(k)$, $k \geqslant 0$ $(1 \leqslant j \leqslant n)$, where $u_j(k+1) = A(k)u_j(k) + f_j(k)$, $k \geqslant 0$; $u_j(0) = u_0^j$.

(i) Define $u(k_1, k_2, \ldots, k_n) := u_1(k_1) + u_2(k_2) + \cdots + u_n(k_n)$ for all $k_1 \geqslant 0, k_2 \geqslant 0, \ldots, k_n \geqslant 0$. Then we have that $u(\cdot)$ is bounded,

$$u(k_1+1, \ldots, k_n+1) = Au(k_1, \ldots, k_n) + F(k_1, \ldots, k_n), \quad (k_1, \ldots, k_n) \in \mathbb{N}_0^n,$$

where $F(k_1, \ldots, k_n) := f_1(k_1) + f_2(k_2) + \cdots + f_n(k_n)$ for all $k_1 \geqslant 0, k_2 \geqslant 0, \ldots, k_n \geqslant 0$ and $u(0, 0, \ldots, 0) = u_0^0 + u_0^1 + \cdots + u_0^n$. Moreover, the solution $u(\cdot)$ is $\mathbb{D}$-asymptotically R_n-multialmost automorphic, where $\mathrm{R}_n = \{(b_1^m, b_2^m, \ldots, b_n^m)_{m\in\mathbb{N}} : (b_j^m)_{m\in\mathbb{N}} \in \mathrm{R}$ for all $j \in \mathbb{N}_n\}$ and $\mathbb{D}$ is any unbounded subset of $\mathbb{N}_0^n$, which enjoys the property that the assumptions $(k_1, \ldots, k_n) \in \mathbb{D}$ and $|(k_1, \ldots, k_n)| \to +\infty$ imply $k_1 \to +\infty, \ldots, k_n \to +\infty$.

(ii) Suppose now that $a \in \mathbb{N}$, R is the collection of all sequences in $\mathbb{Z}$ and $u(k_1, k_2, \ldots, k_n) := u_1(ak_1) + u_2(ak_2) + \cdots + u_n(ak_n)$ for all $k_1 \geqslant 0, k_2 \geqslant 0, \ldots, k_n \geqslant 0$. Inductively, we can simply prove that

$$u_j(k+a) = A^a u_j(k) + \sum_{j=0}^{a} A^j f_j(k+a-1-j), \quad k \geqslant 0,\ j \in \mathbb{N}_n.$$

If $\mathbb{D}$ has the property stated in part (i), then the above simply implies that $u(\cdot)$ is $\mathbb{D}$-asymptotically almost automorphic and

$$u(k_1+1, \ldots, k_n+1) = A^a u(k_1, \ldots, k_n) + F(k_1, \ldots, k_n), \quad (k_1, \ldots, k_n) \in \mathbb{N}_0^n,$$

where

$$\begin{aligned} F(k_1, \ldots, k_n) := {} & A^{a-1}\big[f_1(ak_1) + f_2(ak_2) + \cdots + f_n(ak_n)\big] \\ & + A^{a-2}\big[f_1(ak_1+1) + f_2(ak_2+1) + \cdots + f_n(ak_n+1)\big] \\ & + \cdots \\ & + \big[f_1(ak_1+a-1) + f_2(ak_2+1) + \cdots + f_n(ak_n+1)\big], \end{aligned}$$

for any $k_1 \geqslant 0, k_2 \geqslant 0, \ldots, k_n \geqslant 0$. Moreover, $F(k_1, \ldots, k_n)$ is $\mathbb{D}$-asymptotically almost automorphic and $u(0, 0, \ldots, 0) = u_0^0 + u_0^1 + \cdots + u_0^n$.

Consider now the following semilinear abstract difference equation:

$$u(k+1) = A(k)u(k) + f(k; u(k)), \quad k \geqslant 0; \quad u(0) = u_0.$$

A simple computation shows that the solution $u(\cdot)$ of this problem must satisfy

$$\begin{aligned} u(k) := {} & B_{k-1,0}u_0 + B_{k-1,1}f(0; u(0)) + B_{k-1,2}f(1; u(1)) \\ & + \cdots + B_{k-1,k-1}f(k-2; u(k-2)) + B_{k-1,k}f(k-1; u(k-1)), \end{aligned} \tag{92}$$

for any $k \geqslant 1$, where $B_{k-1,l} := A_{k-1} \cdot A_{k-2} \cdot \cdots \cdot A_l$ for $0 \leqslant l \leqslant k-1$ and $B_{k-1,k} := \mathrm{I}$ $(k \geqslant 1)$. Keeping in mind Theorem 2.3.11 and the formula (92), we can simply formulate and prove an analogue of Theorem 2.3.10 in the nonautonomous case.

2.3.3 On the abstract difference equation $u(k, m) = A(k, m)u(k-1, m-1) + f(k, m)$, $k, m \in \mathbb{N}$

In this subsection, we investigate the abstract difference equation

$$u(k, m) = A(k, m)u(k-1, m-1) + f(k, m), \quad k, m \in \mathbb{N}, \tag{93}$$

subjected with the initial conditions

$$u(k, 0) = u_{k,0}; \quad u(0, m) = u_{0,m}, \quad k, m \in \mathbb{N}_0, \tag{94}$$

and its semilinear analogues.

We will assume that $A(k, m) \in L(X)$ for all $k, m \in \mathbb{N}$. Then we can inductively prove that a unique solution of the problem (93)–(94) is given by

$$\begin{aligned} u(k, m) = {} & A(k, m)A(k-1, m-1) \cdot \cdots \cdot A(k-m-1, 1)u(k-m, 0) \\ & + [f(k, m) + A(k, m)f(k-1, m-1) + A(k, m)A(k-1, m-1)f(k-2, m-2) \\ & + \cdots + A(k, m)A(k-1, m-1) \cdot \cdots \cdot A(k-m+2, 2)f(k-m+1, 1)], \end{aligned}$$

provided that $k \geqslant m$, and

$$\begin{aligned} u(k, m) = {} & A(k, m)A(k-1, m-1) \cdot \cdots \cdot A(1, m-k-1)u(0, m-k) \\ & + [f(k, m) + A(k, m)f(k-1, m-1) + A(k, m)A(k-1, m-1)f(k-2, m-2) \\ & + \cdots + A(k, m)A(k-1, m-1) \cdot \cdots \cdot A(2, m-k+2)f(1, m-k+1)], \end{aligned}$$

provided that $k < m$.

Our first result about the existence and uniqueness of bounded, $\mathbb{D}$-asymptotically R-multialmost automorphic solutions of the problem (93)–(94) reads as follows (we will use the same notion and notation as in [184]).

Theorem 2.3.14. *Suppose that $A(k,m) \equiv A \in L(X)$, $\|A\| < 1$, R is any collection of sequences in $\mathbb{Z}^2$, $g : \mathbb{Z}^2 \to X$ is bounded and* R*-multialmost automorphic, $q \in C_{0,\mathbb{N}_0^2}(\mathbb{N}_0^2 : X)$, $f(k,m) = g(k,m) + q(k,m)$, $k,\ m \in \mathbb{N}_0^2$, $(u_{k,0})_k$ and $(u_{0,m})_m$ are bounded sequences and $\mathbb{D} := \{(k,m) \in \mathbb{N}_0^2 \setminus \{(0,0)\} : \arg(k + i \cdot m) \in [\varepsilon, (\pi/2) - \varepsilon]\}$, where $\varepsilon \in (0, \pi/2)$. Then the unique solution $u(k,m)$ of the problem* (93)–(94) *is bounded and $\mathbb{D}$-asymptotically* R*-multialmost automorphic.*

Proof. The above expressions show that

$$\begin{aligned} u(k,m) &= A^m u_{k-m,0} + [f(k,m) + Af(k-1,m-1) + \cdots + A^{m-1} f(k-m+1,1)] \\ &= \sum_{l=0}^{\infty} A^l g(k-l,m-l) + \left[A^m u_{k-m,0} + \sum_{l=0}^{m-1} A^l q(k-l,m-l) - \sum_{l=m}^{\infty} A^l g(k-l,m-l) \right] \\ &:= G(k,m) + Q(k,m), \quad k,\ m \in \mathbb{N}_0,\ k \geqslant m \end{aligned}$$

and

$$\begin{aligned} u(k,m) &= A^k u_{0,m-k} + [f(k,m) + Af(k-1,m-1) + \cdots + A^{k-1} f(1,m-k+1)] \\ &= \sum_{l=0}^{\infty} A^l g(k-l,m-l) + \left[A^k u_{0,m-k} + \sum_{l=0}^{k-1} A^l q(k-l,m-l) - \sum_{l=k}^{\infty} A^l g(k-l,m-l) \right] \\ &:= G(k,m) + Q(k,m), \quad k,\ m \in \mathbb{N}_0,\ k < m. \end{aligned}$$

The function $G(\cdot;\cdot)$ can be defined for all $k,\ m \in \mathbb{Z}$ and a simple computation involving the discrete version of the dominated convergence theorem shows that $G(\cdot;\cdot)$ is bounded and R-multialmost automorphic. If $(k,m) \in \mathbb{D}$ and $|(k,m)| \to +\infty$, then $k \to +\infty$ and $m \to +\infty$; taking into account the prescribed assumptions, a relatively simple computation yields that $Q \in C_{0,\mathbb{D}}(\mathbb{N}_0^2 : X)$. It can be simply shown that $Q(\cdot;\cdot)$ is bounded as well, which completes the proof. □

In the general case, we assume that $A(k,m) \in L(X)$ for all $k,\ m \in \mathbb{Z}$. Define $B_{k,m} := A(k,m)A(k-1,m-1) \cdot \cdots \cdot A(k-m-1,1)$ if $k \geqslant m$, $B_{k,m} := A(k,m)A(k-1,m-1) \cdot \cdots \cdot A(1,m-k-1)$ if $k < m$ and $A_l(k,m) := A(k,m)A(k-1,m-1) \cdot \cdots \cdot A(k-l+1,m-l+1)$, $l \geqslant 0$. Then we have

$$\begin{aligned} u(k,m) &= \sum_{l=0}^{\infty} A_l(k,m) g(k-l,m-l) \\ &\quad + \left[B_{k,m} u(k-m,0) + \sum_{l=0}^{m-1} A_l(k,m) q(k-l,m-l) - \sum_{l=m}^{\infty} A_l(k,m) g(k-l,m-l) \right], \end{aligned}$$

if $k \geqslant m$, and

$$u(k,m) = \sum_{l=0}^{\infty} A_l(k,m)g(k-l,m-l)$$
$$+ \left[B_{k,m}u(0,m-k) + \sum_{l=0}^{k-1} A_l(k,m)q(k-l,m-l) - \sum_{l=k}^{\infty} A_l(k,m)g(k-l,m-l) \right],$$

if $k < m$. Using these formulae and the argumentation contained in the proofs of Theorem 2.3.11 and Theorem 2.3.14, it is not difficult to deduce the following result.

Theorem 2.3.15. *Suppose that* R *is any collection of sequences in* $\mathbb{Z}^2$*, which satisfies that, for every sequence* $(b_k) \in$ R*, we have that any subsequence of* (b_k) *also belongs to* R*. Suppose further that* $g : \mathbb{Z}^2 \to X$ *is bounded and* R*-multialmost automorphic,* $q \in C_{0,\mathbb{N}_0^2}(\mathbb{N}_0^2 : X)$*,* $f(k,m) = g(k,m) + q(k,m)$*,* $k,\ m \in \mathbb{N}_0^2$*,* $(u_{k,0})_k$ *and* $(u_{0,m})_m$ *are bounded sequences and* $\mathbb{D} := \{(k,m) \in \mathbb{N}_0^2 \setminus \{(0,0)\} : \arg(k + i \cdot m) \in [\varepsilon, (\pi/2) - \varepsilon]\}$*, where* $\varepsilon \in (0, \pi/2)$*. Then the unique solution* $u(k,m)$ *of the problem* (93)–(94) *is bounded and* $\mathbb{D}$*-asymptotically* R*-multialmost automorphic provided that the following conditions hold:*
(i) $\lim_{|(k,m)| \to +\infty; (k,m) \in \mathbb{D}} \|B_{k,m}\| = 0$.
(ii) $\sum_{l=0}^{\infty} \sup_{(k,m) \in \mathbb{Z}^2} \|A_l(k,m)\| < +\infty$.
(iii) *For every* $l \geqslant 0$*, we have that* $(A_l(k,m))_{(k,m) \in \mathbb{Z}^2}$ *is an* R*-multialmost automorphic sequence.*

The following example illustrates an application of Theorem 2.3.15.

Example 2.3.16. Suppose that R satisfies the first requirement from Theorem 2.3.15 as well as that $A \in L(X)$, $\|A\| < 1$ and $A(k,m) := A^{a_{k,m}}$ for all $k,\ m \in \mathbb{Z}$, where $(a_{k,m})_{(k,m) \in \mathbb{Z}^2}$ is an R-multialmost automorphic sequence of positive integers. Then Theorem 2.3.15 can be applied; observe also that, here and in Example 2.3.12, we can use the fractional powers of A provided that this operator is nonnegative (cf. [443] for the notion; this is important for the applications in which R is the collection of all sequences in $\mathbb{Z}^2$).

Consider now the abstract semilinear difference equation

$$u(k,m) = A(k,m)u(k-1,m-1) + f(k,m; u(k,m)), \quad k,\ m \in \mathbb{N}, \tag{95}$$

subjected with the initial conditions (94), where $A(k,m) \in L(X)$ for all $k,\ m \in \mathbb{N}$. Then a unique solution of this problem must satisfy

$$u(k,m) = A(k,m)A(k-1,m-1) \cdot \dots \cdot A(k-m-1,1)u(k-m,0)$$
$$+ \big[f(k,m;u(k,m)) + A(k,m)f(k-1,m-1;u(k-1,m-1))$$
$$+ A(k,m)A(k-1,m-1)f(k-2,m-2;u(k-2,m-2))$$
$$+ \dots + A(k,m)A(k-1,m-1) \cdot \dots \cdot A(k-m+2,2)f(k-m+1,1;u(k-m+1,1)) \big],$$

provided that $k \geqslant m$, and

$$\begin{aligned} u(k,m) &= A(k,m)A(k-1,m-1)\cdot\dots\cdot A(1,m-k-1)u(0,m-k) \\ &\quad + \big[f(k,m;u(k,m)) + A(k,m)f(k-1,m-1;u(k-1,m-1)) \\ &\quad + A(k,m)A(k-1,m-1)f(k-2,m-2;u(k-2,m-2)) \\ &\quad + \dots + A(k,m)A(k-1,m-1)\cdot\dots\cdot A(2,m-k+2)f(1,m-k+1;u(1,m-k+1))\big], \end{aligned}$$

provided that $k < m$.

Keeping in mind Theorem 2.3.14 and the argumentation given in the proof of Theorem 2.3.10, we can simply deduce the following result.

Theorem 2.3.17. *Suppose that $A(k,m) \equiv A \in L(X)$, $\|A\| < 1$, $\mathcal{B}$ denotes the collection of all bounded subsets of X and R is any collection of sequences in $\mathbb{Z}^2$ such that the conditions* [D1]–[D2] *hold. Suppose further that the function $f_0 : \mathbb{Z}^2 \times X \to X$ is $(\mathrm{R},\mathcal{B})$-almost automorphic, $f_1 : \mathbb{N}_0^2 \times X \to X$ satisfies that for each bounded set B of X we have $\lim_{|(k,m)|\to+\infty} \sup_{x\in B} \|f_1(k,m;x)\| = 0$ and $f(k,m;x) = f_0(k,m;x) + f_1(k,m;x)$, $k,\ m \geqslant 0$, $x \in X$. Let the following conditions hold:*

(i) *The set $\{f(k,m;x) : k,\ m \in \mathbb{N}_0,\ x \in B\}$ is bounded for any bounded subset B of X.*

(ii) *There exists a constant $L \in (0,1)$ such that, for every $k,\ m \geqslant 0$ and $x,\ y \in X$, we have*

$$\max\big(\|f(k,m;x) - f(k,m;y)\|,\ \|f_0(k,m;x) - f_0(k,m;y)\|\big) \leqslant L\|x-y\|.$$

(iii) $\|A\| < 1 - L$.

(iv) *$(u_{k,0})_k$ and $(u_{0,m})_m$ are bounded sequences.*

Let $\mathbb{D} := \{(k,m) \in \mathbb{N}_0^2 \setminus \{(0,0)\} : \arg(k + i\cdot m) \in [\varepsilon,(\pi/2)-\varepsilon]\}$, where $\varepsilon \in (0,\pi/2)$. Then there exists a unique bounded, $\mathbb{D}$-asymptotically R-almost automorphic solution of (95).

The interested reader might wonder to formulate a semilinear analogue of Theorem 2.3.15 as well as to consider the higher-dimensional analogues of the equation (95).

2.3.4 Asymptotically almost automorphic sequences and some applications

In this part, we aim to establish some sufficient conditions for the existence of asymptotically almost automorphic solutions of Volterra type difference equations. We employ the fixed-point theorems and work only in the one-dimensional setting. For the setup of our analysis, we need to recall the notion of exponential dichotomy.

Suppose that $X(t)$ is the fundamental matrix solution of the linear homogeneous system

$$x(t+1) = A(t)x(t), \tag{96}$$

where A is an invertible matrix function for all $t \in \mathbb{Z}$. Recall that the problem (96) is said to admit an exponential dichotomy if there exist a projection P and positive constants $\alpha_1, \alpha_2, \beta_1$ and β_2 such that

$$\begin{aligned} \|X(t)PX^{-1}(s)\| &\leqslant \beta_1(1+\alpha_1)^{s-t}, \quad t \geqslant s, \\ \|X(t)(I-P)X^{-1}(s)\| &\leqslant \beta_2(1+\alpha_2)^{t-s}, \quad s \geqslant t. \end{aligned}$$

In this construction, if we specifically choose the projection $P = I$, then the exponential dichotomy turns into the uniform asymptotical stability, that is, there exist real constants $\alpha, \beta > 0$ so that

$$\|\Phi(t,s)\| = \|X(t)X^{-1}(s)\| \leqslant \beta(1+\alpha)^{s-t}, \quad t \geqslant s, \tag{97}$$

where $\Phi(\cdot,\cdot)$ is the state-transition matrix of (96) and

$$X(t)X^{-1}(s) = \begin{cases} \prod_{k=s}^{t-1} A(s+t-1-k), & \text{for all } s \leqslant t \in \mathbb{Z} \\ \prod_{k=t}^{s-1} A^{-1}(k) & \text{for all } s \geqslant t \in \mathbb{Z} \end{cases}$$

(see [19, Theorem 2.4.1] and [536, Theorem 2.10]).

Let us consider now the following nonconvolution type Volterra difference equation:

$$x(t+1) = A(t)x(t) + \sum_{s=-\infty}^{t-1} K(t,s)x(s), \quad t \in \mathbb{N}_0, \tag{98}$$

where $A(t) = [a_{i,j}(t)]$ is an invertible matrix function for all $t \in \mathbb{N}_0$ with bounded inverse and $K(t,s) = [k_{i,j}(t,s)]$.

We suppose that the following conditions hold:

C1 $A(\cdot)$ is an almost automorphic matrix function.

C2 The estimate (97) holds with $\Phi(\cdot,\cdot)$ being the state-transition matrix for the homogeneous part of (98).

C3 $K(t,s)$ is bi-almost automorphic in t and k, that is, for any given integer sequence $(v_n')_{n\in\mathbb{Z}}$, there exist a subsequence $(v_n)_{n\in\mathbb{Z}}$ of $(v_n')_{n\in\mathbb{Z}}$ and a mapping $\overline{K}(\cdot,\cdot)$ such that

$$\lim_{n\to\infty} K(t+v_n, s+v_n) = \bar{K}(t,s) \text{ and } \lim_{n\to\infty} \bar{K}(t-v_n, s-v_n) = K(t,s)$$

for all $(t,s) \in \mathbb{Z} \times \mathbb{Z}$.

C4 There exists a real number $U_K > 0$ such that

$$0 < \sup_{t\in\mathbb{Z}} \sum_{s=-\infty}^{t-1} \|K(t,s)\| \leqslant U_K < \infty.$$

The following auxiliary result can be shown by repeating the procedure from the proof of [536, Theorem 3.1].

Lemma 2.3.18. *Suppose that* **(C1)** *holds. Then the state-transition matrix* $\Phi(\cdot,\cdot)$ *for the homogeneous part of* (98) *is bi-almost automorphic.*

We will include all relevant details of the proof of the following lemma, which is crucial for the formulation and proof of Theorem 2.3.21, which is the main result of this subsection (cf. also [19]).

Lemma 2.3.19 (Variation of constants). *A function* $x(\cdot)$ *is a solution of* (98) *with initial data* $x(0) = x_0$ *if and only if*

$$x(t) = \Phi(t,0)x_0 + \sum_{s=0}^{t-1}\Phi(t,s+1)\sum_{j=-\infty}^{s} K(s,j)x(j), \quad t \geqslant 0.$$

Proof. Let $X(t)$ be the fundamental matrix solution for the homogeneous part of (98), and let the vector-valued function $w(\cdot)$ be such that the function $x(t) := X(t)w(t)$, $t \geqslant 0$ is a solution of (98). Then $x(0) = X(0)w(0) = x_0$ and we must have

$$\begin{aligned} x(t+1) = X(t+1)w(t+1) &= A(t)x(t) + \sum_{s=-\infty}^{t-1} K(t,s)x(s) \\ &= A(t)X(t)w(t) + \sum_{s=-\infty}^{t-1} K(t,s)x(s) \\ &= X(t+1)w(t) + \sum_{s=-\infty}^{t-1} K(t,s)x(s). \end{aligned}$$

This results in

$$X(t+1)\Delta w(t) = \sum_{s=-\infty}^{t-1} K(t,s)x(s),$$

and

$$\Delta w(t) = X^{-1}(t+1)\sum_{s=-\infty}^{t-1} K(t,s)x(s),$$

where Δ stands for the standard forward difference operator. This simply implies

$$w(t) = w(0) + \sum_{s=1}^{t} X^{-1}(s)\sum_{j=-\infty}^{s-1} K(s-1,j)x(j), \quad t \in \mathbb{N}_0.$$

After that, we get

$$x(t) = X(t)w(t) = X(t)\left[w(0) + \sum_{s=1}^{t} X^{-1}(s)\sum_{j=-\infty}^{s-1} K(s-1,j)x(j)\right]$$

$$= X(t)\left[X^{-1}(0)x_0 + \sum_{s=0}^{t-1} X^{-1}(s+1) \sum_{j=-\infty}^{s} K(s,j)x(j)\right],$$

which completes the proof. □

Subsequently, we introduce the mapping $H(\cdot)$ as follows:

$$(Hx)(t) := \Phi(t,0)x_0 + \sum_{s=0}^{t-1} \Phi(t,s+1) \sum_{j=-\infty}^{s} K(s,j)x(j), \quad t \geqslant 0.$$

We need the following result.

Lemma 2.3.20. *Assume that conditions* **(C1–C4)** *hold. Then $H(\cdot)$ maps $AAA(\mathbb{N}_0 : X)$ into itself, where $AAA(\mathbb{N}_0 : X)$ denotes the Banach space of all asymptotically almost automorphic functions endowed with the sup-norm.*

Proof. Suppose that $x \in AAA(\mathbb{N}_0 : X)$ so that $x(t) = x_1(t) + x_2(t)$, where $x_1(t)$ is almost automorphic in $t \in \mathbb{Z}$ and $\lim_{t\to+\infty} \|x_2(t)\| = 0$. First, we rewrite $H(\cdot)$ as

$$\begin{aligned}(Hx)(t) &:= \Phi(t,0)x_0 + \sum_{s=0}^{t-1} \Phi(t,s+1) \sum_{j=-\infty}^{s} K(s,j)(x_1(j) + x_2(j)) \\ &= \Phi(t,0)x_0 + \sum_{s=0}^{t-1} \Phi(t,s+1)[\Lambda(s,x_1(s)) + \Lambda(s,x_2(s))],\end{aligned}$$

where

$$\Lambda(s,x(s)) := \sum_{j=-\infty}^{s} K(s,j)x(j).$$

Clearly, $H(\cdot)$ can be decomposed as $(Hx)(t) = (H_1x)(t) + (H_2x)(t)$, $t \geqslant 0$, where

$$(H_1x)(t) := \sum_{s=-\infty}^{t-1} \Phi(t,s+1)\Lambda(s,x_1(s)), \quad t \in \mathbb{Z}$$

and

$$(H_2x)(t) := \Phi(t,0)x_0 + \sum_{s=0}^{t-1} \Phi(t,s+1)\Lambda(s,x_2(s)) - \sum_{s=-\infty}^{-1} \Phi(t,s+1)\Lambda(s,x_1(s)), \quad t \geqslant 0.$$

First of all, we will prove that the function $\Lambda(\cdot,x_1(\cdot))$ is almost automorphic. If $y_1(\cdot)$ and $y_2(\cdot)$ are almost automorphic, then we have

$$\|\Lambda(s,y_1) - \Lambda(s,y_2)\| \leqslant \sup_{s\in\mathbb{Z}} \sum_{j=-\infty}^{s} \|K(s,j)y_1 - K(s,j)y_2\|$$

$$\leqslant \sup_{s\in\mathbb{Z}} \sum_{j=-\infty}^{s} \|K(s,j)\| \|y_1 - y_2\|$$
$$\leqslant U_K \|y_1 - y_2\|.$$

Then [65, Theorem 2.10] implies that $s \mapsto \Lambda(s, x(s))$, $s \in \mathbb{Z}$ is almost automorphic. Furthermore, for every given integer sequence $(v_k')_{k\in\mathbb{Z}}$, there exist a subsequence $(v_k)_{k\in\mathbb{Z}}$ of $(v_k')_{k\in\mathbb{Z}}$ and a mapping $\bar{\Lambda}(\cdot, \overline{x_1}(\cdot))$ such that, for every $s \in \mathbb{Z}$, we have $\lim_{k\to+\infty} \Lambda(s + v_k, x_1(s + v_k)) = \bar{\Lambda}(s, \overline{x_1}(s))$ and $\lim_{k\to+\infty} \bar{\Lambda}(s - v_k, \overline{x_1}(s - v_k)) = \Lambda(s, x_1(s))$. Then we have

$$(H_1x)(t + v_k) = \sum_{s=-\infty}^{t+v_k-1} \Phi(t + v_k, s + 1)\Lambda(s, x_1(s))$$
$$= \sum_{s=-\infty}^{t-1} \Phi(t + v_k, s + v_k + 1)\Lambda(s + v_k, x_1(s + v_k))$$

for all $t \in \mathbb{Z}$. If we let the limit as $k \to \infty$ and utilize the Lebesgue convergence theorem, we get the limit result

$$(\overline{H_1x})(t) = \sum_{s=-\infty}^{t-1} \bar{\Phi}(t, s + 1)\bar{\Lambda}(s, \overline{x_1}(s)).$$

Using the same procedure, we get $\lim_{k\to\infty} (\overline{H_1x})(t - v_k) = (H_1x)(t)$ for all $t \in \mathbb{Z}$ so that $(H_1x)(t)$ is almost automorphic in $t \in \mathbb{Z}$.

It remains to show that $\lim_{t\to\infty} \|(H_2x)(t)\| = 0$. Since $\lim_{t\to+\infty} \|x_2(t)\| = 0$, for a given $\varepsilon > 0$ there exists $N > 0$ such that $\|x_2(t)\| < \varepsilon$ for all $t > N$. Let U_1 and U_2 be the upper bounds of $\Lambda(\cdot, x_1(\cdot))$ and $\Lambda(\cdot, x_2(\cdot))$, respectively. Additionally, we have $\Lambda(s, 0) = 0$, and consequently, $\|\Lambda(s, x)\| \leqslant U_K\|x\|$. Then, for every $t \geqslant N + 1$, we have

$$\begin{aligned}
\|(H_2x)(t)\| &\leqslant \|\Phi(t, 0)\| \|x_0\| \\
&\quad + \sum_{s=-\infty}^{-1} \|\Phi(t, s + 1)\| \|\Lambda(s, x_1(s))\| + \sum_{s=0}^{t-1} \|\Phi(t, s + 1)\| \|\Lambda(s, x_2(s))\| \\
&\leqslant \beta(1 + \alpha)^{-t}\|x_0\| + U_1\beta(1 + \alpha)^{1-t} \sum_{s=-\infty}^{-1} (1 + \alpha)^s + U_2\beta(1 + \alpha)^{1-t} \sum_{s=0}^{N} (1 + \alpha)^s \\
&\quad + U_K\varepsilon \sum_{s=N+1}^{t-1} \beta(1 + \alpha)^{1+s-t}.
\end{aligned} \tag{99}$$

It is obvious that all terms in (99) converge to 0. Also, a direct calculation results in

$$U_K\varepsilon \sum_{s=N+1}^{t-1} \beta(1 + \alpha)^{1+s-t} < \frac{U_K\varepsilon\beta(1 + \alpha)}{\alpha}.$$

Hence, we have $\lim_{t\to\infty} \|(H_2x)(t)\| = 0$. This completes the proof. □

Now we are able to prove the following result.

Theorem 2.3.21. *In addition to* **(C1–C4)**, *suppose also that*
C5 $\frac{\beta U_K(1+\alpha)}{\alpha} < 1.$

Then the equation (98) *has a unique asymptotically almost automorphic solution.*

Proof. In Lemma 2.3.20, it is shown that $H : AAA(\mathbb{N}_0 : X) \to AAA(\mathbb{N}_0 : X)$. To complete the proof, it is enough to prove that $H(\cdot)$ is a contraction. Take any y_1, $y_2 \in AAA(\mathbb{N}_0 : X)$; then we have

$$\begin{aligned}\|(Hy_1)(t) - (Hy_2)(t)\| &\leqslant \sum_{s=0}^{t-1} \|\Phi(t, s+1)\| \cdot \|\Lambda(s, y_1(s)) - \Lambda(s, y_2(s))\| \\ &\leqslant \sum_{s=0}^{t-1} \beta(1+\alpha)^{1+s-t} U_k \|y_1 - y_2\| \\ &\leqslant \frac{\beta U_K(1+\alpha)}{\alpha} \|y_1 - y_2\|.\end{aligned}$$

As a result, $H(\cdot)$ is a contraction and the Banach fixed-point theorem implies that $H(\cdot)$ has a unique fixed point. Consequently, (98) has a unique asymptotically almost automorphic solution. □

3 Abstract degenerate Volterra difference equations, abstract degenerate fractional difference equations and their almost periodic type solutions

In this chapter, we investigate various classes of the abstract degenerate Volterra difference equations, the abstract degenerate fractional difference equations as well as the existence and uniqueness of their almost periodic type solutions.

3.1 On some classes of abstract degenerate difference equations

In the existing literature, there exist many research articles devoted to the study of the well-posedness and qualitative properties of solutions for the following class of the abstract first-order differential equations:

$$Au'(t) + Bu(t) = f(t), \quad t \in \mathbb{R}, \tag{100}$$

where A and B are closed linear operators on X, and the following class of the abstract higher-order differential equations:

$$A_p u^{(p)}(t) + \sum_{i=0}^{p-1} A_i u^{(i)}(t) = f(t), \quad t \in \mathbb{R} \tag{101}$$

where $A_p, \ldots, A_0$ are closed linear operators on X and $p \in \mathbb{N} \setminus \{1\}$; cf. [445] and references quoted therein. The main aim of this section is to provide certain results about the existence of (almost periodic type) solutions of the following discrete analogues of (100) and (101):

$$Au_{k+1} + Bu_k = f_k, \quad k \in \mathbb{Z} \tag{102}$$

and

$$A_p u_{k+p} + A_{p-1} u_{k+p-1} + \cdots + A_0 u_k = f_k, \quad k \in \mathbb{Z}, \tag{103}$$

respectively.

In [251], T. Diagana and D. Pennequin have considered the case in which A and B $(A_p, \ldots, A_0)$ are singular real matrices and $S_1 = \{\lambda \in \mathbb{C} : |\lambda| = 1\} \subseteq \rho(A + B)$ $(S_1 \subseteq \{\lambda \in \mathbb{C} : (\lambda^p A_p + \lambda^{p-1} A_{p-1} + \cdots + \lambda A_1 + A_0)^{-1} \in L(X)\})$; cf. also the research article [161] by S. L. Campbell. In this section, we reconsider the results established in [251] for general closed linear operators acting on X. We also investigate the existence and uniqueness of almost periodic type solutions to the abstract degenerate difference equations under our consideration (the material is taken from [473]).

https://doi.org/10.1515/9783111689746-005

3.1.1 Abstract degenerate first-order difference equations

In this subsection, we investigate the abstract degenerate difference equation (102). Our first structural result, which is formulated in the general infinite-dimensional setting, reads as follows.

Theorem 3.1.1. *Suppose that A and B are closed linear operators on X, $C \in L(X)$ is injective, $CA \subseteq AC$, $CB \subseteq BC$ and $(A+B)^{-1}C \in L(X)$. Then, for every (polynomially) bounded sequence $(f_k)_{k\in\mathbb{Z}}$, the sequence*

$$u_k := (C-(A+B)^{-1}C)\sum_{v=1}^{+\infty}(-1)^{v+1} \times [A(C-(A+B)^{-1}CA)(A+B)^{-1}C]^{v-1}(A+B)^{-1}Cf_{k-1+v}, \quad k\in\mathbb{Z} \tag{104}$$

is a (polynomially) bounded solution of (102), *provided that*

$$\|A(C-(A+B)^{-1}CA)(A+B)^{-1}C\|_{L(X)} < 1. \tag{105}$$

Moreover, if $(f_k)_{k\in\mathbb{Z}}$ is T-almost periodic, where $T \in L(X)$ commutes with A, B and C, or R*-almost automorphic, then the solution $(u_k)_{k\in\mathbb{Z}}$ has the same property.*

Proof. Due to the closed graph theorem, we have $A(A+B)^{-1}C \in L(X)$ and $A(C-(A+B)^{-1}CA)(A+B)^{-1}C \in L(X)$. Clearly, the series in (104) absolutely converges due to (105). Furthermore, $u_k \in D(A)\cap D(B)$ for all $k\in\mathbb{Z}$ and

$$u_k = \sum_{v=1}^{+\infty}(-1)^{v+1}[(C-(A+B)^{-1}CA)(A+B)^{-1}CA]^{v-1} \times (C-(A+B)^{-1}CA)(A+B)^{-1}C(A+B)^{-1}Cf_{k-1+v},$$

for any $k\in\mathbb{Z}$. Hence,

$$\begin{aligned} Au_{k+1}+Bu_k &= \sum_{v=1}^{+\infty}(-1)^{v+1}A[(C-(A+B)^{-1}CA)(A+B)^{-1}CA]^{v-1} \\ &\quad\times (C-(A+B)^{-1}CA)(A+B)^{-1}C(A+B)^{-1}Cf_{k+v} \\ &\quad+\sum_{v=1}^{+\infty}(-1)^{v+1}B[(C-(A+B)^{-1}CA)(A+B)^{-1}CA]^{v-1} \\ &\quad\times (C-(A+B)^{-1}CA)(A+B)^{-1}C(A+B)^{-1}Cf_{k-1+v} \\ &= \sum_{m=k+1}^{+\infty}(-1)^{k+1-m}A[(C-(A+B)^{-1}CA)(A+B)^{-1}CA]^{k-m-1} \\ &\quad\times (C-(A+B)^{-1}CA)(A+B)^{-1}C(A+B)^{-1}Cf_m \end{aligned}$$

$$
\begin{aligned}
&+\sum_{m=k}^{+\infty}(-1)^{k-m}B\big[(C-(A+B)^{-1}CA)(A+B)^{-1}CA\big]^{k-m}\\
&\times(C-(A+B)^{-1}CA)(A+B)^{-1}C(A+B)^{-1}Cf_m\\
&=\sum_{m=k}^{+\infty}(-1)^{k+1-m}A\big[(C-(A+B)^{-1}CA)(A+B)^{-1}CA\big]^{k-m-1}\\
&\times(C-(A+B)^{-1}CA)(A+B)^{-1}C(A+B)^{-1}Cf_m\\
&+A\big[(C-(A+B)^{-1}CA)^{-1}(A+B)^{-1}CA\big]^{-1}\\
&\times(C-(A+B)^{-1}CA)^{-1}(A+B)^{-1}Cf_k\\
&+\sum_{m=k}^{+\infty}(-1)^{k-m}B\big[(C-(A+B)^{-1}CA)(A+B)^{-1}CA\big]^{k-m}\\
&\times(C-(A+B)^{-1}CA)(A+B)^{-1}C(A+B)^{-1}Cf_m\\
&=\sum_{m=k}^{+\infty}(-1)^{k+1-m}A\big[(C-(A+B)^{-1}CA)(A+B)^{-1}CA\big]^{k-m-1}\\
&\times(C-(A+B)^{-1}CA)(A+B)^{-1}C(A+B)^{-1}Cf_m\\
&+AA^{-1}C^{-1}(A+B)(C-(A+B)^{-1}CA)(C-(A+B)^{-1}CA)^{-1}(A+B)^{-1}Cf_k\\
&+\sum_{m=k}^{+\infty}(-1)^{k-m}B\big[(C-(A+B)^{-1}CA)(A+B)^{-1}CA\big]\\
&\times\big[(C-(A+B)^{-1}CA)(A+B)^{-1}CA\big]^{k-m-1}\\
&\times(C-(A+B)^{-1}CA)(A+B)^{-1}C(A+B)^{-1}Cf_m,
\end{aligned}
$$

for any $k \in \mathbb{Z}$. Since

$$
AA^{-1}C^{-1}(A+B)(C-(A+B)^{-1}CA)(C-(A+B)^{-1}CA)^{-1}(A+B)^{-1}Cf_k = f_k,
$$

for any $k \in \mathbb{Z}$ and

$$
\begin{aligned}
&B\big[(C-(A+B)^{-1}CA)^{-1}(A+B)^{-1}CA\big]x\\
&\quad= B\big[(A+B)^{-1}CB\big]^{-1}(A+B)^{-1}CAx\\
&\quad= BB^{-1}C^{-1}(A+B)(A+B)^{-1}CAx\\
&\quad= BB^{-1}Ax = Ax, \quad x \in X,
\end{aligned}
$$

the first part of result simply follows. Arguing similarly as in the proof of [491, Theorem 2.1], we can simply deduce that the solution $(u_k)_{k\in\mathbb{Z}}$ is T-almost periodic, where $T \in L(X)$ satisfies the prescribed properties, or R-almost automorphic, provided that $(f_k)_{k\in\mathbb{Z}}$ has the same property. □

Remark 3.1.2. It is worth noting that the formula (104) is derived after multiplying (102) with $(A+B)^{-1}C$ and using a relatively simple trick with the multivalued linear operator $\mathcal{A} = -[(A+B)^{-1}CA]^{-1}(C-(A+B)^{-1}CA)$ in [491, Theorem 2.1]. But, arguing in a such a way, we can only prove that the sequence $(u_k)_{k\in\mathbb{Z}}$ satisfies the difference equation $A^2u_{k+1} + ABu_k = Af_k$, $k \in \mathbb{Z}$.

The requirements of Theorem 3.1.1 are really satisfied in many concrete situations, with $C \neq \mathrm{I}$.

Example 3.1.3. Suppose that A and B are higher-order differential operators with constant coefficients in the space $L^p(\mathbb{R}^n)$, $BUC(\mathbb{R}^n)$ or $C_b(\mathbb{R}^n)$, where $1 \leqslant p < +\infty$, the operator $A+B$ generates an exponentially bounded C-regularized semigroup, $0 \in \rho_C(A+B)$ and

$$\|C\| \cdot \|A(A+B)^{-1}C\| + \|A(A+B)^{-1}C\|^2 < 1.$$

Then the requirements of Theorem 3.1.1 are satisfied and the resolvent set of $A+B$ can be empty.

We continue by providing a few useful observations.

Remark 3.1.4. Suppose that

$$\kappa := \|A(C-(A+B)^{-1}CA)(A+B)^{-1}C\|_{L(X)} \geqslant 1$$

and all remaining requirements in Theorem 3.1.1 hold. Then the sequence $(u_k)_{k\in\mathbb{Z}}$, given by (104), is an exponentially bounded solution of (102), provided that there exists $c \in (0,\kappa^{-1})$ such that $\|f(k)\| \leqslant Mc^{-|k|}$, $k \in \mathbb{Z}$ for some real number $M > 0$; more precisely, we have

$$\begin{aligned}
\|u_k\| &\leqslant \|C-(A+B)^{-1}C\| \cdot \|(A+B)^{-1}C\| \cdot \sum_{v=1}^{+\infty} \kappa^{v-1}\|f(k-1+v)\| \\
&\leqslant M\|C-(A+B)^{-1}C\| \cdot \|(A+B)^{-1}C\| \cdot \sum_{v=1}^{+\infty} \kappa^{v-1}c^{|v-1+k|} \\
&\leqslant Mc^{-|k|}\|C-(A+B)^{-1}C\| \cdot \|(A+B)^{-1}C\| \cdot \sum_{v=1}^{+\infty} \kappa^{v-1}c^{v-1} \\
&= M\frac{c^{-|k|}}{1-\kappa c}\|C-(A+B)^{-1}C\| \cdot \|(A+B)^{-1}C\|, \quad k \in \mathbb{Z}.
\end{aligned}$$

Remark 3.1.5. (i) We can similarly analyze the situation in which

$$\kappa_1 := \|B(C-(A+B)^{-1}CB)(A+B)^{-1}C\|_{L(X)} < 1$$

or $\kappa_1 \geqslant 1$ and there exists $c \in (0,\kappa_1^{-1})$ such that $\|f(k)\| \leqslant Mc^{-|k|}$, $k \in \mathbb{Z}$ for some real number $M > 0$, by considering the sequences $(\check{u}(k) \equiv u(-k))_{k\in\mathbb{Z}}$ and $(\check{f}(k) \equiv f(-k))_{k\in\mathbb{Z}}$.

(ii) If A and B are singular real matrices, $(f_k)_{k\in\mathbb{Z}}$ is an almost periodic sequence and the condition

$$S_1 \subseteq \rho_C^{-A}(-B) := \{\lambda \in \mathbb{C} : (\lambda A + B)^{-1}C \in L(X)\}$$

holds with $C = \mathrm{I}$, then [251, Theorem 3.1] shows that there exists a unique almost periodic solution of (102). Concerning the uniqueness of solutions of (102), we would like to emphasize first that the requirements of Theorem 3.1.1 do not imply the uniqueness of almost periodic solutions of (102) with $f(\cdot) \equiv 0$. For example, let us consider the Banach space $X := l^{\infty}(\mathbb{N} : \mathbb{C})$. Define $A\langle x_m\rangle := \langle x_{2m-1}/2\rangle$, $\langle x_m\rangle \in X$ and $B := \mathrm{I} - A$. Then $A \in L(X)$, $\|A\| < 1/2$, A and B are noninvertible, the requirements of Theorem 3.1.1 hold with $C = \mathrm{I}$, and $\rho_{\mathrm{I}}^{-A}(-B) = \{1\}$, as easily approved (cf. [445, Section 2.1] for more details about the (C, B)-resolvents of closed linear operators). But any antiperiodic sequence $(u_k)_{k\in\mathbb{Z}} = (\langle x_m^k\rangle)_{k\in\mathbb{Z}} = (\langle(-1)^k a, 0, 0, \ldots\rangle)_{k\in\mathbb{Z}}$, where $a \in \mathbb{C}$, is a solution of (102). It would be very tempting to formulate a satisfactory result concerning the uniqueness of (almost periodic) solutions of problem (102) in the infinite-dimensional setting.

(iii) In [251, Subsection 3.2], the authors have considered the existence and uniqueness of Besicovitch almost periodic solutions to (102). Besides T-almost periodic, R-almost automorphic and Besicovitch almost periodic solutions, we can also analyze some other type of generalized almost periodic (automorphic) type solutions to (102), like Weyl T-almost periodic solutions.

3.1.2 Abstract degenerate higher-order difference equations

This subsection investigates the abstract degenerate higher-order difference equations with unbounded linear operators and their systems. We will first consider the following system of the abstract degenerate higher-order difference equations:

$$\begin{aligned}(S):\quad & A_{j;1}u_k + (A_{j;2} + B_{j;1})u_{k+1} + \cdots + (A_{j;p} + B_{j;p-1})u_{p-1+k} + B_{j;p}u_{p+k} = f_j(k),\\ & k \in \mathbb{Z} \quad (1 \leqslant j \leqslant p),\end{aligned}$$

where $A_{i;j}$, $B_{i;j}$ are closed linear operators on X and $p \in \mathbb{N} \setminus \{1\}$. The system (S) can be rewritten in the following equivalent matricial form:

$$B\vec{u}_{k+1} + A\vec{u}_k = [f_k\ 0 \ldots 0]^T, \quad k \in \mathbb{Z},$$

where

$$A = \begin{bmatrix} A_{1;1} & A_{1;2} & A_{1;3} & \cdots & A_{1;p} \\ A_{2;1} & A_{2;2} & & \cdots & A_{2;p} \\ \cdot & \cdot & \cdot & \cdots & \cdot \\ A_{p;1} & A_{p;2} & A_{p;3} & \cdots & A_{p;p} \end{bmatrix}, \quad B = \begin{bmatrix} B_{1;1} & B_{1;2} & B_{1;3} & \cdots & B_{1;p} \\ B_{2;1} & B_{2;2} & & \cdots & B_{2;p} \\ \cdot & \cdot & \cdot & \cdots & \cdot \\ B_{p;1} & B_{p;2} & B_{p;3} & \cdots & B_{p;p} \end{bmatrix}$$

and $\vec{u}_k := [u_k\ u_{k+1}\ \cdots\ u_{k+p-1}]^T$, $k \in \mathbb{Z}$; for more details about operator matrices and their applications, we refer the reader to the book manuscript [277] by K. J. Engel. Therefore, as an immediate consequence of Theorem 3.1.1 (cf. also Remark 3.1.4), we have the following result.

Theorem 3.1.6. *Suppose that there exist bounded linear operators $C_{i';j'}$ on X such that the operator matrix*

$$C = \begin{bmatrix} C_{1;1} & C_{1;2} & C_{1;3} & \cdots & C_{1;p} \\ C_{2;1} & C_{2;2} & & \cdots & C_{2;p} \\ \cdot & \cdot & \cdot & \cdots & \cdot \\ C_{p;1} & C_{p;2} & C_{p;3} & \cdots & C_{p;p} \end{bmatrix}$$

is injective on X^p as well as that $C_{i';j'}A_{i;j} \subseteq A_{i;j}C_{i';j'}$, $C_{i';j'}B_{i;j} \subseteq B_{i;j}C_{i';j'}$ for $1 \leqslant i,j,i',j' \leqslant p$ and $(A+B)^{-1}C \in L(X^p)$. Then we have the following:

(i) *If* (105) *holds with the space $L(X)$ replaced therein with the space $L(X^p)$, then there exists a solution $(u_k)_{k\in\mathbb{Z}}$ of system* (S). *Moreover, if all sequences $(f_j(k))_{k\in\mathbb{Z}}$ are T-almost periodic, where $T \in L(X)$ commutes with A, B and C, or* R*-almost automorphic, then the solution $(u_k)_{k\in\mathbb{Z}}$ of system* (S) *has the same property.*
(ii) *If there exists $c \in (0,1)$ such that*

$$c\|A(C-(A+B)^{-1}CA)(A+B)^{-1}C\|_{L(X^p)} < 1$$

and $\|f(k)\| \leqslant Mc^{-|k|}$, $k \in \mathbb{Z}$ for some real number $M > 0$, then there exists a solution $(u_k)_{k\in\mathbb{Z}}$ of system (S)*; moreover, $\|u_k\| \leqslant M_1c^{-|k|}$, $k \in \mathbb{Z}$ for some real number $M_1 > 0$.*

Now we will consider the abstract higher-order difference equation (103), with $A_p,\ldots,A_0$ being closed linear operators on X and $p \in \mathbb{N}\setminus\{1\}$. Our first result concerning the well-posedness of (103) reads as follows.

Theorem 3.1.7. *Suppose that any of the operators $A_p,\ldots,A_0$ belongs to the space $L(X)$, commutes with C and $-R_1 := (A_0+A_1+\cdots+A_p)^{-1}C \in L(X)$. Set*

$$A := \begin{bmatrix} A_0 & 0 & 0 & \cdots & 0 \\ 0 & -I & & \cdots & 0 \\ \cdot & \cdot & \cdot & \cdots & \cdot \\ 0 & 0 & 0 & \cdots -I & 0 \\ 0 & 0 & 0 & \cdots & -I \end{bmatrix}, \quad B := \begin{bmatrix} A_1 & A_2 & A_3 & \cdots & A_p \\ I & 0 & 0 & \cdots & 0 \\ 0 & I & \cdot & \cdots & 0 \\ 0 & 0 & \cdots & I\ 0 & 0 \\ 0 & 0 & 0 & \cdots I & 0 \end{bmatrix},$$

$\mathbf{C} := CI_{X^p}$, *where I_{X^p} denotes the identity operator on X^p, $X_k := -A_0 - A_1 - \cdots - A_{k+1}$, $1 \leqslant k \leqslant p-1$ and $Y_k := A_{k+1} + \cdots + A_p$ for $1 \leqslant k \leqslant p-1$. Then $(A+B)^{-1}\mathbf{C} \in L(X^p)$ and there exists a solution $(u_k)_{k\in\mathbb{Z}}$ of* (103), *which is given by the first projection in the formula* (104) *with the operator C replaced therein with the operator $\mathbf{C}$ and the term f_{k-1+v}*

replaced therein with the term $[f_{k-1+v}\ 0\ \dots\ 0]^T$, *provided that there exists* $c \in (0,1)$ *such that*

$$c\kappa_{\mathbf{C}} := c\|A(\mathbf{C} - (A+B)^{-1}\mathbf{C}A)(A+B)^{-1}\mathbf{C}\|_{L(X^p)} < 1$$

and $\|f(k)\| \leqslant Mc^{-|k|}$, $k \in \mathbb{Z}$ *for some real number* $M > 0$. *Moreover,* $\|u_k\| \leqslant M_1 c^{-|k|}$, $k \in \mathbb{Z}$ *for some real number* $M_1 > 0$.

Proof. The operator matrices A and B are closed. Arguing as in the proof of [621, Proposition 7], with $\lambda = 1$ and the transpose of $A + B$, we may conclude that $(A+B)^{-1}\mathbf{C} \in L(X^p)$ as well as that

$$(A+B)^{-1}\mathbf{C} = -\begin{bmatrix} R_1 & Y_1R_1 & Y_2R_1 & \cdots & Y_{p-1}R_1 \\ R_1 & X_1R_1 & Y_2R_1 & \cdots & Y_{p-1}R_1 \\ R_1 & X_1R_1 & X_2R_1 & Y_3R_1\cdots & Y_{p-1}R_1 \\ R_1 & X_1R_1 & \cdots & X_{p-2}R_1 & Y_{p-1}R_1 \\ R_1 & X_1R_1 & X_2R_1 & \cdots & X_{p-1}R_1 \end{bmatrix}. \tag{106}$$

On the other hand, we can write (103) as

$$B\vec{u}_{k+1} + A\vec{u}_k = [f_k\ 0\ \dots\ 0]^T, \quad k \in \mathbb{Z},$$

where $\vec{u}_k := [u_k\ u_{k+1}\ \dots\ u_{k+p-1}]^T$, $k \in \mathbb{Z}$. Now the final conclusion follows from Theorem 3.1.1 and the consideration from Remark 3.1.4. □

Remark 3.1.8. In contrast to Theorem 3.1.1, where the case $p = 1$ has been considered, we can hardly expect that

$$\|A(\mathbf{C} - (A+B)^{-1}\mathbf{C}A)(A+B)^{-1}\mathbf{C}\|_{L(X^p)} < 1.$$

The second result on the well-posedness of (103) reads as follows.

Theorem 3.1.9. *Suppose that any of the operators* $A_p, \dots, A_0$ *is linear, closed, commutes with* C *and* $R_1' := (A_0 + A_1 + \cdots + A_p)^{-1}C \in L(X)$. *Suppose also that* $A_iA_jx = A_jA_ix$ *for all* $x \in D(A_iA_j) \cap D(A_jA_i)$ *and* $0 \leqslant i, j \leqslant p$. *Set*

$$A' := \begin{bmatrix} A_0R_1' & 0 & 0 & \cdots & 0 \\ 0 & -\mathrm{I} & & \cdots & 0 \\ \cdot & \cdot & \cdot & \cdots & \cdot \\ 0 & 0 & 0 & \cdot\cdot -\mathrm{I} & 0 \\ 0 & 0 & 0 & \cdots & -\mathrm{I} \end{bmatrix}, \quad B' := \begin{bmatrix} A_1R_1' & A_2R_1' & A_3R_1' & \cdots & A_pR_1' \\ \mathrm{I} & 0 & 0 & \cdots & 0 \\ 0 & \mathrm{I} & \cdot & \cdots & 0 \\ 0 & 0 & \cdots & \mathrm{I}\ 0 & 0 \\ 0 & 0 & 0 & \cdot\cdot\,\mathrm{I} & 0 \end{bmatrix},$$

$\mathbf{C} := CI_{X^p}$, $X_k' := -(A_0 + A_1 + \cdots + A_{k+1})R_1'$, $1 \leqslant k \leqslant p-1$ *and* $Y_k' := (A_{k+1} + \cdots + A_p)R_1'$ *for* $1 \leqslant k \leqslant p-1$. *Then* $(A' + B')^{-1}\mathbf{C} \in L(X^p)$ *and there exists a solution* $(u_k)_{k\in\mathbb{Z}}$ *of* (103), *which is given by the first projection in the formula* (104) *with the operators* A, B, C *replaced* *therein with the operator A', B', $\mathbf{C}$ and the term f_{k-1+v} replaced therein with the term $[R_1'f_{k-1+v}\ 0\ \ldots\ 0]^T$, provided that $f_k \in D(A_j)$ for all $k \in \mathbb{Z}$ and $0 \leqslant j \leqslant p$, as well as that there exists $c \in (0,1)$ such that*

$$c\kappa_{\mathbf{C}}' := c\|A'(\mathbf{C} - (A' + B')^{-1}\mathbf{C}A')(A' + B')^{-1}\mathbf{C}\|_{L(X^p)} < 1$$

and $\max(\|R_1'f(k)\|, \|f(k)\|) \leqslant Mc^{-|k|}$, $k \in \mathbb{Z}$ for some real number $M > 0$. Moreover, $\|u_k\| \leqslant M_1c^{-|k|}$, $k \in \mathbb{Z}$ for some real number $M_1 > 0$.

Proof. The requirements of Theorem 3.1.1 are satisfied with the operators $A_p, \ldots, A_0$ and the sequence $(f_k)_{k\in\mathbb{Z}}$ replaced with the operators $A_pR_1', \ldots, A_0R_1'$ and the sequence $(R_1'f_k)_{k\in\mathbb{Z}}$ therein. Hence, there exists a solution $(u_k)_{k\in\mathbb{Z}}$ of problem

$$A_pR_1'u_{k+p} + A_{p-1}R_1'u_{k+p-1} + \cdots + A_0R_1'u_k = R_1'f_k, \quad k \in \mathbb{Z}. \tag{107}$$

Since we have assumed that $A_iA_jx = A_jA_ix$ for all $x \in D(A_iA_j) \cap D(A_jA_i)$ and $0 \leqslant i,j \leqslant p$, it readily follows that $R_1'A_j \subseteq A_jR_1'$ for $0 \leqslant j \leqslant p$ and, since the operator R_1' is injective, we only need to verify that the solution $(u_k)_{k\in\mathbb{Z}}$ of (107) satisfies $u_k \in D(A_j)$ for all $k \in \mathbb{Z}$ and $0 \leqslant j \leqslant p$. In order to see this, let us observe that the argumentation contained in the proof of Theorem 3.1.1 shows that

$$\begin{aligned}
&[u_k\ u_{k+1} \ldots u_{k-1+p}]^T \\
&\quad = (\mathbf{C} - (A' + B')^{-1}\mathbf{C}) \sum_{v=1}^{+\infty} (-1)^{v+1} \\
&\qquad \times [A'(\mathbf{C} - (A' + B')^{-1}\mathbf{C}A')(A' + B')^{-1}\mathbf{C}]^{v-1}(A' + B')^{-1}\mathbf{C}[R_1'f_{k-1+v}\ 0\ \ldots\ 0]^T,
\end{aligned} \tag{108}$$

for any $k \in \mathbb{Z}$. Moreover, using the same notation as in the proof of Theorem 3.1.1, we have

$$(A' + B')^{-1}\mathbf{C} = -\begin{bmatrix} \mathrm{I} & Y_1R_1 & Y_2R_1 & \cdots & Y_{p-1}R_1 \\ \mathrm{I} & X_1R_1 & Y_2R_1 & \cdots & Y_{p-1}R_1 \\ \mathrm{I} & X_1R_1 & X_2R_1 & Y_3R_1 \cdots & Y_{p-1}R_1 \\ \mathrm{I} & X_1R_1 & \cdots & X_{p-2}R_1 & Y_{p-1}R_1 \\ \mathrm{I} & X_1R_1 & X_2R_1 & \cdots & X_{p-1}R_1 \end{bmatrix}.$$

This simply implies that any element of the vector column $(\mathbf{C} - (A' + B')^{-1}\mathbf{C})(A' + B')^{-1}\mathbf{C}[R_1'f_{k-1+v}\ 0\ \ldots\ 0]^T$, obtained by plugging $v = 1$ in the formula (108), belongs to $D(A_j)$ for $0 \leqslant j \leqslant p$. Since $R_1'A_j \subseteq A_jR_1'$ for $0 \leqslant j \leqslant p$ and $f_k \in D(A_j)$ for all $k \in \mathbb{Z}$ and $0 \leqslant j \leqslant p$, we also have

$$\sum_{v=2}^{+\infty} (-1)^{v+1}[A'(\mathbf{C} - (A' + B')^{-1}\mathbf{C}A')(A' + B')^{-1}\mathbf{C}]^{v-1}(A' + B')^{-1}\mathbf{C}[R_1'f_{k-1+v}\ 0\ \ldots\ 0]^T$$

$$= \sum_{v=2}^{+\infty} (-1)^{v+1} (R_1' I)_{X^p} [A'(\mathbf{C} - (A'+B')^{-1} \mathbf{C} A')(A'+B')^{-1} \mathbf{C}]^{v-1} (A'+B')^{-1} \mathbf{C} [f_{k-1+v}\, 0 \dots 0]^T$$

$$= (R_1' I)_{X^p} \sum_{v=2}^{+\infty} (-1)^{v+1} [A'(\mathbf{C} - (A'+B')^{-1} \mathbf{C} A')(A'+B')^{-1} \mathbf{C}]^{v-1} (A'+B')^{-1} \mathbf{C} [f_{k-1+v}\, 0 \dots 0]^T,$$

where we have used the fact that the last series converges on the account of our assumption $\|f(k)\| \leqslant Mc^{-|k|}$, $k \in \mathbb{Z}$ and the closedness of the operator matrix $(R_1' I)_{X^p}$. This simply completes the proof. □

We will illustrate Theorem 3.1.9 with the following example:

Example 3.1.10. Suppose that $A_p = P_j(A)$ for some closed linear operator A and a complex polynomial $P(\cdot)$, $0 \leqslant j \leqslant p$. If there exists an injective operator $C \in L(X)$ which satisfies $CA \subseteq AC$ and $(P_0(A) + \dots + P_p(A))^{-1} C \in L(X)$, then we can apply Theorem 3.1.9. This, in particular, holds if the operator $P_0(A) + \dots + P_p(A)$ generates an exponentially decaying C-regularized semigroup, when the resolvent set of $P_0(A) + \dots + P_p(A)$ can be empty.

Concerning the nonlinear higher-order difference equations in Banach algebras, we refer the reader to the interesting article [687] by G. Sedaghat. Before we close this section, let us observe that our results are not applicable in the study of nonautonomous analogues of (102) and (103). For more details about the well-posedness of the abstract degenerate difference equation,

$$A_k u_{k+1} + B_k u_k = f_k, \quad k \in \mathbb{N}_0,$$

where A_k and B_k are singular complex matrices ($k \in \mathbb{N}_0$), we refer the reader to [58, 319, 545, 546, 576] and references cited therein.

3.2 Abstract Volterra difference inclusions

The organization and main ideas of this section, which is broken down into three subsections, can be summarized as follows [470]. We generalize here the notion of a discrete resolvent family introduced by V. Keyantuo et al. in [429, Definition 3.2]. More precisely, in Definition 3.2.1, we introduce the notion of a discrete (a, k)-regularized C-resolvent family subgenerated by a multivalued linear operator. The first structural result on the discrete (a, k)-regularized C-resolvent families is given in Proposition 3.2.4. We continue by stating our first main result, Theorem 3.2.5. The discrete abstract Cauchy inclusions are analyzed in Section 3.2.2; cf. Proposition 3.2.6, Proposition 3.2.7 and Proposition 3.2.8 for some structural results established in this part. At this place, we would like to emphasize that V. Keyantuo et al. (cf. also the research articles [173] and [550] by Y.-K. Chang and P. Lü) have analyzed the existence and uniqueness of the weighted pseudo asymptotically mild solutions to the following abstract Volterra difference equation:

$$u(k+1) = A\sum_{j=-\infty}^{k} a(k-j)u(j+1) + \sum_{j=-\infty}^{k} b(k-j)f(k), \quad k \in \mathbb{Z}$$

and its semilinear analogue.

In this section, we follow a slightly different approach. We investigate the existence and uniqueness of the asymptotically almost periodic (automorphic) type solutions to the following abstract Volterra difference equation:

$$u(v) \in f(v) + \mathcal{A}(a *_0 u)(v), \quad v \in \mathbb{N}_0,$$

where $\mathcal{A}$ is an MLO in X, $(a(v))_{v\in\mathbb{N}_0}$ and $(f(v))_{v\in\mathbb{N}_0}$ are given sequences in X; this is the abstract Volterra difference inclusion which is spontaneously connected with the notion of a discrete (a,k)-regularized C-resolvent family. In Section 3.2.3, we analyze the existence and uniqueness of asymptotically almost periodic type solutions and asymptotically almost automorphic type solutions to (122) as well as the existence and uniqueness of almost periodic type solutions and almost automorphic type solutions to the following abstract Volterra difference inclusion:

$$u(k+1) \in \mathcal{A}\sum_{j=-\infty}^{k} a(k-j)u(j+1) + \sum_{j=-\infty}^{k} b(k-j)f(k), \quad k \in \mathbb{Z}.$$

For the sake of simplicity and better exposition, we will not consider here the abstract semilinear Volterra difference inclusions. Before we switch to Section 3.2.1, we would like to notice that T. Furumochi, S. Murakami and Y. Nagabuchi have investigated, in [314], the abstract Volterra difference equation

$$u_{k+1} = \sum_{j=-\infty}^{k} Q(k-j)u_j, \quad k \in \mathbb{Z},$$

where $(Q(k))_{k\in\mathbb{N}_0}$ is a family of bounded linear operators on X such that $\|Q(k)\| \leqslant Me^{-\gamma k}$, $k \in \mathbb{N}_0$ for some positive real constant $\gamma > 0$. The authors have also studied the stability property of the zero solution of this equation and its connection with the following abstract differential equation with piecewise continuous delay:

$$u'(t) = Au(t) + \sum_{k=0}^{\infty} B(k)u(\lfloor t-k \rfloor), \quad t \geqslant 0;$$

cf. also [577, 599, 608, 609, 615].

3.2.1 Discrete (a,k)-regularized C-resolvent families

The following important notion generalizes the notion introduced in [3, Definition 3.1], [46, Definition 2.5] and [429, Definition 3.2], where the authors have assumed that $C = \mathrm{I}$

and $\mathcal{A} = A$ is a closed single-valued linear operator, and the notion introduced in Definition 2.1.47, where we have assumed that $a(v) = k(v) = k^a(v)$ for all $v \in \mathbb{N}_0$:

Definition 3.2.1. Suppose that $\mathcal{A}$ is a closed multivalued linear operator, $a : \mathbb{N}_0 \to \mathbb{C}$, $k : \mathbb{N}_0 \to \mathbb{C}$ and $C \in L(X)$ is an injective operator. Then we say that the operator $\mathcal{A}$ is a subgenerator of a discrete (a,k)-regularized C-resolvent family $(S(v))_{v\in\mathbb{N}_0} \subseteq L(X)$ if $\mathcal{A}C \subseteq C\mathcal{A}$, $\mathcal{A}S(v) \subseteq S(v)\mathcal{A}$ and $CS(v) = S(v)C$ for all $v \in \mathbb{N}_0$, and the assumption $(x,y) \in \mathcal{A}$ implies

$$S(v)x = k(v)Cx + \sum_{j=0}^{v} a(v-j)S(j)y, \quad v \in \mathbb{N}_0. \tag{109}$$

We will not consider here the notion of a discrete mild (a,k)-regularized (C_1, C_2)-existence and uniqueness family. Moreover, we will not consider here the discrete solution operator families in sequentially complete locally convex spaces.

The integral generator $\mathcal{A}_{\text{int}}$ of $(S(v))_{v\in\mathbb{N}_0}$ is defined by

$$\mathcal{A}_{\text{int}} := \{(x,y) \in X \times X : (109) \text{ holds}\}.$$

It is a closed MLO, which extends any other subgenerator of $(S(v))_{v\in\mathbb{N}_0}$ and satisfies $\mathcal{A}_{\text{int}} = C^{-1}\mathcal{A}_{\text{int}}C$; furthermore, for each subgenerator $\mathcal{A}$ of $(S(v))_{v\in\mathbb{N}_0}$ we also have that $C^{-1}\mathcal{A}C$ is a subgenerator of $(S(v))_{v\in\mathbb{N}_0}$.

Before proceeding any further, we would like to observe that the notion of a discrete (a,k)-regularized C-resolvent family is not so simply related with the notion of a continuous (a,k)-regularized C-resolvent family defined on the interval $[0,\infty)$; for example, in the case of the usually considered strongly continuous semigroups $(S(t))_{t\geqslant 0}$, we have $a(t) = k(t) = 1$ for all $t \geqslant 0$ and $C = \mathrm{I}$ but the functional identity (109) does not hold for $v = 0$ since the left-hand side of (109) equals x while the left-hand side of (109) equals $x + Ax$ $(x \in D(A))$. In other words, the restriction of an (a,k)-regularized C-resolvent family $(S(t))_{t\geqslant 0}$ to the set $\mathbb{N}_0$ need not be a discrete (a,k)-regularized C-resolvent family with the same subgenerator. But we know that the expression

$$S(v)x := \int_0^{+\infty} e^{-t}\frac{t^v}{v!}T(t)x\,dt, \quad v \in \mathbb{N}_0,\ x \in X,$$

where $(T(t))_{t\geqslant 0}$ is an exponentially bounded C-regularized semigroup subgenerated by $\mathcal{A}$, determines a discrete $(1,1)$-regularized C-resolvent family $(S(v))_{v\in\mathbb{N}_0}$ with the same subgenerator; see [535] for more details about the subject (here we assume that there exist real numbers $M \geqslant 1$ and $\omega < 1$ such that $\|T(t)\| \leqslant Me^{\omega t}$, $t \geqslant 0$). It is natural to ask what is going on if we consider the expression

$$S_{d,\omega}(v)x := \int_0^{+\infty} e^{-\omega t}\frac{(\omega t)^v}{v!}R(t)x\,dt, \quad v \in \mathbb{N}_0,\ x \in X, \tag{110}$$

where $(R(t))_{t\geqslant 0}$ is an exponentially bounded (a,k)-regularized C-resolvent family subgenerated by $\mathcal{A}$. Concerning this question, we would like to note that the argumentation contained in the proof of [3, Theorem 3.5] essentially shows the following (with the exception of the statement [535, Proposition 3.7(ii)], where $\omega < 0$, C. Lizama has always used the value $\omega = 1$ in [535]).

Theorem 3.2.2. *Suppose that $\mathcal{A}$ is a closed subgenerator of an exponentially bounded (a,k)-regularized C-resolvent family $(R(t))_{t\geqslant 0}$ and there exist real numbers $M \geqslant 1$ and $\omega_0 \geqslant 0$ such that $\|R(t)\| + |a(t)| + |k(t)| \leqslant Me^{\omega_0 t}$, $t \geqslant 0$. If $\omega > \omega_0$, then we set*

$$a_{d,\omega}(v) := \int_0^{+\infty} e^{-\omega t}\frac{(\omega t)^v}{v!}a(t)\,dt \quad \text{and} \quad k_{d,\omega}(v) := \int_0^{+\infty} e^{-\omega t}\frac{(\omega t)^v}{v!}k(t)\,dt,\ v \in \mathbb{N}_0.$$

Then the operator family $(S_{d,\omega}(v))_{v\in\mathbb{N}_0}$, given by (110), is a discrete $(a_{d,\omega}, k_{d,\omega})$-regularized C-resolvent family with subgenerator $\mathcal{A}$; we also have that

$$\int_0^{+\infty} \|R(t)\|\,dt < +\infty \quad \text{implies} \quad \sum_{v=0}^{+\infty} \|S_{d,\omega}(v)\| < +\infty.$$

Moreover, if $R(t)x - k(t)Cx \in \mathcal{A}\int_0^t a(t-s)R(s)x\,ds$ for all $x \in X$ and $t \geqslant 0$, then the equation (111) below holds with $a(\cdot)$ and $k(\cdot)$ replaced therein with $a_{d,\omega}(\cdot)$ and $k_{d,\omega}(\cdot)$, respectively.

Proof. We shall content ourselves with sketching the proof. It is clear that the operator family $(S_{d,\omega}(v))_{v\in\mathbb{N}_0}$, given by (110), is well-defined as well as that $CS(v) = S(v)C$ for all $v \in \mathbb{N}_0$ since $R(t)C = CR(t)$, $t \geqslant 0$; owing to the closedness of $\mathcal{A}$ and the inclusion $R(t)\mathcal{A} \subseteq \mathcal{A}R(t)$, $t \geqslant 0$, we can similarly show that $\mathcal{A}S(v) \subseteq S(v)\mathcal{A}$ for all $v \in \mathbb{N}_0$. Suppose now that $(x,y) \in \mathcal{A}$. Then we have

$$\begin{aligned} S_{d,\omega}(v)x &= \int_0^{+\infty} e^{-\omega t}\frac{(\omega t)^v}{v!}R(t)x\,dt \\ &= \int_0^{+\infty} e^{-\omega t}\frac{(\omega t)^v}{v!}\left[k(t)Cx + \int_0^t a(t-s)R(s)y\,ds\right]dt \\ &= k_{d,\omega}(v)Cx + \int_0^{+\infty} e^{-\omega t}\frac{(\omega t)^v}{v!}\int_0^t a(t-s)R(s)y\,ds\,dt \\ &= k_{d,\omega}(v)Cx + \int_0^{+\infty}\int_s^{+\infty} e^{-\omega t}\frac{(\omega t)^v}{v!}a(t-s)R(s)y\,dt\,ds \\ &= k_{d,\omega}(v)Cx + \int_0^{+\infty}\int_0^{+\infty} e^{-\omega(r+s)}\frac{(\omega(r+s))^v}{v!}a(r)R(s)y\,dr\,ds \end{aligned}$$

$$= k_{d,\omega}(v)Cx + \int_0^{+\infty}\int_0^{+\infty} e^{-\omega(r+s)}\frac{1}{v!}\sum_{j=0}^{v}\binom{v}{j}(\omega r)^{v-j}(\omega s)^j a(r)R(s)y\,dr\,ds$$

$$= k_{d,\omega}(v)Cx + \sum_{j=0}^{v}\Bigg[\int_0^{+\infty} e^{-\omega r}\frac{(\omega r)^{v-j}}{(v-j)!}a(r)\,dr\Bigg]\cdot\Bigg[\int_0^{+\infty} e^{-\omega s}\frac{(\omega s)^j}{j!}R(s)y\,ds\Bigg]$$

$$= k_{d,\omega}(v)Cx + \sum_{j=0}^{v} a_{d,\omega}(v-j)S_{d,\omega}(j)y, \quad v \in \mathbb{N}_0,$$

where we have used the change of variables in the (double) integral, the binomial theorem and the Fubini theorem. By the foregoing, it follows that $(S_{d,\omega}(v))_{v\in\mathbb{N}_0}$ is a discrete $(a_{d,\omega}, k_{d,\omega})$-regularized C-resolvent family with subgenerator $\mathcal{A}$. Using a similar argumentation and the closedness of $\mathcal{A}$, we can prove that the assumption $R(t)x - k(t)Cx \in \mathcal{A}\int_0^t a(t-s)R(s)x\,ds$ for all $x \in X$ and $t \geqslant 0$ implies that the equation (111) below holds with $a(\cdot)$ and $k(\cdot)$ replaced therein with $a_{d,\omega}(\cdot)$ and $k_{d,\omega}(\cdot)$, respectively. Suppose, finally, that $\int_0^{+\infty}\|R(t)\|\,dt < +\infty$. Then we have

$$\sum_{v=0}^{+\infty}\|S_{d,\omega}(v)\| \leqslant M\sum_{v=0}^{+\infty}\int_0^{+\infty} e^{-\omega t}\frac{(\omega t)^v}{v!}\|R(t)\|\,dt$$

$$= M\int_0^{+\infty} e^{-\omega t}\sum_{v=0}^{+\infty}\frac{(\omega t)^v}{v!}\|R(t)\|\,dt = M\int_0^{+\infty}\|R(t)\|\,dt < +\infty.$$

This completes the proof. □

We continue by providing the following useful observation.

Remark 3.2.3. Suppose that $\mathcal{A}$ is a closed subgenerator of a local (a,k)-regularized C-resolvent family $(R(t))_{t\in[0,\tau)}$, $0 < \tau_0 < \tau$ and $\omega > 0$. Define

$$S_{d,\omega,\tau_0}(v)x := \int_0^{\tau_0} e^{-\omega t}\frac{(\omega t)^v}{v!}R(t)x\,dt, \quad v \in \mathbb{N}_0,\ x \in X,$$

$$a_{d,\omega,\tau_0}(v) := \int_0^{\tau_0} e^{-\omega t}\frac{(\omega t)^v}{v!}a(t)\,dt \quad\text{and}\quad k_{d,\omega,\tau_0}(v) := \int_0^{\tau_0} e^{-\omega t}\frac{(\omega t)^v}{v!}k(t)\,dt, \quad v \in \mathbb{N}_0.$$

In the local situation, it is not clear whether the operator family $(S_{d,\omega,\tau_0}(v))_{v\in\mathbb{N}_0}$ is a discrete $(a_{d,\omega,\tau_0}, k_{d,\omega,\tau_0})$-regularized C-resolvent family with subgenerator $\mathcal{A}$; let us briefly explain the main problem here. In the third line of the long computation given on [3, p. 12], we can change the order of integration and obtain, with our new notation used, that the corresponding term looks like (for simplicity, we assume here that $\mathcal{A} = A$ is single-valued)

$$k_{d,\omega,\tau_0}(v)Cx + \mathcal{A}\int_0^{\tau_0}\int_s^{\tau_0} e^{-\omega t}\frac{(\omega t)^v}{v!}a(t-s)R(s)x\,dt\,ds,$$

as well as that the corresponding term on [3, p. 12, l. -2] looks like

$$k_{d,\omega,\tau_0}(v)Cx + \mathcal{A}\int_0^{\tau_0} e^{-\omega s}\left[\sum_{j=0}^{v}\frac{(\omega s)^j}{j!}\int_0^{\tau_0-s} e^{-\omega\tau}\frac{(\omega\tau)^{v-j}}{(v-j)!}a(\tau)\,d\tau\right]R(s)x\,ds.$$

But the integral

$$\int_0^{\tau_0-s} e^{-\omega\tau}\frac{(\omega\tau)^{v-j}}{(v-j)!}a(\tau)\,d\tau$$

depends on s in the local setting and differs from the integral

$$\int_0^{+\infty} e^{-\omega\tau}\frac{(\omega\tau)^{v-j}}{(v-j)!}a(\tau)\,d\tau$$

in the global setting. This is the main reason why the further summation and the corresponding result cannot be given in the local setting.

The following results can be deduced by means of the argumentation contained in the proofs of [491, Proposition 2.4, Proposition 2.5].

Proposition 3.2.4. (i) *Suppose that $a(v) \neq 0$ for all $v \in \mathbb{N}_0$ and $\mathcal{A}$ is a subgenerator of a discrete (a,k)-regularized C-resolvent family $(S(v))_{v\in\mathbb{N}_0}$. Then $S(v)x = 0$ for all $x \in \mathcal{A}0$ and $v \in \mathbb{N}_0$.*

(ii) *Suppose that $\mathcal{A}$ is a subgenerator of a discrete (a,C)-resolvent family $(S(v))_{v\in\mathbb{N}_0}$. If $\lambda \in \rho_C(\mathcal{A})$, then we have $C\sum_{j=0}^{v} a(v-j)S(j)x \in D(\mathcal{A})$ for all $x \in X$ and $v \in \mathbb{N}_0$, as well as*

$$S(v)Cx - k(v)C^2x \in C^{-1}\mathcal{A}0 + \mathcal{A}C\sum_{j=0}^{v} a(v-j)S(j)x, \quad v \in \mathbb{N}_0,\ x \in X.$$

In particular, if $\mathcal{A} = A$ is single-valued, then

$$S(v)x - k(v)Cx = C^{-1}AC\sum_{j=0}^{v} a(v-j)S(j)x, \quad v \in \mathbb{N}_0,\ x \in X.$$

Further on, we would like to notice the following facts:

(i) Suppose that $a_0k_0 \neq 0$ and $\mathcal{A}$ is a subgenerator of a discrete (a,k)-regularized C-resolvent family $(S(v))_{v\in\mathbb{N}_0}$ satisfying the equation (111) below. Then $1/a_0 \in \rho_C(\mathcal{A})$. In order to see that, observe first that the equation (111) with $v = 0$ gives $S(0)x \in$

$k(0)Cx + \mathcal{A}[a_0 S(0)x]$, which further implies that $k_0 a_0^{-1}Cx \in (a_0^{-1} - \mathcal{A})S(0)x$ and $(a_0^{-1} - \mathcal{A})^{-1}Cx \in k_0 a_0^{-1}S(0)x$ $(x \in X)$. It remains to be proved that the operator $(a_0^{-1} - \mathcal{A})^{-1}C$ is single-valued. Assume that $y \in (a_0^{-1} - \mathcal{A})^{-1}Cx$. Then we have $(y, a_0^{-1}y - Cx) \in \mathcal{A}$ and, due to (109), we get $S(0)y = k_0 Cy + a_0 S(0)[a_0^{-1}y - Cx]$. This yields $k_0 Cy = a_0 S(0)Cx$ and $y = a_0 k_0^{-1} S(0)x$.

(ii) If $a_0 \neq 0$ and $1/a_0 \in \rho_C(\mathcal{A})$, then there exists a unique discrete (a, k)-regularized C-resolvent family $(S(v))_{v\in\mathbb{N}_0}$, which satisfies (111) and which is subgenerated by $\mathcal{A}$ (cf. also [429, Proposition 3.5]). Indeed, the equation (111) with $v = 0$ implies $S(0)x = a_0^{-1}k_0(a_0^{-1} - \mathcal{A})^{-1}Cx$ for all $x \in X$. If the operator family $(R(v))_{v\in\mathbb{N}_0}$ is a discrete (a, k)-regularized C-resolvent family subgenerated by $\mathcal{A}$ and if the equation (111) holds with $S(\cdot)$ replaced therein with $R(\cdot)$, then we define $T(v)x := S(v)x - T(v)x$, $v \in \mathbb{N}_0$, $x \in X$. By the foregoing, we have $T(0)x = 0$, $x \in X$. Proceeding by induction, we get $T(v)x \in \mathcal{A}[a_0 T(v)x]$, $v \in \mathbb{N}_0$, $x \in X$. Therefore, we have $T(v)x = (a_0^{-1} - \mathcal{A})^{-1}C0 = 0$, $v \in \mathbb{N}_0$, $x \in X$ and $S(v)x = T(v)x$, $v \in \mathbb{N}_0$, $x \in X$.

Let us define now the sequence $(\phi_j(v))$ in the same way as in the formulation of [429, Theorem 3.8]. Our second main result reads as follows.

Theorem 3.2.5. (i) *Suppose that $a_0 a_1 k_0 \neq 0$ and $\mathcal{A}$ is the integral generator of a discrete (a, k)-regularized C-resolvent family $(S(v))_{v\in\mathbb{N}_0}$ satisfying the following functional equality:*

$$S(v)x \in k(v)Cx + \mathcal{A}\sum_{j=0}^{v} a(v-j)S(j)x, \quad x \in X,\ v \in \mathbb{N}_0. \tag{111}$$

Then $C^{-j}[(1-\mathcal{A})^{-1}C]^{v+1} = (1-\mathcal{A})^{-v-1}C \in L(X)$ for all $v \in \mathbb{N}_0$, $S(0) = k_0(1 - a_0\mathcal{A})^{-1}C$ and $(S(v))_{v\in\mathbb{N}_0}$ is given recursively by the equation (112) below. Furthermore, $(1 - a_0\mathcal{A})^{-m}S(v) \in L(X)$ for all m, $v \in \mathbb{N}_0$ and the equation (113) below holds.

(ii) *Suppose that $a_0 \neq 0$, $\mathcal{A}$ is a closed MLO and $(1-a_0\mathcal{A})^{-m}C \in L(X)$ for all $m \in \mathbb{N}$. Define $S(0) := k_0(1 - a_0\mathcal{A})^{-1}C$ and $(S(v))_{v\in\mathbb{N}_0}$ recursively by*

$$S(v)x := k_v a_0^{-1}(a_0^{-1} - \mathcal{A})^{-1}Cx + \sum_{j=0}^{v-1} a(v-j)[-a_0^{-1}S(j)x + a_0^{-2}(a_0^{-1} - \mathcal{A})^{-1}S(j)x], \tag{112}$$

for any $v \in \mathbb{N}$ and $x \in X$. Then the sequence $(S(v))_{v\in\mathbb{N}_0} \subseteq L(X)$ is well-defined and satisfies that $(a_0^{-1} - \mathcal{A})^{-m}S(v) \in L(X)$ for all m, $v \in \mathbb{N}_0$; furthermore, $(S(v))_{v\in\mathbb{N}_0}$ is a unique discrete (a, k)-regularized C-resolvent family with subgenerator $\mathcal{A}$,

$$X_j := C^{-j-1}[a_0^{-1}(1 - a_0\mathcal{A})^{-1}C - C]^j[(1 - a_0\mathcal{A})^{-1}C]^2 \in L(X), \quad j \in \mathbb{N}$$

and the assumption $a(\cdot) \equiv k(\cdot)$ implies

$$S(v) := \sum_{j=1}^{v} \phi_j(v)X_j, \quad v \in \mathbb{N} \setminus \{1\}.$$

In the general case, we have

$$S(v)x = \sum_{j=1}^{v+1} \beta_v(j)(a_0^{-1} - \mathcal{A})^{-j}Cx, \quad v \in \mathbb{N}_0,\ x \in X, \tag{113}$$

where $\beta_0(1) := k_0/a_0$,

$$\beta_v(1) := \frac{k_v}{a_0} - a_0^{-1}\sum_{l=0}^{v-1} a(v-l)\beta_l(1), \quad v \in \mathbb{N}, \tag{114}$$

$$\beta_v(j) := -a_0^{-1}\sum_{l=j}^{v-1} a(v-l)\beta_l(j) + a_0^{-2}\sum_{l=j-1}^{v-1} a(v-l)\beta_l(j-1), \quad v \in \mathbb{N},\ 2 \leqslant j \leqslant v \tag{115}$$

and

$$\beta_v(v+1) := a_0^{-2}a_1\beta_{v-1}(v), \quad v \in \mathbb{N}. \tag{116}$$

(iii) *Suppose that* $a_0 \neq 0$,

$$\sum_{v=1}^{+\infty} |a_v| < |a_0|, \quad \sum_{v=1}^{+\infty} |k_v| < +\infty, \tag{117}$$

$\mathcal{A}$ is a closed MLO *and the mapping* $\lambda \mapsto (\lambda - \mathcal{A})^{-1}C \in L(X)$, $\lambda \in \{z \in \mathbb{C} : |a_0^{-1} - z| < |a_0|^{-1}\}$ *is continuous. Then* $(1 - a_0\mathcal{A})^{-v}C \in L(X)$ *for all* $v \in \mathbb{N}$ *and the discrete* (a,k)*-regularized C-resolvent family* $(S(v))_{v\in\mathbb{N}_0}$*, defined in part* (ii)*, satisfies* $\sum_{v=0}^{+\infty} \|S(v)\| < +\infty$.

Proof. The items (i) and (ii) can be deduced with the help of the argumentation contained in the proofs of [429, Theorem 3.8] and [491, Theorem 2.7(i)–(ii)]; we will only outline here the main details of the proof of (i). Using our assumptions and (111), we can simply show that $S(0) = k_0(1 - a_0\mathcal{A})^{-1}C$ and

$$CS(v)x := k_v a_0^{-1}(a_0^{-1} - \mathcal{A})^{-1}C^2x + \sum_{j=0}^{v-1} a(v-j)\big[-a_0^{-1}S(j)Cx + a_0^{-2}(a_0^{-1} - \mathcal{A})^{-1}S(j)Cx\big], \tag{118}$$

for any $v \in \mathbb{N}$ and $x \in X$. This implies that the formulae $C^{-v}[(1 - \mathcal{A})^{-1}C]^{v+1} = (1 - \mathcal{A})^{-v-1}C \in L(X)$ and (112) hold with $v = 0$. Now we proceed by induction; assume that these formulae hold for any natural number strictly less than $v \geqslant 1$ and let us prove that these formulae hold for v. Our hypothesis imply together with (118) that

$$\begin{aligned} CS(v)x := k_v a_0^{-1}(a_0^{-1} - \mathcal{A})^{-1}C^2x + \sum_{j=0}^{v-1} a(v-j)\Bigg[-a_0^{-1}\sum_{l=1}^{j+1}\beta_j(l)(a_0^{-1} - \mathcal{A})^{-l}C^2x \\ + a_0^{-2}(a_0^{-1} - \mathcal{A})^{-1}\sum_{l=1}^{j+1}\beta_j(l)(a_0^{-1} - \mathcal{A})^{-l}C^2x\Bigg], \quad x \in X. \end{aligned} \tag{119}$$

Since $a_0 a_1 k_0 \neq 0, \beta_0(1) = k_0/a_0$ and (116) holds, we have $\beta_j(j+1) \neq 0$ for all $j \in \mathbb{N}$. Keeping in mind (119), we simply get that $C^{-1}(a_0^{-1} - \mathcal{A})^{-v-1}C^2 \in L(X)$. But we have $C^{-1}\mathcal{A}C = \mathcal{A}$, which simply implies $C^{-1}(a_0^{-1} - \mathcal{A})C = a_0^{-1} - \mathcal{A}$ and, by taking the inverse, $C^{-1}(a_0^{-1} - \mathcal{A})^{-1}C = (a_0^{-1} - \mathcal{A})^{-1}$. Keeping in mind that $(a_0^{-1} - \mathcal{A})^{-v}C \in L(X)$, the above yields

$$\begin{aligned} C^{-1}(a_0^{-1} - \mathcal{A})^{-v-1}C^2 &= [C^{-1}(a_0^{-1} - \mathcal{A})^{-1}C] \cdot (a_0^{-1} - \mathcal{A})^{-v}C \\ &= (a_0^{-1} - \mathcal{A})^{-1} \cdot (a_0^{-1} - \mathcal{A})^{-v}C = (a_0^{-1} - \mathcal{A})^{-v-1}C \in L(X). \end{aligned}$$

This implies $C^{-v}[(1 - \mathcal{A})^{-1}C]^{v+1} = (1 - \mathcal{A})^{-v-1}C \in L(X)$ since $(1 - \mathcal{A})^{-v-1}C \subseteq C^{-v}[(1 - \mathcal{A})^{-1}C]^{v+1}$ and $C^{-v}[(1 - \mathcal{A})^{-1}C]^{v+1}$ is single-valued. As a consequence, we have that the equation (113) holds, which completes the induction step. After that, we can simply prove that $(S(v))_{v \in \mathbb{N}_0}$ is given recursively by (112) as well as that $(1 - a_0\mathcal{A})^{-m}S(v) \in L(X)$ for all $m, v \in \mathbb{N}_0$.

Now we will prove (iii). Due to (117), there exists $r \in (c, |a_0|^{-1})$, where $c := \sum_{v=1}^{+\infty} |a_v|/|a_0^2|$. Arguing as in the proof of [491, Theorem 2.7(iii)], it follows that the mapping $\lambda \mapsto (\lambda - \mathcal{A})^{-1}C \in L(X)$, $\lambda \in \{z \in \mathbb{C} : |a_0^{-1} - z| < |a_0|^{-1}\}$ is analytic as well as that there exists a finite real constant $M_r \geqslant 1$ such that

$$\|(a_0^{-1} - \mathcal{A})^{-j}C\| \leqslant \frac{M_r}{r^j}, \quad j \in \mathbb{N}. \tag{120}$$

On the other hand, we have

$$\begin{aligned} \sum_{v=0}^{+\infty} \|S(v)\| &\leqslant \sum_{v=0}^{+\infty} \sum_{j=1}^{v+1} |\beta_v(j)| \cdot \|(a_0^{-1} - \mathcal{A})^{-j}C\| \\ &= \sum_{j=1}^{+\infty} \left[\sum_{v=j-1}^{+\infty} |\beta_v(j)| \right] \cdot \|(a_0^{-1} - \mathcal{A})^{-j}C\| \\ &:= \sum_{j=1}^{+\infty} m_j \cdot \|(a_0^{-1} - \mathcal{A})^{-j}C\|. \end{aligned} \tag{121}$$

Implementing (114), we get

$$\begin{aligned} \sum_{v=1}^{+\infty} |\beta_v(1)| &\leqslant \frac{1}{|a_0|} \sum_{v=1}^{+\infty} |k_v| + \frac{1}{|a_0|} \sum_{v=1}^{+\infty} \sum_{l=0}^{v-1} |a_{v-l}| \cdot |\beta_l(1)| \\ &= \frac{1}{|a_0|} \sum_{v=1}^{+\infty} |k_v| + \frac{1}{|a_0|} \sum_{v=0}^{+\infty} \left[\sum_{j=1}^{+\infty} |a_j| \right] \cdot |\beta_v(1)|. \end{aligned}$$

Since $\beta_0(1) := k_0/a_0$ and (117) holds, this implies

$$m_1 \leqslant |k_0/a_0| + \frac{|a_0|}{|a_0| - \sum_{v=1}^{+\infty} [|a_v| + |k_v|]} \left[|k_0/a_0^2| \cdot \sum_{v=1}^{+\infty} |a_v| + |1/a_0| \cdot \sum_{v=1}^{+\infty} |k_v| \right].$$

Let $j \geqslant 2$. Then (116) implies $\beta_{j-1}(j) = (a_1/a_0^2)^{j-1} \cdot (k_0/a_0)$; furthermore, we can use (115) in order to see that

$$\begin{aligned}
m_j &= \left|(a_1/a_0^2)^{j-1} \cdot (k_0/a_0)\right| + \sum_{v=j}^{+\infty} |\beta_v(j)| \\
&\leqslant \frac{|k_0|}{|a_0|} \frac{|a_1|^{j-1}}{|a_0|^{2j-2}} + \sum_{v=j}^{+\infty} \left\{ \frac{1}{|a_0|} \sum_{l=j}^{v-1} |a_{v-l}| \cdot |\beta_l(j)| + \frac{1}{|a_0|^2} \sum_{l=j-1}^{v-1} |a_{v-l}| \cdot |\beta_l(j-1)| \right\} \\
&\leqslant \frac{|k_0|}{|a_0|} \frac{|a_1|^{j-1}}{|a_0|^{2j-2}} + \left[\frac{1}{|a_0|} \sum_{v=1}^{+\infty} |a_v| \right] \cdot |\beta_j(j)| + \left[\frac{1}{|a_0^2|} \sum_{v=1}^{+\infty} |a_v| \right] \\
&\quad \times \left[|\beta_{j-1}(j-1)| + |\beta_j(j-1)| + |\beta_{j+1}(j-1)| + \cdots \right] \\
&\leqslant \frac{|k_0|}{|a_0|} \frac{|a_1|^{j-1}}{|a_0|^{2j-2}} + \left[\frac{1}{|a_0|} \sum_{v=1}^{+\infty} |a_v| \right] \cdot |\beta_j(j)| + \left[\frac{1}{|a_0^2|} \sum_{v=1}^{+\infty} |a_v| \right] \cdot m_{j-1} \\
&= \frac{|k_0|}{|a_0|} \frac{|a_1|^{j-1}}{|a_0|^{2j-2}} + \left[\frac{1}{|a_0|} \sum_{v=1}^{+\infty} |a_v| \right] \cdot \left| \frac{a_1}{a_0^2} \right|^j \cdot |\beta_1(1)| + \left[\frac{1}{|a_0^2|} \sum_{v=1}^{+\infty} |a_v| \right] \cdot m_{j-1} \\
&:= x_{j-1} + c \cdot m_{j-1}.
\end{aligned}$$

Since $x_j = |a_1/a_0^2| x_{j-1}$ and $c \geqslant |a_1/a_0^2|$, the above yields

$$m_j \leqslant c^{j-2} x_1 + c^{j-1} m_1,$$

so that (cf. (121)):

$$\sum_{v=0}^{+\infty} \|S(v)\| \leqslant m_1 \left\|(a_0^{-1} - \mathcal{A})^{-1} C\right\| + \sum_{j=2}^{+\infty} \left[c^{j-2} x_1 + c^{j-1} m_1\right] \cdot \left\|(a_0^{-1} - \mathcal{A})^{-j} C\right\|.$$

Keeping in mind this estimate, the final conclusion simply follows from (120), since $r > c$. □

Concerning the statement of [429, Theorem 3.9], we will only emphasize the following (we will use the same notation as in [429]):

(i) It seems very plausible that this statement can be reformulated if $\mathcal{A}$ is a closed MLO, which is a subgenerator of a bounded, analytic C-regularized semigroup; cf. also [445, pp. 316–317] for some observations given in the degenerate setting.
(ii) Let $k,\ g \in L^1_{loc}([0,\infty))$, let $k \neq 0$ and let $k,\ g \geqslant 0$. We relax first the assumption $g \in W^{1,1}([0,\infty))$ here by the assumption $g \in L^1([0,\infty))$. Then the second condition in (117) holds with the sequence $k(\cdot)$ replaced therein by the sequence $b(\cdot)$ since

$$\sum_{v=1}^{\infty} b_v = \sum_{v=1}^{\infty} \int_0^{+\infty} e^{-t} \frac{t^v}{v!} g(t)\,dt \leqslant \int_0^{+\infty} e^{-t}(e^t - 1) g(t)\,dt < \int_0^{+\infty} g(t)\,dt < +\infty.$$

In place of all restrictive conditions on the function $k(t)$, we only assume here that $k \in L^1([0,\infty))$ and

$$\int_0^{+\infty} k(t)\,dt < 2\int_0^{+\infty} e^{-t}k(t)\,dt.$$

Using a similar computation as above, it can be simply shown that the first condition in (117) also holds. Hence, we can apply Theorem 3.2.5(iii) in order to see that the existence and continuity of mapping $\lambda \mapsto (\lambda - \mathcal{A})^{-1}C \in L(X)$, $\lambda \in \{z \in \mathbb{C} : |a_0^{-1} - z| < a_0^{-1}\}$ implies that $\mathcal{A}$ is a subgenerator of a discrete (a,k)-regularized C-resolvent family $(S(v))_{v\in\mathbb{N}_0}$ which satisfies $\sum_{v=0}^{+\infty}\|S(v)\| < +\infty$. In this case, the conditions that $\mathcal{A}$ generates a bounded, analytic strongly continuous semigroup and $0 \in \rho(\mathcal{A})$ seem to be very redundant.

3.2.2 The discrete abstract Cauchy problem

In this subsection, we consider the following discrete abstract Cauchy problem:

$$u(v) \in f(v) + \mathcal{A}(a *_0 u)(v), \quad v \in \mathbb{N}_0, \tag{122}$$

where $\mathcal{A}$ is an MLO in X, $(a(v))_{v\in\mathbb{N}_0}$ and $(f(v))_{v\in\mathbb{N}_0}$ are given sequences in X. This is a special case of the problem

$$\mathcal{B}u(v) \in \mathcal{F}(v) + \mathcal{A}(a *_0 u)(v), \quad v \in \mathbb{N}_0, \tag{123}$$

where $\mathcal{A}$ and $\mathcal{B}$ are given MLOs in X, $(a(v))_{v\in\mathbb{N}_0}$ is a given sequence in X and $(\mathcal{F}(v))_{v\in\mathbb{N}_0}$ is a given sequence in $P(X)$. The concept of (pre)solution of (123) and the concept of strong solution of (123) can be introduced in the same manner as in [445, Definition 3.1.1(i)], where we have considered the corresponding continuous analogue. For simplicity, we will consider here the problem (122); we say that a sequence $u : \mathbb{N}_0 \to X$ is:

(i) A mild solution of (122) if $(a *_0 u)(v) \in D(\mathcal{A})$ for all $v \in \mathbb{N}_0$ and (122) holds.
(ii) A strong solution of (122) if there exists a sequence $u_{\mathcal{A}} : \mathbb{N}_0 \to X$ such that $u_{\mathcal{A}}(v) \in \mathcal{A}u(v)$ for all $v \in \mathbb{N}_0$ and

$$u(v) = f(v) + (a *_0 u_{\mathcal{A}})(v), \quad v \in \mathbb{N}_0.$$

It is clear that any strong solution of (122) is a mild solution of (122) as well as that the converse statement is not true, in general. Immediately from the above definition, we have the following.

Proposition 3.2.6. *Suppose that $\mathcal{A}$ is a closed subgenerator of a discrete (a,k)-regularized C-resolvent family $(S(v))_{v\in\mathbb{N}_0}$ satisfying* (111), *and $g : \mathbb{N}_0 \to X$ is a given sequence. Define $u(v) := (S *_0 g)(v)x$, $x \in X$, $v \in \mathbb{N}_0$. Then the following holds:*

(i) *For every* $x \in X$, *the sequence* $u(\cdot)$ *is a mild solution of* (122) *with* $f(v) := (k *_0 g)(v)Cx$, $v \in \mathbb{N}_0$.
(ii) *For every* $x \in D(\mathcal{A})$, *the sequence* $u(\cdot)$ *is a strong solution of* (122) *with* $f(v) := (k *_0 g)(v)Cx$, $v \in \mathbb{N}_0$.

Concerning the uniqueness of solutions of (122), we will state the following result.

Proposition 3.2.7. (i) *Suppose that* $a_0 = 0$. *Then there exists at most one mild (strong) solution of* (122) *if and only if* $\mathcal{A} = A$ *is single-valued.*
(ii) *Suppose that* $a_0 \neq 0$ *and there exists an injective operator* $C \in L(X)$ *which commutes with* $\mathcal{A}$ *and satisfies* $a_0^{-1} \in \rho_C(\mathcal{A})$. *Then there exists at most one mild (strong) solution of* (122).

Proof. The proof of (i) is almost trivial and therefore omitted. Suppose now that the requirements of (ii) hold; it suffices to show then that there exists at most one mild solution of (122). In actual fact, if $u(\cdot)$ is a mild solution of (122), then we have $Cu(0) = (1 - a_0\mathcal{A})^{-1}Cf(0)$; furthermore, a simple computation tells us that

$$Cu(v) \in (1 - a_0\mathcal{A})^{-1}Cf(v) + (1 - a_0\mathcal{A})^{-1}C\mathcal{A}\big[a(v)u(0) + \cdots + a(1)u(v-1)\big], \quad v \in \mathbb{N},$$

so that $Cu(v)$ is uniquely determined by Lemma 1.1.2. The final conclusion follows from the injectivity of C. □

Now we will state without a proof the following discrete analogue of [445, Proposition 3.2.8(ii)] (the interested reader is advised to state a discrete analogue of [445, Theorem 3.2.9(ii)], as well):

Proposition 3.2.8. *Suppose that* $\mathcal{A}$ *is a closed subgenerator of a discrete* (a,k)*-regularized* C*-resolvent family* $(S(v))_{v\in\mathbb{N}_0}$. *Then any strong solution* $u(\cdot)$ *of* (122) *satisfies* $(S *_0 g)(v) = (kC *_0 u)(v)$ *for all* $v \in \mathbb{N}_0$.

3.2.3 Almost periodic type solutions to abstract Volterra difference inclusions

In this subsection, we will first provide the basic details about the existence and uniqueness of asymptotically almost periodic type solutions and asymptotically almost automorphic type solutions to (122). Our basic assumption will be that $a_0 \neq 0$, (117) holds, $\mathcal{A}$ is a closed MLO and the mapping $\lambda \mapsto (\lambda - \mathcal{A})^{-1}C \in L(X)$, $\lambda \in \{z \in \mathbb{C} : |a_0^{-1} - z| < |a_0|^{-1}\}$ is continuous. Then Theorem 3.2.5(iii) shows that there exists a unique discrete (a,k)-regularized C-resolvent family $(S(v))_{v\in\mathbb{N}_0}$ subgenerated by $\mathcal{A}$ as well as that $\sum_{v=0}^{+\infty} \|S(v)\| < +\infty$. As is well known, the summability of $(S(v))_{v\in\mathbb{N}_0}$ is indispensable for studying the existence and uniqueness of asymptotically almost periodic type solutions and asymptotically almost automorphic type solutions to (122).

For example, suppose that there exist a bounded sequence $h : \mathbb{Z} \to X$ and a sequence $q : \mathbb{N} \to X$ vanishing at plus infinity such that $g(v) = h(v) + q(v)$ for all $v \in \mathbb{N}_0$. We already know that, for every $x \in X$, the function $u(v) := (S *_0 g)(v)x$, $x \in X$, $v \in \mathbb{N}_0$ is a mild solution of (122) with $f(v) := (k *_0 g)(v)Cx$, $v \in \mathbb{N}_0$. Furthermore, we have

$$u(v) = \sum_{k=-\infty}^{v} S(v-k)h(k) - \sum_{k=-\infty}^{-1} S(v-k)h(k) + \sum_{k=0}^{\lfloor v/2 \rfloor} S(v-k)q(k) + \sum_{k=\lfloor v/2 \rfloor}^{v} S(v-k)q(k), \quad v \in \mathbb{N}_0,$$

as well as

$$\left\| \sum_{k=-\infty}^{-1} S(v-k)h(k) \right\| \leqslant \|g\|_{\infty} \sum_{w=v+1}^{+\infty} \|S(w)\| \to 0, \quad v \to +\infty$$

and

$$\left\| \sum_{k=0}^{\lfloor v/2 \rfloor} S(v-k)q(k) \right\| \leqslant \|q\|_{\infty} \sum_{w=\lfloor v/2 \rfloor}^{+\infty} \|S(w)\| \to 0, \quad v \to +\infty.$$

Furthermore, for any $\varepsilon > 0$ we can find $v_0 \in \mathbb{N}$ such that $\|q(v)\| \leqslant \varepsilon$ for $v \geqslant v_0$; if this is the case, then we have

$$\left\| \sum_{k=\lfloor v/2 \rfloor}^{v} S(v-k)q(k) \right\| \leqslant \varepsilon \sum_{k=\lfloor v/2 \rfloor}^{v} \|S(v-k)\| \leqslant \varepsilon/2, \quad v \geqslant v_1,$$

for some integer $v_1 \geqslant v_0$. The function $H(v) := \sum_{k=-\infty}^{v} S(v-k)h(k)$, $v \in \mathbb{Z}$ is bounded and T-almost periodic, where $T \in L(X)$ commutes with $S(\cdot)$, $\mathcal{A}$ and C, or bounded and R-almost automorphic, where R is any collection of sequences in $\mathbb{Z}$, provided that the forcing term $h(\cdot)$ enjoys the same features; therefore, we have the following:

(i) If $h(\cdot)$ is T-almost periodic, where $T \in L(X)$ commutes with $S(\cdot)$, $\mathcal{A}$ and C, then there exist a bounded, T-almost periodic sequence $H : \mathbb{Z} \to X$ and a sequence $Q : \mathbb{N} \to X$ vanishing at plus infinity such that $u(v) = H(v) + Q(v)$ for all $v \in \mathbb{N}_0$.
(ii) If $h(\cdot)$ is R-almost automorphic, then there exist a bounded, R-almost automorphic sequence $H : \mathbb{Z} \to X$ and a sequence $Q : \mathbb{N} \to X$ vanishing at plus infinity such that $u(v) = H(v) + Q(v)$ for all $v \in \mathbb{N}_0$.

Suppose now that $\mathcal{A}$ is a closed subgenerator of an exponentially bounded (a,k)-regularized C-resolvent family $(R(t))_{t\geqslant 0}$, $\int_0^{+\infty} \|R(t)\|\, dt < +\infty$ and there exist real numbers $M \geqslant 1$ and $\omega_0 \geqslant 0$ such that $\|R(t)\| + |a(t)| + |k(t)| \leqslant Me^{\omega_0 t}$, $t \geqslant 0$. If $\omega > \omega_0$, then we define $(S_{d,\omega}(v))_{v\in\mathbb{N}_0}$ by (110), as well as $a_{d,\omega}(\cdot)$ and $k_{d,\omega}(\cdot)$ as in the formulation of Theorem 3.2.2. Then $(S_{d,\omega}(v))_{v\in\mathbb{N}_0}$, given by (110), is a summable discrete $(a_{d,\omega}, k_{d,\omega})$-regularized C-resolvent family with subgenerator $\mathcal{A}$. Due to Proposition 3.2.6, we have that the sequence $u(v) := (S_{d,\omega} *_0 g)(v)x$, $v \in \mathbb{N}_0$ ($x \in X$ is fixed) is a mild solution of (122) with $f(v) := (k_{d,\omega} *_0 g)(v)Cx$, $v \in \mathbb{N}_0$. If $g(\cdot)$ has a certain asymptotically almost periodic (automorphic) behavior, then the solution $u(\cdot)$ of (122) will have the same asymptotically almost periodic (automorphic) behavior, roughly speaking.

Consider now the abstract Volterra difference inclusion:

$$u(k+1) \in \mathcal{A} \sum_{j=-\infty}^{k} a(k-j)u(j+1) + \sum_{j=-\infty}^{k} b(k-j)Cf(k), \quad k \in \mathbb{Z}. \tag{124}$$

It is said that a sequence $u : \mathbb{Z} \to X$ is:

(i) A mild solution of (124) if $u \in l^1(\mathbb{Z} : X)$, $\sum_{j=-\infty}^{k} a(k-j)u(j+1) \in D(\mathcal{A})$ for all $k \in \mathbb{Z}$ and (124) holds.

(ii) A strong solution of (124) if $u \in l^1(\mathbb{Z} : X)$ and there exists a sequence $u_{\mathcal{A}} : \mathbb{Z} \to X$ such that $u_{\mathcal{A}} \in l^1(\mathbb{Z} : X)$, $u_{\mathcal{A}}(k) \in \mathcal{A}u(k)$ for all $k \in \mathbb{Z}$ and

$$u(k+1) = \sum_{j=-\infty}^{k} a(k-j)u_{\mathcal{A}}(j+1) + \sum_{j=-\infty}^{k} b(k-j)Cf(k), \quad k \in \mathbb{Z}.$$

Any strong solution of (124) is a mild solution of (124), while the converse statement is not true in general. Arguing similarly as in the proof of [429, Theorem 3.13], we may deduce the following result.

Theorem 3.2.9. *Suppose that $\mathcal{A}$ is a closed subgenerator of a discrete (a,k)-regularized C-resolvent family and $\sum_{v=0}^{+\infty} \|S(v)\| < +\infty$. Then the following holds:*

(i) *If $f \in l^1(\mathbb{Z} : X)$ and* (111) *holds, then the function $u(\cdot)$, defined by*

$$u(k) := \sum_{j=-\infty}^{k-1} S(k-1-j)f(j), \quad k \in \mathbb{Z}, \tag{125}$$

is a mild solution of (124).

(ii) *If $f \in l^1(\mathbb{Z} : X)$ and there exists a sequence $f_{\mathcal{A}} : \mathbb{Z} \to X$ such that $f_{\mathcal{A}} \in l^1(\mathbb{Z} : X)$ and $f_{\mathcal{A}}(k) \in \mathcal{A}f(k)$ for all $k \in \mathbb{Z}$, then the function $u(\cdot)$, given by* (125), *is a strong solution of* (124).

By the foregoing, we have that the function $u(\cdot)$, given by (125), is bounded and T-almost periodic, where $T \in L(X)$ commutes with $S(\cdot)$, $\mathcal{A}$ and C, respectively, bounded and R-almost automorphic, provided that the forcing term $f(\cdot)$ enjoys the same features.

Before proceeding further, we would like to mention that R. Grau and A. Pereira have recently investigated the class of (a,k)-regularized resolvent families on time scales using the vector-valued Laplace transform [340]. It seems very plausible that their results can be slightly extended in the degenerate setting by using the additional regularizing operator $C \neq \mathrm{I}$ in the whole analysis. We will analyze (a,k)-regularized C-resolvent families on time scales somewhere else.

3.3 Abstract multiterm fractional difference equations

Discrete fractional calculus and discrete fractional equations are rapidly growing fields of research of many authors (cf. the monographs [7] by S. Abbas et al., [59] by M. H. Annaby, Z. S. Mansour, [330] by C. Goodrich, A. C. Peterson and the research articles [3, 4, 46, 74–76, 164, 281, 254, 288, 367, 375, 491, 534, 535] for some results obtained recently in this direction). The main aim of this section is to expand and contemplate the research studies [3, 4, 375, 470, 491, 535] by investigating the abstract multiterm fractional difference equations in Banach spaces.

We consider here the abstract multiterm fractional difference equations with the Riemann–Liouville fractional derivatives and Caputo fractional derivatives, which will be denoted henceforth by Δ^{α} and Δ_C^{α}, respectively ($\alpha > 0$). If $(X, \|\cdot\|)$ is a complex Banach space, $n \in \mathbb{N}$, $0 \leqslant \alpha_1 < \cdots < \alpha_n$ and $A_1, \ldots, A_n$ are closed linear operators on X, then we consider the well-posedness of the following abstract multiterm fractional difference equation with Caputo fractional derivatives:

$$\sum_{i=1}^{n} T_i u(v + v_i) = f(v), \quad v \in \mathbb{N}_0, \tag{126}$$

where $v_1, \ldots, v_n \in \mathbb{N}_0, f : \mathbb{N}_0 \to X$ is a given sequence and, for every $v \in \mathbb{N}_0$ and $1 \leqslant i \leqslant n$, $T_i u(v)$ is the term $T_{i,L} u(v)$ or $T_{i,R} u(v)$, where $T_{i,L} u(v) := A_i \Delta_C^{\alpha_i} u(v)$, if $v \in \mathbb{N}_0$, $1 \leqslant i \leqslant n$, and $T_{i,R} u(v) := \Delta_C^{\alpha_i} A_i u(v)$, if $v \in \mathbb{N}_0, 1 \leqslant i \leqslant n$ ($\Delta_C^0 u \equiv u$). If $m_i = \lceil \alpha_i \rceil$ for $1 \leqslant i \leqslant n$, $\mathcal{I} := \{i \in \mathbb{N}_n : \alpha_i > 0$ and $T_{i,L} u(v)$ appears on the left-hand side of (150)$\}$, $Q := \max \mathcal{I}$, if $\mathcal{I} \neq \emptyset$, $Q := m_Q := v_Q := 0$, if $\mathcal{I} = \emptyset$, $\tau := \max\{m_i + v_i : i \in I\}$, if $\mathcal{I} \neq \emptyset$, and $\tau := 0$, if $\mathcal{I} = \emptyset$, then we endow (126) with the following initial conditions:

$$u(j) = u_j, \quad 0 \leqslant j \leqslant \tau - 1 \quad \text{and} \quad A_i u(j) = u_{i,j} \quad \text{if } i \notin \mathcal{I} \text{ and } v_i + m_i - 1 \geqslant j \geqslant 0. \tag{127}$$

We investigate the analogues of (126)–(127) with the Riemann–Liouville fractional derivatives as well.

In this section, we also analyze some new classes of the abstract higher-order difference equations with integer-order derivatives. In Section 3.3.1, we extend the statement of [320, Theorem 13.2.1] to the closed linear operators (cf. Theorems 3.3.1–3.3.2) by assuming that the operator $(\sum_{s=0}^{m} z^{m-s} A_s)^{-1} C$ exists at a neighborhood of infinity for a certain bounded, possibly noninjective, operator $C \in L(X)$; here, we would like to emphasize that it is not clear how we can reconsider the representation formulae established in these results if the operator $(\sum_{s=0}^{m} z^{m-s} A_s)^{-1} C$ exists on some different region of the complex plane (we follow a pure analytical approach and remove the condition $\limsup_{k\to+\infty} \|f_k\|^{1/k} < r < +\infty$ from the analysis carried out in [320]). For the first time in the existing literature, we construct here an example of a closed linear operator A on X, which is not bounded, and an injective operator $C \in L(X)$ such that

$$\|(\lambda + A)^{-1} C\| \leqslant \frac{M}{|\lambda|}, \quad |\lambda| \geqslant r,$$

for some real numbers $r > 0$ and $M \geqslant 1$; cf. the final part of Section 3.3.2 for more details. If $C = I$, then the assumptions of Theorem 3.3.1 imply that the operators $A_0, \ldots, A_{p-1}$ are bounded (cf. Proposition 3.3.5); in Section 3.3.2, we provide certain applications of Theorem 3.3.1 with $C \neq I$. Further on, in Theorem 3.3.3, we consider the abstract higher-order nonautonomous difference equations by using a matricial approach (cf. also Remark 3.3.4).

The main aim of Section 3.3.3 is to consider a class of the abstract (degenerate) higher-order difference equations by applying the Poisson-like transforms to the solutions of the corresponding abstract (degenerate) higher-order differential equations [535]. Here, we would like to point out that the Poisson transform

$$v \mapsto \int_0^{+\infty} e^{-t} \frac{t^v}{v!} u(t)\, dt, \quad v \in \mathbb{N}_0$$

cannot be defined for the negative values of v. The important novelty of this section is putting forward to consideration the following transform:

$$v \mapsto y_{a,b,c,j,\omega}(v) := \int_0^{+\infty} e^{-b(ct^{-1}+at)^j} \frac{(\omega t)^{v-\frac{1}{2}}}{\Gamma(v+\frac{1}{2})} u(t)\, dt, \quad v \in \mathbb{Z},$$

where $a \in \mathbb{R}$, $b, c, \omega \in \mathbb{R} \setminus \{0\}$ and $j \in \mathbb{N}$. The Poisson transform cannot be defined for not-exponentially bounded functions $u(\cdot)$; for this reason, we also suggest using the following transform:

$$v \mapsto \int_0^{+\infty} e^{-b(at)^j} \frac{(\omega t)^v}{v!} u(t)\, dt, \quad v \in \mathbb{N}_0,$$

where $a \in \mathbb{R}$, $b, \omega \in \mathbb{R} \setminus \{0\}$ and $j \in \mathbb{N} \setminus \{1\}$. The main results established in Section 3.3.3 are Theorem 3.3.7 and Theorem 3.3.8; here, it is worth noticing that Theorem 3.3.8 can be applied in the analysis of the existence of the almost periodic (automorphic) type solutions for certain classes of the abstract higher-order difference equations even if the Poisson transform is applied to the exponentially bounded solution operator families (we do not need exponential decaying or integrability of solution operator families in any sense).

A short review of already known definitions and results about the Riemann–Liouville and Caputo fractional difference operators are given in Section 3.3.4, where we also provide some original results, like Theorem 3.3.9 (cf. also Example 3.3.10 and Example 3.3.11).

Our main results about the abstract multiterm fractional difference equations are given in Section 3.3.5, where we analyze the abstract multiterm fractional difference equations with Caputo fractional derivatives, and Section 3.3.6, where we analyze the abstract multiterm fractional difference equations with Riemann–Liouville fractional derivatives. The main results of Section 3.3.5 are Theorem 3.3.15, Theorem 3.3.16 and Theorem 3.3.17 (Example 3.3.19 is also important since we provide here some applications of results from the theory of C-regularized semigroups of operators). We continue our work by observing that the abstract multiterm fractional difference equations with Caputo fractional derivatives or Riemann–Liouville fractional derivatives are special cases of the following abstract nonautonomous difference equation with integer-order derivatives:

$$A_{v+p,v}u(v+p)+A_{v+p-1,v}u(v+p-1)+\cdots+A_{0,v}u(0)=f(v),\quad v\in\mathbb{N}_0,$$

where $p\in\mathbb{N}$, $(A_{j,v})_{v\in\mathbb{N}_0;0\leqslant j\leqslant v+p-1}$ is a sequence of linear operators and $f(\cdot)$ is given inhomogeneity; this equation generalizes the equation (3.1) considered in [320, Section 13.3], where it has been assumed that $A_{v+p,v}=\mathrm{I}$, $A_{v+p-j,v}=-A_j(v)$ for $1\leqslant j\leqslant p$ and $A_{j,v}=0$ for $0\leqslant j\leqslant v-1$. The main result about the well-posedness of this problem is given in Theorem 3.3.20.

3.3.1 Abstract higher-order difference equations

In this subsection, we will provide several new results about the abstract higher-order difference equations. We will first extend the statement of [320, Theorem 13.2.1] to the closed linear operators (we will use a slightly different notation here; cf. also the representation formula (11), which plays an important role in our analysis).

Theorem 3.3.1. *Suppose that $p\geqslant 2$, $A_0,\dots,A_{p-1}$ are closed linear operators, $C\in L(X)$ is injective and commutes with all operators $A_0,\dots,A_{p-1}$, $r>0$ and the following conditions hold:*

(i) *$(z^p+z^{p-1}A_{p-1}+\cdots+zA_1+A_0)^{-1}C\in L(X)$ for $|z|\geqslant r$ and the mapping $z\mapsto(z^p+z^{p-1}A_{p-1}+\cdots+zA_1+A_0)^{-1}Cx$, $|z|\geqslant r$ is continuous for every $x\in X$.*
(ii) *The mapping $z\mapsto A_j(z^p+z^{p-1}A_{p-1}+\cdots+zA_1+A_0)^{-1}Cx$, $|z|\geqslant r$ is continuous for every $x\in X$ and $j\in\mathbb{N}_{p-1}^0$.*
(iii) *The mapping $z\mapsto(1+zA_{p-1}+\cdots+z^{p-1}A_1+z^pA_0)^{-1}C\in L(X)$, $0<|z|\leqslant 1/r$ is bounded and the mapping $z\mapsto z^{p-j}A_j(1+zA_{p-1}+\cdots+z^{p-1}A_1+z^pA_0)^{-1}C\in L(X)$, $0<|z|\leqslant 1/r$ is bounded for all $j\in\mathbb{N}_{p-1}$.*

Then the equation

$$u(v+p)+A_{p-1}u(v+p-1)+\cdots+A_0u(v)=Cf(v),\quad v\in\mathbb{N}_0;\quad u(v)=0,\ 0\leqslant v\leqslant p-1 \tag{128}$$

has a unique solution $(u(v))_{v\in\mathbb{N}_0}$ given by $u(v):=(G*_0f)(v)$, $v\in\mathbb{N}_0$, where $(G(v))_{v\in\mathbb{N}_0}\subseteq L(X)$ and

$$G(v)x:=\frac{1}{2\pi i}\oint_{|z|=r}z^{v-1}(z^p+z^{p-1}A_{p-1}+\cdots+zA_1+A_0)^{-1}Cx\,dz,\ x\in X,\ v\in\mathbb{N}_0. \tag{129}$$

Proof. It is clear that the sequence $(G(v))_{v\in\mathbb{N}_0}$ is well-defined in $L(X)$. Set

$$P_z^{-1}Cx := (z^p + z^{p-1}A_{p-1} + \cdots + zA_1 + A_0)^{-1}Cx, \quad |z| \geqslant r,\ x \in X.$$

Since we have assumed that (i)–(ii) hold, $C \in L(X)$ is injective and commutes with all operators $A_0, \ldots, A_{p-1}$, we can apply the Hilbert resolvent equation and the argumentation contained in the proof of [445, Lemma 2.6.3] in order to show the following:
(i.1) The mapping $z \mapsto P_z^{-1}Cx$, $|z| > r$ is analytic for every $x \in X$.
(ii.1) The mapping $z \mapsto A_jP_z^{-1}Cx$, $|z| > r$ is analytic for every $x \in X$ and $j \in \mathbb{N}_{p-1}^0$.

If $0 \leqslant v \leqslant p-1$, then we have $G(v) = 0$ and, therefore, $u(v) = 0$; more precisely, the first estimate in (iii) and an elementary argumentation shows that

$$\begin{aligned} G(v)x &= \operatorname{Res}_{z=0}\left[\frac{1}{z^2}z^{1-v}P_{1/z}^{-1}Cx\right] \\ &= \operatorname{Res}_{z=0}\left[z^{p-v-1}(1 + zA_{p-1} + \cdots + z^{p-1}A_1 + z^pA_0)^{-1}Cx\right] = 0, \quad x \in X,\ 0 \leqslant v \leqslant p-1. \end{aligned}$$

Keeping this in mind, we only need to prove that

$$\begin{aligned} &\sum_{j=0}^{v+p} G(v+p-j)f(j) + A_{p-1}\sum_{j=0}^{v+p-1} G(v+p-j-1)f(j) \\ &\quad + \cdots + A_0\sum_{j=0}^{v} G(v-j)f(j) = Cf(v), \quad v \geqslant p, \end{aligned}$$

i. e.,

$$\begin{aligned} &\sum_{j=0}^{v+p}\frac{1}{2\pi i}\oint_{|z|=r} z^{v+p-j-1}P_z^{-1}Cf(j)\,dz + \sum_{j=0}^{v+p-1}\frac{1}{2\pi i}\oint_{|z|=r} z^{v+p-j-2}A_{p-1}P_z^{-1}Cf(j)\,dz \\ &\quad + \cdots + \sum_{j=0}^{v}\frac{1}{2\pi i}\oint_{|z|=r} z^{v-j-1}A_0P_z^{-1}Cf(j)\,dz = Cf(v), \quad v \geqslant p, \end{aligned}$$

since (ii) holds and $A_0, \ldots, A_{p-1}$ are closed. Let us fix an integer $v \geqslant p$. On the left-hand side of the above equality, the summation goes from $j = 0$ to $j = v + p$; moreover, it can be simply shown that the term containing $f(j)$ for $0 \leqslant j < v$ is equal to

$$\frac{1}{2\pi i}\oint_{|z|=r} z^{v-j-1}Cf(j)\,dz = 0$$

by the Cauchy theorem. Similarly, the term containing $f(v)$ equals

$$\frac{1}{2\pi i}\oint_{|z|=r} z^{-1}Cf(v)\,dz = Cf(v)$$

by the residue theorem and it remains to be proved that the term containing $f(j)$ for $v < j \leqslant v + p$ is equal to zero. If $j = v + k$ for some integer $k \in \mathbb{N}_p$, then a simple computation shows that the above mentioned term equals

$$W = \frac{1}{2\pi i} \oint_{|z|=r} z^{v-j-1}[z^p + z^{p-1}A_{p-1} + \cdots + z^k A_k]P_z^{-1}Cf(j)\,dz.$$

Applying the second estimate in (iii) and the equality

$$W = \operatorname{Res}_{z=0}\big[z^{j-1-v}(1 + zA_{p-1} + \cdots + z^{p-k}A_k)(1 + zA_{p-1} + \cdots + z^{p-1}A_1 + z^p A_0)^{-1}Cf(j)\big],$$

obtained similarly as in the computation of value $G(v)$ for $0 \leqslant v \leqslant p-1$, it readily follows that $W = 0$. The proof of the theorem is thereby complete. □

Concerning the proof of Theorem 3.3.1, we would like to observe that the injectivity of C and its commutation with the operators $A_0, \ldots, A_{p-1}$ have been used only for proving that the condition (i), respectively, (ii), implies the condition (i.1), resp. (ii.1). Therefore, we can also state the following result.

Theorem 3.3.2. *Suppose that $p \geqslant 2$, $A_0, \ldots, A_{p-1}$ are closed linear operators, $C \in L(X)$, $r > 0$ and the following conditions hold:*

(i) $(z^p + z^{p-1}A_{p-1} + \cdots + zA_1 + A_0)^{-1}C \in L(X)$ *for* $|z| \geqslant r$ *and the mapping* $z \mapsto (z^p + z^{p-1}A_{p-1} + \cdots + zA_1 + A_0)^{-1}Cx$, $|z| \geqslant r$ *is analytic for every* $x \in X$.

(ii) *The mapping* $z \mapsto A_j(z^p + z^{p-1}A_{p-1} + \cdots + zA_1 + A_0)^{-1}Cx$, $|z| \geqslant r$ *is analytic for every* $x \in X$ *and* $j \in \mathbb{N}_{p-1}^0$.

(iii) *The condition* (iii) *from the formulation of* Theorem 3.3.1 *holds.*

Then the equation (128) *has a unique solution* $(u(v))_{v\in\mathbb{N}_0}$, *given by* $u(v) := (G *_0 f)(v)$, $v \in \mathbb{N}_0$, *where* $(G(v))_{v\in\mathbb{N}_0} \subseteq L(X)$ *is defined through* (129).

The representation formulae established in [320, Theorem 13.2.1] and Theorem 3.3.1 cannot be always applied and our second contribution, formulated only in the nonautonomous setting, reads as follows.

Theorem 3.3.3. (i) *Suppose that* $A_0(v)$ *is a linear operator* ($v \in \mathbb{N}_0$), $u_0 \in D(A_0(v)A_0(v-1)\cdot\cdots\cdot A_0(0))$ *for all* $v \in \mathbb{N}$ *and* $f(v-1-j) \in D(A_0(v-1)A_0(v-2)\cdot\cdots\cdot A_0(v-j))$ *for all* $v \in \mathbb{N}$ *and* $j \in \mathbb{N}_{v-1}$. *Then the abstract difference equation*

$$u(v+1) + A_0(v)u(v) = f(v), \quad v \in \mathbb{N}_0; \quad u(0) = u_0 \tag{130}$$

has a unique solution $(u(v))_{v\in\mathbb{N}_0}$, *which is given by*

$$u(v) := (-1)^v A_0(v)A_1(v-1)\cdot\cdots\cdot A_0(0)u_0 + \sum_{j=1}^{v-1}(-1)^j A_0(v-1)A_0(v-2)\cdot\cdots\cdot A_0(v-j)f_{v-1-j} + f_{v-1}, \quad v \in \mathbb{N}. \tag{131}$$

(ii) *Suppose that* $p \geqslant 2$, $A_0(v), \ldots, A_{p-1}(v)$ *are linear operators* ($v \in \mathbb{N}_0$),

$$A(v) := \begin{bmatrix} A_{p-1}(v) & A_{p-2}(v) & A_{p-3}(v) & \cdots & A_0(v) \\ -\mathrm{I} & 0 & & \cdots & 0 \\ 0 & -\mathrm{I} & 0 & \cdots & 0 \\ \cdot & \cdot & \cdot & \cdots & \cdot \\ 0 & 0 & 0 & \cdots -\mathrm{I} & 0 \end{bmatrix}, \quad v \in \mathbb{N}_0, \tag{132}$$

$[u_{p-1} \ \ldots \ u_0]^T \in D(A(v)A(v-1) \cdot \cdots \cdot A(0))$ *for all* $v \in \mathbb{N}$ *and* $[f_{v-1-j} \ \ldots \ 0]^T \in D(A(v-1)A(v-2) \cdot \cdots \cdot A(v-j))$ *for all* $v \in \mathbb{N}$ *and* $j \in \mathbb{N}_{v-1}$. *Then the abstract difference equation*

$$\begin{aligned} u(v+p) + A_{p-1}(v)u(v+p-1) + \cdots + A_0(v)u(v) = f(v), \quad v \in \mathbb{N}_0; \\ u(v) = u_v, \quad 0 \leqslant v \leqslant p-1 \end{aligned} \tag{133}$$

has a unique solution $(u(v))_{v \in \mathbb{N}_0}$, *which is given by*

$$\begin{aligned} u(v) = \Pr_p\Bigg(& (-1)^v A(v)A(v-1) \cdot \cdots \cdot A(0)[u_{p-1} \ \ldots \ u_0]^T \\ & + \sum_{j=1}^{v-1} (-1)^j A(v-1)A(v-2) \cdot \cdots \cdot A(v-j)[f_{v-1-j} \ \ldots \ 0]^T \Bigg), \quad v \in \mathbb{N}; \end{aligned} \tag{134}$$

here, $\Pr_p([x_{p-1} \ \ldots \ x_0]^T) = x_0$ *if* $x_j \in X$ *for* $0 \leqslant j \leqslant p-1$.

Proof. The proof of (i) is omitted since it follows from a simple induction argument. To prove (ii), we only need to rewrite (133) into the equivalent matricial form

$$[u_{v+p} \ \ldots \ u_{v+1}]^T + A(v)[u_{v+p-1} \ \ldots \ u_v]^T = [f_v \ 0 \ \ldots \ 0]^T, \quad v \in \mathbb{N}_0,$$

and apply the result established in part (i). □

Remark 3.3.4. (i) Suppose that, for every $v \in \mathbb{N}_0$, $A_0(v), \ldots, A_{p-1}(v)$ are bounded linear operators and $\sup_{v \in \mathbb{N}_0}(\|A_0(v)\| + \cdots + \|A_{p-1}(v)\|) < +\infty$. Then a simple calculation involving the formula (134) shows that any solution $(u(v))_{v \in \mathbb{N}_0}$ of (133) is exponentially bounded, provided that the sequence $(f(v))_{v \in \mathbb{N}_0}$ is exponentially bounded; cf. also [320, Theorem 13.3.1].

(ii) The representation formula (131) can be used in the analysis of the existence and uniqueness of the asymptotically almost periodic (automorphic) type solutions of the equation (130).

We continue by noticing that the requirements of Theorem 3.3.1 hold if $A_0, \ldots, A_{p-1}$ are bounded linear operators, $C = \mathrm{I}$ and $(z^p + z^{p-1}A_{p-1} + \cdots + zA_1 + A_0)^{-1} \in L(X)$ for $|z| \geqslant r$. For example, (iii) holds since

$$\|(1 + zA_{p-1} + \cdots + z^{p-1}A_1 + z^p A_0)^{-1}\| \leqslant \frac{1}{1 - \|zA_{p-1} + \cdots + z^{p-1}A_1 + z^p A_0\|} \leqslant 2, \quad |z| \leqslant r_0,$$

for a sufficiently small real number $r_0 > 0$; a similar estimate holds for the term $z^{p-j}A_j(1 + zA_{p-1} + \cdots + z^{p-1}A_1 + z^pA_0)^{-1}$, where $j \in \mathbb{N}_{p-1}$. Therefore, Theorem 3.3.1 provides a proper extension of [320, Theorem 13.2.1] and can be applied even if the condition $\limsup_{k\to+\infty} \|f_k\|^{1/k} < r < +\infty$ is neglected; for example, we can consider the sequence $f(k) = \exp(k^2)$ for all $k \in \mathbb{N}_0$.

If the first condition in part (iii) of Theorem 3.3.1 holds with $C = I$, then we must have $A_0 \in L(X), \ldots, A_{p-1} \in L(X)$. Furthermore, we can clarify the following much more general result.

Proposition 3.3.5. *Suppose that $A_0, \ldots, A_{p-1}$ are closed linear operators and there exist real numbers $r > 0$, $M \geqslant 1$ and $\zeta > 0$ such that $(\lambda^p + \lambda^{p-1}A_{p-1} + \cdots + \lambda A_1 + A_0)^{-1} \in L(X)$ for $|\lambda| \geqslant r$ and*

$$\|(\lambda^p + \lambda^{p-1}A_{p-1} + \cdots + \lambda A_1 + A_0)^{-1}\| \leqslant M(1 + |\lambda|)^{\zeta}, \quad |\lambda| \geqslant r.$$

Then $A_0 \in L(X), \ldots, A_{p-1} \in L(X)$.

Proof. Consider the operator matrix $A(v) \equiv A$ on X^p, given by (132) with $A_{p-1}(v) \equiv A_{p-1}$, $A_{p-2}(v) \equiv A_{p-2}, \ldots, A_0(v) \equiv A_0$. Then the prescribed assumptions easily imply that $(\lambda + A)^{-1} \in L(X^p)$ for $|\lambda| \geqslant r$ and that there exist real numbers $M_1 \geqslant 1$ and $\zeta_1 > 0$ such that $\|(\lambda+A)^{-1}\|_{L(X^p)} \leqslant M_1(1+|\lambda|)^{\zeta_1}$ for $|\lambda| \geqslant r$; cf. also [320, Lemma 13.1.2]. As is well known (see, e. g., [792, pp. 137–138]), this implies $A \in L(X^p)$, which simply completes the proof. □

In the following subsection, we will focus our attention to the case $C \neq I$.

3.3.2 Some applications of Theorems 3.3.1–3.3.2 with $C \neq I$

In this subsection, we will provide certain applications of Theorems 3.3.1–3.3.2 with $C \neq I$. Let us observe first that the assumptions (i) and (ii) in the formulation of Theorem 3.3.1 can be satisfied for unbounded linear operators, with $C = I$.

Example 3.3.6. Suppose that $s > 1$,

$$X := \left\{ f \in C^{\infty}[0,1];\ \|f\| := \sup_{p\geqslant 0} \frac{\|f^{(p)}\|_{\infty}}{p!^s} < \infty \right\}$$

and

$$A := -d/ds, \quad D(A) := \{ f \in X;\ f' \in X,\ f(0) = 0 \}.$$

Suppose further that $P(z) = \sum_{j=0}^{N} a_j z^j$, $z \in \mathbb{C}$, $a_N \neq 0$ is a complex nonzero polynomial and $N = dg(P) \geqslant 2$. We define the operator $P(A)$ in the usual way; then $P(A)$ is not densely defined and there exist real numbers $M > 0$, $b > 0$ and $c > 0$ such that

$$\rho(P(A)) = \mathbb{C} \quad \text{and} \quad \|R(\lambda : P(A))\| \leqslant M e^{b|\lambda|^{1/Ns} + c|\lambda|^{1/N}}, \quad \lambda \in \mathbb{C};$$

see [443, Example 2.6.10] and [445, Example 2.2.18] for the notion and more details.

Suppose now that $P_0(z), \ldots, P_{p-1}(z)$ and $Q_0(z), \ldots, Q_{p-1}(z)$ are complex polynomials, $dg(P_j) \geqslant 2$ for $0 \leqslant j \leqslant p-1$ and

$$z^p + z^{p-1}Q_{p-1}(A) + \cdots + zQ_1(A) + Q_0(A) = (z - P_{p-1}(A)) \cdot \cdots \cdot (z - P_0(A)), \quad z \in \mathbb{C}.$$

Set $A_j := Q_j(A)$ for $0 \leqslant j \leqslant p-1$. Then we have

$$P_z^{-1} = (z - P_{p-1}(A))^{-1} \cdot \cdots \cdot (z - P_0(A))^{-1}, \quad z \in \mathbb{C}.$$

Keeping in mind the usual resolvent equation, the last equality implies the validity of the assumptions (i) and (ii). On the other hand, the first assumption in part (iii) of Theorem 3.3.1 does not hold and we can simply prove this fact in the following way: Suppose the contrary; then we can apply Theorem 3.3.1 to see that $G(0) = 0$. But, in our concrete situation, the mapping $z \mapsto P_z^{-1} \in L(X)$, $z \in \mathbb{C}$ is entire so that the residue theorem implies $G(0) = A_0^{-1} = (-1)^p (P_{p-1}(A))^{-1} \cdot \cdots \cdot (P_0(A))^{-1} \neq 0$, which is a contradiction.

It is also worth noticing that there exists a bounded linear operator $C \in L(X)$ such that the operator $B = P(A)$, considered in the first part of Example 3.3.6, satisfies that, for every $r > 0$, there exists a finite real number $M_r \geqslant 1$ such that

$$\|(\lambda + B)^{-1}C\| \leqslant \frac{M_r}{|\lambda|}, \quad |\lambda| \geqslant r; \tag{135}$$

then a very simple calculation shows that Theorem 3.3.2 can be applied with $A_0 = B$ and $A_j = c_j \mathrm{I}$ for all $j \in \mathbb{N}_{p-1}$, where c_j are arbitrary complex constants.

In actual fact, we know that there exist two finite real constants $M \geqslant 1$ and $c > 0$ such that $\|(\lambda + B)^{-1}\| \leqslant M \exp(c|\lambda|^{1/N})$ for all $\lambda \in \mathbb{C}$. We choose an angle $\psi \in (0, \pi/2)$ and a number $a > 0$ such that $1/N < a < \pi/(2(\pi - \psi))$; after that, we set

$$Cx := \frac{1}{2\pi i} \int_{\Gamma_\psi} e^{-(-z)^a} (z + B)^{-1} x \, dz, \quad x \in X,$$

where Γ_ψ is the upwards oriented boundary of the region $\Omega := \{z \in \mathbb{C} : |z| \leqslant d\} \cup \{z \in \mathbb{C} \setminus \{0\} : |\arg(z)| \leqslant \psi\}$ for a suitable number $d > 0$. Using the resolvent equation, the residue theorem and an elementary argumentation involving the estimate $1/N < a < \pi/(2(\pi - \psi))$, we can prove that $C \in L(X)$ and (135) holds. Furthermore, since the C-resolvent of operator B is polynomially bounded on Ω, we can prove that the operator $C^2 \in L(X)$ is injective so that the operator C is injective and satisfies that, for every $r > 0$, there exists a finite real number $M_r \geqslant 1$ such that (135) holds; then Theorem 3.3.1 can be applied with the same choice of the operators $A_0, \ldots, A_{p-1}$.

Finally, we would like to note that it seems very plausible that, in some concrete choices of the polynomial $P(z)$, the operator B generates a tempered, regular, Beurling ultradistribution semigroup of $(p!^2)$-class so that there exists an injective operator $C \in L(X)$ such that the operator B generates an exponentially bounded C-regularized semigroup [442].

3.3.3 Poisson-like transforms and abstract higher-order differential equations

In this subsection, we follow the ideas of C. Lizama established in [535]. Our first structural result reads as follows.

Theorem 3.3.7. *Suppose that $a \in \mathbb{R}$, $\omega \in \mathbb{R} \setminus \{0\}$, $n \in \mathbb{N}$, $A_0, \dots, A_{n-1}$ are closed linear operators, $u \in C^n([0,\infty) : X)$, the terms $A_j u^{(j)}(t)$ are continuous for $t \geqslant 0$ and $0 \leqslant j \leqslant n$, $f \in C([0,\infty) : X)$ and*

$$A_n u^{(n)}(t) + A_{n-1} u^{(n-1)}(t) + \cdots + A_0 u(t) = f(t), \quad t \geqslant 0. \tag{136}$$

Suppose that there exist $M \geqslant 1$ and $\varepsilon > 0$ such that

$$\max_{j \in \mathbb{N}_n^0} \big(\|u^{(j)}(t)\| + \|A_j u^{(j)}(t)\| \big) \leqslant M e^{a-\omega-\varepsilon}.$$

Set

$$y_{a,\omega}(v) := \int_0^{+\infty} e^{-at} \frac{(\omega t)^v}{v!} u(t)\, dt, \quad v \in \mathbb{N}_0. \tag{137}$$

Then, for every $v \in \mathbb{N}_0$, we have

$$\begin{aligned} &\sum_{s=0}^{n} (-1)^s \binom{n}{s} a^s \omega^{n-s} A_n y_{a,\omega}(v+s) + \sum_{j=0}^{n-1} \sum_{s=0}^{j} (-1)^s \binom{j}{s} a^s \omega^{j-s} A_j y_{a,\omega}(v+n-j+s) \\ &\quad = \int_0^{+\infty} e^{-at} \frac{(\omega t)^{v+n}}{(v+n)!} f(t)\, dt. \end{aligned} \tag{138}$$

Proof. Applying the partial integration, we get

$$\int_0^{+\infty} e^{-at} \frac{(\omega t)^{v+1}}{(v+1)!} u'(t)\, dt = -\omega y_{a,\omega}(v) + a y_{a,\omega}(v+1), \quad v \in \mathbb{N}_0.$$

Since $\max_{j \in \mathbb{N}_n^0} \|u^{(j)}(t)\| \leqslant M e^{a-\omega-\varepsilon}$, we can inductively prove that

$$\int_0^{+\infty} e^{-at} \frac{(\omega t)^{v+j}}{(v+j)!} u^{(j)}(t)\, dt = \sum_{s=0}^{j} (-1)^s \binom{j}{s} a^s \omega^{j-s} y_{a,\omega}(v+s), \quad v \in \mathbb{N}_0,\ j \in \mathbb{N}_n^0. \tag{139}$$

Furthermore, since the operators $A_0, \ldots, A_{n-1}$ are closed and $\max_{j\in\mathbb{N}_n^0} \|A_j u^{(j)}(t)\| \leqslant Me^{a-\omega-\varepsilon}$, (139) implies

$$\int_0^{+\infty} e^{-at}\frac{(\omega t)^{v+j}}{(v+j)!}A_j u^{(j)}(t)\,dt = \sum_{s=0}^{j}(-1)^s\binom{j}{s}a^s\omega^{j-s}A_j y_{a,\omega}(v+s), \quad v\in\mathbb{N}_0.$$

Now the final conclusion simply follows from the last equality and the fact that $u(\cdot)$ solves the equation (136). □

Our second structural result reads as follows.

Theorem 3.3.8. *Suppose that $a \in \mathbb{R}$, $\omega \in \mathbb{R}\setminus\{0\}$, $n \in \mathbb{N}$, $A_0, \ldots, A_{n-1}$ are closed linear operators, $(S(t))_{t\geqslant 0} \subseteq L(X)$, $S(\cdot)x \in C^n([0,\infty) : X)$ for all $x \in X$, the terms $A_j S^{(j)}(t)x$ are continuous for $x \in X$, $t \geqslant 0$, $0 \leqslant j \leqslant n$, $(f_k)_{k\in\mathbb{Z}}$ is a bounded sequence and*

$$A_n S^{(n)}(t)x + A_{n-1}S^{(n-1)}(t)x + \cdots + A_0 S(t)x = 0, \quad t \geqslant 0. \tag{140}$$

Suppose that there exist $M \geqslant 1$ and $\varepsilon > 0$ such that

$$\max_{j\in\mathbb{N}_n^0}\left(\|S^{(j)}(t)\|_{L(X)} + \|A_j S^{(j)}(t)\|_{L(X)}\right) \leqslant Me^{a-\omega-\varepsilon}. \tag{141}$$

Set

$$\mathbf{S}_{a,\omega}(v)x := \int_0^{+\infty} e^{-at}\frac{(\omega t)^v}{v!}S(t)x\,dt, \quad v\in\mathbb{N}_0,\ x\in X$$

and

$$y^*_{a,\omega}(v) := \sum_{l=-\infty}^{v}\mathbf{S}_{a,\omega}(v-l)f(l), \quad v\in\mathbb{Z}.$$

Then $\sum_{v=0}^{\infty}\|\mathbf{S}_{a,\omega}(v)\| \leqslant M/\varepsilon$ and, for every $v \in \mathbb{Z}$, we have

$$\sum_{s=0}^{n}(-1)^s\binom{n}{s}a^s\omega^{n-s}A_n y^*_{a,\omega}(v+s) + \sum_{j=0}^{n-1}\sum_{s=0}^{j}(-1)^s\binom{j}{s}a^s\omega^{j-s}A_j y^*_{a,\omega}(v+n-j+s) = g(v), \tag{142}$$

where

$$g(v) := \sum_{j=1}^{n}\sum_{l=v+j}^{v+n}\Bigg[(-1)^j\binom{n}{j}a^j\omega^{n-j}A_n + \sum_{k=n-j}^{n-1}(-1)^{j+k-n}\binom{k}{j+k-n}a^{j+k-n}\omega^{n-j}A_k\Bigg]\cdot\int_0^{+\infty}e^{-at}\frac{(\omega t)^{v-l+j}}{(v+l+j)!}S(t)g(l)\,dt,$$

for any $v \in \mathbb{Z}$.

Proof. It is clear that

$$\sum_{v=0}^{\infty}\|\mathbf{S}_{a,\omega}(v)\| \leqslant \int_0^{+\infty} e^{-at}e^{\omega t}\|S(t)\|\,dt \leqslant M\int_0^{+\infty} e^{-\varepsilon t}\,dt = M/\varepsilon.$$

Since (140)–(141) hold, we can argue in the same way as in the proof of Theorem 3.3.7 in order to see that, for every $v,\ l \in \mathbb{Z}$ with $v \geqslant l$, (138) holds with $f \equiv 0$, the number v replaced therein with the number $v-l$ and the sequence $y_{a,\omega}(\cdot)$ replaced therein with the sequence $\int_0^{+\infty} e^{-at}\frac{(\omega t)^{\cdot}}{\cdot!}S(t)g(l)\,dt$. Keeping in mind that the equation (138) with $f \equiv 0$ can be rewritten as

$$\sum_{j=1}^{n}\Bigg[(-1)^j\binom{n}{j}a^j\omega^{n-j}A_n+\sum_{k=n-j}^{n-1}(-1)^{j+k-n}\binom{k}{j+k-n}a^{j+k-n}\omega^{n-j}A_k\Bigg]y(v+j)+\big[\omega^nA_n\big]y(v)=0,$$

the above simply yields the required conclusion. □

Since $(f_k)_{k\in\mathbb{Z}}$ is bounded and $\sum_{v=0}^{\infty}\|\mathbf{S}_{a,\omega}(v)\| < +\infty$, we have an open door to consider the existence of almost periodic (automorphic) type solutions of the abstract higher-order difference equation (142), provided that $(f_k)_{k\in\mathbb{Z}}$ enjoys a certain almost periodic (automorphic) behavior. For example, we have the following:

(i) $y^*_{a,\omega}(\cdot)$ will be c-almost periodic (c-uniformly recurrent), where $c \in \mathbb{C}\setminus\{0\}$, provided that $(f_k)_{k\in\mathbb{Z}}$ is T-almost periodic (T-uniformly recurrent).

(ii) $y^*_{a,\omega}(\cdot)$ will be equi-Weyl-$(\mathbb{F},1,c)$-almost periodic [bounded Weyl-$(\mathbb{F},1,c)$-almost periodic; bounded Doss-$(\mathbb{F},1,c)$-almost periodic], provided that $f(\cdot)$ is.

(iii) $y^*_{a,\omega}(\cdot)$ will be Levitan c-almost periodic, provided that $f(\cdot)$ is.

Let us consider now the sequence

$$y_{a,b,j,\omega}(v) := \int_0^{+\infty} e^{-b(at)^j}\frac{(\omega t)^v}{v!}u(t)\,dt, \quad v \in \mathbb{N}_0, \tag{143}$$

where $a \in \mathbb{R}$, $b,\ \omega \in \mathbb{R}\setminus\{0\}$ and $j \in \mathbb{N}\setminus\{1\}$, in place of the sequence $(y_{a,\omega}(v))_{v\in\mathbb{N}_0}$ given by (137). Assuming that $\lim_{t\to+\infty} t^ve^{-b(at)^j}u(t) = 0$ for all $v \in \mathbb{N}$, we can apply the partial integration in order to see that

$$\int_0^{+\infty} e^{-b(at)^j}\frac{(\omega t)^{v+1}}{(v+1)!}u'(t)\,dt = -\omega y_{a,b,j,\omega}(v) + jb\omega^{1-j}a^j(v+j)\cdot\cdots\cdot(v+2)y_{a,b,j,\omega}(v+j),$$

for any $v \in \mathbb{N}_0$. Repeating this procedure, we can compute the value of

$$\int_0^{+\infty} e^{-b(at)^j}\frac{(\omega t)^{v+s}}{(v+s)!}u^{(s)}(t)\,dt \quad (2 \leqslant s \leqslant n;\ v \in \mathbb{N}_0)$$

in terms of $v, y_{a,b,j,\omega}(v), y_{a,b,j,\omega}(v+j), \ldots, y_{a,b,j,\omega}(v+js)$, providing some very mild conditions. Therefore, if $u(t)$ is a (not-exponentially bounded) solution of (136), then we can similarly formulate an analogue of Theorem 3.3.7 with the sequence $(y_{a,\omega}(v))_{v\in\mathbb{N}_0}$ replaced by $(y_{a,b,j,\omega}(v))_{v\in\mathbb{N}_0}$; in contrast to Theorem 3.3.7, the operator coefficients now depend on time $v \in \mathbb{N}_0$.

As already marked, one is in chancery if one wants to define the Poisson transform (cf. (137) with $a = \omega = 1$) for the negative values of v. In order to overcome this difficulty, we suggest using the following transform:

$$v \mapsto y_{a,b,c,j,\omega}(v) := \int_0^{+\infty} e^{-b(ct^{-1}+at)^j} \frac{(\omega t)^{v-\frac{1}{2}}}{\Gamma(v+\frac{1}{2})} u(t)\,dt, \quad v \in \mathbb{Z}, \tag{144}$$

where $a \in \mathbb{R}$, $b,\ c,\ \omega \in \mathbb{R} \setminus \{0\}$ and $j \in \mathbb{N}$. Then an analogue of Theorem 3.3.7 can be formulated for the abstract nonautonomous higher-order difference equations defined on $\mathbb{Z}$. For example, let $a = b = c = j = \omega = 1$; then we can apply the partial integration in order to see that

$$\begin{aligned} y_{1,1,1,1,1}(v) = &- \int_0^{+\infty} e^{-t^{-1}-t} \frac{t^{v+\frac{1}{2}}}{\Gamma(v+\frac{3}{2})} u'(t)\,dt \\ &- y_{1,1,1,1,1}(v+1) + \left(v - \frac{1}{2}\right)\left(v + \frac{1}{2}\right) y_{1,1,1,1,1}(v-1), \quad v \in \mathbb{Z}. \end{aligned}$$

Repeating verbatim this procedure, we can simply compute the terms

$$\int_0^{+\infty} e^{-t^{-1}-t} \frac{t^{v+j-\frac{1}{2}}}{\Gamma(v+j+\frac{1}{2})} u^{(j)}(t)\,dt, \quad v \in \mathbb{Z} \quad (2 \leqslant j \leqslant n)$$

and constitute the abstract nonautonomous higher-order difference equation with a solution $(y_{1,1,1,1,1}(v))_{v\in\mathbb{Z}}$. It is clear that Theorem 3.3.8 can be reformulated for the abstract nonautonomous higher-order difference equations obtained by performing the transforms introduced in (143) and (144).

Unfortunately, it would be very difficult to compute the value of

$$\int_0^{+\infty} e^{-b(ct^{-1}+at)^j} \frac{(\omega t)^{v-\frac{1}{2}}}{\Gamma(v+\frac{1}{2})} \left[\int_{-\infty}^{t} a(t-s)u(s)\,ds \right] dt$$

in terms of $v \in \mathbb{Z}$ and the members of the sequence $(y_{a,b,c,j,\omega}(v))_{v\in\mathbb{Z}}$; because of that, it is not clear how we can apply the transform introduced in (144) to the abstract fractional integrodifferential equations considered by R. Ponce in [646].

3.3.4 On fractional difference operators

Given a one-dimensional sequence (u_k) in X, the Euler forward difference operator Δ is defined by $\Delta u_k := u_{k+1} - u_k$. The operator Δ^m is defined inductively; then, for every integer $m \geqslant 1$, we have

$$\Delta^m u_k = \sum_{j=0}^{m} (-1)^{m-j} \binom{m}{j} u_{k+j}. \tag{145}$$

Further on, if $u : \mathbb{N}_0 \to X$ is a given sequence and $\alpha \in (0, \infty) \smallsetminus \mathbb{N}_0$, then we define the fractional integral $\Delta^{-\alpha} u : \mathbb{N}_0 \to X$ by

$$\Delta^{-\alpha} u(v) := \sum_{j=0}^{v} k^\alpha (v - j) u(j), \quad v \in \mathbb{N}_0.$$

Set $m := \lceil \alpha \rceil$, $k^0(0) := 1$ and $k^0(v) := 0$, $v \in \mathbb{N}$; then $k^\alpha *_0 k^\beta \equiv k^{\alpha+\beta}$ for all $\alpha, \beta \geqslant 0$ and the formula (148) below coincides with the formula (145) for integer values of order $\alpha = m \in \mathbb{N}$. The Riemann–Liouville fractional difference operator of order $\alpha > 0$, Δ^α for short, is defined by

$$\Delta^\alpha u(v) := [\Delta^m (\Delta^{-(m-\alpha)} u)](v), \quad v \in \mathbb{N}_0,$$

while the Caputo fractional difference operator of order $\alpha > 0$, Δ_C^α for short, is defined by

$$\Delta_C^\alpha u(v) := [\Delta^{\alpha-m} (\Delta^m u)](v), \quad v \in \mathbb{N}_0.$$

The Riemann–Liouville fractional difference operator of order $\alpha > 0$ and the Caputo fractional difference operator of order $\alpha > 0$ are not equal: if $0 < \alpha < 1$, then [535, Theorem 2.4] states that

$$\Delta_C^\alpha u(v) = \Delta^\alpha u(v) - k^{1-\alpha}(v + 1) u(0), \quad v \in \mathbb{N}_0; \tag{146}$$

on the other hand, if $1 < \alpha < 2$, then [4, Theorem 2.5] states that

$$\Delta_C^\alpha u(v) = \Delta^\alpha u(v) - k^{2-\alpha}(v + 1)[u(1) - 2u(0)] - k^{2-\alpha}(v + 2) u(0), \quad v \in \mathbb{N}_0.$$

Now we will state and prove the following general result, which connects the fractional difference operators Δ^α and Δ_C^α; this is a generalization of [535, Theorem 2.4] and [4, Theorem 2.5].

Theorem 3.3.9. *Suppose that $\alpha \in (0, \infty) \smallsetminus \mathbb{N}_0$ and $m = \lceil \alpha \rceil$. Then we have*

$$\Delta_C^\alpha u(v) = \Delta^\alpha u(v) + c_0^\alpha(v) u(0) + \cdots + c_{m-1}^\alpha(v) u(m - 1), \quad v \in \mathbb{N}_0,$$

where

$$c_s^a(v) = \sum_{\substack{j+r=s\\ 0\leqslant j\leqslant \min(s,v),0\leqslant r\leqslant s}} (-1)^{m-r}\binom{m}{r}k^{m-a}(v-j) - \sum_{j=\max(0,s-v)}^{m} (-1)^{m-j}\binom{m}{j}k^{m-a}(v+j-s), \quad v \in \mathbb{N}_0,\ s \in \mathbb{N}_{m-1}^0.$$

Proof. Using the corresponding definitions and (145), it readily follows that

$$\Delta^a u(v) = [\Delta^m(\Delta^{-(m-a)}u)](v) = \sum_{j=0}^{m}\sum_{l=0}^{v+j}(-1)^{m-j}\binom{m}{j}k^{m-a}(v+j-l)u(l), \quad v \in \mathbb{N}_0 \tag{147}$$

and

$$\Delta_C^a u(v) = [\Delta^{a-m}(\Delta^m u)](v) = \sum_{j=0}^{v}\sum_{r=0}^{m}(-1)^{m-r}\binom{m}{r}k^{m-a}(v-j)u(j+r), \quad v \in \mathbb{N}_0. \tag{148}$$

This immediately implies the required result if we prove that, for every integers $s \geqslant m$ and $v \in \mathbb{N}_0$ such that $v + m \geqslant s$, we have

$$\sum_{\substack{j+r=s\\ 0\leqslant j\leqslant \min(s,v),0\leqslant r\leqslant m}} (-1)^{m-r}\binom{m}{r}k^{m-a}(v-j) = \sum_{j=\max(0,s-v)}^{m} (-1)^{m-j}\binom{m}{j}k^{m-a}(v+j-s). \tag{149}$$

The first term in (149) equals

$$\sum_{j=s-m}^{\min(s,v)} (-1)^{m-(s-j)}\binom{m}{s-j}k^{m-a}(v-j),$$

while the second term in (149) equals

$$\sum_{j=\max(0,s-v)}^{m} (-1)^{m-j}\binom{m}{j}k^{m-a}(v+j-s) = \sum_{j_1=s-m}^{s-\max(0,s-v)} (-1)^{m-(s-j_1)}\binom{m}{s-j_1}k^{m-a}(v-j_1).$$

The final conclusion simply follows since we have $\min(s,v) = s - \max(0,s-v)$. □

It can be proved the following: if A is a closed linear operator and $u(v) \in D(A)$ for all $v \in \mathbb{N}_0$, then the terms $A(\Delta^a u)$ and $A(\Delta_C^a u)$ are well-defined and we have $A(\Delta^a u) = \Delta^a(Au)$ and $A(\Delta_C^a u) = \Delta_C^a(Au)$. But, if the term $A(\Delta^a u)$ or $A(\Delta_C^a u)$ is well-defined, then we do not necessarily have that $u(v) \in D(A)$ for all $v \in \mathbb{N}_0$ as well as that the above equalities hold.

Example 3.3.10. Suppose that $\alpha > 0$, $m = \lceil\alpha\rceil$, A is a closed linear operator and $u(0),\ldots,u(m-1) \in X \setminus D(A)$. We determine $u(m)$ such that

$$\sum_{r=0}^{m}(-1)^{m-r}\binom{m}{r}k^{m-\alpha}(v)u(r) = 0.$$

After that, we determine $u(m+1), u(m+2), \ldots$ such that

$$\sum_{j=0}^{v}\sum_{r=0}^{m}(-1)^{m-r}\binom{m}{r}k^{m-\alpha}(v-j)u(j+r) = 0, \quad v \in \mathbb{N}_0;$$

cf. (148). Then $A(\Delta_C^\alpha u) \equiv \Delta_C^\alpha u \equiv 0$ and $u(v) \notin D(A)$ for some values of $v \in \mathbb{N}_0$.

Another simple counterexample is given by $u(v) \equiv x \notin D(A)$. Since $\sum_{r=0}^{m}(-1)^{m-r}\binom{m}{r} = 0$, we have $\Delta_C^\alpha u \equiv 0$ and $u(v) \notin D(A)$ for all values of $v \in \mathbb{N}_0$; let us only observe here that (146) implies $[\Delta^\alpha u](v) = k^{1-\alpha}(v+1)x \neq 0$ for all $v \in \mathbb{N}_0$ and $\alpha \in (0,1)$.

Therefore, in the sequel, we must distinguish the terms $A(\Delta^\alpha u)$ $[A(\Delta_C^\alpha u)]$ and $\Delta^\alpha(Au)$ $[\Delta_C^\alpha(Au)]$. Now we will deduce the following identity for the Riemann–Liouville fractional derivatives, which is well known in the continuous framework (see, e. g., [97, p. 11]).

Example 3.3.11. Suppose that $\alpha > 0$; then $\Delta^\alpha k^\alpha = 0$. In actual fact, we have ($m = \lceil\alpha\rceil$):

$$[\Delta^\alpha k^\alpha](v) = \sum_{j=0}^{m}(-1)^{m-j}\binom{m}{j}k^m(v+j), \quad v \in \mathbb{N}_0.$$

Therefore, we only need to prove that

$$\sum_{j=0}^{m}(-1)^{m-j}\binom{m}{j}\binom{v+j+m-1}{m-1} = 0, \quad v \in \mathbb{N}_0.$$

But, this is true since we have (cf. [338, (1.13), p. 2]):

$$\sum_{j=0}^{m}(-1)^{m-j}\binom{m}{j}\binom{v+j+m-1}{s} = 0, \quad v \in \mathbb{N}_0,\ s \in \mathbb{N}_{m-1}^0.$$

Suppose now that $\beta > \alpha > 0$. Then we already know that $\Delta^\alpha k^\beta(\cdot) = k^{\beta-\alpha}(\cdot + \lceil\alpha\rceil)$; see [4, Example 2.3]. This yields

$$\sum_{j=0}^{m}(-1)^{m-j}\binom{m}{j}\frac{\Gamma(v+j+m+\beta-\alpha)}{\Gamma(m+\beta-\alpha)\cdot(v+j)!} = \frac{\Gamma(v+\beta+\lceil\alpha\rceil-\alpha)}{\Gamma(\beta-\alpha)\cdot(v+\lceil\alpha\rceil)!}, \quad v \in \mathbb{N}_0.$$

Keeping in mind our definition of $k^0(\cdot)$, the above implies

$$\Delta^\alpha k^\beta(\cdot) = k^{\beta-\alpha}(\cdot + \lceil\alpha\rceil), \quad \beta \geqslant \alpha > 0.$$

The Riemann–Liouville fractional derivatives of the finite convolution product have been analyzed in [534, Lemma 3.6] and [4, Theorem 2.6], provided that $0 < \alpha \leqslant 1$ and $1 < \alpha \leqslant 2$, respectively. The interested reader may try to extend these statements for the general fractional order $\alpha > 0$ as well as to clarify the corresponding statements for the Caputo fractional derivatives. It could be also worthwhile to state certain relations between the fractional operators $\Delta^{\alpha}\Delta^{\beta}$, $\Delta^{\beta}\Delta^{\alpha}$ and $\Delta^{\alpha+\beta}$ for $\alpha, \beta > 0$; the same problem can be proposed for the Caputo fractional derivatives (see also [329]).

In the remainder of this section, we will always assume that $n \in \mathbb{N}$, $0 \leqslant \alpha_1 < \cdots < \alpha_n$ and $A_1, \ldots, A_n$ are closed linear operators on X. Define $m_i := \lceil \alpha_i \rceil$, $i \in \mathbb{N}_n$.

3.3.5 Abstract multiterm fractional difference equations with Caputo fractional derivatives

We start by introducing the terms $T_{i,L}u(v) := A_i\Delta_C^{\alpha_i}u(v)$, if $v \in \mathbb{N}_0$, $i \in \mathbb{N}_n$ and $T_{i,R}u(v) := \Delta_C^{\alpha_i}A_iu(v)$, if $v \in \mathbb{N}_0$ and $i \in \mathbb{N}_n$. Henceforth it will be assumed that, for every $v \in \mathbb{N}_0$ and $i \in \mathbb{N}_n$, $T_iu(v)$ denotes either $T_{i,L}u(v)$ or $T_{i,R}u(v)$. Of concern is the following abstract degenerate multi-term fractional difference equation with Caputo fractional derivatives:

$$\sum_{i=1}^{n} T_iu(v + v_i) = f(v), \quad v \in \mathbb{N}_0, \tag{150}$$

where $v_1, \ldots, v_n \in \mathbb{N}_0$ and $f : \mathbb{N}_0 \to X$ is a given sequence.

Define $\mathcal{I} := \{i \in \mathbb{N}_n : \alpha_i > 0$ and $T_{i,L}u(v)$ appears on the left-hand side of (150)$\}$, $Q := \max \mathcal{I}$, if $\mathcal{I} \neq \emptyset$, $Q := m_Q := v_Q := 0$, if $\mathcal{I} = \emptyset$, $\tau := \max\{m_i + v_i : i \in I\}$, if $\mathcal{I} \neq \emptyset$, and $\tau := 0$, if $\mathcal{I} = \emptyset$.

In order to endow the equation (150) with the initial conditions, let us observe that the identity (148) implies that (150) is equivalent to

$$\begin{aligned} &\sum_{i\in\mathcal{I}} A_i\left[\sum_{j=0}^{v+v_i}\sum_{r=0}^{m_i}(-1)^{m_i-r}\binom{m_i}{r}k^{m_i-\alpha_i}(v+v_i-j)u(j+r)\right] \\ &+ \sum_{i\in\mathbb{N}_n\setminus\mathcal{I}}\left[\sum_{j=0}^{v+v_i}\sum_{r=0}^{m_i}(-1)^{m_i-r}\binom{m_i}{r}k^{m_i-\alpha_i}(v+v_i-j)A_iu(j+r)\right] = f(v), \quad v \in \mathbb{N}_0. \end{aligned} \tag{151}$$

This suggests us to endow (150) with the following initial conditions:

$$u(j) = u_j, \quad 0 \leqslant j \leqslant \tau - 1 \quad \text{and} \quad A_iu(j) = u_{i,j} \quad \text{if } i \notin \mathcal{I} \text{ and } v_i + m_i - 1 \geqslant j \geqslant 0. \tag{152}$$

If $T_nu(v) = T_{n,L}u(v)$, then (152) reduces to: $u(j) = u_j$, $0 \leqslant j \leqslant \tau - 1$. If this is not the case, then the choice (152) may be nonoptimal and, as in the continuous setting, we cannot expect the existence of solutions of problem [(150); (152)], in general. In order to expect the existence of solutions of problem [(150); (152)], we must also impose the following compatibility condition:

$$A_i u_j = u_{i,j} \quad \text{if } i \notin \mathcal{I} \text{ and } 0 \leqslant j \leqslant \min(\tau - 1, v_i + m_i - 1). \tag{153}$$

In the sequel of this subsection, we will always assume that (153) holds. Now we are ready to introduce the following notion.

Definition 3.3.12. A sequence $u : \mathbb{N}_0 \to X$ is said to be a strong solution of problem [(150); (152)] if the term $T_i u(v+v_i)$ is well-defined for any $v \in \mathbb{N}_0$, $i \in \mathbb{N}_n$, and [(150); (152)] holds.

If $Q > 0$, then we can consider the problem obtained from the problem (150) by replacing some of terms $T_{i,R}(v)$, for $1 \leqslant i \leqslant Q$, with the corresponding terms of form $T_{i,L}(v)$. By the foregoing, we know that any strong solution of problem [(150); (152)] will be a strong solution of such a problem.

We continue with some useful observations.

Remark 3.3.13. (i) Suppose that a sequence $u(\cdot)$ is a strong solution of the following abstract multiterm fractional difference equation

$$\sum_{i=1}^{n} T_{i,R} u(v + v_i) = f(v), \quad v \in \mathbb{N}_0.$$

Let $\Delta^{\alpha_1} g = 0$, $g(v) \notin D(A_1)$ for some $v \in \mathbb{N}_0$, and let $A_2, \ldots, A_n$ be bounded linear operators, for example. Define $u_g(v) := u(v) + g(v)$, $v \in \mathbb{N}_0$. Then we have

$$\sum_{i=1}^{n} T_{1,L} u_g(v + v_1) + \sum_{i=2}^{n} T_{i,R} u_g(v + v_i) = f(v), \quad v \in \mathbb{N}_0;$$

furthermore, the term $\Delta^{\alpha_1}(A_1 u_g)$ is not well-defined (cf. also Example 3.3.21 below).

(ii) Convoluting the equation (150) with $k^{\alpha_n}(\cdot)$, we obtain that any strong solution $v \mapsto u(v)$, $v \in \mathbb{N}_0$ of problem [(150); (152)] satisfies a multiterm Volterra difference equation

$$\sum_{i=1}^{n} \mathcal{T}_i u(v) = (k^{\alpha_n} *_0 f)(v), \quad v \in \mathbb{N}_0, \tag{154}$$

with certain terms $\mathcal{T}_i u(\cdot)$ for $1 \leqslant i \leqslant n-1$, since for each $m \in \mathbb{N}$ with $\alpha_n \geqslant m$ we have

$$\begin{aligned}
(k^{\alpha_n} *_0 \Delta^m u)(v) &= (k^{\alpha_n - m} *_0 (k^m *_0 \Delta^m u))(v) \\
&= \sum_{j=0}^{v} k^{\alpha_n - m}(v - j)(k^m *_0 \Delta^m u)(j) \\
&= \sum_{j=0}^{v} \sum_{s=0}^{j} \sum_{l=0}^{m} (-1)^{m-l} k^{\alpha_n - m}(v - j) \binom{m}{l} \binom{j - s + m - 1}{m - 1} u(s + l), \quad v \in \mathbb{N}_0;
\end{aligned}$$

cf. also [445, Definition 2.3.2], where we have introduced the notion of a $\mathcal{V}$-mild solution for a continuous analogue of (154). In the discrete setting, it would be very unreasonable to consider the well-posedness of problem [(150); (152)] and its relatives by convoluting the equations with $k^{\alpha_n}(\cdot)$, as it has been usually done in the continuous setting.

The values of translation numbers $v_1, \ldots, v_n$ are essentially important sometimes; for example, in some cases, we can trivially show that there exists a unique strong solution of problem [(150); (152)].

Example 3.3.14. Let us consider the following abstract difference equation:

$$(\Delta_C^{\alpha} u)(v) = Au(v) + f(v), \quad v \in \mathbb{N}_0; \quad u(0) = u_0,$$

where $0 < \alpha < 1$ and $A : X \to X$ is an arbitrary function. This equation is equivalent with

$$\sum_{j=0}^{v+1} k^{1-\alpha}(v+1-j)u(j) - \sum_{j=0}^{v} k^{1-\alpha}(v-j)u(j) - k^{1-\alpha}(v+1)u_0 = Au(v) + f(v),$$
$$v \in \mathbb{N}_0; \quad u(0) = u_0.$$

Since $k^{1-\alpha}(0) = 1$, we can simply show by induction that there exists a unique strong solution $u(\cdot)$ of this problem.

Further on, if $\mathcal{I} \neq \emptyset$, then we define

$$\mathcal{I}_{\max} := \{i \in \mathcal{I} : m_i + v_i = \tau\}.$$

Now we are ready to formulate the following structural result.

Theorem 3.3.15. *Suppose that $\mathcal{I} \neq \emptyset$, $m_i + v_i < \tau$ for all $i \in \mathbb{N}_n \setminus \mathcal{I}$ and there exists an injective operator $C \in L(X)$ such that $CA_i \subseteq A_iC$ for all $i \in \mathcal{I}_{\max}$ and*

$$\Big(\sum_{i\in\mathcal{I}_{\max}} A_i\Big)^{-1} C \in L(X).$$

Then there exists at most one strong solution of problem [(150); (152)].

Proof. Due to the linearity of problem [(150); (152)], we may assume that $u_j = 0$ for $0 \leqslant j \leqslant \tau - 1$, $A_i u(j) = 0$ if $i \notin \mathcal{I}$ and $v_i + m_i - 1 \geqslant j \geqslant 0$ and $f \equiv 0$. Since $m_i + v_i < \tau$ for all $i \in \mathbb{N}_n \setminus \mathcal{I}$, the problem [(150); (152)] has the form:

$$\sum_{i\in\mathcal{I}_{\max}} A_i[u(v+\tau) + c_1u(v+\tau-1) + \cdots + c_{v+\tau}u(0)] + \sum_{i\in\mathcal{I}\setminus\mathcal{I}_{\max}} A_i[u(v+m_i+v_i) + \cdots]$$

$$+\sum_{i\in\mathbb{N}_n\smallsetminus\mathcal{I}}\left[\sum_{j=0}^{v+v_i}\sum_{r=0}^{m_i}(-1)^{m_i-r}\binom{m_i}{r}k^{m_i-\alpha}(v+v_i-j)A_iu(j+r)\right]=0,\quad v\in\mathbb{N}_0.$$

The first addend in the above sum, with $v = 0$, equals

$$\sum_{i\in\mathcal{I}_{\max}}A_iu(\tau).$$

It follows that $\sum_{i\in\mathcal{I}_{\max}}A_iu(\tau)$ has a given value; keeping in mind that $CA_i \subseteq A_iC$ for all $i \in \mathcal{I}_{\max}$, we can multiply this term with $(\sum_{i\in\mathcal{I}_{\max}}A_i)^{-1}C$ in order to see that $Cu(\tau)$ is uniquely determined, so that $u(\tau)$ is uniquely determined since C is injective. Furthermore, we have $u(\tau) \in D(A_i)$ for all $i \in \mathcal{I}_{\max}$. Plugging repeatedly $v = 1, \ldots,$ we can simply conclude that the values of $u(\tau+1), \ldots,$ are uniquely determined. This simply completes the proof. □

If $\mathcal{I} \neq \mathbb{N}_n$, then we define $\tau_D := \max\{m_i + v_i : i \in \mathbb{N}_n \smallsetminus \mathcal{I}\}$ and

$$\mathcal{I}_{\max,D} := \{i \in \mathbb{N}_n \smallsetminus \mathcal{I} : m_i + v_i = \tau_D\}.$$

Define also $\tau_b := \max\{m_i + v_i : i \in \mathbb{N}_n\}$, $\mathcal{I}_{\max,R,b} := \{i \in \mathcal{I} : m_i + v_i = \tau_b\}$, $\mathcal{I}_{\max,D,b} := \{i \in \mathbb{N}_n \smallsetminus \mathcal{I} : m_i + v_i = \tau_b\}$ and $\mathcal{I}_{\max,b} := \mathcal{I}_{\max,R,b} \cup \mathcal{I}_{\max,D,b}$. Then we can similarly prove the following results.

Theorem 3.3.16. (i) *Suppose that $\mathcal{I} \neq \mathbb{N}_n$, $m_i + v_i < \tau_D$ for all $i \in \mathcal{I}$ and there exists an injective operator $C \in L(X)$ such that $CA_i \subseteq A_iC$ for all $i \in \mathcal{I}_{\max,D}$ and*

$$\left(\sum_{i\in\mathcal{I}_{\max,D}}A_i\right)^{-1}C \in L(X).$$

Then there exists at most one strong solution of problem (150) *when endowed with the initial conditions*

$$u(j) = u_j,\quad 0 \leqslant j \leqslant \tau_D - 1 \quad \textit{and} \quad A_iu(j) = u_{i,j} \quad \textit{if } i \notin \mathcal{I} \textit{ and } v_i + m_i - 1 \geqslant j \geqslant 0. \tag{155}$$

(ii) *Suppose that $\mathcal{I}_{\max,R,b} \neq \emptyset$, $\mathcal{I}_{\max,D,b} \neq \emptyset$ and there exists an injective operator $C \in L(X)$ such that $CA_i \subseteq A_iC$ for all $i \in \mathcal{I}_{\max,R,b} \cup \mathcal{I}_{\max,D,b}$ and*

$$\left(\sum_{i\in\mathcal{I}_{\max,R,b}\cup\mathcal{I}_{\max,D,b}}A_i\right)^{-1}C \in L(X).$$

Then there exists at most one strong solution of problem [(150); (152)].

Concerning the existence and uniqueness of strong solutions of problem (150), defined in a similar way as in Definition 3.3.12, we will first state the following result with

$\mathcal{I} = \emptyset$ (it is clear that, if the operator $\sum_{i\in\mathcal{I}_{\max,b}} A_i$ is not injective, then we cannot expect the uniqueness of strong solutions of problem (150)).

Theorem 3.3.17. *Suppose that $\mathcal{I} = \emptyset$ and there exists an injective operator $C \in L(X)$ such that $CA_i \subseteq A_iC$ for all $i \in \mathcal{I}_{\max,b}$,*

$$\Big(\sum_{i\in\mathcal{I}_{\max,b}} A_i\Big)^{-1} C \in L(X) \quad \text{and} \quad C^{-1}\Big(\sum_{i\in\mathcal{I}_{\max,b}} A_i\Big)C = \sum_{i\in\mathcal{I}_{\max,b}} A_i. \tag{156}$$

Then there exists a strong solution of problem (150)*, endowed with the initial conditions of the form*

$$\begin{aligned} A_iu(j) &= u_{i,j} \quad \text{if } i \in \mathcal{I}_{\max,b} \text{ and } \tau_b - 1 \geqslant j \geqslant 0, \quad \text{and} \\ A_iu(j) &= u_{i,j} \quad \text{if } i \notin \mathcal{I}_{\max,b} \text{ and } v_i + m_i \geqslant j \geqslant 0, \end{aligned} \tag{157}$$

respectively, there exists a unique strong solution of problem (150)*, endowed with the initial conditions of the form*

$$u(j) = u_j, \quad 0 \leqslant j \leqslant \tau_b - 1, \tag{158}$$

provided that

$$\Big(\sum_{i\in\mathcal{I}_{\max,b}} A_i\Big)^{-1} f(v) \in D(A_l), \quad \Big(\sum_{k\in\mathcal{I}_{\max,b}} A_k\Big)^{-1} u_{i,j} \in D(A_l) \tag{159}$$

for all $v \in \mathbb{N}$, $l \in \mathbb{N}_n \smallsetminus \mathcal{I}_{\max,b}$ and all terms $u_{i,j}$ defined in (157)*, as well as that*

$$\Big(\sum_{i\in\mathcal{I}_{\max,b}} A_i\Big)^{-1} A_j[D(A_1) \cap \cdots \cap D(A_n)] \subseteq D(A_1) \cap \cdots \cap D(A_n), \quad j \in \mathbb{N}_n. \tag{160}$$

Proof. It is clear that the problem [(150); (152)] is equivalent with a problem of the following form:

$$\begin{aligned} &\Big[\sum_{i\in\mathcal{I}_{\max,b}} A_i\Big]u(v + \tau_b) \\ &\quad + \sum_{i\in\mathcal{I}_{\max,b}} [b_{v+v_i+m_i-1}(v)A_iu(v + v_i + m_i - 1) + \cdots + b_0(v)A_iu(0)] \\ &\quad + \sum_{i\in\mathbb{N}_n\smallsetminus\mathcal{I}_{\max,b}} [A_iu(v + v_i + m_i) + \cdots + c_0(v)A_iu(0)] = f(v), \quad v \in \mathbb{N}_0. \end{aligned} \tag{161}$$

Keeping in mind the initial conditions (158), the above implies that the values of terms $A_iu(j)$ in (157) are uniquely determined; therefore, it suffices to show that there exists a sequence $(u(v))_{v\geqslant\tau_b}$ such that the above equality holds. Plugging $v = 0$, we obtain

$$\Big[\sum_{i\in\mathcal{I}_{\max,b}} A_i\Big]u(\tau_b) + \sum_{i\in\mathcal{I}_{\max,b}} [b_{v_i+m_i-1}(0)u_{i,v_i+m_i-1} + \cdots + b_0(0)u_{i,0}] + \sum_{i\in\mathbb{N}_n\setminus\mathcal{I}_{\max,b}} [u_{i,v_i+m_i} + \cdots + c_0(0)u_{i,0}] = f(0).$$

The second equality in (156) implies

$$C^{-1}\Big(\sum_{i\in\mathcal{I}_{\max,b}} A_i\Big)^{-1} C = \Big(\sum_{i\in\mathcal{I}_{\max,b}} A_i\Big)^{-1}.$$

Keeping in mind this equality, the first inclusion in (156) and our assumption (159), it follows that $u(\tau_b)$ is uniquely determined as well as that $u(\tau_b) \in D(A_j)$ for all $j \in \mathbb{N}_n$ and

$$u(\tau_b) = \Big(\sum_{i\in\mathcal{I}_{\max,b}} A_i\Big)^{-1} f(0) - \Big(\sum_{i\in\mathcal{I}_{\max,b}} A_i\Big)^{-1} \sum_{j\in\mathcal{I}_{\max,b}} [b_{v+v_j+m_j-1}(v)A_j u(v+v_j+m_j-1) + \cdots + b_0(v)A_j u(0)] - \Big(\sum_{i\in\mathcal{I}_{\max,b}} A_i\Big)^{-1} \sum_{j\in\mathbb{N}_n\setminus\mathcal{I}_{\max,b}} [u_{j,v_j+m_j} + \cdots + c_0(0)u_{j,0}].$$

Plugging $v = 1$, we obtain

$$\Big[\sum_{i\in\mathcal{I}_{\max,b}} A_i\Big]u(\tau_b+1) + \sum_{i\in\mathcal{I}_{\max,b}} [b_{v_i+m_i}(1)A_i u(\tau_b) + \cdots + b_0(1)u_{i,0}] + \sum_{i\in\mathbb{N}_n\setminus\mathcal{I}_{\max,b}} [A_i u(1+v_i+m_i) + \cdots + c_0(1)u_{i,0}] = f(1).$$

Since we have assumed (160), it follows that $u(\tau_b + 1)$ is uniquely determined as well as that $u(\tau_b + 1) \in D(A_j)$ for all $j \in \mathbb{N}_n$ and

$$u(\tau_b+1) = \Big(\sum_{i\in\mathcal{I}_{\max,b}} A_i\Big)^{-1} f(1) - \Big(\sum_{i\in\mathcal{I}_{\max,b}} A_i\Big)^{-1} \sum_{j\in\mathcal{I}_{\max,b}} [b_{v+v_j+m_j}(v)A_j u(v+v_j+m_j-1) + \cdots + b_0(v)A_j u(0)] - \Big(\sum_{i\in\mathcal{I}_{\max,b}} A_i\Big)^{-1} \sum_{j\in\mathbb{N}_n\setminus\mathcal{I}_{\max,b}} [u_{j,v_j+m_j} + \cdots + c_0(0)u_{j,0}].$$

Taking into account the assumptions (159)–(160), we can continue this procedure by plugging $v = 2, \ldots$, determining in such a way the values of $u(\tau_b + 2), \ldots$. The constructed sequence $(u(v))_{v \geqslant \tau_b}$ clearly satisfies (161). □

It is important to state the following important corollary of Theorem 3.3.17.

Corollary 3.3.18. *Suppose $A_1, \ldots, A_n \in L(X)$ and $(\sum_{i \in \mathcal{I}_{\max,b}} A_i)^{-1} \in L(X)$. Then there exists a strong solution of problem* (150), *endowed with the initial conditions of the form* (157), *respectively, there exists a unique strong solution of problem* (150), *endowed with the initial conditions of the form* (158).

We continue by providing a few illustrative applications of Theorem 3.3.17.

Example 3.3.19. (i) Suppose that $\mathcal{I}_{\max,b} = \mathbb{N}_n$. Then (160) holds and Theorem 3.3.17 can be applied provided that the operator $\sum_{i \in \mathcal{I}_{\max,b}} A_i$ is the integral generator of an exponentially decaying C-regularized semigroup; in this case, the resolvent set of $\sum_{i \in \mathcal{I}_{\max,b}} A_i$ can be empty. For example, Theorem 3.3.17 can be applied to the abstract fractional difference equation of the form

$$\Delta^{\alpha_1} A_1 u(v+1) + \Delta^{\alpha_2} A_2 u(v+1) + A_3 u(v+2) = f(v), \quad v \in \mathbb{N}_0,$$

where $0 < \alpha_1 \leqslant 1$ and $0 < \alpha_2 \leqslant 1$, as well as to the abstract fractional difference equation of the form

$$\Delta_C^{\alpha_1} A_1 u(v+1) + \Delta_C^{\alpha_2} A_2 u(v+2) + \Delta_C^{\alpha_3} A_3 u(v+3) + A_4 u(v+4) = f(v), \quad v \in \mathbb{N}_0,$$

where $2 < \alpha_1 \leqslant 3$, $1 < \alpha_2 \leqslant 2$ and $0 < \alpha_3 \leqslant 1$.

(ii) Suppose that $P_i(z)$ is a nonzero complex polynomial and $A_i := P_i(A)$, where A is a closed linear operator ($i \in \mathbb{N}_n$). Suppose further that the operator $\sum_{i \in \mathcal{I}_{\max,b}} A_i$ is the integral generator of an exponentially decaying C-regularized semigroup and there exists an index $i \in \mathcal{I}_{\max,b}$ such that $\mathrm{dg}(P_i) = \max\{\mathrm{dg}(P_j) : 1 \leqslant j \leqslant n\}$. Then Theorem 3.3.17 can be successfully applied; for example, this result can be applied to the abstract fractional difference equation of the form

$$\Delta_C^{\alpha_1} P_1(A) u(v+3) + \Delta_C^{\alpha_2} P_2(A) u(v+2) + \Delta_C^{\alpha_1} P_3(A) u(v+1) = f(v), \quad v \in \mathbb{N}_0,$$

where $1 < \alpha_1, \alpha_2, \alpha_3 \leqslant 2$ and $\mathrm{dg}(P_1) \geqslant \max(\mathrm{dg}(P_2), \mathrm{dg}(P_3))$.

Further on, it is clear that the problem (150) with $\mathcal{I} = \emptyset$ is a special case of the following problem:

$$A_{v+p,v} u(v+p) + A_{v+p-1,v} u(v+p-1) + \cdots + A_{0,v} u(0) = f(v), \quad v \in \mathbb{N}_0, \tag{162}$$

with $p := \tau_b$, $A_{v+p,v} := \sum_{i \in \mathcal{I}_{\max,b}} A_i$ and a certain sequence of linear operators $(A_{j,v})_{v \in \mathbb{N}_0; 0 \leqslant j \leqslant v+p-1}$. Concerning the well-posedness of this general problem, we will state the following result.

Theorem 3.3.20. *Suppose that $p \in \mathbb{N}$ and $A_{j,v}$ is a linear operator for all $v \in \mathbb{N}_0$ and $0 \leqslant j \leqslant v+p-1$, as well as that there exists an injective operator $C \in L(X)$ such that $CA_{j,v} \subseteq A_{j,v}C$ for all $v \in \mathbb{N}_0$ and $0 \leqslant j \leqslant v+p-1$, $C^{-1}A_{v+p,v}C = A_{v+p,v}$ for all $v \in \mathbb{N}_0$ and $A_{v+p,v}^{-1}C \in L(X)$ for all $v \in \mathbb{N}_0$. Then there exists a unique solution of problem* (162), *endowed with the initial conditions of the form*

$$u(j) = u_j, \quad 0 \leqslant j \leqslant p-1,$$

provided that $f(v) \in R(A_{v+p,v})$ for all $v \in \mathbb{N}_0$, $A_{v+p,v}^{-1}f(v) \in D([A_{p+k+1,k+1}^{-1}A_{p+k,k+1}] \cdot \cdots \cdot [A_{v+p+1,v+1}^{-1}A_{v+p,v+1}])$ for all $k \geqslant v$, $u_0 \in D([A_{p+k+1,k+1}^{-1}A_{p+k,k+1}] \cdot \cdots \cdot [A_{p+1,1}^{-1}A_{p,1}] \cdot [A_{p,0}^{-1}A_{0,0}])$ for all $k \in \mathbb{N}_0, \ldots$, and $u_{p-1} \in D([A_{p+k+1,k+1}^{-1}A_{p+k,k+1}] \cdot \cdots \cdot [A_{p+1,1}^{-1}A_{p,1}] \cdot [A_{p,0}^{-1}A_{p-1,0}])$ for all $k \in \mathbb{N}_0$.

Proof. Multiplying the equation (162) with $A_{v+p,v}^{-1}C$ and using the equality $C^{-1}A_{v+p,v}^{-1}C = A_{v+p,v}^{-1}$ for all $v \in \mathbb{N}_0$, it can be simply proved that the equation (162) is equivalent with the equation

$$u(v+p) + B_{v+p-1,v}u(v+p-1) + \cdots + B_{0,v}u(0) = A_{v+p,v}^{-1}f(v), \quad v \in \mathbb{N}_0;$$
$$u(j) = u_j, \quad 0 \leqslant j \leqslant p-1,$$

where $B_{j,v} := A_{v+p}^{-1}A_{j,v}$ for all $v \in \mathbb{N}_0$ and $0 \leqslant j \leqslant v+p-1$. Proceeding by induction, and using our assumptions, this implies the required statement as in the proof of Theorem 3.3.17. □

3.3.6 Degenerate multiterm fractional difference equations with Riemann–Liouville fractional derivatives

In this subsection, we will briefly consider the abstract multiterm fractional difference equations with Riemann–Liouville fractional derivatives. Set $T_{i,L}^{RL}u(v) := A_i\Delta^{\alpha_i}u(v)$, if $v \in \mathbb{N}_0$ and $i \in \mathbb{N}_n$, and $T_{i,R}^{RL}u(v) := \Delta^{\alpha_i}A_iu(v)$, if $v \in \mathbb{N}_0$ and $i \in \mathbb{N}_n$. Henceforth, it will be assumed that, for every $v \in \mathbb{N}_0$ and $i \in \mathbb{N}_n$, $T_i^{RL}u(v)$ denotes either $T_{i,L}^{RL}u(v)$ or $T_{i,R}^{RL}u(v)$. Of concern is the following analogue of (150):

$$\sum_{i=1}^{n} T_i^{RL}u(v+v_i) = f(v), \quad v \in \mathbb{N}_0, \tag{163}$$

where $v_1, \ldots, v_n \in \mathbb{N}_0$ and $f : \mathbb{N}_0 \to X$ is a given sequence. The equation (163) is equivalent with (see also (147)):

$$\sum_{i \in \mathcal{I}} A_i \left[\sum_{j=0}^{m_i} \sum_{l=0}^{v+v_i+j} (-1)^{m_i-j} \binom{m_i}{j} k^{m_i-\alpha_i}(v+v_i+j-l)u(l) \right]$$

$$+\sum_{i\in\mathbb{N}_n\setminus\mathcal{I}}\left[\sum_{j=0}^{m_i}\sum_{l=0}^{v+v_i+j}(-1)^{m_i-j}\binom{m_i}{j}k^{m_i-a_i}(v+v_i+j-l)A_iu(l)\right]=f(v),\quad v\in\mathbb{N}_0. \tag{164}$$

This suggests us to endow (163) with the initial conditions (152); we assume here that the compatibility condition (153) holds. We introduce the notion of a strong solution of problem [(163); (152)] in the same way as in Definition 3.3.12.

It is worth mentioning that Theorem 3.3.16 and Theorem 3.3.17(ii) can serve one to clarify a sufficient condition for the problem [(163); (152)] to have at most one strong solution, while Theorem 3.3.17(i) can serve one to clarify a sufficient condition for the problem [(163); (155)] to have at most one strong solution. It is also worth noting that Theorem 3.3.17 remains true for the equation (164) endowed with the same initial conditions as in the formulation of this result. The problem (163) with $\mathcal{I}=\emptyset$ is a special case of the problem (162), as well.

We continue with the following example.

Example 3.3.21. Suppose that $a_j-a_1\in\mathbb{N}$ for $j=2,3,\ldots,n$ and

$$\sum_{i=1}^{n}T_{i,R}^{RL}u(v+v_i)=f(v),\quad v\in\mathbb{N}_0.$$

Let $\Delta^{a_1}g=0$ and $g(v)\notin D(A_1)\cup\cdots\cup D(A_n)$ for $v=0,1,\ldots,m_1-1$; the sequence $g(\cdot)$ can be constructed similarly as in Example 3.3.10. Since $\Delta^{a_1+k}g=\Delta^k(\Delta^{a_1}g)=0$ for all $k\in\mathbb{N}$, it follows that $\Delta^{a_j}g=0$ for all $j\in\mathbb{N}_n$. Define $u_g(v):=u(v)+g(v)$, $v\in\mathbb{N}_0$. Then it can be easily shown that

$$\sum_{i=1}^{n}T_{i,L}^{RL}u_g(v+v_i)=f(v),\quad v\in\mathbb{N}_0;$$

furthermore, the term $\Delta^{a_j}(A_ju_g)$ is not defined for any $j\in\mathbb{N}_n$.

Using the recent result of L. Abadias et al. [4, Theorem 5.5], we can simply transfer the statement of Theorem 3.3.7 to the abstract multiterm fractional difference equations with Riemann–Liouville derivatives. More precisely, suppose that

$$A_nD_t^{a_n}u(t)+A_{n-1}D_t^{a_{n-1}}u(t)+\cdots+A_1D_t^{a_1}u(t)=f(t),$$

where $f:[0,\infty)\to X$ is a continuous function. Set $y(v):=\int_0^{+\infty}e^{-t}(t^v/v!)u(t)\,dt$, $v\geqslant 0$. Then, under certain logical assumptions, we have

$$A_n(\Delta^{a_n}y)(v)+\sum_{j=1}^{n-1}A_j(\Delta^{a_j}y)(v+m_n-m_j)=\int_0^{+\infty}e^{-t}\frac{t^{v+m_n}}{(v+m_n)!}f(t)\,dt,\quad v\in\mathbb{N}_0; \tag{165}$$

cf. [443–445] and [792] for many interesting examples which can be used to provide certain applications of this result and the formula established in the equation (166) below.

Furthermore, using the assertions of [4, Theorem 5.5(ii), Theorem 5.5] and the identities [97, (1.22)–(1.23)], we can simply show that, under certain logical assumptions, we have the following formula:

$$\int\limits_{0}^{+\infty} e^{-t} \frac{t^{v+m_n}}{(v+m_n)!} \mathbf{D}_t^{\alpha} u(t)\, dt$$

$$= (\Delta^{\alpha} y)(v) + \frac{(-1)^{v+m+1}}{(v+m)!} \sum_{k=0}^{m-1} u^{(k)}(0)(\alpha-1-k)\cdots\cdots(\alpha-k-v-m), \quad v \in \mathbb{N}_0, \quad (166)$$

where $m = \lceil \alpha \rceil$. This formula can be used to transfer the statement of Theorem 3.3.7 to the abstract multiterm fractional difference equations with Caputo derivatives. The statement of [4, Theorem 5.5] and the equation (166) can be reworded for the Poisson-like transforms given in (137) and (143).

We close this section with the observation that there are many other important questions with regards to the abstract (degenerate) multiterm fractional difference equations with Caputo fractional derivatives or Riemann–Liouville fractional derivatives. For example, C. Lizama and M. Murillo-Arcila have analyzed, in [537], the question of L^p-maximal regularity for a class of the abstract fractional difference equations with Riemann–Liouville fractional derivatives of order $\alpha \in (1,2]$. Their investigation has been carried out on UMD spaces (cf. also the research article [814], where J. Zhang and S. Bu have recently analyzed the same equation for the abstract fractional difference equations of order $\alpha \in (2,3]$).

3.4 Abstract nonscalar Volterra difference equations

Concerning the abstract nonscalar Volterra integral equations, mention should be made of the important monograph [652] by J. Prüss and the monographs [442, 443, 445] by M. Kostić. The main aim of this section is to examine various classes of the abstract nonscalar Volterra difference equations. We use the Poisson-like transforms to make a connection between the solutions of the abstract nonscalar Volterra integrodifferential equations and the abstract nonscalar Volterra difference equations. We consider here the abstract degenerate Volterra difference equation

$$Bu(v) = f(v) + \sum_{j=0}^{v} A(v-j)u(j), \quad v \in \mathbb{N}_0, \quad (167)$$

where $f : \mathbb{N}_0 \to X$, B is a closed linear operator on X and $A : \mathbb{N}_0 \to L(Y,X)$. We assume that Y is a Banach space which is continuously embedded in X and we also consider the nonautonomous generalization of (167). We thoroughly analyze the well-posedness of (167) in the following particular case:

$$A(v) = a_1(v)A_1 + \cdots + a_n(v)A_n, \quad v \in \mathbb{N}_0, \quad (168)$$

where $n \in \mathbb{N}, A_1, \ldots, A_n$ are closed linear operators on X, and $(a_1(v))_{v\in\mathbb{N}_0}, \ldots, (a_n(v))_{v\in\mathbb{N}_0}$ are given sequences in X. It is also worth noticing that the equation (167) is a special case of the equation

$$Bu(v) = f(v) + \sum_{i=1}^{n} \sum_{j=0}^{v+v_i} A_i(v + v_i - j)u(j), \quad v \in \mathbb{N}_0, \tag{169}$$

where $v_1, \ldots, v_n \in \mathbb{N}_0$ and $A_1 : \mathbb{N}_0 \to L(Y,X), \ldots, A_n : \mathbb{N}_0 \to L(Y,X)$. We will not consider the well-posedness of problem (169) in general case and we will only focus our attention here to the case in which the kernel $(A_i(v))_{v\in\mathbb{N}_0}$ can be expressed as $A_i(\cdot) := a_i(\cdot)A_i, 1 \leqslant i \leqslant n$.

In the previous section, we have analyzed the abstract multiterm fractional difference equations with Riemann–Liouville and Caputo fractional derivatives. In this section, we continue this part by investigating the abstract multiterm fractional difference equations with Weyl fractional derivatives. In contrast to the study of fractional difference equations with Riemann–Liouville fractional derivatives and Caputo fractional derivatives, the domain of solutions of the considered fractional difference equations is the whole integer line $\mathbb{Z}$; moreover, we do not accompany the considered fractional difference equations with any initial conditions. Here, we also want to emphasize that the use of sequences $a_1(\cdot) = k^{m_1-a_1}(\cdot), \ldots, a_n(\cdot) = k^{m_n-a_n}(\cdot)$ in (168), where $(k^a(v))_{v\in\mathbb{N}_0}$ is the Cèsaro sequence, plays a crucial role in our analysis; cf. Section 3.4.5 for the notion and more details.

The structure of this section can be briefly described as follows. First of all, we recall the basic facts about the abstract Volterra integral equations of nonscalar type in Section 3.4.1. Discrete (A, k, B)-regularized C-resolvent families are investigated in Section 3.4.2; in Section 3.4.3, we introduce and systematically analyze various notions of the $(k, C, B, (A_i)_{1\leqslant i\leqslant n})$-solution operator families connected with the use of kernel $(A(v))_{v\in\mathbb{N}_0}$ given by (168). The asymptotically almost periodic type solutions of the abstract multiterm discrete abstract Cauchy problem

$$Bu(v) = f(v) + \sum_{i=1}^{n} A_i(a_i *_0 u)(v + v_i), \quad v \in \mathbb{N}_0,$$

where $v_1, \ldots, v_n \in \mathbb{N}_0$, are analyzed in Section 3.4.4.

The abstract multiterm fractional difference equations with Weyl fractional derivatives are investigated in Section 3.4.5; here, we also investigate the well-posedness of the following abstract multiterm difference equation:

$$Bu(v) = A_1 \sum_{l=-\infty}^{v+v_1} a_1(v + v_1 - l)u(l) + \cdots + A_n \sum_{l=-\infty}^{v+v_n} a_n(v + v_n - l)u(l), \quad v \in \mathbb{Z},$$

where B, $A_1, \ldots, A_n$ are closed linear operator on X. In particular, we consider the existence of solutions to the problems

$$\begin{aligned}(\Delta^m Bu)(v) &= A_1(\Delta_W^{\alpha_1}u)(v+v_1)+\cdots\\ &\quad + A_n(\Delta_W^{\alpha_n}u)(v+v_n)+\Delta^m(k\circ Cf)(v)+\Delta^m g(v), \quad v\in\mathbb{Z}\end{aligned}$$

and

$$\begin{aligned}B(\Delta^{m_n}h)(v) &= \sum_{j=1}^{n-1} A_j(\Delta^{m_n-m_j}\Delta_W^{\alpha_j}h)(v+v_j)\\ &\quad + A_n(\Delta_W^{\alpha_n}h)(v+v_n)+(k\circ Cf)(v)+g(v), \quad v\in\mathbb{Z};\end{aligned}$$

cf. Theorem 3.4.17 for the notation and more details. It is also worthwhile to mention that we connect the solutions of the abstract multiterm fractional differential equation

$$A_n D_t^{\alpha_n}u(t)+A_{n-1}D_t^{\alpha_{n-1}}u(t)+\cdots+A_1D_t^{\alpha_1}u(t)=0, \quad t>0,$$

where $0 \leqslant \alpha_1 < \alpha_2 < \cdots < \alpha_n$ and $D_t^{\alpha}u(t)$ denotes the Riemann–Liouville fractional derivative of function $u(t)$ of order $\alpha > 0$, with the solutions of the abstract multi-term fractional difference equation

$$\begin{aligned}&A_n[\Delta_W^{\alpha_n}u](v)+A_{n-1}[\Delta_W^{\alpha_{n-1}}u](v+m_n-m_{n-1})\\ &\quad +\cdots+A_1[\Delta_W^{\alpha_1}u](v+m_n-m_1)=-g(v), \quad v\in\mathbb{Z};\end{aligned}$$

cf. Example 3.4.20 for more details. We also consider certain connections between the abstract multiterm fractional differential equations with Caputo derivatives and the abstract multiterm fractional difference equation of the above form; we particularly analyze the existence of (almost periodic type) solutions to the problem

$$[\Delta_W^{\alpha}u](v)+Au(v+m)=-g_{\omega}(v), \quad v\in\mathbb{Z},$$

where $\alpha > 0$ and $m = \lceil\alpha\rceil$; cf. Example 3.4.22 for more details. All considered Volterra difference equations and fractional difference equations can be unsolvable with respect to the highest derivative.

If the sequences $(a_k)_{k\in\mathbb{N}_0}$ and $(b_k)_{k\in\mathbb{Z}}$ are given, then we define the Weyl convolution product $(a\circ b)(\cdot)$ by

$$(a\circ b)(v) := \sum_{l=-\infty}^{v} a(v-l)b(l), \quad v\in\mathbb{Z}.$$

Due to [429, Theorem 3.12(ii)–(iii)], some logical assumptions ensure that (cf. also the proof of Theorem 3.4.17 below):

$$(f *_0 g)\circ h = g\circ(f\circ h) = f\circ(g\circ h). \tag{170}$$

3.4.1 Abstract nonscalar Volterra integral equations

Let $A(t)$ be a locally integrable function from $[0,\tau)$ into $L(Y,X)$. For the sake of better readability, we will recall the notion introduced in [445, Definitions 2.9.1–2.9.3].

Definition 3.4.1. Let $k \in C([0,\tau))$ and $k \neq 0$, let $\tau \in (0,\infty]$, $f \in C([0,\tau) : X)$, and let $A \in L^1_{loc}([0,\tau) : L(Y,X))$. Of concern is the following degenerate Volterra equation:

$$Bu(t) = f(t) + \int_0^t A(t-s)u(s)\,ds, \quad t \in [0,\tau). \tag{171}$$

Then a function $u \in C([0,\tau) : [D(B)])$ is said to be a strong solution of (171) if $u \in L^\infty_{loc}([0,\tau) : Y)$ and (171) holds on $[0,\tau)$.

In the following definition, we introduce the main solution concepts for dealing with (171).

Definition 3.4.2. Let $\tau \in (0,\infty]$, $k \in C([0,\tau))$, $k \neq 0$ and $A \in L^1_{loc}([0,\tau) : L(Y,X))$. A family $(S(t))_{t\in[0,\tau)}$ in $L(X,[D(B)])$ is said to be an (A,k,B)-regularized C-pseudoresolvent family if the following holds:

$(S1)_c$ The mappings $t \mapsto S(t)x$, $t \in [0,\tau)$ and $t \mapsto BS(t)x$, $t \in [0,\tau)$ are continuous in X for every $x \in X$, $BS(0) = k(0)C$ and $S(t)C = CS(t)$, $t \in [0,\tau)$.

(S2) Set $U(t)x := \int_0^t S(s)x\,ds$, $x \in X$, $t \in [0,\tau)$. Then (S2) means $U(t)Y \subseteq Y$, $U(t)_{|Y} \in L(Y)$, $t \in [0,\tau)$ and $(U(t)_{|Y})_{t\in[0,\tau)}$ is locally Lipschitz continuous in $L(Y)$.

(S3) The resolvent equations

$$BS(t)y = k(t)Cy + \int_0^t A(t-s)dU(s)y, \quad t \in [0,\tau),\ y \in Y, \tag{172}$$

$$BS(t)y = k(t)Cy + \int_0^t S(t-s)A(s)y\,ds, \quad t \in [0,\tau),\ y \in Y, \tag{173}$$

hold; (172), resp. (173), is called the first resolvent equation, respectively, the second resolvent equation.

An (A,k,B)-regularized C-pseudoresolvent family $(S(t))_{t\in[0,\tau)}$ is said to be an (A,k,B)-regularized C-resolvent family if additionally:

(S4) For every $y \in Y$, $S(\cdot)y \in L^\infty_{loc}([0,\tau) : Y)$.

An operator family $(S(t))_{t\in[0,\tau)}$ in $L(X,[D(B)])$ is called a weak (A,k,B)-regularized C-pseudoresolvent family if and only if $(S1)_c$ and (173) hold.

The condition (S3) can be rewritten in the following equivalent form:

(S3)' $$BU(t)y = \Theta(t)Cy + \int_0^t A(t-s)U(s)y\,ds, \quad t \in [0,\tau),\ y \in Y,$$

$$BU(t)y = \Theta(t)Cy + \int_0^t U(t-s)A(s)y\,ds, \quad t \in [0,\tau),\ y \in Y.$$

We also need the following notion.

Definition 3.4.3. Let $\tau \in (0,\infty]$, $k \in C([0,\tau))$, $k \neq 0$ and $A \in L^1_{loc}([0,\tau) : L(Y,X))$. A strongly continuous operator family $(V(t))_{t\in[0,\tau)} \subseteq L(X)$ is said to be an (A,k,B)-regularized C-uniqueness family if

$$V(t)By = k(t)Cy + \int_0^t V(t-s)A(s)y\,ds, \quad t \in [0,\tau),\ y \in Y \cap D(B).$$

3.4.2 Discrete (A,k,B)-regularized C-resolvent families

Suppose that $B(k)$ is a closed linear operator acting in X ($k \in \mathbb{N}_0$), and $A : \mathbb{N}_0 \to L(Y,X)$. We would like to introduce the following notion.

Definition 3.4.4. (i) Let $f : \mathbb{N}_0 \to X$. Then it is said that a sequence $(u_k)_{k\in\mathbb{N}_0}$ is a solution of the abstract degenerate Volterra difference equation

$$B(v)u(v) = f(v) + \sum_{j=0}^{v} A(v-j)u(j), \quad v \in \mathbb{N}_0 \tag{174}$$

if $u(j) \in D(B(j)) \cap Y$ for all j, $v \in \mathbb{N}_0$, and (174) holds.

(ii) Let $k : \mathbb{N}_0 \to \mathbb{C}$ and $k \neq 0$. Then a family $(S(v))_{v\in\mathbb{N}_0}$ in $L(X)$ is said to be a discrete (A,k,B)-regularized C-resolvent family if $S(v)C = CS(v)$, $v \in \mathbb{N}_0$ and the following holds:

(S1) $S(v)Y \subseteq Y$, $S(v)_{|Y} \in L(Y)$, $v \in \mathbb{N}_0$,

$$B(v)S(v)y = k(v)Cy + \sum_{j=0}^{v} A(v-j)S(j)y, \quad v \in \mathbb{N}_0,\ y \in Y, \tag{175}$$

and

$$B(v)S(v)y = k(v)Cy + \sum_{j=0}^{v} S(v-j)A(j)y, \quad v \in \mathbb{N}_0,\ y \in Y; \tag{176}$$

(175), respectively, (176), is called the first resolvent equation, respectively, the second resolvent equation.

An operator family $(S(v))_{v\in\mathbb{N}_0}$ in $L(X)$ is called a discrete weak (A,k,B)-regularized C-resolvent family if $S(v)C = CS(v)$, $v \in \mathbb{N}_0$, $S(v)Y \subseteq Y$, $S(v)_{|Y} \in L(Y)$, $v \in \mathbb{N}_0$ and (176) hold.

(iii) Let $k : \mathbb{N}_0 \to \mathbb{C}$ and $k \neq 0$. Then a strongly continuous operator family $(W(v))_{v\in\mathbb{N}_0} \subseteq L(X)$ is said to be a discrete (A,k,B)-regularized C-uniqueness family if, for every $v \in \mathbb{N}_0$, we have

$$W(v)B(v)y = k(v)Cy + \sum_{j=0}^{v} W(v-j)A(j)y, \quad y \in Y \cap D(B(v)).$$

The notion introduced in Definition 3.4.4(ii)–(iii) has not been considered elsewhere even in the case that $B = I$. Further on, it is clear that the assumptions $S(v)Y \subseteq Y$, $S(v)_{|Y} \in L(Y)$, $v \in \mathbb{N}_0$ and $A(j)S(v)y = S(v)A(j)y$ for all j, $v \in \mathbb{N}_0$ and $y \in Y$ ensure that (176) implies (175). We feel it is our duty to say that the equation (175) is only an unsatisfactory attempt to provide a discrete analogue of the first resolvent equation (172) as well as that the last mentioned equation can be also considered in the nonautonomous setting (with certain difficulties).

In the following proposition, we analyze the uniqueness of solutions to (174).

Proposition 3.4.5. *Suppose that a sequence $(u_k)_{k\in\mathbb{N}_0}$ is a solution of* (174) *and the operator $B(v) - A(0)$ is injective for all $v \in \mathbb{N}_0$. Then $u(\cdot)$ is uniquely determined.*

Proof. Suppose that $f \equiv 0$ in (174). Then it is clear that (174) with $v = 0$ gives $B(0)u(0) = A(0)u(0)$, so that $u(0)$ is uniquely determined. If $v = 1$, then we get $B(1)u(1) = A(1)u(0) + A(0)u(1)$, so that $u(1)$ is uniquely determined, as well. Proceeding in this way and using the injectiveness of the operator $B(v)-A(0)$ for all $v \in \mathbb{N}_0$, we simply obtain the required statement. □

Further on, if $B(v) \equiv B$, $g : \mathbb{N}_0 \to \mathbb{C}$ and $g \neq 0$, then we define $S_g(v)x := (g *_0 S)(v)x$, $v \in \mathbb{N}_0$ and $x \in X$. It can be simply shown that, if $(S(v))_{v\in\mathbb{N}_0} \subseteq L(X)$ is a discrete (weak) (A,k,B)-regularized C-resolvent family [$(W(v))_{v\in\mathbb{N}_0} \subseteq L(X)$ is a discrete (A,k,B)-regularized C-uniqueness family], then $(S_g(v))_{v\in\mathbb{N}_0} \subseteq L(X)$ [$(W_g(v))_{v\in\mathbb{N}_0} \subseteq L(X)$] is a discrete (weak) $(A,k *_0 g,B)$-regularized C-resolvent family [discrete $(A,k *_0 g,B)$-regularized C-uniqueness family].

Concerning the notion of a discrete (A,k,B)-regularized C-uniqueness family, we would like to state the following result (cf. also [445, Proposition 2.9.5(i)]; the interested reader may try to transfer the statement of [445, Proposition 2.9.4(i)] to the discrete solution operator families, as well).

Proposition 3.4.6. *Suppose that $k : \mathbb{N}_0 \to \mathbb{C}$, $k \neq 0$, a sequence $(u_k)_{k\in\mathbb{N}_0}$ is a solution of the abstract degenerate Volterra difference equation* (174) *and a strongly continuous operator family $(W(v))_{v\in\mathbb{N}_0} \subseteq L(X)$ is a discrete (A,k,B)-regularized C-uniqueness family.*

(i) *Then* $W(0)f(0) = k(0)Cu(0)$.
(ii) *If* $B(v) \equiv B$, *then we have* $(kC *_0 u)(v) = (W *_0 f)(v)$ *for all* $v \in \mathbb{N}_0$. *In particular, if* $k(0) \neq 0$ *and* C *is injective, then there exists at most one solution of* (174).

Proof. The given assumptions imply

$$W(0)B(0)u(0) = k(0)Cu(0) + W(0)A(0)u(0) \quad \text{and} \quad B(0)f(0) = f(0) + A(0)u(0).$$

This simply yields $W(0)f(0) = k(0)Cu(0)$. To prove (ii) with $B(v) \equiv B$, observe first that

$$W(1)Bu(0) = k(1)Cu(0) + W(1)A(0)u(0) + W(0)A(1)u(0) \quad \text{and} \quad Bu(0) = f(0) + A(0)u(0).$$

Hence,

$$W(1)f(0) = k(1)Cu(0) + W(0)A(1)u(0). \tag{177}$$

Similarly, we have

$$W(0)Bu(1) = k(0)Cu(1) + W(0)A(0)u(1) \quad \text{and} \quad Bu(1) = f(1) + A(1)u(0) + A(0)u(1).$$

Hence,

$$W(0)f(1) = k(0)Cu(1) - W(0)A(1)u(0). \tag{178}$$

Keeping in mind (177)–(178), we get that $(kC *_0 u)(1) = (W *_0 f)(1)$. Considering the terms $W(v)Bu(0)$, $W(v-1)Bu(1), \ldots,$ and $W(0)Bu(v)$, a similar line of reasoning gives that $(kC *_0 u)(v) = (W *_0 f)(v)$ for all $v \in \mathbb{N}$. It is clear that, if $f \equiv 0$, $k(0) \neq 0$ and C is injective, then $(kC *_0 u)(v) = 0$ for all $v \in \mathbb{N}_0$; inductively, it readily follows that $u \equiv 0$. □

We continue by stating the following result (cf. also Theorem 3.4.17 with $v_1 = \cdots = v_n = 0$).

Theorem 3.4.7. *Suppose that* $k : \mathbb{N}_0 \to \mathbb{C}$, $k \neq 0$ *and* $(S(v))_{v \in \mathbb{N}_0} \subseteq L(X)$ *is an operator family such that* $S(v)Y \subseteq Y$, $S(v)_{|Y} \in L(Y)$, $v \in \mathbb{N}_0$, (175) *holds (this, in particular, holds if* $(S(v))_{v \in \mathbb{N}_0}$ *is a discrete* (A, k, B)*-regularized* C*-resolvent family),* $\sum_{v=0}^{+\infty} \|S(v)_{|Y}\|_{L(Y)} < +\infty$ *and* $\sum_{v=0}^{+\infty} \|[A *_0 S](v)\|_{L(Y,X)} < +\infty$. *Suppose, further, that* (i) *or* (ii) *holds, where:*
(i) $f : \mathbb{Z} \to Y$ *is a bounded sequence,* $k \in l^1(\mathbb{Z} : Y)$ *and* $\sum_{v=0}^{+\infty} \|A(v)\|_{L(Y,X)} < +\infty$.
(ii) $f \in l^1(\mathbb{Z} : Y)$, $k : \mathbb{Z} \to Y$ *is a bounded sequence and* $\sup_{v \geqslant 0} \|A(v)\|_{L(Y,X)} < +\infty$.

Define $u(v) := \sum_{l=-\infty}^{v} S(v-l)f(l)$, $v \in \mathbb{Z}$. *Then* $u(\cdot)$ *is bounded,* $u \in l^1(\mathbb{Z} : Y)$ *provided that* (ii) *holds, and*

$$B(v)u(v) = (k \circ Cf)(v) + \sum_{l=-\infty}^{v} A(v-l)u(l), \quad v \in \mathbb{Z}. \tag{179}$$

Proof. It is clear that the sequence $(k \circ Cf)(\cdot)$ is well-defined, $u(\cdot)$ is well-defined and bounded if (i) holds as well as that $u \in l^1(\mathbb{Z} : X)$ if (b) holds. The functional equality (S1) and the closedness of the linear operator $B(v)$ in combination with the assumptions $\sum_{v=0}^{+\infty} \|S(v)_{|Y}\|_{L(Y)} < +\infty$ and $\sum_{v=0}^{+\infty} \|[A *_0 S](v)\|_{L(Y,X)} < +\infty$ simply imply the following:

$$
\begin{aligned}
&Bu(v) - (k \circ Cf)(v) \\
&= \sum_{l=-\infty}^{v} BS(v-l)f(l) - (k \circ Cf)(v) \\
&= -(k \circ Cf)(v) + \sum_{l=-\infty}^{v} [k(v-l)Cf(l) + [A *_0 S](v-l)f(l)] \\
&= \sum_{l=-\infty}^{v} [A *_0 S](v-l)f(l) = \sum_{s=0}^{+\infty} [A *_0 S](s)f(v-s) \\
&= \sum_{s=0}^{+\infty} A(s) \sum_{r=0}^{+\infty} S(r)f(v-s-r) = \sum_{s=0}^{+\infty} A(s)u(v-s) = \sum_{l=-\infty}^{v} A(v-l)u(l), \quad v \in \mathbb{Z},
\end{aligned}
$$

where we have also used the Fubini theorem in the last line of computation. □

The main result of this subsection reads as follows.

Theorem 3.4.8. *Suppose that $a \in \mathbb{R}$, $\omega \in \mathbb{R} \setminus \{0\}$, $k : \mathbb{N}_0 \to \mathbb{C}$ and $k \neq 0$, $(S(t))_{t\geqslant 0}$ in $L(X, [D(B)])$ is a weak (A, k, B)-regularized C-pseudoresolvent family and $(W(t))_{t\geqslant 0} \subseteq L(X)$ is an (A, k, B)-regularized C-uniqueness family. Suppose further that there exist finite real constants $M \geqslant 1$ and $\varepsilon > 0$ such that $\|A(t)\|_{L(Y,X)} + \|S(t)\|_{L(X,[D(B)])} + \|W(t)\|_{L(X)} \leqslant M\exp(t(a-\varepsilon))$, $t \geqslant 0$. Define*

$$
k_{a,w}(v) := \int_0^{+\infty} e^{-at} \frac{(\omega t)^v}{v!} k(t)\,dt, \quad v \in \mathbb{N}_0,
$$

$$
A_{a,w}(v)y := \int_0^{+\infty} e^{-at} \frac{(\omega t)^v}{v!} A(t)y\,dt, \quad v \in \mathbb{N}_0,\ y \in Y,
$$

$$
S_{a,w}(v)x := \int_0^{+\infty} e^{-at} \frac{(\omega t)^v}{v!} S(t)x\,dt, \quad v \in \mathbb{N}_0,\ x \in X
$$

and

$$
W_{a,w}(v)x := \int_0^{+\infty} e^{-at} \frac{(\omega t)^v}{v!} W(t)x\,dt, \quad v \in \mathbb{N}_0,\ x \in X.
$$

Then $(S_{a,w}(v))_{v\in\mathbb{N}_0}$ is a discrete weak $(A, k_{a,w}, B)$-regularized C-resolvent family and $(W_{a,w}(v))_{v\in\mathbb{N}_0}$ is a discrete $(A, k_{a,w}, B)$-regularized C-uniqueness family. Furthermore, *if $(S(t))_{t\geqslant 0}$ is an (A,k,B)-regularized C-pseudoresolvent family, then for any $v\in\mathbb{N}$ and $y\in Y$, we have*

$$BS_{a,w}(v)y=k_{a,\omega}(v)Cy+a\omega^{v}(A_{a,w}*_0 U)(v)y-\omega^{v}(-1)^{v}(A_{a,w}*_0 U)(v-1)y.$$

Proof. Suppose that $v\in\mathbb{N}_0$ and $y\in Y$ are fixed. Applying the second resolvent equation, the basic operational properties of the vector-valued Laplace transform and the Leibniz rule, it readily follows that (cf. also the proof of [535, Theorem 3.4]):

$$\begin{aligned}
BS_{a,w}(v)y&=\int_0^{+\infty}e^{-at}\frac{(\omega t)^{v}}{v!}S(t)y\,dt=\int_0^{+\infty}e^{-at}\frac{(\omega t)^{v}}{v!}BS(t)y\,dt\\
&=\int_0^{+\infty}e^{-at}\frac{(\omega t)^{v}}{v!}\Bigg[k(t)Cy+\int_0^{t}S(t-s)A(s)y\,ds\Bigg]dt\\
&=k_{a,w}(v)Cy+(-1)^{v}\frac{\omega^{v}}{v!}\big[\hat{S}(\lambda)\hat{A}(\lambda)\big]^{(v)}_{\lambda=a}\\
&=k_{a,w}(v)Cy+(-1)^{v}\frac{\omega^{v-j}\omega^{j}}{v!}\sum_{j=0}^{v}\binom{v}{j}\hat{S}^{(v-j)}(a)\hat{A}^{(j)}(a)\\
&=k_{a,w}(v)Cy+\frac{\omega^{v-j}\omega^{j}}{v!}\sum_{j=0}^{v}\binom{v}{j}\int_0^{+\infty}e^{-at}t^{v-j}S(t)\Bigg[\int_0^{+\infty}e^{-ar}r^{v-j}A(r)y\,dr\Bigg]dt\\
&=k_{a,w}(v)Cy+(S_{a,w}*_0 A_{a,w})(v)y,
\end{aligned}$$

as required. The corresponding statement for $(W_{a,w}(v))_{v\in\mathbb{N}_0}$ follows similarly; suppose now that $(S(t))_{t\geqslant 0}$ is an (A,k,B)-regularized C-pseudoresolvent family. Fix $v\in\mathbb{N}$ and $y\in Y$. Applying the partial integration and (S3)', we get

$$\begin{aligned}
BS_{a,w}(v)y&=\int_0^{+\infty}e^{-at}\bigg[a\frac{(\omega t)^{v}}{v!}-\omega\frac{(\omega t)^{v-1}}{(v-1)!}\bigg]BU(t)y\,dt\\
&=\int_0^{+\infty}e^{-at}\bigg[a\frac{(\omega t)^{v}}{v!}-\omega\frac{(\omega t)^{v-1}}{(v-1)!}\bigg]\cdot\bigg\{\int_0^{t}k(s)\,ds\,Cy+(A*U)(t)y\bigg\}dt\\
&=k_{a,w}(v)Cy+\int_0^{+\infty}e^{-at}\bigg[a\frac{(\omega t)^{v}}{v!}-\omega\frac{(\omega t)^{v-1}}{(v-1)!}\bigg](A*U)(t)y\,dt.
\end{aligned}$$

Then we can argue as in the first part of the proof to complete it. □

It is worth mentioning that Theorem 3.4.8 can be applied to the abstract nonscalar Volterra integral equations considered by J. Prüss in [652, Chapters 7–9] (see, especially, the analysis of viscoelastic Timoshenko beam on p. 240) as well as to the abstract degenerate nonscalar Volterra integral equations considered in [445, Theorem 2.9.7] (see also the applications with the abstract differential operators in L^p-spaces given on pp. 240–241). In such a way, we can consider the well-posedness for certain classes of the abstract semidiscrete nonscalar Volterra difference equations.

Further on, the integrability of (A,k,B)-regularized C-resolvent families has been considered by J. Prüss in [652, Section 10.5, pp. 277–281], provided that $k \equiv 1$ and $B = C = I$. It is worth noting that the condition $\sum_{v=0}^{+\infty} \|S(v)_{|Y}\|_{L(Y)} < +\infty$ in Theorem 3.4.7 holds for the resolvents $(S(v) \equiv S_{1,1}(v))_{v\in\mathbb{N}_0}$ obtained by applying Theorem 3.4.8 to the resolvent operator families used in the analysis of Volterra nonscalar equations of variational type (cf. [652, Corollaries 10.7, 10.8; pp. 280–281]). In general case of exponentially bounded (A,k,B)-regularized C-resolvent families, we must use different values of a, which should be sufficiently large, and $w > 0$ in order to ensure the validity of condition $\sum_{v=0}^{+\infty} \|S(v)_{|Y}\|_{L(Y)} < +\infty$ in Theorem 3.4.7. If this is the case, then we can consider the existence and uniqueness of almost periodic type solutions to (179); the usual almost periodic solutions can be considered if the condition (i) in Theorem 3.4.7 holds, while the (metrically) Weyl almost periodic solutions of (179) can be considered if the condition (ii) in Theorem 3.4.7 holds.

We round off this subsection by providing some observations.

Remark 3.4.9. Define also

$$\Theta_{a,w}(v) := \int_0^{+\infty} e^{-at} \frac{(\omega t)^v}{v!} \int_0^t k(s)\,ds\,dt, \quad v \in \mathbb{N}_0$$

and

$$U_{a,w}(v)y := \int_0^{+\infty} e^{-at} \frac{(\omega t)^v}{v!} U(t)y\,dt, \quad v \in \mathbb{N}_0,\ y \in Y.$$

Then we have

$$BU_{a,w}(v)y = \Theta_{a,w}(v)Cy + (U_{a,w} *_0 A_{a,w})(v)y, \quad v \in \mathbb{N}_0,\ y \in Y$$

and

$$BU_{a,w}(v)y = \Theta_{a,w}(v)Cy + (A_{a,w} *_0 U_{a,w})(v)y, \quad v \in \mathbb{N}_0,\ y \in Y,$$

so that, for every $y \in Y$, the sequence $(u(v) \equiv U_{a,w}(v)y)_{v\in\mathbb{N}_0}$ is a strong solution of problem (174) with $B(v) \equiv B$ and $f(v) \equiv \Theta_{a,w}(v)Cy$.

Remark 3.4.10. We have already analyzed the following Poisson type transform:

$$v \mapsto S_{a,w,j}(v)x := \int_0^{+\infty} e^{-(at)^j} \frac{(\omega t)^v}{v!} S(t)x\,dt, \quad v \in \mathbb{N}_0,\ x \in X,$$

where $j \in \mathbb{N}$ and $j \geqslant 2$. It would be very difficult to state an analogue of Theorem 3.4.8 for this transform because we cannot so simply compute the term

$$\int_0^{+\infty} e^{-(at)^j} \frac{(\omega t)^v}{v!} \left[\int_0^t S(t-s)A(s)y \right] dt, \quad v \in \mathbb{N}_0, \; x \in X$$

in this framework.

3.4.3 $(k, C, B, (A_i)_{1\leqslant i\leqslant n}, (v_i)_{1\leqslant i\leqslant n})$-solution operator families

In this subsection, we analyze various classes of $(k, C, B, (A_i)_{1\leqslant i\leqslant n}, (v_i)_{1\leqslant i\leqslant n})$-solution operator families connected with the use of kernel $(A_i(v) \equiv a_i(v)A_i)_{v\in\mathbb{N}_0}$ in (169). If $v_1, \ldots, v_n \in \mathbb{N}_0$, then we define $v_{\max} := \max(v_1, \ldots, v_n)$ and $M := \{i \in \mathbb{N}_n : v_i = v_{\max}\}$. We will always assume henceforth that the following condition holds:

$$a_i : \mathbb{N}_0 \to \mathbb{C} \quad \text{for all } i \in \mathbb{N}_n \quad \text{and} \quad a_i(0) \neq 0, \quad i \in M. \tag{180}$$

We continue by introducing the following notion.

Definition 3.4.11. Suppose that $B, A_1, \ldots, A_n$ are closed linear operators on X, $C \in L(X)$, $v_1, \ldots, v_n \in \mathbb{N}_0$, $\mathcal{I} \subseteq \mathbb{N}_n$, $k : \mathbb{N}_0 \to \mathbb{C}$, $k \neq 0$ and (180) holds. Then we say that the operator family $(S(v))_{v\in\mathbb{N}_0} \subseteq L(X)$ is a discrete:

(i) $(k, C, B, (A_i)_{1\leqslant i\leqslant n}, (v_i)_{1\leqslant i\leqslant n})$-existence family if the mapping $x \mapsto A_i(a_i *_0 S)(v + v_i)x$, $x \in X$ belongs to $L(X)$ for $v \in \mathbb{N}_0, 1 \leqslant i \leqslant n$ and

$$BS(v)x = k(v)Cx + \sum_{i=1}^{n} A_i(a_i *_0 S)(v + v_i)x, \quad v \in \mathbb{N}_0, \; x \in X. \tag{181}$$

(ii) $(k, C, B, (A_i)_{1\leqslant i\leqslant n}, (v_i)_{1\leqslant i\leqslant n}, \mathcal{I})$-existence family if $(S(v))_{v\in\mathbb{N}_0}$ is $(k, C, B, (A_i)_{1\leqslant i\leqslant n}, (v_i)_{1\leqslant i\leqslant n})$-existence family and $S(v)A_i \subseteq A_iS(v)$ for all $v \in \mathbb{N}_0$ and $i \in \mathbb{N}_n \setminus \mathcal{I}$.

Any discrete $(k, C, B, (A_i)_{1\leqslant i\leqslant n}, (v_i)_{1\leqslant i\leqslant n})$-existence family $(S(v))_{v\in\mathbb{N}_0}$ satisfies $(S(v))_{v\in\mathbb{N}_0} \subseteq L(X, [D(B)])$. In the case that $v_1 = v_2 = \cdots = v_n = 0$, we omit the term "$(v_i)_{1\leqslant i\leqslant n}$" from the notation. If this is not the case, then we cannot expect the uniqueness of $(k, C, B, (A_i)_{1\leqslant i\leqslant n}, (v_i)_{1\leqslant i\leqslant n})$-existence family when all terms of this tuple are given in advance. For example, if $n = 1$, $B = C = \mathrm{I}$, $A_1 = 2\mathrm{I}$, $a_1 \equiv k \equiv 1$ and $v_1 = 1$, then we have $S(0)x + 2S(1)x + Cx = 0$, so that neither $S(0)$ nor $S(1)$ can be uniquely determined. The notion of a discrete $(k, C, B, (A_i)_{1\leqslant i\leqslant n})$-existence family looks similar but it is not exactly the same as the notion of a weak discrete (A, k, B)-regularized C-resolvent family with $B(\cdot) \equiv B$, $Y := X$ and $A(v)y := \sum_{i=1}^{n} a_i(v)A_iy$ for all $y \in Y$ and $v \in \mathbb{N}_0$.

We continue by stating the following result.

Proposition 3.4.12. *Suppose that $B, A_1, \ldots, A_n$ are closed linear operators on X, $C \in L(X)$, $v_1, \ldots, v_n \in \mathbb{N}_0$, $k : \mathbb{N}_0 \to \mathbb{C}$, $k \neq 0$, (180) holds and $(S(v))_{v\in\mathbb{N}_0} \subseteq L(X)$ is a discrete $(k, C, B, (A_i)_{1\leqslant i\leqslant n}, (v_i)_{1\leqslant i\leqslant n})$-existence family. If $x \in X$, $i \in \mathbb{N}_n$ and $v_i = 0$, then $S(v)x \in D(A_i)$ for all $v \in \mathbb{N}_0$ and $i \in \mathbb{N}_n$; the same holds for each $i \in \mathbb{N}_n$ with $v_i > 0$, provided that $S(0)x \in D(A_i), \ldots, S(v_i - 1)x \in D(A_i)$.*

Proof. We will prove the statement in case $i \in \mathbb{N}_n$ and $v_i > 0$. Then we have $(a_i *_0 S)(v_i)x \in D(A_i)$. Since we have assumed $a_i(0) \neq 0$ and $S(0)x \in D(A_i), \ldots, S(v_i-1)x \in D(A_i)$, this simply implies $S(v_i)x \in D(A_i)$. Since $(a_i *_0 S)(1 + v_i)x \in D(A_i)$, we similarly obtain $S(v_i + 1)x \in D(A_i)$. Proceeding by induction, we get $S(v)x \in D(A_i)$ for all $v \in \mathbb{N}_0$. □

Now we will state and prove the following result (cf. also [470, Theorem 2.5]).

Theorem 3.4.13. *Suppose that $B, A_1, \ldots, A_n$ are closed linear operators on X, $C \in L(X)$ is injective, $k : \mathbb{N}_0 \to \mathbb{C}$, $k(0) \neq 0$ and (180) holds.*

(i) *Suppose further that $(S(v))_{v\in\mathbb{N}_0} \subseteq L(X)$ is a discrete $(k, C, B, (A_i)_{1\leqslant i\leqslant n})$-existence family such that $S(0)Bx = BS(0)x$ and $S(0)A_ix = A_iS(0)x$ for all $x \in D(B) \cap D(A_1) \cap \cdots \cap D(A_n)$. Then $(B - \sum_{i=0}^{n} a_i(0)A_i)^{-1}C \in L(X)$, $S(0) = k(0)(B - \sum_{i=0}^{n} a_i(0)A_i)^{-1}C$,*

$$S(v)x = \left(B - \sum_{i=0}^{n} a_i(0)A_i\right)^{-1} \left[k(v)Cx + \sum_{i=1}^{n} A_i \sum_{j=0}^{v-1} a_i(v-j)S(j)x\right], \quad v \in \mathbb{N},\ x \in X, \tag{182}$$

and $A_iS(v) \in L(X)$ for all $i \in \mathbb{N}_n$ and $v \in \mathbb{N}_0$.

(ii) *Suppose that $C \in L(X)$ is injective, $(B - \sum_{i=0}^{n} a_i(0)A_i)^{-1}C \in L(X)$ and, for every $l \in \mathbb{N}$ and for every choice of integers $a_j \in \mathbb{N}_n$ for $1 \leqslant j \leqslant l$, we have*

$$\left[\prod_{j=1}^{l}\left(B - \sum_{i=0}^{n} a_i(0)A_i\right)^{-1} A_{a_j}\right] \cdot \left(B - \sum_{i=0}^{n} a_i(0)A_i\right)^{-1} C \in L(X). \tag{183}$$

Define $S(0) := k(0)(B - \sum_{i=0}^{n} a_i(0)A_i)^{-1}C$ and $S(v)$, $v \in \mathbb{N}$, recursively by (182). Then $(S(v))_{v\in\mathbb{N}_0} \subseteq L(X)$ is well-defined, $A_iS(v) \in L(X)$ for all $i \in \mathbb{N}_n$, $v \in \mathbb{N}_0$ and $(S(v))_{v\in\mathbb{N}_0}$ is a unique discrete $(k, C, B, (A_i)_{1\leqslant i\leqslant n})$-existence family. Furthermore, if $\mathcal{I} \subseteq \mathbb{N}_n$ and

$$\begin{aligned}
&CB \subseteq BC, \quad CA_i \subseteq A_iC \quad \text{for all } i \in \mathbb{N}_n \setminus \mathcal{I},\\
&(\forall i \in \mathbb{N}_n \setminus \mathcal{I})\,(\forall x \in D(A_i) \cap D(B))\, Bx \in D(A_i),\\
&A_ix \in D(B) \quad \text{and} \quad A_iBx = BA_ix,\\
&(\forall i \in \mathbb{N}_n \setminus \mathcal{I})(\forall j \in \mathbb{N}_n)\,(\forall x \in D(A_i) \cap D(A_j)),\\
&A_jx \in D(A_i), \quad A_ix \in D(A_j) \quad \text{and} \quad A_iA_jx = A_jA_ix,
\end{aligned} \tag{184}$$

respectively, there exist a closed linear operator A and the complex polynomials $P_B(\cdot)$, $P_1(\cdot), \ldots, P_n(\cdot)$ such that $CA \subseteq AC$ and $B = P_B(A)$, $A_1 = P_1(A), \ldots, A_n = P_n(A)$, then

$(S(v))_{v\in\mathbb{N}_0}$ is a discrete $(k, C, B, (A_i)_{1\leqslant i\leqslant n}, \mathcal{I})$-existence family, respectively, $(S(v))_{v\in\mathbb{N}_0}$ is a discrete $(k, C, B, (A_i)_{1\leqslant i\leqslant n}, \emptyset)$-existence family.

(iii) *Suppose that $C = \mathrm{I}$, $(B - \sum_{j=0}^{n} a_j(0)A_j)^{-1} \in L(X)$, $\sum_{v=0}^{+\infty} |a_i(v)| < +\infty$ for $1 \leqslant i \leqslant n$, $\sum_{v=0}^{\infty} |k(v)| < +\infty$, and* (a) *or* (b) *holds, where:*

(a) *$A_i \in L(X)$ for $1 \leqslant i \leqslant n$ and*

$$1 > \sum_{i=1}^{n}\sum_{v=1}^{+\infty} |a_i(v)| \cdot \left\| \left(B - \sum_{j=0}^{n} a_j(0)A_j\right)^{-1} A_i \right\|. \tag{185}$$

(b) *Suppose that $C = \mathrm{I}$,* (184) *holds or there exist a closed linear operator A and the complex polynomials $P_B(\cdot), P_1(\cdot), \ldots, P_n(\cdot)$ such that $B = P_B(A)$, $A_1 = P_1(A), \ldots, A_n = P_n(A)$ and*

$$1 > \sum_{i=1}^{n}\sum_{v=1}^{+\infty} |a_i(v)| \cdot \left\| A_i \left(B - \sum_{j=0}^{n} a_j(0)A_j\right)^{-1} \right\|. \tag{186}$$

Then the requirements in (ii) *hold and we have*

$$\sum_{v=0}^{+\infty} \|S(v)\| < +\infty \quad \text{and} \quad \sum_{v=0}^{+\infty} \|A_i(a_i *_0 S)(v)\| < +\infty \quad (1 \leqslant i \leqslant n), \tag{187}$$

provided that (a) *holds, respectively, we have* (187) *and*

$$\sum_{v=0}^{+\infty} \|A_i S(v)\| < +\infty \quad (1 \leqslant i \leqslant n), \tag{188}$$

provided that (b) *holds.*

Proof. Suppose that the assumptions in (i) hold and $x \in X$. Then (181) with $v = 0$ implies $BS(0)x = k(0)Cx + \sum_{i=1}^{n} a_i(0)A_iS(0)x$ so that $(B - \sum_{i=0}^{n} a_i(0)A_i)S(0)x = k(0)Cx$ and $S(0)x \in k(0)(B - \sum_{i=0}^{n} a_i(0)A_i)^{-1}Cx$. Now we will prove that the multivalued linear operator $(B - \sum_{i=0}^{n} a_i(0)A_i)^{-1}C$ is single-valued so that $(B - \sum_{i=0}^{n} a_i(0)A_i)C \in L(X)$ and $S(0) = k(0)(B - \sum_{i=0}^{n} a_i(0)A_i)^{-1}C$. In actual fact, let us assume that $y \in (B - \sum_{i=0}^{n} a_i(0)A_i)^{-1}C0$ for some $y \in X$. This implies $(B - \sum_{i=0}^{n} a_i(0)A_i)y = 0$ and $y \in D(B) \cap D(A_1) \cap \cdots \cap D(A_n)$ since $a_1(0) \neq 0, \ldots, a_n(0) \neq 0$; now the prescribed assumption implies that $0 = S(0)(B - \sum_{i=0}^{n} a_i(0)A_i)y = (B - \sum_{i=0}^{n} a_i(0)A_i)S(0)y = k(0)Cy$. Since $k(0) \neq 0$ and C is injective, this simply implies $y = 0$, the required conclusions and $A_iS(0) \in L(X)$ for all $i \in \mathbb{N}_n$ by the closed graph theorem.

Since $BS(v) \in L(X)$, $A_i(a_i *_0 S)(v) \in L(X)$ for $v \in \mathbb{N}_0$, $1 \leqslant i \leqslant n$,

$$BS(v)x = k(v)Cx + \sum_{i=1}^{n} A_i[a_i(0)S(v)x + \cdots + a_i(v)S(0)x], \quad v \in \mathbb{N}_0,\ x \in X \tag{189}$$

and $a_1(0) \neq 0, \ldots, a_n(0) \neq 0$, we can inductively prove that $A_iS(v) \in L(X)$ for all $i \in \mathbb{N}_n$ and $v \in \mathbb{N}_0$. Moreover, (189) easily implies (182) after a simple calculation.

In order to prove (ii), let us observe first that the mappings $S(0)$, $BS(0)$ and $x \mapsto A_i(a_i *_0 S)(0)x$, $x \in X$ belong to $L(X)$ for $v = 0$, $1 \leqslant i \leqslant n$ as well as that (181) holds with $v = 0$ and $A_jS(0) \in L(X)$ all $j \in \mathbb{N}_0$, since we have assumed that $a_i \neq 0$ for all $i \in \mathbb{N}_n$. Now we proceed by induction. Assume that the mappings $S(v')$, $BS(v')$ and $x \mapsto A_i(a_i *_0 S)(v')x$, $x \in X$ belong to $L(X)$ for all $v' < v \in \mathbb{N}$, $1 \leqslant i \leqslant n$ as well as that (181) holds for all $v' < v$ and $A_jS(v') \in L(X)$ for all $j \in \mathbb{N}_0$. The induction hypothesis together with the representation formula (182), the assumption (183) and the closed graph theorem imply that $S(v) \in L(X)$ and $BS(v) \in L(X)$. Since

$$A_i(a_1 *_0 S)(v)x = A_i\left[a_i(0)S(v)x + \sum_{j=1}^{v-1} a_i(v-j)S(j)x\right], \quad x \in X,\ 1 \leqslant i \leqslant n,$$

the representation formula (182), the closed graph theorem and the assumption $a_i \neq 0$ for all $i \in \mathbb{N}_0$ imply that $A_iS(v) \in L(X)$ and the mapping $x \mapsto A_i(a_i *_0 S)(v)x$, $x \in X$ belongs to $L(X)$ for $1 \leqslant i \leqslant n$. Keeping in mind (182), the above simply implies (181), so that the conclusions stated in the first part of theorem hold true. Suppose now that $\mathcal{I} \subseteq \mathbb{N}_n$ and the commutation relations in (184) hold. We will prove that $(S(v))_{v \in \mathbb{N}_0}$ is a discrete $(k, C, B, (A_i)_{1 \leqslant i \leqslant n}, \mathcal{I})$-existence family, i. e., that for each fixed $x \in D(A_i)$ we have $A_ix \in D(S(v))$ and $A_iS(v)x = S(v)A_ix$ for all $v \in \mathbb{N}_0$ and $i \in \mathbb{N}_n \setminus \mathcal{I}$. If $v = 0$, then we need to prove that

$$A_i\left(B - \sum_{j=0}^{n} a_j(0)A_j\right)^{-1} Cx = \left(B - \sum_{j=0}^{n} a_j(0)A_j\right)^{-1} CA_ix,$$

i. e.,

$$\left(B - \sum_{i=0}^{n} a_j(0)A_j\right)A_i\left(B - \sum_{j=0}^{n} a_j(0)A_j\right)^{-1} Cx = CA_ix.$$

Bringing into play the equalities in the second line and the fourth line of (184), the above is equivalent to

$$A_i\left(B - \sum_{i=0}^{n} a_j(0)A_j\right)\left(B - \sum_{j=0}^{n} a_j(0)A_j\right)^{-1} Cx = CA_ix,$$

i. e., with $A_iCx = CA_ix$, which it has been assumed in the first line of (184). Now we proceed by induction. Suppose that for each $v' < v$, $i \in \mathbb{N}_n \setminus \mathcal{I}$ and $x \in D(A_i)$ we have $A_ix \in D(S(v'))$ and $A_iS(v')x = S(v')A_ix$. Let us prove that the last equality holds with $v' = v$. Using (182), it suffices to show that

$$\begin{aligned}
&A_i\left(B - \sum_{j=0}^{n} a_j(0)A_j\right)^{-1}\left[k(v)Cx + \sum_{j=1}^{n} A_j \sum_{l=0}^{v-1} a_j(v-l)S(l)x\right] \\
&= \left(B - \sum_{j=0}^{n} a_j(0)A_j\right)^{-1}\left[k(v)C + \sum_{j=1}^{n} A_j \sum_{l=0}^{v-1} a_j(v-l)S(l)\right]A_ix.
\end{aligned}$$

The equality of terms containing $k(v)Cx$ is clear, while the equality of the remaining parts follows from the induction hypothesis and the assumption made in the third line of (184) and the fourth line of (184), since $\sum_{l=0}^{v-1} a_j(v-l)S(l)x \in D(A_j)$ for all $j \in \mathbb{N}_n$, $v \in \mathbb{N}$ and $x \in X$. It is quite easy to show that, if there exist a closed linear operator A and the complex polynomials $P_B(\cdot), P_1(\cdot), \ldots, P_n(\cdot)$ such that $CA \subseteq AC$ and $B = P_B(A)$, $A_1 = P_1(A), \ldots, A_n = P_n(A)$, then $(S(v))_{v \in \mathbb{N}_0}$ is a discrete $(k, C, B, (A_i)_{1 \leqslant i \leqslant n}, \emptyset)$-existence family.

If $C = \mathrm{I}$ and $(B - \sum_{j=0}^{n} a_j(0)A_j)^{-1} \in L(X)$, then the equation (183) automatically holds by the closed graph theorem. Suppose now that the assumptions in (iii)(a) hold. Then we have

$$\begin{aligned}\sum_{v=0}^{+\infty}\|S(v)\| &\leqslant \left\|\left(B - \sum_{i=0}^{n} a_i(0)A_i\right)^{-1}\right\| \cdot \sum_{v=0}^{+\infty}|k(v)| \\ &\quad + \sum_{i=1}^{n}\left\|\left(B - \sum_{j=0}^{n} a_j(0)A_j\right)^{-1} A_i\right\| \cdot \sum_{v=0}^{+\infty}\sum_{j=0}^{v-1}|a_i(v-j)| \cdot \|S(j)\| \\ &= \left\|\left(B - \sum_{i=0}^{n} a_i(0)A_i\right)^{-1}\right\| \cdot \sum_{v=0}^{+\infty}|k(v)| \\ &\quad + \sum_{i=1}^{n}\sum_{v=1}^{+\infty}|a_i(v)| \cdot \left\|\left(B - \sum_{j=0}^{n} a_j(0)A_j\right)^{-1} A_i\right\| \cdot \sum_{v=0}^{+\infty}\|S(v)\|.\end{aligned}$$

Keeping in mind (185), we simply get the first estimate in (187). The second estimate in (187) holds since $A_i \in L(X)$ for $1 \leqslant i \leqslant n$ and

$$\sum_{v=0}^{+\infty}\|A_i(a_i *_0 S)(v)\| \leqslant \|A_i\| \cdot \sum_{v=0}^{+\infty}|a_i(v)| \cdot \sum_{v=0}^{+\infty}\|S(v)\| \quad (1 \leqslant i \leqslant n).$$

Suppose now that the assumptions in (iii)(b) hold. Then we similarly obtain

$$\begin{aligned}\sum_{v=0}^{+\infty}\|S(v)\| &\leqslant \left\|\left(B - \sum_{i=0}^{n} a_i(0)A_i\right)^{-1}\right\| \cdot \sum_{v=0}^{+\infty}|k(v)| \\ &\quad + \sum_{i=1}^{n}\left\|A_i\left(B - \sum_{j=0}^{n} a_j(0)A_j\right)^{-1}\right\| \cdot \sum_{v=0}^{+\infty}\sum_{j=0}^{v-1}|a_i(v-j)| \cdot \|S(j)\| \\ &= \left\|\left(B - \sum_{i=0}^{n} a_i(0)A_i\right)^{-1}\right\| \cdot \sum_{v=0}^{+\infty}|k(v)| \\ &\quad + \sum_{i=1}^{n}\sum_{v=1}^{+\infty}|a_i(v)| \cdot \left\|A_i\left(B - \sum_{j=0}^{n} a_j(0)A_j\right)^{-1}\right\| \cdot \sum_{v=0}^{+\infty}\|S(v)\|.\end{aligned}$$

Now we can use (186) to deduce the first estimate in (187). Let $1 \leqslant i \leqslant n$; then the second estimate in (187) can be proved as follows: Clearly, we have

$$\sum_{v=0}^{\infty}\left\|A_i\sum_{j=0}^{v}a_i(v-j)S(j)\right\| \leqslant \sum_{v=0}^{+\infty}|a_i(v)|\cdot\sum_{v=0}^{+\infty}\|A_iS(v)\|.$$

Therefore, it suffices to show (188). But, arguing as above and using our commuting assumptions (184), it follows that

$$\begin{aligned}\sum_{v=0}^{+\infty}\|A_iS(v)\| &\leqslant \left\|A_i\left(B-\sum_{j=0}^{n}a_j(0)A_j\right)^{-1}\right\|\cdot\sum_{v=0}^{+\infty}|k(v)|\\
&\quad+\sum_{v=0}^{+\infty}\sum_{j=1}^{n}\left\|A_i\left(B-\sum_{j=0}^{n}a_j(0)A_j\right)^{-1}A_j\sum_{l=0}^{v-1}a_j(v-l)S(l)\right\|\\
&=\left\|A_i\left(B-\sum_{j=0}^{n}a_j(0)A_j\right)^{-1}\right\|\cdot\sum_{v=0}^{+\infty}|k(v)|\\
&\quad+\sum_{v=0}^{+\infty}\sum_{j=1}^{n}\left\|A_j\left(B-\sum_{j=0}^{n}a_j(0)A_j\right)^{-1}\sum_{l=0}^{v-1}a_j(v-l)A_iS(l)\right\|\\
&\leqslant\left\|A_i\left(B-\sum_{j=0}^{n}a_j(0)A_j\right)^{-1}\right\|\cdot\sum_{v=0}^{+\infty}|k(v)|\\
&\quad+\sum_{j=1}^{n}\sum_{v=1}^{+\infty}|a_j(v)|\cdot\left\|A_j\left(B-\sum_{j=0}^{n}a_j(0)A_j\right)^{-1}\right\|\cdot\sum_{v=0}^{+\infty}\|A_iS(v)\|.\end{aligned}$$

Now the required conclusion simply follows from (186). □

We proceed with some observations.

Remark 3.4.14. (i) As already clarified, the equation (183) automatically holds if $C = I$; moreover, in general case, we have $C^lD_l \in L(X)$, where D_l denotes the operator appearing on the left-hand side of (183).

(ii) The assumptions made in (184) do not imply that the operators $A_1, \dots, A_n$ are bounded, in general. A simple counterexample can be given with $B = C = I$, the operators A_i being bounded for $i \in \mathbb{N}_n \setminus \mathcal{I}$, by assuming also that $R(A_i) \subseteq D(A_j)$ for all $j \in \mathbb{N}_n$ and $A_iA_jx = A_jA_ix$ for all $j \in \mathbb{N}_n$ and $x \in D(A_j)$.

(iii) It would be very tempting to deduce the estimates (187) and (188) using the argumentation contained in the proofs of parts (ii)–(iii) of [470, Theorem 2.5].

Now we will consider the situation in which $v_{\max} > 0$. Set $M := \{i \in \mathbb{N}_n : v_i = v_{\max}\}$; then we have the following analogue of Theorem 3.4.13.

Theorem 3.4.15. (i) *Suppose that $B, A_1, \dots, A_n$ are closed linear operators on X, $C \in L(X)$, $k : \mathbb{N}_0 \to \mathbb{C}$, $k(0) \neq 0$, (180) holds, $(S(v))_{v\in\mathbb{N}_0} \subseteq L(X)$ is a discrete $(k, C, B, (A_i)_{1\leqslant i\leqslant n}, (v_i)_{1\leqslant i\leqslant n})$-existence family, and for every $i \in M$ and $x \in X$, we have $S(0)x \in D(A_i), \dots, S(v_{\max}-1)x \in D(A_i)$. Then $S(v)x \in D(A_i)$ for all $v \in \mathbb{N}_0$, $x \in X$, $i \in M$,*

$$BS(0)x = k(0)Cx + \sum_{i=1}^{n} A_i(a_i *_0 S)(v_i)x, \quad x \in X \tag{190}$$

and $S(v)$ *is a uniquely determined for* $v > v_{max}$, *provided that the operator* $\sum_{i\in M} a_i(0)A_i$ *is injective.*

Furthermore, let us assume that (a)–(c) *hold, where:*

(a) $[\sum_{i\in M} a_i(0)A_i]^{-1}B \in L(X)$ *and* $[\sum_{i\in M} a_i(0)A_i]^{-1}C \in L(X)$.

(b) $[\sum_{i\in M} a_i(0)A_i]^{-1}A_j(a_i *_0 S)(v + v_i) \in L(X)$ *for all* $v \in \mathbb{N}_0$ *and* $j \in \mathbb{N}_n \setminus M$.

(c) $[\sum_{i\in M} a_i(0)A_i]^{-1}A_jS(v) \in L(X)$ *for all* $v \in \mathbb{N}_0$ *and* $j \in M$.

Then, for every $v \in \mathbb{N}$ *and* $x \in X$, *we have*

$$S(v + v_{max})x = \Big[\sum_{i\in M} a_i(0)A_i\Big]^{-1}\Big\{BS(v)x - k(v)Cx - \sum_{i\in\mathbb{N}_n\setminus M} A_i(a_i *_0 S)(v + v_i)x - \sum_{i\in M} A_i\big[(a_i *_0 S)(v + v_i)x - a_i(0)S(v + v_{max})x\big]\Big\}, \tag{191}$$

and $A_iS(v) \in L(X)$ *for all* $i \in M$ *and* $v \in \mathbb{N}_0$.

(ii) *Suppose that* $B, A_1, \ldots, A_n$ *are closed linear operators on* X, $C \in L(X)$, $k : \mathbb{N}_0 \to \mathbb{C}$, $k(0) \neq 0$, (180) *holds, and the operators* $S(0) \in L(X), \ldots, S(v_{max}) \in L(X)$ *satisfy* (190). *Suppose, further, that:*

(a) $[\sum_{i\in M} a_i(0)A_i]^{-1} \in L(X)$ *and* $[\sum_{i\in M} a_i(0)A_i]^{-1}B \in L(X)$.

(b) $A_j \in L(X)$ *for all* $j \in \mathbb{N}_n \setminus M$.

(c) $[\sum_{i\in M} a_i(0)A_i]^{-1}A_jS(v) \in L(X)$ *for all* $v \in \mathbb{N}^0_{v_{max}}$ *and* $j \in M$.

Define, for every $v \in \mathbb{N}$, *the operator* $S(v+v_{max})$ *through* (191). *Then* $(S(v))_{v\in\mathbb{N}_0} \subseteq L(X)$ *is a discrete* $(k, C, B, (A_i)_{1\leqslant i\leqslant n}, (v_i)_{1\leqslant i\leqslant n})$*-existence family and* $A_iS(v) \in L(X)$ *for all* $i \in M$ *and* $v \in \mathbb{N}_0$. *Furthermore, if there exist a closed linear operator* A *and the complex polynomials* $P_B(\cdot), P_1(\cdot), \ldots, P_n(\cdot)$ *such that* $CA \subseteq AC$, $B = P_B(A)$, $A_1 = P_1(A), \ldots, A_n = P_n(A)$, *and conditions* (a)–(c) *hold, then* $(S(v))_{v\in\mathbb{N}_0}$ *is a discrete* $(k, C, B, (A_i)_{1\leqslant i\leqslant n}, \emptyset, (v_i)_{1\leqslant i\leqslant n})$*-existence family provided that* $S(v)A \subseteq AS(v)$ *for* $0 \leqslant v \leqslant v_{max}$.

(iii) *Suppose that the assumptions in* (ii) *hold,* $\sum_{v=0}^{+\infty} |a_i(v)| < +\infty$ *for* $1 \leqslant i \leqslant n$, $\sum_{v=0}^{\infty} |k(v)| < +\infty$, *and* (a) *or* (b) *holds, where:*

(a) $A_i \in L(X)$ *for all* $i \in \mathbb{N}_n$, *and*

$$1 > \Big\|\Big[\sum_{j\in M} a_j(0)A_j\Big]^{-1}B\Big\| + \sum_{v=0}^{+\infty}|a_i(v)| \cdot \sum_{i\in\mathbb{N}_n\setminus M}\Big\|\Big[\sum_{j\in M} a_j(0)A_j\Big]^{-1}A_i\Big\| + \sum_{v=1}^{+\infty}|a_i(v)| \cdot \sum_{i\in M}\Big\|\Big[\sum_{j\in M} a_j(0)A_j\Big]^{-1}A_i\Big\|.$$

(b) *Suppose that there exist a closed linear operator* A *and the complex polynomials* $P_B(\cdot), P_i(\cdot)$ $(i \in M)$ *such that* $B = P_B(A)$, $A_i = P_i(A)$ *for* $i \in M$, $A_j \in L(X)$ *for all* $j \in \mathbb{N} \setminus M$, *and*

$$1 > \left\| \left[\sum_{j \in M} a_j(0) A_j \right]^{-1} B \right\| + \sum_{v=0}^{+\infty} |a_i(v)| \cdot \sum_{i \in \mathbb{N}_n \setminus M} \left\| \left[\sum_{j \in M} a_j(0) A_j \right]^{-1} A_i \right\|$$
$$+ \sum_{v=1}^{+\infty} |a_i(v)| \cdot \sum_{i \in M} \left\| A_i \left[\sum_{j \in M} a_j(0) A_j \right]^{-1} \right\|.$$

Then we have

$$\sum_{v=0}^{+\infty} \|S(v)\| < +\infty \quad \text{and} \quad \sum_{v=0}^{+\infty} \|A_i(a_i *_0 S)(v + v_i)\| < +\infty \quad (1 \leqslant i \leqslant n), \tag{192}$$

provided that (a) *holds, respectively, we have* (192) *and*

$$\sum_{v=0}^{+\infty} \|A_i S(v)\| < +\infty \quad (i \in M),$$

provided that (b) *holds and* $A_j A_i \subseteq A_i A_j$ *for every* $i \in M$ *and* $j \in \mathbb{N}_n \setminus M$.

Proof. It is clear that (190) holds (cf. (181) with $v = 0$) as well as that Proposition 3.4.12 implies $S(v)x \in D(A_i)$ for all $v \in \mathbb{N}_0$, $x \in X$, $i \in M$. Plugging $v = 1$ in (181) and using this fact, we get

$$\left[\sum_{i \in M} a_i(0) A_i \right] S(v_{\max} + 1)x = BS(1)x - k(1)Cx - \sum_{i \in \mathbb{N}_n \setminus M} A_i(a_i *_0 S)(1 + v_i)x$$
$$- \sum_{i \in M} A_i \left[(a_i *_0 S)(1 + v_i)x - a_i(0)S(1 + v_{\max})x \right].$$

If the operator $\sum_{i \in M} a_i(0) A_i$ is injective, the above implies that the value of $S(v_{\max} + 1)$ is uniquely determined; moreover, if (a)–(c) hold, then we obtain that the formula (191) is valid with $v = 1$ as well as that $A_i S(v_{\max} + 1) \in L(X)$ for all $i \in M$. Repeating this argumentation with $v = 2$, $v = 3, \ldots$, we obtain the part (i). The first conclusion in (ii) follows similarly, while the remaining parts can be deduced as in the proof of Theorem 3.4.13. □

It is also worth noting that, in Definition 3.4.11, we can consider arbitrary sequence $(h(v; x))_{v \in \mathbb{N}_0}$ in X as a substitute of the already considered sequence $(k(v)Cx)_{v \in \mathbb{N}_0}$, especially in the case that $v_{\max} > 0$.

3.4.4 The corresponding discrete abstract Cauchy problems

In this subsection, we will briefly consider the following discrete abstract Cauchy problem:

$$Bu(v) = f(v) + \sum_{i=1}^{n} A_i(a_i *_0 u)(v + v_i), \quad v \in \mathbb{N}_0, \tag{193}$$

where $v_1, \ldots, v_n \in \mathbb{N}_0$. We say that a mapping $u : \mathbb{N}_0 \to X$ is:

(i) A mild solution of (193) if $u(v) \in D(B)$ and $(a_i *_0 u)(v + v_i) \in D(A_i)$ for all $v \in \mathbb{N}_0$, $i \in \mathbb{N}_n$ and (193) holds.
(ii) A strong solution of (193) if $u(v) \in D(A_i)$ for all $v \in \mathbb{N}_0$, $i \in \mathbb{N}_n$ and

$$Bu(v) = f(v) + \sum_{i=1}^{n}(a_i *_0 A_i u)(v + v_i), \quad v \in \mathbb{N}_0.$$

It is obvious that any strong solution of (193) is a mild solution of (193). If $(S(v))_{v\in\mathbb{N}_0}$ is a discrete $(k, C, B, (A_i)_{1\leqslant i\leqslant n}, (v_i)_{1\leqslant i\leqslant n})$-existence family, then it is also clear that the function $u(v) := S(v)x$, $v \in \mathbb{N}_0$, where $x \in X$, is a mild solution of (193). Since we have assumed (180), it readily follows that any mild solution of (193) is also a strong solution of (193), if for each $i \in \mathbb{N}_n$ with $v_i > 0$ we have $u(0) \in D(A_i), \ldots, u(v_i - 1) \in D(A_i)$. Furthermore, the problem (193) has at most one mild (strong) solution provided that for each $i \in \mathbb{N}_n$ with $v_i > 0$ we have $u(0) \in D(A_i), \ldots, u(v_i - 1) \in D(A_i)$ as well as that the operator $B - \sum_{i=1}^{n} a_i(0)A_i$ is injective; cf. also the proof of Proposition 3.4.12. The interested reader may attempt to prove some discrete variation of parameters formula for the problem (193); cf. also [470, Proposition 3.3].

In the remainder of this subsection, we will consider the case $v_1 = v_2 = \cdots = v_n = 0$. The proof of following result is simple and therefore omitted.

Proposition 3.4.16. *Suppose that $B, A_1, \ldots, A_n$ are closed linear operators on X, $C \in L(X)$, $k : \mathbb{N}_0 \to \mathbb{C}$, $k \neq 0$, $g : \mathbb{N}_0 \to \mathbb{C}$, $g \neq 0$, and $(S(v))_{v\in\mathbb{N}_0} \subseteq L(X)$ is a discrete $(k, C, B, (A_i)_{1\leqslant i\leqslant n})$-existence family. If $x \in X$, then the sequence $u(v) := (g *_0 S)(v)x$, $v \in \mathbb{N}_0$ is a strong solution of* (193) *with $f(v) = (k *_0 g)(v)$, $v \in \mathbb{N}_0$.*

Now we will explain how Proposition 3.4.16 can be applied in the analysis of the existence and uniqueness of asymptotically almost periodic type solutions of the discrete abstract Cauchy problem (193), with $v_1 = v_2 = \cdots = v_n = 0$. Suppose that there exist a bounded sequence $h : \mathbb{Z} \to X$ and a sequence $q : \mathbb{N}_0 \to X$ vanishing at plus infinity such that $g(v) = h(v) + q(v)$ for all $v \in \mathbb{N}_0$. If the requirements of Proposition 3.4.16 hold, then we know that the function $u(v) := (g *_0 S)(v)x$, $x \in X$, $v \in \mathbb{N}_0$ is a strong solution of (193) with $f(v) = (k *_0 g)(v)$, $v \in \mathbb{N}_0$. If, additionally, $\sum_{v=0}^{+\infty} \|S(v)\| < +\infty$, then the argumentation given at the end of [470, Subsection 3.1] shows that there exist a bounded sequence $H : \mathbb{Z} \to X$ and a sequence $Q : \mathbb{N}_0 \to X$ vanishing at plus infinity such that $u(v) = H(v) + Q(v)$ for all $v \in \mathbb{N}_0$; moreover, $H(\cdot)$ is bounded and c-almost periodic, where $c \in \mathbb{C} \setminus \{0\}$, or bounded and R-almost automorphic, where R is any collection of sequences in $\mathbb{Z}$, provided that the forcing term $h(\cdot)$ has the same property (cf. [470] for the notion).

3.4.5 The abstract multiterm fractional difference equations with Weyl fractional derivatives

Let us recall that, if $\alpha > 0$, $m = \lceil \alpha \rceil$ and $u : \mathbb{Z} \to X$ satisfies $\sum_{v=-\infty}^{\infty} \|u(v)\| \cdot (1 + |v|)^{m-\alpha-1} < +\infty$, then the Weyl fractional derivative $[\Delta_W^{\alpha} u](\cdot)$ is defined by

$$[\Delta_W^a u](v) := [\Delta^m(\Delta_W^{-(m-a)}u)](v), \quad v \in \mathbb{Z},$$

where

$$(\Delta_W^{-(m-a)}u)(v) := \sum_{l=-\infty}^{v} k^{m-a}(v-l)u(l), \quad v \in \mathbb{Z}.$$

Hence,

$$[\Delta_W^a u](v) = \sum_{j=0}^{m} \sum_{l=-\infty}^{v+j} (-1)^{m-j}\binom{m}{j} k^{m-a}(v+j-l)u(l), \quad v \in \mathbb{Z}. \tag{194}$$

Due to [3, Remark 2.4], the assumption $\sum_{v=-\infty}^{\infty} \|u(v)\| \cdot (1+|v|)^{m-a-1} < +\infty$ implies

$$[\Delta_W^a u](v) = [\Delta^m(\Delta_W^{-(m-a)}u)](v) = [\Delta_W^{-(m-a)}(\Delta^m u)](v), \quad v \in \mathbb{Z}; \tag{195}$$

furthermore, it can be simply proved that, for every real number a, the sequence $h(\cdot) = u(\cdot + a)$ satisfies $\sum_{v=-\infty}^{\infty} \|h(v)\| \cdot (1+|v|)^{m-a-1} < +\infty$ as well as that (cf. (194)):

$$[\Delta_W^a h](v) = [\Delta_W^a u](v+a), \quad v \in \mathbb{Z}. \tag{196}$$

In the sequel, we also write $\Delta_W^a u(v)$ in place of $[\Delta_W^a u](v)$.

If A is a closed linear operator, then we must distinguish the terms $A(\Delta_W^a u)$ and $\Delta_W^a(Au)$ for $a > 0$. For example, if $u(v) \equiv x \notin D(A)$, then we have

$$\begin{aligned} \Delta_W^a u(v) &= \sum_{j=0}^{m} \sum_{l=-\infty}^{v+j} (-1)^{m-j}\binom{m}{j} k^{m-a}(v+j-l)x \\ &= \sum_{j=0}^{m} (-1)^{m-j}\binom{m}{j}\left[\sum_{s=0}^{\infty} k^{m-a}(s)x\right] = 0, \quad v \in \mathbb{Z}, \end{aligned} \tag{197}$$

but the term $[\Delta_W^a(Au)](v)$ is not well-defined for any $v \in \mathbb{Z}$.

Define now $T_{i,L}^W u(v) := A_i \Delta_W^{a_i} u(v)$, if $v \in \mathbb{Z}$ and $i \in \mathbb{N}_n$, and $T_{i,R}^W u(v) := \Delta_C^{a_i} A_i u(v)$, if $v \in \mathbb{Z}$ and $i \in \mathbb{N}_n$. We assume that, for every $v \in \mathbb{N}_0$ and $i \in \mathbb{N}_n$, $T_i^W u(v)$ denotes either $T_{i,L}^W u(v)$ or $T_{i,R}^W u(v)$. Of concern is the following abstract multiterm fractional difference equation without initial conditions:

$$\sum_{i=1}^{n} T_i^W u(v+v_i) = f(v), \quad v \in \mathbb{Z}, \tag{198}$$

where $v_1, \ldots, v_n \in \mathbb{Z}$ and $f : \mathbb{Z} \to X$ is a given sequence. A strong solution of problem (198) is any sequence $u : \mathbb{Z} \to X$ such that the term $T_i^W u(\cdot + v_i)$ is well-defined for any $i \in \mathbb{N}_n$ as well as that (198) identically holds for any $v \in \mathbb{Z}$.

Set now $v_{\min} := \min\{v_1, \dots, v_n\}$ and $y(v) := u(v + v_{\min})$, $v \in \mathbb{Z}$. Then it is clear that the problem (198) is equivalent with the problem (cf. (196)):

$$\sum_{i=1}^{n} T_i^W y(v + [v_i - v_{\min}]) = f(v), \quad v \in \mathbb{Z},$$

as well as that $v_i - v_{\min} \geqslant 0$ for all $i \in \mathbb{N}_n$. Therefore, in the analysis of the existence and uniqueness of solutions of the problem (198), we may assume without loss of generality that $v_1 \geqslant 0, \dots, v_n \geqslant 0$. This will be our standing assumption henceforth.

Due to (194), the problem (198) is equivalent with

$$\sum_{i\in\mathcal{I}} A_i \left[\sum_{j=0}^{m_i} \sum_{l=-\infty}^{v+j} (-1)^{m_i-j} \binom{m_i}{j} k^{m_i-a}(v+j-l)u(l) \right]$$
$$+ \sum_{i\notin\mathcal{I}} \left[\sum_{j=0}^{m_i} \sum_{l=-\infty}^{v+j} (-1)^{m_i-j} \binom{m_i}{j} k^{m_i-a}(v+j-l)A_i u(l) \right] = f(v), \quad v \in \mathbb{Z},$$

where $\mathcal{I} := \{i \in \mathbb{N}_n : a_i > 0$ and $T_{i,L}^W u(v)$ appears on the left-hand side of (198)$\}$. We cannot expect the uniqueness of (periodic/almost periodic) strong solutions of (198) in any sense: if $u(\cdot)$ is a strong solution of (198), then the sequence $u_x(\cdot) := u(\cdot) + x$, where $x \in D(A_i)$ for all $i \notin \mathcal{I}$, is likewise a strong solution of (198); cf. (197).

The statement of [470, Theorem 3.4] can be helpful in the analysis of the existence of strong solutions to some classes of the abstract multiterm fractional difference equations. This result is far from being satisfactory in the general analysis of problem (198), when we need the notion introduced in Definition 3.4.11.

The following result seems to be extremely important in the study of the abstract multiterm fractional difference equations with Weyl fractional derivatives.

Theorem 3.4.17. *Suppose that $v_1 \geqslant 0, \dots, v_n \geqslant 0$, $(S(v))_{v\in\mathbb{N}_0} \subseteq L(X)$ is a discrete $(k, C, B, (A_i)_{1\leqslant i\leqslant n}, (v_i)_{1\leqslant i\leqslant n})$-existence family, (187) and the following hold:*

(a) *$f : \mathbb{Z} \to X$ is a bounded sequence, $k \in l^1(\mathbb{N}_0 : Y)$ and $\sum_{v=0}^{+\infty} |a_i(v)| < +\infty$ for $1 \leqslant i \leqslant n$, or*

(b) *$f \in l^1(\mathbb{Z} : X)$, $k : \mathbb{N}_0 \to X$ is a bounded sequence and $a_i : \mathbb{Z} \to \mathbb{C}$ is a bounded sequence for $1 \leqslant i \leqslant n$.*

Define

$$u(v) := \sum_{l=-\infty}^{v} S(v-l)f(l), \quad v \in \mathbb{Z}$$

and

$$g(v) := A_1 \sum_{l=v+1}^{v+v_1} (a_1 *_0 S)(v + v_1 - l)f(l) + \cdots$$

$$+ A_n \sum_{l=v+1}^{v+v_n} (a_n *_0 S)(v + v_n - l)f(l), \quad v \in \mathbb{Z}. \tag{199}$$

Then $u(\cdot)$ is bounded if (a) *holds, $u \in l^1(\mathbb{Z} : X)$ if* (b) *holds, and we have*

$$\begin{aligned} Bu(v) &= A_1 \sum_{l=-\infty}^{v+v_1} a_1(v + v_1 - l)u(l) + \cdots \\ &+ A_n \sum_{l=-\infty}^{v+v_n} a_n(v + v_n - l)u(l) + (k \circ Cf)(v) + g(v), \quad v \in \mathbb{Z}. \end{aligned} \tag{200}$$

Especially, the following holds:

(i) *Suppose that $f \in l^1(\mathbb{Z} : X)$ and $g(v)$ is given by* (199), *with $a_i(v) = k^{m_i - \alpha_i}(v)$ for all $v \in \mathbb{N}_0$ and $i \in \mathbb{N}$, where $m = m_i \in \mathbb{N}$ for all $i \in \mathbb{N}_n$. Then we have*

$$\begin{aligned} (\Delta^m Bu)(v) &= A_1(\Delta_W^{\alpha_1} u)(v + v_1) + \cdots \\ &+ A_n(\Delta_W^{\alpha_n} u)(v + v_n) + \Delta^m (k \circ Cf)(v) + \Delta^m g(v), \quad v \in \mathbb{Z}. \end{aligned}$$

(ii) *Suppose that $f \in l^1(\mathbb{Z} : X)$ and $u = \Delta^{m_n} h$ for a certain sequence $h : \mathbb{Z} \to X$. Then we have*

$$\begin{aligned} B(\Delta^{m_n} h)(v) &= \sum_{j=1}^{n-1} A_j(\Delta^{m_n - m_j} \Delta_W^{\alpha_j} h)(v + v_j) \\ &+ A_n(\Delta_W^{\alpha_n} h)(v + v_n) + (k \circ Cf)(v) + g(v), \quad v \in \mathbb{Z}. \end{aligned}$$

Proof. It is clear that $u(\cdot)$ is well-defined and bounded if (a) holds as well as $u \in l^1(\mathbb{Z} : X)$ if (b) holds; furthermore, the sequence $(k \circ Cf)(\cdot)$ is well-defined. The functional equality (181) and the closedness of the linear operator B together imply with the prescribed assumptions and the Fubini theorem that

$$\begin{aligned} &Bu(v) - (k \circ Cf)(v) \\ &= \sum_{l=-\infty}^{v} BS(v - l)f(l) - (k \circ Cf)(v) = -(k \circ Cf)(v) \\ &\quad + \sum_{l=-\infty}^{v} [k(v - l)Cf(l) + A_1(a_1 *_0 S)(v + v_1 - l)f(l) + \cdots + A_n(a_n *_0 S)(v + v_n - l)f(l)] \\ &= \sum_{l=-\infty}^{v} [A_1(a_1 *_0 S)(v + v_1 - l)f(l) + \cdots + A_n(a_n *_0 S)(v + v_n - l)f(l)] \\ &= A_1 \sum_{l=-\infty}^{v} (a_1 *_0 S)(v + v_1 - l)f(l) + \cdots + A_n \sum_{l=-\infty}^{v} (a_n *_0 S)(v + v_n - l)f(l) \qquad (201) \\ &= A_1 \sum_{l=-\infty}^{v+v_1} (a_1 *_0 S)(v + v_1 - l)f(l) + \cdots + A_n \sum_{l=-\infty}^{v+v_n} (a_n *_0 S)(v + v_n - l)f(l) - g(v) \end{aligned}$$

$$
\begin{aligned}
&= \sum_{j=1}^{n} A_j \sum_{s=0}^{+\infty} [a_j(s)S(0) + \cdots + a_j(0)S(s)]f(v + v_j - s) - g(v) \\
&= \sum_{j=1}^{n} A_j \sum_{s=0}^{+\infty} a_j(s) \sum_{r=0}^{+\infty} S(r)f(v + v_j - s - r) - g(v) \qquad (202) \\
&= \sum_{j=1}^{n} A_j \sum_{l=-\infty}^{v+v_j} a_j(v + v_j - l)u(l) - g(v), \quad v \in \mathbb{Z},
\end{aligned}
$$

so that (200) holds true; here, (201) follows from the assumption $\sum_{v=0}^{+\infty} \|A_i(a_i *_0 S)(v+v_i)\| < +\infty$ for $1 \leqslant i \leqslant n$ and the closedness of the linear operators $A_1, \ldots, A_n$, while (202) follows from an application of the discrete Fubini theorem, the assumption $\sum_{v=0}^{+\infty} \|S(v)\| < +\infty$ and the assumption (a) or (b). The proofs of (i) and (ii) follow simply from this general result with condition (b) being satisfied, the definitions of Weyl fractional derivatives and the equation (195). □

We continue be providing certain observations.

Remark 3.4.18. (i) It is clear that $g(\cdot) \equiv 0$, if $v_1 = v_2 = \cdots = v_n = 0$.
(ii) The assumption $u(v) = \Delta^{m_n} h(v)$, $v \in \mathbb{Z}$ in (ii) implies that $h(v) = u_1(v) + c_0 + c_1 v + \cdots + c_{m_n-1} v^{m_n-1}$ for all $v \in \mathbb{Z}$, for some sequence $(u_1(\cdot))$ depending on $(u(\cdot))$ and arbitrary complex constants $c_0, \ldots, c_{m_n-1}$. For more details about the calculus of finite differences and its applications, we refer the reader to the monographs [410] and [719]; cf. also the research articles [83, 84].
(iii) Suppose that $c_i \in \mathbb{C}$, $v_1 = v_2 = \cdots = v_n = 0$ and $h_i \in \mathbb{Z}$ for $1 \leqslant i \leqslant m$. Define $y(v) := \sum_{i=1}^{m} c_i u(v + h_i)$, $v \in \mathbb{Z}$; then (200) implies after the substitution $s = v - l \geqslant 0$ that

$$
\begin{aligned}
By(v) = A_1 \sum_{l=-\infty}^{v} a_1(v - l)y(l) + \cdots + A_n \sum_{l=-\infty}^{v} a_n(v - l)y(l) \\
+ \sum_{i=i}^{m} c_i(k \circ Cf)(v + h_i), \quad v \in \mathbb{Z}.
\end{aligned}
$$

We can similarly prove the following general result (the strong solutions of problems under our consideration are obtained by plugging $\mathcal{I} = \emptyset$).

Theorem 3.4.19. *Suppose that $v_1 \geqslant 0, \ldots, v_n \geqslant 0$, $\mathcal{I} \subseteq \mathbb{N}_n$, $(S(v))_{v \in \mathbb{N}_0} \subseteq L(X)$ is a discrete $(k, C, B, (A_i)_{1\leqslant i \leqslant n}, (v_i)_{1 \leqslant i \leqslant n}, \mathcal{I})$-existence family, $\sum_{v=0}^{+\infty} \|S(v)\| < +\infty$, $\sum_{v=0}^{+\infty} \|A_i(a_i *_0 S)(v + v_i)\| < +\infty$ for $i \in \mathcal{I}$ and the following holds:*
(a) *$f : \mathbb{Z} \to X$ is a bounded sequence, $k \in l^1(\mathbb{N}_0 : X)$ and $\sum_{v=0}^{+\infty} |a_i(v)| < +\infty$ for $i \in \mathcal{I}$, or*
(b) *$f \in l^1(\mathbb{Z} : X)$, $k : \mathbb{N}_0 \to X$ is a bounded sequence and $a_i : \mathbb{Z} \to \mathbb{C} \setminus \{0\}$ is a bounded sequence for $i \in \mathcal{I}$*

as well as

(c) $A_i f : \mathbb{Z} \to X$ *is a bounded sequence,* $\sum_{v=0}^{+\infty} |a_i(v)| < +\infty$ *for* $i \in \mathbb{N}_n \setminus \mathcal{I}$ *and* $(S(v))_{v\in\mathbb{N}_0} \subseteq L(X)$ *is a discrete* $(k, C, B, (A_i)_{1\leqslant i\leqslant n}, (v_i)_{1\leqslant i\leqslant n}, \mathcal{I})$*-existence family, or*

(d) $f \in l^1(\mathbb{Z} : X)$, $\sum_{v=0}^{+\infty} \|A_i S(v)\| < +\infty$ *for all* $i \in \mathbb{N}_n \setminus \mathcal{I}$ *and* $a_i : \mathbb{N}_0 \to \mathbb{C} \setminus \{0\}$ *is a bounded sequence for* $i \in \mathbb{N}_n \setminus \mathcal{I}$*, or*

(e) $f \in l^1(\mathbb{Z} : [D(A_i)])$ *for all* $i \in \mathbb{N}_n \setminus \mathcal{I}$, $a_i : \mathbb{N}_0 \to \mathbb{C} \setminus \{0\}$ *is a bounded sequence for* $i \in \mathbb{N}_n \setminus \mathcal{I}$ *and* $(S(v))_{v\in\mathbb{N}_0} \subseteq L(X)$ *is a discrete* $(k, C, B, (A_i)_{1\leqslant i\leqslant n}, (v_i)_{1\leqslant i\leqslant n}, \mathcal{I})$*-existence family.*

Define $u(\cdot)$ *and* $g(\cdot)$ *in the same way as in the proof of* Theorem 3.4.17. *Then we have*

$$Bu(v) = \sum_{i\in\mathcal{I}} A_i \sum_{l=-\infty}^{v+v_i} a_i(v + v_i - l)u(l) + \sum_{i\in\mathbb{N}_n\setminus\mathcal{I}} \sum_{l=-\infty}^{v+v_i} a_i(v + v_i - l)A_i u(l) + (k \circ Cf)(v) + g(v), \quad v \in \mathbb{Z}.$$

Especially, the following holds for a function $f \in l^1(\mathbb{Z} : X)$:

(i) *Suppose that* $a_i(v) = k^{m_i - a_i}(v)$ *for all* $v \in \mathbb{N}_0$ *and* $i \in \mathbb{N}$*, where* $m = m_i \in \mathbb{N}$ *for all* $i \in \mathbb{N}_n$*. If* $\sum_{v=0}^{+\infty} \|A_i S(v)\| < +\infty$ *for all* $i \in \mathbb{N}_n \setminus \mathcal{I}$ *or* $f \in l^1(\mathbb{Z} : [D(A_i)])$ *for all* $i \in \mathbb{N}_n \setminus \mathcal{I}$ *and* $(S(v))_{v\in\mathbb{N}_0} \subseteq L(X)$ *is a discrete* $(k, C, B, (A_i)_{1\leqslant i\leqslant n}, (v_i)_{1\leqslant i\leqslant n}, \mathcal{I})$*-existence family, then we have*

$$(\Delta^m Bu)(v) = \sum_{i\in\mathcal{I}} A_i(\Delta_W^{a_i} u)(v + v_i) + \sum_{i\in\mathbb{N}_n\setminus\mathcal{I}} (\Delta_W^{a_i} A_i u)(v + v_i) + \Delta^m(k \circ Cf)(v) + \Delta^m g(v), \quad v \in \mathbb{Z}.$$

(ii) *Suppose that* $\sum_{v=0}^{+\infty} \|A_i S(v)\| < +\infty$ *for all* $i \in \mathbb{N}_n \setminus \mathcal{I}$ *or* $f \in l^1(\mathbb{Z} : [D(A_i)])$ *for all* $i \in \mathbb{N}_n \setminus \mathcal{I}$ *and* $(S(v))_{v\in\mathbb{N}_0} \subseteq L(X)$ *is a discrete* $(k, C, B, (A_i)_{1\leqslant i\leqslant n}, (v_i)_{1\leqslant i\leqslant n}, \mathcal{I})$*-existence family as well as* $u = \Delta^{m_n} h$ *for a certain sequence* $h : \mathbb{Z} \to \bigcap_{i\in\mathbb{N}_n\setminus\mathcal{I}} D(A_i)$*. Then we have*

$$B(\Delta^{m_n} h)(v) = \sum_{i\in\mathcal{I}} A_i(\Delta^{m_n - m_i}\Delta_W^{a_i} h)(v + v_i) + \sum_{i\in\mathbb{N}_n\setminus\mathcal{I}} (\Delta^{m_n - m_i}\Delta_W^{a_j} A_i h)(v + v_i) + (k \circ Cf)(v) + g(v), \quad v \in \mathbb{Z}.$$

It is clear that Theorem 3.4.17 and Theorem 3.4.19 can be illustrated with many concrete examples in which the corresponding discrete $(k, 0, 0, (A_i)_{1\leqslant i\leqslant n}, (v_i)_{1\leqslant i\leqslant n}, \mathcal{I})$-existence families are integrable on account of Theorem 3.4.13(iii) or Theorem 3.4.15(iii). Now we will provide one more important application of Theorem 3.4.19.

Example 3.4.20. Suppose that $(T(t))_{t\geqslant 0} \subseteq L(X)$ is a strongly continuous operator family such that $\int_0^{+\infty} \|T(t)\|\,dt < +\infty$ and, for every $x \in X$, the mapping $t \mapsto T(t)x$, $t \geqslant 0$ is a mild solution of the abstract Cauchy problem

$$A_n D_t^{\alpha_n} T(t)x + A_{n-1} D_t^{\alpha_{n-1}} T(t)x + \cdots + A_1 D_t^{\alpha_1} T(t)x = 0, \quad t > 0,$$

accompanied with certain initial conditions (cf. [445, pp. 161–163] for more details in this direction), where $0 \leqslant \alpha_1 < \alpha_2 < \cdots < \alpha_n$ and the operators A_j are polynomials of a certain closed linear operator A; the interested reader may consult [443–445] for more details and many interesting examples of differential operators, which can be used to provide certain applications of the conclusions established here (we can also consider the exponentially bounded operator families $(T(t))_{t\geqslant 0}$ using the Poisson like transform from Theorem 3.4.8 since the statement of [4, Theorem 5.5] admits an extension in this framework). Set

$$S(v)x := \int_0^{+\infty} e^{-t}(t^v/v!)T(t)x\,dt, \quad v \in \mathbb{N}_0,\ x \in X.$$

Then we have $\sum_{v=0}^{+\infty} \|S(v)\| < +\infty$ and

$$A_n(\Delta^{\alpha_n} S(\cdot)x)(v) + \sum_{s=1}^{n-1} A_s(\Delta^{\alpha_s} S(\cdot)x)(v + m_n - m_s) = 0, \quad v \in \mathbb{N}_0,\ x \in X.$$

In other words,

$$\begin{aligned} &A_n\left[\sum_{j=0}^{m_n}\sum_{l=0}^{v+j}(-1)^{m_n-j}\binom{m_n}{j}k^{m_n-\alpha_n}(v+j-l)S(l)x\right] \\ &+\sum_{s=1}^{n-1}A_s\left[\sum_{j=0}^{m_s}\sum_{l=0}^{v+m_n-m_s+j}(-1)^{m_s-j}\binom{m_s}{j}k^{m_s-\alpha_s}(v+m_n-m_s+j-l)S(l)x\right]=0, \end{aligned}$$

for any $v \in \mathbb{N}_0$ and $x \in X$. Now it can be easily shown that Theorem 3.4.19 is applicable with $\mathcal{I} = \emptyset$, $B = C = 0$ and arbitrary nontrivial sequence $f \in l^1(\mathbb{Z} : [D(A_i)])$ for all $i \in \mathbb{N}_n$; in our concrete situation, we have that $(S(v))_{v\in\mathbb{N}_0} \subseteq L(X)$ is a discrete $(k, 0, 0, (A_i)_{1\leqslant i\leqslant n}, (v_i)_{1\leqslant i\leqslant n}, \mathcal{I})$-existence family with $v_i = m_i$ for $1 \leqslant i \leqslant n$ and

$$a_s(v) := \sum_{j=0}^{m_s}(-1)^{m_s-j}\binom{m_s}{j}k^{m_s-\alpha_s}(v + m_n + j - 2m_s), \quad v \in \mathbb{N}_0 \quad (1 \leqslant s \leqslant n),$$

where we have put $k^{m_s-\alpha_s}(v) := 0$ if $v > 0$ $(1 \leqslant s \leqslant n)$. Next, let us observe that we have $a_s(0) = 1$, $i \in M$ and $\sum_{v=0}^{+\infty} |a_s(v)| < +\infty$ for $s \in \mathbb{N}_n$ since $a_s(v) \sim c_s v^{-2+m_s-\alpha_s}$ as $v \to +\infty$; this follows form a relatively simple analysis involving the identity $\sum_{j=0}^{m_s}(-1)^{m_s-j}\binom{m_s}{j} = 0$, the Lagrange mean value theorem and the asymptotic formula $k^{\alpha}(v) \sim g_{\alpha}(v)$ as $v \to +\infty$

($a > 0$). Consequently, we have that there do not exist an integer $m_s \in \mathbb{N}$, the complex constants $\beta_{1;s}, \ldots, \beta_{m;s}$ and the translation numbers $v'_{1;s} \in \mathbb{N}_0, \ldots, v'_{m;s} \in \mathbb{N}_0$ such that $k^{m_s - a_s}(v) = \beta_{1;s} a_s(v + v'_{1;s}) + \cdots + \beta_{m;s} a_s(v + v'_{m;s})$.

In our concrete situation, we are solving the equation

$$\sum_{l=-\infty}^{v+v_n} a_n(v + v_n - l) A_n u(l) + \cdots + \sum_{l=-\infty}^{v+v_1} a_1(v + v_1 - l) A_1 u(l) = -g(v), \quad v \in \mathbb{Z}, \tag{203}$$

where $g(\cdot)$ is given by (199). Keeping in mind the representation of $a_s(\cdot)$ for $1 \leqslant s \leqslant n$, we get

$$\begin{aligned} &\sum_{l=-\infty}^{v+m_n} \sum_{j=0}^{m_n} (-1)^{m_n - j} \binom{m_n}{j} k^{m_n - a_n}(v + j - l) A_n u(l) \\ &\quad + \sum_{s=1}^{n-1} \sum_{l=-\infty}^{v+m_s} \sum_{j=0}^{m_s} (-1)^{m_s - j} \binom{m_s}{j} k^{m_s - a_s}(v + m_n - m_s + j - l) A_s u(l) = -g(v), \end{aligned}$$

for any $v \in \mathbb{Z}$, i. e.,

$$\begin{aligned} &\sum_{j=0}^{m_n} (-1)^{m_n - j} \binom{m_n}{j} \sum_{l=-\infty}^{v+m_n} k^{m_n - a_n}(v + j - l) A_n u(l) \\ &\quad + \sum_{s=1}^{n-1} \sum_{j=0}^{m_s} (-1)^{m_s - j} \binom{m_s}{j} \sum_{l=-\infty}^{v+m_s} k^{m_s - a_s}(v + m_n - m_s + j - l) A_s u(l) = -g(v), \end{aligned}$$

for any $v \in \mathbb{Z}$. Taking into account the formula (194), we finally get

$$\begin{aligned} &[\Delta_W^{a_n} A_n u](v) + [\Delta_W^{a_{n-1}} A_{n-1} u](v + m_n - m_{n-1}) \\ &\quad + \cdots + [\Delta_W^{a_1} A_1 u](v + m_n - m_1) = -g(v), \quad v \in \mathbb{Z}. \end{aligned}$$

The term $g(\cdot)$ can be directly computed for small values of a_j ($1 \leqslant j \leqslant n$); for example, if $a_n \leqslant 1$, then we have

$$g(v) = \sum_{i \in M} A_i \int_0^{+\infty} e^{-t} T(t) f(v + 1)\, dt, \quad v \in \mathbb{Z}.$$

The kernels $a(v) = k^{m-a}(v)$, $v \in \mathbb{N}_0$ are not integrable provided that $a \in (0, \infty) \smallsetminus \mathbb{N}$ and $m = \lceil a \rceil$. Therefore, we cannot use Theorem 3.4.13(iii) or Theorem 3.4.15(iii) to deduce the integrability of the corresponding discrete $(k, B, C, (A_i)_{1 \leqslant i \leqslant n}, (v_i)_{1 \leqslant i \leqslant n}, \mathcal{I})$-existence families. We can overcome this difficulty by applying the following trick.

Theorem 3.4.21. *Suppose that $v_1 \geqslant 0, \ldots, v_n \geqslant 0$, $\mathcal{I} \subseteq \mathbb{N}_n$, $(S(v))_{v \in \mathbb{N}_0} \subseteq L(X)$ is a discrete $(k, C, B, (A_i)_{1 \leqslant i \leqslant n}, (v_i)_{1 \leqslant i \leqslant n}, \mathcal{I})$-existence family, $\omega > 0$, $\sum_{v=0}^{+\infty} \| e^{-\omega v} S(v) \| < +\infty$, $\sum_{v=0}^{+\infty} \| A_i (a_i *_0 e^{-\omega \cdot} S(\cdot))(v + v_i) \| < +\infty$ for $i \in \mathcal{I}$ and the following holds:*

(a) $e^{-\omega \cdot}f : \mathbb{Z} \to X$ *is a bounded sequence,* $e^{-\omega \cdot}k \in l^1(\mathbb{N}_0 : X)$ *and* $\sum_{v=0}^{+\infty} |a_i(v)e^{-\omega(\cdot - v_i)}| < +\infty$ *for* $i \in \mathcal{I}$, *or*

(b) $e^{-\omega \cdot}f \in l^1(\mathbb{Z} : X)$, $e^{-\omega \cdot}k : \mathbb{N}_0 \to X$ *is a bounded sequence and* $e^{-\omega(\cdot - v_i)}a_i : \mathbb{Z} \to \mathbb{C} \setminus \{0\}$ *is a bounded sequence for* $i \in \mathcal{I}$

as well as

(c) $e^{-\omega \cdot}A_i f : \mathbb{Z} \to X$ *is a bounded sequence,* $\sum_{v=0}^{+\infty} |a_i(v)e^{-\omega(\cdot - v_i)}| < +\infty$ *for* $i \in \mathbb{N}_n \setminus \mathcal{I}$ *and* $(S(v))_{v \in \mathbb{N}_0} \subseteq L(X)$ *is a discrete* $(k, C, B, (A_i)_{1 \leqslant i \leqslant n}, (v_i)_{1 \leqslant i \leqslant n}, \mathcal{I})$*-existence family, or*

(d) $e^{-\omega \cdot}f \in l^1(\mathbb{Z} : X)$, $\sum_{v=0}^{+\infty} \|A_i e^{-\omega v}S(v)\| < +\infty$ *for all* $i \in \mathbb{N}_n \setminus \mathcal{I}$ *and* $e^{-\omega(\cdot - v_i)}a_i : \mathbb{N}_0 \to \mathbb{C} \setminus \{0\}$ *is a bounded sequence for* $i \in \mathbb{N}_n \setminus \mathcal{I}$, *or*

(e) $e^{-\omega \cdot}f \in l^1(\mathbb{Z} : [D(A_i)])$ *for all* $i \in \mathbb{N}_n \setminus \mathcal{I}$, $e^{-\omega(\cdot - v_i)}a_i : \mathbb{N}_0 \to \mathbb{C} \setminus \{0\}$ *is a bounded sequence for* $i \in \mathbb{N}_n \setminus \mathcal{I}$ *and* $(S(v))_{v \in \mathbb{N}_0} \subseteq L(X)$ *is a discrete* $(k, C, B, (A_i)_{1 \leqslant i \leqslant n}, (v_i)_{1 \leqslant i \leqslant n}, \mathcal{I})$*-existence family.*

Define

$$u(v) := e^{\omega v} \sum_{l=-\infty}^{v} [e^{-\omega(v-l)}S(v-l)] \cdot [e^{-\omega l}f(l)], \quad v \in \mathbb{Z}$$

and $g_\omega(\cdot)$ *in the same way as in the proof of* Theorem 3.4.17 *with the terms* $g(\cdot)$, $a_i(\cdot)$, $S(\cdot)$ *and* $f(\cdot)$ *replaced therein with the terms* $g_\omega(\cdot)$, $e^{-\omega(\cdot - v_i)}a_i(\cdot)$, $e^{-\omega \cdot}S(\cdot)$ *and* $e^{-\omega \cdot}f(\cdot)$, *respectively. Then we have:*

$$\begin{aligned} Bu(v) &= \sum_{i \in \mathcal{I}} A_i \sum_{l=-\infty}^{v+v_i} a_i(v + v_i - l)u(l) \\ &\quad + \sum_{i \in \mathbb{N}_n \setminus \mathcal{I}} \sum_{l=-\infty}^{v+v_i} a_i(v + v_i - l)A_i u(l) + e^{-\omega v}(k \circ Cf)(v) + g_\omega(v), \quad v \in \mathbb{Z}. \end{aligned} \tag{204}$$

Especially, the following holds for a function $e^{-\omega \cdot}f \in l^1(\mathbb{Z} : X)$:

(i) *Suppose that* $a_i(v) = k^{m_i - a_i}(v)$ *for all* $v \in \mathbb{N}_0$ *and* $i \in \mathbb{N}$, *where* $m = m_i \in \mathbb{N}$ *for all* $i \in \mathbb{N}_n$. *If* $\sum_{v=0}^{+\infty} \|A_i e^{-\omega v}S(v)\| < +\infty$ *for all* $i \in \mathbb{N}_n \setminus \mathcal{I}$ *or* $e^{-\omega \cdot}f \in l^1(\mathbb{Z} : [D(A_i)])$ *for all* $i \in \mathbb{N}_n \setminus \mathcal{I}$ *and* $(S(v))_{v \in \mathbb{N}_0} \subseteq L(X)$ *is a discrete* $(k, C, B, (A_i)_{1 \leqslant i \leqslant n}, (v_i)_{1 \leqslant i \leqslant n}, \mathbb{N}_n \setminus \mathcal{I})$*-existence family, then we have*

$$\begin{aligned} (\Delta^m Bu)(v) &= \sum_{i \in \mathcal{I}} A_i (\Delta_W^{a_i} u)(v + v_i) \\ &\quad + \sum_{i \in \mathbb{N}_n \setminus \mathcal{I}} (\Delta_W^{a_i} A_i u)(v + v_i) + \Delta^m e^{-\omega v}(k \circ Cf)(v) + \Delta^m g_\omega(v), \quad v \in \mathbb{Z}. \end{aligned}$$

(ii) *Suppose that* $\sum_{v=0}^{+\infty} \|A_i e^{-\omega v}S(v)\| < +\infty$ *for all* $i \in \mathbb{N}_n \setminus \mathcal{I}$ *or* $e^{-\omega \cdot}f \in l^1(\mathbb{Z} : [D(A_i)])$ *for all* $i \in \mathbb{N}_n \setminus \mathcal{I}$ *and* $(S(v))_{v \in \mathbb{N}_0} \subseteq L(X)$ *is a discrete* $(k, C, B, (A_i)_{1 \leqslant i \leqslant n}, (v_i)_{1 \leqslant i \leqslant n}, \mathbb{N}_n \setminus \mathcal{I})$*-*

existence family as well as $u = \Delta^{m_n} h$ *for a certain sequence* $h : \mathbb{Z} \to \bigcap_{i \in \mathbb{N}_n \setminus \mathcal{I}} D(A_i)$. *Then we have*

$$B(\Delta^{m_n} h)(v) = \sum_{i \in \mathcal{I}} A_i(\Delta^{m_n - m_i} \Delta_W^{a_i} h)(v + v_i) + \sum_{i \in \mathbb{N}_n \setminus \mathcal{I}} (\Delta^{m_n - m_i} \Delta_W^{a_j} A_i h)(v + v_i) + e^{-\omega v}(k \circ Cf)(v) + g_\omega(v), \quad v \in \mathbb{Z}.$$

Proof. The proof simply follows by applying Theorem 3.4.19, after observing that $(e^{-\omega v} S(v))_{v \in \mathbb{N}_0}$ is a discrete $(e^{-\omega \cdot} k(\cdot), C, B, (A_i)_{1 \leqslant i \leqslant n}, (v_i)_{1 \leqslant i \leqslant n}, \mathcal{I})$-existence family with the kernels $a_i(\cdot)$ replaced therein with the kernels $a_i(\cdot) e^{-\omega(\cdot - v_i)}$ for $1 \leqslant i \leqslant n$. □

As in many research articles published by now, this enables us to consider the almost periodic features of the function $e^{-\omega \cdot} u(\cdot)$, where $u(\cdot)$ solves (204); cf. [447] for more details on the subject. The exponential boundedness of discrete $(k, B, C, (A_i)_{1 \leqslant i \leqslant n}, (v_i)_{1 \leqslant i \leqslant n}, \mathcal{I})$-existence family can be proved in many concrete situations; for example, if $v_1 = \cdots = v_n = 0$, $k(\cdot)$ is exponentially bounded, the kernels $a_i(\cdot)$ are bounded for $1 \leqslant i \leqslant n$, $C = \mathrm{I}$ and there exists a closed linear operator A and the complex polynomials $P_B(\cdot)$, $P_i(\cdot)$ $(1 \leqslant i \leqslant n)$ such that $B = P_B(A)$ and $A_i = P_i(A)$ for $1 \leqslant i \leqslant n$.

Finally, we will present the following illustrative applications of Theorem 3.4.21.

Example 3.4.22. (i) Suppose for simplicity that $\alpha > 0$, $m = \lceil \alpha \rceil$ and a closed linear operator $-A$ is a subgenerator of (g_α, g_α)-regularized C-resolvent family $(T(t))_{t \geqslant 0}$ such that there exist two real constants $M \geqslant 1$ and $c \in (0, 1)$ such that $\|T(t)\| \leqslant Me^{(1-c)t}$, $t \geqslant 0$; see [445] for the notion and more details. Then it is not difficult to show that, for every $x \in X$, the function $t \mapsto T(t)x$, $t \geqslant 0$ is a mild solution of the abstract Cauchy problem (cf. [428] for the case $1 < \alpha \leqslant 2$; if $0 < \alpha \leqslant 1$, then we recover some known results about the problem (125) with $\mathcal{A} = -A$):

$$D_t^\alpha u(t) + Au(t) = 0, \quad t > 0.$$

We define the sequence $(S(v))_{v \in \mathbb{N}_0}$ as in Example 3.4.20. Then we have $\|S(v)\| \leqslant c^{-1}(c^{-1})^v$, $v \in \mathbb{N}_0$ so that $\sum_{v=0}^{+\infty} \|e^{-\omega v} S(v)\| < +\infty$ for $\omega > \ln(c^{-1})$. Let us assume that the sequence $f : \mathbb{Z} \to X$ satisfies the assumptions (a) and (c) from the formulation of Theorem 3.4.21. If we define $u(\cdot)$ as in Example 3.4.20, then the equation (203) holds with the sequence $g(\cdot)$ replaced with the sequence $g_\omega(\cdot)$, obtained by replacing the forcing term $f(\cdot)$ in (199) by $e^{-\omega \cdot} f(\cdot)$. Arguing in the same way as in Example 3.4.20, we get that

$$[\Delta_W^\alpha u](v) + Au(v + m) = -g_\omega(v), \quad v \in \mathbb{Z}.$$

(ii) In this part, we will transfer the conclusions established in (i) for the equations with the Caputo fractional derivatives. Suppose that $\alpha > 0$, $m = \lceil \alpha \rceil$ and a closed linear operator $-A$ is a subgenerator of a global (g_α, C)-regularized resolvent family $(T(t))_{t \geqslant 0}$

such that there exist two real constants $M \geqslant 1$ and $c \in (0,1)$ with $\|T(t)\| \leqslant Me^{(1-c)t}$, $t \geqslant 0$; see [97], [443] and [445] for the notion and the corresponding examples. Then we know that, for every $x \in X$, the function $t \mapsto T(t)x$, $t \geqslant 0$ is a mild solution of the abstract Cauchy problem

$$\mathbf{D}_t^{\alpha} u(t) + Au(t) = 0, \quad t > 0;\ u(0) = Cx,\ u^{(j)}(0) = 0,\ 1 \leqslant j \leqslant m-1.$$

Using the formula (7.2.4), we can similarly prove that

$$0 = (\Delta^{\alpha} S)(v)x + AS(v+m)x + k(v)Cx, \quad v \in \mathbb{N}_0,\ x \in X,$$

where

$$k(v) = \frac{(-1)^{v+m+1}}{(v+m)!}(\alpha-1)\cdot\dots\cdot(\alpha-v-m), \quad v \in \mathbb{N}_0.$$

Define the sequence $(S(v))_{v\in\mathbb{N}_0}$ as in (i). Then $(S(v))_{v\in\mathbb{N}_0}$ is a discrete $(k, 0, C, (I, A), (0, m), \emptyset)$-existence family, with

$$a_2(v) = \sum_{j=0}^{m} (-1)^{m-j}\binom{m}{j} k^{m-\alpha}(v+j-m) \quad \text{and} \quad a_1(v) = k^0(v), \quad v \in \mathbb{N}_0.$$

Arguing as in Example 3.4.20, with the same notion of the term $g_\omega(\cdot)$, we obtain that

$$[\Delta_W^{\alpha} u](v) + Au(v+m) = -g_\omega(v) - e^{\omega v}(k \circ Cf)(v), \quad v \in \mathbb{Z},$$

provided that the sequence $f : \mathbb{Z} \to X$ satisfies the assumptions (a) and (c) from the formulation of Theorem 3.4.21.

(iii) Suppose finally that $0 \leqslant \alpha_1 < \alpha_2 < \dots < \alpha_n$, $k \in \mathbb{N}^0_{m_n-1}$ and the resolvent operator family $(T(t))_{t\geqslant 0}$ satisfies that there exist two real constants $M \geqslant 1$ and $c \in (0,1)$ with $\|T(t)\| \leqslant Me^{(1-c)t}$, $t \geqslant 0$ as well as that, for every $x \in X$, the function $t \mapsto T(t)x$, $t \geqslant 0$ is a mild solution of the abstract Cauchy problem

$$A_n \mathbf{D}_t^{\alpha_n} u(t) + \dots + A_1 \mathbf{D}_t^{\alpha_1} u(t) = 0, \quad t > 0; \quad u^{(j)}(0) = \delta_{jk} Cx, \quad 0 \leqslant j \leqslant m_n - 1;$$

see [443] and [445] for the notion and the corresponding examples. Set $A_k := \{j \in \mathbb{N}^0_{n-1} : m_j - 1 \geqslant k\}$ and

$$k(v) := \sum_{j\in A_k} \frac{(-1)^{v+m_j+1}}{(v+m_j)!}(\alpha-1-k)\cdot\dots\cdot(\alpha-1-k-m_j).$$

We define the sequence $(S(v))_{v\in\mathbb{N}_0}$ as in Example 3.4.20. Then we have

$$A_n(\Delta^{\alpha_n} S(\cdot)x)(v) + \sum_{s=1}^{n-1} A_s(\Delta^{\alpha_s} S(\cdot)x)(v+m_n-m_s) = -k(v)Cx, \quad v \in \mathbb{N}_0,\ x \in X,$$

so that $(S(v))_{v\geqslant 0}$ is a discrete $(k, 0, C, (A_i)_{1\leqslant i\leqslant n}, (m_n - m_1, \ldots, m_n - m_{n-1}, 0), \emptyset)$-existence family. If the sequence $f : \mathbb{Z} \to X$ satisfies the assumptions (a) and (c) from the formulation of Theorem 3.4.21, then the foregoing argumentation shows that

$$\begin{aligned}&[\Delta_W^{a_n} A_n u](v) + [\Delta_W^{a_{n-1}} A_{n-1} u](v + m_n - m_{n-1})\\&\quad + \cdots + [\Delta_W^{a_1} A_1 u](v + m_n - m_1) = -g_\omega(v) - e^{\omega v}(k \circ Cf)(v), \quad v \in \mathbb{Z},\end{aligned}$$

where we define $g_\omega(\cdot)$ as in (ii).

For more details about discrete fractional calculus, we also refer the reader to the monograph [301] by R. A. C. Ferreira and the doctoral dissertation [384] by M. T. Holm; cf. also [302].

3.5 Abstract nonscalar Volterra difference equations of several variables

In the previous section, we have analyzed various classes of discrete (A, k, B)-regularized C-solution operator families for the abstract Volterra nonscalar difference equation

$$B(v)u(v) = f(v) + \sum_{j=0}^{v} A(v - j)u(j), \quad v \in \mathbb{N}_0,$$

where $B(k)$ is a closed linear operator acting in X ($k \in \mathbb{N}_0$) and $A : \mathbb{N}_0 \to L(Y, X)$; here, Y is any Banach space which is continuously embedded into X. We can similarly analyze the well-posedness of the abstract Volterra nonscalar difference equation

$$B(\mathrm{v})u(\mathrm{v}) = f(\mathrm{v}) + \sum_{j\in\mathbb{N}_0^n; j\leqslant \mathrm{v}} A(\mathrm{v} - j)u(j), \quad \mathrm{v} \in \mathbb{N}_0^n, \tag{205}$$

where $B(k)$ is a closed linear operator acting in X ($k \in \mathbb{N}_0^n$) and $A : \mathbb{N}_0^n \to L(Y, X)$. The notion of a discrete (weak) (A, k, B)-regularized C-resolvent family for (205) and the notion of a discrete (A, k, B)-regularized C-uniqueness family for (205) can be introduced in the same way as in the one-dimensional setting. After that, we can simply transfer the statements of Proposition 3.4.5, Proposition 3.4.6 and Theorem 3.4.7 to the higher-dimensional setting. It is also worth noting that the notion introduced in Definition 3.2.1 can be reconsidered in the multidimensional setting.

If $\mathrm{j} = (j_1, \ldots, j_n) \in \mathbb{N}_0^n$ and $\mathrm{k} = (k_1, \ldots, k_n) \in \mathbb{N}_0^n$, then we write $\mathrm{j} \leqslant \mathrm{k}$ if $j_m \leqslant k_m$ for all $1 \leqslant m \leqslant n$. If the sequences $(a_{\mathrm{k}})_{\mathrm{k}\in\mathbb{N}_0^n}$ and $(b_{\mathrm{k}})_{\mathrm{k}\in\mathbb{N}_0^n}$ are given, then we define $(a *_0 b)(\cdot)$ by

$$(a *_0 b)(\mathrm{k}) := \sum_{j\in\mathbb{N}_0^n;\, j\leqslant \mathrm{k}} a_{\mathrm{k}-j} b_j, \quad \mathrm{k} \in \mathbb{N}_0^n.$$

It can be simply proved that the convolution product $*_0$ is commutative and associative. If the sequences $(a_{\mathrm{k}})_{\mathrm{k}\in\mathbb{N}_0^n}$ and $(b_{\mathrm{k}})_{\mathrm{k}\in\mathbb{Z}^n}$ are given, then we define the Weyl convolution product $(a\circ b)(\cdot)$ by

$$(a\circ b)(\mathrm{v}) := \sum_{l\in\mathbb{Z}^n;l\leqslant\mathrm{v}} a(\mathrm{v}-l)b(l), \quad v\in\mathbb{Z}^n.$$

Under certain assumptions, the equalities stated in (170) hold true.

The main purpose of this section is to analyze some classes of the discrete $(k, C, B, (A_i)_{1\leqslant i\leqslant m}, (\mathrm{v}_i)_{1\leqslant i\leqslant m})$-solution operator families in the multidimensional setting as well as to provide certain applications of the introduced notion to the abstract non-scalar Volterra difference equations of several variables. The following notion plays an essential role in our study.

Definition 3.5.1. Suppose that $B, A_1, \ldots, A_m$ are closed linear operators on X, $C\in L(X)$, $\mathrm{v}_1,\ldots,\mathrm{v}_m\in\mathbb{N}_0^n$, $\mathcal{I}\subseteq\mathbb{N}_m$, $k:\mathbb{N}_0^n\to\mathbb{C}$ and $k\neq 0$. Then we say that the operator family $(S(\mathrm{v}))_{\mathrm{v}\in\mathbb{N}_0^n}\subseteq L(X)$ is a discrete:

(i) $(k, C, B, (A_i)_{1\leqslant i\leqslant m}, (\mathrm{v}_i)_{1\leqslant i\leqslant m})$-existence family if the mapping $x\mapsto A_i(a_i *_0 S)(\mathrm{v}+\mathrm{v}_i)x$, $x\in X$ belongs to $L(X)$ for $\mathrm{v}\in\mathbb{N}_0^n$, $1\leqslant i\leqslant m$ and

$$BS(\mathrm{v})x = k(\mathrm{v})Cx + \sum_{i=1}^{m} A_i(a_i *_0 S)(\mathrm{v}+\mathrm{v}_i)x, \quad \mathrm{v}\in\mathbb{N}_0^n,\ x\in X.$$

(ii) $(k, C, B, (A_i)_{1\leqslant i\leqslant m}, (\mathrm{v}_i)_{1\leqslant i\leqslant m}, \mathcal{I})$-existence family if $(S(\mathrm{v}))_{\mathrm{v}\in\mathbb{N}_0^n}$ is $(k, C, B, (A_i)_{1\leqslant i\leqslant m}, (\mathrm{v}_i)_{1\leqslant i\leqslant m})$-existence family and $S(\mathrm{v})A_i\subseteq A_iS(\mathrm{v})$ for all $v\in\mathbb{N}_0^n$ and $i\in\mathbb{N}_m\setminus\mathcal{I}$.

If $\mathrm{v}_1=\mathrm{v}_2=\cdots=\mathrm{v}_m=0$, then we omit the term "$(\mathrm{v}_i)_{1\leqslant i\leqslant m}$" from the notation. The proofs of the following results can be given as in the one-dimensional setting.

Proposition 3.5.2. *Suppose that $B, A_1, \ldots, A_m$ are closed linear operators on X, $C\in L(X)$, $\mathrm{v}_1,\ldots,\mathrm{v}_m\in\mathbb{N}_0^n$, $k:\mathbb{N}_0^n\to\mathbb{C}$, $k\neq 0$, $1\leqslant i\leqslant m$, $a_i(0)\neq 0$ and $(S(\mathrm{v}))_{\mathrm{v}\in\mathbb{N}_0^n}\subseteq L(X)$ is a discrete $(k, C, B, (A_i)_{1\leqslant i\leqslant m}, (\mathrm{v}_i)_{1\leqslant i\leqslant m})$-existence family. If $x\in X$ and $\mathrm{v}_i=0$, then $S(\mathrm{v})x\in D(A_i)$ for all $\mathrm{v}\in\mathbb{N}_0^n$; the same holds provided that $S(\mathrm{j})x\in D(A_i)$ for all $\mathrm{j}\in\mathbb{N}_0^n\setminus(\mathrm{v}_i+\mathbb{N}_0^n)$.*

Theorem 3.5.3. *Suppose that $B, A_1, \ldots, A_m$ are closed linear operators on X, $C\in L(X)$ is injective, $k:\mathbb{N}_0^n\to\mathbb{C}$, $k(0)\neq 0$ and $a_i(0)\neq 0$ for $1\leqslant i\leqslant m$.*

(i) *Suppose further that $(S(\mathrm{v}))_{\mathrm{v}\in\mathbb{N}_0^n}\subseteq L(X)$ is a discrete $(k, C, B, (A_i)_{1\leqslant i\leqslant m})$-existence family such that $S(0)Bx = BS(0)x$ and $S(0)A_ix = A_iS(0)x$ for all $x\in D(B)\cap D(A_1)\cap\cdots\cap D(A_m)$. Then $(B-\sum_{i=0}^{m} a_i(0)A_i)^{-1}C\in L(X)$, $S(0) = k(0)(B-\sum_{i=0}^{m} a_i(0)A_i)^{-1}C$,*

$$S(\mathrm{v})x = \left(B-\sum_{i=0}^{m} a_i(0)A_i\right)^{-1}\left[k(\mathrm{v})Cx + \sum_{i=1}^{m} A_i\sum_{j\in A_{\mathrm{v}}} a_i(\mathrm{v}-j)S(j)x\right],$$

$$\mathrm{v}\in\mathbb{N}_0^n\setminus\{0\},\ x\in X, \tag{206}$$

where

$$A_{v} := \{j \in \mathbb{N}_0^n : j \leqslant v,\ j \neq v\}, \quad v \in \mathbb{N}_0^n \setminus \{0\},$$

and $A_i S(v) \in L(X)$ *for all* $i \in \mathbb{N}_m$ *and* $v \in \mathbb{N}_0^n$.

(ii) *Suppose that* $C \in L(X)$ *is injective,* $(B - \sum_{i=0}^{m} a_i(0)A_i)^{-1}C \in L(X)$ *and, for every* $l \in \mathbb{N}$ *and for every choice of integers* $a_j \in \mathbb{N}_m$ *for* $1 \leqslant j \leqslant l$, *we have*

$$\left[\prod_{j=1}^{l}\left(B - \sum_{i=0}^{m} a_i(0)A_i\right)^{-1} A_{a_j}\right] \cdot \left(B - \sum_{i=0}^{m} a_i(0)A_i\right)^{-1} C \in L(X).$$

Define $S(0) := k(0)(B - \sum_{i=0}^{m} a_i(0)A_i)^{-1}C$ *and* $S(v)$, $v \in \mathbb{N}_0^n \setminus \{0\}$, *recursively by* (206). *Then* $(S(v))_{v\in\mathbb{N}_0^n} \subseteq L(X)$ *is well-defined,* $A_i S(v) \in L(X)$ *for all* $i \in \mathbb{N}_m$, $v \in \mathbb{N}_0^n$ *and* $(S(v))_{v\in\mathbb{N}_0^n}$ *is a unique discrete* $(k, C, B, (A_i)_{1\leqslant i\leqslant m})$*-existence family. Furthermore, if* $\mathcal{I} \subseteq \mathbb{N}_m$ *and* (184) *holds, respectively, there exist a closed linear operator* A *and the complex polynomials* $P_B(\cdot)$, $P_1(\cdot), \ldots, P_m(\cdot)$ *such that* $CA \subseteq AC$ *and* $B = P_B(A)$, $A_1 = P_1(A), \ldots, A_m = P_m(A)$, *then* $(S(v))_{v\in\mathbb{N}_0^n}$ *is a discrete* $(k, C, B, (A_i)_{1\leqslant i\leqslant m}, \mathcal{I})$*-existence family, respectively,* $(S(v))_{v\in\mathbb{N}_0^n}$ *is a discrete* $(k, C, B, (A_i)_{1\leqslant i\leqslant m}, \emptyset)$*-existence family.*

(iii) *Suppose that* $C = I$, $(B - \sum_{j=0}^{m} a_j(0)A_j)^{-1} \in L(X)$, $\sum_{v\in\mathbb{N}_0^n\setminus\{0\}} |a_i(v)| < +\infty$ *for* $1 \leqslant i \leqslant m$, $\sum_{v\in\mathbb{N}_0^n} |k(v)| < +\infty$ *and* (a) *or* (b) *holds, where:*

(a) $A_i \in L(X)$ *for* $1 \leqslant i \leqslant m$ *and*

$$1 > \sum_{i=1}^{m} \sum_{v\in\mathbb{N}_0^n\setminus\{0\}} |a_i(v)| \cdot \left\|\left(B - \sum_{j=0}^{m} a_j(0)A_j\right)^{-1} A_i\right\|.$$

(b) *Suppose that* $C = I$, (184) *holds or there exist a closed linear operator* A *and the complex polynomials* $P_B(\cdot)$, $P_1(\cdot), \ldots, P_m(\cdot)$ *such that* $B = P_B(A)$, $A_1 = P_1(A), \ldots, A_m = P_m(A)$ *and*

$$1 > \sum_{i=1}^{m} \sum_{v\in\mathbb{N}_0^n\setminus\{0\}} |a_i(v)| \cdot \left\|A_i\left(B - \sum_{j=0}^{m} a_j(0)A_j\right)^{-1}\right\|.$$

Then the requirements in (ii) *hold and we have*

$$\sum_{v\in\mathbb{N}_0^n} \|S(v)\| < +\infty \quad \text{and} \quad \sum_{v\in\mathbb{N}_0^n} \|A_i(a_i *_0 S)(v)\| < +\infty \quad (1 \leqslant i \leqslant m), \tag{207}$$

provided that (a) *holds, respectively, we have* (207) *and*

$$\sum_{v\in\mathbb{N}_0^n} \|A_i S(v)\| < +\infty \quad (1 \leqslant i \leqslant m),$$

provided that (b) *holds.*

Regrettably, the statement of [472, Theorem 3.5] cannot be so straightforwardly transferred to the higher-dimensional setting. The extension is straightforward only in the case that there exists a tuple $\mathrm{v}_i =: \mathrm{v}_{\max} \in \mathbb{N}_0^n$, for some $i \in \mathbb{N}_m$, such that $\mathrm{v}_{i;j} \geqslant \mathrm{v}_{l;j}$ for all $l \in \mathbb{N}_m$ and $j \in \mathbb{N}_n$, with the meaning clear; in this case, we define $M \subseteq \mathbb{N}_m$ as a set of all indexes $i \in \mathbb{N}_m$ with the above property. Then the result of [472, Theorem 3.5] can be simply extended to the higher-dimensional setting with almost the same notation used; for example, in part (i) of this result, we have to assume that the compatibility condition

$$BS(0)x = k(0)Cx + \sum_{i=1}^{m} A_i[a_i(\mathrm{v}_i)S(0)x + \cdots + a_i(0)S(\mathrm{v}_i)x], \quad x \in X$$

holds, so that the value of $S(\mathrm{v})$ will be uniquely determined for any $\mathrm{v} \in (\mathrm{v}_{\max} + \mathbb{N}_0^n) \setminus \{\mathrm{v}_{\max}\}$. This always happens if $m = 1$; all other details can be left to the interested readers.

If there does not exist a tuple $\mathrm{v}_i \in \mathbb{N}_0^n$ with the above described property, then there is no easy way to generalize [472, Theorem 3.5] to the higher-dimensional setting; the main problem lies in the fact that the partial order relation $\sim \subseteq \mathbb{N}_0^n \times \mathbb{N}_0^n$, defined by

$$\mathrm{v} = (v_1, \ldots, v_n) \sim \mathrm{v}' = (v_1', \ldots, v_n') \Leftrightarrow v_i \leqslant v_i', \quad i \in \mathbb{N}_n,$$

is not a total order if $n \geqslant 2$.

We continue by providing some useful observations about the abstract multi-term Volterra difference equation:

$$Bu(\mathrm{v}) = f(\mathrm{v}) + \sum_{i=1}^{m} A_i(a_i *_0 u)(\mathrm{v} + \mathrm{v}_i), \quad \mathrm{v} \in \mathbb{N}_0^n, \tag{208}$$

where $\mathrm{v}_1, \ldots, \mathrm{v}_n \in \mathbb{N}_0^m$.

Remark 3.5.4. All established results about the well-posedness of problem (208) continue to hold in the multidimensional setting; concerning the existence and uniqueness of asymptotically almost periodic type solutions of (208) and similar problems, we would like to note that we must require some additional conditions on the solution operator family $(S(\mathrm{v}))_{\mathrm{v} \in \mathbb{N}_0^n}$, besides its uniform integrability, in order to see that the sequence $u(\mathrm{v}) := (g *_0 S)(\mathrm{v})$, $\mathrm{v} \in \mathbb{N}_0^n$ is $\mathbb{D}$-asymptotically almost periodic (in the sense that there exist an almost periodic sequence $H : \mathbb{Z}^n \to X$ and a continuous function $Q : \mathbb{N}_0^n \to X$ such that $u = H + Q$ on $\mathbb{N}_0^n$ and $\lim_{|\mathrm{v}| \to +\infty; \mathrm{v} \in \mathbb{D}} \|Q(\mathrm{v})\| = 0$, where $\mathbb{D}$ is a certain nonempty subset of $\mathbb{N}_0^n$), provided that the function $g(\cdot)$ is $\mathbb{D}$-asymptotically almost periodic. In the multidimensional setting, the main problem is the $\mathbb{D}$-asymptotical vanishing of the function

$$\mathrm{v} \mapsto \sum_{j \leqslant \mathrm{v}; \neg(0 \leqslant j)} S(\mathrm{v} - j)h(j), \quad \mathrm{v} \in \mathbb{N}_0^n$$

as $|\mathrm{v}| \to +\infty$, where $h(\cdot)$ is the almost periodic part of $g(\cdot)$.

The following results can be proved in the same way as in the corresponding parts of the proofs of [472, Theorem 4.1, Theorem 4.3, Theorem 4.5].

Theorem 3.5.5. (i) *Suppose that* $\mathrm{v}_1 \in \mathbb{N}_0^n, \ldots, \mathrm{v}_m \in \mathbb{N}_0^n$, $(S(\mathrm{v}))_{\mathrm{v}\in\mathbb{N}_0^n} \subseteq L(X)$ *is a discrete* $(k, C, B, (A_i)_{1\leqslant i\leqslant m}, (\mathrm{v}_i)_{1\leqslant i\leqslant m})$*-existence family,* $\sum_{\mathrm{v}\in\mathbb{N}_0^n} \|S(\mathrm{v})\| < +\infty$ *and the following holds:*

(a) $f : \mathbb{Z}^n \to X$ *is a bounded sequence,* $k \in l^1(\mathbb{N}_0^n : Y)$ *and* $\sum_{\mathrm{v}\in\mathbb{N}_0^n} |a_i(\mathrm{v})| < +\infty$ *for* $1 \leqslant i \leqslant m$*, or*

(b) $f \in l^1(\mathbb{Z}^n : X)$, $k : \mathbb{N}_0^n \to X$ *is a bounded sequence and* $a_i : \mathbb{Z}^n \to \mathbb{C}$ *is a bounded sequence for* $1 \leqslant i \leqslant m$.

Define

$$u(\mathrm{v}) := \sum_{l\in\mathbb{Z}^n; l\leqslant \mathrm{v}} S(\mathrm{v} - l)f(l), \quad \mathrm{v} \in \mathbb{Z}^n \tag{209}$$

and

$$\begin{aligned} g(\mathrm{v}) := A_1\Big(\sum_{l\leqslant \mathrm{v}+\mathrm{v}_1} - \sum_{l\leqslant \mathrm{v}}\Big)(a_1 *_0 S)(\mathrm{v} + \mathrm{v}_1 - l)f(l) + \cdots \\ + A_m\Big(\sum_{l\leqslant \mathrm{v}+\mathrm{v}_m} - \sum_{l\leqslant \mathrm{v}}\Big)(a_m *_0 S)(\mathrm{v} + \mathrm{v}_m - l)f(l), \quad v \in \mathbb{Z}^n. \end{aligned} \tag{210}$$

Then $u(\cdot)$ *is bounded if* (a) *holds,* $u \in l^1(\mathbb{Z}^n : X)$ *if* (b) *holds, and we have*

$$\begin{aligned} Bu(\mathrm{v}) = A_1 \sum_{l\in\mathbb{Z}^n; l\leqslant \mathrm{v}+\mathrm{v}_1} a_1(\mathrm{v} + \mathrm{v}_1 - l)u(l) + \cdots \\ + A_m \sum_{l\in\mathbb{Z}^n; l\leqslant \mathrm{v}+\mathrm{v}_m} a_1(\mathrm{v} + \mathrm{v}_m - l)u(l) + g(\mathrm{v}), \quad v \in \mathbb{Z}^n. \end{aligned}$$

(ii) *Suppose that* $\mathrm{v}_1 \in \mathbb{N}_0^n, \ldots, \mathrm{v}_m \in \mathbb{N}_0^n$, $\mathcal{I} \subseteq \mathbb{N}_m$, $(S(\mathrm{v}))_{\mathrm{v}\in\mathbb{N}_0^n} \subseteq L(X)$ *is a discrete* $(k, C, B, (A_i)_{1\leqslant i\leqslant m}, (\mathrm{v}_i)_{1\leqslant i\leqslant m}, \mathcal{I})$*-existence family,* $\sum_{\mathrm{v}\in\mathbb{N}_0^n} \|S(\mathrm{v})\| < +\infty$, $\sum_{\mathrm{v}\in\mathbb{N}_0^n} \|A_i(a_i *_0 S)(\mathrm{v} + \mathrm{v}_i)\| < +\infty$ *for* $i \in \mathcal{I}$ *and the following holds:*

(a) $f : \mathbb{Z}^n \to X$ *is a bounded sequence,* $k \in l^1(\mathbb{N}_0^n : X)$ *and* $\sum_{\mathrm{v}\in\mathbb{N}_0^n} |a_i(\mathrm{v})| < +\infty$ *for* $i \in \mathcal{I}$*, or*

(b) $f \in l^1(\mathbb{Z}^n : X)$, $k : \mathbb{N}_0^n \to X$ *is a bounded sequence and* $a_i : \mathbb{Z}^n \to \mathbb{C} \smallsetminus \{0\}$ *is a bounded sequence for* $i \in \mathcal{I}$

as well as

(c) $A_i f : \mathbb{Z}^n \to X$ *is a bounded sequence,* $\sum_{\mathrm{v}\in\mathbb{N}_0^n} |a_i(\mathrm{v})| < +\infty$ *for* $i \in \mathbb{N}_m \smallsetminus \mathcal{I}$ *and* $(S(\mathrm{v}))_{\mathrm{v}\in\mathbb{N}_0} \subseteq L(X)$ *is a discrete* $(k, C, B, (A_i)_{1\leqslant i\leqslant m}, (\mathrm{v}_i)_{1\leqslant i\leqslant m}, \mathcal{I})$*-existence family, or*

(d) $f \in l^1(\mathbb{Z}^n : X)$, $\sum_{\mathrm{v}\in\mathbb{N}_0^n} \|A_i S(\mathrm{v})\| < +\infty$ *for all* $i \in \mathbb{N}_m \smallsetminus \mathcal{I}$ *and* $a_i : \mathbb{N}_0^n \to \mathbb{C} \smallsetminus \{0\}$ *is a bounded sequence for* $i \in \mathbb{N}_m \smallsetminus \mathcal{I}$*, or*

(e) $f \in l^1(\mathbb{Z}^n : [D(A_i)])$ *for all* $i \in \mathbb{N}_m \smallsetminus \mathcal{I}$, $a_i : \mathbb{N}_0^n \to \mathbb{C} \smallsetminus \{0\}$ *is a bounded sequence for* $i \in \mathbb{N}_m \smallsetminus \mathcal{I}$ *and* $(S(\mathrm{v}))_{\mathrm{v}\in\mathbb{N}_0^n} \subseteq L(X)$ *is a discrete* $(k, C, B, (A_i)_{1\leqslant i\leqslant m}, (\mathrm{v}_i)_{1\leqslant i\leqslant m}, \mathcal{I})$*-existence family.*

Define $u(\cdot)$ and $g(\cdot)$ in the same way as in part (i). *Then we have*

$$\begin{aligned} Bu(\mathrm{v}) = {} & \sum_{i\in\mathcal{I}} A_i \sum_{l\in\mathbb{Z}^n;l\leqslant \mathrm{v}+\mathrm{v}_i} a_i(\mathrm{v}+\mathrm{v}_i-l)u(l) \\ & + \sum_{i\in\mathbb{N}_m\smallsetminus\mathcal{I}} \sum_{l\in\mathbb{Z}^n;l\leqslant \mathrm{v}+\mathrm{v}_i} a_i(\mathrm{v}+\mathrm{v}_i-l)A_i u(l) + (k\circ Cf)(\mathrm{v}) + g(\mathrm{v}), \quad v\in\mathbb{Z}^n. \end{aligned}$$

(iii) *Suppose that $\omega > 0$, $\mathrm{v}_1 \in \mathbb{N}_0^n, \ldots, \mathrm{v}_m \in \mathbb{N}_0^n$, $\mathcal{I} \subseteq \mathbb{N}_m$, $(S(\mathrm{v}))_{\mathrm{v}\in\mathbb{N}_0^n} \subseteq L(X)$ is a discrete $(k, C, B, (A_i)_{1\leqslant i\leqslant m}, (\mathrm{v}_i)_{1\leqslant i\leqslant m}, \mathcal{I})$-existence family, $\sum_{\mathrm{v}\in\mathbb{N}_0^n} \|e^{-\omega[v_1+\cdots+v_n]}S(\mathrm{v})\| < +\infty$, $\sum_{\mathrm{v}\in\mathbb{N}_0^n} \|A_i(e^{-\omega[(\cdot_1-\mathrm{v}_{i;1})+\cdots+(\cdot_n-\mathrm{v}_{i;n})]}a_i *_0 [e^{-\omega[\cdot_1+\cdots+\cdot_n]}S])(\mathrm{v}+\mathrm{v}_i)\| < +\infty$ for $i \in \mathcal{I}$ and the following holds:*

(a) $e^{-\omega[\cdot_1+\cdots+\cdot_n]}f : \mathbb{Z}^n \to X$ *is a bounded sequence,* $e^{-\omega[\cdot_1+\cdots+\cdot_n]}k \in l^1(\mathbb{N}_0^n : X)$ *and* $\sum_{\mathrm{v}\in\mathbb{N}_0^n} |e^{-\omega[(v_1-\mathrm{v}_{i;1})+\cdots+(v_n-\mathrm{v}_{i;n})]}a_i(\mathrm{v})| < +\infty$ *for* $i \in \mathcal{I}$, *or*

(b) $e^{-\omega[\cdot_1+\cdots+\cdot_n]}f \in l^1(\mathbb{Z}^n : X)$, $e^{-\omega[\cdot_1+\cdots+\cdot_n]}k : \mathbb{N}_0^n \to X$ *is a bounded sequence and* $e^{-\omega[(\cdot_1-\mathrm{v}_{i;1})+\cdots+(\cdot_n-\mathrm{v}_{i;n})]}a_i : \mathbb{Z}^n \to \mathbb{C}\smallsetminus\{0\}$ *is a bounded sequence for* $i \in \mathcal{I}$

as well as

(c) $e^{-\omega[\cdot_1+\cdots+\cdot_n]}A_i f : \mathbb{Z}^n \to X$ *is a bounded sequence,* $\sum_{\mathrm{v}\in\mathbb{N}_0^n} |e^{-\omega[(v_1-\mathrm{v}_{i;1})+\cdots+(v_n-\mathrm{v}_{i;n})]}a_i(\mathrm{v})| < +\infty$ *for* $i \in \mathbb{N}_m \smallsetminus \mathcal{I}$ *and* $(S(\mathrm{v}))_{\mathrm{v}\in\mathbb{N}_0} \subseteq L(X)$ *is a discrete* $(k, C, B, (A_i)_{1\leqslant i\leqslant m}, (\mathrm{v}_i)_{1\leqslant i\leqslant m}, \mathcal{I})$*-existence family, or*

(d) $e^{-\omega[\cdot_1+\cdots+\cdot_n]}f \in l^1(\mathbb{Z}^n : X)$, $\sum_{\mathrm{v}\in\mathbb{N}_0^n} \|e^{-\omega[v_1+\cdots+v_n]}A_i S(\mathrm{v})\| < +\infty$ *for all* $i \in \mathbb{N}_m\smallsetminus\mathcal{I}$ *and* $e^{-\omega[(\cdot_1-\mathrm{v}_{i;1})+\cdots+(\cdot_n-\mathrm{v}_{i;n})]}a_i : \mathbb{N}_0^n \to \mathbb{C}\smallsetminus\{0\}$ *is a bounded sequence for* $i \in \mathbb{N}_m\smallsetminus\mathcal{I}$, *or*

(e) $e^{-\omega[\cdot_1+\cdots+\cdot_n]}f \in l^1(\mathbb{Z}^n : [D(A_i)])$ *for all* $i \in \mathbb{N}_m \smallsetminus \mathcal{I}$, $e^{-\omega[(\cdot_1-\mathrm{v}_{i;1})+\cdots+(\cdot_n-\mathrm{v}_{i;n})]}a_i : \mathbb{N}_0^n \to \mathbb{C}\smallsetminus\{0\}$ *is a bounded sequence for* $i \in \mathbb{N}_m\smallsetminus\mathcal{I}$ *and* $(S(\mathrm{v}))_{\mathrm{v}\in\mathbb{N}_0} \subseteq L(X)$ *is a discrete* $(k, C, B, (A_i)_{1\leqslant i\leqslant m}, (\mathrm{v}_i)_{1\leqslant i\leqslant m}, \mathcal{I})$*-existence family.*

Define

$$u(\mathrm{v}) := e^{\omega[v_1+\cdots+v_n]} \sum_{l\in\mathbb{Z}^n;l\leqslant\mathrm{v}} [e^{-\omega[(v_1-l_1)+\cdots+(v_n-l_n)]}S(\mathrm{v}-l)][e^{-\omega[l_1+\cdots+l_n]}f(l)], \tag{211}$$

for any $\mathrm{v} \in \mathbb{Z}^n$ *and* $g_\omega(\cdot)$ *in the same way as in part* (i), *with the operator family* $S(\cdot)$, *the kernels* $a_i(\cdot)$ *and the function* $f(\cdot)$ *replaced therein with the operator family* $e^{-\omega[\cdot_1+\cdots+\cdot_n]}S(\cdot)$, *the kernels* $e^{-\omega[(\cdot_1-\mathrm{v}_{i;1})+\cdots+(\cdot_n-\mathrm{v}_{i;n})]}a_i(\cdot)$ *and the function* $e^{-\omega[\cdot_1+\cdots+\cdot_n]}f(\cdot)$, *respectively,* $(1 \leqslant i \leqslant m)$. *Then we have*

$$\begin{aligned} Bu(\mathrm{v}) = {} & \sum_{i\in\mathcal{I}} A_i \sum_{l\in\mathbb{Z}^n;l\leqslant \mathrm{v}+\mathrm{v}_i} a_i(\mathrm{v}+\mathrm{v}_i-l)u(l) \\ & + \sum_{i\in\mathbb{N}_m\smallsetminus\mathcal{I}} \sum_{l\in\mathbb{Z}^n;l\leqslant \mathrm{v}+\mathrm{v}_i} a_i(\mathrm{v}+\mathrm{v}_i-l)A_i u(l) \\ & + e^{-\omega[v_1+\cdots+v_n]}(k\circ Cf)(\mathrm{v}) + g_\omega(\mathrm{v}), \end{aligned} \tag{212}$$

for any $\mathrm{v} \in \mathbb{Z}^n$.

Keeping in mind the representation formulae (209) and (211), we are in a position to consider the existence and uniqueness of almost periodic and almost automorphic type solutions to the abstract multiterm problems (210) and (212), respectively. For example, in the concrete situation of Theorem 3.5.5(i), the almost periodicity of the inhomogeneity of $f(\cdot)$ implies the almost periodicity of the solution $u(\cdot)$.

3.6 Notes and Appendices to Part I

In this section, we will present several new results and remarks about the discrete dynamical systems and some classes of the (abstract) Volterra difference equations, which are not considered so far. First of all, we will provide a brief description of some recent results obtained in a collaboration with Professors H. C. Koyuncuoğlu/Y. N. Raffoul [488, 487] and H. C. Koyuncuoğlu/V. E. Fedorov [484].

Affine-periodic solutions of discrete dynamical systems: Massera's criterion and affine-periodic Floquet decomposition

Let us recall that a matrix-valued function $A : \mathbb{Z} \to \mathbb{R}^{n\times n}$ is said to be (Q, T)-affine periodic if there is $T \in \mathbb{N}$ such that

$$A(t+T) = QA(t)Q^{-1} \quad \text{for } Q \in GL(\mathbb{R}^n), \quad t \in \mathbb{Z},$$

where $GL(\mathbb{R}^n)$ stands for the n-dimensional linear group over $\mathbb{R}$. Besides that, the linear discrete dynamical system

$$x(t+1) = A(t)x(t), \quad t \in \mathbb{Z} \tag{213}$$

is called (Q, T)-affine periodic if A is a (Q, T)-affine periodic matrix function. Furthermore, a vector-valued function $f : \mathbb{Z} \times \mathbb{R}^n \to \mathbb{R}^n$ is said to be (Q, T)-affine symmetric if there are $Q \in GL(\mathbb{R}^n)$ and $T \in \mathbb{N}$ such that

$$f(t+T, x) = Qf(t, Q^{-1}x) \quad \text{for all } t \in \mathbb{Z} \text{ and } x \in \mathbb{R}^n. \tag{214}$$

The nonlinear system

$$x(t+1) = f(t, x(t)), \quad t \in \mathbb{Z}$$

is called (Q, T)-affine periodic if $f(\cdot, \cdot)$ is (Q, T)-affine symmetric. Finally, a solution $x(\cdot)$ of the system (213) (or (214)) is called (Q, T)-affine periodic if $x(t+T) = Qx(t)$ for all $t \in \mathbb{Z}$.

In [488, Theorem 2], we have proved a generalized discrete Massera type condition, which states that the discrete dynamical system

$$x(t+1) = A(t)x(t) + f(t), \quad t \in \mathbb{Z} \tag{215}$$

has a (Q,T)-affine periodic solution if and only if it has an $m_{(Q,T)}$-bounded solution. By that, we mean the following: Let $T \in \mathbb{N}$ be fixed, let $Q \in GL(\mathbb{R}^n)$ and let

$$m_T^{(t)} = \begin{cases} \max\{n \in \mathbb{N}_0 : t \geqslant nT\}, & t \in \mathbb{N}_0, \\ \max\{n \in \mathbb{Z}_- : t \geqslant nT\}, & t \in -\mathbb{N}. \end{cases}$$

Then we say that $x : \mathbb{Z} \to \mathbb{R}^n$ is $m_{(Q,T)}$-bounded if

$$\sup_{t \in \mathbb{Z}} \left| Q^{-m_T^{(t)}} x(t) \right| < +\infty.$$

In [488, Theorem 3], we have proved an affine periodic Floquet decomposition type theorem: Let X be the principal fundamental matrix of the affine periodic linear discrete dynamical system

$$x(t+1) = A(t)x(t), \quad t \in \mathbb{Z}. \tag{216}$$

Then $Z(t) = Q^{-1}X(t+T)$ is also a fundamental matrix for (216) and

$$Z(t) = X(t)C \quad \text{for all } t \in \mathbb{Z},$$

where

$$C = Q^{-1}A(T-1)A(T-2) \cdot \cdots \cdot A(0).$$

In addition to the above, there exist a (Q,T)-affine periodic matrix function R and a nonsingular matrix B so that

$$X(t) = R(t)QB^t.$$

The existence of affine-periodic solutions of discrete dynamical systems and some applications have been also considered in [488]; in particular, we have analyzed there the existence of affine-periodic solutions to the Keynesian cross-economic model with lagged income

$$\begin{aligned} D(t) &= C(t) + I(t-1) + G(t-1), \\ C(t) &= cx(t), \\ x(t+1) &= \delta D(t+1) + x(t)(1-\delta), \end{aligned}$$

where c is a nonnegative constant, D is aggregate demand, x is aggregate income, C is aggregate consumption, I is aggregate investment, G is government spent and $\delta < 1$ is speed of adjustment term.

We also refer the reader to [188] for further information concerning the Massera theorem for asymptotically periodic scalar differential equations.

Positive periodic solutions for certain kinds of delayed q-difference equations with biological background

By a quantum domain, we mean any closed subset $\overline{q^{\mathbb{Z}}}$ of $\mathbb{R}$ given by

$$\overline{q^{\mathbb{Z}}} := \{q^t : t \in \mathbb{Z}\} \cup \{0\}, \quad q > 1; \quad q^{\mathbb{Z}} := \{q^t : t \in \mathbb{Z}\}.$$

We refer the reader to the pioneering book [414] for more details regarding the quantum calculus. It is obvious that $\overline{q^{\mathbb{Z}}}$ is not translation invariant since it is not closed under addition. Roughly speaking, a q-difference equation is a difference equation involving a q-derivative

$$f^{\Delta}(t) = \frac{f(qt) - f(t)}{qt - t}, \quad t \in q^{\mathbb{Z}}$$

of its unknown. It is clear that the q-derivative f^{Δ} turns into the ordinary derivative f' if $q \to 1$. It should be also emphasized that the theory of q-difference equations is an alternative and a practical tool for the discretization of continuous time mathematical models.

Motivated by the results established in some earlier research articles, we have recently analyzed the q-difference equation

$$x^{\Delta}(t) = x(t)[a(t) - g(t, x(\delta_1(t)), x(\delta_2(t)), \ldots, x(\delta_n(t)))], \quad t \in q^{\mathbb{Z}} \tag{217}$$

with delay terms δ_i; cf. [487]. We have replaced the ordinary derivative with q-derivative in the equation

$$x'(t) = x(t)[a(t) - g(t, x(t - \tau_1(t)), x(t - \tau_2(t)), \ldots, x(t - \tau_n(t)))], \quad t \in \mathbb{R}, \tag{218}$$

proposing thus an alternative direct discretization for (218), which covers many important biological models of single species. It is worthwhile to study (217) under certain conditions due to the linkage between (217) and the equations

$$x^{\Delta}(t) = x(t)\left[a(t) - \sum_{i=1}^{n} a_i(t)x(\delta_i(t))\right], \quad t \in q^{\mathbb{Z}},$$

$$x^{\Delta}(t) = a(t)x(t)\left[1 - \prod_{i=1}^{n} \frac{x(\delta_i(t))}{K(t)}\right], \quad t \in q^{\mathbb{Z}},$$

and

$$x^{\Delta}(t) = a(t)x(t)\left[1 - \sum_{i=1}^{n} \frac{a_i(t)x(\delta_i(t))}{1 + c_i(t)x(\delta_i(t))}\right], \quad t \in q^{\mathbb{Z}},$$

which can be regarded as certain q-analogues of the logistic equation. The outcomes of [487] can be implemented in the study of population dynamics of single species; in our analysis, we use the coincidence degree theory and investigate sufficient conditions for the existence of positive periodic solutions to (217). See also the research articles [28] by B. Ahmad, K. Ntouyas, [43] by B. Alqahtani et al., [680] by A. Salim et al., [743] by J. Tariboon, S. Ntouyas, P. Agarwal and references quoted therein as well as the research article [145] by T. Brikshavana and T. Sitthiwiratthambrick.

Periodic solutions of Kolmogorov systems on quantum time scales

The Kolmogorov systems are extremely important in modeling various biological processes. Here, we concentrate on the coupled functional dynamic equations as Kolmogorov systems on quantum time scales and study the existence of their periodic solutions. As already mentioned, quantum time scales are not translation invariant, which results in using multiplicative periodicity perceptions as an alternative to conventional periodicity. We focus our attention to the Kolmogorov systems with/without delay term and employ the coincidence degree theory and the fixed-point theory for investigating the sufficient conditions for the existence of positive periodic solutions. This is a part of joint research study [499] with H. C. Koyuncuoğlu, Ö. Ö. Kaymak and T. Katican.

Dynamic equations of the form

$$x_i' = x_i h_i(x_1, x_2, \dots, x_n),\ 1 \leqslant i \leqslant n$$

are called Kolmogorov equations. These equations are often used in real life models in which the per unit change x_i'/x_i is explained by the functions $h_i(x_1, x_2, \dots, x_n)$. This model can be considered as a unified abstract model in mathematical biology since it turns into logistic, Volterra prey–predator, and Lotka–Volterra competition model in particular cases (concerning some problems in mathematical biology, we can also recommend reading the monograph [710] by Hal L. Smith and Horst R. Thieme). Inspired by [421], we consider here the following 2-D Kolmogorov system on quantum domains:

$$\begin{cases} x^\Delta(t) = x(t)\hat{f}_1(t, x(t), y(t)), \\ y^\Delta(t) = y(t)\hat{f}_2(t, x(t), y(t)), \end{cases} \quad t \in q^{\mathbb{Z}}, \tag{219}$$

where $q > 1$ and Δ stands for the q-difference operator, i. e.,

$$x^\Delta(t) = \frac{x(qt) - x(t)}{(q-1)t}, \quad t \in q^{\mathbb{Z}}.$$

Obviously, the q-difference system (219) can be used to describe the dynamics of prey–predator interaction in mathematical biosciences. We rewrite the system (219) as follows:

$$\begin{cases} x^\Delta(t) = x(t)[a(t) - f_1(t, x(t), y(t))], \\ y^\Delta(t) = y(t)[b(t) - f_2(t, x(t), y(t))], \end{cases} \quad t \in q^{\mathbb{Z}}; \tag{220}$$

cf. also [220]. Obviously, when (220) is interpreted as a prey–predator system then $x(\cdot)$ can be considered as the population of prey and $y(\cdot)$ indicates the population of the predator. For exhibiting the functional q-difference system (220) as a Kolmogorov prey–predator system, we impose the following conditions (see [421]):

(i) There is a carrying capacity τ_1 for the prey population. That is, there exists $\tau_1 > 0$ so that $f_1(t, \tau_1, 0) = a(t)$ and $a(t) - f_1(t, x, y) < 0$ whenever $x > \tau_1$.
(ii) We naturally assume that predation effects growth in prey population negatively, and this means $f_1(\cdot)$ is increasing in y.
(iii) There is a minimum value for the prey population to promote the population of predators. Equivalently, there exists $\tau_2 > 0$ so that $b(t) - f_2(t, \tau_2, 0) = 0$.
(iv) The function $f_2(\cdot)$ is decreasing with respect to x, and increasing in y.

Clearly, it is reasonable to focus on nonnegative solutions of the quantum Kolmogorov system (220). Since we are interested in the existence of periodic solutions of (220), it is natural to make some periodicity assumptions on the model. For a fixed $T \in q^{\mathbb{N}}$, we suppose that:

(v) $a(\cdot)$ and $b(\cdot)$ are positive valued multiplicatively T-periodic functions

$$a(tT) = a(t) \quad \text{and} \quad b(tT) = b(t) \quad \text{for all } t \in q^{\mathbb{Z}}.$$

(vi) $f_1(\cdot)$ and $f_2(\cdot)$ are positive valued multiplicatively T-periodic functions in t; that is,

$$f_{1,2}(tT, x, y) = f_{1,2}(t, x, y) \quad \text{for all } t \in q^{\mathbb{Z}}.$$

We will always assume that conditions (v–vi) hold. Fix now $T \in q^{\mathbb{N}}$ and consider the set θ_T, which is the set of all T-periodic couples (x, y) where $x,\ y : q^{\mathbb{Z}} \longrightarrow [0, \infty)$. Then $(\theta_T, \|\cdot\|)$ is a Banach space with the norm

$$\|(x, y)\| := \max_{[1,T]\cap q^{\mathbb{Z}}} |x(t)| + \max_{[1,T]\cap q^{\mathbb{Z}}} |y(t)|.$$

Let X and Z be two normed spaces and the mapping $L : domL \subseteq X \longrightarrow Z$ be linear. Let us recall that L is called a Fredholm mapping of index zero if $\dim KerL < \infty$ and $ImL \subseteq Z$ is closed with $co \dim ImL < \infty$. Furthermore, if L is a Fredholm mapping of index zero and there exist continuous projections $P : Z \longrightarrow Z$ and $Q : Z \longrightarrow Z$ so that $ImP = KerL$, $ImL = KerQ = \mathrm{Im}(I - Q)$, then the mapping $L_{DomL\cap KerP} : (I - P)X \longrightarrow ImL$ has the inverse $K_P : ImL \longrightarrow (I - P)X$. Also, if Ω is an open bounded subset of X, then the continuous mapping $N : X \longrightarrow Z$ is said to be L-compact on $\bar{\Omega}$ whenever $QN(\bar{\Omega})$ is bounded, and $K_P(I - Q)N : \bar{\Omega} \longrightarrow X$ is compact. Also, there exists an isomorphism $J : ImQ \longrightarrow KerL$ since $\dim ImQ = co \dim ImL$.

The following result is extremely important for us.

Theorem 3.6.1 (Continuation theorem). *Let L be a Fredholm mapping of index zero, and N be L-compact on $\bar{\Omega}$. Suppose*

(a) $L_Z = \lambda N_Z$ *for each* $\lambda \in (0,1)$ *and* $Z \notin \partial\Omega$.
(b) $QN_Z \neq 0$ *for each* $Z \in \partial\Omega \cap KerL$, *and Brouwer degree* $\deg\{JQN, \Omega \cap KerL, 0\} \neq 0$.

Then the equation $L_Z = N_Z$ *has at least one solution in* $DomL \cap \bar{\Omega}$.

Define now:
- $\Psi := \min\{[0,\infty) \cap q^{\mathbb{Z}}\}$.
- $I_T := [\Psi, \Psi T] \cap q^{\mathbb{Z}}$.
- $\bar{g} := \frac{1}{\Psi(T-1)} \int_{I_T} g(s)\Delta s$.
- $\theta_T^0 := \{(x,y) \in \theta_T : \bar{x} = \bar{y} = 0\}$.
- $\theta_T^{k_1,k_2} := \{(x,y) \in \theta_T : \bar{x} = k_1 \text{ and } \bar{y} = k_2,\ k_{1,2} \geqslant 0, \text{ for all } t \in q^{\mathbb{Z}}\}$.

Then θ_T^0 and $\theta_T^{k_{1,2}}$ are both closed and linear subspaces of θ_T. Moreover, we can simply prove that $\theta_T = \theta_T^0 \oplus \theta_T^{k_{1,2}}$ following the lines of the proof of [421, Lemma 4.1].

We will use the following auxiliary result.

Lemma 3.6.2. *Suppose that the pair* (x,y) *is a nonnegative T-periodic solution of the quantum Kolmogorov system* (220). *Then we have*

$$\min_{t \in I_T} x(t) \geqslant e_{\ominus a}(\Psi T, \Psi)\|x\|,$$
$$\min_{t \in I_T} y(t) \geqslant e_{\ominus b}(\Psi T, \Psi)\|y\|.$$

Proof. As an implementation of q-derivative, we write

$$x(qt) = x(t) + (q-1)tx^{\Delta}(t),$$

and we reconstruct the Kolmogorov system (220) as

$$x^{\Delta}(t) = a(t)(x(qt) - (q-1)tx^{\Delta}(t)) - x(t)f_1(t,x,y),$$
$$y^{\Delta}(t) = b(t)(y(qt) - (q-1)ty^{\Delta}(t)) - y(t)f_2(t,x,y).$$

Then we get

$$\begin{cases} x^{\Delta}(t)(1 + (q-1)ta(t)) - a(t)x(qt) = -x(t)f_1(t,x,y), \\ y^{\Delta}(t)(1 + (q-1)tb(t)) - b(t)y(qt) = -y(t)f_2(t,x,y). \end{cases} \tag{221}$$

The functions a and b are tacitly assumed to be regressive, namely $1 + (q-1)ta(t) \neq 0$ and $1 + (q-1)tb(t) \neq 0$ for all $t \in q^{\mathbb{Z}}$. Consequently, (221) turns into the coupled system

$$\begin{cases} x^{\Delta}(t) - \frac{a(t)}{1+(q-1)ta(t)}x(qt) = \frac{-x(t)}{1+(q-1)ta(t)}f_1(t,x,y), \\ y^{\Delta}(t) - \frac{b(t)}{1+(q-1)tb(t)}y(qt) = \frac{-y(t)}{1+(q-1)tb(t)}f_2(t,x,y). \end{cases} \tag{222}$$

We multiply the equations in (222) with $e_{\ominus a}(t,\Psi)$ and $e_{\ominus b}(t,\Psi)$, respectively. In such a way, we get

$$\begin{cases} (e_{\ominus a}(t,\Psi)x(t))^{\Delta} = \frac{-e_{\ominus a}(t,\Psi)x(t)}{1+(q-1)ta(t)}f_1(t,x,y), \\ (e_{\ominus b}(t,\Psi)y(t))^{\Delta} = \frac{-e_{\ominus b}(t,\Psi)y(t)}{1+(q-1)tb(t)}f_2(t,x,y). \end{cases} \tag{223}$$

We integrate the equations in (223) from t to tT, and obtain

$$\begin{cases} x(tT)e_{\ominus a}(tT,\Psi) - x(t)e_{\ominus a}(t,\Psi) = \int\limits_t^{tT} \frac{-e_{\ominus a}(u,\Psi)x(u)}{1+(q-1)ua(u)}f_1(u,x,y)\Delta u, \\ y(tT)e_{\ominus b}(tT,\Psi) - y(t)e_{\ominus b}(t,\Psi) = \int\limits_t^{tT} \frac{-e_{\ominus b}(u,\Psi)y(u)}{1+(q-1)ub(u)}f_2(u,x,y)\Delta u. \end{cases} \tag{224}$$

By T-periodicity of the couple (x,y), (224) can be alternatively represented as

$$\begin{cases} x(t)e_{\ominus a}(t,\Psi)(e_{\ominus a}(tT,t) - 1) = \int\limits_t^{tT} \frac{-e_{\ominus a}(u,\Psi)x(u)}{1+(q-1)ua(u)}f_1(u,x,y)\Delta u, \\ y(t)e_{\ominus b}(t,\Psi)(e_{\ominus b}(tT,t) - 1) = \int\limits_t^{tT} \frac{-e_{\ominus b}(u,\Psi)y(u)}{1+(q-1)ub(u)}f_2(u,x,y)\Delta u. \end{cases}$$

At this stage, we write

$$e_{\ominus a}(tT,t) = e_{\ominus a}(\Psi T,\Psi) \quad \text{and} \quad e_{\ominus b}(tT,t) = e_{\ominus b}(\Psi T,\Psi).$$

This yields to

$$x(t) = \int\limits_t^{tT} \frac{-x(u)}{1+(q-1)ua(u)} \frac{e_{\ominus a}(u,t)}{e_{\ominus a}(\Psi T,\Psi) - 1} f_1(u,x,y)\Delta u,$$

$$y(t) = \int\limits_t^{tT} \frac{-y(u)}{1+(q-1)ub(u)} \frac{e_{\ominus b}(u,t)}{e_{\ominus b}(\Psi T,\Psi) - 1} f_2(u,x,y)\Delta u.$$

Set

$$G_1(t,u) := \frac{e_{\ominus a}(u,t)}{1 - e_{\ominus a}(\Psi T,\Psi)}$$

and

$$G_2(t,u) := \frac{e_{\ominus b}(u,t)}{1 - e_{\ominus b}(\Psi T,\Psi)},$$

for all $u \in [t,tT] \cap q^{\mathbb{Z}}$. We have the following inequalities:

$$\frac{e_{\ominus a}(\Psi T,\Psi)}{1 - e_{\ominus a}(\Psi T,\Psi)} \leqslant G_1(t,u) \leqslant \frac{1}{1 - e_{\ominus a}(\Psi T,\Psi)}$$

and

$$\frac{e_{\ominus b}(\Psi T, \Psi)}{1 - e_{\ominus b}(\Psi T, \Psi)} \leqslant G_2(t, u) \leqslant \frac{1}{1 - e_{\ominus b}(\Psi T, \Psi)},$$

for all $u \in [t, tT] \cap q^{\mathbb{Z}}$. Hence, we obtain

$$\begin{cases} \|x\| \leqslant \frac{1}{1-e_{\ominus a}(\Psi T,\Psi)} \int\limits_t^{tT} \frac{x(u)}{1+(q-1)ua(u)} f_1(u,x,y)\Delta u, \\ \|y\| \leqslant \frac{1}{1-e_{\ominus b}(\Psi T,\Psi)} \int\limits_t^{tT} \frac{y(u)}{1+(q-1)ub(u)} f_2(u,x,y)\Delta u, \end{cases} \tag{225}$$

and

$$\begin{cases} \min_{t\in I_T} x(t) \geqslant \frac{e_{\ominus a}(\Psi T,\Psi)}{1-e_{\ominus a}(\Psi T,\Psi)} \int\limits_t^{tT} \frac{x(u)}{1+(q-1)ua(u)} f_1(u,x,y)\Delta u, \\ \min_{t\in I_T} y(t) \geqslant \frac{e_{\ominus b}(\Psi T,\Psi)}{1-e_{\ominus b}(\Psi T,\Psi)} \int\limits_t^{tT} \frac{y(u)}{1+(q-1)ub(u)} f_2(u,x,y)\Delta u. \end{cases} \tag{226}$$

Considering (225) and (226) together implies that the assertion is correct. □

We need the following conditions:

C1: There exist constants $M_2 > M_1 > 0$ so that if $x(t) \geqslant M_2$ for all $t \in I_T$, then

$$f_1(t, x(t), y(t)) > a(t), \quad t \in I_T$$

and if $0 < x(t) \leqslant M_1$ for all $t \in I_T$, then

$$f_1(t, x(t), y(t)) < a(t), \quad t \in I_T.$$

C2: There exist constants $M_4 > M_3 > 0$ so that if $y(t) \geqslant M_4$ for all $t \in I_T$, then

$$f_2(t, x(t), y(t)) > b(t), \quad t \in I_T$$

and if $0 < y(t) \leqslant M_3$ for all $t \in I_T$, then

$$f_2(t, x(t), y(t)) < b(t), \quad t \in I_T.$$

Theorem 3.6.3. *Suppose that* (C1–C2) *hold. Then the Kolmogorov system* (220) *has a T-periodic solution.*

Proof. In order to use the continuation theorem in coincidence degree theory, we fix $X = Z = \theta_T$, and present the operators as

$$L\begin{pmatrix} x \\ y \end{pmatrix} = \begin{pmatrix} x^{\Delta} \\ y^{\Delta} \end{pmatrix},$$

$$N\begin{pmatrix} x \\ y \end{pmatrix} = \begin{pmatrix} x(t)[a(t) - f_1(t, x(t), y(t))] \\ y(t)[b(t) - f_2(t, x(t), y(t))] \end{pmatrix},$$

where $\operatorname{Ker}(L) = \theta_T^{k_1,k_2}$ and $\operatorname{Im}(L) = \theta_T^0$. Then L is a Fredholm operator of index zero since $\operatorname{Im}(L)$ is closed and $\dim \mathit{KerL} = \operatorname{codim} \mathit{ImL} = 1$. Besides that, there exist projection $P : X \longrightarrow X$ and $Q : Z \longrightarrow Z$ so that

$$P\begin{pmatrix} x \\ y \end{pmatrix} = Q\begin{pmatrix} x \\ y \end{pmatrix} = \begin{pmatrix} \bar{x} \\ \bar{y} \end{pmatrix}.$$

It is worth emphasizing that $\mathit{ImP} = \mathit{KerL}$ and $\mathit{ImL} = \mathit{KerQ} = \operatorname{Im}(I - Q)$. Accordingly, the inverse of L exists and it is represented by $K_P : \mathit{ImL} \longrightarrow \mathit{KerP} \cap \mathit{DomL}$ so that

$$K_P\begin{pmatrix} x \\ y \end{pmatrix} = \begin{pmatrix} \hat{x} - \bar{\hat{x}} \\ \hat{y} - \bar{\hat{y}} \end{pmatrix},$$

where

$$\hat{x} = \int_{\Psi}^{t} x(s)\Delta s.$$

Now we get

$$QN\begin{pmatrix} x \\ y \end{pmatrix} = \begin{pmatrix} \frac{1}{\Psi T - \Psi}\int_{\Psi}^{\Psi T} x(s)[a(s) - f_1(s, x(s), y(s))]\Delta s \\ \frac{1}{\Psi T - \Psi}\int_{\Psi}^{\Psi T} y(s)[b(s) - f_2(s, x(s), y(s))]\Delta s \end{pmatrix}$$

and

$$K_P(I - Q)N\begin{pmatrix} x \\ y \end{pmatrix}$$
$$= \begin{pmatrix} \int_{\Psi}^{t} \mathcal{K}_1(s)\Delta s - \frac{1}{\Psi T - \Psi}\int_{\Psi}^{\Psi T}\int_{\Psi}^{t} \mathcal{K}_1(s)\Delta s\Delta t - (t - \Psi - \frac{1}{\Psi T - \Psi}\int_{\Psi}^{\Psi T} (t - \Psi)\Delta t)\overline{\mathcal{K}_1(t)} \\ \int_{\Psi}^{t} \mathcal{K}_2(s)\Delta s - \frac{1}{\Psi T - \Psi}\int_{\Psi}^{\Psi T}\int_{\Psi}^{t} \mathcal{K}_2(s)\Delta s\Delta t - (t - \Psi - \frac{1}{\Psi T - \Psi}\int_{\Psi}^{\Psi T} (t - \Psi)\Delta t)\overline{\mathcal{K}_2(t)} \end{pmatrix},$$

where $\mathcal{K}_1$ and $\mathcal{K}_2$ are chosen as

$$\begin{pmatrix} \mathcal{K}_1 \\ \mathcal{K}_2 \end{pmatrix} = \begin{pmatrix} x(t)[a(t) - f_1(t, x(t), y(t))] \\ y(t)[b(t) - f_2(t, x(t), y(t))] \end{pmatrix}$$

for the sake of brevity.

Noting that X is a Banach space, $\overline{P(I-Q)N(\bar{\Omega})}$ is compact for any open subset $\Omega \subseteq X$ by Arzela–Ascoli theorem. On the other hand, $QN(\bar{\Omega})$ is bounded and this yields to L-compactness of N for any open subset $\Omega \subseteq X$.

For $\alpha \in (0,1)$, consider the equation $Lx = \alpha Nx$, that is,

$$\begin{pmatrix} x^{\Delta} \\ y^{\Delta} \end{pmatrix} = \begin{pmatrix} \alpha x(t)[a(t) - f_1(t, x(t), y(t))] \\ \alpha y(t)[b(t) - f_2(t, x(t), y(t))] \end{pmatrix} \tag{227}$$

and suppose that the pair (x, y) is an arbitrary solution of (227). Then we have

$$\begin{aligned} \min_{t \in I_T} x(t) &\geqslant e_{\ominus \alpha a}(\Psi T, \Psi)\|x\| \geqslant e_{\ominus a}(\Psi T, \Psi)\|x\|, \\ \min_{t \in I_T} y(t) &\geqslant e_{\ominus \alpha b}(\Psi T, \Psi)\|y\| \geqslant e_{\ominus b}(\Psi T, \Psi)\|y\| \end{aligned}$$

as a consequence of Lemma 3.6.2. Additionally, we integrate (219) from Ψ to ΨT and get

$$\int_{\Psi}^{\Psi T} \alpha x(s)[a(s) - f_1(s, x(s), y(s))]\Delta s = 0 \tag{228}$$

and

$$\int_{\Psi}^{\Psi T} \alpha y(s)[b(s) - f_2(s, x(s), y(s))]\Delta s = 0. \tag{229}$$

We pursue this part of the proof by establishing contradictions. First, we aim to show that

$$\|x\| < \frac{M_2}{e_{\ominus a}(\Psi T, \Psi)} \quad \text{and} \quad \|y\| < \frac{M_4}{e_{\ominus b}(\Psi T, \Psi)}.$$

Assume the opposite. If

$$\|x\| \geqslant \frac{M_2}{e_{\ominus a}(\Psi T, \Psi)} \quad \text{and} \quad \|y\| \geqslant \frac{M_4}{e_{\ominus b}(\Psi T, \Psi)},$$

then

$$\min_{t \in q^{Z_1}} x(t) = \min_{t \in I_T} x(t) \geqslant e_{\ominus a}(\Psi T, \Psi)\|x\| \geqslant M_2$$

and

$$\min_{t \in q^{Z_1}} y(t) = \min_{t \in I_T} y(t) \geqslant e_{\ominus b}(\Psi T, \Psi)\|y\| \geqslant M_4.$$

This indicates

$$f_1(t, x(t), y(t)) > a(t), \quad t \in I_T$$

and

$$f_2(t, x(t), y(t)) > b(t), \quad t \in I_T,$$

which contradicts with (228) and (229), respectively. In a similar fashion, suppose that

$$\min_{t \in I_T} x(t) < e_{\ominus a}(\Psi T, \Psi) M_1$$

and

$$\min_{t \in I_T} y(t) < e_{\ominus b}(\Psi T, \Psi) M_3.$$

Then we obtain the inequalities

$$M_1 e_{\ominus a}(\Psi T, \Psi) > \min_{t \in I_T} x(t) \geqslant e_{\ominus a}(\Psi T, \Psi)\|x\|$$

and

$$M_3 e_{\ominus b}(\Psi T, \Psi) > \min_{t \in I_T} y(t) \geqslant e_{\ominus b}(\Psi T, \Psi)\|y\|$$

in the light of Lemma 3.6.2. These contradict with (228) and (229) again. Consequentially, we deduce that

$$e_{\ominus a}(\Psi T, \Psi) M_1 < x(t) < \frac{M_2}{e_{\ominus a}(\Psi T, \Psi)} \tag{230}$$

and

$$e_{\ominus b}(\Psi T, \Psi) M_3 < y(t) < \frac{M_4}{e_{\ominus b}(\Psi T, \Psi)} \tag{231}$$

for all $t \in I_T$.

Next, we introduce the set Ω from the couples $(x, y) \in X$ so that (230)–(231) hold. Let us underline that (a) of continuation theorem, Theorem 3.6.1, is satisfied. Also, suppose that $(x, y) \in \partial\Omega \cap KerL$. Then, x and y must be constant with

$$QN\begin{pmatrix} x \\ y \end{pmatrix} = \begin{pmatrix} \frac{1}{\Psi T - \Psi} \int_{\Psi}^{\Psi T} x(s)[a(s) - f_1(s, x(s), y(s))]\Delta s \\ \frac{1}{\Psi T - \Psi} \int_{\Psi}^{\Psi T} y(s)[b(s) - f_2(s, x(s), y(s))]\Delta s \end{pmatrix} \neq \begin{pmatrix} 0 \\ 0 \end{pmatrix}.$$

Moreover, $J : ImQ \longrightarrow KerL$ and $ImQ = KerL$, thus $J = L$. We introduce the homotopy

$$H_{\lambda^*}\begin{pmatrix} x \\ y \end{pmatrix} := \alpha^* \begin{pmatrix} \frac{1}{2}(e_{\ominus a}(\Psi T, \Psi) M_1 + \frac{M_2}{e_{\ominus a}(\Psi T, \Psi)}) - x \\ \frac{1}{2}(e_{\ominus b}(\Psi T, \Psi) M_3 + \frac{M_4}{e_{\ominus b}(\Psi T, \Psi)}) - y \end{pmatrix} + (1 - \alpha^*) QN\begin{pmatrix} x \\ y \end{pmatrix}$$

for $\alpha^* \in [0,1]$. Obviously,

$$H_{\lambda^*}\begin{pmatrix} x \\ y \end{pmatrix} \neq \begin{pmatrix} 0 \\ 0 \end{pmatrix}$$

for any $\alpha^* \in [0,1]$ and $(x,y) \in \partial\Omega \cap KerL$. Then we evaluate the degree

$$\begin{aligned}\deg\{JQN, \Omega \cap KerL, 0\} &= \deg\{QN, \Omega \cap KerL, 0\} \\ &= \deg\left\{\begin{pmatrix} \frac{1}{2}(e_{\ominus a}(\Psi T, \Psi)M_1 + \frac{M_2}{e_{\ominus a}(\Psi T,\Psi)}) - x \\ \frac{1}{2}(e_{\ominus b}(\Psi T, \Psi)M_3 + \frac{M_4}{e_{\ominus b}(\Psi T,\Psi)}) - y, \end{pmatrix} \Omega \cap KerL, 0\right\} \\ &\neq 0.\end{aligned}$$

To conclude, the Kolmogorov system (220), which is constructed on quantum domain, has at least one positive T-periodic solution by Theorem 3.6.1. □

Consider now the following conditions:

C3: There exist constants $M_6 > M_5 > 0$ such that if $x(t) \geqslant M_6$ for all $t \in I_T$, then

$$f_1(t, x(t), y(t)) < a(t), \quad t \in I_T$$

and if $0 < x(t) \leqslant M_5$ for all $t \in I_T$, then

$$f_1(t, x(t), y(t)) > a(t), \quad t \in I_T.$$

C4: There exist constants $M_8 > M_7 > 0$ such that if $y(t) \geqslant M_8$ for all $t \in I_T$, then

$$f_2(t, x(t), y(t)) < b(t), \quad t \in I_T$$

and if $0 < y(t) \leqslant M_7$ for all $t \in I_T$, then

$$f_2(t, x(t), y(t)) > b(t), \quad t \in I_T.$$

The following result can be proved by exactly repeating the same steps in the proof of Theorem 3.6.3, therefore it is trivial.

Theorem 3.6.4. *Suppose that* (C3) *and* (C4) *are satisfied. Then the Kolmogorov system* (220) *has a T-periodic solution.*

In [499], we have also considered the existence and uniqueness of periodic solutions of the following delayed Kolmogorov system on quantum domains:

$$\begin{cases} x^{\Delta}(t) = x(t)[a(t) - f_1(t, x(\delta(t)), y(t))], \\ y^{\Delta}(t) = y(t)[b(t) - f_2(t, x(t), y(\delta(t)))], \end{cases} \quad t \in q^{\mathbb{Z}}.$$

(h,k)-Dichotomies on time scales and its application to Volterra integro-dynamic systems

In our recent joint paper with Professor H. C. Koyuncuoğlu and Y. Raffoul [489], we have investigated the notion of (h,k)-dichotomy on time scales and provided certain applications to the Volterra integro-dynamic system

$$x^{\Delta}(t) = A(t)x(t), \quad x(t_0) = x_0, \quad t \in \mathbb{T}, \tag{232}$$

where A is an $n \times n$, invertible, regressive matrix valued function.

Suppose that $h,\ k : \mathbb{T} \to (0,\infty)$. The linear, time-varying dynamical system (96) is said to admit an (h,k)-dichotomy if there exist positive constants $\beta_{1,2}$, α and projection P such that

$$\begin{aligned}|X(t)PX^{-1}(s)| &\leqslant \beta_1 \frac{h(t)}{h(s)} e_{\ominus\alpha}(t,s), \quad s,t \in \mathbb{T},\ t \geqslant s,\\ |X(t)(I-P)X^{-1}(s)| &\leqslant \beta_2 \frac{k(t)}{k(s)} e_{\ominus\alpha}(s,t), \quad s,t \in \mathbb{T},\ t \leqslant s,\end{aligned}$$

where $X(t)$ is principal fundamental matrix solution of (232). Further on, the functions $h,\ k : \mathbb{T} \to (0,\infty)$ satisfy the compensation law if there exists a constant $C > 0$ such that

$$\frac{h(t)}{h(s)} \leqslant C\frac{k(t)}{k(s)} \quad \text{for all } s,\ t \in \mathbb{T} \text{ with } t \geqslant s.$$

In the sequel, we define the Green function of (96) as

$$G(t,s) := \begin{cases} X(t)PX^{-1}(\sigma(s)), & t \geqslant \sigma(s);\\ -X(t)(I-P)X^{-1}(\sigma(s)), & t < \sigma(s)\end{cases} \tag{233}$$

for a projection P. The linear, homogeneous, time-varying dynamical system (232) satisfies integrability condition with pair (γ, P) if

$$\sup_{t\in\mathbb{T}} \int_{-\infty}^{\infty} |G(t,s)|\Delta s \leqslant \gamma < \infty. \tag{234}$$

If the linear system (232) admits an (h,k)-dichotomy, then it satisfies the integrability condition whenever

$$\sup_{t\in\mathbb{T}} \left(\beta_1 \int_{-\infty}^{t} \frac{h(t)}{h(s)} e_{\ominus\alpha}(t,s)\Delta s + \beta_2 \int_{t}^{\infty} \frac{k(t)}{k(s)} e_{\ominus\alpha}(s,t)\Delta s \right) \leqslant \gamma < \infty. \tag{235}$$

In [489], we have established the following results.

Proposition 3.6.5. *Suppose that* (234) *holds. If the linear system* (232) *admits an* (h,k)*-dichotomy, then* $x(t) = 0$ *is a unique bounded solution of* (232).

Theorem 3.6.6. *Suppose that the conditions*

(A1) $\mathbb{T}$ *is an unbounded time scale from below and above.*

(A2) $\mathbb{T}$ *is a periodic time scale in shifts associated with the initial point* t_0.

(A3) *The matrix function* A *is* Δ T*-periodic in shifts, that is,* $A(\delta_+^T(t))\delta_+^{\Delta T}(t) = A(t)$.

Hold and the linear, homogeneous, time-varying system (232) *admits an* (h,k)*-dichotomy with integrability condition* (235). *Then the Green function of* (232) *is unique up to period* T *in shifts, i. e.,*

$$G(\delta_+^T(t), \delta_+^T(s)) = G(t,s).$$

For more details, we refer the reader to [489].

Almost automorphic solutions to a class of nonlinear difference equations

Let us consider the following abstract nonlinear difference equation:

$$x(t+1) = a(t)x(t) + \sum_{j=-\infty}^{t-1} \Lambda_1(t,j,x(j)) + \sum_{j=t}^{\infty} \Lambda_2(t,j,x(j)), \tag{236}$$

where $a : \mathbb{Z} \to \mathbb{C}$, $a(t) \neq 0$ for all $t \in \mathbb{Z}$, and $\Lambda_{1,2} : \mathbb{Z} \times \mathbb{Z} \times X \to X$.

Then we know that a function $x(\cdot)$ is a solution of (236) with the initial data $x(t_0) = x_0$ if and only if

$$x(t) = x_0 \prod_{s=t_0}^{t-1} a(s) + \sum_{k=t_0}^{t-1} \left(\prod_{s=k+1}^{t-1} a(s) \right) \left(\sum_{j=-\infty}^{k} \Lambda_1(k,j,x(j)) + \sum_{j=k+1}^{\infty} \Lambda_2(k,j,x(j)) \right).$$

In [484], we have assumed that the following conditions are satisfied:

C1n The function $a(\cdot)$ is discrete almost automorphic.

C2n $\Lambda_{1,2}$ are discrete bi-almost automorphic in t and s, uniformly for x.

C3n For $u_{1,2} \in X$, the Lipschitz inequalities

$$\|\Lambda_1(t,s,u_1) - \Lambda_1(t,s,u_2)\| \leqslant m_1(t,s)\|u_1 - u_2\|$$

and

$$\|\Lambda_2(t,s,u_1) - \Lambda_2(t,s,u_2)\| \leqslant m_2(t,s)\|u_1 - u_2\|$$

hold, as well as

$$\sup_{t\in\mathbb{Z}} \sum_{j=-\infty}^{t-1} m_1(t,j) = M_1 < \infty \quad \text{and} \quad \sup_{t\in\mathbb{Z}} \sum_{j=t}^{\infty} m_2(t,j) = M_2 < \infty.$$

C4n For every integer sequence $(v_n')_{n\in\mathbb{Z}}$, there exists a subsequence $(v_n)_{n\in\mathbb{Z}}$ of $(v_n')_{n\in\mathbb{Z}}$ such that

$$\lim_{n\to\infty} x(t_0 \pm v_n) = x(t_0) = x_0.$$

After that, we have introduced the mapping $H : X \to X$ by

$$(Hx)(t) := x_0 \prod_{s=t_0}^{t-1} a(s) + \sum_{k=t_0}^{t-1} \left(\prod_{s=k+1}^{t-1} a(s) \right) (S_1(k, x(k)) + S_2(k, x(k))),$$

where

$$S_1(k, x(k)) := \sum_{j=-\infty}^{k} \Lambda_1(k, j, x(j))$$

and

$$S_2(k, x(k)) := \sum_{j=k+1}^{\infty} \Lambda_2(k, j, x(j)).$$

Then we have the following: If $x(\cdot)$ is almost automorphic, then $S_1(\cdot, x(\cdot))$ and $S_2(\cdot, x(\cdot))$ are almost automorphic and H maps $\mathcal{AA}(X)$ into $\mathcal{AA}(X)$, where $\mathcal{AA}(X)$ stands for the vector space of all almost automorphic sequences from $\mathbb{Z}$ into X, equipped with the sup-norm. This is the crucial for further applications of the Banach contraction principle; in [484], we have proved the following results:

(i) Suppose also

C5n

$$\sup_{t\in\mathbb{Z}} \sum_{k=t_0}^{t-1} \left\| \prod_{s=k+1}^{t-1} a(s) \right\| (M_1 + M_2) = \kappa < 1.$$

Then the abstract difference equation (236) has a unique discrete almost automorphic solution.

(ii) Assume that the conditions **C1n–C5n** hold. For a positive constant γ, we define the set

$$W_\gamma := \{x \in \mathcal{AA}(X) : \|x - x^0\|_{\mathcal{AA}(X)} \leqslant \gamma\},$$

where

$$x^0(t) := \sum_{k=t_0}^{t-1} \left(\prod_{s=k+1}^{t-1} a(s) \right) (S_1(k, 0) + S_2(k, 0)).$$

Let $\|x\|_{\mathcal{AA}(X)} \leqslant \gamma$ and

C6n $\|\prod_{s=t_0}^{t-1} a(s)\|_{\mathcal{A}\mathcal{A}(X)} \leqslant \psi$ for all t.
If

$$\|x_0\|\psi + \kappa\gamma \leqslant \gamma,$$

then the nonlinear difference equation (236) has a unique discrete almost automorphic solution in W_γ.

(iii) Suppose that the conditions **C1n–C5n** are satisfied. Then a bounded solution of nonlinear abstract difference equation (236) is almost automorphic if and only if it has a relatively compact range.

Fractional economic cobweb models with Hilfer nabla fractional differences

In the field of business economics, the cobweb model is a well-known important model used for describing the equilibrium price between demand and supply function for non-storable goods. The cobweb model appeared already in the eighteenth century, but the original name "Cobweb" is given to this concept by N. Kaldor in [415]; cf. also M. Ezekiel [282]. The pioneering book of G. Gandolfo [315] provides a systematic study of the cobweb model defined as

$$\begin{cases} D(t) = a + bp(t+1), \\ S(t) = a_1 + b_1 p(t), \qquad t \in \mathbb{N}_0, \\ D(t) = S(t), \end{cases} \tag{237}$$

where a, b, a_1, $b_1 \in \mathbb{R}$, $b \neq 0$, $b \neq b_1$, $p(t)$ is the market price, $D(t)$ is the demand function and $S(t)$ is the supply function. The solution of (237) is given by

$$p(t) = (p_0 - p^*)\left(\frac{b_1}{b}\right)^t + p^*,$$

where $p_0 = p(0)$ and p^* is the equilibrium value.

Recently, the fractional analogues of cobweb models are studied in a series of papers; see, e. g., the research articles [124–126, 193, 194, 616, 679, 721] and references therein. In this part, we will provide a brief description of the main results established by H. C. Koyuncuoğlu, Ö. Ö. Kaymak and J. M. Jonnalagadda in [498].

If $\alpha > 0$, $m = \lceil \alpha \rceil$, $0 \leqslant \beta \leqslant 1$ and $u : \mathbb{N}_0 \to X$, then we define the Hilfer fractional derivative $\Delta^{\alpha,\beta} u$ through

$$[\Delta^{\alpha,\beta} u](v) := [\Delta^{-\beta(m-\alpha)} \Delta^m \Delta^{(\beta-1)(m-\alpha)}](v), \quad v \in \mathbb{N}_0.$$

If $\beta = 0$, respectively, $\beta = 1$, then the Hilfer fractional derivative $\Delta^{\alpha,\beta} u$ reduces to the Riemann–Liouville fractional derivative Δ^α, respectively, the Caputo fractional derivative Δ_C^α; see, e. g., [13, 54, 356, 377, 400, 401, 770] for more details on the subject.

Let us consider the following cobweb model with Hilfer nabla fractional difference in the demand function:

$$\begin{cases} D(t) = a + b[p(t) + (\Delta^{\alpha,\beta} p)(t)], \\ S(t) = a_1 + b_1 p(t), & t \in \mathbb{N}_0, \\ D(t) = S(t), \end{cases} \tag{238}$$

where $0 < \alpha \leqslant 1$, $0 \leqslant \beta \leqslant 1$, a, b, a_1, $b_1 \in \mathbb{R}$ and $b \neq b_1$. Due to the equilibrium condition in (238), we get the following scalar fractional difference equation:

$$(\Delta^{\alpha,\beta} p)(t) + \lambda p(t) = \theta,$$

where $\lambda = \frac{b-b_1}{b}$ and $\theta = \frac{a_1-a}{b}$.

Suppose that $|p| < 1$, $\alpha > 0$ and $\beta \in \mathbb{R}$. Then the nabla Mittag-Leffler function is defined by

$$E_{p,\alpha,\beta}(t,a) := \sum_{k=0}^{\infty} p^k H_{\alpha k+\beta}(t,a), \quad t \in \mathbb{N}_a \equiv a + \mathbb{N}_0,$$

where the μth-order nabla fractional Taylor monomial is defined by

$$H_\mu(t,a) := \frac{(t-a)^{\bar{\mu}}}{\Gamma(\mu+1)} = \frac{\Gamma(t-a+\mu)}{\Gamma(t-a)\Gamma(\mu+1)}, \quad \mu \in \mathbb{R}\setminus\{\dots,-2,-1\}.$$

We have the following results.

Theorem 3.6.7. (i) *The unique solution of* (238) *is given by*

$$p(t) = (\lambda p_0 - \theta) E_{-\lambda,\alpha,\alpha-1}(t,\rho(0)) + p_0 E_{-\lambda,\alpha,\gamma-1}(t,\rho(0)) + \frac{\theta}{\lambda}\big(1 - E_{-\lambda,\alpha,0}(t,\rho(0))\big),$$
$$t \in \mathbb{N}_0,$$

where $p_0 = p(0)$ and $\gamma = \alpha + \beta - \alpha\beta$.

(ii) *Assume that $b_1 < b$. Then the solution $p(\cdot)$ of* (238) *converges to its equilibrium*

$$p^* = \frac{a_1 - a}{b - b_1}.$$

The authors also suggested the use of Hilfer nabla fractional difference in the supply function; they have analyzed the model

$$\begin{cases} D(t) = a + bp(t), \\ S(t) = a_1 + b_1[p(t) + d(\Delta^{\alpha,\beta} p)(t)], & t \in \mathbb{N}_0, \\ D(t) = S(t), \end{cases} \tag{239}$$

where $0 < \alpha \leqslant 1$, $0 \leqslant \beta \leqslant 1$, a, b, a_1, b_1, $d \in \mathbb{R}$, $b \neq 0$ and $b \neq b_1$. Now we have

$$(\Delta^{\alpha,\beta}p)(t) + \delta p(t) = \eta,$$

where $\delta = \frac{b_1-b}{b_1 d}$ and $\eta = \frac{a-a_1}{b_1 d}$.

The following results can be achieved.

Theorem 3.6.8. *The cobweb model* (239) *has the unique solution*

$$p(t) = (\delta p_0 - \eta)E_{-\delta,\alpha,\alpha-1}(t,\rho(0)) + p_0 E_{-\delta,\alpha,\gamma-1}(t,\rho(0)) + \frac{\eta}{\delta}(1 - E_{-\delta,\alpha,0}(t,\rho(0))), \quad t \in \mathbb{N}_0,$$

where $p_0 = p(0)$ and $\gamma = \alpha + \beta - \alpha\beta$. Moreover, the solution converges to the equilibrium point $p^* = \frac{a_1-a}{b-b_1}$, whenever $0 < \frac{b-b_1}{b_1 d} < 1$.

Almost periodic points and minimal sets in ω-regular spaces

Topological dynamics is still a very active field of research. For the basic source of information about topological dynamics, we refer the reader to the research monographs [118, 119, 237, 271, 335, 394, 420, 767]. In this part, we will first describe the results established by J.-H. Mai and W.-H. Sun in [567], which regards the almost periodic points and minimal sets in ω-regular spaces. The interested reader may also consult the paper [795], where J. C. Xiong has analyzed the set of almost periodic points of a continuous self-map of an interval, and the paper [376], where W. H. He, J. D. Yin and Z. L. Zhou have analyzed the notion of a quasiweakly almost periodic point; cf. also [323, 423, 583, 688] and the research studies of R. A. Johnson [403–407].

If (X,τ) is a topological space and $f : X \to X$ is continuous, then a point $x \in X$ is said to be a periodic point of T if there exists $m \in \mathbb{N}$ such that $f^m x = x$; further on, it is said that x is:

(i) A recurrent point of $f(\cdot)$ if for each neighborhood U of x and for each $N \in \mathbb{N}$ there exists $k > N$ such that $f^k(x) \in U$.
(ii) An almost periodic point of $f(\cdot)$ if for each neighborhood U of x there exists $N \in \mathbb{N}$ such that $\{f^{k+i}(x) : 0 \leqslant i \leqslant N\} \cap U \neq \emptyset$ for all $k \in \mathbb{N}_0$.

Further on, a nonempty subset W of X is called a minimal set of $f(\cdot)$ if W is closed, f-invariant (i. e., $f(W) \subseteq W$) and no proper subset of W enjoys these properties. We know that every point in a minimal set of $f(\cdot)$ is a recurrent point of $f(\cdot)$.

Before proceeding any further, we would like to note that, if X is a complex Banach space, $f \in L(X)$ and $x \in X$ is a recurrent vector of $f(\cdot)$, respectively, a frequently hypercyclic vector of $f(\cdot)$, then x is a recurrent point of $f(\cdot)$, respectively, an almost periodic point of $f(\cdot)$; cf. [96, 344, 446] for the notion and more details about the linear topological dynamics of operators and its applications.

In [567, Definition 2.1], the authors have introduced the following notion: A topological space X is said to be ω-regular if for any closed subset W of X, any point $x \in X \setminus W$

and any countable set $A \subseteq W$, there exist disjoint open sets U and V in X such that $x \in U$ and $A \subseteq V$. Every ω-regular topological space is regular (i. e., for any closed subset W of X and any point $x \in X \setminus W$ there exist disjoint open sets U and V in X such that $x \in U$ and $W \subseteq V$), while the converse statement does not hold in general. The notion of an ω-regular space is extremely important because several results established by one of the founders of topological dynamics, W. H. Gottschalk, admits the extensions from regular spaces to ω-regular spaces. For example, in [567, Theorem 2.3], the authors have shown that, if X is an ω-regular topological space and $f : X \to X$ is continuous, then the closure of every almost periodic orbit of $f(\cdot)$ is a minimal set. Furthermore, the authors have proved the following results:

(i) If X is a locally compact topological space and $f : X \to X$ is continuous, then all points in any minimal set of $f(\cdot)$ are almost periodic points of $f(\cdot)$; cf. [567, Theorem 3.1].
(ii) Let X be the Banach space of all square-summable complex sequences equipped with the usual norm. Then there exists an isometric homeomorphism $f(\cdot)$ of X such that any point $x \in X$ is a recurrent point of $f(\cdot)$ and no point $x \in X$ is an almost periodic point of $f(\cdot)$; cf. [567, Example 3.3].
(iii) If X is a topological space, $f : X \to X$ is continuous and W is a finite minimal set of $f(\cdot)$, then all points in W are almost periodic points of $f(\cdot)$; cf. [567, Corollary 3.6].
(iv) If X is an ω-regular topological space and $f : X \to X$ is continuous, then all points in the closure of any almost periodic orbit of $f(\cdot)$ are almost periodic points of $f(\cdot)$.

We continue by noticing that we can consider the following notion for general binary relations, as well (in place of T, we can also consider the general binary relations on X).

Definition 3.6.9. Suppose that (X, τ) is a topological space, $T : X \to X$ and $\rho \subseteq X \times X$. Then a point $x \in D_\infty(\rho) := \bigcap_{n\in\mathbb{N}} D(\rho^n)$ is said to be:
(i) T-recurrent point of $\rho(\cdot)$ if for each neighborhood U of Tx and for each $N \in \mathbb{N}$ there exists $k > N$ such that $\rho^k(x) \subseteq U$;
(ii) T-almost periodic point of $\rho(\cdot)$ if for each neighborhood U of Tx there exists $N \in \mathbb{N}$ such that $\{\rho^{k+i}(x) : 0 \leqslant i \leqslant N\} \cap U \neq \emptyset$ for all $k \in \mathbb{N}_0$.

In [446], we have investigated the unbounded recurrent linear operators and the unbounded frequently hypercyclic linear operators. Our previous observation on recurrent points of $f(\cdot)$ and almost periodic points of $f(\cdot)$ continues to hold for unbounded linear operators.

The notion introduced in Definition 3.6.9(ii) can be further generalized by taking into consideration various kinds of lower and upper (Banach) densities; cf. [446, Sections 1.2–1.4] for some details about the class of $\mathcal{F}$-hypercyclic operators on Banach spaces. We will analyze such notion somewhere else.

Further on, the notion of an almost periodic point of a continuous mapping $f : T \to X$ is a little bit strange because it has been assumed there that for each neighborhood

U of x there exists $N \in \mathbb{N}$ such that $\{f^{k+i}(x) : 0 \leqslant i \leqslant N\} \cap U \neq \emptyset$ for all $k \in \mathbb{N}_0$; in the pseudometric spaces, we can also assume that for each $\varepsilon > 0$ there exists $N \in \mathbb{N}$ such that $\inf(\{d(f^{k+i}(x), f^k(x)) : 0 \leqslant i \leqslant N\}) \leqslant \varepsilon$ for all $k \in \mathbb{N}_0$, i. e., that the sequence $(f^k x)_{k\in\mathbb{N}}$ is almost periodic. Concerning this issue (cf. also Definition 3.6.9(ii)), we would like to introduce the following notion.

Definition 3.6.10. Suppose that (X, d) is a pseudometric space and $\rho \subseteq X \times X$. Let $\mathcal{F}$-almost periodicity, respectively, $\mathcal{F}$-almost automorphy, be any type of almost periodicity (almost automorphy) considered by now (we assume, if necessary, that X has a linear vector structure). Then a point $x \in D_\infty(\rho)$ is said to be:
(i) A regular $\mathcal{F}$-almost periodic point of $\rho(\cdot)$ if there exists an $\mathcal{F}$-almost periodic sequence $(x_k)_{k\in\mathbb{N}}$ such that $x_k \in \rho^k x$ for all $k \in \mathbb{N}$.
(ii) A regular $\mathcal{F}$-almost automorphic point of $\rho(\cdot)$ if there exists an $\mathcal{F}$-almost automorphic sequence $(x_k)_{k\in\mathbb{N}}$ such that $x_k \in \rho^k x$ for all $k \in \mathbb{N}$.

For example, a point $x \in D_\infty(\rho)$ is a regular T-almost periodic point of function $\rho : X \to X$, where $T : X \to X$, if the sequence $(\rho^k x)_{k\in\mathbb{N}}$ is T-almost periodic, i. e., if for each $\varepsilon > 0$ there exists $N \in \mathbb{N}$ such that $\inf(\{d(f^{k+i}(x), Tf^k(x)) : 0 \leqslant i \leqslant N\}) \leqslant \varepsilon$ for all $k \in \mathbb{N}_0$.

The notion introduced in Definition 3.6.10 seems to be new even for bounded linear operators on complex Banach spaces; but this notion also clarifies what is a regular $\mathcal{F}$-almost periodic (automorphic) point of a digraph $\rho \subseteq X \times X$. More details will appear somewhere else.

For further information about almost automorphic points in topological dynamics, we refer the reader to the articles [107, 311, 317, 373, 569, 572, 603, 636].

Discrete fractional operators

The first type of fractional difference was introduced in 1832 by J. Liouville. Besides the generalized Hilfer (a, b, α)-fractional derivatives, we can also analyze many other types of fractional difference operators, like:

(i) The sequential delta fractional difference

$$\Delta^\alpha f(m) := \sum_{k=0}^{+\infty} (-1)^k \frac{\Gamma(\alpha+1)}{k! \cdot \Gamma(\alpha+1-k)} f(m+k),$$

where $\alpha \in \mathbb{R} \smallsetminus \mathbb{Z}$ and $m \in \mathbb{N}_0$ were introduced and analyzed by S. Chapman in 1911 [178].

(ii) In [253], L. Díaz and T. J. Osler have introduced the following fractional difference operator of order $\alpha > 0$:

$$\Delta^\alpha f(z) := \sum_{k=0}^{+\infty} (-1)^k \binom{\alpha}{k} f(z+\alpha-k) \tag{240}$$

and proposed the following Leibniz rule

$$\Delta^{a}[fg](z)=\sum_{k=0}^{+\infty}\binom{a}{k}\Delta^{a-k}f(z)\Delta^{k}g(z+a-k).$$

The notion given in the formula (240) has several serious drawbacks; for example, the semigroup law $\Delta^{a+\beta}=\Delta^{a}\Delta^{\beta}$ does not hold in this framework. For this reason, C. W. J. Granger and R. Joyeux have proposed the following notion:

$$\Delta_{1}^{a}f(z):=\sum_{k=0}^{+\infty}(-1)^{k}\binom{a}{k}f(z-k),$$

where a is arbitrary real number. See also the research article [341] by H. L. Gray and N. fan Zhang.

(iiii) The discrete Prabhakar fractional differences and sums have been introduced and analyzed in [601].

(iv) In a recent research article [742] by V. E. Tarasov, the interested readers can find a short review of exact finite-differences calculus of integer orders. The proposed exact finite-differences are exact algebraic analogues of derivatives; that is, these differences satisfy the same characteristic relations as the standard derivatives in the space of entire functions.

(v) The various types of linear fractional-order discrete systems have been discussed by D. Mozyrska and M. Wyrwas in [606]. In order to achieve their aims, the authors perform the Z-transform. The authors also analyze the discrete Mittag-Leffler functions, some new types of fractional difference sums and operators and compute their Z-transforms.

The discrete fractional derivatives based on the bilinear transformation have been introduced and analyzed in [628], while the proportional fractional discrete operators of the Riemann–Liouville and Caputo type have been introduced and analyzed in [539]. For more details about discrete fractional derivatives, we also recommend the survey articles [627] by M. D. Ortigueira and [771] by Q. Wang, R. Xu.

Let us finally note that we have not analyzed applications of the generalized almost periodic type sequences to the (abstract) Volterra integrodifferential equations with piecewise constant argument here. We will consider this topic in our follow-up research studies.

Part II: **Generalized metrical almost periodicity and generalized metrical almost automorphy**

In this part, we analyze various classes of generalized metrically almost periodic functions and generalized metrically almost automorphic functions, and provide certain applications to the abstract Volterra integrodifferential equations. We follow the notion and notation used in the previous part of the book.

We start our exposition by investigating the metrical Bochner criterion and the notion of metrical Stepanov almost periodicity. In Chapter 4, we also analyze metrically piecewise continuous ρ-almost periodic functions and various classes of generalized vectorially almost periodic functions. Chapter 5 investigates the notions of metrical almost automorphy in the sense of Stepanov, Weyl and Besicovitch approaches. We also analyze the notions of Stepanov-p-almost periodicity and Stepanov-p-almost automorphy with a general exponent $p > 0$.

https://doi.org/10.1515/9783111689746-006

4 Generalized almost periodic functions, generalized metrically almost periodic functions and applications

4.1 Metrical Bochner criterion and metrical Stepanov almost periodicity

In this section, we continue our previous analysis of multidimensional metrically Stepanov p-almost periodic type functions [458]. We introduce the corresponding classes of Stepanov almost periodic type functions in a Bochner-like manner and clarify their main structural properties.

The organization and main ideas of this section can be briefly summarize as follows. In Section 4.1.1, we first recall the basic definitions of metrically almost periodic type functions considered in [448]; the main structural result of this section is Theorem 4.1.3, which is completely new and provide a generalization of the well-known Bochner criterion (we call this result *metrical Bochner criterion*). In Corollary 4.1.5, we provide an important application of Theorem 4.1.3 to Doss-p-almost periodic functions. The main aim of Section 4.1.2 is to consider the notion of Stepanov $(\Omega, \mathrm{R}, \mathcal{B}, \mathcal{P}_Z)$-multialmost periodicity; in Definition 4.1.7, we introduce the class of (strongly) Stepanov $(\Omega, \mathrm{R}, \mathcal{B}, \mathcal{P}_Z)$-multialmost periodic functions. The Bochner type criterion for Stepanov $(\Omega, \mathrm{R}, \mathcal{B}, \mathcal{P}_Z)$-multialmost periodic functions is established in Theorem 4.1.9 (cf. also Proposition 4.1.13, where we consider the invariance of Stepanov $(\Omega, \mathrm{R}, \mathcal{B}, \mathcal{P}_Z)$-multialmost periodicity under the actions of convergence in the pseudometric space $\mathcal{P}_Z$). The notion of Stepanov $(\Omega, \mathcal{B}, \mathcal{P}_Z)$-boundedness and the notion of Stepanov $(\Omega, \mathcal{B}, Z)$-boundedness are introduced in Definition 4.1.11, and later examined in Proposition 4.1.12. In Section 4.1.3, we analyze the extensions of almost periodic sequences. The material of this section is taken from our joint paper [481] with V. E. Fedorov and H. C. Koyuncuoğlu.

Before proceeding further, we would like to note that the following weighted function space has not been used in the previous part of book. Suppose that $\nu : I \to [0, \infty)$ is any nontrivial function. Then we define the vector space $C_{0,\nu}(I : Y)$ $[C_{b,\nu}(I : Y)]$ as above; equipped with the pseudometric $d(u, v) := \sup_{\mathbf{t} \in I} \|\nu(\mathbf{t})[u(\mathbf{t}) - v(\mathbf{t})]\|_Y$, $(C_{0,\nu}(I : Y), d)$ $[(C_{b,\nu}(I : Y), d)]$ becomes a pseudometric space.

4.1.1 Metrical Bochner criterion

In this subsection, we will always assume that $\emptyset \neq I \subseteq \mathbb{R}^n$, $P \subseteq Y^I$, the space of all functions from I into Y, the zero function belongs to P and $\mathcal{P} = (P, d)$ is a pseudometric space; if $f \in P$, then we designate $\|f\|_P := d(f, 0)$. We need the following notion from [448].

https://doi.org/10.1515/9783111689746-007

Definition 4.1.1. Suppose that $\emptyset \neq I \subseteq \mathbb{R}^n$, $F : I \times X \to Y$ is a given function and the following holds:

$$\text{If} \quad \mathbf{t} \in I,\ \mathbf{b} \in \mathrm{R} \text{ and } l \in \mathbb{N}, \quad \text{then we have } \mathbf{t} + \mathbf{b}(l) \in I.$$

Then we say that the function $F(\cdot;\cdot)$ is $(\mathrm{R}, \mathcal{B}, \mathcal{P})$-multialmost periodic, respectively, strongly $(\mathrm{R}, \mathcal{B}, \mathcal{P})$-multialmost periodic in the case that $I = \mathbb{R}^n$, if for every $B \in \mathcal{B}$ and for every sequence $(\mathbf{b}_k = (b_k^1, b_k^2, \ldots, b_k^n)) \in \mathrm{R}$ there exist a subsequence $(\mathbf{b}_{k_l} = (b_{k_l}^1, b_{k_l}^2, \ldots, b_{k_l}^n))$ of $(\mathbf{b}_k)$ and a function $F^* : I \times X \to Y$ such that, for every $l \in \mathbb{N}$ and $x \in B$, we have $F(\cdot + (b_{k_l}^1, \ldots, b_{k_l}^n); x) - F^*(\cdot; x) \in P$ and

$$\lim_{l \to +\infty} \sup_{x \in B} \| F(\cdot + (b_{k_l}^1, \ldots, b_{k_l}^n); x) - F^*(\cdot; x) \|_P = 0, \tag{241}$$

respectively, $F(\cdot + (b_{k_l}^1, \ldots, b_{k_l}^n); x) - F^*(\cdot; x) \in P$, $F^*(\cdot - (b_{k_l}^1, \ldots, b_{k_l}^n); x) - F(\cdot; x) \in P$, (241) holds and

$$\lim_{l \to +\infty} \sup_{x \in B} \| F^*(\cdot - (b_{k_l}^1, \ldots, b_{k_l}^n); x) - F(\cdot; x) \|_P = 0.$$

Definition 4.1.2. Suppose that $\emptyset \neq I' \subseteq \mathbb{R}^n$, $\emptyset \neq I \subseteq \mathbb{R}^n$, $F : I \times X \to Y$ is a given function, ρ is a binary relation on Y, $R(F) \subseteq D(\rho)$ and $I + I' \subseteq I$. Then we say that $F(\cdot;\cdot)$ is Bohr $(\mathcal{B}, I', \rho, \mathcal{P})$-almost periodic if for every $B \in \mathcal{B}$ and $\varepsilon > 0$ there exists $l > 0$ such that for each $\mathbf{t}_0 \in I'$ there exists $\tau \in B(\mathbf{t}_0, l) \cap I'$ such that, for every $\mathbf{t} \in I$ and $x \in B$, there exists an element $y_{\mathbf{t};x} \in \rho(F(\mathbf{t}; x))$ such that $F(\cdot + \tau; x) - y_{\cdot;x} \in P$ for all $x \in B$ and

$$\sup_{x \in B} \| F(\cdot + \tau; x) - y_{\cdot;x} \|_P \leqslant \varepsilon.$$

The proof of the following structural result, where we relate the notion from the above two definitions, can be given using the equivalence between the (conditional) sequential compactness and the total boundedness in the complete pseudometric spaces as well as the argumentation contained in the proof of Bochner criterion (see, e. g., [805, pp. 28–29]).

Theorem 4.1.3 (Metrical Bochner criterion). *Suppose that $F : \mathbb{R}^n \times X \to Y$, any set B of the collection $\mathcal{B}$ is compact, $F(\cdot; x) \in P$ for all $x \in X$ and the following conditions hold:*

(i) *For every $\varepsilon > 0$ and $B \in \mathcal{B}$, there exists $\delta > 0$ such that, for every τ', $\tau'' \in \mathbb{R}^n$, the assumption $|\tau' - \tau''| < \delta$ implies $\|F(\cdot + \tau'; x) - F(\cdot + \tau''; x)\|_P \leqslant \varepsilon$ for all $x \in B$.*
(ii) *P is a complete pseudometric vector space and the assumption f, $g \in P$ implies $f \pm g \in P$ and $\|f \pm g\|_P \leqslant \|f\|_P + \|g\|_P$.*
(iii) *There exists $c > 0$ such that, for every $\tau \in \mathbb{R}^n$ and $f \in P$, we have $f(\cdot + \tau) \in P$ and $\|f(\cdot + \tau)\|_P \leqslant c\|f\|_P$.*
(iv) *For every $B \in \mathcal{B}$, there exists $L_B > 0$ such that $\|F(\cdot; x) - F(\cdot; y)\|_P \leqslant L_B \|x - y\|_Y$ for all x, $y \in B$.*

Then $F(\cdot;\cdot)$ is Bohr $(\mathcal{B}, \mathbb{R}^n, \mathrm{I}, \mathcal{P})$-almost periodic if and only if $F(\cdot;\cdot)$ is $(\mathrm{R}, \mathcal{B}, \mathcal{P})$-multialmost periodic with R being the collection of all sequences in $\mathbb{R}^n$.

Proof. We shall content ourselves with sketching the proof. Suppose first that $F(\cdot;\cdot)$ is Bohr $(\mathcal{B}, \mathbb{R}^n, \mathrm{I}, \mathcal{P})$-almost periodic. Let $\varepsilon > 0$ and $B \in \mathcal{B}$ be given. Then there exists a finite subset $\{x_i : 1 \leqslant i \leqslant m\}$ of B such that $B \subseteq \bigcup_{1\leqslant i\leqslant m} L(x_i, \varepsilon)$, where $L(\cdot, \varepsilon)$ denotes the open ball of radius ε in X. Keeping in mind the prescribed assumptions (i)–(iii), we can replace the sup-norm $\|\cdot\|_\infty$ with the pseudonorm $\|\cdot\|_P$ throughout the proof of the Bochner criterion in order to see that the set $\overline{\{F(\cdot + \tau; x_i) : \tau \in \mathbb{R}^n, 1 \leqslant i \leqslant m\}}$ is complete and totally bounded in P; therefore, this set is sequentially compact in P and from any sequence (b_k) in $\mathbb{R}^n$ we can extract a subsequence (b_{k_l}) of (b_k) such that the sequence $(F(\cdot + b_{k_l}; x_i))_l$ is convergent in P for $1 \leqslant i \leqslant m$. Using this fact, (ii), the Lipschitz type assumption (iv) and the following decomposition ($x \in B$; $l, l' \in \mathbb{N}$):

$$\begin{aligned}
&\|F(\cdot + b_{k_l}; x) - F(\cdot + b_{k_{l'}}; x)\|_P \\
&\quad \leqslant \|F(\cdot + b_{k_l}; x) - F(\cdot + b_{k_l}; x_i)\|_P + \|F(\cdot + b_{k_l}; x_i) - F(\cdot + b_{k_{l'}}; x)\|_P \\
&\qquad + \|F(\cdot + b_{k_{l'}}; x_i) - F(\cdot + b_{k_{l'}}; x)\|_P, \qquad (242)
\end{aligned}$$

we simply get that $F(\cdot;\cdot)$ is $(\mathrm{R}, \mathcal{B}, \mathcal{P})$-multialmost periodic. The converse statement is much simpler and principally follows from the fact that the set $\{F(\cdot + \tau; x_i) : \tau \in \mathbb{R}^n, 1 \leqslant i \leqslant m\}$ is totally bounded in P. Keeping in mind the proof of the Bochner criterion and the assumptions (ii)–(iii), we get the existence of a sufficiently large real number $l > 0$ such that for each $\mathbf{t}_0 \in \mathbb{R}^n$ there exists $\boldsymbol{\tau} \in B(\mathbf{t}_0, l)$ such that, for every $\mathbf{t} \in \mathbb{R}^n$ and $i \in \mathbb{N}_m$, we have $\|F(\cdot + \tau; x_i) - F(\cdot; x_i)\|_P \leqslant \varepsilon$. Using this estimate, the estimate (242), (ii) and (iv), we simply get that $F(\cdot;\cdot)$ is Bohr $(\mathcal{B}, \mathbb{R}^n, \mathrm{I}, \mathcal{P})$-almost periodic. □

Remark 4.1.4. (i) It is almost impossible to state any satisfactory analogue of Theorem 4.1.3 in the case that $\rho \neq \mathrm{I}$. If R is not a collection of all sequences in $\mathbb{R}^n$ and $\rho = \mathrm{I}$, then the situation is not trivial for consideration; for example, the multidimensional analogue of Haraux's criterion states that a bounded continuous function $F : \mathbb{R}^n \to Y$ is almost periodic if and only if the set of translations $\{F(\cdot + d) : d \in D\}$ is relatively compact in $C_b(\mathbb{R}^n : Y)$ for some (any) relatively dense subset D of $\mathbb{R}^n$ (cf. the recent research article [205] by P. Cieutat for more details about the Bochner almost periodicity criterion and its generalizations). In this section, we will not consider the metrical analogues of Haraux's criterion.

(ii) The general situation $F : I \times X \to Y$, where $\emptyset \neq I \subsetneq \mathbb{R}^n$, is also not trivial for consideration. For example, we know that a bounded continuous function $F : [0,\infty)^n \to Y$ is asymptotically almost periodic if and only if the set of translations $\{F(\cdot + \tau) : \tau \in [0,\infty)^n\}$ is relatively compact in $C_b([0,\infty)^n : Y)$; see, e. g., the proofs of [447, Propositions 6.1.34, 6.1.35]. Concerning this subject, we will only note that various classes of asymptotically ρ-almost periodic type functions in general metric have recently been introduced and analyzed in [448, Chapter 5] as well as that it is not difficult to profile the class of I-asymptotically $(\mathcal{B}, I, \mathrm{I}, \mathcal{P})$-almost periodic functions (of type 1) in terms of a metrical Bochner type criterion.

(iii) It is worth noting that Theorem 4.1.3 remains true if Y is a complete pseudometric space (the notion under our consideration can be simply understood in the newly arisen situation); let us only note that, in the formulation of condition (iv), we should require that $\|F(\cdot;x) - F(\cdot;y)\|_P \leqslant L_B d_Y(x,y)$ for all $x,\ y \in B$.

It is clear that Theorem 4.1.3 can be applied in many different ways; for example, we can use Theorem 4.1.3 to deduce several new Bochner type criteria for the multidimensional Hölder (Lipschitz) almost periodic type functions and the multidimensional almost periodic functions in variation because the metric spaces considered in their definitions satisfy conditions (ii) and (iii) from the formulation of Theorem 4.1.3 [448]. We will apply here Theorem 4.1.3 to the multidimensional Doss-p-almost periodic functions. In order to do that, let us recall first that the notion of Besicovitch-p-boundedness and the notion of Besicovitch-p-continuity for a p-locally integrable function $F : \mathbb{R}^n \to Y$ are special cases of the notion introduced in [448, Definition 3.2.1], with $\mathrm{F}(t) \equiv t^{-n/p}$, $\phi(x) \equiv x$, $\Lambda' = \Lambda = \mathbb{R}^n$ and $\rho = \mathrm{I}$. We say that a p-locally integrable function $F(\cdot)$ is Doss-p-almost periodic if for each $\varepsilon > 0$ there exists a relatively dense set $R \subseteq \mathbb{R}^n$, with the meaning clear, such that for each $\tau \in R$ we have

$$\limsup_{t\to+\infty}\left[\frac{1}{(2t)^n}\int_{[-t,t]^n}\|F(x+\tau)-F(x)\|_Y^p\,dx\right]^{1/p}\leqslant\varepsilon.$$

Clearly, $F(\cdot)$ is Doss-p-almost periodic if and only if $F(\cdot)$ is Bohr $(I', \rho, \mathcal{P})$-almost periodic with $I' = \mathbb{R}^n$, $\rho = \mathrm{I}$ and P being the pseudometric space consisting of all Besicovitch-p-bounded functions, i. e., those p-locally integrable functions $f : \mathbb{R}^n \to Y$ such that its Besicovitch-p-seminorm

$$\|f\|_P := \limsup_{t\to+\infty}\left[\frac{1}{(2t)^n}\int_{[-t,t]^n}\|f(x)\|_Y^p\,dx\right]^{1/p}$$

is finite ($1 \leqslant p < +\infty$). We equip this space with the Besicovitch p-pseudodistance, which is given by

$$d(f,g) := \limsup_{t\to+\infty}\left[\frac{1}{(2t)^n}\int_{[-t,t]^n}\|f(x)-g(x)\|_Y^p\,dx\right]^{1/p},\quad f,\ g\in P.$$

We know that (P, d) is a complete pseudometric space due to the famous result of J. Marcinkiewicz (cf. [508, Theorem 10.5.1, pp. 249–252] for the one-dimensional setting, and [448, Theorem 3.3.6] for the multidimensional setting). Condition (iii) from the formulation of Theorem 4.1.3 holds with $c = 1$ and we can therefore state the following corollary.

Corollary 4.1.5. *Suppose that $1 \leqslant p < +\infty$ as well as $F : \mathbb{R}^n \to Y$ is Besicovitch-p-bounded and Besicovitch-p-continuous. Then $F(\cdot)$ is Doss-p-almost periodic if and only if for every sequence (b_k) in $\mathbb{R}^n$ there exist a subsequence (b_{k_l}) of (b_k) and a Lebesgue measurable function $F^* : \mathbb{R}^n \to Y$ such that*

$$\lim_{l\to+\infty} \limsup_{t\to+\infty} \left[\frac{1}{(2t)^n} \int_{[-t,t]^n} \|F(x+b_{k_l}) - F^*(x)\|_Y^p \, dx \right]^{1/p} = 0.$$

We close this subsection with the following observation.

Remark 4.1.6. If $F : \mathbb{R}^n \to \mathbb{R}$ is Bohr $(\mathbb{R}^n, \mathrm{I}, \mathcal{P})$-almost periodic with $\mathcal{P}$ being defined as above, then we do not necessarily have that $F(\cdot) \in P$; cf. the formulation of Theorem 4.1.3. For example, if $n = 1$, then the function $f(\cdot)$ considered in [447, Example 3.2.13] is uniformly continuous, Doss-p-almost periodic and not Besicovitch-p-bounded ($1 \leqslant p < +\infty$).

4.1.2 Stepanov $(\Omega, \mathrm{R}, \mathcal{B}, \mathcal{P}_Z)$-multialmost periodicity

The main aim of this subsection is to introduce and analyze the class of (strongly) Stepanov $(\Omega, \mathrm{R}, \mathcal{B}, \mathcal{P}_Z)$-multialmost periodic functions in a Bochner-like manner. We will assume henceforth that $\emptyset \neq \Lambda \subseteq \mathbb{R}^n$, $\emptyset \neq Z \subseteq Y^\Omega$ and $\Lambda + \Omega \subseteq \Lambda$, where $\emptyset \neq \Omega \subseteq \mathbb{R}^n$ is a fixed compact set with positive Lebesgue measure. Let $P_Z \subseteq Z^\Lambda$, $0 \in P_Z$, let (P_Z, d_{P_Z}) be a pseudometric space, and let $\|f\|_{P_Z} := d_{P_Z}(f, 0)$, $f \in P_Z$. If $F : \Lambda \times X \to Y$, then we introduce the multidimensional Bochner transform $\hat{F}_\Omega : \Lambda \times X \to Y^\Omega$ by

$$[\hat{F}_\Omega(\mathbf{t}; x)](u) := F(\mathbf{t} + \mathbf{u}; x), \quad \mathbf{t} \in \Lambda,\ \mathbf{u} \in \Omega,\ x \in X.$$

The following notion will be essential in our further analysis.

Definition 4.1.7. Suppose that $\emptyset \neq \Lambda \subseteq \mathbb{R}^n$, $F : \Lambda \times X \to Y$ is a given function and the assumptions $\mathbf{t} \in \Lambda$, $\mathbf{b} \in \mathrm{R}$ and $l \in \mathbb{N}$ imply $\mathbf{t} + \mathbf{b}(l) \in \Lambda$. Then we say that the function $F(\cdot;\cdot)$ is Stepanov $(\Omega, \mathrm{R}, \mathcal{B}, \mathcal{P}_Z)$-multialmost periodic, respectively, strongly Stepanov $(\Omega, \mathrm{R}, \mathcal{B}, \mathcal{P}_Z)$-multialmost periodic in the case that $\Lambda = \mathbb{R}^n$, if for every $B \in \mathcal{B}$ and for every sequence $(\mathbf{b}_k = (b_k^1, b_k^2, \ldots, b_k^n)) \in \mathrm{R}$ there exist a subsequence $(\mathbf{b}_{k_l} = (b_{k_l}^1, b_{k_l}^2, \ldots, b_{k_l}^n))$ of $(\mathbf{b}_k)$ and a function $F_\Omega^* : \Lambda \times X \to Z$ such that, for every $l \in \mathbb{N}$ and $x \in B$, we have $\hat{F}_\Omega(\cdot + (b_{k_l}^1, \ldots, b_{k_l}^n); x) - F_\Omega^*(\cdot; x) \in P_Z$ and

$$\lim_{l\to+\infty} \sup_{x\in B} \|\hat{F}_\Omega(\cdot + (b_{k_l}^1, \ldots, b_{k_l}^n); x) - F_\Omega^*(\cdot; x)\|_{P_Z} = 0, \tag{243}$$

respectively, $\hat{F}(\cdot + (b_{k_l}^1, \ldots, b_{k_l}^n); x) - F^*(\cdot; x) \in P_Z$, $F^*(\cdot - (b_{k_l}^1, \ldots, b_{k_l}^n); x) - \hat{F}(\cdot; x) \in P_Z$, (243) holds and

$$\lim_{l\to+\infty} \sup_{x\in B} \|F_\Omega^*(\cdot - (b_{k_l}^1, \ldots, b_{k_l}^n); x) - \hat{F}_\Omega(\cdot; x)\|_{P_Z} = 0.$$

The usual notion, obtained by plugging $Z = L^{p(\mathbf{u})}(\Omega : Y)$ and $P_Z = C_b(\Lambda : Z)$ can be further generalized using the spaces $Z = L_\nu^{p(\mathbf{u})}(\Omega : Y)$, where $p \in \mathcal{P}(\Omega)$ and $\nu : \Omega \to (0, \infty)$ is a Lebesgue measurable function, and $P_Z = C_{b,\zeta}(\Lambda : Z)$, where $\zeta : \Lambda \to [0, \infty)$ is any nonzero function. If $\Lambda = \mathbb{R}^n$ and $P_Z = C_{b,\zeta}(\mathbb{R}^n : Z)$, where $\zeta : \mathbb{R}^n \to (0, \infty)$, then it can be simply proved that any strongly Stepanov $(\Omega, \mathrm{R}, \mathcal{B}, \mathcal{P}_Z)$-multialmost periodic function has to be Stepanov $(\Omega, \mathrm{R}, \mathcal{B}, Z^{\mathcal{P}})$-multi-almost automorphic in the sense of Definition 5.1.3 below.

Suppose now that, for every sequence which belongs to R, any its subsequence also belongs to R. Then it is very simple to show that, if $F(\cdot;\cdot)$ and $G(\cdot;\cdot)$ are (strongly) Stepanov $(\Omega, \mathrm{R}, \mathcal{B}, \mathcal{P}_Z)$-multialmost periodic functions, $\alpha, \beta \in \mathbb{C}$, Z and P_Z are vector spaces and there exist two finite real constants $c_\alpha > 0$ and $c_\beta > 0$ such that the assumption $f, g \in P_Z$ implies $\|\alpha f + \beta g\|_{P_Z} \leqslant c_\alpha \|f\|_{P_Z} + c_\beta \|g\|_{P_Z}$, then the function $[\alpha F + \beta G](\cdot;\cdot)$ is likewise (strongly) Stepanov $(\Omega, \mathrm{R}, \mathcal{B}, \mathcal{P}_Z)$-multialmost periodic.

The subsequent result follows directly from the corresponding definitions.

Proposition 4.1.8. *Suppose that $\emptyset \neq \Lambda \subseteq \mathbb{R}^n$, $F : \Lambda \times X \to Y$ and the assumptions $\mathbf{t} \in \Lambda$, $\mathbf{b} \in \mathrm{R}$ and $l \in \mathbb{N}$ imply $\mathbf{t} + \mathbf{b}(l) \in \Lambda$. If $\hat{F}_\Omega : \Lambda \times X \to Z$, then the function $F(\cdot;\cdot)$ is Stepanov $(\Omega, \mathrm{R}, \mathcal{B}, \mathcal{P}_Z)$-multialmost periodic if and only if the function $\hat{F}_\Omega(\cdot;\cdot)$ is $(\mathrm{R}, \mathcal{B}, \mathcal{P}_Z)$-multialmost periodic.*

Keeping in mind Proposition 4.1.8, we can immediately apply Theorem 4.1.3 (cf. also Remark 4.1.4(iii)) in order to deduce the following Bochner type criterion.

Theorem 4.1.9. *Suppose that $F : \mathbb{R}^n \times X \to Y$, $\hat{F}_\Omega : \Lambda \times X \to Z$, any set B of the collection $\mathcal{B}$ is compact, $\hat{F}_\Omega(\cdot; x) \in P_Z$ for all $x \in X$ and the following conditions hold:*

(i) *For every $\varepsilon > 0$ and $B \in \mathcal{B}$, there exists $\delta > 0$ such that, for every $\tau', \tau'' \in \mathbb{R}^n$, the assumption $|\tau' - \tau''| < \delta$ implies $\|\hat{F}_\Omega(\cdot + \tau'; x) - \hat{F}_\Omega(\cdot + \tau''; x)\|_{P_Z} \leqslant \varepsilon$ for all $x \in B$.*
(ii) *P_Z is a complete pseudometric vector space and the assumption $f, g \in P_Z$ implies $f \pm g \in P_Z$ and $\|f \pm g\|_{P_Z} \leqslant \|f\|_{P_Z} + \|g\|_{P_Z}$.*
(iii) *There exists $c > 0$ such that, for every $\tau \in \mathbb{R}^n$ and $f \in P_Z$, we have $f(\cdot + \tau) \in P_Z$ and $\|f(\cdot + \tau)\|_{P_Z} \leqslant c\|f\|_{P_Z}$.*
(iv) *For every $B \in \mathcal{B}$, there exists $L_B > 0$ such that $\|\hat{F}_\Omega(\cdot; x) - \hat{F}_\Omega(\cdot; y)\|_{P_Z} \leqslant L_B \|x - y\|_Y$ for all $x, y \in B$.*

Then $F(\cdot;\cdot)$ is Stepanov $(\Omega, \mathrm{R}, \mathcal{B}, \mathcal{P}_Z)$-multialmost periodic if and only if $\hat{F}_\Omega(\cdot;\cdot)$ is Bohr $(\mathcal{B}, \mathbb{R}^n, \mathrm{I}, \mathcal{P}_Z)$-almost periodic, i. e., for every $B \in \mathcal{B}$ and $\varepsilon > 0$, there exists $l > 0$ such that for each $\mathbf{t}_0 \in \mathbb{R}^n$ there exists $\tau \in B(\mathbf{t}_0, l)$ such that $\hat{F}_\Omega(\cdot + \tau; x) - \hat{F}_\Omega(\cdot; x) \in P_Z$ for all $x \in B$ and

$$\sup_{x \in B} \big\|\hat{F}_\Omega(\cdot + \tau; x) - \hat{F}_\Omega(\cdot; x)\big\|_{P_Z} \leqslant \varepsilon.$$

Remark 4.1.10. The notion of Stepanov ρ-almost periodicity in general metric has recently been analyzed in [458]. We would like to note that it is not clear how we can relate the notion introduced in [458, Definitions 2.1–2.3], given in a Bohr-like manner, with the notion introduced in Definition 4.1.7. It is also worth noting that Theorem 4.1.9 provides a metrical generalization of [447, Theorem 6.2.13].

Condition $\hat{F}_\Omega(\cdot;x) \in P_Z$ for all $x \in X$ and condition (i) from the formulation of Theorem 4.1.9 holds in the usually considered framework (cf. [447] and [508] for more details; the interested reader should make an attempt to formulate the metrical generalizations of [508, Theorem 5.2.3, Corollary 1; pp. 201–202]). Now we will introduce the following notion, which generalizes the notion of Stepanov $(\Omega, p(u))$-boundedness introduced in [447, Definition 6.2.1].

Definition 4.1.11. Suppose that $\emptyset \neq \Lambda \subseteq \mathbb{R}^n$, $F : \Lambda \times X \to Y$ is a given function and $\hat{F}_\Omega : \Lambda \times X \to Z$.

(i) Then we say that the function $F(\cdot;\cdot)$ is Stepanov $(\Omega, \mathcal{B}, \mathcal{P}_Z)$-bounded if $\hat{F}_\Omega(\cdot;x) \in P_Z$ for all $x \in X$ and for each set $B \in \mathcal{B}$ we have $\sup_{x\in B} \|\hat{F}_\Omega(\cdot;x)\|_{P_Z} < +\infty$.
(ii) Suppose that (Z, d_Z) is a pseudometric space and $0 \in Z$. Then we say that the function $F(\cdot;\cdot)$ is Stepanov $(\Omega, \mathcal{B}, Z)$-bounded if for each set $B \in \mathcal{B}$ we have $\sup_{\mathbf{t}\in\Lambda;x\in B} \|\hat{F}_\Omega(\mathbf{t};x)\|_Z < +\infty$, where $\|\cdot\|_Z = d_Z(0,\cdot)$.

It is clear that any Stepanov $(\Omega, \mathcal{B}, \mathcal{P}_Z)$-bounded function $F(\cdot;\cdot)$ is Stepanov $(\Omega, \mathcal{B}, Z)$-bounded if there exists a finite constant $c_1 > 0$ such that, for every $f \in P_Z$, we have $\sup_{\mathbf{t}\in\Lambda} \|f(t)\|_Z \leqslant c_1\|f\|_{P_Z}$ as well as that any Stepanov $(\Omega, \mathcal{B}, Z)$-bounded function $F(\cdot;\cdot)$ is Stepanov $(\Omega, \mathcal{B}, \mathcal{P}_Z)$-bounded if $\hat{F}_\Omega(\cdot;x) \in P_Z$ for all $x \in X$ and there exists a finite constant $c_2 > 0$ such that, for every $f \in P_Z$, we have $\|f\|_{P_Z} \leqslant c_2 \sup_{\mathbf{t}\in\Lambda} \|f(t)\|_Z$. Using the argumentation contained in the proof of [184, Proposition 2.5(i)] and the assumptions (i)–(ii) stated below, we can simply deduce the following result.

Proposition 4.1.12. *Suppose that (Z, d_Z) is a complete pseudometric space and $0 \in Z$. Suppose further that $F : \mathbb{R}^n \times X \to Y$ is Stepanov $(\Omega, \mathrm{R}, \mathcal{B}, \mathcal{P}_Z)$-multi-almost periodic with R being the collection of all sequences in $\mathbb{R}^n$ and $\mathcal{B}$ being a collection of some compact subsets of X, $\hat{F}_\Omega : \mathbb{R}^n \times X \to Z$ and* (i)–(ii) *hold, where:*

(i) *There exist a finite real constant $c > 0$ and a point $\mathbf{t}_0 \in \mathbb{R}^n$ such that, for every $f \in P_Z$, we have $\|f(\mathbf{t}_0)\|_Z \leqslant c\|f\|_{P_Z}$.*
(ii) *For each set $B \in \mathcal{B}$, there exists a finite real constant $L_B > 0$ such that*

$$\left\|\hat{F}_\Omega(\mathbf{t};x) - \hat{F}_\Omega(\mathbf{t};y)\right\|_Z \leqslant L_B\|x-y\|, \quad \mathbf{t} \in \mathbb{R}^n; \quad x, y \in B.$$

Then the set $\{\hat{F}_\Omega(\mathbf{t};x) : \mathbf{t} \in \mathbb{R}^n,\ x \in B\}$ is relatively compact in Z ($B \in \mathcal{B}$); especially, for each set $B \in \mathcal{B}$ we have $\sup_{\mathbf{t}\in\mathbb{R}^n;x\in B} \|\hat{F}_\Omega(\mathbf{t};x)\|_Z < +\infty$ and the function $F(\cdot,\cdot)$ is Stepanov $(\Omega, \mathcal{B}, Z)$-bounded.

We continue by noting that the statements of [447, Propositions 6.2.22–6.2.25] can be simply formulated for the metrical Stepanov almost periodicity; this is not the case with the Bochner type result [508, Theorems 5.2.7–5.2.9, pp. 205–209]. Details can be left to the curious reader.

Using Proposition 4.1.8, we can immediately clarify many other structural results for the (strongly) Stepanov $(\Omega, \mathrm{R}, \mathcal{B}, \mathcal{P}_Z)$-multialmost periodic functions. For example, the statements of [452, Propositions 2.3(i), 2.4, Theorems 2.5, 2.7] and the conclusions (i)–(iii) established on [448, pp. 234–235] can be almost directly reformulated in our new framework (concerning possible applications, Proposition 2.4, Theorem 2.5 and Theorem 2.7 are most important since these statements can be applied in the analysis of metrically Stepanov almost periodic solutions of the heat equation in $\mathbb{R}^n$ and a large class of abstract fractional semilinear inclusions in Banach spaces; cf. [448, Section 4] for more details). For the sake of brevity, we will only state here the following result, which is a direct consequence of Proposition 4.1.8 and [452, Proposition 2.6].

Proposition 4.1.13. *Suppose that $\emptyset \neq \Lambda \subseteq \mathbb{R}^n$, the assumptions $\mathbf{t} \in \Lambda$, $\mathbf{b} \in \mathrm{R}$ and $l \in \mathbb{N}$ imply $\mathbf{t} + \mathbf{b}(l) \in \Lambda$, P_Z has a linear vector structure, P_Z is complete and the metric d_Z is translation invariant in the sense that $d_Z(f+g, h+g) = d_Z(f,h)$ whenever f, h, $f+g$, $h+g \in P_Z$. Suppose further that, for each integer $j \in \mathbb{N}$ the function $F_j : \Lambda \times X \to Y$ is Stepanov $(\Omega, \mathrm{R}, \mathcal{B}, \mathcal{P}_Z)$-multialmost periodic, $\hat{F_{j;\Omega}} : \Lambda \times X \to Z$, $\hat{F_{j;\Omega}}(\cdot + b_k; x) \in P_Z$ and $\hat{F_{j;\Omega}}^*(\cdot; x) \in P_Z$ $(x \in B; (b_k) \in \mathrm{R})$, with the meaning clear, as well as that, for every sequence which belongs to R, any its subsequence also belongs to R. If $F : \Lambda \times X \to Y$ and, for every $B \in \mathcal{B}$ and $(b_k) \in \mathrm{R}$, we have*

$$\lim_{(l,l)\to+\infty} \sup_{x\in B} \big\| F_l(\cdot + b_{k_l}; x) - F(\cdot + b_{k_l}; x) \big\|_{P_Z} = 0,$$

then the function $F(\cdot;\cdot)$ is Stepanov $(\Omega, \mathrm{R}, \mathcal{B}, \mathcal{P}_Z)$-multialmost periodic, $\hat{F_\Omega}(\cdot + b_k; x) \in P_Z$ and $\hat{F_\Omega}^*(\cdot; x) \in P_Z$ $(x \in B; (b_k) \in \mathrm{R})$.

It could be also interesting to reconsider the statements of [176, Proposition 2.10, Theorem 2.11], which concerns the convolution invariance of metrical Stepanov almost periodicity, in our new framework and provide certain applications.

4.1.3 Extensions of almost periodic sequences

Extensions of almost periodic sequences and extensions of almost automorphic sequences have been analyzed in many research articles by now. In [176], we have recently shown that the extensions of almost automorphic sequences belong to some special classes of compactly almost automorphic functions. In this subsection, we will follow a similar approach and observe that the notion of metrical Stepanov almost periodicity can be useful in the study of extensions of almost periodic sequences; more precisely, we will see that the extensions of almost periodic sequences belong to some special classes of (metrically) almost periodic functions. For simplicity, we will consider here the one-dimensional setting.

Suppose that $F : \mathbb{Z} \to Y$ is an almost periodic sequence. If $t \in [m, m+1)$ for some $m \in \mathbb{Z}$, then we define $\tilde{F}(t) := F(m) + (t-m)[F(m+1) - F(m)]$. It can be simply shown that $\tilde{F} : \mathbb{R} \to Y$ is an almost periodic function and $\tilde{F}(k) = F(k)$ for all $k \in \mathbb{Z}$. Suppose further that $\Omega = [0,1]$ and (Z_0, d_{Z_0}) is a metric vector space such that any function from Z_0 is essentially bounded, $\|f \pm g\|_{Z_0} \leqslant \|f\|_{Z_0} + \|g\|_{Z_0}$ for all $f,\ g \in Z_0$ and the following condition holds:

(Q1) There exists a finite real number $c > 0$ such that the function $u \mapsto h(u) \equiv y_1' + (a_1 + u)y_1$, $u \in [0, u_1)$; $u \mapsto h(u) \equiv y_2' + (a_2 + u)y_2$, $u \in [u_1, 1]$ belongs to Z_0 and $\|h\|_{Z_0} \leqslant c(\|y_1\|_Y + \|y_2\|_Y + \|y_1'\|_Y + \|y_2'\|_Y)$ for all $u_1 \in [0,1]$, $a_1,\ a_2 \in [-1,1)$ and $y_1, y_2, y_1', y_2' \in Y$.

Arguing as in the proofs of Proposition 5.1.20 and Section 5.1 below, we can deduce the following results.

Theorem 4.1.14. *Suppose that $\Omega = [0,1]$, condition (Q1) holds, (Z_0, d_{Z_0}) is a metric vector space such that any function from Z_0 is essentially bounded and $\|f \pm g\|_{Z_0} \leqslant \|f\|_{Z_0} + \|g\|_{Z_0}$ for all $f,\ g \in Z_0$. Then the following holds:*

(i) *The function $\tilde{F}(\cdot)$ is Stepanov $(\mathrm{R}, \mathcal{P}_Z)$-almost periodic, where R denotes the collection of all sequences in $\mathbb{Z}$, (Z, d_Z) is the metric vector space equipped with the distance $d_Z(f,g) := \|f - g\|_\infty + \|f - g\|_{Z_0}$, $f,\ g \in Z$ and $P_Z = C_b(\mathbb{R} : Z)$.*

(ii) *Assume, in addition to the above requirements, that the following condition holds:*

  (Q2) *There exists a finite real number $d > 0$ such that, if $f : \mathbb{R} \to Y$ is a Lipschitz continuous function and $f(t + \cdot) \in Z_0$ for all $t \in \mathbb{R}$, then*

$$\left\| f(t+\cdot) - f(t'+\cdot) \right\|_{Z_0} \leqslant d\left|t - t'\right|, \quad t,\ t' \in \mathbb{R}.$$

  Then the function $\tilde{F}(\cdot)$ is Stepanov $(\mathrm{R}, \mathcal{P}_Z)$-almost periodic, where R denotes the collection of all sequences in $\mathbb{R}$, (Z, d_Z) and P_Z have the same meaning as above.

In particular, we can use the metric $d_{Z_0}(\cdot;\cdot)$ of the Banach space of k-times piecewise continuously differentiable functions on the interval $[0,1]$, equipped with the usual metric; cf. [176] for more details.

Let us finally mention two topics, which have not been investigated by now:

1. As already mentioned, we have followed here the idea from our recent research article [176], where we have considered the values of multidimensional Bochner transform in a general pseudometric space Z. This idea can be also used to further generalize the notion of:
   (i) Generalized Weyl classes of metrically almost periodic type functions introduced in [448, Definitions 4.3.19–4.3.21].
   (ii) Stepanov weighted ergodic components in general metric (see [448, Definition 5.2.1]).

(iii) Weyl weighted ergodic components in general metric (see [448, Definitions 5.2.8–5.2.10]).
(iv) Weighted pseudoergodic components in general metric (see [448, Definition 5.2.15]).
2. Asymptotically Stepanov almost periodic type functions in general metric can be also introduced and analyzed (cf. [447, Subsection 6.2.3] for the classical concept).

4.2 Metrically piecewise continuous ρ-almost periodic functions and applications

The main aim of this section is to introduce and analyze several new classes of metrically piecewise continuous ρ-almost periodic type functions as well as to present certain applications to the abstract Volterra integrodifferential equations in Banach spaces.

Assume that $(t_k)_{k\in\mathbb{Z}}$ $[(t_k)_{k\in\mathbb{N}}]$ is a sequence in $\mathbb{R}$ [in $(0,\infty)$] such that $\delta_0 := \inf_{k\in\mathbb{Z}}(t_{k+1}-t_k) > 0$ $[\delta_0 := \inf_{k\in\mathbb{N}}(t_{k+1}-t_k) > 0]$; this will be our standing assumption henceforth. Set $t_k^j := t_{k+j} - t_k$, $j, k \in \mathbb{Z}$ $[j, k \in \mathbb{N}]$. Let us recall that the sequence $(t_k)_{k\in\mathbb{Z}}$ $[(t_k)_{k\in\mathbb{N}}]$ is said to be uniformly almost periodic if, for every $\varepsilon > 0$, there exists a relatively dense set Q_ε in $\mathbb{Z}$ [in $\mathbb{N}$] such that

$$\left|t_{i+q}^j - t_i^j\right| < \varepsilon, \quad i, j \in \mathbb{Z} \quad [i, j \in \mathbb{N}], \quad q \in Q_\varepsilon.$$

It is well known that, if the sequence $(t_k)_{k\in\mathbb{Z}}$ $[(t_k)_{k\in\mathbb{N}}]$ is uniformly almost periodic, then the family of sequences $(t_k^j)_{k\in\mathbb{Z}}$ $[(t_k^j)_{k\in\mathbb{N}}]$, $j \in \mathbb{Z}$ $[j \in \mathbb{N}]$ is equipotentially almost periodic. It is also well known that the family of sequences $(t_k^j)_{k\in\mathbb{Z}}$, $j \in \mathbb{Z}$ is equipotentially almost periodic if and only if there exist a nonzero real number ζ and an almost periodic sequence $(a_k)_{k\in\mathbb{Z}}$ such that $t_k = \zeta k + a_k$ for all $k \in \mathbb{Z}$.

Now we would like to recall the following well-known notion.

Definition 4.2.1. Suppose that the function $f : \mathbb{R} \to X$ $[f : [0,\infty) \to X]$ is piecewise continuous with the possible first kind discontinuities at the points of a fixed sequence $(t_k)_{k\in\mathbb{Z}}$ $[(t_k)_{k\in\mathbb{N}}]$. Then $f(\cdot)$ is said to be (t_k)-piecewise continuous almost periodic if the following holds:
(i) The family of sequences $(t_k^j)_{k\in\mathbb{Z}}$ $[(t_k^j)_{k\in\mathbb{N}}]$, $j \in \mathbb{Z}$ $[j \in \mathbb{N}]$ is equipotentially almost periodic, i. e., (t_k) is a Wexler sequence.
(ii) For every $\varepsilon > 0$, there exists $\delta > 0$ such that, if the points t_1 and t_2 belong to (t_i, t_{i+1}) for some $i \in \mathbb{Z}$ $[i \in \mathbb{N}_0;\ t_0 \equiv 0]$ and $|t_1 - t_2| < \delta$, then $\|f(t_1) - f(t_2)\| < \varepsilon$.
(iii) For every $\varepsilon > 0$, there exists a relatively dense set S in $\mathbb{R}$ [in $[0,\infty)$] such that, if $\tau \in S$, then $\|f(t+\tau) - f(t)\| < \varepsilon$ for all $t \in \mathbb{R}$ such that $|t - t_k| > \varepsilon$, $k \in \mathbb{Z}$ $[k \in \mathbb{N}]$. Such a point τ is called an ε-almost period of $f(\cdot)$.

The space of all (t_k)-piecewise continuous almost periodic functions will be denoted by $PCAP_{(t_k)}(\mathbb{R} : X)$ $[PCAP_{(t_k)}([0,\infty) : X)]$.

Any almost periodic function $f(\cdot)$ is (t_k)-piecewise continuous almost periodic, while the converse statement is not true; see, e. g., Example 4.2.4 below.

The organization and main ideas of section can be briefly outlined as follows. We introduce and analyze various classes of metrically piecewise continuous ρ-almost periodic type functions in Definition 4.2.2. We continue our exposition by stating Proposition 4.2.3; an illustrative example is given after that. Section 4.2.1 investigates the relations between metrically piecewise continuous ρ-almost periodic type functions and metrically Stepanov ρ-almost periodic type functions. The main results of this subsection are Theorem 4.2.5 and Theorem 4.2.6, which slightly generalize the corresponding results established in our recent research article [477]; cf. also Example 4.2.7. In Section 4.2.2, we analyze some applications to the abstract Volterra integrodifferential equations.

First of all, if $0 < \varepsilon < \delta_0/2$ [$0 < \varepsilon < \delta_0/2$ and $0 < \varepsilon < t_1$], then we set $A_\varepsilon := I \setminus \bigcup_{k\in\mathbb{Z}} L(t_k, \varepsilon)$ [$A_\varepsilon := I \setminus \bigcup_{k\in\mathbb{N}} L(t_k, \varepsilon)$] and assume that $P_\varepsilon \subseteq Y^{A_\varepsilon}$, $0 \in P_\varepsilon$ and $(P_\varepsilon, d_\varepsilon)$ is a pseudometric space. Set $\|f\|_\varepsilon := d_\varepsilon(f, 0)$, $f \in P_\varepsilon$. If we take $d_\varepsilon(f,g) = \sup_{t\in A_\varepsilon} \|f(t)-g(t)\|$, then the following notion generalizes the notion introduced in Definition 4.2.1 and [267, Definition 6] (this notion also generalizes the notion introduced in [289, Definition 2.1] and [448, Definition 4.1.19], provided that $I = [0,\infty)$ or $I = \mathbb{R}$, $I' = I$, the pseudometric spaces P_ε satisfy certain extra assumptions and (t_k) is an arbitrary sequence obeying our general requirements).

Definition 4.2.2. Suppose that ρ is a binary relation on Y and the function $F : \mathbb{R}\times X \to Y$ [$F : [0,\infty) \times X \to Y$] satisfies that, for every $x \in X$, the function $t \mapsto F(t; x)$ is piecewise continuous with the possible first kind discontinuities at the points of a fixed sequence $(t_k)_{k\in\mathbb{Z}}$ [$(t_k)_{k\in\mathbb{N}}$]. Then we say that the function $F(\cdot)$ is:

(i) pre-$(\mathcal{B}, \rho, (t_k), \mathcal{P})$-piecewise continuous almost periodic if, for every $\varepsilon > 0$ such that $0 < \varepsilon < \delta_0/2$ [$0 < \varepsilon < \delta_0/2$ and $0 < \varepsilon < t_1$], $\sigma > 0$ and $B \in \mathcal{B}$, there exists a relatively dense set S in $\mathbb{R}$ [in $[0,\infty)$] such that, if $\tau \in S$, $x \in B$ and $t \in \mathbb{R}$ ($t \geqslant 0$) satisfies $|t - t_k| > \varepsilon$ for all $k \in \mathbb{Z}$ [$k \in \mathbb{N}$], then there exists $y_{t,x} \in \rho(F(t; x))$ such that $F(\cdot + \tau; x) - y_{\cdot,x} \in P_\varepsilon$ and $\|F(\cdot + \tau; x) - y_{\cdot,x}\|_\varepsilon < \sigma$.

(ii) $(\mathcal{B}, \rho, (t_k), \mathcal{P})$-piecewise continuous almost periodic if $F(\cdot;\cdot)$ is pre-$(\mathcal{B}, \rho, (t_k), \mathcal{P})$-piecewise continuous almost periodic, (t_k) is a Wexler sequence and (QUC) holds, where:

    (QUC) For every $\varepsilon > 0$ and $B \in \mathcal{B}$, there exists $\delta > 0$ such that, if $x \in B$ and the points t_1 and t_2 belong to (t_i, t_{i+1}) for some $i \in \mathbb{Z}$ [$i \in \mathbb{N}_0$; $t_0 \equiv 0$] and $|t_1 - t_2| < \delta$, then $\|F(t_1; x) - F(t_2; x)\|_Y < \varepsilon$.

(iii) pre-$(\mathcal{B}, \rho, (t_k), \mathcal{P})$-piecewise continuous uniformly recurrent if there exists a strictly increasing sequence (α_l) of positive real numbers tending to plus infinity and satisfying that, for every $\varepsilon > 0$ with $0 < \varepsilon < \delta_0/2$ [$0 < \varepsilon < \delta_0/2$ and $0 < \varepsilon < t_1$], $\sigma > 0$ and $B \in \mathcal{B}$, there exists an integer $l_0 \in \mathbb{N}$ such that, if $x \in B$, $l \geqslant l_0$ and $t \in \mathbb{R}$ satisfies $|t - t_k| > \varepsilon$ for all $k \in \mathbb{Z}$ [$k \in \mathbb{N}$], then there exists $y_{t,x} \in \rho(F(t; x))$ such that $F(\cdot + \alpha_l; x) - y_{\cdot,x} \in P_\varepsilon$ and $\|F(\cdot + \alpha_l; x) - y_{\cdot,x}\|_\varepsilon < \sigma$.

(iv) $(\mathcal{B}, \rho, (t_k), \mathcal{P})$-piecewise continuous uniformly recurrent if $F(\cdot; \cdot)$ is pre-$(\mathcal{B}, \rho, (t_k), \mathcal{P})$-piecewise continuous uniformly recurrent and the condition (QUC) holds.

We say that the function $F(\cdot; \cdot)$ is (pre-)$(\mathcal{B}, \rho, \mathcal{P})$-piecewise continuous almost periodic [(pre-)$(\mathcal{B}, \rho, \mathcal{P})$-piecewise continuous uniformly recurrent] if $F(\cdot; \cdot)$ is (pre-)$(\mathcal{B}, \rho, (t_k), \mathcal{P})$-piecewise continuous almost periodic [(pre-)$(\mathcal{B}, \rho, (t_k), \mathcal{P})$-piecewise continuous uniformly recurrent] for a certain sequence $(t_k)_{k\in\mathbb{Z}}$ [$(t_k)_{k\in\mathbb{N}}$] obeying our general requirements. If $\rho = c\mathrm{I}$ for some $c \in \mathbb{C}$, then we also say that $F(\cdot; \cdot)$ is (pre-)piecewise continuous $(c, \mathcal{P})$-almost periodic [(pre-)piecewise continuous $(c, \mathcal{P})$-uniformly recurrent]; furthermore, if $c = -1$, then we also say that $F(\cdot; \cdot)$ is (pre-)piecewise continuous $\mathcal{P}$-almost anti-periodic [(pre-)piecewise continuous $\mathcal{P}$-uniformly antirecurrent].

As already marked in [267, Remark 2(i)], the condition (QUC) is primarily intended for the analysis of case in which $\rho = \mathrm{I}$; the interested reader is advised to formulate an analogue of [267, Proposition 1] for metrically piecewise continuous ρ-almost periodic type functions. The following simple result, which discuss the use of two different parameters $\varepsilon > 0$ and $\sigma > 0$ in Definition 4.2.2, can be formulated for all other classes of functions introduced therein (cf. also [267, Remark 1]); in particular, this holds provided that $P_\varepsilon = C_{b,\nu}(A_\varepsilon : Y)$, where $\nu : I \to [0, \infty)$ is an arbitrary function:

Proposition 4.2.3. *Suppose that the general requirements from* Definition 4.2.2 *and the following condition hold:*
(C) *If* $0 < \varepsilon' < \varepsilon$, *then* $\|\cdot\|_\varepsilon \leqslant \|\cdot\|_{\varepsilon'}$.

Then $F(\cdot)$ is pre-$(\mathcal{B}, \rho, (t_k), \mathcal{P})$-piecewise continuous almost periodic if and only if, for every $\varepsilon > 0$ such that $0 < \varepsilon < \delta_0/2$ [$0 < \varepsilon < \delta_0/2$ and $0 < \varepsilon < t_1$] and $B \in \mathcal{B}$, there exists a relatively dense set S in $\mathbb{R}$ [in $[0, \infty)$] such that, if $\tau \in S$, $x \in B$ and $t \in \mathbb{R}$ ($t \geqslant 0$) satisfies $|t - t_k| > \varepsilon$ for all $k \in \mathbb{Z}$ [$k \in \mathbb{N}$], then there exists $y_{t,x} \in \rho(F(t; x))$ such that $F(\cdot + \tau; x) - y_{\cdot,x} \in P_\varepsilon$ and $\|F(\cdot + \tau; x) - y_{\cdot,x}\|_\varepsilon < \varepsilon$.

We continue with the following simple example.

Example 4.2.4. Suppose that $p : \mathbb{R} \to \mathbb{R}$ is a periodic trigonometric polynomial, $\nu : \mathbb{R} \to [0, \infty)$ is an arbitrary function and $P_\varepsilon = C_{b,\nu}(A_\varepsilon : Y)$ for every sufficiently small number $\varepsilon > 0$. Then there exists a unique $\mathcal{P}$-piecewise continuous almost periodic function $f(\cdot)$ such that $f(t) = sign(p(t))$ for all real values of t, which are not zeros of $p(\cdot)$.

If the choice of pseudometric spaces P_ε is the same as in the last example, then we extend the function spaces introduced in [267, Definition 6]; on the other hand, if $P_\varepsilon = C_{b,\nu}(A_\varepsilon : Y)$ for all sufficiently small numbers $\varepsilon > 0$, with some function $\nu : I \to (0, \infty)$ such that $(1/\nu)(\cdot)$ is a bounded function, then the function spaces introduced in Definition 4.2.2 are subspaces of the corresponding spaces introduced in [267, Definition 6]. Further on, the statements [267, Propositions 3, 4, 8] can be reformulated in the metrical framework.

4.2.1 Relations with metrically Stepanov ρ-almost periodic type functions

In this subsection, we will briefly analyze the relations between metrically ρ-piecewise continuous almost periodic type functions and metrically Stepanov ρ-almost periodic type functions. The following result is a metrical extension of [477, Theorem 1].

Theorem 4.2.5. *Suppose that $\rho = T \in L(Y)$, $p > 0$, $F : \Lambda \times X \to Y$ is pre-$(\mathcal{B}, T, (t_k), \mathcal{P})$-piecewise continuous almost periodic, where $\Lambda = \mathbb{R}$ or $\Lambda = [0, \infty)$, and for every $B \in \mathcal{B}$, we have $\|F\|_{\infty,B} \equiv \sup_{t\in\Lambda,x\in B} \|F(t; x)\| < +\infty$. Suppose further that for every sufficiently small $\varepsilon > 0$, we have $P_\varepsilon = C_{b,\eta}(\Lambda : Y)$ with some positive function $\eta : \mathbb{R} \to (0, \infty)$, $\nu : \mathbb{R} \to (0, \infty)$, the function $(\nu/\eta)(\cdot)$ is Stepanov-p-bounded and the following condition holds:*

(LQ0) *For every $\varepsilon > 0$, there exist $d > 0$ and $\varepsilon_0 > 0$ such that, for every $x \in \mathbb{R}$ and for every Lebesgue measurable set $\Omega \subseteq [x, x + 1]$ such that $m(\Omega) < \varepsilon_0$, we have $\int_d^{+\infty} y^{p-1} m(\{x \in \Omega : \nu(x) > y\})\, dy < \varepsilon$.*

Then $F \in S^{(\mathbb{F},T,\mathcal{P}_t,\mathcal{P})}_{\Omega,\Lambda'}(\Lambda : Y)$, where $\Lambda = \Lambda' = \mathbb{R}$, $\mathbb{F}(\cdot\cdot) \equiv 1$, $\Omega = [0, 1]$, $P = C_b(\mathbb{R} : \mathbb{C})$ and $P_t = L^p_\nu(t + [0, 1] : \mathbb{C})$ for all $t \in \mathbb{R}$, i. e., for each $\varepsilon > 0$ and $B \in \mathcal{B}$ there exists a relatively dense set S' in $\mathbb{R}$ such that for each $\tau \in S'$ we have

$$\int_x^{x+1} \|F(t + \tau; b) - cF(t; b)\|^p \nu^p(t)\, dt \leqslant \varepsilon, \quad x \in \mathbb{R},\ b \in B.$$

Proof. We will only outline the main details of the proof, with $\Lambda = \mathbb{R}$ and $T = cI$ for some $c \in \mathbb{C}$. Let a number $\varepsilon > 0$ and a set $B \in \mathcal{B}$ be given. Suppose $x \in \mathbb{R}$ and the interval $[x, x + 1]$ contains the possible first kind discontinuities of functions $F(\cdot; b)$ at the points $\{t_m, \ldots, t_{m+k}\} \subseteq [x, x+1]$ $(b \in X)$; then $k \leqslant \lceil 1/\delta_0 \rceil$. Let the numbers $d > 0$ and $\varepsilon_0 > 0$ satisfy (LQ0), with the number ε replaced therein with the number $\varepsilon/(2((1 + |c|)\|F\|_{\infty,B})^p p)$. We know that there exists a relatively dense set S in $\mathbb{R}$ such that, if $\tau \in S$ and $b \in B$, then $\|F(t+\tau; x) - cF(t; x)\|\eta(t) < \varepsilon_1$ for all $t \in \mathbb{R}$ such that $|t - t_k| > \varepsilon_1$, $k \in \mathbb{Z}$, where the number $\varepsilon_1 \in (0, \varepsilon_0/2\lceil 1/\delta_0 \rceil)$ is defined in the same way as in the proof of [477, Theorem 1], with the function $\nu(\cdot)$ replaced by the function $(\nu/\eta)(\cdot)$ in the last line of the proof.

The function $t \mapsto F(t + \tau; b) - cF(t; b)$, $t \in [x, x + 1]$ is not greater than $\varepsilon_1/\eta(t)$ if $t \in A_x := [x, t_m - \varepsilon_1] \cup (t_m + \varepsilon_1, t_{m+1} - \varepsilon_1] \cup \cdots \cup (t_{m+k}, x+1]$; otherwise, $\|F(t+\tau; b) - cF(t; b)\| \leqslant (1 + |c|)\|F\|_{\infty,B}$. Keeping in mind [477, Lemma 1], this implies

$$\int_x^{x+1} \|F(t + \tau; b) - cF(t; b)\|^p \nu^p(t)\, dt$$
$$\leqslant \varepsilon_1^p \int_{A_x} \nu^p(t)\eta^{-p}(t)\, dt + ((1 + |c|)\|F\|_{\infty,B})^p \int_{[x,x+1]\smallsetminus A_x} \nu^p(t)\, dt$$

$$\leqslant \varepsilon_1^p \int_x^{x+1} \frac{\nu^p(t)}{\eta^p(t)}\,dt + ((1+|c|)\|F\|_{\infty,B})^p p \int_0^\infty y^{p-1} m(\{s \in [x,x+1] \smallsetminus A_X : \nu(s) > y\})\,dy$$

$$\leqslant \varepsilon_1^p \|\nu/\eta\|_{S^p} + 2((1+|c|)\|F\|_{\infty,B})^p d^p \lceil 1/\delta_0 \rceil \varepsilon_1$$

$$+ ((1+|c|)\|F\|_{\infty,B})^p p \int_d^\infty y^{p-1} m(\{s \in [x,x+1] \smallsetminus A_X : \nu(s) > y\})\,dy, \quad b \in B,$$

which simply completes the proof of result. □

The following result is a metrical analogue of the first part of [477, Theorem 2].

Theorem 4.2.6. *Suppose that $\Lambda = \mathbb{R}$ or $\Lambda = [0,\infty)$, $\nu \in PC(\Lambda : [0,\infty))$ is bounded, has the possible first kind discontinuities at the points of the sequence (t_k) and satisfies the condition $(\mathrm{QUC})_\nu$, where:*

$(\mathrm{QUC})_\nu$ *For every $\varepsilon > 0$, there exists $\delta > 0$ such that, if the points t_1 and t_2 belong to (t_i, t_{i+1}) for some $i \in \mathbb{Z}$ [$i \in \mathbb{N}_0$; $t_0 \equiv 0$] and $|t_1 - t_2| < \delta$, then $|\nu(t_1) - \nu(t_2)| < \varepsilon$.*

Suppose further that $\rho = T \in L(Y)$, $p > 0$ and $F \in S_{\Omega,\Lambda'}^{(\mathbb{F},T,\mathcal{P}_t,\mathcal{P})}(\Lambda : Y)$, where $\Lambda = \Lambda'$, $\mathbb{F}(\cdot\cdot) \equiv 1$, $\Omega = [0,1]$, $P = C_b(\Lambda : \mathbb{C})$ and $P_t = L_\nu^p(t + [0,1] : \mathbb{C})$ for all $t \in \Lambda$. If, for every $B \in \mathcal{B}$, we have $\|F\|_{\infty,B} < +\infty$, and $F(\cdot;\cdot)$ satisfies that, for every $x \in X$, the function $t \mapsto F(t;x)$, $t \in \Lambda$ is piecewise continuous with the possible first kind discontinuities at the points of the sequence (t_k) and the condition (QUC) holds, then $F(\cdot;\cdot)$ is pre-$(\mathcal{B}, T, (t_k), \mathcal{P})$-piecewise continuous almost periodic with $P_\varepsilon = C_{b,\nu}(A_\varepsilon : Y)$ for each $\varepsilon > 0$ sufficiently small.

Proof. We will provide all relevant details of the proof, which basically follows from the argumentation contained in the proof of [267, Theorem 2] and the prescribed assumptions on the function $\nu(\cdot)$. For simplicity, we will assume that $\Lambda = \mathbb{R}$, $T = cI$ for some $c \in S_1$ and $X = \{0\}$. Let $\varepsilon > 0$ be fixed; then there exists $\delta \in (0, \min\{\varepsilon/2, \delta_0/4\})$ such that, if the points t_1 and t_2 belong to the same interval (t_i, t_{i+1}) of the continuity of functions $F(\cdot)$ and $\nu(\cdot)$ and $|t_1 - t_2| < \delta$, then $\|F(t_1) - F(t_2)\| + |\nu(t_1) - \nu(t_2)| < \varepsilon/8(1 + \|\nu\|_\infty + \|F\|_\infty)$. Suppose that $\eta_k \in (0, \varepsilon\delta^{1/p}/4)$ for all $k \in \mathbb{N}$ and $\lim_{k\to+\infty} \eta_k = 0$. We will prove that there exists $k_0 \in \mathbb{N}$ such that, for every $\tau \in \mathbb{R}$ with $\int_t^{t+1} \|F(s+\tau) - cF(s)\|^p \nu^p(s)\,ds \leqslant \eta_{k_0}^p$, $t \in \mathbb{R}$, we have $\|F(t+\tau) - cF(t)\|\nu(t) \leqslant \varepsilon$ for all $t \notin \bigcup_{l\in\mathbb{Z}}(t_l - \varepsilon, t_l + \varepsilon)$. If we assume the contrary, then for each $k \in \mathbb{N}$ there exist points $s_k \notin \bigcup_{l\in\mathbb{Z}}(t_l - \varepsilon, t_l + \varepsilon)$ and $\tau_k \in \mathbb{R}$ such that $\int_t^{t+1} \|F(s+\tau_k) - cF(s)\|^p \nu^p(s)\,ds \leqslant \eta_k^p$, $t \in \mathbb{R}$ and $\|F(s_k + \tau_k) - cF(s_k)\|\nu(s_k) > \varepsilon$. Since the functions $F(\cdot)$ and $\nu(\cdot)$ are continuous from the left side, for each $k \in \mathbb{N}$ there exist points $s_k' \notin \bigcup_{l\in\mathbb{Z}}(t_l - (3\varepsilon/4), t_l + (3\varepsilon/4))$ and $\tau_k \in \mathbb{R}$ such that $\int_t^{t+1} \|F(s+\tau_k) - cF(s)\|^p \nu^p(s)\,ds \leqslant \eta_k^p$, $t \in \mathbb{R}$, $\|F(s_k' + \tau_k) - cF(s_k')\|\nu(s_k') > 3\varepsilon/4$ and $s_k' + \tau_k \notin \{t_l : l \in \mathbb{Z}\}$. Since $\delta < \varepsilon/2$, it follows that, for every $k \in \mathbb{N}$, the interval $(s_k' - \delta, s_k' + \delta)$ belongs to the same interval (t_j, t_{j+1}) of continuity of function $f(\cdot)$, for some $j \in \mathbb{Z}$. On the other hand, at least one of the intervals $(s_k' + \tau_k, s_k' + \tau_k + \delta)$ and $(s_k' + \tau_k - \delta, s_k' + \tau_k)$ belongs to the same interval (t_p, t_{p+1}) of continuity of function $F(\cdot)$, for some $p \in \mathbb{Z}$. For fixed $k \in \mathbb{N}$, we may assume

w. l. o. g. that the above holds for the interval $(s'_k+\tau_k, s'_k+\tau_k+\delta)$; since $|c|=1$, this readily implies

$$\begin{aligned}
&\|\nu(s+s'_k)[F(s+s'_k+\tau_k)-cF(s+s'_k)]-\nu(s'_k)[F(s'_k+\tau_k)-cF(s'_k)]\| \\
&\quad\leqslant \|\nu(s+s'_k)F(s+s'_k+\tau_k)-\nu(s'_k)F(s'_k+\tau_k)\|+\|\nu(s+s'_k)F(s+s'_k)-\nu(s'_k)F(s'_k)\| \\
&\quad\leqslant \nu(s+s'_k)\|F(s+s'_k+\tau_k)-F(s'_k+\tau_k)\|+\|F(s'_k+\tau_k)\|\cdot|\nu(s+s'_k)-\nu(s'_k)| \\
&\qquad+\nu(s+s'_k)\|F(s+s'_k)-F(s'_k)\|+\|F(s'_k)\|\cdot|\nu(s+s'_k)-\nu(s'_k)|\leqslant \varepsilon/2, \quad \text{a. e. } s\in[0,\delta].
\end{aligned}$$

Since $\nu(\cdot)$ and $F(\cdot)$ are bounded, the above implies

$$\|F(s+s'_k+\tau_k)-cF(s+s'_k)\|\nu(s+s'_k)\geqslant \varepsilon/4, \quad \text{a. e. } s\in[0,\delta],\ k\in\mathbb{N}$$

and

$$\begin{aligned}
\eta_k^p &\geqslant \int_{t_k}^{t_k+1}\|F(s+s'_k+\tau_k)-cF(s+s'_k)\|^p\nu^p(s+s'_k)\,ds \\
&\geqslant \int_{t_k}^{t_k+\delta}\|F(s+s'_k+\tau_k)-cF(s+s'_k)\|^p\nu^p(s+s'_k)\,ds\geqslant(\varepsilon/4)^p\delta, \quad k\in\mathbb{N}.
\end{aligned}$$

This is a contradiction and the proof of theorem is therefore completed. □

We continue with the following useful observation.

Remark 4.2.7. Suppose that $p:\mathbb{R}\to\mathbb{R}$ is a nonperiodic trigonometric polynomial, $\nu:\mathbb{R}\to(0,\infty)$ is a Lebesgue measurable function and the condition (LT0) holds, where:

(LT0) For every $\varepsilon>0$, there exists a sufficiently large integer $d>0$ such that $\sup_{t\in\mathbb{R}}\int_d^{+\infty}y^{p-1}m(\{x\in[t,t+1]:\nu(x)>y\})\,dy<\varepsilon$, where $m(\cdot)$ denotes the Lebesgue measure.

Then we have shown, in [477, Example 3], that the function $F(t):=\operatorname{sign}(p(t))$, $t\in\mathbb{R}$ belongs to the class $S_{\Omega,\Lambda'}^{(\mathbb{F},\rho,\mathcal{P}_t,\mathcal{P})}(\Lambda:Y)$ with $\Lambda=\Lambda'=\mathbb{R}$, $\mathbb{F}(\cdot\cdot)\equiv 1$, $\rho=\mathrm{I}$, $\Omega=[0,1]$, $P=C_b(\mathbb{R}:\mathbb{C})$ and $P_t=L_\nu^p(t+[0,1]:\mathbb{C})$ for all $t\in\mathbb{R}$, where $p>0$ is arbitrary. Disappointingly, the set consisting of all zeroes of $p(\cdot)$ can have infinitely many accumulation points in $\mathbb{R}$ (cf. also [267, Remark 5(ii)]).

We know that the pre-$(\mathcal{B},(t_k))$-piecewise continuous almost periodic functions form, under certain logical assumptions, the vector space with the usual operations; see, e. g., [267, Theorem 4]. It is not clear how we can extend the above mentioned results to the metrically pre-$(\mathcal{B},(t_k))$-piecewise continuous almost periodic functions.

4.2.2 Applications to the abstract Volterra integrodifferential equations

In this subsection, we will present some applications of our results to the abstract Volterra integrodifferential equations in Banach spaces.

1. The results obtained in [267] and [477], which have been slightly extended in Section 4.2.1, and the composition principles for Stepanov-p-almost periodic functions (cf. [444] for more details) can be successfully applied in the analysis of the existence and uniqueness of (t_k)-piecewise continuous almost periodic solutions of the following semilinear Volterra integral equation:

$$u(t) = g(t) + \int_{-\infty}^{t} a(t-s)F(s;u(s))\,ds, \quad t \in \mathbb{R}. \tag{244}$$

Let us assume that the following conditions hold:
(i) $1 \leqslant p < \infty$, $1/p + 1/q = 1$ and $\mathcal{B}$ denotes the collection of all sets in X with relatively compact range.
(ii) $F(\cdot;\cdot)$ is pre-$(\mathcal{B},(t_k))$-piecewise continuous almost periodic and, for every $B \in \mathcal{B}$, we have $\|F\|_{\infty,B} < +\infty$.
(iii) There exists a finite real constant $L > 0$ such that $\|F(t;x) - F(t;y)\| \leqslant L\|x-y\|$ for all $t \in \mathbb{R}$ and $x, y \in X$.
(iv) $g \in PCAP_{(t_k)}(\mathbb{R} : X)$.
(v) $\sum_{k=0}^{\infty} \|a(\cdot)\|_{L^q[k,k+1]} < +\infty$ and $L\int_0^{\infty} |a(t)|\,dt < 1$.

Since $PCAP_{(t_k)}(\mathbb{R} : X)$ is a Banach space equipped with the sup-norm, any function $f \in PCAP_{(t_k)}(\mathbb{R} : X)$ has relatively compact range (see [267, Proposition 2]) and the conditions (i)–(v) are satisfied, we can apply Theorem 4.2.5 with $\nu \equiv \eta \equiv 1$ and [444, Proposition 2.6.11, Theorem 2.7.2] in order to see that the mapping $\Psi : u \mapsto g(\cdot) + \int_{-\infty}^{\cdot} a(\cdot - s)F(s;u(s))\,ds$, $u \in PCAP_{(t_k)}(\mathbb{R} : X)$ is a well-defined contraction. Using the Banach contraction principle, it readily follows that there exists a unique (t_k)-piecewise continuous almost periodic solution of (244).

2. It would be very difficult to say something more about the existence and uniqueness of metrically (t_k)-piecewise continuous almost periodic solutions of (244); cf. also [267, Example 3] and [458, Theorem 2.1, Propositions 2.2, 2.3]. On the other hand, Theorem 4.2.5 can be simply reworded for the corresponding classes of piecewise continuous T-uniformly recurrent functions. Keeping this in mind, we can provide some applications to the abstract degenerate semilinear fractional differential equations considered in [458, Section 3]; see, especially, [458, Theorem 3.1].

3. Suppose that $c \in \mathbb{C}$, $|c| = 1$, $1 \leqslant p < +\infty$, $1/p + 1/q = 1$, $f : \mathbb{R} \to X$ is bounded and pre-$(c,(t_k),\mathcal{P})$-piecewise continuous almost periodic, where for every sufficiently small $\varepsilon > 0$, we have $P_\varepsilon = C_{b,\eta}(\mathbb{R} : X)$ with some positive function $\eta : \mathbb{R} \to (0,\infty)$ such that the function $(\nu/\eta)(\cdot)$ is Stepanov-p-bounded for some function $\nu : \mathbb{R} \to (0,\infty)$ satisfying

(LQ0). Due to Theorem 4.2.5, for each $\varepsilon > 0$ there exists a relatively dense set S' in $\mathbb{R}$ such that for each $\tau \in S'$, we have

$$\int_x^{x+1} \|f(t+\tau) - cf(t)\|^p v^p(t)\,dt \leqslant \varepsilon^p, \quad x \in \mathbb{R}. \tag{245}$$

Suppose now that $(R(t))_{t>0} \subseteq L(X,Y)$ is a strongly continuous operator family such that $\sum_{k=0}^{\infty} \|R(\cdot)\|_{L_v^q[k,k+1]} < \infty$ and the function $(1/v)(\cdot)$ is bounded. Applying (245), we get

$$\begin{aligned}
&\int_{x-k-1}^{x-k} \|f(t+\tau) - cf(t)\|^p v^{-p}(x-t)\,dt \\
&\quad \leqslant \int_{x-k-1}^{x-k} \|f(t+\tau) - cf(t)\|^p v^p(t) v^{-p}(x-t) v^{-p}(t)\,dt \\
&\quad \leqslant \int_{x-k-1}^{x-k} \|f(t+\tau) - cf(t)\|^p v^p(t)\,dt \cdot \|1/v(\cdot)\|_\infty^{2p} \leqslant \varepsilon^p \|1/v(\cdot)\|_\infty^{2p}, \quad x \in \mathbb{R},\ k \in \mathbb{Z}.
\end{aligned}$$

Now we can apply [458, Proposition 2.3], with $\sigma \equiv 1$ and $\Lambda = \Lambda' = \mathbb{R}$, in order to see that the function

$$t \mapsto F(t) := \int_{-\infty}^{t} R(t-s)f(s)\,ds, \quad t \in \mathbb{R}$$

is c-almost periodic in the usual sense. This can be applied in the analysis of the existence and uniqueness of c-almost periodic solutions for a large class of the abstract (fractional) Volterra integrodifferential inclusions without initial conditions; see [444] for many applications of this type.

At the end of this section, we would like to emphasize that it would be curious to find some new applications of metrically piecewise continuous ρ-almost periodic functions to the abstract impulsive Volterra integrodifferential equations (cf. also [266, 267] and references quoted therein for more details on the subject). The metrically piecewise continuous almost automorphic functions and their relations with metrically Stepanov almost automorphic type functions will be investigated somewhere else (cf. also [756]).

4.3 Generalized vectorial almost periodicity

In this section, we investigate several new classes of vectorially Stepanov-p-almost periodic type functions and vectorially (equi)-Weyl-p-almost periodic type functions, where $p > 0$. The novelty of our approach lies in the fact that we use the vector-valued integration in the analysis of these classes of functions; unfortunately, we cannot consider the general value of exponent $p \neq 1$ in the pure vector-valued setting. The introduced classes of functions extend the usually considered classes of Stepanov-p-almost periodic type functions and (equi)-Weyl-p-almost periodic type functions in the scalar-valued setting as well as the usually considered classes of Stepanov-1-almost periodic type functions and (equi)-Weyl-1-almost periodic type functions in the vector-valued setting. We also analyze Σ-almost periodic type functions and the invariance of generalized vectorial almost periodicity under the actions of convolution products, which is incredibly important for applications to the abstract Volterra integrodifferential equations.

The structure of this section can be briefly depicted as follows [474]. We analyze the notion of vectorial Stepanov almost periodicity in Section 4.3.1. The class of Σ-almost periodic functions is analyzed in Section 4.3.2, while the class of vectorially Stepanov almost periodic functions in general metric is analyzed in Section 4.3.3. Vectorially Weyl almost periodic type functions are analyzed in Section 4.3.4, while the invariance of vectorial Stepanov almost periodicity and vectorial Weyl almost periodicity under the actions of convolution products is analyzed in Section 4.3.5; for simplicity, we will not consider here the extensions of vectorially generalized almost periodic functions and the corresponding composition principles as well as the vectorially Weyl almost periodic type sequences.

4.3.1 Vectorial Stepanov almost periodicity

We open this subsection by introducing the following notion.

Definition 4.3.1. (i) A locally integrable function $F : \mathbb{R}^n \to X$ is vectorially Stepanov almost periodic if for every $\varepsilon > 0$ there exists $l > 0$ such that for each $\mathbf{t}_0 \in \mathbb{R}^n$ there exists $\boldsymbol{\tau} \in B(\mathbf{t}_0, l)$ with

$$\left\| \int_{[0,1]^n} [F(\mathbf{t} + \boldsymbol{\tau} + \mathbf{u}) - F(\mathbf{t} + \mathbf{u})]\, d\mathbf{u} \right\| \leqslant \varepsilon, \quad \mathbf{t} \in \mathbb{R}^n. \tag{246}$$

(ii) A p-locally integrable function $F : \mathbb{R}^n \to \mathbb{C}$ is vectorially Stepanov-p-almost periodic if for every $\varepsilon > 0$ there exists $l > 0$ such that for each $\mathbf{t}_0 \in \mathbb{R}^n$ there exists $\boldsymbol{\tau} \in B(\mathbf{t}_0, l)$ with

$$\left| \int_{[0,1]^n} [F(\mathbf{t} + \boldsymbol{\tau} + \mathbf{u}) - F(\mathbf{t} + \mathbf{u})]^p\, d\mathbf{u} \right| \leqslant \varepsilon^p, \quad \mathbf{t} \in \mathbb{R}^n. \tag{247}$$

It is clear that the notion of vectorial Stepanov-p-almost periodicity cannot be easily introduced if $p \neq 1$ and $X \neq \mathbb{C}$.

Immediately from the equations (246) and (247), it follows that any Stepanov almost periodic function is vectorially Stepanov almost periodic as well as that any scalar-valued Stepanov-p-almost periodic function is vectorially Stepanov-p-almost periodic.

The relations between Stepanov-p-almost periodic functions in norm and vectorial Stepanov-p-almost periodic functions are quite nontrivial ($p = 1$, general X; $p > 0$ and $p \neq 1$, $X = \mathbb{C}$). Furthermore, the class of scalar-valued vectorially Stepanov-p-almost periodic functions behaves very badly if $p \neq 1$ and we will not examined this class in more detail henceforth. The use of vector-valued integration in Definition 4.3.1(i) may cause some unpleasant difficulties in many concrete situations, as well; for example, it is not simple to consider the pointwise products of vectorially Stepanov almost periodic functions and the existence of Bohr–Fourier coefficients of vectorially Stepanov almost periodic functions (see, e. g., [444, Theorem 2.1.1] for the notion and more details in the one-dimensional setting).

We continue by providing the following illustrative example.

Example 4.3.2. Let Y be one of the spaces $L^p(\mathbb{R}^n)$, $C_0(\mathbb{R}^n)$ or $BUC(\mathbb{R}^n)$, where $1 \leqslant p < \infty$. Then we know that the Gaussian semigroup

$$(G(t)F)(x) := (4\pi t)^{-(n/2)} \int_{\mathbb{R}^n} F(x-y)e^{-\frac{|y|^2}{4t}}\,dy, \quad t > 0,\ f \in Y,\ x \in \mathbb{R}^n,$$

can be extended to a bounded analytic C_0-semigroup of angle $\pi/2$, generated by the Laplacian Δ_Y acting with its maximal distributional domain in Y. Let $F(\cdot)$ be bounded and vectorially Stepanov almost periodic, and let $t_0 > 0$ be fixed. Then the function $\mathbb{R}^n \ni x \mapsto u(x,t_0) \equiv (G(t_0)F)(x) \in \mathbb{C}$ is likewise bounded and vectorially Stepanov almost periodic, which essentially follows from the next simple computation involving the Fubini theorem ($x \in \mathbb{R}^n$; $\tau \in \mathbb{R}^n$ satisfies the necessary requirements):

$$\begin{aligned}
&\Big| \int_{[0,1]^n} \big[(G(t_0)F)(x+\tau+u) - (G(t_0)F)(x+u)\big]\,du \Big| \\
&= \Big| \int_{[0,1]^n} \Big[\int_{\mathbb{R}^n} F(x+\tau+u-y)e^{-\frac{|y|^2}{4t_0}}\,dy - \int_{\mathbb{R}^n} F(x+u-y)e^{-\frac{|y|^2}{4t_0}}\,dy \Big]\,du \Big| \\
&= \Big| \int_{[0,1]^n} \Big[\int_{\mathbb{R}^n} [F(x+\tau+u-y) - F(x+u-y)]e^{-\frac{|y|^2}{4t_0}}\,dy \Big]\,du \Big| \\
&= \Big| \int_{\mathbb{R}^n} e^{-\frac{|y|^2}{4t_0}} \Big[\int_{[0,1]^n} [F(x+\tau+u-y) - F(x+u-y)]\,du \Big]\,dy \Big| \\
&\leqslant \int_{\mathbb{R}^n} e^{-\frac{|y|^2}{4t_0}} \Big| \int_{[0,1]^n} [F(x+\tau+u-y) - F(x+u-y)]\,du \Big|\,dy \leqslant \varepsilon \int_{\mathbb{R}^n} e^{-\frac{|y|^2}{4t_0}}\,dy.
\end{aligned}$$

We can also consider here vectorially Stepanov c-almost periodic functions; see Definition 4.3.9 below with $\rho = cI$ and $c \in \mathbb{C} \setminus \{0\}$.

The proof of the following result is simple and, therefore, omitted.

Proposition 4.3.3. *A locally integrable function $F : \mathbb{R}^n \to X$ is vectorially Stepanov almost periodic if and only if the function $G = \Sigma(F) : \mathbb{R}^n \to X$, given by $G(\mathbf{t}) := \int_{[0,1]^n} F(\mathbf{t}+\mathbf{u})\, d\mathbf{u}$, $\mathbf{t} \in \mathbb{R}^n$, is almost periodic.*

In the one-dimensional setting, we have that $G(t) = F^{[1]}(t+1) - F^{[1]}(t)$, $t \in \mathbb{R}$, where $F^{[1]}(t) := \int_0^t F(s)\, ds$, $t \in \mathbb{R}$ is the first antiderivative of function $F(\cdot)$. In particular, if $F^{[1]}(\cdot)$ is almost periodic, then $F(\cdot)$ is vectorially Stepanov almost periodic (see also the formulation of Kadets's theorem [69, Theorem 4.6.11] and the research articles [57] by J. Andres, D. Pennequin, [149] by C. Budde, J. Kreulich, [262] by H.-S. Ding et al., and [684] by A. M. Samoilenko, S. I. Trofimchuk for further information concerning the integration of almost periodic functions). Now we would like to present the following example, which justifies the introduction of notion in Definition 4.3.1.

Example 4.3.4. Let us recall that B. Basit and H. Güenzler have constructed, in [92, Example 3.2], a bounded continuous function $f : \mathbb{R} \to \mathbb{R}$, which satisfies that the function $f^{[1]}(\cdot)$ is almost periodic and the function $f(\cdot)$ is not Stepanov almost periodic. Since $f^{[1]}(\cdot)$ is almost periodic, we have that $f(\cdot)$ is vectorially Stepanov almost periodic; see also [447, p. 62] for more details about this function.

It is also worth noticing that Proposition 4.3.3 gives a rise for the introduction of space consisting of all locally integrable functions $f : \mathbb{R} \to X$ such that its first antiderivative $f^{[1]}(\cdot)$ is Stepanov-p-almost periodic; for example, the first derivative of function $f(t) = \sin(1/(2 + \cos t + \cos(\sqrt{2}t)))$, $t \in \mathbb{R}$ belongs to this space but it is not vectorially Stepanov almost periodic [448].

We continue by observing that Proposition 4.3.3 implies that the collection of all vectorially Stepanov almost periodic functions forms a vector space with the usual operations as well as that any vectorially Stepanov almost periodic function $F : \mathbb{R}^n \to X$ is vectorially Stepanov bounded in the sense that

$$\sup_{\mathbf{t}\in\mathbb{R}^n} \left\| \int_{t+[0,1]^n} F(\mathbf{u})\, d\mathbf{u} \right\| < +\infty.$$

The notion of vectorial Stepanov boundedness and the usually considered Stepanov boundedness are different, as the following example shows.

Example 4.3.5. Suppose that $X := \mathbb{C}$ and $F(t) := t\sin(t^2)$, $t \in \mathbb{R}$. Then we have

$$\int_t^{t+1} |s \cdot \sin(s^2)|\, ds = \frac{1}{2} \int_{t^2}^{(t+1)^2} |\sin v|\, dv \geqslant c|t|, \quad t \in \mathbb{R},$$

for some positive real constant $c > 0$. Hence,

$$\sup_{t\in\mathbb{R}} \left[(1+|t|)^{-\sigma} \int_t^{t+1} |f(s)|\, ds \right] = +\infty, \quad \sigma \in (0,1).$$

A similar argumentation shows that

$$\left|\int_t^{t+1} f(s)\,ds\right| = \frac{1}{2}\left|\cos(t^2) - \cos((t+1)^2)\right| \leqslant 1, \quad t \in \mathbb{R},$$

which yields the required conclusion. Further on, we have

$$\int_t^{t+1} [f(s+\tau) - f(s)]\,ds$$
$$= \frac{1}{2}\big[(\cos((t+\tau)^2) - \cos((t+\tau+1)^2)) - (\cos(t^2) - \cos((t+1)^2))\big], \quad t,\ \tau \in \mathbb{R},$$

which shows that $f(\cdot)$ cannot be vectorially Stepanov almost periodic.

Without going into further details, we will only emphasize here that the vector-valued integration can be useful in the studies of vectorial weighted ergodic components and vectorial weighted ergodic components in general metric (see [447, Section 6.4] and [448, Section 5.2] for more details on the subject).

4.3.2 Σ-almost periodicity

In the remainder of this section, we will always assume that $\emptyset \neq I \subseteq \mathbb{R}^n$, $P \subseteq Y^I$, the zero function belongs to P and $\mathcal{P} = (P, d)$ is a pseudometric space; if $f \in P$, then we set $\|f\|_P := d(f, 0)$.

We refer the reader to [452, Definition 2.1] for the notion of a (strongly) $(\mathrm{R}, \mathcal{B}, \mathcal{P}, L)$-multialmost periodic function. Keeping in mind Proposition 4.3.3, we would like to introduce the following general notion.

Definition 4.3.6. Suppose that $\emptyset \neq I \subseteq \mathbb{R}^n$, $\Sigma : Y^{I\times X} \to Y^{I\times X}$, R is an arbitrary collection of sequences in $\mathbb{R}^n$, $F : I \times X \to Y$ is a given function, for each $B \in \mathcal{B}$ and $\mathbf{b} \in \mathrm{R}$ the set $L(B; \mathbf{b})$ is a collection of subsets of B, and the following holds:

$$\text{If} \quad \mathbf{t} \in I,\ \mathbf{b} \in \mathrm{R} \text{ and } l \in \mathbb{N}, \quad \text{then we have } \mathbf{t} + \mathbf{b}(l) \in I. \tag{248}$$

Then we say that the function $F(\cdot;\cdot)$ is $(\Sigma, \mathrm{R}, \mathcal{B}, \mathcal{P}, L)$-multialmost periodic, respectively, strongly $(\Sigma, \mathrm{R}, \mathcal{B}, \mathcal{P}, L)$-multialmost periodic in the case that $I = \mathbb{R}^n$, if the function $\Sigma(F)$ is $(\mathrm{R}, \mathcal{B}, \mathcal{P}, L)$-multialmost periodic, respectively, strongly $(\mathrm{R}, \mathcal{B}, \mathcal{P}, L)$-multialmost periodic.

If for each $B \in \mathcal{B}$ and $\mathbf{b} \in \mathrm{R}$, we have $L(B; \mathbf{b}) = \{B\}$, then we omit the term “L” from the notation.

We can similarly define the notion of a (strong) $(\Sigma, \mathrm{R}_{\mathrm{X}}, \mathcal{B}, \mathcal{P})$-multialmost periodic function (of type 1), where R_{X} is an arbitrary collection of sequences in $\mathbb{R}^n \times X$; cf. [452, Definition 2.2]. Once it is done, we can formulate the following slight extension of [452, Proposition 2.6].

Proposition 4.3.7. *Suppose that P is a vector structure with the usual operations, P is complete and the metric d is translation invariant, i. e., $d(f+g,h+g)=d(f,h)$ iff, h, $f+g$, $h+g\in P$. Let us assume further that, for every $j\in\mathbb{N}$, the function $F_j : I\times X\to Y$ is $(\Sigma,\mathrm{R}_X,\mathcal{B},\mathcal{P},L)$-multialmost periodic of type 1 as well as that, for every sequence which belongs to R_X, any its subsequence also belongs to R_X. If $F : I\times X\to Y$ and for each $B\in\mathcal{B}$, $(\mathbf{b};\mathbf{x})=((b_k;x_k))=((b_k^1,b_k^2,\ldots,b_k^n);x_k)\in\mathrm{R}_X$, $B'\in L(B;(\mathbf{b};\mathbf{x}))$ and we have*

$$\lim_{(i,l)\to(+\infty,+\infty)}\sup_{x\in B'}\big\|F_i(\cdot+b_{k_l};x+x_{k_l})-F(\cdot+b_{k_l};x+x_{k_l})\big\|_P=0,$$

then the function $F(\cdot;\cdot)$ is $(\Sigma,\mathrm{R}_X,\mathcal{B},\mathcal{P},L)$-multialmost periodic of type 1, provided that the following conditions hold true:

(i) *$\Sigma(f-g)=\Sigma f-\Sigma g$ for all f, $g\in P$.*
(ii) *There exists a finite real constant $c>0$ such that $\|\Sigma f\|_P\leqslant c\|f\|_P$ for all $f\in P$.*

For example, this condition holds if $I=\mathbb{R}^n$, $P=C_b(\mathbb{R}^n:Y)$ and $\Sigma(\cdot)$ is given as in the formulation of Proposition 4.3.3. Suppose now that P has a linear vector structure, the metric d is translation invariant and, for every $c\in\mathbb{C}$, one has $cf\in P$, $f\in P$ and the existence of a real number $\phi(c)>0$ such that $\|cf\|_P\leqslant\phi(c)\|f\|_P$ for all $f\in P$. If we assume that for each sequence of collection R $[\mathrm{R}_X]$ any its subsequence also belongs to R $[\mathrm{R}_X]$ as well as that the mapping Σ is linear, then the space consisting of all (strongly) $(\Sigma,\mathrm{R},\mathcal{B},\mathcal{P},L)$-multialmost periodic [(strongly) $(\Sigma,\mathrm{R}_X,\mathcal{B},\mathcal{P},L)$-multialmost periodic] functions is a vector space; see also [452, Remark(ii), pp. 234–235].

We refer the reader to [452, Definition 3.1] for the notions of a Bohr $(\mathcal{B},I',\rho,\mathcal{P})$-almost periodic function and a $(\mathcal{B},I',\rho,\mathcal{P})$-uniformly recurrent function. The following notion could be also introduced (we can similarly introduce and analyze many other classes of (metrically) generalized Σ-almost periodic type functions, which have been considered in [448, Chapters 4–7] in the case that Σ is the identity mapping).

Definition 4.3.8. Suppose that $\emptyset\neq I'\subseteq\mathbb{R}^n$, $\emptyset\neq I\subseteq\mathbb{R}^n$, $\Sigma : Y^{I\times X}\to Y^{I\times X}$, $F : I\times X\to Y$ is a given function, ρ is a binary relation on Y, $R(\Sigma(F))\subseteq D(\rho)$ and $I+I'\subseteq I$. Then we say that:

(i) $F(\cdot;\cdot)$ is Bohr $(\Sigma,\mathcal{B},I',\rho,\mathcal{P})$-almost periodic if the function $\Sigma(F)$ is Bohr $(\mathcal{B},I',\rho,\mathcal{P})$-almost periodic.
(ii) $F(\cdot;\cdot)$ is $(\Sigma,\mathcal{B},I',\rho,\mathcal{P})$-uniformly recurrent if the function $\Sigma(F)$ is $(\mathcal{B},I',\rho,\mathcal{P})$-uniformly recurrent.

It is clear that the structural results established for Bohr $(\mathcal{B},I',\rho,\mathcal{P})$-almost periodic functions and $(\mathcal{B},I',\rho,\mathcal{P})$-uniformly recurrent functions can be straightforwardly generalized for Bohr $(\Sigma,\mathcal{B},I',\rho,\mathcal{P})$-almost periodic functions and $(\Sigma,\mathcal{B},I',\rho,\mathcal{P})$-uniformly recurrent functions, provided that the mapping $\Sigma(\cdot)$ has suitable properties; we can slightly extend the statements of [452, Proposition 3.7, Corollary 3.8, Proposition 3.10] in this manner, for example. In order to avoid any repeating and plagiarism, we will not formulate such results here.

4.3.3 Vectorially Stepanov almost periodic functions in general metric

As in all previous research studies of generalized multidimensional almost periodic type functions, we will also denote here the region I by Λ and the region I' by Λ'. We assume that the following condition holds true:

(SM1-1v): $\emptyset \neq \Lambda \subseteq \mathbb{R}^n$, $\emptyset \neq \Lambda' \subseteq \mathbb{R}^n$, $\emptyset \neq \Omega \subseteq \mathbb{R}^n$ is a Lebesgue measurable set such that $m(\Omega) > 0$, $\Lambda' + \Lambda \subseteq \Lambda$ and $\Lambda + \Omega \subseteq \Lambda$.

The following notion extends the notion introduced in Definition 4.3.9(i), with $\rho = \mathrm{I}$ and $P = C_b(\mathbb{R}^n : Y)$; see Proposition 4.3.3.

Definition 4.3.9. Suppose that (SM1-1v) holds, ρ is a binary relation on Y, $F : \Lambda \times X \to Y$, for each $x \in X$ the mapping $F(\cdot; x)$ is locally integrable,

$$[\Sigma(F)](\mathbf{t}; x) := \int_{\Omega} F(\mathbf{t} + \mathbf{u}; x)\, d\mathbf{u}, \quad \mathbf{t} \in \Lambda,\ x \in X \tag{249}$$

and $R(\Sigma(F)) \subseteq D(\rho)$. Then the function $F(\cdot;\cdot)$ is said to be vectorially Stepanov $(\mathcal{B}, \Lambda', \rho, \mathcal{P})$-almost periodic, respectively, vectorially Stepanov $(\mathcal{B}, \Lambda', \rho, \mathcal{P})$-uniformly recurrent, if $F(\cdot;\cdot)$ is Bohr $(\Sigma, \mathcal{B}, \Lambda', \rho, \mathcal{P})$-almost periodic, respectively, $(\Sigma, \mathcal{B}, \Lambda', \rho, \mathcal{P})$-uniformly recurrent, with $\Sigma(\cdot)$ being defined through (249).

If R is an arbitrary collection of sequences in $\mathbb{R}^n$ and (248) holds, then we say that the function $F(\cdot;\cdot)$ is vectorially Stepanov $(\mathrm{R}, \mathcal{B}, \mathcal{P}, L)$-multialmost periodic, respectively, strongly vectorially Stepanov $(\mathrm{R}, \mathcal{B}, \mathcal{P}, L)$-multialmost periodic in the case that $\Lambda = \mathbb{R}^n$, if the function $\Sigma(F)$ is $(\mathrm{R}, \mathcal{B}, \mathcal{P}, L)$-multialmost periodic, respectively, strongly $(\mathrm{R}, \mathcal{B}, \mathcal{P}, L)$-multialmost periodic, with $\Sigma(\cdot)$ being defined through (249).

If $F : \Lambda \times X \to Y$, $G : \Lambda \times X \to Y$ and for each $x \in X$ the mappings $F(\cdot; x)$ and $G(\cdot; x)$ are locally integrable, then we have $\Sigma(\alpha F + \beta G) = \alpha\Sigma(F) + \beta\Sigma(G)$ for all scalars α, $\beta \in \mathbb{C}$, so that the collection of all vectorially Stepanov $(\mathcal{B}, \Lambda', \rho, \mathcal{P})$-almost periodic functions is a linear vector space if the corresponding space consisting of all $(\mathcal{B}, \Lambda', \rho, \mathcal{P})$-almost periodic functions is a linear vector space (we cannot expect the linearity of space of all vectorially Stepanov $(\mathcal{B}, \Lambda', \rho, \mathcal{P})$-uniformly recurrent functions in any sense; see [447] for more details). Furthermore, if $\Lambda = \Lambda' = \mathbb{R}^n$, $\rho = \mathrm{I}$, R is the collection of all sequences in $\mathbb{R}^n$ and $\mathcal{B}$ is the collection of all compact subsets of X, then the metrical Bochner criterion immediately clarifies the coincidence of the class consisting of all vectorially Stepanov $(\mathcal{B}, \Lambda', \rho, \mathcal{P})$-almost periodic functions and the class consisting of all vectorially Stepanov $(\mathrm{R}, \mathcal{B}, \mathcal{P})$-multialmost periodic functions, provided that the pseudometric space P has the properties stated in this result.

Now we are in a position to state the following result.

Proposition 4.3.10. *Suppose that* (SM1-1v) *holds,* $\rho = T \in L(Y)$, $F_k : \Lambda \times X \to Y$ *and the function* $F_k(\cdot;\cdot)$ *is vectorially Stepanov* $(\mathcal{B}, \Lambda', T, \mathcal{P})$*-almost periodic, respectively, vec-*

torially Stepanov $(\mathcal{B}, \Lambda', T, \mathcal{P})$-uniformly recurrent ($k \in \mathbb{N}$), where $P = C_b(\Lambda : Y)$. If $F : \Lambda \times X \to Y$, for each $x \in X$ the mapping $F(\cdot; x)$ is locally integrable, and for each $B \in \mathcal{B}$,

$$\lim_{k \to +\infty} \sup_{\mathbf{t} \in \Lambda, x \in B} \left\| \int_{\Omega} [F_k(\mathbf{t} + \mathbf{u}; x) - F(\mathbf{t} + \mathbf{u}; x)]\, d\mathbf{u} \right\| = 0,$$

then the function $F(\cdot;\cdot)$ is vectorially Stepanov $(\mathcal{B}, \Lambda', T, \mathcal{P})$-almost periodic, respectively, vectorially Stepanov $(\mathcal{B}, \Lambda', T, \mathcal{P})$-uniformly recurrent.

Proof. We will consider the class of vectorially Stepanov $(\mathcal{B}, \Lambda', T, \mathcal{P})$-almost periodic functions, only. Let $B \in \mathcal{B}$ and $\varepsilon > 0$ be fixed. Then there exist $k_0 \in \mathbb{N}$ and $l > 0$ such that for each $\mathbf{t}_0 \in \Lambda'$ there exists $\tau \in B(\mathbf{t}_0, l) \cap \Lambda'$ such that

$$\sup_{\mathbf{t} \in \Lambda, x \in B} \left\| \int_{\mathbf{t}+\Omega} [F_{k_0}(\mathbf{u} + \tau; x) - TF_{k_0}(\mathbf{u}; x)]\, d\mathbf{u} \right\| \leqslant (\varepsilon/3).$$

Then the final conclusion simply follows from the next computation ($\mathbf{t} \in \Lambda$, $x \in B$):

$$\begin{aligned}
&\left\| \int_{\mathbf{t}+\Omega} [F(\mathbf{u} + \tau; x) - TF(\mathbf{u}; x)]\, d\mathbf{u} \right\| \\
&\quad \leqslant \left\| \int_{\mathbf{t}+\Omega} [F(\mathbf{u} + \tau; x) - F_{k_0}(\mathbf{u} + \tau; x)]\, d\mathbf{u} \right\| + \left\| \int_{\mathbf{t}+\Omega} [F_{k_0}(\mathbf{u} + \tau; x) - TF_{k_0}(\mathbf{u}; x)]\, d\mathbf{u} \right\| \\
&\qquad + \left\| \int_{\mathbf{t}+\Omega} [TF_{k_0}(\mathbf{u}; x) - TF(\mathbf{u}; x)]\, d\mathbf{u} \right\| \\
&\quad \leqslant \left\| \int_{\mathbf{t}+\Omega} [F(\mathbf{u} + \tau; x) - F_{k_0}(\mathbf{u} + \tau; x)]\, d\mathbf{u} \right\| + \left\| \int_{\mathbf{t}+\Omega} [F_{k_0}(\mathbf{u} + \tau; x) - TF_{k_0}(\mathbf{u}; x)]\, d\mathbf{u} \right\| \\
&\qquad + \|T\| \cdot \left\| \int_{\mathbf{t}+\Omega} [F_{k_0}(\mathbf{u}; x) - F(\mathbf{u}; x)]\, d\mathbf{u} \right\|.
\end{aligned}$$

□

Suppose now that $\emptyset \neq \Lambda \subseteq \mathbb{R}^n$, $\emptyset \neq \Omega \subseteq \mathbb{R}^n$ is a Lebesgue measurable set such that $m(\Omega) > 0$ and $\Lambda + \Omega \subseteq \Lambda$. Denote by $S_v^1(\Lambda \times X : Y)$ the collection of all functions $F : \Lambda \times X \to Y$ such that for each $x \in X$ the function $F(\cdot; x)$ is locally integrable and, for every $B \in \mathcal{B}$,

$$\|F\|_{S^1_{v,B,\Omega}} := \sup_{\mathbf{t} \in \Lambda, x \in B} \frac{1}{m(\Omega)} \left\| \int_{\mathbf{t}+\Omega} F(\mathbf{s}; x)\, d\mathbf{s} \right\| < +\infty. \tag{250}$$

For the functions $F : \Lambda \to Y$, we omit the term "B" from the notation; moreover, we omit the term "Ω" from the notation if $\Omega = [0,1]^n$.

Let us observe the following facts:

(i) The assumption $\|F\|_{S^1_v} = 0$ does not imply $F(\mathbf{t}) = 0$ for a. e. $\mathbf{t} \in \Lambda$; for example, this is not true for the function $F(t) = \cos(2\pi t)$, $t \in \mathbb{R}$.

(ii) If $\lambda \in \mathbb{C}$, $B \in \mathcal{B}$ and $\|F\|_{S^1_{\nu,B,\Omega}} < +\infty$, then we have $\|\lambda \cdot F\|_{S^1_{\nu,B,\Omega}} = |\lambda| \cdot \|F\|_{S^1_{\nu,B,\Omega}}$.

(iii) If $B \in \mathcal{B}$, $\|F\|_{S^1_{\nu,B,\Omega}} < +\infty$, the function $G : \Lambda \times X \to Y$ satisfies that for each $x \in X$ the function $G(\cdot; x)$ is locally integrable and $\|G\|_{S^1_{\nu,B,\Omega}} < +\infty$, then we have

$$\|F + G\|_{S^1_{\nu,B,\Omega}} \leqslant \|F\|_{S^1_{\nu,B,\Omega}} + \|G\|_{S^1_{\nu,B,\Omega}}.$$

If $B \in \mathcal{B}$, then the symbol $S^1_{\nu,B}(\Lambda \times B : Y)$ denotes the collection of all functions $F : \Lambda \times B \to Y$ such that for each $x \in B$ the function $F(\cdot; x)$ is locally integrable and (250) holds. Then $S^1_{\nu,B}(\Lambda \times B : Y)$ is a vector space and (i)–(iii) simply yield that $\|\cdot\|_{S^1_{\nu,B,\Omega}}$ is a seminorm on $S^1_{\nu,B}(\Lambda \times B : Y)$.

Concerning the completeness of space $S^1_{\nu,B}(\Lambda \times B : Y)$, we will only state and prove the following simple result (cf. also [447, p. 376, l. -1] and [508, Theorem 5.2.1, pp. 199–200]).

Proposition 4.3.11. *Suppose that $B \in \mathcal{B}$, $(F_k(\cdot; \cdot))_{k \in \mathbb{N}}$ is a sequence of functions in $S^1_{\nu,B}(\Lambda \times B : Y)$, and the following two conditions hold:*

(i) *For every $\varepsilon > 0$, there exists $k_0 \in \mathbb{N}$ such that, for every $k, m \in \mathbb{N}$ with $\min(k, m) \geqslant k_0$, we have $\|F_k - F_m\|_{S^1_{\nu,B,\Omega}} \leqslant \varepsilon$.*

(ii) *There exists a function $F : \Lambda \times B \to Y$ such that for each $x \in B$ the function $F(\cdot; x)$ is locally integrable and $\lim_{k \to +\infty} \|F_k(\mathbf{t}; x) - F(\mathbf{t}; x)\| = 0$ for a. e. $\mathbf{t} \in \Lambda$.*

(iii) *For every compact set $K \subseteq \Lambda$ and for every $x \in B$, there exists a function $g_x \in L^1(K : Y)$ such that $\|F_k(\mathbf{t}; x)\| \leqslant g_x(\mathbf{t})$ a. e. on K for all $k \in \mathbb{N}$.*

Then we have $\lim_{k \to +\infty} \|F_k - F\|_{S^1_{\nu,B,\Omega}} = 0$.

Proof. If $\varepsilon > 0$, $x \in B$ and $\mathbf{t} \in \Lambda$, then the dominated convergence theorem [69, Theorem 1.1.8] and (ii)–(iii) together imply that

$$\lim_{m \to +\infty} \int_{\mathbf{t}+\Omega} F_m(\mathbf{s}; x)\, d\mathbf{s} = \int_{\mathbf{t}+\Omega} F(\mathbf{s}; x)\, d\mathbf{s}. \tag{251}$$

On the other hand, (i) implies that there exists $k_0 \in \mathbb{N}$ such that, for every $k, m \in \mathbb{N}$ with $\min(k, m) \geqslant k_0$, we have

$$\left\| \int_{\mathbf{t}+\Omega} [F_k(\mathbf{s}; x) - F_m(\mathbf{s}; x)]\, d\mathbf{s} \right\| \leqslant \varepsilon, \quad \mathbf{t} \in \Lambda,\ x \in B.$$

The final conclusion follows by letting $m \to +\infty$ in the above estimate and using (251). □

4.3.4 Vectorially Weyl almost periodic type functions

Here, we would like to propose the following notion (we will not consider here the metrical generalizations of this definition; see [448, Section 4.3] for more details).

Definition 4.3.12. Suppose that $p > 0$.

(i) A locally integrable function $F : \mathbb{R}^n \to X$ is said to be vectorially equi-Weyl-almost periodic if, for every $\varepsilon > 0$, there exist two finite real numbers $l > 0$ and $L > 0$ such that for each $\mathbf{t}_0 \in \mathbb{R}^n$ there exists $\tau \in B(\mathbf{t}_0, L)$ with

$$\sup_{\mathbf{t}\in\mathbb{R}^n}\left\| \int_{\mathbf{t}+l[0,1]^n} [F(\tau + \mathbf{u}) - F(\mathbf{u})]\, d\mathbf{u} \right\| < \varepsilon l^n.$$

(ii) A locally integrable function $F : \mathbb{R}^n \to X$ is said to be vectorially Weyl-almost periodic if, for every $\varepsilon > 0$, there exists a finite real number $L > 0$ such that for each $\mathbf{t}_0 \in \mathbb{R}^n$ there exists $\tau \in B(\mathbf{t}_0, L)$ with

$$\limsup_{l\to+\infty} \sup_{\mathbf{t}\in\mathbb{R}^n}\left[l^{-n} \left\| \int_{\mathbf{t}+l[0,1]^n} [F(\tau + \mathbf{u}) - F(\mathbf{u})]\, d\mathbf{u} \right\| \right] < \varepsilon.$$

It is clear that any (equi-)Weyl-1-almost periodic function is vectorially (equi-)Weyl-almost periodic. Furthermore, it can be simply proved that any vectorially Stepanov almost periodic function $F : \mathbb{R}^n \to X$ is vectorially equi-Weyl-almost periodic.

We continue with the following example (cf. also [69, Example 4.6.5] and Example 5.2.18 below).

Example 4.3.13. Suppose that $X := c_0$, the Banach space of all numerical sequences vanishing at plus infinity, equipped with the sup-norm. Define $f(t) := ((1/k)\cos(t/k))_k$, $t \in \mathbb{R}$. Then $f : \mathbb{R} \to X$ is almost periodic but its first anti-derivative $F(t) := (\sin(t/k))_k$, $t \in \mathbb{R}$ is bounded, uniformly continuous but not Stepanov-p-almost automorphic for any exponent $p > 0$; furthermore, we will prove later that $F(\cdot)$ is vectorially Weyl almost automorphic as well as that $F(\cdot)$ is not vectorially Weyl almost automorphic of type 1 nor jointly vectorially Weyl almost automorphic (see Section 4.3.4 for the notion and more details).

Now we will prove that $F(\cdot)$ is vectorially Weyl-almost periodic, i. e., the vectorially Weyl-1-almost periodic. In order to show this, let us notice that (t, $x \in \mathbb{R}$, $\tau \in \mathbb{R}$):

$$F(t + x + \tau) - F(t + x) = 2\left(\sin\frac{\tau}{2k} \cdot \cos\frac{2t + 2x + 2\tau}{2k}\right)_k,$$

so that ($t \in \mathbb{R}$, $\tau \in \mathbb{R}$; $l > 0$)

$$\begin{aligned}
&\frac{1}{l}\int_t^{t+l} [F(x + \tau) - F(x)]\, dx \\
&= \frac{2}{l}\left(k\sin\frac{\tau}{2k} \cdot \left[\sin\frac{2t + 2l + \tau}{2k} - \sin\frac{2t + \tau}{2k}\right]\right)_k \\
&= \frac{4}{l}\left(k\sin\frac{\tau}{2k} \cdot \cos\frac{2t + l + \tau}{2k} \cdot \sin\frac{l}{2k}\right)_k.
\end{aligned} \tag{252}$$

This immediately implies the required statement since $|k\sin(\tau/2k)| \leqslant |\tau|/2$, $k \in \mathbb{N}$ and

$$\left|\cos\frac{4t+2l+2\tau}{4k}\cdot\sin\frac{l}{2k}\right| \leqslant 1, \quad k \in \mathbb{N}.$$

Now we will prove that $F(\cdot)$ is not Weyl-p-almost periodic for any finite exponent $p > 0$. Suppose that $0 < \varepsilon < 32^{-1}\cdot(\cos(7\pi/16)\cdot\sqrt{2}/2)^p$. Then there exists $L > 0$ such that, for every $t_0 \in \mathbb{R}$, we can find $\tau \in [t_0 - L, t_0 + L]$ such that there exists a finite real number $l_0(\tau) > 0$ with

$$\int_0^l \left\|\left(\cos\frac{2x+\tau}{2k}\cdot\sin\frac{\tau}{2k}\right)_k\right\|^p dx \leqslant \varepsilon\cdot l, \quad l \geqslant l_0(\tau). \tag{253}$$

Suppose that $k_0 \in \mathbb{N}$ is sufficiently large and $2k_0\pi/4 \leqslant \tau < (2k_0+2)\pi/4$. Then $\pi/4 \leqslant \tau/2k_0 < 3\pi/8$ and, therefore, $\sin(\tau/2k_0) \geqslant \sqrt{2}/2$. Suppose further that $x_0/k_0 = \pi/16$; then $\cos((2x+\tau)/2k_0) \geqslant \cos(7\pi/16)$, $x \in [2m\pi k_0, 2m\pi k_0 + x_0]$ ($m \in \mathbb{N}_0$) and taking into account (253) with $l = 2m\pi k_0 + x_0$, where m is sufficiently large, we get

$$\begin{aligned}\varepsilon(2m\pi k_0 + x_0) &\geqslant \int_0^{2m\pi k_0+x_0} \left\|\left(\cos\frac{2x+\tau}{2k}\cdot\sin\frac{\tau}{2k}\right)_k\right\|^p dx \\ &\geqslant \int_0^{2m\pi k_0+x_0} \left|\cos\frac{2x+\tau}{2k_0}\cdot\sin\frac{\tau}{2k_0}\right|^p dx \geqslant (m+1)x_0\left(\frac{\sqrt{2}}{2}\cdot\cos\frac{7\pi}{16}\right)^p, \quad l \geqslant l_0(\tau).\end{aligned}$$

Dividing by m and letting $m \to +\infty$, we get

$$2\pi\varepsilon(k_0/x_0) = 32\varepsilon \geqslant \left(\frac{\sqrt{2}}{2}\cdot\cos\frac{7\pi}{16}\right)^p,$$

which is a contradiction.

Now we will prove that $F(\cdot)$ is not vectorially equi-Weyl-almost periodic. Suppose that $0 < \varepsilon < \sqrt{2}/2$. Then there exist $l > 0$ and $L > 0$ such that, for every $t_0 \in \mathbb{R}$, there exists $\tau \in [t_0 - L, t_0 + L]$ such that

$$\left\|\frac{4}{l}\left(k\sin\frac{\tau}{2k}\cdot\cos\frac{2t+l+\tau}{2k}\cdot\sin\frac{l}{2k}\right)_k\right\| \leqslant \varepsilon, \quad t \in \mathbb{R}; \tag{254}$$

see (252). It is clear that there exists $k_0(\varepsilon, l) \in \mathbb{N}$ such that

$$4k\sin\frac{l}{2k} \geqslant l, \quad k \geqslant k_0(\varepsilon, l). \tag{255}$$

After that, take any $t_0 > L + \pi k_0(\varepsilon, l)/2$. Then there exists $k_0 \geqslant k_0(\varepsilon, l)$ such that $2k_0\pi/4 \leqslant \tau < (2k_0+2)\pi/4$. Then, as above, $\pi/4 \leqslant \tau/2k_0 < 3\pi/8$ and $\sin(\tau/2k_0) \geqslant \sqrt{2}/2$. Since (254) holds for all $t \in \mathbb{R}$, we can take $t = -(l+\tau)/2$ in order to see that $\cos((2t+l+\tau)/2k_0) = 1$ so that (255) and the above estimates enable one to see that the right-hand side of (254) is greater than or equal to $\sqrt{2}/2$, which is a contradiction.

Let us finally observe that for each $l > 0$ we have

$$\sup_{t\in\mathbb{R}} \frac{1}{l}\left\| \int_t^{t+l} F(s)\,ds \right\| = 1, \tag{256}$$

so that the vectorial Weyl seminorm of $F(\cdot)$, defined through

$$\|F\|_{W,v} := \lim_{l\to+\infty} \sup_{t\in\mathbb{R}} \frac{1}{l}\left\| \int_t^{t+l} F(s)\,ds \right\|, \tag{257}$$

exists and equals 1. In order to see that (256) holds, notice that a simple computation shows that

$$\frac{1}{l}\left\| \int_t^{t+l} F(s)\,ds \right\| = \sup_{k\in\mathbb{N}}\left[\frac{2k}{l} \cdot \left| \sin\frac{2t+l}{2k} \cdot \sin\frac{l}{2k} \right| \right], \quad t \in \mathbb{R}.$$

This implies that for each $l > 0$ we have

$$\begin{aligned}\frac{1}{l}\left\| \int_t^{t+l} F(s)\,ds \right\| &\geqslant \sup_{t\in(\pi\mathbb{Z}-l)/2}\ \sup_{k\in\mathbb{N}}\left[\frac{2k}{l} \cdot \left| \sin\frac{2t+l}{2k} \cdot \sin\frac{l}{2k} \right| \right] \\ &\geqslant \sup_{k\in\mathbb{N}}\left[\frac{2k}{l} \cdot \left| \sin\frac{l}{2k} \right| \right] = 1,\end{aligned}$$

since $\lim_{k\to+\infty}(2k/l)\cdot|\sin(l/2k)| = 1$.

If the function $F(\cdot)$ is locally integrable, then the existence of vectorial Weyl seminorm of $F(\cdot)$ in $[0,+\infty]$ cannot be proved with the help of the argumentation given on [447, pp. 375–376]. Moreover, this seminorm does not necessarily exist, as the following counterexample shows.

Example 4.3.14. Define the function $F : \mathbb{R} \to \mathbb{R}$ by $F(t) := -2k$ if $t \in [2k, 2k+1)$ for some $k \in \mathbb{Z}$ and $F(t) := 2k+2$ if $t \in [2k+1, 2k+2)$ for some $k \in \mathbb{Z}$. Then we have

$$\limsup_{l\to+\infty} \sup_{t\in\mathbb{R}} \frac{1}{l}\left| \int_t^{t+l} F(s)\,ds \right| = +\infty \quad \text{and} \quad \liminf_{l\to+\infty} \sup_{t\in\mathbb{R}} \frac{1}{l}\left| \int_t^{t+l} F(s)\,ds \right| = 1$$

and, therefore, $\|F\|_{W,v}$ does not exist in $[0,+\infty]$. In order to see this, it suffices to show that

$$\sup_{t\in\mathbb{R}}\left|\int_t^{t+2k} F(s)\,ds\right| = 2k \quad \text{and} \quad \sup_{t\in\mathbb{R}}\left|\int_t^{t+2k+1} F(s)\,ds\right| = +\infty \quad (k \in \mathbb{N}). \tag{258}$$

By our construction, we have $\int_{2m+1}^{2m+1+2k} F(s)\,ds = 0$ for all $m,\ k \in \mathbb{Z}$. Hence,

$$\sup_{t\in\mathbb{R}}\left|\int_t^{t+2k+1} F(s)\,ds\right| \geqslant \sup_{t\in 2\mathbb{Z}+1}\left|\int_t^{t+2k+1} F(s)\,ds\right|.$$

If $t = 2m+1$ for some $m \in \mathbb{Z}$, then we have

$$\int_t^{t+2k+1} F(s)\,ds = \int_{2m+2k+1}^{2m+2k+2} F(s)ds = 2m+2k+2,$$

which simply implies the second equality in (258). On the other hand, we have

$$\sup_{t\in\mathbb{R}}\left|\int_t^{t+2k} F(s)\,ds\right| \geqslant \int_0^{2k} F(s)\,ds = 2k \quad (k \in \mathbb{N}). \tag{259}$$

If $t \in [2m, 2m+1)$ for some $m \in \mathbb{Z}$, then we have

$$\begin{aligned}\int_t^{t+2k} F(s)\,ds &= \int_t^{2m+1} F(s)\,ds - \int_{t+2k}^{2m+2k+1} F(s)\,ds \\ &= (-2)(2m+1-t)m + (2m+1-t)(2m+2k) = 2(2m+1-t)k,\end{aligned}$$

so that

$$\left|\int_t^{t+2k} F(s)\,ds\right| \leqslant 2k. \tag{260}$$

Similarly, if $t \in [2m+1, 2m+2)$ for some $m \in \mathbb{Z}$, then we have

$$\begin{aligned}\int_t^{t+2k} F(s)\,ds &= \int_t^{2m+1} F(s)\,ds + \int_{2m+2k+1}^{t+2k} F(s)\,ds \\ &= (2m+1-t)(2m+2) - (2m+1-t)(2m+2k+2) = 2(2m+1-t)k,\end{aligned}$$

so that (260) again holds. Keeping in mind (259)–(260), we simply get the first equality in (258).

The uniform convergence of a sequence of locally integrable functions implies the convergence of this sequence in the vectorial Weyl seminorm; furthermore, the standard evidence shows that, if the sequence of vectorially (equi)-Weyl-almost periodic functions converges in the vectorial Weyl seminorm, then the limit function is likewise vectorially (equi)-Weyl-almost periodic. These results can be formulated in the multidimensional setting, as well.

It is not so simple to deduce a proper analogue of Proposition 4.3.3 for vectorially Weyl almost periodic type functions. After clarifying this fact, we will generalize the notion introduced in Definition 4.3.12 as follows (cf. also [448, Definitions 3.1.1–3.1.6]).

Definition 4.3.15. Suppose that $F : \Lambda \times X \to Y$, ρ is a binary relation on Y and the following condition holds:

(WM1-1v): $\emptyset \neq \Lambda \subseteq \mathbb{R}^n$, $\emptyset \neq \Lambda' \subseteq \mathbb{R}^n$, $\emptyset \neq \Omega \subseteq \mathbb{R}^n$ is a Lebesgue measurable set such that $m(\Omega) > 0$, $\Lambda' + \Lambda \subseteq \Lambda$ and $\Lambda + l\Omega \subseteq \Lambda$ for $l > 0$.

(i) It is said that $F(\cdot;\cdot)$ is vectorially equi-Weyl-$(\mathcal{B}, \Lambda', \rho, \Omega)$-almost periodic if, for every $\varepsilon > 0$ and $B \in \mathcal{B}$, there exist two finite real numbers $l > 0$ and $L > 0$ such that for each $\mathbf{t}_0 \in \Lambda'$ there exists $\tau \in B(\mathbf{t}_0, L) \cap \Lambda'$ such that for each $\mathbf{t} \in \Lambda$ and $x \in B$ there exists an integrable mapping $y_{\mathbf{t};x} : \mathbf{t} + l\Omega \to Y$ such that $y_{\mathbf{t};x}(\mathbf{u}) \in \rho(F(\mathbf{u}; x))$ for a. e. $\mathbf{u} \in \mathbf{t} + l\Omega$ and
$$\sup_{\mathbf{t}\in\Lambda;x\in B} \left\| \int_{\mathbf{t}+l\Omega} [F(\tau + \mathbf{u}; x) - y_{\mathbf{t};x}(\mathbf{u})]\, d\mathbf{u} \right\| < \varepsilon l^n. \tag{261}$$

(ii) It is said that $F(\cdot;\cdot)$ is vectorially Weyl-$(\mathcal{B}, \Lambda', \rho, \Omega)$-almost periodic if, for every $\varepsilon > 0$ and $B \in \mathcal{B}$, there exists a finite real number $L > 0$ such that for each $\mathbf{t}_0 \in \Lambda'$ there exists $\tau \in B(\mathbf{t}_0, L) \cap \Lambda'$ such that for each $\mathbf{t} \in \Lambda$ and $x \in B$ there exists an integrable mapping $y_{\mathbf{t};x} : \mathbf{t} + l\Omega \to Y$ such that $y_{\mathbf{t};x}(\mathbf{u}) \in \rho(F(\mathbf{u}; x))$ for a. e. $\mathbf{u} \in \mathbf{t} + l\Omega$ and
$$\limsup_{l\to+\infty} \sup_{\mathbf{t}\in\Lambda;x\in B} \left[l^{-n} \left\| \int_{\mathbf{t}+l\Omega} [F(\tau + \mathbf{u}; x) - y_{\mathbf{t};x}(\mathbf{u})]\, d\mathbf{u} \right\| \right] < \varepsilon.$$

Now we will formulate and prove the following result (cf. also [448, pp. 23–24]).

Proposition 4.3.16. *Suppose that $m \in \mathbb{N}$, $F : \Lambda \times X \to Y$, $\rho = T \in L(Y)$ and (WM1-1v) holds. If $F(\cdot;\cdot)$ is vectorially (equi-)Weyl-$(\mathcal{B}, \Lambda', T, \Omega)$-almost periodic, then $F(\cdot;\cdot)$ is vectorially (equi-)Weyl-$(\mathcal{B}, m\Lambda', T^m, \Omega)$-almost periodic, as well.*

Proof. We will consider the class of vectorially equi-Weyl-$(\mathcal{B}, \Lambda', T, \Omega)$-almost periodic functions, only. Let $\varepsilon > 0$ and $B \in \mathcal{B}$ be fixed. Then there exist two finite real numbers $l > 0$ and $L > 0$ such that for each $\mathbf{t}_0 \in \Lambda'$ there exists $\tau \in B(\mathbf{t}_0, L) \cap \Lambda'$ such that for each $\mathbf{t} \in \Lambda$ and $x \in B$ we have that (261) holds with $y_{\mathbf{t};x} = TF(\mathbf{u}; x)$ for a. e. $\mathbf{u} \in \mathbf{t} + l\Omega$. Let $\mathbf{t} \in \Lambda$

and $x \in B$ be fixed, and let $\mathbf{t}_0 \in \Lambda'$ and $\tau \in B(\mathbf{t}_0, L) \cap \Lambda'$ be as above. Then it is clear that $s\Lambda' + \Lambda \subseteq \Lambda$, $s \in \mathbb{N}$ and

$$F(\mathbf{u} + m\tau; x) - T^m F(\mathbf{u}; x) = \sum_{j=0}^{m-1} T^j \big[F(\mathbf{u} + (m-j)\tau; x) - TF(\mathbf{u} + (m-j-1)\tau; x)\big],$$

for any $\mathbf{u} \in \mathbf{t} + l\Omega$. This implies

$$\begin{aligned}
&\Big\| \int_{\mathbf{t}+l\Omega} [F(\mathbf{u} + m\tau; x) - T^m F(\mathbf{u}; x)]\, d\mathbf{u} \Big\| \\
&\quad \leqslant \sum_{j=0}^{m-1} \|T\|^j \cdot \Big\| \int_{\mathbf{t}+l\Omega} [F(\mathbf{u} + (m-j)\tau; x) - T^m F(\mathbf{u} + (m-j-1)\tau; x)]\, d\mathbf{u} \Big\| \\
&\quad = \sum_{j=0}^{m-1} \|T\|^j \cdot \Big\| \int_{\mathbf{t}+(m-j-1)\tau+l\Omega} [F(\mathbf{u} + \tau; x) - T^m F(\mathbf{u}; x)]\, d\mathbf{u} \Big\| \leqslant \sum_{j=0}^{m-1} \|T\|^j \varepsilon,
\end{aligned}$$

which simply yields the required conclusion. □

We conclude this subsection by observing that Proposition 4.3.16 can be also formulated for the corresponding classes of vectorially Stepanov almost periodic functions.

4.3.5 Invariance of vectorial Stepanov almost periodicity and vectorial Weyl almost periodicity under the actions of convolution products

In this subsection, we investigate the invariance of generalized vectorial almost periodicity under the actions of convolution products. Our first result reads as follows.

Proposition 4.3.17. *Suppose that $\nu : \mathbb{R} \to (0, +\infty)$ satisfies that the function $1/\nu(\cdot)$ is locally bounded and there exists a Lebesgue measurable function $\psi : \mathbb{R} \to (0, +\infty)$ such that $\nu(x+y) \leqslant \psi(x)\nu(y)$ for all $x, y \in \mathbb{R}$. Suppose further that $f : \mathbb{R} \to X$ is a bounded, vectorially $(\Lambda', T, \mathcal{P})$-almost periodic function, respectively, a bounded, vectorially $(\Lambda', T, \mathcal{P})$-uniformly recurrent function, where $\rho = T \in L(X)$ and $P = C_{b,\nu}(\mathbb{R} : X)$. If $(R(t))_{t>0}$ is any strongly continuous operator family in $L(X)$ such that $R(t)T = TR(t)$ for all $t > 0$ and $\int_0^{+\infty} \|R(r)\| \cdot (1 + \psi(r))\, dr < +\infty$, then the function $F : \mathbb{R} \to X$, given by*

$$F(t) := \int_{-\infty}^{t} R(t-s) f(s)\, ds, \quad t \in \mathbb{R}, \tag{262}$$

is bounded, continuous and vectorially $(\Lambda', T, \mathcal{P})$-almost periodic, respectively, bounded, continuous and vectorially $(\Lambda', T, \mathcal{P})$-uniformly recurrent.

Proof. Suppose that $\varepsilon > 0$. Then there exists $l > 0$ such that, for every $t_0 \in \Lambda'$ there exists $\tau \in [t_0 - l, t_0 + l] \cap \Lambda'$ such that

$$\left\| \left[\int_0^1 f(t+\tau+s)\,ds - T \int_0^1 f(t+s)\,ds \right] \cdot \nu(t) \right\| \leqslant \varepsilon, \quad t \in \mathbb{R}. \tag{263}$$

It can be simply proved that the function $F(\cdot)$ is well-defined, bounded and continuous as well as that $F(t) = \int_0^{+\infty} R(r)f(t-r)\,dr$ for all $t \in \mathbb{R}$. Furthermore, we can use (263), the Fubini theorem and the prescribed assumptions in order to see that the following holds:

$$\begin{aligned}
&\left\| \left[\int_0^1 F(t+\tau+s)\,ds - T \int_0^1 F(t+s)\,ds \right] \cdot \nu(t) \right\| \\
&= \left\| \left[\int_0^1 \int_0^{+\infty} R(r)f(t+\tau+s-r)\,dr\,ds - T \int_0^1 \int_0^{+\infty} R(r)f(t+s-r)\,dr\,ds \right] \cdot \nu(t) \right\| \\
&= \left\| \left[\int_0^1 \int_0^{+\infty} R(r)[f(t+\tau+s-r) - Tf(t+s-r)]\,dr\,ds \right] \cdot \nu(t) \right\| \\
&= \left\| \int_0^{+\infty} R(r) \left[\int_0^1 [f(t+\tau+s-r) - Tf(t+s-r)]\,ds \right] dr \cdot \nu(t) \right\| \\
&\leqslant \int_0^{+\infty} \|R(r)\| \cdot \left\| \int_0^1 [f(t+\tau+s-r) - Tf(t+s-r)]\,ds \right\| dr \cdot \nu(t) \\
&\leqslant \int_0^{+\infty} \|R(r)\| \cdot \frac{\varepsilon}{\nu(t-r)} \cdot \nu(t)\,dr \leqslant \varepsilon \int_0^{+\infty} \|R(r)\| \psi(r)\,dr.
\end{aligned}$$

This simply implies the required conclusion. □

The statement of [444, Proposition 2.6.11] cannot be reconsidered for the vectorial Stepanov almost periodicity. Concerning the vectorial Weyl almost periodicity, we will clarify the following result; the proof is almost completely the same as the proof of [444, Proposition 2.11.1(i)] and, therefore, omitted.

Proposition 4.3.18. *Suppose that $(R(t))_{t>0} \subseteq L(X)$ is a strongly continuous operator family satisfying that $\int_0^\infty \|R(s)\|\,ds < \infty$, $T \in L(X)$ and $R(t)T = TR(t)$ for all $t > 0$. If $f : \mathbb{R} \to X$ is bounded and vectorially (equi-)Weyl-$(\Lambda', T, \mathcal{P})$-almost periodic, where $P = C_b(\mathbb{R} : X)$, then the function $F(\cdot)$, given by (262), is bounded, continuous and vectorially (equi-)Weyl-$(\Lambda', T, \mathcal{P})$-almost periodic.*

It is clear that Proposition 4.3.17 and Proposition 4.3.18 can be applied in the qualitative analysis of solutions for various classes of the abstract Volterra integro-differential equations without initial conditions; for example, we can consider the existence and uniqueness of vectorially Stepanov (Weyl) almost periodic solutions to the abstract fractional Poisson heat equation without initial conditions (see [444–448] for more details in this direction).

The statement of [444, Theorem 2.11.4] cannot be reconsidered for the vectorial Weyl almost periodicity, unfortunately. Let us also note that we can use two different pivot spaces X and Y in the formulations of Proposition 4.3.17 and Proposition 4.3.18, provided that $T = cI$ for some $c \in \mathbb{C} \setminus \{0\}$.

Conclusions and final remarks

The notion introduced in Definition 4.3.9(ii) can be extended in the following way (the metrical generalizations of this definition can be also introduced; see [448, Section 4.2] for more details).

Definition 4.3.19. Suppose that (SM1-1v) holds, $p > 0$, ρ is a binary relation on Y, $F : \Lambda \times X \to Y$ and $R(F) \subseteq D(\rho)$. The function $F(\cdot;\cdot)$ is said to be vectorially Stepanov $(\mathcal{B}, \Lambda', \rho, p)$-almost periodic if for each $x \in X$ the mapping $F(\cdot; x)$ is p-locally integrable as well as, for every $\varepsilon > 0$ and $B \in \mathcal{B}$, there exists $l > 0$ such that for each $\mathbf{t}_0 \in \Lambda$ there exists a point $\tau \in B(\mathbf{t}_0, l) \cap \Lambda'$ such that for each $\mathbf{t} \in \Lambda$ and $x \in B$ there exists a p-locally integrable function $y_{\mathbf{t};x}(\cdot)$ on Ω such that $y_{\mathbf{t};x}(\mathbf{u}) \in \rho(F(\mathbf{t} + \mathbf{u}; x))$ for a. e. $\mathbf{u} \in \Omega$ and

$$\left| \int_{\Omega} [F(\mathbf{t} + \tau + \mathbf{u}; x) - y_{\mathbf{t};x}(\mathbf{u})]^p \, d\mathbf{u} \right| \leqslant \varepsilon^p, \quad \mathbf{t} \in \Lambda, \ x \in B.$$

We can similarly define the notion of a vectorially Stepanov $(\mathcal{B}, \Lambda', \rho, p)$-uniformly recurrent function in the scalar-valued setting.

Let us finally note that we can also consider the class of scalar-valued vectorially (equi-)Weyl-p-almost periodic functions and the class of scalar-valued vectorially (equi-)Weyl-$(\mathcal{B}, \Lambda', \rho, p)$-almost periodic functions following our approaches from Definition 4.3.12 and Definition 4.3.15 as well as some other classes of generalized vectorially almost periodic functions, like vectorially Doss almost periodic type functions (see [448, Section 3.2]). More details will appear somewhere else.

5 Generalized almost automorphic functions, generalized metrically almost automorphic functions and applications

In this chapter, we analyze some new classes of generalized almost automorphic functions, generalized metrically almost automorphic functions and provide certain applications.

5.1 Metrical Stepanov almost automorphy and applications

In this section, we analyze various classes of multidimensional Stepanov almost automorphic type functions in general metric. We clarify the main structural properties for the introduced classes of metrically Stepanov almost automorphic-type functions, providing also some applications to the abstract Volterra integro-differential equations. Our basic idea is to consider the values of multidimensional Bochner transform in a general pseudometric space Z, not only in the space $Z = L^p(\Omega : Y)$, which has been followed in all research studies of Stepanov almost automorphy by now.

The organization of the section can be briefly summarized as follows (the material is taken from [176]). First of all, we present an illustrative example concerning the existence and uniqueness of metrically Stepanov almost automorphic solutions of the wave equations in $\mathbb{R}^3$. In Definition 5.1.3, which opens the first subsection, we introduce the notion of a Stepanov $(\Omega, \mathrm{R}, \mathcal{B}, Z^{\mathcal{P}})$-multialmost automorphic function $F : \mathbb{R}^n \times X \to Y$ as well as the notion of Stepanov $(\Omega, \mathrm{R}, \mathcal{B}, Z^{\mathcal{P}}, W_{\mathcal{B},\mathrm{R}})$-multialmost automorphy [Stepanov $(\Omega, \mathrm{R}, \mathcal{B}, Z^{\mathcal{P}}, P_{\mathcal{B},\mathrm{R}})$-multialmost automorphy] for $F(\cdot;\cdot)$. The main structural characterizations for the introduced classes of functions are given in Theorem 5.1.6, Propositions 5.1.7, 5.1.9 and 5.1.10 (see also Remark 5.1.5 and Example 5.1.8). Convolution invariance of metrical Stepanov almost automorphy and some applications are considered in Section 5.1.2; Section 5.1.3 investigates some subspaces of compactly almost automorphic functions, which are very special cases of the function spaces introduced in Definition 5.1.3.

In order to motivate our research, we would like to present the following illustrative example based on our analyses from [494, Section 2.1].

Example 5.1.1. Consider the following wave equation in $\mathbb{R}^3$:

$$u_{tt}(t,x) = d^2 \Delta_x u(t,x), \quad x \in \mathbb{R}^3,\ t > 0; \quad u(0,x) = g(x), \quad u_t(0,x) = h(x), \tag{264}$$

where $d > 0$, $g \in C^3(\mathbb{R}^3 : \mathbb{R})$ and $h \in C^2(\mathbb{R}^3 : \mathbb{R})$. By the Kirchhoff formula (see e. g., [681, Theorem 5.4, pp. 277–278]), we know that the function

$$u(t,x) := \frac{\partial}{\partial t}\left[\frac{1}{4\pi d^2 t} \int\limits_{\partial B_{dt}(x)} g(\boldsymbol{\sigma})\, d\sigma\right] + \frac{1}{4\pi d^2 t} \int\limits_{\partial B_{dt}(x)} h(\boldsymbol{\sigma})\, d\sigma$$

https://doi.org/10.1515/9783111689746-008

$$= \frac{1}{4\pi} \int_{\partial B_1(0)} g(x + dt\omega)\, d\omega + \frac{dt}{4\pi} \int_{\partial B_1(0)} \nabla g(x + dt\omega) \cdot \omega\, d\omega$$
$$+ \frac{t}{4\pi} \int_{\partial B_1(0)} h(x + dt\omega)\, d\omega, \quad t \geqslant 0,\ x \in \mathbb{R}^3,$$

is a unique solution of problem (264), which belongs to the class $C^2([0,\infty) \times \mathbb{R}^3)$. Fix now a number $t_0 > 0$ and assume that the functions $g(\cdot)$, $\nabla g(\cdot)$ and $h(\cdot)$ are bounded and Stepanov $([0,1]^3, \mathrm{R}, Z^{\mathcal{P}})$-multialmost automorphic in the sense of Definition 5.1.3 below, where $Z = L^1_\nu([0,1]^3)$ with $\nu : [0,1]^3 \to (0,\infty)$ being an arbitrary Lebesgue integrable function; here, R is any collection of sequences in $\mathbb{R}^3$ which satisfies that, for every sequence $(\mathbf{b}_k) \in \mathrm{R}$, any subsequence $(\mathbf{b}_{k_l})$ of $(\mathbf{b}_k)$ also belongs to R. Using the dominated convergence theorem and the Fubini theorem, it readily follows that the function $x \mapsto u(t_0, x)$, $x \in \mathbb{R}^3$ is bounded and Stepanov $([0,1]^3, \mathrm{R}, Z^{\mathcal{P}})$-multialmost automorphic. We can similarly consider the wave equation in $\mathbb{R}^2$ [$\mathbb{R}$], whose unique solution is given by the Poisson formula [d'Alembert formula].

In this section, we will always assume that Ω is a fixed compact subset of $\mathbb{R}^n$ with positive Lebesgue measure and $p \in \mathcal{P}(\Omega)$. If $F : \mathbb{R}^n \times X \to Y$, then we introduce the multidimensional Bochner transform $\hat{F}_\Omega : \mathbb{R}^n \times X \to Y^\Omega$ by

$$[\hat{F}_\Omega(\mathbf{t}; x)](u) := F(\mathbf{t} + \mathbf{u}; x), \quad \mathbf{t} \in \mathbb{R}^n,\ \mathbf{u} \in \Omega,\ x \in X.$$

Since no confusion seems likely, we also write $\hat{F}$ in place of $\hat{F}_\Omega$. Finally, we recall the following notion (see [494, Definition 2.1]).

Definition 5.1.2. Suppose $p \in \mathcal{P}(\Omega)$ and the Bochner transform $\hat{F}_\Omega : \mathbb{R}^n \times X \to L^{p(\mathbf{u})}(\Omega : Y)$ is continuous. Then we say that $F(\cdot;\cdot)$ is:

(i) Stepanov $(\Omega, p(\mathbf{u}))$-$(\mathrm{R}, \mathcal{B})$-multialmost automorphic if $\hat{F} : \mathbb{R}^n \times X \to L^{p(\mathbf{u})}(\Omega : Y)$ is $(\mathrm{R}, \mathcal{B})$-multialmost automorphic.
(ii) Stepanov $(\Omega, p(\mathbf{u}))$-$(\mathrm{R}, \mathcal{B}, W_{\mathcal{B},\mathrm{R}})$-multialmost automorphic [Stepanov $(\Omega, p(\mathbf{u}))$-$(\mathrm{R}, \mathcal{B}, P_{\mathcal{B},\mathrm{R}})$-multialmost automorphic] if $\hat{F} : \mathbb{R}^n \times X \to L^{p(\mathbf{u})}(\Omega : Y)$ is $(\mathrm{R}, \mathcal{B}, W_{\mathcal{B},\mathrm{R}})$-multialmost automorphic [$(\mathrm{R}, \mathcal{B}, P_{\mathcal{B},\mathrm{R}})$-multialmost automorphic].

5.1.1 Metrical Stepanov almost automorphy

Suppose that X and Y are complex Banach spaces as well as that $\emptyset \neq Z \subseteq Y^\Omega$, where $\emptyset \neq \Omega \subseteq \mathbb{R}^n$ is a fixed compact set with positive Lebesgue measure. Let $Z \subseteq Y^\Omega$, $0 \in Z$ and let (Z, d_Z) be a pseudometric space. Set $\|f\|_Z := d_Z(f, 0)$, $f \in Z$.

We start this subsection by introducing the following notion.

Definition 5.1.3. Suppose that $F : \mathbb{R}^n \times X \to Y$ is a given function and R is a collection of sequences in $\mathbb{R}^n$. Then we say that the function $F(\cdot;\cdot)$ is Stepanov $(\Omega, \mathrm{R}, \mathcal{B}, Z^{\mathcal{P}})$-multialmost automorphic if, for every $B \in \mathcal{B}$ and for every sequence $(\mathbf{b}_k = (b_k^1, b_k^2, \ldots, b_k^n)) \in \mathrm{R}$, there exist a subsequence $(\mathbf{b}_{k_l} = (b_{k_l}^1, b_{k_l}^2, \ldots, b_{k_l}^n))$ of $(\mathbf{b}_k)$ and a function $F_B^* : \mathbb{R}^n \times X \to Z$ such that, for every $\mathbf{t} \in \mathbb{R}^n$, $l \in \mathbb{N}$ and $x \in B$, we have $F(\mathbf{t} + \cdot + (b_{k_l}^1, \ldots, b_{k_l}^n); x) - [F_B^*(\mathbf{t}; x)](\cdot) \in Z$, $[F_B^*(\mathbf{t} - (b_{k_l}^1, \ldots, b_{k_l}^n); x)](\cdot) - F(\mathbf{t} + \cdot; x) \in Z$,

$$\lim_{l \to +\infty} \left\| F(\mathbf{t} + \cdot + (b_{k_l}^1, \ldots, b_{k_l}^n); x) - [F_B^*(\mathbf{t}; x)](\cdot) \right\|_Z = 0, \tag{265}$$

and

$$\lim_{l \to +\infty} \left\| [F_B^*(\mathbf{t} - (b_{k_l}^1, \ldots, b_{k_l}^n); x)](\cdot) - F(\mathbf{t} + \cdot; x) \right\|_Z = 0. \tag{266}$$

Furthermore, if for each $x \in B$ the convergence in (265)–(266) is uniform in $\mathbf{t}$ for any element of the collection $W_{B;(\mathbf{b}_k)}(x)$ [the convergence in (265)–(266) is uniform in $(\mathbf{t}; x)$ for any set of the collection $P_{B;(\mathbf{b}_k)}$], then we say that $F(\cdot; \cdot)$ is Stepanov $(\Omega, \mathrm{R}, \mathcal{B}, Z^{\mathcal{P}}, W_{\mathcal{B},\mathrm{R}})$-multialmost automorphic [Stepanov $(\Omega, \mathrm{R}, \mathcal{B}, Z^{\mathcal{P}}, P_{\mathcal{B},\mathrm{R}})$-multi-almost automorphic].

The usual notion, obtained by plugging $Z = L^{p(\mathbf{u})}(\Omega : Y)$, can be further generalized by plugging $Z = L_\nu^{p(\mathbf{u})}(\Omega : Y)$, where $p \in \mathcal{P}(\Omega)$ and $\nu : \Omega \to (0, \infty)$ is a Lebesgue measurable function, which can be used for consideration of some subspaces [extensions] of usually considered classes of Stepanov almost automorphic type functions [494]; here, we identify the functions that are equal almost everywhere on Ω, which is a little bit inappropriate because we have assumed that $Z \subseteq Y^\Omega$. We can also use here the space $Z = C_{0,\nu}(\Omega : Y)$ $[C_{b,\nu}(\Omega : Y)]$, where $\nu : \Omega \to [0, \infty)$ is any nontrivial function. Furthermore, if there exist two finite real constants $m > 0$ and $M > 0$ such that $m \leqslant \|\nu(\mathbf{u})\| \leqslant M$ for all $\mathbf{u} \in \Omega$, then we recover the notion of compact almost automorphy, which will be further examined in Section 5.1.3; if $m \leqslant \|\nu(\mathbf{u})\|$ for all $\mathbf{u} \in \Omega$ $[\|\nu(\mathbf{u})\| \leqslant M$ for all $\mathbf{u} \in \Omega]$, then we consider some subspaces [extensions] of compactly almost automorphic functions.

We continue by providing some useful remarks.

Remark 5.1.4. It is worth noting that the notion of Stepanov $(\Omega, \mathrm{R}, \mathcal{B}, Z^{\mathcal{P}})$-multialmost automorphy can be further generalized in the following sense (cf. also [448, Definitions 6.1.5, 7.1.35]): We say that a function $F : \mathbb{R}^n \times X \to Y$ is Stepanov $(\Omega, \mathrm{R}, \mathcal{B}, Z^{\mathcal{P}})$-normal if, for every $B \in \mathcal{B}$ and for every sequence $(\mathbf{b}_k = (b_k^1, b_k^2, \ldots, b_k^n)) \in \mathrm{R}$, there exists a subsequence $(\mathbf{b}_{k_l} = (b_{k_l}^1, b_{k_l}^2, \ldots, b_{k_l}^n))$ of $(\mathbf{b}_k)$ such that, for every $\mathbf{t} \in \mathbb{R}^n$ and $\varepsilon > 0$, there exists $l_0 \in \mathbb{N}$ such that, for every $x \in B$ and l, $l' \geqslant l_0$, we have $F(\mathbf{t} + \cdot + (b_{k_l}^1, \ldots, b_{k_l}^n); x) - F(\mathbf{t} + \cdot + (b_{k_{l'}}^1, \ldots, b_{k_{l'}}^n); x) \in Z$ and

$$\left\| F(\mathbf{t} + \cdot + (b_{k_l}^1, \ldots, b_{k_l}^n); x) - F(\mathbf{t} + \cdot + (b_{k_{l'}}^1, \ldots, b_{k_{l'}}^n); x) \right\|_Z \leqslant \varepsilon.$$

If Z is a metric vector space such that the assumption f, $g \in Z$ implies $f \pm g \in Z$ and $\|f \pm g\|_Z \leqslant \|f\|_Z + \|g\|_Z$, then the first limit equation (265) shows that any Stepanov $(\Omega, \mathrm{R}, \mathcal{B}, Z^{\mathcal{P}})$-multialmost automorphic function $F(\cdot; \cdot)$ has to be Stepanov $(\Omega, \mathrm{R}, \mathcal{B}, Z^{\mathcal{P}})$-normal. We will not consider the notion of Stepanov $(\Omega, \mathrm{R}, \mathcal{B}, Z^{\mathcal{P}})$-normality henceforth.

Remark 5.1.5. It is worth noting that, if the convergence of a sequence $(f_k(\cdot))$ in Z implies the pointwise convergence of the sequence $(f_k(\mathbf{u}))$ in Y for some $\mathbf{u} \in \Omega$, then any continuous, Stepanov $(\Omega, \mathrm{R}, \mathcal{B}, Z^{\mathcal{P}})$-multialmost automorphic function $F : \mathbb{R}^n \times X \to Y$ is $(\mathrm{R}, \mathcal{B})$-multialmost automorphic. On the other hand, the dominated convergence theorem in Lebesgue spaces with variable exponent (see, e. g., [494, Lemma 1.2(iv)]) shows that any $(\mathrm{R}, \mathcal{B})$-multialmost automorphic function $F : \mathbb{R}^n \times X \to Y$, which satisfies $\sup_{\mathbf{t}\in\mathbb{R}^n;x\in B} \|F(\mathbf{t};x)\|_Y < +\infty$ for any set $B \in \mathcal{B}$ has to be Stepanov $(\Omega, \mathrm{R}, \mathcal{B}, Z^{\mathcal{P}})$-multialmost automorphic, where $Z = L_\nu^{p(\mathbf{u})}(\Omega : Y)$ with any $p \in D_+(\Omega)$ and $\nu \in L^{p(\mathbf{u})}(\Omega)$, that provides a proper generalization of [494, Proposition 2.6]. It is logical to ask whether we can construct, for some concrete examples of almost automorphic functions $F : \mathbb{R}^n \to Y$ which are not compactly almost automorphic, the pseudometric spaces (Z, d_Z) of weighted continuous functions such that $F(\cdot)$ is Stepanov $(\Omega, \mathrm{R}, Z^{\mathcal{P}})$-multialmost automorphic with R being the collection of all sequences in $\mathbb{R}^n$; see, e. g., [184, Example, p. 809].

In the proof of the following structural result, we will not use the dominated convergence theorem (see also [452, Subsection 2.1] for more details about the interplay between the metrical almost periodicity and (Stepanov) almost automorphy; the results established in this section are not so easily comparable to the results established in [452]).

Theorem 5.1.6. *Suppose that* $F : \mathbb{R}^n \times X \to Y$ *is* $(\mathrm{R}, \mathcal{B}, W_{\mathcal{B},\mathrm{R}})$*-multialmost automorphic, respectively,* $(\mathrm{R}, \mathcal{B}, P_{\mathcal{B},\mathrm{R}})$*-multialmost automorphic. Define*

$$W'_{B;(b_k)}(x) := \{D' \subseteq \mathbb{R}^n : (\exists D \in W_{B;(b_k)}(x))(\forall \mathbf{t} \in D')(\forall \mathbf{u} \in \Omega)\ \mathbf{t} + \mathbf{u} \in D\},$$

provided $B \in \mathcal{B}$, $x \in B$ *and* $(b_k) \in \mathrm{R}$*, respectively,*

$$P'_{B;(b_k)} := \{D' \subseteq \mathbb{R}^n \times X : (\exists D \in P_{B;(b_k)})(\forall (\mathbf{t};x) \in D')(\forall \mathbf{u} \in \Omega)\ (\mathbf{t} + \mathbf{u};x) \in D\},$$

provided $B \in \mathcal{B}$ *and* $(b_k) \in \mathrm{R}$. *Then* $F(\cdot;\cdot)$ *is Stepanov* $(\Omega, \mathrm{R}, \mathcal{B}, Z^{\mathcal{P}}, W'_{\mathcal{B},\mathrm{R}})$*-multi-almost automorphic, respectively, Stepanov* $(\Omega, \mathrm{R}, \mathcal{B}, Z^{\mathcal{P}}, P'_{\mathcal{B},\mathrm{R}})$*-multialmost automorphic, whenever there exists a finite real number* $c_2 > 0$ *such that* $\|f\|_Z \leqslant c_2\|f\|_{L^\infty(\Omega:Y)}$ *for all essentially bounded functions* $f \in Z$.

Proof. We will consider $(\mathrm{R}, \mathcal{B}, W_{\mathcal{B},\mathrm{R}})$-multialmost automorphic functions, only. Let $\varepsilon > 0$, $(b_k) \in \mathrm{R}$, $B \in \mathcal{B}$ and $x \in B$ be given. Let $D' \in W'_{B;(b_k)}(x)$ and let $D \in W_{B;(b_k)}(x)$ be as in the above definition. Then there exists $l_0 \in \mathbb{N}$ such that, for every $\mathbf{t} \in D$, $\mathbf{u} \in \Omega$ and $l \geqslant l_0$, we have $\|F(\mathbf{t} + \mathbf{u} + b_{k_l};x) - F_B(\mathbf{t} + \mathbf{u};x)\|_Y \leqslant \varepsilon$, where $F_B : \mathbb{R}^n \times X \to Y$ denotes the corresponding limit function. Define $[F_B^*(\mathbf{t};x)](\mathbf{u}) := F_B(\mathbf{t}+\mathbf{u};x)$, $\mathbf{u} \in \Omega$, $\mathbf{t} \in \mathbb{R}^n$. Using the last estimate and the last given assumption, we get

$$\|F(\mathbf{t} + \cdot + b_{k_l};x) - F_B(\mathbf{t} + \cdot;x)\|_Z \leqslant c_2\|F(\mathbf{t} + \cdot + b_{k_l};x) - F_B(\mathbf{t} + \cdot;x)\|_{L^\infty(\Omega:Y)} \leqslant c_2\varepsilon,$$

for all $l \geqslant l_0$ and $\mathbf{t} \in D'$. This simply completes the proof. □

In particular, Theorem 5.1.6 holds for the choice $Z = L^{p(\cdot)}_{\nu}(\Omega : Y)$ with some $p \in \mathcal{P}(\Omega)$ and $\nu \in L^{p(\mathbf{u})}(\Omega)$; cf. also [447, Lemma 1.1.7(iii)]. Suppose further that $k \in \mathbb{N}$ and $F_i : \mathbb{R}^n \times X \to Y_i$ $(1 \leqslant i \leqslant k)$. We define $(F_1, \ldots, F_k) : \mathbb{R}^n \times X \to Y_1 \times \cdots \times Y_k$ by

$$(F_1, \ldots, F_k)(\mathbf{t}; x) := (F_1(\mathbf{t}; x), \ldots, F_k(\mathbf{t}; x)), \quad \mathbf{t} \in \mathbb{R}^n,\ x \in X.$$

If the pseudometric spaces (Z_i, d_{Z_i}) are given, where $Z_i \subseteq Y_i^{\Omega}$, then we define Z to be the set of all functions $F : \Omega \to Y_1 \times \cdots \times Y_k$, which have the form $F(\mathbf{u}) := (F_1(\mathbf{u}), \ldots, F_k(\mathbf{u}))$, $u \in \Omega$ for some functions $F_i \in Z_i$ $(1 \leqslant i \leqslant k)$. The pseudometric $d_Z(\cdot;\cdot)$ on Z is given by

$$d_Z(F, G) := \sum_{i=1}^{k} d_{Z_i}(F_i, G_i), \quad F_i,\ G_i \in Z_i \quad (1 \leqslant i \leqslant k).$$

We continue by stating the following analogue of [494, Proposition 2.2]; the proof is simple and, therefore, omitted.

Proposition 5.1.7. *Suppose that $k \in \mathbb{N}$, for every sequence (b_k), which belongs to R, we have that any its subsequence also belongs to R and the following condition holds:*

(S1)' *For each set $B \in \mathcal{B}$, for each sequence $(b_k) \in \mathrm{R}$ and for every subsequence $(b_k)'$ of (b_k) we have $W_{B;(b_k)}(x) \subseteq W_{B;(b_k)'}(x)$ for all $x \in B$ and $\mathrm{P}_{B;(b_k)} \subseteq \mathrm{P}_{B;(b_k)'}$.*

If $F_i : \mathbb{R}^n \times X \to Y_i$ is Stepanov $(\Omega, \mathrm{R}, \mathcal{B}, Z_i^{\mathcal{P}})$-multialmost automorphic [Stepanov $(\Omega, \mathrm{R}, \mathcal{B}, Z_i^{\mathcal{P}}, W_{\mathcal{B},\mathrm{R}})$-multialmost automorphic; Stepanov $(\Omega, \mathrm{R}, \mathcal{B}, Z_i^{\mathcal{P}}, \mathrm{P}_{\mathcal{B},\mathrm{R}})$-multi-almost automorphic] for $1 \leqslant i \leqslant k$, then $(F_1, \ldots, F_k)(\cdot;\cdot)$ is Stepanov $(\Omega, \mathrm{R}, \mathcal{B}, Z^{\mathcal{P}})$-multialmost automorphic [Stepanov $(\Omega, \mathrm{R}, \mathcal{B}, Z^{\mathcal{P}}, W_{\mathcal{B},\mathrm{R}})$-multialmost automorphic; Stepanov $(\Omega, \mathrm{R}, \mathcal{B}, Z^{\mathcal{P}}, \mathrm{P}_{\mathcal{B},\mathrm{R}})$-multialmost automorphic].

The statement of [494, Proposition 2.3(i)] cannot be so simply formulated in our new framework. Further on, the conclusions obtained in [494, Examples 2.4, 2.5] can be slightly strengthened using the notion introduced in Definition 5.1.3; for example, we have the following.

Example 5.1.8. Let $\varphi : \mathbb{R} \to \mathbb{C}$ be an almost periodic function, let $\Omega := [0,1]^2$ and let $(T(t))_{t \in \mathbb{R}} \subseteq L(X, Y)$ be an operator family, which is strongly locally integrable and not strongly continuous at zero. Suppose that there exist a finite real number $M \geqslant 1$ and a real number $\gamma \in (0, 1)$ such that

$$\|T(t)\|_{L(X,Y)} \leqslant \frac{M}{|t|^{\gamma}}, \quad t \in \mathbb{R} \setminus \{0\},$$

R is the collection of all sequences in $\Delta_2 \equiv \{(t,t) : t \in \mathbb{R}\}$ and $\mathcal{B}$ is the collection of all bounded subsets of X. Define $F : \mathbb{R}^2 \times X \to Y$ by

$$F(t, s; x) := e^{\int_s^t \varphi(\tau)\, d\tau} T(t - s)x, \quad (t, s) \in \mathbb{R}^2,\ x \in X.$$

Suppose further that for each bounded subset B of X and for each sequence $(\mathbf{b}_k = (b_k, b_k))$ in R the collection $P_{B;(\mathbf{b}_k)}$ is constituted of all sets of form $\{(t,s) \in \mathbb{R}^2 : |t-s| \leqslant L\} \times B$, where $L > 0$. In [494, Example 2.4], we have shown that the function $F(\cdot, \cdot; \cdot)$ is Stepanov $(\Omega, 1)$-$(\mathrm{R}, \mathcal{B}, \mathrm{P}_{\mathcal{B},\mathrm{R}})$-multialmost automorphic. Arguing in the same way, we can prove a slightly stronger result: the function $F(\cdot, \cdot; \cdot)$ is Stepanov $(\Omega, \mathrm{R}, \mathcal{B}, Z^{\mathcal{P}})$-multialmost automorphic, where $Z = L^1_\nu([0,1]^2 : Y)$ with $\nu(\mathbf{u}) = |\mathbf{u}|^\sigma$ for some $\sigma \in (\gamma - 1, 0)$.

The existence of a function $F_B : \mathbb{R}^n \times X \to Y$ such that $[F^*_B(\mathbf{t}; x)](\mathbf{u}) = F_B(\mathbf{t} + \mathbf{u}; x)$ for all $\mathbf{t} \in \mathbb{R}^n$, $x \in X$ and $\mathbf{u} \in \Omega$ can be proved in some specific cases (see Definition 5.1.3); for example, in the case that $\Omega = [0,1]^n$. The statement of [494, Proposition 2.9] can be simply reformulated in our new framework; details can be left to the interested readers. Further on, the statements of [494, Lemma 2.12, Theorem 2.13] can be straightforwardly extended to metrically Stepanov almost automorphic type functions by replacing the Banach space $L^{p(\mathbf{u})}(\Omega : Y)$ in their formulations with a general complete metric vector space $Z = (P_Z, d_Z)$. Arguing as in the proof of [184, Proposition 2.10], we can deduce the following supremum formula for Stepanov $(\Omega, \mathrm{R}, \mathcal{B}, Z^{\mathcal{P}})$-multialmost automorphic functions.

Proposition 5.1.9. *Let $F : \mathbb{R}^n \times X \to Y$ be Stepanov $(\Omega, \mathrm{R}, \mathcal{B}, Z^{\mathcal{P}})$-multi-almost automorphic, let $a > 0$ and let the assumption $f,\ g \in Z$ imply $f \pm g \in Z$ and $\|f \pm g\|_Z \leqslant \|f\|_Z + \|g\|_Z$. Suppose that there exists a sequence $b(\cdot)$ in R whose any subsequence is unbounded. Then we have $F(\mathbf{t} + \cdot) \in Z$ for all $\mathbf{t} \in \mathbb{R}^n$ and*

$$\sup_{\mathbf{t}\in\mathbb{R}^n; x\in X} \|F(\mathbf{t} + \cdot)\|_Z = \sup_{\mathbf{t}\in\mathbb{R}^n, |\mathbf{t}|\geqslant a; x\in X} \|F(\mathbf{t} + \cdot)\|_Z.$$

The results established in [184, Proposition 2.14] can be simply reformulated in our new framework as well. Now we will state and provide the main details of the proof of the following result.

Proposition 5.1.10. *Suppose that $F_m : \mathbb{R}^n \times X \to Y$ is Stepanov $(\Omega, \mathrm{R}, \mathcal{B}, Z^{\mathcal{P}})$-multialmost automorphic [Stepanov $(\Omega, \mathrm{R}, \mathcal{B}, Z^{\mathcal{P}}, W_{\mathcal{B},\mathrm{R}})$-multialmost automorphic; Stepanov $(\Omega, \mathrm{R}, \mathcal{B}, Z^{\mathcal{P}}, P_{\mathcal{B},\mathrm{R}})$-multialmost automorphic] for all $m \in \mathbb{N}$, $F : \mathbb{R}^n \times X \to Y$ and $\lim_{m\to+\infty} F_m(\mathbf{t} + \cdot; x) = F(\mathbf{t} + \cdot; x)$ for the topology of Z, uniformly on the set $\mathbb{R}^n \times B$ for each $B \in \mathcal{B}$. Suppose further that Z is a complete pseudometric space, the assumption $f,\ g \in Z$ implies $f \pm g \in Z$ and $\|f \pm g\|_Z \leqslant \|f\|_Z + \|g\|_Z$ and, for every sequence $b \in \mathrm{R}$, all of its subsequences also belong to R. Then $F(\cdot; \cdot)$ is likewise Stepanov $(\Omega, \mathrm{R}, \mathcal{B}, Z^{\mathcal{P}})$-multialmost automorphic [Stepanov $(\Omega, \mathrm{R}, \mathcal{B}, Z^{\mathcal{P}}, W_{\mathcal{B},\mathrm{R}})$-multialmost automorphic; Stepanov $(\Omega, \mathrm{R}, \mathcal{B}, Z^{\mathcal{P}}, P_{\mathcal{B},\mathrm{R}})$-multi-almost automorphic].*

Proof. Let $B \in \mathcal{B}$ be given. Using the prescribed assumptions and the well-known 3-ε argument, we can prove that the sequence $([F^*_{B,m}(\mathbf{t}; x)](\cdot))_{m\in\mathbb{N}}$ is a Cauchy sequence in Z and, therefore, convergent for each fixed pair $(\mathbf{t}; x) \in \mathbb{R}^n \times B$. If $[F^*_B(\mathbf{t}; x)](\cdot)$ is the corresponding limit function in Z, then $([F^*_{B,m}(\mathbf{t}; x)](\cdot))_{m\in\mathbb{N}}$ converges uniformly to $[F^*_B(\mathbf{t}; x)](\cdot)$ in Z on the set $\mathbb{R}^n \times B$, which follows from the estimate

$$\begin{aligned}&\big\|[F^*_{B,m}(\mathbf{t};x)](\cdot)-[F^*_{B,m'}(\mathbf{t};x)](\cdot)\big\|_Z\\&\quad\leqslant \limsup_{l\to+\infty}\big\|[F_m(\mathbf{t}+\cdot+b_{k_l};x)-F_{m'}(\mathbf{t}+\cdot+b_{k_l};x)]\big\|_Z\leqslant\varepsilon,\end{aligned}$$

for all $(\mathbf{t};x)\in\mathbb{R}^n\times B$ and $m,\ m'\geqslant m_0(\varepsilon)$ sufficiently large, by letting $m'\to+\infty$. The remainder of proof is standard and, therefore, omitted. □

The composition principle for $(\mathrm{R},\mathcal{B})$-multialmost automorphic functions has recently been established in [184, Theorem 2.20]. This result can be formulated for Stepanov $(\Omega,\mathrm{R},\mathcal{B},Z^{\mathcal{P}})$-multialmost automorphic [Stepanov $(\Omega,\mathrm{R},\mathcal{B},Z^{\mathcal{P}},W_{\mathcal{B},\mathrm{R}})$-multi-almost automorphic; Stepanov $(\Omega,\mathrm{R},\mathcal{B},Z^{\mathcal{P}},P_{\mathcal{B},\mathrm{R}})$-multi-almost automorphic] functions; for the sake of brevity, we will only formulate here the following composition principle for Stepanov $(\Omega,\mathrm{R},\mathcal{B},Z^{\mathcal{P}})$-multialmost automorphic functions (the proof is based on the argumentation contained in the proof of the above mentioned theorem and can be omitted).

Theorem 5.1.11. *Suppose that $L,\ L'>0$, $F:\mathbb{R}^n\times X\to Y$ is Stepanov $(\Omega,\mathrm{R},\mathcal{B},Z^{\mathcal{P}})$-multialmost automorphic and $G:\mathbb{R}^n\times Y\to E$ is Stepanov $(\Omega,\mathrm{R}',\mathcal{B}',M^{\mathcal{P}_G})$-multialmost automorphic, where $(E,\|\cdot\|_E)$ is a complex Banach space, R' is a collection of all sequences from R and all their subsequences, $\mathcal{B}':=\{B':B\in\mathcal{B}\}$ with*

$$\begin{aligned}B'&:=\{F(\mathbf{t};x):\mathbf{t}\in\mathbb{R}^n,\ x\in B\}\\&\quad\cup\{[F^*_B(\mathbf{t};x)](\mathbf{u}):\mathbf{t}\in\mathbb{R}^n,\ x\in B,\ \mathbf{u}\in\Omega,\ (b_k)\in\mathrm{R}\}\quad(B\in\mathcal{B}),\end{aligned}$$

$M\subseteq E^{\Omega}$ and (M,d_M^G) is a pseudometric space. Suppose that the assumption $f,\ g\in M$ implies $f\pm g\in M$ and $\|f\pm g\|_M\leqslant\|f\|_M+\|g\|_M$,

$$\|G(\mathbf{t};x)-G(\mathbf{t};y)\|_E\leqslant L\|x-y\|,\quad \mathbf{t}\in\mathbb{R}^n,\ x,\ y\in Y,$$

and the assumptions $f\in M$, $g\in P$ and $\|f(\mathbf{u})\|_E\leqslant L\|g(\mathbf{u})\|_Y$ for all $\mathbf{u}\in\Omega$ imply $\|f\|_M\leqslant L'\|g\|_P$. Then the Nemytskii operator $W:\mathbb{R}^n\times X\to E$, given by $W(\mathbf{t};x):=G(\mathbf{t};F(\mathbf{t};x))$, $\mathbf{t}\in\mathbb{R}^n$, $x\in X$, is Stepanov $(\Omega,\mathrm{R},\mathcal{B},M^{\mathcal{P}_G})$-multialmost automorphic.

5.1.2 Convolution invariance of metrical Stepanov almost automorphy and applications

Concerning the convolution invariance of metrical Stepanov almost automorphy introduced in Definition 5.1.3, we will state and prove the following result (cf. also Remark 5.1.5, [447, Proposition 8.1.17] and [494, Proposition 2.11]).

Theorem 5.1.12. *Suppose that $F:\mathbb{R}^n\times X\to Y$ is a Stepanov $(\Omega,\mathrm{R},\mathcal{B},Z^{\mathcal{P}})$-multialmost automorphic function, $h\in L^1(\mathbb{R}^n)$ and the following conditions hold:*

(i) *For every $B\in\mathcal{B}$, we have $\sup_{\mathbf{t}\in\mathbb{R}^n;x\in B}\|F(\mathbf{t};x)\|_Y<+\infty$.*

(ii) *The mapping* $\sigma \mapsto F(\mathbf{t}-\sigma;x)$, $\sigma \in \mathbb{R}^n$ *is Lebesgue measurable for all* $\mathbf{t} \in \mathbb{R}^n$ *and* $x \in X$.
(iii) *The convergence of a sequence* $(f_k(\cdot))$ *in* Z *implies the pointwise convergence of a sequence* $(f_k(\mathbf{u}))$ *in* Y *for all* $\mathbf{u} \in \Omega$.
(iv) *For every sequence* $(\mathbf{b}_k) \in \mathrm{R}$ *and for any subsequence* $(\mathbf{b}_{k_l})$ *of* $(\mathbf{b}_k)$, *there is a finite real number* $c_1 > 0$ *such that the mappings*

$$\int_{\mathbb{R}^n} h(\sigma) \cdot \{F(\mathbf{t} + \cdot + \mathbf{b}_{k_l} - \sigma; x) - [G(\mathbf{t} - \sigma; x)](\cdot)\}\, d\sigma$$

and

$$\int_{\mathbb{R}^n} h(\sigma) \cdot \{[G(\mathbf{t} - \mathbf{b}_{k_l} - \sigma; x)](\cdot) - F(\mathbf{t} + \cdot - \sigma; x)\}\, d\sigma$$

belongs to Z *for every function* $G : \mathbb{R}^n \times X \to Z$ *such that, for every* $B \in \mathcal{B}$, $\sup_{\mathbf{t}\in\mathbb{R}^n,\mathbf{u}\in\Omega;x\in B} \|[G(\mathbf{t};x)](\mathbf{u})\|_Y < +\infty$,

$$\begin{aligned}&\left\| \int_{\mathbb{R}^n} h(\sigma) \cdot \{F(\mathbf{t} + \cdot + \mathbf{b}_{k_l} - \sigma; x) - [G(\mathbf{t} - \sigma; x)](\cdot)\}\, d\sigma \right\|_Z \\ &\quad \leqslant c_1 \int_{\mathbb{R}^n} |h(\sigma)| \cdot \|F(\mathbf{t} + \cdot + \mathbf{b}_{k_l} - \sigma; x) - [G(\mathbf{t} - \sigma; x)](\cdot)\|_Z\, d\sigma\end{aligned}$$

and

$$\begin{aligned}&\left\| \int_{\mathbb{R}^n} h(\sigma) \cdot \{[G(\mathbf{t} - \mathbf{b}_{k_l} - \sigma; x)](\cdot) - F(\mathbf{t} + \cdot - \sigma; x)\}\, d\sigma \right\|_Z \\ &\quad \leqslant c_1 \int_{\mathbb{R}^n} |h(\sigma)| \cdot \|[G(\mathbf{t} - \mathbf{b}_{k_l} - \sigma; x)](\cdot) - F(\mathbf{t} + \cdot - \sigma; x)\|_Z\, d\sigma.\end{aligned}$$

(v) *There is a finite real number* $c_2 > 0$ *such that* $\|f\|_Z \leqslant c_2 \|f\|_{L^\infty(\Omega:Y)}$ *for all essentially bounded functions* $f \in Z$.

Then the function $(h * F)(\cdot;\cdot)$, *given by*

$$(h * F)(\mathbf{t};x) := \int_{\mathbb{R}^n} h(\sigma) F(\mathbf{t} - \sigma; x)\, d\sigma, \quad \mathbf{t} \in \mathbb{R}^n,\ x \in X,$$

is Stepanov $(\Omega, \mathrm{R}, \mathcal{B}, Z^{\mathcal{P}})$*-multialmost automorphic and, for every* $B \in \mathcal{B}$, *we have* $\sup_{\mathbf{t}\in\mathbb{R}^n;x\in B} \|(h * F)(\mathbf{t};x)\|_Y < +\infty$.

Proof. Let $B \in \mathcal{B}$ and $(\mathbf{b}_k) \in \mathrm{R}$ be given. Then we know that there exist a subsequence $(\mathbf{b}_{k_l})$ of $(\mathbf{b}_k)$ and a function $F_B^* : \mathbb{R}^n \times X \to Z$ such that, for every $\mathbf{t} \in \mathbb{R}^n$ and $x \in B$, we have (265)–(266), with the meaning clear. Due to (i) and (iii), we have

$$\sup_{\mathbf{t}\in\mathbb{R}^n;x\in B} \|[F_B^*(\mathbf{t};x)](\mathbf{u})\|_Y \leqslant \sup_{\mathbf{t}\in\mathbb{R}^n;x\in B} \|F(\mathbf{t};x)\|_Y < +\infty$$

for all $\mathbf{u} \in \Omega$. Due to (ii) and the above estimate, the mapping $(h * F)(\cdot;\cdot)$ is well-defined and satisfies $\sup_{\mathbf{t}\in\mathbb{R}^n;x\in B} \|(h * F)(\mathbf{t};x)\|_Y < +\infty$; the same arguments show that the function $(h * F)^*_B(\cdot;\cdot)$, given by

$$[(h * F)^*_B(\mathbf{t};x)](\mathbf{u}) := \int_{\mathbb{R}^n} h(\sigma)[F^*_B(\mathbf{t}-\sigma;x)](\mathbf{u})\,d\sigma, \quad \mathbf{t} \in \mathbb{R}^n,\ x \in B,$$

is well-defined. Therefore, it suffices to show that, for every fixed $\mathbf{t} \in \mathbb{R}^n$ and $x \in B$, we have

$$\lim_{l\to+\infty} \|(h * F)(\mathbf{t} + \cdot + (b^1_{k_l},\dots,b^n_{k_l});x) - [(h * F)^*_B(\mathbf{t};x)](\cdot)\|_Z = 0, \tag{267}$$

and

$$\lim_{l\to+\infty} \|[(h * F)^*_B(\mathbf{t} - (b^1_{k_l},\dots,b^n_{k_l});x)](\cdot) - (h * F)(\mathbf{t} + \cdot;x)\|_Z = 0.$$

The proofs of both limit equalities are based on the same arguments and will prove here (267), only. Due to (iv), we simply get

$$\begin{aligned}&\|(h * F)(\mathbf{t} + \cdot + (b^1_{k_l},\dots,b^n_{k_l});x) - [(h * F)^*_B(\mathbf{t};x)](\cdot)\|_Z\\ &\quad\leqslant \int_{\mathbb{R}^n} |h(\sigma)| \cdot \|F(\mathbf{t} + \cdot + \mathbf{b}_{k_l} - \sigma;x) - [F^*_B(\mathbf{t}-\sigma;x)](\cdot)\|_Z\,d\sigma.\end{aligned}$$

Applying the dominated convergence theorem and (v), we obtain the required limit equality. □

We feel it is our duty to say that the result established in Theorem 5.1.12 is unsatisfactory to a certain extent (e. g., in (iii), we have assumed that the pointwise convergence in Y holds for all $\mathbf{u} \in \Omega$, not only for a. e. $\mathbf{u} \in \Omega$). The most natural choice for possible applications is the use of Banach space $Z = C_{b,\nu}(\Omega : Y)$, where $0 < \nu(\mathbf{u}) \leqslant M$, $\mathbf{u} \in \Omega$ for some finite real constant $M > 0$; in particular, we obtain that the mapping $(h * F)(\cdot)$ is compactly almost automorphic provided that the mapping $F(\cdot)$ is (this result can be applied in the analysis of the existence and uniqueness of compactly almost automorphic solutions of the heat equation in $\mathbb{R}^n$; see also [447, pp. 543–545] and [184, Subsection 3.3]). It is also worth noting that we can apply Theorem 5.1.12 in the analysis of the existence and uniqueness of metrically Stepanov almost automorphic solutions of the semilinear Hammerstein integral equations of convolution type on $\mathbb{R}^n$; see, e. g., [184, p. 348].

The following result can be also formulated in the multidimensional setting (the proof is very similar to the proof of Theorem 5.1.12 and, therefore, omitted).

Theorem 5.1.13. *Suppose that* $f : \mathbb{R} \to X$ *is an essentially bounded, Stepanov* $(\Omega, \mathrm{R}, Z^{\mathcal{P}})$-*almost automorphic function,* $(R(t))_{t>0} \subseteq L(X,Y)$ *is a strongly continuous operator family such that* $\int_0^\infty \|R(t)\|\,dt < +\infty$ *and the following conditions hold:*

(i) *The convergence of a sequence $(f_k(\cdot))$ in Z implies the pointwise convergence of a sequence $(f_k(\mathbf{u}))$ in Y for all $\mathbf{u} \in \Omega$.*

(iv) *For every sequence $(b_k) \in \mathrm{R}$ and for any subsequence (b_{k_l}) of (b_k), there is a finite real number $c_1 > 0$ such that the mappings*

$$\int_0^{\infty} R(s) \cdot \{f(t + \cdot + b_{k_l} - s) - [g(t - s)](\cdot)\} \, ds$$

and

$$\int_0^{\infty} R(s) \cdot \{[g(t - b_{k_l} - s)](\cdot) - F(t + \cdot - s)\} \, ds$$

belongs to Z for every function $g : \mathbb{R} \to Z$ such that $\sup_{t \in \mathbb{R}, u \in \Omega} \|[g(t)](u)\|_Y < +\infty$, and we have

$$\left\| \int_0^{\infty} R(s) \cdot \{f(t + \cdot + b_{k_l} - s) - [g(t - s)](\cdot)\} \, ds \right\|_Z \leqslant c_1 \int_0^{\infty} \|R(s)\| \cdot \|f(t + \cdot + b_{k_l} - s) - [g(t - s)](\cdot)\|_Z \, ds$$

as well as

$$\left\| \int_0^{\infty} R(s) \cdot \{[g(t - b_{k_l} - s)](\cdot) - F(t + \cdot - s)\} \, ds \right\|_Z \leqslant c_1 \int_0^{\infty} \|R(s)\| \cdot \|[g(t - b_{k_l} - s)](\cdot) - F(t + \cdot - s)\|_Z \, ds.$$

(v) *There is a finite real number $c_2 > 0$ such that $\|f\|_Z \leqslant c_2 \|f\|_{L^{\infty}(\Omega:Y)}$ for all essentially bounded functions $f \in Z$.*

Then the function $F(\cdot)$, given by

$$F(t) := \int_{-\infty}^{t} R(t - s) f(s) \, ds, \quad t \in \mathbb{R},$$

is essentially bounded and Stepanov $(\Omega, \mathrm{R}, Z^{\mathcal{P}})$-multialmost automorphic.

It is well known that Theorem 5.1.13 can be successfully applied in the analysis of the existence and uniqueness of metrically Stepanov almost automorphic solutions for various classes of the abstract Volterra integrodifferential inclusions; cf. [444] for more details. Concerning the semilinear abstract Cauchy problems, we will only state here the following result in the one-dimensional setting (the proof follows by applying Proposition 5.1.10, Theorem 5.1.11, Theorem 5.1.13 and the Banach contraction principle, by plugging $X = Y = E$ and $Z = M = C_{b,\nu}(\Omega : Y)$).

Theorem 5.1.14. *Suppose that $(R(t))_{t>0} \subseteq L(X)$ is a strongly continuous operator family, $\int_0^{\infty} \|R(t)\|\,dt < +\infty$ and $0 < \nu(u) \leqslant M$, $u \in \Omega$ for some finite real constant $M > 0$. Suppose further, for every sequence $b \in \mathrm{R}$, all of its subsequences also belong to R, the functions $F : \mathbb{R} \times X \to X$ and $G : \mathbb{R} \times X \to X$ are Stepanov $(\Omega, \mathrm{R}, \mathcal{B}, Z^{\mathcal{P}})$-almost automorphic, where $X \in \mathcal{B}$, and*

$$\sup_{t\in\mathbb{R};x\in X} \big[\|F(t;x)\| + \|G(t;x)\|\big] < +\infty.$$

If there exist two finite real constants $L_F > 0$ and $L_G > 0$ such that $\|F(t;x) - F(t;y)\| \leqslant L_F\|x-y\|$ and $\|G(t;x)-G(t;y)\| \leqslant L_G\|x-y\|$ for all $t \in \mathbb{R}$ and $x, y \in X$, then the assumption $L_F + L_G \int_0^{\infty} \|R(t)\|\,dt < 1$ implies that the semilinear integral equation

$$u(t) = F(t;u(t)) + \int_{-\infty}^{t} R(t-s)G(s;u(s))\,ds, \quad t \in \mathbb{R}$$

has a unique bounded, Stepanov $(\Omega, \mathrm{R}, \mathcal{B}, Z^{\mathcal{P}})$-almost automorphic solution.

5.1.3 On some subspaces of compactly almost automorphic functions

The main aim of this subsection is to introduce and briefly analyze some subspaces of compactly almost automorphic functions.

1. In this part, we will introduce the class of Hölder-α-almost automorphic functions (cf. the pioneering papers [725–727] by S. Stoínski, [457] and [448, Section 7.1] for more details about Hölder-α-almost periodic functions).

Suppose that $\alpha \in (0,1]$. If $B \in \mathcal{B}$ and $\mathbf{u} \in \Omega \equiv [0,1]^n$, then we define

$$L_{\alpha,B}(\mathbf{u},\delta;F) := \sup_{x\in B}\ \sup_{\mathbf{u}_1,\mathbf{u}_2\in\Omega\cap B(\mathbf{u},\delta);\mathbf{u}_1\neq\mathbf{u}_2} \frac{\|F(\mathbf{u}_1;x) - F(\mathbf{u}_2;x)\|_Y}{|\mathbf{u}_1 - \mathbf{u}_2|^{\alpha}}.$$

We say that a point $\mathbf{u} \in \Omega$ is an α-regular point of function $F(\cdot;\cdot)$ if for every set $B \in \mathcal{B}$ there exists a finite real number $\delta > 0$ such that $L_{\alpha,B}(\mathbf{u},\delta;F) < +\infty$. Define after that $P^{1,\mathcal{B}}_{0,\alpha}(\Omega\times X : Y)$ as the set of all functions $F : \Omega\times X \to Y$ satisfying that each point $\mathbf{u} \in \Omega$ is an α-regular point of function $F(\cdot;\cdot)$. Assuming $F \in P^{1,\mathcal{B}}_{0,\alpha}(\Omega \times X : Y)$ and $B \in \mathcal{B}$, we define

$$L_{\alpha,B}(\mathbf{u};F) := \lim_{\delta\to 0+} L_{\alpha,B}(\mathbf{u},\delta;F), \quad \mathbf{u} \in \Omega.$$

By $P^{\mathcal{B}}_{0,\alpha}(\Omega \times X : Y)$, we denote the set of all functions $F \in P^{1,\mathcal{B}}_{0,\alpha}(\Omega \times X : Y)$ satisfying that for each set $B \in \mathcal{B}$ the function $\mathbf{u} \mapsto L_{\alpha,B}(\mathbf{u}, 1; F)$, $\mathbf{u} \in \Omega$ is locally bounded. We know that the space $Z = \mathcal{P}_\alpha = (P_\alpha, d_\alpha)$ is a metric space if $P_\alpha = C_b(\Omega : Y) \cap P_{0,\alpha}(\Omega : Y)$ and

$$d_\alpha(f,g) := \sup_{\mathbf{u}\in\Omega}\Big(\|f(\mathbf{u}) - g(\mathbf{u})\|_Y + \lim_{\delta\to 0+} L_\alpha(\mathbf{u},\delta; f-g)\Big). \tag{268}$$

Now we are ready to introduce the following notion.

Definition 5.1.15. We say that a function $F : \mathbb{R}^n \times X \to Y$ is Hölder-$(\mathcal{B},\alpha)$-almost automorphic [$\mathcal{B}$-Lipschitz almost automorphic, if $\alpha = 1$] if $F(\cdot;\cdot)$ is Stepanov $(\Omega, \mathrm{R}, \mathcal{B}, Z^{\mathcal{P}})$-multialmost automorphic with $\Omega = [0,1]^n$, R being the collection of all sequences in $\mathbb{R}^n$ and Z being defined as above. If we replace the set P_α with the set $P'_\alpha = P_{0,\alpha}(\Omega : Y)$ and the distance $d_\alpha(f,g)$ in (268) with the distance

$$d'_\alpha(f,g) \equiv \sup_{\mathbf{u}\in\Omega} \lim_{\delta\to 0+} L_\alpha(\mathbf{u},\delta; f-g),$$

then we say that $F(\cdot;\cdot)$ is weakly Hölder-$(\mathcal{B},\alpha)$-almost automorphic [weakly $\mathcal{B}$-Lipschitz almost automorphic, if $\alpha = 1$].

It is clear that the assumption $0 < \alpha' \leqslant \alpha'' \leqslant 1$ implies that any (weakly) Hölder-$(\mathcal{B},\alpha'')$-almost automorphic function $F : \mathbb{R}^n \times X \to Y$ is (weakly) Hölder-$(\mathcal{B},\alpha')$-almost automorphic. Any Hölder-α-almost automorphic function $F : \mathbb{R}^n \times X \to Y$ is compactly $(\mathrm{R},\mathcal{B})$-almost automorphic with R being the collection of all sequences in $\mathbb{R}^n$, while the converse statement seems not to be true. Without entering into all the details about Hölder-$(\mathcal{B},\alpha)$-almost automorphic functions, which will be given somewhere else, we would like to note that the argumentation contained in the proof of [448, Theorem 7.1.15] (cf. also [448, Corollary 7.1.17]) implies the following.

Theorem 5.1.16. *Suppose that $F : \mathbb{R}^n \to \mathbb{R}$ is continuously differentiable and all partial derivatives of first order of $F(\cdot)$ are compactly almost automorphic. Then $F(\cdot)$ is Lipschitz almost automorphic.*

For more details about one-dimensional $C^{(n)}$-almost automorphic functions and their applications, we refer the reader to the references quoted in the research article [602] by G. P. Mophou and G. M. N'Guérékata; we can similarly analyze multidimensional $C^{(n)}$-almost automorphic functions and provide certain applications (see, e. g., the third part of [448, Subsection 7.1.5]).

Example 5.1.17. Suppose that $1 \leqslant p < \infty$. Then we know that the function $f(\cdot)$, given by (72), is almost automorphic, Stepanov-p-almost periodic (S^p-almost periodic) but not compactly almost automorphic; see [444] for the notion and more details. In [448, Example 6.2.4], we have recently proved that the function $f(\cdot)$ is not Lipschitz S^p-almost periodic, i. e., that the Bochner transform $\hat{f} : \mathbb{R} \to L^p([0,1] : \mathbb{C})$ is not Lipschitz almost periodic. We would like to observe here that $f(\cdot)$ is not weakly Lipschitz S^p-almost automorphic, i. e., that the Bochner transform $\hat{f} : \mathbb{R} \to L^p([0,1] : \mathbb{C})$ is not weakly Lipschitz almost automorphic. In actual fact, the argumentation contained in the first part of the above mentioned example shows that the converse statement would imply that the function $t \mapsto g(t) \equiv (\zeta'(t)/\zeta(t)) \cos(1/\zeta(t))$, $t \in \mathbb{R}$ is Stepanov-p-almost automorphic, where we have put $\zeta(t) \equiv 2 + \cos t + \cos(\sqrt{2}t)$. But this cannot be true since the function $g(\cdot)$ is not Stepanov-p-bounded, as shown in the second part of the same example. We conclude this example by noticing that for each number $\alpha \in (0,1)$ the Bochner transform $\hat{f} : \mathbb{R} \to L^p([0,1] : \mathbb{C})$ is both Hölder-α-almost periodic and Hölder-α-almost automorphic, which follows from the analyses carried out in [448, Remark 7.1.13, Example 7.1.14].

In connection with the last conclusion, one may ask whether, for a given exponent $\alpha \in (0,1]$, any Hölder-α-almost periodic function $F : \mathbb{R}^n \to Y$ is Hölder-α-almost automorphic. The answer is affirmative for continuously differentiable Hölder-α-almost periodic functions $F : \mathbb{R}^n \to Y$. This is clear provided that $\alpha < 1$ (see [448, Remark 7.1.13]); if $\alpha = 1$, then the statement follows from the facts that the mapping $\tau \mapsto F(\cdot + \tau) \in P_1$, $\tau \in \mathbb{R}^n$ is continuous (see, e. g., the proof of [448, Theorem 7.1.28]) and the set of functions $\{F(\cdot + \tau) : \tau \in \mathbb{R}^n\}$ is sequentially compact in P_1 (equivalently, totally bounded in P_1), which can be shown following the lines of the proof of the Bochner criterion given on [805, pp. 28–29], with the norm $\| \cdot \|_\infty$ replaced therein by the norm $\| \cdot \|_{P_1}$.

2. The best possible way to introduce the class of one-dimensional compactly almost automorphic functions in variation is to use the following metric space $Z = (P_Z, d_Z)$, where P_Z is the space of all essentially bounded functions on $\Omega \equiv [0,1]$ such that $V_0^1(f) \equiv \sup_\Pi \sum_{j=0}^{k-1} \|f(x_{k+1}) - f(x_k)\|$ is finite, and

$$d_Z(f,g) := \operatorname{ess\,sup}_{u\in\Omega} \|f(u) - g(u)\|_Y + \sup_\Pi \sum_{j=0}^{k-1} \|[f(x_{k+1}) - g(x_{k+1})] - [f(x_k) - g(x_k)]\|, \quad f,\ g \in P_Z, \tag{269}$$

where the summations are taken over all partition $\Pi = \{0 = x_0 < \cdots < x_k = 1\}$ of the interval $[0,1]$.

Definition 5.1.18. We say that a continuous function $F : \mathbb{R} \times X \to Y$ is compactly $(\mathrm{R}, \mathcal{B})$-almost automorphic in variation if $F(\cdot;\cdot)$ is Stepanov $(\Omega, \mathrm{R}, \mathcal{B}, Z^{\mathcal{P}})$-multialmost automorphic with $\Omega = [0,1]$ and Z being defined as above. Furthermore, we say that $F(\cdot;\cdot)$ is compactly $\mathcal{B}$-almost automorphic in variation if $F(\cdot;\cdot)$ is compactly $(\mathrm{R}, \mathcal{B})$-almost automorphic in variation with R being the collection of all sequences in $\mathbb{R}$.

Using the argumentation from [448, Remark 7.1.5], we can simply deduce the following result.

Proposition 5.1.19. *Suppose that $F : \mathbb{R}\times X \to Y$ is $\mathcal{B}$-Lipschitz almost automorphic. Then $F(\cdot;\cdot)$ is compactly $\mathcal{B}$-almost automorphic in variation.*

The situation is much more complicated in the multidimensional setting, when we can follow several different ways for the introduction of multivariate functions bounded in variation. In [448, Section 7.3, pp. 417–419], we have presented the main details about the class of multidimensional functions bounded in variation and multidimensional almost periodic functions in variation. The traditional approaches of G. Vitali (1908), G. H. Hardy (1905) and C. Arzelá (1905) can be straightforwardly generalized to the vector-valued setting, so that Definition 5.1.18 can be extended to the multidimensional setting using the metric space $Z = (P_Z, d_Z)$, where P_Z is the space of all essentially bounded functions on $\Omega \equiv [0,1]^n$ such that its Vitali, Hardy or Arzelá variation on $[0,1]^n$, denoted by a common symbol $V(f; [0,1]^n)$ here, is finite and

$$d_Z(f,g) := \operatorname*{ess\,sup}_{\mathbf{u}\in\Omega} \| f(\mathbf{u}) - g(\mathbf{u})\|_Y + V(f(\cdot) - g(\cdot); [0,1]^n), \quad f,\ g \in P_Z.$$

The modern definition of a multivariate function bounded in variation is based on the distributional techniques; for simplicity, we will skip all details concerning multidimensional compactly almost automorphic functions in variation here.

In [482, Theorem 2.5, Corollary 2.6], we have recently analyzed the extensions of multidimensional almost automorphic sequences $F : \mathbb{Z}^n \to Y$. It is logical to ask whether the compactly almost automorphic function $\tilde{F} : \mathbb{R}^n \to Y$, constructed in the proof of Theorem 2.5 mentioned above, belongs to some specific subspaces of compactly almost automorphic functions. Concerning this issue, we would like to introduce the following function spaces (for simplicity, we will consider here the one-dimensional setting).

3. Suppose that $F : \mathbb{Z} \to Y$ is an almost automorphic sequence. If $t \in [m, m+1)$ for some $m \in \mathbb{Z}$, then we define $\tilde{F}(t) := F(m) + (t-m)[F(m+1) - F(m)]$. We know that $\tilde{F} : \mathbb{R} \to Y$ is a compactly almost automorphic function and $\tilde{F}(k) = F(k)$ for all $k \in \mathbb{Z}$; cf. [305] and [482]. Suppose further that $\Omega = [0,1]$ and (Z_0, d_{Z_0}) is a metric vector space such that any function from Z_0 is essentially bounded, $\|f \pm g\|_{Z_0} \leqslant \|f\|_{Z_0} + \|g\|_{Z_0}$ for all $f,\ g \in Z_0$ as well as the condition (Q1) holds. Suppose further that (b_l') is any sequence of integer numbers. Then there exist a subsequence (b_l) of (b_l') and a function $G : \mathbb{Z} \to Y$ such that $\lim_{l\to+\infty} F(k+b_l) = G(k)$ and $\lim_{l\to+\infty} G(k-b_l) = F(k)$ for all $k \in \mathbb{Z}$. If $t \in [m, m+1)$ for some $m \in \mathbb{Z}$, then we define $\tilde{G}(t) := G(m) + (t-m)[G(m+1) - G(m)]$. Fix now a number $t \in \mathbb{R}$; then we know that $\lim_{l\to+\infty} \tilde{F}(t+b_l) = \tilde{G}(t)$ and $\lim_{l\to+\infty} \tilde{G}(t-b_l) = \tilde{F}(t)$ [482]. Furthermore, there exist a unique number $u_1 \in [0,1]$ and a unique integer $m \in \mathbb{Z}$ such that $t+u \in [m, m+1)$ for $u \in [0, u_1)$ and $t+u \in [m+1, m+2)$ for $u \in [u_1, 1]$. Then $t - m \in [-1, 1)$,

$$\begin{aligned}&\tilde{F}(t+u+b_l) - \tilde{G}(t+u)\\&\quad= [F(m+b_l) - G(m)] + (t+u-m)\{[F(m+1+b_l) - G(m+1)] - [F(m+b_l) - G(m)]\},\end{aligned}$$

for any $u \in [0, u_1)$, and

$$\tilde{F}(t+u+b_l) - \tilde{G}(t+u)$$

$$= [F(m+1+b_l) - G(m+1)] + (t+u-m-1)\{[F(m+2+b_l) - G(m+2)] - [F(m+1+b_l) - G(m+1)]\},$$

for any $u \in [u_1, 1]$. Keeping in mind condition (Q1) and the above two equalities, it simply follows that the subsequent result holds true.

Proposition 5.1.20. *Suppose that* $\Omega = [0,1]$, *condition* (Q1) *holds,* (Z_0, d_{Z_0}) *is a metric vector space such that any function from* Z_0 *is essentially bounded and* $\|f \pm g\|_{Z_0} \leqslant \|f\|_{Z_0} + \|g\|_{Z_0}$ *for all* $f, g \in Z_0$. *Then the function* $\tilde{F}(\cdot)$ *is* $(\Omega, \mathrm{R}, Z^{\mathcal{P}}, W_{\mathrm{R}})$*-almost automorphic, where* R *denotes the collection of all sequences in* $\mathbb{Z}$, $W_{(b_l')}$ *denotes the collection of all compact subsets of* $\mathbb{R}$ *for any sequence* (b_l') *in* R *and* (Z, d_Z) *is the metric vector space equipped with the distance* $d_Z(f,g) := \|f-g\|_\infty + \|f-g\|_{Z_0}$, $f, g \in Z$.

In particular, we can use the metric $d_Z(\cdot;\cdot)$ given by (269). Suppose now that (a_l') is any sequence of real numbers, $b_l' = \lfloor a_l' \rfloor$ for all $l \in \mathbb{N}$, (b_l) is given as above and $\lim_{l\to+\infty}(a_l - b_l) = c_1 \in [0,1]$. By the proof of [482, Theorem 2.5], we have $\lim_{l\to+\infty} \tilde{F}(t + a_l) = \tilde{G}(t+c_1)$ and $\lim_{l\to+\infty} \tilde{G}(t + c_1 - a_l) = \tilde{F}(t)$, uniformly on compact subsets of $\mathbb{R}$ containing the point $t \in \mathbb{R}$. Assume now that (Q1) and (Q2) hold. Since the function $\tilde{F}(\cdot)$ is Lipschitz continuous, as easily approved, and

$$\|\tilde{F}(t+a_l+\cdot) - \tilde{G}(t+c_1+\cdot)\|_{Z_0} \leqslant \|\tilde{F}(t+a_l+\cdot) - \tilde{F}(t+b_l+c_1+\cdot)\|_{Z_0} + \|\tilde{F}(t+b_l+c_1+\cdot) - \tilde{G}(t+c_1+\cdot)\|_{Z_0},$$

it readily follows that the subsequent result holds true.

Theorem 5.1.21. *Suppose that the requirements of* Proposition 5.1.20 *and condition* (Q2) *hold. Then the function* $\tilde{F}(\cdot)$ *is* $(\Omega, \mathrm{R}, Z^{\mathcal{P}}, W_{\mathrm{R}})$*-almost automorphic, where* R *denotes the collection of all sequences in* $\mathbb{R}$ *as well as* $W_{(b_l')}$ *and* (Z, d_Z) *have the same meaning as above.*

In particular, we can use the metric $d_{Z_0}(\cdot;\cdot)$ of the Banach space of k-times piecewise continuously differentiable functions on the interval $[0,1]$, equipped with the usual metric, which is not equivalent with the metric induced by the sup-norm for $k \geqslant 1$; see, e. g., [266, Definition 4]. Condition (Q2) does not hold for the metric spaces of functions bounded in variation and it is logical to ask whether the function $\tilde{F}(\cdot)$ has to be compactly almost automorphic in variation. In particular, this can be asked for the extension of the almost automorphic sequence $(F(k) \equiv \mathrm{sign}(\cos(2\pi k a)))_{k\in\mathbb{Z}}$, where a is an irrational number [122].

We close this section with some concluding remarks and observations. First of all, we would like to note that it is very difficult to introduce the notion of a metrically Stepanov almost automorphic sequence $F : \mathbb{Z}^n \times X \to Y$ in a proper way. Of course, this can be done in the same way as for functions, with the pointwise convergence for all $\mathbf{t} \in \mathbb{Z}^n$ and the corresponding limit function $F_B^* : \mathbb{Z}^n \times X \to Z$; but, in any reasonable situation, the introduced notion is equivalent with the usual notion of $(\mathrm{R}, \mathcal{B})$-multialmost automorphy of the corresponding sequence.

The following topics have not been investigated in the previous part of the section:

1. We have not considered here the $\mathbb{D}$-asymptotically Stepanov multialmost automorphic type functions in general metric and their applications. See also [184, Subsection 2.3] and [494, p. 19] for more details about the subject.

2. We have recently considered, in [458], the notion of a metrical Stepanov ρ-almost periodicity in a Bohr like manner. Suppose now that X and Y are complex Banach spaces as well as that $\emptyset \neq Z \subseteq Y^{\Omega}$, where $\emptyset \neq \Omega \subseteq \mathbb{R}^n$ is a fixed compact set with positive Lebesgue measure [$\emptyset \neq \Omega \subseteq \mathbb{Z}^n$ is a fixed bounded set]. Let $P_Z \subseteq Z^{\Lambda}$, $0 \in P_Z$ and let (P_Z, d_{P_Z}) be a pseudometric space. Set $\|f\|_{P_Z} := d_{P_Z}(f, 0), f \in P_Z$. The following notion can be also considered.

Definition 5.1.22. Suppose that $\emptyset \neq \Lambda \subseteq \mathbb{R}^n$, $F : \Lambda \times X \to Y$ is a given function and the assumptions $\mathbf{t} \in \Lambda$, $\mathbf{b} \in \mathrm{R}$ and $l \in \mathbb{N}$ imply $\mathbf{t} + \mathbf{b}(l) \in \Lambda$. Then we say that the function $F(\cdot;\cdot)$ is Stepanov $(\mathrm{R}, \mathcal{B}, \mathcal{P}_Z, L)$-multialmost periodic, respectively, strongly Stepanov $(\Omega, \mathrm{R}, \mathcal{B}, \mathcal{P}_Z, L)$-multialmost periodic in the case that $\Lambda = \mathbb{R}^n$ [$\Lambda = \mathbb{Z}^n$] if for every $B \in \mathcal{B}$ and for every sequence $(\mathbf{b}_k = (b_k^1, b_k^2, \ldots, b_k^n)) \in \mathrm{R}$ there exist a subsequence $(\mathbf{b}_{k_l} = (b_{k_l}^1, b_{k_l}^2, \ldots, b_{k_l}^n))$ of $(\mathbf{b}_k)$ and a function $F^* : \Lambda \times X \to Z$ such that, for every $l \in \mathbb{N}$ and $x \in B$, we have $\hat{F}(\cdot + (b_{k_l}^1, \ldots, b_{k_l}^n); x) - F^*(\cdot; x) \in P_Z$ and, for every $B' \in L(B; \mathbf{b})$,

$$\lim_{l \to +\infty} \sup_{x \in B'} \left\| \hat{F}(\cdot + (b_{k_l}^1, \ldots, b_{k_l}^n); x) - F^*(\cdot; x) \right\|_{P_Z} = 0, \tag{270}$$

respectively, $\hat{F}(\cdot + (b_{k_l}^1, \ldots, b_{k_l}^n); x) - F^*(\cdot; x) \in P_Z$, $F^*(\cdot - (b_{k_l}^1, \ldots, b_{k_l}^n); x) - \hat{F}(\cdot; x) \in P_Z$, (270) holds and, for every $B' \in L(B; \mathbf{b})$,

$$\lim_{l \to +\infty} \sup_{x \in B'} \left\| F^*(\cdot - (b_{k_l}^1, \ldots, b_{k_l}^n); x) - \hat{F}(\cdot; x) \right\|_{P_Z} = 0.$$

The usual notion, obtained by plugging $Z = L^{p(\mathbf{u})}(\Omega : Y)$ and $P_Z = C_b(\Lambda : Z)$ can be further generalized by plugging $Z = L_{\nu}^{p(\mathbf{u})}(\Omega : Y)$, where $p \in \mathcal{P}(\Omega)$ and $\nu : \Omega \to (0, \infty)$ is a Lebesgue measurable function, and $P_Z = C_{b,\zeta}(\Lambda : Z)$, where $\zeta : \Lambda \to [0, \infty)$ is any nonzero function. If $\Lambda = \mathbb{R}^n$ and $P_Z = C_{b,\zeta}(\mathbb{R}^n : Z)$, where $\zeta : \mathbb{R}^n \to (0, \infty)$, then it can be simply proved that any strongly Stepanov $(\Omega, \mathrm{R}, \mathcal{B}, \mathcal{P}_Z, L)$-multialmost periodic function has to be Stepanov $(\Omega, \mathrm{R}, \mathcal{B}, Z^{\mathcal{P}})$-multialmost automorphic. We close this part with the observation that the notion introduced in Definition 5.1.22 is not so easily comparable to the notion introduced in [458, Definitions 2.1–2.3].

3. It could be interesting to reconsider a Bochner type result [494, Theorem 2.16] for certain classes of metrically Stepanov almost automorphic functions, as well.

5.2 Metrical Weyl almost automorphy and applications

The main aim of this section is to investigate several new classes of metrically Weyl almost automorphic functions of the form $F : \mathbb{R}^n \times X \to Y$ and metrically Weyl almost automorphic sequences of the form $F : \mathbb{Z}^n \times X \to Y$, where X and Y are complex Banach spaces [10]. We provide many illustrative examples, useful remarks and applications to the abstract Volterra integrodifferential equations and the abstract Volterra difference equations.

The organization of this section, which is written in a semiheuristical manner, can be briefly described as follows. Various classes of metrically Weyl almost automorphic type functions and metrically Weyl almost automorphic type sequences are introduced and thoroughly analyzed in Section 5.2.1. The main purpose of this subsection is to clarify the metrical generalizations of the structural results presented in our recent research article [461] (Proposition 5.2.5 and Proposition 5.2.7 are new; see also Example 5.2.8 for the slight improvements of the conclusions established in [461] as well as Remark 5.2.6 and Remark 5.2.9, where we provide several observations about the notion under our consideration).

Section 5.2.2, where we provide many supporting examples, investigates the extensions of metrically Weyl almost automorphic sequences. The main structural results established in this subsection are Theorem 5.2.11 and Theorem 5.2.12. The main purpose of Section 5.2.3 is to present some applications of the obtained theoretical results to the abstract Volterra integrodifferential equations and the abstract Volterra difference equations. In this subsection, we first examine the convolution invariance of joint Weyl almost automorphy and Weyl almost automorphy of type 2; see Proposition 5.2.13 and Theorem 5.2.14 for more details. The second part of this subsection pertains to the study of applications to the abstract (fractional) difference equations. We also consider vectorially Weyl almost automorphic type functions here.

5.2.1 Metrically Weyl almost automorphic functions and metrically Weyl almost automorphic sequences

The main aim of this subsection is to introduce and analyze various classes of metrically Weyl almost automorphic functions and metrically Weyl almost automorphic sequences as well as to slightly generalize the notion introduced recently in [447, Section 8.3] (we will continue citing [447] since the paper [461] is still not published in the final form). Unless stated otherwise, we will always assume here that $\Omega := [-1,1]^n \subseteq \mathbb{R}^n$ [$\Omega := [-1,1]^n \cap \mathbb{Z}^n \subseteq \mathbb{Z}^n$], $\mathbb{F} : (0,\infty) \times \mathbb{R}^n \to (0,\infty)$ [$\mathbb{F} : (0,\infty) \times \mathbb{Z}^n \to (0,\infty)$] as well as that, for every $l > 0$, (P_l, d_l) is a pseudometric space, where $P_l \subseteq Y^{l\Omega}$ [$P_l \subseteq Y^{l\Omega \cap \mathbb{Z}^n}$] is closed under the addition and subtraction of functions, containing the zero-function: $0 \in P_l$. Set $\|f\|_l := d_l(f,0)$ for all $f \in P_l$ ($l > 0$). We will always assume henceforth that R is a collection of sequences in $\mathbb{R}^n$ [$\mathbb{Z}^n$]; for simplicity and better understanding, we will not consider here the corresponding classes of functions with the collections R_X of sequences in $\mathbb{R}^n \times X$ [$\mathbb{Z}^n \times X$].

We will first introduce the following notion, which generalizes the corresponding notion from [447, Definitions 8.3.17, 8.3.18, 8.3.28].

Definition 5.2.1. Suppose that $F : \mathbb{R}^n \times X \to Y$ [$F : \mathbb{Z}^n \times X \to Y$] satisfies that for each $x \in X$, $l > 0$ and $\mathbf{t} \in \mathbb{R}^n$ [$\mathbf{t} \in \mathbb{Z}^n$] we have $F(\mathbf{t} + \cdot; x) \in P_l$. Let for every $l > 0$, $B \in \mathcal{B}$ and $(\mathbf{b}_k = (b_k^1, b_k^2, \ldots, b_k^n)) \in \mathrm{R}$ there exist a subsequence $(\mathbf{b}_{k_m} = (b_{k_m}^1, b_{k_m}^2, \ldots, b_{k_m}^n))$ of $(\mathbf{b}_k)$ and a function $F^* : \mathbb{R}^n \times X \to P_l$ [$F^* : \mathbb{Z}^n \times X \to P_l$] such that for each $x \in B$, $l > 0$ and $\mathbf{t} \in \mathbb{R}^n$ [$\mathbf{t} \in \mathbb{Z}^n$] we have

(i)

$$\lim_{m \to +\infty} \lim_{l \to +\infty} \mathbb{F}(l, \mathbf{t}) \| F(\mathbf{t} + \cdot + (b_{k_m}^1, \ldots, b_{k_m}^n); x) - [F^*(\mathbf{t}; x)](\cdot) \|_l = 0 \tag{271}$$

and

$$\lim_{m \to +\infty} \lim_{l \to +\infty} \mathbb{F}(l, \mathbf{t}) \| [F^*(\mathbf{t} - (b_{k_m}^1, \ldots, b_{k_m}^n); x)](\cdot) - F(\mathbf{t} + \cdot; x) \|_l = 0, \tag{272}$$

pointwise for all $x \in B$ and $\mathbf{t} \in \mathbb{R}^n$ [$\mathbf{t} \in \mathbb{Z}^n$], or

(ii)

$$\lim_{l \to +\infty} \lim_{m \to +\infty} \mathbb{F}(l, \mathbf{t}) \| F(\mathbf{t} + \cdot + (b_{k_m}^1, \ldots, b_{k_m}^n); x) - [F^*(\mathbf{t}; x)](\cdot) \|_l = 0 \tag{273}$$

and

$$\lim_{l \to +\infty} \lim_{m \to +\infty} \mathbb{F}(l, \mathbf{t}) \| [F^*(\mathbf{t} - (b_{k_m}^1, \ldots, b_{k_m}^n); x)](\cdot) - F(\mathbf{t} + \cdot; x) \|_l = 0, \tag{274}$$

pointwise for all $x \in B$ and $\mathbf{t} \in \mathbb{R}^n$ [$\mathbf{t} \in \mathbb{Z}^n$], or

(iii)

$$\lim_{(l,m) \to +\infty} \mathbb{F}(l, \mathbf{t}) \| F(\mathbf{t} + \cdot + (b_{k_m}^1, \ldots, b_{k_m}^n); x) - [F^*(\mathbf{t}; x)](\cdot) \|_l = 0$$

and

$$\lim_{(l,m) \to +\infty} \mathbb{F}(l, \mathbf{t}) \| [F^*(\mathbf{t} - (b_{k_m}^1, \ldots, b_{k_m}^n); x)](\cdot) - F(\mathbf{t} + \cdot; x) \|_l = 0,$$

pointwise for all $x \in B$ and $\mathbf{t} \in \mathbb{R}^n$ [$\mathbf{t} \in \mathbb{Z}^n$].

In the case that (i), respectively, [(ii); (iii)], holds, then we say that $F(\cdot; \cdot)$ is Weyl-$(\mathbb{F}, \mathcal{P}, \mathrm{R}, \mathcal{B})$-multialmost automorphic, respectively, [Weyl-$(\mathbb{F}, \mathcal{P}, \mathrm{R}, \mathcal{B})$-multialmost automorphic of type 1; jointly Weyl-$(\mathbb{F}, \mathcal{P}, \mathrm{R}, \mathcal{B})$-multialmost automorphic].

Definition 5.2.2. Suppose that $\emptyset \neq W \subseteq \mathbb{R}^n$ $[\emptyset \neq W \subseteq \mathbb{Z}^n]$, $\mathbb{F} : (0,\infty) \times \mathbb{R}^n \to (0,\infty)$ $[\mathbb{F} : (0,\infty) \times \mathbb{Z}^n \to (0,\infty)]$ and $F : \mathbb{R}^n \times X \to Y$ $[F : \mathbb{Z}^n \times X \to Y]$ satisfies that for each $x \in X$, $l > 0$ and $\mathbf{t} \in \mathbb{R}^n$ $[\mathbf{t} \in \mathbb{Z}^n]$ we have $F(\mathbf{t} + \cdot; x) \in P_l$. If for every $B \in \mathcal{B}$ and $(\mathbf{b}_k = (b_k^1, b_k^2, \ldots, b_k^n)) \in \mathrm{R}$, there exists a subsequence $(\mathbf{b}_{k_m} = (b_{k_m}^1, b_{k_m}^2, \ldots, b_{k_m}^n))$ of $(\mathbf{b}_k)$ such that for each $\varepsilon > 0$, $x \in B$ and $\mathbf{t} \in \mathbb{R}^n$ $[\mathbf{t} \in \mathbb{Z}^n]$ there exists $m_0 \in \mathbb{N}$ such that, for every m, $m' \in \mathbb{N}$ with $m \geqslant m_0$ and $m' \geqslant m_0$, there exists $l_0 > 0$ such that, for every $l \geqslant l_0$ and $w \in lW$, we have

$$\left\|F(\mathbf{t} + \cdot + (b_{k_m}^1, \ldots, b_{k_m}^n) - w; x) - F(\mathbf{t} + \cdot + (b_{k_{m'}}^1, \ldots, b_{k_{m'}}^n) - w; x)\right\|_l < \varepsilon / \mathbb{F}(l, \mathbf{t} - w),$$

then we say that $F(\cdot;\cdot)$ is Weyl-$(\mathbb{F}, \mathcal{P}, \mathrm{R}, \mathcal{B}, W)$-multialmost automorphic of type 2.

Definition 5.2.3. Suppose that $\emptyset \neq W \subseteq \mathbb{R}^n$ $[\emptyset \neq W \subseteq \mathbb{Z}^n]$, $\mathbb{F} : (0,\infty) \times \mathbb{R}^n \to (0,\infty)$ $[\mathbb{F} : (0,\infty) \times \mathbb{Z}^n \to (0,\infty)]$ and $F : \mathbb{R}^n \times X \to Y$ $[F : \mathbb{Z}^n \times X \to Y]$ satisfies that for each $x \in X$, $l > 0$ and $\mathbf{t} \in \mathbb{R}^n$ $[\mathbf{t} \in \mathbb{Z}^n]$ we have $F(\mathbf{t} + \cdot; x) \in P_l$. Let for every $l > 0$, $B \in \mathcal{B}$ and $(\mathbf{b}_k = (b_k^1, b_k^2, \ldots, b_k^n)) \in \mathrm{R}$ there exist a subsequence $(\mathbf{b}_{k_m} = (b_{k_m}^1, b_{k_m}^2, \ldots, b_{k_m}^n))$ of $(\mathbf{b}_k)$ and a function $F^* : \mathbb{R}^n \times X \to P_l$ $[F^* : \mathbb{Z}^n \times X \to P_l]$ such that for each $\varepsilon > 0$, $x \in B$ and $\mathbf{t} \in \mathbb{R}^n$ $[\mathbf{t} \in \mathbb{Z}^n]$, there exists $p > 0$ such that, for every $l \in [p, +\infty)$, $m \in \mathbb{N}$ with $m \geqslant p$ and $w \in lW$, we have

$$\mathbb{F}(l, \mathbf{t} - w)\left\|F(\mathbf{t} + \cdot + (b_{k_m}^1, \ldots, b_{k_m}^n) - w; x) - [F^*(\mathbf{t} - w; x)](\cdot)\right\|_l < \varepsilon$$

and

$$\mathbb{F}(l, \mathbf{t} - w)\left\|[F^*(\mathbf{t} - (b_{k_m}^1, \ldots, b_{k_m}^n) - w; x)](\cdot) - F(\mathbf{t} + \cdot - w; x)\right\|_l < \varepsilon,$$

then we say that $F(\cdot;\cdot)$ is jointly Weyl-$(\mathbb{F}, \mathcal{P}, \mathrm{R}, \mathcal{B}, W)$-multialmost automorphic.

In the recent research paper [464], we have considered the Besicovitch multidimensional almost automorphic type functions and their applications. This is the first paper in the existing literature dealing with the notion of generalized almost automorphy, which additionally involves the growth order of limit function $F^*(\cdot;\cdot)$; cf. [464, Definition 3.1]. We will consider henceforth the notion in which the limit function $F^*(\cdot;\cdot)$ from Definition 5.2.1, respectively, Definition 5.2.3, is bounded by the function $\omega : \mathbb{R}^n \to (0,\infty)$ $[\omega : \mathbb{Z}^n \to (0,\infty)]$ in the sense that there exists $M > 0$ such that, for every $x \in B$, $l > 0$ and $\mathbf{u} \in l\Omega$, we have $\|[F^*(\mathbf{t}; x)](\mathbf{u})\|_Y \leqslant M\omega(|\mathbf{t}|)$, $\mathbf{t} \in \mathbb{R}^n$ $[\|[F^*(\mathbf{t}; x)](\mathbf{u})\|_Y \leqslant M\omega(|\mathbf{t}|)$, $\mathbf{t} \in \mathbb{Z}^n]$. If this is the case, then we say that the function $F(\cdot;\cdot)$ is Weyl-$(\mathbb{F}, \mathcal{P}, \mathrm{R}, \mathcal{B}, \omega)$-multialmost automorphic [Weyl-$(\mathbb{F}, \mathcal{P}, \mathrm{R}, \mathcal{B}, \omega)$-multialmost automorphic of type 1; jointly Weyl-$(\mathbb{F}, \mathcal{P}, \mathrm{R}, \mathcal{B}, \omega)$-multi-almost automorphic], respectively, jointly Weyl $(\mathbb{F}, \mathcal{P}, \mathrm{R}, \mathcal{B}, W, \omega)$-multialmost automorphic; furthermore, if $\omega(\cdot) \equiv 1$, then we write "b" in place of "ω."

If $1 \leqslant p < \infty$, then the notion of a (jointly) Weyl-p-almost automorphic function (of type 1) [(jointly) Weyl-(p, b)-multialmost automorphic (of type 1)] $F : \mathbb{R}^n \to Y$ is obtained with $\Omega = [-1, 1]^n$, R being the collection of all sequences in $\mathbb{R}^n$, $\mathbb{F}(l, \mathbf{t}) \equiv l^{-n/p}$

and $P_l = L^p([-l,l]^n : Y)$; the notion of a (jointly) Weyl-p-almost automorphic sequence (of type 1) [(jointly) Weyl-(p,b)-multialmost automorphic (of type 1)] $F : \mathbb{Z}^n \to Y$ is new and can be obtained with $\Omega = [-1,1]^n$, R being the collection of all sequences in $\mathbb{Z}^n$, $\mathbb{F}(l,\mathbf{t}) \equiv l^{-n/p}$ and $P_l = L^p([-l,l]^n \cap \mathbb{Z}^n : Y)$ [cf. the first term in the equation (276) below with $p_l \equiv p$ and $\nu_l \equiv 1$]. We similarly define the notion of (joint) Weyl-(p,R)-almost automorphy (of type 1) [(joint) Weyl-(p,R,b)-almost automorphy (of type 1)], where R is a general collection of sequences obeying our requirements.

We continue by introducing the corresponding notion in which $0 < p < 1$ (cf. also [448, Subsection 4.3.1] for the notion of metrical Weyl distance; if $1 \leqslant p < \infty$, then we have $\mathbb{F}(l,\mathbf{t}) \equiv l^{-n/p}$).

Definition 5.2.4. Suppose that $0 < p < 1$ and $F : \mathbb{R}^n \to Y$ $[F : \mathbb{Z}^n \to Y]$. Then we say that $F(\cdot)$ is (jointly) Weyl-p-almost automorphic function (of type 1) [(jointly) Weyl-(p,b)-almost automorphic sequence (of type 1)] if $F(\cdot)$ is (jointly) Weyl-$(\mathbb{F},\mathcal{P},\mathrm{R})$-multialmost automorphic (of type 1) [(jointly) Weyl-$(\mathbb{F},\mathcal{P},\mathrm{R},b)$-multialmost automorphic (of type 1)], where $\Omega = [-1,1]^n$, R is the collection of all sequences in $\mathbb{R}^n$ $[\mathbb{Z}^n]$, $\mathbb{F}(l,\mathbf{t}) \equiv l^{-n}$ and $P_l = L^p([-l,l]^n : Y)$ $[P_l = l^p([-l,l]^n \cap \mathbb{Z}^n : Y)$; cf. the second term in the equation (276) below with $\nu_l \equiv 1]$. If R is a general collection of sequences obeying our requirements, then we also say that $F(\cdot)$ is (jointly) Weyl-(p,R)-almost automorphic (of type 1) [(jointly) Weyl-(p,R,b)-almost automorphic (of type 1)].

We can further generalize the notion introduced in the above three definitions following our approach from [447, Definition 8.1.2]; cf. also [447, Remark 8.3.19(i)] and [464, Example 2.7]. The notion in which $P_l = L^p([-l,l]^n : Y)$ $[P_l = l^p([-l,l]^n \cap \mathbb{Z}^n : Y)]$ and $\mathbb{F}(l,\mathbf{t}) \equiv l^{-n/p}$, if $1 \leqslant p < +\infty$, respectively, $\mathbb{F}(l,\mathbf{t}) \equiv l^{-n}$, if $0 < p < 1$, is extremely important; if this is the case, then we have the following result, which can be simply formulated for the general function spaces introduced in Definitions 5.2.1–5.2.3 as well.

Proposition 5.2.5. *Let P_l and $\mathbb{F}(l,\cdot)$ be as above ($l > 0$). Then the following holds:*

(i) *Suppose that $0 < p \leqslant q < +\infty$. Then any (jointly) Weyl-q-almost automorphic function $F : \mathbb{R}^n \to Y$ $[F : \mathbb{Z}^n \to Y]$ (of type 1) is (jointly) Weyl-p-almost automorphic (of type 1); furthermore, the same holds for the corresponding classes of (jointly) Weyl-(q,b)-almost automorphic functions (of type 1) and (jointly) Weyl-(p,b)-almost automorphic functions (of type 1).*
(ii) *Suppose that $0 < p \leqslant q < +\infty$. Then $F : \mathbb{R}^n \to Y$ $[F : \mathbb{Z}^n \to Y]$ is essentially bounded, (jointly) Weyl-(q,b)-almost automorphic (of type 1) if and only if $F(\cdot)$ is essentially bounded, (jointly) Weyl-(p,b)-almost automorphic (of type 1).*

Proof. The proof of (i) is very simple and follows from an elementary application of the Hölder inequality. Keeping this in mind, the statement (ii) follows immediately if we prove that any essentially bounded, (jointly) Weyl-(p,b)-almost automorphic function $F : \mathbb{R}^n \to Y$ $[F : \mathbb{Z}^n \to Y]$ (of type 1) is (jointly) Weyl-(q,b)-almost automorphic (of type 1). For the sake of brevity, we will consider here the essentially bounded, jointly Weyl-(p,b)-

almost automorphic functions $F : \mathbb{R}^n \to Y$, only. Let (b_k) be a given sequence. Then, for every $l > 0$, there exist a subsequence (b_{k_m}) of (b_k), a function $F^* : \mathbb{R}^n \to L^p([-l,l]^n : Y)$ and a finite real number $M > 0$ such that $\|[F^*(\mathbf{t})](x)\|_Y \leqslant M$ for all $\mathbf{t} \in \mathbb{R}^n$ and $x \in [-l,l]^n$ as well as that, for every $\mathbf{t} \in \mathbb{R}^n$ and $\varepsilon > 0$, there exists $s > 0$ such that, for every $l \geqslant s$ and for every $m \in \mathbb{N}$ with $m \geqslant s$, we have

$$l^{-n} \int\limits_{[-l,l]^n} \|F(\mathbf{t}+x+b_{b_m}) - [F^*(\mathbf{t})](x)\|_Y^p \, dx < \varepsilon$$

and

$$l^{-n} \int\limits_{[-l,l]^n} \|[F^*(\mathbf{t}-b_{b_m})](x) - F(\mathbf{t}+x)\|_Y^p \, dx < \varepsilon. \tag{275}$$

Then the final conclusion simply follows from the estimate

$$\begin{aligned} l^{-n} & \int\limits_{[-l,l]^n} (\|F(\mathbf{t}+x+b_{b_m}) - [F^*(\mathbf{t})](x)\|_Y/(\|F\|_\infty + M))^q \, dx \\ & \leqslant l^{-n} \int\limits_{[-l,l]^n} (\|F(\mathbf{t}+x+b_{b_m}) - [F^*(\mathbf{t})](x)\|_Y/(\|F\|_\infty + M))^p \, dx \end{aligned}$$

and the corresponding estimate for the term appearing in (275). □

We continue by providing some useful observations.

Remark 5.2.6. (i) As multiple times before, we omit the term "R" if R denotes the collection of all sequences in $\mathbb{R}^n$ $[\mathbb{Z}^n]$, and we omit the term "-multi" if $n = 1$.

(ii) If we consider the continuous notion from the above three definitions, then the very natural extension of the notion introduced in [447, Definitions 8.3.17, 8.3.18, 8.3.28] can be obtained by setting $\|f\|_l \equiv \|f\|_{L^{p_l(\mathbf{t})}_{\nu_l}(l\Omega:Y)}$, where $p_l \in \mathcal{P}(l\Omega)$ $[p_l \equiv p \in (0,1)$ for all $l > 0]$ and $\nu_l : l\Omega \to (0,\infty)$ is a Lebesgue measurable function, or $\|f\|_l \equiv \|f\|_{C_{b,\nu_l}(l\Omega:Y)}$, where $\nu_l : l\Omega \to [0,\infty)$ is any nonzero function $(l > 0)$. Concerning the discrete notion from the above mentioned definitions, the natural choices for $\|f\|_l$ can be obtained by setting

$$\|f\|_l \equiv \Big[\sum_{j \in l\Omega \cap \mathbb{Z}^n} \|\nu_l(j) f(j)\|^{p_l}\Big]^{1/p_l} \quad \Big[\|f\|_l \equiv \sum_{j \in l\Omega \cap \mathbb{Z}^n} \|\nu_l(j) f(j)\|^{p}\Big], \\ \text{or} \quad \|f\|_l \equiv \sup_{j \in l\Omega \cap \mathbb{Z}} \|\nu_l(j) f(j)\|, \tag{276}$$

where $1 \leqslant p_l < +\infty$ $[p_l \equiv p \in (0,1)$ for all $l > 0]$ and $\nu_l : l\Omega \cap \mathbb{Z}^n \to [0,\infty)$ is any nonzero function $(l > 0)$.

(iii) The notions introduced in the above definitions provide a very general approach to Weyl-p-almost automorphy. For example, if we assume that for each $l > 0$ we have $P_l = L^{\infty}(l\Omega : Y)$ or $P_l = L^p_{\nu}(l\Omega : Y)$ with $\nu \in L^p(\mathbb{R}^n : (0,\infty))$ and $p > 0$ as well as that for each $x \in X$ the function $F(\cdot; x)$ is bounded and for each $t \in \mathbb{R}^n$ we have $\lim_{l\to+\infty} \mathbb{F}(l,\mathbf{t}) = 0$, then the function $F(\cdot;\cdot)$ is (jointly) Weyl-$(\mathbb{F},\mathcal{P},\mathrm{R},\mathcal{B})$-multialmost automorphic [Weyl-$(\mathbb{F},\mathcal{P},\mathrm{R},\mathcal{B})$-multialmost automorphic of type 1, provided that the limits $\lim_{m\to+\infty}\cdot$ in the equations (273) and (274) exist], which can be shown by plugging $F^* \equiv F$.

In [461], we have observed that any Stepanov-p-almost automorphic function $f : \mathbb{R} \to Y$ is Weyl-p-almost automorphic of type 1. This result can be simply extended as follows. Suppose that the requirements clarified directly before Definition 5.1.3 hold with $\Omega = [-1,1]^n$ and $G : (0,\infty) \to (0,\infty)$. Let for each $l > 0$ the nonempty set $Z_l \subseteq Y^{l\Omega}$ be defined as the set of all functions $f : [-l,l]^n \to Y$ such that, for every cube of the form $k + [-1,1]^n$, which belongs to $[-l,l]^n$, where $k \in (2\mathbb{Z}+1)^n$, we have that the restriction of function $f(\cdot - k)$ to the set $[-1,1]^n$ belongs to P; suppose further that the set Z_l is equipped with any pseudometric $d_l(\cdot;\cdot)$ such that

$$d_l(f,g) \leqslant G(l)\sum_{k} d(f(\cdot - k)_{|[-1,1]^n}, g(\cdot - k)_{|[-1,1]^n}), \quad f,\, g \in Z_l,\ l > 0,$$

where the summation is taken over all points $k \in (2\mathbb{Z}+1)^n$ such that $k+[-1,1]^n \subseteq [-l,l]^n$. If for each $l > 0$ there exists a finite $c_l > 0$ such that $\mathbb{F}(l,\mathbf{t})G(l) \leqslant c_l$ for all $\mathbf{t} \in \mathbb{R}^n$, then the second limits in (273)–(274) are equal to zero for every fixed $l > 0$ and, therefore, we can clarify our next structural result.

Proposition 5.2.7. *Let $F : \mathbb{R}^n \times X \to Y$ be Stepanov $([-1,1]^n, \mathrm{R}, \mathcal{B}, Z^{\mathcal{P}})$-multialmost automorphic and let for each $l > 0$ we have the existence of a finite $c_l > 0$ such that $\mathbb{F}(l,\mathbf{t})G(l) \leqslant c_l$ for all $\mathbf{t} \in \mathbb{R}^n$. Then $F(\cdot;\cdot)$ is Weyl-$(\mathbb{F},\mathcal{P},\mathrm{R},\mathcal{B})$-multialmost automorphic of type* 1, *with $\mathcal{P}$ being defined as above.*

The class of Stepanov-p-almost periodic functions with a general exponent $p > 0$ has been systematically analyzed in the joint paper of the second named author and W.-S. Du [477]. With the help of the metrical Bochner criterion, we can simply prove that any Stepanov-p-almost periodic function $F : \mathbb{R}^n \to Y$, where $p > 0$, is Weyl-p-almost automorphic, Weyl-p-almost automorphic of type 1, as well as jointly Weyl-p-almost automorphic in the usual sense, with the limit function $F^* \equiv F$; the converse statement is not true since there exists a jointly Weyl-p-almost automorphic function $f \in L^p_{\mathrm{loc}}(\mathbb{R} : \mathbb{R})$, which is not Stepanov-$p$-almost automorphic [461]. Furthermore, the joint Weyl-p-almost automorphy of $F : \mathbb{R}^n \times X \to Y$ [$F : \mathbb{Z}^n \times X \to Y$] implies its Weyl-$p$-almost automorphy provided that for each $k \in \mathbb{N}$ the both limits in the equations (271)–(272) exist as $l \to +\infty$; a similar comment can be given for the Weyl-p-almost automorphy of type 1.

We continue by providing some illustrative examples.

Example 5.2.8. (i) In [447, Theorem 8.3.8], we have considered the function $f(x) := |x|^{\sigma}$, $x \in \mathbb{R}$, where $\sigma \in (0,1)$, $p \in [1,\infty)$ and $(1-\sigma)p < 1$. Among many other conclusions, we have deduced there that the function $f(\cdot)$ is Weyl-p-almost automorphic, not Weyl-p-almost automorphic of type 1 nor joint Weyl-p-almost automorphic. Let us consider now case in which $\sigma > 0, p \in (0,1)$ and $a > 1-(1-\sigma)p > 0$. Concerning the Weyl-p-almost automorphy of $f(\cdot)$, we would like to observe here that the same argumentation as in the proof of the above-mentioned theorem shows that $f(\cdot)$ is Weyl-$(\mathbb{F}, \mathcal{P}, \mathrm{R})$-multialmost automorphic, where $\mathbb{F}(l,t) \equiv l^{-a}$, $P_l \equiv L^p([-l,l] : Y)$ and R is the collection of all real sequences. Furthermore, if $\sigma \geqslant 1$, $\nu \in L^{\infty}[-l,l]$ for all $l > 0$ and $\lim_{l \to +\infty} l^{\sigma-1}\mathbb{F}(l) = 0$, then the function $f(\cdot)$ is Weyl-$(\mathbb{F}, \mathcal{P}, \mathrm{R})$-multialmost automorphic, where $\mathbb{F}(l,t) \equiv \mathbb{F}(l)$, $P_l \equiv C_{b,\nu}([-l,l] : Y)$ for all $l > 0$ and R is the collection of all real sequences; in order to see this, we can apply the Lagrange mean value theorem and the following simple computation ($w \in \mathbb{R}$ and $l > 0$ are arbitrary, $f^* \equiv f$):

$$
\begin{aligned}
&\mathbb{F}(l) \sup_{x \in [-l,l]} \big|f(t+w+x) - f(t+x)\big| \cdot \nu(x) \\
&= \mathbb{F}(l) \sup_{x \in [-l,l]} \big||t+w+x|^{\sigma} - |t+x|^{\sigma}\big| \cdot \nu(x) \\
&\leqslant \sigma|w|\mathbb{F}(l) \sup_{x \in [-l,l]} \sup_{y \in [|t+x|,|t+x+\omega|] \cup [|t+x+\omega|,|t+x|]} y^{\sigma-1} \cdot \nu(x) \\
&\leqslant \sigma|w|\mathbb{F}(l) \sup_{x \in [-l,l]} \big[|t+x|^{\sigma-1} + |t+x+w|^{\sigma-1}\big] \cdot \nu(x) \\
&\leqslant \sigma|w|\mathbb{F}(l) \sup_{x \in [-l,l]} \big[2|t|^{\sigma-1} + |w|^{\sigma-1} + 2|x|^{\sigma-1}\big] \cdot \nu(x).
\end{aligned}
$$

(ii) Let $p > 0$. Arguing as in the proof of [447, Theorem 8.3.10], we have that the Heaviside function $\chi_{[0,\infty)}(\cdot)$ is not jointly Weyl-p-almost automorphic as well as that $\chi_{[0,\infty)}(\cdot)$ is both Weyl-p-almost automorphic and Weyl-p-almost automorphic of type 1; furthermore, we can similarly prove the following:
   (a) The Heaviside function $\chi_{[0,\infty)}(\cdot)$ is not jointly Weyl-$(\mathbb{F}, \mathcal{P})$-almost automorphic if $\mathbb{F}(l) \equiv 1/l$ and $P_l \equiv L^p_{\nu}([-l,l] : \mathbb{C})$ for all $l > 0$, where $\nu : \mathbb{R} \to (0,\infty)$ is any Lebesgue measurable function such that $\limsup_{l \to +\infty}(1/l)\int_{-l}^{0} \nu^p(x)\,dx > 0$.
   (b) The Heaviside function $\chi_{[0,\infty)}(\cdot)$ is Weyl-$(\mathbb{F}, \mathcal{P})$-almost automorphic if $\lim_{l \to +\infty} \mathbb{F}(l) = 0$ and $P_l \equiv L^p_{\nu}([-l,l] : \mathbb{C})$ for all $l > 0$, where $\nu : \mathbb{R} \to (0,\infty)$ is any p-locally integrable function.

   We can similarly consider the multidimensional analogue of this example and provide the basic information about the metrical Weyl-p-almost automorphic properties of the function $\chi_K(\cdot)$, where K is a nonempty compact subset of $\mathbb{R}^n$; cf. [447, Example 8.3.21] for more details in this direction.

(iii) In [447, Example 8.3.20], we have reconsidered the well-known example proposed by J. Stryja [732, pp. 42–47]; see also [55, Example 4.28]: Define $f : \mathbb{R} \to \mathbb{R}$ by $f(x) := 0$ for $x \leqslant 0$, $f(x) := \sqrt{n/2}$ if $x \in (n-2, n-1]$ for some $n \in 2\mathbb{N}$ and $f(x) := -\sqrt{n/2}$ if

$x \in (n-1, n]$ for some $n \in 2\mathbb{N}$. We have shown that the function $f(\cdot)$ is not (jointly) Weyl 1-almost automorphic (of type 1; of type 2). On the other hand, we have $|f(x)| \leqslant 2^{-1/2}\sqrt{|x|}$, $x \in \mathbb{R}$ so that a simple computation shows that the function $f(\cdot)$ is $(\mathbb{F}, \mathcal{P})$-almost automorphic, provided that for each $l > 0$ we have $P_l = L_\nu^p([-l, l])$ with some $p > 0$ and a Lebesgue measurable function $\nu(\cdot)$ such that

$$\lim_{l \to +\infty} \mathbb{F}(l) \int_{-l}^{l} (1 + |x|^{p/2}) \nu^p(x)\, dx = 0.$$

Furthermore, we can prove that the function $f(\cdot)$ is $(\mathbb{F}, \mathcal{P}, \mathrm{R})$-almost automorphic, provided that R is the collection of all real sequences (a_m) satisfying that $a_m \in 2\mathbb{N}$ for all $m \in \mathbb{N}$ as well as that for each $l > 0$ we have $P_l = L_\nu^p([-l, l])$ with some $p > 0$ and a Lebesgue measurable function $\nu(\cdot)$ such that $\lim_{l \to +\infty} \mathbb{F}(l) \int_{-l}^{l} \nu^p(x)\, dx = 0$.

(iv) In [464, Example 2.3] (cf. also [448, Example 3.4.4]), we have analyzed the Weyl-p-almost automorphic properties of the function $f : \mathbb{R} \to l_\infty$, given by $f(t) := (e^{-|t|/k})_{k \in \mathbb{N}}$, $t \in \mathbb{R}$. Since this function is slowly oscillating, it can be shown that the function $f(\cdot)$ is Weyl-$(\mathbb{F}, \mathcal{P})$-almost automorphic provided that $\lim_{l \to +\infty} \mathbb{F}(l) = 0$, there exist $l_0 > 0$ and $M > 0$ such that, for every $l \geqslant l_0$, we have $\mathbb{F}(l) \int_{-l}^{l} \nu^p(x)\, dx \leqslant M$ and $P_l = L_\nu^p([-l, l])$ for some $p > 0$ $(l > 0)$ and a p-locally integrable function $\nu : \mathbb{R} \to (0, \infty)$. Moreover, we can prove that the function $f(\cdot)$ is Weyl-$(\mathbb{F}, \mathcal{P})$-almost automorphic of type 1 provided that $\nu : \mathbb{R} \to (0, \infty)$ is a p-locally integrable function and $P_l = L_\nu^p([-l, l])$ for some $p > 0$ $(l > 0)$. In connection with this example, we would like to stress the following:
   (a) Clearly, $\|f(t)\| = 1$ for all $t \in \mathbb{R}$ so that $\lim_{t \to +\infty} t^{-1} \int_0^t \|f(s)\|\, ds = 1 \neq 0$, as it has been mistakenly written.
   (b) Concerning the Weyl-p-almost automorphy of the function $f(\cdot)$, we should use the limit function $f^* \equiv f$, if (b_k) is a sequence tending to plus infinity or minus infinity (not $f^* \equiv 0$).
   (c) The corresponding statement for the joint Weyl-p-almost automorphy of the function $f(\cdot)$ is not true.

Remark 5.2.9. (i) The result of [447, Proposition 8.3.9] remains true in the higher-dimensional setting so that any Weyl-p-almost automorphic function of type 1 (jointly Weyl-p-almost automorphic function) $F : \mathbb{R}^n \to Y$ must be Stepanov-p-bounded, i. e., $\sup_{\mathbf{t} \in \mathbb{R}^n} \int_{\mathbf{t}+[0,1]^n} \|F(\mathbf{t})\|^p\, d\mathbf{t} < +\infty$ $(p > 0$; cf. also Proposition 5.2.5(i)). Furthermore, in the discrete setting, a similar argumentation shows that any Weyl-p-almost automorphic sequence of type 1 (jointly Weyl-p-almost automorphic sequence) $F : \mathbb{Z}^n \to Y$, where $p > 0$, must be bounded.

(ii) The statement of [447, Proposition 8.3.13] can be also formulated in the higher-dimensional setting, with the usage of spaces $L_\nu^p(l[-1, 1]^n : Y)$, where $p > 0$ and $\nu : \mathbb{R}^n \to (0, \infty)$ is any Lebesgue measurable function. Details can be left to the interested readers (cf. also [447, Questions 8.3.14, 8.3.15]).

Concerning [447, Example 8.3.16], we would like to recall that A. Haraux and P. Souplet have considered the function $f : \mathbb{R} \to \mathbb{R}$ given by

$$f(t) := \sum_{m=1}^{\infty} \frac{1}{m} \sin^2\left(\frac{t}{2^m}\right), \quad t \in \mathbb{R}. \tag{277}$$

We already know that this function is (Besicovitch) unbounded and Weyl p-almost automorphic for any finite exponent $p \geqslant 1$ as well as that for each $\tau \in \mathbb{R}$ the function $f(\cdot + \tau) - f(\cdot)$ is almost antiperiodic (see [444] for the notion). Due to Proposition 5.2.5(i), this clearly implies that the function $f(\cdot)$ is Weyl-p-almost automorphic for any finite exponent $p > 0$. Here, we would like to note that the function $f(\cdot)$ has bounded differences, i. e., for each $\tau \in \mathbb{R}$ we have that the function $f(\cdot + \tau) - f(\cdot)$ is bounded. Every such a Lebesgue measurable function has to be Weyl-$(\mathbb{F}, \mathcal{P})$-almost automorphic, where $p > 0$ and $P_l = L^p_\nu([-l, l])$ for some p-locally integrable function $\nu(\cdot)$ satisfying that

$$\lim_{l \to +\infty} \mathbb{F}(l, t) \int_{-l}^{l} \nu^p(x)\, dx = 0, \quad t \in \mathbb{R};$$

cf. also [447, p. 63] for more details about functions with bounded differences.

Before proceeding to the next subsection, we would like to observe that the statements of [447, Propositions 8.3.23, 8.3.24] can be formulated for the function spaces introduced in this section, like the statement of [447, Proposition 8.3.30] concerning the pointwise products of metrically Weyl almost automorphic type functions.

5.2.2 Extensions of metrically Weyl almost automorphic sequences

This subsection aims to provide the basic information about the extensions of metrically Weyl almost automorphic sequences. At the very beginning, we would like to emphasize that the statements of [483, Theorems 2.3, 2.5] cannot be so simply formulated for the general classes of metrically Weyl almost automorphic sequences and functions. Discretization of Weyl almost automorphic type functions is a completely new topic and here we will first present some positive results in this direction, which can be approved similarly as in [461] (we will also consider noncontinuous functions; for more details about discretization of generalized almost periodic type functions and generalized almost automorphic type functions, we also refer the reader to [56, 494, 483]).

Example 5.2.10. Suppose that $p \geqslant 1$.

(i) If K is any nonempty compact subset of $\mathbb{R}^n$, then the function $\chi_K(\cdot)$ is (jointly) Weyl-p-almost automorphic (of type 1). The same holds for the sequence $(\chi_K(\mathbf{t}))_{\mathbf{t} \in \mathbb{Z}^n}$.

(ii) We already know that the function $\chi_{[0,\infty)^n}(\cdot)$ is not jointly Weyl-p-almost automorphic, not Weyl-p-almost automorphic of type 1 as well as that the function $\chi_{[0,\infty)^n}(\cdot)$ is Weyl-p-almost automorphic. The same holds for the sequence $(\chi_{[0,\infty)^n}(\mathbf{t}))_{\mathbf{t} \in \mathbb{Z}^n}$.

(iii) All conclusions established for the function $x \mapsto f(x) \equiv |x|^{\sigma}$, $x \in \mathbb{R}$, where $\sigma \in (0,1)$ and $(1-\sigma)p < 1$, remain true for the sequence $(f(k))_{k\in\mathbb{Z}}$; cf. [447, Theorem 8.3.8] for more details.

Concerning some negative results in this direction, we will only recall once more that J. Andres and D. Pennequin have constructed, in [56, Example 4], an infinitely differentiable Stepanov-1-almost periodic function $f : \mathbb{R} \to \mathbb{R}$ such that the sequence $(f(k))_{k\in\mathbb{Z}}$ is not bounded. Therefore, the function $f(\cdot)$ is (jointly) Weyl-1-almost automorphic (of type 1), but the sequence $(f(k))_{k\in\mathbb{Z}}$ is not bounded and, therefore, not (jointly) Weyl-1-almost automorphic (of type 1); see also Remark 5.2.9(i) and [476, Remark 4].

In connection with [483, Theorem 2.5], we will state and prove the following result (cf. Definition 5.3.2 for the notion).

Theorem 5.2.11. *Suppose that $p > 0$ and $F : \mathbb{Z}^n \to Y$ is bounded, jointly Weyl-(p,R)-multialmost automorphic sequence, where R is any collection of sequences in $\mathbb{Z}^n$ such that the assumption $(b_k) \in \mathrm{R}$ implies that any subsequence of (b_k) also belongs to R. Let R' be the collection of all sequences (a_k) in $\mathbb{R}^n$ satisfying that there exists a sequence $(b_k) \in \mathrm{R}$ such that $\sup_{k\in\mathbb{N}} |a_k - b_k| < +\infty$. Then there exists a bounded, uniformly continuous, jointly Weyl-(p,R')-multialmost automorphic function $\tilde{F} : \mathbb{R}^n \to Y$ such that $R(\tilde{F}(\cdot)) \subseteq CH(\overline{R(F)})$, $\|\tilde{F}\|_{\infty} = \|F\|_{\infty}$ and $\tilde{F}(k) = F(k)$ for all $k \in \mathbb{Z}^n$.*

Proof. We will consider the one-dimensional setting for the sake of brevity (in general case, the result follows from a similar argumentation but the proof is much more technically complicated). If $t \in [k,k+1)$ for some $k \in \mathbb{Z}$, then we define $\tilde{F}(t) := F(k) + (t-k)\cdot [F(k+1) - F(k)]$. Since $F(\cdot)$ is bounded, it readily follows that $\tilde{F}(\cdot)$ is bounded, uniformly continuous as well as that $R(\tilde{F}(\cdot)) \subseteq CH(\overline{R(F)})$, $\|\tilde{F}\|_{\infty} = \|F\|_{\infty}$ and $\tilde{F}(k) = F(k)$ for all $k \in \mathbb{Z}$. It remains to be proved that $\tilde{F}(\cdot)$ is jointly Weyl-(p,R')-almost automorphic. Let $(a_k) \in \mathrm{R}'$ be fixed. Then there exists a sequence $(b_k) \in \mathrm{R}$ such that $\sup_{k\in\mathbb{N}} |a_k - b_k| < +\infty$. Due to our assumptions, for every $l > 0$, we can find a subsequence $(b_{k_m}) \in \mathrm{R}$ of (b_k) and a function $F^* : \mathbb{Z} \to L^p([-l,l] \cap \mathbb{Z})$ such that, for every $t \in \mathbb{Z}$, we have

$$\lim_{(m,l)\to+\infty} l^{-1} \sum_{j\in[-l,l]\cap\mathbb{Z}} \|F(t + b_{k_m} + j) - [F^*(t)](j)\|_Y^p = 0 \tag{278}$$

and

$$\lim_{(m,l)\to+\infty} l^{-1} \sum_{j\in[-l,l]\cap\mathbb{Z}} \|[F^*(t - b_{k_m})](j) - F(t+j)\|_Y^p = 0. \tag{279}$$

Since $F(\cdot)$ is bounded, it simply follows that for each $\varepsilon > 0$ there exist real numbers $c_p > 0$ and $l_0 > 0$ such that

$$\sum_{j\in[-l,l]\cap\mathbb{Z}} \|[F^*(0)](j)\|_Y^p \leqslant c_p\big((2l+2)\|F\|_{\infty}^p + \varepsilon l\big), \quad l \geqslant l_0.$$

Keeping this in mind, it is not difficult to prove that (278)–(279) imply

$$\lim_{(m,l)\to+\infty} l^{-1} \sum_{j\in[-l,l]\cap\mathbb{Z}} \|F(t+b_{k_m}+j)-[F^*(0)](t+j)\|_Y^p = 0$$

and

$$\lim_{(m,l)\to+\infty} l^{-1} \sum_{j\in[-l,l]\cap\mathbb{Z}} \|[F^*(0)](t-b_{k_m}+j)-F(t+j)\|_Y^p = 0. \tag{280}$$

Furthermore, we may assume without loss of generality that $\lim_{k\to\infty}(a_k-b_k)=c\in\mathbb{R}$. In order to complete the proof, it suffices to prove that, for every fixed number $t\in\mathbb{R}$, we have

$$\lim_{(m,l)\to+\infty} l^{-1} \int_{-l}^{l} \|\tilde{F}(t+a_{k_m}+x)-G(t+x+c)\|_Y^p\,dx = 0 \tag{281}$$

and

$$\lim_{(m,l)\to+\infty} l^{-1} \int_{-l}^{l} \|G(t-a_{k_m}+x+c)-F(t+x)\|_Y^p\,dx = 0, \tag{282}$$

where $G(t):=[F^*(0)](k)+(t-k)\cdot\{[F^*(0)](k+1)-[F^*(0)](k)\}$ if $t\in[k,k+1)$ for some $k\in\mathbb{Z}$. We will only prove the limit equation (281) since the limit equation (282) can be proved analogously, with the help of the limit equation (280). Let $\varepsilon>0$ be fixed. Since $\tilde{F}(\cdot)$ is uniformly continuous, (281) follows automatically if we prove that there exists a sufficiently large real number $s>0$ such that, for every $l\geqslant s$ and $m\geqslant s$, we have

$$l^{-1} \int_{-l}^{l} \|\tilde{F}(t+x+c+b_{k_m})-G(t+x+c)\|_Y^p\,dx \leqslant \varepsilon. \tag{283}$$

To deduce the validity of (283), we divide the segment $[-l,l]$ into disjoint intervals depending on the belonging of the number $t+c+x$ to some interval of the form $[k,k+1)$, where $k\in\mathbb{Z}$. Keeping in mind the definitions of $\tilde{F}(\cdot)$, $G(\cdot)$ and such a division of the interval $[-l,l]$, we easily get that there exists a finite real number $c_p>0$ such that

$$\int_{-l}^{l} \|\tilde{F}(t+x+c+b_{k_m})-G(t+x+c)\|_Y^p\,dx \leqslant c_p \sum_{j=-\lceil l\rceil-1}^{\lceil l\rceil+1} \|F(\lfloor t\rfloor+j+b_{k_m})-G(\lfloor t\rfloor+j)\|_Y^p,$$

for all sufficiently large numbers $l>0$ and $m\in\mathbb{N}$, which simply implies the required. □

We can similarly prove the following result for jointly Weyl-$(\mathbb{F},\mathcal{P},\mathrm{R},W)$-multialmost automorphic sequences of type 2.

Theorem 5.2.12. *Suppose that $p > 0$ and $F : \mathbb{Z}^n \to Y$ is a bounded, jointly Weyl-$(\mathbb{F},\mathcal{P},\mathrm{R},W)$-multialmost automorphic sequence of type 2, where $\mathbb{F}(l,\mathbf{t}) \equiv l^{-n/p}$ if $p \geqslant 1$ [$\mathbb{F}(l,\mathbf{t}) \equiv l^{-1}$ if $p \in (0,1)$], $\emptyset \neq W \subseteq \mathbb{Z}^n$, for each $l > 0$ and $f \in P_l$ we have $\|f\|_l \equiv [\sum_{j\in[-l,l]^n\cap\mathbb{Z}^n} \|f(j)\|_Y^p]^{1/p}$ if $p \geqslant 1$ [$\|f\|_l \equiv \sum_{j\in[-l,l]^n\cap\mathbb{Z}^n} \|f(j)\|_Y^p$ if $p \in (0,1)$], R is any collection of sequences in $\mathbb{Z}^n$ such that the assumption $(b_k) \in \mathrm{R}$ implies that any subsequence of (b_k) also belongs to R. Let R' be the collection of all sequences (a_k) in $\mathbb{R}^n$ satisfying that there exists a sequence $(b_k) \in \mathrm{R}$ such that $\sup_{k\in\mathbb{N}} |a_k - b_k| < +\infty$. Then there exists a bounded, uniformly continuous, Weyl-$(\mathbb{F},\mathcal{P}',\mathrm{R}',W)$-multialmost automorphic function $\tilde{F} : \mathbb{R}^n \to Y$ of type 2, where for each $l > 0$ we have $P_l' = L^p([-l,l]^n : Y)$; furthermore, we have $R(\tilde{F}(\cdot)) \subseteq CH(\overline{R(F)})$, $\|\tilde{F}\|_\infty = \|F\|_\infty$ and $\tilde{F}(k) = F(k)$ for all $k \in \mathbb{Z}^n$.*

It is worthwhile to mention that Theorem 5.2.12 can be simply formulated for the class of jointly Weyl-$(\mathbb{F},\mathcal{P},\mathrm{R},W)$-multialmost automorphic sequences, as well. Various classes of Stepanov-p-almost periodic functions in norm and Stepanov-p-almost automorphic functions in norm ($p > 0$) will be consider a little bit later; we will not consider here the corresponding classes of (metrically) Weyl-p-almost automorphic functions in norm ($p > 0$).

5.2.3 Applications to the abstract Volterra integrodifferential equations and the abstract Volterra difference equations

In this subsection, we provide some applications of the introduced notion:

1. Convolution invariance of joint Weyl-$(\mathbb{F},\mathcal{P},\mathrm{R},\mathcal{B},\omega)$-multialmost automorphy and Weyl-$(\mathbb{F},\mathcal{P},\mathrm{R},\mathcal{B},W)$-multialmost automorphy of type 2

The introduced classes of metrically Weyl almost automorphic type functions, especially the class of Weyl-$(\mathbb{F},\mathcal{P},\mathrm{R},\mathcal{B})$-multialmost automorphic functions (of type 1), behave very badly with respect to the invariance under the actions of infinite convolution products. The main problem is the existence of the second limits in the equations (271)–(272), respectively, (273)–(274). Concerning this question, we will only emphasize that the corresponding classes of metrically Besicovitch almost automorphic functions and sequences, which will be considered in the next section, behave much better with respect to this matter as well as that the certain results can be obtained provided that the second limits in these equations are equal to zero or that the limit function $F^*(\cdot;\cdot)$ satisfies $F^* \equiv F$.

Concerning [447, Proposition 8.3.6, Question 8.3.7], we would like to present the following result, which can be deduced by means of the argumentation contained in the proof of [464, Proposition 3.2] (observe only that the function $F(\cdot)$ appearing on [464, l. -2, p. 47] is bounded in the newly arisen situation; taking into account the corresponding definition of joint Weyl-$(\mathbb{F}, \mathcal{P}, \mathrm{R}, \mathcal{B}, \omega)$-multialmost automorphy, the remainder of proof can be simply copied).

Proposition 5.2.13. *Suppose that the operator family $(R(t))_{t>0} \subseteq L(X, Y)$ satisfies that there exist finite real constants $M > 0$, $\beta \in (0,1]$ and $\gamma > \beta$ such that*

$$\|R(t)\|_{L(X,Y)} \leqslant M\frac{t^{\beta-1}}{1+t^{\gamma}}, \quad t > 0. \tag{284}$$

Suppose further that $a > 0$, $\alpha > 0$, $1 \leqslant p < +\infty$, $ap \geqslant 1$, $\alpha p \geqslant 1$, $ap(\beta-1)/(ap-1) > -1$ if $ap > 1$ and $\beta = 1$ if $ap = 1$. If $b \in [0, \gamma-\beta)$, $w(t) := (1+|t|)^b$, $t \in \mathbb{R}$ and the function $f : \mathbb{R} \to X$ is both essentially bounded and jointly Weyl-$(\mathbb{F}, \mathcal{P}, \mathrm{R}, \omega)$-multialmost automorphic with $\mathbb{F}(l) \equiv l^{-a/\alpha}$ and $P_l = L^{\alpha p}([-l, l])$ for all $l > 0$, then the function $F(\cdot)$, given by (262), is continuous, bounded and jointly Weyl-$(\mathbb{F}, \mathcal{P}, \mathrm{R}, \omega)$-multialmost automorphic.

It is clear that Proposition 5.2.13 can be applied in the analysis of the existence and uniqueness of jointly Weyl-$(\mathbb{F}, \mathcal{P}, \mathrm{R}, \omega)$-multialmost automorphic-type solutions for a large class of the abstract fractional integrodifferential inclusions without initial conditions. For example, we can apply this result in the study of the fractional Poisson heat equation $D_{t,+}^{\gamma}[m(x)v(t,x)] = (\Delta - b)v(t,x) + f(t,x)$, $t \in \mathbb{R}$, $x \in \Omega$; $v(t,x) = 0$, $v(t,x) \in [0,\infty) \times \partial\Omega$ in the space $X := L^p(\Omega)$, where Ω is a bounded domain in $\mathbb{R}^n$ and some extra assumptions are satisfied. See [444] for more details about applications of this type.

The proof of following result is very similar to the proof of [447, Theorem 8.3.25] and, therefore, omitted.

Theorem 5.2.14. *Suppose that $h \in L^1(\mathbb{R}^n)$, $p \in \mathcal{P}(\mathbb{R}^n)$ and $F : \mathbb{R}^n \times X \to Y$ is Weyl-$(\mathbb{F}, \mathcal{P}, \mathrm{R}, \mathcal{B}, (2\mathbb{Z}+1)^n)$-multialmost automorphic of type 2, where $P_l = L_{\nu}^{p(\mathbf{u})}(l\Omega : Y)$ for all $l > 0$ and some Lebesgue measurable function $\nu : \mathbb{R}^n \to (0, +\infty)$. Let $p_1, q \in \mathcal{P}(\mathbb{R}^n)$, let $1/p(\mathbf{u}) + 1/q(\mathbf{u}) \equiv 1$ and let $\mathbb{F}_1 : (0,\infty) \times \mathbb{R}^n \to (0,\infty)$. Suppose further that, for every $x \in X$, one has $\sup_{\mathbf{t}\in\mathbb{R}^n} \|F(\mathbf{t}; x)\|_Y < \infty$, as well as that $\emptyset \neq W_2 \subseteq (2\mathbb{Z})^n$ and for every $\mathbf{t} \in \mathbb{R}^n$ there exists $l_1 > 0$ such that, for every $l \geqslant l_1$ and $w \in lW_2$, we have*

$$\int_{l\Omega} \varphi_{p_1(\mathbf{u})}\left(2\mathbb{F}_1(l, \mathbf{t}+w)\nu_1(\mathbf{u}) \sum_{k\in l(2\mathbb{Z}+1)^n} \frac{\|h(\mathbf{u}+k-\mathbf{v})/\nu(\mathbf{v})\|_{L^{q(\mathbf{v})}(l\Omega)}}{\mathbb{F}(l, \mathbf{t}-k+w)}\right) du \leqslant 1.$$

Then the function $h * F : \mathbb{R}^n \times X \to Y$, defined by

$$(h * F)(\mathbf{t}; x) := \int_{\mathbb{R}^n} h(\sigma)F(\mathbf{t}-\sigma; x)\, d\sigma, \quad \mathbf{t} \in \mathbb{R}^n,\ x \in X, \tag{285}$$

is Weyl $(\mathbb{F}_1, \mathcal{P}_1, \mathrm{R}, \mathcal{B}, W_2)$-multialmost automorphic of type 2, where $p_1 \in \mathcal{P}(\mathbb{R}^n)$ and $P_l^1 = L_{\nu_1}^{p_1(\mathbf{u})}(l\Omega : Y)$ for all $l > 0$ and some Lebesgue measurable function $\nu_1 : \mathbb{R}^n \to (0, +\infty)$.

We can similarly consider the invariance of Weyl-$(\mathbb{F}, \mathcal{P}, \mathrm{R}, \mathcal{B}, W)$-multialmost automorphy of type 2 under the actions of the infinite convolution product (262) as well as the convolution invariance of joint Weyl-$(\mathbb{F}, \mathcal{P}, \mathrm{R}, \mathcal{B}, W)$-multialmost automorphy under the actions of convolution product (285) and the convolution product (262); cf. also [447, Theorems 8.3.27, 8.3.29]. The possible applications can be given to the heat equation in $\mathbb{R}^n$ and the evolution systems generated by the family of operators $(A(t) \equiv \Delta + a(t)\mathrm{I})_{t\geqslant 0}$, where Δ denotes the Dirichlet Laplacian on $L^r(\mathbb{R}^n)$ for some $r \geqslant 1$ and $a \in L^\infty([0,\infty))$; cf. also the second application and the third application given in [447, Subsection 8.3.5].

2. Some applications to the abstract fractional difference equations

The class of jointly equi-Weyl-p-normal functions, where $p \geqslant 1$, have been introduced in the final section of [461]. This class and its metrical generalizations are important because they are stable, in a certain sense, under the actions of convolution products. This enables one to take up the study of the asymptotically Weyl almost automorphic type solutions for a class of the abstract impulsive first-order differential inclusions (cf. [266, Subsections 4.1, 4.2] for more details in this direction).

For our next application, we need the following notion.

Definition 5.2.15. Suppose that $\mathbb{F} : \mathbb{R} \to (0,\infty)$, $\nu : \mathbb{Z} \to (0,\infty)$ and $p \in [1,\infty)$. Then we say that a sequence $f : \mathbb{Z} \to X$ is jointly equi-Weyl-$(\mathbb{F}, p, \nu, p_b)$-almost automorphic if there exist positive real numbers $M \geqslant 1$ and $s \geqslant 0$ such that $\|f(k)\| \leqslant M(1+|k|)^s$, $k \in \mathbb{Z}$ as well as that for any integer sequence (s_r) there exist a subsequence (s_{r_m}) of (s_r), a sequence $f^* : \mathbb{Z} \to X$ and positive real numbers $M' \geqslant 1$ and $s' \geqslant 0$ such that $\|f^*(k)\| \leqslant M'(1+|k|)^{s'}$, $k \in \mathbb{Z}$ and

$$\lim_{(m,l)\to+\infty} \sup_{k\in\mathbb{R}} \mathbb{F}(l)\left[\sum_{j=-l}^{l} \|f(j+k+s_{r_m}) - f^*(k+j)\|^p \nu^p(j)\right]^{1/p} = 0.$$

In the following result, we analyze the existence and uniqueness of jointly equi-Weyl-$(\mathbb{F}, p, \nu, p_b)$-almost automorphic solutions of the first-order difference equation (41), where $A \in L(X)$ and $(f_k \equiv f(k))_{k\in\mathbb{Z}}$ is a jointly equi-Weyl-$(\mathbb{F}, p, \nu, p_b)$-almost automorphic sequence; cf. also [65, Section 3] and [476, Theorem 7].

Theorem 5.2.16. *Suppose that $\|A\| < 1$, $\mathbb{F} : \mathbb{R} \to (0,\infty)$, $\nu : \mathbb{Z} \to (0,\infty)$, $p \in [1,\infty)$ and $f : \mathbb{Z} \to X$ is a jointly equi-Weyl-$(\mathbb{F}, p, \nu, p_b)$-almost automorphic sequence. Then there exists a unique polynomially bounded solution $u(\cdot)$ of (41) and $u(\cdot)$ is jointly equi-Weyl-$(\mathbb{F}, p, \nu, p_b)$-almost automorphic, provided that there exist a sequence $\psi : \mathbb{Z} \to (0,\infty)$ and a number $\sigma > 0$ such that $\nu(x+y) \leqslant \nu(x)\psi(y)$ for all $x, y \in \mathbb{Z}$ and $\sum_{v=0}^{+\infty}[\psi(v+1)]^p/(1+v^\sigma)^p < +\infty$.*

Proof. It is well known that a unique polynomially bounded solution $u(\cdot)$ of (41) is given by

$$u(k) = \sum_{v=0}^{+\infty} A^v f(k-v-1), \quad k \in \mathbb{Z};$$

observe that the above series converges since there exist positive real numbers $M \geqslant 1$ and $s \geqslant 0$ such that $\|f(k)\| \leqslant M(1+|k|)^s$, $k \in \mathbb{Z}$. Let $\varepsilon > 0$ be fixed. Then we know that for any integer sequence (s_r) there exist a subsequence (s_{r_m}) of (s_r), a sequence $f^* : \mathbb{Z} \to X$ and positive real numbers $M' \geqslant 1$ and $s' \geqslant 0$ such that $\|f^*(k)\| \leqslant M'(1+|k|)^{s'}$, $k \in \mathbb{Z}$ as well as that there exists $w > 0$ such that, for every $k \in \mathbb{Z}$, $l \geqslant w$ and for every $m \in \mathbb{N}$ with $m \geqslant w$, we have

$$\mathbb{F}(l)\left[\sum_{j=-l}^{l} \|f(j+k+s_{r_m}) - f^*(k+j)\|^p \nu^p(j)\right]^{1/p} \leqslant \varepsilon.$$

Set $u^*(k) := \sum_{v=0}^{+\infty} A^v f^*(k-v-1)$, $k \in \mathbb{Z}$; obviously, this series is convergent due to the polynomial boundedness of the sequence $f^*(\cdot)$. Observing that $\nu(j) \leqslant \nu(j-v-1)\psi(v+1)$ for all $j \in \mathbb{Z}$, $v \in \mathbb{N}_0$ and $\sum_{v=0}^{+\infty} [\psi(v+1)]^p/(1+v^\sigma)^p < +\infty$, we can repeat verbatim the argumentation contained in the proof of [476, Theorem 7] in order to see that there exists an absolute constant $c_A^p > 0$ such that, for every $k \in \mathbb{Z}$, $l \geqslant w$ and for every $m \in \mathbb{N}$ with $m \geqslant w$, we have

$$\mathbb{F}(l)\left[\sum_{j=-l}^{l} \|u(j+k+s_{r_m}) - u^*(k+j)\|^p \nu^p(j)\right]^{1/p} \leqslant c_A^p \varepsilon.$$

□

We can also formulate an analogue of Theorem 5.2.16 for the abstract first-order difference inclusion (47). Furthermore, the statements of [476, Theorems 8, 9] can be formulated for jointly equi-Weyl-$(\mathbb{F}, p, \nu, p_b)$-almost automorphic sequences as well. We can similarly analyze the existence and uniqueness of jointly equi-Weyl-$(\mathbb{F}, p, \nu, b)$-almost automorphic solutions of the already considered equation

$$\Delta^\alpha u(k) = Au(k+1) + f(k), \quad k \in \mathbb{Z},$$

where A is a closed linear operator on X, $0 < \alpha < 1$ and $\Delta^\alpha u(k)$ denotes the Caputo fractional difference operator of order α. We would like to note that we can similarly analyze the existence and uniqueness of jointly equi-Weyl-$(\mathbb{F}, p, \nu, b)$-almost automorphic solutions of this equation (cf. Definition 5.2.15 with $s = s' = 0$). Furthermore, the existence and uniqueness of $\mathbb{D}$-asymptotically jointly equi-Weyl-$(\mathbb{F}, p, \nu, p_b)$-almost automorphic type solutions for the difference equation

$$u(k+1) = Au(k) + f(k), \quad k \geqslant 0; \quad u(0) = u_0$$

as well as the difference equation

$$u(k, m) = A(k, m)u(k-1, m-1) + f(k, m), \quad k,\ m \in \mathbb{N},$$

subjected with the initial conditions

$$u(k, 0) = u_{k,0}; \quad u(0, m) = u_{0,m}, \quad k,\ m \in \mathbb{N}_0,$$

can be considered similarly as before.

5.2.4 Vectorially Weyl almost automorphic type functions

Concerning the various notions of (metrical) Weyl almost automorphy considered in [6, 461] and this section, we would like to emphasize that we can also consider the following classes of functions (here, the assumption $p = 1$ is almost inevitable).

Definition 5.2.17. Let $f \in L^1_{\mathrm{loc}}(\mathbb{R} : X)$. Then we say that $f(\cdot)$ is vectorially Weyl almost automorphic, respectively, vectorially Weyl almost automorphic of type 1, if for every real sequence (s_k), there exist a subsequence (s_{k_m}) and a function $f^* \in L^1_{\mathrm{loc}}(\mathbb{R} : X)$ such that

$$\lim_{m\to\infty} \lim_{l\to+\infty} \frac{1}{2l} \int_{-l}^{l} [f(t + s_{k_m} + x) - f^*(t + x)]\, dx = 0, \tag{286}$$

respectively,

$$\lim_{l\to+\infty} \lim_{m\to\infty} \frac{1}{2l} \int_{-l}^{l} [f(t + s_{k_m} + x) - f^*(t + x)]\, dx = 0, \tag{287}$$

and

$$\lim_{m\to\infty} \lim_{l\to+\infty} \frac{1}{2l} \int_{-l}^{l} [f^*(t - s_{k_m} + x) - f(t + x)]\, dx = 0, \tag{288}$$

respectively,

$$\lim_{l\to+\infty} \lim_{m\to\infty} \frac{1}{2l} \int_{-l}^{l} [f^*(t - s_{k_m} + x) - f(t + x)]\, dx = 0, \tag{289}$$

for each $t \in \mathbb{R}$. We similarly introduce the class of jointly Weyl almost automorphic functions in the vectorial sense.

It is worth noting that we use the vector-valued integration in the equations (286)–(289). This can be essentially in some concrete situations, as the following well-known example indicates.

Example 5.2.18. Let $X := c_0$ and let $f(t) := ((1/n)\cos(t/n))_n$, $t \in \mathbb{R}$. Let us reconsider its first antiderivative $F(t) := (\sin(t/n))_n$, $t \in \mathbb{R}$ from Example 4.3.13.

It is clear that $F(\cdot)$ is vectorially Weyl almost automorphic since for every fixed integer $k_m \in \mathbb{N}$ we have that the second limits in (286) and (288) are equal to zero, with $F^* \equiv F$. In order to see this, notice that we have ($t,\ x \in \mathbb{R}$, $s_{k_m} \in \mathbb{R}$, $k_m \in \mathbb{N}$):

$$F(t + x + s_{k_m}) - F(t + x) = 2\left(\sin \frac{s_{k_m}}{2n} \cdot \cos \frac{2t + 2x + 2s_{k_m}}{2n}\right)_n,$$

which simply implies that, for every $l > 0$, we have

$$\begin{aligned}&\left\| \frac{1}{2l} \int_{-l}^{l} [F(t+x+s_{k_m}) - F(t+x)]\, dx \right\| \\ &= \left\| \frac{4}{l} \left(n \cdot \sin \frac{s_{k_m}}{2n} \cdot \sin \frac{l}{n} \cdot \cos \frac{2t + s_{k_m}}{2n} \right)_n \right\| \leqslant \frac{4}{l} \frac{|s_{k_m}|}{2};\end{aligned}$$

the second limit equation in (288) can be considered similarly. Moreover, $F(\cdot)$ is not vectorially Weyl almost automorphic of type 1, which can be proved as follows. If we suppose the contrary, then a simple computation shows that for each sequence (s_k) there exist a subsequence (s_{k_m}) of (s_k) and a mapping $F^* : \mathbb{R} \to c_0$ such that, for every $t \in \mathbb{R}$, we have

$$\lim_{l \to +\infty} \lim_{m \to +\infty} l^{-1} \left[\left(n \cdot \sin \frac{t + s_{k_m}}{n} \cdot \sin \frac{l}{n} \right)_n - \frac{1}{2} \int_{-l}^{l} F^*(t+x)\, dx \right] = 0.$$

This, in particular, implies that the second limit $\lim_{m \to +\infty} \cdot$ in the above expression always exists if l is an irrational multiple of π. After a simple calculation, we get that the sequence $(\sin((t + s_{k_m})/n))_n$ exists in c_0 as $m \to +\infty$, whence we may conclude by the proof of [184, Proposition 2.5] that the range of function $F(\cdot)$ is relatively compact in c_0. This is false and yields the contradiction.

For the continuation, let us observe that an application of Proposition 5.2.5 shows the following:

(HP) For each exponent $p > 0$, the function $F(\cdot)$ is (jointly) Weyl-p-almost automorphic (of type 1) if and only if $F(\cdot)$ is (jointly) Weyl-1-almost automorphic (of type 1).

We will prove that $F(\cdot)$ is neither (jointly) Weyl-p-almost automorphic (of type 1) nor jointly Weyl almost automorphic in the vectorial sense. Due to (HP), it suffices to consider the case $p = 1$ in the sequel.

We will first prove that $F(\cdot)$ is not Weyl-1-almost automorphic of type 1. Let $\varepsilon \in (0, 2^{-1} \sin 1)$ be fixed. If we suppose the contrary, then for each sequence (s_k) there exist a subsequence (s_{k_m}) of (s_k), a mapping $F^* : \mathbb{R} \to c_0$ and a number $l_0 > 0$ such that for each $l \geqslant l_0$ there exists $m_l \in \mathbb{N}$ such that for each $m \geqslant m_l$ we have (put $t = 0$):

$$\frac{1}{2l} \int_{-l}^{l} \|F(s_{k_m} + x) - F^*(x)\|\, dx \leqslant \varepsilon/2.$$

In particular, for each $l \geqslant l_0$ there exists $m_l \in \mathbb{N}$ such that for each $m,\ m' \geqslant m_l$ we have

$$\frac{1}{2l} \left\| \int_{-l}^{l} [F(s_{k_m} + x) - F(x + s_{k_{m'}})]\, dx \right\| \leqslant \varepsilon. \tag{290}$$

Applying the Newton–Leibniz formula and a few elementary transformations, we get for each $l \geqslant l_0$ there exists $m_l \in \mathbb{N}$ such that for each $m,\ m' \geqslant m_l$ we have

$$\sup_{n\in\mathbb{N}}\left|\frac{2n}{l}\cdot\cos\frac{s_{k_m}+s_{k_{m'}}}{2n}\cdot\sin\frac{l}{n}\cdot\sin\frac{s_{k_m}-s_{k_{m'}}}{2n}\right|\leqslant\varepsilon.$$

Plug now $s_k \equiv k$, $l = l_0$ and $m' = m_0$. Since $[n/l\sin(l/n)] \to 1$ as $n \to +\infty$, we obtain that there exist integers $n_0,\ m_0 \in \mathbb{N}$ such that, for every $n \geqslant n_0$ and $m \geqslant m_0$, we have

$$\sup_{n\geqslant n_0}\left|\cos\frac{k_m+k_{m_0}}{2n}\cdot\sin\frac{k_m-k_{m_0}}{2n}\right|\leqslant\varepsilon.$$

But, if $k_m \geqslant k_{m_0} + n_0$, then

$$\begin{aligned}&\sup_{n\geqslant n_0}\left|\cos\frac{k_m+k_{m_0}}{2n}\cdot\sin\frac{k_m-k_{m_0}}{2n}\right|\\&\geqslant\left|\cos\frac{k_m+k_{m_0}}{2(k_m-k_{m_0})}\cdot\sin\frac{k_m-k_{m_0}}{2(k_m-k_{m_0})}\right|\to 2^{-1}\sin 1,\quad m\to+\infty,\end{aligned}$$

which is a contradiction. We can similarly prove that $F(\cdot)$ is not jointly Weyl almost automorphic in the vectorial sense and, therefore, not jointly Weyl almost automorphic.

It remains to be proved that $F(\cdot)$ is not Weyl-1-almost automorphic. If we suppose the contrary, then there exists $m_0 \in \mathbb{N}$ such that, for every $m,\ m' \geqslant m_0$, there exists $l_{m,m'} > 0$ such that for each $l \geqslant l_{m,m'}$ we have (290). Plug now $s_k \equiv k$, $m' = m_0$ and minorize the norm of the sequence in (290) in c_0 by the absolute value of its $(k_m - k_{m_0})$-th element. This yields that for each integer $m \geqslant m_0$ there exists $l_m > 0$ such that for each $l \geqslant l_m$ we have

$$\frac{1}{2l}\int_{-l}^{l}\left|\cos\frac{2x+k_m+k_{m_0}}{2(k_m-k_{m_0})}\cdot\sin\frac{k_m-k_{m_0}}{2(k_m-k_{m_0})}\right|dx\leqslant\varepsilon. \tag{291}$$

Substituting $y = \frac{2x+k_m+k_{m_0}}{2(k_m-k_{m_0})}$ in (291), we simply get

$$\frac{k_m-k_{m_0}}{2l}\int_{\frac{-2l+k_m+k_{m_0}}{2(k_m-k_{m_0})}}^{\frac{2l+k_m+k_{m_0}}{2(k_m-k_{m_0})}}|\cos y|\,dy\leqslant\varepsilon/\sin(1/2),\quad l\geqslant l_m.$$

This is impossible since the term on the left-hand side of the above estimate behaves like

$$\frac{k_m-k_{m_0}}{2l}\cdot 2\cdot\frac{\frac{2l+k_m+k_{m_0}}{2(k_m-k_{m_0})}-\frac{-2l+k_m+k_{m_0}}{2(k_m-k_{m_0})}}{\pi}=\frac{2}{\pi}.$$

Conclusions and final remarks

It is worth noting that the approach obeyed in this section and the previous one can be useful for the introduction and further analysis of some other classes of generalized vectorially almost automorphic functions. For example, we can introduce Σ-almost automorphic type functions and vectorially Stepanov almost automorphic functions in the following way: First of all, we refer the reader to [447, Definitions 8.1.1, 8.1.8] for the notion of a (compactly) $(\mathrm{R}, \mathcal{B})$-multialmost automorphic [(compactly) $(\mathrm{R}_X, \mathcal{B})$-multialmost automorphic] function $F : \mathbb{R}^n \times X \to Y$, where R $[\mathrm{R}_X]$ is a certain collection of sequences in $\mathbb{R}^n$ $[\mathbb{R}^n \times X]$; cf. also [447, Definition 8.1.2 and p. 519].

Now we are ready to introduce the following analogue of Definition 4.3.6.

Definition 5.2.19. Suppose that $\Sigma : Y^{\mathbb{R}^n \times X} \to Y^{\mathbb{R}^n \times X}$ and $F : \mathbb{R}^n \times X \to Y$ is a given function. Then we say that the function $F(\cdot;\cdot)$ is (compactly) $(\Sigma, \mathrm{R}, \mathcal{B})$-multialmost automorphic [(compactly) $(\Sigma, \mathrm{R}_X, \mathcal{B})$-multialmost automorphic] if the function $\Sigma(F)$ is (compactly) $(\mathrm{R}, \mathcal{B})$-multialmost automorphic [(compactly) $(\mathrm{R}_X, \mathcal{B})$-multialmost automorphic].

The following analogue of Definition 4.3.9 can be also analyzed.

Definition 5.2.20. Suppose that $F : \mathbb{R}^n \times X \to Y$, for each $x \in X$ the mapping $F(\cdot; x)$ is locally integrable, Ω is a compact subset of $\mathbb{R}^n$ with positive Lebesgue measure and $\Sigma : Y^{\mathbb{R}^n \times X} \to Y^{\mathbb{R}^n \times X}$ is given through (249), with $\Lambda = \mathbb{R}^n$. Then the function $F(\cdot;\cdot)$ is said to be vectorially (compactly) Stepanov $(\mathrm{R}, \mathcal{B})$-multialmost automorphic [vectorially (compactly) Stepanov $(\mathrm{R}_X, \mathcal{B})$-multialmost automorphic] if the function $F(\cdot;\cdot)$ is (compactly) $(\Sigma, \mathrm{R}, \mathcal{B})$-multialmost automorphic [(compactly) $(\Sigma, \mathrm{R}_X, \mathcal{B})$-multialmost automorphic].

Any Stepanov $(\Omega, 1)$-$(\mathrm{R}, \mathcal{B})$-multialmost automorphic [Stepanov $(\Omega, 1)$-$(\mathrm{R}_X, \mathcal{B})$-multialmost automorphic] function, defined in the sense of [447, Definition 8.2.1(i), (iii)], is vectorially Stepanov $(\mathrm{R}, \mathcal{B})$-multialmost automorphic [vectorially Stepanov $(\mathrm{R}_X, \mathcal{B})$-multialmost automorphic], which can be simply proved in the following way. Let us consider, for simplicity, a Stepanov $(\Omega, 1)$-$(\mathrm{R}, \mathcal{B})$-multialmost automorphic function $F : \mathbb{R}^n \times X \to Y$. If $B \in \mathcal{B}$ and $(b_k)_{k \in \mathbb{N}} \in \mathrm{R}$ are given, then we can find a subsequence $(b_{k_l})_{l \in \mathbb{N}}$ of $(b_k)_{k \in \mathbb{N}}$ and a limit function $F^* : \mathbb{R}^n \times X \to L^1(\Omega : Y)$ such that for each $\mathbf{t} \in \mathbb{R}^n$ and $x \in B$ we have that the mapping $\mathbf{u} \mapsto [F^*(\mathbf{t}; x)](\mathbf{u})$, $\mathbf{u} \in \Omega$ is locally integrable,

$$\lim_{l \to +\infty} \int_{\Omega} \left\| F(\mathbf{t} + \mathbf{u} + b_{k_l}; x) - [F^*(\mathbf{t}; x)](\mathbf{u}) \right\| d\mathbf{u} = 0$$

and

$$\lim_{l \to +\infty} \int_{\Omega} \left\| [F^*(\mathbf{t} + \mathbf{u} - b_{k_l}; x)](\mathbf{u}) - F(\mathbf{t} + \mathbf{u}; x) \right\| d\mathbf{u} = 0.$$

Define $F_1^* : \mathbb{R}^n \times X \to Y$ by $F_1^*(\mathbf{t}; x) := \int_{\Omega} [F^*(\mathbf{t}; x)](\mathbf{u})\, d\mathbf{u}$, $\mathbf{t} \in \mathbb{R}^n$, $x \in B$. Then it readily follows that, for every $\mathbf{t} \in \mathbb{R}^n$ and $x \in B$, we have

$$\lim_{l\to+\infty}\int_{\Omega} F(\mathbf{t}+\mathbf{u}+b_{k_l};x)=F_1^*(\mathbf{t};x)$$

and

$$\lim_{l\to+\infty} F_1^*(\mathbf{t}-b_{k_l};x)=\int_{\Omega} F(\mathbf{t}+\mathbf{u};x)\,d\mathbf{u},$$

as required.

An attempt should be made to reconsider Proposition 4.3.7 for Σ-almost automorphic type functions and the structural results presented in [447, Section 8.2] for vectorially Stepanov almost automorphic type functions.

5.3 Metrical Besicovitch almost automorphy and applications

In this section, we reconsider and slightly generalize various classes of Besicovitch almost automorphic functions [447, 464]. More precisely, we consider here various classes of metrically Besicovitch almost automorphic functions of the form $F : \mathbb{R}^n \times X \to Y$ and metrically Besicovitch almost automorphic sequences of the form $F : \mathbb{Z}^n \times X \to Y$; the class of vectorially Besicovitch almost automorphic type functions and the class of vectorially Besicovitch almost automorphic type functions in general metric can be also introduced (cf. [448, Section 3.4] and the subsequent section for more details). The main structural characterizations for the introduced classes of metrically Besicovitch almost automorphic functions and sequences are established. In addition to the above, we provide some applications of our results to the abstract Volterra integrodifferential equations.

Unless stated otherwise, we will always assume henceforth that $\Omega := [-1,1]^n \subseteq \mathbb{R}^n$ $[\Omega := [-1,1]^n \cap \mathbb{Z}^n \subseteq \mathbb{Z}^n]$, $\mathbb{F} : (0,\infty)\times\mathbb{R}^n \to (0,\infty)$ $[\mathbb{F} : (0,\infty)\times\mathbb{Z}^n \to (0,\infty)]$ as well as that, for every $l > 0$, (P_l, d_l) is a pseudometric space, where $P_l \subseteq Y^{l\Omega}$ $[P_l \subseteq Y^{l\Omega\cap\mathbb{Z}^n}]$ is closed under the addition and subtraction of functions, containing the zero-function. Define $\|f\|_l := d_l(f,0)$ for all $f \in P_l$ $(l > 0)$. We will always assume henceforth that R is a collection of sequences in $\mathbb{R}^n$ $[\mathbb{Z}^n]$; for simplicity and better understanding, we will not consider here the corresponding classes of functions with the collections R_X of sequences in $\mathbb{R}^n \times X$ $[\mathbb{Z}^n \times X]$.

The following notion generalizes the notion introduced recently in [464, Definition 2.1] (for simplicity, we will consider the case $\phi(x) \equiv x$ here).

Definition 5.3.1. Suppose that $F : \mathbb{R}^n \times X \to Y$ $[F : \mathbb{Z}^n \times X \to Y]$ satisfies that for each $x \in X$, $l > 0$ and $\mathbf{t} \in \mathbb{R}^n$ $[\mathbf{t} \in \mathbb{Z}^n]$ we have $F(\mathbf{t}+\cdot;x) \in P_l$. Let for every $l > 0$, $B \in \mathcal{B}$ and $(\mathbf{b}_k = (b_k^1, b_k^2, \ldots, b_k^n)) \in \mathrm{R}$ there exist a subsequence $(\mathbf{b}_{k_m} = (b_{k_m}^1, b_{k_m}^2, \ldots, b_{k_m}^n))$ of $(\mathbf{b}_k)$ and a function $F^* : \mathbb{R}^n \times X \to P_l$ $[F^* : \mathbb{Z}^n \times X \to P_l]$ such that for each $x \in B$, $l > 0$ and $\mathbf{t} \in \mathbb{R}^n$ $[\mathbf{t} \in \mathbb{Z}^n]$ we have

(i)

$$\lim_{m\to+\infty}\limsup_{l\to+\infty}\mathbb{F}(l,\mathbf{t})\big\|F(\mathbf{t}+\cdot+(b_{k_m}^1,\ldots,b_{k_m}^n);x)-[F^*(\mathbf{t};x)](\cdot)\big\|_l=0 \tag{292}$$

and

$$\lim_{m\to+\infty}\limsup_{l\to+\infty}\mathbb{F}(l,\mathbf{t})\big\|[F^*(\mathbf{t}-(b_{k_m}^1,\ldots,b_{k_m}^n);x)](\cdot)-F(\mathbf{t}+\cdot;x)\big\|_l=0, \tag{293}$$

pointwise for all $x\in B$ and $\mathbf{t}\in\mathbb{R}^n$ [$\mathbf{t}\in\mathbb{Z}^n$], or

(ii)

$$\lim_{m\to+\infty}\liminf_{l\to+\infty}\mathbb{F}(l,\mathbf{t})\big\|F(\mathbf{t}+\cdot+(b_{k_m}^1,\ldots,b_{k_m}^n);x)-[F^*(\mathbf{t};x)](\cdot)\big\|_l=0 \tag{294}$$

and

$$\lim_{m\to+\infty}\liminf_{l\to+\infty}\mathbb{F}(l,\mathbf{t})\big\|[F^*(\mathbf{t}-(b_{k_m}^1,\ldots,b_{k_m}^n);x)](\cdot)-F(\mathbf{t}+\cdot;x)\big\|_l=0, \tag{295}$$

pointwise for all $x\in B$ and $\mathbf{t}\in\mathbb{R}^n$ [$\mathbf{t}\in\mathbb{Z}^n$],

(iii)

$$\lim_{l\to+\infty}\limsup_{m\to+\infty}\mathbb{F}(l,\mathbf{t})\big\|F(\mathbf{t}+\cdot+(b_{k_m}^1,\ldots,b_{k_m}^n);x)-[F^*(\mathbf{t};x)](\cdot)\big\|_l=0$$

and

$$\lim_{l\to+\infty}\limsup_{m\to+\infty}\mathbb{F}(l,\mathbf{t})\big\|[F^*(\mathbf{t}-(b_{k_m}^1,\ldots,b_{k_m}^n);x)](\cdot)-F(\mathbf{t}+\cdot;x)\big\|_l=0,$$

pointwise for all $x\in B$ and $\mathbf{t}\in\mathbb{R}^n$ [$\mathbf{t}\in\mathbb{Z}^n$], or

(iv)

$$\lim_{l\to+\infty}\liminf_{m\to+\infty}\mathbb{F}(l,\mathbf{t})\big\|F(\mathbf{t}+\cdot+(b_{k_m}^1,\ldots,b_{k_m}^n);x)-[F^*(\mathbf{t};x)](\cdot)\big\|_l=0$$

and

$$\lim_{l\to+\infty}\liminf_{m\to+\infty}\mathbb{F}(l,\mathbf{t})\big\|[F^*(\mathbf{t}-(b_{k_m}^1,\ldots,b_{k_m}^n);x)](\cdot)-F(\mathbf{t}+\cdot;x)\big\|_l=0,$$

pointwise for all $x\in B$ and $\mathbf{t}\in\mathbb{R}^n$ [$\mathbf{t}\in\mathbb{Z}^n$].

In the case that (i), respectively, (ii) holds, respectively, [(iii), resp., (iv) holds], then we say that $F(\cdot;\cdot)$ is Besicovitch-$(\mathbb{F},\mathcal{P},\mathrm{R},\mathcal{B})$-multialmost automorphic, respectively, weakly Besicovitch-$(\mathbb{F},\mathcal{P},\mathrm{R},\mathcal{B})$-multialmost automorphic [Besicovitch-$(\mathbb{F},\mathcal{P},\mathrm{R},\mathcal{B})$-multialmost automorphic of type 1, respectively, weakly Besicovitch-$(\mathbb{F},\mathcal{P},\mathrm{R},\mathcal{B})$-multialmost automorphic of type 1].

We will also consider the notion in which the limit function $F^*(\cdot;\cdot)$ from Definition 5.3.1 is bounded by the function $\omega : \mathbb{R}^n \to (0,\infty)$ $[\omega : \mathbb{Z}^n \to (0,\infty)]$ in the sense that there exists $M > 0$ such that, for every $x \in B$, $l > 0$ and $\mathbf{u} \in l\Omega$, we have $\|[F^*(\mathbf{t};x)](\mathbf{u})\|_Y \leqslant M\omega(|\mathbf{t}|)$, $\mathbf{t} \in \mathbb{R}^n$ $[\|[F^*(\mathbf{t};x)](\mathbf{u})\|_Y \leqslant M\omega(|\mathbf{t}|)$, $\mathbf{t} \in \mathbb{Z}^n]$. If this is the case, then we say that the function $F(\cdot;\cdot)$ is Besicovitch-$(\mathbb{F},\mathcal{P},\mathrm{R},\mathcal{B},\omega)$-multi-almost automorphic, e. g.; furthermore, if $\omega(\cdot) \equiv 1$, then we write "b" in place of "ω". For simplicity, we will not consider here the notion in which the first limits in the above equations are replaced by $\limsup_{m\to}\cdot$, $\liminf_{m\to}\cdot$ and the corresponding limits with the variable $l > 0$.

If $1 \leqslant p < \infty$, then the notion of a (weakly) Besicovitch-p-almost automorphic function (of type 1) $F : \mathbb{R}^n \to Y$ is obtained with $\Omega = [-1,1]^n$, R being the collection of all sequences in $\mathbb{R}^n$, $\mathbb{F}(l,\mathbf{t}) \equiv l^{-n/p}$ and $P_l = L^p([-l,l]^n : Y)$; the notion of a (weakly) Besicovitch-p-almost automorphic sequence (of type 1) $F : \mathbb{Z}^n \to Y$ is new and can be obtained with $\Omega = [-1,1]^n$, R being the collection of all sequences in $\mathbb{Z}^n$, $\mathbb{F}(l,\mathbf{t}) \equiv l^{-n/p}$ and $P_l = L^p([-l,l]^n \cap \mathbb{Z}^n : Y)$, where

$$\|f\|_l \equiv \left[\sum_{j\in[-l,l]^n\cap\mathbb{Z}^n} \|f(j)\|_Y^p\right]^{1/p}, \quad \text{if } p \geqslant 1, \tag{296}$$

$$\text{respectively, } \|f\|_l \equiv \sum_{j\in[-l,l]^n\cap\mathbb{Z}^n} \|f(j)\|_Y^p, \quad \text{if } p \in (0,1). \tag{297}$$

We similarly define the notion of (weak) Besicovitch-(p,R)-almost automorphy (of type 1) [(weak) Besicovitch-(p,R,b)-almost automorphy (of type 1)], where R is a general collection of sequences obeying our requirements. The important metrical generalizations of (296)–(297) are given by

$$\|f\|_l \equiv \left[\sum_{j\in[-l,l]^n\cap\mathbb{Z}^n} \|f(j)\|_Y^p \nu^p(j)\right]^{1/p}, \quad \text{if } p \geqslant 1,$$

$$\text{resp., } \|f\|_l \equiv \sum_{j\in[-l,l]^n\cap\mathbb{Z}^n} \|f(j)\|_Y^p \nu^p(j), \quad \text{if } p \in (0,1),$$

where $\nu : \mathbb{Z}^n \to (0,\infty)$ is an arbitrary sequence.

Now we will introduce the corresponding notion with $0 < p < 1$.

Definition 5.3.2. Suppose that $0 < p < 1$ and $F : \mathbb{R}^n \to Y$ $[F : \mathbb{Z}^n \to Y]$. Then we say that $F(\cdot)$ is (weakly) Besicovitch-p-almost automorphic function (of type 1) [(weakly) Besicovitch-(p,b)-almost automorphic sequence (of type 1)] if $F(\cdot)$ is (weakly) Besicovitch-$(\mathbb{F},\mathcal{P},\mathrm{R})$-multialmost automorphic (of type 1) [(weakly) Besicovitch-$(\mathbb{F},\mathcal{P},\mathrm{R},b)$-multialmost automorphic (of type 1)], where $\Omega = [-1,1]^n$, R is the collection of all sequences in $\mathbb{R}^n$ $[\mathbb{Z}^n]$, $\mathbb{F}(l,\mathbf{t}) \equiv l^{-n}$ and $P_l = L^p([-l,l]^n : Y)$ $[P_l = l^p([-l,l]^n \cap \mathbb{Z}^n : Y)]$. If R is a general collection of sequences obeying our requirements, then we also say that $F(\cdot)$ is (weakly) Besicovitch-(p,R)-almost automorphic (of type 1) [(weakly) Besicovitch-(p,R,b)-almost automorphic (of type 1)].

We can further generalize the notion introduced in the above definitions following our approach from [464, Definition 2.6]. The notion in which $P_l = L^p([-l,l]^n : Y)$ $[P_l = l^p([-l,l]^n \cap \mathbb{Z}^n : Y)]$ and $\mathbb{F}(l,\mathbf{t}) \equiv l^{-n/p}$, if $1 \leqslant p < +\infty$, respectively, $\mathbb{F}(l,\mathbf{t}) \equiv l^{-n}$, if $0 < p < 1$, is the most important, when we have the following result; the proof is almost the same as the corresponding proof for the Weyl classes of metrically almost automorphic functions and, therefore, omitted.

Proposition 5.3.3. *Let P_l and $\mathbb{F}(l,\cdot)$ be as above ($l > 0$). Then the following holds:*

(i) *Suppose that $0 < p \leqslant q < +\infty$. Then any (weakly) Besicovitch-q-almost automorphic function $F : \mathbb{R}^n \to Y$ $[F : \mathbb{Z}^n \to Y]$ (of type 1) is (weakly) Besicovitch-p-almost automorphic (of type 1); furthermore, the same holds for the corresponding classes of (weakly) Besicovitch-(q,b)-almost automorphic functions (of type 1) and (weakly) Besicovitch-(p,b)-almost automorphic functions (of type 1).*
(ii) *Suppose that $0 < p \leqslant q < +\infty$. Then $F : \mathbb{R}^n \to Y$ $[F : \mathbb{Z}^n \to Y]$ is essentially bounded, (weakly) Besicovitch-(q,b)-almost automorphic (of type 1) if and only if $F(\cdot)$ is essentially bounded, (weakly) Besicovitch-(p,b)-almost automorphic (of type 1).*

The statement of [464, Proposition 2.5] can be simply formulated for a class of metrically Besicovich almost automorphic functions of type 1, with a general exponent $p > 0$; for example, if $p > 0$, $\sigma > 0$, $\mathbb{F}(l) \equiv l^{-\sigma}$, $f \in L^p_{\mathrm{loc}}(\mathbb{R} : X)$ and there exist a strictly increasing sequence (l_k) of positive real numbers tending to plus infinity, a sequence (b_k) of real numbers and a positive real number $\varepsilon_0 > 0$ such that, for every $k \in \mathbb{N}$ and for every subsequence of (b_{k_m}) of (b_k), we have

$$\lim_{m\to+\infty} l_k^{-\sigma} \int_{b_{k_m}-l_k}^{b_{k_m}+l_k} \|f(x)\|^p \nu^p(x)\,dx = +\infty,$$

where $\nu : \mathbb{R} \to (0,\infty)$ is a Lebesgue measurable function, then $f(\cdot)$ cannot be Besicovitch-$(\mathbb{F},\mathcal{P})$-multialmost automorphic, where $P_l = L^p_\nu([-l,l] : X)$ for all $l > 0$. A similar statement can be formulated for the sequences.

Furthermore, the statement of [464, Proposition 2.8], which concerns the pointwise multiplication of Besicovitch almost automorphic-type functions, can be slightly generalized with the usage of the Banach spaces $P_l = L^p_\nu([-l,l]^n : Y)$ for all $l > 0$, provided that a Lebesgue measurable function $\nu : \mathbb{R}^n \to Y$ satisfies that there exists a bounded function $\varphi : \mathbb{R}^n \to (0,\infty)$ such that $\nu(x+y) \leqslant \nu(x)\varphi(y)$ for all $x, y \in \mathbb{R}^n$. This can be simply employed for the construction of multidimensional Besicovitch almost periodic type functions in general metric; cf. [464, Example 2.9].

Concerning the composition principle for a class of Besicovich almost automorphic functions of type 1, established in [464, Theorem 2.10], we would like to note that the same proof shows the validity of this result for all exponents p, $q > 0$ such that $p = aq$. Moreover, we can similarly deduce the following result in which we consider the metrical Besicovitch-p-almost automorphy of the multidimensional Nemytskii operator $W : \mathbb{R}^n \times X \to Z$, given by

$$W(\mathbf{t};x) := G(\mathbf{t};F(\mathbf{t};x)), \quad \mathbf{t} \in \mathbb{R}^n,\ x \in X, \tag{298}$$

where $F : \mathbb{R}^n \times X \to Y$ and $G : \mathbb{R}^n \times Y \to Z$ (for the notion of uniform $(\mathrm{R}, \mathcal{B}')$-almost automorphy, we refer the reader to [184]; the result of [464, Theorem 2.10(ii)] can be generalized in a similar fashion).

Theorem 5.3.4. *Suppose that* $0 < p,\ q < +\infty$, $\alpha > 0$, $p = \alpha q$, $\mathbb{F}(l,t) \equiv l^{-n/p}$, $F(\cdot;\cdot)$ *is Besicovitch-*$(\mathbb{F}, \mathcal{P}, \mathrm{R}, \mathcal{B})$*-multialmost automorphic of type* 1, *where* $P_l = L^p_\nu([-l,l]^n : Y)$ *for all* $l > 0$ *and some p-locally integrable function* $\nu : \mathbb{R}^n \to (0,\infty)$, *and for every* $B \in \mathcal{B}$ *and* $(\mathbf{b}_k) \in \mathrm{R}$, *the subsequence* $(\mathbf{b}_{k_m})$ *of* $(\mathbf{b}_k)$ *and the function* $F^* : \mathbb{R}^n \times X \to Y$ *from the corresponding definition satisfy* $F^*(\mathbf{t};x) \in \overline{\bigcup_{\mathbf{s}\in\mathbb{R}^n} F(\mathbf{s};x)}$, $\mathbf{t} \in \mathbb{R}^n$, $x \in X$. *Define* $B' := \overline{\bigcup_{\mathbf{t}\in\mathbb{R}^n} F(\mathbf{t};B)}$ *for each set* $B \in \mathcal{B}$, *and* $\mathcal{B}' := \{B' : B \in \mathcal{B}\}$. *Assume, additionally, that, for every sequence from* R, *any its subsequence also belongs to* R, $G : \mathbb{R}^n \times Y \to Z$ *is uniformly* $(\mathrm{R}, \mathcal{B}')$*-almost automorphic and there exists a finite real constant* $a > 0$ *such that*

$$\|G(\mathbf{t};y) - G(\mathbf{t};y')\|_Z \leqslant a\|y - y'\|_Y^{\alpha}, \quad \mathbf{t} \in \mathbb{R}^n,\ y,\ y' \in Y.$$

Then the function $W(\cdot;\cdot)$, *given by* (298), *is Besicovitch-*$(\mathbb{F}^{p/q}, \mathcal{P}_q, \mathrm{R}, \mathcal{B})$*-multi-almost automorphic of type* 1, *where* $P_l = L^q_{\nu^{p/q}}([-l,l]^n : Y)$ *for all* $l > 0$.

We can apply such results in the analysis of the existence and uniqueness of Besicovitch-$(\mathbb{F}^{p/q}, \mathcal{P}_q, \mathrm{R}, \mathcal{B})$-multialmost automorphic solutions for various classes of the abstract (fractional) semilinear Cauchy problems; see the third application in [464, Section 3] and the third application in [455, Section 4].

Concerning the extensions of metrically Besicovitch almost automorphic sequences, we will only state the following result; the proof is much the same as that of Theorem 5.2.12 and, therefore, omitted.

Theorem 5.3.5. *Suppose that* $p > 0$ *and* $F : \mathbb{Z}^n \to Y$ *is bounded, (weakly) Besicovitch-*(p, R)*-multialmost automorphic sequence (of type* 1*), where* R *is any collection of sequences in* $\mathbb{Z}^n$ *such that the assumption* $(b_k) \in \mathrm{R}$ *implies that any subsequence of* (b_k) *also belongs to* R. *Let* R$'$ *be the collection of all sequences* (a_k) *in* $\mathbb{R}^n$ *satisfying that there exists a sequence* $(b_k) \in \mathrm{R}$ *such that* $\sup_{k\in\mathbb{N}} |a_k - b_k| < +\infty$. *Then there exists a bounded, uniformly continuous, (weakly) Besicovitch-*(p, R)*-multialmost automorphic function* $\tilde{F} : \mathbb{R}^n \to Y$ *(of type* 1*) such that* $R(\tilde{F}(\cdot)) \subseteq CH(\overline{R(F)})$, $\|\tilde{F}\|_\infty = \|F\|_\infty$ *and* $\tilde{F}(k) = F(k)$ *for all* $k \in \mathbb{Z}^n$.

Now we will provide certain applications; we will first analyze the convolution invariance of metrical Besicovitch almost type automorphy. Let the operator family $(R(t))_{t>0} \subseteq L(X,Y)$ satisfy that there exist finite real constants $M > 0$, $\beta \in (0,1]$ and $\gamma > \beta$ such that (284) holds. If this is the case, then we are in a position to state and prove the following slight generalization of [464, Proposition 3.2].

Proposition 5.3.6. *Suppose that (284) holds, $a > 0$, $\alpha > 0$, $1 \leqslant p < +\infty$, $p \geqslant 1$, $p(\beta-1)/(p-1) > -1$ if $p > 1$, and $\beta = 1$ if $p = 1$. Suppose further that $b \in [0, \gamma - \beta)$, $w(t) := (1+|t|)^b$, $t \in \mathbb{R}$ and the function $f : \mathbb{R} \to X$ is Besicovitch-$(t^{-a}, \mathcal{P}, \mathrm{R}, w)$-almost automorphic, where $P_l = L^p_\nu([-l,l] : X)$ for all $l > 0$ and some Lebesgue measurable function $\nu : \mathbb{R} \to (0,\infty)$ satisfying that*

$$\limsup_{l\to+\infty}\left[l^{-ap}\int_{-l}^{l}\nu^p(s)\,ds\right] < +\infty \tag{299}$$

and there exists a Lebesgue measurable function $\varphi : \mathbb{R} \to (0,\infty)$ such that $\nu(x) \leqslant \nu(y)\varphi(x-y)$ for all $x, y \in \mathbb{R}$ and there exists $\zeta \in ((1/p)+b, (1/p)+\gamma-\beta)$ with

$$\int_{-\infty}^{+\infty}\frac{\varphi^p(s)}{(1+|s|^\zeta)^p}\,ds < +\infty. \tag{300}$$

If there exists a finite real constant $M' > 0$ such that $\|f(t)\| \leqslant M'w(t)$, $t \in \mathbb{R}$, then the function $F(\cdot)$, given by (262), is continuous and Besicovitch-$(t^{-a}, \mathcal{P}_Y, \mathrm{R}, w)$-almost automorphic, where $P_{l;Y} = L^p_\nu([-l,l] : Y)$ for all $l > 0$, and there exists a finite real constant $M'' > 0$ such that $\|F(t)\|_Y \leqslant M''w(t)$, $t \in \mathbb{R}$.

Proof. We shall content ourselves with sketching the proof. Arguing as in the proof of [455, Proposition 4.2], we can show that the function $F(\cdot)$ is well-defined, continuous and there exists a finite real constant $M'' > 0$ such that $\|F(t)\|_Y \leqslant M''w(t)$, $t \in \mathbb{R}$. Let a sequence $(b_k) \in \mathrm{R}$ be given and let $l > 0$. Then we know that there exist a subsequence (b_{k_m}) of (b_k), a function $f^* : \mathbb{R} \to X$ and a finite real constant $M > 0$ such that $\|[f^*(t)](s)\| \leqslant Mw(t)$, $t \in \mathbb{R}$, $s \in [-l,l]$,

$$\lim_{m\to+\infty}\limsup_{l\to+\infty}\left[l^{-ap}\int_{-l}^{l}\big\|f(t+s+b_{k_m}) - [f^*(t)](s)\big\|^p\nu^p(s)\,ds\right] = 0$$

and

$$\lim_{m\to+\infty}\limsup_{l\to+\infty}\left[l^{-ap}\int_{-l}^{l}\big\|[f^*(t-b_{k_m})](s) - f(t+s)\big\|^p\nu^p(s)\,ds\right] = 0.$$

Define $F^* : \mathbb{R} \to Y$ by $[F^*(t)](u) := \int_{-\infty}^{t} R(t-s)[f^*(s)](u)\,ds$, $t \in \mathbb{R}$, $u \in [-l,l]$. Take now any real number $\zeta \in ((1/p)+b, (1/p)+\gamma-\beta)$ such that (300) holds. Then we have

$$\frac{1}{2l^{ap}}\int_{-l}^{l}\big\|F(s+b_{k_m}+t) - [F^*(t)](s)\big\|^p\nu^p(s)\,ds$$

$$
\begin{aligned}
&\leqslant \frac{M_1}{2l^{ap}} \int_{-l}^{l} \int_{-\infty}^{0} \frac{1}{(1+|z|^{\alpha\zeta})^p} \|F(s+b_{k_m}+t+z)-F^*(s+t+z)\|^p \nu^p(s)\,dz\,ds \\
&= \frac{M_1}{2l^{ap}} \int_{-l}^{l} \int_{z-s}^{l} \frac{1}{(1+|z-s|^{\alpha\zeta})^p} \|F(b_{k_m}+t+z)-F^*(t+z)\|^p \nu^p(s)\,ds\,dz \\
&+ \frac{M_1}{2l^{ap}} \int_{-\infty}^{-l} \int_{-l}^{l} \frac{1}{(1+|z-s|^{\alpha\zeta})^p} \|F(b_{k_m}+t+z)-F^*(t+z)\|^p \nu^p(s)\,ds\,dz \\
&\leqslant \frac{M_1}{l^{ap}} \int_{-l}^{l} \|F(b_{k_m}+t+z)-F^*(t+z)\|^p \nu^p(z) \cdot \int_{-\infty}^{+\infty} \frac{\varphi^p(s-z)\,ds}{(1+|s-z|^{\zeta})^p}\,dz \\
&+ \frac{M_1}{2l^{ap}} \int_{-\infty}^{-3l} \int_{-l}^{l} \frac{1}{(1+|z-s|^{\alpha\zeta})^p} \|F(b_{k_m}+t+z)-F^*(t+z)\|^p \nu^p(z)\varphi^p(s-z)\,ds\,dz \\
&+ \frac{M_1}{2l^{ap}} \int_{-3l}^{3l} \int_{-l}^{l} \frac{1}{(1+|z-s|^{\alpha\zeta})^p} \|F(b_{k_m}+t+z)-F^*(t+z)\|^p \nu^p(z)\varphi^p(s-z)\,ds\,dz \\
&\leqslant \frac{M_1}{l^{ap}} \int_{-l}^{l} \|F(b_{k_m}+t+z)-F^*(t+z)\|^p \nu^p(z) \cdot \int_{-\infty}^{+\infty} \frac{\varphi^p(s-z)\,ds}{(1+|s-z|^{\zeta})^p}\,dz \\
&+ \frac{M_1}{l^{ap}} \int_{-3l}^{3l} \|F(b_{k_m}+t+z)-F^*(t+z)\|^p \nu^p(z) \cdot \int_{-\infty}^{+\infty} \frac{\varphi^p(s-z)\,ds}{(1+|s-z|^{\zeta})^p}\,dz \\
&+ \frac{cM_1}{2} \int_{-\infty}^{-3l} \frac{1}{(1+|z/2|^{\alpha\zeta})^p} \|F(b_{k_m}+t+z)-F^*(t+z)\|^p \nu^p(z)\,dz \cdot \left[l^{-ap} \int_{-l}^{l} \nu^p(s)\,ds \right],
\end{aligned}
$$

for any $t \in \mathbb{R}$; here, we have applied the Hölder inequality as well as the Fubini theorem and an elementary change of variables in the double integral. The first limit equation follows by applying (299); the second limit equation can be shown analogously. □

It is clear that Proposition 5.3.6 can be applied in the analysis of the existence and uniqueness of Besicovitch-$(\mathbb{F}, \mathcal{P}, \mathrm{R}, \omega)$-almost automorphic type solutions for a substantially large class of the abstract Volterra integrodifferential inclusions without initial conditions; see [444] for more details about applications of this type. We can similarly analyze the invariance of Besicovitch-$(\mathbb{F}, \mathcal{P}, \mathrm{R}, \omega)$-multi-almost automorphy under the actions of the usual convolution product

$$
F \mapsto (h * F)(\mathbf{t}) \equiv \int_{\mathbb{R}^n} h(\mathbf{t}-\mathbf{s})F(\mathbf{s})\,d\mathbf{s}, \quad \mathbf{t} \in \mathbb{R}^n
$$

and provide certain applications to the heat equation in $\mathbb{R}^n$; cf. also [464, Theorem 3.3].

Concerning certain applications to the abstract Volterra difference equations, we will only note that we can consider the existence and uniqueness of metrically Besicovitch almost automorphic type solutions of the first-order difference equation $u(k+1) = Au(k) + f(k)$, $k \in \mathbb{Z}$, where $A \in L(X)$ and $(f(k))_{k\in\mathbb{Z}}$ satisfies certain assumptions. Here, we must impose the additional condition on the sequence $(f(k))_{k\in\mathbb{Z}}$ by requiring that the limits in Definition 5.3.1 are uniform with respect to the point $t \in \mathbb{Z}$; this is almost inevitable since the discrete version of the dominated convergence theorem is no longer applicable in our new framework. Keeping in mind this extra condition, we can similarly analyze the already considered classes of the abstract difference equations.

5.4 Stepanov-p-almost periodicity and Stepanov-p-almost automorphy ($p > 0$)

In this section, we analyze various classes of Stepanov-p-almost periodic functions and Stepanov-p-almost automorphic functions ($p > 0$). We also examine the class of Stepanov-p-almost periodic (automorphic) functions in norm ($p > 0$). The main structural properties for the introduced classes of functions are clarified; we also provide several important theoretical examples, useful remarks and some new applications of Stepanov-p-almost periodic functions.

The organization and main ideas of this section can be briefly explained as follows. One of the main novelties of this section is the analysis of Stepanov-p-almost periodic functions and Stepanov-p-almost automorphic functions, where the exponent p has the value between 0 and 1; here, we would like to notice that the class of complex-valued Stepanov-p-almost periodic functions and the class of Stepanov-p-normal functions $f : \mathbb{R} \to \mathbb{C}$, where $p > 0$, was introduced for the first time by H. D. Ursell in [755], the paper which has been cited only six times from 1931 onwards. A Stepanov-p-almost periodic (automorphic) function $f : \mathbb{R} \to [0,\infty)$ need not be locally integrable for $0 < p < 1$. For example, the function

$$f(t) := \begin{cases} |\sin t|^{-1}, & t \notin \mathbb{Z}\pi \\ 0, & t \in \mathbb{Z}\pi, \end{cases}$$

is not locally integrable and, therefore, not Stepanov-p-almost periodic (automorphic) for any finite exponent $p \geqslant 1$; on the other hand, it can be simply shown that this function is Stepanov-p-almost periodic for any exponent $p \in (0,1)$.

In Section 5.4.1, we introduce and analyze the notion of Stepanov-$(p,\mathcal{B})$-almost periodicity, where $p > 0$ (cf. Definition 5.4.1 for the notions of Stepanov-$(p,\rho,\mathcal{B},\Lambda')$-almost periodicity, (strong) Stepanov-$(p,\mathrm{R},\mathcal{B})$-almost periodicity and their particular cases, the notions of Stepanov-p-almost periodicity and Bochner–Stepanov-p-almost periodicity, where $p > 0$). If $0 < p < 1$, then the notion of a (Bochner–)Stepanov-p-almost periodic function $F : \Lambda \to Y$ is new even in the case that $\Lambda = \mathbb{R}$ and $Y = \mathbb{C}$. The main structural results of Section 5.4.1 are Proposition 5.4.4 and Corollary 5.4.5; we also propose here many illustrative examples (without any doubt, the most important are Example 5.4.7 and Example 5.4.8, where we provide the proper extensions of the well-known statement [508, Theorem 5.3.1, p. 210] and some conclusions established on [508, Theorem p. 212]); an interesting open problem is also proposed in this context. Section 5.4.2 investigates the relations between piecewise continuous almost periodic functions and metrically Stepanov-p-almost periodic functions ($p > 0$). In this subsection, we provide proper generalizations of [266, Theorems 1, 2] concerning the relations between the class of pre-$(\mathcal{B}, T, (t_k))$-piecewise continuous almost periodic functions and the class $S^{(\mathbb{F},T,\mathcal{P}_t,\mathcal{P})}_{\Omega,\Lambda'}(\Lambda : Y)$ with $\Lambda = \Lambda' = \mathbb{R}$, $\mathbb{F}(\cdot\cdot) \equiv 1$, $\Omega = [0,1]$, $P = C_b(\mathbb{R} : \mathbb{C})$ and $P_t = L^p_\nu(t + [0,1] : \mathbb{C})$ for all $t \in \mathbb{R}$ (cf. Theorem 5.4.10). The main aim of Theorem 5.4.11 is to show that, if $\rho = T \in L(Y)$, $p > 0$ and $F : \Lambda \times X \to Y$ is a Stepanov-$(\Omega, p, \rho, \mathcal{B}, \Lambda')$-almost periodic function, where $\Lambda = \mathbb{R}$ or $\Lambda = [0, \infty)$, $\Omega = [0,1]$ and $\Lambda' = \Lambda$, then the validity of a quasiuniformly convergent type condition (QUC) implies that the considered function $F(\cdot;\cdot)$ is pre-$(\mathcal{B}, T, (t_k))$-piecewise continuous almost periodic. In Corollary 5.4.12(ii), we particularly show that any uniformly continuous, Stepanov-p-almost periodic function $F : \mathbb{R}^n \to Y$ is almost periodic ($p > 0$), extending thus the well-known Bochner theorem. The main aim of Section 5.4.3 is to investigate the invariance of Stepanov-p-almost periodicity under the actions of the infinite convolution products ($0 < p < 1$) and provide certain applications of the introduced notion to the abstract Volterra integrodifferential equations; the main results of this subsection are Theorem 5.4.13 and Proposition 5.4.14 (concerning applications, we feel it is our duty to emphasize at the very beginning that the situation is very complicated in the case that $0 < p < 1$ since the reverse Hölder inequality is valid in our new framework; see, e. g., [670, Proposition 3]).

The main purpose of Section 5.4.4 is to analyze various classes of Stepanov-p-almost automorphic type functions ($p > 0$). In Definition 5.1.3 and Definition 5.4.17, we introduce the notions of Stepanov $(\Omega, p, \mathrm{R}, \mathcal{B}, \nu)$-almost automorphy, Stepanov $(\Omega, p, \mathrm{R}, \mathcal{B}, \nu, W_{\mathcal{B},\mathrm{R}})$-almost automorphy, Stepanov $(\Omega, p, \mathrm{R}, \mathcal{B}, \nu, P_{\mathcal{B},\mathrm{R}})$-almost automorphy and some special subnotions of them. After that, we quote some statements established recently for the general classes of metrically Stepanov almost automorphic functions, which can be reformulated in our special framework (see, e. g., Proposition 5.4.18). Section 5.4.5 investigates certain relations between piecewise continuous almost automorphic functions and metrically Stepanov-p-almost automorphic functions ($p > 0$); the main structural results presented here are Proposition 5.4.21, Theorem 5.4.24 and Theorem 5.4.25. In Section 5.4.6, we analyze the notion of Stepanov-p-almost periodicity in norm and the notion of Stepanov-p-almost automorphy in norm ($p > 0$). The considered function spaces are introduced in Definition 5.4.26 and Definition 5.4.27. The main result established in this section, where we also propose some open problems for our readers, is Theorem 5.4.28; cf. also Remark 5.4.29.

The main aim of Section 5.4.7 is to present applications of the obtained results to the abstract (impulsive) Volterra integrodifferential inclusions in Banach spaces. This section contains two separate parts: the first part considers certain applications to the abstract impulsive first-order differential inclusions, while the second part considers certain applications to the abstract fractional differential inclusions.

Suppose now that $T > 0$. We will deal later on with the space of X-valued piecewise continuous functions on $[0, T]$, which is defined by

$$PC([0,T] : X) \equiv \{u : [0,T] \to X : u \in C((t_i, t_{i+1}] : X),\ u(t_i-) = u(t_i) \text{ exist for any } i \in \mathbb{N}_l,\ u(t_i+) \text{ exist for any } i \in \mathbb{N}_l^0 \text{ and } u(0) = u(0+)\},$$

where $0 \equiv t_0 < t_1 < t_2 < \cdots < t_l < T \equiv t_{l+1}$ and the symbols $u(t_i-)$ and $u(t_i+)$ denote the left and the right limits of the function $u(t)$ at the point $t = t_i$, $i \in \mathbb{N}_{l-1}^0$, respectively. Let us recall that $PC([0,T] : X)$ is a Banach space endowed with the norm $\|u\| := \max\{\sup_{t\in[0,T)} \|u(t+)\|, \sup_{t\in(0,T]} \|u(t-)\|\}$. The space of X-valued piecewise continuous functions on $[0,\infty)$, denoted by $PC([0,\infty) : X)$, if defined as the union of those functions $f : [0,\infty) \to X$ such that the discontinuities of $f(\cdot)$ form a discrete set and that for each $T > 0$ we have $f_{|[0,T]}(\cdot) \in PC([0,T] : X)$. We similarly define the space $PC(\mathbb{R} : X)$. If $\omega \in \mathbb{R}$, then $C_\omega([0,\infty) : X)$ denotes the space of all continuous functions $f : [0,\infty) \to X$ such that the function $t \mapsto e^{-\omega t}\|f(t)\|$, $t \geqslant 0$ is bounded; the space $PC_\omega([0,\infty) : X)$ denotes the space of all piecewise continuous functions $f : [0,\infty) \to X$ such that the function $t \mapsto e^{-\omega t}\|f(t)\|$, $t \geqslant 0$ is bounded.

5.4.1 Stepanov-p-almost periodic type functions ($p > 0$)

In this subsection, we will consider the following special kinds of Stepanov-p-almost periodic type functions ($p > 0$).

Definition 5.4.1. (i) Suppose that $p > 0$, (SM-1) holds true and $\nu_{\mathbf{t}} : \mathbf{t} + \Omega \to (0,\infty)$ is a Lebesgue measurable function ($\mathbf{t} \in \Lambda$). Then we say that a function $F : \Lambda \times X \to Y$ is Stepanov-$(\Omega, p, \rho, \mathcal{B}, \Lambda', \nu_\cdot)$-almost periodic if $F(\cdot;\cdot)$ belongs to the class $S_{\Omega,\Lambda',\mathcal{B}}^{(\mathbb{F},\rho,\mathcal{P}_{\mathbf{t}},\mathcal{P})}(\Lambda \times X : Y)$ with $\mathbb{F}(\cdot\cdot) \equiv 1$, $P = C_b(\Lambda : \mathbb{C})$ and $P_{\mathbf{t}} = L^p_{\nu_{\mathbf{t}}}(\mathbf{t} + \Omega : Y)$ for all $\mathbf{t} \in \Lambda$. If there exists a Lebesgue measurable function $\nu : \Lambda \to (0,\infty)$ such that $\nu_{\mathbf{t}}(\cdot) \equiv \nu_{|\mathbf{t}+\Omega}(\cdot)$ for all $\mathbf{t} \in \Lambda$, then we also say that $F(\cdot;\cdot)$ is Stepanov-$(\Omega, p, \rho, \mathcal{B}, \Lambda', \nu)$-almost periodic; furthermore, if $\nu_{\mathbf{t}}(\cdot) \equiv 1$ for all $\mathbf{t} \in \Lambda$, then we omit the term "$\nu_\cdot$" from the notation.

(ii) Suppose that $p > 0$, $\emptyset \neq \Lambda \subseteq \mathbb{R}^n$, $F : \Lambda \times X \to Y$ is a given function and the assumptions $\mathbf{t} \in \Lambda$, $\mathbf{b} \in \mathrm{R}$ and $l \in \mathbb{N}$ imply $\mathbf{t} + \mathbf{b}(l) \in \Lambda$. If $\nu : \Omega \to (0,\infty)$ is a Lebesgue measurable function, then we say that the function $F(\cdot;\cdot)$ is (strongly) Stepanov-$(\Omega, p, \mathrm{R}, \mathcal{B}, \nu)$-almost periodic if $F(\cdot;\cdot)$ is (strongly) Stepanov $(\Omega, \mathrm{R}, \mathcal{B}, \mathcal{P}_Z)$-multialmost periodic with $Z = L^p_\nu(\Omega : Y)$ and $P_Z = C_b(\Lambda : Z)$; if $\nu(\cdot) \equiv 1$, then we omit the term "ν" from the notation (this will be our general rule henceforth).

(iii) Suppose that $p > 0$, (SM-1) holds true, $\nu_{\mathbf{t}} : \mathbf{t} + \Omega \to (0,\infty)$ is a Lebesgue measurable function ($\mathbf{t} \in \Lambda$) and $\nu : \Lambda \to (0,\infty)$ is a Lebesgue measurable function. Then we say that a function $F : \Lambda \to Y$ is Stepanov-$(\Omega, p, \nu_.)$-almost periodic [Stepanov-(Ω, p, ν)-almost periodic] if $F(\cdot)$ is Stepanov-$(\Omega, p, \rho, \Lambda', \nu_.)$-almost periodic [Stepanov-$(\Omega, p, \rho, \Lambda', \nu)$-almost periodic] with $\rho \equiv \mathrm{I}$ and $\Lambda' \equiv \{\tau \in \mathbb{R}^n : \tau + \mathbf{t} \in \Lambda,\ \mathbf{t} \in \Lambda\}$.

(iv) Suppose that $p > 0$, $\emptyset \neq \Lambda \subseteq \mathbb{R}^n$, $F : \Lambda \to Y$ is a given function, $\nu : \Omega \to (0, \infty)$ is a Lebesgue measurable function and R denotes the collection of all sequences in $\mathbb{R}^n$ such that the assumptions $\mathbf{t} \in \Lambda$, $\mathbf{b} \in \mathrm{R}$ and $l \in \mathbb{N}$ imply $\mathbf{t} + \mathbf{b}(l) \in \Lambda$. Then we say that the function $F(\cdot)$ is Bochner–Stepanov-(Ω, p, ν)-almost periodic if $F(\cdot)$ is Stepanov-$(\Omega, p, \mathrm{R}, \nu)$-almost periodic.

In all above definitions, we omit the term "Ω" from the notation if $\Omega = [0,1]^n$. If we denote by $A_{X,Y}$ any of the above introduced classes of function spaces, $c_1 \in \mathbb{R} \setminus \{0\}$, $\tau \in \mathbb{R}^n$, c, $c_2 \in \mathbb{C} \setminus \{0\}$ and $x_0 \in X$, then it is not difficult to find some sufficient conditions ensuring that the function $cF(\cdot;\cdot)$, $F(c_1\cdot; c_2\cdot)$, $\|F(\cdot;\cdot)\|_Y$ or $F(\cdot + \tau; \cdot + x_0)$ belongs to $A_{X,Y}$ if $F(\cdot;\cdot)$ belongs to $A_{X,Y}$. Using [481, Proposition 3.2], the statements of [452, Propositions 2.3(i), 2.4, Theorems 2.5, 2.7] and the conclusions (i)–(iii) established on [452, pp. 234–235] can be simply rephrased for the (strongly) Stepanov-$(\Omega, p, \mathrm{R}, \mathcal{B}, \nu)$-almost periodic functions. We also have the following consequence of [481, Proposition 3.7].

Proposition 5.4.2. *Suppose that $\emptyset \neq \Lambda \subseteq \mathbb{R}^n$ and the assumptions $\mathbf{t} \in \Lambda$, $\mathbf{b} \in \mathrm{R}$ and $l \in \mathbb{N}$ imply $\mathbf{t}+\mathbf{b}(l) \in \Lambda$. Suppose further that for each integer $j \in \mathbb{N}$ the function $F_j : \Lambda \times X \to Y$ is Stepanov-$(\Omega, p, \mathrm{R}, \mathcal{B}, \nu)$-almost periodic, $\hat{F_{j;\Omega}} : \Lambda \times X \to Z$, $\hat{F_{j;\Omega}}(\cdot + b_k; x) \in C_b(\Lambda : L_\nu^p(\Omega : Y))$ and $\hat{F_{j;\Omega}}^*(\cdot; x) \in C_b(\Lambda : L_\nu^p(\Omega : Y))$ ($x \in B$; $(b_k) \in \mathrm{R}$), with the meaning clear, as well as that, for every sequence which belongs to R, any its subsequence also belongs to R. If $F : \Lambda \times X \to Y$ and, for every $B \in \mathcal{B}$ and $(b_k) \in \mathrm{R}$, we have*

$$\lim_{(i,l)\to+\infty} \sup_{x\in B} \|F_i(\cdot + b_{k_l}; x) - F(\cdot + b_{k_l}; x)\|_{C_b(\Lambda : L_\nu^p(\Omega:Y))} = 0,$$

then the function $F(\cdot;\cdot)$ is Stepanov-$(\Omega, p, \mathrm{R}, \mathcal{B}, \nu)$-almost periodic, $\hat{F_\Omega}(\cdot + b_k; x) \in C_b(\Lambda : L_\nu^p(\Omega : Y))$ and $\hat{F_\Omega}^*(\cdot; x) \in C_b(\Lambda : L_\nu^p(\Omega : Y))$ ($x \in B$; $(b_k) \in \mathrm{R}$).

Let us recall from the introductory part that a Stepanov-p-almost periodic function $F : \mathbb{R}^n \to Y$ need not be locally integrable for $0 < p < 1$. Moreover, if we assume that $F : \mathbb{R}^n \to Y$ is both locally integrable and Stepanov-p-almost periodic for some $p \in (0,1)$, then it is not clear (cf. the proof of [448, Proposition 3.5.9]) whether the expression

$$\varphi \mapsto T(\varphi) \equiv \int_{\mathbb{R}^n} \varphi(\mathbf{t}) F(\mathbf{t})\, d\mathbf{t}, \quad \varphi \in \mathcal{D}(\mathbb{R}^n)$$

determines a regular almost periodic distribution; here, $\mathcal{D}(\mathbb{R}^n)$ stands for the space of all smooth test functions $\varphi : \mathbb{R}^n \to \mathbb{C}$ with compact support. We will examine generalized almost periodic functions of this type somewhere else.

We continue by providing the following illustrative example.

Example 5.4.3. In [448, Example 9.2.7], we have constructed a piecewise continuous almost periodic function $f : \mathbb{R} \to Y$, which is not continuous and satisfies that for each $\varepsilon > 0$ there exists a relatively dense subset R of $\mathbb{R}$ such that for each $\tau \in R$ and $t \in \mathbb{R}$ we have $\|f(t+\tau) - f(t)\|_Y \leqslant \varepsilon$. This simply implies that, for every $p > 0$ and for every Stepanov-p-bounded function $\nu : \mathbb{R} \to (0, \infty)$, the function $f(\cdot)$ is Stepanov-(p, ν)-almost periodic. Of course, by the Stepanov-p-boundedness of $\nu(\cdot)$, we mean that $\sup_{t\in\mathbb{R}} \int_t^{t+1} |\nu(s)|^p\, ds < +\infty$.

Now we will state the following simple result (cf. also [131, Theorem 1, Corollary, p. 62]).

Proposition 5.4.4. *Suppose that $p > 0$, $q > 0$ and (SM-1) holds true.*
(i) *If $F : \Lambda \times X \to Y$ is Stepanov-$(\Omega, q, \rho, \mathcal{B}, \Lambda')$-almost periodic and $p \leqslant q$, then $F(\cdot;\cdot)$ is Stepanov-$(\Omega, p, \rho, \mathcal{B}, \Lambda')$-almost periodic.*
(ii) *Suppose that $F : \Lambda \times X \to Y$, $\rho = T \in L(X)$ and, for every set $B \in \mathcal{B}$, we have $\|F\|_{B,\infty} := \sup_{x\in B; \mathbf{t}\in\Lambda} \|F(\mathbf{t}; x)\|_Y < +\infty$. Then $F(\cdot;\cdot)$ is Stepanov-$(\Omega, p, \rho, \mathcal{B}, \Lambda')$-almost periodic if and only if $F(\cdot;\cdot)$ is Stepanov-$(\Omega, q, \rho, \mathcal{B}, \Lambda')$-almost periodic.*

Proof. We will prove only (ii) because (i) follows almost directly from the definition of Stepanov-$(\Omega, p, \rho, \mathcal{B})$-almost periodicity and an easy application of the Hölder inequality. Due to (i), it suffices to show that the assumption $F(\cdot;\cdot)$ is Stepanov-$(\Omega, p, \rho, \mathcal{B})$-almost periodic implies that $F(\cdot;\cdot)$ is Stepanov-$(\Omega, q, \rho, \mathcal{B})$-almost periodic. Let $B \in \mathcal{B}$ and $\varepsilon > 0$ be given. Then we know there exists $l > 0$ such that for each $\mathbf{t}_0 \in \Lambda'$ there exists $\tau \in B(\mathbf{t}_0, l) \cap \Lambda'$ such that, for every $\mathbf{t} \in \Lambda$ and $x \in B$, we have

$$\int_{\mathbf{t}+\Omega} \|F(\tau + \mathbf{s}; x) - TF(\mathbf{s}; x)\|^p\, ds \leqslant \varepsilon^p.$$

Since $x^p \geqslant x^q$ for all $x \in [0, 1]$, we have

$$\int_{\mathbf{t}+\Omega} \left(\frac{\|F(\tau + \mathbf{s}; x) - TF(\mathbf{s}; x)\|}{\|F\|_{B,\infty}(1 + \|T\|)} \right)^q ds$$
$$\leqslant \int_{\mathbf{t}+\Omega} \left(\frac{\|F(\tau + \mathbf{s}; x) - TF(\mathbf{s}; x)\|}{\|F\|_{B,\infty}(1 + \|T\|)} \right)^p ds \leqslant \varepsilon^p \big[\|F\|_{B,\infty}(1 + \|T\|)\big]^{-p} m(\Omega).$$

This simply completes the proof. □

Before proceeding any further, we would like to notice that the conclusion established in [444, Example 2.2.3(i)] directly follows from the conclusion established in [444, Example 2.2.2] and Proposition 5.4.4. Now we will state the following important corollary of Proposition 5.4.4.

Corollary 5.4.5. (i) *Suppose that $0 < p \leqslant q < \infty$ and $F \in L^p_{loc}(\mathbb{R}^n : Y)$. If $F(\cdot)$ is Stepanov-q-almost periodic, then $F(\cdot)$ is Stepanov-p-almost periodic.*

(ii) *Suppose that $F \in L^{\infty}(\mathbb{R}^n : Y)$ and $0 < p \leqslant q < \infty$. Then $F(\cdot)$ is Stepanov-p-almost periodic if and only if $F(\cdot)$ is Stepanov-q-almost periodic.*
(iii) *Suppose that $F \in BUC(\mathbb{R}^n : Y)$ and $p > 0$. Then $F(\cdot)$ is almost periodic if and only if $F(\cdot)$ is Stepanov-p-almost periodic.*

We continue with the following illustrative example, which shows that the class of equi-Weyl-almost periodic functions is essentially larger than the union of all classes of Stepanov-p-almost periodic functions with the exponent $p > 0$; cf. [444–447] for the notion and more details about the Weyl almost periodic type functions.

Example 5.4.6. (i) Suppose that K is any compact subset of $\mathbb{R}^n$ with a positive Lebesgue measure. Then we know that the function $F(\mathbf{t}) := \chi_K(\mathbf{t})$, $\mathbf{t} \in \mathbb{R}^n$ is equi-Weyl-p-almost periodic for any exponent $p \geqslant 1$; see also [447, Example 6.3.8] for a slightly stronger result. Arguing similarly as in [55, Example 4.27], we may conclude that $F(\cdot)$ cannot be Stepanov-p-almost periodic for any exponent $p > 0$; moreover, this function cannot be Stepanov-p-almost periodic in norm for any exponent $p > 0$ (cf. Section 5.4.6 for the notion).

(ii) Denote by $E_\alpha(\cdot)$ the Mittag-Leffler function. If $\alpha \in (0,2) \setminus \{1\}$ and $r \in \mathbb{R} \setminus \{0\}$, then the function $t \mapsto E_\alpha((ir)^\alpha t^\alpha)$, $t \in \mathbb{R}$ is not Stepanov-p-almost periodic for any exponent $p > 0$, which follows from the argumentation contained in the proof of [444, Lemma 2.6.9]. On the other hand, if $1 < \alpha < 2$, then the asymptotic expansion formula for the Mittag-Leffler functions (see, e. g., [444, Theorem 1.4.1 and the formulae (16)–(18)]) shows that the formula stated on the fourth line of the proof of the above mentioned lemma continues to hold for $t \leqslant -1$, as well as that the term $|\varepsilon_\alpha((ir)^\alpha t^\alpha)|$ is bounded by Const.$\cdot|t|^{-\alpha}$ for $|t| \geqslant 1$; throughout the proof, we have mistakenly used the constant β, we actually have $\beta = 1$ here. Keeping in mind the argumentation given on [448, pp. 407–408] and the fact that equi-Weyl-p-almost periodic functions form a vector space with the usual operations, it readily follows that the function $t \mapsto E_\alpha((ir)^\alpha t^\alpha)$, $t \in \mathbb{R}$ is equi-Weyl-p-almost periodic for any exponent $p \geqslant 1$. The situation is a little bit complicated if $0 < \alpha < 1$, because in this case, the asymptotic expansion formula for the Mittag-Leffler functions shows that there exists a continuous function $q : \mathbb{R} \to \mathbb{C}$ such that $\lim_{|t| \to +\infty} q(t) = 0$, $E_\alpha((ir)^\alpha t^\alpha) = \alpha^{-1}(ir)^{1-\beta} e^{irt} + q(t)$ for $t \geqslant 0$ and $E_\alpha((ir)^\alpha t^\alpha) = q(t)$ for $t < 0$. Using again the argumentation given on [448, pp. 407–408], it readily follows that the function $t \mapsto E_\alpha((ir)^\alpha t^\alpha)$, $t \in \mathbb{R}$ is equi-Weyl-p-almost periodic if and only if the function $t \mapsto F(t)$, $t \in \mathbb{R}$, given by

$$F(t) := \alpha^{-1} e^{irt} \varepsilon_\alpha((ir)^\alpha t^\alpha), \quad t \geqslant 0; \quad F(t) := 0,\ t < 0,$$

is equi-Weyl-p-almost periodic ($p \geqslant 1$). But the last statement is not true on account of the formula given on [447, l. 8, p. 425] so that the function $t \mapsto E_\alpha((ir)^\alpha t^\alpha)$, $t \in \mathbb{R}$ is not equi-Weyl-p-almost periodic ($0 < \alpha < 1$; $p \geqslant 1$).

The subsequent examples, in which we provide the proper extensions of the conclusions established in [508, Theorem 5.3.1, p. 210 and p. 212], indicate the importance of the notion introduced in this section (without going into specifics, we will only mention in passing that the conclusions established here can be also clarified in the multidimensional setting; see, e. g., [447, Example 6.2.9]).

Example 5.4.7. Suppose that $f : \mathbb{R} \to \mathbb{R}$ is an almost periodic function and there exist a finite real number $c > 0$ and an analytic function $g : \{z \in \mathbb{C} : |\operatorname{Im} z| < c\} \to \mathbb{C}$ such that $g(t) = f(t)$ for all $t \in \mathbb{R}$. Then we know that the function $F(t) := \operatorname{sign}(f(t))$, $t \in \mathbb{R}$ is Stepanov-p-almost periodic for any $p \geqslant 1$. Now we will improve this result by showing that the function $F(\cdot)$ belongs to the class $S_{\Omega,\Lambda'}^{(\mathbb{F},\rho,\mathcal{P}_t,\mathcal{P})}(\Lambda : Y)$ with $\Lambda = \Lambda' = \mathbb{R}$, $\mathbb{F}(\cdot\cdot) \equiv 1$, $\rho = \mathrm{I}$, $\Omega = [0,1]$, $P = C_b(\mathbb{R} : \mathbb{C})$ and $P_t = L_\nu^p(t + [0,1] : \mathbb{C})$ for all $t \in \mathbb{R}$, where $p > 0$ is an arbitrary exponent and $\nu : \mathbb{R} \to (0,\infty)$ is a Lebesgue measurable function satisfying the following condition:

(LT) For every $\varepsilon > 0$, there exists a sufficiently large integer $d > 0$ such that $\sup_{t\in\mathbb{R}} \int_d^{+\infty} y^{p-1} m(\{x \in [t,t+1] : \nu(x) > y\})\, dy < \varepsilon$, where $m(\cdot)$ denotes the Lebesgue measure.

In order to do that, set $E_\alpha := \{x \in \mathbb{R} : |f(x)| > \alpha\}$ $(\alpha > 0)$ and fix a number $\varepsilon > 0$. Let $d > 0$ be an integer such that

$$\sup_{t\in\mathbb{R}} \int_d^{+\infty} y^{p-1} m(\{x \in [t,t+1] : \nu(x) > y\})\, dy < 2^{-p} p^{-1} 2^{-1} \varepsilon.$$

If $\tau \in \mathbb{R}$ is an α-almost period of the function $f(\cdot)$, then we know that $F(x+\tau) = F(x)$ for all $x \in E_\alpha$ and, because of that, we have (the equation (301) is a consequence of (1.7)):

$$\begin{aligned}
\int_t^{t+1} |F(x+\tau) - F(x)|^p \nu^p(x)\, dx &\leqslant 2^p \int_{E_\alpha^c \cap [t,t+1]} \nu^p(x)\, dx \\
&= 2^p p \int_0^{+\infty} y^{p-1} m(\{x \in E_\alpha^c \cap [t,t+1] : \nu(x) > y\})\, dy \\
&= 2^p p \int_0^{d} y^{p-1} m(\{x \in E_\alpha^c \cap [t,t+1] : \nu(x) > y\})\, dy \\
&\quad + 2^p p \int_d^{+\infty} y^{p-1} m(\{x \in E_\alpha^c \cap [t,t+1] : \nu(x) > y\})\, dy \\
&\leqslant 2^p p \int_0^{d} y^{p-1} m(\{x \in E_\alpha^c \cap [t,t+1] : \nu(x) > y\})\, dy
\end{aligned}$$

$$
\begin{aligned}
&+ 2^p p \int_d^{+\infty} y^{p-1} m(\{x \in [t,t+1] : \nu(x) > y\})\, dy \\
&\leqslant 2^p p \int_0^d y^{p-1} m(\{x \in E_\alpha^c \cap [t,t+1] : \nu(x) > y\})\, dy + (\varepsilon/2) \\
&\leqslant 2^p p \cdot m(E_\alpha^c \cap [t,t+1]) \int_0^d y^{p-1}\, dy + (\varepsilon/2) \\
&= 2^p d^p m(E_\alpha^c \cap [t,t+1]) + (\varepsilon/2) \qquad (301)
\end{aligned}
$$

for any $t \in \mathbb{R}$. By the proof of [508, Theorem 5.3.1], we have $\lim_{\alpha\to 0+} m(E_\alpha^c \cap [t,t+1]) = 0$, uniformly in $t \in \mathbb{R}$, which simply implies the required conclusion.

Concerning the condition (LT), we would like to emphasize the following facts:

(LT1): It is clear that condition (LT) holds provided that there exists a finite real constant $M > 0$ such that $\nu(x) \leqslant M$ for all $x \in \mathbb{R}$. But condition (LT) also holds for some unbounded functions $\nu(x)$; for example, set $\nu(x) := |k|$ if $x \in [k, k+2^{-|k|}]$ for some $k \in \mathbb{Z}$ and $\nu(x) := 1$, otherwise. Then the validity of (LT) simply follows from the fact that $m(\{x \in [t,t+1] : \nu(x) > y\}) \leqslant 2^{-|k|}$, $t \in \mathbb{R}$, provided that $y \in [k, k+1]$ for some $k \in \mathbb{Z}$.

(LT2): Suppose that $\nu(t) = c + \zeta(t)$, $t \in \mathbb{R}$, where $c \geqslant 0$, $\zeta : \mathbb{R} \to (0,+\infty)$ is a Lebesgue measurable function and $\int_{-\infty}^{+\infty} \zeta^p(x)\, dx < +\infty$. Then condition (LT) also holds, which can be shown as follows. Let a number $\varepsilon > 0$ be fixed; then it is clear that there exists a sufficiently large number $d > 1$ such that $c + d \in \mathbb{N}$ and $\int_d^{+\infty} y^{p-1} m(\{x \in [t,t+1] : \zeta(x) > y\})\, dy < \varepsilon$. We will prove that (LT) holds with the integer $c + d$ in place of d. In actual fact, we have

$$
\begin{aligned}
&\int_{c+d}^{+\infty} y^{p-1} m(\{x \in [t,t+1] : \nu(x) > y\})\, dy \\
&\quad = \int_d^{+\infty} (y+c)^{p-1} m(\{x \in [t,t+1] : \zeta(x) > y\})\, dy.
\end{aligned}
$$

If $0 < p \leqslant 1$, then we have

$$
\begin{aligned}
&\int_d^{+\infty} (y+c)^{p-1} m(\{x \in [t,t+1] : \zeta(x) > y\})\, dy \\
&\quad \leqslant \int_d^{+\infty} y^{p-1} m(\{x \in [t,t+1] : \zeta(x) > y\})\, dy \\
&\quad \leqslant \int_d^{+\infty} y^{p-1} m(\{x \in \mathbb{R} : \zeta(x) > y\})\, dy < \varepsilon;
\end{aligned}
$$

on the other hand, if $p > 1$, then there exists a real number $c_p > 0$ such that

$$\int_d^{+\infty} (y+c)^{p-1} m(\{x \in [t,t+1] : \zeta(x) > y\})\, dy$$
$$\leqslant c_p \Bigg[\int_d^{+\infty} y^{p-1} m(\{x \in [t,t+1] : \zeta(x) > y\})\, dy + \int_d^{+\infty} m(\{x \in [t,t+1] : \zeta(x) > y\})\, dy \Bigg]$$
$$\leqslant 2c_p \int_d^{+\infty} y^{p-1} m(\{x \in [t,t+1] : \zeta(x) > y\})\, dy < 2c_p \varepsilon.$$

This simply implies the required conclusion. We can similarly prove that condition (LT) holds for the function $c + \nu(\cdot)$ if (LT) holds for the function $\nu(\cdot)$.

Further on, it can be simply shown that condition (LT) does not hold if $\nu(t) = e^{\sigma|t|}$, $t \in \mathbb{R}$ or $\nu(t) = (1+|t|)^{\sigma}$, $t \in \mathbb{R}$ for some real number $\sigma > 0$. The following condition is similar to (LT) but it holds for any p-locally integrable function $\nu(\cdot)$, as easily explained:

(LT-K) For every $\varepsilon > 0$ and for every compact set $K \subseteq \mathbb{R}$, there exists a sufficiently large integer $d > 0$ such that, for every $t \in K$, we have $\int_d^{+\infty} y^{p-1} m(\{x \in [t,t+1] : \nu(x) > y\})\, dy < \varepsilon$.

If (LT-K) holds, then we can similarly prove that the function $F(\cdot)$ is Stepanov–Levitan-(N,ν)-almost periodic in the following sense:

(SL-1) For every $\varepsilon > 0$ and $N > 0$, there exists a relatively dense subset $R_{\varepsilon,N} \subseteq \mathbb{R}$ of Stepanov–Levitan-(N,ν)-almost periods of $F(\cdot)$, which means that if, $\tau \in R_{\varepsilon,N}$ and $|t| \leqslant N$, then $\int_t^{t+1} |F(x+\tau) - F(x)|^p \nu^p(x)\, dx \leqslant \varepsilon$.

(SL-2) For every $\varepsilon > 0$ and $N > 0$, there exist a number $\delta > 0$ and a relatively dense subset $R_{\delta,N} \subseteq \mathbb{R}$ of Stepanov–Levitan-(N,ν,δ)-almost periods of $F(\cdot)$ such that $R_{\delta,N} \pm R_{\delta,N} \subseteq R_{\varepsilon,N}$.

We will consider this class of generalized Levitan N-almost periodic functions somewhere else (cf. [448] for further information in this direction).

Example 5.4.8. Suppose that α, $\beta \in \mathbb{R}$ and $\alpha\beta^{-1}$ is a well-defined irrational number. Then we know that the functions

$$f(t) = \sin\left(\frac{1}{2 + \cos \alpha t + \cos \beta t}\right), \quad t \in \mathbb{R}$$

and

$$g(t) = \cos\left(\frac{1}{2 + \cos \alpha t + \cos \beta t}\right), \quad t \in \mathbb{R}$$

are Stepanov-p-almost periodic but not almost periodic ($1 \leqslant p < \infty$). Suppose now that $p > 0$ is an arbitrary exponent as well as that the function $\nu : \mathbb{R} \to (0, \infty)$ is Stepanov-p-bounded and satisfies (LT); for example, the function $\nu(\cdot)$ constructed in the final part of the previous example enjoys these features. Then the functions $f(\cdot)$ and $g(\cdot)$ belong to the class $S_{\Omega,\Lambda'}^{(\mathbb{F},\rho,\mathcal{P}_t,\mathcal{P})}(\Lambda : Y)$ with $\Lambda = \Lambda' = \mathbb{R}$, $\mathbb{F}(\cdot\cdot) \equiv 1$, $\rho = \mathrm{I}$, $\Omega = [0,1]$, $P = C_b(\mathbb{R} : \mathbb{C})$ and $P_t = L_\nu^p(t + [0,1] : \mathbb{C})$ for all $t \in \mathbb{R}$.

We will prove this fact only for the function $f(\cdot)$, with $\alpha = 1$ and $\beta = \sqrt{2}$. In order to do that, set $E_a := \{x \in \mathbb{R} : |2 + \cos x + \cos(\sqrt{2}x)| > a\}$ ($a > 0$) and fix a number $\varepsilon > 0$. Let $a > 0$ and $\delta > 0$ be arbitrary real numbers such that $a > \delta$. Arguing as in [508] (see p. 212), it readily follows that, for every δ-almost period $\tau \in \mathbb{R}$ of the function $2 + \cos(\cdot) + \cos(\sqrt{2}\cdot)$ and for every $t \in \mathbb{R}$, we have

$$\int_t^{t+1} |f(x+\tau) - f(x)|^p \nu^p(x)\,dx \leqslant \left[\frac{\delta}{a(a-\delta)}\right]^p \|\nu\|_{S^p} + 2^p \int_{E_a^c \cap [t,t+1]} \nu^p(x)\,dx,$$

where $\|\nu\|_{S^p} := \sup_{t\in\mathbb{R}} \int_t^{t+1} \nu^p(x)\,dx$. Then the final conclusion follows similarly as in [508] and the previous example (if the function $\nu(\cdot)$ is Stepanov-p-bounded, then the functions $f(\cdot)$ and $g(\cdot)$ are Stepanov–Levitan-(N,ν)-almost periodic).

5.4.2 Relations between piecewise continuous almost periodic functions and metrically Stepanov-p-almost periodic functions ($p > 0$)

In our joint research article [266] with W.-S. Du and D. Velinov, we have recently analyzed certain relations between piecewise continuous almost periodic functions (piecewise continuous uniformly recurrent functions) and Stepanov almost periodic functions (Stepanov uniformly recurrent functions). We start this subsection by recalling the following notion.

Definition 5.4.9 (cf. [266, Definition 6(i)]). Suppose that ρ is a binary relation on Y, the function $F : \mathbb{R} \times X \to Y$ [$F : [0,\infty) \times X \to Y$] satisfies that, for every $x \in X$, the function $t \mapsto F(t;x)$, $t \in \mathbb{R}$ is piecewise continuous with the possible first kind discontinuities at the points of a fixed sequence $(t_k)_{k\in\mathbb{Z}}$ [$(t_k)_{k\in\mathbb{N}}$]. Suppose further that $(t_k)_{k\in\mathbb{Z}}$ [$(t_k)_{k\in\mathbb{N}}$] satisfies that $\delta_0 := \inf_{k\in\mathbb{Z}}(t_{k+1} - t_k) > 0$ [$\delta_0 := \inf_{k\in\mathbb{N}}(t_{k+1} - t_k) > 0$]. Then we say that the function $F(\cdot;\cdot)$ is pre-$(\mathcal{B},\rho,(t_k))$-piecewise continuous almost periodic if, for every $\varepsilon > 0$ and $B \in \mathcal{B}$, there exists a relatively dense set S in $\mathbb{R}$ [in $[0,\infty)$] such that, if $\tau \in S$, $x \in B$ and $t \in \mathbb{R}$ satisfies $|t - t_k| > \varepsilon$ for all $k \in \mathbb{Z}$ [$k \in \mathbb{N}$], then there exists $y_{t,x} \in \rho(F(t;x))$ such that $\|F(t+\tau;x) - y_{t,x}\| < \varepsilon$.

The following result provides, even for the usually considered exponents $p \geqslant 1$, an extension of [266, Theorem 1] for pre-$(\mathcal{B},T,(t_k))$-piecewise continuous almost periodic functions [the extension for pre-$(\mathcal{B}, T, (t_k))$-piecewise continuous uniformly recurrent functions can be deduced in a similar manner].

Theorem 5.4.10. *Suppose that $\rho = T \in L(Y)$, $p > 0$, $F : \Lambda \times X \to Y$ is pre-$(\mathcal{B}, T, (t_k))$-piecewise continuous almost periodic, where $\Lambda = \mathbb{R}$ or $\Lambda = [0, \infty)$, and, for every $B \in \mathcal{B}$, $\|F\|_{\infty,B} \equiv \sup_{t\in\Lambda, x\in B} \|F(t; x)\| < +\infty$. Suppose further that the function $\nu : \mathbb{R} \to (0, \infty)$ is Stepanov-p-bounded and satisfies the following condition:*

(LQ) *For every $\varepsilon > 0$, there exist $d > 0$ and $\varepsilon_0 > 0$ such that, for every $x \in \mathbb{R}$ and for every Lebesgue measurable set $\Omega \subseteq [x, x + 1]$ such that $m(\Omega) < \varepsilon_0$, we have $\int_d^{+\infty} y^{p-1} m(\{x \in \Omega : \nu(x) > y\})\, dy < \varepsilon$.*

Then the function $F(\cdot; \cdot)$ belongs to the class $S^{(\mathbb{F},T,\mathcal{P}_t,\mathcal{P})}_{\Omega,\Lambda'}(\Lambda : Y)$ with $\Lambda = \Lambda' = \mathbb{R}$, $\mathbb{F}(\cdot\cdot) \equiv 1$, $\Omega = [0, 1]$, $P = C_b(\mathbb{R} : \mathbb{C})$ and $P_t = L^p_\nu(t + [0, 1] : \mathbb{C})$ for all $t \in \mathbb{R}$.

Proof. Without loss of generality, we may assume that $\Lambda = \mathbb{R}$ and $T = cI$ for some $c \in \mathbb{C}$. Let a number $\varepsilon > 0$ and a set $B \in \mathcal{B}$ be given. Suppose that a point $x \in \mathbb{R}$ is fixed and the interval $[x, x + 1]$ contains the possible first kind discontinuities of functions $F(\cdot; b)$ at the points $\{t_m, \ldots, t_{m+k}\} \subseteq [x, x + 1]$ $(b \in X)$; then we clearly have $k \leqslant \lceil 1/\delta_0 \rceil$. Let the numbers $d > 0$ and $\varepsilon_0 > 0$ be determined from condition (LQ), with the number ε replaced therein with the number $\varepsilon/(2((1+|c|)\|F\|_{\infty,B})^p p)$. Further on, let S be a relatively dense set in $\mathbb{R}$ such that, if $\tau \in S$ and $b \in B$, then $\|F(t + \tau; x) - cF(t; x)\| < \varepsilon_1$ for all $t \in \mathbb{R}$ such that $|t - t_k| > \varepsilon_1$, $k \in \mathbb{Z}$, where the number $\varepsilon_1 \in (0, \varepsilon_0/2\lceil 1/\delta_0 \rceil)$ will be precisely clarified a bit later. The function $t \mapsto F(t + \tau; b) - cF(t; b)$, $t \in [x, x + 1]$ is not greater than ε_1 if $t \in [x, t_m - \varepsilon_1] \cup (t_m + \varepsilon_1, t_{m+1} - \varepsilon_1] \cup \cdots \cup (t_{m+k}, x + 1]$; otherwise, $\|F(t + \tau; b) - cF(t; b)\| \leqslant (1 + |c|)\|F\|_{\infty,B}$. Using Lemma 1.1.11, the above implies

$$
\begin{aligned}
&\int_x^{x+1} \|F(t + \tau; b) - cF(t; b)\|^p \nu^p(t)\, dt \\
&\quad \leqslant \varepsilon_1^p \int_{A_x} \nu^p(t)\, dt + ((1 + |c|)\|F\|_{\infty,B})^p \int_{[x,x+1]\smallsetminus A_x} \nu^p(t)\, dt \\
&\quad \leqslant \varepsilon_1^p \int_x^{x+1} \nu^p(t)\, dt + ((1 + |c|)\|F\|_{\infty,B})^p p \int_0^{\infty} y^{p-1} m(\{s \in [x, x + 1] \smallsetminus A_x : \nu(s) > y\})\, dy \\
&\quad \leqslant \varepsilon_1^p \|\nu\|_{S^p} + ((1 + |c|)\|F\|_{\infty,B})^p p \int_0^{d} y^{p-1} m(\{s \in [x, x + 1] \smallsetminus A_x : \nu(s) > y\})\, dy \\
&\qquad + ((1 + |c|)\|F\|_{\infty,B})^p p \int_d^{\infty} y^{p-1} m(\{s \in [x, x + 1] \smallsetminus A_x : \nu(s) > y\})\, dy \\
&\quad \leqslant \varepsilon_1^p \|\nu\|_{S^p} + 2((1 + |c|)\|F\|_{\infty,B})^p d^p \lceil 1/\delta_0 \rceil \varepsilon_1
\end{aligned}
$$

$$+ ((1+|c|)\|F\|_{\infty,B})^p p \int_d^\infty y^{p-1} m(\{s \in [x,x+1] \setminus A_X : v(s) > y\})\, dy$$

$$\leqslant \varepsilon_1^p \|v\|_{S^p} + 2((1+|c|)\|F\|_{\infty,B})^p d^p \lceil 1/\delta_0 \rceil \varepsilon_1 + \frac{\varepsilon}{2}, \quad b \in B,$$

where $A_x := [x, t_m - \varepsilon_0] \cup (t_m + \varepsilon_0, t_{m+1} - \varepsilon_0] \cup \cdots \cup (t_{m+k}, x+1]$. This simply completes the proof of theorem since we can always find a sufficiently small number $\varepsilon_1 > 0$ such that

$$\varepsilon_1^p \|v\|_{S^p} + 2((1+|c|)\|F\|_{\infty,B})^p d^p \lceil 1/\delta_0 \rceil \varepsilon_1 < \frac{\varepsilon}{2}. \qquad \square$$

It is clear that condition (LT) implies condition (LQ), so that we can use the weight function $v(\cdot)$ constructed in Example 5.4.8 here.

The subsequent result follows from [266, Theorem 3] and the argumentation contained in the proof of [266, Theorem 2]; the only thing worth noting is that, if $0 < p < 1$, then we should replace the number η_k^p with the number η_k throughout the proof of Theorem 2 and assume that $\eta_k \in (0, (\varepsilon/4)^p \delta)$ for all $k \in \mathbb{N}$.

Theorem 5.4.11. *Suppose that $\rho = T \in L(Y)$, $p > 0$ and $F : \Lambda \times X \to Y$ is a Stepanov-$(\Omega, p, \rho, \mathcal{B}, \Lambda')$-almost periodic function, where $\Lambda = \mathbb{R}$ or $\Lambda = [0,\infty)$, $\Omega = [0,1]$ and $\Lambda' = \Lambda$. Suppose further that $F(\cdot;\cdot)$ satisfies that, for every $x \in X$, the function $t \mapsto F(t;x)$, $t \in \mathbb{R}$ is piecewise continuous with the possible first kind discontinuities at the points of a fixed sequence $(t_k)_{k\in\mathbb{Z}}$ $[(t_k)_{k\in\mathbb{N}}]$ and $\delta_0 := \inf_{k\in\mathbb{Z}}(t_{k+1} - t_k) > 0$ $[\delta_0 := \inf_{k\in\mathbb{N}}(t_{k+1} - t_k) > 0]$. If the condition (QUC) holds, then $F(\cdot;\cdot)$ is pre-$(\mathcal{B}, T, (t_k))$-piecewise continuous almost periodic. Furthermore, if $T = \mathrm{I}$, $F(\cdot;\cdot)$ is continuous and any set of collection $\mathcal{B}$ is compact, then $F(\cdot;\cdot)$ is Bohr $\mathcal{B}$-almost periodic, i. e., for every $B \in \mathcal{B}$ and $\varepsilon > 0$, there exists $l > 0$ such that for each $t_0 \in \Lambda$ there exists $\tau \in B(t_0, l) \cap \Lambda$ such that, for every $t \in \Lambda$ and $x \in B$, we have $\|F(t+\tau;x) - F(t;x)\| \leqslant \varepsilon$.*

If $p \geqslant 1$, then it is well known that any uniformly continuous, Stepanov-p-almost periodic function $F : \mathbb{R}^n \to Y$ is almost periodic and, therefore, bounded. Furthermore, we know that there exists a Stepanov-1-almost periodic function $F : \mathbb{R} \to \mathbb{R}$, which is not uniformly continuous (bounded); see, e. g., [386]. In connection with this problem, we would like to state the following result; the proof can be deduced using the argumentation, which is very similar to the argumentation used for proving [266, Theorems 2, 3], given in the one-dimensional setting and, therefore, omitted.

Corollary 5.4.12. (i) *Suppose that $F : \mathbb{R}^n \times X \to Y$ is continuous, any set of collection $\mathcal{B}$ is compact and $F(\cdot;\cdot)$ is uniformly continuous on the set $\mathbb{R}^n \times B$ $(B \in \mathcal{B})$. If $F(\cdot;\cdot)$ is Stepanov-$(\Omega, p, \mathrm{I}, \mathcal{B}, \mathbb{R}^n)$-almost periodic function, where $p > 0$ and $\Omega = [0,1]^n$, then $F(\cdot;\cdot)$ is Bohr $\mathcal{B}$-almost periodic.*

(ii) *Suppose that $p > 0$. Then any uniformly continuous, Stepanov-p-almost periodic function $F : \mathbb{R}^n \to Y$ is almost periodic.*

5.4.3 The invariance of Stepanov-p-almost periodicity under the actions of the infinite convolution products ($0 < p < 1$)

In the series of our recent research articles, we have examined the invariance of Stepanov-p-almost periodicity under the actions of the infinite convolution products and provide various applications of the usually considered classes of Stepanov-p-almost periodic (automorphic) type functions with the exponent $p \geqslant 1$; in this subsection, we will analyze the invariance of Stepanov-p-almost periodicity under the action of the infinite convolution product (262), where $(R(t))_{t>0} \subseteq L(X,Y)$ is a strongly continuous operator family satisfying certain extra conditions ($0 < p < 1$). For the sake of brevity, we will analyze here the one-dimensional setting.

The consideration from [444, Proposition 2.6.11], where we have analyzed the case $p \geqslant 1$, is essential but cannot be replicated or modified in our new framework since the reverse Hölder inequality is valid for $0 < p < 1$; we feel it is our duty to say aloud that the last mentioned result has been previously considered by M. Maqbul and D. Bahuguna [578] in the scalar-valued setting, for the nondegenerate strongly continuous semigroups of operators. In order to overcome this difficulty, we must impose some new unpleasant conditions; for example, we can prove the following result.

Theorem 5.4.13. *Suppose that $\nu : \mathbb{R} \to (0,\infty)$ is a Lebesgue measurable function and there exists a function $\omega : \mathbb{R} \to [0,\infty)$ such that $\nu(x+y) \leqslant \nu(x)\omega(y)$ for all $x, y \in \mathbb{R}$. Suppose further that $f : \mathbb{R} \to X$ is Stepanov-(p,T,ν)-almost periodic, where $\rho = T \in L(Y)$, and $(R(t))_{t>0} \subseteq L(X,Y)$ is a strongly continuous operator family commuting with T and satisfying the following conditions:*

(i) *There exists a finite real number $s \geqslant 1$ such that $\sum_{k=0}^{\infty} \|\omega(\cdot)R(\cdot)\|_{L^{s'}[k,k+1]} < +\infty$, where* $1/s + 1/s' = 1$.

(ii) *There exists a finite real number $M > 0$ such that, for every $t \in \mathbb{R}$ and $\tau \in \mathbb{R}$, we have*

$$\sum_{k=0}^{\infty} \int_{k}^{k+1} \|R(r)\| \cdot \|f(t-r)\|\, dr + \sum_{k=0}^{\infty} \max\{[\|R(r)\| \cdot \|f(t+\tau-r) - Tf(t-r)\|\nu(t-r)\omega(r)] : r \in [k,k+1],\ \|f(t+\tau-r) - Tf(t-r)\|\nu(t-r) > 1\} \leqslant M.$$

Then the function $F(\cdot)$, given by (262), *is (T,ν)-almost periodic provided that*

(iii) $\lim_{\delta\to 0} \sup_{t\in\mathbb{R}} \int_t^{t+1} \|f(s+\delta) - Tf(s)\|^p \nu^p(s)\, ds = 0$;

here, by (T,ν)-almost periodicity of $F(\cdot)$, we mean that $F(\cdot)$ is continuous and for each $\varepsilon > 0$ there exists a relatively dense set R in $\mathbb{R}$ such that for each $\tau \in R$ and $t \in \mathbb{R}$ we have $\|F(t+\tau) - TF(t)\|_Y \nu(t) \leqslant \varepsilon$. Furthermore, the condition (iii) *holds provided that the function $\omega(\cdot)$ is bounded and $T = \mathrm{I}$.*

Proof. It can be simply shown that the function $F(\cdot)$ is well-defined and the integral, which defines $F(\cdot)$, is absolutely convergent since we have

$$\left\| \int_{-\infty}^{t} R(t-s)f(s)\,ds \right\|_Y = \left\| \int_0^{\infty} R(s)f(t-s)\,ds \right\|_Y \leqslant \int_0^{\infty} \|R(s)\| \cdot \|f(t-s)\|\,ds \leqslant M;$$

cf. (ii). Let R be a relatively dense subset of $\mathbb{R}$ such that for each $\tau \in R$ and $t \in \mathbb{R}$ we have $\int_t^{t+1} \|f(s+\tau) - Tf(s)\|^p \nu^p(s)\,ds \leqslant \varepsilon$. Fix a number $t \in \mathbb{R}$. Then the Lebesgue measure of the set $B_t \equiv \{s \in [t,t+1] : \|f(s+\tau) - Tf(s)\| \cdot \nu(s) > 1\}$ is less than or equal to ε; hence, we have ($\tau \in R$):

$$\begin{aligned}
&\|F(t+\tau) - TF(t)\|_Y \nu(t) \\
&\leqslant \sum_{k=0}^{\infty} \int_k^{k+1} \|R(r)\| \cdot \|f(t+\tau-r) - Tf(t-r)\| \nu(t)\,dr \\
&= \sum_{k=0}^{\infty} \int_{B_{t-k-1}} \|R(r)\| \cdot \|f(t+\tau-r) - Tf(t-r)\| \nu(t-r)\omega(r)\,dr \\
&\quad + \sum_{k=0}^{\infty} \int_{[k,k+1] \setminus B_{t-k-1}} \|R(r)\| \cdot \|f(t+\tau-r) - Tf(t-r)\| \nu(t-r)\omega(r)\,dr \\
&\leqslant \sum_{k=0}^{\infty} \int_{B_{t-k-1}} \|R(r)\| \cdot \|f(t+\tau-r) - Tf(t-r)\| \nu(t-r)\omega(r)\,dr \\
&\quad + \sum_{k=0}^{\infty} \int_{[k,k+1] \setminus B_{t-k-1}} \|R(r)\| \cdot \|f(t+\tau-r) - Tf(t-r)\|^{p/s} \nu^{p/s}(t-r)\omega(r)\,dr \\
&\leqslant \sum_{k=0}^{\infty} \int_{B_{t-k-1}} \|R(r)\| \cdot \|f(t+\tau-r) - Tf(t-r)\| \nu(t-r)\omega(r)\,dr \\
&\quad + \sum_{k=0}^{\infty} \int_{[k,k+1]} \|R(r)\| \cdot \|f(t+\tau-r) - Tf(t-r)\|^{p/s} \nu^{p/s}(t-r)\omega(r)\,dr \\
&\leqslant \sum_{k=0}^{\infty} \int_{B_{t-k-1}} \|R(r)\| \cdot \|f(t+\tau-r) - Tf(t-r)\| \nu(t-r)\omega(r)\,dr \\
&\quad + \sum_{k=0}^{\infty} \|\omega(\cdot)R(\cdot)\|_{L^{s'}[k,k+1]} \left(\int_k^{k+1} \|f(t-r+\tau) - Tf(t-r)\|^p \nu^p(t-r)\,dr \right)^{1/s} \\
&\leqslant \sum_{k=0}^{\infty} \int_{B_{t-k-1}} \|R(r)\| \cdot \|f(t+\tau-r) - Tf(t-r)\| \nu(t-r)\omega(r)\,dr + \varepsilon \sum_{k=0}^{\infty} \|\omega(\cdot)R(\cdot)\|_{L^{s'}[k,k+1]} \\
&\leqslant \varepsilon \sum_{k=0}^{\infty} \max\{[\|R(r)\| \cdot \|f(t+\tau-r) - Tf(t-r)\| \nu(t-r)\omega(r)] :
\end{aligned}$$

$$r \in [k,k+1],\ \|f(t+\tau-r)-Tf(t-r)\|v(t-r)>1\}+\varepsilon\sum_{k=0}^{\infty}\|\omega(\cdot)R(\cdot)\|_{L^{s'}[k,k+1]}$$

$$\leqslant\left(M+\sum_{k=0}^{\infty}\|\omega(\cdot)R(\cdot)\|_{L^{s'}[k,k+1]}\right)\varepsilon.$$

The continuity of function $F(\cdot)$ can be proved as above, using the condition (iii) and replacing the number τ with the number δ throughout the above computation. It remains to be proved that (ii) holds provided that the function $\omega(\cdot)$ is bounded and $T=\mathrm{I}$. Clearly, for every $t \in \mathbb{R}$, $\tau \in \mathbb{R}$ and $\delta \in (-1,1)$, we have

$$\begin{aligned}
&\int_t^{t+1}\|f(s+\delta)-f(s)\|^p v^p(s)\,ds\\
&\quad\leqslant\int_t^{t+1}\|f(s+\delta)-f(s+\delta+\tau)\|^p v^p(s)\,ds\\
&\qquad+\int_t^{t+1}\|f(s+\delta+\tau)-f(s+\tau)\|^p v^p(s)\,ds+\int_t^{t+1}\|f(s+\tau)-f(s)\|^p v^p(s)\,ds\\
&\quad\leqslant\int_t^{t+1}\|f(s+\delta)-f(s+\delta+\tau)\|^p v^p(s+\delta)\omega^p(-\delta)\,ds\\
&\qquad+\int_t^{t+1}\|f(s+\delta+\tau)-f(s+\tau)\|^p v^p(s+\tau)\omega^p(-\tau)\,ds+\int_t^{t+1}\|f(s+\tau)-f(s)\|^p v^p(s)\,ds\\
&\quad\leqslant(1+\|\omega\|_\infty)\Bigg[\int_t^{t+1}\|f(s+\delta)-f(s+\delta+\tau)\|^p v^p(s+\delta)\,ds\\
&\qquad+\int_t^{t+1}\|f(s+\delta+\tau)-f(s+\tau)\|^p v^p(s+\tau)\,ds+\int_t^{t+1}\|f(s+\tau)-f(s)\|^p v^p(s)\,ds\Bigg].
\end{aligned}$$

For a given $\varepsilon>0$, we can find a real number $l>3$ such that any interval $I\subseteq\mathbb{R}$ contains a number $\tau\in I$ such that $\int_t^{t+1}\|f(s+\tau)-f(s)\|^p v^p(s)\,ds\leqslant\varepsilon/3$ for all $t\in\mathbb{R}$. Fix now a real number t. Then we can always find a number $\tau\in\mathbb{R}$ such that the last inequality holds and $s+\tau\subseteq[0,l]$ for all $s\in[t,t+1]$. The first addend and the third addend in the above sum can be simply estimated by $\varepsilon/3$; this can be also done for the second addend in the above sum since we can argue as in the proof of [444, Proposition 3.5.3], by choosing a sequence of infinitely differentiable functions (φ_k), which converges to $[f(\cdot+\tau)-f(\cdot)]v(\cdot)$ in $L^1[0,l]$ and apply the Hölder inequality and the same procedure after that. □

We continue by stating the following result, which is not so easily comparable to Theorem 5.4.13 and [458, Proposition 2.3].

Proposition 5.4.14. *Suppose that $p \in [1,\infty)$, $1/p + 1/q = 1$, there exists a finite real constant $c > 0$ such that $\nu(t) \geqslant c > 0$, $t \in \mathbb{R}$ as well as $\nu : \mathbb{R} \to (0,\infty)$ is a Lebesgue measurable function and there exists a function $\omega : \mathbb{R} \to [0,\infty)$ such that $\nu(x+y) \leqslant \nu(x)\omega(y)$ for all x, $y \in \mathbb{R}$. Suppose further that $f : \mathbb{R} \to X$ is Stepanov-(p, T, ν)-almost periodic, where $\rho = T \in L(Y)$, and $(R(t))_{t>0} \subseteq L(X,Y)$ is a strongly continuous operator family commuting with T and satisfying the following conditions:*

(i) $\sum_{k=0}^{\infty} \|[1 + \omega(\cdot)]R(\cdot)\|_{L^q[k,k+1]} < +\infty$.

(ii) *For every $t \in \mathbb{R}$, we have*

$$\sum_{k=0}^{\infty} \int_{k}^{k+1} \|R(r)\| \cdot \|f(t-r)\| \, dr < +\infty.$$

Then the function $F(\cdot)$, given by (262), is (T,ν)-almost periodic; in particular, the function $F(\cdot)$ is both T-almost periodic and almost periodic.

Proof. We will provide the main details of the proof since it can be given with the help of the argumentation employed for proving [444, Proposition 2.6.11] and [458, Proposition 2.3]. Due to (ii), the function $F(\cdot)$ is well-defined. Observe that the condition $\nu(t) \geqslant c > 0$, $t \in \mathbb{R}$ implies that the function $f(\cdot)$ is Stepanov-(p,T)-almost periodic and, therefore, Stepanov-p-almost periodic in the usual sense. Since $\sum_{k=0}^{\infty} \|R(\cdot)\|_{L^q[k,k+1]} < +\infty$, the continuity and almost periodicity of $F(\cdot)$ follows directly from an application of [444, Proposition 2.6.11]. The remainder of proof follows similarly as in the proof of the above mentioned result and, therefore, omitted. □

Remark 5.4.15. It is worth noting that many statements for Stepanov-p-almost periodic functions with the exponent $p \geqslant 1$, which can be deduced without the help of the Hölder inequality, continue to hold for Stepanov-p-almost periodic functions with the exponent $p \in (0,1)$; for example, this is the case with the statements of [444, Theorem 2.6.17(i)] and [447, Theorem 6.2.15, Proposition 6.2.22, Theorem 6.2.30, Corollary 6.2.31]. On the other hand, the statements of [447, Propositions 6.2.18, 6.2.19], where the Hölder inequality is essentially employed in the proofs, cannot be clarified for Stepanov-p-almost periodic functions with the exponent $p \in (0,1)$. Especially, we would like to emphasize that the composition principles established in [548, Theorem 2.2], [458, Theorem 2.2] and [447, Theorems 6.2.32, 6.2.33] cannot be clarified for Stepanov-p-almost periodic functions with the exponent $p \in (0,1)$.

Using a similar argumentation as in the proof of Theorem 5.4.13, we may deduce certain results concerning the invariance of Stepanov p-almost periodicity (automorphy) under the actions of the infinite convolution product

$$t \mapsto H(t) \equiv \int_{-\infty}^{+\infty} h(t-s)f(s)\, ds, \quad t \in \mathbb{R}, \tag{302}$$

where $h \in L^1(\mathbb{R})$. This can be applied in the analysis of the existence and uniqueness of the Stepanov-p-almost periodic (automorphic) solutions to the inhomogenous heat equation (cf. [447, Subsection 6.2.6 and pp. 558–559]).

Finally, we would like to note that it is almost impossible to state any relevant result concerning the invariance of Stepanov p-almost periodicity (automorphy) in norm under the actions of the infinite convolution products (262) and (302); cf. the next two subsections for the notion.

5.4.4 Stepanov-p-almost automorphic type functions ($p > 0$)

The main aim of this subsection is to introduce and analyze the following classes of Stepanov-p-almost automorphic functions.

Definition 5.4.16. Suppose that $p > 0$, $\nu : \Omega \to (0, \infty)$ is a Lebesgue measurable function, $F : \mathbb{R}^n \times X \to Y$ is a given function and R is a certain collection of sequences in $\mathbb{R}^n$. Then we say that the function $F(\cdot;\cdot)$ is Stepanov $(\Omega, p, \mathrm{R}, \mathcal{B}, \nu)$-almost automorphic [Stepanov $(\Omega, p, \mathrm{R}, \mathcal{B}, \nu, W_{\mathcal{B},\mathrm{R}})$-almost automorphic; Stepanov $(\Omega, p, \mathrm{R}, \mathcal{B}, \nu, P_{\mathcal{B},\mathrm{R}})$-almost automorphic] if and only if $F(\cdot;\cdot)$ is Stepanov $(\Omega, \mathrm{R}, \mathcal{B}, Z^{\mathcal{P}})$-multialmost automorphic [Stepanov $(\Omega, \mathrm{R}, \mathcal{B}, Z^{\mathcal{P}}, W_{\mathcal{B},\mathrm{R}})$-multialmost automorphic; Stepanov $(\Omega, \mathrm{R}, \mathcal{B}, Z^{\mathcal{P}}, P_{\mathcal{B},\mathrm{R}})$-multialmost automorphic] with $Z = L^p_\nu(\Omega : Y)$.

Any strongly Stepanov-$(\Omega, p, \mathrm{R}, \mathcal{B}, \nu)$-almost periodic function $F : \mathbb{R}^n \times X \to Y$ is Stepanov $(\Omega, p, \mathrm{R}, \mathcal{B}, \nu)$-almost automorphic.

Definition 5.4.17. Suppose that $p > 0$ and $\nu : \Omega \to (0, \infty)$ is a Lebesgue measurable function. Then we say that a function $F : \mathbb{R}^n \to Y$ is Stepanov-(Ω, p, ν)-almost automorphic [Stepanov-(Ω, p, ν, W)-almost automorphic] if $F(\cdot)$ is Stepanov-$(\Omega, p, \mathrm{R}, \nu)$-almost automorphic [Stepanov-$(\Omega, p, \mathrm{R}, \nu, W_{\mathrm{R}})$-almost automorphic] with R being the collection of all sequences in $\mathbb{R}^n$; $F(\cdot)$ is Stepanov-p-almost automorphic if $F(\cdot)$ is Stepanov-(Ω, p, ν)-almost automorphic with $\Omega = [0,1]^n$ and $\nu(\cdot) \equiv 1$.

The statements of [176, Theorem 2.4, Propositions 2.5, 2.7, 2.8, Theorem 2.9], established recently for general classes of metrically Stepanov almost automorphic functions, can be simply reformulated for the special classes of Stepanov $(\Omega, p, \mathrm{R}, \mathcal{B}, \nu)$-almost automorphic functions since the metric space $Z = L^p_\nu(\Omega : Y)$ satisfies all necessary requirements for applications of these results. For example, we have the following (Proposition 2.8).

Proposition 5.4.18. *Suppose that $F_m : \mathbb{R}^n \times X \to Y$ is Stepanov $(\Omega, p, \mathrm{R}, \mathcal{B}, \nu)$-almost automorphic [Stepanov $(\Omega, p, \mathrm{R}, \mathcal{B}, \nu, W_{\mathcal{B},\mathrm{R}})$-almost automorphic; Stepanov $(\Omega, p, \mathrm{R}, \mathcal{B}, \nu, P_{\mathcal{B},\mathrm{R}})$-almost automorphic] for all $m \in \mathbb{N}$, $F : \mathbb{R}^n \times X \to Y$ and $\lim_{m\to+\infty} F_m(\mathbf{t}+\cdot; x) = F(\mathbf{t}+\cdot; x)$ for the topology of $L^p_\nu(\Omega : Y)$, uniformly on the set $\mathbb{R}^n \times B$ for each $B \in \mathcal{B}$. Suppose further that, for every sequence $b \in \mathrm{R}$, all of its subsequences also belong to R. Then $F(\cdot;\cdot)$ is likewise*

Stepanov $(\Omega, p, \mathrm{R}, \mathcal{B}, \nu)$*-almost automorphic [Stepanov* $(\Omega, p, \mathrm{R}, \mathcal{B}, \nu, W_{\mathcal{B},\mathrm{R}})$*-almost automorphic; Stepanov* $(\Omega, p, \mathrm{R}, \mathcal{B}, \nu, P_{\mathcal{B},\mathrm{R}})$*-almost automorphic].*

The following analogue of Corollary 5.4.12 can be deduced using the argumentation contained in the proof of [399, Proposition 3.1] and a relatively simple argumentation involving the compactness of sets of the collection $\mathcal{B}$ and the condition (LB) clarified below (cf. also the proof of Theorem 5.4.25).

Corollary 5.4.19. (i) *Suppose that* $p > 0$, $\Omega = [0,1]^n$, $\nu : \Omega \to (0, \infty)$ *is a Lebesgue measurable function,* R *is any collection of sequences in* $\mathbb{R}^n$*, which satisfies that for each sequence in* R *any its subsequence also belongs to* R, $F : \mathbb{R}^n \times X \to Y$ *is continuous, any set of collection* $\mathcal{B}$ *is compact,* $F(\cdot; x)$ *is uniformly continuous on* $\mathbb{R}^n$ *for every fixed element* $x \in X$ *and the following condition holds:*

(LB) *For every set* $B \in \mathcal{B}$*, there exists a finite real number* $L_B > 0$ *such that* $\|F(\mathbf{t}; x) - F(\mathbf{t}; y)\|_Y \leqslant L_B \|x - y\|$, $\mathbf{t} \in \mathbb{R}^n$, $x, y \in B$.

If $F(\cdot;\cdot)$ *is Stepanov* $(\Omega, p, \mathrm{R}, \mathcal{B}, \nu)$*-almost automorphic, then* $F(\cdot;\cdot)$ *is compactly* $(\mathrm{R}, \mathcal{B})$*-multialmost automorphic.*

(ii) *Suppose that* $p > 0$, $\Omega = [0,1]^n$ *and* $\nu : \Omega \to (0, \infty)$ *is a Lebesgue measurable function. Then any uniformly continuous, Stepanov-*(Ω, p, ν)*-almost automorphic function* $F : \mathbb{R}^n \to Y$ *is almost automorphic.*

Before proceeding to the next subsection, we would like to note that Corollary 5.4.5 continues to hold for Stepanov-p-almost automorphic functions ($p > 0$). This essentially follows from the Hölder inequality and the fact that, for every fixed number $\mathbf{t} \in \mathbb{R}^n$, the assumptions $\lim_{l\to+\infty} \int_{[0,1]^n} \|F(\mathbf{t}+b_{k_l}+\mathbf{s}) - [F^*(\mathbf{t})](\mathbf{s})\|^p \, d\mathbf{s} = 0$ and $\lim_{l\to+\infty} \int_{[0,1]^n} \|[F^*(\mathbf{t}-b_{k_l})](\mathbf{s}) - F(\mathbf{t}+\mathbf{s})\|^p d\mathbf{s} = 0$ imply the existence of a set $N \subseteq [0,1]^n$ with the Lebesgue zero measure such that $\lim_{l\to+\infty} F(\mathbf{t}+b_{k_l}+\mathbf{s}) = [F^*(\mathbf{t})](\mathbf{s})$ and $\lim_{l\to+\infty} [F^*(\mathbf{t}-b_{k_l})](\mathbf{s}) = F(\mathbf{t}+\mathbf{s})$ for all $\mathbf{s} \in [0,1]^n \setminus N$; after that, we may conclude that the essential boundedness of $F(\cdot)$ implies the essential boundedness of $F^*(\cdot)$ and argue as in the proof of Proposition 5.4.4(ii). This argumentation also provides the affirmative answer to the problem proposed on page 9 of our joint research study [249] with T. Diagana.

5.4.5 Relations between piecewise continuous almost automorphic functions and metrically Stepanov-p-almost periodic functions ($p > 0$)

We start this subsection by recalling that the notion of a Bochner spatially almost automorphic sequence $(t_k)_{k\in\mathbb{Z}}$ has recently been introduced by L. Qi and R. Yuan in [756, Definition 3.1], who proved that any Wekler sequence $(t_k)_{k\in\mathbb{Z}}$ is Bochner spatially almost automorphic as well as that the converse statement is not true in general. The authors have analyzed the classes of Bohr, Bochner and Levitan piecewise continuous almost automorphic functions; in [756, Theorem 4.8], the authors have proved that these classes coincide (cf. also the research article [260] by W. Dimbour and V. Valmorin for the notion of $\mathbb{S}$-almost automorphy). Moreover, in [756, Theorem 8.2], the essential relationship between piecewise continuous almost automorphic functions and Stepanov-p-almost automorphic functions has been clarified ($p \geqslant 1$). In order to further study the relations between piecewise continuous almost automorphic functions and metrically Stepanov-p-almost automorphic functions ($p > 0$), we need to introduce the following, rather general, notion (cf. also condition (iii) in [756, Definition 4.2]).

Definition 5.4.20. Suppose that R is a certain collection of real sequences and the function $F : \mathbb{R} \times X \to Y$ satisfies that, for every $x \in X$, the function $t \mapsto F(t;x)$, $t \in \mathbb{R}$ is piecewise continuous with the possible first kind discontinuities at the points of a fixed sequence $(t_k)_{k\in\mathbb{Z}}$. Let (t_k') be any real sequence. Then we say that the function $F(\cdot;\cdot)$ is pre-(R, $\mathcal{B}$, (t_k), (t_k'))-piecewise continuous almost automorphic if, for every $B \in \mathcal{B}$ and $(b_k) \in$ R, there exist a subsequence (b_{k_l}) of (b_k) and a function $F_B^* : \mathbb{R}\times X \to Y$ such that, for every $x \in B$, the function $t \mapsto F_B^*(t;x)$, $t \in \mathbb{R}$ is piecewise continuous with the possible first kind discontinuities at the points of sequence $(t_k')_{k\in\mathbb{Z}}$, $\lim_{l\to+\infty} F(t+b_{k_l};x) = F_B^*(t;x)$ for all $t \in \mathbb{R} \smallsetminus \{t_{k_l}' : l \in \mathbb{Z}\}$ and $\lim_{l\to+\infty} F_B^*(t - b_{k_l};x) = F(t;x)$ for all $t \in \mathbb{R} \smallsetminus \{t_{k_l} : l \in \mathbb{Z}\}$.

The next simple result follows directly by applying the dominated convergence theorem with $[F_B^*(t;x)](u) \equiv F_B^*(t+u;x)$.

Proposition 5.4.21. *Suppose that $p > 0$, R is a certain collection of real sequences and the function $F : \mathbb{R} \times X \to Y$ satisfies that, for every $x \in X$, the function $t \mapsto F(t;x)$, $t \in \mathbb{R}$ is piecewise continuous with the possible first kind discontinuities at the points of a fixed sequence $(t_k)_{k\in\mathbb{Z}}$. Let (t_k') be any real sequence and let $\sup_{t\in\mathbb{R};x\in B}\|F(t;x)\|_Y < +\infty$ ($B \in \mathcal{B}$). If the function $F(\cdot;\cdot)$ is pre-(R, $\mathcal{B}$, (t_k), (t_k'))-piecewise continuous almost automorphic, then $F(\cdot;\cdot)$ is Stepanov (Ω, p, R, $\mathcal{B}$, ν)-almost automorphic for any function $\nu \in L^p(\Omega)$.*

Remark 5.4.22. In [260, Definition 2.3], the authors have introduced the notion of $\mathbb{S}$-almost automorphy for a function $F : \mathbb{R} \to Y$, where $\mathbb{S}$ is any subset of $\mathbb{R}$. In this slightly different approach, the authors have not used the assumption that the limit function $F^*(\cdot)$ is piecewise continuous; without entering into further details, we will only note here that an analogue of Proposition 5.4.21 can be simply formulated for $\mathbb{S}$-almost automorphic functions, provided that the Lebesgue measure of the set $\mathbb{S}$ is equal to zero and R is the collection of all sequences with values in $\mathbb{S}$.

Further on, the notion of a Levitan piecewise continuous almost automorphic function has been introduced in [756, Definition 4.6]; now we would like to extend this notion by introducing the following general class of functions (we use the condition (iii) from this definition, in a slightly modified form, only).

Definition 5.4.23. Suppose that $F : \mathbb{R}\times X \to Y$ satisfies that, for every $x \in X$, the function $t \mapsto F(t;x)$, $t \in \mathbb{R}$ is piecewise continuous with the possible first kind discontinuities at the points of a fixed sequence $(t_k)_{k\in\mathbb{Z}}$. If ρ is a binary relation on Y, then we say that the function $F(\cdot;\cdot)$ is pre-($\mathcal{B}$, ρ, (t_k))-Levitan piecewise continuous almost automorphic if, for every $B \in \mathcal{B}$, $\varepsilon > 0$ and $N > 0$, there exists a relatively dense subset S of $\mathbb{R}$ such that, if $\tau \in S$, $x \in B$, $|t| \leqslant N$ and $|t - \tau_j| > \varepsilon$ for all $j \in \mathbb{Z}$, then there exists $y_{t;x} \in \rho(F(t;x))$ such that $\|F(t + \tau; x) - y_{t,x}\| \leqslant \varepsilon$.

The introduced notion is important for our purposes because a slight modification of the proof of Theorem 5.4.10 shows that the following result holds true.

Theorem 5.4.24. *Suppose that* $p > 0$, $\rho = T \in L(Y)$ *and* $F : \mathbb{R} \times X \to Y$ *satisfies that, for every* $x \in X$, *the function* $t \mapsto F(t;x)$, $t \in \mathbb{R}$ *is piecewise continuous with the possible first kind discontinuities at the points of a fixed sequence* $(t_k)_{k \in \mathbb{Z}}$, *which is strictly monotonically increasing and satisfies that there exists* $\delta_0 > 0$ *such that* $\inf_{k \in \mathbb{Z}}(t_{k+1} - t_k) > \delta_0$. *If* $\nu : \mathbb{R} \to (0, \infty)$ *is Stepanov-p-bounded and satisfies condition* (LQ), $F(\cdot;\cdot)$ *is pre-*$(\mathcal{B}, \rho, (t_k))$*-Levitan piecewise continuous almost automorphic and* $\sup_{t \in \mathbb{R}; x \in B} \|F(t;x)\|_Y < +\infty$ ($B \in \mathcal{B}$), *then we have the following: For every* $B \in \mathcal{B}$, $\varepsilon > 0$ *and* $t \in \mathbb{R}$, *there exists a relatively dense subset* S *of* $\mathbb{R}$ *such that, if* $\tau \in S$ *and* $x \in B$, *then*

$$\int_t^{t+1} \|F(s + \tau; x) - TF(s;x)\|^p \nu^p(s)\, ds \leqslant \varepsilon.$$

Now we would like to note that Proposition 5.4.21 and Theorem 5.4.24 can serve one to provide several different extensions of [756, Theorem 8.2]. For example, the statement of this result holds for any exponent $p > 0$; moreover, we have the following (for the notion of a Levitan s. a. a. sequence, we refer the reader to [756, Definition 3.12]).

Theorem 5.4.25. *Suppose that* $p > 0$, $\Omega = [0,1]$, $\nu : \Omega \to (0, \infty)$ *is a p-integrable function,* $F : \mathbb{R} \to Y$ *is piecewise continuous with possible discontinuities at the points of a subset of a Levitan s. a. a. sequence* $(t_k)_{k \in \mathbb{Z}}$ *and the following condition:*

(QUC1) *For every* $\varepsilon > 0$, *there exists* $\delta > 0$ *such that, if the points* t' *and* t'' *belong to* (t_i, t_{i+1}) *for some* $i \in \mathbb{Z}$ *and* $|t' - t''| < \delta$, *then* $\|F(t') - F(t'')\| < \varepsilon$.

Holds. Then $F(\cdot)$ *is Stepanov-*(Ω, p, ν)*-almost automorphic if and only if* $F(\cdot)$ *is piecewise continuous almost automorphic.*

Proof. It is clear that Proposition 5.4.21 implies that, if $F : \mathbb{R}^n \to Y$ is piecewise continuous almost automorphic, then $F(\cdot)$ is Stepanov-(Ω, p, ν)-almost automorphic. To prove the converse statement, it suffices to show that for each number $\sigma \in (0,1]$ the Steklov function $F_\sigma := \sigma^{-1} \int_0^\sigma F(t + s)\, ds$, $t \in \mathbb{R}$ is compactly almost automorphic (see the proof of the above mentioned result). The uniform continuity of $F_\sigma(\cdot)$ can be simply proved using the condition (QUC1) and we will only prove here that $F_\sigma(\cdot)$ is almost automorphic. If (b_k) is a real sequence, then we can always find a subsequence (b_{k_l}) of (b_k) and a function $F^* : \mathbb{R} \to Y$ such that, for every fixed number $t \in \mathbb{R}$, we have $\lim_{l \to +\infty} \int_0^1 \|F(t + b_{k_l} + s) - [F^*(t)](s)\|^p \nu^p(s)\, ds = 0$ and $\lim_{l \to +\infty} \int_0^1 \|[F^*(t - b_{k_l})](s) - F(t + s)\|^p \nu^p(s)\, ds = 0$. This simply implies the existence of a set $N \subseteq [0,1]$ with the Lebesgue zero measure such that $\lim_{l \to +\infty} F(t + b_{k_l} + s) = [F^*(t)](s)$ and $\lim_{l \to +\infty} [F^*(t - b_{k_l})](s) = F(t + s)$

for all $s \in [0,1] \setminus N$. Applying the dominated convergence theorem, we simply get that $\lim_{l\to+\infty} F_\sigma(t + b_{k_l}) = \sigma^{-1} \int_0^\sigma [F^*(t)](s)\,ds$ and $\lim_{l\to+\infty} \sigma^{-1} \int_0^\sigma [F^*(t - b_{k_l})](s)\,ds = F_\sigma(t)$, which implies the required conclusion. □

We will consider two-dimensional analogues of Theorem 5.4.25 somewhere else. The interested reader may also try to prove certain analogues of Theorem 5.4.13 and Proposition 5.4.14 for Stepanov-p-almost automorphic type functions ($0 < p < 1$).

5.4.6 Stepanov-p-almost periodicity in norm and Stepanov-p-almost automorphy in norm ($p > 0$)

In this subsection, we will consider the following notion.

Definition 5.4.26. (i) Suppose that $p > 0$, (SM-1) holds true and $\nu_{\mathbf{t}} : \mathbf{t} + \Omega \to (0, \infty)$ is a Lebesgue measurable function ($\mathbf{t} \in \Lambda$). Then we say that a function $F : \Lambda \times X \to Y$ is Stepanov-$(\Omega, p, \rho, \mathcal{B}, \Lambda', \nu_\cdot)$-almost periodic in norm [Stepanov-$(\Omega, p, \rho, \mathcal{B}, \Lambda', \nu)$-almost periodic in norm, where a Lebesgue measurable function $\nu : \Lambda \to (0, \infty)$ satisfies that $\nu_{\mathbf{t}}(\cdot) \equiv \nu_{|\mathbf{t}+\Omega}(\cdot)$ for all $\mathbf{t} \in \Lambda$] if the function $\|F(\cdot;\cdot)\|_Y^p : \Lambda \times X \to \mathbb{C}$ is Stepanov-$(\Omega, 1, \rho_N, \mathcal{B}, \Lambda', \nu_\cdot^p)$-almost periodic [Stepanov-$(\Omega, 1, \rho_N, \mathcal{B}, \Lambda', \nu^p)$-almost periodic], where $\rho_N := \{(\|x\|_Y^p, \|y\|_Y^p); (x, y) \in \rho\}$; furthermore, if $\nu_{\mathbf{t}}(\cdot) \equiv 1$ for all $\mathbf{t} \in \Lambda$, then we omit the term "$\nu_\cdot$" from the notation.

(ii) Suppose that $p > 0$, $\emptyset \neq \Lambda \subseteq \mathbb{R}^n$, $F : \Lambda \times X \to Y$ is a given function and the assumptions $\mathbf{t} \in \Lambda$, $\mathbf{b} \in \mathrm{R}$ and $l \in \mathbb{N}$ imply $\mathbf{t} + \mathbf{b}(l) \in \Lambda$. If $\nu : \Omega \to (0, \infty)$ is a Lebesgue measurable function, then we say that the function $F(\cdot;\cdot)$ is (strongly) Stepanov-$(\Omega, p, \mathrm{R}, \mathcal{B}, \nu)$-almost periodic in norm if $\|F(\cdot;\cdot)\|_Y^p$ is (strongly) Stepanov $(\Omega, \mathrm{R}, \mathcal{B}, \mathcal{P}_Z)$-multialmost periodic with $Z = L^1_{\nu^p}(\Omega : \mathbb{C})$ and $\mathcal{P}_Z = C_b(\Lambda : Z)$; if $\Omega = [0,1]^n$, then we omit the term "Ω" from the notation.

(iii) Suppose that $p > 0$, (SM-1) holds true, $\nu_{\mathbf{t}} : \mathbf{t} + \Omega \to (0, \infty)$ is a Lebesgue measurable function ($\mathbf{t} \in \Lambda$) and $\nu : \Lambda \to (0, \infty)$ is a Lebesgue measurable function. Then we say that a function $F : \Lambda \to Y$ is Stepanov-$(\Omega, p, \nu_\cdot)$-almost periodic in norm [Stepanov-(Ω, p, ν)-almost periodic in norm] if $\|F(\cdot)\|_Y^p$ is Stepanov-$(\Omega, 1, \rho, \Lambda', \nu_\cdot^p)$-almost periodic [Stepanov-$(\Omega, 1, \rho, \Lambda', \nu^p)$-almost periodic] with $\rho \equiv \mathrm{I}$ and $\Lambda' \equiv \{\tau \in \mathbb{R}^n : \tau + \mathbf{t} \in \Lambda,\ \mathbf{t} \in \Lambda\}$.

(iv) Suppose that $p > 0$, $\emptyset \neq \Lambda \subseteq \mathbb{R}^n$, $F : \Lambda \to Y$ is a given function, $\nu : \Omega \to (0, \infty)$ is a Lebesgue measurable function and R denotes the collection of all sequences in $\mathbb{R}^n$ such that the assumptions $\mathbf{t} \in \Lambda$, $\mathbf{b} \in \mathrm{R}$ and $l \in \mathbb{N}$ imply $\mathbf{t} + \mathbf{b}(l) \in \Lambda$. Then we say that the function $F(\cdot)$ is Bochner–Stepanov-(Ω, p, ν)-almost periodic in norm if $\|F(\cdot)\|_Y^p$ is Stepanov-$(\Omega, 1, \mathrm{R}, \nu^p)$-almost periodic.

In the almost automorphic setting, we will consider the following notion.

Definition 5.4.27. (i) Suppose that $p > 0$, $\nu : \Omega \to (0, \infty)$ is a Lebesgue measurable function, $F : \mathbb{R}^n \times X \to Y$ is a given function and R is a certain collection of sequences in $\mathbb{R}^n$. Then we say that the function $F(\cdot;\cdot)$ is Stepanov $(\Omega, p, \mathrm{R}, \mathcal{B}, \nu)$-almost automorphic in norm [Stepanov $(\Omega, p, \mathrm{R}, \mathcal{B}, \nu, W_{\mathcal{B},\mathrm{R}})$-almost automorphic in norm; Stepanov $(\Omega, p, \mathrm{R}, \mathcal{B}, \nu, P_{\mathcal{B},\mathrm{R}})$-almost automorphic in norm] if $\|F(\cdot;\cdot)\|_Y^p$ is Stepanov $(\Omega, \mathrm{R}, \mathcal{B}, Z^{\mathcal{P}})$-multialmost automorphic [Stepanov $(\Omega, \mathrm{R}, \mathcal{B}, Z^{\mathcal{P}}, W_{\mathcal{B},\mathrm{R}})$-multialmost automorphic; Stepanov $(\Omega, \mathrm{R}, \mathcal{B}, Z^{\mathcal{P}}, P_{\mathcal{B},\mathrm{R}})$-multialmost automorphic] with $Z = L^1_{\nu^p}(\Omega : Y)$.

(ii) Suppose that $p > 0$ and $\nu : \Omega \to (0, \infty)$ is a Lebesgue measurable function. Then we say that a function $F : \mathbb{R}^n \to Y$ is Stepanov-(Ω, p, ν)-almost automorphic in norm [Stepanov-(Ω, p, ν, W)-almost automorphic in norm] if $\|F(\cdot)\|_Y^p$ is Stepanov-$(\Omega, 1, \mathrm{R}, \nu^p)$-almost automorphic [Stepanov-$(\Omega, 1, \mathrm{R}, \nu^p, W_{\mathrm{R}})$-almost automorphic] with R being the collection of all sequences in $\mathbb{R}^n$; $F(\cdot)$ is Stepanov-p-almost automorphic in norm if $\|F(\cdot)\|_Y^p$ is Stepanov-$(\Omega, 1, \nu^p)$-almost automorphic with $\Omega = [0,1]^n$ and $\nu(\cdot) \equiv 1$.

If $0 < p \leqslant 1$, then we have the following inequality:

$$\|x - y\|_Y^p \geqslant \big|\|x\|_Y^p - \|y\|_Y^p\big|, \quad x, y \in Y. \tag{303}$$

Using (303), it can be simply verified that any Stepanov-$(\Omega, p, \rho, \mathcal{B}, \Lambda', \nu_\cdot)$-almost periodic function is Stepanov-$(\Omega, p, \rho, \mathcal{B}, \Lambda', \nu_\cdot)$-almost periodic in norm; this holds for all other classes of functions introduced in Definition 5.4.26 and Definition 5.4.27, provided that $0 < p \leqslant 1$. In particular, if $F : \mathbb{R}^n \to Y$ is a Stepanov-p-almost periodic (automorphic) function and $0 < p \leqslant 1$, then the function $\|F(\cdot)\|_Y^p$ is Stepanov-1-almost periodic (automorphic).

If $p \geqslant 1$, then we cannot expect the existence of a finite real constant $c_p > 0$ such that the inequality

$$\|x - y\|_Y^p \leqslant c_p \big|\|x\|_Y^p - \|y\|_Y^p\big|, \quad x, y \in Y$$

holds true (consider the case in which $\|x\|_Y = \|y\|_Y$ but $x \neq y$). But the inequality

$$|x - y|^p \leqslant \big||x|^p - |y|^p\big|, \quad x, y \geqslant 0$$

is true, as easily shown with the help of the elementary differential calculus and, therefore, any Stepanov-$(\Omega, p, \rho, \mathcal{B}, \Lambda', \nu_\cdot)$-almost periodic function in norm with the nonnegative real values is Stepanov-$(\Omega, p, \rho, \mathcal{B}, \Lambda', \nu_\cdot)$-almost periodic; this holds for all other classes of functions introduced in Definition 5.4.26 and Definition 5.4.27, provided that $p \geqslant 1$. In particular, if $F : \mathbb{R}^n \to [0, \infty)$, $p \geqslant 1$ and $F^p(\cdot)$ is Stepanov-1-almost periodic (automorphic), then $F(\cdot)$ is Stepanov-p-almost periodic (automorphic).

We continue by recalling that, for every exponent $p \geqslant 1$, H. Bohr and E. Fölner have constructed, in [131, Main example II c], a Stepanov-p-almost periodic function $g : \mathbb{R} \to [0, \infty)$, which is not Stepanov-$q$-bounded and, therefore, not Stepanov-q-almost periodic (automorphic) for any exponent $q > p$. We will use this important example to prove the following result.

Theorem 5.4.28. *Suppose that $0 < p < 1$. Then there exists a function $f : \mathbb{R} \to [0, \infty)$, which is Stepanov-p-almost periodic in norm, not Stepanov-q-bounded and, therefore, not Stepanov-q-almost periodic (automorphic) for any exponent $q > p$.*

Proof. By the foregoing, there exists a Stepanov-1-almost periodic function $g : \mathbb{R} \to [0, \infty)$, which is not Stepanov-$q$-bounded and, therefore, not Stepanov-q-almost periodic (automorphic) for any exponent $q > 1$. Define $f(t) := g^{1/p}(t)$, $t \in \mathbb{R}$. Then it is clear that $f(\cdot)$ is Stepanov-p-almost periodic in norm. If we assume that $f(\cdot)$ is Stepanov-q-bounded for some exponent $q > p$, then $g(\cdot)$ must be Stepanov-(q/p)-bounded, which is a contradiction. This simply implies that $f(\cdot)$ is not Stepanov-q-almost periodic (automorphic) for any exponent $q > 1$, which can be also directly shown as follows: Assume the contrary; then the function $g^{q/p}(\cdot)$ must be Stepanov-1-almost periodic (automorphic) due to (303). But this would imply that the function $g(\cdot)$ is Stepanov-(q/p)-almost periodic (automorphic), which is a contradiction. □

Remark 5.4.29. Since $f(\cdot)$ is not Stepanov-q-bounded, we also have that $f(\cdot)$ is not Stepanov-q-almost periodic (automorphic) in norm for any exponent $q > p$.

Further on, we know that any Stepanov-p-almost periodic function $f : \mathbb{R} \to [0, \infty)$ is Stepanov-p-almost periodic in norm ($0 < p < 1$); hence, it is logical to ask whether the function $f(\cdot)$, considered in the proof of Theorem 5.4.28, is Stepanov-p-almost periodic? If this is the case, then for each exponent $p > 0$, we would have an example of a Stepanov-p-almost periodic function, which is not Stepanov-q-bounded and, therefore, not Stepanov-q-almost periodic (automorphic) for any exponent $q > p$.

Concerning [131, Main example II c] and Theorem 5.4.28, we would like to present the following examples as well.

Example 5.4.30. This example is a simple modification of the example already considered in the introductory part of this section. Suppose that $p > 0$ and

$$f(t) := \begin{cases} |\sin t|^{-(1/p)}, & t \notin \mathbb{Z}\pi, \\ 0, & t \in \mathbb{Z}\pi. \end{cases}$$

Then the function $F(\cdot)$ is not p-locally integrable and, therefore, not Stepanov-q-almost periodic (automorphic) for any exponent $q \geqslant p$. On the other hand, it can be simply shown that the function $f(\cdot)$ is Stepanov-p_0-almost periodic for any exponent $p_0 \in (0, p)$.

In connection with this example, it is also worth noting that H. D. Ursell has constructed many nontrivial examples of functions $f : \mathbb{R} \to [0, \infty)$, which are Stepanov-$p_0$-almost periodic for any exponent $p_0 \in (0, 1)$ but not Stepanov p_1-bounded for any exponent $p > 1$ (in this case, the number 1 is said to be the critical index of $f(\cdot)$; cf. [755, pp. 430–440] for more details on the subject).

Example 5.4.31. Suppose that $0 < p < 1$. Let us observe that the function $g(\cdot)$ from the proof of Theorem 5.4.28 has the form $g(t) = \sum_{k=1}^{+\infty} g_k(t)$, $t \in \mathbb{R}$, where $g_k(\cdot)$ is an essentially bounded, periodic function with the nonnegative values ($k \in \mathbb{N}$) and the above series is convergent in the Stepanov S^1-norm. Consider now the following function:

$$h(t) := \sum_{k=1}^{\infty} g_k^{1/p}(t), \quad t \in \mathbb{R}.$$

The function $h(\cdot)$ is well-defined and Stepanov-p-almost periodic because the sequence of functions $\sum_{k=1}^{\infty} g_k^{1/p}(\cdot)$ is Cauchy and, therefore, convergent in the Stepanov S^p-norm, which can be shown using the following simple calculation ($k,\ m \in \mathbb{N}$, $k < m$):

$$\sup_{t\in\mathbb{R}} \int_t^{t+1} \left[\sum_{i=k}^{m} g_i^{1/p}(s)\right]^p ds \leqslant \sup_{t\in\mathbb{R}} \int_t^{t+1} \left[\sum_{i=k}^{m} g_i(s)\right] ds.$$

It is not clear whether the function $h(\cdot)$ is locally integrable or Stepanov-q-almost periodic for some exponent $q > p$. Let us also emphasize that there is no simple theoretical explanation, which would imply that the above conclusions hold in a general situation of this example and that the function $h(\cdot)$ will be Stepanov-q-almost periodic for every exponent $q > p$ in the case that $g_k(t) \geqslant 1$ for all $t \in \mathbb{R}$ and $k \in \mathbb{N}$; this simply follows from the inequality

$$\left[\sum_{i=k}^{m} g_i^{1/q}(s)\right]^q \leqslant \left[\sum_{i=k}^{m} g_i^{1/p}(s)\right]^q.$$

Finally, we would like to recall that, for every almost periodic function $f : \mathbb{R} \to Y$ and for every positive real number $p > 0$, the function $\|f(\cdot)\|^p : \mathbb{R} \to [0,\infty)$ is almost periodic [509], whence we may conclude that $f(\cdot)$ is Stepanov-p-almost periodic in norm. In connection with Theorem 5.4.28 and this observation, we would like to raise the following issue.

Problem 5.4.32. *Suppose that $p > 0$. Can we find some sufficient conditions ensuring that the assumptions $f : \mathbb{R} \to Y$ is an unbounded Stepanov-p-almost periodic (automorphic) function, $r > 0$ and $q > 0$ imply that the function $\|f(\cdot)\|^r : \mathbb{R} \to [0,\infty)$ is Stepanov-q-almost periodic (automorphic)?*

Let us also observe that the class of Stepanov-p-almost periodic (automorphic) functions in norm is extremely nontrivial. For example, it seems very plausible that the Stepanov-p-almost periodic (automorphic) functions in norm do not form a vector space with the usual operations; furthermore, it is very difficult to state some satisfactory analogues of Proposition 5.4.4 and Corollary 5.4.5 for this class of functions.

5.4.7 Applications to the abstract Volterra integrodifferential inclusions

The main aim of this subsection is to provide certain applications of our results to the abstract (impulsive) Volterra integrodifferential inclusions in Banach spaces.

We will first present some applications of Theorem 5.4.13 and Proposition 5.4.14 concerning the invariance of Stepanov-p-almost periodicity under the actions of the infinite convolution products ($0 < p < 1$). It is clear that Theorem 5.4.13 can be applied in the analysis of the existence and uniqueness of (T, ν)-almost periodic solutions for various classes of the abstract fractional integrodifferential inclusions without initial conditions; cf. [444] for more details about applications of this type (here, the main problem is to find the inhomogeneity $f(\cdot)$, which is not Stepanov-1-almost periodic, such that the requirements of Theorem 5.4.13 hold). For example, we can analyze the existence and uniqueness of (T, ν)-almost periodic solutions of the fractional Poisson heat equation

$$\begin{cases} D_{t,+}^{\gamma}[m(x)v(t,x)] = (\Delta - b)v(t,x) + f(t,x), & t \in \mathbb{R},\ x \in \Omega; \\ v(t,x) = 0, & (t,x) \in [0,\infty) \times \partial\Omega, \end{cases}$$

in the space $X := L^p(\Omega)$, where Ω is a bounded domain in $\mathbb{R}^n$, $b > 0$, $m(x) \geqslant 0$ a. e. $x \in \Omega$, $m \in L^{\infty}(\Omega)$, $\gamma \in (0,1)$ and $1 < p < \infty$.

Concerning Proposition 5.4.14, we would like to note that the possible applications of this result can be always made to the abstract differential first-order inclusions provided that the operator family $(R(t))_{t>0}$ is a (degenerate) strongly continuous semigroup of operators satisfying that $\|R(t)\| \leqslant Me^{-ct}t^{\beta-1}$, $t > 0$ for some real constants $M > 0$, $c > 0$ and $\beta \in (0,1]$ as well as the function $\nu : \mathbb{R} \to (0,\infty)$ is an admissible weight function with the property that $\nu(t) \leqslant M'e^{\omega|s|}\nu(t+s)$, $t, s \in \mathbb{R}$ for some real constants $M' > 0$ and $\omega < c$. The applications can be simply given in the analysis of the existence and uniqueness of ν-almost periodic solutions of the abstract Poisson heat equation

$$\begin{cases} \frac{\partial}{\partial t}[m(x)v(t,x)] = (\Delta - b)v(t,x) + f(t,x), & t \in \mathbb{R},\ x \in \Omega; \\ v(t,x) = 0, & (t,x) \in [0,\infty) \times \partial\Omega, \end{cases}$$

in the space $X := L^p(\Omega)$; cf. also the first application made in [267, Subsection 4.4], where we have analyzed the almost periodic type solutions to the abstract higher-order impulsive Cauchy problems. In particular, we can use the inhomogeneities from Example 5.4.8 with $\nu(t) = a + \zeta(t)$, $t \in \mathbb{R}$, where $a \geqslant 0$ and an admissible weight function $\zeta : \mathbb{R} \to (0,\infty)$ satisfies $\zeta(t) \leqslant M'e^{\omega|s|}\zeta(t+s)$, $t, s \in \mathbb{R}$ and $\int_{-\infty}^{+\infty} \zeta^p(x)\,dx < +\infty$ (e. g., we can take $\zeta(t) = e^{-\omega|t|}$, $t \in \mathbb{R}$ with $0 < \omega < c$); many other examples of such admissible weight functions can be given following the recent investigations of chaotic translation semigroups on weighted L^p-spaces (see, e. g., [443, Chapter 3] for more details).

Before dividing the remainder of this subsection into two separate parts, we would like to note that the argumentation similar to that one contained in the proofs of Theorem 5.4.13 and Proposition 5.4.14 can be useful to deduce certain results concerning the invariance of Stepanov p-almost periodicity under the actions of the infinite convolution product

$$t \mapsto H(t) \equiv \int_{-\infty}^{+\infty} h(t-s)f(s)\,ds, \quad t \in \mathbb{R},$$

where $h \in L^1(\mathbb{R})$. This can be applied in the analysis of the existence and uniqueness of the Stepanov-p-almost periodic type solutions to the inhomogeneous heat equation (cf. [447, Subsection 6.2.6 and pp. 558–559]).

1. Applications to the abstract impulsive first-order differential inclusions

In this part, we will analyze the existence and uniqueness of ν-almost periodic type solutions to the abstract impulsive differential inclusions of first order. Of concern is the following abstract impulsive Cauchy inclusion

$$(ACP)_{1;1}: \begin{cases} u'(t) \in \mathcal{A}u(t) + f(t), & t \in [0,\infty) \setminus \{t_1, \dots, t_l, \dots\}, \\ (\Delta u)(t_k) = u(t_k+) - u(t_k-) = Cy_k, & k \in \mathbb{N}_l, \\ u(0) = Cu_0. \end{cases}$$

We refer the reader to [266] for the notion of a (pre)solution of $(ACP)_{1;1}$. We need the following result from this paper.

Lemma 5.4.33. *Suppose that $\mathcal{A}$ is a closed subgenerator of a global C-regularized semigroup $(R(t))_{t\geqslant 0}$. Suppose further that $0 < t_1 < \cdots < t_l < \cdots < +\infty$, the sequence $(t_l)_l$ has no accumulation point, the functions $C^{-1}f(\cdot)$ and $f_{\mathcal{A}}(\cdot)$ are continuous on the set $[0,\infty) \setminus \{t_1, \dots, t_l, \dots\}$, $f_{\mathcal{A}}(t) \in \mathcal{A}C^{-1}f(t)$ for all $t \in [0,\infty) \setminus \{t_1, \dots, t_l, \dots\}$, as well as the right limits and the left limits of the functions $C^{-1}f(\cdot)$ and $f_{\mathcal{A}}(\cdot)$ exist at any point of the set $\{t_1, \dots, t_l, \dots\}$. Define the functions $u(t)$ and $\omega(t)$ for $t \geqslant 0$ by*

$$u(t) := R(t)u_0 + \int_0^t R(t-s)(C^{-1}f)(s)\,ds + \omega(t), \quad t \geqslant 0$$

and

$$\omega(t) := \begin{cases} 0, & t \in [0, t_1], \\ \sum_{p=1}^{k} R(t-t_p)y_0^p, & \text{if } t \in (t_k, t_{k+1}] \text{ for some } k \in \mathbb{N}_{l-1}^0, \end{cases} \tag{304}$$

respectively. Then the function $u(t)$ is a unique solution of the problem $(ACP)_{1;1}$, provided that $u_0 \in D(\mathcal{A})$ and $y_k \in D(\mathcal{A})$ for all $k \in \mathbb{N}$.

In order to formulate our main result, we need to impose the following conditions:

(AS1) $\mathcal{A}$ is a closed subgenerator of a global C-regularized semigroup $(R(t))_{t\geqslant 0}$ and $\|R(t)\| \leqslant Me^{\omega_0 t}$, $t \geqslant 0$ for some real numbers $M > 0$ and $\omega_0 < 0$.

(AS2) $m \in \mathbb{N}$, $f_i : [0,\infty) \to \mathbb{C}$ is an almost periodic function ($1 \leqslant i \leqslant m$), $f_1(0) \cdot \dots \cdot f_m(0) \neq 0$, and the sequence $\{t_1, \dots, t_l, \dots\}$ of all possible zeroes of functions $f_0(\cdot), \dots, f_m(\cdot)$ has no accumulation point.

(AS3) $(x_i, y_i) \in \mathcal{A}$ for $1 \leqslant i \leqslant m$, $t \mapsto f_0(t)$, $t \geqslant 0$ is a piecewise continuous function uniquely determined by the function $t \mapsto \sum_{i=1}^m \operatorname{sign}(f_i(t))Cx_i$, $t \geqslant 0$ and $t \mapsto f_{\mathcal{A}}(t)$, $t \geqslant 0$ is a piecewise continuous function uniquely determined by the function $t \mapsto \sum_{i=1}^m \operatorname{sign}(f_i(t))y_i$, $t \geqslant 0$.

(AS4) $p \geqslant 1, \zeta : \mathbb{R} \to (0,\infty)$ is an admissible weight function, $\zeta(t) \leqslant Me^{\omega|s|}\zeta(t+s), t, s \in \mathbb{R}$ for some real constant $\omega < \omega_0$, $\int_{-\infty}^{+\infty} \zeta^p(x)\, dx < +\infty, c > 0$ and $\nu(t) = c+\zeta(t), t \in \mathbb{R}$.

(AS5) $\sum_{k\geqslant 1} e^{-\omega_0 t_k}\|y_k\| < +\infty$, $u_0 \in D(\mathcal{A}), y_k \in D(\mathcal{A})$ for all $k \in \mathbb{N}$ and $q : [0,\infty) \to X$ is a locally integrable function such that $\|q(t)\| \leqslant Me^{\omega_1 t}$, $t \geqslant 0$ for some real number $\omega_1 < \omega_0$.

(AS6) The functions $q(\cdot)$ and $f_{\mathcal{A},q}(\cdot)$ are continuous on the set $[0,\infty) \smallsetminus \{t_1,\ldots,t_l,\ldots\}$, $f_{\mathcal{A},q}(t) \in \mathcal{A}q(t)$ for all $t \in [0,\infty) \smallsetminus \{t_1,\ldots,t_l,\ldots\}$, as well as the right limits and the left limits of the functions $q(\cdot)$ and $f_{\mathcal{A},q}(\cdot)$ exist at any point of the set $\{t_1,\ldots,t_l,\ldots\}$.

Then we have the following result.

Theorem 5.4.34. *Suppose that conditions* (AS1)–(AS6) *hold good. Then there exists a unique solution of the problem* $(ACP)_{1;1}$, *which can be written as a sum of a* ν*-almost periodic function, a function from the space* $PC_{\omega_0}([0,\infty) : X)$ *and a function from the space* $C_{\omega_0+\omega_1}([0,\infty) : X)$.

Proof. It can be simply proved that we have $\nu(t) \leqslant Me^{\omega|s|}\nu(t+s)$, t, $s \in \mathbb{R}$. Keeping in mind the first condition in (AS5) and the consideration from [266, Application 1, Subsection 4.1], we have that the function $\omega(\cdot)$, given by (304), belongs to the space $PC_{\omega_0}([0,\infty) : X)$. Moreover, due to the consideration from Example 5.4.7 (cf. (LT2)), we have that $C^{-1}f_0 \in S_{\Omega,\Lambda'}^{(\mathbb{F},\rho,\mathcal{P}_t,\mathcal{P})}(\Lambda : \mathbb{C})$ with $\Lambda = \Lambda' = \mathbb{R}$, $\mathbb{F}(\cdot\cdot) \equiv 1$, $\rho = \mathrm{I}$, $\Omega = [0,1]$, $P = C_b(\mathbb{R} : \mathbb{C})$ and $P_t = L_\nu^p(t+[0,1] : \mathbb{C})$ for all $t \in \mathbb{R}$. Since $p \geqslant 1$ and $\nu(t) \geqslant c > 0, t \in \mathbb{R}$, we can apply Proposition 5.4.14 in order to see that the function $t \mapsto \int_{-\infty}^{t} R(t-s)(C^{-1}f_0(s))\, ds$, $t \in \mathbb{R}$ is ν-almost periodic. Furthermore, it is clear that the function $t \mapsto R(t)u_0, t \geqslant 0$ belongs to the space $C_{\omega_0}([0,\infty) : X)$. Due to [69, Proposition 1.3.4], the mapping $t \mapsto \int_0^t R(t-s)(C^{-1}f(s))\, ds, t \geqslant 0$ is continuous so that the mapping $t \mapsto \int_t^{+\infty} R(t-s)(C^{-1}f_0(s))\, ds, t \geqslant 0$ is likewise continuous due to the representation formula given in Lemma 5.4.33 and the formula (305) given below. Then the final conclusion simply follows from a simple computation and the decomposition

$$u(t) = \int_{-\infty}^{t} R(t-s)(C^{-1}f_0(s))\, ds + \left[R(t)u_0 + \int_{t}^{+\infty} R(t-s)(C^{-1}f_0(s))\, ds + \omega(t)\right] + \int_0^t R(t-s)q(s)\, ds, \quad t \geqslant 0. \tag{305}$$

□

Before proceeding further, we would like to note that Theorem 5.4.34 can be applied in the analysis of the existence and uniqueness of asymptotically ν-almost periodic type solutions for a class of the abstract impulsive first-order (degenerate) differential equations involving the (noncoercive) differential operators with constant coefficients in L^p-spaces (cf. [443], [444] and [445] for more details).

2. Applications to the abstract fractional integrodifferential inclusions

In this part, we will use the following lemma (cf. [445, Proposition 3.2.15(i)]; for the notion of solutions, cf. [445, Definition 3.1.1(ii)]).

Lemma 5.4.35. *Suppose $\alpha \in (0,\infty) \smallsetminus \mathbb{N}$, $x \in D(\mathcal{A})$ as well as $C^{-1}f, f_{\mathcal{A}} \in C([0,\infty) : X)$, $f_{\mathcal{A}}(t) \in \mathcal{A}C^{-1}f(t)$, $t \in [0,\infty)$ and $\mathcal{A}$ is a closed subgenerator of a global (g_α, C)-regularized resolvent family $(R(t))_{t\geqslant 0}$. Set $v(t) := (g_{\lceil\alpha\rceil-\alpha} * f)(t)$, $t \in [0,\infty)$. If $v \in C^{\lceil\alpha\rceil-1}([0,\infty) : X)$ and $v^{(k)}(0) = 0$ for $1 \leqslant k \leqslant \lceil\alpha\rceil - 2$, then the function $u(t) := R(t)x + (R * C^{-1}f)(t)$, $t \geqslant 0$ is a unique solution of the following abstract time-fractional inclusion:*

$$(ACP)_\alpha^f : \begin{cases} u \in C^{\lceil\alpha\rceil}((0,\infty) : X) \cap C^{\lceil\alpha\rceil-1}([0,\infty) : X), & \\ \mathbf{D}_t^\alpha u(t) \in \mathcal{A}u(t) + \frac{d^{\lceil\alpha\rceil-1}}{dt^{\lceil\alpha\rceil-1}}(g_{\lceil\alpha\rceil-\alpha} * f)(t), & t \geqslant 0, \\ u(0) = Cx, \quad u^{(k)}(0) = 0, & 1 \leqslant k \leqslant \lceil\alpha\rceil - 1. \end{cases}$$

Now, we will use the following conditions:

(AS1)' $\alpha \in (0,2) \smallsetminus \{1\}$, $\mathcal{A}$ is a closed subgenerator of an exponentially bounded (g_α, C)-regularized resolvent family $(R(t))_{t\geqslant 0}$ and $\|R(t)\| \leqslant Me^{\omega_0 t}$, $t \geqslant 0$ for some real numbers $M > 0$ and $\omega_0 \geqslant 0$.

(AS2)' $\omega > \omega_0$, $m \in \mathbb{N}$, $f_i : [0,\infty) \to \mathbb{C}$ is identically equal to the function $\sin(\cdot)$ or $\cos(\cdot)$, α_i and β_i are nonzero real numbers such that $\alpha_i\beta_i^{-1} \notin \mathbb{Q}$ $(1 \leqslant i \leqslant m)$.

(AS3)' $q : [0,\infty) \to X$ is a locally integrable function such that $\|q(t)\| \leqslant Me^{\omega_1 t}$, $t \geqslant 0$ for some real number $\omega_1 < \omega - \omega_0$.

(AS4)' $(x,y) \in \mathcal{A}$, $(x_i,y_i) \in \mathcal{A}$ for $1 \leqslant i \leqslant m$, and $e^{-\omega t}(C^{-1}f)(t) = \sum_{i=1}^m f_i(1/(2 + \cos(\alpha_i t) + \cos(\beta_i t)))x_i + q(t)$, $t \geqslant 0$.

(AS5)' The functions $q(\cdot)$ and $f_{\mathcal{A},q}(\cdot)$ are continuous on $[0,\infty)$, $f_{\mathcal{A},q}(t) \in \mathcal{A}q(t)$ for all $t \geqslant 0$, and $g_{2-\alpha} * e^{\omega\cdot}q(\cdot) \in C^1([0,\infty) : X)$ if $1 < \alpha < 2$.

(AS6)' $p \geqslant 1$, $\zeta : \mathbb{R} \to (0,\infty)$ is an admissible weight function, $\zeta(t) \leqslant Me^{\omega|s|}\zeta(t+s)$, $t, s \in \mathbb{R}$ for some real constant $\omega < \omega_0$, $\int_{-\infty}^{+\infty} \zeta^p(x)\,dx < +\infty$, $c > 0$ and $\nu(t) = c + \zeta(t)$, $t \in \mathbb{R}$.

Then we have the following result, which can be proved in a similar fashion as Theorem 5.4.34.

Theorem 5.4.36. *Suppose that conditions* (AS1)'–(AS6)' *hold good. Then there exists a unique solution $u(\cdot)$ of the problem $(ACP)_\alpha^f$ with $f(t) = f_0(t) + Cq(t)$, $t \geqslant 0$; furthermore, the function $e^{-\omega\cdot}u(\cdot)$ can be written as a sum of a ν-almost periodic function, a function from the space $C_{\omega_0-\omega}([0,\infty) : X)$ and a function from the space $C_{\omega_0-\omega+\omega_1}([0,\infty) : X)$.*

It is clear that Theorem 5.4.36 can be applied in the analysis of the existence and uniqueness of asymptotically ν-almost periodic type solutions for a class of the abstract fractional (degenerate) differential equations involving the (noncoercive) differential operators with constant coefficients in L^p-spaces. Furthermore, we can apply this result in the study of the existence and uniqueness of asymptotically ν-almost periodic type solutions for a class of the abstract fractional (degenerate) differential equations considered in [445, Section 3.5] and a class of the abstract fractional integrodifferential inclusions with impulsive effects considered in [267, Section 5] (cf. [267, Theorem 10]).

If a sufficiently small real number $\sigma > 0$ is given, then it would be interesting to construct an example of a Lebesgue measurable function $\nu : \mathbb{R} \to (0, \infty)$ satisfying the following three conditions:

(i) There exists a finite real number $M_\sigma \geqslant 1$ such that $\nu(x+y) \leqslant M_\sigma(1+|y|)^\sigma \nu(x)$ for all $x, y \in \mathbb{R}$.
(ii) $\nu(\cdot)$ is unbounded and satisfies (LT).
(iii) There exists a finite real number $c > 0$ such that $\nu(t) \geqslant c$ for all $t \in \mathbb{R}$.

If such a function exists, then we can consider the case $\omega = 0$ in Theorem 5.4.36 and apply Proposition 5.4.14, Lemma 5.4.35 and our conclusion established in Example 5.4.8 in the study of the existence and uniqueness of asymptotically ν-almost periodic type solutions of the abstract fractional Cauchy inclusion $(ACP)^f_a$ (cf. also [444, Remarks 2.6.12, 2.6.14(ii)]).

Before closing this section by quoting some topics not considered in our previous work, we would like to point out that L. Gao and X. Sun have analyzed, in [316], the almost periodic solutions to impulsive stochastic delay differential equations driven by fractional Brownian motion with the Hurst index $H \in (1/2, 1)$; cf. also [132–135, 268–270, 361, 643, 672] for some other important references concerning the impulsive Volterra integrodifferential equations:

1. The class of $\mathbb{D}$-asymptotically Stepanov-p-almost periodic (automorphic) functions has not been considered here in great detail. We will analyze this class and the corresponding classes of Weyl, Besicovitch and Doss $\mathbb{D}$-asymptotically p-almost periodic (automorphic) functions in our forthcoming research studies ($p > 0$).
2. Suppose that $p \geqslant 1$. Then any Stepanov-p-almost periodic function $F : \mathbb{R}^n \to Y$ has the mean value
$$M(F) := \lim_{T\to+\infty} \frac{1}{(2T)^n} \int\limits_{\mathbf{s}+K_T} F(\mathbf{t})\, d\mathbf{t},$$
which does not depend on $s \in \mathbb{R}^n$; here, $K_T := \{\mathbf{t} = (t_1, t_2, \ldots, t_n) \in \mathbb{R}^n : |t_i| \leqslant T \text{ for } 1 \leqslant i \leqslant n\}$. The Bohr–Fourier coefficient $F_\lambda \in X$ is defined by
$$F_\lambda := M(e^{-i\langle\lambda,\cdot\rangle} F(\cdot)), \quad \lambda \in \mathbb{R}^n,$$
where $\langle\lambda, \cdot\rangle$ denotes the usual inner product in $\mathbb{R}^n$. We know that the Bohr spectrum of $F(\cdot)$, defined by $\sigma(F) := \{\lambda \in \mathbb{R}^n : F_\lambda \neq 0\}$, is at most a countable set.
We have not considered here the existence of Bohr–Fourier coefficients for locally integrable Stepanov-p-almost periodic functions with the exponent $p \in (0, 1)$. Moreover, we have not reconsidered here the important theoretical results [444, Theorems 2.6.8, 2.6.10] for Stepanov-p-almost periodic functions with the exponent $p \in (0, 1)$ as well as Stepanov approximations by trigonometric polynomials in L^p-spaces (cf. [448, Subsection 6.2.1] for more details on the subject).

6 Almost periodic type functions in general measure, almost automorphic type functions in general measure and applications

6.1 Measure theoretical approach to almost periodicity

The class of almost periodic functions in the sense of the Lebesgue measure (also called the class of m-almost periodic functions) was introduced by W. Stepanov in 1926 [724] and further analyzed by S. Stoiński in [726] (1994) and [728] (1999); see also the research monograph [729]. A Lebesgue measurable function $f : \mathbb{R} \to X$ is said to be m-almost periodic (almost periodic in view of the Lebesgue measure m) if for each real numbers $\varepsilon,\ \eta > 0$ the set

$$\Big\{\tau \in \mathbb{R} : \sup_{x\in\mathbb{R}} m(\{t \in [x, x+1] : \|f(t+\tau) - f(t)\| \geqslant \eta\}) \leqslant \varepsilon\Big\}$$

is relatively dense in $\mathbb{R}$. In this definition, we do not need two real numbers $\varepsilon,\ \eta > 0$: a very simple argumentation shows that a Lebesgue measurable function $f : \mathbb{R} \to X$ is m-almost periodic if and only if for each real number $\varepsilon > 0$ the set

$$\Big\{\tau \in \mathbb{R} : \sup_{x\in\mathbb{R}} m(\{t \in [x, x+1] : \|f(t+\tau) - f(t)\| \geqslant \varepsilon\}) \leqslant \varepsilon\Big\}$$

is relatively dense in $\mathbb{R}$. Further on, we know that, if $g : \mathbb{R} \to \mathbb{C}$ is an almost periodic function, which has a bounded analytical extension in a strip around the real axis, then the function $f : \mathbb{R} \to \mathbb{C}$, given by $f(x) := 1/g(x)$, if $g(x) \neq 0$, and $f(x) := 0$, if $g(x) = 0$, is m-almost periodic. In particular, the function $f(\cdot)$, given by (72), is not bounded but $f(\cdot)$ is m-almost periodic; furthermore, we know that $f(\cdot)$ is not Stepanov almost periodic since it is not Stepanov bounded (see [151, Example 6] given by D. Bugajewski and A. Nawrocki), as well as that $f(\cdot)$ is Stepanov (1/4)-almost periodic (see [151, Example 7, Theorem 8]).

It is worth noting that P. Kasprzaka, A. Nawrocki and J. Signerska-Rynkowska have analyzed, in [419], the integrate-and-fire models with an almost periodic type input function. In [419, Example 3.3], the authors have constructed a continuous m-almost periodic function $f : \mathbb{R} \to \mathbb{R}$ such that the mean value

$$M(f) := \lim_{t\to+\infty} \frac{1}{t} \int_0^t f(s)\, ds$$

does not exist; in particular, this implies that $f(\cdot)$ cannot be Besicovitch almost periodic in the sense of [444] (cf. also [419, Theorem 3.10]). It is well known that there exists a bounded, uniformly continuous Levitan almost periodic function $f : \mathbb{R} \to \mathbb{R}$, which is

https://doi.org/10.1515/9783111689746-009

not m-almost periodic as well as that there exists a bounded continuous function $f : \mathbb{R} \to \mathbb{R}$ that is m-almost periodic but not Levitan almost periodic; see, e. g., [618, Examples 3.1, 3.3].

For more details about the class of one-dimensional m-almost periodic functions and their applications, we refer the reader to the research articles [151] by D. Bugajewski, A. Nawrocki, [150] by D. Bugajewski, K. Kasprzak, A. Nawrocki and the doctoral dissertation [620] of A. Nawrocki. Some extensions of the class of m-almost periodic functions have been analyzed by A. Michalowicz and S. Stoiński in [589] following the approach of M. Levitan; it is also worth noting that L. I. Danilov has considered, in [226], a class of the Lebesgue measurable functions $f : \mathbb{R} \to X$ such that for each real number $\varepsilon > 0$ there exists a strictly increasing sequence (τ_k) of positive real numbers such that $\lim_{k\to+\infty} \tau_k = +\infty$ and

$$\lim_{k\to+\infty} \sup_{x\in\mathbb{R}} m(\{t \in [x, x+1] : \|f(t+\tau_k) - f(t)\| \geqslant \varepsilon\}) = 0.$$

The class of c-almost periodic functions in view of the Lebesgue measure has not been analyzed so far, even in the one-dimensional setting, and this fact strongly influenced us to write the paper [490], from which we have taken the material of this section ($c \in \mathbb{C} \setminus \{0\}$). Here, we introduce and analyze various classes of multi-dimensional ρ-almost periodic type functions in general measure; furthermore, in place of the usually considered bounded linear operators $\rho = c\mathrm{I}$, where $c \in \mathbb{C} \setminus \{0\}$, we consider here the general binary relations ρ.

If the governing solution family $(R(t))_{t>0}$ of bounded linear operators is only norm integrable and $f(\cdot)$ is Stepanov-p-almost periodic, then we consider here, for the first time in the existing literature, the almost periodicity of function $u(\cdot)$, given by $u(t) = \int_{-\infty}^{t} R(t-s)f(s)\,ds$, $t \in \mathbb{R}$. It should be also emphasized that the existence and uniqueness of m-almost periodic solutions of the semilinear Volterra integral equations have not been analyzed before.

In order to achieve our aims, we essentially employ the ideas developed by S. Stoiński. We can freely say that these ideas are extremely important in the deeper analysis of Stepanov-p-almost periodic functions because a very simple argumentation shows that any Stepanov-p-almost periodic function $f : \mathbb{R} \to Y$ is m-almost periodic ($p > 0$) as well as that any bounded m-almost periodic function $f : \mathbb{R} \to Y$ is Stepanov-p-almost periodic ($p > 0$). We have already constructed an example of an unbounded m-almost periodic function $f : \mathbb{R} \to \mathbb{R}$, which is not locally p-integrable ($p > 0$).

The section is structurally organized as follows. Our main contributions are given in Section 6.1.1, where we introduce several new classes of multidimensional ρ-almost periodic type functions in general measure. We explore the convolution invariance of multidimensional ρ-almost periodicity in general measure in Section 6.1.3; the applications to the semilinear Volterra integral equations are given in Section 6.1.4.

6.1.1 Multidimensional ρ-almost periodic type functions in general measure

We will always assume henceforth that $\emptyset \neq \Lambda \subseteq \mathbb{R}^n$, $\nu : \Lambda \to [0, \infty)$, $m' : P(\mathbb{R}^n) \to [0, \infty]$, $m'(\emptyset) = 0$, $\emptyset \neq \Omega \subseteq \mathbb{R}^n$ is a nonempty compact set and $\Lambda + \Omega \subseteq \Lambda$. For every $\varepsilon > 0$ and for every two functions $f : \Lambda \to Y$ and $g : \Lambda \to Y$, we define

$$d_\varepsilon(f, g) := \sup_{\mathbf{t}\in\Lambda} m'(\{\mathbf{s} \in \mathbf{t} + \Omega : \|f(\mathbf{s}) - g(\mathbf{s})\|_Y \cdot \nu(\mathbf{s}) \geqslant \varepsilon\}); \tag{306}$$

set also $\|f\|_{P_\varepsilon} := d_\varepsilon(0, f)$. Then we have $0 \leqslant d_\varepsilon(f, g)$ and $d_\varepsilon(f, f) = 0$, so that $d_\varepsilon(\cdot;\cdot)$ is a premetric on the space of all functions from Λ into Y; furthermore, we have $d_\varepsilon(f, g) = d_\varepsilon(g, f)$ and $d_\varepsilon(f, g) = d_\varepsilon(f + h, g + h)$ so that $d_\varepsilon(\cdot;\cdot)$ is a translation invariant pseudo-semimetric on the space of all functions from Λ into Y (according to M. M. Deza and M. Laurent [244], these features are sufficiently enough to call $d_\varepsilon(\cdot;\cdot)$ a distance). Furthermore, we have the following:

(i) If $f : \Lambda \to Y$, $g : \Lambda \to Y$, $h : \Lambda \to Y$ and the assumptions A, B, $C \subseteq \mathbb{R}^n$ and $A \subseteq B \cup C$ imply $m'(A) \leqslant m'(B) + m'(C)$, then we have

$$d_\varepsilon(f, h) \leqslant d_{\varepsilon/2}(f, g) + d_{\varepsilon/2}(g, h), \quad \varepsilon > 0. \tag{307}$$

(ii) Suppose that $\emptyset \neq \Lambda' \subseteq \mathbb{R}^n$, $\Lambda + \Lambda' \subseteq \Lambda$, $\tau \in \Lambda'$, $M > 0$, the assumption $\mathbf{v} \in \Lambda + \Omega + \tau$ implies $\nu(\mathbf{v} - \tau) \leqslant M\nu(\mathbf{v})$ and the assumption $A \subseteq B \subseteq \mathbb{R}^n$ implies $m'(A) \leqslant m'(B)$. Then we have

$$d_\varepsilon(f(\cdot + \tau), g(\cdot + \tau)) \leqslant d_{\varepsilon/M}(f, g), \tag{308}$$

for any two functions $f : \Lambda \to Y$ and $g : \Lambda \to Y$.

(iii) Suppose that $T \in L(Y)$, $f : \Lambda \to Y$ and $g : \Lambda \to Y$. Then we have

$$d_\varepsilon(Tf, Tg) \leqslant d_{\varepsilon/\|T\|}(f, g), \tag{309}$$

where $d_{\varepsilon/\|T\|}(f, g) = 0$ for $T = 0$.

(iv) Suppose that $f : \Lambda \to Y$ and $g : \Lambda \to Y$. If the assumption $A \subseteq B \subseteq \mathbb{R}^n$ implies $m'(A) \leqslant m'(B)$, then for each $\varepsilon' \in (0, \varepsilon)$ we have

$$d_\varepsilon(f, g) \leqslant d_{\varepsilon'}(f, g) \quad \text{and} \quad \|f\|_{P_\varepsilon} \leqslant \|f\|_{P_{\varepsilon'}}. \tag{310}$$

(v) The triangle inequality

$$d_\varepsilon(f, h) \leqslant d_\varepsilon(f, g) + d_\varepsilon(g, h)$$

does not hold, in general (choose, e. g., $f \equiv \varepsilon$, $g \equiv \varepsilon/2$, $h \equiv 0$ and $\nu \equiv 1$).

(vi) The assumption $d_\varepsilon(f, g) = 0$ does not imply $f = g$ a. e. (choose, e. g., $f \equiv \varepsilon/2$, $g \equiv \varepsilon/4$ and $\nu \equiv 1$).

Now we would like to introduce the following notion.

Definition 6.1.1. Suppose that $\emptyset \neq \Lambda' \subseteq \mathbb{R}^n$, $\emptyset \neq \Lambda \subseteq \mathbb{R}^n$, $F : \Lambda \times X \to Y$ is a given function, ρ is a binary relation on Y, $R(F) \subseteq D(\rho)$ and $\Lambda + \Lambda' \subseteq \Lambda$. Then we say that:

(i) $F(\cdot;\cdot)$ is Bohr $(\mathcal{B}, \Lambda', \rho, \Omega, m', \nu)$-almost periodic if for every $B \in \mathcal{B}$ and $\varepsilon > 0$ there exists $l > 0$ such that for each $\mathbf{t}_0 \in \Lambda'$ there exists $\tau \in B(\mathbf{t}_0, l) \cap \Lambda'$ such that, for every $\mathbf{t} \in \Lambda$, $x \in B$ and $\mathbf{s} \in \mathbf{t} + \Omega$, there exists an element $y_{\mathbf{s};x} \in \rho(F(\mathbf{s}; x))$ such that
$$\sup_{x \in B} \|F(\cdot + \tau; x) - y_{\cdot;x}\|_{P_\varepsilon} \leqslant \varepsilon.$$

(ii) $F(\cdot;\cdot)$ is $(\mathcal{B}, \Lambda', \rho, \Omega, m', \nu)$-uniformly recurrent if for every $B \in \mathcal{B}$ and $\varepsilon > 0$ there exists a sequence (τ_k) in Λ' such that $\lim_{k \to +\infty} |\tau_k| = +\infty$ and that, for every $\mathbf{t} \in \Lambda$, $x \in B$ and $\mathbf{s} \in \mathbf{t} + \Omega$, there exists an element $y_{\mathbf{s};x} \in \rho(F(\mathbf{s}; x))$ such that
$$\lim_{k \to +\infty} \sup_{x \in B} \|F(\cdot + \tau_k; x) - y_{\cdot;x}\|_{P_\varepsilon} = 0.$$

Remark 6.1.2. In our approach, the function $m'(\cdot)$ is defined for all subsets in $\mathbb{R}^n$. If $m'(\cdot)$ is defined only for subsets belonging to a certain σ-algebra on $\mathbb{R}^n$, e. g., then the notion introduced in Definition 6.1.1 can be understood only for the functions, which are measurable in a certain sense. For example, in the case of consideration of the usual Lebesgue measure $m(\cdot)$, we must additionally assume that for each $x \in X$ the function $F(\cdot; x)$ is Lebesgue measurable as well as that the function $\nu : \Lambda \to [0, \infty)$ is Lebesgue measurable. In our later investigations of m-almost periodic functions, we will tacitly assume these conditions.

The subsequent result follows directly from the corresponding definitions.

Proposition 6.1.3. *Suppose that $\emptyset \neq \Lambda' \subseteq \mathbb{R}^n$, $\emptyset \neq \Lambda \subseteq \mathbb{R}^n$, $F : \Lambda \times X \to Y$ is a given function, ρ is a binary relation on Y and $\Lambda + \Lambda' \subseteq \Lambda$. If $F(\cdot;\cdot)$ is Bohr $(\mathcal{B}, \Lambda', \rho, \nu)$-almost periodic ($(\mathcal{B}, \Lambda', \rho, \nu)$-uniformly recurrent), then $F(\cdot;\cdot)$ is Bohr $(\mathcal{B}, \Lambda', \rho, \Omega, m', \nu)$-almost periodic ($(\mathcal{B}, \Lambda', \rho, \Omega, m', \nu)$-uniformly recurrent).*

Suppose that the requirements sufficient for the validity of the estimates (307)–(308) hold, $\Lambda + \Lambda' \subseteq \Lambda$, $\tau + \Lambda = \Lambda$ for all $\tau \in \Lambda'$, $F : \Lambda \times X \to Y$, $R(F) \subseteq D(\rho)$ and $\rho(x)$ is singleton for every $x \in R(F)$. Arguing as in the proof of [289, Proposition 2.2], we may conclude that $\Lambda + (\Lambda' - \Lambda') \subseteq \Lambda$ as well as that the Bohr $(\mathcal{B}, \Lambda', \rho, \Omega, m', \nu)$-almost periodicity ($(\mathcal{B}, \Lambda', \rho, \Omega, m', \nu)$-uniform recurrence) implies the Bohr $(\mathcal{B}, \Lambda' - \Lambda', \mathrm{I}, \Omega, m', \nu)$-almost periodicity ($(\mathcal{B}, \Lambda' - \Lambda', \mathrm{I}, \Omega, m', \nu)$-uniform recurrence) of $F(\cdot;\cdot)$; the statement of [289, Corollary 2.3] can be generalized in this manner, as well. Furthermore, if the requirements sufficient for the validity of the estimates (307)–(309) hold, then the conclusions established in [289, Example 2.8] can be formulated in our new framework; this can be also done with the statements established in [289, Theorem 2.11, (i)–(iv); Proposition 2.12] and [430, Proposition 2.7].

The following result can be also formulated for the corresponding classes of uniformly recurrent functions in measure; the proof is rather simple and can be left to the interested readers (here, $L^p_{m',\nu}(\mathbf{t}+\Omega : Y) = \{u : \mathbf{t}+\Omega \to Y; \|u(\cdot)\|_Y \nu(\cdot) \in L^p_{m'}(\mathbf{t}+\Omega : Y)\}$ is equipped with the norm $\|u\|_{L^p_{m',\nu}(\mathbf{t}+\Omega:Y)} = \|\|u(\cdot)\|_Y \nu(\cdot)\|_{L^p_{m'}(\mathbf{t}+\Omega:Y)}$ for all $p > 0$ and $\mathbf{t} \in \Lambda$).

Proposition 6.1.4. *Suppose that $\emptyset \neq \Lambda' \subseteq \mathbb{R}^n$, $\emptyset \neq \Lambda \subseteq \mathbb{R}^n$, $F : \Lambda \times X \to Y$ is a given function, ρ is a binary relation on Y and $\Lambda + \Lambda' \subseteq \Lambda$.*

(i) *If $F \in S^{(1,\rho,\mathcal{P}_{\mathbf{t}},\mathcal{P})}_{\Omega,\Lambda',\mathcal{B}}(\Lambda \times X : Y)$, where $P = C_b(\Lambda : \mathbb{C})$ and $P_t = L^p_{m',\nu}(\mathbf{t}+\Omega : Y)$ for some measure $m'(\cdot)$ on $\mathbb{R}^n$ and $p > 0$ ($\mathbf{t} \in \Lambda$), then $F(\cdot;\cdot)$ is Bohr $(\mathcal{B}, \Lambda', \rho, \Omega, m', \nu)$-almost periodic.*

(ii) *If $\rho = T \in L(Y)$, $\sup_{x \in B, \mathbf{t} \in \Lambda} \|F(\mathbf{t}; x)\|_Y < \infty$ for all $B \in \mathcal{B}$ and the function $F(\cdot;\cdot)$ is Bohr $(\mathcal{B}, \Lambda', \rho, \Omega, m', \nu)$-almost periodic for some bounded function $\nu(\cdot)$ and measure $m'(\cdot)$ on $\mathbb{R}^n$ satisfying $\sup_{\mathbf{t}\in\Lambda} m'(\mathbf{t}+\Omega) < +\infty$, then $F \in S^{(1,\rho,\mathcal{P}_{\mathbf{t}},\mathcal{P})}_{\Omega,\Lambda',\mathcal{B}}(\Lambda \times X : Y)$ with $P = C_b(\Lambda : \mathbb{C})$ and $P_t = L^p_{m',\nu}(\mathbf{t}+\Omega : Y)$ for all $\mathbf{t} \in \Lambda$.*

It is worth noting that Proposition 6.1.4(i) can be applied to the functions considered in [477, Examples 3, 4], with $\nu(\cdot) \not\equiv 1$, as well as that Proposition 6.1.4(ii) extends [726, Theorem 2.7]. We have already shown that any uniformly continuous, Stepanov-p-almost periodic function $F : \mathbb{R}^n \to Y$ is almost periodic ($p > 0$). This result can be further extended in the following way (we define the notion of m-almost periodicity of $F(\cdot)$ as in the one-dimensional setting; cf. also [618, Remark 3.4], where the author has also imposed the boundedness of function $F(\cdot)$ to derive its almost periodicity).

Proposition 6.1.5. *Suppose that $F : \mathbb{R}^n \to Y$ is m-almost periodic and uniformly continuous. Then $F(\cdot)$ is almost periodic.*

Proof. Keeping in mind Proposition 6.1.4(ii) and [477, Corollary 2], it suffices to show that $F(\cdot)$ is bounded. Toward this end, fix a number $\varepsilon > 0$. Then we can find three finite real numbers $c > 0$, $l > 0$ and $0 < \delta < \varepsilon/2$ such that, for every $\mathbf{t}_0 \in \mathbb{R}^n$, there exists $\tau \in B(\mathbf{t}_0, l)$ such that $m(\{\mathbf{s} \in \mathbf{t} + [0,1]^n : \|F(\mathbf{s}+\tau) - F(\mathbf{s})\|_Y < \delta\}) \geqslant 1 - c\delta^n$ for all $\mathbf{t} \in \mathbb{R}^n$, $m((\mathbf{t} + [0,1]^n) \cap B(\mathbf{s}, \delta)) > c\delta^n$, provided $\mathbf{t} \in \mathbb{R}^n$ and $\mathbf{s} \in \mathbf{t} + [0,1]^n$ and that the assumption $|x - y| < \delta$ for some $x, y \in \mathbb{R}^n$ implies $\|F(x) - F(y)\|_Y \leqslant \varepsilon/2$. If $\mathbf{t} \in \mathbb{R}^n$ is arbitrary and $\mathbf{s} \in \mathbf{t} + [0,1]^n$, then it can be simply shown that there exists $x \in B(\mathbf{s}, \delta)$ such that $\|F(x+\tau) - F(x)\| < \delta$. Since $F(\cdot)$ is continuous, we have that there exists a finite real number $M > 0$ such that $\|F(\mathbf{t})\|_Y \leqslant M$ for all $\mathbf{t} \in B(0, 2l)$. Pick up $\tau \in \mathbb{R}^n$ such that $x + \tau \in B(0, 2l)$ and $m(\{\mathbf{s} \in \mathbf{t} + [0,1]^n : \|F(\mathbf{s}+\tau) - F(\mathbf{s})\|_Y < \delta\}) \geqslant 1 - c\delta^n$ for all $\mathbf{t} \in \mathbb{R}^n$. Then we have $\|F(x+\tau) - F(x)\|_Y < \delta$ and $\|F(x) - F(\mathbf{s})\| \leqslant \varepsilon/2$ so that $\|F(\mathbf{s})\| \leqslant M + \delta + \varepsilon/2$. This implies the required result. □

We will illustrate Proposition 6.1.5 with the following example.

Example 6.1.6. The function $f : \mathbb{R} \to \mathbb{R}$, given by (277), is (Besicovitch) unbounded, uniformly continuous and uniformly recurrent (see [371, Theorem 1.1] and [447, Theorem 2.4.2]). Due to Proposition 6.1.5, $f(\cdot)$ cannot be m-almost periodic; on the other hand, Proposition 6.1.3 yields that $f(\cdot)$ is m-uniformly recurrent, with the meaning clear.

The following extension of an old result of S. Stoiński (see, e.g., [618, Theorem 2.7]) is applicable if $\Lambda = [0,\infty)^n$ or $\Lambda = \mathbb{R}^n$; albeit the proof is very similar to that of [448, Proposition 2.1.8(i)], we will present all relevant details for the sake of completeness.

Proposition 6.1.7. *Suppose that* $\emptyset \neq \Lambda \subseteq \mathbb{R}^n$, $\Lambda + \Lambda \subseteq \Lambda$, $\rho = T \in L(Y)$, $F : \Lambda \times X \to Y$ *is Bohr* $(\mathcal{B}, \Lambda, T, \Omega, m')$*-almost periodic,* $\sup_{x\in B, \mathbf{t}\in\Lambda\cap K} \|F(\mathbf{t};x)\|_Y < +\infty$ *for each compact* $K \subseteq \mathbb{R}^n$ *and the assumption* $A \subseteq B \subseteq \mathbb{R}^n$ *implies* $m'(A) \leqslant m'(B)$*. If*

$$(\forall l > 0)\,(\exists \mathbf{t}_0 \in \Lambda)\,(\exists k > 0)\,(\forall \mathbf{t} \in \Lambda)(\exists \mathbf{t}_0' \in \Lambda)$$
$$(\forall \mathbf{t}_0'' \in B(\mathbf{t}_0', l) \cap \Lambda)\, \mathbf{t} - \mathbf{t}_0'' \in B(\mathbf{t}_0, kl) \cap \Lambda, \tag{311}$$

then for each $B \in \mathcal{B}$ *and each sequence of positive real numbers* (λ_k) *tending to zero, we have*

$$\lim_{k\to+\infty} \sup_{\mathbf{t}\in\Lambda; x\in B} m'(\{\mathbf{s} \in \mathbf{t} + \Omega : \lambda_k \|F(\mathbf{s};x)\|_{Y \geqslant 1}\}) = 0. \tag{312}$$

Proof. Let $B \in \mathcal{B}$ and $\varepsilon > 0$ be given. Then we can find a finite number $l > 0$ such that for each $\mathbf{t}_0 \in \Lambda$ there exists $\tau \in B(\mathbf{t}_0, l) \cap \Lambda$ such that

$$\sup_{x\in B} m'(\{\mathbf{s} \in \mathbf{t} + \Omega : \|F(\mathbf{s} + \tau; x) - TF(\mathbf{s};x)\|_Y \geqslant \varepsilon\}) \leqslant \varepsilon, \quad \mathbf{t} \in \Lambda. \tag{313}$$

Suppose that $\mathbf{t}_0 \in \Lambda$ and $k > 0$ are chosen such that (311) holds. The prescribed assumption implies that the set $\{F(\mathbf{s};x) : \mathbf{s} \in B(\mathbf{t}_0, kl) \cap \Lambda,\ x \in B\}$ is bounded in Y. Let $\mathbf{t} \in \Lambda$ be fixed. Then there exists $\mathbf{t}_0' \in \Lambda$ such that, for every $\mathbf{t}_0'' \in B(\mathbf{t}_0', l) \cap \Lambda$, we have $\mathbf{t} \in \mathbf{t}_0'' + [B(\mathbf{t}_0, kl) \cap \Lambda]$. On the other hand, there exists $\tau = \mathbf{t}_0'' \in B(\mathbf{t}_0', l) \cap \Lambda$ such that (313) holds. Clearly, $\mathbf{s} = \mathbf{t} - \tau \in B(\mathbf{t}_0, kl) \cap \Lambda$, which simply implies that

$$\sup_{x\in B} m'(\{\mathbf{s} \in \mathbf{t} - \tau + \Omega : \|F(\mathbf{s} + \tau; x) - TF(\mathbf{s};x)\|_Y \geqslant \varepsilon\}) \leqslant \varepsilon, \quad \mathbf{t} \in \Lambda$$

and

$$\sup_{x\in B} m'(\{\mathbf{s} \in \mathbf{t} + \Omega : \|F(\mathbf{s}; x) - TF(\mathbf{s} - \tau;x)\|_Y \geqslant \varepsilon\}) \leqslant \varepsilon, \quad \mathbf{t} \in \Lambda.$$

Since $\mathbf{s} - \tau \in \Omega + [B(\mathbf{t}_0, kl) \cap \Lambda]$ and $T \in L(Y)$, there exists a finite constant $M_\varepsilon > 0$ such that $\|TF(\mathbf{s} - \tau; x)\|_Y \leqslant M_\varepsilon$ for all $x \in B$ and $\mathbf{s} \in \mathbf{t} + \Omega$. The final conclusion (312) follows from this estimate, the given assumption on $m'(\cdot)$ and the inclusions

$$\{\mathbf{s} \in \mathbf{t} + \Omega : \lambda_k \|F(\mathbf{s};x)\|_Y \geqslant 1\} \subseteq \{\mathbf{s} \in \mathbf{t} + \Omega : \|F(\mathbf{s};x)\|_Y > M_\varepsilon + \varepsilon\}$$
$$\subseteq \{\mathbf{s} \in \mathbf{t} - \tau + \Omega : \|F(\mathbf{s};x) - TF(\mathbf{s} - \tau;x)\|_Y \geqslant \varepsilon\},$$

which hold for all $x \in B$ and all sufficiently large integers $k \geqslant k_0(\varepsilon)$. □

The condition [151, (13)] and the statement of [151, Theorem 8] can be straightforwardly extended to the multidimensional setting. But it seems much more complicated to provide some sufficient conditions, which would ensure that an m-almost periodic function $F : \mathbb{R}^n \to Y$ is equi-Weyl-p-almost periodic for some (all) exponents $p \geqslant 1$.

We continue with the following illustrative application of the d'Alembert formula.

Example 6.1.8. Let $a > 0$ and $|c| = 1$; then a unique regular solution of the wave equation $u_{tt} = a^2 u_{xx}$ in domain $\{(x,t) : x \in \mathbb{R},\ t > 0\}$, equipped with the initial conditions $u(x,0) = f(x) \in C^2(\mathbb{R})$ and $u_t(x,0) = g(x) \in C^1(\mathbb{R})$, is given by

$$u(x,t) = \frac{1}{2}[f(x-at) + f(x+at)] + \frac{1}{2a}\int_{x-at}^{x+at} g(s)\,ds, \quad x \in \mathbb{R},\ t > 0.$$

If we assume that the function $x \mapsto (f(x), g^{[1]}(x))$, $x \in \mathbb{R}$ is $(c\mathrm{I}, [0,1], m)$-almost periodic, where $g^{[1]}(\cdot) \equiv \int_0^{\cdot} g(s)\,ds$, then the solution $u(x,t)$ can be extended to the whole real line in the time variable and this solution is $(c\mathrm{I}, [0,1]^2, m_2)$-almost periodic in $(x,t) \in \mathbb{R}^2$; in order not to make any confusion, $m(\cdot)$ denotes here the Lebesgue measure in $\mathbb{R}$ and $m_2(\cdot)$ denotes the Lebesgue measure in $\mathbb{R}^2$. To show this, fix a positive real number $\varepsilon > 0$. Then there exists a finite real number $l > 0$ such that any subinterval I of $\mathbb{R}$ of length l contains a point $\tau \in I$ such that

$$\sup_{t\in\mathbb{R}} m(\{x \in [t,t+1] : |f(x+\tau) - cf(x)| < \varepsilon\}) \geqslant 1 - \varepsilon, \quad t \in \mathbb{R} \tag{314}$$

and (314) holds with the function $f(\cdot)$ replaced by the function $g^{[1]}(\cdot)$ therein. Furthermore, one has ($x,\ t,\ \tau_1,\ \tau_2 \in \mathbb{R}$):

$$\begin{aligned} &|u(x+\tau_1, t+\tau_2) - cu(x,t)| \\ &\quad \leqslant \frac{1}{2}|f((x-at) + (\tau_1 - a\tau_2)) - cf(x-at)| \\ &\qquad + \frac{1}{2}|f((x+at) + (\tau_1 + a\tau_2)) - cf([x+at+(\tau_1+a\tau_2)] - (\tau_1 + a\tau_2))| \\ &\qquad + \frac{1}{2a}|g^{[1]}((x-at) + (\tau_1 - a\tau_2)) - cg^{[1]}(x-at)| \\ &\qquad + \frac{1}{2a}|g^{[1]}((x+at) + (\tau_1 + a\tau_2)) - cg^{[1]}(x+at)|. \end{aligned} \tag{315}$$

Set now $X := x - at$ and $Y := x + at$. Then $x = (X+Y)/2$, $y = (-X+Y)/2a$, the linear transformation $T : \mathbb{R}^2 \to \mathbb{R}^2$, given by $T(X,Y) = ((X+Y)/2, (-X+Y)/2a)$ for all $(X,Y) \in \mathbb{R}^2$, is an isomorphism and $\det([T]) = 1/2a$. Suppose now that $(x,t) \in [s_1, s_1+1]\times[s_2, s_2+1]$ for some $s_1,\ s_2 \in \mathbb{R}$. Then it can be simply shown that the transformation $T^{-1}(\cdot)$ maps the rectangle $[s_1, s_1+1] \times [s_2, s_2+1]$ onto the closed quadrilateral P with the vertices $A(s_1 - as_2, s_1 + as_2)$, $B(s_1 - as_2 - a, s_1 + as_2 + a)$, $C(s_1 + 1 - as_2, s_1 + 1 + as_2)$ and $D(s_1 + 1 - as_2 - a, s_1 + 1 + as_2 + a)$; clearly, $m_2(P) = 2a$. Define $X_\varepsilon := \{x \in [s_1 - as_2 - a, s_1 - as_2 + 1] :$

$\max(|f(x+\tau)-cf(x)|, |g^{[1]}(x+\tau)-cg^{[1]}(x)|)<\varepsilon\}$ and $Y_\varepsilon := \{x \in [s_1+as_2, s_1+as_2+a+1] : \max(|f(x+\tau)-cf(x)|, |g^{[1]}(x+\tau)-cg^{[1]}(x)|)<\varepsilon\}$. Then $m(X_\varepsilon) \geqslant a+1-\lceil a+1\rceil\varepsilon$, $m(Y_\varepsilon) \geqslant a+1-\lceil a+1\rceil\varepsilon$ and $m_2(\{(X,Y) \in P : X \in X_\varepsilon,\ Y \in Y_\varepsilon\}) \geqslant 2a-(\lceil a+1\rceil\varepsilon)^2$. Therefore,

$$m_2(\{(x,t) \in [s_1,s_1+1]\times[s_2,s_2+1] : X \in X_\varepsilon,\ Y \in Y_\varepsilon\}) \geqslant \frac{1}{2a}\big(2a-(\lceil a+1\rceil\varepsilon)^2\big).$$

Keeping in mind this estimate and (315), the final conclusion follows similarly as in [447, Example 2, pp. XXXIV–XXXV]. Let us finally note that the possible applications can be also given to the Kirchhoff formula and the Poisson formula; see [447] for more details.

Before proceeding further, let us only mention that [447, Examples 6.1.13, 6.1.16] can be formulated in our new framework; these examples show the importance of considerations of general regions Λ and Λ' in our analysis. Now we will state and prove the following extension of [726, Theorem 6] (cf. also [448, Theorem 2.1.12(v)], the notion of metric space $\mathcal{X}$ introduced in [726], and [151, Definition 7]).

Theorem 6.1.9. *Suppose that $M>0$, $\emptyset \neq \Lambda' \subseteq \mathbb{R}^n$, $\emptyset \neq \Lambda \subseteq \mathbb{R}^n$, $\rho = T \in L(Y)$ and $\Lambda+\Lambda' \subseteq \Lambda$. Suppose further that for each $k \in \mathbb{N}$ we have that $F_k : \Lambda\times X \to Y$ is a Bohr $(\mathcal{B},\Lambda',\rho,\Omega,m',\nu)$-almost periodic ($(\mathcal{B},\Lambda',\rho,\Omega,m',\nu)$-uniformly recurrent) function and, for every $\varepsilon>0$ and $B \in \mathcal{B}$, we have*

$$\lim_{k\to+\infty}\sup_{x\in B}\|F_k(\cdot;x)-F(\cdot;x)\|_{P_\varepsilon}=0.$$

Then $F(\cdot;\cdot)$ is Bohr $(\mathcal{B},\Lambda',\rho,\Omega,m',\nu)$-almost periodic ($(\mathcal{B},\Lambda',\rho,\Omega,m',\nu)$-uniformly recurrent), provided that the assumptions A, B, $C \subseteq \mathbb{R}^n$ and $A \subseteq B\cup C$ imply $m'(A) \leqslant m'(B)+m'(C)$, and the assumption $\mathbf{v} \in \Lambda+\Omega+\tau$ for some $\tau \in \Lambda'$ implies $\nu(\mathbf{v}-\tau) \leqslant M\nu(\mathbf{v})$.

Proof. The proof basically follows from the next estimates ($k \in \mathbb{N}$, $x \in X$, $\tau \in \Lambda'$):

$$\begin{aligned}
&\|F(\cdot+\tau;x)-TF(\cdot;x)\|_{P_\varepsilon}\\
&\quad\leqslant \|F(\cdot+\tau;x)-F_k(\cdot+\tau;x)\|_{P_{\varepsilon/2}}+\|F_k(\cdot+\tau;x)-TF(\cdot;x)\|_{P_{\varepsilon/2}}\\
&\quad\leqslant \|F(\cdot+\tau;x)-F_k(\cdot+\tau;x)\|_{P_{\varepsilon/2}}+\|F_k(\cdot+\tau;x)-TF_k(\cdot;x)\|_{P_{\varepsilon/4}}+\|TF_k(\cdot;x)-TF(\cdot;x)\|_{P_{\varepsilon/4}}\\
&\quad\leqslant \|F(\cdot;x)-F_k(\cdot;x)\|_{P_{\varepsilon/2M}}+\|F_k(\cdot+\tau;x)-TF_k(\cdot;x)\|_{P_{\varepsilon/4}}+\|F_k(\cdot;x)-F(\cdot;x)\|_{P_{\varepsilon/4\|T\|}}.
\end{aligned}$$

Keeping this in mind, we can apply the estimates (307)–(309) to complete the proof. □

We proceed further with the observation that it would be very difficult to state a satisfactory analogue of Theorem 6.1.9 for the class of (strongly) $(\mathrm{R},\mathcal{B},\Omega,L,m',\nu)$-multialmost periodic functions; cf. also the proof of [452, Proposition 2.6] and the estimates given in (310).

In connection with the statement of [618, Theorem 2.9], we will present the following illustrative example; here, we do not use the boundedness of the analytical extension on the strips around the real axes.

Example 6.1.10. Suppose that $G : \mathbb{R}^n \to \mathbb{R}$ is an almost periodic function, $G(\mathbf{t}) \neq 0$ for all $\mathbf{t} \in \mathbb{R}^n$ and there exist real numbers a and b such that $a < 0 < b$ and the function $G(\cdot)$ can be analytically extended to the region $\{(z_1, \ldots, z_n) \in \mathbb{C}^n : \operatorname{Re} z_i \in (a, b) \text{ for } 1 \leqslant i \leqslant n\}$. Then the argumentation contained in the proof of [508, Theorem 5.3.1] (cf. also [447, Example 6.2.9]) shows that $\lim_{\delta \to 0+} m(\{\mathbf{s} \in \mathbf{t} + [0,1]^n : |G(\mathbf{s})| < \delta\}) = 0$, uniformly in $\mathbf{t} \in \mathbb{R}^n$, so that

$$(\forall \varepsilon > 0)(\exists \delta_0 > 0)(\forall \delta \in (0, \delta_0))(\forall \mathbf{t} \in \mathbb{R}^n)\, m(\{\mathbf{s} \in \mathbf{t} + [0,1]^n : |G(\mathbf{s})| \geqslant \delta\}) \geqslant 1 - \varepsilon.$$

Suppose now that $\varepsilon > 0$ is given, $\delta \in (0, \min(1/\varepsilon, \delta_0/2))$ and $\tau \in \mathbb{R}^n$ is a $(\delta/2)$-almost period of the function $G(\cdot)$. If $\mathbf{s} \in \mathbf{t} + [0,1]^n$ for some $\mathbf{t} \in \mathbb{R}^n$ and $|G(\mathbf{s})| \geqslant \delta$, then we cannot have $|F(\mathbf{s} + \tau) - F(\mathbf{s})| \geqslant \varepsilon$ because this would imply

$$\delta/2 > \left|G(\mathbf{s} + \tau) - G(\mathbf{s})\right| \geqslant \varepsilon \cdot \left|G(\mathbf{s})\right| \cdot \left|G(\mathbf{s} + \tau)\right| \geqslant \varepsilon \cdot \delta \cdot (\delta/2),$$

which is a contradiction. Hence,

$$m(\{\mathbf{s} \in \mathbf{t} + [0,1]^n : \left|F(\mathbf{s} + \tau) - F(\mathbf{s})\right| < \varepsilon\}) \geqslant 1 - \varepsilon, \quad \mathbf{t} \in \mathbb{R}^n$$

and

$$m(\{\mathbf{s} \in \mathbf{t} + [0,1]^n : \left|F(\mathbf{s} + \tau) - F(\mathbf{s})\right| \geqslant \varepsilon\}) \leqslant \varepsilon, \quad \mathbf{t} \in \mathbb{R}^n.$$

This implies that $F(\cdot)$ is m-almost periodic. In particular, the function

$$F(t_1, \ldots, t_n) = \frac{1}{2 + \cos t_1 + \cos(\sqrt{2} t_1)} \cdot \cdots \cdot \frac{1}{2 + \cos t_n + \cos(\sqrt{2} t_n)}, \quad \mathbf{t} = (t_1, \ldots, t_n) \in \mathbb{R}^n$$

is m-almost periodic.

In connection with this example and the function $f(\cdot)$ given by (72), we would like to recall that we have already asked in [448, Example 6.4.10(i)] whether the function $f(\cdot)$ is Besicovitch-p-bounded for some finite exponent $p \geqslant 1$. It is our strong belief that this is not the case as well as that this is a simple example of a continuous, m-almost periodic function $f : \mathbb{R} \to \mathbb{R}$ such that $M(f)$ does not exist; cf. also [619].

Further on, if $\delta > 0$, then we set $\Lambda(\delta) := \{h \in \mathbb{R}^n : |h| \leqslant \delta \text{ and } \mathbf{t} + h \in \Lambda \text{ for all } \mathbf{t} \in \Lambda\}$. We say that a function $F : \Lambda \times X \to Y$ is $(\mathcal{B}, \Omega, m')$-continuous if, for every $\varepsilon > 0$ and $B \in \mathcal{B}$, there exists $\delta > 0$ such that, for every $h \in \Lambda(\delta)$, we have

$$\sup_{\mathbf{t} \in \Lambda; x \in B} m'(\{\mathbf{s} \in \mathbf{t} + \Omega : \|F(\mathbf{s} + h; x) - F(\mathbf{s}; x)\|_Y \geqslant \varepsilon\}) \leqslant \varepsilon.$$

Immediately from definition, it follows that any function $F : \Lambda \times X \to Y$, which is uniformly continuous on the sets of the form $\Lambda \times B$, where $B \in \mathcal{B}$, is $(\mathcal{B}, \Omega, m')$-continuous. Moreover, if the function $F : \Lambda \times X \to Y$ is continuous on the sets of the form $\Lambda \times B$, where

$B \in \mathcal{B}$, the region Λ has certain geometrical properties (we can always take $\Lambda = \mathbb{R}^n$ here) and $F(\cdot;\cdot)$ is Bohr $(\mathcal{B}, \Lambda, \mathrm{I}, \Omega, m')$-almost periodic, then $F(\cdot;\cdot)$ is $(\mathcal{B}, \Omega, m')$-continuous. To the best knowledge of the author, it is not clear how one can prove that the statements of [620, Twierdzenie 1.8, Lemat 1.8, Lemat 1.9, Uwaga 1.9; pp. 23–24] hold if the function under consideration is not continuous; because of that we will not reconsider these statements in the multi-dimensional setting (it is also worth noticing that the argumentation of H. Bohr, which has particularly been used in the proof of [620, Lemat 1.9], is inapplicable in the multidimensional setting).

6.1.2 Strongly $(\mathcal{B}, \Omega, m', \nu)$-almost periodic type functions

In this subsection, we assume that the general requirements made at the beginning of this section hold. We start by introducing the following notion (cf. also [448, Definition 6.1.1], with $\phi(x) \equiv x$).

Definition 6.1.11. Suppose that $F : \Lambda \times X \to Y$ is a given function. Then we say that $F(\cdot;\cdot)$ is strongly $(\mathcal{B}, \Omega, m', \nu)$-almost periodic if for each $B \in \mathcal{B}$ and $\varepsilon > 0$ there exists a sequence $(P_k(\cdot;\cdot))$ of trigonometric polynomials such that

$$\lim_{k \to +\infty} \sup_{x \in B; \mathbf{t} \in \Lambda} m'\big(\{\mathbf{s} \in \mathbf{t} + \Omega : \|P_k(\mathbf{s};x) - F(\mathbf{s};x)\|_Y \cdot \nu(\mathbf{s}) \geqslant \varepsilon\}\big) = 0.$$

We can similarly consider the notion in which the sequence $(P_k(\cdot;\cdot))$ of trigonometric polynomials is replaced by a sequence of ρ-periodic functions or a sequence of $(\rho_j)_{j \in \mathbb{N}_n}$-periodic functions; cf. [289] for the notion used.

The following analogue of [448, Proposition 6.1.9(ii)] holds true; for simplicity, we assume here that $\Omega = [0,1]^n$, $\nu(\cdot) \equiv 1$ and $m' = m$ is the Lebesgue measure.

Proposition 6.1.12. *Suppose that $F : \Lambda \times X \to Y$, $\Lambda + \Lambda' \subseteq \Lambda$, $\Omega = [0,1]^n$ and $F(\cdot;\cdot)$ is strongly $(\mathcal{B}, \Omega, m)$-almost periodic. Then $F(\cdot;\cdot)$ is Bohr $(\mathcal{B}, \Lambda', \Omega, m', \nu)$-almost periodic.*

Proof. Let $B \in \mathcal{B}$ and $\varepsilon > 0$ be given. Then we know that there exists $k_0 \in \mathbb{N}$ such that, for every $k \geqslant k_0$, we have

$$\sup_{x \in B; \mathbf{t} \in \Lambda} m\big(\{\mathbf{s} \in \mathbf{t} + \Omega : \|P_k(\mathbf{s};x) - F(\mathbf{s};x)\|_Y \geqslant \varepsilon/3\}\big) \leqslant \varepsilon/2.$$

This implies

$$\sup_{x \in B; \mathbf{t} \in \Lambda} m\big(\{\mathbf{s} \in \mathbf{t} + \Omega : \|P_k(\mathbf{s};x) - F(\mathbf{s};x)\|_Y < \varepsilon/3\}\big) \geqslant 1 - (\varepsilon/2).$$

Denote $A_{\mathbf{t},\varepsilon/3,k} := \{\mathbf{s} \in \mathbf{t} + \Omega : \|P_k(\mathbf{s};x) - F(\mathbf{s};x)\|_Y \geqslant \varepsilon/3\}$. Then for each points $\mathbf{t}$, $\tau \in \mathbb{R}^n$ we have $m(A_{\mathbf{t}+\tau,\varepsilon/3,k} - \tau) \geqslant 1 - (\varepsilon/2)$ and this simply implies $m(A_{\mathbf{t},\varepsilon/3,k} \cap (A_{\mathbf{t}+\tau,\varepsilon/3,k} - \tau)) \geqslant 1 - \varepsilon$

for all $\mathbf{t},\ \tau \in \mathbb{R}^n$. Let an integer $k \geqslant k_0$ be fixed. Then the final conclusion simply follows from the estimate

$$\begin{aligned}
&\|F(\mathbf{s}+\tau;x) - F(\mathbf{s};x)\|_Y \\
&\quad \leqslant \|F(\mathbf{s}+\tau;x) - P_k(\mathbf{s}+\tau;x)\|_Y + \|P_k(\mathbf{s}+\tau;x) - P_k(\mathbf{s};x)\|_Y + \|P_k(\mathbf{s};x) - F(\mathbf{s};x)\|_Y \\
&\quad \leqslant 3 \cdot (\varepsilon/3),
\end{aligned}$$

provided that $s \in A_{\mathbf{t},\varepsilon/3,k} \cap (A_{\mathbf{t}+\tau,\varepsilon/3,k} - \tau)$ and $\tau \in \mathbb{R}^n$ is a $(B,\varepsilon/3)$-almost period of the trigonometric polynomial $P_k(\cdot;\cdot)$, with the meaning clear. □

Let us finally notice that we can also introduce and analyze the notion of $(\mathrm{R},\mathcal{B},m',\nu)$-normality (cf. [448, Definition 6.1.5], with $\phi(x) \equiv x$). We will skip all details concerning this topic here.

Applications

In the continuation of this section, we will provide several applications of the introduced notion to the abstract Volterra integrodifferential inclusions and present some new important contributions to the theory of Stepanov almost periodic functions. We will divide the material into two separate subsections.

6.1.3 Convolution invariance of measure almost periodicity

The convolution invariance of Bohr $(\mathcal{B},\Lambda',\rho,\Omega,m',\nu)$-almost periodicity and $(\mathcal{B},\Lambda',\rho,\Omega,m',\nu)$-uniform recurrence is an extremely delicate theme. We start this subsection by recalling that the convolution of an m-almost periodic function $f : \mathbb{R}^n \to Y$ with a function $h \in L^1(\mathbb{R}^n)$ does not have to exists; furthermore, if the value $(h*f)(\mathbf{t}) = \int_{-\infty}^{\infty} h(\mathbf{t}-\mathbf{s})f(\mathbf{s})\,d\mathbf{s}$ exists for a. e. $\mathbf{t} \in \mathbb{R}^n$, then the convolution $(h * f)(\cdot)$ is not m-almost periodic, in general (cf. [151, Examples 3, 4] for the one-dimensional setting).

Now we will state and prove the following result (for simplicity, we consider here the case in which $\nu(\cdot) \equiv 1$; $\mathbf{t} > \mathbf{0}$ means than any component of tuple $\mathbf{t} \in \mathbb{R}^n$ is positive).

Theorem 6.1.13. (i) *Let $(R(\mathbf{t}))_{\mathbf{t}>\mathbf{0}} \subseteq L(X)$ be a strongly continuous operator family such that $\int_{(0,\infty)^n} \|R(\mathbf{t})\|\,d\mathbf{t} < \infty$. If $f : \mathbb{R}^n \to X$ is bounded and Bohr $(\Lambda',\rho,[0,1]^n,m)$-almost periodic ($(\Lambda',\rho,[0,1]^n,m)$-uniformly recurrent), where $\rho = T \in L(X)$, then the function $F : \mathbb{R}^n \to X$, given by*

$$F(\mathbf{t}) := \int_{-\infty}^{t_1}\int_{-\infty}^{t_2}\cdots\int_{-\infty}^{t_n} R(\mathbf{t}-\mathbf{s})f(\mathbf{s})\,d\mathbf{s}, \quad \mathbf{t} \in \mathbb{R}^n, \tag{316}$$

is bounded, continuous and Bohr $(\Lambda',\rho,[0,1]^n)$-almost periodic ($(\Lambda',\rho,[0,1]^n)$-uniformly recurrent), provided that $R(\mathbf{t})T = TR(\mathbf{t})$, $\mathbf{t} > \mathbf{0}$.

(ii) *Let* $a(\cdot)$ *be Lebesgue measurable and let* $\int_{(0,\infty)^n} |a(\mathbf{t})|\,d\mathbf{t} < \infty$. *If* $f : \mathbb{R}^n \to X$ *is bounded and Bohr* $(\Lambda', \rho, [0,1]^n, m)$*-almost periodic* ($(\Lambda', \rho, [0,1]^n, m)$*-uniformly recurrent), where* $\rho = T \in L(X)$, *then the function* $F : \mathbb{R}^n \to X$, *given by*

$$F(\mathbf{t}) := \int_{-\infty}^{t_1}\int_{-\infty}^{t_2}\cdots\int_{-\infty}^{t_n} a(\mathbf{t}-\mathbf{s})f(\mathbf{s})\,d\mathbf{s}, \quad \mathbf{t} \in \mathbb{R}^n, \tag{317}$$

is bounded, uniformly continuous and Bohr $(\Lambda', \rho, [0,1]^n)$*-almost periodic* ($(I', \rho, [0,1]^n)$*-uniformly recurrent).*

Proof. We will prove part (i) only for the class of one-dimensional, bounded, Bohr $(\mathcal{B}, \Lambda', T, [0,1], m)$-almost periodic functions. It is clear that the function $F(\cdot)$ is well-defined and bounded; the continuity is a simple consequence of the dominated convergence theorem, the boundedness of the function $f(\cdot)$ and the strong continuity of $(R(t))_{t>0}$. It is clear that the function $F(\cdot)$ is well-defined and bounded; the continuity is a simple consequence of the dominated convergence theorem, the boundedness and continuity of the function $f(\cdot)$. Let $\varepsilon > 0$ be fixed. Then there exists $k \in \mathbb{N}$ such that $\int_k^{+\infty} \|R(r)\|\,dr \leqslant \varepsilon/(2(1+\|T\|)\|f\|_\infty)$. This implies

$$\int_k^{+\infty} \|R(r)\| \cdot \|f(s+\tau-r) - Tf(s-r)\|\,dr \leqslant \varepsilon/2, \quad s \in \mathbb{R}. \tag{318}$$

Let $0 < \varepsilon' < \varepsilon/(4k(1+\int_0^\infty \|R(r)\|\,dr))$ and let $0 < \varepsilon' < \delta/2$, where $\delta > 0$ satisfies that the assumption $m(A) < \delta$ for some Lebesgue measurable set $A \subseteq (0,\infty)$ implies $\int_A \|R(r)\|\,dr < \varepsilon/4k$. Then we know that there exists $l > 0$ such that for each $t_0 \in \Lambda'$ there exists $\tau \in B(t_0, l) \cap \Lambda'$ such that, for every $t \in \mathbb{R}$ and $s \in [t, t+1]$, we have $\|f(\cdot+\tau) - Tf(\cdot)\|_{P_{\varepsilon'}} \leqslant \varepsilon'$. Suppose that $s \in \mathbb{R}$ is arbitrary. We will show that $\|F(s+\tau) - TF(s)\|_Y < \varepsilon$. Assume the contrary, i. e., $\|F(s+\tau) - TF(s)\|_Y \geqslant \varepsilon$. Then an elementary argumentation involving (318) and the commutativity assumption $R(t)T = TR(t)$, $t > 0$ shows that

$$\begin{aligned}&\sum_{j=0}^{k-1}\int_j^{j+1} \|R(r)\| \cdot \|f(s+\tau-r) - Tf(s-r)\|\,dr\\ &\quad = \int_0^k \|R(r)\| \cdot \|f(s+\tau-r) - Tf(s-r)\|\,dr \geqslant \varepsilon/2.\end{aligned} \tag{319}$$

If $j \in \mathbb{N}_0$, $0 \leqslant j \leqslant k-1$ and $r \in [j, j+1]$, then $s - r \in [s-j-1, s-j]$ so that

$$m(\{r \in [j, j+1] : \|f(s+\tau-r) - Tf(s-r)\| \geqslant \varepsilon'\}) \leqslant \varepsilon'$$

and

$$\int_j^{j+1} \|R(r)\| \cdot \|f(s+\tau-r) - Tf(s-r)\| \, dr < \varepsilon'(1+\|T\|)\|f\|_\infty \varepsilon/4k + \varepsilon' \int_j^{j+1} \|R(r)\| \, dr.$$

The last estimate and the choice of $\varepsilon' > 0$ together imply

$$\int_0^k \|R(r)\| \cdot \|f(s+\tau-r) - Tf(s-r)\| \, dr < \varepsilon/2,$$

which contradicts (319). Concerning the part (ii), we will only note that the argumentation contained in the proof of [69, Proposition 1.3.2 c)] shows that the resulting function $F(\cdot)$ is uniformly continuous. □

Similarly, we can prove the following result (by Id we denote the identity operator on X).

Theorem 6.1.14. (i) *Let $(R(\mathbf{t}))_{\mathbf{t}>\mathbf{0}} \subseteq L(X,Y)$ be a strongly continuous operator family such that $\int_{(0,\infty)^n} \|R(\mathbf{t})\| \, d\mathbf{t} < \infty$. If $f : \mathbb{R}^n \to X$ is bounded and Bohr $(\Lambda', \rho, [0,1]^n, m)$-almost periodic ($(\Lambda', \rho, [0,1]^n, m)$-uniformly recurrent), where $\rho = cId \in L(X)$, then the function $F : \mathbb{R}^n \to Y$, given by* (316)*, is bounded, continuous and Bohr $(\Lambda', \rho_Y, [0,1]^n)$-almost periodic ($(\Lambda', \rho_Y, [0,1]^n)$-uniformly recurrent), where $\rho_Y = c\mathrm{I} \in L(Y)$.*

(ii) *Let $a(\cdot)$ be Lebesgue measurable and let $\int_{(0,\infty)^n} |a(\mathbf{t})| \, d\mathbf{t} < \infty$. If $f : \mathbb{R}^n \to Y$ is bounded and Bohr $(\Lambda', \rho, [0,1]^n, m)$-almost periodic ($(\Lambda', \rho, [0,1]^n, m)$-uniformly recurrent), where $\rho = c\mathrm{I} \in L(Y)$, then the function $F : \mathbb{R}^n \to Y$, given by* (317)*, is bounded, uniformly continuous and Bohr $(\Lambda', \rho, [0,1]^n)$-almost periodic ($(I', \rho, [0,1]^n)$-uniformly recurrent).*

Now we will provide several important observations about the last two results.

Remark 6.1.15. If the function $f(\cdot)$ is uniformly continuous in part (i) of Theorem 6.1.13 or Theorem 6.1.14, then the resulting function $F(\cdot)$ will be also uniformly continuous, which directly follows from [69, Proposition 1.3.5 c)]. But, in the operator-valued setting, it is not clear how to prove that the boundedness of $f(\cdot)$ implies the uniform continuity of $F(\cdot)$.

Remark 6.1.16. (i) Suppose that the requirements of Theorem 6.1.13(i) or Theorem 6.1.14(i) hold. If the function $f(\cdot)$ has relatively compact range, then the function $F(\cdot)$ has relatively compact range as well.

(ii) Suppose that the requirements of Theorem 6.1.13(ii) or Theorem 6.1.14(ii) hold. If the function $f(\cdot)$ has relatively compact range, then the function $F(\cdot)$ has relatively compact range as well.

In order to see that (i) holds, observe that $F(\mathbf{t}) := \int_{(0,\infty)^n} R(\mathbf{s}) f(\mathbf{t}-\mathbf{s}) \, d\mathbf{s}$, $\mathbf{t} \in \mathbb{R}^n$. If $\varepsilon > 0$ is given, then there exists a finite set $\{x_i : 1 \leqslant i \leqslant k\} \subseteq X$ such that $\overline{R(f)} \subseteq \bigcup_{1\leqslant i\leqslant k} L(x_i, \varepsilon)$, where $L(x_i, \varepsilon)$ denotes the open ball in X with the center x_i and the radius ε. Then we have

$$\overline{R(F)} \subseteq \bigcup_{1\leqslant i\leqslant k} L\Big(\int_{(0,\infty)^n} R(\mathbf{s})x_i\, d\mathbf{s}, \varepsilon \int_{(0,\infty)^n} \|R(\mathbf{s})\|\, d\mathbf{s} \Big),$$

which simply the required. We can similarly show (ii).

Remark 6.1.17. It is not clear whether we can replace the boundedness of function $f(\cdot)$ in the formulation of Theorem 6.1.13 or Theorem 6.1.14 by some weaker conditions, for example, by its Stepanov p-boundedness for some $p > 0$. In connection with this issue, we would like to stress that such attempts for the usual infinite convolution product have been analyzed already by G. Bruno and A. Pankov in [146, Lemma 2]. Disappointingly, the proof of this result is not completely correct because the authors have not proved the (absolute) convergence of the integral $\int_{\mathbb{R}} \varphi(t)u(x-t)\, dt$ in a proper way ($x \in \mathbb{R}$); here, we use the same notation as in [146].

Remark 6.1.18. As the referee of the old version of paper [490] has noticed, the assumption $R(\mathbf{t})T = TR(\mathbf{t})$, $\mathbf{t} > \mathbf{0}$ is crucial and the proof of Theorem 6.1.13 does not work if this condition is disregarded. For example, suppose that $n = 1$, $X := L^1(\mathbb{R})$, $T \in L(X)$ is given by $[Tg](t) := g(-t)$, $t \in \mathbb{R}$, $g \in L^1(\mathbb{R})$ and $f(\cdot)$ is the 2-periodic extension of the function $f_0 : [0,2) \to \mathbb{R}$, given by $f_0(t) := 1$ for $0 \leqslant t < 1$ and $f_0(t) := 0$ for $1 \leqslant t < 2$, to the whole real line. Then we have $f(t+1) = Tf(t)$ for all $t \in \mathbb{R}$ and an arduous computation yields that $\|F(t+1) - TF(t)\|_{L^1(\mathbb{R})} \geqslant 2/(1+e)$ for all $t \in [0,1]$.

In order to provide certain applications of Theorem 6.1.13(i) to the abstract Volterra integrodifferential equations in Banach spaces, we will first revisit the paper [593] by R. K. Miller, R. L. Wheeler and our previous analysis from [447, Example 3.3.32].

Example 6.1.19. Let $Y = H$ be an infinite-dimensional Hilbert space with inner product $\langle \cdot, \cdot \rangle$. In [593], R. K. Miller and R. L. Wheeler have analyzed the well-posedness of the following abstract Cauchy problem of nonscalar type:

$$x'(t) = Ax(t) + \int_0^t b(t-s)(A + aI)x(s)\, ds + f(t), \quad x(0) = x_0, \tag{320}$$

where $b(t)$ is a scalar-valued kernel, $b \in C^1([0,\infty))$, $a \in \mathbb{C}$, $f : [0,\infty) \to H$ is continuous and A is a densely defined, self-adjoint closed linear operator in H. We know that the validity of assumptions [593, (A1)–(A5)] with the coefficients $\alpha = \beta_0 = \beta_1 = 0$ and the validity of assumption [593, (A6)] with $B\sigma(L) \neq \emptyset$ (cf. [593, p. 273] for the notion) imply by [593, Theorem 8] that there exists a unique residual resolvent $(R(t))_{t\geqslant 0}$ for (320) such that $\|R(\cdot)\| \in L^q([0,\infty))$ for $1 \leqslant q < \infty$; then [593, Theorem 2] yields that the unique solution of (320) for all $x_0 \in D(A)$ and $f \in C^1([0,\infty) : X)$ is given by

$$x(t) = R(t)x_0 + \int_0^t R(t-s)f(s)\, ds, \quad t \geqslant 0.$$

For the sequel, let us emphasize that the assumption $\|R(\cdot)\| \in L^q([0,\infty))$ for $1 \leqslant q < \infty$ on the resolvent solution family $(R(t))_{t\geqslant 0}$ does not directly imply that

$$\sum_{k=0}^{\infty} \|R(\cdot)\|_{L^q[k,k+1]} < +\infty \quad \text{for some } q \in (1,+\infty],$$

which would be also very difficult to prove using the methods proposed in [593] (it should be also said that we have not been able to locate any relevant result examining the issue whether the last condition holds for the operator solution families appearing in the theory of abstract ill-posed Cauchy problems). If we assume that the forcing term $f \in C^1([0,\infty) : X)$ satisfies that there exists a function

$$f_1 \in C^1(\mathbb{R} : H) \cap \Big[C_b(\mathbb{R} : H) \bigcap_{p\geqslant 1} S^p AP(\mathbb{R} : H) \Big]$$

such that $f_1(t) = f(t)$ for all $t \geqslant 0$, then a simple argumentation involving the decomposition

$$x(t) = R(t)x_0 + \int_{-\infty}^{t} R(t-s)f(s)\,ds - \int_{-\infty}^{0} R(t-s)f(s)\,ds, \quad t \geqslant 0,$$

Theorem 6.1.13(i) and Proposition 6.1.4 shows that the solution $x(\cdot)$ belongs to the space

$$\bigcap_{p\geqslant 1} L^p([0,\infty) : H) + C_0([0,\infty) : H) + AP(\mathbb{R} : H).$$

We continue by observing that the integrability of resolvent operator families for the abstract Volterra integral equations has been considered by J. Prüss in [652, Part III, Section 10]; see, e. g., [652, Theorem 10.1, Corollary 10.1, pp. 262–263] for some sufficient conditions ensuring that $\int_0^{+\infty} \|R(s)\|\,ds < +\infty$. Therefore, the asymptotical almost periodicity of the corresponding abstract Volterra integral equations, with the forcing terms of the same type, follows from the consideration carried out in Example 6.1.19 and Theorem 6.1.13(i); [444, Proposition 2.6.11] is generally not applicable here.

Now we will provide the following simple application of Theorem 6.1.13(ii).

Example 6.1.20. We can simply construct a great number of kernels $a \in L^1((0,\infty)^n)$, which are not locally q-integrable at zero for any exponent $q \in (1,\infty]$. If the function $f(\cdot)$ is bounded and Stepanov-p-almost periodic for some $p \in [1,\infty)$, then we cannot apply [447, Theorem 6.2.36] in order to see that the resulting function $F(\cdot)$ will be almost periodic. But the almost periodicity of $F(\cdot)$ can be proved with the help of Theorem 6.1.13(ii).

6.1.4 Semilinear Volterra integral equations

Let us define $S^{\infty}(\mathbb{R}^n : X) := C_b(\mathbb{R}^n : X) \bigcap_{p\geqslant 1} S^p AP(\mathbb{R}^n : X)$; equipped with the sup-norm, $S^{\infty}(\mathbb{R}^n : X)$ is a Banach space. By $S^{\infty}(\mathbb{R}^n \times X : Y)$, we denote the set of all continuous functions $F : \mathbb{R}^n \times X \to Y$ such that $\sup_{\mathbf{t}\in\mathbb{R}^n, x\in K} \|F(\mathbf{t}; x)\|_Y < +\infty$ for each compact set $K \subseteq \mathbb{R}^n$ and the Bochner transform $\hat{F}(\cdot;\cdot) : \mathbb{R}^n \times X \to L^p([0,1]^n : Y)$ is Bohr $\mathcal{B}$-almost periodic for every finite exponent $p \geqslant 1$, where $\mathcal{B}$ denotes the collection of all compact subsets of X.

Consider now the following integral equation:

$$u(\mathbf{t}) = f(\mathbf{t}) + \int_{-\infty}^{t_1} \int_{-\infty}^{t_2} \cdots \int_{-\infty}^{t_n} a(\mathbf{t} - \mathbf{s}) F(\mathbf{s}; u(\mathbf{s}))\, d\mathbf{s}, \quad \mathbf{t} \in \mathbb{R}^n, \tag{321}$$

where $X = Y$ is a finite-dimensional complex Banach space, $a \in L^1((0,\infty)^n)$ and $f \in S^{\infty}(\mathbb{R}^n : X)$. The mapping $\Psi : S^{\infty}(\mathbb{R}^n : X) \to S^{\infty}(\mathbb{R}^n : X)$, given by

$$(\Psi u)(\mathbf{t}) := f(\mathbf{t}) + \int_{-\infty}^{t_1} \int_{-\infty}^{t_2} \cdots \int_{-\infty}^{t_n} a(\mathbf{t} - \mathbf{s}) F(\mathbf{s}; u(\mathbf{s}))\, d\mathbf{s}, \quad \mathbf{t} \in \mathbb{R}^n, \ \mathbf{u} \in S^{\infty}(\mathbb{R}^n : X),$$

is well-defined due to [185, Theorem 4.4] and Theorem 6.1.13(ii). Moreover, this mapping is a contraction provided that there exists a finite real constant $L > 0$ such that $\|F(\mathbf{t}; x) - F(\mathbf{t}; y)\| \leqslant L\|x - y\|$ for all $\mathbf{t} \in \mathbb{R}^n$, $x, y \in X$ and $L \int_{(0,\infty)^n} |a(\mathbf{s})|\, d\mathbf{s} < 1$ so that the integral equation (321) has a unique solution, which belongs to the space $S^{\infty}(\mathbb{R}^n : X)$. Let us finally notice that the solution $u(\cdot)$ will be almost periodic if the function $f(\cdot)$ is almost periodic.

Let us also emphasize that Theorem 6.1.13 can be reformulated for the usual convolution product

$$(\mathbf{t}, x) \mapsto \int_{\mathbb{R}^n} h(\mathbf{t} - \mathbf{s}) F(\mathbf{s}; x)\, d\mathbf{s}, \quad \mathbf{t} \in \mathbb{R}^n, \ x \in X,$$

where $h \in L^1(\mathbb{R}^n)$ and $F(\cdot;\cdot)$ is bounded, Bohr $(\mathcal{B}, \Lambda', \rho, [0,1]^n, m)$-almost periodic ($(\mathcal{B}, \Lambda', \rho, [0,1]^n, m)$-uniformly recurrent), with $\rho = T \in L(Y)$. The resulting function has the same properties as $F(\cdot;\cdot)$ and this can be applied in the analysis of the existence and uniqueness of bounded, Bohr $(\Lambda', \rho, [0,1]^n, m)$-almost periodic ($(\Lambda', \rho, [0,1]^n, m)$-uniformly recurrent) solutions of the abstract semilinear integral equation

$$u(\mathbf{t}) = f(\mathbf{t}) + \int_{\mathbb{R}^n} h(\mathbf{t} - \mathbf{s}) F(\mathbf{s}; u(\mathbf{s}))\, d\mathbf{s}, \quad \mathbf{t} \in \mathbb{R}^n,$$

where $f(\cdot)$ is bounded, Bohr $(\Lambda', \rho, [0,1]^n, m)$-almost periodic ($(\Lambda', \rho, [0,1]^n, m)$-uniformly recurrent). Regrettably, the applications to the heat equation in $\mathbb{R}^n$ are really confined because the Gaussian kernel rapidly decays at infinity and the constructed solutions are always almost periodic [447]; the applications to the abstract ill-posed Cauchy problems of first order will be given in the next section.

Finally, we would like to point out the following: Suppose that $\emptyset \neq \Lambda \subseteq \mathbb{R}^n$, $P_\varepsilon \subseteq Y^\Lambda$, the space of all functions from Λ into Y, the zero function belongs to P_ε and $\mathcal{P}_\varepsilon = (P_\varepsilon, d_\varepsilon)$ is a premetric space ($\varepsilon > 0$); if $f \in P_\varepsilon$, then we set $\|f\|_{P_\varepsilon} := d_\varepsilon(f, 0)$. It is worth noticing that the properties of pseudo-semimetric given by (306) can give us the idea to generalize the notion from [452, Definitions 2.1, 2.2, 3.1] following the method proposed for the introduction of notion in Definition 6.1.1. In general case, the quantities $\|\cdot\|_{P_\varepsilon}$ and $\|\cdot\|_{P_\eta}$ are not comparable if $\varepsilon < \eta$, as in the case of our previous consideration of m-almost periodicity, so that we can also introduce the following notion: $F(\cdot;\cdot)$ is said to be Bohr $(\mathcal{B}, \Lambda', \rho, \mathcal{P})_{\varepsilon,\eta}$-almost periodic if for every $B \in \mathcal{B}$ and ε, $\eta > 0$ there exists $l > 0$ such that for each $\mathbf{t}_0 \in \Lambda'$ there exists $\tau \in B(\mathbf{t}_0, l) \cap \Lambda'$ such that, for every $\mathbf{t} \in \Lambda$ and $x \in B$, there exists an element $y_{\mathbf{t};x} \in \rho(F(\mathbf{t}; x))$ such that $F(\cdot + \tau; x) - y_{\cdot;x} \in P_\eta$ for all $x \in B$ and

$$\sup_{x\in B} \|F(\cdot + \tau; x) - y_{\cdot;x}\|_{P_\eta} \leqslant \varepsilon.$$

It is very difficult to say anything relevant about the introduced notion if the premetric spaces under our consideration are not pseudometric spaces. Because of that, we can freely say that the validity of triangle inequality is almost inevitable in any serious research of the metrical almost periodicity and its generalizations. However, some statements like [452, Propositions 2.3, 3.14], the statements (i) and (iii) clarified on pp. 234–235 of [452] and the statements (i)–(ii) clarified on p. 246 of [452] can be formulated with general premetric spaces; on the other hand, it seems that the statements of [452, Proposition 2.6, Theorem 2.7, Proposition 3.7] cannot be properly reformulated if the triangle inequality does not hold in our framework.

We close the section by proposing the following open problem (cf. also [447, Theorem 2.1.26], [185, Theorem 2.15] and [289, Theorem 2.28] for some results obtained in this direction).

PROBLEM. Suppose that $(\mathbf{v}_1, \ldots, \mathbf{v}_n)$ is a basis of $\mathbb{R}^n$,

$$\Lambda' = \Lambda = \{\alpha_1\mathbf{v}_1 + \cdots + \alpha_n\mathbf{v}_n : \alpha_i \geqslant 0 \text{ for all } i \in \mathbb{N}_n\}$$

is a convex polyhedral in $\mathbb{R}^n$ and $F : \Lambda \to Y$ is an unbounded Bohr (Ω, m)-almost periodic function, where $\Omega = [-1, 1]^n \cap \Lambda$. Is there a (unique) Bohr (Ω, m)-almost periodic function $\tilde{F} : \mathbb{R}^n \to Y$ such that $\tilde{F}(\mathbf{t}) = F(\mathbf{t})$ for all $\mathbf{t} \in \Lambda$?

6.2 Almost periodic functions and almost automorphic functions in general measure

The main purpose of this section is to continue the study raised in the previous section by examining several new classes of multidimensional almost automorphic type functions in general measure and several new classes of multidimensional almost periodic type functions in general measure defined in a Bochner like manner [469]. In the existing literature concerning almost automorphic functions, we have not been able to find any relevant reference treating the notion of almost automorphy in the Lebesgue measure and because of that we can freely say that our results seem to be completely new even for the real-valued functions defined on $\mathbb{R}$; concerning this problem, we would like to note that the class of μ-normal functions was introduced by S. Stoiński in 1999 [728] following the approach of S. Bochner.

In Section 6.2.1, we analyze the convolution invariance of measure almost periodicity and measure almost automorphy. Some applications of our theoretical results are given in the next subsection.

We assume that $\nu : \mathbb{R}^n \to [0,\infty)$, $m' : P(\mathbb{R}^n) \to [0,\infty]$, $m'(\emptyset) = 0$ and $\emptyset \neq \Omega \subseteq \mathbb{R}^n$ is a nonempty compact set. Furthermore, for every $\varepsilon > 0$, for every two functions $F : \mathbb{R}^n \times X \to Y$ and $F^* : \mathbb{R}^n \times X \to Y^{\Omega}$ and for every sequence $(\mathbf{b}_k)$ in $\mathbb{R}^n$, we define

$$d^1_{\varepsilon,\mathbf{t},\mathbf{b}_k,x}(F,F^*) := m'(\{\mathbf{u} \in \Omega : \|F(\mathbf{t}+\mathbf{b}_k+\mathbf{u};x) - [F^*(\mathbf{t};x)](\mathbf{u})\|_Y \cdot \nu(\mathbf{u}) \geqslant \varepsilon\})$$

and

$$d^2_{\varepsilon,\mathbf{t},\mathbf{b}_k,x}(F,F^*) := m'(\{\mathbf{u} \in \Omega : \|[F^*(\mathbf{t}-\mathbf{b}_k;x)](\mathbf{u}) - F(\mathbf{t}+\mathbf{u};x)\|_Y \cdot \nu(\mathbf{u}) \geqslant \varepsilon\}).$$

We have already considered several new classes of multidimensional ρ-almost periodic type functions in general measure. The following classes of functions, introduced in a Bochner like manner, have not been analyzed so far.

Definition 6.2.1. Suppose that, for every $B \in \mathcal{B}$, $\varepsilon > 0$ and $\mathbf{b} \in \mathrm{R}$, $L(B,\varepsilon;\mathbf{b})$ denotes a nonempty collection of certain subsets of B. Suppose further that $\emptyset \neq \Lambda \subseteq \mathbb{R}^n$, $F : \Lambda \times X \to Y$ is a given function and (248) holds with the region I replaced therein with Λ. It is said that the function $F(\cdot;\cdot)$ is $(\mathrm{R},\mathcal{B},\Omega,L,m',\nu)$-multialmost periodic, respectively, strongly $(\mathrm{R},\mathcal{B},\Omega,L,m',\nu)$-multialmost periodic in the case that $\Lambda = \mathbb{R}^n$, if, for every $B \in \mathcal{B}$ and for every sequence $(\mathbf{b}_k = (b_k^1, b_k^2, \ldots, b_k^n)) \in \mathrm{R}$, there exist a subsequence $(\mathbf{b}_{k_l} = (b_{k_l}^1, b_{k_l}^2, \ldots, b_{k_l}^n))$ of $(\mathbf{b}_k)$ and a function $F^* : \Lambda \times X \to Y$ such that, for every $\varepsilon > 0$, $l \in \mathbb{N}$, $x \in B$ and $B' \in L(B,\varepsilon;\mathbf{b})$,

$$\lim_{l\to+\infty} \sup_{x\in B'} \|F(\cdot + (b_{k_l}^1,\ldots,b_{k_l}^n);x) - F^*(\cdot;x)\|_{P_\varepsilon} = 0, \tag{322}$$

respectively, (322) holds and, for every $B' \in L(B,\varepsilon;\mathbf{b})$,

$$\lim_{l\to+\infty} \sup_{x\in B'} \|F^*(\cdot - (b_{k_l}^1,\ldots,b_{k_l}^n);x) - F(\cdot;x)\|_{P_\varepsilon} = 0.$$

Definition 6.2.2. Suppose that, for every $B \in \mathcal{B}$, $\varepsilon > 0$, $\mathbf{b} \in \mathrm{R}$ and $\mathbf{t} \in \mathbb{R}^n$, $L(B,\varepsilon;\mathbf{b},\mathbf{t})$ denotes a nonempty collection of certain subsets of B. Suppose further that $F : \mathbb{R}^n \times$

$X \to Y$ is a given function and P is a collection of nonempty subsets of $\mathbb{R}^n$. Then it is said that the function $F(\cdot;\cdot)$ is $(\mathrm{R}, \mathcal{B}, \Omega, L, m', \nu)$-multialmost automorphic if for every $B \in \mathcal{B}$ and for every sequence $(\mathbf{b}_k = (b_k^1, b_k^2, \ldots, b_k^n)) \in \mathrm{R}$ there exist a subsequence $(\mathbf{b}_{k_l} = (b_{k_l}^1, b_{k_l}^2, \ldots, b_{k_l}^n))$ of $(\mathbf{b}_k)$ and a function $F^* : \mathbb{R}^n \times X \to Y^\Omega$ such that, for every $\varepsilon > 0$, $\mathbf{t} \in \mathbb{R}^n$ and $B' \in L(B, \varepsilon; \mathbf{b}, \mathbf{t})$, we have

$$\lim_{l\to+\infty} \sup_{x\in B'} d^1_{\varepsilon,\mathbf{t},\mathbf{b}_k,x}(F, F^*) = 0 \quad \text{and} \quad \lim_{l\to+\infty} \sup_{x\in B'} d^2_{\varepsilon,\mathbf{t},\mathbf{b}_k,x}(F, F^*) = 0.$$

Furthermore, we say that $F(\cdot;\cdot)$ is $(\mathrm{R}, \mathcal{B}, \Omega, L, P, m', \nu)$-multialmost automorphic if, additionally, for every set $I \in P$, we have

$$\lim_{l\to+\infty} \sup_{\mathbf{t}\in I; x\in B'} d^1_{\varepsilon,\mathbf{t},\mathbf{b}_k,x}(F, F^*) = 0 \quad \text{and} \quad \lim_{l\to+\infty} \sup_{\mathbf{t}\in I; x\in B'} d^2_{\varepsilon,\mathbf{t},\mathbf{b}_k,x}(F, F^*) = 0.$$

It would be very difficult to clarify a satisfactory Bochner type criterion for the notion introduced in Definition 6.2.1 and Definition 6.2.2. The following results are straightforward and the proofs are therefore omitted.

Proposition 6.2.3. (i) *Let $F : \mathbb{R}^n \times X \to Y$ be a compactly $(\mathrm{R}, \mathcal{B})$-multialmost automorphic function and let, for every $B \in \mathcal{B}$, $\varepsilon > 0$, $\mathbf{b} \in \mathrm{R}$ and $\mathbf{t} \in \mathbb{R}^n$, $L(B, \varepsilon; \mathbf{b}, \mathbf{t})$ denote the collection of all singletons $\{x\}$ when $x \in B$. Then $F(\cdot;\cdot)$ is $(\mathrm{R}, \mathcal{B}, \Omega, L, P, m', \nu)$-multialmost automorphic, where P denotes the collection of all compact subsets of $\mathbb{R}^n$.*

(ii) *Let $F : \mathbb{R}^n \times X \to Y$ be an $(\mathrm{R}, \mathcal{B}, P_{\mathcal{B},\mathrm{R}})$-multialmost automorphic function, where, for every $B \in \mathcal{B}$ and $(\mathbf{b}_k) \in \mathrm{R}$, $P_{B;\mathbf{b}_k}$ denotes the collection of all sets of the form $K \times P'_{B;\mathbf{b}_k}$ with K being a compact subset of $\mathbb{R}^n$ and $P'_{B;\mathbf{b}_k}$ a certain collection of subsets of B. Let for every $B \in \mathcal{B}$, $\varepsilon > 0$, $\mathbf{b} \in \mathrm{R}$ and $\mathbf{t} \in \mathbb{R}^n$, $L(B, \varepsilon; \mathbf{b}, \mathbf{t})$ be equal to $P'_{B;\mathbf{b}_k}$. Then $F(\cdot;\cdot)$ is $(\mathrm{R}, \mathcal{B}, \Omega, L, m', \nu)$-multialmost automorphic.*

(iii) *Suppose that $F : \mathbb{R}^n \times X \to Y$ is Stepanov $(\Omega, \mathrm{R}, \mathcal{B}, Z^{\mathcal{P}}, W_{\mathcal{B},\mathrm{R}})$-multi-almost automorphic function, where $Z = C_{b,\nu}(\Omega : Y)$ with some function $\nu : \Omega \to [0, \infty)$ and, for every $B \in \mathcal{B}$, $x \in B$ and $(\mathbf{b}_k) \in \mathrm{R}$, $W_{B;(\mathbf{b}_k)}(x)$ denotes the collection of all compact subsets in $\mathbb{R}^n$. Then $F(\cdot;\cdot)$ is $(\mathrm{R}, \mathcal{B}, \Omega, L, m', \nu)$-multialmost automorphic, where for every $B \in \mathcal{B}$, $\varepsilon > 0$, $\mathbf{b} \in \mathrm{R}$ and $\mathbf{t} \in \mathbb{R}^n$, $L(B, \varepsilon; \mathbf{b}, \mathbf{t})$ denotes the collection of all singletons $\{x\}$ when $x \in B$.*

(iv) *Suppose that $F : \mathbb{R}^n \times X \to Y$ is Stepanov $(\Omega, \mathrm{R}, \mathcal{B}, Z^{\mathcal{P}}, P_{\mathcal{B},\mathrm{R}})$-multi-almost automorphic function, where $Z = C_{b,\nu}(\Omega : Y)$ with some function $\nu : \Omega \to [0, \infty)$ and, for every $B \in \mathcal{B}$ and $(\mathbf{b}_k) \in \mathrm{R}$, $P_{B;(\mathbf{b}_k)}$ denotes the collection of all sets of the form $K \times P'_{B;\mathbf{b}_k}$ with K being a compact subset of $\mathbb{R}^n$ and $P'_{B;\mathbf{b}_k}$ a certain collection of subsets of B. Then $F(\cdot;\cdot)$ is $(\mathrm{R}, \mathcal{B}, \Omega, L, m', \nu)$-multialmost automorphic, where for every $B \in \mathcal{B}$, $\varepsilon > 0$, $\mathbf{b} \in \mathrm{R}$ and $\mathbf{t} \in \mathbb{R}^n$, $L(B, \varepsilon; \mathbf{b}, \mathbf{t})$ is equal to $P'_{B;\mathbf{b}_k}$.*

Moreover, we can simply prove the following results (the only nontrivial thing that should be explained is the essential boundedness of function $F^*(\mathbf{t}; x) : \Omega \to Y$ in part (iii); this can be deduced as in the proof of Theorem 6.2.6).

Proposition 6.2.4. (i) *Suppose that, for every $B \in \mathcal{B}$, $\varepsilon > 0$ and $\mathbf{b} \in \mathrm{R}$, $L(B, \varepsilon; \mathbf{b})$ denotes a nonempty collection of certain subsets of B. If $F : \mathbb{R}^n \times X \to Y$ is a strongly $(\mathrm{R}, \mathcal{B}, \Omega, L, m', \nu)$-multialmost periodic function and there exists a function $\varphi : \mathbb{R}^n \to (0, \infty)$ such that $\nu(\mathbf{u}) \leqslant \nu(\mathbf{t} + \mathbf{u})\varphi(\mathbf{t})$ for all $\mathbf{u} \in \Omega$ and $\mathbf{t} \in \mathbb{R}^n$, then $F(\cdot; \cdot)$ is $(\mathrm{R}, \mathcal{B}, \Omega, L, m', \nu)$-multialmost automorphic with $L(B, \varepsilon; \mathbf{b}, \mathbf{t}) = L(B, \varepsilon; \mathbf{b})$ for all $B \in \mathcal{B}$, $\varepsilon > 0$, $\mathbf{b} \in \mathrm{R}$ and $\mathbf{t} \in \mathbb{R}^n$, provided that the assumption $A \subseteq B \subseteq \mathbb{R}^n$ implies $m'(A) \leqslant m'(B)$.*

(ii) *Suppose that the function $F : \mathbb{R}^n \times X \to Y$ is Stepanov $(\Omega, \mathrm{R}, \mathcal{B}, Z^{\mathcal{P}})$-multialmost automorphic, where $Z = L^p_{m',\nu}(\Omega : Y)$ with some $p > 0$, function $\nu : \Omega \to [0, \infty)$ and measure $m'(\cdot)$ on Ω. Then $F(\cdot; \cdot)$ is $(\mathrm{R}, \mathcal{B}, \Omega, L, m', \nu)$-multialmost automorphic, where for every $B \in \mathcal{B}$, $\varepsilon > 0$, $\mathbf{b} \in \mathrm{R}$ and $\mathbf{t} \in \mathbb{R}^n$, $L(B, \varepsilon; \mathbf{b}, \mathbf{t})$ denotes the collection of all singletons $\{x\}$ when $x \in B$.*

(iii) *Suppose that the function $F(\cdot; \cdot)$ is $(\mathrm{R}, \mathcal{B}, \Omega, L, m', \nu)$-multialmost automorphic, where for every $B \in \mathcal{B}$, $\varepsilon > 0$, $\mathbf{b} \in \mathrm{R}$ and $\mathbf{t} \in \mathbb{R}^n$, $L(B, \varepsilon; \mathbf{b}, \mathbf{t})$ denotes the collection of all singletons $\{x\}$ when $x \in B$. Suppose further that $\nu(\cdot)$ is bounded, $m'(\cdot)$ is a finite measure on Ω and $Z = L^p_{m',\nu}(\Omega : Y)$. If $\sup_{\mathbf{t} \in \mathbb{R}^n} \|F(\mathbf{t}; x)\|_Y < +\infty$ for all $x \in X$ and the function $F^*(\mathbf{t}; x) : \Omega \to Y$ from* Definition 6.2.2 *is essentially bounded for every $\mathbf{t} \in \mathbb{R}^n$ and $x \in X$ (this always holds provided that there exists $c > 0$ such that $\nu(\mathbf{u}) \geqslant c > 0$ for a. e. $\mathbf{u} \in \Omega$), then $F(\cdot; \cdot)$ is Stepanov $(\Omega, \mathrm{R}, \mathcal{B}, Z^{\mathcal{P}})$-multi-almost automorphic.*

The interested reader may try to prove some analogues of Proposition 6.1.7, Theorem 6.1.9 and the conclusion established in [184, Example, p. 819] for almost automorphic type functions in general measure. Before proceeding to the next subsection, we will present the following example.

Example 6.2.5. Suppose that $a > 0$; then a unique regular solution of the wave equation $u_{tt} = a^2 u_{xx}$ in domain $\{(x, t) : x \in \mathbb{R},\ t > 0\}$, equipped with the initial conditions $u(x, 0) = f(x) \in C^2(\mathbb{R})$ and $u_t(x, 0) = g(x) \in C^1(\mathbb{R})$, is given by the d'Alembert formula already considered in Example 6.1.8. If we assume that the function $x \mapsto (f(x), g^{[1]}(x))$, $x \in \mathbb{R}$ is $(\mathrm{R}, [0, 1], m)$-almost automorphic, where $g^{[1]}(\cdot) \equiv \int_0^{\cdot} g(s)\, ds$, then the solution $u(x, t)$ can be extended to the whole real line in the time variable and this solution is $(\mathrm{R}_2, [0, 1]^2, m_2)$-almost automorphic in $(x, t) \in \mathbb{R}^2$; here, $m(\cdot)$ and R denote the Lebesgue measure in $\mathbb{R}$ and the collection of all sequences in $\mathbb{R}$, respectively, while $m_2(\cdot)$ and R_2 denote the Lebesgue measure in $\mathbb{R}^2$ and the collection of all sequences in $\mathbb{R}^2$, respectively. This can be verified as in the above mentioned example.

6.2.1 Convolution invariance of measure almost periodicity and measure almost automorphy

In this subsection, we will first state the following result.

Theorem 6.2.6. *Let $(R(\mathbf{t}))_{\mathbf{t}>\mathbf{0}} \subseteq L(X,Y)$ be a strongly continuous operator family such that $\int_{(0,\infty)^n} \|R(\mathbf{t})\|\, d\mathbf{t} < \infty$. If $f : \mathbb{R}^n \to X$ is bounded and $(\mathrm{R}, [0,1]^n, m)$-multialmost periodic, respectively, strongly $(\mathrm{R}, [0,1]^n, m)$-multialmost periodic, then the function $F : \mathbb{R}^n \to Y$, given by*

$$F(\mathbf{t}) := \int_{-\infty}^{t_1}\int_{-\infty}^{t_2}\cdots\int_{-\infty}^{t_n} R(\mathbf{t}-\mathbf{s})f(\mathbf{s})\, d\mathbf{s}, \quad \mathbf{t} \in \mathbb{R}^n,$$

is well-defined, bounded, continuous and $(\mathrm{R}, [0,1]^n, m)$-multialmost periodic, respectively, strongly $(\mathrm{R}, [0,1]^n, m)$-multialmost periodic.

Proof. We will provide the main details of the proof for (i), considering only the class of one-dimensional, bounded $(\mathrm{R}, [0,1], m)$-multialmost periodic functions. It is clear that the function $F(\cdot)$ is well-defined and bounded; the continuity is a simple consequence of the dominated convergence theorem. Let $(b_k) \in \mathrm{R}$. Then we know that there exist a subsequence (b_{k_l}) of (b_k) and a function $f^* : \mathbb{R} \to Y^{[0,1]}$ such that, for every $\varepsilon > 0$,

$$\lim_{l\to+\infty} \sup_{t\in\mathbb{R}} m(\{u \in [0,1] : \|f(t+u+b_{k_l}) - [f^*(t)](u)\|_Y \geqslant \varepsilon\}) = 0. \tag{323}$$

Using (323), we can construct a strictly increasing sequence (l_s) of positive integers such that

$$m(\{u \in [0,1] : \|f(t+u+b_{k_l}) - [f^*(t)](u)\|_Y < \varepsilon\}) \geqslant 1 - (\varepsilon/2^l), \quad l \geqslant l_s,\ t \in \mathbb{R}.$$

Set $A := \bigcup_{t\in\mathbb{R}, s\in\mathbb{N}}\{u \in [0,1] : \|f(t+u+b_{k_{l_s}}) - [f^*(t)](u)\|_Y < \varepsilon\}$. Then we have $m(A) = 1$ and $\|[f^*(t)](u)\|_Y \leqslant \|f\|_\infty + \varepsilon$, $t \in \mathbb{R}$, $u \in A$. Therefore, we can define $[F^*(t)](u) := \int_{-\infty}^{t} R(t-s)[f^*(s)](u)\, ds$, $t \in \mathbb{R}$, $u \in A$ and $[F^*(t)](u) := 0$, $t \in \mathbb{R}$, $u \in [0,1] \setminus A$. It is simple to show that (323) holds with $f(\cdot)$ and $f^*(\cdot)$ replaced therein with $F(\cdot)$ and $F^*(\cdot)$. □

It is also not clear whether the resulting function $F(\cdot)$ has to be R-almost periodic in the usual sense (cf. also [447, Theorem 6.1.54]). Now we will state the following result; now it is not clear why the resulting function $F(\cdot)$ has to be R-almost automorphic in the usual sense.

Theorem 6.2.7. *Let $(R(\mathbf{t}))_{\mathbf{t}>\mathbf{0}} \subseteq L(X,Y)$ be a strongly continuous operator family such that $\int_{(0,\infty)^n} \|R(\mathbf{t})\|\, d\mathbf{t} < \infty$ and let P be the collection of all compact subsets of $\mathbb{R}^n$. If $f : \mathbb{R}^n \to X$ is bounded and $(\mathrm{R}, [0,1]^n, P, m)$-multialmost automorphic, then the function $F : \mathbb{R}^n \to Y$, given by (316), is well-defined, bounded, continuous and $(\mathrm{R}, [0,1]^n, P, m)$-multialmost automorphic.*

Proof. The proof is almost the same as the proof of Theorem 6.2.6. The only difference is the existence of a finite real number $M > 0$ and a Lebesgue measurable set $A \subseteq [0,1]$ such that $m(A) = 1$ and $\|[f^*(t)](u)\|_Y \leqslant M$, $t \in \mathbb{R}$, $u \in A$. Arguing as in the proof of

Theorem 6.2.6, for each integer $k \in \mathbb{Z}$ we get the existence of a Lebesgue measurable set $A_k \subseteq [0,1]$ such that $m(A_k) = 1$ and $\|[f^*(t)](u)\|_Y \leqslant \|f\|_\infty + \varepsilon$, $t \in [k, k+1]$, $u \in A_k$. This simply implies the required result by setting $A := \bigcap_{k\in\mathbb{Z}} A_k$. □

Remark 6.2.8. As before, the range of $F(\cdot)$ is relatively compact provided that the range of $f(\cdot)$ is relatively compact, in both statements, Theorem 6.2.6 and Theorem 6.2.7.

In connection with the assumption made in the formulation of Theorem 6.2.7, we would like to present the following result.

Proposition 6.2.9. *Suppose that $f \in L^\infty(\mathbb{R}^n : Y)$. Then the following statements are equivalent:*

(i) *$f(\cdot)$ is $(\mathrm{R}, [0,1]^n, m)$-multialmost automorphic.*
(ii) *$f(\cdot)$ is $(\mathrm{R}, [0,1]^n, P, m)$-multialmost automorphic, where P denotes the collection of all compact subsets of $\mathbb{R}^n$.*
(iii) *For every (some) $p > 0$, $f(\cdot)$ is Stepanov $([0,1]^n, \mathrm{R}, Z^{\mathcal{P}})$-multialmost automorphic, where $Z = L^p([0,1]^n : Y)$.*

Proof. The equivalence of (i) and (iii) follows from Proposition 6.2.4(ii)–(iii). It is clear that (ii) implies (i) and we only need to prove that (iii) implies (ii). We will do that only in the one-dimensional setting because the proof is quite similar in the higher-dimensional setting. Let $(b_k) \in \mathrm{R}$. Then there exist a subsequence (b_{k_l}) of (b_k) and a locally integrable function $f^* : \mathbb{R} \to Y$ such that, for every $\varepsilon > 0$ and $t \in \mathbb{R}$, we have $(p = 1)$: $\lim_{l\to+\infty} \int_t^{t+1} \|f(s + b_{k_l}) - f^*(s)\|\, ds = 0$ and $\lim_{l\to+\infty} \int_t^{t+1} \|f^*(s - b_{k_l}) - f(s)\|\, ds = 0$. This implies that for each $\delta > 0$ there exists $l_0(\delta, t) \in \mathbb{N}$ such that for each $l \geqslant l_0(\delta, t)$ we have $\int_t^{t+1} \|f(s + b_{k_l}) - f^*(s)\|^p\, ds \leqslant \delta/2$ and $\int_t^{t+1} \|f^*(s - b_{k_l}) - f(s)\|\, ds < \delta/2$. Using the absolute continuity of the Lebesgue integral, we can find $\sigma(\delta, t) > 0$ such that the assumption $|v| < \sigma(\delta, t)$ implies $\int_{t+v}^{t+1+v} \|f(s + b_{k_l}) - f^*(s)\|\, ds \leqslant \delta$ and $\int_{t+v}^{t+1+v} \|f^*(s - b_{k_l}) - f(s)\|\, ds < \delta$. Suppose now that K is a compact set. Then we can find a finite subset $\{t_1, \dots, t_s\}$ of K such that $K \subseteq \bigcup_{1\leqslant i\leqslant s} L(t_i, \sigma(\delta, t_i))$. Plugging $l_0 := \max\{l_0(\delta, t_1), \dots, l_0(\delta, t_s)\}$, we easily get $\int_t^{t+1} \|f(s + b_{k_l}) - f^*(s)\|\, ds \leqslant \delta$ and $\int_t^{t+1} \|f^*(s - b_{k_l}) - f(s)\|\, ds < \delta$ for all $t \in K$ and $l \geqslant l_0$. This simply completes the proof. □

Furthermore, we have the following.

Theorem 6.2.10. *Suppose that $F : \mathbb{R}^n \to Y$ is uniformly continuous and $(\mathrm{R}, [0,1]^n, m)$-multialmost automorphic with R being the collection of all sequences in $\mathbb{R}^n$. Then $F(\cdot)$ is compactly almost automorphic.*

Proof. Since any uniformly continuous, Stepanov p-almost automorphic function $F : \mathbb{R}^n \to Y$ is compactly almost automorphic [447], Proposition 6.2.9 yields that we only need to show that $F \in L^\infty(\mathbb{R}^n : Y)$. We will prove that $R(F)$ is relatively compact in Y; so, let (b_k) be any given sequence in $\mathbb{R}^n$. Then there exist a subsequence (b_{k_l}) of (b_k) and a function $F^* : \mathbb{R}^n \to Y^{[0,1]^n}$ such that, for every $\varepsilon > 0$,

$$\lim_{l\to+\infty} m(\{\mathbf{u} \in [0,1]^n : \|f(\mathbf{u}+b_{k_l}) - [f^*(0)](\mathbf{u})\|_Y \geqslant \varepsilon\}) = 0. \tag{324}$$

Using (324), we can construct a Lebesgue measurable subset A of $[0,1]^n$ such that $\lim_{l\to+\infty} F(\mathbf{u} + b_{k_l}) = [F^*(0)](\mathbf{u})$ for all $\mathbf{u} \in A$. Then it is very simple to prove that the mapping $\mathbf{u} \mapsto [F^*(0)](\mathbf{u})$, $\mathbf{u} \in [0,1]^n$ is uniformly continuous as well as that

$$[F^*(0)](0) = \lim_{m\to+\infty} [F^*(0)](\mathbf{u}_m) = \lim_{m\to+\infty} \lim_{l\to+\infty} F(\mathbf{u}_m + b_{k_l}) = \lim_{l\to+\infty} F(b_{k_l}),$$

where $(\mathbf{u}_m)$ is any sequence in A tending to zero. This completes the proof. □

The following slightly generalizes the conclusion from our previous section: The function $f : \mathbb{R} \to \mathbb{R}$, given by (277), is uniformly continuous and unbounded but not Stepanov-1-almost automorphic [447]. Due to Proposition 6.2.10, $f(\cdot)$ cannot be m-almost automorphic, i. e., (R, $[0,1], m$)-almost automorphic with R being the collection of all sequences in $\mathbb{R}$.

6.2.2 Some applications to Volterra integrodifferential equations

In this subsection, we will provide certain applications of the obtained results.

1. As mentioned before, there exists a large class of kernels $a \in L^1((0,\infty)^n)$, which are not locally q-integrable at zero for any exponent $q \in (1,\infty]$. If the function $f(\cdot)$ is bounded and Stepanov-p-almost automorphic for some $p \in [1,\infty)$, then we cannot apply [447, Proposition 3.5.3] in order to see that the resulting function $F(\cdot)$ will be compactly almost automorphic. On the other hand, the compact almost automorphy of $F(\cdot)$ can be proved with the help of Theorem 6.2.6; cf. also Remark 6.1.15.

2. It is worth noting that the statements of Theorem 6.2.7 and [469, Theorem 4.1(i)] can be simply reformulated for the usual convolution product

$$(h * F)(\mathbf{t}) := \int_{\mathbb{R}^n} h(\mathbf{t}-\mathbf{s})F(\mathbf{s})\, d\mathbf{s}, \quad \mathbf{t} \in \mathbb{R}^n,$$

where $h \in L^1(\mathbb{R}^n)$. Keeping this in mind, we can provide the following important application to the abstract ill-posed Cauchy problems of first order (cf. [184, Subsection 3.3] for more details and applications of similar type).

Example 6.2.11. Suppose that $k \in \mathbb{N}$, $a_\alpha \in \mathbb{C}$, $0 \leqslant |\alpha| \leqslant k$, $a_\alpha \neq 0$ for some α with $|\alpha| = k$, $P(x) = \sum_{|\alpha|\leqslant k} a_\alpha i^{|\alpha|} x^\alpha$, $x \in \mathbb{R}^n$, $P(\cdot)$ is an elliptic polynomial, i. e., there exist $C > 0$ and $L > 0$ such that $|P(x)| \geqslant C|x|^k$, $|x| \geqslant L$, $\omega := \sup_{x\in\mathbb{R}^n} \mathrm{Re}(P(x)) < \infty$ and $X := C_b(\mathbb{R}^n)$. Set

$$P(D) := \sum_{|\alpha|\leqslant k} a_\alpha f^{(\alpha)} \quad \text{and} \quad Dom(P(D)) := \{f \in X : P(D)f \in X \text{ distributionally}\}.$$

Then the operator $P(D)$ generates an exponentially bounded r-times integrated semigroup $(S_r(t))_{t\geqslant 0}$ in X for any $r > n/2$; moreover, we know that for each $t \geqslant 0$ there exists a function $f_t \in L^1(\mathbb{R}^n)$ such that

$$[S_r(t)f](x) := (f_t * f)(x), \quad x \in \mathbb{R}^n, f \in X.$$

Fix now a number $t_0 \geqslant 0$ and assume that the function $f \in X$ is (strongly) $(\mathrm{R}, [0,1]^n, m)$-almost periodic. By the foregoing, the function $x \mapsto [S_r(t_0)f](x)$, $x \in \mathbb{R}^n$ will be bounded, continuous and (strongly) $(\mathrm{R}, [0,1]^n, m)$-almost periodic. Concerning the corresponding abstract first-order Cauchy problem, the above implies that there exists a unique X-valued continuous function $t \mapsto u(t)$, $t \geqslant 0$ such that $\int_0^t u(s)\,ds \in Dom(P(D))$ for every $t \geqslant 0$ and

$$u(t) = P(D)\int_0^t u(s)\,ds - \frac{t^r}{\Gamma(r+1)}f, \quad t \geqslant 0; \tag{325}$$

furthermore, the solution $t \mapsto u(t)$, $t \geqslant 0$ of (325) has the property that its orbit consists of bounded, continuous and (strongly) $(\mathrm{R}, [0,1]^n, m)$-almost periodic functions.

3. We continue with the observation that Theorem 6.2.6 and Theorem 6.2.7 can be successfully applied in the qualitative analysis of solutions for a large class of the abstract Volterra integrodifferential equations whose solution operator family is only norm integrable. See the previous section for more details.

4. Keeping in mind Proposition 6.2.9, Theorem 6.2.7 and the fact that the composition result established in [444, Theorem 3.2.3] can be formulated in the multi-dimensional setting (with $p = 1$, as well, if the function $f(\cdot;\cdot)$ is Lipschitz in the second variable), we can simply consider the existence and uniqueness of a unique solution of the semilinear integral equation (here, $X = Y$ is finite-dimensional)

$$u(\mathbf{t}) = f(\mathbf{t}) + \int_{-\infty}^{t_1}\int_{-\infty}^{t_2}\cdots\int_{-\infty}^{t_n} a(\mathbf{t}-\mathbf{s})F(\mathbf{s}; u(\mathbf{s}))\,ds, \quad \mathbf{t} \in \mathbb{R}^n,$$

which belongs to the space $S_A^\infty(\mathbb{R}^n : X) := C_b(\mathbb{R}^n : X) \bigcap_{p\geqslant 1} S^p AA(\mathbb{R}^n : X)$; equipped with the sup-norm, $S_A^\infty(\mathbb{R}^n : X)$ is a Banach space. We assume here that $f \in S_A^\infty(\mathbb{R}^n : X)$, $a \in L^1((0,\infty)^n)$, $F : \mathbb{R}^n \times X \to Y$ is a continuous function such that $\sup_{\mathbf{t}\in\mathbb{R}^n, x\in K} \|F(\mathbf{t}; x)\|_Y < +\infty$ for each compact set $K \subseteq \mathbb{R}^n$; here, $S^p AA(\mathbb{R}^n : X)$ denotes the space of all Stepanov-p-almost automorphic functions ($p \geqslant 1$).

6.3 Measure theoretical approach to generalized almost periodicity

In this section, we will further generalize the class of m-almost periodic functions and the class of (equi-)Weyl-p-almost periodic functions [Doss-p-almost periodic functions] by considering a new class of (equi-)Weyl-p-almost periodic functions [Doss-p-almost periodic functions] in general measure, with a general exponent $p > 0$; at this place, it should be also worthwhile to mention that A. Michalowicz and S. Stoiński have extended the class of Levitan N-almost periodic functions in a similar manner, by considering the class of Levitan N-almost periodic functions in the Lebesgue measure [589].

The section is structurally organized as follows. First of all, we recall the basic definitions and results about multidimensional Weyl ρ-almost periodic type functions in general metric. In Section 6.3.1, we introduce and analyze the class of (equi-)Weyl $(\mathbb{F}, \mathcal{B}, \Lambda', \rho, \Omega, m', \nu)$-almost periodic functions; see Definition 6.3.4. Our first structural result is given in Proposition 6.3.6; in particular, this results shows that any (equi)-Weyl-p-almost periodic function $F : \mathbb{R}^n \to X$ is (equi)-Weyl-m-almost periodic as well as that any bounded, (equi)-Weyl-m-almost periodic function $F : \mathbb{R}^n \to X$ is (equi)-Weyl-p-almost periodic ($p \geqslant 1$). Section 6.3.2 investigates the multidimensional Doss ρ-almost periodic type functions in general measure. The class of Doss $(\mathrm{F}, \mathcal{B}, \Lambda', \rho, m', \nu)$-almost periodic functions is introduced in Definition 6.3.8 and our first structural result is given in Proposition 6.3.9. At the end of section, we provide several useful remarks, observations and perspectives for the further explorations of generalized almost periodicity and generalized almost automorphy in measure.

Generalized ρ-almost periodic type functions and their metrical generalizations

In this part, we recall the basic definitions and facts about generalized ρ-almost periodic type functions and their metrical generalizations.

Assume that the following conditions hold true:

(WM1-1): $\emptyset \neq \Lambda \subseteq \mathbb{R}^n$, $\emptyset \neq \Lambda' \subseteq \mathbb{R}^n$, $\emptyset \neq \Omega \subseteq \mathbb{R}^n$ is a Lebesgue measurable set such that $m(\Omega) > 0$, $p \in \mathcal{P}(\Lambda)$, the collection of all Lebesgue measurable functions from Λ into $[1, +\infty]$, $\Lambda' + \Lambda \subseteq \Lambda$, $\Lambda + l\Omega \subseteq \Lambda$ for all $l > 0$, $\phi : [0, \infty) \to [0, \infty)$ and $\mathbb{F} : (0, \infty) \times \Lambda \to (0, \infty)$.

(WM1-2): For every $\mathbf{t} \in \Lambda$ and $l > 0$, $\mathcal{P}_{\mathbf{t},l} = (P_{\mathbf{t},l}, d_{\mathbf{t},l})$ is a pseudometric space of functions from $\mathbb{C}^{\mathbf{t}+l\Omega}$ containing the zero function. Define $\|f\|_{P_{\mathbf{t},l}} := d_{\mathbf{t},l}(f, 0)$ for all $f \in P_{\mathbf{t},l}$; $\mathcal{P} = (P, d)$ is a pseudometric space of functions from $\mathbb{C}^{\Lambda}$ containing the zero function and $\|f\|_P := d(f, 0)$ for all $f \in P$. The argument from Λ will be denoted by $\cdot\cdot$ and the argument from $\mathbf{t} + l\Omega$ will be denoted by $\cdot$.

We recall the following notion [448].

Definition 6.3.1. (i) By $e - W^{(\phi,\mathbb{F},\rho,\mathcal{P}_{\mathbf{t},l},\mathcal{P})_1}_{\Omega,\Lambda',\mathcal{B}}(\Lambda \times X : Y)$ we denote the set consisting of all functions $F : \Lambda \times X \to Y$ such that, for every $\varepsilon > 0$ and $B \in \mathcal{B}$, there exist two finite real numbers $l > 0$ and $L > 0$ such that for each $\mathbf{t}_0 \in \Lambda'$ there exists $\tau \in B(\mathbf{t}_0, L) \cap \Lambda'$ such that, for every $x \in B$, the mapping $\mathbf{u} \mapsto G_x(\mathbf{u}) \in \rho(F(\mathbf{u}; x))$, $\mathbf{u} \in \bigcup_{l>0;\mathbf{t}\in\Lambda}(\mathbf{t} + l\Omega)$ is well-defined, and

$$\sup_{x\in B}\big\|\mathbb{F}(l, \cdot\cdot)\phi(\|F(\tau + \cdot; x) - G_x(\cdot)\|_{P_{\cdot\cdot,l}})\big\|_P < \varepsilon.$$

(ii) By $W^{(\phi,\mathbb{F},\rho,\mathcal{P}_{t,l},\mathcal{P})_1}_{\Omega,\Lambda',\mathcal{B}}(\Lambda\times X : Y)$, we denote the set consisting of all functions $F : \Lambda\times X \to Y$ such that, for every $\varepsilon > 0$ and $B \in \mathcal{B}$, there exists a finite real number $L > 0$ such that for each $\mathbf{t}_0 \in \Lambda'$ there exists $\tau \in B(\mathbf{t}_0, L) \cap \Lambda'$ such that, for every $x \in B$, the mapping $\mathbf{u} \mapsto G_x(\mathbf{u}) \in \rho(F(\mathbf{u}; x))$, $\mathbf{u} \in \bigcup_{l>0;\mathbf{t}\in\Lambda}(\mathbf{t} + l\Omega)$ is well-defined, and

$$\limsup_{l\to+\infty} \sup_{x\in B} \|\mathbb{F}(l, \cdot\cdot)\phi(\|F(\tau + \cdot; x) - G_x(\cdot)\|_{P_{\cdot\cdot,l}})\|_P < \varepsilon.$$

The multidimensional Weyl ρ-almost periodic functions introduced in the first three definitions of [456, Section 3] are special cases of the above introduced classes of functions, with $P_{\mathbf{t},l} = L^{p(\cdot)}(\mathbf{t}+l\Omega : \mathbb{C})$, the metric $d_{\mathbf{t},l}$ induced by the norm of this Banach space ($\mathbf{t} \in \Lambda$, $l > 0$), $P = L^{\infty}(\Lambda : \mathbb{C})$ and the metric d induced by the norm of this Banach space. If $P_{\mathbf{t},l} = L^{p(\cdot)}_{\nu}(\mathbf{t} + l\Omega : \mathbb{C})$ ($\mathbf{t} \in \Lambda$, $l > 0$) and $P = L^{\infty}(\Lambda : \mathbb{C})$, then the corresponding space will be denoted by $(e-)W^{p(\mathbf{u}),\phi,\mathbb{F},\nu}_{\Omega,\Lambda',\mathcal{B}}(\Lambda \times X : Y)$; similarly, if $p \in (0,1)$, $P_{\mathbf{t},l} = L^{p}_{\nu}(\mathbf{t} + l\Omega : \mathbb{C})$ ($\mathbf{t} \in \Lambda$, $l > 0$) and $P = L^{\infty}(\Lambda : \mathbb{C})$, then the corresponding space will be denoted by $(e-)W^{p,\phi,\mathbb{F},\nu}_{\Omega,\Lambda',\mathcal{B}}(\Lambda \times X : Y)$. In the sequel, we will use the same notion and notation as in [447], where we have considered various classes of generalized almost periodic functions in the Lebesgue spaces with variable exponent $L^{p(x)}$.

We also need the following notion (see [448, Definition 3.2.1(ii)(b)] with $\phi(x) \equiv x$, and [448, Definition 6.2.11(ii)(b)]; here and hereafter, $\Lambda_l := \Lambda \cap B(0, l)$).

Definition 6.3.2. Suppose that $\emptyset \neq \Lambda \subseteq \mathbb{R}^n$, $\emptyset \neq \Lambda' \subseteq \mathbb{R}^n$, $\Lambda + \Lambda' \subseteq \Lambda$, $\nu : \Lambda \to [0, +\infty)$, $p \in \mathcal{P}(\Lambda)$ $[0 < p < 1]$ and the function $F : \Lambda\times X \to Y$ satisfies that $\|F(\cdot + \tau; x) - y_{\cdot;x}\| \cdot \nu(\cdot) \in L^{p(\cdot)}(\Lambda_l)$ $[L^p(\Lambda_l)]$ for all $l > 0$, $x \in X$, $\tau \in \Lambda'$ and $y_{\cdot;x} \in \rho(F(\cdot; x))$. Then it is said that $F(\cdot;\cdot)$ is Doss-$(p(\cdot), \mathrm{F}, \mathcal{B}, \Lambda', \rho, \nu)$-almost periodic [Doss-$(p, \mathrm{F}, \mathcal{B}, \Lambda', \rho, \nu)$-almost periodic] if, for every $B \in \mathcal{B}$ and $\varepsilon > 0$, there exists $L > 0$ such that for each $\mathbf{t}_0 \in \Lambda'$ there exists a point $\tau \in B(\mathbf{t}_0, L) \cap \Lambda'$ such that, for every $l > 0$, $x \in B$ and $\cdot \in \Lambda_l$, we have the existence of an element $y_{\cdot;x} \in \rho(F(\cdot; x))$ such that

$$\limsup_{l\to+\infty} \mathrm{F}(l) \sup_{x\in B} [\|F(\cdot + \tau; x) - y_{\cdot;x}\|_Y \cdot \nu(\cdot)]_P < \varepsilon,$$

where $P = L^{p(\cdot)}(\Lambda_l)$ $[P = L^p(\Lambda_l)]$.

The usual class of Doss-p-almost periodic functions is obtained by plugging $\Lambda = \Lambda' = \mathbb{R}^n$, $\rho = \mathrm{I}$, $\nu(\cdot) \equiv 1$, $\Omega = [-1,1]^n$ and $\mathrm{F}(l) \equiv l^{-n/p}$ ($p > 0$). A very simple argumentation shows that a p-locally integrable function $F : \mathbb{R}^n \to Y$ is Doss-p-almost periodic if and only if, for every $\varepsilon > 0$, there exists $L > 0$ such that for each $\mathbf{t}_0 \in \mathbb{R}^n$ there exists a point $\tau \in B(\mathbf{t}_0, L)$ such that, for every $\mathbf{t} \in \mathbb{R}^n$, there exists $l_{\mathbf{t}} > 0$ such that, for every $l \geqslant l_{\mathbf{t}}$, we have

$$\left[(2l)^{-n} \int_{\mathbf{t}+l[-1,1]^n} \|F(\mathbf{s} + \tau) - F(\mathbf{s})\|_Y^p \, d\mathbf{s}\right] \leqslant \varepsilon.$$

6.3.1 Multidimensional Weyl ρ-almost periodic type functions in general measure

We will always assume here that $\emptyset \neq \Lambda \subseteq \mathbb{R}^n$, $\nu : \Lambda \to [0,\infty)$, $m' : P(\mathbb{R}^n) \to [0,\infty]$, $m'(\emptyset) = 0$, $\emptyset \neq \Omega \subseteq \mathbb{R}^n$ is a nonempty compact set, $\mathbb{F} : (0,\infty)\times\Lambda \to (0,\infty)$ and $\Lambda + l\Omega \subseteq \Lambda$ for all $l > 0$. For every $\varepsilon > 0$, $l > 0$ and for every two functions $f : \Lambda \to Y$ and $g : \Lambda \to Y$, we define

$$d_{\varepsilon,l,\mathbb{F},\nu}(f,g) := \sup_{\mathbf{t}\in\Lambda}\big[\mathbb{F}(l,\mathbf{t})\cdot m'(\{\mathbf{s}\in\mathbf{t}+l\Omega : \|f(\mathbf{s})-g(\mathbf{s})\|_Y\cdot\nu(\mathbf{s})\geqslant\varepsilon\})\big]$$

and $\|f\|_{P_{\varepsilon,l,\mathbb{F},\nu}} := d_{\varepsilon,l,\mathbb{F},\nu}(0,f)$. Then we have $0 \leqslant d_{\varepsilon,l,\mathbb{F},\nu}(f,g) \leqslant +\infty$, $d_{\varepsilon,l,\mathbb{F},\nu}(f,f) = 0$, $d_{\varepsilon,l,\mathbb{F},\nu}(f,g) = d_{\varepsilon,l,\mathbb{F},\nu}(g,f)$ and $d_{\varepsilon,l,\mathbb{F},\nu}(f,g) = d_{\varepsilon,l,\mathbb{F},\nu}(f+h,g+h)$ so that $d_{\varepsilon,l,\mathbb{F},\nu}(\cdot;\cdot)$ is a translation invariant pseudo-semimetric on the space of all functions from Λ into Y, provided that for each $l > 0$ we have $\sup_{\mathbf{t}\in\Lambda}\mathbb{F}(l,\mathbf{t}) < +\infty$. Moreover, the following holds:

(i) If $f : \Lambda \to Y$, $g : \Lambda \to Y$, $h : \Lambda \to Y$ and the assumptions A, B, $C \subseteq \mathbb{R}^n$ and $A \subseteq B \cup C$ imply $m'(A) \leqslant m'(B) + m'(C)$, then we have

$$d_{\varepsilon,l,\mathbb{F},\nu}(f,h) \leqslant d_{\varepsilon/2,l,\mathbb{F},\nu}(f,g) + d_{\varepsilon/2,l,\mathbb{F},\nu}(g,h), \quad \varepsilon > 0,\ l > 0. \tag{326}$$

(ii) Suppose that $\emptyset \neq \Lambda' \subseteq \mathbb{R}^n$, $\Lambda + \Lambda' \subseteq \Lambda$, $\Lambda + l\Omega \subseteq \Lambda$ for all $l > 0$, $\tau \in \Lambda'$, $M > 0$, the assumption $\mathbf{v} \in \Lambda + l\Omega + \tau$ implies $\nu(\mathbf{v}-\tau) \leqslant M\nu(\mathbf{v})$ and the assumption $A \subseteq B \subseteq \mathbb{R}^n$ implies $m'(A) \leqslant m'(B)$. Then we have

$$d_{\varepsilon,l,\mathbb{F},\nu}(f(\cdot+\tau),g(\cdot+\tau)) \leqslant d_{\varepsilon/M,l,\mathbb{F},\nu}(f,g), \tag{327}$$

for any two functions $f : \Lambda \to Y$ and $g : \Lambda \to Y$.

(iii) Suppose that $T \in L(Y)$, $f : \Lambda \to Y$ and $g : \Lambda \to Y$. Then we have

$$d_{\varepsilon,l,\mathbb{F},\nu}(Tf,Tg) \leqslant d_{\varepsilon/\|T\|,l,\mathbb{F},\nu}(f,g), \tag{328}$$

where $d_{\varepsilon/\|T\|,l,\mathbb{F}}(f,g) = 0$ for $T = 0$.

(iv) Suppose that $f : \Lambda \to Y$ and $g : \Lambda \to Y$. If the assumption $A \subseteq B \subseteq \mathbb{R}^n$ implies $m'(A) \leqslant m'(B)$, then for each $\varepsilon' \in (0,\varepsilon)$ we have

$$d_{\varepsilon,l,\mathbb{F},\nu}(f,g) \leqslant d_{\varepsilon',l,\mathbb{F},\nu}(f,g) \quad\text{and}\quad \|f\|_{P_{\varepsilon,l,\mathbb{F},\nu}} \leqslant \|f\|_{P_{\varepsilon',l,\mathbb{F},\nu}}.$$

(v) The triangle inequality

$$d_{\varepsilon,l,\mathbb{F},\nu}(f,h) \leqslant d_{\varepsilon,l,\mathbb{F},\nu}(f,g) + d_{\varepsilon,l,\mathbb{F},\nu}(g,h)$$

does not hold, in general, and the assumption $d_{\varepsilon,l,\mathbb{F},\nu}(f,g) = 0$ does not imply $f = g$ a. e., in general.

In particular, the property (v) shows that $d_{\varepsilon,l,\mathbb{F},\nu}(\cdot;\cdot)$ is not a pseudometric on the space of all functions from Λ into Y; therefore, we cannot apply [462, Theorem 2.1] in order to see that $\lim_{l\to+\infty} d_{\varepsilon,l,\mathbb{F},\nu}(f,g)$ exists for fixed $\varepsilon > 0$, $\mathbb{F}(\cdot;\cdot)$ and $f(\cdot)$, $g(\cdot)$. Concerning this issue, we will state and prove the following result.

Theorem 6.3.3. *Suppose that $\Lambda = [0,\infty)^n$ or $\Lambda = \mathbb{R}^n$, $\Omega = [0,1]^n$, $F : \Lambda \times X \to Y$ and $G : \Lambda \times X \to Y$. If $B \subseteq X$ is an arbitrary nonempty set, then we define*

$$d^B_{\varepsilon,l,\mathbb{F},\nu}(F,G) := \sup_{x\in B;\mathbf{t}\in\Lambda} \big[\mathbb{F}(l)\cdot m'\big(\{\mathbf{s}\in \mathbf{t}+l\Omega : \|F(\mathbf{s};x)-G(\mathbf{s};x)\|_Y \cdot \nu(\mathbf{s}) \geqslant \varepsilon\}\big)\big].$$

Then $\lim_{l\to+\infty} d^B_{\varepsilon,l,\mathbb{F},\nu}(F,G)$ exists in $[0,\infty]$, provided that the following conditions hold:
(i) *If A, $B \subseteq \mathbb{R}^n$, then $m'(A\cup B) \leqslant m'(A)+m'(B)$.*
(ii) *For every $l_1 > 0$, we have $\limsup_{l_2\to+\infty}[\frac{\mathbb{F}(l_2)}{\mathbb{F}(l_1)}\cdot\lceil\frac{l_2}{l_1}\rceil^n] \leqslant 1$.*

In particular, (ii) *holds with $\mathbb{F}(l)\equiv l^{-n}$.*

Proof. If $l_2 > l_1 > 0$, then (i) easily implies

$$\begin{aligned}&\sup_{x\in B;\mathbf{t}\in\Lambda} \big[\mathbb{F}(l_2)\cdot m'\big(\{\mathbf{s}\in \mathbf{t}+l_2\Omega : \|F(\mathbf{s};x)-G(\mathbf{s};x)\|_Y \cdot \nu(\mathbf{s}) \geqslant \varepsilon\}\big)\big]\\ &\quad\leqslant \left[\frac{\mathbb{F}(l_2)}{\mathbb{F}(l_1)}\cdot\left\lceil\frac{l_2}{l_1}\right\rceil^n\right]\cdot \sup_{x\in B;\mathbf{t}\in\Lambda} \big[\mathbb{F}(l_1)\cdot m'\big(\{\mathbf{s}\in \mathbf{t}+l_1\Omega : \|F(\mathbf{s};x)-G(\mathbf{s};x)\|_Y \cdot \nu(\mathbf{s}) \geqslant \varepsilon\}\big)\big].\end{aligned}$$

Applying (ii), we get

$$\limsup_{l_2\to+\infty} d^B_{\varepsilon,l_2,\mathbb{F},\nu}(F,G) \leqslant d^B_{\varepsilon,l_1,\mathbb{F},\nu}(F,G)$$

and

$$\limsup_{l_2\to+\infty} d^B_{\varepsilon,l_2,\mathbb{F},\nu}(F,G) \leqslant \liminf_{l_1\to+\infty} d^B_{\varepsilon,l_1,\mathbb{F},\nu}(F,G),$$

which simply yields the final conclusion. □

Now we would like to introduce the following notion.

Definition 6.3.4. Suppose that $\emptyset \neq \Lambda' \subseteq \mathbb{R}^n$, $\emptyset \neq \Lambda \subseteq \mathbb{R}^n$, $F : \Lambda \times X \to Y$ is a given function, ρ is a binary relation on Y, $R(F) \subseteq D(\rho)$, $\mathbb{F} : (0,\infty)\times\Lambda \to (0,\infty)$, $\Lambda + l\Omega \subseteq \Lambda$ for all $l > 0$ and $\Lambda + \Lambda' \subseteq \Lambda$. Then we say that:
(i) $F(\cdot;\cdot)$ is equi-Weyl $(\mathbb{F},\mathcal{B},\Lambda',\rho,\Omega,m',\nu)$-almost periodic if for every $B\in\mathcal{B}$ and $\varepsilon > 0$ there exist $l > 0$ and $L > 0$ such that for each $\mathbf{t}_0 \in \Lambda'$ there exists $\tau \in B(\mathbf{t}_0,L)\cap\Lambda'$ such that, for every $\mathbf{t}\in\Lambda$, $x\in B$ and $\mathbf{s}\in\mathbf{t}+l\Omega$, there exists an element $y_{\mathbf{s};x}\in\rho(F(\mathbf{s};x))$ such that

$$\sup_{x\in B}\|F(\cdot+\tau;x)-y_{\cdot;x}\|_{P_{\varepsilon,l,\mathbb{F},\nu}} \leqslant \varepsilon. \tag{329}$$

(ii) $F(\cdot;\cdot)$ is Weyl $(\mathbb{F},\mathcal{B},\Lambda',\rho,\Omega,m',\nu)$-almost periodic if for every $B\in\mathcal{B}$ and $\varepsilon>0$ there exists $L>0$ such that for each $\mathbf{t}_0\in\Lambda'$ there exists $\tau\in B(\mathbf{t}_0,L)\cap\Lambda'$ such that there exists $l(\tau)>0$ such that, for every $\mathbf{t}\in\Lambda$, $x\in B$, $l\geqslant l(\tau)$ and $\mathbf{s}\in\mathbf{t}+l\Omega$, there exists an element $y_{\mathbf{s};x}\in\rho(F(\mathbf{s};x))$ such that (329) holds.

We would like to observe the following fact.

Remark 6.3.5. In place of (329), we can also consider a slightly stronger condition:

$$\sup_{l_2\geqslant l;x\in B}\big\|F(\cdot+\tau;x)-y_{\cdot;x}\big\|_{P_{\varepsilon,l_2,\mathbb{F},\nu}}\leqslant\varepsilon.$$

If $\mathbb{F}(l,\mathbf{t})\equiv l^{-n}$, then Theorem 6.3.3 shows that this condition is equivalent to (329).

Any Bohr $(\mathcal{B},\Lambda',\rho,\Omega,m',\nu)$-almost periodic function has to be equi-Weyl $(1,\mathcal{B},\Lambda',\rho,\Omega,m',\nu)$-almost periodic, provided that $\Lambda+l\Omega\subseteq\Lambda$ for all $l>0$. It is also clear that any equi-Weyl $(\mathbb{F},\mathcal{B},\Lambda',\rho,\Omega,m',\nu)$-almost periodic function is Weyl $(\mathbb{F},\mathcal{B},\Lambda',\rho,\Omega,m',\nu)$-almost periodic.

If $\phi(x)\equiv x$, then the notion introduced in Definition 6.3.4 generalizes the notion introduced in [480, Definition 3], where we have considered the case in which $\rho=\mathrm{I}$. For simplicity, we will consider here the case in which $\phi(x)\equiv x$, only, and we will not generalize here the notion introduced in [480, Definitions 4–8] and the corresponding classes of Weyl ρ-uniformly recurrent type functions in general measure. The notion of (equi)-Weyl-(Doss)-m-almost periodicity for a Lebesgue measurable function $F:\mathbb{R}^n\to Y$ is obtained by plugging $\mathbb{F}(\mathbf{t},l)\equiv l^{-n/p}$, $\Lambda'=\mathbb{R}^n$, $\rho=\mathrm{I}$, $\Omega=[0,1]^n$ and $\nu(\cdot)\equiv 1$.

We continue by stating the following result; we can also consider general measures here.

Proposition 6.3.6. *Suppose that* (WM1-1) *holds. Then we have the following:*
(i) *Suppose that* $F\in(e-)W^{p(\mathbf{u}),x,\mathbb{F},\nu}_{\Omega,\Lambda',\mathcal{B}}(\Lambda\times X:Y)$ *or* $F\in(e-)W^{p,x,\mathbb{F},\nu}_{\Omega,\Lambda',\mathcal{B}}(\Lambda\times X:Y)$ *for some* $p\in(0,1)$. *Then* $F(\cdot;\cdot)$ *is (equi-)Weyl* $(\mathbb{F},\mathcal{B},\Lambda',\rho,\Omega,m,\nu)$*-almost periodic.*
(ii) *Suppose that* $\rho=T\in L(Y)$, $F(\cdot;\cdot)$ *is (equi-)Weyl* $(\mathbb{F},\mathcal{B},\Lambda',\rho,\Omega,m,\nu)$*-almost periodic, there exists* $M>0$ *such that* $\nu(\mathbf{t})\leqslant M$ *for all* $\mathbf{t}\in\Lambda$, *there exists* $l_0>0$ *such that* $\sup_{l\geqslant l_0,\mathbf{t}\in\Lambda}[\mathbb{F}(l,\mathbf{t})l^{n/p}]<+\infty$ *and for each set* $B\in\mathcal{B}$ *we have* $\sup_{x\in B;\mathbf{t}\in\Lambda}\|F(\mathbf{t};x)\|_Y<+\infty$. *Then, for every* $p\geqslant 1$, *we have* $F\in(e-)W^{p,x,\mathbb{F},\nu}_{\Omega,\Lambda',\mathcal{B}}(\Lambda\times X:Y)$.

Proof. The proof of (i) is almost trivial and, therefore, omitted. To deduce (ii), define

$$A_{\varepsilon,l,\mathbf{t},x}:=\{\mathbf{s}\in\mathbf{t}+l\Omega:\|F(\mathbf{s}+\tau;x)-F(\mathbf{s};x)\|_Y\cdot\nu(\mathbf{s})\geqslant\varepsilon\},\quad\varepsilon>0,\ l>0,\ \mathbf{t}\in\Lambda,\ x\in X.$$

Then the final conclusion simply follows from the prescribed assumptions and the estimate $(B\in\mathcal{B};\tau\in\Lambda')$:

$$\sup_{x\in B;\mathbf{t}\in\Lambda}\mathbb{F}(l,\mathbf{t})\Biggl(\int_{\mathbf{t}+l\Omega}\|F(\mathbf{s}+\tau;x)-TF(\mathbf{s};x)\|_Y^p\cdot\nu^p(\mathbf{s})\,d\mathbf{s}\Biggr)^{1/p}$$
$$\leqslant\sup_{x\in B;\mathbf{t}\in\Lambda}\mathbb{F}(l,\mathbf{t})\Bigl(m(A_{\varepsilon,l,\mathbf{t},x})\Bigl[(1+\|T\|)\cdot\sup_{x\in B;\mathbf{t}\in\Lambda}\|F(\mathbf{t};x)\|_Y\Bigr]^p+\varepsilon^pl^nM^p\Bigr)^{1/p}.\qquad\square$$

We continue by stating the following result.

Theorem 6.3.7. *Suppose that* $M>0$, $\emptyset\neq\Lambda'\subseteq\mathbb{R}^n$, $\emptyset\neq\Lambda\subseteq\mathbb{R}^n$, $F:\Lambda\times X\to Y$ *is a given function,* $\rho=T\in L(Y)$, $\mathbb{F}:(0,\infty)\to(0,\infty)$, $\nu:\Lambda\to[0,\infty)$ *and* $\Lambda+\Lambda'\subseteq\Lambda$. *Suppose further that, for every* $k\in\mathbb{N}$, *the function* $F_k:\Lambda\times X\to Y$ *is (equi-)Weyl-*$(\mathbb{F},\mathcal{B},\Lambda',\rho,\Omega,m',\nu)$*-almost periodic and, for every* $\varepsilon>0$, $l>0$ *and* $B\in\mathcal{B}$, *we have*

$$\lim_{k\to+\infty}\sup_{x\in B}\|F_k(\cdot;x)-F(\cdot;x)\|_{P_{\varepsilon,l,\mathbb{F},\nu}}=0.$$

Then $F(\cdot;\cdot)$ *is (equi-)Weyl-*$(\mathbb{F},\mathcal{B},\Lambda',\rho,\Omega,m',\nu)$*-almost periodic, provided that the assumptions* A, B, $C\subseteq\mathbb{R}^n$ *and* $A\subseteq B\cup C$ *imply* $m'(A)\leqslant m'(B)+m'(C)$, *and the assumption* $\mathbf{v}\in\Lambda+l\Omega+\tau$ *for some* $\tau\in\Lambda'$ *and* $l>0$ *implies* $\nu(\mathbf{v}-\tau)\leqslant M\nu(\mathbf{v})$.

Proof. If $k\in\mathbb{N}$, $x\in X$ and $\tau\in\Lambda'$, then the estimates (326)–(328) imply

$$\begin{aligned}&\|F(\cdot+\tau;x)-TF(\cdot;x)\|_{P_{\varepsilon,l,\mathbb{F},\nu}}\\&\leqslant\|F(\cdot+\tau;x)-F_k(\cdot+\tau;x)\|_{P_{\varepsilon/2,l,\mathbb{F},\nu}}+\|F_k(\cdot+\tau;x)-TF(\cdot;x)\|_{P_{\varepsilon/2,l,\mathbb{F},\nu}}\\&\leqslant\|F(\cdot+\tau;x)-F_k(\cdot+\tau;x)\|_{P_{\varepsilon/2,l,\mathbb{F},\nu}}+\|F_k(\cdot+\tau;x)-TF_k(\cdot;x)\|_{P_{\varepsilon/4,l,\mathbb{F},\nu}}\\&\quad+\|TF_k(\cdot;x)-TF(\cdot;x)\|_{P_{\varepsilon/4,l,\mathbb{F},\nu}}\\&\leqslant\|F(\cdot;x)-F_k(\cdot;x)\|_{P_{\varepsilon/2M,l,\mathbb{F},\nu}}+\|F_k(\cdot+\tau;x)-TF_k(\cdot;x)\|_{P_{\varepsilon/4,l,\mathbb{F},\nu}}\\&\quad+\|F_k(\cdot;x)-F(\cdot;x)\|_{P_{\varepsilon/4\|T\|,l,\mathbb{F},\nu}}.\end{aligned}$$

This simply completes the proof. □

6.3.2 Multidimensional Doss ρ-almost periodic type functions in general measure

In this subsection, we will extend the class of Doss ρ-almost periodic functions from Definition 6.3.2, provided that the function $\mathbb{F}(\cdot;\cdot)$ does not depend on the second argument, by considering the class of Doss $(\mathbb{F},\mathcal{B},\Lambda',\rho,m',\nu)$-almost periodic functions. The precise definition goes as follows.

Definition 6.3.8. Suppose that $\emptyset\neq\Lambda'\subseteq\mathbb{R}^n$, $\emptyset\neq\Lambda\subseteq\mathbb{R}^n$, $F:\Lambda\times X\to Y$ is a given function, ρ is a binary relation on Y, $R(F)\subseteq D(\rho)$, $\mathbb{F}:(0,\infty)\to(0,\infty)$, $\nu:\Lambda\to[0,\infty)$ and $\Lambda+\Lambda'\subseteq\Lambda$. Then we say that $F(\cdot;\cdot)$ is Doss $(\mathbb{F},\mathcal{B},\Lambda',\rho,m',\nu)$-almost periodic if for every $B\in\mathcal{B}$ and $\varepsilon>0$ there exists $L>0$ such that for each $\mathbf{t}_0\in\Lambda'$ there exists $\tau\in B(\mathbf{t}_0,L)\cap\Lambda'$

such that, for every $\mathbf{s} \in \Lambda$ and $x \in B$, there exists an element $y_{\mathbf{s};x} \in \rho(F(\mathbf{s};x))$ such that

$$\limsup_{l\to+\infty}\Big[\mathrm{F}(l)\sup_{x\in B} m'\big(\{\mathbf{s}\in\Lambda_l : \|F(\mathbf{s}+\tau;x)-y_{\mathbf{s};x}\|_Y\cdot\nu(\mathbf{s})\geqslant\varepsilon\}\big)\Big]\leqslant\varepsilon.$$

The proof of the following result is simple and, therefore, omitted.

Proposition 6.3.9. *Suppose that $\emptyset \neq \Lambda' \subseteq \mathbb{R}^n$, $\emptyset \neq \Lambda \subseteq \mathbb{R}^n$, $F : \Lambda\times X \to Y$ is a given function, ρ is a binary relation on Y, $R(F) \subseteq D(\rho)$, $\mathrm{F} : (0,\infty) \to (0,\infty)$, $\nu : \Lambda \to [0,\infty)$ and $\Lambda + \Lambda' \subseteq \Lambda$.*

(i) *If $F(\cdot;\cdot)$ is Doss-$(p, \mathrm{F}, \mathcal{B}, \Lambda', \rho, \nu)$-almost periodic and $p \in D_+(\Lambda)$, then $F(\cdot;\cdot)$ is Doss $(\mathrm{F}, \mathcal{B}, \Lambda', \rho, m, \nu)$-almost periodic.*
(ii) *If $\rho = T \in L(Y)$, there exists $M > 0$ such that $\nu(\mathbf{s}) \leqslant M$ for a. e. $\mathbf{s} \in \Lambda$, $\limsup_{l\to+\infty}[\mathrm{F}(l)l^{n/p}] < +\infty$, $F(\cdot;\cdot)$ is Doss $(\mathrm{F}, \mathcal{B}, \Lambda', \rho, m, \nu)$-almost periodic and $\sup_{x\in B;\mathbf{t}\in\Lambda}\|F(\mathbf{t};x)\|_Y < +\infty$ for each set $B \in \mathcal{B}$, then $F(\cdot;\cdot)$ is Doss-$(p, \mathrm{F}, \mathcal{B}, \Lambda', \rho, \nu)$-almost periodic for each $p \geqslant 1$.*
(iii) *If $0 \in \Lambda$, $l\Omega = \Lambda_l$ for all $l > 0$ and $F(\cdot;\cdot)$ is Weyl $(\mathbb{F}, \mathcal{B}, \Lambda', \rho, \Omega, m', \nu)$-almost periodic with $\mathbb{F}(l,\mathbf{t}) \equiv \mathbb{F}(l)$, then $F(\cdot;\cdot)$ is Doss $(\mathrm{F}, \mathcal{B}, \Lambda', \rho, m', \nu)$-almost periodic with $\mathrm{F} = \mathbb{F}$.*

The class of Doss $(\mathrm{F}, \mathcal{B}, \Lambda', \rho, m, \nu)$-uniformly recurrent functions can be introduced in the same way as above. Keeping this observation in mind, Proposition 6.3.9(ii) and the estimate established on [448, p. 113, l. 1] can be used to prove that there does not exist a nonempty subset Λ' of $\mathbb{R}$ such that the function $f(\cdot)$ analyzed in [448, Example 3.2.3] is Doss $(l^{-1}, \Lambda', \rho, m, 1)$-uniformly recurrent with $\rho(t) = 1$ for all $t \in \mathbb{R}$. The interested reader can make an effort to reformulate Theorem 6.3.7 and the conclusions established in [448, Example 3.2.4] for multi-dimensional Doss almost periodic type functions in general measure; cf. also [448, Example 3.2.7].

6.3.3 Applications

In this subsection, we will present some applications of Weyl and Doss almost periodic type functions to the abstract Volterra integro-differential equations in Banach spaces. We will divide this subsection into two separate parts.

1. Invariance of generalized almost periodicity in measure under the actions of convolution products

In this part, we will examine the invariance of Weyl and Doss almost periodicity in measure under the actions of convolution products; for simplicity, we will consider the one-dimensional setting only. We start by stating the following result.

Theorem 6.3.10. *Suppose that $\emptyset \neq \Lambda' \subseteq \mathbb{R}$, $f : \mathbb{R} \to Y$ is a bounded function, $\rho = T \in L(Y)$, $\mathbb{F} : (0,\infty) \times \mathbb{R} \to (0,\infty)$, $\mathbb{F}_1 : (0,\infty) \times \mathbb{R} \to (0,\infty)$ and $\Omega = [0,1]$. Suppose further that $(R(t))_{t>0} \subseteq L(X,Y)$ is a strongly continuous operator family such that $R(t)T = TR(t)$ for all $t > 0$ and $\int_{(0,\infty)} \|R(t)\|\,dt < \infty$. If $f(\cdot)$ is (equi-)Weyl $(\mathbb{F},\Lambda',\rho,\Omega,m)$-almost periodic, then the function $F : \mathbb{R} \to Y$, given by (316), is bounded, continuous and (equi-)Weyl $(\mathbb{F}_1,\Lambda',\rho,\Omega,m)$-almost periodic, provided that the following condition holds true:*

(Q1w) *For every $\varepsilon > 0$, there exists $\varepsilon' \in (0, \varepsilon/[2(1+\int_{(0,\infty)} \|R(t)\|\,dt)])$ such that, for every $l > 0$ and $t \in \mathbb{R}$, we have*

$$m\left(\left\{s \in [t,t+l] : (1+\|T\|)\|f\|_\infty \sum_{k=0}^{+\infty} \frac{\|R(\cdot)\|_{L^\infty[kl,(k+1)l]}}{\mathbb{F}(l,s-(k+1)l)} < \frac{\varepsilon}{2\varepsilon'}\right\}\right) \geqslant l - \frac{\varepsilon}{\mathbb{F}_1(l,\mathbf{t})}.$$

Proof. We will consider the class of equi-Weyl $(\mathbb{F},\Lambda',\rho,[0,1],m)$-almost periodic functions. We can simply prove that $F(\cdot)$ is well-defined, bounded and continuous. Let $\varepsilon > 0$ be given, and let $\varepsilon' > 0$ be determined from condition (Q1w). Then we know that there exist $l > 0$ and $L > 0$ such that for each $t_0 \in \Lambda'$ there exists $\tau \in B(t_0,L) \cap \Lambda'$ such that

$$\|f(\cdot+\tau) - Tf(\cdot)\|_{P_{\varepsilon',l,\mathbb{F}}} \leqslant \varepsilon'.$$

Furthermore, we have

$$\begin{aligned}
&\|F(s+\tau) - TF(s)\|_Y \\
&\leqslant \int_0^{+\infty} \|R(r)\| \cdot \|f(s+\tau-r) - Tf(s-r)\|_Y\,dr \\
&= \sum_{k=0}^{+\infty} \int_{kl}^{(k+1)l} \|R(r)\| \cdot \|f(s+\tau-r) - Tf(s-r)\|_Y\,dr \\
&\leqslant (1+\|T\|)\|f\|_\infty \cdot \varepsilon' \cdot \sum_{k=0}^{+\infty} \frac{\|R(\cdot)\|_{L^\infty[kl,(k+1)l]}}{F(l,s-(k+1)l)} + \varepsilon' \cdot \sum_{k=0}^{+\infty} \int_{kl}^{(k+1)l} \|R(r)\|\,dr.
\end{aligned}$$

Since $\varepsilon/ < 1/[2(1+\int_0^\infty \|R(r)\|\,dr)]$, the above calculation implies

$$\begin{aligned}
&\left\{s \in [t,t+l] : \sum_{k=0}^{+\infty} \frac{\|R(\cdot)\|_{L^\infty[kl,(k+1)l]}}{F(l,s-(k+1)l)} < \frac{\varepsilon}{2\varepsilon'\|f\|_\infty(1+\|T\|)}\right\} \\
&\quad \subseteq \{s \in [t,t+l] : \|F(s+\tau) - TF(s)\|_Y < \varepsilon\},
\end{aligned}$$

so that the final conclusion simply follows from condition (Q1w). □

We continue with the following example (compare with [447, Example 6.3.4]).

Example 6.3.11. Suppose that the function $f \in C^2(\mathbb{R}^3)$ has a compact support. Then we know that the function

$$u(x) = \frac{1}{4\pi} \int_{\mathbb{R}^3} \frac{f(x-y)}{|y|}\, dy, \quad x \in \mathbb{R}^3,$$

is a unique classical solution of the partial differential equations $\Delta u = -f$. Since for each $T \in L(\mathbb{R}^3)$, $l > 0$ and $\varepsilon' > 0$ we have

$$\begin{aligned} |u(x+\tau) - Tu(x)| &\leqslant \frac{1}{4\pi} \int_{\mathbb{R}^3} \frac{|f(x-y+\tau) - Tf(x-y)|}{|y|}\, dy, \\ &\leqslant \frac{\varepsilon'}{4\pi} \int_{\mathbb{R}^3} \frac{dy}{|y| \cdot \mathbb{F}(l, x-y)}, \quad x \in \mathbb{R}^3, \end{aligned}$$

it readily follows that the (equi-)Weyl $(\mathbb{F}, \Lambda', T, \Omega, m)$-almost periodicity of $f(\cdot)$ implies the (equi-)Weyl $(\mathbb{F}_1, \Lambda', T, \Omega, m)$-almost periodicity of $u(\cdot)$, provided that the following condition holds:

(D) For every $\varepsilon > 0$, there exists $\varepsilon' > 0$ such that, for every $l > 0$ and $\mathbf{t} \in \mathbb{R}^3$, we have

$$m\Bigg(\Bigg\{x \in \mathbf{t} + l\Omega : \int_{\mathbb{R}^3} \frac{dy}{|y| \cdot \mathbb{F}(l, x-y)} < \frac{\varepsilon}{4\pi\varepsilon'}\Bigg\}\Bigg) \geqslant m(l\Omega) - \frac{\varepsilon}{\mathbb{F}_1(l, \mathbf{t})}.$$

Remark 6.3.12. The invariance of (equi-)Weyl-p-almost periodicity under the actions of the infinite convolution products has been investigated in [444, Proposition 2.11.1, Theorem 2.11.4]; these results can be also formulated with a general operator $\rho = T \in L(Y)$ and a general set Λ' of the corresponding Weyl-ε-almost periods. Because of that, the case in which $\mathbb{F}(l, \mathbf{t}) \equiv l^{-n/p}$ will not occupy our attention here.

For the sake of completeness, we will provide the main details of the proof of the following result.

Theorem 6.3.13. *Suppose that $\emptyset \neq \Lambda' \subseteq \mathbb{R}$, $f : \mathbb{R} \to Y$ is a bounded function, $\rho = T \in L(Y)$, $\mathrm{F} : (0,\infty) \to (0,\infty)$, $\mathrm{F}_1 : (0,\infty) \to (0,\infty)$ and $(R(t))_{t>0} \subseteq L(X,Y)$ is a strongly continuous operator family such that $R(t)T = TR(t)$ for all $t > 0$ and $\int_{(0,\infty)} \|R(t)\|\, dt < \infty$. If $f(\cdot)$ is Doss $(\mathrm{F}, \Lambda', T, m)$-almost periodic, then the function $F : \mathbb{R} \to Y$, given by (316), is bounded, continuous and Doss $(\mathrm{F}_1, \Lambda', T, m)$-almost periodic, provided that the following condition holds true:*

(Q2w) *For every $\varepsilon > 0$, there exist $\varepsilon' \in (0, \varepsilon/[2(1 + \int_{(0,\infty)} \|R(t)\|\, dt)])$ and $l_0 > 0$ such that, for every $l \geqslant l_0$, we have*

$$m\Bigg(\Bigg\{s \in [-l, l] : (1 + \|T\|)\|f\|_\infty \cdot \varepsilon' \cdot \limsup_{v \to +\infty} \frac{\|R(\cdot)\|_{L^\infty[0,v]}}{\mathrm{F}(v + |s|)} < \frac{\varepsilon}{2}\Bigg\}\Bigg) \geqslant 2l - \frac{\varepsilon}{\mathrm{F}_1(l)}.$$

Proof. Let $\varepsilon > 0$ be given, and let $\varepsilon' > 0$ and $l_0 > 0$ be determined from condition (Q1w). Then we know that there exists $L > 0$ such that for each $t_0 \in \Lambda'$ there exists $\tau \in B(t_0, L) \cap \Lambda'$ such that

$$\mathrm{F}(l)m(\{s \in [-l,l] : \|f(s+\tau) - Tf(\cdot)\|_Y \geqslant \varepsilon'\}) \leqslant \varepsilon'.$$

Furthermore, we have

$$\begin{aligned}
&\|F(s+\tau) - TF(s)\|_Y \\
&\quad \leqslant \int_0^{+\infty} \|R(r)\| \cdot \|f(s+\tau-r) - Tf(s-r)\|_Y \, dr \\
&\quad = \lim_{v\to+\infty} \int_0^{l} \|R(r)\| \cdot \|f(s+\tau-r) - Tf(s-r)\|_Y \, dr \\
&\quad \leqslant (1+\|T\|)\|f\|_\infty \cdot \varepsilon' \cdot \limsup_{v\to+\infty} \frac{\|R(\cdot)\|_{L^\infty[0,v]}}{\mathrm{F}(v+|s|)} + \varepsilon' \cdot \sum_{k=0}^{+\infty} \int_0^{+\infty} \|R(r)\| \, dr,
\end{aligned}$$

so that

$$\begin{aligned}
&\left\{s \in [-l,l] : (1+\|T\|)\|f\|_\infty \cdot \varepsilon' \cdot \limsup_{v\to+\infty} \frac{\|R(\cdot)\|_{L^\infty[0,v]}}{\mathrm{F}(v+|s|)} < \frac{\varepsilon}{2}\right\} \\
&\quad \subseteq \{s \in [-l,l] : \|F(s+\tau) - TF(s)\|_Y < \varepsilon\}.
\end{aligned}$$

The final conclusion now follows from condition (Q2w). □

It is clear that Theorem 6.3.10 [Theorem 6.3.13] can be applied in the analysis of Weyl [Doss] almost periodic solutions in the Lebesgue measure for a large class of the abstract Volterra integrodifferential inclusions without boundary conditions; cf. [444] for more details about this problematic.

2. Abstract semilinear Cauchy inclusions

In this part, we will analyze the existence and uniqueness of (equi-)Weyl almost periodic solutions for the following abstract semilinear Cauchy inclusion:

$$D_{\gamma,+}^t u(t) \in \mathcal{A}u(t) + f(t, u(t)), \quad t \in \mathbb{R}, \tag{330}$$

where $D_{\gamma,+}^t u(t)$ denotes the Weyl–Liouville fractional derivative of order $\gamma \in (0,1)$, $\mathcal{A}$ is a multivalued linear operator and $f(\cdot,\cdot)$ has some extra features. It is well known that there exists a large class of multivalued linear operators $\mathcal{A}$ for which the solution operator family $(R(t))_{t>0}$ for (330) has the growth of type (284), for some finite real constants $M \geqslant 1$, $\beta \in (0,1]$ and $\gamma > \beta$; cf. [444] and [445] for more details. Moreover, a unique mild solution of (330) is given by

$$u(t) = \int_{-\infty}^{t} R(t-s)f(s,u(s))\,ds, \quad t \in \mathbb{R}.$$

Here, we will consider the space $e - W^{\infty}(\mathbb{R} : X) := C_b(\mathbb{R} : X) \bigcap_{p \geqslant 1} e - W^p AP(\mathbb{R} : X)$. It is clear that $e - W^{\infty}(\mathbb{R} : X) := C_b(\mathbb{R} : X) \bigcap_{p > 1/\beta} e - W^p AP(\mathbb{R} : X)$; equipped with the sup-norm, $e - W^{\infty}(\mathbb{R} : X)$ is a Banach space. Observe also that Proposition 6.3.6 yields that $e - W^{\infty}(\mathbb{R}^n : X)$ is the Banach space of all bounded continuous functions $f : \mathbb{R} \to X$, which are equi-Weyl almost periodic in the Lebesgue measure. By $e - W^{\infty,rc}(\mathbb{R} : X)$, we denote the Banach subspace of $e - W^{\infty}(\mathbb{R} : X)$, which contains the functions with relatively compact range.

Keeping in mind the Banach contraction principle, [444, Proposition 2.11.1, Theorem 2.11.4] and the composition principles established in [450, Theorems 2.2, 2.3], it readily follows that there exists a unique mild solution of (330), which belongs to the space $e - W^{\infty,rc}(\mathbb{R} : X)$, provided that the following conditions hold:

(i) There exists $L > 0$ such that $L \cdot \int_0^{\infty} \|R(r)\|\,dr < 1$ and $\|f(t,x) - f(t,y)\| \leqslant L\|x - y\|$ for all $t \in \mathbb{R}$ and $x, y \in X$.
(ii) For every relatively compact set $K \in X$, the set $\{f(t,x) : t \in \mathbb{R},\ x \in K\}$ is relatively compact in X.
(iii) The conditions (i) and (iii) given in the formulation of [450, Theorem 2.2] hold for any exponent $p > 1/\beta$.

We can similarly analyze the existence and uniqueness of (equi-)Weyl almost periodic solutions for some other classes of the semilinear integral equations. Let us finally mention some topics, which are not considered here:

1. In [444, Definition 2.13.2], we have introduced the class of one-dimensional Besicovitch–Doss-p-almost periodic functions. The conditions (ii), (iii) [already done for Doss-p-almost periodic functions in measure] and (iv) in this definition can be further extended by considering the same condition in view of the general measure. In such a way, we can extend the class of Besicovitch–Doss-p-almost periodic functions ($p \geqslant 1$). We can also consider the case in which $0 < p < 1$ and some multidimensional analogues of it; see, e. g., [447, Section 3.3].
2. We can analyze various classes of Stepanov quasiasymptotically ρ-almost periodic type functions in general measure; see, e. g., [447, Subsection 3.3.3].
3. We can also consider Weyl and Besicovitch almost automorphic type functions in general measure.

6.4 Notes and Appendices to Part II

Between Bohr ρ-almost periodicity and Stepanov ρ-almost periodicity

In this part, we will reconsider the notion of Riemann integrability and introduce some new classes of generalized ρ-almost periodic type functions (the introduced classes are new even in the one-dimensional setting, with $\rho = \mathrm{I}$, and these classes should not be mistakenly identified with the corresponding classes of generalized (real-valued) almost periodic type functions considered by R. Doss in [263], where the author has followed a completely different approach with the use of the upper Lebesgue integrals and the lower Lebesgue integrals).

If $\Omega \subseteq \mathbb{R}^n$ is a Jordan measurable set, then $m_J(\Omega)$ denotes henceforth its Jordan measure. Further on, if $\Omega_1, \ldots, \Omega_k$ are Jordan measurable sets, $\Omega = \Omega_1 \cup \cdots \Omega_k$ and $\Omega_i \cap \Omega_j = \emptyset$ for $1 \leqslant i, j \leqslant k$, then we say that $P = \{\Omega_1, \ldots, \Omega_k\}$ is a partition of Ω. By $d(P) := \max\{d(\Omega_1), \ldots, d(\Omega_k)\}$, where $d(\Omega_i)$ denotes the diameter of set Ω_i for $1 \leqslant i \leqslant k$, we denote the diameter of partition P; if the function $f : \Omega \to \mathbb{R}$ and the points $\xi_i \in \Omega_i$ are given $(1 \leqslant i \leqslant k)$, then we form the Riemann sum

$$\sigma(f; P, (\xi_i)_{1 \leqslant i \leqslant k}) := \sum_{i=1}^{k} m_J(\Omega_i) f(\xi_i). \tag{331}$$

Now we are ready to introduce the following notion.

Definition 6.4.1. Suppose that $p > 0$, $\emptyset \neq \Lambda' \subseteq \mathbb{R}^n$, $\emptyset \neq \Lambda \subseteq \mathbb{R}^n$, $F : \Lambda \times X \to Y$ is a given function, ρ is a binary relation on Y and $\Lambda + \Lambda' \subseteq \Lambda$. Suppose also that for each $\mathbf{t} \in \Lambda$ the set $\Lambda(\mathbf{t}) := \Lambda \cap (\mathbf{t} + [0,1]^n)$ is Jordan measurable with positive Jordan measure. Then we say that:

(i) $F(\cdot;\cdot)$ is summable $(\mathcal{B}, \Lambda', \rho, p)$-almost periodic if for every $B \in \mathcal{B}$ and $\varepsilon > 0$ there exists $l > 0$ such that for each $\mathbf{t}_0 \in \Lambda'$ there exists $\tau \in B(\mathbf{t}_0, l) \cap \Lambda'$ such that, for every $\mathbf{t} \in \Lambda$, there exists $k_0 \in \mathbb{N}$ such that, for every $k \geqslant k_0$, every partition $P = \{\Lambda_1(\mathbf{t}), \ldots, \Lambda_k(\mathbf{t})\}$ of $\Lambda(\mathbf{t})$, every choice of the points $\xi_i \in \Lambda_i(\mathbf{t})$ for $1 \leqslant i \leqslant k$, and every $x \in B$, there exists an element $y_{i;x} \in \rho(F(\xi_i; x))$ for $1 \leqslant i \leqslant k$ such that

$$\sup_{x \in B} \sum_{i=1}^{k} \|F(\xi_i + \tau; x) - y_{i;x}\|_Y^p \cdot m_J(\Lambda_i(\mathbf{t})) \leqslant \varepsilon. \tag{332}$$

(ii) $F(\cdot;\cdot)$ is summable $(\mathcal{B}, \Lambda', \rho, p)$-uniformly recurrent if for every $B \in \mathcal{B}$ there exists a sequence (τ_m) in Λ' such that $\lim_{m \to +\infty} |\tau_m| = +\infty$ and, for every $\mathbf{t} \in \Lambda$, there exists $k_0 \in \mathbb{N}$ such that, for every $k \geqslant k_0$, every partition $P = \{\Lambda_1(\mathbf{t}), \ldots, \Lambda_k(\mathbf{t})\}$ of $\Lambda(\mathbf{t})$, every choice of the points $\xi_i \in \Lambda_i(\mathbf{t})$ for $1 \leqslant i \leqslant k$, and every $x \in B$, there exists an element $y_{i;x} \in \rho(F(\xi_i; x))$ for $1 \leqslant i \leqslant k$ such that

$$\lim_{m \to +\infty} \sup_{x \in B} \sum_{i=1}^{k} \|F(\xi_i + \tau_m; x) - y_{i;x}\|_Y^p \cdot m_J(\Lambda_i(\mathbf{t})) = 0.$$

It can be simply shown that any summable $(\mathcal{B}, \Lambda', \rho, p)$-uniformly recurrent is summable $(\mathcal{B}, \Lambda', \rho, p)$-almost periodic provided that the set Λ' is unbounded as well as that every Bohr $(\mathcal{B}, \Lambda', \rho)$-almost periodic function ($(\mathcal{B}, I', \rho)$-uniformly recurrent function) is summable $(\mathcal{B}, \Lambda', \rho, p)$-almost periodic (summable $(\mathcal{B}, \Lambda', \rho, p)$-uniformly recurrent) for every exponent $p > 0$. If $\rho = \mathrm{I}$, then the sum in (332) is nothing else but $\sigma(f; P, (\xi_i)_{1\leqslant i\leqslant k})$ with $f(\cdot) \equiv \|F(\cdot + \tau; x) - F(\cdot; x)\|_Y^p$; cf. (331).

The following structural result is relatively simple and states that, under certain logical assumptions, the class of summable $(\mathcal{B}, \Lambda', \rho, p)$-almost periodic functions lies between the class of Bohr $(\mathcal{B}, \Lambda', \rho)$-almost periodic functions and the class of Stepanov $([0,1]^n, p)$-$(\mathcal{B}, \Lambda', \rho)$-almost periodic functions.

Proposition 6.4.2. *Suppose that $p > 0$, $\emptyset \neq \Lambda' \subseteq \mathbb{R}^n$, $\emptyset \neq \Lambda \subseteq \mathbb{R}^n$, $F : \Lambda \times X \to Y$ is a given function, ρ is a binary relation on Y, $\Lambda + \Lambda' \subseteq \Lambda$, $\Lambda + [0,1]^n \subseteq \Lambda$ and the function $\hat{F}_{[0,1]^n} : \Lambda \times X \to L^p([0,1]^n : Y)$ is well-defined and continuous. Suppose further that the condition (C_ρ) holds, where:*

(C_ρ) *For each $\varepsilon > 0$, there exists $\delta > 0$ such that, for every $y_1, y_2 \in Y$ with $\|y_1 - y_2\|_Y < \delta$, we have $\|z_1 - z_2\|_Y < \varepsilon/3$ for every $z_1 \in \rho(y_1)$ and $z_2 \in \rho(y_2)$.*

If $F(\cdot;\cdot)$ is summable $(\mathcal{B}, \Lambda', \rho)$-almost periodic (summable $(\mathcal{B}, \Lambda', \rho)$-uniformly recurrent) and for each $x \in X$ the function $F(\cdot; x)$ is locally bounded and has at most countably many discontinuities, then $F(\cdot;\cdot)$ is Stepanov $([0,1]^n, p)$-$(\mathcal{B}, \Lambda', \rho)$-almost periodic (Stepanov $([0,1]^n, p)$-$(\mathcal{B}, \Lambda', \rho)$-uniformly recurrent).

Proof. We will consider here the corresponding classes of generalized almost periodic functions with $p \geqslant 1$. Let $B \in \mathcal{B}$ and $\varepsilon > 0$ be fixed. Then there exists $l > 0$ such that for each $\mathbf{t}_0 \in \Lambda'$ there exists $\boldsymbol{\tau} \in B(\mathbf{t}_0, l) \cap \Lambda'$ such that, for every $\mathbf{t} \in \Lambda$, there exists $k_0 \in \mathbb{N}$ such that, for every $k \geqslant k_0$, every partition $P = \{\Lambda_1(\mathbf{t}), \ldots, \Lambda_k(\mathbf{t})\}$ of set $\Lambda(\mathbf{t}) = \mathbf{t} + [0,1]^n$ (since $\Lambda + [0,1]^n \subseteq \Lambda$), every choice of the points $\xi_i \in \Lambda_i(\mathbf{t})$ for $1 \leqslant i \leqslant k$, and every $x \in B$, there exists an element $y_{i;x} \in \rho(F(\xi_i; x))$ for $1 \leqslant i \leqslant k$ such that (332) holds with the number ε replaced by the number $(\varepsilon/2)^p$ therein. On the other hand, we know that, due to our assumptions and condition (C_ρ), for each $x \in X$ and $\tau \in \Lambda'' \equiv \{\mathbf{t} \in \mathbb{R}^n : \mathbf{t} + \Lambda \subseteq \Lambda\}$ the mapping $\cdot \mapsto \|F(\cdot + \tau; x) - y_{\cdot;x}\|_Y^p$ is locally Riemann integrable for each function $\cdot \mapsto y_{\cdot;x} \in \rho(F(\cdot; x))$, which means that we can find a sufficiently small number $\delta > 0$ such that, for every partition $P = \{\Lambda_1(\mathbf{t}), \ldots, \Lambda_k(\mathbf{t})\}$ of set $\Lambda(\mathbf{t})$ with parameter $d(P) < \delta$ and for every choice of points $\xi_i \in \Lambda_i(\mathbf{t})$ for $1 \leqslant i \leqslant k$, we have that the absolute value of the difference between

$$\|F(\mathbf{t} + \boldsymbol{\tau} + \mathbf{u}; x) - z_{\mathbf{t},\mathbf{u},x}\|_{L^p([0,1]^n:Y)} = \Bigg(\int\limits_{[0,1]^n} \|F(\mathbf{t} + \boldsymbol{\tau} + \mathbf{u}; x) - z_{\mathbf{t},\mathbf{u},x}\|^p \, d\mathbf{u}\Bigg)^{1/p}$$
$$= \Bigg(\int\limits_{\mathbf{t}+[0,1]^n} \|F(\mathbf{v} + \boldsymbol{\tau}; x) - z_{\mathbf{t},\mathbf{v}-\mathbf{t},x}\|^p \, d\mathbf{v}\Bigg)^{1/p}$$

and

$$\Bigg(\sum_{i=1}^{k} \|F(\xi_i + \tau; x) - y_{i;x}\|_Y^p \cdot m_J(\Lambda_i(\mathbf{t}))\Bigg)^{1/p} \tag{333}$$

is smaller than $\varepsilon/2$. The value of term in (333), with $y_{i;x} = z_{\mathbf{t},\xi_i-\mathbf{t},x}$ is smaller than $\varepsilon/2$ if $d(P) < \delta$ and the number k_0 is sufficiently large so that $m_J(\Lambda_1(\mathbf{t})) = \cdots = m_J(\Lambda_k(\mathbf{t})) = 1/k \leqslant \delta$ for $k \geqslant k_0$, which simply implies the required. □

It is not clear whether a Stepanov $([0,1]^n, p)$-$(\mathcal{B}, \Lambda', \rho)$-almost periodic (Stepanov $([0,1]^n, p)$-$(\mathcal{B}, \Lambda', \rho)$-uniformly recurrent) function has to be summable $(\mathcal{B}, \Lambda', \rho)$-almost periodic (summable $(\mathcal{B}, \Lambda', \rho)$-uniformly recurrent). This is primarily caused by the fact that the value of Riemann sum $\sigma(f; P, (\xi_i)_{1\leqslant i\leqslant k})$ and the value of Riemann sum $\sigma(f; P', (\xi_i')_{1\leqslant i\leqslant m})$, where the partition P' is finer than P and the set $\{\xi_i' : 1 \leqslant i \leqslant m\}$ contains the set $\{\xi_i : 1 \leqslant i \leqslant k\}$, cannot be easily compared (if we do not use the points $(\xi_i)_{1\leqslant i\leqslant k}$ in our analysis, then we compare the upper and lower Darboux sums, which will not be employed here). The corresponding concepts of the Lebesgue summable $(\mathcal{B}, \Lambda', \rho, p)$-almost periodic function and the Lebesgue summable $(\mathcal{B}, \Lambda', \rho, p)$-uniformly recurrent function can be also introduced, with many nontrivial technical details.

It could be interesting to see what features of Bohr $(\mathcal{B}, \Lambda', \rho)$-almost periodic type functions continue to hold for the function spaces introduced in Definition 6.4.1.

On almost periodic solutions of KdV equation

In the research articles [239, 240], P. Deift has asked whether for the KdV equation

$$q_t(t,x) - 6q(t,x)q_x(t,x) + q_{xxx}(t,x) = 0 \tag{334}$$

with almost periodic initial data $q(0,x) = V(x)$, the solution evolves almost periodically in time, i. e., whether for each $t \in \mathbb{R}$ the solution $x \mapsto q(t,x)$, $x \in \mathbb{R}$ is almost periodic. The negative answer to this conjecture has recently been given by D. Damanik et al. [224] and A. Chapouto et al. [179]. In the last mentioned paper, the authors have introduced the notion of distributional and bounded solutions to (334) (cf. [179, Definition 1.1]) and shown the uniqueness of solutions to (334) that are merely bounded, without any further assumptions (cf. [179, Theorem 1.2]). The main result of paper is [179, Theorem 1.3], where the authors have shown that there is a bounded solution $q : [-T,T] \times \mathbb{R} \to \mathbb{R}$ of (334) with almost periodic initial data $V(x)$ for which the solution $x \mapsto q(t,x)$, $x \in \mathbb{R}$ is not almost periodic at some time $t_0 \in [-T,T]$.

Piecewise continuous Hölder-α-almost periodic type functions of one real variable

In the monograph [447], we have considered various classes of multidimensional Hölder-α-almost periodic type functions and provided certain applications to the abstract Volterra integrodifferential inclusions (cf. also Section 5.1.3). In this part, we will briefly explain how one can introduce and analyze several new classes of piecewise continuous Hölder-α-almost periodic type functions of one real variable.

Suppose that $\Lambda = [0,\infty)$ or $\Lambda = \mathbb{R}$, $F : \Lambda \to Y$ is a piecewise continuous mapping and $0 < \alpha \leqslant 1$. If $t \in \Lambda$ is a point of continuity of $F(\cdot)$, then we define

$$L_\alpha(t,\delta;F) := \sup_{u_1,u_2\in\Lambda\cap B(t,\delta);u_1\neq u_2} \frac{\|F(u_1)-F(u_2)\|_Y}{|u_1-u_2|^\alpha};$$

on the other hand, if the function $F(\cdot)$ is continuous from the left side at the point t, only, then we define

$$L_\alpha(t,\delta;F) := \sup_{u_1,u_2\in\Lambda\cap[t-\delta,t];u_1\neq u_2} \frac{\|F(u_1)-F(u_2)\|_Y}{|u_1-u_2|^\alpha}.$$

A point $t\in\Lambda$ is said to be an α-regular point of function $F(\cdot)$ if there exists a finite real number $\delta>0$ such that $L_\alpha(t,\delta;F)<+\infty$. Then we define $P^1_{0,\alpha}(\Lambda:Y)$ as the collection of all functions $F:\Lambda\to Y$ satisfying that each point $t\in\Lambda$ is an α-regular point of function $F(\cdot)$. If $F\in P^1_{0,\alpha}(\Lambda:Y)$, then we define $L_\alpha(t;F) := \lim_{\delta\to 0+} L_\alpha(t,\delta;F)$, $t\in\Lambda$. Finally, by $P_{0,\alpha}(\Lambda:Y)$ we denote the set of all functions $F\in P^1_{0,\alpha}(\Lambda:Y)$ satisfying that the function $t\mapsto L_\alpha(t,1;F)$, $t\in\Lambda$ is locally bounded. It is simply shown that $P^1_{0,\alpha}(\Lambda:Y)$ and $P_{0,\alpha}(\Lambda:Y)$ are vector spaces with the usual operations as well as that these spaces are translation invariant in the usual sense.

It is not difficult to prove that the space $\mathcal{P}_\alpha=(P_\alpha,d_\alpha)$ is a metric space if $P_\alpha = PC(\Lambda:Y)\cap P_{0,\alpha}(\Lambda:Y)$ and $d_\alpha(f,g) := \sup_{t\in\Lambda}(\|f(t)-g(t)\|_Y + \lim_{\delta\to 0+} L_\alpha(t,\delta;f-g))$. Moreover, it can be simply proved that the space $\mathcal{P}^1_\alpha=(P^1_\alpha,d^1_\alpha)$ is a pseudometric space if $P^1_\alpha = P_{0,\alpha}(\Lambda:Y)$ and $d_\alpha(f,g) := \sup_{t\in\Lambda}\lim_{\delta\to 0+} L_\alpha(t,\delta;f-g)$.

Keeping in mind the above notion, we can consider the notion of a piecewise continuous strongly ($\mathbb{F}$)-Hölder-α-almost periodic (semi-($\rho,\mathbb{F}$)-Hölder-α-periodic) function (of type 1), a piecewise continuous ($\mathbb{F},\Lambda',\rho$)-Hölder-α-almost periodic function (of type 1) and a piecewise continuous ($\mathbb{F},\Lambda',\rho$)-Lipschitz-uniformly recurrent function (of type 1). The corresponding classes of piecewise continuous Lipschitz almost periodic type functions (of type 1) can be defined by plugging $\alpha=1$; cf. also [448, Section 7.1]. More details about the above mentioned classes of almost periodic type functions will appear somewhere else.

Almost periodic type solutions to nonlinear evolution equations of first order

There are many research articles devoted to the study of almost periodic type solutions to nonlinear evolution equations of first order. The main problem in the analysis of the existence and uniqueness of almost periodic type solutions for nonlinear evolution equations presents the fact that the variation of parameters formula cannot be applied, or can be applied with certain obvious difficulties (cf. [284, 392, 549, 626, 776] and references quoted there for more details in this direction). Besides the papers mentioned in our previously published monographs [444–448], we would like to mention here the following research articles concerning this problem: [36, 196, 198, 582, 585, 634, 763, 764, 797].

In this part, we will briefly summarize the main results of the recent research article [800] by L. Ye and Y. Liu, where the authors have considered the existence and uniqueness of pseudo-almost periodic C^0-solutions for evolution inclusions with mixed nonlocal plus local initial conditions. The starting point is that $(X, \|\cdot\|)$ is a real Banach space and A is an m-dissipative operator, which generates a nonlinear semigroup of contractions $(S(t))_{t\geqslant 0}$ by the exponential formula. If a real constant $\tau \geqslant 0$ and a continuous function $f : [0,\infty) \times C([-\tau, 0] : \overline{D(A)}) \to X$ are given, the authors consider the following evolution inclusion with mixed nonlocal plus local initial conditions:

$$u'(t) \in Au(t) + f(t, u_t), \quad t \geqslant 0; \quad u(t) = p(u)(t) + \psi(t), \quad t \in [-\tau, 0]. \tag{335}$$

It has been assumed that the function $p : C([-\tau, +\infty) : \overline{D(A)}) \to C([-\tau, 0] : \overline{D(A)})$ is nonexpansive, and the local initial condition $\psi : [-\tau, 0] \to X$ is continuous such that $p(u) + \psi \in C([-\tau, 0] : \overline{D(A)})$ for every $u \in C_b([-\tau, +\infty) : \overline{D(A)})$. The function $p(\cdot)$ is usually called a feedback operator; if $\psi \equiv 0$, then the problem (335) reduces to

$$u'(t) \in Au(t) + f(t, u_t), \quad t \geqslant 0; \quad u(t) = p(u)(t), \quad t \in [-\tau, 0]. \tag{336}$$

A sufficient condition for the unique C^0-solution of problem (336) to be almost periodic has been obtained by I. I. Vrabie in [764]; furthermore, under certain reasonable assumptions, the existence and boundedness of the C^0-solution to problem (335) have been considered by the same author in [765] with the help of the Schaefer fixed-point theorem (let us also mention here the important research monograph [763], which considers the compactness methods for nonlinear evolution equations).

The following special case of (335) is also of concern:

$$u'(t) \in Au(t) + f(t), \quad t \geqslant 0; \quad u(t) = p(u)(t) + \psi(t), \quad t \in [-\tau, 0]. \tag{337}$$

In [585], B. Meknani, J. Zhang and T. Abdelhamid have proved that the problem (337) has a unique pseudo-almost periodic C^0-solution.

The main results of paper [800] are given in its third section. In the final section of this paper, the authors consider the existence and uniqueness of pseudo-almost periodic C^0-solution for the following problem:

$$\begin{aligned} &u_t(t,x) = -au_x(t,x) - \omega u(t,x) + f(t,u_t)(x), \quad (t,x) \in [0,\infty) \times \mathbb{R}; \\ &u(t,x) = u(t, x+\pi), \quad t \geqslant 0; \\ &u(t,x) = [p(u)(t)](x) + \psi(t,x), \quad t \in [-\tau, 0],\ x \in (0,\pi). \end{aligned}$$

A note on the generation of (degenerate) linear semigroups of contractions

In the theory of nonlinear semigroups of operators, it is crucial to assume that the semigroup $(S(t))_{t\geqslant 0}$ under the consideration has a growth of type $\|S(t)\| \leqslant e^{\omega t}$, $t \geqslant 0$ for some real number $\omega \in \mathbb{R}$; in contrast to the theory of linear semigroups of operators, the results concerning the generation of nonlinear semigroups $(S(t))_{t\geqslant 0}$ of the growth $\|S(t)\| \leqslant Me^{\omega t}$, $t \geqslant 0$ do not exist if $M \neq 1$.

In this part, we will provide some new observations and results on the generation of (degenerate) linear semigroups of contractions. Let us assume first that a densely defined multivalued linear operator $\mathcal{A}$ is the integral generator of an $(a,1)$-regularized I-resolvent family $(R(t))_{t\geqslant 0} \subseteq L(X)$ of contractions as well as that the mapping $\tilde{a}$: $(0,\infty) \to (0,\infty)$ is a bijection. Then an application of [445, Theorem 3.2.5] yields that

$$\left\| \left(\frac{1}{\tilde{a}(\lambda)} - \mathcal{A} \right)^{-1} x \right\| \leqslant |\tilde{a}(\lambda)| \cdot \|x\|, \quad \lambda > 0,\ x \in X.$$

This simply implies $\|(z-\mathcal{A})^{-1}x\| \leqslant z^{-1}\|x\|$ for all $z > 0$ and $x \in X$. Now we can apply [445, Proposition 1.2.6] in order to see that

$$\left\| \frac{d^{n-1}}{dz^{n-1}} [(z-\mathcal{A})^{-1}x] \right\| \leqslant (n-1)! \cdot z^{-n} \cdot \|x\|, \quad n \in \mathbb{N},\ z > 0,\ x \in X.$$

Keeping in mind [445, Theorem 3.2.12], we get that $\mathcal{A}$ is the integral generator of a linear semigroup $(S(t))_{t\geqslant 0}$ of contractions. This, in particular, holds if $a(t) = g_\alpha(t)$ for some $\alpha > 0$. Concerning this special case, we would like to state the following result.

Theorem 6.4.3. *Suppose that $\mathcal{A}$ is a densely defined multivalued linear operator. Then the following statements are equivalent:*
(i) *$\mathcal{A}$ is the integral generator of a linear semigroup $(S(t))_{t\geqslant 0} \subseteq L(X)$ of contractions.*
(ii) *For every $\alpha \in (0,1)$, $\mathcal{A}$ is the integral generator of a $(g_\alpha,1)$-regularized* I-*resolvent family $(R(t))_{t\geqslant 0} \subseteq L(X)$ of contractions.*
(iii) *For some $\alpha \in (0,1)$, $\mathcal{A}$ is the integral generator of a $(g_\alpha,1)$-regularized* I-*resolvent family $(R(t))_{t\geqslant 0} \subseteq L(X)$ of contractions.*

Proof. Using the argumentation contained in the proof of the subordination principle [97, Theorem 3.1], we can simply prove that (i) implies (ii). It is clear that (ii) implies (iii); the above analysis shows that (iii) implies (i), completing the proof. □

Asymptotically almost periodic solutions to nonlinear Volterra integral equations

The existence and uniqueness of almost periodic solutions to nonlinear Volterra integral equations have been sought in many research articles by now; cf. the list of references quoted in [279] and [444–448] for further information in this direction. On the other hand, in the existing literature concerning the abstract nonlinear Volterra integrodifferential equations (see also [89, 214, 342, 343, 547, 626] and the references quoted therein), we have not been able to locate any relevant result with regards to the question of the existence and uniqueness of asymptotically almost periodic solutions.

In this part, we will present a new interesting result concerning the existence and uniqueness of asymptotically almost periodic solutions to the abstract nonlinear Volterra integral equations with Lipschitz continuous operators (Theorem 6.4.4; cf. also the results established in [280]); the proof of this result is relatively simple and it is based on the use of the Banach contraction principle. The operators under our consideration fail to be m-accretive in any sense and our results cannot be formulated for the class of asymptotically periodic functions or the class of pseudo-almost periodic functions. We also introduce and analyze here the class of mild (a, k, C, B)-regularized resolvent families in the nonlinear setting and provide the basic details about the well-posedness of abstract nonlinear Volterra inclusions [467].

Suppose that $B : X \to X$ and $\|Bx - By\| \leqslant L\|x - y\|$, $x, y \in X$ for some finite real constant $L > 0$. Of concern is the following abstract nonlinear integral equation:

$$u(t) = b(t) + \int_0^t a(t-s)Bu(s)\, ds, \quad t \geqslant 0. \tag{338}$$

It is worth noting that the Banach contraction principle can be successfully applied in the analysis of the existence of a unique asymptotically almost periodic solution of problem (338), provided that $a \in L^1([0,\infty))$ and $b(\cdot)$ is asymptotically almost periodic.

To explain this in more detail, denote by $AAA([0,\infty) : X)$ the Banach space of all asymptotically almost periodic functions $f : [0,\infty) \to X$, equipped with the sup-norm. Let us recall that a necessary and sufficient condition for a continuous function $f : [0,\infty) \to +\infty$ to be asymptotically almost periodic is that for each sequence (b_k) in $[0,\infty)$ there exist a subsequence (b_{k_l}) of (b_k) and a function $f^* : [0,\infty) \to X$ such that $\lim_{l\to+\infty} f(t + b_{k_l}) = f^*(t)$, uniformly for $t \geqslant 0$. Using this criterion and the Lipschitz continuity of the operator B, it readily follows that the mapping $Bf(\cdot)$ is asymptotically almost periodic, provided that $f(\cdot)$ is asymptotically almost periodic. Using now the Bochner criterion and the Lipschitz continuity of the operator B together imply that the mapping $Bf(\cdot)$ is asymptotically almost periodic, provided that $f(\cdot)$ is asymptotically almost periodic. Further on, we have

$$\begin{aligned}\int_0^t a(t-s)f(s)\, ds &= \int_0^t a(t-s)[g(s) + q(s)]\, ds = \int_{-\infty}^t a(t-s)g(s)\, ds \\ &\quad + \left[\int_0^{t/2} a(t-s)q(s)\, ds + \int_{t/2}^t a(t-s)q(s)\, ds - \int_{-\infty}^0 a(t-s)g(s)\, ds\right] \\ &:= G(t) + Q(t), \quad t \geqslant 0.\end{aligned}$$

Using the integrability of $a(\cdot)$ and the standard argumentation, it follows that the mapping $t \mapsto G(t)$, $t \in \mathbb{R}$ is almost periodic, the function $Q(\cdot)$ is continuous and $\lim_{t\to+\infty} Q(t) = 0$. Therefore, the function

$$t \mapsto b(t) + \int_0^t a(t-s)f(s)\, ds, \quad t \geqslant 0$$

is asymptotically almost periodic provided that $f(\cdot)$ is asymptotically almost periodic.

By the foregoing, we have that the operator $\Psi : AAA([0,\infty) : X) \to AAA([0,\infty) : X)$, given by

$$[\Psi(f)](t) := b(t) + \int_0^t a(t-s)Bf(s)\,ds, \quad t \geqslant 0,\ f \in AAA([0,\infty) : X),$$

is well-defined for every $x \in X$. Furthermore, the assumption $L\|a\|_{L^1([0,\infty))} < 1$ implies that $\Psi(\cdot)$ is a contraction and an application of the Banach contraction principle yields that the following result holds true.

Theorem 6.4.4. *Suppose that $B : X \to X$, $\|Bx - By\| \leqslant L\|x-y\|$, $x, y \in X$ for some finite real constant $L > 0$, $a \in L^1([0,\infty))$, $L\|a\|_{L^1([0,\infty))} < 1$ and $b(\cdot)$ is asymptotically almost periodic. Then there exists a unique asymptotically almost periodic solution $u(t)$ of* (338).

We can similarly analyze the existence and uniqueness of asymptotically almost automorphic solutions of (338), provided that the function $b(\cdot)$ is asymptotically almost automorphic.

In connection with the above analysis, we would like to introduce the following notion (we can also generalize the notion from [445, Definitions 3.2.1, 3.2.2] in a similar manner).

Definition 6.4.5. Suppose $0 < \tau \leqslant \infty$, $k \in C([0,\tau))$, $k \neq 0$, $a \in L^1_{loc}([0,\tau))$, $a \neq 0$, $C : X \to X$ and $B : D(B) \subseteq X \to P(X)$ is a given function. Then we say that B is a subgenerator of [the generator, if $C = I$] a (local, if $\tau < \infty$) mild (a,k,C,B)-regularized resolvent family $(R(t))_{t\in[0,\tau)}$ if and only if $R(t) : X \to X$ is continuous for every $t \in [0,\tau)$, $(R(t))_{t\in[0,\tau)}$ is strongly continuous and, for every $x \in D(B)$, there exists a locally integrable mapping $t \mapsto r_B(t;x)$, $t \in [0,\tau)$ such that $r_B(t;x) \in BR(t)x$, $t \in [0,\tau)$ and

$$\int_0^t a(t-s)r_B(s)\,ds = R(t)x - k(t)Cx, \quad t \in [0,\tau). \tag{339}$$

If $C = I$, then we omit the term "C" from the notation.

In the nonlinear setting, we would like to emphasize the following:

(i) It is redundant to assume that $R(t)B \subseteq BR(t)$, $t \in [0,\tau)$ or $R(t)x - k(t)Cx \in B\int_0^t a(t-s)R(s)x\,ds$, $t \in [0,\tau)$, $x \in X$.

(ii) The use of function $k(\cdot)$ is also discussable because we cannot prove that B generates a mild $(a*b,k,B)$-regularized resolvent family $((b*R)(t))_{t\in[0,\tau)}$ provided that B generates a mild (a,k,B)-regularized resolvent family $(R(t))_{t\in[0,\tau)}$; here, $b \in L^1_{loc}([0,\tau))$.

If $\tau = \infty$, then we say that $(R(t))_{t\geqslant 0}$ is exponentially nonexpansive (nonexpansive) if there exists $\omega \in \mathbb{R}$ ($\omega = 0$) such that

$$\|R(t)x - R(t)y\| \leqslant e^{\omega t}\|x-y\|;$$

the infimum of such numbers is said to be the exponential type of $(R(t))_{t\geqslant 0}$. In the local setting, the notion from Definition 6.4.5 can be also introduced for the strongly continuous operator families defined on the closed interval $[0,\tau]$, where $0<\tau<\infty$.

Let us consider now the Banach space $C([0,\tau] : X)$, if $0<\tau<\infty$, and the Banach space $AAA([0,\infty) : X)$, if $\tau=\infty$. Applying the Banach contraction principle and the Grönwall inequality, we can prove the following result.

Theorem 6.4.6. (i) *Suppose that* $k\in C[0,\tau]$, $k\geqslant 0$, $a\in L^1[0,\tau]$, $B: X\to X$ *is a Lipschitz continuous operator,* $\|Bx-By\|\leqslant L\|x-y\|$, $x, y\in X$ *for some finite real constant* $L>0$, *and* $L\int_0^\tau |a(s)|\,ds<1$. *Then there exists a unique mild* (a,k,B,C)*-regularized resolvent family* $(R(t))_{t\in[0,\tau]}$ *subgenerated by* B *and the following holds:*

$$\|R(t)x-R(t)y\|\leqslant\left[k(t)+\int_0^t k(s)|a(t-s)|\exp\left(\int_s^t |a(t-r)|\,dr\right)ds\right]\cdot\|Cx-Cy\|, \tag{340}$$

for any $t\in[0,\tau]$ *and* $x, y\in X$. *Moreover, if* $k(\cdot)$ *is nondecreasing, then*

$$\|R(t)x-R(t)y\|\leqslant k(t)\exp\left(\int_0^t |a(s)|\,ds\right)\cdot\|Cx-Cy\|, \tag{341}$$

for any $t\in[0,\tau]$ *and* $x, y\in X$.

(ii) *Suppose that* $k\in C([0,\infty))$, $k\geqslant 0$, $a\in L^1([0,\infty))$, $B: X\to X$ *is a Lipschitz continuous operator,* $\|Bx-By\|\leqslant L\|x-y\|$, $x, y\in X$ *for some finite real constant* $L>0$, *and* $L\int_0^\infty |a(s)|\,ds<1$. *Then there exists a unique mild* (a,k,B,C)*-regularized resolvent family* $(R(t))_{t\geqslant 0}$ *subgenerated by* B *and* (340) *holds for any* $\tau>0$, *respectively,* (341) *holds for any* $\tau>0$, *provided that* $k(\cdot)$ *is nondecreasing. Furthermore, if* $k(\cdot)$ *is asymptotically almost periodic, then the mapping* $t\mapsto R(t)x$, $t\geqslant 0$ *is asymptotically almost periodic for every element* $x\in X$.

If the operator $B : D(B)\subseteq X\to X$ is not Lipschitz continuous, then we must apply some other types of fixed-point theorems in order to prove the existence of a local mild (a,k,B,C)-regularized resolvent family $(R(t))_{t\in[0,\tau]}$ subgenerated by B. It would be very tempting to apply the structural results about mild (a,k,B,C)-regularized resolvent families in the analysis of the abstract nonlinear Volterra integrodifferential equations, which do not involve Lipschitz continuous operators.

Suppose now that $0<\tau\leqslant\infty$, $a\in L^1_{loc}([0,\tau))$ and $f:[0,\tau)\to X$. By a solution of the abstract Volterra integral inclusion,

$$u(t)\in f(t)+\int_0^t a(t-s)Bu(s)\,ds,\quad t\in[0,\tau), \tag{342}$$

we mean any continuous function $t \mapsto u(t)$, $t \in [0,\tau)$ such that there exists a locally integrable mapping $t \mapsto u_B(t)$, $t \in [0,\tau)$ such that $u_B(t) \in Bu(t)$, $t \in [0,\tau)$ and

$$u(t) = f(t) + \int_0^t a(t-s)u_B(s)\,ds, \quad t \in [0,\tau).$$

Hence, if B subgenerates a mild (a,k,C,B)-regularized resolvent family $(R(t))_{t\in[0,\tau)}$, then for each $x \in D(B)$ the function $t \mapsto R(t)x$, $t \in [0,\tau)$ is a solution of the problem (342) with $f(t) \equiv k(t)Cx$. On the other hand, it is said that any $(m-1)$-times continuously differentiable function $t \mapsto u(t)$, $t \in [0,\tau)$ is a solution of the abstract fractional Cauchy inclusion

$$\mathbf{D}_t^{\alpha}u(t) \in Bu(t) + h(t), \quad t \in [0,\tau); \quad u^{(j)}(0) = u_j, \quad 0 \leqslant j \leqslant m-1, \tag{343}$$

where $h : [0,\tau) \to X$ is a continuous mapping, $\alpha \in (0,\infty) \setminus \mathbb{N}$ and $m = \lceil \alpha \rceil$, if the initial conditions are satisfied and there exists a continuous mapping $t \mapsto u_B(t)$, $t \in [0,\tau)$ such that $u_B(t) \in Bu(t)$, $t \in [0,\tau)$ and $\mathbf{D}_t^{\alpha}u(t) = u_B(t) + h(t)$, $t \in [0,\tau)$. The following statement can be simply proved as in the linear case ([445]; cf. also the identity [97, (1.21)]).

Proposition 6.4.7. *Suppose that the mapping $t \mapsto u(t)$, $t \in [0,\tau)$ is $(m-1)$-times continuously differentiable. Then $u(\cdot)$ is a solution of the abstract fractional Cauchy inclusion* (343) *if and only if $u(\cdot)$ is a solution of the abstract Volterra integral inclusion* (342) *with $a(t) \equiv g_\alpha(t)$ and $f(t) \equiv \sum_{j=0}^{m-1} g_{j+1}(t)u_j + (g_\alpha * h)(t)$, $t \in [0,\tau)$.*

Furthermore, we have the following.

Proposition 6.4.8. *Suppose that $B : X \to X$, $\|Bx - By\| \leqslant L\|x-y\|$, $x, y \in X$ for some finite real constant $L > 0$ and for each $x \in X$ there exists a solution of the abstract fractional Cauchy inclusion* (343) *with $h(t) \equiv 0$, $u_0 = Cx$ and $u^{(j)}(0) = 0$ for $1 \leqslant j \leqslant m-1$. Then B subgenerates a mild $(a,1,C,B)$-regularized resolvent family on $[0,\tau)$.*

Proof. We define $R(t)x := u(t;x)$, $x \in X$, $t \in [0,\tau)$, where $u(\cdot;x)$ is a solution of the abstract fractional Cauchy inclusion (343) with $h(t) \equiv 0$, $u_0 = Cx$ and $u^{(j)}(0) = 0$ for $1 \leqslant j \leqslant m-1$. Then it is clear that the family $(R(t))_{t\in[0,\tau)}$ is strongly continuous as well as that Proposition 6.4.7 shows that, for every $x \in D(B)$, there exists a locally integrable mapping $t \mapsto r_B(t;x)$, $t \in [0,\tau)$ such that $r_B(t;x) \in BR(t)x$, $t \in [0,\tau)$ and (339) holds with $k(t) \equiv 1$. It remains to be proved that the mapping $R(t) : X \to X$ is continuous for every fixed number $t \in [0,\tau)$. But this simply follows from an application of the Gronwall inequality, since we have

$$\|R(t)x - R(t)y\| \leqslant \|Cx - Cy\| + L\int_0^t g_\alpha(t-s)\|R(s)x - R(s)y\|\,ds, \quad x, y \in X.$$

□

We will analyze mild (a,k,C,B)-regularized resolvent families somewhere else. For more details about the abstract nonlinear Volterra integrodifferential equations, we also refer the reader to [89, 214, 342, 343, 547, 626] and references quoted therein. Let us also note that we can similarly analyze some classes of the abstract nonlinear functional Volterra equations with Lipschitz continuous operators, involving the bounded or unbounded delays.

Almost periodicity on time scales

Time scale calculus is an important, relatively new, branch of mathematics, which deals with systems that evolve over time and whose behavior changes at different rates. The idea of working with hybrid domains like time scales was proposed in the research article [383] by S. Hilger (1990). In time scales calculus, the authors treat noncontinuous functions that change at different rates in different parts of the time domain. It is an interesting combination of differential equations, difference equations and dynamic systems theory. Time scale calculus has many important applications in various fields like physics, engineering, economics and biology; especially, we would like to emphasize here the applications of time scale calculus in population dynamics, epidemics and control systems.

As already mentioned, in [202], M. F. Choquehuanca, J. G. Mesquita and A. Pereira have investigated the existence and uniqueness of the almost automorphic solutions to the second-order equations with boundary conditions on time scales. Weyl almost periodic functions on time scales and their Fourier series have recently been analyzed by Y. Li and X. Huang in [514].

In our recent joint paper with D. Agarwal, S. Dhama and S. Abbas [18], we have investigated the periodicity, stability and synchronization of solutions for a class of hybrid coupled dynamic equations with multiple delays; cf. also [15, 27, 129, 137, 222, 225, 245, 309, 357, 422, 698, 699] and references quoted therein. More precisely, in [18], we have considered general coupled dynamic equations on the time scale with multiple delays and nonlinear feedback. The special cases of this model describe the mathematical model of Nicholson blowflies, Lasota–Wazewska model, and Mackey glass. These models and their modifications have been subjected to extensive and rigorous analysis, yielding various results of their stability, persistence, attractivity, periodic solutions, almost periodic solutions, and other related aspects. Many authors have studied the oscillation of the discrete Nicholson's blowflies model, the periodicity of the generalized delayed Nicholson blowflies model using the fixed-point theorem and the existence of positive periodic solutions to the discrete Lasota–Wazewska model with impulses using the coincidence degree theory.

The investigation of the periodicity of Nicholson blowflies, Lasota–Wazewska and Mackey–Glass models on time scales has been initiated in [18] with the help of the coincidence degree theory. Furthermore, synchronization is a novel analysis for this type of model. Synchronization refers to the phenomenon where individuals within a population exhibit similar behavior patterns, such as reproduction or migration, at the same time. This can occur due to various factors, such as environmental cues or interactions between individuals. Synchronization can play an important role in determining the dynamics of the population. For example, if individuals synchronize their reproduction, it can lead to a boom-and-bust cycle in the population, where there are periods of high reproductive success followed by periods of low reproductive success. On the other hand, if individuals stagger their reproduction, it can lead to a more stable population size. Synchronization can also have implications for the conservation of endangered species, as it can affect the viability of small populations. If individuals in a small population do not synchronize their behavior, it can lead to a loss of genetic diversity and an increased risk of inbreeding, which can reduce the population's ability to adapt to changing environment. In [18], we have analyzed the problem of the exponential synchronization for the equations under our consideration by means of feedback control.

Concerning almost periodic fractional fuzzy dynamic equations on time scales, it is worthwhile to mention the survey article [768] by C. Wang, Y. Tan and R. P. Agarwal.

Katznelson–Tzafriri type theorems

Before proceeding further, we would like to note that the Katznelson–Tzafriri type theorems for linear evolution equations have been established by B. X. Dieu, S. Siegmund and N. V. Minh in [258], while the Katznelson–Tzafriri type theorems for linear difference equations have been established by N. V. Minh, H. Matsunaga, N. D. Huy and V. T. Luong in [597]; cf. also [598], where the authors have considered spectral results for polynomially bounded sequences and the Katznelson–Tzafriri type theorems for linear fractional difference equations. Concerning the oscillatory solutions of linear difference equations (with several delays), we can warmly recommend the research articles [72, 105, 180, 181, 204] and references cited therein, as well.

On the solvability of a fourth-order differential evolution equation on singular cylindrical domain of $\mathbb{R}^4$

In this part, we will present the main details of our joint research study [175] with B. Chaouchi. In [175], we have investigated the following fourth-order evolution equation

$$\frac{\partial^4 u}{\partial t^4} = -\Delta^2 u + h, \tag{344}$$

where Δ^2 is the biharmonic operator on $\mathbb{R}^3$ defined by

$$\Delta^2 := \frac{\partial^4}{\partial z^4} + 2\left(\frac{\partial^4}{\partial z^2 \partial x^2} + \frac{\partial^4}{\partial z^2 \partial y^2}\right) + \left(\frac{\partial^2}{\partial x^2} + \frac{\partial^2}{\partial y^2}\right)^2.$$

Due to some realistic physical reasons, the solvability of the problem (344) is considered in a nonsmooth cylindrical domain $Q := \mathbb{R}^+ \times \Pi$, where $\Pi = \Omega \times \mathbb{R}^+$ and Ω the base of Π is a planar cusp domain defined by

$$\Omega := \{(x,y) \in \mathbb{R}^2 : 0 < x < x_0,\ -\psi(x) < y < \psi(x)\};$$

here, $\mathbb{R}_+ = [0,\infty)$,
(1) $\psi(x) = x^\alpha$ with $1 < \alpha \leqslant 2$.
(2) x_0 is a given point near the cuspidal point $\{0\}$.

We have assumed that

$$h \in C^{4\sigma}(\mathbb{R}^+; C^{4\sigma}(\Pi)), \quad 0 < 4\sigma < 1,$$

with

$$f|_{\partial Q} = 0. \tag{345}$$

In our study, we have dealt with the space $C^{4\sigma}(\mathbb{R}^+; C^{4\sigma}(\Pi))$ consisting of all (4σ)-Hölder continuous functions $f : \mathbb{R}^+ \to C^{4\sigma}(\Pi)$ such that

$$\|f\|_{C^{4\sigma}(\mathbb{R}^+;C^{4\sigma}(\Pi))} := \|f\|_{BUC(\mathbb{R}^+;C^{4\sigma}(\Pi))} + \sup_{t>\tau\geqslant 0} \frac{\|f(t) - f(\tau)\|_{C^{4\sigma}(\Pi)}}{|t-\tau|^{4\sigma}} < \infty.$$

Evolution differential equations involving the biharmonic operator arise in several models describing various phenomena in the applied sciences. Among the most famous, we quote the Kardar–Parisi–Zhang equation

$$\frac{\partial u}{\partial t} = \mu\Delta^2 + \kappa\Delta|\nabla|^2 + f,$$

arising in crystallography; here, μ and κ are some physical parameters. The second famous equation with biharmonic operator is the equation

$$\frac{\partial^2 u}{\partial t^2} = \Delta^2 + f,$$

arising in statics and dynamics of a suspension bridge. In description of the instability and the structural behavior of suspension bridges under the action of dead and live loads, the suggested model is given by the following fourth-order evolution equation:

$$\frac{\partial^2 u}{\partial t^2} = -u_{xxxx} - \gamma u^+ + h, \quad x > 0,\ t > 0,$$

where $u = u(x,t)$ denotes the vertical displacement of the beam in the downward direction, $u^+ = \max(u,0)$, γu^+ represents the force due to the hangers, and h represents the forcing term acting on the bridge, including its own weight per unit length.

We accompany the problem (344) with the periodic type conditions

$$u|_{\{t\}\times\Pi} - u|_{\{t+T\}\times\Pi} = 0, \tag{346}$$

where $T > 0$ is a positive real number. We also suppose that a solution u of (344) satisfies

$$\begin{cases} u|_{z=z} - u|_{z=z+\varpi} = 0, \\ u|_{\mathbb{R}^+\times\partial\Pi} = 0, \end{cases} \tag{347}$$

where $\varpi > 0$ a positive real number.

The solvability of problem (344)–(346)–(347) is discussed using the transformation of the problem (344)–(346)–(347) set in the nonregular domain Q into a new transformed problem equation set in a regular domain. The change of variables

$$\Psi : (t, x, y, z) \mapsto (t, \xi, \eta, \lambda) = \left(t, \frac{x^{1-\alpha}}{\alpha - 1}, \frac{y}{\psi(x)}, z\right),$$

transforms the cylinder Q into a new one, given by

$$Q_\infty = \mathbb{R}^+ \times D \times \mathbb{R}^+,$$

where

$$D := (\xi_0, +\infty) \times (-1, 1) \quad \text{and} \quad \xi_0 := \frac{1}{\alpha - 1} x_0^{1-\alpha} > 0.$$

Set

$$u(t, x, y, z) := v(t, \xi, \eta, \lambda) \quad \text{and} \quad h(t, x, y, z) := g(t, \xi, \eta, \lambda).$$

In such a way, the problem (344) is transformed in

$$\frac{\partial^4 v}{\partial t^4} + \Delta^2 v + \frac{1}{\xi}[Lv] = \theta^{-4}\xi^{-4\beta} g, \quad \text{in } Q_\infty, \tag{348}$$

where

$$\beta := \alpha/(\alpha - 1), \quad \theta := (\alpha - 1)^\beta,$$

and L is a fourth differential operator with C^∞-bounded coefficients on Q_∞. The new boundary conditions are given by

$$\begin{gathered} v|_{\xi=\xi_0} = 0, \quad v|_{\xi\to+\infty} = 0, \\ v|_{\eta=1} = 0, \quad v|_{\eta=-1} = 0, \\ v|_{\lambda=\lambda} = v|_{\lambda=\lambda+\varpi}, \\ v|_{\{t\}\times\Pi} - v|_{\{t+T\}\times\Pi} = 0. \end{gathered} \tag{349}$$

After a translation to the origin with a respect to the variable ξ, we focus on the study of the following problem in Q_∞, given by

$$\frac{\partial^4 v}{\partial t^4} + \Delta^2 v = f, \tag{350}$$

with

$$\begin{aligned} v|_{\xi=0} = 0, \quad v|_{\xi\to+\infty} = 0, \\ v|_{\eta=1} = 0, \quad v|_{\eta=-1} = 0, \\ v|_{\lambda=\lambda} = v|_{\lambda=\lambda+\varpi}, \\ v|_{\{t\}\times\Pi} - v|_{\{t+T\}\times\Pi} = 0, \end{aligned} \tag{351}$$

and

$$f = \theta^{-4}\xi^{-4\beta}g.$$

Set $E := C^{4\sigma}(D \times \mathbb{R}^+)$ and define the functions:

$$\begin{aligned} v : \mathbb{R}^+ \to E;\ t \longrightarrow v(t); \quad v(t)(\xi,\eta,\lambda) := v(t,\xi,\eta,\lambda), \\ f : \mathbb{R}^+ \to E;\ t \longrightarrow f(t); \quad f(t)(\xi,\eta,\lambda) := f(t,\xi,\eta,\lambda). \end{aligned}$$

Let A denote the biharmonic operator Δ^2 with domain consisting of those functions

$$v \in C^4(D \times \mathbb{R}^+) \cap W^{4,p}(D \times \mathbb{R}^+), \quad 1 < p < +\infty,$$

such that

$$v|_{\xi=\xi_0} = 0, \quad v|_{\xi\to+\infty} = 0, \quad v|_{\eta=1} = 0, \quad v|_{\eta=-1} = 0, \quad v|_{\lambda=\lambda} = v|_{\lambda=\lambda+\varpi}.$$

Summing up, the abstract version of problem (350)–(351) is given in the complex Banach space E by the following fourth-order abstract differential equation:

$$\frac{d^4v(t)}{dt^4} + Av(t) = f(t), \tag{352}$$

under periodic boundary conditions

$$v(t+T) = v(t). \tag{353}$$

Consider now the following problem:

$$\Delta^2 w = \phi, \tag{354}$$

associated with the following conditions:

$$\begin{aligned} w|_{\xi=0} = 0, \quad w|_{\xi\to+\infty} = 0, \\ w|_{\eta=1} = 0, \quad w|_{\eta=-1} = 0, \\ w|_{\lambda=\lambda} = w|_{\lambda=\lambda+\varpi}. \end{aligned} \tag{355}$$

The study of (354)–(355) is performed in the Banach space $C_0^{4\sigma}(D \times \mathbb{R}^+)$, where

$$C_0^{4\sigma}(D \times \mathbb{R}^+) := \{\varphi \in C^{4\sigma}(D \times \mathbb{R}^+) : \varphi|_{\partial D} = 0\}, \quad 0 < 4\sigma < 1.$$

We rewrite the problem (354)–(355) in the form of the following abstract differential equation:

$$\frac{d^4 w(\lambda)}{d\lambda^4} + 2M \frac{d^2 w(\lambda)}{d\lambda^4} + M^2 w(\lambda) = \phi(\lambda), \quad \lambda \in \mathbb{R}^+, \tag{356}$$

subjected with the periodic condition

$$w(\lambda) = w(\lambda + \varpi),$$

where

$$\begin{cases} D(M) := \{w \in W^{4,p}(D) \cap C^{4\sigma}(D) : w|_{\xi=0} = 0,\ w|_{\xi \to +\infty} = 0,\ w|_{\eta=1} = 0,\ w|_{\eta=-1} = 0\}, \\ M\varphi := \Delta\varphi, \end{cases}$$

and the vector valued functions w and ϕ are defined as follows:

$$\begin{aligned} w &: \mathbb{R}^+ \to X;\ \lambda \longrightarrow w(\lambda); \quad w(\lambda)(\xi,\eta) := w(\xi,\eta,\lambda), \\ \phi &: \mathbb{R}^+ \to X;\ \lambda \longrightarrow \phi(\lambda); \quad \phi(\lambda)(\xi,\eta) := \phi(\xi,\eta,\lambda), \end{aligned}$$

where

$$X = C^{4\sigma}(D), \quad 0 < 4\sigma < 1.$$

The following results hold true.

Proposition 6.4.9. *The operator M is closed, nondensely defined and satisfies the Krein-ellipticity property, that is, $\mathbb{R}^+ \subseteq \rho(M)$ and there exists $c > 0$ such that for all $\mu \geqslant 0$ we have*

$$\|(M - \mu I)^{-1}\|_{\mathcal{L}(X)} \leqslant \frac{c}{1 + |\mu|}. \tag{357}$$

Moreover, (357) *holds in a sector of the form*

$$S_{\varepsilon_0,\delta_0} := \{z \in \mathbb{C} : |z| \geqslant \varepsilon_0 \text{ and } |\arg z| \leqslant \delta_0\} \cup \{z = \varepsilon_0 e^{iv} : -\delta_0 \leqslant v \leqslant \delta_0\},$$

with some small $\varepsilon_0 > 0$ and $\delta_0 \in (\pi/2, \pi)$.

Proposition 6.4.10. *The operator A is closed, nondensely defined and satisfies the Krein-ellipticity in a sector of the form $S_{\varepsilon_0,\delta_0}$ for some $\varepsilon_0 > 0$ and $\delta_0 \in (\pi/2, \pi)$.*

By a strict solution v to (352), we mean a vectorial function v such that

$$v \in C^4(\mathbb{R}^+; E) \cap C(\mathbb{R}^+; D(A))$$

and (353) is satisfied. The natural representation of the solution of (352)–(353), in the abstract case, is given by the sum of the Dunford integrals

$$v(t) = v_1(t) + v_2(t) + v_3(t), \tag{358}$$

where, for every $t \geqslant 0$,

$$v_1(t) := -\frac{1}{2\pi i} \int_{\gamma_1} \int_t^{t+T} \frac{e^{-r(\mu)(s-t)}}{c_\mu(T)} (A - \mu I)^{-1} f(s)\, ds,$$

$$v_2(t) := -\frac{1}{2\pi i} \int_{\gamma_1} \int_t^{t+T} \frac{e^{-r(\mu)(s-t-T)}}{c_\mu(T)} (A - \mu I)^{-1} f(s)\, ds,$$

$$v_3(t) := -\frac{1}{2\pi i} \int_{\gamma_1} \int_t^{t+T} \frac{\cosh(-r(\mu)(t - s + \frac{T}{2}))}{s_\mu(T)} (A - \mu I)^{-1} f(s)\, ds.$$

Here,

$$r(\mu) := \sqrt[4]{-\mu},$$
$$c_\mu(T) := 1 - e^{-r(\mu)T},$$
$$s_\mu(T) := 4r^2(\mu) \sinh\left(\frac{r(\mu)}{2} T\right),$$

with $r(\mu)$ is the analytic determination defined by $\mathrm{Re}(r(\mu)) > 0$ while γ_1 is a suitable curve in the complex plane.

The following results are needed in order to study the optimal regularity of the solution (358).

Proposition 6.4.11. *Let $f \in C^{4\sigma}(\mathbb{R}^+; E)$ with $0 < 4\sigma < 1$, and let $f(0) = 0$. Then we have:*
(1) $v_i(\cdot) \in D(A)$, $i \in \{1, 2, 3\}$.
(2) $f(\cdot) - Av_i(\cdot) \in C^{4\sigma}(\mathbb{R}^+; E)$, $i \in \{1, 2, 3\}$.
(3) $\frac{d^k v_i(\cdot)}{dt^4} \in C^{4\sigma}(\mathbb{R}^+; E)$, $i \in \{1, 2, 3\}$ *and* $k \in \{1, 2, 3, 4\}$.

Theorem 6.4.12. *Let $f \in C^{4\sigma}(\mathbb{R}^+, E)$ with $0 < 4\sigma < 1$, and let $f(0) = 0$. Then, the unique strict solution of* (352)–(353) *given by* (358) *satisfies the property of maximal regularity, that is,*

$$Av(\cdot), \quad \frac{d^4 v(\cdot)}{dt^4} \in C^{4\sigma}(\mathbb{R}^+, E).$$

At this level, we are able to state our main result concerning the transformed problem given by (350)–(351).

Proposition 6.4.13. *Let* $f \in C^{4\sigma}(\mathbb{R}^+, C^{4\sigma}(D \times \mathbb{R}^+))$ *with* $0 < 4\sigma < 1$. *Then problem* (350)–(351) *has a unique strict solution*

$$v \in C^4(Q_\infty).$$

Moreover, $v(\cdot)$ *satisfies the following maximal regularity property:*

$$\frac{\partial^4 v}{\partial t^4}, \quad \Delta^2 v \in C^{4\sigma}(\mathbb{R}^+, C^{4\sigma}(D \times \mathbb{R}^+)).$$

By a classical argument of perturbation, the same results hold true for the complete transformed problem (348)–(349).

Recall that our change of variables is defined as follows:

$$\Psi : Q \to Q_\infty, \quad (t, x, y, z) \mapsto (t, \xi, \eta, \lambda) = \left(t, \frac{x^{1-\alpha}}{\alpha - 1}, \frac{y}{\psi(x)}, z\right).$$

Then, in order to justify the conclusion for problem (344)–(346)–(347) set in the singular cylinder Q, we need to use the inverse change of variables

$$\Psi^{-1} : Q_\infty \to Q, \quad (t, \xi, \eta, \lambda) \mapsto (t, x, y, z) = \left(t, e^{\frac{\ln(\alpha-1)\xi}{1-\alpha}}, \eta e^{\frac{\alpha \ln(\alpha-1)\xi}{1-\alpha}}, \lambda\right).$$

Our main result reads as follows.

Theorem 6.4.14. *Let* $h \in C^{4\sigma}(\mathbb{R}^+; C^{4\sigma}(\Pi))$, *let* $0 < 4\sigma < 1$ *and let* (345) *hold. Then problem* (344)–(346)–(347) *admits a unique strict solution* u *satisfying*

$$\frac{\partial^4 u}{\partial t^4} \quad \text{and} \quad \Delta^2 u \in C^{4\sigma}(\mathbb{R}^+, C^{4\sigma,\infty}(\Pi)),$$

where

$$C^{4\sigma,\infty}(\Pi) = \{\varphi \in C^{4\sigma,\infty}(\Pi) : x^{4\sigma(2\alpha-1)}\varphi \in C^{4\sigma}(\Pi)\}.$$

Almost periodic solutions of singular integrodifferential equations and singularly perturbed integrodifferential equations

The existence and uniqueness of periodic solutions to singular ordinary differential equations and their systems have attracted the considerable attention of many authors so far; see, e. g., [199, 203, 505, 515, 520, 523, 524, 532, 552, 668, 794] and references quoted therein.

It is worth noticing that A. C. Lazer and S. Solimini were the first who investigated the positive periodic solutions for singular ordinary differential equations ([505], 1987). In this seminal paper, the authors analyzed the nonlinear ordinary differential equation

$$u'' - u^{-\alpha} = p(t), \quad u > 0, \tag{359}$$

where $\alpha > 0$ and $p : \mathbb{R} \to \mathbb{R}$ is a continuous 2π-periodic function; it was proved that the equation (359) with $\alpha \geqslant 1$ has a positive periodic solution if and only if the function $p(\cdot)$ has the negative mean value as well as that the above is no longer true if $\alpha \in (0, 1)$. The particular case $\alpha = 1/2$ has recently been studied by D. Rojas in [667], where the author proved that this equation corresponds to a perturbed isochronous oscillator and resonance conditions on the forcing term $p(t)$ are given; see also the research article [668] by D. Rojas and P. J. Torres, where the authors have examined the existence of harmonic and subharmonic bouncing solutions of (359) in the case that $\alpha \in (0, 1)$.

Concerning some other results established in the aforementioned papers, we would like to emphasize the following:

(i) In [203], J. Chu and Z. Zhang have analyzed the existence of positive periodic solutions to second-order singular differential equation

$$x'' + a(t)x = f(t, x) + e(t),$$

with the sign-changing potential; here, the potential $f(t, x)$ is continuous in (t, x), T-periodic in the variable t and may be singular at the point $x = 0$. The authors have studied both, the repulsive case and the attractive case, using essentially the Schauder fixed-point theorem.

(ii) In [520], J. Liang, Y. Liu and C. Gao have analyzed the existence and uniqueness of almost periodic solution to singular systems of the form

$$Ex'(t) = Ax(t) + f(t),$$

where E and A are real-valued matrices of format $n \times n$, $f : \mathbb{R} \to \mathbb{R}^n$ is an almost periodic function and some extra conditions are satisfied. It has been proved that the homogeneous linear counterpart of the above differential equation has an almost periodic solution if and only if there exists $\lambda \in i\mathbb{R}$ such that $\det(\lambda E - A) = 0$; cf. also the second main result [520, Theorem 2]. It could be worthwhile to mention that, in this paper, we have located a new research monograph "Almost Periodic Differential Equations" by C. Y. He (Beijing: Higher Education Press, 1992; in Chinese).

(iii) In [552], S. Lu has investigated the existence of a periodic solution for the second-order differential equation with a singularity of repulsive type

$$x''(t) + f(x, t)x'(t) - g(x(t)) + \varphi(t)x(t) = h(t),$$

where $g(x)$ is singular at $x = 0$ and the functions $\phi(\cdot)$ and $h(\cdot)$ are T-periodic. The author has employed the continuation theorem of Manásevich and Mawhin in order to achieve a new result on the existence of positive periodic solution of the above equation. See also the research article [532] by Y. Liu et al. for some application of the Mawhin continuation theorem of coincidence degree theory.

(iv) The existence of a positive periodic solution for second-order singular semipositone differential equation

$$x'' + h(t)x' + a(t)x = f(t,x,x'),$$

where $\lim_{x\to 0+} f(t,x,y) = +\infty$ uniformly in $(t,y) \in \mathbb{R}^2$, has been analyzed by X. Xing, in [794], using a nonlinear alternative principle of Leray–Schauder.

See also the research article [250] by T. Diagana and M. M. Mbaye, where the authors have examined the existence of square-mean almost periodic solutions to some singular stochastic differential equations with square-mean almost periodic coefficients, and the research article [656] by S. Radchenko, V. Samoilenko and P. Samusenko, where the authors have examined the asymptotic solutions of singularly perturbed linear differential-algebraic equations with periodic coefficients.

A note on the generalized multidimensional almost periodic functions

Let $\emptyset \neq \Lambda \subseteq \mathbb{R}^n$ and let $F : \Lambda \times X \to Y$. In a series of our recent research studies, we have analyzed the generalized almost periodicity of function $F(\cdot;\cdot)$ taking into account a fixed compact subset Ω of $\mathbb{R}^n$ with positive Lebesgue measure such that $\Lambda + \Omega \subseteq \Lambda$; if $0 \in \Omega$, then this condition is clearly equivalent with $\Lambda + \Omega = \Lambda$. Here, we would like to emphasize that there exist some real situations when we cannot find the set Ω with the above described properties; for example, this happens if $\Lambda = \{(x,y) \in \mathbb{R}^2 : |x-y| \leqslant L\}$ for some positive real constant $L > 0$. Even for such regions Λ, we can analyze the generalized multidimensional almost periodicity of function $F(\cdot;\cdot)$ in the following way.

Assume that $p : \Omega \to [1,\infty]$ belongs to the space $\mathcal{P}(\Omega)$ and $\mathcal{B}$ denotes a certain collection of nonempty subsets of X. We will first extend [185, Definition 2.1] as follows.

Definition 6.4.15. Suppose that $\emptyset \neq \Lambda \subseteq \mathbb{R}^n$ and $F : \Lambda \times X \to Y$ satisfies that for each $\mathbf{t} \in \Lambda$ and $x \in X$, the function $F(\mathbf{t} + \mathbf{u}; x)$ belongs to the space $L^{p(\mathbf{u})}(\Omega_{\mathbf{t}} : Y)$, where $\Omega_{\mathbf{t}} := \{\mathbf{u} \in \Omega : \mathbf{t} + \mathbf{u} \in \Omega\}$ $(\mathbf{t} \in \Lambda)$. Then we say that $F(\cdot;\cdot)$ is Stepanov $(\Omega, p(\mathbf{u}))$-bounded on $\mathcal{B}$ if and only if for each $B \in \mathcal{B}$ we have

$$\sup_{\mathbf{t}\in\Lambda; x\in B} \|F(\mathbf{t}+\mathbf{u};x)\|_{L^{p(\mathbf{u})}(\Omega_{\mathbf{t}}:Y)} < \infty.$$

We can similarly introduce the notion of Stepanov distance $D_{S_\Omega}^{p(\cdot)}(F,G)$ of functions $F(\cdot)$ and $G(\cdot)$. Unfortunately, in a new concept, we cannot so simply proved the existence of Weyl distance $D_W^{p(\cdot)}(F,G)$ of functions $F(\cdot)$ and $G(\cdot)$, i. e., the existence of limit $\lim_{l\to+\infty} D_{S_\Omega}^{p(\cdot)}(F,G)$ with $p(\cdot) \equiv p \in [1,+\infty)$ and $\Omega = [0,1]^n$; see [185, pp. 3688–3690] for more details.

It could be worthwhile to introduce the following extension of [185, Definition 2.7], as well (besides the condition $\Lambda + \Omega \subseteq \Lambda$, we also remove here the conditions $\Lambda' \subseteq \Lambda$ and $\Lambda + \Lambda' \subseteq \Lambda$ from the notion).

Definition 6.4.16. Suppose that $\emptyset \neq \Lambda' \subseteq \mathbb{R}^n$ and $F : \Lambda \times X \to Y$.

(i) Then we say that $F(\cdot;\cdot)$ is Stepanov $(\Omega, p(\mathbf{u}))$-$(\mathcal{B}, \Lambda')$-almost periodic (Stepanov $(\Omega, p(\mathbf{u}))$-$\mathcal{B}$-almost periodic, if $\Lambda' = \Lambda$) if and only if for every $B \in \mathcal{B}$ and $\varepsilon > 0$ there exists $l > 0$ such that for each $\mathbf{t}_0 \in \Lambda'$ there exists $\tau \in B(\mathbf{t}_0, l) \cap \Lambda'$ such that

$$\|F(\mathbf{t} + \tau + \mathbf{u}; x) - F(\mathbf{t} + \mathbf{u}; x)\|_{L^{p(\mathbf{u})}(\Omega_{\mathbf{t},\tau}:Y)} \leqslant \varepsilon, \quad \mathbf{t} \in \Lambda,\ x \in B,$$

where $\Omega_{\mathbf{t},\tau} := \{\mathbf{u} \in \Omega : \mathbf{t} + \mathbf{u} \in \Lambda,\ \mathbf{t} + \tau + \mathbf{u} \in \Lambda\}$.

(ii) Then we say that $F(\cdot;\cdot)$ is Stepanov $(\Omega, p(\mathbf{u}))$-$(\mathcal{B}, \Lambda')$-uniformly recurrent (Stepanov $(\Omega, p(\mathbf{u}))$-$\mathcal{B}$-uniformly recurrent, if $\Lambda' = \Lambda$) if and only if for every $B \in \mathcal{B}$ there exists a sequence (τ_k) in Λ' such that $\lim_{k\to+\infty} |\tau_k| = +\infty$ and

$$\lim_{k\to+\infty} \sup_{\mathbf{t}\in\Lambda; x\in B} \|F(\mathbf{t} + \tau_k + \mathbf{u}; x) - F(\mathbf{t} + \mathbf{u}; x)\|_{L^{p(\mathbf{u})}(\Omega_{\mathbf{t},\tau_k}:Y)} = 0.$$

If R is a nonempty collection of sequences in $\mathbb{R}^n$ and R_X is a nonempty collection of sequences in $\mathbb{R}^n \times X$, then we can similarly extend the notion of Stepanov $(\Omega, p(\mathbf{u}))$-$(\mathrm{R}_X, \mathcal{B})$-multialmost periodicity and the notion of Stepanov $(\Omega, p(\mathbf{u}))$-$(\mathrm{R}_X, \mathcal{B})$-multialmost periodicity introduced in [185, Definitions 2.4, 2.5]. We can use a similar idea to analyze some new classes of metrically Stepanov almost periodic functions and (metrically) Weyl almost periodic functions. All details can be left to the enthusiastic readers.

Almost periodic solutions to the semilinear diffusion equations with rough coefficients

In a recent research article [677], L. T. Sac and P. T. Xuan have analyzed the existence, uniqueness and stability of the pseudo-almost periodic mild solutions to the following semilinear diffusion equations with rough coefficients

$$u'(t) - b\Delta u = g(t, u), \quad (t, x) \in \mathbb{R} \times \mathbb{R}^d, \tag{360}$$

where $b \in L^\infty(\mathbb{R}^d)$, $\operatorname{Re} b \geqslant \delta > 0$ for some positive real constant $\delta > 0$ and $g(t, u) = |u(t)|^{m-1}u(t) + F(t)$, where $F(\cdot)$ is a given bounded function on the real line. The authors have considered the following abstract version of problem (360):

$$u'(t) + Au(t) = BG(u)(t), \quad t \in \mathbb{R},$$

where $-A$ generates a strongly continuous semigroup $(\exp(-tA))_{t\geqslant 0}$ on some interpolation spaces and B is a connection operator between the various spaces involved.

In the research study [677], the authors allow the zero number to belong to the spectrum of A, so that the semigroup $(\exp(-tA))_{t\geqslant 0}$ is not exponentially decaying; the crucial thing in their analysis is the polynomial decay of the B-regularized semigroup $(\exp(-tA)B)_{t\geqslant 0}$. The authors deal with the Lorentz spaces, which are defined as follows: Let $\emptyset \neq \Omega \subseteq \mathbb{R}^d$, let $1 < p < +\infty$ and let $1 \leqslant q \leqslant +\infty$. Then the Lorentz space $L^{p,q}(\Omega)$ is defined by

$$L^{p,q}(\Omega) := \{u \in L^1_{\text{loc}}(\Omega) : \|u\|_{p,q} < +\infty\},$$

where

$$\|u\|_{p,q} := \left(\int_0^{+\infty} \left(sm(\{x \in \Omega : |u(x)| > s\})^{1/p}\right)^q \frac{ds}{s} \right)^{1/q} \quad (1 \leqslant q < +\infty)$$

and

$$\|u\|_{p,+\infty} := \sup_{s>0} \left[sm(\{x \in \Omega : |u(x)| > s\})^{1/p} \right].$$

Using the real-interpolation functor $(\cdot,\cdot)_{\theta,q}$, we have

$$(L^{p_0}(\Omega), L^{p_1}(\Omega))_{\theta,q} = L^{p,q}(\Omega),$$

where $1 < p_0 < p < p_1 < +\infty$ and $\theta \in (0,1)$ is such that

$$\frac{1-\theta}{p_0} + \frac{\theta}{p_1} = \frac{1}{p}, \quad 1 \leqslant q \leqslant +\infty.$$

The authors employ the general interpolation result [108, Theorem 3.11.2] to achieve their aims; for more details about the interpolation theory and its applications, we also refer the reader to the research monographs [557] and [748].

The authors first consider the following inhomogeneous linear evolution equation of the form:

$$u'(t) + Au(t) = Bf(t), \quad t \in \mathbb{R},$$

where the unknown function $u(\cdot)$ is Y-valued, $-A$ generates a strongly continuous semigroup $(\exp(-tA))_{t\geqslant 0}$ on the interpolation space $Y_1, f(\cdot)$ is a function from $\mathbb{R}$ to X and B is the connection operator between X and Y such that $\exp(-tA)B \in L(X, Y_i)$ for $i = 1,2$ and $t \geqslant 0$. See also the results established in [622] and [543, 544].

ρ-Almost periodic type ultradistributions in $\mathbb{R}^n$

The notion of a bounded distribution and the notion an almost periodic distribution were introduced in the pioneering papers by L. Schwartz (see, e.g., [686]), where the author analyzed the scalar-valued case. The bounded and almost periodic distributions with values in general Banach spaces were introduced by I. Cioranescu in [207] (1990); see also [90, 91]. Further on, the class of scalar-valued almost periodic ultradistributions was introduced by I. Cioranescu [208] (1992) and the class of vector-valued almost periodic ultradistributions was introduced in [449] (2018).

In this part, we consider various classes of ρ-almost periodic type ultradistributions in $\mathbb{R}^n$; for more details about one-dimensional almost periodic ultradistributions, we also refer the reader to the list of references quoted in [444]. In such a way, we continue the research study of ρ-almost periodic type distributions in $\mathbb{R}^n$ (see [465] for more details in this direction) as well as the research studies [459] and [478] (a joint work with S. Pilipović, D. Velinov and V. Fedorov), where we have analyzed some classes of one-dimensional c-almost periodic (ultra)distributions.

Concerning our topic, we would like to recall that the scalar-valued almost periodic ultradistributions in $\mathbb{R}^n$ was analyzed by M. C. Gómez-Collado in [324] within the theory of ω-ultradistributions (2000); furthermore, all structural results established in [324] hold in the vector-valued setting. The main result of this research article, [324, Theorem 4.2], holds for the almost periodic ultradistributions of (M_p)-class and the almost periodic ultradistributions of $\{M_p\}$-class, provided that (M_p) is a sequence of positive real numbers satisfying $M_0 = 1$ as well as the conditions (M.1), (M.2) and (M.3) clarified below. Our main results are formulated within the Komatsu theory of ultradistributions, in the concrete situation in which the sequence (M_p) does not necessarily satisfy (M.3) but a slightly weaker condition (M.3'); the material is taken from [475].

Suppose that $\emptyset \neq I' \subseteq \mathbb{R}^n$, $\emptyset \neq I \subseteq \mathbb{R}^n$, $F : I \to Y$ is a continuous function, ρ is a binary relation on X and $I + I' \subseteq I$. We refer the reader to [448, Definition 2.1.1] for the notion of a Bohr (I',ρ)-almost periodic function ((I',ρ)-uniformly recurrent function) $F : I \times X \to Y$. Denote by $AP_{I',\rho}(I : X)$ $[UR_{I',\rho}(I : X)]$ the collection of all Bohr (I',ρ)-almost periodic functions [(I',ρ)-uniformly recurrent functions].

We refer the reader to [438–440] for more details about the theory of ultradistributions. Let (M_p) be a sequence of positive real numbers satisfying $M_0 = 1$ and the following conditions:

(M.1): $M_p^2 \leqslant M_{p+1}M_{p-1}$, $p \in \mathbb{N}$.

(M.2): $M_p \leqslant AH^p \sup_{0\leqslant i\leqslant p} M_i M_{p-i}$, $p \in \mathbb{N}$, for some A, $H > 1$.

(M.3'): $\sum_{p=1}^{\infty} \frac{M_{p-1}}{M_p} < \infty$.

Any use of the condition

(M.3): $\sup_{p\in\mathbb{N}} \sum_{q=p+1}^{\infty} \frac{M_{q-1}M_{p+1}}{pM_pM_q} < \infty$,

which is slightly stronger than (M.3′), will be explicitly emphasized. If $s > 1$, then we know that the Gevrey sequence $(p!^s)$ satisfies the above conditions. The space of Beurling, respectively, Roumieu ultradifferentiable functions, is defined by $\mathcal{D}^{(M_p)}(\mathbb{R}^n) :=$

$\operatorname{indlim}_{K \Subset \mathbb{R}^n} \mathcal{D}_K^{(M_p)}$, respectively, $\mathcal{D}^{\{M_p\}}(\mathbb{R}^n) := \operatorname{indlim}_{K \Subset \mathbb{R}^n} \mathcal{D}_K^{\{M_p\}}$, where $\mathcal{D}_K^{(M_p)} := \operatorname{projlim}_{h\to\infty} \mathcal{D}_K^{M_p,h}$, respectively, $\mathcal{D}_K^{\{M_p\}} := \operatorname{indlim}_{h\to 0} \mathcal{D}_K^{M_p,h}$,

$$\mathcal{D}_K^{M_p,h} := \{\phi \in C^\infty(\mathbb{R}^n) : \operatorname{supp}\phi \subseteq K,\ \|\phi\|_{M_p,h,K} < \infty\}$$

and

$$\|\phi\|_{M_p,h,K} := \sup\left\{\frac{h^{|\alpha|}|\phi^{(\alpha)}(\mathbf{t})|}{M_{|\alpha|}} : \mathbf{t} \in K,\ \alpha \in \mathbb{N}_0^n\right\}.$$

The asterisk $*$ is used to denote both, the Beurling case (M_p) or the Roumieu case $\{M_p\}$. The space consisted of all continuous linear functions from $\mathcal{D}^*(\mathbb{R}^n)$ into X, denoted by $\mathcal{D}'^*(\mathbb{R}^n : X)$, is said to be the space of n-dimensional X-valued ultradistributions of $*$-class.

We say that the operator of infinite differentiation $P(D) = \sum_{\alpha\in\mathbb{N}_0^n} a_\alpha D^\alpha$ is an ultradifferential operator of class (M_p), respectively, of class $\{M_p\}$, if there exist $l > 0$ and $C > 0$, respectively, for every $l > 0$ there exists a constant $C > 0$, such that $|a_\alpha| \leqslant C l^{|\alpha|}/M_{|\alpha|}$, $\alpha \in \mathbb{N}_0^n$; see [438] for further information. We introduce the space $\mathcal{E}^*(\mathbb{R}^n : X)$ and the convolution of an n-dimensional X-valued ultradistribution of $*$-class and an n-dimensional scalar-valued ultradifferentiable function in the same way as on pages 671 and 685 in [440]. If $T \in \mathcal{D}'^*(\mathbb{R}^n : X)$ and $\varphi \in \mathcal{D}^*(\mathbb{R}^n)$, then we define $(T * \varphi)(t) := \langle T, \varphi(\mathbf{t} - \cdot)\rangle$, $\mathbf{t} \in \mathbb{R}^n$; then we know that $T * \varphi \in \mathcal{E}^*(\mathbb{R}^n : X)$. Set also $\langle T_{\mathbf{t}}, \varphi\rangle := \langle T, \varphi(\cdot - \mathbf{t})\rangle$ for $\mathbf{t} \in \mathbb{R}^n$.

The tempered ultradistributions of Beurling, respectively, Roumieu type, are defined by S. Pilipović [641] as duals of the corresponding test spaces

$$\mathcal{S}^{(M_p)}(\mathbb{R}^n) := \operatorname*{projlim}_{h\to\infty} \mathcal{S}^{M_p,h}(\mathbb{R}^n), \quad \text{resp.,}\ \mathcal{S}^{\{M_p\}}(\mathbb{R}^n) := \operatorname*{indlim}_{h\to 0} \mathcal{S}^{M_p,h}(\mathbb{R}^n),$$

where

$$\mathcal{S}^{M_p,h}(\mathbb{R}^n) := \{\phi \in C^\infty(\mathbb{R}^n) : \|\phi\|_{M_p,h} < \infty\} \quad (h > 0),$$

$$\|\phi\|_{M_p,h} := \sup\left\{\frac{h^{|\alpha|+|\beta|}}{M_{|\alpha|}M_{|\beta|}}(1 + |\mathbf{t}|^2)^{|\beta|/2}|\phi^{(\alpha)}(\mathbf{t})| : \mathbf{t} \in \mathbb{R}^n,\ \alpha,\ \beta \in \mathbb{N}_0^n\right\}.$$

A continuous linear mapping $\mathcal{S}^{(M_p)}(\mathbb{R}^n) \to X$, resp., $\mathcal{S}^{\{M_p\}}(\mathbb{R}^n) \to X$, is said to be an n-dimensional X-valued tempered ultradistribution of Beurling, respectively, Roumieu type.

For any $h > 0$, we define

$$\mathcal{D}_{L^1}(\mathbb{R}^n, (M_p), h) := \left\{f \in \mathcal{D}'^*(\mathbb{R}^n : X)\ ;\ \|f\|_{1,h} := \sup_{\alpha\in\mathbb{N}_0^n} \frac{h^{|\alpha|}\|f^{(\alpha)}\|_1}{M_{|\alpha|}} < \infty\right\},$$

where $\|\cdot\|_1$ denotes the norm in $L^1(\mathbb{R}^n)$. Then $(\mathcal{D}_{L^1}(\mathbb{R}^n, (M_p), h), \|\cdot\|_{1,h})$ is a Banach space and the space of all n-dimensional X-valued bounded Beurling ultradistributions of class (M_p), respectively, n-dimensional X-valued bounded Roumieu ultradistributions of class $\{M_p\}$, is defined as the space consisting of all linear continuous mappings from $\mathcal{D}_{L^1}(\mathbb{R}^n,(M_p))$, respectively, $\mathcal{D}_{L^1}(\mathbb{R}^n,\{M_p\})$, into X, where

$$\mathcal{D}_{L^1}(\mathbb{R}^n,(M_p)) := \operatorname*{projlim}_{h\to+\infty} \mathcal{D}_{L^1}(\mathbb{R}^n,(M_p),h),$$

respectively,

$$\mathcal{D}_{L^1}(\mathbb{R}^n,\{M_p\}) := \operatorname*{indlim}_{h\to0+} \mathcal{D}_{L^1}(\mathbb{R}^n,(M_p),h).$$

We assume that these spaces are equipped with the strong topologies and we denote them by $\mathcal{D}'_{L^1}(\mathbb{R}^n,(M_p):X)$, respectively, $\mathcal{D}'_{L^1}(\mathbb{R}^n,\{M_p\}:X)$. We can prove that $\mathcal{S}^{(M_p)}(\mathbb{R}^n)$, respectively, $\mathcal{S}^{\{M_p\}}(\mathbb{R}^n)$, is a dense subspace of $\mathcal{D}_{L^1}(\mathbb{R}^n,(M_p))$, respectively, $\mathcal{D}_{L^1}(\mathbb{R}^n,\{M_p\})$ and that $f_{|\mathcal{S}^{(M_p)}(\mathbb{R}^n)} : \mathcal{S}^{(M_p)}(\mathbb{R}^n) \to X$, respectively, $f_{|\mathcal{S}^{\{M_p\}}(\mathbb{R}^n)} : \mathcal{S}^{\{M_p\}}(\mathbb{R}^n) \to X$, is a tempered X-valued ultradistribution of class (M_p), respectively, of class $\{M_p\}$; cf. [444] for the notion and more details about the tempered vector-valued ultradistributions.

For our further work, we will introduce the following spaces of vector-valued bounded ultradistributions of $*$-class. For any $h > 0$, we define

$$\begin{aligned}
&\mathcal{D}_{L^1}(\mathbb{R}^n,(M_p),h,s)\\
&\quad := \Big\{ f \in C^\infty(\mathbb{R}^n : X);\ \sup_{s\in B(\cdot,1)} \big|f^{(\alpha)}(s)\big| \in L^1(\mathbb{R}^n : X) \text{ for all}\\
&\qquad \alpha \in \mathbb{N}_0^n \text{ and } \|f\|_{1,h,s} := \sup_{\alpha\in\mathbb{N}_0^n} \frac{h^{|\alpha|}\,|\sup_{s\in B(\cdot,1)} |f^{(\alpha)}(s)||_1}{M_{|\alpha|}}\\
&\quad = \sup_{\alpha\in\mathbb{N}_0^n} \frac{h^{|\alpha|}\int_{\mathbb{R}^n} |\sup_{s\in B(x,1)} |f^{(\alpha)}(s)||\,dx}{M_{|\alpha|}} < \infty \Big\}.
\end{aligned}$$

Then $(\mathcal{D}_{L^1}(\mathbb{R}^n,(M_p),h,s), \|\cdot\|_{1,h,s})$ is a Banach space, which is continuously embedded into $(\mathcal{D}_{L^1}(\mathbb{R}^n,(M_p),h), \|\cdot\|_{1,h})$ for all $h > 0$. The space of all n-dimensional X-valued strongly bounded Beurling ultradistributions of class (M_p), respectively, n-dimensional X-valued strongly bounded Roumieu ultradistributions of class $\{M_p\}$, is defined as the space consisting of all linear continuous mappings from $\mathcal{D}_{L^1}(\mathbb{R}^n,(M_p),s)$, respectively, $\mathcal{D}_{L^1}(\mathbb{R}^n,\{M_p\},s)$, into X, where

$$\mathcal{D}_{L^1}(\mathbb{R}^n,(M_p),s) := \operatorname*{projlim}_{h\to+\infty} \mathcal{D}_{L^1}(\mathbb{R}^n,(M_p),h,s),$$

respectively,

$$\mathcal{D}_{L^1}(\mathbb{R}^n,\{M_p\},s) := \operatorname*{indlim}_{h\to0+} \mathcal{D}_{L^1}(\mathbb{R}^n,(M_p),h,s).$$

As above, we assume that these spaces are equipped with the strong topologies; we denote them by $\mathcal{D}'_{L^1}(\mathbb{R}^n,(M_p),s : X)$ and $\mathcal{D}'_{L^1}(\mathbb{R}^n,\{M_p\},s : X)$, respectively. We will not further analyze the topological properties and the structural theorems for the spaces $\mathcal{D}_{L^1}(\mathbb{R}^n, *, s)$ and their duals here; cf. also [640]. We will only note that a very simple argumentation shows that $\mathcal{S}^*(\mathbb{R}^n)$ is a linear subspace of $\mathcal{D}_{L^1}(\mathbb{R}^n, *, s)$.

The space of all continuous linear mappings from $\mathcal{D}_{L^1}(\mathbb{R}^n, *)$ into X, equipped with the topology of uniform convergence over bounded subsets of $\mathcal{D}_{L^1}(\mathbb{R}^n, *, s)$, will be denoted by $\mathcal{D}'_{L^1,s}(\mathbb{R}^n, * : X)$. This space is locally convex since the family of seminorms $(\sup_{\varphi\in B} \|\langle\cdot,\varphi\rangle\|)_{B\in\mathcal{B}d}$, where $\mathcal{B}d$ denotes the collection of all bounded subsets of $\mathcal{D}'_{L^1}(\mathbb{R}^n, *, s : X)$, satisfies the conditions (1) and (2) from [584, Lemma 22.4], as easily approved; cf. also [584, Lemma 22.5]. It is not clear whether the spaces $\mathcal{D}'_{L^1,s}(\mathbb{R}^n, * : X)$ and $\mathcal{D}'_{L^1}(\mathbb{R}^n, *)$ are topologically equivalent.

We continue by introducing the following notion (cf. [465, Definition 2.1] for the distributional analogue).

Definition 6.4.17. Suppose that $I' \subseteq \mathbb{R}^n$, ρ is a binary relation on X and $T \in \mathcal{D}'^*(\mathbb{R}^n : X)$. Then we say that T is an (I',ρ)-almost periodic ultradistribution of $*$-class [(I',ρ)-uniformly recurrent ultradistribution of $*$-class], if $T * \varphi \in AP_{I',\rho}(\mathbb{R}^n : X)$ for all $\varphi \in \mathcal{D}^*(\mathbb{R}^n)$ [$T * \varphi \in UR_{I',\rho}(\mathbb{R}^n : X)$ for all $\varphi \in \mathcal{D}^*(\mathbb{R}^n)$]. If $I' = I$ and $\rho = \mathrm{I}$, then we also say that T is an almost periodic ultradistribution of $*$-class [uniformly recurrent ultradistribution of $*$-class].

It is clear that the structural characterizations of Bohr (I',ρ)-almost periodic functions [(I',ρ)-uniformly recurrent functions] can be used to provide certain results about (I',ρ)-almost periodic ultradistributions of $*$-class [(I',ρ)-uniformly recurrent ultradistributions of $*$-class]. For example, using [448, Corollary 2.1.4], we can immediately clarify the following result.

Proposition 6.4.18. *Suppose that $I' \subseteq \mathbb{R}^n$ and $\rho : X \to X$. If T is an (I',ρ)-almost periodic ultradistribution of $*$-class [(I',ρ)-uniformly recurrent ultradistribution of $*$-class], then T is an $(I'-I',\mathrm{I})$-almost periodic ultradistribution of $*$-class [$(I'-I',\mathrm{I})$-uniformly recurrent ultradistribution of $*$-class].*

We will omit such results in the sequel. Now we will reconsider the result established in [465, Theorem 2.2].

Theorem 6.4.19. *Suppose that $\rho = A \in L(X)$, $\emptyset \neq I' \subseteq \mathbb{R}^n$, there exist an integer $k \in \mathbb{N}$ and (I',A)-almost periodic ((I',A)-uniformly recurrent) functions $F_j : \mathbb{R}^n \to X$ $(0 \leqslant j \leqslant k)$ such that the function*

$$\mathbf{t} \mapsto F(\mathbf{t}) \equiv \big(F_0(\mathbf{t}), \ldots, F_k(\mathbf{t})\big), \quad t \in \mathbb{R}^n$$

is (I',A_{k+1})-almost periodic ((I',A_{k+1})-uniformly recurrent), where $A_{k+1} \in L(X^{k+1})$ is given by $A_{k+1}(x_0, x_1, \ldots, x_k) := (Ax_0, Ax_1, \ldots, Ax_k)$, $(x_0, x_1, \ldots, x_k) \in X^{k+1}$. Set

$$T := \sum_{j=0}^{k} \sum_{\alpha\in\mathbb{N}_0^n} a_{\alpha,j} F_j^{(\alpha)}$$

and suppose that there exist $l > 0$ and $C > 0$, respectively, for every $l > 0$ there exists a constant $C > 0$, such that $|a_{\alpha,j}| \leqslant Cl^{|\alpha|}/M_{|\alpha|}$ for all $\alpha \in \mathbb{N}_0^n$ and $0 \leqslant j \leqslant k$. Then T is an (I',ρ)-almost periodic ultradistribution of $*$-class [(I',ρ)-uniformly recurrent ultradistribution of $*$-class].

Proof. We will provide all details of proof in the case of consideration of (I',ρ)-almost periodic ultradistribution of Beurling class. First of all, [440, Theorem 7.7] implies that $T \in \mathcal{D}'^{*}(\mathbb{R}^n : X)$; further on, for each $\varphi \in \mathcal{D}^{*}(\mathbb{R}^n)$ and $\mathbf{t} \in \mathbb{R}^n$ we have

$$(T * \varphi)(\mathbf{t}) = \langle T, \varphi(\mathbf{t} - \cdot)\rangle = \sum_{j=0}^{k}\sum_{\alpha\in\mathbb{N}_0^n}\int_{\mathbb{R}^n} \varphi^{(\alpha)}(\mathbf{t} - v)F_j(v)\,dv$$
$$= \sum_{j=0}^{k}\sum_{\alpha\in\mathbb{N}_0^n}\int_{\mathbb{R}^n} \varphi^{(\alpha)}(v)F_j(\mathbf{t} - v)\,dv.$$

Let $\varepsilon > 0$ be given. Then there exists $l > 0$ such that for each $\mathbf{t}_0 \in I'$ there exists $\tau \in B(\mathbf{t}_0, l) \cap I'$ such that, for every $\mathbf{t} \in \mathbb{R}^n$, we have

$$\|F_j(\mathbf{t} + \tau) - AF_j(\mathbf{t})\|_Y \leqslant \varepsilon, \quad 0 \leqslant j \leqslant k.$$

Suppose that $\varphi \in \mathcal{D}^{*}(\mathbb{R}^n)$ and $\operatorname{supp}\varphi \subseteq K$. Then there exists $h > l$ such that

$$\int_{\mathbb{R}^n} |\varphi^{(\alpha)}(v)|\,dv = \int_{K} |\varphi^{(\alpha)}(v)|\,dv \leqslant m(K)\|\varphi\|_{M_p,h,K}\frac{M_{|\alpha|}}{h^{|\alpha|}}, \quad \alpha \in \mathbb{N}_0^n.$$

Therefore, for every $\mathbf{t} \in \mathbb{R}^n$, we have

$$\|(T * \varphi)(t + \tau) - A(T * \varphi)(\mathbf{t})\|$$
$$\leqslant \sum_{j=0}^{k}\sum_{\alpha\in\mathbb{N}_0^n}\int_{\mathbb{R}^n} |\varphi^{(\alpha)}(v)| \cdot \|F_j(\mathbf{t} + \tau - v) - AF_j(\mathbf{t} - v)\|\,dv$$
$$\leqslant \varepsilon\sum_{j=0}^{k}\sum_{\alpha\in\mathbb{N}_0^n}\frac{l^{|\alpha|}}{M_{|\alpha|}}\int_{\mathbb{R}^n} |\varphi^{(\alpha)}(v)|\,dv \leqslant \sum_{j=0}^{k}\sum_{\alpha\in\mathbb{N}_0^n}\frac{l^{|\alpha|}}{M_{|\alpha|}}m(K)\|\varphi\|_{M_p,h,K}\frac{M_{|\alpha|}}{h^{|\alpha|}},$$

which simply implies the required statement. □

As already mentioned in the introductory part, if the sequence (M_p) additionally satisfies (M.3), $A = \mathrm{I}$ and T is an almost periodic ultradistribution of $*$-class, then there exist two almost periodic functions $F : \mathbb{R}^n \to X$, $G : \mathbb{R}^n \to X$ and an ultradifferential operator $P(D)$ of $*$-class such that $T = P(D)F + G$; see the formulation of [324, Theorem 4.2] and the statements of [324, Corollary 2.6] and [640, Lemma 5], which are the main auxiliary results needed for the proof of this result. Unfortunately, it is not clear how one can prove an analogue of this result for c-almost periodic ultradistributions of $*$-class (i. e.,

ρ-almost periodic ultradistributions of $*$-class with $\rho = c\mathrm{I}$ for some $c \in \mathbb{C} \setminus \{0\}$); see also [459, p. 18] for more details given in the one-dimensional setting.

Suppose now that $I' = I$ and $\rho = \mathrm{I}$. Then the requirements of Theorem 6.4.19 imply that the following condition holds:

(BC) The set of all translations $\{T_{\mathbf{t}} : \mathbf{t} \in \mathbb{R}^n\}$ is relatively compact in $\mathcal{D}'_{L^1}(\mathbb{R}^n, *)$.

Furthermore, the validity of (BC) implies that T is an almost periodic ultradistribution of $*$-class; see the proofs of [208, Theorem 2] and [324, Theorem 4.2].

Define now

$$\mathcal{E}^*_{I',\rho,AP}(\mathbb{R}^n : X) := \mathcal{E}^*(\mathbb{R}^n : X) \cap AP_{I',\rho}(\mathbb{R}^n : X)$$

and

$$\mathcal{E}^*_{I',\rho,UR}(\mathbb{R}^n : X) := \mathcal{E}^*(\mathbb{R}^n : X) \cap UR_{I',\rho}(\mathbb{R}^n : X).$$

If $\varphi : \mathbb{R}^n \to \mathbb{C}$, then we define $\check{\varphi}(x) := \varphi(-x)$, $x \in \mathbb{R}^n$; let us recall that $(e_1, \ldots, e_n)$ denotes the standard basis of $\mathbb{R}^n$. We continue by stating the following result.

Theorem 6.4.20. *Suppose that T is an n-dimensional X-valued bounded ultradistribution of $*$-class, $I' \subseteq \mathbb{R}^n$, ρ is a binary relation on X, $D(\rho)$ is a closed subset of Y and C_ρ holds. Then T is an (I', ρ)-almost periodic ultradistribution of $*$-class [(I', ρ)-uniformly recurrent ultradistribution of $*$-class] if and only if there exists a sequence $(\psi_k)_{k\in\mathbb{N}}$ in $\mathcal{E}^*_{I',\rho,AP}(\mathbb{R}^n : X)$ [$\mathcal{E}^*_{I',\rho,UR}(\mathbb{R}^n : X)$] such that $\lim_{k\to+\infty} \psi_k = T$ for the topology of $\mathcal{D}'_{L^1,s}(\mathbb{R}^n, * : X)$.*

Proof. We will consider the (I', ρ)-almost periodic ultradistributions of $\{M_p\}$-class, only. Suppose that there exists a sequence $(\psi_k)_{k\in\mathbb{N}}$ in $\mathcal{E}^*_{I',\rho,AP}(\mathbb{R}^n : X)$ with the prescribed properties. First of all, we will prove that for each fixed test function $\varphi \in \mathcal{D}^{\{M_p\}}$ the set of all translations $\{\varphi(\cdot - \mathbf{t}) : \mathbf{t} \in \mathbb{R}^n\}$ is bounded in $\mathcal{D}_{L^1}(\mathbb{R}^n, \{M_p\}, s : X)$. We know that there exist a compact set $K \subseteq \mathbb{R}^n$ and two real numbers $h > 0$ and $c > 0$ such that $|\varphi^{(\alpha)}(x)| \leqslant cM_{|\alpha|}/h^{|\alpha|}$ for all $\alpha \in \mathbb{N}_0^n$ and $x \in \mathbb{R}^n$. Let K_1 denotes the compact set in $\mathbb{R}^n$ given by $K_1 = K + B(0,1)$. Then we have

$$\begin{aligned} &\sup_{\mathbf{t}\in\mathbb{R}^n;\alpha\in\mathbb{N}_0^n} \frac{h^{|\alpha|}\int_{\mathbb{R}^n}\sup_{\mathbf{s}\in B(x,1)}|\varphi^{(\alpha)}(\mathbf{s}-\mathbf{t})|\,dx}{M_{|\alpha|}} \\ &\leqslant \sup_{\mathbf{t}\in\mathbb{R}^n;\alpha\in\mathbb{N}_0^n} \frac{h^{|\alpha|}\int_{\mathbb{R}^n}\sup_{\mathbf{s}\in B(x-\mathbf{t},1)}|\varphi^{(\alpha)}(\mathbf{s})|\,dx}{M_{|\alpha|}} \\ &= \sup_{\mathbf{t}\in\mathbb{R}^n;\alpha\in\mathbb{N}_0^n} \frac{h^{|\alpha|}\int_{\mathbf{t}+K_1} cM_{|\alpha|}/h^{|\alpha|}\,dx}{M_{|\alpha|}} = cm(K_1). \end{aligned}$$

Keeping in mind this fact, the required conclusion almost immediately from the argumentation contained in the first part of proof of [140, Proposition 7], since we have assumed that $D(\rho)$ is a closed subset of Y and ρ satisfies (C_ρ); cf. also [448, Theorem 2.1.12(v)]. Assume now that T is an (I',ρ)-almost periodic ultradistribution of $*$-class. Let $(\delta_k)_{k\in\mathbb{N}}$ be a sequence of infinitely ultradifferentiable functions of $\{M_p\}$-class such that $\operatorname{supp}\delta_k \subseteq [-(1/k),1/k]^n$ and $\int_{\mathbb{R}^n}\delta_k(\mathbf{t})\,d\mathbf{t}=1$ for all $k\in\mathbb{N}$. Set $\psi_k := T*\delta_k$ for all $k\in\mathbb{N}$. Then $(\psi_k)_{k\in\mathbb{N}}$ is a sequence in $\mathcal{E}^*_{I',\rho,AP}(\mathbb{R}^n:X)$ and we only need to prove that $\lim_{k\to+\infty}\psi_k = T$ for the topology of $\mathcal{D}'_{L^1,s}(\mathbb{R}^n,\{M_p\}:X)$. Let $B\in\mathcal{B}d$ be fixed. Then there exists $h>0$ such that B is contained and bounded in $\mathcal{D}_{L^1}(\mathbb{R}^n,(M_p),h_0H,s)$ for all $h_0\in(0,h]$, where H denotes the constant from (M.2). We may assume without loss of generality that there exists $c>0$ such that $\|\langle T,\varphi\rangle\|\leqslant c\|\varphi\|_{1,h}$, $\varphi\in\mathcal{D}_{L^1}(\mathbb{R}^n,(M_p),h)$. Now we will estimate the term $\|[\check{\psi}_k*\varphi-\varphi]^{(\alpha)}\|_{L^1(\mathbb{R}^n)}$ $(k\in\mathbb{N},\ \alpha\in\mathbb{N}_0^n)$. The mean value theorem implies that for each $x\in\mathbb{R}^n$ and $y\in\mathbb{R}^n$ there exists a constant $c_{x,y}\in(0,1)$ such that, for every $k\geqslant 1+\sqrt{n}$, we have

$$\begin{aligned}
\|[\check{\psi}_k*\varphi-\varphi]^{(\alpha)}\|_{L^1(\mathbb{R}^n)} &= \int_{\mathbb{R}^n}\Big|\int_{\mathbb{R}^n}\check{\psi}_k(y)[\varphi^{(\alpha)}(x-y)-\varphi^{(\alpha)}(x)]\,dy\Big|\,dx\\
&\leqslant \int_{\mathbb{R}^n}\check{\psi}_k(y)\int_{\mathbb{R}^n}|\varphi^{(\alpha)}(x-y)-\varphi^{(\alpha)}(x)|\,dx\,dy\\
&\leqslant \sum_{i=1}^n\int_{\mathbb{R}^n}|y_i|\check{\psi}_k(y)\int_{\mathbb{R}^n}|\varphi^{(\alpha+e_i)}(x-y+c_{x,y}y)|\,dx\,dy\\
&\leqslant \sum_{i=1}^n\int_{\mathbb{R}^n}|y_i|\check{\psi}_k(y)\int_{\mathbb{R}^n}\sup_{\mathbf{s}\in[x-y,x]}|\varphi^{(\alpha+e_i)}(\mathbf{s})|\,dx\,dy\\
&\leqslant \frac{n}{k}\sum_{i=1}^n\int_{\mathbb{R}^n}\sup_{\mathbf{s}\in[x-y,x]}|\varphi^{(\alpha+e_i)}(\mathbf{s})|\,dx \leqslant \frac{n}{k}\int_{\mathbb{R}^n}\sup_{\mathbf{s}\in B(x,1)}|\varphi^{(\alpha+e_i)}(\mathbf{s})|\,dx,
\end{aligned}$$

where we have also used the Fubini theorem. Hence, we have the following:

$$\begin{aligned}
\sup_{\varphi\in B}\|\langle T*\psi_k-T,\varphi\rangle\| &= \sup_{\varphi\in B}\|\langle T,\check{\psi}_k*\varphi-\varphi\rangle\|\\
&\leqslant cd\sup_{\varphi\in B}\sup_{\alpha\in\mathbb{N}_0^n}\frac{h^{|\alpha|}\|[\check{\psi}_k*\varphi-\varphi]^{(\alpha)}\|_{L_1(\mathbb{R}^n)}}{M_{|\alpha|}}\\
&\leqslant \frac{cnd}{k}\sup_{\varphi\in B}\sup_{\alpha\in\mathbb{N}_0^n}\frac{h^{|\alpha|}\sum_{i=1}^n\int_{\mathbb{R}^n}\sup_{\mathbf{s}\in B(x,1)}|\varphi^{(\alpha+e_i)}(\mathbf{s})|\,dx}{M_{|\alpha|}}\\
&\leqslant \frac{nc}{kh}AM_1\sup_{\alpha\in\mathbb{N}_0^n}\frac{(hH)^{|\alpha|+1}\sum_{i=1}^n\int_{\mathbb{R}^n}\sup_{\mathbf{s}\in B(x,1)}|\varphi^{(\alpha+e_i)}(\mathbf{s})|\,dx}{M_{|\alpha|+1}}\\
&\leqslant \frac{cn^2d}{kh}AM_1\sup_{\varphi\in B}\|\varphi\|_{1,hH,s},
\end{aligned}$$

which simply completes the proof. □

Remark 6.4.21. The argumentation contained in the second part of the proof of [140, Proposition 7] (cf. also [449, Lemma 1], where we have made the same mistake, and [141, Proposition 10]) is a little bit misleading since the equality

$$\int_{\mathbb{R}^n} |\varphi^{(a+e_i)}(x - y + c_{x,y}y)| \, dx = \int_{\mathbb{R}^n} |\varphi^{(a+e_i)}(x)| \, dx$$

is not true because the value of $c_{x,y}$ strongly depends on $x, y \in \mathbb{R}^n$ and it is not a constant so that the change of variable $x \mapsto x - y + c_{x,y}y$ cannot be done without further information on the Jacobian of $c_{x,y}$ for fixed $y \in \mathbb{R}^n$. Moreover, the inequality

$$\int_{\mathbb{R}^n} |\varphi^{(a)}(x - y) - \varphi^{(a)}(x)| \, dx \leqslant \sum_{i=1}^{n} |y_i| \int_{\mathbb{R}^n} |\varphi^{(a+e_i)}(x)| \, dx$$

can be wrong, as the following counterexample shows: Suppose that $b > a > 0, y > 0$, $n = 1$, $\alpha = 0$ and $f(x) = x^2$, $x \geqslant 0$. Then

$$\begin{aligned}\int_a^b |f(x+y) - f(y)| \, dx &= \int_a^b [y^2 + 2xy] \, dx = y \int_a^b [y + 2x] \, dx = (b^3 - a^3)/3 + y \int_a^b 2x \, dx \\ &= (b^3 - a^3)/3 + y \int_a^b |f'(x)| \, dx > y \int_a^b |f'(x)| \, dx.\end{aligned}$$

For the test functions, we can consider the sequence of smooth functions, which sufficiently good approximates the function $f(\cdot)$ in the Sobolev space $W^{1,1}((a, b + y))$.

Suppose now that there exists a sequence $(\psi_k)_{k\in\mathbb{N}}$ in $\mathcal{E}^*_{I',\rho,AP}(\mathbb{R}^n : X)$ such that $\lim_{k\to+\infty} \psi_k = T$ for the topology of $\mathcal{D}'_{L^1}(\mathbb{R}^n, * : X)$. Then T is an (I', ρ)-almost periodic ultradistribution of $*$-class [(I', ρ)-uniformly recurrent ultradistribution of $*$-class]; especially, in the case that $I' = I$ and $\rho = \mathrm{I}$, then T is an almost periodic ultradistribution of $*$-class [uniformly recurrent ultradistribution of $*$-class] and it can be approximated by trigonometric polynomials in the space of bounded X-valued ultradistributions of $*$-class, which has been used in [208] and [449] for the definition of an almost periodic ultradistribution of $*$-class. If the last condition holds, then we have $T * \varphi \in AP(\mathbb{R}^n : X)$, respectively, $T * \varphi \in UR(\mathbb{R}^n : X)$, for all $\varphi \in \mathcal{D}^*(\mathbb{R}^n)$; see the proof of implication (i) $\Leftrightarrow$ (iii) in [208, Theorem 2]. We can similarly reformulate the statements of [449, Theorem 1] and [449, Theorem 2] for almost periodic ultradistributions of $*$-class in $\mathbb{R}^n$. The statement of [459, Theorem 2.2] can be simply reformulated for c-almost periodic ultradistributions of $*$-class in $\mathbb{R}^n$ by replacing the space $\mathcal{D}'_{L^1,s}(\mathbb{R}^n, * : X)$ in its formulation with the space $\mathcal{D}'_{L^1}(\mathbb{R}^n, * : X)$; the same modification has to be done in the one-dimensional setting.

We will omit here all details concerning the existence of Bohr–Fourier coefficients of almost periodic ultradistributions of $*$-class; cf. [324] and the final part of [449, Section 2] for further information in this direction.

If $\emptyset \neq \mathbb{A} \subseteq C^{\infty}(\mathbb{R}^n : X)$, let us denote by $B'^{*}_{\mathbb{A}}(\mathbb{R}^n : X)$ the space of all vector-valued ultradistributions $T \in \mathcal{D}'^{*}(\mathbb{R}^n : X)$ such that $T * \varphi \in \mathbb{A}$ for all $\varphi \in \mathcal{D}^{*}(\mathbb{R}^n)$. The interested reader should reconsider the statements established in [449, Section 3] in the higher-dimensional setting. We close this part by observing that it could be also interesting to introduce and analyze the Colombeau ρ-almost periodic generalized functions in $\mathbb{R}^n$ and the Fourier ρ-almost periodic hyperfunctions in $\mathbb{R}^n$. The corresponding classes of almost automorphic generalized functions, with $\rho = \mathrm{I}$, can be also investigated.

Stepanov-like pseudo-S-asymptotically (ω, c)-periodic solutions for a class of stochastic Volterra integrodifferential equations

The study of periodicity and asymptotic behavior of solutions to the stochastic Volterra integrodifferential equations has become a crucial subject of investigation due to its wide range of applications in various scientific and engineering fields. Particularly, the pseudo-S-asymptotic periodicity and the (ω, c)-periodicity have attracted significant attention in the recent literature. Let us also notice that P. Bezandry [113] and P. Bezandry, T. Diagana [114] have analyzed almost periodic and S^2-almost periodic solutions for stochastic evolution equations driven by fractional Brownian motion; inspired by their results, we examine here the Stepanov-like pseudo-S-asymptotically (ω, c)-periodic solutions for a class of stochastic Volterra integrodifferential equations [492]. For more details about almost periodic solutions of stochastic Volterra integrodifferential equations, we refer the reader to the list of references cited in [444], [447] and [448].

Of concern is the following stochastic Volterra integrodifferential equation:

$$\Psi'(\tau) = A\Psi(\tau)+g(\tau,\Psi(\tau))+\int_{-\infty}^{\tau} A_1(\tau-s)f(s,\Psi(s))\,ds+\int_{-\infty}^{\tau} A_2(\tau-s)h(s,\Psi(s))\,dW(s), \quad (361)$$

where $A : D(A) \subseteq H \to H$ is the infinitesimal generator of a C_0-semigroup $(T(\tau))_{\tau\geqslant 0}$, A_1, A_2 are convolution type kernels in $L^1(0,\infty)$ and $L^2(0,\infty)$, respectively, $f : \mathbb{R} \times L^2(\Omega : H) \to L^2(\Omega : H)$, $g : \mathbb{R} \times L^2(\Omega : H) \to L^2(\Omega : H)$, $h : \mathbb{R} \times L^2(\Omega : H) \to L^2(\mathbb{R} : L^2(\Omega : H))$ and W is a Brownian motion, with covariance operator Q, such that $tr\, Q < \infty$. Here, we assume that $(\Omega, \mathcal{F}, p)$ is a probability space and H is a complex separable Hilbert space. We will use the notation $\|\cdot\|$ to represent the norm on H; the expectation $E(\cdot)$ is defined by $E\|\Psi\|^2 = \int_{\Omega} \|\Psi\|^2\, dp$.

For a stochastic process $\Psi : \mathbb{R} \to L^2(\Omega : H)$ is said to be stochastically bounded if

$$E\|\Psi(\tau)\|^2 = \int_{\Omega} \|\Psi(\tau)\|^2\, dp < M,$$

for some positive constant M and all $\tau \in \mathbb{R}$; furthermore, a stochastic process $\Psi : \mathbb{R} \to L^2(\Omega : H)$ is said to be stochastically continuous if

$$\lim_{\tau\to s} E\|\Psi(\tau) - \Psi(s)\| = 0.$$

The space of all bounded and continuous stochastic processes $\Psi : \mathbb{R} \to L^2(\Omega : H)$ will be denoted by $BC(\mathbb{R} : L^2(\Omega : H))$. The space $BC(\mathbb{R} : L^2(\Omega : H))$ equipped with the norm $\|\Psi\| = (\sup_{\tau\in\mathbb{R}} E\|\Psi(\tau)\|^2)^{1/2}$ becomes a Banach space. Further on, the space of all Stepanov bounded stochastic processes $\Psi : \mathbb{R} \to L^2(\Omega : H)$, such that $\Psi^*(t,s) = \Psi(t+s) \in L^\infty(\mathbb{R} : L^2([0,1] : L^2(\Omega : H)))$ will be denoted by $\mathcal{B}SS(\mathbb{R} : L^2(\Omega : H))$. When endowed with the norm

$$\|\Psi\|_{BSS} := \sup_{\tau\in\mathbb{R}}\left(\int_0^1 E\|\Psi(\tau+s)\|^2\,ds\right)^{\frac{1}{2}} = \sup_{\tau\in\mathbb{R}}\left(\int_\tau^{\tau+1} E\|\Psi(r)\|^2\,dr\right)^{\frac{1}{2}}$$

this space becomes a Banach space. Finally, let us recall that, if $c \in \mathbb{C}\setminus\{0\}$ and $\omega > 0$, then a continuous function $u : \mathbb{R} \to X$ is called (ω, c)-periodic if $u(\tau+\omega) = cu(\tau)$ for all $\tau \in \mathbb{R}$. In the sequel of this part, we will always assume that $|c| = 1$; for more details regarding the general case $|c| \neq 1$, we refer the reader to [492].

We need the following notion (cf. also [779]).

Definition 6.4.22. For a stochastic process $\Psi \in BC(\mathbb{R} : L^2(\Omega : H))$, it is said that is mean-square pseudo-S-asymptotically (ω, c)-periodic if

$$\lim_{\mu\to\infty}\frac{1}{2\mu}\int_{-\mu}^{\mu} E\|\Psi(s+\omega) - c\Psi(s)\|^2\,ds = 0.$$

By $\mathcal{P}SAP_{\omega,c}(\mathbb{R} : L^2(\Omega : H))$ will be denoted the set of all square-mean pseudo-S-asymptotically (ω, c)-periodic stochastic processes.

We have the following simple results:

Theorem 6.4.23. *Let* Ψ, Ψ_1, $\Psi_2 \in \mathcal{P}SAP_{\omega,c}(\mathbb{R} : L^2(\Omega : H))$. *Then:*
(i) $\Psi_1 + \Psi_2 \in \mathcal{P}SAP_{\omega,c}(\mathbb{R} : L^2(\Omega : H))$.
(ii) $a\Psi \in \mathcal{P}SAP_{\omega,c}(\mathbb{R} : L^2(\Omega : H))$, *for each* $a \in \mathbb{C}$.
(iii) $\Psi_a(\tau) = \Psi(\tau + a) \in \mathcal{P}SAP_{\omega,c}(\mathbb{R} : L^2(\Omega : H))$, *for each* a, $\tau \in \mathbb{R}$.
(iv) $\mathcal{P}SAP_{\omega,c}(\mathbb{R} : L^2(\Omega : H))$ *endowed with the norm* $\|\Psi\|_\infty = (\sup_{\tau\in\mathbb{R}} E\|\Psi(\tau)\|^2)^{1/2}$, $\Psi \in \mathcal{P}SAP_{\omega,c}(\mathbb{R} : L^2(\Omega : H))$ *is a Banach space.*

Definition 6.4.24. For a stochastic process $\Psi \in \mathcal{B}SS(\mathbb{R} : L^2(\Omega : H))$, it is said to be mean-square Stepanov-like pseudo-S-asymptotically (ω, c)-periodic if

$$\lim_{\mu\to\infty}\frac{1}{2\mu}\int_{-\mu}^{\mu}\left(\int_\tau^{\tau+1} E\|\Psi(s+\omega) - c\Psi(s)\|^2\,ds\right)^{\frac{1}{2}} d\tau = 0.$$

By $\mathcal{S}PSAP_{\omega,c}$, we denote the collection of all square-mean Stepanov-like pseudo-S-asymptotically (ω, c)-periodic stochastic processes.

We will prove the following results.

Theorem 6.4.25. (i) *$SPSAP_{\omega,c}(\mathbb{R} : L^2(\Omega : H))$ is a Banach space with the norm $\|\Psi\|_{BSS} := \sup_{\tau\in\mathbb{R}}(\int_\tau^{\tau+1} E\|\Psi(s)\|^2\,ds)^{1/2}$, $\Psi \in SPSAP_{\omega,c}(\mathbb{R} : L^2(\Omega : H))$.*
(ii) *$PSAP_{\omega,c}(\mathbb{R} : L^2(\Omega : H)) \subseteq SPSAP_{\omega,c}(\mathbb{R} : L^2(\Omega : H))$.*

Proof. (i) Let (Ψ_k) be a sequence in $SPSAP_{\omega,c}(\mathbb{R} : L^2(\Omega : H))$ and $\Psi_k \to \Psi$ as $k \to \infty$. Hence, for every $\varepsilon > 0$ there exist constants k_0 and $M > 0$ such that for $k \geqslant k_0$, $\|\Psi_k - \Psi\|_{BSS} \leqslant \frac{\varepsilon}{3|c|}$ and for $\mu > M$,

$$\frac{1}{2\mu}\int_{-\mu}^{\mu}\left(\int_{\tau}^{\tau+1} E\|\Psi_k(s+\omega) - c\Psi(s)\|\,ds\right)^{\frac{1}{2}} d\tau \leqslant \frac{\varepsilon}{3}.$$

Now,

$$\begin{aligned}
&\frac{1}{2\mu}\int_{-\mu}^{\mu}\left(\int_{\tau}^{\tau+1} E\|\Psi(s+\omega) - c\Psi(s)\|^2\,ds\right)^{\frac{1}{2}} d\tau \\
&= \frac{1}{2\mu}\int_{-\mu}^{\mu}\left(\int_{\tau}^{\tau+1} E\|\Psi(s+\omega) - \Psi_k(s+\omega) + \Psi_k(s+\omega) - c\Psi_k(s) + c\Psi_k(s) - c\Psi(s)\|^2\,ds\right)^{\frac{1}{2}} d\tau \\
&\leqslant \frac{1}{2\mu}\int_{-\mu}^{\mu}\left(\int_{\tau}^{\tau+1} E\|\Psi(s+\omega) - \Psi_k(s+\omega)\|^2\,ds\right)^{\frac{1}{2}} d\tau \\
&\quad + \frac{1}{2\mu}\int_{-\mu}^{\mu}\left(\int_{\tau}^{\tau+1} E\|\Psi_k(s+\omega) - c\Psi_k(s)\|^2\,ds\right)^{\frac{1}{2}} d\tau \\
&\quad + \frac{1}{2\mu}\int_{-\mu}^{\mu}\left(\int_{\tau}^{\tau+1} E\|c\Psi_k(s) - c\Psi(s)\|^2\,ds\right)^{\frac{1}{2}} d\tau \\
&\leqslant \|\Psi_k - \Psi\|_{BSS} + \frac{\varepsilon}{3} + |c|\|\Psi_k - \Psi\|_{BSS} \leqslant \frac{\varepsilon}{3} + \frac{\varepsilon}{3} + \frac{\varepsilon}{3} = \varepsilon.
\end{aligned}$$

Therefore, the space $SPSAP_{\omega,c}(\mathbb{R} : L^2(\Omega : H))$ is a closed subspace of $BSS(\mathbb{R} : L^2(\Omega : H))$, so this space is a Banach space, when equipped with the $\|\cdot\|_{BSS}$ norm.

(ii) Let $\Psi \in PSAP_{\omega,c}(\mathbb{R} : L^2(\Omega : H))$. Hence, $\Psi(\cdot + s) \in PSAP_{\omega,c}(\mathbb{R} : L^2(\Omega : H))$ for $s \in [0,1]$. Now,

$$\begin{aligned}
&\frac{1}{2\mu}\int_{-\mu}^{\mu}\left(\int_{\tau}^{\tau+1} E\|\Psi(s+\omega) - c\Psi(s)\|^2\,ds\right)^{\frac{1}{2}} d\tau \\
&\leqslant \frac{(2\mu)^{\frac{1}{2}}}{2\mu}\left(\int_{-\mu}^{\mu}\int_{\tau}^{\tau+1} E\|\Psi(s+\omega) - c\Psi(s)\|^2\,ds\,d\tau\right)^{\frac{1}{2}}
\end{aligned}$$

$$= \frac{1}{(2\mu)^{\frac{1}{2}}} \left(\int_{-\mu}^{\mu} \int_{0}^{1} E\|\Psi(\tau + s + \omega) - c\Psi(\tau + s)\|^2 \, ds \, d\tau \right)^{\frac{1}{2}}$$

$$= \left(\int_{0}^{1} \frac{1}{2\mu} \int_{-\mu}^{\mu} E\|\Psi(\tau + s + \omega) - c\Psi(\tau + s)\|^2 \, d\tau \, ds \right) \to 0,$$

when $\mu \to +\infty$. Hence, $\Psi \in \mathcal{S}PSAP_{\omega,c}(\mathbb{R} : L^2(\Omega : H))$. □

Let $f, g, h \in \mathcal{B}SS(\mathbb{R} \times L^2(\Omega : H) : L^2(\Omega : H))$. We will use the following conditions:

(A0s) Let $(\tau, x) \in \mathbb{R} \times L^2(\Omega : H)$. Then (A0s) means that

$$\lim_{\mu\to\infty} \frac{1}{2\mu} \int_{-\mu}^{\mu} \left(\int_{\tau}^{\tau+1} E\|f(s + \omega, x) - cf(s, c^{-1}x)\|^2 \, ds \right)^{\frac{1}{2}} d\tau = 0,$$

$$\lim_{\mu\to\infty} \frac{1}{2\mu} \int_{-\mu}^{\mu} \left(\int_{\tau}^{\tau+1} E\|g(s + \omega, x) - cg(s, c^{-1}x)\|^2 \, ds \right)^{\frac{1}{2}} d\tau = 0,$$

$$\lim_{\mu\to\infty} \frac{1}{2\mu} \int_{-\mu}^{\mu} \left(\int_{\tau}^{\tau+1} E\|h(s + \omega, x) - ch(s, c^{-1}x)\|^2 \, ds \right)^{\frac{1}{2}} d\tau = 0,$$

uniformly on any bounded set of $L^2(\Omega : H)$.

(A1s) There exist constants L_1, L_2, $L_3 > 0$ such that

$$E\|g(\tau, \Phi) - g(\tau, \Psi)\|^2 \leqslant L_1 E\|\Phi - \Psi\|^2,$$
$$E\|f(\tau, \Phi) - f(\tau, \Psi)\|^2 \leqslant L_2 E\|\Phi - \Psi\|^2,$$
$$E\|h(\tau, \Phi) - h(\tau, \Psi)\|^2_{L(\mathbb{R}:L^2(\Omega:H))} \leqslant L_3 E\|\Phi - \Psi\|^2,$$

for all $\tau \in \mathbb{R}$ and every Φ, $\Psi \in L^2(\Omega : H)$.

(A2s) The semigroup $(T(\tau))_{\tau\geqslant 0}$ is compact and exponentially stable, meaning that there are constants K, $\theta > 0$ such that $\|T(\tau)\| \leqslant Ke^{-\theta t}$, $\tau \geqslant 0$.

(A3s) The functions f, g, h are uniformly continuous on every bounded set $B \subseteq L^2(\Omega : H)$ for every $\tau \in \mathbb{R}$. Additionally, for every bounded set $B \subseteq L^2(\Omega : H)$, the sets $g(\mathbb{R}, B)$, $f(\mathbb{R}, B)$ and $h(\mathbb{R}, B)$ are bounded. Furthermore, there exists $b > 0$ such that $\Lambda_b \leqslant \frac{\theta b}{20\nu K^2}$, where

$$\Lambda_b = \max\left\{ \sup_{\tau\in\mathbb{R},\|\Psi\|\leqslant b} \|g(\tau, \Psi)\|, \sup_{\tau\in\mathbb{R},\|\Psi\|\leqslant b} \|f(\tau, \Psi)\|, \sup_{\tau\in\mathbb{R},\|\Psi\|\leqslant b} \|h(\tau, \Psi)\| \right\},$$

and $\nu = \max\{1, \|A_1\|^2_{L^1(0,\infty)}, 4\|A_2\|^2_{L^2(0,\infty)}\}$.

(A4s) There exist measurable functions m_f, m_g and m_h from $\mathbb{R}$ to $[0,\infty)$ such that

$$E\|g(\tau,\Phi)-g(\tau,\Psi)\|^2 \leqslant m_g(\tau)\cdot E\|\Phi-\Psi\|^2,$$
$$E\|f(\tau,\Phi)-f(\tau,\Psi)\|^2 \leqslant m_f(\tau)\cdot E\|\Phi-\Psi\|^2,$$
$$E\|h(\tau,\Phi)-h(\tau,\Psi)\|^2_{L(\mathbb{R},L^2(\Omega:H))} \leqslant m_h(\tau)\cdot E\|\Phi-\Psi\|^2,$$

for all $\tau\in\mathbb{R}$ and every Φ, $\Psi\in L^2(\Omega:H)$.

(A5s) Let $(\Psi_k)_k\in\mathcal{BSS}(\mathbb{R}:L^2(\Omega:H))$ be uniformly bounded and uniformly convergent on every compact subset of $\mathbb{R}$. Then $g(\cdot,\Psi_k(\cdot))$, $f(\cdot,\Psi_k(\cdot))$ and $h(\cdot,\Psi_k)$ are relatively compact in $\mathcal{BSS}(\mathbb{R}:L^2(\Omega:H))$.

Then we can state the following results.

Theorem 6.4.26. *Let $u\in\mathcal{BSS}(\mathbb{R}\times L^2(\Omega:H):L^2(\Omega:H))$ fulfill the assumption* (A0), *let us suppose the existence of a positive real number L such that for any Ψ_1, $\Psi_2\in L^2(\Omega:H)$ we have $E\|u(\tau,\Psi_1)-u(\tau,\Psi_2)\|^2\leqslant L\|\Psi_1-\Psi_2\|^2$ uniformly for all $\tau\in\mathbb{R}$, and let $\Psi\in SPSAP_{\omega,c}(\mathbb{R}:L^2(\Omega:H))$. Then $u(\cdot,\Psi(\cdot))\in SPSAP_{\omega,c}(\mathbb{R},L^2(\Omega:H))$.*

Theorem 6.4.27. *Let for every $\varepsilon>0$ and for every bounded subset $B\subseteq L^2(\Omega:H)$, there are constants $C_{\varepsilon,B}$ and $\delta_{\varepsilon,B}>0$ such that*

$$E\|w(\tau,\Psi_1)-w(\tau,\Psi_2)\|^2\leqslant\varepsilon,$$

for all Ψ_1, $\Psi_2\in B$ with $E\|\Psi_1-\Psi_2\|^2\leqslant\delta_{\varepsilon,B}$ and $\tau\geqslant C_{\varepsilon,B}$. Let $u\in\mathcal{BSS}(\mathbb{R}\times L^2(\Omega:H):L^2(\Omega:H))$ fulfill the assumption (A0) *and let $\Psi\in SPSAP_{\omega,c}(\mathbb{R}:L^2(\Omega:H))$. Then $u(\cdot,\Psi(\cdot))\in SPSAP_{\omega,c}(\mathbb{R}:L^2(\Omega:H))$.*

In [492], we have also introduced the notion of a $\mathcal{F}_\tau$-progressively measurable process $(\Psi(\tau))_{\tau\in\mathbb{R}}$ and said that $(\Psi(\tau))_{\tau\in\mathbb{R}}$ is a mild solution of (361) if the following stochastic integral equation is satisfied:

$$\begin{aligned}\Psi(\tau)&=T(\tau-\tau_0)\Psi(\tau_0)+\int_{\tau_0}^{\tau}T(\tau-s)g(s,\Psi(s))\,ds\\&\quad+\int_{\tau_0}^{\tau}T(\tau-\sigma)\int_{\tau_0}^{\sigma}A_1(\sigma-s)f(s,\Psi(s))\,ds\,d\sigma\\&\quad+\int_{\tau_0}^{\tau}T(\tau-\sigma)\int_{\tau_0}^{\sigma}A_2(\sigma-s)h(s,\Psi(s))\,dW(s)\,d\sigma\end{aligned}$$

for all $\tau\geqslant\tau_0$. Here, the third integral is interpreted in the sense of Itô approach.

We define the integral operator Θ by

$$(\Theta\Psi)(\tau) := \int_{-\infty}^{\tau} T(\tau-s)g(s,\Psi(s))\,ds + \int_{-\infty}^{\tau} T(\tau-\sigma)\int_{-\infty}^{\sigma} A_1(\sigma-s)f(s,\Psi(s))\,ds\,d\sigma$$
$$+ \int_{-\infty}^{\tau} T(\tau-\sigma)\int_{-\infty}^{\sigma} A_2(\sigma-s)h(s,\Psi(s))\,dW(s)\,d\sigma.$$

Then we have the following results.

Lemma 6.4.28. *Let* (A2) *and* (A3) *be satisfied. Then* $\Theta : BC(\mathbb{R} : L^2(\Omega : H)) \to BC(\mathbb{R} : L^2(\Omega : H))$ *is well-defined and continuous.*

Theorem 6.4.29. *Let* (A0), (A2)–(A5) *hold and let* $\varphi : [0,\infty) \to [0,\infty)$ *be a nonincreasing function with* $\sum_{k=1}^{\infty}\varphi(k) < \infty$ *such that* $\|T(\tau)\|^2 \leqslant \varphi(\tau)$ *for all* $\tau \in [0,\infty)$. *If* f, $g \in \mathcal{S}PSAP_{\omega,c}(\mathbb{R}\times L^2(\Omega : H) : L^2(\Omega : H))$ *and* $h \in \mathcal{S}PSAP_{\omega,c}(\mathbb{R}\times L^2(\Omega : H) : L(\mathbb{R}, L^2(\Omega : H))$, *then the stochastic equation* (361) *has at least one Stepanov-like pseudo-S-asymptotically* (ω, c)*-periodic solution mild solution, provided that*

$$L_g = \sup_{\tau\in\mathbb{R}} \int_{-\infty}^{\tau} e^{-\theta(\tau-s)} m_g(s)\,ds < \infty,$$

$$L_f = \sup_{\tau\in\mathbb{R}} \int_{-\infty}^{\tau} \|A_1(\tau-s)\|^2 m_f(s)\,ds < \infty,$$

$$L_h = \sup_{\tau\in\mathbb{R}} \int_{-\infty}^{\tau} \|A_2(\tau-s)\|^2 m_h(s)\,ds < \infty,$$

and

$$\rho = 6\frac{K}{\theta^2}\big(\theta L_g + L_f\|A_1\|_{L^1(0,\infty)} + L_h\big) < 1.$$

Let us finally consider the following example.

Example 6.4.30. Consider the following stochastic evolution integrodifferential equation:

$$\frac{\partial\Psi(\tau,x)}{\partial\tau} = \frac{\partial^2}{\partial x^2}\Psi(\tau,x) + g(\tau,\Psi(\tau,x))$$
$$+ \int_{-\infty}^{\tau} e^{-\pi^2(\tau-s)} f(s,\Psi(x,\tau))\,ds + \int_{-\infty}^{\tau} e^{-\pi^2(\tau-s)} h(s,\Psi(s,x))\,dW(s), \tag{362}$$

with $\Psi(t,0) = \Psi(t,1) = 0$, $t \in \mathbb{R}$ and W is a Brownian motion, with covariance operator Q, such that $tr\, Q < \infty$. The forcing terms are given by

$$g(\tau,\Psi)(s) = a(\tau)\sigma_1(\Psi(s)),$$
$$f(\tau,\Psi)(s) = a(\tau)\sigma_2(\Psi(s)),$$
$$h(\tau,\Psi)(s) = a(\tau)\sigma_3(\Psi(s)),$$

where some extra conditions are satisfied. Now,

$$f(\tau+\omega,\Psi)(s) = a(\tau+\omega)\sigma_1(\Psi) = a(\tau)c\sigma_1(c^{-1}\Psi) = cf(\tau,c^{-1}\Psi)(s),$$
$$g(\tau+\omega,\Psi)(s) = a(\tau+\omega)\sigma_2(\Psi) = a(\tau)c\sigma_2(c^{-1}\Psi) = cg(\tau,c^{-1}\Psi)(s),$$
$$h(\tau+\omega,\Psi)(s) = a(\tau+\omega)\sigma_3(\Psi) = a(\tau)c\sigma_3(c^{-1}\Psi) = ch(\tau,c^{-1}\Psi)(s),$$

so that the Lebesgue dominated convergence theorem yields

$$\lim_{\mu\to\infty}\frac{1}{2\mu}\int_{-\mu}^{\mu}\left(\int_{\tau}^{\tau+1} E\|f(s+\omega,\Psi(s+\omega)) - cf(s,c^{-1}\Psi(s+\omega))\|^2\,ds\right)^{\frac{1}{2}} d\tau = 0,$$

$$\lim_{\mu\to\infty}\frac{1}{2\mu}\int_{-\mu}^{\mu}\left(\int_{\tau}^{\tau+1} E\|g(s+\omega,\Psi(s+\omega)) - cg(s,c^{-1}\Psi(s+\omega))\|^2\,ds\right)^{\frac{1}{2}} d\tau = 0,$$

$$\lim_{\mu\to\infty}\frac{1}{2\mu}\int_{-\mu}^{\mu}\left(\int_{\tau}^{\tau+1} E\|h(s+\omega,\Psi(s+\omega)) - ch(s,c^{-1}\Psi(s+\omega))\|^2\,ds\right)^{\frac{1}{2}} d\tau = 0,$$

and (A0) is therefore satisfied.

Set $H := L^2(0,1)$. Let us consider the linear operator $A : D(A) \subseteq H \to H$ defined by $D(A) := H^2(0,1)\cap H_0^1(0,1)$ and $A\phi(s) := \Delta\psi$ for $s \in (0,1)$ and $\phi \in D(A)$. Note that $H^1(0,1) = \{u \in L^2(0,1) : u' \in L^2(0,1)\}$, $H^2(0,1) = \{u \in L^2(0,1) : u',\ u'' \in L^2(0,1)\}$ and $H_0^1(0,1) = \{u \in H^1(0,1) : u(0) = u(1) = 0\}$, where the derivatives are taken in a weak sense. Then we know that the operator A generates a strongly continuous semigroup $(T(\tau))_{\tau\geqslant 0}$ given by

$$(T(\tau)\phi)(z) = \sum_{k=1}^{\infty} e^{-k^2\pi^2\tau}\langle\phi, e_k\rangle_{L^2} e_k(z),$$

where $e_k(z) = \sqrt{2}\sin(k\pi z)$, for $k = 1,2,\ldots$ and $\|T(\tau)\| \leqslant e^{-\pi^2\tau}$ for all $\tau \geqslant 0$. In [492], we have proved that there exists a unique Stepanov-like pseudo-S-asymptotically (ω,c)-periodic mild solution of (362).

Poincaré–Perron problem for ordinary differential equations of higher order in the class of almost periodic type functions

The first result about the existence of solutions to the following nonautonomous linear differential equation of higher order:

$$y^{(n)}(t) + \sum_{i=0}^{n-1}(a_i + r_i(t))y^{(i)}(t) = 0, \tag{363}$$

where $a_0, \ldots, a_{n-1} \in \mathbb{C}$ and the functions $r_i(\cdot)$ are small in a certain sense for $i \geqslant 2$, was proved by H. Poincaré in 1885 and later improved by O. Perron in 1909. More precisely, H. Poincaré proved the existence of a solution to (363) whose logarithmic derivative $y'(t)/y(t)$ converges as $t \to +\infty$, provided that the roots $\lambda_1, \ldots, \lambda_n \in \mathbb{C}$ of the associated polynomial equation $x^n + \sum_{i=0}^{n-1} a_i x^i = 0$ have distinct real parts, the functions $r_i(\cdot)$ are continuous for $t \geqslant t_0$ and vanish at plus infinity. Under the same assumptions, O. Perron proved the existence of n linearly independent solutions $y_1(\cdot), \ldots, y_n(\cdot)$ of (363) such that their logarithmic derivatives $y_i'(t)/y_i(t)$ converges to λ_i as $t \to +\infty$, for $1 \leqslant i \leqslant n$. From then on, the Poincaré–Perron type problems for ordinary differential equations of higher order have been analyzed by many authors, especially in the case that the perturbations $r_i(\cdot)$ belong to a weighted L^p-space ($p \geqslant 1; 1 \leqslant i \leqslant n$).

The first results about the Poincaré problem in the class of almost periodic type functions were given by P. Figueroa and M. Pinto in [303] (2015), where the authors investigated the second-order differential equation (363). In [154], H. Bustos, P. Figueroa and M. Pinto obtained the important scientific results in the case that $n \geqslant 3$, where the explicit formulae for the solutions of (363) are given by studying a Riccati type equation associated with the logarithmic derivatives of solutions. Furthermore, the authors provided some sufficient conditions ensuring the existence of a fundamental system of solution, using also the Banach fixed-point theorem for findings of a unique almost periodic and asymptotically almost periodic solutions to this Riccati type equation. The results are well illustrated for the third-order linear differential equation (363).

Before we close this part, let us note that M. Cheng and Z. Liu have analyzed, in [168], the second Bogolyubov theorem and applied this result to the global averaging principle for SPDEs with monotone coefficients.

Part III: Abstract Volterra integrodifferential functional inclusions and their almost periodic type solutions

Functional differential equations, specifically referred to as neutral differential equations, are encountered in a variety of phenomena, particularly in the analysis of oscillatory systems and the modeling of various physical problems. For further details and in-depth exploration, the interested readers are encouraged to refer to [278, 673] and the references provided therein. The problem of existence and uniqueness of almost periodic solutions of integrodifferential equations and neutral integrodifferential equations is very popular since they play a crucial role in modeling dynamic systems with delayed interactions, reflecting scenarios encountered in various scientific and engineering areas.

The exploration of generalized periodicity in the solutions of diverse classes of neutral differential equations was initially prompted by considerations of periodic behavior in their solutions. Key contributions to this line of inquiry can be found in works such as [99, 390, 624, 635, 657, 685, 690, 793] and [769]. Researchers E. Ait Dads and K. Ezzinbi [31], as well as A. Fink and J. Gatica [304], extended this exploration by examining almost periodic solutions for specific classes of nonlinear neutral integral equations. Subsequent investigations by S. Abbas and D. Bahuguna [8] and X. Chen and F. Lin [195] further expanded the scope, examining almost periodicity in more general neutral functional differential equations within Banach spaces. Notably, the pursuit of almost periodic solutions for a class of nonlinear integrodifferential equations with neutral delay is presented in [827]. Additionally, in a related but distinct conceptualization of almost periodicity, studies on positive pseudo almost periodicity [30, 248], and investigations into (μ, ν)-pseudo S-asymptotic ω-periodicity [790] for solutions of various classes of neutral differential equations have been conducted.

The main aim of this part is to structurally analyze several new classes of Volterra integrodifferential functional inclusions as well as to contribute to the theory of fractional calculus. We also examine the existence and uniqueness of almost periodic type solutions to the Volterra integrodifferential functional inclusions.

https://doi.org/10.1515/9783111689746-010

7 (F, G, C)-resolvent operator families, abstract fractional differential inclusions and some classes of abstract nonlinear fractional differential equations with delay

7.1 (F, G, C)-resolvent operator families and applications

The theory of abstract Volterra integrodifferential equations is still a very active field of research of many authors (cf. the monographs [443, 445, 652] and references cited therein for more details on the subject). In a joint research paper with V. E. Fedorov [292], which contains the main results of this section, we have continued our recent investigations of the abstract Volterra integrodifferential inclusions by examining several new classes of the abstract fractional differential-difference inclusions and the abstract Volterra integrodifference inclusions.

The use of vector-valued Laplace transform and (a, k)-regularized C-resolvent families is almost inevitable in any important research of the abstract nonscalar Volterra integrodifferential equations. The main purpose of this section is to extend the notion of an exponentially equicontinuous (a, k)-regularized C-resolvent family [mild (a, k)-regularized C_1-existence family, mild (a, k)-regularized C_2-uniqueness family] by introducing the concept of an (F, G, C)-regularized resolvent family [mild (F, G, C_1)-regularized existence family, mild (F, G, C_2)-regularized uniqueness family]; cf. Definition 7.1.1 and Proposition 7.1.2. The introduced solution operator families are subgenerated by multivalued linear operators in locally convex spaces.

The organization and main ideas of this section can be briefly summarized as follows. After explaining the main notation and terminology used, we introduce and analyze various classes of (F, G, C)-regularized resolvent operator type families in Section 7.1.1. Many structural characterizations of (a, k)-regularized C-resolvent operator type families continue to hold for (F, G, C)-regularized resolvent operator type families; for the sake of brevity, we will only quote the corresponding results and explain the basic differences in our new framework (cf. [445, Chapter 3] and [460] for more details on the subject; before proceeding further, we would like to say that it is far from being clear how we can introduce our new concepts in the local framework). We introduce the notion of the integral generator of a mild (F, G, C_2)-regularized uniqueness family [(F, G, C)-regularized resolvent family] and clarify the main structural features of subgenerators of (F, G, C)-regularized resolvent operator type families. In Theorem 7.1.4 [Theorem 7.1.5], we prove the existence of a very specific example of an (F, G)-regularized resolvent family $(R(t))_{t \geqslant 0}$, which cannot be (a, k)-regularized resolvent family for any choice of a Laplace transformable function $a \in L^{\infty}_{loc}([0, \infty))$ [$a \in L^{1}_{loc}([0, \infty))$] and a continuous Laplace transformable function $k(\cdot)$. The notion of an (exponentially equicontinuous) analytic (F, G, C)-regularized resolvent family of angle α is introduced in Defi-

https://doi.org/10.1515/9783111689746-011

nition 7.1.6. After that, we examine the subordination principles, the generation results for (exponentially equicontinuous, analytic) (F, G, C)-regularized resolvent families and the smoothing properties of (F, G, C)-regularized resolvent families.

We continue by recalling that S. Bochner [122] has analyzed the linear difference-differential operators

$$L_h = \sum_{i=0}^{p} \sum_{j=0}^{q} a_{ij} \frac{d^i}{dx^i} h(\cdot + h_j),$$

where a_{ij} are complex numbers ($0 \leqslant i \leqslant p$, $0 \leqslant j \leqslant q$) and $h = (h_j)_{0 \leqslant j \leqslant q} \subseteq \mathbb{R}^q$; some results about the existence and uniqueness of generalized almost periodic solutions of the above equation can be found in [444]. Concerning the applications of (F, G, C)-regularized resolvent operator families to the abstract fractional differential-difference inclusions (cf. Section 7.1.2), we would like to emphasize first that C. Lizama and his coauthors have examined, in a series of important research papers, the well-posedness of the abstract fractional difference equation (50), where A is a closed linear operator on Banach space X, $0 < \alpha < 1$ and $\Delta^\alpha u(k)$ denotes the Caputo fractional difference operator of order α; cf. [3, 45, 46] and references cited therein for more details about this problematic. The results established in the aforementioned research papers have strongly influenced us to examine what happens with the well-posedness of the following abstract degenerate fractional differential-difference inclusions:

$$(\mathrm{DFP})_{R,a,b}: \begin{cases} \mathbf{D}_t^\alpha Bu(t+a) \in \mathcal{A}u(t+b) + f(t), & t \geqslant 0, \\ (Bu)(t) = Bu_0(t), & 0 \leqslant t \leqslant c \end{cases}$$

and

$$(\mathrm{DFP})_{L,a,b}: \begin{cases} \mathcal{B}\mathbf{D}_t^\alpha u(t+a) \subseteq \mathcal{A}u(t+b) + f(t), & t \geqslant 0, \\ u(t) = u_0(t), & 0 \leqslant t \leqslant c, \end{cases}$$

where $0 \leqslant a, b < \infty$, $c = \max(a, b)$, $f : [0, \infty) \to X$ is Lebesgue measurable, B is a closed linear operator, $\mathcal{A}$ and $\mathcal{B}$ are closed MLOs. In Example 7.1.7, which opens Section 7.1.2, we perceive that the solutions of problems $(\mathrm{DFP})_{R,a,b}$ and $(\mathrm{DFP})_{L,a,b}$ can be simply calculated step by step in many concrete situations as well as that it is almost impossible to apply the vector-valued Laplace transform in the deeper analysis of the abstract fractional Cauchy inclusions $(\mathrm{DFP})_{R,a,b}$ and $(\mathrm{DFP})_{L,a,b}$, provided that $0 \leqslant a < b$; on the other hand, if $0 \leqslant b < a$, then (a, k)-regularized C-resolvent operator families can be employed for giving the simple form of solutions of these problems (it would be very difficult to summarize here the basic results concerning the Volterra integrofunctional equations and their applications, even in the scalar-valued setting; for further information in this direction, we refer the reader to the lectures of H. Brunner [147], the research articles [211] by C. Corduneanu, [283] by K. Ezzinbi and M. A. Hachimi, [355] by J. R. Haddock, M. N. Nkashama, J. H. Wu, the important research monographs [78] by N. V. Azbelev, V. P. Maksimov, L. F.

Rakhmatullina, [278] by L. Erbe, Q. Kong, B. G. Zhang, [366] by J. K. Hale, [437] by V. B. Kolmanovskii, A. Myshkis as well as the research articles [359, 360] by R. Hakl, [570] by P. Martinez-Amores, [700, 702, 703, 706] by V. E. Slyusarchuk and references cited therein).

In the continuation of Section 7.1.2, we analyze the well-posedness of the following abstract Volterra initial value problem:

$$u(t) \in f(t) + \sum_{i=1}^{m} (a_i * \mathcal{A}u(\cdot + t_i))(t), \quad t \geqslant 0; \; u(t) = u_0(t), \; 0 \leqslant t \leqslant \tau,$$

where $m \in \mathbb{N}$, $0 \leqslant t_1, t_2, \ldots, t_m < +\infty$, $\tau := \max(t_1, t_2, \ldots, t_m)$, $f : [0, \infty) \to X$ is Lebesgue measurable, $\mathcal{A}$ is a closed MLO on X and $a_1(t), \ldots, a_m(t)$ are locally integrable scalar-valued functions defined for $t \geqslant 0$; in contrast to many other recent research papers concerning the abstract delay integrodifferential equations, the terms $u(\cdot + t_i)$ considered here are under the action of the multivalued linear operator $\mathcal{A}$ (the problem seems to be new even for the single-valued linear operators).

The solution of the above problem can be very difficultly calculated step by step, with the exception of some extremely peculiar situations (e. g., this can be done provided that $a_1(t) = a_2(t) = \cdots = a_m(t) = g_\alpha(t)$ for some $\alpha > 0$ and $0 \in \rho(\mathcal{A})$; details can be left to the interested readers). Because of that, we will follow here the abstract theoretical approach based on the use of vector-valued Laplace transform and (F, G, C)-regularized resolvent families; in many concrete situations, the theory of (a, k)-regularized C-resolvent families can be used, as well. We also consider the following abstract Volterra initial value problem:

$$u(t) \in f(t) + \sum_{i=1}^{m} (a_i * \mathcal{A}u(\cdot + t_i))(t), \quad t \in \mathbb{R}; \; u(t) = u_0(t), \; t \in [\tau_1, \tau_2],$$

where $m \in \mathbb{N}$, $t_1, t_2, \ldots, t_m \in \mathbb{R}$ and there exists an index $i \in \mathbb{N}_m \equiv \{1, \ldots, m\}$ with $t_i < 0$, $f : [0, \infty) \to X$ is Lebesgue measurable, $\mathcal{A}$ is a closed MLO on X and $a_1(t), \ldots, a_m(t)$ are locally integrable scalar-valued functions defined for $t \in \mathbb{R}$, $\tau_1 := \min(t_1, t_2, \ldots, t_m)$, $\tau_2 := \max(t_1, t_2, \ldots, t_m)$ if there exists an index $j \in \mathbb{N}_m$ with $t_j > 0$ and $\tau_2 := 0$, otherwise.

Before proceeding any further, we would like to emphasize that there exist some unpleasant situations in which the unique solution of the above mentioned Cauchy inclusions exists only if the multivalued linear operator $\mathcal{A}$ satisfies certain very exceptional spectral conditions; for example, the C-resolvent set of $\mathcal{A}$ must exist outside a compact set $K \subseteq \mathbb{C}$ sometimes. Concerning this issue, we would like to recall that, if $\mathcal{A} = A$ is single-valued with polynomially bounded resolvent existing outside a compact set $K \subseteq \mathbb{C}$ and X is a Banach space, then A must be bounded (see, e. g., the proof of [792, Theorem 6.5, pp. 135–138]). On the other hand, there exists a nondensely defined operator A on a Banach space X with ultra-polynomially bounded resolvent existing on $\mathbb{C}$, which will be extremely important for some applications (e. g., we can employ this operator in the analysis of problem (367) since the term $\|\lambda^{\alpha-1}e^{-\lambda}(\lambda^\alpha e^{-\lambda} - A)^{-1}\|$ is ultra-polynomially bounded for $\operatorname{Re}\lambda \geqslant 0$; this cannot be done in the analysis of problem (368) below).

Notation and terminology

In this section, X denotes a Hausdorff sequentially complete locally convex space over the field of complex numbers, SCLCS for short. If Y is also an SCLCS over the same field of scalars as X, then the shorthand $L(X, Y)$ denotes the vector space consisting of all continuous linear mappings from X into Y; $L(X) \equiv L(X,X)$. By $\circledast$ ($\circledast_Y$), we denote the fundamental system of seminorms, which defines the topology of X (Y). The symbol I denotes the identity operator on X. Let $0 < \tau \leqslant \infty$; then a strongly continuous operator family $(W(t))_{t\in[0,\tau)} \subseteq L(X, Y)$ is said to be locally equicontinuous if, for every $T \in (0, \tau)$ and $p \in \circledast_Y$, there exist $q_p \in \circledast$ and $c_p > 0$ such that $p(W(t)x) \leqslant c_p q_p(x)$, $x \in X$, $t \in [0, T]$; the notions of equicontinuity of $(W(t))_{t\in[0,\tau)}$ and the (exponential) equicontinuity of $(W(t))_{t\geqslant 0}$ are defined similarly.

By $\mathcal{B}$, we denote the family consisting of all bounded subsets of X. Define $p_{\mathbb{B}}(T) := \sup_{x\in\mathbb{B}} p(Tx)$, $p \in \circledast_Y$, $\mathbb{B} \in \mathcal{B}$, $T \in L(X, Y)$. Then $p_{\mathbb{B}}(\cdot)$ is a seminorm on $L(X, Y)$ and the system $(p_{\mathbb{B}})_{(p,\mathbb{B})\in\circledast_Y\times\mathcal{B}}$ induces the Hausdorff locally convex topology on $L(X, Y)$.

For more details about multivalued linear operators in locally convex spaces, (a, k)-regularized C-resolvent solution operator families subgenerated by multivalued linear operators and the Laplace transform of functions with values in SCLCSs, we refer the reader to [445] as well as to [69, 233, 287, 792]; we will use the same notion and notation as in the easily accessible monograph [445]. Concerning fractional calculus and fractional differential equations, we will only recommend here the research monographs [257, 431, 682] and the doctoral dissertation [97]. Basic source of information about abstract degenerate Volterra integrodifferential equations can be obtained by consulting monographs [287, 445, 734] and references cited therein.

7.1.1 (F, G, C)-Resolvent operator families: definitions and main results

We will always assume here that $\mathcal{A} : X \to P(X)$ is an MLO, $C_1 \in L(Y,X)$, $C_2 \in L(X)$ is injective, $C \in L(X)$ is injective and $C\mathcal{A} \subseteq \mathcal{A}C$.

The following notion will be essentially important in our further work.

Definition 7.1.1. Suppose that $\omega \in \mathbb{R}$, $F : \{\lambda \in \mathbb{C} \mid \operatorname{Re}\lambda > \omega\} \to \mathbb{C}$ and $G : \{\lambda \in \mathbb{C} \mid \operatorname{Re}\lambda > \omega\} \to \mathbb{C}$.

(i) A strongly continuous operator family $(R_1(t))_{t\geqslant 0} \subseteq L(Y,X)$ is said to be a mild (F, G, C_1)-regularized existence family with a subgenerator $\mathcal{A}$ if for each $y \in Y$ the mapping $t \mapsto R_1(t)y$, $t \geqslant 0$ is Laplace transformable, $abs(R_1(\cdot)y) \leqslant \omega$ and

$$F(\lambda)C_1 y \in (G(\lambda) - \mathcal{A}) \int_0^\infty e^{-\lambda t} R_1(t)y\,dt, \quad y \in Y,\ \operatorname{Re}\lambda > \omega.$$

(ii) A strongly continuous operator family $(R_2(t))_{t\geqslant 0} \subseteq L(X)$ is said to be a mild (F, G, C_2)-regularized uniqueness family with a subgenerator $\mathcal{A}$ if for each $x \in D(\mathcal{A}) \cup R(\mathcal{A})$ the mapping $t \mapsto R_2(t)x$, $t \geqslant 0$ is Laplace transformable, $abs(R_2(\cdot)x) \leqslant \omega$ and

$$F(\lambda)C_2y = G(\lambda)\int_0^\infty e^{-\lambda t}R_2(t)x\,dt - \int_0^\infty e^{-\lambda t}R_2(t)y\,dt, \tag{364}$$

whenever $(x,y) \in \mathcal{A}$ and $\operatorname{Re}\lambda > \omega$.

(iii) A strongly continuous operator family $(R(t))_{t\geqslant 0} \subseteq L(X)$ is said to be an (F, G, C)-regularized resolvent family with a subgenerator $\mathcal{A}$ if for each $x \in X$ the mapping $t \mapsto R(t)x$, $t \geqslant 0$ is Laplace transformable, $abs(R(\cdot)x) \leqslant \omega$, $G(\lambda) \in \rho_C(\mathcal{A})$ for $\operatorname{Re}\lambda > \omega$, and

$$F(\lambda)(G(\lambda) - \mathcal{A})^{-1}Cx = \int_0^\infty e^{-\lambda t}R(t)x\,dt, \quad x \in X,\ \operatorname{Re}\lambda > \omega.$$

If $C = \mathrm{I}$, then we omit the term "C" from the notation, with the meaning clear.

Keeping in mind the introduced notion, we can almost immediately clarify the following result.

Proposition 7.1.2. *Suppose $\mathcal{A}$ is a closed* MLO *in X, $C_1 \in L(Y,X)$, $C_2 \in L(X)$, C_2 is injective, $\omega_0 \geqslant 0$, $\omega \geqslant \max(\omega_0, abs(|a|), abs(k))$ and $\tilde{a}(\lambda) \neq 0$ for* $\operatorname{Re}\lambda > \omega$. *Then the following holds:*

(i) *If $(R_1(t))_{t\geqslant 0} \subseteq L(Y,X)$ is a mild (a,k)-regularized C_1-existence family with a subgenerator $\mathcal{A}$ and the family $\{e^{-\omega t}R_1(t) : t \geqslant 0\} \subseteq L(Y,X)$ is equicontinuous, then $(R_1(t))_{t\geqslant 0}$ is a mild (F, G, C_1)-regularized existence family with a subgenerator $\mathcal{A}$, where*

$$F(\lambda) := \tilde{k}(\lambda)/\tilde{a}(\lambda), \quad \operatorname{Re}\lambda > \omega \quad \textit{and} \quad G(\lambda) := 1/\tilde{a}(\lambda), \quad \operatorname{Re}\lambda > \omega. \tag{365}$$

(ii) *If $(R_2(t))_{t\geqslant 0} \subseteq L(X)$ is a mild (a,k)-regularized C_2-uniqueness family with a subgenerator $\mathcal{A}$ and the family $\{e^{-\omega t}R_2(t) : t \geqslant 0\} \subseteq L(X)$ is equicontinuous, then $(R_2(t))_{t\geqslant 0}$ is a mild (F, G, C_2)-regularized uniqueness family with a subgenerator $\mathcal{A}$, where $F(\cdot)$ and $G(\cdot)$ are given through* (365).

(iii) *If $(R(t))_{t\geqslant 0} \subseteq L(X)$ is an (a,k)-regularized C-resolvent family with a subgenerator $\mathcal{A}$ and the family $\{e^{-\omega t}R(t) : t \geqslant 0\} \subseteq L(X)$ is equicontinuous, then $(R(t))_{t\geqslant 0}$ is an (F, G, C)-regularized resolvent family with a subgenerator $\mathcal{A}$, where $F(\cdot)$ and $G(\cdot)$ are given through* (365).

The case in which $G(\lambda) = \tilde{a}(\lambda)$ for some Laplace transformable function $a(\cdot)$ (see also (365)) is not important for us since, in this case, the corresponding abstract Volterra problem is equivalent to the problem [445, (262)] with $\mathcal{B} = A$ and $A = \mathrm{I}$ therein.

If $(R_1(t))_{t\geqslant 0} \subseteq L(Y,X)$ is a mild (F, G, C_1)-regularized existence family with a subgenerator $\mathcal{A}$, then it is clear that any extension of $\mathcal{A}$ is also a subgenerator of $(R_1(t))_{t\geqslant 0}$. The integral generator $\mathcal{A}_{\mathrm{int}}$ of a mild (F, G, C_2)-regularized uniqueness family $(R_2(t))_{t\geqslant 0}$ is defined by

$$\mathcal{A}_{\mathrm{int}} := \{(x,y) \in X \times X : (364) \text{ holds for } \operatorname{Re}\lambda > \omega\}.$$

It is clear that the integral generator $\mathcal{A}_{\text{int}}$ of $(R_2(t))_{t\geqslant 0}$ is an extension of any subgenerator of $(R_2(t))_{t\geqslant 0}$ as well that $\mathcal{A}_{\text{int}}$ is likewise a subgenerator of $(R_2(t))_{t\geqslant 0}$. After dividing (364) with λ, it is not difficult to prove that $\mathcal{A}_{\text{int}}$ must be closed. Further on, if $(R(t))_{t\geqslant 0} \subseteq L(X)$ is an (F,G,C)-regularized resolvent family with a closed subgenerator $\mathcal{A}$, then we can prove that $R(t)\mathcal{A} \subseteq \mathcal{A}R(t)$ for all $t \geqslant 0$, so that $(R(t))_{t\geqslant 0} \subseteq L(X)$ is a mild (F,G,C)-regularized uniqueness family with subgenerator $\mathcal{A}$. In this case, we define the integral generator $\mathcal{A}_{\text{int}}$ as above. The interested reader may try to clarify the main properties of subgenerators of (F,G,C)-regularized resolvent families.

The following simple example shows that it is not necessarily true that $F \equiv F_1$ and $G \equiv G_1$ if $(R(t))_{t\geqslant 0}$ is both an (F,G,C)-regularized resolvent family with the integral generator $\mathcal{A}$ and an (F_1,G_1,C)-regularized resolvent family with the integral generator $\mathcal{A}$.

Example 7.1.3. Suppose that $\mathcal{A} = cI$, where $c \neq 1$, $C = I$ and $R(t)x := (1/(1-c))e^{c/(1-c)t}$, $t \geqslant 0$, $x \in X$. Then $(R(t))_{t\geqslant 0}$ is both $(1/(\lambda+1), \lambda/(\lambda+1), C)$-resolvent family with the integral generator $\mathcal{A}$ and $((2\lambda-c)/((1-c)\lambda-c), 2\lambda, C)$-resolvent family with the integral generator $\mathcal{A}$, as easily approved.

We continue by stating the following result.

Theorem 7.1.4. *There exist a real number $\omega \geqslant 0$, an analytic function $F : \{\lambda \in \mathbb{C} \mid \operatorname{Re}\lambda > \omega\} \to \mathbb{C}$, a function $G : \{\lambda \in \mathbb{C} \mid \operatorname{Re}\lambda > \omega\} \to \mathbb{C}$, an infinite-dimensional Banach space X and an exponentially bounded (F,G)-regularized resolvent family $(R(t))_{t\geqslant 0}$ with the integral generator $\mathcal{A} = A \in L(X)$, having the empty point spectrum, such that there do not exist a Laplace transformable function $a \in L^{\infty}_{\text{loc}}([0,\infty))$ and a continuous Laplace transformable function $k(\cdot)$ such that $(R(t))_{t\geqslant 0}$ is an exponentially bounded (a,k)-regularized resolvent family generated by A.*

Proof. It is well known that there exist an infinite-dimensional Banach space X and an operator $\mathcal{A} = A \in L(X)$ having the empty point spectrum. Then it is very simple to show the following:

(i1) If $\emptyset \neq \Omega \subseteq \mathbb{C}$ and $c_i : \Omega \to \mathbb{C}$ are given functions $(1 \leqslant i \leqslant 4)$ such that $c_1(\lambda)I - c_2(\lambda)A = c_3(\lambda)I - c_4(\lambda)A$ for all $\lambda \in \Omega$, then $c_1(\lambda) = c_3(\lambda)$ and $c_2(\lambda) = c_4(\lambda)$ for all $\lambda \in \Omega$.

(i2) If $\emptyset \neq \Omega \subseteq \rho(A)$ and $c_i : \Omega \to \mathbb{C}$ are given functions $(1 \leqslant i \leqslant 4)$, then the assumption $c_1(\lambda)(c_2(\lambda) - A)^{-1} = c_3(\lambda)(c_4(\lambda) - A)^{-1}$ for all $\lambda \in \Omega$ implies $c_1(\lambda) = c_3(\lambda)$ and $c_2(\lambda) = c_4(\lambda)$ for all $\lambda \in \Omega$.

Further on, due to the famous counterexample by W. Desch and J. Prüss [242, Proposition 4.1], we know that there exist an analytic function $H : \{\lambda \in \mathbb{C} : \operatorname{Re}\lambda > 0\} \to \mathbb{C}$ and a finite real constant $M > 0$ such that $|H(\lambda)| \leqslant M/(1+|\lambda|)$, $\operatorname{Re}\lambda > 0$ and there does not exist a Laplace transformable function $a \in L^{\infty}_{\text{loc}}([0,\infty))$ such that $\tilde{a}(\lambda) = H(\lambda)$ for $\operatorname{Re}\lambda > 0$. Put $F(\lambda) := \lambda^{-1}H(\lambda)$, $\operatorname{Re}\lambda > 0$, $G(\lambda) := 1/H(\lambda)$ if $\operatorname{Re}\lambda > 0$ and $H(\lambda) \neq 0$, and $G(\lambda) := \lambda_0 \in \rho(A)$, if $\operatorname{Re}\lambda > 0$ and $H(\lambda) = 0$. Suppose that there exist two functions $a(\cdot)$ and $k(\cdot)$ with the prescribed assumptions such that A generates the exponentially bounded (a,k)-regularized resolvent family $(R(t))_{t\geqslant 0}$. Then we have two possibilities:

(1) $H(\lambda) \neq 0$ for all $\operatorname{Re}\lambda > 0$. Then a very simple analysis involving the inequality $\|(\lambda - A)^{-1}\| \leq 1/(|\lambda| - \|A\|)$ for $|\lambda| > \|A\|$ implies that the function $\lambda \mapsto F(\lambda)(G(\lambda) - A)^{-1}$, $\operatorname{Re}\lambda > 0$ is analytic and bounded by $M'/|\lambda|^2$ on the right half-plane, where $M' > 0$ is finite. Using the complex characterization theorem for the Laplace transform, we get the existence of an exponentially bounded (F, G)-regularized resolvent family $(R(t))_{t\geqslant 0}$ with the integral generator A. Due to (i1)–(i2), we obtain that $H(\lambda) = \tilde{a}(\lambda)$, $\operatorname{Re}\lambda > 0$, which is a contradiction.
(2) There exists a discrete sequence $(\lambda_k)_{k\in\mathbb{N}}$ in the right half-plane such that $H(\lambda_k) = 0$ for all $k \in \mathbb{N}$. Then the mapping $\lambda \mapsto F(\lambda)(G(\lambda) - A)^{-1}$, $\operatorname{Re}\lambda > 0$, $\lambda \neq \lambda_k$ for all $k \in \mathbb{N}$; $\lambda \mapsto 0$, $\lambda = \lambda_k$ for some $k \in \mathbb{N}$, is analytic and bounded by $M''/|\lambda|^2$ on some right half-plane, where $M'' > 0$ is finite, as easily approved. Using the complex characterization theorem for the Laplace transform, we get the existence of an exponentially bounded (F, G)-regularized resolvent family $(R(t))_{t\geqslant 0}$ with the integral generator A. Then the final conclusion follows similarly as in part (i), with the help of the uniqueness theorem for the analytic functions and the issues (i1)–(i2). □

The proof of [242, Proposition 4.1] is not given in a constructive way and it is not clear whether the function $H(\cdot)$ has infinitely many zeroes in the right half-plane. Furthermore, if the Laplace transform of a function $b \in L^1_{\text{loc}}([0, \infty))$ is equal to $H(\lambda)$, then the complex characterization theorem for the Laplace transform implies that the function $t \mapsto (g_a * b)(t)$, $t \geqslant 0$ must be exponentially bounded for each $a > 0$; but this does not imply the local essential boundedness for $b(\cdot)$ as a class of very simple counterexamples shows (consider, e. g., the function $b(t) = |t - 1|^{-1/2}$, $t \neq 1$). In the following extension of Theorem 7.1.4, we will use the well-known identities from the theory of Laplace transform of distributions:

$$[\mathcal{L}(\delta(\cdot - a))] = e^{-a\lambda} \quad \text{and} \quad [\mathcal{L}(\delta'(\cdot - a))] = \lambda e^{-a\lambda}, \tag{366}$$

where $\delta(\cdot - a)$ denotes the Dirac delta distribution centered at the point $t = a > 0$.

Theorem 7.1.5. *There exist a real number $\omega \geqslant 0$, an analytic function $F : \{\lambda \in \mathbb{C} \mid \operatorname{Re}\lambda > \omega\} \to \mathbb{C}$, a function $G : \{\lambda \in \mathbb{C} \mid \operatorname{Re}\lambda > \omega\} \to \mathbb{C}$, an infinite-dimensional Banach space X and an exponentially bounded (F, G)-regularized resolvent family $(R(t))_{t\geqslant 0}$ with the integral generator $\mathcal{A} = A \in L(X)$, having the empty point spectrum, such that the conclusions of* Theorem 7.1.4 *hold with a Laplace transformable function $a \in L^1_{\text{loc}}([0, \infty))$ and a continuous Laplace transformable function $k(\cdot)$.*

Proof. Let $\omega = 0$ and let $X := c_0$, the Banach space of all numerical sequences vanishing at infinity, equipped with the sup-norm. Define

$$A\langle x_k\rangle := \langle 0, e^{-e^1}x_1, e^{-e^2}x_2, \ldots, e^{-e^k}x_k, \ldots\rangle, \quad \langle x_k\rangle \in c_0.$$

Then $A \in L(X)$, the spectrum of A is equal to $\{0\}$ and the point spectrum of A is empty, which can be simply shown. Let $\sigma < -1$, $F(\lambda) := \lambda^{1+\sigma}e^{-\lambda}$ and $G(\lambda) := \lambda e^{-\lambda}$ $(\operatorname{Re}\lambda > 0)$. Since there exists a finite real constant $M > 0$ such that $\|(\lambda e^{-\lambda} - A)^{-1}\| \leqslant M$ if $|\lambda e^{-\lambda}| \leqslant 2\|A\|$ and

$$\|(\lambda e^{-\lambda} - A)^{-1}\| \leqslant \frac{|\lambda e^{-\lambda}|}{|\lambda e^{-\lambda}| - \|A\|} \leqslant M$$

if $|\lambda e^{-\lambda}| > 2\|A\|$, the complex characterization theorem for the Laplace transform implies that A is the integral generator of an exponentially bounded (F, G)-regularized resolvent family $(R(t))_{t\geqslant 0}$. Keeping in mind (366) and the uniqueness theorem for the Laplace transform of distributions (see, e. g., [586] and [649]), the argumentation contained in the proof of Theorem 7.1.4 shows that there do not exist a Laplace transformable function $a \in L^1_{loc}([0,\infty))$ and a continuous Laplace transformable function $k(\cdot)$ such that $(R(t))_{t\geqslant 0}$ is an exponentially bounded (a, k)-regularized resolvent family generated by A. □

We continue by introducing the following notion.

Definition 7.1.6. (i) Suppose that $\mathcal{A}$ is an MLO in X. Let $\alpha \in (0, \pi]$, and let $(R(t))_{t\geqslant 0}$ be an (F, G, C)-regularized resolvent family with subgenerator $\mathcal{A}$. Then it is said that $(R(t))_{t\geqslant 0}$ is an analytic (F, G, C)-regularized resolvent family of angle α if there exists a function $\mathbf{R} : \Sigma_\alpha \to L(X)$, which satisfies that, for every $x \in X$, the mapping $z \mapsto \mathbf{R}(z)x$, $z \in \Sigma_\alpha$ is analytic as well as that:
(a) $\mathbf{R}(t) = R(t)$, $t > 0$ and
(b) $\lim_{z\to 0, z\in\Sigma_\gamma} \mathbf{R}(z)x = k(0)Cx$ for all $\gamma \in (0, \alpha)$ and $x \in X$.

(ii) Let $(R(t))_{t\geqslant 0}$ be an analytic (F, G, C)-regularized resolvent family of angle $\alpha \in (0, \pi]$. Then it is said that $(R(t))_{t\geqslant 0}$ is an exponentially bounded, analytic (F, G, C)-regularized resolvent family of angle α, respectively, bounded analytic (F, G, C)-regularized resolvent family of angle α if for every $\gamma \in (0, \alpha)$, there exists $\omega_\gamma \geqslant 0$, respectively, $\omega_\gamma = 0$, such that the family $\{e^{-\omega_\gamma \operatorname{Re} z}\mathbf{R}(z) : z \in \Sigma_\gamma\} \subseteq L(X)$ is equicontinuous. Since there is no risk for confusion, we will identify in the sequel $R(\cdot)$ and $\mathbf{R}(\cdot)$.

The statements of [445, Theorems 3.2.19, 3.2.25, 3.2.26] can be straightforwardly formulated in our new framework. Further on, the subordination principle for abstract time-fractional inclusions clarified in [445, Theorem 3.1.8] can be extended to (F, G, C)-regularized families; for example, if $\mathcal{A}$ is a closed subgenerator of an (F, G, C)-regularized resolvent family $(R(t))_{t\geqslant 0}$ satisfying certain extra properties, then for each number $\gamma \in (0, 1)$ we can define the operator family $(R_\gamma(t))_{t\geqslant 0}$ by $R_\gamma(0) := R(0)$ and

$$R_\gamma(t) := \int_0^\infty t^{-\gamma}\Phi_\gamma(st^{-\gamma})R(s)x\, ds, \quad x \in X,\ t > 0,$$

where $\Phi_\gamma(\cdot)$ denotes the Wright function. Then we have $[\widetilde{R_\gamma(t)}](\lambda) = \lambda^{\gamma-1}[\widetilde{R(t)}](\lambda^\gamma)$ for $\operatorname{Re}\lambda > \omega$ suff. large, so that $(R_\gamma(t))_{t\geqslant 0}$ is a $(\cdot^{\gamma-1}F(\cdot^\gamma), G(\cdot^\gamma), C)$-regularized resolvent family with subgenerator $\mathcal{A}$. Moreover, it can be proved that $(R_\gamma(t))_{t\geqslant 0}$ is an exponentially equicontinuous, analytic $(\cdot^{\gamma-1}F(\cdot^{\gamma}), G(\cdot^{\gamma}), C)$-regularized resolvent family of certain angle (for more details, see the issues (i)–(iii) in the formulation of the last mentioned result).

In [445, Proposition 3.1.8(i)], we have proved that the equality $C_2R_1(t) = R_2(t)C_1$, $t \in [0, \tau)$ holds for the local (a, k)-regularized (C_1, C_2)-existence and uniqueness families; in our new framework, we must assume certain extra conditions ensuring the validity of this equality. An attempt should be made to extend the real characterization theorem [445, Theorem 3.2.12] for (F, G, C)-regularized resolvent families, as well; on the other hand, the statement of [445, Proposition 3.2.3] and the complex characterization theorem [445, Theorem 3.2.10] can be straightforwardly extended to (F, G, C)-regularized resolvent families.

7.1.2 Applications to the abstract Volterra integrodifference inclusions

In this subsection, we provide some applications of (F, G, C)-regularized resolvent families and (a, k)-regularized C-resolvent families to the abstract Volterra integrodifference inclusions. We mainly use the vector-valued Laplace transform here.

First of all, we will provide the following illustrative example with $C = \mathrm{I}$.

Example 7.1.7. (i) Suppose that $0 < \alpha < 1$ and $\mathcal{A}$ is a closed MLO. Consider the following abstract fractional Cauchy inclusion:

$$\mathbf{D}_t^{\alpha} u(t) \in \mathcal{A}u(t+1) + f(t), \quad t \geqslant 0; \quad u(t) = u_0(t), \quad 0 \leqslant t < 1. \tag{367}$$

In some very simple situations (e. g., if $\mathcal{A} = c\mathrm{I}$ for some $c \in \mathbb{C} \setminus \{0\}$), the unique solution of (367) can be directly calculated step by step; on the other hand, the theory of vector-valued Laplace transform is completely inapplicable here. To explain this in more detail, let us try to apply the Laplace transform identity [97, (1.23)] and Lemma 1.1.10(i); then we get

$$\lambda^{\alpha}\tilde{u}(\lambda) - \lambda^{\alpha-1}u(0) \in \mathcal{A}\left[e^{\lambda}\left\{\tilde{u}(\lambda) - \int_0^1 e^{-\lambda t}u_0(t)\,dt\right\}\right] + \tilde{f}(\lambda), \quad \operatorname{Re}\lambda > 0 \text{ suff. large,}$$

provided that all terms are well-defined and exponentially bounded, with the meaning clear. This simply implies

$$u(t) = \mathcal{L}^{-1}\left[(\lambda^{\alpha} - e^{\lambda}\mathcal{A})^{-1}\left\{\lambda^{\alpha-1}u(0) + \tilde{f}(\lambda) - e^{\lambda}\mathcal{A}\int_0^1 e^{-\lambda t}u_0(t)\,dt\right\}\right](t), \quad t \geqslant 0,$$

provided that all terms are well-defined. Consider now the case $u_0(t) \equiv 0$, with $\mathcal{A} = c\mathrm{I}$ for some $c \in \mathbb{C} \setminus \{0\}$; then we obtain

$$u(t) = (\mathcal{L}^{-1}[\lambda^{\alpha-1}(\lambda^{\alpha} - ce^{\lambda})^{-1}] * f)(t), \quad t \geqslant 0.$$

But there does not exist an exponentially bounded (F, G)-regularized resolvent family $(R(t))_{t\geqslant 0}$ with the integral generator cI such that

$$\lambda^{a-1}(\lambda^a - ce^\lambda)^{-1}x = \int_0^\infty e^{-\lambda t}R(t)x\,dt, \quad \operatorname{Re}\lambda > 0 \text{ suff. large}, x \in X.$$

In actual fact, the mapping $\lambda \mapsto \lambda^a - ce^\lambda$ has infinitely many zeros in any right half-plane and the function $\lambda \mapsto \lambda^{a-1}(\lambda^a - ce^\lambda)^{-1}$ cannot be analytic there (the situation in which there exists a complex number $c \in \mathbb{C} \setminus \{0\}$ such that $c \notin \rho(\mathcal{A})$ cannot be considered, as well; on the other hand, the situation in which $\sigma(A) = \{0\}$ can be simply considered following the lines of the proof of Theorem 7.1.5).
As mentioned in the introductory part of this section, the above example indicates that it is very difficult to apply the vector-valued Laplace transform in the analysis of the abstract fractional Cauchy inclusions $(\mathrm{DFP})_{R,a,b}$ and $(\mathrm{DFP})_{L,a,b}$, where $0 \leqslant a < b$, $f : [0, \infty) \to X$ is exponentially bounded, B is a closed linear operator, $\mathcal{A}$ and $\mathcal{B}$ are closed MLOs [we can use the substitution $t \mapsto t + a$ and a similar calculation].

(ii) Suppose now that $0 < a < 1$ and consider the following abstract fractional Cauchy inclusion:

$$\mathbf{D}_t^a u(t+1) \in \mathcal{A}u(t) + f(t), \quad t \geqslant 0; \quad u(0) = x, \quad \mathbf{D}_t^a u(t) = u_0(t), \quad 0 \leqslant t < 1. \quad (368)$$

The unique solution of (368) can be directly calculated step by step in some situations but much better results can be established by applying the vector-valued Laplace transform and the theory of (a, k)-regularized C-resolvent families. In fact, the unique solution $u(t)$ of (368) is given by

$$u(t) = \mathcal{L}^{-1}\left[(\lambda^a e^\lambda - \mathcal{A})^{-1}\left\{\tilde{f}(\lambda) + e^\lambda \int_0^1 e^{-\lambda t}u_0(t)\,dt + \lambda^{a-1}e^\lambda u(0)\right\}\right](t), \quad t \geqslant 0.$$

If $u_0(t) \equiv 0$ and $x = 0$, then it can be simply verified that we have $u(t) = (R * f)(t)$, $t \geqslant 0$, where $(R(t))_{t\geqslant 0}$ is an exponentially equicontinuous (a, a)-regularized resolvent family generated by $\mathcal{A}$ with $a(t)$ being equal to zero for $0 \leqslant t \leqslant 1$ and $g_{2-a}(t-1)$ for $t > 1$. But some serious problems occur again since the complement of the set $\{\lambda^a e^\lambda : \lambda \in \mathbb{C}, \operatorname{Re}\lambda > \omega\}$ is compact for any $\omega \geqslant 0$, which implies that the resolvent set of $\mathcal{A}$ must contain the set $\mathbb{C} \setminus K$, for some compact set $K \subseteq \mathbb{C}$. If $\mathcal{A} = A$ is single-valued with polynomially bounded resolvent and X is a Banach space, then the above implies that A must be bounded, when some applications can be given.

Usually, in order to study the abstract (multiterm) fractional differential inclusions, we convert them into the corresponding abstract (multiterm) Volterra integrodifferential inclusions. The abstract multiterm Volterra integrofunctional inclusions will be our main subject in the remainder of this section, which will be broken down into two individual parts:

1. In the first part, we consider the following abstract Volterra initial value problem:

$$u(t) \in f(t) + \sum_{i=1}^{m} (a_i * \mathcal{A}u(\cdot + t_i))(t), \quad t \geqslant 0; \quad u(t) = u_0(t), \quad 0 \leqslant t \leqslant \tau, \tag{369}$$

where $m \in \mathbb{N}$, $0 \leqslant t_1, t_2, \ldots, t_m < +\infty$, $\tau := \max(t_1, t_2, \ldots, t_m)$, $f : [0, \infty) \to X$ is Lebesgue measurable, $u_0 \in L^1([0, \tau] : X)$, $\mathcal{A}$ is a closed MLO on X and for each $i \in \mathbb{N}_m$ we have that $a_i(\cdot)$ is a sum of a locally integrable scalar-valued function defined for $t \geqslant 0$ and a scalar-valued distribution $a_i(\cdot) = c_i\delta(\cdot - b_i)$, where $c_i \in \mathbb{C}$ and $0 \leqslant b_i < +\infty$ [if $a_i(\cdot) = c_i\delta(\cdot - b_i)$, then according to Lemma 1.1.10 and (366) we define $(a_i * u)(t) := c_i u(t - b_i)$, $t > b_i$ and $(a_i * u)(t) := 0$, $0 \leqslant t \leqslant b_i$].

We will use the following concepts of solutions.

Definition 7.1.8. (i) By a mild solution of (369), we mean any continuous function $u : [0, \infty) \to X$ such that $(a_i * u(\cdot + t_i))(t) \in D(\mathcal{A})$ for all $t \geqslant 0$, $i \in \mathbb{N}_m$ and

$$u(t) \in f(t) + \sum_{i=1}^{m} \mathcal{A}(a_i * u(\cdot + t_i))(t), \quad t \geqslant 0; \quad u(t) = u_0(t), \quad 0 \leqslant t \leqslant \tau.$$

(ii) By an LT-mild solution of (369), we mean any Laplace transformable mild solution of (369).

(iii) By a strong solution of (369), we mean any continuous function $u : [0, \infty) \to X$ of (369), which satisfies that, for every $i \in \mathbb{N}_m$, there exists a continuous function $u_i : [0, \infty) \to X$ such that $u_i(t) \in \mathcal{A}(u(t + t_i))$ for all $t \geqslant 0$ and

$$u(t) = f(t) + \sum_{i=1}^{m} (a_i * u_i(\cdot))(t), \quad t \geqslant 0; \quad u(t) = u_0(t), \quad 0 \leqslant t \leqslant \tau. \tag{370}$$

(iv) By a strong LT-solution of (369), we mean any continuous Laplace transformable function $u : [0, \infty) \to X$ of (369), which satisfies that, for every $i \in \mathbb{N}_m$, there exists a continuous Laplace transformable function $u_i : [0, \infty) \to X$ such that $u_i(t) \in \mathcal{A}(u(t + t_i))$ for all $t \geqslant 0$ and (370) holds.

Since $\mathcal{A}$ is closed, any strong (LT-)solution of (369) is also a mild (LT-)solution of (369); on the other hand, it is clear that any LT-mild solution of (369) is a mild solution of (369) and any strong LT-solution of (369) is a strong solution of (369). Unfortunately, it is very difficult to say anything relevant about the existence and uniqueness of mild (strong) solutions of the problem (369).

In the sequel, we assume that the following condition holds:

(Q): There exists $\omega \in \mathbb{R}$ such that $\omega \geqslant \max(abs(|a_1|), \ldots, abs(|a_m|))$ and

$$\widetilde{a_1}(\lambda)e^{\lambda t_1} + \cdots + \widetilde{a_m}(\lambda)e^{\lambda t_m} \neq 0, \quad \operatorname{Re}\lambda > \omega,$$

where we put $abs(|a_i|) := -\infty$ if $a_i(\cdot) = c_i\delta(\cdot - b_i)$ with some $c_i \in \mathbb{C}$ and $0 \leqslant b_i < +\infty$ as well as $abs(|a_i|) := abs(|b_i|)$ if $a_i(\cdot) = b_i(\cdot) + c_i\delta(\cdot - b_i)$ with some Laplace transformable function $b_i(\cdot)$, $c_i \in \mathbb{C}$ and $0 \leqslant b_i < +\infty$ $(1 \leqslant i \leqslant m)$.

Define

$$G(\lambda) := \frac{1}{\widetilde{a_1}(\lambda)e^{\lambda t_1} + \cdots + \widetilde{a_m}(\lambda)e^{\lambda t_m}}, \quad \operatorname{Re}\lambda > \omega.$$

The main result concerning the well-posedness of problem (369) reads as follows.

Theorem 7.1.9. *Suppose that $\mathcal{A}$ is a closed subgenerator of an exponentially equicontinuous $(\tilde{k}(\cdot)G, G, C)$-regularized resolvent family $(R(t))_{t\geqslant 0}$, where $k(\cdot)$ is Laplace transformable or $k(\cdot) = \delta(\cdot)$, the Dirac delta distribution.*
(i) *If $f(\cdot)$ is Laplace transformable and $k \neq 0$, then the solutions of* (369) *are unique.*
(ii) *If there exist a function $v_0 \in L^1([0,\tau] : X)$ such that $v_0(t) \in \mathcal{A}u_0(t)$ for a. e. $t \in [0,\tau]$ and a Laplace transformable function $u_1 : [0,\infty) \to X$ such that*

$$\int_0^\infty e^{-\lambda t}u_1(t)\,dt = \sum_{i=1}^m \widetilde{a_i}(\lambda)e^{\lambda t_i}\int_0^{t_i} e^{-\lambda t}v_0(t)\,dt, \quad \operatorname{Re}\lambda > 0 \textit{ suff. large},$$

then $u(\cdot)$ is a strong LT *solution of* (369) *if and only if*

$$(k * Cu)(t) = (R * [f - u_1])(t), \quad t \geqslant 0. \tag{371}$$

(iii) *Suppose that there exists a Laplace transformable function $u_2 : [0,\infty) \to X$ such that*

$$\int_0^\infty e^{-\lambda t}u_2(t)\,dt = G(\lambda)\sum_{i=1}^m \widetilde{a_i}(\lambda)e^{\lambda t_i}\int_0^{t_i} e^{-\lambda t}u_0(t)\,dt, \quad \operatorname{Re}\lambda > 0 \textit{ suff. large}.$$

Then $u(\cdot)$ is a strong LT *solution of* (369) *if and only if*

$$(k * Cu)(t) = (R * [f - u_2])(t) + (kC * u_2)(t), \quad t \geqslant 0.$$

Proof. We have

$$\tilde{k}(\lambda)\left[I - \sum_{i=1}^m \widetilde{a_i}(\lambda)e^{\lambda t_i}\mathcal{A}\right]^{-1} Cx = \int_0^\infty e^{-\lambda t}R(t)x\,dt, \quad x \in X,\ \operatorname{Re}\lambda > \omega_0 > \omega \text{ suff. large};$$

cf. also (366). Suppose first that $u(\cdot)$ is a strong LT solution of (369). Then $\tilde{u}_i(\lambda) \in e^{\lambda t_i}\mathcal{A}[\tilde{u}(\lambda) - \int_0^{t_i} e^{-\lambda t}u_0(t)\,dt]$, $\operatorname{Re}\lambda > \omega_0$ $(1 \leqslant i \leqslant m)$. Applying the Laplace transform to (370), we get

$$\tilde{u}(\lambda) \in \tilde{f}(\lambda) + \sum_{i=1}^{m} \widetilde{a_i}(\lambda)e^{\lambda t_i}\mathcal{A}\left[\tilde{u}(\lambda) - \int_0^{t_i} e^{-\lambda t}u_0(t)\,dt\right], \quad \operatorname{Re}\lambda > \omega_0.$$

The prescribed assumptions imply

$$\left[I - \sum_{i=1}^{m} \widetilde{a_i}(\lambda)e^{\lambda t_i}\mathcal{A}\right]C\tilde{u}(\lambda) \in C\tilde{f}(\lambda) - C\mathcal{A}\sum_{i=1}^{m} \widetilde{a_i}(\lambda)e^{\lambda t_i}\int_0^{t_i} e^{-\lambda t}u_0(t)\,dt, \quad \operatorname{Re}\lambda > \omega_0 \tag{372}$$

and

$$\begin{aligned} C\tilde{u}(\lambda) \in &\left[I - \sum_{i=1}^{m} \widetilde{a_i}(\lambda)e^{\lambda t_i}\mathcal{A}\right]^{-1} C\tilde{f}(\lambda) \\ &- \left[I - \sum_{i=1}^{m} \widetilde{a_i}(\lambda)e^{\lambda t_i}\mathcal{A}\right]^{-1} C\mathcal{A}\sum_{i=1}^{m} \widetilde{a_i}(\lambda)e^{\lambda t_i}\int_0^{t_i} e^{-\lambda t}u_0(t)\,dt, \quad \operatorname{Re}\lambda > \omega_0. \end{aligned}$$

Since the operator $[I - \sum_{i=1}^{m} \widetilde{a_i}(\lambda)e^{\lambda t_i}\mathcal{A}]^{-1}C\mathcal{A}$ is single-valued on account of [445, Theorem 1.2.4(i)], we get that

$$\begin{aligned} \tilde{k}(\lambda)C\tilde{u}(\lambda) = &\tilde{k}(\lambda)G(\lambda)(G(\lambda) - \mathcal{A})^{-1}C\tilde{f}(\lambda) \\ &- \tilde{k}(\lambda)G(\lambda)[G(\lambda)(G(\lambda) - \mathcal{A})^{-1}C - C]\sum_{i=1}^{m} \widetilde{a_i}(\lambda)e^{\lambda t_i}\int_0^{t_i} e^{-\lambda t}u_0(t)\,dt, \quad \operatorname{Re}\lambda > \omega_0. \end{aligned} \tag{373}$$

If $u_0 \equiv 0$, then the Titchmarsh convolution theorem, the injectivity of C and the above arguments together imply (371) with $u_1 \equiv 0$, completing the proof of (i). On the other hand, a simple argumentation involving [445, Theorem 1.2.3] and (372) shows that

$$\begin{aligned} (\widetilde{k * Cu)(\cdot})(\lambda) = &\tilde{k}(\lambda)\left[I - \sum_{i=1}^{m} \widetilde{a_i}(\lambda)e^{\lambda t_i}\mathcal{A}\right]^{-1} C\tilde{f}(\lambda) \\ &- \tilde{k}(\lambda)\left[I - \sum_{i=1}^{m} e^{\lambda t_i}\mathcal{A}\right]^{-1} C\sum_{i=1}^{m} \widetilde{a_i}(\lambda)e^{\lambda t_i}\int_0^{t_i} e^{-\lambda t}v_0(t)\,dt, \quad \operatorname{Re}\lambda > \omega_0. \end{aligned}$$

Performing the inverse Laplace transform, we get (371). The proof of sufficiency in part (ii) can be deduced by following the inverse procedure. The proof of (iii) follows from a similar argumentation involving the equation (373) and, therefore, can be omitted. □

We continue by providing some illustrative examples.

Example 7.1.10. (i) Suppose that $m = 1$, $t_1 = 1$ and $a_1(t) \equiv 1$. If $\mathcal{A} = A$ is the operator from the formulation of Theorem 7.1.5, then the theory of (a, k)-regularized

C-resolvent families cannot be applied in the analysis of the well-posedness of problem (369). On the other hand, we can apply Theorem 7.1.9, with $F(\lambda) = \lambda^{1+\sigma}e^{-\lambda}$ for arbitrary $\sigma < -1$ and $G(\lambda) = \lambda e^{-\lambda}$, in the analysis of the well-posedness of problem (369).

(ii) Suppose that $a(\cdot)$ and $k(\cdot)$ are Laplace transformable, $k(\cdot)$ is continuous, $\tilde{a}(\cdot)$ does not vanish on some right half-plane and $\mathcal{A}$ is a subgenerator of an exponentially equicontinuous (a,k)-regularized C-resolvent operator family $(R(t))_{t\geqslant 0}$. If $m = 1$ and $t_1 > 0$, then we define $a_1(t) := a(t - t_1)$, $t \geqslant t_1$ and $a_1(t) := 0$, $0 \leqslant t < t_1$. Then $\widetilde{a_1}(\lambda) = e^{-\lambda t_1}\tilde{a}(\lambda)$ for $\operatorname{Re}\lambda > 0$ suff. large and Theorem 7.1.9 can be applied with $G(\lambda) = [\tilde{a}(\lambda)]^{-1}$, $u_1(\cdot) = (a * v_0\chi_{[0,t_1]})(\cdot)$ and $u_2(\cdot) = (u_0\chi_{[0,t_1]})(\cdot)$, which is the most symptomatic and easiest way to use this result. If this is the case and $u(\cdot)$ is a strong LT-solution of (369), then we have

$$\int_0^t a_1(s)u_1(t + t_1 - s)\,ds = \int_{t_1}^t a_1(s)u_1(t + t_1 - s)\,ds = \int_{t_1}^t a(s - t_1)u_1(t - s + t_1)\,ds$$
$$= \chi_{[t_1,+\infty)}(t)\int_0^{t-t_1} a(s)u_1(t - s)\,ds, \quad t \geqslant 0,$$

so that $u_0(t) = f(t)$, $0 \leqslant t \leqslant t_1$ and $u(\cdot)$ solves, in a certain sense, the abstract Cauchy inclusion

$$v(t) \in f(t) + \mathcal{A}\int_0^{t-t_1} a(s)v(t - s)\,ds, \quad t > t_1; \quad v(t) = f(t), \quad 0 \leqslant t \leqslant t_1. \tag{374}$$

Moreover, if the requirement from Theorem 7.1.9(i) holds and $u_{cl}(\cdot)$ is a strong solution of the abstract Cauchy inclusion

$$u(t) \in f(t) + \mathcal{A}\int_0^t a(t - s)u(s)\,ds, \quad t \geqslant 0,$$

then the use of (371) and [445, Proposition 2.3.8(ii)] imply

$$(k * C[u_{cl} - u])(t) = (R * a * v_0\chi_{[0,t_1]})(t), \quad t \geqslant 0. \tag{375}$$

As an application, we can consider the fractional-functional Poisson heat equations of the form (374), for example; cf. [287, 445] for more details about the subject.

(iii) The situation in which there exist positive real numbers $\alpha_i > 0$ such that $\widetilde{a_i}(\lambda)e^{\lambda t_i} = \lambda^{-\alpha_i}$ for $1 \leqslant i \leqslant m$ is also important. In this situation, $\mathcal{A}$ needs to be a subgenerator of an exponentially equicontinuous (a,k)-regularized C-resolvent operator family $(R(t))_{t\geqslant 0}$, where $a(t) = g_{\alpha_1}(t) + \cdots + g_{\alpha_m}(t)$; also, in place of kernels $g_{\alpha_1}(t), \ldots, g_{\alpha_m}(t)$, we can consider the general Laplace transformable kernels $a_1(t), \ldots, a_m(t)$.

In many concrete situations, which are not similar to the situations explored in Example 7.1.10(ii)–(iii), the complement of the set $\{G(\lambda) : \operatorname{Re}\lambda > \omega\}$ is compact in $\mathbb{C}$; without going into further details concerning this question and many other important questions concerning the range and zeroes of complex exponential polynomials (see, e. g., [73, 604] and references cited therein for more details about these issues), it seems very plausible that the above happens if

$$G(\lambda) = \frac{1}{\lambda^{-a_1}e^{\lambda t_1'} + \cdots + \lambda^{-a_m}e^{\lambda t_m'}}, \quad \operatorname{Re}\lambda > \omega,$$

for some positive real numbers $a_i > 0$ and nonzero real numbers $t_i' \neq 0$ $(1 \leqslant i \leqslant m)$. Therefore, the operators from Example 3.3.6 are sometimes essentially important in the analysis of the existence and uniqueness of convoluted solutions of the abstract Cauchy problem (369).

Example 7.1.11. Suppose that A is the operator from Example 3.3.6, $\omega_0 \geqslant 0$, $a_1(\cdot) = \cdots = a_m(\cdot) = a(\cdot)$ satisfy the general assumptions of Theorem 7.1.9, $0 < t_1 < t_2 < \cdots < t_m$, the numbers $t_2, \ldots, t_m$ are integer multiples of t_1 and there exists a complex polynomial $S(\cdot)$ such that $S(z) \neq 0$ for $\operatorname{Re} z \geqslant \omega_0$ and

$$e^{\lambda t_1} + \cdots + e^{\lambda t_m} = S(e^{\lambda t_1}), \quad \operatorname{Re}\lambda > \omega_0.$$

Then there exists $\omega \geqslant \omega_0$ such that condition (Q) holds. Suppose further that there exist $\omega_1 \geqslant \omega_0$, $M > 0$ and $\beta \in (0, N)$ such that $|\tilde{a}(\lambda)|^{-\sigma} \leqslant M|\lambda|^{\beta}$ for $\operatorname{Re}\lambda \geqslant \omega_1$. Since

$$\left|e^{\lambda t_1} + \cdots + e^{\lambda t_m}\right| \geqslant e^{\operatorname{Re}(\lambda)t_m} - e^{\operatorname{Re}(\lambda)t_{m-1}} - \cdots - e^{\operatorname{Re}(\lambda)t_1}, \quad \lambda \in \mathbb{C},$$

there exists a sufficiently large real number $\omega \geqslant \omega_1$ such that the assumptions of Theorem 7.1.9 are satisfied with a kernel $k(\cdot)$ whose Laplace transform decays ultra-polynomially on some right half-space. For example, Theorem 7.1.9(ii) can be applied provided that $a(\cdot) = c\delta(\cdot - t_{m+1})$, where $c \in \mathbb{C}$ and $+\infty > t_{m+1} > t_m$ while Theorem 7.1.9(iii) can be applied, with $u_2(\cdot) \equiv u_0(\cdot)$, provided that $u_0(t) = 0$, $t > t_1$; concerning the corresponding abstract Cauchy problems, the main applications can be given in the case that the function $(f - u_1)(\cdot)$ $[(f - u_2)(\cdot)]$ belongs to the range of the convolution transform $k * \cdot$, which can really come off.

We will only note that Example 3.3.6 can be reconsidered in the degenerate setting as well as that certain applications can be given to the fractional-functional analogues of the linearized Benney–Luke type equation

$$(\lambda u - \Delta u)_t = \alpha\Delta u - \beta\Delta^2 u \quad (\alpha > 0,\ \beta > 0;\ \lambda \in \mathbb{R});$$

see [287, Theorem 1.14, p. 28] and [445, Example 2.2.18] for further information in this direction.

2. In the second part, we consider the following abstract Volterra initial value problem:

$$u(t) \in f(t) + \sum_{i=1}^{m} (a_i * \mathcal{A}u(\cdot + t_i))(t), \quad t \in \mathbb{R}; \quad u(t) = u_0(t), \quad t \in [\tau_1, \tau_2], \tag{376}$$

where $m \in \mathbb{N}$, $t_1, t_2, \ldots, t_m \in \mathbb{R}$ and there exists an index $i \in \mathbb{N}_m$ with $t_i < 0$, $f : [0, \infty) \to X$ is Lebesgue measurable, $\mathcal{A}$ is a closed MLO on X and $a_1(\cdot), \ldots, a_m(\cdot)$ are locally integrable scalar-valued functions defined for $t \in \mathbb{R}$ such that the functions $a_{i|[0,\infty)}(\cdot)$ and $\check{a}_{i|[0,\infty)}(\cdot)$ are Laplace transformable for $1 \leqslant i \leqslant m$ [for the sake of simplicity, we will not consider the case in which some $a_i(\cdot)$ contains a nonvanishing distribution part], $\tau_1 := \min(t_1, t_2, \ldots, t_m)$, $\tau_2 := \max(t_1, t_2, \ldots, t_m)$ if there exists an index $j \in \mathbb{N}_m$ with $t_j > 0$ and $\tau_2 := 0$, otherwise.

We define a mild (strong) solution of (376) in the same way as in Definition 7.1.8. By a mild LT-solution of (376), we mean any mild solution of (376) such that the functions $u_{[0,\infty)}(\cdot)$ and $u_{[0,\infty)}\check{}(\cdot)$ are Laplace transformable; by a strong LT solution of (376), we mean any continuous function $u : \mathbb{R} \to X$ such that the functions $u_{[0,\infty)}(\cdot)$ and $u_{[0,\infty)}\check{}(\cdot)$ are Laplace transformable, which satisfies that, for every $i \in \mathbb{N}_m$, there exists a continuous function $u_i : \mathbb{R} \to X$ such that the functions $u_{i;[0,\infty)}(\cdot)$ and $u_{i;[0,\infty)}\check{}(\cdot)$ are Laplace transformable, $u_i(t) \in \mathcal{A}(u(t + t_i))$ for all $t \in \mathbb{R}$ and

$$u(t) = f(t) + \sum_{i=1}^{m} (a_i * u_i(\cdot))(t), \quad t \in \mathbb{R}; \quad u(t) = u_0(t), \quad \tau_1 \leqslant t \leqslant \tau_2.$$

Keeping in mind the introduced notion, we have

$$\begin{aligned}
\check{u}(t) = u(-t) &\in f(-t) + \sum_{i=1}^{m} \int_0^{-t} a_i(s)\mathcal{A}u(-t - s + t_i)\, ds \\
&\in f(-t) + \sum_{i=1}^{m} \int_0^{-t} a_i(s)\mathcal{A}\check{u}(t + s - t_i)\, ds \in f(-t) - \sum_{i=1}^{m} \int_0^{t} a_i(-s)\mathcal{A}\check{u}(t - s - t_i)\, ds \\
&\in f(-t) + \sum_{i=1}^{m} \int_t^{0} a_i(s - t)\mathcal{A}\check{u}(s - t_i)\, ds \in f(-t) - \sum_{i=1}^{m} \int_0^{t} \check{a}_i(t - s)\mathcal{A}\check{u}(s - t_i)\, ds, \quad t \in \mathbb{R},
\end{aligned}$$

and

$$\begin{aligned}
\check{u}(t) = u(-t) &= f(-t) + \sum_{i=1}^{m} \int_0^{-t} a_i(s)u_i(-t - s)\, ds \\
&= f(-t) + \sum_{i=1}^{m} \int_0^{-t} a_i(s)\check{u}_i(t + s)\, ds = f(-t) - \sum_{i=1}^{m} \int_0^{t} a_i(-s)\check{u}_i(t - s)\, ds
\end{aligned}$$

$$= f(-t) - \sum_{i=1}^{m} \int_0^t \check{a}_i(-s)\check{u}_i(t-s)\,ds = f(-t) - \sum_{i=1}^{m} \int_0^t \check{a}_i(t-s)\check{u}_i(t-s)\,ds, \quad t \in \mathbb{R}.$$

Since $\check{u}_i(t) \in \mathcal{A}\check{u}(t-t_i)$, $t \in \mathbb{R}$, $1 \leqslant i \leqslant m$, this implies the following.

Proposition 7.1.12. *Suppose that $u : \mathbb{R} \to X$ is a continuous function. Then $u(\cdot)$ is a mild (strong) solution of* (376) *[mild* LT *(strong* LT*) solution of* (376)*] if and only if the function $v(\cdot) = \check{u}(\cdot)$ is a mild (strong) solution of* (377) *[mild* LT *(strong* LT*) solution of* (377)*], where*

$$v(t) \in \check{f}(t) + \sum_{i=1}^{m} (\check{a}_i * (-\mathcal{A})u(\cdot - t_i))(t), \quad t \in \mathbb{R}; \quad v(t) = v_0(t) = \check{u}_0(t), \quad t \in [-\tau_2, -\tau_1]. \tag{377}$$

If the functions $a(\cdot)$ and $\check{a}(\cdot)$ are Laplace transformable, then we have

$$\int_0^\infty e^{-\lambda t}\check{a}(t)\,dt = \int_{-\infty}^0 e^{\lambda t}a(t)\,dt, \quad \operatorname{Re}\lambda > 0 \text{ suff. large.}$$

Keeping this in mind, Lemma 1.1.10(i), Proposition 7.1.12, Theorem 7.1.9 and its proof together imply, after considering the functions $u_{|[0,\infty)}$ and $u_{|(-\infty,0]}$ separately, the following result.

Theorem 7.1.13. *Suppose that $\mathcal{A}$ is a closed subgenerator of an exponentially equicontinuous $(\tilde{k}(\cdot)G, G, C)$-regularized resolvent family $(R(t))_{t\geqslant 0}$, where $k(\cdot)$ is Laplace transformable or $k(\cdot) = \delta(\cdot)$, the Dirac delta distribution. Suppose further that $-\mathcal{A}$ is a subgenerator of an exponentially equicontinuous $(\tilde{k_1}(\cdot)G_1, G_1, C')$-regularized resolvent family $(R_1(t))_{t\geqslant 0}$, where $C' \in L(X)$ is injective and commutes with $\mathcal{A}$ as well as $k_1(\cdot)$ is Laplace transformable or $k_1(\cdot) = \delta(\cdot)$, and*

$$G_1(\lambda) = \frac{1}{[\int_{-\infty}^0 e^{\lambda t}a_1(t)\,dt]e^{-\lambda t_1} + \cdots + [\int_{-\infty}^0 e^{\lambda t}a_m(t)\,dt]e^{-\lambda t_m}},$$

where we assume that $G_1(\lambda)$ is well-defined on some right half-plane.

(i) *If $f(\cdot)$ and $\check{f}(\cdot)$ are Laplace transformable, $k \neq 0$ and $k_1 \neq 0$, then the solutions of* (376) *are unique.*

(ii) *If there exists a function $v_0 \in L^1([\tau_1, \tau_2] : X)$ such that $v_0(t) \in \mathcal{A}u_0(t)$ for a. e. $t \in [\tau_1, \tau_2]$, there exists a Laplace transformable function $u_1 : [0, \infty) \to X$ such that*

$$\int_0^\infty e^{-\lambda t}u_1(t)\,dt = \sum_{i=1}^{m} \tilde{a}_i(\lambda)e^{\lambda t_i}\int_0^{t_i} e^{-\lambda t}v_0(t)\,dt, \quad \operatorname{Re}\lambda > 0 \textit{ suff. large,}$$

and there exists a Laplace transformable function $u_2 : [0, \infty) \to X$ such that

$$\int_0^\infty e^{-\lambda t}u_2(t)\,dt = \sum_{i=1}^{m}\left[\int_{-\infty}^0 e^{\lambda t}a_i(t)\,dt\right]e^{-\lambda t_i}\int_0^{-t_i} e^{-\lambda t}v_0(-t)\,dt, \quad \operatorname{Re}\lambda > 0 \textit{ suff. large.}$$

Then $u(\cdot)$ is a strong LT *solution of* (376) *if and only if*

$$(k * Cu)(t) = (R * [f - u_1])(t), \quad t \geqslant 0 \quad \textit{and} \quad (k_1 * C'\check{u})(t) = (R_1 * [\check{f} - u_2])(t), \quad t \geqslant 0.$$

(iii) *Suppose that there exists a Laplace transformable function $u_3 : [0, \infty) \to X$ such that*

$$\int_0^\infty e^{-\lambda t} u_3(t)\, dt = G(\lambda) \sum_{i=1}^m \widetilde{a_i}(\lambda) e^{\lambda t_i} \int_0^{t_i} e^{-\lambda t} C u_0(t)\, dt, \quad \operatorname{Re} \lambda > 0 \textit{ suff. large,}$$

and there exists a Laplace transformable function $u_4 : [0, \infty) \to X$ such that

$$\int_0^\infty e^{-\lambda t} u_4(t)\, dt = G_1(\lambda) \sum_{i=1}^m \left[\int_{-\infty}^0 e^{\lambda t} a_i(t)\, dt \right] e^{-\lambda t_i} \int_0^{-t_i} e^{-\lambda t} C' u_0(t)\, dt, \quad \operatorname{Re} \lambda > 0 \textit{ suff. large.}$$

Then $u(\cdot)$ is a strong LT *solution of* (376) *if and only if*

$$(k * Cu)(t) = (R * [f - u_3])(t) + (kC * u_2)(t), \quad t \geqslant 0$$

and

$$(k_1 * C_1\check{u})(t) = (R_1 * [\check{f} - u_4])(t) + (k_1 C' * u_4)(t), \quad t \geqslant 0.$$

The illustrative applications made in Example 7.1.10 and Example 7.1.11 can be repeated for the abstract Volterra initial value problem (376); for example, we have the following.

Example 7.1.14. Suppose that $m = 1$, $t_1 < 0$, $a(\cdot)$ and $k(\cdot)$ are Laplace transformable, $k(\cdot)$ is continuous, $\tilde{a}(\cdot)$ does not vanish on some right half-plane and $\pm\mathcal{A}$ are subgenerators of exponentially equicontinuous (a, k)-regularized C-resolvent operator families $(R_\pm(t))_{t\geqslant 0}$. Define $a(-t) := a(t)$, $t \geqslant 0$. If $\widetilde{a_1}(\lambda) = e^{-\lambda t_1}\tilde{a}(\lambda)$ for $\operatorname{Re} \lambda > 0$ suff. large, then Theorem 7.1.13 can be applied. Furthermore, an analogue of the formula (375) can be given in our new framework.

We close the section by quoting some topics not considered here:

1. In [445, Subsection 3.2.2], we have investigated the case in which some of the operators C_2 or C is not injective. We will skip all related details concerning the corresponding classes of (F, G, C_2)-regularized uniqueness families and (F, G, C)-regularized resolvent families.

2. Let $\mathcal{A}_i$ be a closed MLO ($1 \leqslant i \leqslant m$). The abstract multiterm Volterra integrodifference inclusion

$$u(t) \in f(t) + \sum_{i=1}^m (a_i * \mathcal{A}_i u(\cdot + t_i))(t), \quad t \geqslant 0; \quad u(t) = u_0(t), \quad 0 \leqslant t \leqslant \tau$$

and the abstract multiterm Volterra integrodifference inclusion

$$u(t) \in f(t) + \sum_{i=1}^{m} (a_i * \mathcal{A}_i u(\cdot + t_i))(t), \quad t \in \mathbb{R}; \quad u(t) = u_0(t), \quad t \in [\tau_1, \tau_2]$$

will be considered somewhere else.

3. The abstract degenerate higher-order Cauchy problems of the form

$$Bu^{(n)}(t) = A_{n-1}u^{(n-1)}(t + t_{n-1}) + \cdots + A_0 u(t + t_0); \quad u(t) \equiv u_0(t), \quad 0 \leqslant t \leqslant \tau,$$

deserve a special attention and will be considered somewhere else ($0 \leqslant t_0, \ldots, t_{n-1} < +\infty$, $\tau \equiv \max(t_0, \ldots, t_{n-1})$; B, $A_{n-1}, \ldots, A_0$ are closed linear operators).

Finally, we would like to draw the attention of readers to the research article [12], where A. Abilassan, J. E. Restrepo and D. Suragann have introduced a new class of the multivariate Mittag-Leffler functions and compute the inverse Laplace transform of terms like

$$\lambda^{-a}\left(1 - \frac{\omega_1}{\lambda^{a_1}} - \cdots - \frac{\omega_m}{\lambda^{a_m}}\right)^{-\beta}, \quad \operatorname{Re}\lambda > 0,$$

where $\operatorname{Re} a > 0$ and $\operatorname{Re} a_i > 0$ for $1 \leqslant i \leqslant m$; cf. [12, Theorem 2.3].

7.2 Abstract fractional differential inclusions with Hilfer derivatives

It is well known that R. Hilfer has introduced a fractional derivative with two parameters, which generalizes the Riemann–Liouville fractional derivatives and the Caputo fractional derivatives ([381]; cf. also the important research articles [553, 554, 738, 739, 741], where the authors have analyzed the general Sonin and Luchko conditions in fractional calculus). In this section, we analyze the abstract fractional differential inclusions with Hilfer derivatives; in such a way, we continue the analyses raised in [762], where A. R. Volkova, V. E. Fedorov and D. M. Gordievskikh have analyzed the solvability of some classes of the abstract fractional Cauchy problems with Hilfer derivatives, and [290], where V. E. Fedorov, Y. Apakov and A. Skorynin have analyzed the analytic resolving families of operators for the abstract fractional differential equations with Hilfer derivatives [293]; cf. also [605].

We would like to emphasize that some results from the existing theory of (degenerate) (a, k)-regularized C-resolvent operator families can be successfully applied in the analysis of the abstract fractional differential inclusions with Hilfer derivatives; we also provide some new applications of (F, G, C)-regularized resolvent families here (cf. Theorem 7.2.8 and Remark 7.2.9 below). This is the first research study of this topic in which the regularizing operators C, different from the identity operator, are used; moreover, the multivalued linear operators approach is followed here, which enables us to study the well-posedness of the abstract fractional Poisson heat equation with Hilfer derivative, for example. It is also worth noting that this is probably the first serious research study of the abstract fractional differential equations with Hilfer derivatives whose solutions are not analytical in the time variable.

It seems very plausible that our results can be also reconsidered for the abstract fractional Cauchy inclusions with Ψ-Hilfer fractional derivatives. Concerning this theme, we would like to notice that K. Karthikeyan et al. have recently analyzed the well-posedness and the existence results for the Ψ-Hilfer fractional impulsive integrodifferential equations involving almost sectorial operators [418]; let us also emphasize that M. S. Abdo, S. T. Thabet and B. Ahmad have analyzed, in [11], the existence and Ulam-stability results for Ψ-Hilfer fractional integrodifferential equations (cf. also [243, 312, 337, 347, 382, 398, 417, 502, 718, 766] for some related results established recently in this direction). We would like to notice that, in all above mentioned papers, the authors have assumed that the corresponding closed linear operator A generates a strongly continuous semigroup or that it is almost sectorial, which seems to be very redundant in the study of the abstract fractional differential equations with Hilfer derivatives of order $\alpha \in (0,1)$.

In the remainder of book, we assume that X is a complex Banach space. In Section 7.2.1, we emphasize the unimportance of type $\beta \in (0,1)$ in the analysis of a class of the abstract degenerate relaxation differential equations with Hilfer derivatives; the existence and uniqueness of asymptotically almost periodic solutions of the abstract fractional Cauchy problems with Hilfer derivatives are considered in Section 7.2.2.

In this section, we define the Hilfer fractional derivative $D_t^{\alpha,\beta}u(t)$ of order $\alpha > 0$ and type $\beta \in [0,1]$ for any locally integrable function $u : [0,\infty) \to X$ for which the function $v(t) := J_t^{(1-\beta)(m-\alpha)}u(t)$, $t \geqslant 0$ is $(m-1)$-continuously differentiable and $v^{(m-1)} \in AC_{\mathrm{loc}}([0,\infty) : X)$, i. e., the function $v^{(m-1)}(\cdot)$ is locally absolutely continuous on $[0,\infty)$, through

$$D_t^{\alpha,\beta}u(t) := \left[J_t^{\beta(m-\alpha)}\frac{d^m}{dt^m}J_t^{(1-\beta)(m-\alpha)}u\right](t), \quad t > 0, \tag{378}$$

where $m = \lceil \alpha \rceil$. This is a slightly different definition from the definition of a regular Hilfer fractional derivative $D_t^{\alpha,\beta}u(t)$ given by the equation [290, (2.1)]; for the sequel, we would like to note that the Laplace transform identity established in [290, Lemma 2.1] continues to hold for the Hilfer fractional derivative $D_t^{\alpha,\beta}u(t)$ introduced here, under certain mild assumptions. The Hilfer fractional derivative $D_t^{\alpha,\beta}u(t)$ reduces to the Riemann–Liouville fractional derivative $D_t^{\alpha}u(t)$, respectively, the Caputo fractional derivative $\mathbf{D}_t^{\alpha}u(t)$, when $\beta = 0$, respectively, $\beta = 1$. For the notion and more details about the abstract (multiterm) fractional differential equations with the Riemann–Liouville fractional derivatives and the Caputo fractional derivatives, we refer the reader to [443, 445] and the list of references quoted therein. In the sequel, we will always assume that the following condition holds:

$$\beta \in (0,1) \quad \text{and} \quad \alpha \in (0,+\infty) \setminus \mathbb{N};$$

the function $v(\cdot)$ will be always defined through

$$v(t) := J_t^{(1-\beta)(m-\alpha)} u(t), \quad t \geqslant 0. \tag{379}$$

We start by introducing the following notion (cf. (379)).

Definition 7.2.1. Suppose that $\mathcal{A}$ is an MLO in X and $f \in L^1_{loc}([0,\infty) : X)$. Then we say that a function $u \in L^1_{loc}([0,\infty) : X)$ is a solution of the abstract fractional inclusion

$$D_t^{\alpha,\beta} u(t) \in \mathcal{A}u(t) + f(t) \quad \text{for a. e. } t > 0 \tag{380}$$

if the function $v(\cdot)$ is $(m-1)$-continuously differentiable, the function $v^{(m-1)}(\cdot)$ is locally absolutely continuous on $[0,\infty)$, and (380) holds for a. e. $t > 0$.

Our first structural result reads as follows.

Theorem 7.2.2. *Suppose that $C_2 \in L(X)$ is injective and $\mathcal{A}$ is a closed subgenerator of a global (g_α, k)-regularized C_2-uniqueness family $(R_2(t))_{t\geqslant 0}$, where $k(t)$ is a continuous nonzero function for $t \geqslant 0$. Then there exists a unique solution $u(\cdot)$ of* (380) *with the given initial values $v_j := v^{(j-1)}(0)$ for $2 \leqslant j \leqslant m$.*

Proof. Suppose that $u(\cdot)$ is a solution of (380) with the initial values $v_j = 0$ for $2 \leqslant j \leqslant m$ and $f \equiv 0$. Define $u_{\mathcal{A}}(t) := J_t^{\beta(m-\alpha)} v^{(m)}(t)$ for a. e. $t \geqslant 0$. Then $u_{\mathcal{A}} \in L^1_{loc}([0,\infty) : X)$ and $u_{\mathcal{A}}(t) \in \mathcal{A}u(t)$ for a. e. $t > 0$. Keeping in mind the closedness of $\mathcal{A}$, Lemma 1.1.1 and the definition of $D_t^{\alpha,\beta} u(t)$, we can convolute both sides of (380) with $g_{1-\beta(m-\alpha)}(\cdot)$ in order to see that

$$(g_1 *_0 v^{(m)})(t) = (g_{1-\beta(m-\alpha)} *_0 u_{\mathcal{A}})(t), \quad \text{for a. e. } t \geqslant 0.$$

After integration, the above implies

$$v^{(m-1)}(t) - v^{(m-1)}(0) = (g_{1-\beta(m-\alpha)} *_0 u_{\mathcal{A}})(t), \quad t \geqslant 0.$$

Inductively, we get

$$v(t) - \sum_{j=2}^{m} g_j(t) v^{(j-1)}(0) = v(t) = (g_{m-\beta(m-\alpha)} *_0 u_{\mathcal{A}})(t) = (g_\alpha *_0 v_{\mathcal{A}})(t), \quad t \geqslant 0,$$

where $v_{\mathcal{A}}(t) := (g_{m-\alpha-\beta(m-\alpha)} *_0 u_{\mathcal{A}})(t)$, $t \geqslant 0$. Using Lemma 1.1.1, it follows that $v_{\mathcal{A}}(t) \in \mathcal{A}v(t)$ for a. e. $t \geqslant 0$ so that

$$(g_\alpha *_0 v_{\mathcal{A}})(t) \in \mathcal{A}(g_\alpha *_0 v)(t), \quad t \geqslant 0, \tag{381}$$

$v(\cdot) = (g_a *_0 v_{\mathcal{A}})(\cdot) \in C([0,\infty) : X)$ and $v(\cdot)$ is a solution of (1.1) with $\mathcal{B} = \mathrm{I}$, $a(t) = g_a(t)$ and $\mathcal{F} \equiv 0$. Applying [445, Proposition 3.2.8(ii)], we get $v \equiv 0$, which simply yields $u \equiv 0$ since

$$u(t) = \frac{d}{dt}(g_{1-(1-\beta)(m-a)} *_0 v)(t), \quad t \geqslant 0. \qquad \square$$

We also have the following analogue of the Ljubich uniqueness type result [445, Theorem 3.1.6].

Theorem 7.2.3. *Suppose that $\mathcal{A}$ is a closed* MLO, $C\mathcal{A} \subseteq \mathcal{A}C$, $\lambda > 0$, $\{(m\lambda)^a : m \in \mathbb{N}\} \subseteq \rho_C(\mathcal{A})$ *and, for every $\sigma > 0$ and $x \in X$, we have*

$$\lim_{m \to +\infty} \frac{((m\lambda)^a - \mathcal{A})^{-1}C}{e^{m\lambda\sigma}} = 0.$$

Then there exists a unique solution $u(\cdot)$ of (380) *with the given initial values $v_j := v^{(j-1)}(0)$ for $2 \leqslant j \leqslant m$.*

Proof. We will use the same notation as in the proof of the former theorem. Suppose that $u(\cdot)$ is a solution of (380) with the initial values $v_j = 0$ for $2 \leqslant j \leqslant m$ and $f \equiv 0$. Then we have that the function $t \mapsto v_{\mathcal{A}}^{[1]}(t) \equiv (g_1 *_0 v_{\mathcal{A}})(t)$, $t \geqslant 0$ is continuous and

$$v^{[1]}(t) \equiv \int_0^t v(s)\,ds = (g_a *_0 v_{\mathcal{A}}^{[1]})(t), \quad t \geqslant 0.$$

It is clear that $((d^j/dt^j)v^{[1]}(t))_{t=0} = 0$ for $0 \leqslant j \leqslant m$. Keeping in mind [445, Theorem 3.1.6], it suffices to show that $\mathbf{D}_t^a v^{[1]}(t) \in \mathcal{A}v^{[1]}(t)$, $t \geqslant 0$, i. e.,

$$\frac{d^m}{dt^m}[g_{m-a} *_0 v^{[1]}](t) \in \mathcal{A}v^{[1]}(t), \quad t \geqslant 0.$$

But this is equivalent with

$$\frac{d^m}{dt^m}[g_{m-a} *_0 g_a *_0 v_{\mathcal{A}}^{[1]}](t) \in \mathcal{A}v^{[1]}(t), \quad t \geqslant 0, \text{ i. e., with } v_{\mathcal{A}}^{[1]}(t) \in \mathcal{A}v_{\mathcal{A}}^{[1]}(t), \quad t \geqslant 0.$$

Since $\mathcal{A}$ is closed and $v_{\mathcal{A}}(t) \in \mathcal{A}v_{\mathcal{A}}(t)$ for all $t \geqslant 0$ (cf. the proof of Theorem 7.2.2), the last inclusion simply follows by integration. $\square$

In particular, the (ultra-)polynomial boundedness of C_2-resolvent of $\mathcal{A}$ on the region $\{\lambda^a : \operatorname{Re}\lambda > \omega\}$, where $\omega \geqslant 0$, $C_2 \in L(X)$ is injective and commutes with $\mathcal{A}$, implies that there exists a unique solution $u(\cdot)$ of (380) with the given initial values $v_j := v^{(j-1)}(0)$ for $2 \leqslant j \leqslant m$; cf. [445, Theorems 3.2.4, 3.2.5]. Concerning the uniqueness of solutions of (380), this is a much better result compared with the results established in [762, Teorema 1], where the authors have assumed that $\mathcal{A} = A \in L(X)$, and [290, Corollary 4.1,

Theorem 5.1], where the authors have assumed that the region $\{\lambda^{a} : \lambda \in \omega + \Sigma_{\theta}\}$ belongs to the resolvent set of $\mathcal{A} = A$, which is a closed linear operator, with some $\omega > 0$ and $\theta \in (\pi/2, \pi]$; here, also, $C_2 = I$.

Now, we will state and prove the following result.

Theorem 7.2.4. *Suppose that $\mathcal{A}$ is a closed* MLO *in X and $f \in C([0,\infty) : X)$. Then the following holds:*

(i) *If a function $u \in L^1_{\text{loc}}([0,\infty) : X)$ is a solution of the abstract fractional inclusion* (380), *then the function $v(\cdot)$ is a solution of the following integral inclusion:*

$$v(t) - \sum_{j=2}^{m} g_j(t)v^{(j-1)}(0) \in (g_a *_0 \mathcal{A}v)(t) + (g_{m-\beta(m-a)} *_0 f)(t), \quad t \geqslant 0; \tag{382}$$

furthermore, if $v \in C^m([0,\infty) : X)$, then the function $v(\cdot)$ is a strong solution of (382).

(ii) *Suppose that $u \in L^1_{\text{loc}}([0,\infty) : X)$, $v(\cdot)$ is given by* (379), *there exist a function $v_{\mathcal{A}} \in L^1_{\text{loc}}([0,\infty) : X)$ and the elements $v_j \in X$ $(2 \leqslant j \leqslant m)$ such that $v_{\mathcal{A}}(t) \in \mathcal{A}v(t)$ for a. e. $t > 0$ and*

$$v(t) - \sum_{j=2}^{m} g_j(t)v_j = (g_a *_0 v_{\mathcal{A}})(t) + (g_{m-\beta(m-a)} *_0 f)(t), \quad t \geqslant 0. \tag{383}$$

Then $v \in C^{m-1}([0,\infty) : X)$ and $v_j = v^{(j-1)}(0)$ for $2 \leqslant j \leqslant m$; furthermore, if the function $v^{(m-1)}(\cdot)$ is locally absolutely continuous on $[0,\infty)$, then the function $u(\cdot)$ is a solution of the abstract fractional inclusion (380).

(iii) *Suppose that $u \in L^1_{\text{loc}}([0,\infty) : X)$, $v(\cdot)$ is given by* (379), *$v \in C^{m-1}([0,\infty) : X)$ and the function $v^{(m-1)}(\cdot)$ is locally absolutely continuous on $[0,\infty)$. Suppose further that the function $t \mapsto (g_{1-\beta(m-a)} *_0 f)(t)$, $t \geqslant 0$ is locally absolutely continuous on $[0,\infty)$, as well as that there exists a function $v_{a,\mathcal{A}} \in C([0,\infty) : X)$ such that $v_{a,\mathcal{A}}(t) \in \mathcal{A}[g_a *_0 v](t)$ for a. e. $t > 0$ and*

$$v(t) - \sum_{j=2}^{m} g_j(t)v^{(j-1)}(0) = v_{a,\mathcal{A}}(t) + (g_{m-\beta(m-a)} *_0 f)(t), \quad t \geqslant 0. \tag{384}$$

Then the function $u(\cdot)$ is a solution of the abstract fractional inclusion (380).

Proof. If $u \in L^1_{\text{loc}}([0,\infty) : X)$ solves the abstract fractional inclusion (380), then we define $u_{\mathcal{A}}(t) := J_t^{\beta(m-a)} v^{(m)}(t) - f(t)$, $t \geqslant 0$ and $v_{\mathcal{A}}(t) := (g_{m-a-\beta(m-a)} *_0 u_{\mathcal{A}})(t)$, $t \geqslant 0$. The argumentation contained in the proof of Theorem 7.2.2 shows that

$$v(t) - \sum_{j=2}^{m} g_j(t)v^{(j-1)}(0) = (g_a *_0 v_{\mathcal{A}})(t) + (g_{m-\beta(m-a)} *_0 f)(t), \quad t \geqslant 0, \tag{385}$$

$v_{\mathcal{A}}(t) \in \mathcal{A}v(t)$ for all $t \geqslant 0$, and (381) holds. Since $f \in C([0,\infty) : X)$, we have $(g_{m-\beta(m-a)} *_0 f)(\cdot) \in C([0,\infty) : X)$ and, therefore, $(g_a *_0 v_{\mathcal{A}})(\cdot) \in C([0,\infty) : X)$, so that $v(\cdot)$ is a solution of (382); furthermore, if $v \in C^m([0,\infty) : X)$, then the functions $u_{\mathcal{A}}(t)$ and $v_{\mathcal{A}}(t)$ are continuous for $t \geqslant 0$ so that $v(\cdot)$ is a strong solution of (382). This completes the proof of (i). In order to prove (ii), let us suppose that there exist a function $v_{\mathcal{A}}(\cdot)$ and the elements $v_j \in X$ $(2 \leqslant j \leqslant m)$ with the prescribed properties. Then it is clear that $v \in C^{m-1}([0,\infty) : X)$, $v_j = v^{(j-1)}(0)$ for $2 \leqslant j \leqslant m$ and that the differentiation of (383) yields

$$v^{(m-1)}(t) - v^{(m-1)}(0) = (g_{1+a-m} *_0 v_{\mathcal{A}})(t) + (g_{1-\beta(m-a)} *_0 f)(t) \quad \text{for a. e. } t \geqslant 0; \tag{386}$$

see, e. g., [69, Proposition 1.3.6]. If, additionally, the function $v^{(m-1)}(\cdot)$ is locally absolutely continuous on $[0,\infty)$, then the last equality simply implies that

$$(g_{1+\beta(m-a)} *_0 v^{(m)})(t) = (g_{1+a-m+\beta(m-a)} *_0 v_{\mathcal{A}})(t) + (g_1 *_0 f)(t) \quad \text{for a. e. } t \geqslant 0,$$

which further implies that the function $(g_{1+a-m+\beta(m-a)} *_0 v_{\mathcal{A}})(\cdot)$ is locally absolutely continuous on $[0,\infty)$ and

$$J_t^{\beta(m-a)} v^{(m)}(t) = \frac{d}{dt}(g_{1+a-m+\beta(m-a)} *_0 v_{\mathcal{A}})(t) + f(t) \quad \text{for a. e. } t \geqslant 0.$$

It remains to be proved that $(d/dt)(g_{1+a-m+\beta(m-a)} *_0 v_{\mathcal{A}})(t) \in \mathcal{A}u(t)$ for a. e. $t > 0$. Keeping in mind [69, Proposition 1.2.2(i)], it suffices to show that $(g_{1+a-m+\beta(m-a)} *_0 v_{\mathcal{A}})(t) \in \mathcal{A}\int_0^t u(s)\,ds$ for a. e. $t > 0$. But this simply follows from the definition of $v(\cdot)$, the closedness of $\mathcal{A}$ and the assumption $v_{\mathcal{A}}(t) \in \mathcal{A}v(t)$ for a. e. $t > 0$. Hence, the function $u(\cdot)$ is a solution of the abstract fractional inclusion (380). Suppose finally that the assumptions in part (iii) hold. Then the differentiation of (384) yields $v_{a,\mathcal{A}} \in C^{m-1}([0,\infty) : X)$, $v_{a,\mathcal{A}}^{(j)}(0) = 0$ for $0 \leqslant j \leqslant m-1$, and

$$v^{(m-1)}(t) - v^{(m-1)}(0) = v_{a,\mathcal{A}}^{(m-1)}(t) + (g_{1-\beta(m-a)} *_0 f)(t) \quad \text{for all } t \geqslant 0.$$

Since we have assumed that the functions $v^{(m-1)}(\cdot)$ and $t \mapsto (g_{1-\beta(m-a)} *_0 f)(t)$, $t \geqslant 0$ are locally absolutely continuous on $[0,\infty)$, the last equality easily implies that the function $v_{a,\mathcal{A}}^{(m-1)}(\cdot)$ enjoys the same feature. Furthermore, we can simply show that

$$J_t^{\beta(m-a)} v^{(m)}(t) = (g_{\beta(m-a)} *_0 v_{a,\mathcal{A}}^{(m)})(t) + f(t) \quad \text{for a. e. } t \geqslant 0.$$

Therefore, it remains to be proved that $(g_{\beta(m-a)} *_0 v_{a,\mathcal{A}}^{(m)})(t) \in \mathcal{A}u(t)$ for a. e. $t > 0$. As in part (ii), we will prove that $(g_{1+\beta(m-a)} *_0 v_{a,\mathcal{A}}^{(m)})(t) \in \mathcal{A}\int_0^t u(s)\,ds$ for a. e. $t > 0$, which simply implies the required property. Due to the closedness of $\mathcal{A}$ and Lemma 1.1.1, we have

$$(g_{1+\beta(m-a)} *_0 v_{a,\mathcal{A}})(t) \in \mathcal{A}[g_{a+1+\beta(m-a)} *_0 v](t) = \mathcal{A}[g_{m+1} *_0 u](t) \quad \text{for a. e. } t \geqslant 0.$$

Keeping in mind that $v^{(j)}_{a,\mathcal{A}}(0) = 0$ for $0 \leqslant j \leqslant m-1$, we simply get after differentiation that

$$(g_{1+\beta(m-\alpha)} *_0 v^{(m)}_{a,\mathcal{A}})(t) \in \mathcal{A} \int_0^t u(s)\,ds \quad \text{for a.e. } t \geqslant 0,$$

as claimed. This completes the proof of theorem. □

Remark 7.2.5. Due to (386), the assumptions that $v_{\mathcal{A}}(\cdot)$ and $f(\cdot)$ are locally absolutely continuous on $[0,\infty)$ in part (ii) imply that the function $v^{(m-1)}(\cdot)$ is locally absolutely continuous on $[0,\infty)$ as well.

Our first result concerning the existence and uniqueness of solutions to the abstract fractional Cauchy inclusion (380) reads as follows.

Theorem 7.2.6. *Suppose that $p \in \mathbb{N}$, $\mathcal{A}$ is the integral generator of a global (g_α, g_p)-regularized C-resolvent family $(R(t))_{t\geqslant 0}$ and the equation (1.2) holds with $a(t) \equiv g_\alpha(t)$, $k(t) \equiv g_p(t)$, $R_1(t) \equiv R(t)$ and $C_1 = C$. Suppose further that $f \in C^p([0,\infty) : X)$, the function $f^{(p)}(\cdot)$ is locally absolutely continuous on $[0,\infty)$ and*

$$f(0) = f'(0) = \cdots = f^{(p-1)}(0) = 0. \tag{387}$$

Then the function

$$v(t) := \left(R *_0 \left[\sum_{j=2}^{m} g_{j-p}(\cdot)x_{j-1} + (g_{m-\beta(m-\alpha)} *_0 f^{(p)})(\cdot)\right]\right)(t), \quad t \geqslant 0 \tag{388}$$

belongs to the class $C^m([0,\infty) : X)$. Furthermore, the function

$$u(t) := (g_{1-(1-\beta)(m-\alpha)} *_0 v')(t), \quad t \geqslant 0 \tag{389}$$

is a unique solution of the abstract fractional Cauchy problem

$$D_t^{\alpha,\beta} u(t) \in \mathcal{A}u(t) + Cf(t), \quad t \geqslant 0; \quad (J_\cdot^{(1-\beta)(m-\alpha)} u(\cdot))^{(j-1)}(0) = Cx_{j-1}, \quad 2 \leqslant j \leqslant m. \tag{390}$$

Proof. It is clear that the function $v(\cdot)$ is well-defined as well as that it belongs to the class $C^m([0,\infty) : X)$; cf. [69, Proposition 1.3.4]. Since $f^{(m)}(\cdot)$ is locally absolutely continuous on $[0,\infty)$ and (387) holds, a simple calculation yields that

$$\left(R *_0 \left[\sum_{j=2}^{m} g_j(\cdot)Cx_{j-1} + (g_{m-\beta(m-\alpha)} *_0 Cf)(\cdot)\right]\right)(t) = (g_p C *_0 v)(t), \quad t \geqslant 0.$$

Applying [445, Theorem 3.2.9(i)], we get that there exists a continuous function $t \mapsto v_{a,\mathcal{A}}(t)$, $t \geqslant 0$ such that $v_{a,\mathcal{A}}(t) \in \mathcal{A}[g_a * v](t)$ for all $t \geqslant 0$ and

$$v(t) = v_{\alpha,\mathcal{A}}(t) + \sum_{j=2}^{m} g_j(t)Cx_{j-1} + (g_{m-\beta(m-\alpha)} *_0 Cf)(t), \quad t \geqslant 0.$$

It is clear that the function $t \mapsto (g_{1-\beta(m-\alpha)} *_0 f)(t)$, $t \geqslant 0$ is locally absolutely continuous on $[0, \infty)$. Keeping in mind Theorem 7.2.4(iii), a simple continuation argument shows that the function $u(\cdot)$, given by (389), is a solution of (390). The uniqueness of solutions is a consequence of Theorem 7.2.2. □

It is clear that Theorem 7.2.6 with $C = I$ can be applied in the analysis of the abstract fractional Poisson heat type equations with Hilfer derivatives, as we have already marked (cf. [287, 445] for more details); furthermore, the possible applications of Theorem 7.2.6 with $C \neq I$ can be given to the abstract fractional Cauchy–Hilfer problems with the abstract (noncoercive) differential operators with the constant coefficients in L^p-spaces, for example (cf. [443, Section 2.5] and [445, Subsection 2.2.3] for more details).

Remark 7.2.7. In [290, Definition 3.1], the authors have introduced the notion of an l-resolving solution operator family for the equation of form (380); here, $0 \leqslant l \leqslant m-1$. The existence of a 0-resolving solution operator family for (380) has been proved for the first time in [290, Theorem 4.3(i), Corollary 4.1], where the analytical solutions of (380) have been analyzed. In the nonanalytical setting, the existence of l-resolving solution operator family, for some $l \in \{0, \ldots m-1\}$, has not been proved (cf. the formulations of [290, Lemma 3.1, Lemma 3.2, Theorem 3.1]). Hence, we can freely say that Theorem 7.2.6 is the first important research result concerning the existence of nonanalytical solutions of (380).

Now we would like to present some applications of the Laplace transform identity established in [445, Lemma 2.1]. Suppose that $u(\cdot)$ is a solution of (390); then we can perform the Laplace transform to see that

$$(\lambda^\alpha - \mathcal{A})\tilde{u}(\lambda) = \sum_{j=2}^{m} \lambda^{m-j-\beta(m-\alpha)} Cx_{j-1} + C\tilde{f}(\lambda), \quad \operatorname{Re}\lambda > \omega \text{ suff. large}.$$

Therefore, under certain logical assumptions, we have

$$u(t) = \left[\mathcal{L}^{-1}\left(\sum_{j=2}^{m} \lambda^{m-j-\beta(m-\alpha)}(\lambda^\alpha - \mathcal{A})^{-1}Cx_{j-1} + (\lambda^\alpha - \mathcal{A})^{-1}C\tilde{f}(\lambda)\right)\right](t), \quad t > 0.$$

If there exists a strongly continuous operator family $(R(t))_{t>0} \subseteq L(X)$ integrable at zero such that

$$\lambda^{m-2-\beta(m-\alpha)}(\lambda^\alpha - \mathcal{A})^{-1}Cx = \int_0^{+\infty} e^{-\lambda t}R(t)x\,dt, \quad \operatorname{Re}\lambda > \omega \text{ suff. large}, x \in X \qquad (391)$$

and if there exists a strongly continuous operator family $(P(t))_{t>0} \subseteq L(X)$ integrable at zero such that

$$(\lambda^{\alpha} - \mathcal{A})^{-1}Cx = \int_0^{+\infty} e^{-\lambda t}P(t)x\,dt, \quad \operatorname{Re}\lambda > \omega \text{ suff. large}, x \in X, \tag{392}$$

then the solution of (380) should be looked in the following form:

$$u(t) = R(t)x_1 + (g_1 *_0 R)(t)x_2 + \cdots + (g_{m-2} *_0 R)(t)x_{m-1} + \int_0^t P(t-s)f(s)\,ds, \tag{393}$$

for a. e. $t > 0$. The main problem is how to prove that the function $v(\cdot)$, given by (379), is $(m-1)$-continuously differentiable and the function $v^{(m-1)}(\cdot)$ is locally absolutely continuous on $[0,\infty)$. If $f \equiv 0$, this occurs if there exists a strongly continuous operator family $(W(t))_{t>0} \subseteq L(X)$ integrable at zero such that

$$\lambda^{m+\alpha-2}(\lambda^{\alpha} - \mathcal{A})^{-1}Cx = \int_0^{+\infty} e^{-\lambda t}W(t)x\,dt, \quad \operatorname{Re}\lambda > \omega \text{ suff. large}, x \in X, \tag{394}$$

since in this case, we can use the Laplace transform in order to see that $(g_m *_0 W)(t)x = (g_{(1-\beta)(m-\alpha)} *_0 R)(t)x$ for all $t \geqslant 0$ and $x \in X$. By the foregoing, we have the following result.

Theorem 7.2.8. *Suppose that there exists a strongly continuous operator family $(W(t))_{t>0} \subseteq L(X)$ integrable at zero such that* (394) *holds. Then, for every $x \in X$, the mapping $t \mapsto W(t)x$, $t \geqslant 0$ is $(m-1)$-continuously differentiable and its $(m-1)$-derivative is locally absolutely continuous on $[0,\infty)$. If $x \in X$ is fixed, then we set $R(t)x := (d^m/dt^m)W(t)x$, a. e. $t > 0$. Then the function $u(\cdot)$, given by* (393)*, is a solution of* (380) *with $f \equiv 0$.*

The interested reader could make an effort to find some sufficient conditions ensuring that the function $t \mapsto \int_0^t P(t-s)f(s)\,ds$, $t \geqslant 0$ is $(m-1)$-continuously differentiable and its $(m-1)$-derivative is locally absolutely continuous on $[0,\infty)$.

Remark 7.2.9. If the operator family $(R(t))_{t>0}$ defined through (391) is strongly continuous at zero, then it is the exponentially bounded $(g_{\alpha}, g_{\alpha+2+\beta(m-\alpha)-m})$-regularized C-resolvent family subgenerated by $\mathcal{A}$ since $\alpha + 2 + \beta(m-\alpha) - m \geqslant 1$; furthermore, if the operator family $(P(t))_{t>0}$ defined through (392) is strongly continuous at zero and $\alpha \geqslant 1$, then it is the exponentially bounded (g_{α}, g_{α})-regularized C-resolvent family subgenerated by $\mathcal{A}$. But, if the operator family $(W(t))_{t>0}$ defined through (393) is strongly continuous at zero, then we cannot find a continuous function $k : [0,\infty) \to \mathbb{C}$ such that $(W(t))_{t\geqslant 0}$ is an exponentially bounded (g_{α}, k)-regularized C-resolvent family subgenerated by $\mathcal{A}$. Strictly speaking, $(W(t))_{t\geqslant 0}$ is an exponentially bounded (F, G, C)-regularized resolvent family subgenerated by $\mathcal{A}$, with $F(\lambda) \equiv \lambda^{m+\alpha-2}$ and $G(\lambda) \equiv \lambda^{\alpha}$. Without going into all the details, we will only note here that the notion of an (F, G, C)-regularized resolvent family can be further extended for the operator families that are strongly continuous for $t > 0$ and only locally integrable at zero.

In order to satisfactory formulate the subordination principle for the abstract fractional Cauchy inclusions with Hilfer derivatives, we will first introduce the following notion:

Definition 7.2.10. A function $u \in L^1_{\mathrm{loc}}([0,\infty) : X)$ is said to be a mild solution of (380) if the function $v(\cdot)$, given by (379), is $(m-1)$-continuously differentiable and there exists a locally integrable function $v_{\mathcal{A}} \in L^1_{\mathrm{loc}}([0,\infty) : X)$ such that $v_{\mathcal{A}}(t) \in \mathcal{A}v(t)$ for a. e. $t \geqslant 0$ and (385) holds.

In the first part of proof of Theorem 7.2.4(i), we have shown that any solution of (380) is a mild solution of (380); due to Theorem 7.2.4(ii), we have that any mild solution of (380) is a solution of (380) provided that the function $v^{(m-1)}(\cdot)$ is absolutely continuous on any finite interval $[0, a]$, where $a > 0$.

Now we would like to formulate the following result.

Theorem 7.2.11. *Suppose that $0 < \alpha_1 < \alpha$, the functions $u(\cdot)$, $f(\cdot)$ and $f_1(\cdot)$ are Laplace transformable, the functions $f(\cdot)$ and $f_1(\cdot)$ are continuous,*

$$\tilde{f}_1(\lambda) = \lambda^{\frac{\alpha_1}{\alpha}[1+\beta(m-\alpha)-m]-1-\beta_1(m_1-\alpha_1)+m_1}\tilde{f}(\lambda^{\alpha_1/\alpha}), \quad \operatorname{Re}\lambda > \omega \textit{ suff. large}, \tag{395}$$

and the function $u_{\mathcal{A}}(\cdot) := J_t^{\beta(m-\alpha)}v^{(m)}(\cdot) - f(\cdot)$ is Laplace transformable. Suppose further that the function $v(\cdot)$, given by (379), is a mild solution of the abstract fractional Cauchy inclusion (380) and $v^{(j-1)}(0) = 0$ for $2 \leqslant j \leqslant m$. Define

$$v_1(t) := t^{-\alpha_1/\alpha}\int_0^{+\infty}\Phi_{\alpha_1/\alpha}(st^{-\alpha_1/\alpha})v(s)\,ds, \quad t > 0; \quad v_1(0) := 0,$$

and suppose that we can define the locally integrable function $u_1(\cdot)$ through

$$u_1(t) := \frac{d}{dt}(g_{1-(1-\beta_1)(m_1-\alpha_1)} *_0 v_1)(t), \quad t \geqslant 0,$$

where $m_1 = \lceil \alpha_1 \rceil$. Then the function $u_1(\cdot)$ is a mild solution of the abstract fractional Cauchy inclusion

$$D_t^{\alpha_1,\beta_1}u_1(t) \in \mathcal{A}u_1(t) + f_1(t) \quad \textit{for a. e. } t > 0.$$

Proof. Arguing as in the proof of [97, Theorem 3.1], we can simply prove that the function $v_1(t)$ is continuous for $t \geqslant 0$. Define

$$v_{1,\mathcal{A}}(t) := t^{-\alpha_1/\alpha}\int_0^{+\infty}\Phi_{\alpha_1/\alpha}(st^{-\alpha_1/\alpha})v_{\mathcal{A}}(s)\,ds, \quad t > 0; \quad v_{1,\mathcal{A}}(0) := 0.$$

Then we have $v_{\mathcal{A}}(t) \in \mathcal{A}v(t)$ for a. e. $t \geqslant 0$, and due to the closedness of $\mathcal{A}$, $v_{1,\mathcal{A}}(t) \in \mathcal{A}v_1(t)$ for a. e. $t \geqslant 0$. Since the Laplace transform of $v_1(\cdot)$ can be computed by

$$\widetilde{v_1}(\lambda) := \lambda^{(\alpha_1/\alpha)-1}\tilde{v}(\lambda^{\alpha_1/\alpha}), \quad \operatorname{Re}\lambda > \omega \text{ suff. large},$$

it readily follows from our assumption (395) that (cf. also [445, Theorem 3.1.8]):

$$v_1(t) = (g_{\alpha_1} *_0 v_{1,\mathcal{A}})(t) + (g_{m_1-\beta_1(m_1-\alpha_1)} *_0 f_1)(t), \quad t \geqslant 0,$$

which simply completes the proof of theorem. □

Remark 7.2.12. The function $u_1(t)$ can be defined and it is always continuous for $t \geqslant 0$ if $\alpha_1 > 1$.

The theory of abstract degenerate Volterra integrodifferential equations is pretty complicated and the multivalued linear operators approach is not the only existing approach in this field; see, e. g., [734] and [445]. It could be worthwhile to extend the statements of [445, Theorems 2.2.20, 2.2.21, pp. 91–95], where we have not followed the multivalued linear operators approach, to the abstract degenerate fractional differential equations with Hilfer derivatives.

7.2.1 On a class of abstract semilinear fractional Cauchy problems

In this subsection, we will continue our recent study from [445, Subsection 2.2.5] concerning the existence and uniqueness of mild solutions to the following semilinear degenerate relaxation equation:

$$(\mathrm{DFP})_{sl}: \begin{cases} \mathbf{D}_t^{\alpha}\overline{P_2(A)}u(t) = \overline{P_1(A)}u(t) + f(t,u(t)), & t \geqslant 0, \\ u(0) = x, \end{cases}$$

where $0 < \alpha < 1$, the function $f(\cdot,\cdot)$ satisfies certain properties and iA_j, $1 \leqslant j \leqslant n$ are commuting generators of bounded C_0-groups on a Banach space E. We refer the reader to [445] for more details about the functional calculus for commuting generators of bounded C_0-groups and the definitions of operators $\overline{P_1(A)}$ and $\overline{P_2(A)}$.

We will consider the problem

$$(ACP): D_t^{\alpha,\beta}Bu(t) = Au(t) + f(t,u(t)) \quad \text{for a. e. } t \geqslant 0,$$

without any initial conditions ($0 < \alpha < 1$, $0 < \beta < 1$; A and B are closed linear operators on X). Applying [445, Lemma 2.1] again, we get the following equalities, which hold under certain very mild assumptions (here, we will use the same notation as in [290]):

$$\lambda^{\alpha}B\tilde{u}(\lambda) - \lambda^{\beta(\alpha-1)}D^{(\beta-1)(m-\alpha)}Bu(0) = \lambda^{\alpha}B\tilde{u}(\lambda) = A\tilde{u}(\lambda) + \tilde{f}(\lambda), \tag{396}$$

for $\operatorname{Re}\lambda > \omega$ suff. large. Therefore, the order $\beta \in (0,1)$ should not play any role in the definition of a mild solution of (ACP).

For the sequel, we need to introduce the following notion (cf. also [445, Definition 2.2.28]).

Definition 7.2.13. Suppose that A and B are closed linear operators on X. Let $0 < \alpha < 1$, let $C \in L(X)$ be injective and let $C^{-1}AC = A$, $C^{-1}BC = B$. A strongly continuous operator family $(P_\alpha(t))_{t>0} \subseteq L(X)$ is said to be an $(\alpha, \alpha, A, B, C)$-resolvent family if there exist $M \geqslant 1$ and $\omega \geqslant 0$ such that the mapping $t \mapsto \|t^{1-\alpha}P_\alpha(t)\|$, $t \in (0,1]$ is bounded, $\|P_\alpha(t)\| \leqslant Me^{\omega t}$, $t \geqslant 1$ and

$$(\lambda^\alpha B - A)^{-1}Cx = \int_0^\infty e^{-\lambda t}P_\alpha(t)x\,dt, \quad \operatorname{Re}\lambda > \omega,\ x \in X.$$

Keeping in mind the equation (396), it seems reasonable to define a mild solution of (ACP) as any locally integrable function $t \mapsto u(t)$, $t > 0$ for which the mapping $t \mapsto C^{-1}f(t, u(t))$, $t > 0$ is well-defined, locally integrable and the following integral equation holds:

$$u(t) = \int_0^t P_\alpha(t-s)C^{-1}f(s, u(s))\,ds \quad \text{for a. e. } t > 0;$$

the formula (393) can serve one to introduce the notion of a mild solution of the semilinear analogues of (380) with $\alpha > 1$.

For some important examples of $(\alpha, \alpha, \overline{P_1(A)}, \overline{P_2(A)}, C)$-resolvent families, we refer the reader to [445, Theorem 2.2.29], which can be used to provide certain applications in the study of the existence and uniqueness of mild solutions of problem (ACP); cf. also [445, Theorem 2.2.30].

7.2.2 Asymptotically almost periodic solutions of (380)

The representation formula (388) can serve one to provide certain results about the asymptotically almost periodic properties of the function $v(\cdot)$ in some concrete situations.

Suppose now that the requirements of Theorem 7.2.6 hold with the operator family $(R(t))_{t\geqslant 0}$ satisfying $\int_0^{+\infty} \|R(t)\|\,dt < +\infty$ and $t \cdot \|R(t)\| \to 0$ as $t \to +\infty$; this is always the case provided that there exist finite real constants $M > 0$, $\beta \in (0,1]$ and $\gamma > \beta$ such that (284) holds, which is the usual case in the existing literature concerning this problem. Concerning the initial values x_{j-1} for $2 \leqslant j \leqslant m$ and the forcing term $f(\cdot)$, we assume that $x_{j-1} = 0$ if $j - p \geqslant 2$ and the function $t \mapsto (g_{m-\beta(m-\alpha)} *_0 f^{(p)})(t)$, $t \geqslant 0$ is asymptotically almost periodic. Arguing as in the proof of [444, Lemma 2.9.3], it readily follows that the function $v(\cdot)$ is asymptotically almost periodic. In some similar situations, we can prove that the function $v'(\cdot)$ is asymptotically almost periodic; but even if this is the case, it would be very difficult to prove that the solution $u(\cdot)$ of (390) is asymptotically almost periodic in a certain generalized sense (cf. the formula (389) and [447, pp. 87–88] for some recent results obtained in this direction).

The interested reader may also try to provide certain applications of Theorem 7.2.8 in the study of the existence and uniqueness of asymptotically almost periodic solutions to (380). Let us finally notice that the statements of [290, Theorem 4.3, Corollary 4.1], which concern the existence and uniqueness of analytical solutions of (380), can be further extended by using the additional regularizing operator $C \in L(X)$ in the whole analysis. This is important to be stressed because many abstract noncoercive differential operators with constant coefficients in L^p-spaces can have the empty resolvent set.

7.3 Abstract fractional differential inclusions with generalized Laplace derivatives

Traditionally, the fractional derivatives of functions defined on the nonnegative real axis are introduced as certain combinations of the operators of fractional integration and the operators of integer-order differentiation. The main aim of this section is to explain how one can employ the vector-valued Laplace transform to introduce and systematically analyze the notion of a new type of fractional derivative, called here the generalized Laplace fractional derivative (albeit the Laplace transform has been used in the research article [235] by E. C. de Oliveira, S. Jarosz and J. Vaz Jr., we follow a completely different approach here; cf. also [427, Theorem 3.1] and [521, Theorems 4.1–4.3, 5.6], where the authors have investigated the exponential boundedness of the solutions to the abstract fractional differential equations with Caputo derivatives and the rationality of solving the abstract fractional differential equations by the Laplace transform method). For the Laplace transformable functions, the generalized Laplace fractional derivative extends several types of fractional derivatives of convolution type known in the existing literature. In this section, we investigate the abstract fractional differential inclusions with generalized Laplace derivatives with the help of our recent results about (F, G, C)-regularized resolvent operator families. We introduce the notion of a generalized Hilfer (a, b, α)-fractional derivative and analyze after that the abstract fractional differential inclusions with generalized Hilfer (a, b, α)-derivatives and Prabhakar derivatives. It should be noted that this is probably the first research study of the abstract fractional differential equations with Prabhakar derivatives (the material is taken from [479]; cf. also [85]).

The structure and main ideas of this section, which contributes to the theories of vector-valued Laplace transform, fractional calculus and abstract Volterra integro-differential equations, can be briefly outlined as follows. After collecting some preliminaries, we introduce the notion of a generalized Laplace fractional derivative $D_L^W u$ in

Definition 7.3.2. After that, we list a great number of known types of fractional derivatives, which are very special cases of the notion introduced here. The basic properties of the vector-valued Laplace transform enable one to clarify the most important structural results for the generalized Laplace fractional derivatives; see Proposition 7.3.4, Theorem 7.3.5 and Theorem 7.3.6 below. Section 7.3.1 is devoted to the study of Laplace transform of Marchaud right-sided fractional derivatives. In Section 7.3.2, we investigate the Laplace transform identities for the pointwise product of functions (see Proposition 7.3.10 and Remark 7.3.11); the main result of Section 7.3.3, where we analyze Leibniz rules for generalized Laplace fractional derivatives, is Theorem 7.3.12.

The research studies of Leibniz rules for fractional derivatives date back to 1832, when J. Liouville [531, p. 117] proposed the formula

$$D_z^{\alpha}[u(z)v(z)] = \sum_{k=0}^{+\infty} \binom{\alpha}{k} D_z^{\alpha-k} u(z) v^{(k)}(z) \tag{397}$$

for some kinds of fractional derivatives with infinitely differentiable functions $v(\cdot)$. This formula has been reconsidered and slightly generalized in many research articles by now, starting presumably with those of T. J. Osler [629–631]. In the existing literature on the fractional calculus, we have not been able to find a satisfactory analogue of the formula (397) concerning the Leibniz rule for the Riemann–Liouville, Caputo and Hilfer fractional derivatives in the case that the function $v(\cdot)$ is not infinitely differentiable. This fact has strongly influenced us to analyze the Leibniz type rules for these kinds of fractional derivatives and the introduced class of generalized Hilfer (a, b, α)-fractional derivatives; see Sections 7.3.4–7.3.6, where we deal with the functions, which are not Laplace transformable and the functions $v(\cdot)$, which are not infinitely differentiable, in general. We aim to show that the use of usual Leibniz rule and some elementary transformations can be much better for clarifying the Leibniz type rules for fractional derivatives than the use of vector-valued Laplace transform, the result established in Theorem 7.3.12 and the results established in the known generalizations of formula (397). In the formulations of our results, we do not use the pointwise products of fractional derivatives, only, but also the convolutions of fractional derivatives with some other functions.

In Section 7.3.7, we investigate the abstract fractional Cauchy inclusions with generalized Laplace derivatives. The main results of this subsection are Proposition 7.3.21, Theorem 7.2.8, Theorem 7.3.24, Theorem 7.3.25 and Theorem 7.3.27; Section 7.3.8 is devoted to the study of the abstract fractional Cauchy inclusions with generalized Hilfer (a, b, α)-fractional derivatives, while Section 7.3.9 pertains to the study of the abstract fractional Cauchy inclusions with Prabhakar fractional derivatives.

Let us recall that, if $f \in L^1_{\mathrm{loc}}([0,\infty) : X)$, then $f(\cdot)$ is Laplace transformable if and only if the function $t \mapsto \int_0^t f(s)\,ds$, $t \geqslant 0$ is exponentially bounded [69, 445]. Define $\mathbb{L} := \{f : [0,\infty) \to X : f(\cdot)$ is Laplace transformable$\}$; we identify here the functions that are equal almost everywhere on $[0,\infty)$.

We will use the following well-known result.

Lemma 7.3.1. *Suppose that* $m \in \mathbb{N}$, $f \in C^{m-1}([0,\infty) : X)$, *the function* $f^{(m-1)}(\cdot)$ *is locally absolutely continuous on* $[0,\infty)$ *and the functions* $f(\cdot),\dots,f^{(m-1)}(\cdot)$ *are exponentially bounded. Then the function* $f^{(m)}(\cdot)$ *is Laplace transformable and we have*

$$\int_0^{+\infty} e^{-\lambda t} f^{(m)}(t)\,dt = \lambda^m(\mathcal{L}f)(\lambda) - \lambda^{m-1}f(0) - \lambda^{m-2}f'(0) - \cdots - f^{(m-1)}(0),$$

for $\operatorname{Re}\lambda > 0$ *suff. large.*

The generalized Mittag-Leffler function $E^{\gamma}_{\alpha,\beta}(z)$ was introduced by T. R. Prabhakar in [648]:

$$E^{\gamma}_{\alpha,\beta}(z) := \sum_{k=0}^{\infty} \frac{\Gamma(\gamma+k)}{\Gamma(\gamma)\cdot\Gamma(\alpha k+\beta)}\frac{z^k}{k!}, \quad z \in \mathbb{C}.$$

We will use the following Laplace transform identity [648]:

$$\int_0^{+\infty} e^{-\lambda t} t^{\beta-1} E^{\gamma}_{\alpha,\beta}(\eta t^{\alpha})\,dt = \frac{\lambda^{\gamma\alpha-\beta}}{(\lambda^{\alpha}-\eta)^{\gamma}}, \quad \operatorname{Re}\lambda > 0,\ |\eta/\lambda^{\alpha}| < 1, \tag{398}$$

which holds whenever $\alpha,\beta,\gamma,\eta \in \mathbb{C}$ and $\operatorname{Re}\beta > 0$.

In the sequel, we will always assume that the following condition holds good:

(L): $s \in \mathbb{N}_0$, $\omega \geqslant 0$, $w : \{\lambda \in \mathbb{C} : \operatorname{Re}\lambda > \omega\} \to X$ and $w_k : \{\lambda \in \mathbb{C} : \operatorname{Re}\lambda > \omega\} \to X$ are given functions, D_k is a nonempty subset of $\mathbb{L}$ and $x_k : D_k \to X$ are given mappings $(0 \leqslant k \leqslant s)$. Set

$$W := \big(s, w(\cdot), (w_k(\cdot))_{0\leqslant k\leqslant s}, (D_k)_{0\leqslant k\leqslant s}, (x_k(\cdot))_{0\leqslant k\leqslant s}\big). \tag{399}$$

The following notion plays a crucial role in our investigation.

Definition 7.3.2. The generalized Laplace fractional derivative $D^W_L u$ is defined for any function $u \in D_0 \cap \cdots \cap D_s$ such that there exist a number $\omega' > \omega$ and a Laplace transformable function $h : [0,\infty) \to X$ such that

$$\tilde{h}(\lambda) = w(\lambda)\tilde{u}(\lambda) - \sum_{k=0}^{s} w_k(\lambda)x_k(u), \quad \operatorname{Re}\lambda > \omega'. \tag{400}$$

In this case, we set $(D^W_L u)(t) := h(t)$ for a. e. $t > 0$.

Let $m := \lceil \alpha \rceil$. For the Laplace transformable functions, the notion introduced in Definition 7.3.2 generalizes the notion of a great number of different types of fractional derivatives whose Laplace transform can be intrinsically computed (the precise characterizations of sets $D_0,\dots,D_s$ and the mappings $x_0(\cdot),\dots,x_s(\cdot)$ can be given in a very simply and concise way; for example, in the case of the Caputo fractional derivatives of order α considered below, we have that D_k is the vector space of all functions $u(\cdot)$, which are $(k-1)$-times continuously differentiable on some segment $[0,\delta)$ satisfying that $u^{(k)}(0)$ exists, and $x_k(u) = u^{(k)}(0)$, $u \in D_k$, $0 \leqslant k \leqslant s$):

(i) Riemann–Liouville fractional derivatives of order α, with $w(\lambda) = \lambda^{\alpha}$, $s = m-1$, $x_k(u) = (g_{m-\alpha} *_0 u)^{(k)}(0)$ and $w_k(\lambda) = \lambda^{m-1-k}$, see [97, (1.22)].
(ii) Caputo fractional derivatives of order α, with $w(\lambda) = \lambda^{\alpha}$, $s = m-1$, $x_k(u) = u^{(k)}(0)$ and $w_k(\lambda) = \lambda^{\alpha-1-k}$, see [97, (1.23)].
(iii) Hilfer fractional derivatives of order α and type $\beta \in (0,1)$, with $w(\lambda) = \lambda^{\alpha}$, $s = m-1$, $x_k(u) = D^{k-(1-\beta)(m-\alpha)}u(0)$ and $w_k(\lambda) = \lambda^{m-1-k-\beta(m-\alpha)}$, see [290, Lemma 2.1].
(iv) Dzhrbashyan–Nersesyan fractional derivatives of order $\alpha = \sigma_n$, with $w(\lambda) = \lambda^{\alpha}$, $s = n-1$, $x_k(u) = D^{\sigma_k}u(0)$ and $w_k(\lambda) = \lambda^{\alpha-\sigma_k-1}$, see [294, (10)] and [295, 296].
(v) Prabhakar fractional derivative given by [321, (5.9)], with $w(\lambda) = \lambda^{\beta-\alpha\gamma}(\lambda^{\alpha}-\lambda_0)^{\gamma}$, $s = \lceil\beta\rceil - 1$, $x_k(u) = (\mathcal{J}^{-\gamma}_{\alpha,m-\beta-k,\lambda;0})u(0)$ and $w_k(\lambda) = \lambda^{m-\sigma_k-1}$, see [321, (5.13)] and [322].
(vi) Prabhakar fractional derivative given by [321, (5.11)], with $w(\lambda) = \lambda^{\beta-\alpha\gamma}(\lambda^{\alpha}-\lambda_0)^{\gamma}$, $s = \lceil\beta\rceil - 1$, $x_k(u) = u^{(k)}(0)$ and $w_k(\lambda) = \lambda^{\beta-\alpha\gamma}(\lambda^{\alpha}-\lambda_0)^{\gamma}\lambda^{k-1}$, see [321, p. 28, l. 1];
(vii) Erdélyi–Kober fractional derivatives, see [555, p. 259, l. -1].
(viii) Grünwald–Letnikov fractional derivative of order $\alpha \in (0,1)$ with $w(\lambda) = \lambda^{\alpha}$, $s = 0$ and $w_k(\lambda) = 0$, see [645, (2.255), p. 107].
(ix) Miller–Ross sequential fractional derivatives, see [645, Subsection 2.8.5] for more details.
(x) Modified Riemann–Liouville fractional derivatives, see [411, (2.9)–(2.12)], and so on and so forth.

Before proceeding any further, we would like to note the following.

Remark 7.3.3. It is a little bit inappropriate to call $D_L^W u$ the generalized Laplace fractional derivative since the notion introduced in Definition 7.3.2 also generalizes the notion of several known fractional integrals whose Laplace transform can be intrinsically computed. For the Laplace transformable functions, the notion introduced in Definition 7.3.2 extends the notion of Riemann–Liouville fractional integrals, Weyl fractional integrals and k-Hilfer fractional integrals; see [236, (35), (46)]. On the other hand, it is very difficult to precisely compute the Laplace transform of Hadamard fractional integrals, Erdélyi fractional integrals, Kober fractional integrals and Riemann–Liouville integrals of variable fractional order; see [236, (18), (24)–(28), (34), (39)–(42), (44)–(45)]. In the sequel, we will consider only fractional derivatives and we will continue to call $D_L^W u$ the generalized Laplace fractional derivative.

It is clear that the function $h(\cdot)$ in Definition 7.3.2 can be found from $u(\cdot)$ and W by means of the Post-–Widder inversion formula for the Laplace transform or the Phragmén–Doetsch inversion formula for the Laplace transform; see, e. g., [445, Theorems 1.4.3, 1.4.4]. Furthermore, the function $(D_L^W u)(t)$ can be analytically extended to the sector Σ_α and for each $\gamma \in (0,\alpha)$ there exists a finite constant $M_\gamma \geqslant 1$ such that $\|(D_L^W u)(\lambda)\| \leqslant M_\gamma e^{\omega' \operatorname{Re}\lambda}$, $\lambda \in \Sigma_\gamma$ if and only if the function $q(\lambda) := w(\lambda)\tilde{u}(\lambda) - \sum_{k=0}^{s} w_k(\lambda)x_k(u)$, $\operatorname{Re}\lambda > \omega'$ can be analytically extended to the sector $\omega' + \Sigma_{\alpha+(\pi/2)}$ and for each $\gamma \in (0,\alpha)$ we have that the set $\{(\lambda-\omega')q(\lambda) : \lambda \in \omega' + \Sigma_{\gamma+(\pi/2)}\}$ is bounded $(0 < \alpha \leqslant \pi/2)$; see [445, Theorem 1.4.10(i)].

Since the Laplace transform is linear, it readily follows that, if the generalized Laplace fractional derivatives $D_L^W u_1$ and $D_L^W u_2$ exist and $\alpha_1, \alpha_2 \in \mathbb{C}$, then the generalized Laplace fractional derivative $D_L^W[\alpha_1 u_1 + \alpha_2 u_2]$ also exists and we have

$$D_L^W[\alpha_1 u_1 + \alpha_2 u_2] = \alpha_1 D_L^W u_1 + \alpha_2 D_L^W u_2,$$

provided that $D_0, \ldots, D_s$ are vector spaces with the usual operations and the mappings $x_k : D_k \to X$ are linear $(0 \leqslant k \leqslant s)$.

The operational properties of the vector-valued Laplace transform enable one to formulate many statements for the generalized Laplace fractional derivatives. For example, we have the following.

Proposition 7.3.4. *Suppose that $m \in \mathbb{N}$ and the generalized Laplace fractional derivative $D_L^W u$ exists with $(w_j(\cdot))_{0\leqslant j\leqslant k}$ being a set of analytic functions on some right half-plane and $w(\lambda) \equiv \lambda^a$. Set $g(t) := t^m u(t)$, $t \geqslant 0$ and*

$$\overline{W} := (\overline{s}, \overline{w}(\cdot), (\overline{w_k}(\cdot))_{0\leqslant k\leqslant s}, (\overline{D}_k)_{0\leqslant k\leqslant s}, (\overline{x}_k(\cdot))_{0\leqslant k\leqslant s}),$$

where $\overline{s} := s$, $\overline{w}(\lambda) \equiv w(\lambda)$,

$$\overline{w}_k(\lambda) \equiv (-1)^{m+1}\sum_{j=0}^{m}(-1)^j\binom{m}{j}(-a)(-a-1)\cdot\dots\cdot(-a-j+1)\lambda^{-j}\omega_k^{(m-j)}(\lambda), \quad 0 \leqslant k \leqslant s,$$

$\overline{D} := \{t^{m\cdot}h(\cdot) : h \in D_k\}$ and $\overline{x_k}(t^{m\cdot}h(\cdot)) := x_k(h(\cdot))$ for $0 \leqslant k \leqslant s$ and $h \in D_k$. Then the generalized Laplace fractional derivative $D_L^{\overline{W}} g$ exists and we have

$$(D_L^{\overline{W}} g)(t) = \sum_{j=0}^{m}(-1)^j\binom{m}{j}(-a)(-a-1)\cdot\dots\cdot(-a-j+1)(g_j *_0 \cdot^{m-j} D_L^W u(\cdot))(t), \quad t \geqslant 0.$$

Proof. Keeping in mind the corresponding definitions, (400) and [445, Theorem 1.4.2(vi)], we only need to prove that

$$\begin{aligned}
&\sum_{j=0}^{m}(-1)^j\binom{m}{j}(-a)(-a-1)\cdot\dots\cdot(-a-j+1)\lambda^{-j}[\mathcal{L}(\cdot^{m-j}D_L^W h(\cdot))](\lambda)\\
&\quad = \lambda^a[\mathcal{L}(\cdot^m u(\cdot))](\lambda)\\
&\qquad - \sum_{k=0}^{s}(-1)^{m+1}\sum_{j=0}^{m}(-1)^j\binom{m}{j}(-a)(-a-1)\cdot\dots\cdot(-a-j+1)\lambda^{-j}w_k^{(m-j)}(\lambda)x_k(u), \quad \operatorname{Re}\lambda > \omega'.
\end{aligned}$$

Using the Laplace transform identity [445, (34), p. 58], the result follows if we prove that

$$\sum_{j=0}^{m}(-1)^j\binom{m}{j}(-a)(-a-1)\cdot\dots\cdot(-a-j+1)\lambda^{-j}(-1)^{m-j}\left(\cdot^a\tilde{u}(\cdot)-\sum_{k=0}^{s}w_k(\cdot)x_k(u)\right)^{(m-j)}(\lambda)$$
$$=\lambda^a(-1)^m\tilde{u}^{(m)}(\lambda)-\sum_{k=0}^{s}(-1)^{m+1}\sum_{j=0}^{m}(-1)^j\binom{m}{j}$$
$$\times(-a)(-a-1)\cdot\dots\cdot(-a-j+1)\lambda^{-j}w_k^{(m-j)}(\lambda)x_k(u),\quad \operatorname{Re}\lambda>\omega',$$

i. e.,

$$\sum_{j=0}^{m}(-1)^j\binom{m}{j}(-a)(-a-1)\cdot\dots\cdot(-a-j+1)\lambda^{-j}\times(-1)^{m-j}(\cdot^a\tilde{u}(\cdot))^{(m-j)}(\lambda)$$
$$=\lambda^a(-1)^m\tilde{u}^{(m)}(\lambda),\quad \operatorname{Re}\lambda>\omega'.$$

The last equality simply follows by applying the product rule and the next computation:

$$\tilde{u}^{(m)}(\lambda)=([\cdot^a\tilde{u}(\cdot)]\cdot^{-a})^{(m)}(\lambda)$$
$$=\sum_{j=0}^{m}\binom{m}{j}[\cdot^a\tilde{u}(\cdot)]^{(m-j)}(\lambda)(-1)^{m-j}\lambda^{-a-j}(-a)(-a-1)\cdot\dots\cdot(-a-j+1),\quad \operatorname{Re}\lambda>\omega'.$$

This completes the proof of proposition. □

For the proof of Theorem 7.3.6 below, we need the following lemma, which seems to be new in the existing literature (if the function $u(\cdot)$ is exponentially bounded, then the proof essentially follows from the argumentation contained in the proof of [97, Theorem 3.1]).

Lemma 7.3.5. *Suppose that $\gamma\in(0,1)$ and the function $u:[0,\infty)\to X$ is Laplace transformable. Define*

$$u_\gamma(t):=t^{-\gamma}\int_0^{+\infty}\Phi_\gamma(st^{-\gamma})u(s)\,ds,\quad t>0. \tag{401}$$

Then the function $u_\gamma(\cdot)$ is Laplace transformable and

$$\widetilde{u_\gamma}(\lambda)=\lambda^{\gamma-1}\tilde{u}(\lambda^\gamma),\quad \operatorname{Re}\lambda>0 \textit{ suff. large.} \tag{402}$$

Proof. We can simply show that the function $u_\gamma(\cdot)$ is Lebesgue measurable. Furthermore, we know that there exist two finite real constants $M\geqslant 1$ and $\omega>0$ such that $\|\int_0^t u(s)\,ds\|\leqslant Me^{\omega t}$, $t\geqslant 0$ and it suffices to show that there exists a finite real constant $M_1\geqslant 1$ such that $\|\int_0^t u_\gamma(s)\,ds\|\leqslant M_1e^{\omega^{1/\gamma}t}$, $t\geqslant 0$. Toward this end, observe that the asymptotic expansion formula [97, (1.33)] and the partial integration together imply that

$$
\begin{aligned}
u_\gamma(t) &= t^{-\gamma} \int_0^{+\infty} \Phi_\gamma(st^{-\gamma})u(s)\,ds \\
&= t^{-\gamma}\left[\Phi_\gamma(st^{-\gamma}) \int_0^s u(r)\,dr\right]_{s=0}^{+\infty} - t^{-\gamma}\int_0^{+\infty} \Phi_\gamma'(st^{-\gamma})\left[\int_0^s u(r)\,dr\right] ds \\
&= -t^{-\gamma}\int_0^{+\infty} \Phi_\gamma'(st^{-\gamma})\left[\int_0^s u(r)\,dr\right] ds.
\end{aligned}
$$

Since $\Phi_\gamma(\cdot)$ is an entire function and [97, (1.33)] holds, we can simply prove that there exists a finite real constant $M' > 0$ such that $|\Phi_\gamma'(t)| \leqslant M\Phi_\gamma(t)$ for all $t \geqslant 0$. Arguing similarly as in the proof of [97, Theorem 3.1], we get $\|u_\gamma(t)\| \leqslant M_2 t^{-\gamma} e^{\omega^{1/\gamma} t}$, $t > 0$ for some $M_2 > 0$, which simply implies

$$
\left\|\int_0^t u_\gamma(s)\,ds\right\| \leqslant M_2 \int_0^t s^{-\gamma} e^{\omega^{1/\gamma} s}\,ds \leqslant M_2(1-\gamma)^{-1} t^{1-\gamma} e^{\omega^{1/\gamma} t}, \quad t > 0.
$$

The equality (402) follows similarly as in [97]. □

Now we are able to formulate the following result.

Theorem 7.3.6. *Suppose that the generalized Laplace fractional derivative $D_L^W u$ is defined, $\gamma \in (0,1)$ and the function $u_\gamma(\cdot)$ is given by (401). Let*

$$
W_\gamma := (s, w_\gamma(\cdot), (w_{k,\gamma}(\cdot))_{0\leqslant k\leqslant s}, (D_{k,\gamma})_{0\leqslant k\leqslant s}, (x_{k,\gamma}(\cdot))_{0\leqslant k\leqslant s}),
$$

where $w_\gamma(\lambda) \equiv \omega(\lambda^\gamma)$, $w_{k,\gamma}(\lambda) \equiv \lambda^{\gamma-1} w_k(\lambda^\gamma)$ for $0 \leqslant k \leqslant s$, $D_{k,\gamma} := \{h_\gamma(\cdot) : h \in D_k\}$ and $x_{k,\gamma}(h_\gamma(\cdot)) \equiv x_k(h(\cdot))$ for $0 \leqslant k \leqslant s$ and $h \in D_k$; here, $h_\gamma(\cdot)$ is defined by replacing the function $u(\cdot)$ in (401) with the function $h(\cdot)$. Then the generalized Laplace fractional derivative $D_L^{W_\gamma} u_\gamma$ is defined and we have $(D_L^{W_\gamma} u_\gamma)(t) = (D_L^W u)_\gamma(t)$ for a. e. $t > 0$, where $(D_L^W u)_\gamma(\cdot)$ is defined by replacing the function $u(\cdot)$ in (401) with the function $(D_L^W u)(\cdot)$.

Proof. Since (400) holds, Lemma 7.3.5 gives

$$
\begin{aligned}
\widetilde{h_\gamma}(\lambda) &= \lambda^{\gamma-1}\tilde{h}(\lambda^\gamma) \\
&= \lambda^{\gamma-1}\left[w(\lambda^\gamma)\tilde{u}(\lambda^\gamma) - \sum_{k=0}^{s} w_k(\lambda^\gamma) x_k(u)\right], \quad \operatorname{Re}\lambda > \omega' \text{ suff. large.}
\end{aligned}
$$

Keeping in mind the definition of W_γ, the above simply implies the required. □

Similarly, we can prove the following statements (cf. [445, Theorem 1.4.2] for more details).

Theorem 7.3.7. (i) *Suppose that the generalized Laplace fractional derivative $D_L^W u$ is defined, $z \in \mathbb{C}$ and $g(t) := e^{-zt}u(t)$, $t \geqslant 0$. Define*

$$W_z := \big(s, w(\cdot + z), (w_k(\cdot + z))_{0\leqslant k\leqslant s}, (D_{k,z})_{0\leqslant k\leqslant s}, (x_{k,z}(\cdot))_{0\leqslant k\leqslant s}\big),$$

where $D_{k,z} := \{e^{-z\cdot}h(\cdot) : h \in D_k\}$ and $x_{k,z}(e^{-z\cdot}h(\cdot)) \equiv x_k(h(\cdot))$ for $0 \leqslant k \leqslant s$ and $h \in D_k$. Then the generalized Laplace fractional derivative $D_L^{W_z} g$ is defined and we have $(D_L^{W_z})g(t) = e^{-zt}(D_L^W u)(t)$ for a. e. $t > 0$.

(ii) *Suppose that the generalized Laplace fractional derivative $D_L^W u$ is defined, $T \in L(X, Y)$ and $g(t) := Tu(t)$, $t \geqslant 0$. Define*

$$W_T := \big(s, Tw(\cdot), (Tw_k(\cdot))_{0\leqslant k\leqslant s}, (D_{k,T})_{0\leqslant k\leqslant s}, (x_{k,T}(\cdot))_{0\leqslant k\leqslant s}\big),$$

where $D_{k,T} := \{Th(\cdot) : h \in D_k\}$ and $x_{k,T}(Th(\cdot)) \equiv x_k(h(\cdot))$ for $0 \leqslant k \leqslant s$ and $h \in D_k$. Then the generalized Laplace fractional derivative $D_L^{W_T} g$ is defined and we have $(D_L^{W_T})g(t) = T(D_L^W u)(t)$ for a. e. $t > 0$.

(iii) *Suppose that the generalized Laplace fractional derivative $D_L^W u$ is defined, $b > 0$ and $g(t) := u(t + b)$, $t \geqslant 0$. Define*

$$W_b := \big(s + 1, w(\cdot), (w_{k,b}(\cdot))_{0\leqslant k\leqslant s+1}, (D_{k,b})_{0\leqslant k\leqslant s+1}, (x_{k,b}(\cdot))_{0\leqslant k\leqslant s+1}\big),$$

where $w_{k,b}(\lambda) \equiv e^{\lambda s} w_k(\lambda)$,

$$w_{s+1,b}(\lambda) \equiv -w(\lambda)e^{\lambda b}\int_0^b e^{-\lambda t}u(t)\,dt + e^{\lambda b}\int_0^b e^{-\lambda t}(D_L^W u)(t)\,dt,$$

$D_{k,b} := \{h(\cdot + b) : h \in D_k\}$, $x_{k,b}(h(\cdot + b)) := x_k(h(\cdot))$ for $0 \leqslant k \leqslant s$ and $h \in D_k$, $D_{s+1,b} := \mathbb{L}$ and $x_{s+1,b} \equiv 1$. Then the generalized Laplace fractional derivative $D_L^{W_b} g$ is defined and we have $(D_L^{W_b} g)(t) = (D_L^W u)(t + b)$ for a. e. $t > 0$.

(iv) *Suppose that the generalized Laplace fractional derivative $D_L^W u$ is defined, $b > 0$, $g(t) := 0$, $0 \leqslant t < b$ and $g(t) := u(t - b)$, $t \geqslant b$. Define*

$$W_{b;1} := \big(s, w(\cdot), (w_{k,b,1}(\cdot))_{0\leqslant k\leqslant s}, (D_{k,b,1})_{0\leqslant k\leqslant s}, (x_{k,b,1}(\cdot))_{0\leqslant k\leqslant s}\big),$$

where $w_{k,b,1}(\lambda) \equiv e^{-\lambda s} w_k(\lambda)$, $D_{k,b,1} := \{\chi_{[b,+\infty)}(\cdot)h(\cdot - b) : h \in D_k\}$ and $x_{k,b,1}(\chi_{[b,+\infty)}(\cdot)h(\cdot - b)) := x_k(h(\cdot))$ for $0 \leqslant k \leqslant s$ and $h \in D_k$. Then the generalized Laplace fractional derivative $D_L^{W_{b;1}} g$ is defined and we have $(D_L^{W_{b;1}} g)(t) = \chi_{[b,+\infty)}(\cdot)(D_L^W u)(t - b)$ for a. e. $t > 0$.

(v) *Suppose that the generalized Laplace fractional derivative $D_L^W u$ is defined, $a \in L^1_{\mathrm{loc}}([0, \infty))$, $\mathrm{abs}(|a|) < +\infty$ and $g(t) := (a *_0 u)(t)$, $t \geqslant 0$. Define*

$$W_a := \big(s, w(\cdot), (w_{k,a}(\cdot))_{0\leqslant k\leqslant s}, (D_{k,a})_{0\leqslant k\leqslant s}, (x_{k,a}(\cdot))_{0\leqslant k\leqslant s}\big),$$

where $w_{k,a}(\lambda) \equiv \tilde{a}(\lambda)w_k(\lambda)$, $D_{k,a} := \{(a *_0 h)(\cdot) : h \in D_k\}$ *and* $x_{k,a}((a *_0 h)(\cdot)) := x_k(h(\cdot))$ *for* $0 \leqslant k \leqslant s$ *and* $h \in D_k$. *Then the generalized Laplace fractional derivative* $D_L^{W_a} g$ *is defined and we have* $(D_L^{W_a} g)(t) = (a *_0 (D_L^W u))(t)$ *for a. e.* $t > 0$.

7.3.1 The Laplace transform identity for the Marchaud right-sided fractional derivatives

In this subsection, we will specifically consider the Marchaud right-sided derivatives of order $\alpha \in (0,1)$. We start by recalling that the Marchaud right-sided derivative of order $\alpha \in (0,1)$, denoted shortly by $[D_-^\alpha \cdot]$, is defined for those Lebesgue measurable functions $u : [0,\infty) \to X$ for which the integral $\int_0^{+\infty} \frac{u(t)-u(t+x)}{x^{1+\alpha}}\,dx$ has no singularity at the point zero (plus infinity), through

$$[D_-^\alpha u](t) := \frac{\alpha}{\Gamma(1-\alpha)} \int_0^{+\infty} \frac{u(t) - u(t+x)}{x^{1+\alpha}}\,dx, \quad t > 0;$$

cf. also the equation [236, (13)]. The following result holds true.

Proposition 7.3.8. *Suppose that the Marchaud right-sided derivative* $[D_-^\alpha u](\cdot)$ *is well-defined. Let*

$$w(\lambda) \equiv \frac{1}{\Gamma(1-\alpha)}\left[1 + \int_0^1 \frac{1-e^{\lambda x}}{x^{1+\alpha}}\,dx\right],$$

let D_0 *be the set of those functions* $f \in \mathbb{L}$ *such that there exists* $\omega' > 0$ *with*

$$\int_0^{+\infty}\int_0^{+\infty} e^{-t\operatorname{Re}\lambda} \frac{\|f(t) - f(x+t)\|}{x^{1+\alpha}}\,dx\,dt < +\infty, \quad \operatorname{Re}\lambda > \omega' \tag{403}$$

and the integrals

$$\int_0^1 \frac{\int_0^x e^{\lambda(x-t)} f(t)\,dt}{x^{1+\alpha}}\,dx \quad \text{and} \quad \int_1^{+\infty} \frac{\int_x^{+\infty} e^{\lambda(x-t)} f(t)\,dt}{x^{1+\alpha}}\,dx \tag{404}$$

are convergent, as well as

$$x_0(\lambda, f) := \frac{\alpha}{\Gamma(1-\alpha)}\left[\int_0^1 \frac{\int_0^x e^{\lambda(x-t)} f(t)\,dt}{x^{1+\alpha}} - \int_1^{+\infty} \frac{\int_x^{+\infty} e^{\lambda(x-t)} f(t)\,dt}{x^{1+\alpha}}\,dx\right], \quad f \in D_0.$$

If $u \in D_0$, *then* $D_-^\alpha u$ *is Laplace transformable and we have*

$$\widetilde{D_-^\alpha u}(\lambda) = \omega(\lambda)\tilde{u}(\lambda) + x_0(\lambda, u).$$

Proof. Since (403) holds, we can apply the Fubini theorem and [445, Theorem 1.4.2(ii)] after that in order to see that

$$
\begin{aligned}
&\frac{\alpha}{\Gamma(1-\alpha)}\int_0^{+\infty} e^{-\lambda t}\int_0^{+\infty}\frac{u(t)-u(x+t)}{x^{1+\alpha}}\,dx\,dt\\
&=\frac{\alpha}{\Gamma(1-\alpha)}\int_0^{+\infty}\frac{(1-e^{\lambda x})\tilde{u}(\lambda)+e^{\lambda x}\int_0^x e^{-\lambda t}u(t)\,dt}{x^{1+\alpha}}\,dx\\
&=\frac{\alpha}{\Gamma(1-\alpha)}\left[\int_0^1\frac{(1-e^{\lambda x})\tilde{u}(\lambda)+e^{\lambda x}\int_0^x e^{-\lambda t}u(t)\,dt}{x^{1+\alpha}}\,dx+\int_1^{+\infty}\frac{\tilde{u}(\lambda)-e^{\lambda x}\int_x^{+\infty} e^{-\lambda t}u(t)}{x^{1+\alpha}}\,dx\right].
\end{aligned}
\tag{405}
$$

The estimate (405) simply yields

$$
\frac{\alpha}{\Gamma(1-\alpha)}\int_0^{+\infty} e^{-\lambda t}\int_0^{+\infty}\frac{u(t)-u(x+t)}{x^{1+\alpha}}\,dx\,dt=w(\lambda)\tilde{u}(\lambda)+x_0(\lambda,u),\quad u\in D_0,
$$

as claimed. □

Remark 7.3.9. It can be easily shown that the estimate $\|f(t)\| = O(t^{\sigma+\alpha})$, $t \in (0,1]$ for some $\sigma > -1$ implies the absolute convergence of the first integral in (404) on the right half-plane. This follows from the next computation:

$$
\int_0^1\left\|\frac{\int_0^x e^{\lambda(x-t)}f(t)\,dt}{x^{1+\alpha}}\right\|dx\leqslant M\int_0^1 e^{x\operatorname{Re}\lambda}\frac{x^{1+\sigma+\alpha}}{x^{1+\alpha}}\,dx\leqslant Me^{\operatorname{Re}\lambda}\int_0^1 x^{\sigma}\,dx,
$$

where $M > 0$ is a finite real constant. On the other hand, the estimate $\|f(t)\| = O(t^{\xi})$, $t > 1$ for some $\xi \in (0,\alpha)$ implies the absolute convergence of the second integral in (404) on the right half-plane. This follows from the next computation:

$$
\begin{aligned}
&\int_1^{+\infty}\left\|\frac{\int_x^{+\infty} e^{\lambda(x-t)}f(t)\,dt}{x^{1+\alpha}}\right\|dx\\
&\leqslant\int_1^{+\infty}\frac{\int_0^{+\infty} e^{-v\operatorname{Re}\lambda}\|f(v+x)\|\,dv}{x^{1+\alpha}}\,dx\leqslant M\int_1^{+\infty}\frac{\int_0^{+\infty} e^{-v\operatorname{Re}\lambda}[v^{\xi}+x^{\xi}]\,dv}{x^{1+\alpha}}\,dx,
\end{aligned}
$$

where $M > 0$ is a finite real constant.

We can similarly prove that, if $f \in \mathbb{L}$ is differentiable for $t > 0$, the function $t \mapsto \max_{s\in[t,t+1]}\|f'(s)\|$, $t > 0$ is exponentially bounded and integrable at zero as well as that $\|f(t)\| = O(t^{\xi})$, $t > 1$ for some $\xi \in (0,\alpha)$, then (403) holds.

The formula obtained in Proposition 7.3.8 still does not enable one to show that there exists a tuple W, given through (399), such that the generalized Laplace fractional derivative $D_L^W u$ is well-defined and $D_-^\alpha u = D_L^W u$. In order to consider $D_-^\alpha u$ as a generalized Laplace type fractional derivative, one needs to further generalize the notion introduced in Definition 7.3.2 by allowing that the terms $\omega_k(\lambda)x_k(u)$ can have a general form $W_k(\lambda, u)$, where $W_k : \{\lambda \in \mathbb{C} : \operatorname{Re}\lambda > \omega'\} \times D_k \to X$ are given functions ($0 \leqslant k \leqslant s$). Regrettably, it is very difficult to consider the abstract fractional inclusions in this general framework (see also Theorem 7.3.22 below).

For more details about the Marchaud right-sided fractional derivatives, we refer the reader to the research articles [300, 666] and references quoted therein. We conclude this subsection by observing that the notion of Davidson-Essex, Canavati and Jumarie fractional derivatives (see the equations [236, (15), (19), (20)], respectively) are also very special cases of the notion introduced in Definition 7.3.2 for the Laplace transformable functions.

The material of the next five subsections is taken from our recent research article [468].

7.3.2 Laplace transform of pointwise product of functions

It is clear that the product of Laplace transformable functions $u(\cdot)$ and $v(\cdot)$, where $u(t) = v(t) = g_{1/4}(t)$, $t > 0$ is not a Laplace transformable function since it is not locally integrable at zero. It can be simply proved that the pointwise product $t \mapsto u(t)v(t)$, $t > 0$ is Laplace transformable provided that $u(\cdot)$ is exponentially bounded and $\|v\|(\cdot)$ is Laplace transformable or that $|u|(\cdot)$ is Laplace transformable and $v(\cdot)$ is exponentially bounded.

In Table 14.1 given on page 385 of the monograph [142] by R. N. Bracewell, we have found the following formula:

$$[\mathcal{L}(u(t) \cdot v(t))](\lambda) = \frac{1}{2\pi i} \int_{c-i\infty}^{c+i\infty} (\mathcal{L}u)(z)(\mathcal{L}v)(\lambda - z)\, dz, \tag{406}$$

where $c > 0$ is sufficiently large. This formula can be also found at Wikipedia, and it is only stated without any argumentation and the sufficient conditions on the functions $u(\cdot)$ and $v(\cdot)$ such that (406) holds are not precisely determined. Concerning this issue, we will state and prove the following result.

Proposition 7.3.10. *Suppose that* $\varepsilon > 0$ *and* (i) *or* (ii) *holds, where:*

(i) $u(\cdot)$ *is a scalar-valued Laplace transformable function,* $v(\cdot)$ *is a Lebesgue measurable and there exist finite real constants* $M > 0$, $\omega > 0$ *and* $\sigma > 0$ *such that* $\|v(t)\| \leqslant Me^{\omega t}(1 + t^{\sigma-1})$, $t > 0$.

(ii) *There exist finite real constants* $M > 0$, $\omega > 0$ *and* $\sigma > 0$ *such that* $|u(t)| \leqslant Me^{\omega t}(1 + t^{\sigma-1})$, $t > 0$ *and* $v(\cdot)$ *is a vector-valued Laplace transformable function.*

If (i) *holds, then we have:*

(I) *The function* $t \mapsto (g_{2+\varepsilon} *_0 u)(t)v(t)$, $t > 0$ *is Laplace transformable,* (406) *holds with the function* $u(\cdot)$ *replaced therein with the function* $(g_{2+\varepsilon} *_0 u)(\cdot)$ *and the integral on the right-hand side of* (406) *is absolutely convergent when the function* $u(\cdot)$ *replaced therein with the function* $(g_{2+\varepsilon} *_0 u)(\cdot)$.

Furthermore, if (ii) *holds, then we have:*

(II) *The function* $t \mapsto u(t)(g_{2+\varepsilon} *_0 v)(t)$, $t > 0$ *is Laplace transformable,* (406) *holds with the function* $v(\cdot)$ *replaced therein with the function* $(g_{2+\varepsilon} *_0 v)(\cdot)$ *and the integral on the right-hand side of* (406) *is absolutely convergent when the function* $v(\cdot)$ *replaced therein with the function* $(g_{2+\varepsilon} *_0 v)(\cdot)$.

Proof. We will only consider the items (i) and (I). It is clear that there exist two finite real constants $M' > 0$ and $\omega' \geqslant \max(\operatorname{abs}(u), 0)$ such that $\| \int_0^t u(s)\,ds\| \leqslant M'e^{\omega' t}$, $t \geqslant 0$. Applying the partial integration, we obtain

$$\|\tilde{u}(\lambda)\| \leqslant M' \frac{|\lambda|}{\operatorname{Re}\lambda - \omega'}, \quad \operatorname{Re}\lambda > \omega',$$

which simply implies

$$\left\| \int_0^{+\infty} e^{-\lambda t}(g_{2+\varepsilon} *_0 u)(t)\,dt \right\| \leqslant \frac{M'}{|\lambda|^{1+\varepsilon} \cdot (\operatorname{Re}\lambda - \omega')}, \quad \operatorname{Re}\lambda > \omega'. \tag{407}$$

Keeping in mind the complex inversion theorem for the vector-valued Laplace transform and the uniqueness theorem for the Laplace transform, the above yields

$$(g_{2+\varepsilon} *_0 u)(t) = \frac{1}{2\pi i} \int_{c-i\infty}^{c+i\infty} e^{zt} z^{-2-\varepsilon} \tilde{u}(z)\,dz, \quad t \geqslant 0 \quad (c > \omega') \tag{408}$$

as well as that the function $(g_{2+\varepsilon} *_0 u)(\cdot)$ is exponentially bounded. The growth order of function $v(\cdot)$ implies that the function $t \mapsto (g_{2+\varepsilon} *_0 u)(t)v(t)$, $t > 0$ is Laplace transformable as well as that the Laplace transform of $v(\cdot)$ is bounded on the right half-plane $\{z : \operatorname{Re} z > \omega_1\}$, where $\omega_1 > \omega$ is a fixed number. Let $c > \omega'$ and $\operatorname{Re}\lambda - c > \omega$. Then the estimate (407) and the above mentioned fact imply that the integral on the right-hand side of (406) is absolutely convergent when the function $u(\cdot)$ is replaced therein with the function $(g_{2+\varepsilon} *_0 u)(\cdot)$. It remains to be proved that the formula (406) holds. In actual fact, we have (cf. (408)):

$$\int_0^{+\infty} e^{-\lambda t}(g_{2+\varepsilon} *_0 u)(t)v(t)\,dt = \int_0^{+\infty} e^{-\lambda t}\left[\frac{1}{2\pi i} \int_{c-i\infty}^{c+i\infty} e^{zt} z^{-2-\varepsilon} \tilde{u}(z)\,dz\right] v(t)\,dt \tag{409}$$

$$= \frac{1}{2\pi i} \int_{c-i\infty}^{c+i\infty} z^{-2-\varepsilon} \tilde{u}(z)\tilde{v}(\lambda - z)\,dz, \quad \operatorname{Re}\lambda - c > \omega, \tag{410}$$

where we have applied the Fubini theorem and rearranged the order of integration; the absolute convergence of the double integral on the right-hand side of (409) can be simply proved with the use of estimate (407) and the growth order of $v(\cdot)$. □

Remark 7.3.11. Let $c > \mathrm{abs}(u)$. In general case, the computation carried out in (409)–(410) shows that the formula (406) holds for every complex number λ such that $\operatorname{Re}\lambda - c > \mathrm{abs}(v)$, provided that the function $t \mapsto u(t)v(t)$, $t > 0$ is Laplace transformable, the double integral

$$\int_0^{+\infty} e^{-\lambda t}\left[\frac{1}{2\pi i}\int_{c-i\infty}^{c+i\infty} e^{zt}\tilde{u}(z)\,dz\right]v(t)\,dt$$

is absolutely convergent and

$$u(t) = \frac{1}{2\pi i}\int_{c-i\infty}^{c+i\infty} e^{zt}\tilde{u}(z)\,dz, \quad t \geqslant 0.$$

In particular, the function $t \mapsto u(t)v(t)$, $t > 0$ is Laplace transformable and (406) holds if there exist $M > 0$, $\omega \geqslant \max(\mathrm{abs}(u), 0)$ and $\varepsilon > 0$ such that

$$|\tilde{u}(\lambda)| \leqslant M|\lambda|^{-1-\varepsilon}, \quad \operatorname{Re}\lambda > \omega,$$

$v(\cdot)$ is Lebesgue measurable and has the same growth order as in (i). A similar statement can be formulated by interchanging the roles of functions $u(\cdot)$ and $v(\cdot)$.

We want also to mention the recent result of R. AlAhmad [38], who proved the following identities for the scalar-valued functions $u(\cdot)$ and $v(\cdot)$:

$$\begin{aligned}[\mathcal{L}(u(t)\cdot v(t))](\lambda) &= \int_{\lambda}^{+\infty} (\mathcal{L}u)(t)\cdot[\mathcal{L}^{-1}v](t-\lambda)\,dt \\ &= \int_0^{+\infty} (\mathcal{L}u)(t+\lambda)\cdot[\mathcal{L}^{-1}v](t)\,dt, \quad \lambda > 0 \text{ suff. large}, \end{aligned} \tag{411}$$

under certain reasonable assumptions. We will only mention here that the argumentation contained in the proof of [38, Theorem 2.1] shows that the formula (411) continues to hold if $v(\cdot)$ is a vector-valued function.

7.3.3 Leibniz rule for generalized Laplace fractional derivatives

The Leibniz rules for generalized fractional derivatives have been investigated by many authors so far; see [223, 232, 310, 365, 600, 629–631, 722, 737, 747, 753, 778] and references cited therein. In the next subsections, we analyze the Leibniz rules for various types of fractional derivatives. We start by examining the Leibniz rule for the generalized fractional Laplace derivatives.

Suppose that W is given through (399) and

$$W_1 := (s_1, W(\cdot), (W_k(\cdot))_{0\leqslant k\leqslant s_1}, (D_k^1)_{0\leqslant k\leqslant s_1}, (X_k(\cdot))_{0\leqslant k\leqslant s_1}).$$

Suppose also that the generalized Laplace fractional derivative $D_L^W u$ is defined for any scalar-valued function $u \in D_0 \cap \cdots \cap D_s$, (400) holds, the generalized Laplace fractional derivative $D_L^{W_1} v$ is defined for any function $v \in D_0^1 \cap \cdots \cap D_{s_1}^1$ and

$$\tilde{H}(\lambda) = W(\lambda)\tilde{v}(\lambda) - \sum_{k=0}^{s_1} W_k(\lambda)X_k(v), \quad \operatorname{Re}\lambda > \omega'. \tag{412}$$

In this subsection, we consider the problem of finding a tuple W_{uv} of the form (399) such that the generalized Laplace fractional derivative $D_L^{W_{uv}}(uv)$ is well-defined; furthermore, we want to precisely compute $D_L^{W_{uv}}(uv)$.

Our main result in this direction is given as follows.

Theorem 7.3.12. *Suppose that the generalized Laplace fractional derivative $D_L^W u$ is defined for any scalar-valued function $u \in D_0 \cap \cdots \cap D_s$, (400) holds, the generalized Laplace fractional derivative $D_L^{W_1} v$ is defined for any function $v \in D_0^1 \cap \cdots \cap D_{s_1}^1$ and (412) holds. Suppose further that the product $t \mapsto u(t)v(t)$, $t > 0$ is Laplace transformable and the following conditions hold:*

(i) *There exist a function $Z : \{\lambda \in \mathbb{C} : \operatorname{Re}\lambda > \omega'\} \to \mathbb{C}$, two Lebesgue measurable functions $a : [0,\infty) \to \mathbb{C}$ and $b : [0,\infty) \to \mathbb{C}$ such that $abs(|a|) < +\infty$, $abs(|b|) < +\infty$ and a Laplace transformable function $Y : [0,\infty) \to X$ such that $\tilde{a}(\lambda) = 1/\omega(\lambda)$ and $\tilde{b}(\lambda) = 1/W(\lambda)$ for $\operatorname{Re}\lambda > \omega'$, the product $t \mapsto (a *_0 h)(t) \cdot (b *_0 H)(t)$, $t > 0$ is Laplace transformable and*

$$\tilde{Y}(\lambda) = Z(\lambda)\mathcal{L}((a *_0 h)(t) \cdot (b *_0 H)(t))(\lambda), \quad \operatorname{Re}\lambda > \omega'. \tag{413}$$

(ii) *There exist Laplace transformable functions $W_{k,a}(\cdot)$ such that $\widetilde{W_{k,a}}(\lambda) = \tilde{a}(\lambda)w_k(\lambda)$, $\operatorname{Re}\lambda > \omega'$ suff. large $(0 \leqslant k \leqslant s)$.*

(iii) *There exist Laplace transformable functions $W_{k,b}(\cdot)$ such that $\widetilde{W_{k,b}}(\lambda) = \tilde{b}(\lambda)W_k(\lambda)$, $\operatorname{Re}\lambda > \omega'$ suff. large $(0 \leqslant k \leqslant s_1)$.*

(iv) *The products $t \mapsto (a *_0 h)(t) \cdot W_{k,b}(t)$, $t > 0$, $t \mapsto W_{k,a}(t)(b *_0 H)(t)$, $t > 0$, $t \mapsto (a *_0 h)(t) \cdot W_{k,b}(t)$, $t > 0$ and $t \mapsto W_{j,a}(t) \cdot W_{j_1,b}(t)$, $t > 0$ $(0 \leqslant j \leqslant s$, $0 \leqslant j_1 \leqslant s_1)$ are Laplace transformable.*

Define $s_{uv} := 1 + s + s_1 + (1+s)(1+s_1)$, $Z_{uv}(\cdot) := Z(\cdot)$ as well as $Z_k(\cdot)$, D_k^{uv}, P_k $(0 \leqslant k \leqslant s_{uv})$ and W_{uv} in the following way:

(a) *If* $0 \leqslant k \leqslant s_1$, *then we set* $D_k^{uv} := \{q_1(\cdot)q_2(\cdot) : q_1 \in D_0 \cap \cdots \cap D_s,\ q_2 \in D_k^1\}$,

$$Z_k(\lambda) := Z(\lambda) \cdot \mathcal{L}((a *_0 h)(t) \cdot W_{k,b}(t))(\lambda), \quad \operatorname{Re}\lambda > \omega'$$

and $P_k(q_1q_2) := X_k(q_2)$ *for any* $q_1 \in D_0 \cap \cdots \cap D_s$ *and* $q_2 \in D_k^1$.

(b) *If* $s_1 + 1 \leqslant s_1 + s + 1$, *then we set* $D_k^{uv} := \{q_1(\cdot)q_2(\cdot) : q_1 \in D_{k-s_1-1},\ q_2 \in D_0^1 \cap \cdots \cap D_{s_1}^1\}$,

$$Z_k(\lambda) := Z(\lambda) \cdot \mathcal{L}(W_{k,a}(t) \cdot (b *_0 H)(t))(\lambda), \quad \operatorname{Re}\lambda > \omega'$$

and $P_k(q_1q_2) := x_{k-s_1-1}(q_1)$ *for any* $q_1 \in D_{k-s_1-1}$ *and* $q_2 \in D_0^1 \cap \cdots \cap D_{s_1}^1$.

(c) *If* $s_1+s+2 \leqslant k \leqslant 2s_1+s+2$, *then we set* $D_k^{uv} := \{q_1(\cdot)q_2(\cdot) : q_1 \in D_{k-s_1-1},\ q_2 \in D_0^1 \cap \cdots \cap D_{s_1}^1\}$,

$$Z_k(\lambda) := Z(\lambda) \cdot \mathcal{L}(W_{k,a}(t) \cdot (b *_0 H)(t))(\lambda), \quad \operatorname{Re}\lambda > \omega'$$

and $P_k(q_1q_2) := x_{k-s_1-1}(q_1)$ *for any* $q_1 \in D_{k-s_1-1}$ *and* $q_2 \in D_0^1 \cap \cdots \cap D_{s_1}^1$.

(d) *If* $2s_1 + s + 3 \leqslant k \leqslant 3s_1 + s + 3$, *then we set* $D_k^{uv} := \{q_1(\cdot)q_2(\cdot) : q_1 \in D_0,\ q_2 \in D_{k-(2s_1+s+3)}^1\}$,

$$Z_k(\lambda) := Z(\lambda) \cdot \mathcal{L}(W_{k,0}(t) \cdot W_{k,b}(t))(\lambda), \quad \operatorname{Re}\lambda > \omega'$$

and $P_k(q_1q_2) := x_0(q_1)X_{k-(2s_1+s+3)}(q_2)$ *for any* $q_1 \in D_0$ *and* $q_2 \in D_{k-(s_{uv}-s_1)}^1, \ldots,$ *and*

(e) *If* $s_{uv} - s_1 \leqslant k \leqslant s_{uv}$, *then we set* $D_k^{uv} := \{q_1(\cdot)q_2(\cdot) : q_1 \in D_s,\ q_2 \in D_{k-(s_{uv}-s_1)}^1\}, \ldots,$

$$Z_k(\lambda) := Z(\lambda) \cdot \mathcal{L}(W_{s,a}(t) \cdot W_{k,b}(t))(\lambda), \quad \operatorname{Re}\lambda > \omega'$$

and $P_k(q_1q_2) := x_s(u)x_{k-s_1-s-2}(q_2)$ *for any* $q_1 \in D_s$ *and* $q_2 \in D_{k-(s_{uv}-s_1)}^1$.

In the above enumeration between the parts (d) *and* (e), *we first collect all remaining terms containing* $x_0(q_1)$, *after that we collect all terms containing* $x_1(q_1)$, *and so on. Then the generalized Laplace fractional derivative* $D_L^{W_{uv}}(uv)$ *is defined and we have* $D_L^{W_{uv}}(uv) = Y(t)$ *for a. e.* $t > 0$.

Proof. It is clear that the prescribed assumptions imply that, for every $\lambda \in \mathbb{C}$ with $\operatorname{Re}\lambda > \omega'$, we have

$$\int_0^{+\infty} e^{-\lambda t} u(t)v(t)\,dt = \int_0^{+\infty} e^{-\lambda t}\Big[\mathcal{L}^{-1}\Big(\frac{\tilde{h}(\lambda) + \sum_{k=0}^{s} \omega_k(\lambda)x_k(u)}{\omega(\lambda)}\Big)\Big](t) \cdot \Big[\mathcal{L}^{-1}\Big(\frac{\tilde{H}(\lambda) + \sum_{k=0}^{s_1} W_k(\lambda)X_k(v)}{W(\lambda)}\Big)\Big](t)\,dt$$

$$= \int_0^{+\infty} e^{-\lambda t}\left[\mathcal{L}^{-1}\left(\tilde{a}(\lambda)\left[\tilde{h}(\lambda) + \sum_{k=0}^{s} \omega_k(\lambda)x_k(u)\right]\right)\right](t)$$

$$\cdot \left[\mathcal{L}^{-1}\left(\tilde{b}(\lambda)\left[\tilde{H}(\lambda) + \sum_{k=0}^{s_1} W_k(\lambda)X_k(v)\right]\right)\right](t)\,dt$$

$$= \int_0^{+\infty} e^{-\lambda t}\left[\mathcal{L}^{-1}\left(\widetilde{(a *_0 h)}(\lambda) + \sum_{k=0}^{s} \widetilde{W_{k,a}}(\lambda)x_k(u)\right)\right](t)$$

$$\cdot \left[\mathcal{L}^{-1}\left(\widetilde{(b *_0 H)}(\lambda) + \sum_{k=0}^{s_1} \widetilde{W_{k,b}}(\lambda)X_k(v)\right)\right](t)\,dt$$

$$= \int_0^{+\infty} e^{-\lambda t}\left[(a *_0 h)(t) + \sum_{k=0}^{s} W_{k,a}(t)x_k(u)\right] \cdot \left[(b *_0 H)(t) + \sum_{k=0}^{s_1} W_{k,b}(t)X_k(v)\right] dt.$$

Taking into account the equation (413), we get

$$Z(\lambda)\int_0^{+\infty} e^{-\lambda t}u(t)v(t)\,dt = \tilde{Y}(\lambda) + Z(\lambda)\int_0^{+\infty} e^{-\lambda t}(a *_0 h)(t)\cdot(b *_0 H)(t)\,dt$$
$$+ Z(\lambda)\sum_{k=0}^{s_1}\int_0^{+\infty} e^{-\lambda t}(a *_0 h)(t)\cdot W_{k,b}(t)\,dt$$
$$+ Z(\lambda)\sum_{k=0}^{s}\int_0^{+\infty} e^{-\lambda t}W_{k,a}(t)\cdot(b *_0 H)(t)\,dt$$
$$+ Z(\lambda)\int_0^{+\infty} e^{-\lambda t}\sum_{k=0}^{s}\sum_{j=0}^{s_1} W_{k,a}(t)W_{j,b}(t)x_k(u)X_k(v)\,dt.$$

Keeping in mind our definitions given in parts (a)–(e), the result simply follows. □

A simple analysis shows that the usual Leibniz rule for n-times differentiable, Laplace transformable functions can be viewed, under certain logical assumptions, as a special case of the Laplace transform identity:

$$\widetilde{(uv)^{(n)}}(\lambda) = \lambda^{-n}\mathcal{L}((g_n *_0 f^{(n)})(t)\cdot(g_n *_0 g^{(n)})(t))(\lambda), \quad \operatorname{Re}\lambda > \omega \text{ suff. large}; \tag{414}$$

see also Definition 7.3.2 and Theorem 7.3.12 with $w(\lambda) = \lambda^{\alpha}$ and $s = m-1$. It is not so simply to consider the fractional analogues of (414) using the vector-valued Laplace transform. In the next subsections, we will be dealing with the functions, which are not Laplace transformable, in general; our plan is to investigate the Leibniz rules for some special kinds of fractional derivatives by means of the usual Leibniz rule, which seems to be much more powerful than the use of vector-valued Laplace transform.

7.3.4 Leibniz rule for Caputo fractional derivatives

Suppose that $\alpha > 0$ and $m = \lceil\alpha\rceil$. Let us recall that the Caputo fractional derivative $\mathbf{D}_t^\alpha u(t)$ is usually defined for any function $u \in C^{m-1}([0,\infty) : X)$ for which $g_{m-\alpha} *_0 (u - \sum_{k=0}^{m-1} u_k g_{k+1}) \in C^m([0,\infty) : X)$, by (1.6). Recall, if the Caputo fractional derivative $\mathbf{D}_t^\alpha u(t)$ is well-defined and $0 < \gamma < \alpha$, then we know that the Caputo fractional derivative $\mathbf{D}_t^\gamma u(t)$ is also well-defined [445].

The main result of this subsection reads as follows.

Theorem 7.3.13. *Suppose that $u : [0,\infty) \to \mathbb{C}$, $v : [0,\infty) \to X$, $\alpha > 0$ and $m = \lceil\alpha\rceil$. If the Caputo fractional derivatives $\mathbf{D}_t^\alpha u(\cdot)$, $\mathbf{D}_t^\alpha v(\cdot)$ and $\mathbf{D}_t^\alpha [uv](\cdot)$ are well-defined, then we have*

$$\sum_{j=0}^{m-1} \binom{m-1}{j} [u^{(m-1-j)}(0) + (g_{\alpha+1-m} *_0 \mathbf{D}_t^{\alpha-j} u)(t)][v^{(j)}(0) + (g_{\alpha+1-m} *_0 \mathbf{D}_t^{\alpha-m+j+1} v)(t)]$$
$$= (uv)^{(m-1)}(0) + (g_{\alpha+1-m} *_0 \mathbf{D}_t^\alpha [uv])(t), \quad t \geqslant 0. \tag{415}$$

Proof. A simple calculation with the help of partial integration shows that

$$\mathbf{D}_t^\alpha [uv](\cdot) = \frac{d}{dt} [g_{m-\alpha} *_0 ([uv]^{(m-1)}(\cdot) - [uv]^{(m-1)}(0))](t), \quad t \geqslant 0.$$

After the convoluting this equation with $g_{\alpha+1-m}(\cdot)$, we get

$$[uv]^{(m-1)}(t) - [uv]^{(m-1)}(0) = (g_{\alpha+1-m} *_0 \mathbf{D}_t^\alpha [uv](\cdot))(t), \quad t \geqslant 0. \tag{416}$$

By the Leibniz rule, the above equality yields

$$\sum_{j=0}^{m-1} \binom{m-1}{j} u^{(m-1-j)}(t) v^{(j)}(t) - [uv]^{(m-1)}(0) = (g_{\alpha+1-m} *_0 \mathbf{D}_t^\alpha [uv](\cdot))(t), \quad t \geqslant 0. \tag{417}$$

If $m-1 < \alpha \leqslant m$, then we have $m-1-j < \alpha - j \leqslant m-j$ for $0 \leqslant j \leqslant m-1$. Keeping this in mind, we can apply (416), with $v(\cdot) \equiv 1$ and $u(\cdot) \equiv 1$, respectively, in order to see that

$$u^{(m-1-j)}(t) = u^{(m-1-j)}(0) + (g_{\alpha+1-m} *_0 \mathbf{D}_t^{\alpha-j} u)(t), \quad t \geqslant 0$$

and

$$v^{(j)}(t) = v^{(j)}(0) + (g_{\alpha+1-m} *_0 \mathbf{D}_t^{\alpha-m+j+1} v)(t), \quad t \geqslant 0.$$

Inserting this formulae in (417), we simply get the required equality (415). □

Remark 7.3.14. If $\alpha = m \in \mathbb{N}$, then the existence of derivatives $\mathbf{D}_t^\alpha u(\cdot) = u^{(m)}(\cdot)$ and $\mathbf{D}_t^\alpha v(\cdot) = v^{(m)}(\cdot)$ implies that the derivative $\mathbf{D}_t^\alpha [uv](\cdot) = [uv]^{(m)}(\cdot)$ is well-defined as is well known. This is no longer true in the pure fractional case: Suppose that $\alpha \in (0,+\infty) \smallsetminus \mathbb{N}$,

$\mathbf{D}_t^{\alpha}u(\cdot)$ and $\mathbf{D}_t^{\alpha}v(\cdot)$ are well-defined, the functions $u(\cdot)$ and $v(\cdot)$ are m-times continuously differentiable for $t \geqslant 0$, $[uv]^{(m)}(0) = 0$ and $[uv]^{(m-1)}(0) \neq 0$. Then the Caputo fractional derivative $\mathbf{D}_t^{\alpha}[uv](\cdot)$ is not well-defined in the sense of (1.6); in actual fact, if we suppose the contrary, then we would have (see also the proof of [69, Proposition 1.3.6]):

$$\begin{aligned}\mathbf{D}_t^{\alpha}[uv](\cdot) &= \frac{d}{dt}[g_{m-\alpha} *_0 ([uv]^{(m-1)}(\cdot) - [uv]^{(m-1)}(0))](t)\\ &= (g_{m-\alpha} *_0 [uv]^{(m)})(t) - g_{m-\alpha}(t) \cdot [uv]^{(m-1)}(0), \quad t > 0.\end{aligned}$$

This is a contradiction, since the first function on the right-hand side of the last equality is continuous for $t \geqslant 0$, while the second function on the right-hand side of the last equality is not bounded as $t \to 0+$.

7.3.5 Leibniz type rules for Hilfer fractional derivatives and Riemann–Liouville fractional derivatives

Suppose that $\alpha > 0$ and $\beta \in [0,1)$. Let us recall that the Hilfer fractional derivative $D_t^{\alpha,\beta}u(t)$ of order $\alpha > 0$ and type $\beta \in [0,1)$ is defined for any locally integrable function $u : [0,\infty) \to X$ for which the function $v(t) := J_t^{(1-\beta)(m-\alpha)}u(t)$, $t \geqslant 0$ is $(m-1)$-continuously differentiable and the function $v^{(m-1)}(\cdot)$ is locally absolutely continuous on $[0,\infty)$ through (378).

We continue our exposition with the following observation.

Remark 7.3.15. If $\alpha > 0$, then the notion of Caputo fractional derivative $\mathbf{D}_t^{\alpha}u(t)$ considered in Section 7.3.4 is not completely equivalent with the notion of Hilfer fractional derivative $D_t^{\alpha,1}u(t)$. For example, if the Hilfer fractional derivatives $D_t^{\alpha,1}u(t)$ and $D_t^{\alpha,1}v(t)$ are well-defined, then we can use the Leibniz rule and the fact that the pointwise product of a scalar-valued locally absolutely continuous function on $[0,\infty)$ and a vector-valued locally absolutely continuous function on $[0,\infty)$ is again a locally absolutely continuous function on $[0,\infty)$ in order to see that the Hilfer fractional derivative $D_t^{\alpha,1}[uv](t)$ is also well-defined. As we have already clarified in Remark 7.3.14, this is not true for the Caputo fractional derivatives $\mathbf{D}_t^{\alpha}u(t)$, $\mathbf{D}_t^{\alpha}v(t)$ and $\mathbf{D}_t^{\alpha}[uv](t)$; see also [97, pp. 11–12] for more details. The interested reader may try to reconsider the statement of Theorem 7.3.13 for the Hilfer fractional derivatives of type $\beta = 1$.

In contrast to the Caputo fractional derivatives, if the Hilfer fractional derivative $D_t^{\alpha,\beta}u(t)$ is well-defined, then a relatively simple argumentation shows that the Hilfer fractional derivative $D_t^{\gamma,\beta}u(t)$ need not be defined for all $\gamma \in (0,\alpha)$. But, if $k \in \mathbb{N}$ and $\alpha - k > 0$, then the Hilfer fractional derivative $D_t^{\alpha-k,\beta}u(t)$ is well-defined and we have

$$D_t^{\alpha-k,\beta}u(t) = [g_{\beta(m-\alpha)} *_0 v^{(m-k)}](t)$$

$$= \left[g_{\beta(m-\alpha)} *_0 \left\{ [g_k *_0 v^{(m)}](\cdot) + \sum_{j=1}^{k} g_j(\cdot) v^{(m-k+j-1)}(0) \right\} \right](t)$$

$$= (g_k *_0 D_t^{\alpha,\beta} u)(t) + \sum_{j=1}^{k} g_{\beta(m-\alpha)+j}(t) v^{(m-k+j-1)}(0), \quad t > 0. \tag{418}$$

If the Hilfer fractional derivative $D_t^{\alpha,\beta} u(t)$ is well-defined, then we can convolute the both sides of (378) with $g_{1-\beta(m-\alpha)}(\cdot)$ and use the Newton–Leibniz formula after that in order to see that

$$(g_{(1-\beta)(m-\alpha)} *_0 u)^{(m-1)}(t) - (g_{(1-\beta)(m-\alpha)} *_0 u)^{(m-1)}(0) = (g_{1-\beta(m-\alpha)} *_0 D_t^{\alpha,\beta} u)(t), \tag{419}$$

for any $t \geqslant 0$. Set $h(t) := (g_{(1-\beta)(m-\alpha)} *_0 u)(t)$, $t \geqslant 0$. Due to (419), we have

$$h^{(m-1)}(t) - h^{(m-1)}(0) = (g_{1-\beta(m-\alpha)} *_0 D_t^{\alpha,\beta} u)(t), \quad t \geqslant 0,$$

i. e.,

$$h^{(m-1)}(t) - h^{(m-1)}(0) = \left(g_{1-\beta(m-\alpha)} *_0 D_t^{\alpha,\beta} \left[\frac{d}{dt} (g_{1-(1-\beta)(m-\alpha)} *_0 h) \right] \right)(t), \quad t \geqslant 0. \tag{420}$$

If the functions $h_1 : [0,\infty) \to \mathbb{C}$ and $h_2 : [0,\infty) \to X$ are $(m-1)$-times continuously differentiable and $h(t) = h_1(t)h_2(t)$ for all $t \geqslant 0$, then the last formula implies

$$\sum_{j=0}^{m-1} \binom{m-1}{j} h_1^{(m-1-j)}(t) h_2^{(j)}(t) - [h_1 h_2]^{(m-1)}(0)$$
$$= \left(g_{1-\beta(m-\alpha)} *_0 D_t^{\alpha,\beta} \left[\frac{d}{dt} (g_{1-(1-\beta)(m-\alpha)} *_0 [h_1 h_2]) \right] \right)(t), \quad t \geqslant 0. \tag{421}$$

Suppose now that the Hilfer fractional derivatives

$$D_t^{\alpha,\beta} \left[\frac{d}{dt} (g_{1-(1-\beta)(m-\alpha)} *_0 u) \right](\cdot)$$

and

$$D_t^{\alpha,\beta} \left[\frac{d}{dt} (g_{1-(1-\beta)(m-\alpha)} *_0 v) \right](\cdot)$$

are well-defined. Then the Hilfer fractional derivatives

$$D_t^{\alpha-j,\beta} \left[\frac{d}{dt} (g_{1-(1-\beta)(m-\alpha)} *_0 u) \right](\cdot)$$

and

$$D_t^{\alpha-(m-j)+1,\beta}\left[\frac{d}{dt}(g_{1-(1-\beta)(m-\alpha)} *_0 v)\right](\cdot)$$

are well-defined for any $j \in \mathbb{N}_{m-1}^0$; see (418). Keeping in mind this fact and (420)–(421), we can simply prove the following analogue of Theorem 7.3.13 for the Hilfer fractional derivatives.

Theorem 7.3.16. *Suppose that $\alpha > 0$, $\beta \in [0,1)$, $m = \lceil \alpha \rceil$ as well as the functions $u : [0,\infty) \to \mathbb{C}$ and $v : [0,\infty) \to X$ are $(m-1)$-times continuously differentiable. If the Hilfer fractional derivatives*

$$D_t^{\alpha,\beta}\left[\frac{d}{dt}(g_{1-(1-\beta)(m-\alpha)} *_0 u)\right](\cdot),$$

$$D_t^{\alpha,\beta}\left[\frac{d}{dt}(g_{1-(1-\beta)(m-\alpha)} *_0 v)\right](\cdot)$$

and

$$D_t^{\alpha,\beta}\left[\frac{d}{dt}(g_{1-(1-\beta)(m-\alpha)} *_0 [uv])\right](\cdot)$$

are well-defined, then we have

$$\sum_{j=0}^{m-1}\binom{m-1}{j}\left[u^{(m-1-j)}(0) + \left(g_{1-\beta(m-1-\alpha)} *_0 D_t^{\alpha-j,\beta}\left[\frac{d}{dt}(g_{1-(1-\beta)(m-\alpha)} *_0 u)\right]\right)(t)\right]$$
$$\times\left[v^{(j)}(0) + \left(g_{1-\beta(m-\alpha)} *_0 D_t^{\alpha-(m-j)+1,\beta}\left[\frac{d}{dt}(g_{1-(1-\beta)(m-\alpha)} *_0 v)\right]\right)(t)\right]$$
$$= \left(g_{1-\beta(m-\alpha)} *_0 D_t^{\alpha,\beta}\left[\frac{d}{dt}(g_{1-(1-\beta)(m-\alpha)} *_0 [uv])\right]\right)(t), \quad t \geqslant 0.$$

The corresponding formula for the Riemann–Liouville fractional derivatives can be obtained by plugging $\beta = 0$ in Theorem 7.3.16.

7.3.6 Leibniz type rules for generalized Hilfer (a, b, α)-fractional derivatives

If $\delta(t)$ denotes the Dirac delta distribution, then we accept the formal convention $\int_0^t \delta(t-s)f(s)\,ds \equiv f(t)$. Suppose now that $u : [0,\infty) \to X$ is locally integrable, $\alpha > 0$, $m = \lceil \alpha \rceil$, $a \in L^1_{loc}([0,\infty))$ or $a(t) = \delta(t)$, and $b \in L^1_{loc}([0,\infty))$ or $b(t) = \delta(t)$. Set

$$v_a(t) := \int_0^t a(t-s)u(s)\,ds, \quad t \geqslant 0.$$

Now we would like to introduce the following extension of the Hilfer fractional derivative $D_t^{\alpha,\beta}u(t)$; cf. also [12, Section 3].

Definition 7.3.17. The generalized Hilfer (a, b, α)-fractional derivative of function $u(\cdot)$, denoted shortly by $D^{\alpha}_{a,b}u$, is defined for any locally integrable function $u(\cdot)$ such that the function $v_a^{(m-1)}(t)$ is locally absolutely continuous for $t \geqslant 0$, by

$$D^{\alpha}_{a,b}u(t) := (b *_0 v_a^{(m)})(t), \quad \text{a. e. } t \geqslant 0. \tag{422}$$

In the case of consideration of the Hilfer fractional derivative $D_t^{\alpha,\beta}u(t)$, we have $a *_0 b = g_{m-\alpha}$. In Definition 7.3.17, we have not imposed any similar condition and maybe the generalized Hilfer (a, b, α)-fractional derivative should be called the generalized Hilfer (a, b, α)-integrodifferential operator (cf. also the research articles [553, 554], where Y. Luchko has analyzed the importance of the general Sonin conditions in fractional calculus; it is clear that we cannot expect the validity of Sonin conditions and many other properties of the usually considered Hilfer fractional derivatives in the general framework of Definition 7.3.17). It is also worth noting that the notion introduced in Definition 7.3.17 generalizes the notion of the integrodifferential Gerasimov-type operator $D^{K,m}u$, defined by $(D^{K,m}u)(t) := (K *_0 u^{(m)})(t)$ for a. e. $t \geqslant 0$; the integrodifferential equations with Gerasimov type operators have recently been analyzed by V. E. Fedorov and A. D. Godova in [291]. Let us also observe that the notion introduced in Definition 7.3.17 can be further extended following the approach used for the introduction of Ψ-Hilfer fractional derivatives; see the references quoted in [293] for more details about this subject (see also the research monographs [617], [653] and [754] for some contributions of Russian mathematicians in the field of fractional calculus and fractional differential equations).

The notion introduced there can be further extended following the approach used for the introduction of Ψ-Caputo fractional derivatives and Ψ-Hilfer fractional derivatives; see the references quoted in [41] and [293] for more details about this subject. Furthermore, the notion introduced in Definition 7.3.17 can be further extended following the approach used for the introduction of proportional Caputo fractional derivatives introduced in [661, Definition 4]; we will not consider such an extension of generalized Hilfer (a, b, α)-fractional derivatives in this book.

If $k \in \mathbb{N}$ and $\alpha - k > 0$, then the generalized Hilfer fractional derivative $D^{\alpha-k}_{a,b}u(t)$ is well-defined and the argumentation used for proving (418) also shows that we have

$$D^{\alpha-k}_{a,b}u(t) = (g_k *_0 D^{\alpha}_{a,b}u)(t) + \sum_{j=1}^{k}(b *_0 g_j)(t)v_a^{(m-k+j-1)}(0), \quad t > 0. \tag{423}$$

Now we proceed as in the previous subsection. Our basic assumptions will be that there exist two integers $p \in \mathbb{N}$ and $s \in \mathbb{N}$ as well as two locally integrable functions $c(\cdot)$ and $d(\cdot)$ such that

$$b *_0 c = g_p \quad \text{and} \quad a *_0 d = g_s. \tag{424}$$

Suppose now that the generalized Hilfer fractional derivative $D^{\alpha}_{a,b}u(t)$ is well-defined. Convoluting the both sides of (422) with $c(\cdot)$ and using the equalities given in (424), we get

$$(g_p *_0 v_a^{(m)})(t) = \left(c *_0 D^{\alpha}_{a,b}\left[\frac{d^s}{d\cdot^s}(d *_0 v_a)\right]\right)(t), \quad t \geqslant 0.$$

Then the partial integration implies

$$v_a^{(m-p)}(t) - g_p(t)v_a^{(m-1)}(0) - \cdots - g_1(t)v_a^{(m-p)}(0) = \left(c *_0 D^{\alpha}_{a,b}\left[\frac{d^s}{d\cdot^s}(d *_0 v_a)\right]\right)(t),$$

for any $t \geqslant 0$, where we have put (we will accept the same convention henceforth):

$$v_a^{(j)}(t) := (g_{-j} *_0 v_a)(t), \quad t \geqslant 0 \quad \text{and} \quad v_a^{(j)}(0) := 0, \quad \text{if } j \in -\mathbb{N}.$$

If the functions $h_1 : [0,\infty) \to \mathbb{C}$ and $h_2 : [0,\infty) \to X$ are $(m-1)$-times continuously differentiable and $v_a(t) = h_1(t)h_2(t)$ for all $t \geqslant 0$, then the last formula implies

$$\begin{aligned}&[h_1h_2]^{(m-p)}(t) - g_p(t)[h_1h_2]^{(m-1)}(0) - \cdots - g_1(t)[h_1h_2]^{(m-p)}(0)\\ &\quad = \left(c *_0 D^{\alpha}_{a,b}\left[\frac{d^s}{d\cdot^s}(d *_0 [h_1h_2])\right]\right)(t), \quad t \geqslant 0. \end{aligned} \tag{425}$$

Keeping in mind (423) and (425), we can simply deduce the following proper extension of Theorem 7.3.16.

Theorem 7.3.18. *Suppose that $\alpha > 0$, $m = \lceil \alpha \rceil$, there exist two integers $p \in \mathbb{N}$ and $s \in \mathbb{N}$ as well as two locally integrable functions $c(\cdot)$ and $d(\cdot)$ such that (424) holds. Suppose further that the functions $u : [0,\infty) \to \mathbb{C}$ and $v : [0,\infty) \to X$ are $(m-1)$-times continuously differentiable. If the generalized Hilfer fractional derivatives*

$$D^{\alpha}_{a,b}\left[\frac{d^s}{d\cdot^s}(d *_0 u)\right](\cdot), \quad D^{\alpha}_{a,b}\left[\frac{d^s}{d\cdot^s}(d *_0 v)\right](\cdot) \quad \text{and} \quad D^{\alpha}_{a,b}\left[\frac{d^s}{d\cdot^s}(d *_0 [uv])\right](\cdot),$$

are well-defined, then we have

$$\begin{aligned}&\sum_{j=0}^{m-p}\binom{m-p}{j}\left[-\sum_{l=0}^{p-1} g_{p-j}(t)u^{(m-1-j-l)}(0) + \left(c *_0 D^{\alpha-j}_{a,b}\left[\frac{d^s}{d\cdot^s}(d *_0 u)\right]\right)(t)\right]\\ &\quad \times \left[v^{(j)}(0) + \frac{d^{p-1}}{dt^{p-1}}\left(c *_0 D^{\alpha-(m-j)+1}_{a,b}\left[\frac{d^s}{d\cdot^s}(d *_0 v)\right]\right)(t)\right]\\ &= \sum_{j=0}^{p-1} g_{p-j}(t)[uv]^{(m-1-j)}(0) + \left(c *_0 D^{\alpha}_{a,b}\left[\frac{d^s}{d\cdot^s}(d *_0 [uv])\right]\right)(t), \quad t \geqslant 0,\end{aligned} \tag{426}$$

provided that the second addend in (426) is well-defined (this holds provided that $p = 1$).

The Prabhakar fractional derivative of Riemann–Liouville type ${}^{RL}\mathcal{D}^{\gamma}_{\beta,\alpha,\eta;0}$, given by the formula [321, (5.9)], is a special case of the notion introduced in Definition 7.3.17 with $a(t) = t^{m-\alpha-1}E^{-\gamma}_{\beta,m-\alpha}(\eta t^{\beta})$ and $b(t) = \delta(t)$ as well as that the Prabhakar fractional derivative of Caputo type ${}^{C}\mathcal{D}^{\gamma}_{\beta,\alpha,\eta;0}$, given by the formula [321, (5.10)], is a special case of the notion introduced in Definition 7.3.17 with $a(t) = \delta(t)$ and $b(t) = t^{m-\alpha-1}E^{-\gamma}_{\beta,m-\alpha}(\eta t^{\beta})$; here, we have interchanged the role of parameters α and β since, in [321], we have $m = \lceil\beta\rceil$. In the case of the Prabhakar fractional derivative of Riemann–Liouville type, we have $c = g_p(t)$ for any $p \in \mathbb{N}$ and

$$d(t) = \mathcal{L}^{-1}\left(\frac{\lambda^{\gamma\beta-s+m-\alpha}}{(\lambda^{\beta}-\eta)^{\gamma}}\right)(t), \quad t \geqslant 0 \tag{427}$$

for any $s \in \mathbb{N}$; similarly, in the case of the Prabhakar fractional derivative of Caputo type, we can choose $d = g_s(t)$ for any $s \in \mathbb{N}$ and $c(t)$ to be equal with $d(t)$ in (427), with the number s replaced therein with any number $p \in \mathbb{N}$; see, e. g., (398). Therefore, Theorem 7.3.18 provides a Leibniz rule for the Prabhakar fractional derivatives; for example, we have the following result for the Prabhakar fractional derivatives of Riemann–Liouville type.

Theorem 7.3.19. *Suppose that $\alpha > 0$, $m = \lceil\alpha\rceil$, the complex numbers β, γ and η are given, $a(t) = t^{m-\alpha-1}E^{-\gamma}_{\beta,m-\alpha}(\eta t^{\beta})$, $b(t) = \delta(t)$, $c(t) = 1$, $d(\cdot)$ is given by (427), $s \in \mathbb{N}$ is arbitrary as well as that the functions $u : [0,\infty) \to \mathbb{C}$ and $v : [0,\infty) \to X$ are $(m-1)$-times continuously differentiable. If the Prabhakar fractional derivatives of Riemann–Liouville type*

$${}^{RL}\mathcal{D}^{\gamma}_{\beta,\alpha,\eta;0}\left[\frac{d^s}{d\cdot^s}(d *_0 u)\right](\cdot), \quad {}^{RL}\mathcal{D}^{\gamma}_{\beta,\alpha,\eta;0}\left[\frac{d^s}{d\cdot^s}(d *_0 v)\right](\cdot)$$

and

$${}^{RL}\mathcal{D}^{\gamma}_{\beta,\alpha,\eta;0}\left[\frac{d^s}{d\cdot^s}(d *_0 [uv])\right](\cdot),$$

are well-defined, then we have

$$\begin{aligned}
&\sum_{j=0}^{m-p}\binom{m-p}{j}\left[-\sum_{l=0}^{p-1} g_{p-j}(t)u^{(m-1-j-l)}(0) + \left(c *_0^{RL} \mathcal{D}^{\gamma}_{\beta,\alpha-j,\eta;0}\left[\frac{d^s}{d\cdot^s}(d *_0 u)\right]\right)(t)\right] \\
&\quad\times\left[v^{(j)}(0) + \frac{d^{p-1}}{dt^{p-1}}\left(c *_0^{RL} \mathcal{D}^{\gamma}_{\beta,\alpha-(m-j)+1,\eta;0}\left[\frac{d^s}{d\cdot^s}(d *_0 v)\right]\right)(t)\right] \\
&= \sum_{j=0}^{p-1} g_{p-j}(t)[uv]^{(m-1-j)}(0) + \left(c *_0^{RL} \mathcal{D}^{\gamma}_{\beta,\alpha,\eta;0}\left[\frac{d^s}{d\cdot^s}(d *_0 [uv])\right]\right)(t), \quad t \geqslant 0.
\end{aligned}$$

We can similarly formulate an analogue of Theorem 7.3.19 for the Prabhakar fractional derivatives of Caputo type. For more details about fractional differential equations with Prabhakar derivatives, we refer the reader to the recent research article [740] by V. E. Tarasov and the list of references cited therein.

7.3.7 Abstract fractional Cauchy inclusions with generalized Laplace derivatives

We start this subsection by introducing the following notion.

Definition 7.3.20. Suppose that $\mathcal{A}$ is an MLO and $f : [0,\infty) \to X$. Then we say that a locally integrable function $u : [0,\infty) \to X$ is a solution of the abstract fractional Cauchy inclusion

$$(D_L^W u)(t) \in \mathcal{A}u(t) + f(t) \quad \text{for a. e. } t > 0 \tag{428}$$

if the term $D_L^W u$ is well-defined and (428) holds.

In our recent paper [293], we have recently analyzed the abstract fractional Cauchy inclusions with Hilfer derivatives. We will first prove the following simple result concerning the uniqueness of solutions of the abstract fractional Cauchy inclusion (37); cf. also [293, Theorems 2.2, 2.3].

Proposition 7.3.21. *Suppose that $\mathcal{A}$ is a closed* MLO, *$D_0, \ldots, D_s$ are vector spaces with the usual operations and the mappings $x_k : D_k \to X$ are linear ($0 \leqslant k \leqslant s$). If there exists $\omega' > \omega$ such that the operator $w(\lambda) - \mathcal{A}$ is injective for all $\lambda \in \mathbb{C}$ with* $\operatorname{Re}\lambda > \omega'$, *then there exists at most one solution of the abstract fractional Cauchy inclusion* (428) *with the given initial values $x_j(u) := x_j$ for all $u \in D_j$ ($0 \leqslant j \leqslant s$).*

Proof. Suppose that $u_1(\cdot)$ and $u_2(\cdot)$ are solutions of (428) with the given properties. Set $u(t) := u_1(t) - u_2(t)$, $t > 0$. Then the prescribed assumptions imply that $D_L^W u = D_L^W u_1 - D_L^W u_2$ and $(D_L^W u)(t) \in \mathcal{A}u(t)$ for a. e. $t > 0$. Since $\mathcal{A}$ is closed, we can apply the Laplace transform in order to see that $w(\lambda)\tilde{u}(\lambda) \in \mathcal{A}\tilde{u}(\lambda)$ for $\operatorname{Re}\lambda > \omega'$. The final conclusion simply follows from the injectiveness of the operator $w(\lambda) - \mathcal{A}$ for $\operatorname{Re}\lambda > \omega'$ and the uniqueness theorem for the Laplace transform. □

Now we will reconsider the statement of [293, Theorem 2.8] in our general framework.

Theorem 7.3.22. *Suppose that $C \in L(X)$ is injective, $x_0 \in X, \ldots, x_s \in X$ are given elements, $\omega \geqslant 0$ and W is given by* (399), *with $x_j(u) := Cx_j$ for all $u \in D_j$ ($0 \leqslant j \leqslant s$). Suppose further that $\mathcal{A}$ is a closed* MLO, *$C\mathcal{A} \subseteq \mathcal{A}C$ and there exists a strongly continuous operator family $(R(t))_{t>0} \subseteq L(X)$ integrable at zero such that $(w(\lambda) - \mathcal{A})^{-1}C \in L(X)$ for* $\operatorname{Re}\lambda > \omega$,

$$w_0(\lambda)(w(\lambda) - \mathcal{A})^{-1}Cx = \int_0^{+\infty} e^{-\lambda t}R(t)x\,dt, \quad \operatorname{Re}\lambda > \omega,\ x \in X, \tag{429}$$

there exists a strongly continuous operator family $(P(t))_{t>0} \subseteq L(X)$ integrable at zero such that

$$(w(\lambda) - \mathcal{A})^{-1}Cx = \int_0^{+\infty} e^{-\lambda t}P(t)x\,dt, \quad \operatorname{Re}\lambda > \omega,\ x \in X, \tag{430}$$

as well as that the function $|f|(\cdot)$ *and the functions* $|a_j|(\cdot)$ *are Laplace transformable and*

$$w_j(\lambda) = w_0(\lambda)\tilde{a_j}(\lambda), \quad \operatorname{Re}\lambda > \omega \quad (1 \leqslant j \leqslant s). \tag{431}$$

Define

$$u(t) := R(t)x_0 + (a_1 *_0 R)(t)x_1 + \cdots + (a_s *_0 R)(t)x_s + \int_0^t P(t-s)f(s)\,ds, \quad t > 0. \tag{432}$$

If $u \in D_0 \cap \cdots \cap D_s$ *(this particularly holds if* $D_j = \mathbb{L}$ *for* $0 \leqslant j \leqslant s$*), then* $u(\cdot)$ *is a unique solution of* (428) *with the function* $f(\cdot)$ *replaced therein with the function* $Cf(\cdot)$.

Proof. We will outline all relevant details of the proof for the sake of completeness. It is clear that the function $u(\cdot)$ is Laplace transformable as well as that (cf. (429), (430) and (431)):

$$\begin{aligned}\tilde{u}(\lambda) &= \tilde{R}(\lambda)x_0 + \tilde{a_1}(\lambda)\tilde{R}(\lambda)x_1 + \cdots + \tilde{a_s}(\lambda)\tilde{R}(\lambda)x_s + \tilde{P}(\lambda)\tilde{f}(\lambda)\\ &= w_0(\lambda)(w(\lambda) - \mathcal{A})^{-1}Cx_0 + w_1(\lambda)(w(\lambda) - \mathcal{A})^{-1}Cx_1 + \cdots\\ &\quad + w_s(\lambda)(w(\lambda) - \mathcal{A})^{-1}Cx_s + (w(\lambda) - \mathcal{A})^{-1}C\tilde{f}(\lambda),\end{aligned} \tag{433}$$

for $\operatorname{Re}\lambda > \omega$. If $u \in D_0 \cap \cdots \cap D_s$, then the generalized Laplace fractional derivative $(D_L^W u)(t)$ is well-defined; in this case, we can apply the Laplace transform to the both sides of (428) and [445, Proposition 1.4.7] (cf. also [69, Proposition 1.2.2]) to conclude that $u(\cdot)$ is a solution of (428) if the following holds:

$$\omega(\lambda)\tilde{u}(\lambda) - \sum_{k=0}^{s} \omega_k(\lambda)Cx_k \in \mathcal{A}\tilde{u}(\lambda) + C\tilde{f}(\lambda), \quad \operatorname{Re}\lambda > \omega.$$

But the last inclusion is a consequence of (433) and a simple calculation. If $u_1(\cdot)$ and $u_2(\cdot)$ are solutions of (37) with the prescribed properties, then the function $u(t) := u_1(t) - u_2(t)$, $t > 0$ is Laplace transformable and we have $0 \in (w(\lambda) - \mathcal{A})\tilde{u}(\lambda)$ for $\operatorname{Re}\lambda > \omega$. This implies $\tilde{u}(\lambda) = (w(\lambda) - \mathcal{A})^{-1}C0 = 0$ for $\operatorname{Re}\lambda > \omega$, so that the uniqueness theorem for the Laplace transform implies $u(t) = 0$ for a. e. $t > 0$. □

If $u(\cdot)$ is given by (432) and $u \notin D_0 \cap \cdots \cap D_s$, then we say that $u(\cdot)$ is a mild solution of (428). As in many research studies carried out so far, Theorem 7.3.22 enables one to examine the existence and uniqueness of asymptotically almost periodic type solutions to the abstract fractional differential inclusions with generalized Laplace derivatives (see, e. g., [444] and [293] for more details); Theorem 7.3.22 also suggests us to introduce a mild solution of the abstract semilinear fractional Cauchy inclusion

$$(D_L^W u)(t) \in \mathcal{A}u(t) + f(t, u(t)) \quad \text{for a. e. } t > 0$$

as any locally integrable function $u : [0,\infty) \to X$, which satisfies the equation

$$u(t) = R(t)x_0 + (a_1 *_0 R)(t)x_1 + \cdots + (a_s *_0 R)(t)x_s + \int_0^t P(t-s)f(s,u(s))\,ds,$$

for a. e. $t > 0$. We leave all details to the interested readers.

7.3.8 Abstract fractional Cauchy inclusions with generalized Hilfer (a,b,α)-derivatives

Suppose that $\alpha > 0$, $m = \lceil \alpha \rceil$, $a \in L^1_{loc}([0,\infty))$ or $a(t) = \delta(t)$, and $b \in L^1_{loc}([0,\infty))$ or $b(t) = \delta(t)$. Let us recall that $v_a(t) = \int_0^t a(t-s)u(s)\,ds$, $t \geqslant 0$.

We will use the following notion.

Definition 7.3.23. Suppose that $\mathcal{A}$ is an MLO in X and $f \in L^1_{loc}([0,\infty) : X)$. Then we say that a function $u \in L^1_{loc}([0,\infty) : X)$ is a solution of the abstract fractional inclusion

$$D^{\alpha}_{a,b}u(t) \in \mathcal{A}u(t) + f(t) \quad \text{for a. e. } t > 0 \tag{434}$$

if the function $v_a(\cdot)$ is $(m-1)$-continuously differentiable, the function $v_a^{(m-1)}(\cdot)$ is locally absolutely continuous on $[0,\infty)$, and (434) holds for a. e. $t > 0$.

It is very difficult to rephrase the results established in [293, Theorem 2.4] for the generalized Laplace fractional derivatives. But this can be done for the generalized Hilfer (a,b,α)-fractional derivatives; more precisely, we have the following result.

Theorem 7.3.24. *Suppose that $\mathcal{A}$ is a closed* MLO *in X, $f \in C([0,\infty) : X)$ and there exist locally integrable functions $c_1(\cdot)$ and $d_1(\cdot)$ such that $b *_0 c_1 = g_m$ and $c_1 = d_1 *_0 a$. Then the following holds:*

(i) *If a function $u \in L^1_{loc}([0,\infty) : X)$ is a solution of the abstract fractional inclusion* (434), *then the function $v_a(\cdot)$ is a solution of the following integral inclusion:*

$$v_a(t) - \sum_{j=1}^{m} g_j(t)v_a^{(j-1)}(0) \in (d_1 *_0 \mathcal{A}v)(t) + (c_1 *_0 f)(t), \quad t \geqslant 0; \tag{435}$$

furthermore, if $v_a \in C^m([0,\infty) : X)$, then the function $v_a(\cdot)$ is a strong solution of (435).

(ii) *Suppose that $u \in L^1_{loc}([0,\infty) : X)$, there exist a function $v_{a,\mathcal{A}} \in L^1_{loc}([0,\infty) : X)$ and the elements $v_{a,j} \in X$ $(1 \leqslant j \leqslant m)$ such that $v_{a,\mathcal{A}}(t) \in \mathcal{A}v_a(t)$ for a. e. $t > 0$ and*

$$v_a(t) - \sum_{j=1}^{m} g_j(t)v_{a,j} = (d_1 *_0 v_{a,\mathcal{A}})(t) + (c_1 *_0 f)(t), \quad t \geqslant 0. \tag{436}$$

Suppose, moreover, that the following condition holds:

(q) *There exist real numbers* $\xi_1 > m-1$ *and* $\xi_2 > m-1$ *and locally integrable functions* $d_{1,1}(\cdot)$ *and* $c_{1,1}(\cdot)$ *such that* $d_1 = g_{\xi_1} *_0 d_{1,1}$ *and* $c_1 = g_{\xi_2} *_0 c_{1,1}$.

Then $v_a \in C^{m-1}([0,\infty) : X)$ *and* $v_{a,j} = v_a^{(j-1)}(0)$ *for* $1 \leqslant j \leqslant m$*; furthermore, if the function* $v_a^{(m-1)}(\cdot)$ *is locally absolutely continuous on* $[0,\infty)$*, then the function* $u(\cdot)$ *is a solution of the abstract fractional inclusion* (434).

(iii) *Suppose that* $u \in L^1_{loc}([0,\infty) : X)$, $v_a \in C^{m-1}([0,\infty) : X)$*, the function* $v_a^{(m-1)}(\cdot)$ *is locally absolutely continuous on* $[0,\infty)$ *and* (q) *holds. Suppose further that the function* $t \mapsto (g_{\xi_2-(m-1)} *_0 c_{1,1} *_0 f)(t)$, $t \geqslant 0$ *is locally absolutely continuous on* $[0,\infty)$*, as well as that there exists a function* $v_{d_1,a,\mathcal{A}} \in C([0,\infty) : X)$ *such that* $v_{d_1,a,\mathcal{A}}(t) \in \mathcal{A}[d_1 *_0 v_a](t)$ *for a. e.* $t > 0$ *and*

$$v_a(t) - \sum_{j=1}^{m} g_j(t) v_a^{(j-1)}(0) = v_{d_1,a,\mathcal{A}}(t) + (c_1 *_0 f)(t), \quad t \geqslant 0. \tag{437}$$

Then the function $u(\cdot)$ *is a solution of the abstract fractional inclusion* (434).

Proof. We will provide all relevant details of parts (i) and (ii) for the sake of completeness; the proof of part (iii) is much the same as that of the corresponding part of [293, Theorem 2.4] and, therefore, omitted. If $u \in L^1_{loc}([0,\infty) : X)$ is a solution of the abstract fractional inclusion (434), then we define $u_{a,\mathcal{A}}(t) := (b *_0 v^{(m)})(t) - f(t)$, $t \geqslant 0$ and $v_{a,\mathcal{A}}(t) := (a *_0 u_{\mathcal{A}})(t)$, $t \geqslant 0$. We have

$$v_a(t) - \sum_{j=1}^{m} g_j(t) v_a^{(j-1)}(0) = (d_1 *_0 v_{\mathcal{A}})(t) + (c_1 *_0 f)(t), \quad t \geqslant 0, \tag{438}$$

$v_{a,\mathcal{A}}(t) \in \mathcal{A}v(t)$ for all $t \geqslant 0$ and (435) holds. The assumption $f \in C([0,\infty) : X)$ implies $(c_1 *_0 f)(\cdot) \in C([0,\infty) : X)$ and, therefore, $(d_1 *_0 v_{a,\mathcal{A}})(\cdot) \in C([0,\infty) : X)$, so that $v_a(\cdot)$ is a solution of (382); furthermore, if $v_a \in C^m([0,\infty) : X)$, then the functions $u_{a,\mathcal{A}}(t)$ and $v_{a,\mathcal{A}}(t)$ are continuous for $t \geqslant 0$ so that $v_a(\cdot)$ is a strong solution of (435). In order to prove (ii), let us assume that there exist a function $v_{a,\mathcal{A}}(\cdot)$ and the elements $v_{a,j} \in X$ $(1 \leqslant j \leqslant m)$ with the given properties. Then it is clear that $v_a \in C^{m-1}([0,\infty) : X)$, $v_{a,j} = v_a^{(j-1)}(0)$ for $1 \leqslant j \leqslant m$ and that the derivation of (436) yields

$$v_a^{(m-1)}(t) - v_a^{(m-1)}(0) = (g_{\xi_1-(m-1)} *_0 d_{1,1} *_0 v_{\mathcal{A}})(t) + (g_{\xi_2-(m-1)} *_0 c_{1,1} *_0 f)(t),$$

for a. e. $t \geqslant 0$. If the function $v_a^{(m-1)}(\cdot)$ is locally absolutely continuous on $[0,\infty)$, then the last equality implies

$$(g_1 *_0 b *_0 v^{(m)})(t) = (b *_0 g_{\xi_1-(m-1)} *_0 d_{1,1} *_0 v_{\mathcal{A}})(t) + (b *_0 g_{\xi_2-(m-1)} *_0 c_{1,1} *_0 f)(t),$$

for a. e. $t \geqslant 0$. This implies that the function $(b *_0 g_{\xi_1-(m-1)} *_0 d_{1,1} *_0 v_{\mathcal{A}})(\cdot)$ is locally absolutely continuous on $[0,\infty)$ and

$$(b *_0 v^{(m)})(t) = \frac{d}{dt}(b *_0 g_{\xi_1-(m-1)} *_0 d_{1,1} *_0 v_{\mathcal{A}})(t) + f(t) \quad \text{for a. e. } t \geqslant 0,$$

since $b *_0 g_{\xi_2-(m-1)} *_0 c_{1,1} \equiv 1$. It remains to be proved that $(d/dt)(b *_0 g_{\xi_1-(m-1)} *_0 d_{1,1} *_0 v_{\mathcal{A}})(t) \in \mathcal{A}u(t)$ for a. e. $t > 0$. Keeping in mind [69, Proposition 1.2.2(i)], it suffices to show that $(b *_0 g_{\xi_1-(m-1)} *_0 d_{1,1} *_0 v_{\mathcal{A}})(t) \in \mathcal{A} \int_0^t u(s)\, ds$ for a. e. $t > 0$. This is a simple consequence of the definition of $v_a(\cdot)$, the closedness of $\mathcal{A}$ and the assumption $v_{a,\mathcal{A}}(t) \in \mathcal{A}v_a(t)$ for a. e. $t > 0$. Hence, the function $u(\cdot)$ is a solution of the abstract fractional inclusion (434). □

We continue by observing that the statement of [293, Theorem 2.6] cannot be formulated for the generalized Laplace fractional derivatives; it would be very tempting to state a satisfactory analogue of this result for the generalized Hilfer (a, b, α)-fractional derivatives.

In the remainder of this subsection, we will accept the convention $\tilde{\delta}(\lambda) := 1$ and assume that $\alpha > 0$, $m = \lceil \alpha \rceil$, $a \in L^1_{loc}([0,\infty))$, $|a|(\cdot)$ is Laplace transformable or $a(t) = \delta(t)$, and $b \in L^1_{loc}([0,\infty))$, $|b|(\cdot)$ is Laplace transformable or $b(t) = \delta(t)$. Suppose also that $s = m-1$, $D_0 = \cdots = D_{m-1} = D$, where D is equal to the set of all Laplace transformable functions $u : [0,\infty) \to X$ such that $a *_0 u \in C^{m-1}([0,\infty) : X)$, the function $(a *_0 u)^{(m-1)}(\cdot)$ is locally absolutely continuous on $[0,\infty)$ and the functions $(a *_0 u)(\cdot), \ldots, (a *_0 u)^{(m-1)}(\cdot)$ are exponentially bounded. Then Lemma 7.3.1 implies that the function $(a *_0 u)^{(m)}(\cdot)$ is Laplace transformable as well as that

$$\int_0^{+\infty} e^{-\lambda t}(a *_0 u)^{(m)}(t)\, dt$$
$$= \lambda^m \widetilde{a *_0 u}(\lambda) - \lambda^{m-1}(a *_0 u)(0) - \lambda^{m-2}(a *_0 u)'(0) - \cdots - (a *_0 u)^{(m-1)}(0),$$

for $\operatorname{Re}\lambda > 0$ suff. large. In this case, the notion of the generalized Hilfer (a, b, α)-fractional derivative $D^{\alpha}_{a,b}u$, introduced in Definition 7.3.17, is a special case of Definition 7.3.2 with $s = m-1$, $w_k(\lambda) = \tilde{b}(\lambda)\lambda^{m-1-k}$, $x_k(u) = (a *_0 u)^{(k)}(0)$ for $0 \leqslant k \leqslant s$ and $\operatorname{Re}\lambda > \omega$ suff. large ($u \in D$).

Applying Theorem 7.3.22 with $a_j(t) = g_j(t)$ for $t \geqslant 0$ and $1 \leqslant j \leqslant s$, we immediately obtain the following result.

Theorem 7.3.25. *Suppose that $C \in L(X)$ is injective, $x_0 \in X, \ldots, x_s \in X$ are given elements, $\omega \geqslant 0$ and W is given by the above given data, with $(a *_0 u)^{(j)}(0) := Cx_j$ for all $u \in D$. Suppose further that $\mathcal{A}$ is a closed* MLO, *$C\mathcal{A} \subseteq \mathcal{A}C$ and there exists a strongly continuous operator family $(R(t))_{t>0} \subseteq L(X)$ integrable at zero such that $(\lambda^m \tilde{b}(\lambda)\tilde{a}(\lambda) - \mathcal{A})^{-1}C \in L(X)$ for* $\operatorname{Re}\lambda > \omega$,

$$\lambda^{m-1}\tilde{b}(\lambda)(\lambda^m \tilde{b}(\lambda)\tilde{a}(\lambda) - \mathcal{A})^{-1}Cx = \int_0^{+\infty} e^{-\lambda t}R(t)x\, dt, \quad \operatorname{Re}\lambda > \omega,\ x \in X, \qquad (439)$$

there exists a strongly continuous operator family $(P(t))_{t>0} \subseteq L(X)$ *integrable at zero such that*

$$(\lambda^m \tilde{b}(\lambda)\tilde{a}(\lambda) - \mathcal{A})^{-1} Cx = \int_0^{+\infty} e^{-\lambda t} P(t)x \, dt, \quad \operatorname{Re} \lambda > \omega, \ x \in X, \tag{440}$$

as well as that the function $|f|(\cdot)$ *is Laplace transformable. Define*

$$u(t) := R(t)x_0 + (g_1 *_0 R)(t)x_1 + \cdots + (g_s *_0 R)(t)x_s + \int_0^t P(t-s)f(s)\, ds, \quad t > 0. \tag{441}$$

If $u \in D$, *then* $u(\cdot)$ *is a unique solution of* (428) *with* $D_L^W u = D_{a,b}^a u$ *and the function* $f(\cdot)$ *replaced therein with the function* $Cf(\cdot)$.

We continue with some observations.

Remark 7.3.26. (i) The operator family $R(\cdot)$, respectively, $P(\cdot)$, satisfying (439), respectively, (440), is not generally an exponentially bounded (a_1, k)-regularized C-resolvent family, respectively, an exponentially bounded (a_2, k)-regularized C-resolvent family, if the function $1/[\lambda\tilde{a}(\lambda)]$ or the function $1/[\lambda^m \tilde{a}(\lambda)\tilde{b}(\lambda)]$ is not in the range of the Laplace transform, respectively, if the function $1/[\lambda^m \tilde{a}(\lambda)\tilde{b}(\lambda)]$ is not in the range of the Laplace transform. Therefore, it seems that the notion of an (F, G, C)-regularized resolvent family is inevitable for a better understanding of the problem under our consideration.

(ii) If $f \neq 0$, then it is very difficult to clarify certain conditions ensuring that the function $u(\cdot)$, given by (441), belongs to D; if this is not the case, then we say that $u(\cdot)$ is a mild solution of (428) with $D_L^W u = D_{a,b}^a u$. Furthermore, if $f = 0$, then a relatively simple argumentation shows that the function $u(\cdot)$, given by (441), belongs to D provided that there exists a strongly continuous operator family $(W(t))_{t>0} \subseteq L(X)$ integrable at zero such that

$$\lambda^{2m-1}\tilde{a}(\lambda)\tilde{b}(\lambda)(\lambda^m \tilde{b}(\lambda)\tilde{a}(\lambda) - \mathcal{A})^{-1} Cx = \int_0^{+\infty} e^{-\lambda t} W(t)x \, dt, \quad \operatorname{Re} \lambda > \omega, \ x \in X, \tag{442}$$

because then we have $(g_m *_0 W)(t)x = (a *_0 R)(t)x$ for all $t \geqslant 0$ and $x \in X$.

7.3.9 Abstract fractional Cauchy inclusions with Prabhakar derivatives

Let us recall that the Prabhakar fractional derivative of Riemann–Liouville type ${}^{RL}\mathcal{D}^{\gamma}_{\beta,\alpha,\eta;0}$, given by the formula [321, (5.9)], is a special case of the notion introduced in Definition 7.3.17 with $a(t) = t^{m-\alpha-1}E^{-\gamma}_{\beta,m-\alpha}(\eta t^{\beta})$ and $b(t) = \delta(t)$ as well as that the

Prabhakar fractional derivative of Caputo type ${}^{C}\mathcal{D}^{\gamma}_{\beta,\alpha,\eta;0}$, given by the formula [321, (5.10)], is a special case of the notion introduced in Definition 7.3.17 with $a(t) = \delta(t)$ and $b(t) = t^{m-\alpha-1}E^{-\gamma}_{\beta,m-\alpha}(\eta t^{\beta})$. For simplicity, we assume here that $\alpha > 0$ and $m = \lceil \alpha \rceil$; the parameters β, γ, η can take arbitrary complex values.

Using the Laplace transform identity (398) and Theorem 7.3.25, we immediately get the following result.

Theorem 7.3.27. *Suppose that $C \in L(X)$ is injective, $x_0 \in X, \ldots, x_s \in X$ are given elements, $\omega \geqslant 0$ and W is given by the above given data, with $(a *_0 u)^{(j)}(0) := Cx_j$ for all $u \in D$. Suppose further that $\mathcal{A}$ is a closed* MLO, *$C\mathcal{A} \subseteq \mathcal{A}C$ and there exists a strongly continuous operator family $(R(t))_{t>0} \subseteq L(X)$ integrable at zero such that $(\lambda^{\alpha-\beta\gamma}(\lambda^{\beta}-\eta)^{\gamma}-\mathcal{A})^{-1}C \in L(X)$ for* Re $\lambda > \omega$ *as well as that, for every $x \in X$,*

$$\lambda^{m-1}\big(\lambda^{\alpha-\beta\gamma}(\lambda^{\beta}-\eta)^{\gamma}-\mathcal{A}\big)^{-1}Cx = \int_0^{+\infty} e^{-\lambda t}R(t)x\,dt, \quad \operatorname{Re}\lambda > \omega, \tag{443}$$

in the case of consideration of the Prabhakar fractional derivative of Riemann–Liouville type ${}^{RL}\mathcal{D}^{\gamma}_{\beta,\alpha,\eta;0}$, respectively,

$$\lambda^{\alpha-\beta\gamma-1}(\lambda^{\beta}-\eta)^{\gamma}\big(\lambda^{\alpha-\beta\gamma}(\lambda^{\beta}-\eta)^{\gamma}-\mathcal{A}\big)^{-1}Cx = \int_0^{+\infty} e^{-\lambda t}R(t)x\,dt, \quad \operatorname{Re}\lambda > \omega, \tag{444}$$

in the case of consideration of the Prabhakar fractional derivative of Caputo type ${}^{C}\mathcal{D}^{\gamma}_{\beta,\alpha,\eta;0}$, there exists a strongly continuous operator family $(P(t))_{t>0} \subseteq L(X)$ integrable at zero such that

$$\big(\lambda^{\alpha-\beta\gamma}(\lambda^{\beta}-\eta)^{\gamma}-\mathcal{A}\big)^{-1}Cx = \int_0^{+\infty} e^{-\lambda t}P(t)x\,dt, \quad \operatorname{Re}\lambda > \omega, \tag{445}$$

as well as that the function $|f|(\cdot)$ is Laplace transformable. Define $u(\cdot)$ through (441). If, for every $x \in X$, the mapping $t \mapsto (a * R)(t)x$, $t \geqslant 0$ is $(m-1)$-times continuously differentiable and its $(m-1)$-derivative is locally absolutely continuous on $[0,\infty)$, then $u(\cdot)$ is a unique solution of (428) *with $D_L^W u = {}^{RL}\mathcal{D}^{\gamma}_{\beta,\alpha,\eta;0}u$, respectively, $D_L^W u = {}^{C}\mathcal{D}^{\gamma}_{\beta,\alpha,\eta;0}$, and the function $f(\cdot)$ replaced therein with the function $Cf(\cdot)$.*

Especially, (443), respectively, (444), holds if $(R(t))_{t\geqslant 0}$ is an exponentially bounded (a_1,k)-regularized C-resolvent family with subgenerator $\mathcal{A}$, while (445) holds if $(P(t))_{t\geqslant 0}$ is an exponentially bounded (a_1,a_1)-regularized C-resolvent family with subgenerator $\mathcal{A}$, where

$$a_1(t) = t^{\alpha-1}E^{\gamma}_{\beta,\alpha}(\eta t^{\beta}) \quad \text{and} \quad k(t) = t^{\alpha-m}E^{\gamma}_{\beta,\alpha+1-m}(\eta t^{\beta}), \qquad \text{respectively, } k(t) = 1;$$

see (398) and [445, Theorem 3.2.5]. In our concrete situation, we must use (F,G,C)-resolvent operator families in the analysis of the Laplace transform equality (442); if

$u(\cdot)$ is given by (441) and there exists $x \in X$ such that the mapping $t \mapsto (a * R)(t)x$, $t \geqslant 0$ is not $(m-1)$-times continuously differentiable or its $(m-1)$-derivative is not locally absolutely continuous on $[0, \infty)$, then we say that $u(\cdot)$ is a mild solution of (428) with $D_L^W u = {}^{RL}\mathcal{D}^{\gamma}_{\beta,\alpha,\eta;0}u$, respectively, $D_L^W u = {}^{C}\mathcal{D}^{\gamma}_{\beta,\alpha,\eta;0}$, and the function $f(\cdot)$ replaced therein with the function $Cf(\cdot)$.

Suppose now that the following condition holds:

(Qa): $0 < \alpha < \alpha' \leqslant 2$, $C \in L(X)$ is injective, $\mathcal{A}$ is a closed MLO, $C\mathcal{A} \subseteq \mathcal{A}C$, there exist real numbers $\omega' \geqslant 0$, $M \geqslant 1$ and $\xi \in (0, 1]$ such that the mapping $\lambda \to (\lambda - \mathcal{A})^{-1}Cx$, $\lambda \in \omega' + \Sigma_{\alpha'\pi/2}$ is analytic $(x \in X)$, and

$$\|(\lambda - \mathcal{A})^{-1}C\| \leqslant \frac{M}{(1 + |\lambda|)^{\beta}}, \quad \lambda \in \omega' + \Sigma_{\alpha'\pi/2}. \tag{446}$$

Set $\gamma := \min(\pi/2, (\alpha' - \alpha)\pi/(2\alpha))$. Since $\lambda^{\alpha-\beta\gamma}(\lambda^{\beta} - \eta)^{\gamma} \sim \lambda^{\alpha}$ as $|\lambda| \to +\infty$, a simple analysis shows that there exists a sufficiently large real number $\omega > 0$ such that the assumptions $\lambda \in \mathbb{C}$ and $\operatorname{Re}\lambda > \omega$ imply $\lambda^{\alpha-\beta\gamma}(\lambda^{\beta} - \eta)^{\gamma} \in (\omega' + \Sigma_{\alpha'\pi/2}) \cap \Sigma_{\alpha\pi/2}$. Then we can use [445, Theorem 3.2.19] and the estimate (446) in order to see that the following holds:

(i) If there exists $l \in \mathbb{N}_0$ such that $l + 1 \leqslant s$, $x_0 = \cdots = x_l = 0$ in (441) and $m < \alpha\xi + l + 1$, then there exists a strongly continuous operator family $(R(t))_{t \geqslant 0} \subseteq L(X)$ such that

$$\lambda^{m-l-2}\left(\lambda^{\alpha-\beta\gamma}(\lambda^{\beta} - \eta)^{\gamma} - \mathcal{A}\right)^{-1}Cx = \int_0^{+\infty} e^{-\lambda t}R(t)x\,dt, \quad \operatorname{Re}\lambda > \omega,$$

and the mapping $t \mapsto R(t)$, $t > 0$ has an extension to the sector Σ_{γ}, denoted by the same symbol, such that for every $\gamma' \in (0, \gamma)$, there exist finite real constants $M_{\gamma'} \geqslant 1$ and $\omega_{\gamma'} \geqslant 0$ such that $\lim_{z \to 0, z \in \Sigma_{\gamma'}} R(z) = R(0)$ and $\|R(z)\| \leqslant M_{\gamma'}e^{\omega_{\gamma'}|z|}$, $z \in \Sigma_{\gamma'}$.

(ii) If $\mathcal{A}$ is densely defined, then there exists a strongly continuous operator family $(R(t))_{t \geqslant 0} \subseteq L(X)$ such that (444) holds and the mapping $t \mapsto R(t)$, $t > 0$ has an extension to the sector Σ_{γ} satisfying the same properties as in part (i). Furthermore, if $\mathcal{A}$ is not densely defined, then the above holds provided that the numbers β, γ and η are real as well as that there exists an operator $D \in L(X)$ such that $\lim_{\lambda \to +\infty} \lambda(\lambda - \mathcal{A})^{-1}Cx = Dx$ for all $x \in X \setminus \overline{D(\mathcal{A})}$.

(iii) If $\alpha\xi > 1$, then there exists a strongly continuous operator family $(P(t))_{t \geqslant 0} \subseteq L(X)$ such that (445) holds and the mapping $t \mapsto P(t)$, $t > 0$ has an extension to the sector Σ_{γ} satisfying the same properties as the operator family $(R(t))_{t \geqslant 0}$ from part (i).

For many important applications of the statements clarified in (i)–(iii), with a general regularizing operator $C \neq I$, we refer the reader to [443] and [445]; we will only note here that these results can be applied in the analysis of the abstract fractional Poisson heat equation in L^p-spaces, with Prabhakar derivatives.

We will not analyze the semilinear degenerate fractional equations with abstract differential operators and Prabhakar derivatives; see [445, Subsection 2.2.5] for some results established for the Caputo fractional derivatives. For more details about the Prabhakar fractional derivatives and the fractional differential equations with Prabhakar derivatives, we also refer the reader to the research articles [632, 663, 740] and references quoted therein.

It is worth noting that we can similarly analyze the abstract fractional differential inclusions with Davidson-Essex, Canavati and Jumarie derivatives [236]. The representation formula (441) can be helpful in the analysis of the existence and uniqueness of the asymptotically almost periodic properties of solutions to (428) in some concrete situations. All this has been seen many times and we will skip all details concerning this issue here.

We will consider the abstract multiterm fractional differential-difference inclusions with generalized Hilfer (a, b, α)-derivatives somewhere else.

7.4 Asymptotic constancy for solutions of abstract nonlinear fractional equations with delay and generalized Hilfer (a, b, α)-derivatives

The asymptotic behavior of solutions to the functional integrodifferential equations with the help of fixed-point theorems has been analyzed by many authors so far, starting presumably with the paper [153] by T. A. Burton and T. Furumochi; cf. also [70, 103, 152, 355, 14, 501, 658]. The main aim of this section is to consider the asymptotic behavior of solutions to the abstract nonlinear fractional functional differential (difference) equations with generalized Hilfer (a, b, α)-derivatives; see [486]. Our results on the asymptotic behavior of the following abstract fractional functional equation with generalized Hilfer (a, b, α)-derivatives (details and notation will be explained later):

$$D^{\alpha}_{a,b}[u(t) - g(t, u_t)] = \sum_{j=1}^{w} B_j(t)u(t + t_j) + f(t, u_t), \quad t \geqslant 0; \quad u_0 = \xi,$$

can be applied to the equations with Riemann–Liouville, Caputo, Hilfer and Prabhakar derivatives.

Section 7.4.1 investigates the asymptotic behavior of solutions to the abstract nonlinear fractional functional differential equations with generalized Hilfer (a, b, α)-derivatives. The main results of this subsection are Theorem 7.4.2, where we work with the generalized Hilfer (a, b, α)-fractional derivatives of Caputo type, and Theorem 7.4.4, where we work with the generalized Hilfer (a, b, α)-fractional derivatives of Riemann–Liouville type. Section 7.4.2 investigates the asymptotic behavior of solutions to the abstract nonlinear fractional functional difference equation

$$D^{\alpha}_{a,b}[u(v) - g(v, u_v)] = \sum_{j=1}^{w} B_j(v)u(v + v_j) + f(v, u_v), \quad v \in \mathbb{N}_0; \quad u_0 = \xi;$$

cf. Theorem 7.4.10 and Theorem 7.4.11, which present the discrete analogues of Theorem 7.4.2 and Theorem 7.4.4.

7.4.1 Asymptotic constancy for solutions of abstract nonlinear fractional differential equations with delay

Suppose that $a > 0$, $m = \lceil a \rceil$, $\tau > 0$, $w \in \mathbb{N}$, $t_j \in [0, +\infty)$ for all $j \in \mathbb{N}_w$, $f : [0, \infty) \times C([-\tau, 0] : X) \to X$ and $g : [0, \infty) \times C([-\tau, 0] : X) \to X$ are given mappings, $\xi \in C([-\tau, 0] : X)$ and $B_j(t) : X \to X$ are Lipschitz continuous mappings, which satisfy that there exist finite real constants $L_j(t) > 0$ such that $\|B_j(t)x - B_j(t)y\| \leqslant L_j(t)\|x - y\|$ for all $t \geqslant 0$, $x, y \in X$ and $j \in \mathbb{N}_w$. The history function $u_t \in C([-\tau, 0] : X)$ is defined by $u_t(s) := u(t + s)$ for all $s \in [-\tau, 0]$.

In this subsection, we consider the abstract behavior of solutions to the following abstract nonlinear fractional differential equation with finite delay:

$$D_{a,b}^{a}[u(t) - g(t, u_t)] = \sum_{j=1}^{w} B_j(t)u(t + t_j) + f(t, u_t), \quad t \geqslant 0; \quad u_0 = \xi. \tag{447}$$

It is worthwhile to mention that the equation (447) generalizes the equation [26, (1)] for the Caputo fractional derivatives of order $a \in (0, 1)$, where we have the absence of term $\sum_{j=1}^{w} B_j(t)u(t + t_j)$. In [26, Theorem 3.1], R. P. Agarwal, Y. Zhou and Y. He have proved the local existence of solutions for such a problem using the Krasnoselskii fixed point theorem; here, we will not try to extend this result for the equation (447).

In this subsection, we will always assume that the following condition holds true:

(C-c) There exists a function $c \in L^1_{\mathrm{loc}}([0, \infty))$ such that $b *_0 c = g_m$.

Set $v_{a,g} := a *_0 [u(\cdot) - g(\cdot, u_\cdot)]$. Convoluting the both sides of (447) with $c(\cdot)$ and applying the partial integration after that, we get

$$\begin{aligned} & v_{a,g}(t) - g_m(t)v_{a,g}^{(m-1)}(0) - \cdots - g_1(t)v_{a,g}(0) \\ & = \left(c *_0 \left[\sum_{j=1}^{w} B_j(\cdot)u(\cdot + t_j) + f(\cdot, u_\cdot) \right] \right)(t), \quad t \geqslant 0, \end{aligned}$$

i. e.,

$$\begin{aligned} & (a *_0 u)(t) \\ & = (a *_0 g(\cdot, u_\cdot))(t) + g_m(t)[a *_0 [u(\cdot) - g(\cdot, u_\cdot)]]^{(m-1)}(0) + \cdots \\ & + g_1(t)[a *_0 [u(\cdot) - g(\cdot, u_\cdot)]](0) + \left(c *_0 \left[\sum_{j=1}^{w} B_j(\cdot)u(\cdot + t_j) + f(\cdot, u_\cdot) \right] \right)(t), \quad t \geqslant 0. \end{aligned} \tag{448}$$

We will use the following notion.

Definition 7.4.1. Any continuous function $u : [-\tau, \infty) \to X$ satisfying (448) and $u_0 = \xi$ is said to be a mild solution of (447).

In general case, it would be difficult to say when a mild solution of (447) satisfies that the function $v_{a,g}^{(m-1)}(t)$ is well-defined and locally absolutely continuous for $t \geqslant 0$.

Our first structural result reads as follows.

Theorem 7.4.2. *Suppose that $a(t) \equiv \delta(t)$, (C-c) holds, there exist a real constant $L > 0$ and a function $p_f : [0, \infty) \to [0, \infty)$ such that*

$$\|g(t, \xi_1) - g(t, \xi_2)\| \leqslant L \sup_{s\in[-\tau,0]} \|\xi_1(s) - \xi_2(s)\| \tag{449}$$

and

$$\|f(t, \xi_1) - f(t, \xi_2)\| \leqslant p_f(t) \sup_{s\in[-\tau,0]} \|\xi_1(s) - \xi_2(s)\| \tag{450}$$

for all ξ_1, $\xi_2 \in C([-\tau, 0] : X)$ and $t \geqslant 0$. Suppose further that the following conditions hold true:

(C-a) *The functions $g(\cdot, \zeta)$ and $f(\cdot, \zeta)$ are continuous for any fixed function $\zeta \in C([-\tau, 0] : X)$, the mapping $L_j(\cdot)$ is locally bounded on $[0, \infty)$ for every fixed $j \in \mathbb{N}_w$ and the mapping $t \mapsto B_j(t)x$, $t \geqslant 0$ is continuous for every fixed $x \in X$ and $j \in \mathbb{N}_w$;*

(D-a) *There exists a unique element $\beta \in X$ such that the constant function $\beta(s) := \beta$, $s \in [-\tau, 0]$, denoted by the same symbol, satisfies*

$$\beta = \xi(0) - g(0, \xi) + g(t, \beta) + \left(c *_0 \left[\sum_{j=1}^{w} B_j(\cdot)\beta + f(\cdot, \beta) \right] \right)(t) \quad \text{for all } t \geqslant 0. \tag{451}$$

Then there exists a unique mild solution of problem (447) satisfying $\lim_{t\to+\infty} u(t) = \beta$ and $[u(\cdot) - g(\cdot, u_\cdot)]^{(j)}(0) = 0$ for $1 \leqslant j \leqslant m - 1$, provided that there exists a function $z : [0, \infty) \to [0, \infty)$ such that $0 < z(t) < t$ for $t > 0$, $\lim_{t\to+\infty} z(t) = +\infty$,

$$\lim_{t\to+\infty} \int_0^{z(t)} |c(t - s)| \cdot \left[\sum_{j=1}^{w} L_j(s) + p_f(s) \right] ds = 0 \tag{452}$$

and

$$L + \sup_{t\geqslant 0} \int_0^{t} |c(t - s)| \cdot \left[\sum_{j=1}^{w} L_j(s) + p_f(s) \right] ds < 1. \tag{453}$$

In particular, (D-a) holds provided that, for every constant function $v(\cdot) \equiv v$, the function $t \mapsto g(t, v) + (c *_0 [\sum_{j=1}^{w} B_j(\cdot)v + f(\cdot, v)])(t)$, $t \geqslant 0$ is also constant.

Proof. Set

$$P := \left\{ h \in C([-\tau, +\infty) : X) : h_{[-\tau,0]}(\cdot) \equiv \xi(\cdot) \text{ and } \lim_{t\to+\infty} h(t) = \beta \right\},$$

as well as $d(h_1, h_2) := \sup_{t\geqslant -\tau} \|h_1(t) - h_2(t)\|$ for any $h_1,\ h_2 \in P$. Then (P, d) is a complete metric space, as easily approved. Now we will prove that the mapping $\Psi : P \to P$, given by $(\Psi h)(t) := \xi(t)$, $t \in [-\tau, 0]$ and

$$(\Psi h)(t) := \xi(0) - g(0, \xi) + g(t, h_t) + \left(c *_0 \left[\sum_{j=1}^{w} B_j(\cdot)h(\cdot + t_j) + f(\cdot, h_\cdot) \right] \right)(t), \quad t > 0,$$

is a well-defined contraction. First of all, let us prove that the mappings $t \mapsto g(t, h_t)$, $t \geqslant 0$ and $t \mapsto (c *_0 [\sum_{j=1}^{w} B_j(\cdot)h(\cdot + t_j) + f(\cdot, h_\cdot)])(t)$, $t \geqslant 0$ are continuous. Let a number $t \geqslant 0$ and a function $h \in P$ be fixed. Then we have

$$\begin{aligned} \|g(t', h_{t'}) - g(t, h_t)\| &\leqslant \|g(t', h_{t'}) - g(t', h_t)\| + \|g(t', h_t) - g(t, h_t)\| \\ &\leqslant L \sup_{s\in[-\tau,0]} \|h(t + s) - h(t' + s)\| + \|g(t', h_t) - g(t, h_t)\|, \quad t' \geqslant 0. \end{aligned} \tag{454}$$

Since the function $g(\cdot, h_t)$ is continuous by (C-a), the above simply implies that the function $g(\cdot, h_\cdot)$ is continuous at the point t; we can similarly prove that the function $f(\cdot, h_\cdot)$ is continuous at the point t. Since we have assumed that the mapping $L_j(\cdot)$ is locally bounded on $[0, \infty)$ for every fixed $j \in \mathbb{N}_w$ and the mapping $t \mapsto B_j(t)x$, $t \geqslant 0$ is continuous for every fixed $x \in X$ and $j \in \mathbb{N}_w$, a relatively simple argumentation involving the decomposition of type (454) shows that the mapping $t \mapsto \sum_{j=1}^{w} B_j(t)u(t + t_j)$, $t \geqslant 0$ is continuous. This implies that the mapping $(c *_0 [\sum_{j=1}^{w} B_j(\cdot)h(\cdot + t_j) + f(\cdot, h_\cdot)])(\cdot)$ is continuous at the point t; see also the proof of [69, Proposition 1.3.4]. By the foregoing, we have that the mapping $t \mapsto (\Psi h)(t)$, $t \geqslant -\tau$ is continuous.

Let us prove now that $\lim_{t\to+\infty}(\Psi h)(t) = \beta$. In actual fact, we have that $\lim_{t\to+\infty} g(t, h_t) = \beta$ since the mapping $g(\cdot, \cdot)$ is Lipschitz with respect to the second variable and $\lim_{t\to+\infty} h(t) = \beta$; furthermore,

$$\begin{aligned} &\left\| \left(c *_0 \left[\sum_{j=1}^{w} B_j(\cdot)h(\cdot + t_j) + f(\cdot, h_\cdot) \right] \right)(t) - \left(c *_0 \left[\sum_{j=1}^{w} B_j(\cdot)\beta + f(\cdot, \beta) \right] \right)(t) \right\| \\ &\leqslant \int_0^{z(t)} |c(t - s)| \cdot \left\| \left[\sum_{j=1}^{w} B_j(s)h(s + t_j) + f(s, h_s) \right] - \left[\sum_{j=1}^{w} B_j(s)\beta + f(s, \beta) \right] \right\| ds \\ &+ \int_{z(t)}^{t} |c(t - s)| \cdot \left\| \left[\sum_{j=1}^{w} B_j(s)h(s + t_j) + f(s, h_s) \right] - \left[\sum_{j=1}^{w} B_j(s)\beta + f(s, \beta) \right] \right\| ds \\ &\leqslant \int_0^{z(t)} |c(t - s)| \cdot \left[p_f(s) + \sum_{j=1}^{w} L_j(s) \right] (\|h\|_\infty + \|\beta\|)\, ds \end{aligned}$$

$$+\int_{z(t)}^{t}|c(t-s)|\cdot\left[\sum_{j=1}^{w}L_j(s)\|h(s+t_j)-\beta\|+p_f(s)\sup_{r\in[-\tau,0]}\|h(s+r)-\beta\|\right]ds$$
$$\to 0,\quad t\to+\infty,$$

because we have assumed (452) and (453). Our choice of β in (451) finally yields that $\lim_{t\to+\infty}(\Psi h)(t)=\beta$, as claimed.

Therefore, the mapping $\Phi : P \to P$ is well-defined; keeping in mind the estimate (453), a simple calculation yields that the mapping $\Phi : P \to P$ is a contraction. Since $a(t) = \delta(t)$, $[u(\cdot)-g(\cdot,u_\cdot)]^{(j)}(0) = 0$ for $1 \leqslant j \leqslant m-1$ and $u(0) = \xi(0)$, the equation (448) is equivalent to $u = \Psi u$, so that the Banach contraction principle implies that there exists a unique mild solution of problem (447), which satisfies the prescribed properties. It remains to be proved that there exists a unique element $\beta \in X$ such that the constant function $\beta(s) := \beta$, $s \in [-\tau,0]$ satisfies (451), provided that (452) holds. Since, for every constant function $v(\cdot)\equiv v$, the function $t\mapsto g(t,v)+(c *_0 [\sum_{j=1}^{w}B_j(\cdot)v+f(\cdot,v)])(t)$, $t\geqslant 0$ is also constant, we can apply the Banach contraction principle once more in order to see that there exists a unique element $\beta\in X$ such that (D-a) holds; see also (453). □

Before proceeding any further, we would like to emphasize that Theorem 7.4.2 can be successfully applied in the analysis of asymptotic constancy of problem (447) with Caputo derivatives and Prabhakar fractional derivatives of Caputo type; more precisely, for Caputo fractional derivatives, we have $a(t) = \delta(t)$, $b(t) = g_{m-\alpha}(t)$ and $c(t) = g_\alpha(t)$, while for Prabhakar fractional derivatives of Caputo type, we have $a(t) = \delta(t)$, $b(t) = t^{m-\alpha-1}E^{-\gamma}_{\beta,m-\alpha}(\eta t^\beta)$, $t>0$ and

$$c(t)=\mathcal{L}^{-1}\left(\frac{\lambda^{\gamma\beta-\alpha}}{(\lambda^\beta-\eta)^\gamma}\right)(t)=t^{\alpha-1}E^{\gamma}_{\beta,\alpha}(\eta t^\beta),\quad t>0;$$

see, e. g., the formulae [321, (5.10)] and (398).

Remark 7.4.3. Keeping in mind the result established in [500, Lemma 2], Theorem 7.4.2 provides a proper extension of [500, Theorem 1]. It seems plausible that the result established in [500, Theorem 2] can be extended for the equation (447); we will not consider this question here.

Further on, assume that $a\in L^1_{\mathrm{loc}}([0,\infty))$, there exist $s\in\mathbb{N}$ and $d(t)=\delta(t)$ or $d\in L^1_{\mathrm{loc}}([0,\infty))$ such that $a *_0 d = g_s$. After the convolution with $d(\cdot)$, it can been easily shown that (448) is equivalent with

$$\begin{aligned}(g_s *_0 u)(t) &= (g_s *_0 g(\cdot,u_\cdot))(t)+(d *_0 g_m)(t)[a *_0 [u(\cdot)-g(\cdot,u_\cdot)]]^{(m-1)}(0)+\cdots\\ &\quad+(d *_0 g_1)(t)[a *_0 [u(\cdot)-g(\cdot,u_\cdot)]](0)\\ &\quad+\left(c *_0 d *_0\left[\sum_{j=1}^{w}B_j(\cdot)u(\cdot+t_j)+f(\cdot,u_\cdot)\right]\right)(t),\quad t\geqslant 0.\end{aligned}\tag{455}$$

Our second structural result reads as follows.

Theorem 7.4.4. *Suppose that $a \in L^1_{\mathrm{loc}}([0,\infty))$, (C-c) holds, there exist $s \in \mathbb{N}$ and $d(t) = \delta(t)$ or $d \in L^1_{\mathrm{loc}}([0,\infty))$ such that $a *_0 d = g_s$. Suppose further that there exists $e(t) = \delta(t)$ or $e \in L^1_{\mathrm{loc}}([0,\infty))$ such that $c *_0 d = g_s *_0 e$, there exist a real constant $L > 0$ and a function $p_f : [0,\infty) \to [0,\infty)$ such that (449) and (450) hold for all $\xi_1, \xi_2 \in C([-\tau,0] : X)$ and $t \geqslant 0$, (C-a) and the following condition hold:*

(D1) *There exists a unique element $\beta \in X$ such that the constant function $\beta(s) := \beta$, $s \in [-\tau,0]$ satisfies*

$$\beta = g(t,\beta) + \left(e *_0 \left[\sum_{j=1}^{w} B_j(\cdot)\beta + f(\cdot,\beta)\right]\right)(t) \quad \text{for all } t \geqslant 0.$$

Then there exists a unique mild solution of problem (447) satisfying $\lim_{t\to+\infty} u(t) = \beta$ and $[a *_0 (u(\cdot) - g(\cdot,u_\cdot))]^{(j)}(0) = 0$ for $1 \leqslant j \leqslant m-1$, provided that there exists a function $z : [0,\infty) \to [0,\infty)$ such that $0 < z(t) < t$ for $t > 0$, $\lim_{t\to+\infty} z(t) = +\infty$ and the equations (452)–(453) hold with the function $c(\cdot)$ replaced therein with the function $e(\cdot)$, in the case that $e \in L^1_{\mathrm{loc}}([0,\infty)$, respectively,

$$L + \sup_{t\geqslant 0}\left[\sum_{j=1}^{w} L_j(t) + p_f(t)\right] < 1,$$

in the case that $e(t) = \delta(t)$. In particular, (D1) holds provided that, for every constant function $v(\cdot) \equiv v$, the function $t \mapsto g(t,v) + (e *_0 [\sum_{j=1}^{w} B_j(\cdot)v + f(\cdot,v)])(t)$, $t \geqslant 0$ is also constant.

Proof. In our concrete situation, the equation (448) is equivalent with the equation (455). Since we have assumed that there exists $e(t) = \delta(t)$ or $e \in L^1_{\mathrm{loc}}([0,\infty))$ such that $c *_0 d = g_s *_0 e$, the equation (455) is equivalent with the equation

$$u(t) = g(t,u_t) + \left(e *_0 \left[\sum_{j=1}^{w} B_j(\cdot)\beta + f(\cdot,\beta)\right]\right)(t), \quad t \geqslant 0.$$

Now we can argue in the same way as in the proof of Theorem 7.4.2 in order to obtained the required conclusions (observe only that the assumption $a \in L^1_{\mathrm{loc}}([0,\infty))$ implies here that $[a *_0 (u(\cdot) - g(\cdot,u_\cdot))](0) = 0$). □

Remark 7.4.5. If the functions $B_j(\cdot)$ are Laplace transformable for $1 \leqslant j \leqslant w$, the function $|e|(\cdot)$ is Laplace transformable, if $e \in L^1_{\mathrm{loc}}([0,\infty))$, and for every constant function $v(\cdot) \equiv v$, the functions $g(\cdot,v)$ and $f(\cdot,v)$ are Laplace transformable, then the function $t \mapsto g(t,v) + (e *_0 [\sum_{j=1}^{W} B_j(\cdot)v + f(\cdot,v)])(t)$, $t \geqslant 0$ is constant for every constant function $v(\cdot) \equiv v$, provided that, for every constant function $v(\cdot) \equiv v$, there exists a complex constant $c(v) \in X$ such that

$$\int_0^{+\infty} e^{-\lambda t} g(t,v)\,dt + \left[\int_0^{+\infty} e^{-\lambda t} e(t)\,dt\right] \cdot \left[\sum_{j=1}^{w} \int_0^{+\infty} e^{-\lambda t} B_j(t)v\,dt + \int_0^{+\infty} e^{-\lambda t} f(t,v)\,dt\right] = \frac{c(v)}{\lambda}, \tag{456}$$

for every $\lambda \in \mathbb{C}$ such that $\operatorname{Re}\lambda > 0$ is sufficiently large. A similar comment can be given for the last statement in the formulation of Theorem 7.4.2.

In the case of consideration of the usual Hilfer fractional derivative $D_t^{\alpha,\beta}u(t)$ of order $\alpha > 0$ and type $\beta \in [0,1)$, Theorem 7.4.2 can be successfully applied with $a(t) = g_{(1-\beta)(m-\alpha)}(t)$, $b(t) = g_{\beta(m-\alpha)}(t)$, $c(t) = g_{m-\beta(m-\alpha)}(t)$, $d(t) = g_{s-(1-\beta)(m-\alpha)}(t)$ and $e(t) = g_\alpha(t)$, with $s \in \mathbb{N}$ arbitrary. It is also worth noting that Theorem 7.4.2 can be successfully applied in the analysis of the asymptotic constancy of problem (447) with the Prabhakar fractional derivatives of Riemann–Liouville type; in this concrete situation, we have $a(t) = t^{m-\alpha-1}E^{-\gamma}_{\beta,m-\alpha}(\eta t^\beta)$, $b(t) = \delta(t)$, $c(t) = g_m(t)$,

$$d(t) = \mathcal{L}^{-1}\left(\frac{\lambda^{\gamma\beta-s+m-\alpha}}{(\lambda^\beta-\eta)^\gamma}\right)(t) = t^{s-m+\alpha-1}E^{\gamma}_{\beta,s-m+\alpha}(\eta t^\beta), \quad t > 0,$$

and

$$e(t) = \mathcal{L}^{-1}\left(\frac{\lambda^{\gamma\beta-\alpha}}{(\lambda^\beta-\eta)^\gamma}\right)(t) = t^{\alpha-1}E^{\gamma}_{\beta,\alpha}(\eta t^\beta), \quad t > 0,$$

with $s \in \mathbb{N}$ arbitrary.

Now we will illustrate Theorem 7.4.2 and Theorem 7.4.4 with two concrete examples.

Example 7.4.6. (i) Suppose that $1 \leqslant p < +\infty$, Ω is a nonempty Lebesgue measurable subset of $\mathbb{R}^n$, $X := L^p(\Omega)$ and

$$[B_j(t)f](x) := b_j(t)\cdot f(x), \quad t \geqslant 0,\ x \in \Omega,\ f \in X,$$

where $b_j(\cdot)$ are continuous, complex-valued, Laplace transformable functions ($1 \leqslant j \leqslant w$). Then $B_j(t)$ is a linear bounded operator and we have $L_j(t) = |b_j(t)|$ for every $t \geqslant 0$ and $j \in \mathbb{N}_w$. Set $f_1(t) := -\sum_{j=1}^{w} b_j(t)$ and $f(t,v) := f_1(t)v$ for all $t \geqslant 0$ and $v \in X$; assume further that $g(t,v) = g(v)$, where $g : X \to X$ satisfies $\|g(v_1) - g(v_2)\| \leqslant L\|v_1 - v_2\|$ for some $L \in (0,1)$ and all $v_1,\ v_2 \in X$. Then, for every Laplace transformable function $c(\cdot)$, the equation (456) holds with $c(v) = g(v)$ and, therefore, the function $t \mapsto g(t,v) + (e *_0 [\sum_{j=1}^{w} B_j(\cdot)v + f(\cdot,v)])(t)$, $t \geqslant 0$ is constant for every constant function $v(\cdot) \equiv v$; cf. Remark 7.4.5. Therefore, Theorem 7.4.2 can be applied provided that the equations (452) and (453) hold with $L_j(t)$, $1 \leqslant j \leqslant w$ and $p_f(t) = |f_1(t)|$ for all $t \geqslant 0$. In particular, this holds if there exists a real number $p > 1$ such that $b_j \in L^p([0,\infty))$ for all $j \in \mathbb{N}_w$ and $|c|(\cdot)$ is a suitable multiple of a function $c_1 \in L^q([0,\infty))$, where $1/p + 1/q = 1$; cf. [69, Proposition 1.3.5 b)].

(ii) Suppose that $1 \leqslant p < +\infty$, Ω is a nonempty Lebesgue measurable subset of $\mathbb{R}^n$, $X := L^p(\Omega)$ and

$$[B_j(t)f](x) := b_j(t) \cdot b_{j,1}(f(x)), \quad t \geqslant 0,\ x \in \Omega,\ f \in X,$$

where $b_j(\cdot)$ are continuous, complex-valued, Laplace transformable functions ($1 \leqslant j \leqslant w$) and $b_{j,1} : \mathbb{C} \to \mathbb{C}$ satisfies that $\int_\Omega |b_{j,1}(f(x))|^p\,dx < +\infty$ for every $f \in X$ and

$$|b_{j,1}(z) - b_{j,1}(z')| \leqslant L_j|z - z'|, \quad z,\ z' \in \mathbb{C}$$

for some real constant $L_j > 0$. Then $B_j(t)$ is a Lipschitz continuous operator and we have $L_j(t) = |b_j(t)|L_j$ for every $t \geqslant 0$ and $j \in \mathbb{N}_w$. Set $f_1(t) := -\sum_{j=1}^{w} b_j(t)$ and $[f(t,v)](x) := f_1(t)b_{j,1}(v(x))$ for all $t \geqslant 0$, $x \in \Omega$ and $v \in X$; assume further that $g(t,v) = g(v)$ enjoys the same features as in part (i). Arguing as above, we can show that Theorem 7.4.4 can be applied provided that the equations (452) and (453) hold with $L_j(t)$, $1 \leqslant j \leqslant w$, $p_f(t) = |f_1(t)|L_j$ for all $t \geqslant 0$ and the function $c(\cdot)$ replaced therein with $e(\cdot)$. In particular, this holds if $\lim_{t\to+\infty} b_j(t) = 0$ for all $j \in \mathbb{N}_w$ and $|e|(\cdot)$ is a suitable multiple of a function $e_1 \in L^1([0,\infty))$; cf. [69, Proposition 1.3.5 d)].

7.4.2 Asymptotic constancy for solutions of abstract nonlinear fractional difference equations with delay

In the discrete setting, the role of Dirac delta distribution $\delta(t)$ takes the sequence $\delta = k^0 : \mathbb{N}_0 \to \mathbb{N}_0$, which has been defined by $\delta(0) = 1$ and $\delta(v) = 0$, $v \in \mathbb{N}$. Then, for every sequence $a : \mathbb{N}_0 \to X$, we have $\delta *_0 a = a$.

Suppose now that $u : \mathbb{N}_0 \to X$, $\alpha > 0$, $m = \lceil \alpha \rceil$, $a : \mathbb{N}_0 \to X$ and $b : \mathbb{N}_0 \to X$. The following notion is a discrete version of the notion considered in Definition 7.3.17.

Definition 7.4.7. The generalized Hilfer (a,b,α)-fractional derivative of sequence $u(\cdot)$, denoted shortly by $D^\alpha_{a,b}u$, is defined by

$$D^\alpha_{a,b}u(v) := (b *_0 \Delta^m(a *_0 u))(v), \quad v \in \mathbb{N}_0.$$

If $0 \leqslant \beta \leqslant 1$, then the usual Hilfer fractional derivative $D^{\alpha,\beta}u$ of order α and type β is defined as the generalized Hilfer (a,b,α)-fractional derivative of $u(\cdot)$, with $a(v) = k^{(1-\beta)(m-\alpha)}(v)$ and $b(v) = k^{\beta(1-\alpha)}(v)$.

Suppose now that $\tau \in \mathbb{N}$, $w \in \mathbb{N}$, $v_j \in \mathbb{N}_0$ for all $j \in \mathbb{N}_w$, $f : [0,\infty) \times X^{-\mathbb{N}^0_\tau} \to X$ and $g : [0,\infty) \times X^{-\mathbb{N}^0_\tau} \to X$ are given mappings, $\xi \in X^{-\mathbb{N}^0_\tau}$ and $B_j(v) : X \to X$ are Lipschitz continuous mappings, which satisfy that there exist finite real constants $L_j(v) > 0$ such that $\|B_j(v)x - B_j(v)y\| \leqslant L_j(v)\|x - y\|$ for all $v \in \mathbb{N}_0$, $x, y \in X$ and $j \in \mathbb{N}_w$. The history function $u_v \in X^{-\mathbb{N}^0_\tau}$ is defined by $u_v(s) := u(v+s)$ for all $s \in -\mathbb{N}^0_\tau$; of course, here and hereafter, we have $-\mathbb{N}^0_\tau = \{-\tau, -\tau+1, \ldots, 0\}$.

In this subsection, we consider the abstract behavior of solutions to the following abstract nonlinear fractional difference equation with finite delay:

$$D^{\alpha}_{a,b}[u(v) - g(v, u_v)] = \sum_{j=1}^{w} B_j(v)u(v + v_j) + f(v, u_v), \quad v \in \mathbb{N}_0; \quad u_0 = \xi. \tag{457}$$

We need the following auxiliary lemma.

Lemma 7.4.8. *Suppose that $u : \mathbb{N}_0 \to X$ and $l \in \mathbb{N}$. Then, for every $v \in \mathbb{N}_0$, we have*

$$(k^l *_0 \Delta^l u)(v) = u(l + v) + \sum_{s=0}^{l-1}\Bigg[\sum_{\substack{0\leqslant j\leqslant v;\ 0\leqslant r\leqslant l\\ j+r=s}} (-1)^{l-r}\binom{l+v-j-1}{v-j}\binom{l}{r}\Bigg]u(s). \tag{458}$$

Proof. Since $k^1 \equiv 1$, we have $(k^1 *_0 \Delta^1 u)(v) = u(v + 1) - u(0)$, $v \in \mathbb{N}_0$. This implies

$$\begin{aligned}(k^2 *_0 \Delta^2 u)(v) &= (k^1 *_0 k^1 *_0 \Delta(\Delta u))(v)\\ &= (k^1 *_0 [\Delta u(\cdot + 1) - \Delta u(0)])(v) = u(v + 2) - u(1) - (v + 1)[u(1) - u(0)]\\ &= u(v + 2) - (v + 2)u(1) + (v + 1)u(0), \quad v \in \mathbb{N}_0.\end{aligned}$$

Proceeding inductively, we can prove that there exist functions $c_s : \mathbb{N}_0 \to \mathbb{Z}$, $0 \leqslant s \leqslant l-1$, such that

$$(k^l *_0 \Delta^l u)(v) = u(l + v) - \sum_{s=0}^{l-1} c_s(v)u(s), \quad v \in \mathbb{N}_0. \tag{459}$$

Keeping in mind (459) and (145), we immediately obtain the required conclusion since

$$\begin{aligned}(k^l *_0 \Delta^l u)(v) &= \sum_{j=0}^{v}\binom{l+v-j-1}{v-j}(\Delta^l u)(j)\\ &= \sum_{j=0}^{v}\binom{l+v-j-1}{v-j}\sum_{r=0}^{l}(-1)^{l-r}\binom{l}{r}u(j+r)\\ &= \sum_{j=0}^{v}\sum_{r=0}^{l}(-1)^{l-r}\binom{l+v-j-1}{v-j}\binom{l}{r}u(j+r), \quad v \in \mathbb{N}_0.\end{aligned}$$

□

Remark 7.4.9. By the foregoing, the following combinatorial identity holds true:

$$\sum_{\substack{0\leqslant j\leqslant v;\ 0\leqslant r\leqslant l\\ j+r=s}} (-1)^{l-r}\binom{l+v-j-1}{v-j}\binom{l}{r} = 0, \quad \text{if } l \leqslant s \leqslant l + v - 1.$$

This identity is probably known in the existing literature.

In the sequel, we will use the equality (459). Then, by (458), we have

$$c_0(v) = 1, \quad \text{if } m = 1 \quad \text{and} \quad c_0(v) = (-1)^{m+1}\binom{m+v-1}{v}, \quad \text{if } m > 1. \tag{460}$$

In this subsection, we assume that the following condition holds true:

(C-d) There exists a sequence $c(\cdot)$ such that $b *_0 c = k^m$.

Then $b(0)c(0) = 1$ and, therefore, $c(0) \neq 0$, which simply implies that for every two sequences $v_1 : \mathbb{N}_0 \to X$ and $v_2 : \mathbb{N}_0 \to X$, we have

$$c *_0 v_1 = c *_0 v_2 \Rightarrow v_1 = v_2. \tag{461}$$

Set $v_{a,g} := a *_0 [u(\cdot) - g(\cdot, u.)]$. Convoluting the both sides of (457) with $c(\cdot)$ and using Lemma 7.4.8 after that, we get that (457) is equivalent with (cf. also (461)):

$$\begin{aligned}(a *_0 u)(v+m) &= (a *_0 g(\cdot, u.))(v+m) + c_m(v)v_{a,g}(m-1) + \cdots + c_0(v)v_{a,g}(0)\\ &+ \left(c *_0 \left[\sum_{j=1}^{w} B_j(\cdot)u(\cdot + v_j) + f(\cdot, u.)\right]\right)(v), \quad v \in \mathbb{N}_0.\end{aligned} \tag{462}$$

If $a = \delta$, then the last equality is equivalent with

$$\begin{aligned}u(v+m) &= g(\cdot, u.)(v+m) + c_m(v)v_{a,g}(m-1) + \cdots + c_0(v)v_{a,g}(0)\\ &+ \left(c *_0 \left[\sum_{j=1}^{w} B_j(\cdot)u(\cdot + v_j) + f(\cdot, u.)\right]\right)(v), \quad v \in \mathbb{N}_0;\end{aligned} \tag{463}$$

if this is not the case, then we assume that there exist an integer $s \in \mathbb{N}$ and a sequence $d : \mathbb{N}_0 \to \mathbb{C}$ such that $a *_0 d = k^s$ as well as that there exists a sequence $e : \mathbb{N}_0 \to \mathbb{C}$ such that $c *_0 d = k^s *_0 e$. Clearly, $d(0) \neq 0$ and, therefore, (462) is equivalent with

$$\begin{aligned}(k^s *_0 u)(v+m) &= (k^s *_0 g(\cdot, u.))(v+m)\\ &+ (d *_0 c_m)(v)v_{a,g}(m-1) + \cdots + (d *_0 c_0)(v)v_{a,g}(0)\\ &+ \left(k^s *_0 e *_0 \left[\sum_{j=1}^{w} B_j(\cdot)u(\cdot + v_j) + f(\cdot, u.)\right]\right)(v), \quad v \in \mathbb{N}_0.\end{aligned} \tag{464}$$

Keeping in mind the equations (460)–(464) and the argumentation given in the continuous setting, we can simply prove the following analogues of Theorem 7.4.2 and Theorem 7.4.4, respectively (it is worth noting that Theorem 7.4.10 and Theorem 7.4.11 provide a proper extension of the main result of paper [485], Theorem 1; the interested readers may endeavor to extend the stability result established in [485, Theorem 2] to the abstract fractional difference equations with generalized Hilfer (a, b, α)-derivatives).

Theorem 7.4.10. *Suppose that $a = \delta$, (C-d) holds, there exist a real constant $L > 0$ and a function $p_f : \mathbb{N}_0 \to [0, \infty)$ such that*

$$\|g(v, \xi_1) - g(v, \xi_2)\| \leqslant L \sup_{s \in -\mathbb{N}_\tau^0} \|\xi_1(s) - \xi_2(s)\| \tag{465}$$

and

$$\|f(v, \xi_1) - f(v, \xi_2)\| \leqslant p_f(v) \sup_{s \in -\mathbb{N}_\tau^0} \|\xi_1(s) - \xi_2(s)\| \tag{466}$$

for all ξ_1, $\xi_2 \in X^{-\mathbb{N}_\tau^0}$ and $v \in \mathbb{N}_0$. Suppose further that the following condition holds true:

(D-d) *There exists a unique element $\beta \in X$ such that the constant function $\beta(s) := \beta$, $s \in -\mathbb{N}_\tau^0$ satisfies*

$$\beta = g(v + m, \beta) + \left(c *_0 \left[\sum_{j=1}^{w} B_j(\cdot)\beta + f(\cdot, \beta) \right] \right)(v) \quad \text{for all } v \in \mathbb{N}_0,$$

if $m \geqslant 2$, respectively,

$$\beta = g(v + 1, \beta) + \xi(0) - g(0, \xi) + \left(c *_0 \left[\sum_{j=1}^{w} B_j(\cdot)\beta + f(\cdot, \beta) \right] \right)(v) \quad \text{for all } v \in \mathbb{N}_0,$$

if $m = 1$.

Then there exists a unique mild solution of problem (457) *satisfying $\lim_{v \to +\infty} u(v) = \beta$ and $[u(\cdot) - g(\cdot, u_\cdot)](j) = 0$, $j \in \mathbb{N}_{m-1}^0$, if $m \geqslant 2$, provided that there exists a sequence $z : \mathbb{N}_0 \to \mathbb{N}_0$ such that $0 < z(v) < v$ for $v \in \mathbb{N}$, $\lim_{v \to +\infty} z(v) = +\infty$,*

$$\lim_{v \to +\infty} \sum_{j=0}^{z(v)} |c(v - j)| \cdot \left[\sum_{r=1}^{w} L_r(j) + p_f(j) \right] = 0 \tag{467}$$

and

$$L + \sup_{v \in \mathbb{N}_0} \sum_{j=0}^{v} |c(v - s)| \cdot \left[\sum_{r=1}^{w} L_r(j) + p_f(j) \right] < 1. \tag{468}$$

In particular, (D-d) *holds provided that, for every constant function $v(\cdot) \equiv v$, the sequence $v \mapsto g(v + m, v) + (c *_0 [\sum_{j=1}^{w} B_j(\cdot)v + f(\cdot, v)])(v)$, $v \in \mathbb{N}_0$ is also constant.*

Theorem 7.4.11. *Suppose that $a \neq \delta$, (C-d) holds, there exist $s \in \mathbb{N}$ and $d(\cdot)$ such that $a *_0 d = g_s$. Suppose further that there exists $e(\cdot)$ such that $c *_0 d = g_s *_0 e$, there exist a real constant $L > 0$ and a function $p_f : \mathbb{N}_0 \to [0, \infty)$ such that* (465) *and* (466) *hold for all ξ_1, $\xi_2 \in X^{-\mathbb{N}_\tau^0}$ and $v \in \mathbb{N}_0$, and the following condition holds:*

(D1-d) *There exists a unique element $\beta \in X$ such that the constant function $\beta(s) := \beta$, $s \in -\mathbb{N}_\tau^0$ satisfies*

$$\beta = g(v+m,\beta) + \xi(0) - g(0,\xi) + \left(e *_0 \left[\sum_{j=1}^{w} B_j(\cdot)\beta + f(\cdot,\beta)\right]\right)(v) \quad \text{for all } v \in \mathbb{N}_0$$

provided that $d = \delta$ and $m = 1$ or $d \neq \delta$ and $d *_0 c_0 = 1$, respectively,

$$\beta = g(v+m,\beta) + \left(e *_0 \left[\sum_{j=1}^{w} B_j(\cdot)\beta + f(\cdot,\beta)\right]\right)(v) \quad \text{for all } v \in \mathbb{N}_0,$$

otherwise.

Then there exists a unique mild solution of problem (457) *satisfying $\lim_{v\to+\infty} u(v) = \beta$, $[a *_0 (u(\cdot) - g(\cdot,u_\cdot))](j) = 0$ for $1 \leqslant j \leqslant m-1$, and*

$$u(0) = g(0,u_0), \quad \text{i. e., } \xi(0) = g(0,\xi), \quad \text{provided that}$$
$$d = \delta \text{ and } m \geqslant 2, \quad \text{or} \quad d \neq \delta \text{ and } d *_0 c_0 \neq 1,$$

and there exists a sequence $z : \mathbb{N}_0 \to \mathbb{N}_0$ such that $0 < z(v) < v$ for $v \in \mathbb{N}$, $\lim_{v\to+\infty} z(v) = +\infty$, and (467)–(468) *hold with the sequence $c(\cdot)$ replaced therein with the sequence $e(\cdot)$.*

In particular, (D1-d) *holds provided that, for every constant function $v(\cdot) \equiv v$, the sequence $v \mapsto g(v+m,v) + (e *_0 [\sum_{j=1}^{w} B_j(\cdot)v + f(\cdot,v)])(v)$, $v \in \mathbb{N}_0$ is also constant.*

In the case of consideration of Theorem 7.4.10, we have that the condition (D-a) holds provided that, for every constant function $v(\cdot) \equiv v$, the usual Caputo fractional derivative of the sequence $v \mapsto D^{\alpha,1}u(v,v)$, $v \in \mathbb{N}_0$, where $0 < \alpha < 1$, is equal to $-[\sum_{j=1}^{w} B_j(v)v + f(v,v)]$ for all $v \in \mathbb{N}_0$. This is no longer true for the Hilfer fractional derivatives $D^{\alpha,\beta}u$ of order $\beta < 1$ because the equality $D^{\alpha,\beta}u = 0$ does not imply that the sequence $u(\cdot)$ is constant; for example, we have $[D^{\alpha,\beta}k^{1-(1-\beta)(1-\alpha)}](v) = 0$ for all $v \in \mathbb{N}_0$.

Let us finally emphasize that Example 7.4.6 can be reworded in the discrete setting as well as that the analysis carried out in Remark 7.4.5 can be given in the discrete setting using the Z-transform of sequences in place of the Laplace transform of functions.

Asymptotic behavior of the solutions to the systems of partial linear difference equations depending on two variables has been analyzed by K. Konstaninidis, G. Papaschinopoulos and C. J. Schinas in [441]. We close this section by observing that the asymptotic behavior of solutions of delay dynamic equations on time scales has been analyzed by J. Čermak and M. Urbánek in [166]; we will consider the fractional analogues of the results established in [166] somewhere else.

7.5 Asymptotically almost periodic type solutions of abstract nonlinear fractional differential equations with delay and generalized Hilfer (a, b, α)-derivatives

In this section, we investigate the existence and uniqueness of asymptotically almost periodic type solutions for a class of the abstract nonlinear fractional differential equations with delay and generalized Hilfer (a, b, α)-derivatives. We essentially apply the Banach contraction principle to achieve our aims.

Let us recall that a bounded continuous function $f : [0, \infty) \to X$ is said to be asymptotically almost periodic (a. a. p.) if for every $\varepsilon > 0$ we can find numbers $l > 0$ and $M > 0$ such that every subinterval of $[0, \infty)$ of length l contains, at least, one number τ such that $\|f(t+\tau)-f(t)\| \leqslant \varepsilon$ for all $t \geqslant M$. The space consisted of all asymptotically almost periodic functions from $[0, \infty)$ into X will be denoted by $AAP([0, \infty) : X)$. It is well known that, for every continuous function $f : [0, \infty) \to X$, the following statements are equivalent:

(i) $f \in AAP([0, \infty) : X)$.

(ii) There exist uniquely determined functions $g \in AP(\mathbb{R} : X)$ and $\phi \in C_0([0, \infty) : X)$ such that $f(t) = g(t) + \phi(t)$ for all $t \geqslant 0$.

(iii) The set $H(f) := \{f(\cdot + s) : s \geqslant 0\}$ is relatively compact in $C_b([0, \infty) : X)$, which means that for any sequence (b_k) of nonnegative real numbers there exists a subsequence (a_k) of (b_k) such that $(f(\cdot + a_k))$ converges in $C_b([0, \infty) : X)$.

Further on, by $C_0([0, \infty) \times X : Y)$ we denote the vector space of all continuous functions $q : [0, \infty) \times X \to Y$ such that, for every relatively compact set $K \subseteq X$, we have $\lim_{t\to+\infty} \sup_{x\in K} \|q(t, x)\| = 0$. In this section, we will use the following notion.

Definition 7.5.1. (i) A function $f : \mathbb{R} \times X \to Y$ is called almost periodic if $f(\cdot, \cdot)$ is continuous as well as for every $\varepsilon > 0$ and every relatively compact $K \subseteq X$ there exists $l(\varepsilon, K) > 0$ such that every subinterval $J \subseteq \mathbb{R}$ of length $l(\varepsilon, K)$ contains a number τ with the property that $\|f(t + \tau, x) - f(t, x)\| \leqslant \varepsilon$ for all $t \in \mathbb{R}$, $x \in K$. Denote by $AP(\mathbb{R} \times X : Y)$ the vector space consisting of all such functions.

(ii) A function $f : [0, \infty) \times X \to Y$ is said to be asymptotically almost periodic if $f(\cdot, \cdot)$ is continuous and admits a decomposition $f(t, x) = g(t, x) + q(t, x)$, $t \geqslant 0$, $x \in X$, where $g : \mathbb{R} \times X \to Y$ is almost periodic and $q \in C_0([0, \infty) \times X : Y)$. Denote by $AAP([0, \infty) \times X : Y)$ the vector space consisting of all such functions.

The composition principles clarified below can be deduced with the help of the argumentation contained in the proofs of [246, Theorems 3.30, 3.49], where T. Diagana has used bounded sets in place of relatively compact sets.

Lemma 7.5.2. (i) *Let $f \in AP(\mathbb{R} \times X : Y)$ and $h \in AP(\mathbb{R} : X)$. If there exists a finite real constant $L > 0$ such that*

$$\|f(t, x) - f(t, y)\| \leqslant L\|x - y\|, \quad t \in \mathbb{R},\ x, y \in X,$$

then the mapping $t \mapsto f(t, h(t))$, $t \in \mathbb{R}$ belongs to the space $AP(\mathbb{R} : Y)$.

(ii) *Let $f \in AAP([0,\infty) \times X : Y)$ and $h \in AAP([0,\infty) : X)$. If $f = g + q$, where $g(\cdot,\cdot)$ and $q(\cdot,\cdot)$ satisfy the assumptions from* Definition 7.5.1(ii), *and there exist finite real constants $L_1 > 0$ and $L_2 > 0$ such that*

$$\|g(t,x) - g(t,y)\| \leqslant L_1\|x - y\|, \quad t \in \mathbb{R},\ x,\ y \in X,$$

and

$$\|q(t,x) - q(t,y)\| \leqslant L_2\|x - y\|, \quad t \geqslant 0,\ x,\ y \in X,$$

then the mapping $t \mapsto f(t,h(t))$, $t \geqslant 0$ belongs to the space $AAP([0,\infty) : Y)$.

In this section, we examine the existence and uniqueness of asymptotically almost periodic type solutions to the equation (447), provided that condition (C) holds (we will use the same notation as in the previous section). The main result reads as follows.

Theorem 7.5.3. *Suppose that $a(t) \equiv \delta(t)$, (C) holds with $c \in L^1([0,\infty))$, g_{ap}, $f_{ap} \in AP(\mathbb{R} \times \mathbb{X} : X)$, g_q, $f_q \in C_0([0,\infty) \times \mathbb{X} : X)$, $g = g_{ap} + g_q$ and $f = f_{ap} + f_q$ on $[0,\infty) \times \mathbb{X}$, there exist real constants $L_1 > 0$, $L_2 > 0$, $L_3 > 0$ and $L_4 > 0$ such that*

$$\|g_{ap}(t,\xi_1) - g_{ap}(t,\xi_2)\| \leqslant L_1 \sup_{s\in[-\tau,0]} \|\xi_1(s) - \xi_2(s)\|, \quad t \in \mathbb{R},\ \xi_1,\ \xi_2 \in \mathbb{X}, \tag{469}$$

and

$$\|g_q(t,\xi_1) - g_q(t,\xi_2)\| \leqslant L_2 \sup_{s\in[-\tau,0]} \|\xi_1(s) - \xi_2(s)\|, \quad t \geqslant 0,\ \xi_1,\ \xi_2 \in \mathbb{X}, \tag{470}$$

as well as

$$\|f_{ap}(t,\xi_1) - f_{ap}(t,\xi_2)\| \leqslant L_3 \sup_{s\in[-\tau,0]} \|\xi_1(s) - \xi_2(s)\|, \quad t \in \mathbb{R},\ \xi_1,\ \xi_2 \in \mathbb{X}, \tag{471}$$

and

$$\|f_q(t,\xi_1) - f_q(t,\xi_2)\| \leqslant L_4 \sup_{s\in[-\tau,0]} \|\xi_1(s) - \xi_2(s)\|, \quad t \geqslant 0,\ \xi_1,\ \xi_2 \in \mathbb{X}. \tag{472}$$

Suppose further that the following conditions hold true:
(i) *The mapping $t \mapsto B_j(t)x$, $t \geqslant 0$ is continuous for every $x \in X$ and $j \in \mathbb{N}_w$.*
(ii) *For every $j \in \mathbb{N}_w$, for every relatively compact set $K \subseteq X$ and for every sequence (b_n) of nonnegative real numbers, there exists a subsequence (a_n) of (b_n) such that, for every $\varepsilon > 0$, there exists $n_0 \in \mathbb{N}_0$ such that, for every m, $n \in \mathbb{N}$ with $\min(m,n) \geqslant n_0$, we have*

$$\sup_{x\in K;t\geqslant 0} \|B_j(t + a_n)x - B_j(t + a_m)x\| \leqslant \varepsilon. \tag{473}$$

Then there exists a unique asymptotically almost periodic mild solution of problem (447) *satisfying* $[u(\cdot) - g(\cdot, u_\cdot)]^{(j)}(0) = 0$ *for* $1 \leqslant j \leqslant m-1$, *provided that*

$$L_1 + L_2 + (L_3 + L_4 + L_{B,1} + \cdots + L_{B,w}) \cdot \int_0^{+\infty} |c(t)|\, dt < 1. \tag{474}$$

Proof. Set

$$P := \{h \in C([-\tau, +\infty) : X) : h_{|[-\tau,0]}(\cdot) \equiv \xi(\cdot) \text{ and } h_{|[0,\infty)} \in AAP([0,\infty) : X)\},$$

as well as $d(h_1, h_2) := \sup_{t \geqslant -\tau} \|h_1(t) - h_2(t)\|$ for any $h_1,\ h_2 \in P$. Then (P, d) is a complete metric space, as easily approved. Now we will prove that the mapping $\Psi : P \to P$, given by $(\Psi h)(t) := \xi(t)$, $t \in [-\tau, 0]$ and

$$(\Psi h)(t) := \xi(0) - g(0, \xi) + g(t, h_t) + \left(c *_0 \left[\sum_{j=1}^{w} B_j(\cdot) h(\cdot + t_j) + f(\cdot, h_\cdot)\right]\right)(t), \quad t > 0,$$

is a well-defined contraction. First of all, the assumption (i) and the assumptions on $f(\cdot,\cdot)$, $g(\cdot,\cdot)$ and $B_j(\cdot)$ imply that the condition (C1) from the formulation of [486, Theorem 2.2] holds. Then we can repeat verbatim the argumentation contained in the proof of the aforementioned result in order to see that the mapping $t \mapsto (\Psi h)(t)$, $t \geqslant -\tau$ is continuous.

Let us prove now that $\Psi h_{|[0,\infty)} \in AAP([0,\infty) : X)$. First of all, the set $\{h_t : t \geqslant 0\}$ is relatively compact in $\mathbb{X}$ on account of the Arzela–Ascoli theorem. Since the mapping $t \mapsto h_t \in \mathbb{X}$ is bounded, continuous and $h_{|[0,\infty)} \in AAP([0,\infty) : X)$, we can simply prove that the mapping $t \mapsto h_t \in \mathbb{X}$ is a. a. p. The assumption on $g(\cdot,\cdot)$ and Lemma 7.5.2(ii) together imply that the mapping $t \mapsto g(t, u_t)$, $t \geqslant 0$ is likewise a. a. p. Similarly, the mapping $t \mapsto f(t, u_t)$, $t \geqslant 0$ is a. a. p. and the well-known argumentation shows that the mapping $t \mapsto (c *_0 f(\cdot, u_\cdot))(t)$, $t \geqslant 0$ is a. a. p. as well [444]. Basically, it only remains to be proved that the mapping $t \mapsto B_j(t)h(t + t_j)$, $t \geqslant 0$ is a. a. p. Toward this end, let us observe first that this mapping is continuous due to the assumption (i) and the Lipschitz continuity of operators $B_j(\cdot)$. Suppose now that (b_n') is a given sequence of nonnegative real numbers. Then there exists a subsequence (b_n) of (b_n') such that $(u(\cdot + t_j + b_n))_{n\in\mathbb{N}}$ converges in $C_b([0,\infty) : X)$ as $n \to +\infty$. Now we can extract a subsequence (a_n) of (b_n) such that, for every $\varepsilon > 0$, there exists $n_0 \in \mathbb{N}_0$ such that, for every $m,\ n \in \mathbb{N}$ with $\min(m, n) \geqslant n_0$, we have (473). This simply implies that the sequence $(B_j(\cdot + a_n)u(\cdot + t_j + a_n))_{n\in\mathbb{N}}$ converges in $C_b([0,\infty) : X)$ as $n \to +\infty$ since it is a Cauchy one; in actual fact, we have

$$\begin{aligned}
&\|B_j(\cdot + a_n)u(\cdot + t_j + a_n) - B_j(\cdot + a_m)u(\cdot + t_j + a_m)\| \\
&\quad \leqslant \|B_j(\cdot + a_n)u(\cdot + t_j + a_n) - B_j(\cdot + a_n)u(\cdot + t_j + a_m)\| \\
&\qquad + \|B_j(\cdot + a_n)u(\cdot + t_j + a_m) - B_j(\cdot + a_m)u(\cdot + t_j + a_m)\| \\
&\quad \leqslant L_{B,j}\|u(\cdot + t_j + a_n) - u(\cdot + t_j + a_m)\| + \sup_{x\in K; t\geqslant 0} \|B_j(\cdot + a_n)x - B_j(\cdot + a_m)x\|,
\end{aligned}$$

where $K = R(u)$ is a relatively compact set in X. Therefore, it suffices to apply (ii) to obtain the required conclusion.

Therefore, the mapping $\Phi : P \to P$ is well-defined; keeping in mind the estimate (474), a simple calculation yields that the mapping $\Phi : P \to P$ is a contraction. Since $a(t) = \delta(t)$, $[u(\cdot) - g(\cdot, u_\cdot)]^{(j)}(0) = 0$ for $1 \leqslant j \leqslant m-1$ and $u(0) = \xi(0)$, the equation (448) is equivalent to $u = \Psi u$, so that the Banach contraction principle implies that there exists a unique asymptotically almost periodic mild solution of problem (447), which satisfies the given properties. □

Keeping in mind the above argumentation and the argumentation contained in the proof of Theorem 7.4.4, we can similarly prove the following result.

Theorem 7.5.4. *Suppose that $a \in L^1_{loc}([0,\infty))$, (C) holds, there exist $s \in \mathbb{N}$ and $d(t) = \delta(t)$ or $d \in L^1_{loc}([0,\infty))$ such that $a *_0 d = g_s$. Suppose further that there exists $e(t) = \delta(t)$ or $e \in L^1([0,\infty))$ such that $c *_0 d = g_s *_0 e$, g_{ap}, $f_{ap} \in AP(\mathbb{R} \times \mathbb{X} : X)$, g_q, $f_q \in C_0([0,\infty) \times \mathbb{X} : X)$, $g = g_{ap} + g_q$ and $f = f_{ap} + f_q$ on $[0,\infty) \times \mathbb{X}$, and there exist real constants $L_1 > 0$, $L_2 > 0$, $L_3 > 0$ and $L_4 > 0$ such that the equations (469)–(472) hold. Then there exists a unique asymptotically almost periodic mild solution of problem (447) satisfying $[a *_0 (u(\cdot) - g(\cdot, u_\cdot))]^{(j)}(0) = 0$ for $1 \leqslant j \leqslant m-1$, provided that the equation (474) holds with the function $c(\cdot)$ replaced therein with the function $e(\cdot)$.*

Possible applications of our results can be given to the Lipschitz continuous operators considered in Example 7.4.6. In part (i) of this example, it suffices to assume additionally that the functions $b_j(t)$ are asymptotically almost periodic for $t \geqslant 0$ $(1 \leqslant j \leqslant w)$; in part (ii), it suffices to assume additionally the above condition and the condition that, for every relatively compact set K in $L^p(\Omega)$, we have $\sup_{f \in K} \int_\Omega |b_{j,1}(f(x))|^p \, dx < +\infty$. The existence and uniqueness of asymptotically almost periodic type solutions to the equation (457) can be considered in a similar manner.

8 Multidimensional fractional calculus, multidimensional Poisson transform and applications

8.1 Multidimensional fractional calculus: theory and applications

The partial fractional derivatives of functions have not attracted much attention of the authors working in the field of fractional calculus at this time. With the exception of the structural theory developed in Chapter 5 of the fundamental research monograph [682] by S. G. Samko, A. A. Kilbas, O. I. Marichev and some structural results about the partial fractional differential equations given in Chapter 7 in the fundamental research monograph [431] by A. A. Kilbas, M. Srivastava and J. J. Trujillo, we can freely say that almost all established results regarding partial fractional derivatives of functions and partial fractional differential equations given at this time are rather fragmentary and concern very a special kind of functions and partial fractional differential equations.

For example, H. M. Srivastava, R. C. Singh Chandel and P. K. Vishwakarma have analyzed in [720] the partial fractional derivatives of certain generalized hypergeometric functions of several variables (see also [170]); the partial fractional differential equations with Riesz space-fractional derivatives of positive real order (see [682, Section 25, p. 357] for the notion and more details) have been analyzed by H. Jiang et al. in [402] (see also [511]). It is also worth mentioning the recent research article [639] by V. Pilipauskaitė and D. Surgailis, where the authors have analyzed certain fractional operators and fractionally integrated random fields on $\mathbb{Z}^n$. Further on, M. O. Mamchuev [568] and A. V. Pshku [654] have considered the systems of multidimensional fractional partial differential equations containing the terms of form $D_{x_j}^{a_j}u(x_1,\ldots,x_n)$ with just one index $j \in \mathbb{N}_n$ and not the general forms of partial fractional derivatives introduced in this paper. More precisely, A. V. Pshku has considered in [654] the well-posedness of the following multidimensional fractional partial differential equation:

$$\sum_{k=1}^{n} a_k \frac{\partial^{\sigma_k}}{\partial x_k^{\sigma_k}} u(x) + \lambda u(x) = f(x), \quad x \in [0,\infty)^n, \tag{475}$$

where $(\partial^{\sigma_k}/\partial x_k^{\sigma_k} u)$ denotes the fractional partial derivative of order σ_k with respect to the variable x_k with origin $x_k = 0$ (in the sense of Riemann–Liouville, Caputo or Dzhrbashyan–Nersesyan approach). We also refer the reader to the works mentioned on [682, pp. 623–624] and some recent results about nonlinear fractional partial differential equations obtained in [44, 155, 234, 354, 391, 651, 660].

The structure and main ideas of this section can be briefly summarized as follows. We first introduce the notion of multidimensional generalized Hilfer fractional derivative $\mathbb{D}_{\mathbf{a},\mathbf{b}}^{a}u$ for a class of locally integrable functions $u : [0,\infty)^n \to X$. After that, we introduce the multidimensional generalized Hilfer fractional discrete derivative $\mathbb{D}_{\mathbf{a},\mathbf{b}}^{a}u$,

https://doi.org/10.1515/9783111689746-012

for any sequence $u : \mathbb{N}_0^n \to X$. It seems that the notion introduced in this section seems to be not considered elsewhere in the existing literature, even for the Riemann–Liouville or Caputo fractional derivatives. In the next two subsections, we examine the multidimensional generalized Weyl fractional derivatives and differences. The first subsection investigates the generalized Weyl fractional derivatives and differences in the one-dimensional setting. In Definition 8.1.4, we introduce the notion of a generalized Weyl fractional derivative $D_W^{\alpha,a}u$ of function $u(\cdot)$. After that, we examine the basic structural properties of the introduced fractional derivatives. If $a : \mathbb{N}_0 \to \mathbb{C}$ and $f : \mathbb{Z} \to X$ are given sequences, then we define the Weyl fractional difference operator $\Delta_{W,a,m}f$. We show that the approach of R. Hilfer [381] is meaningless for the definitions of Weyl fractional derivatives and differences.

The second subsection investigates the generalized Weyl fractional derivatives and differences in the multidimensional setting (concerning some predecessors of this work, we would like to mention here the research articles [182] by V. B. L. Chaurasia, R. S. Dubey, [339] by S. P. Goyal, Trilok Mathur, [397] by B. B. Jaimini, H. Nagar and [662] by R. K. Raina; see also the lists of references quoted therein). We first introduce the notion of a generalized Weyl $(\alpha, \mathbf{a})$-fractional derivative $\mathbb{D}_W^{\alpha,\mathbf{a}}u$; a very special case of the partial fractional derivative $\mathbb{D}_W^{\alpha,\mathbf{a}}u$ is the generalized Weyl $(\alpha, \mathbf{a})$-fractional derivative $\mathbb{D}_W^{\alpha}u$. After that, if the sequences $a_j : \mathbb{N}_0 \to \mathbb{C}$ and $u : \mathbb{Z}^n \to X$ are given and if $m_j \in \mathbb{N}$ is a given integer $(1 \leqslant j \leqslant n)$, then we introduce the multidimensional Weyl fractional difference operator $\mathbb{D}_{W,\mathbf{a},\mathbf{m}}u$. We investigate the law of exponents for generalized Weyl derivatives and integrals and provide an interesting open problem about the generation of C-regularized solution operator families by the Weyl fractional differential operators with constant coefficients. Furthermore, we reconsider the well-known Clairaut's theorem on equality of mixed partial derivatives (sometimes also called Schwartz's theorem or Young's theorem) in the fractional setting and prove that it is not valid for the Riemann–Liouville and Caputo fractional derivatives (see [682, p. 342] for the first results established in this direction) as well as that it is valid for the Weyl fractional derivatives, under certain reasonable assumptions.

In Section 8.1.4, we introduce and analyze the partial fractional derivatives of functions defined on some special regions in $\mathbb{R}^n$ and the partial fractional differences of sequences defined on some special subsets of $\mathbb{Z}^n$. Further on, the investigation of two-dimensional scalar-valued Laplace transform starts probably with the works of D. L. Bernstein [109, 110] and J. C. Jaeger [395] (1939–1941); for more details about the multidimensional scalar-valued Laplace transform and its applications to (fractional) partial integrodifferential equations, we refer the reader to the research articles [111, 230, 231, 275, 276, 676] and the doctoral dissertations [29, 79, 607]. For the purpose of our investigations of the partial fractional integrodifferential inclusions, we provide the basic details and results about the multidimensional vector-valued Laplace transform in Section 8.1.5 (we will systematically analyze multidimensional vector-valued Laplace transform in our follow-up research studies). Our main structural result established in this section is Theorem 8.1.7, where we clarify the complex inversion theorem for the multidimensional vector-valued Laplace transform.

Fractional partial differential inclusions with generalized Hilfer derivatives are investigated in Section 8.1.6, whose main results are Theorem 8.1.9 and Theorem 8.1.10 (cf. also Remark 8.1.11 and Remark 8.1.12); Section 8.1.7, whose main result is Theorem 8.1.14, investigates the abstract multiterm fractional partial differential equations with Riemann–Liouville and Caputo derivatives, while Section 8.1.8 investigates the fractional partial difference equations with generalized Weyl derivatives.

If $u \in L^1_{\mathrm{loc}}([0,\infty)^n)$, $j \in \mathbb{N}_n$ and $a_j > 0$, then we define

$$J_{t_j}^{a_j} u(x_1,\ldots,x_{j-1},x_j,x_{j+1},\ldots,x_n) := \int_0^{x_j} g_{a_j}(x_j - s)u(x_1,\ldots,x_{j-1},s,x_{j+1},\ldots,x_n)\,ds,$$

$$\mathbf{x} = (x_1,\ldots,x_{j-1},x_j,x_{j+1},\ldots,x_n) \in [0,\infty)^n.$$

Further on, if $a(\cdot)$ is a given sequence in X, which depends on the variables $v_1,\ldots,v_n$, then we define

$$\Delta_{v_i} a(v_1,\ldots,v_i,\ldots,v_n) := a(v_1,\ldots,v_i+1,\ldots,v_n) - a(v_1,\ldots,v_i,\ldots,v_n).$$

After that, we set $\Delta^2_{v_i v_j} a := \Delta_{v_i}\Delta_{v_j} a$ and $\Delta^2_{v_i v_i} a := \Delta_{v_i}\Delta_{v_i} a$; the terms

$$\Delta^m_{v_{i_1}\ldots v_{i_m}} a \quad \text{and} \quad \Delta^{|\alpha|}_{v_1^{\alpha_1}\ldots\ldots v_n^{\alpha_n}} a$$

are defined recursively, as for the partial derivatives of functions ($\alpha_i \in \mathbb{N}_0$; $|\alpha| = \alpha_1 + \cdots + \alpha_n$). It is worth noting that, for every permutation $\sigma : \mathbb{N}_n \to \mathbb{N}_n$, we have

$$\Delta^{|\alpha|}_{v_1^{\alpha_1}\ldots\ldots v_n^{\alpha_n}} a = \Delta^{|\alpha|}_{v_{\sigma(1)}^{\alpha_{\sigma(1)}}\ldots\ldots v_{\sigma(n)}^{\alpha_{\sigma(n)}}} a, \tag{476}$$

as easily approved.

8.1.1 Multidimensional generalized Hilfer fractional derivatives and differences

Suppose that $0 < T_j < +\infty$ and $I_j = [0, T_j)$, $I_j = [0, T_j]$ or $I_j = [0, +\infty)$ for $1 \leqslant j \leqslant n$. Set $I := I_1 \times I_2 \times \cdots \times I_n$. Suppose that $u : I \to X$ is a locally integrable function and, for every $j \in \mathbb{N}_n$, $a_j \in L^1_{\mathrm{loc}}(I_j)$ or $a_j(t) = \delta(t)$, and $b_j \in L^1_{\mathrm{loc}}(I_j)$ or $b_j(t) = \delta(t)$. Suppose further that $\alpha_j \geqslant 0$ for all $j \in \mathbb{N}_n$. Define $\alpha := (\alpha_1,\ldots,\alpha_n)$ and

$$\mathbb{D}^{\alpha}_{\mathbf{a},\mathbf{b}} u(x_1,\ldots,x_n) := [D^{\alpha_1}_{a_1,b_1}(D^{\alpha_2}_{a_2,b_2}(\ldots(D^{\alpha_n}_{a_n,b_n} u(\cdot,\ldots,\cdot))\ldots))](x_1,\ldots,x_n), \tag{477}$$

for a. e. $(x_1,\ldots,x_n) \in I$, provided that the right hand side of (477) is well-defined. Here, we assume that the variables $x_1, x_2,\ldots,x_{n-1}$ are fixed in the computation of the term

$D^{a_n}_{a_n,b_n}u(x_1,\ldots,x_n),\ldots,$ as well as that the variables $x_2,\ x_3,\ldots,x_n$ are fixed in the computation of the final term on the right hand side of (477). We call $\mathbb{D}^{\alpha}_{\mathbf{a},\mathbf{b}}u$ the multidimensional generalized Hilfer $(\mathbf{a},\mathbf{b},\alpha)$-fractional derivative of the function $u(\cdot)$. If for each $j \in \mathbb{N}_n$ we have $D^{\alpha_j}_{a_j,b_j} = D^{\alpha_j}_{R}$, respectively, for each $j \in \mathbb{N}_n$ we have $D^{\alpha_j}_{a_j,b_j} = \mathbf{D}^{\alpha_j}_{C}$, then the corresponding partial fractional derivative $\mathbb{D}^{\alpha}_{\mathbf{a},\mathbf{b}}$ is called the multidimensional Riemann–Liouville fractional operator (cf. also [682, pp. 340–342]), respectively, the multidimensional Caputo fractional operator, and it is denoted by $\mathbb{D}^{\alpha}_{\mathbf{a},\mathbf{b}} = \mathbb{D}^{\alpha}_{R}$, respectively, $\mathbb{D}^{\alpha}_{\mathbf{a},\mathbf{b}} = \mathbb{D}^{\alpha}_{C}$.

In the discrete setting, we assume that $u : \mathbb{N}_0^n \to X$, $a_j : \mathbb{N}_0 \to \mathbb{C}$ and $b_j : \mathbb{N}_0 \to \mathbb{C}$ are given sequences $(1 \leqslant j \leqslant n)$. We define

$$\mathbb{D}^{\alpha}_{\mathbf{a},\mathbf{b}}u(v_1,\ldots,v_n) := \left[D^{\alpha_1}_{a_1,b_1}\left(D^{\alpha_2}_{a_2,b_2}\left(\ldots\left(D^{\alpha_n}_{a_n,b_n}u(\cdot,\ldots,\cdot)\right)\ldots\right)\right)\right](v_1,\ldots,v_n), \tag{478}$$

for any $(v_1,\ldots,v_n) \in \mathbb{N}_0^n$; note that the right-hand side of (478) is always well-defined. We call $\mathbb{D}^{\alpha}_{\mathbf{a},\mathbf{b}}u$ the multidimensional generalized Hilfer $(\mathbf{a},\mathbf{b},\alpha)$-fractional derivative of the sequence $u(\cdot)$; the multidimensional Riemann–Liouville fractional difference operator $\mathbb{D}^{\alpha}_{R}$ and the multidimensional Caputo fractional difference operator $\mathbb{D}^{\alpha}_{C}$ are defined similarly.

We continue by providing certain illustrative examples.

Example 8.1.1. (i) Suppose that $\emptyset \neq D \subseteq [0,+\infty)^n$ is a finite set, $c_\beta \in \mathbb{C}$ for all $\beta = (\beta_1,\ldots,\beta_n) \in D$ and

$$u(x_1,\ldots,x_n) := \sum_{\beta\in D} c_\beta g_{\beta_1}(x_1)\cdot\cdots\cdot g_{\beta_n}(x_n), \quad x_1 \geqslant 0,\ldots,x_n \geqslant 0.$$

Suppose further that $a_j \geqslant 0$, $a_j(t) = g_{\gamma_j}(t)$ and $b_j(t) = g_{\delta_j}(t)$ for some nonnegative numbers $\gamma_j \geqslant 0$ and $\delta_j \geqslant 0$ such that $\gamma_j + \beta_j \geqslant m_j$ $(1 \leqslant j \leqslant n)$. Set $f_j(t) := g_{\delta_j+\gamma_j+\beta_j-m_j}(t)$, $t > 0$, if $\gamma_j + \beta_j > m_j$ and $f_j(t) := 0$, $t \geqslant 0$, if $\gamma_j + \beta_j = m_j$. Then we have

$$\mathbb{D}^{\alpha}_{\mathbf{a},\mathbf{b}}u(x_1,\ldots,x_n) = \sum_{\beta\in D} c_\beta f_1(x_1)\cdot\cdots\cdot f_n(x_n), \quad x_1 \geqslant 0,\ldots,x_n \geqslant 0.$$

This formula enables one to clarify a great number of various partial fractional differential equations, which do have the function $u(x_1,\ldots,x_n)$ as its solution; for example, we have

$$\begin{aligned}&\mathbb{D}^{\alpha}_{\mathbf{a},\mathbf{b}}u(x_1,\ldots,x_n)\\ &= \left[\sum_{\beta\in D} c_\beta \frac{x_1^{\delta_1+\gamma_1-m_1}}{\Gamma(\delta_1+\beta_1+\gamma_1-m_1)}\cdot\cdots\cdot\frac{x_n^{\delta_n+\gamma_n-m_n}}{\Gamma(\delta_n+\beta_n+\gamma_n-m_n)}\right]\cdot u(x_1,\ldots,x_n),\end{aligned}$$

for any $x_1 \geqslant 0,\ldots,x_n \geqslant 0$, provided that $\delta_j + \gamma_j > m_j$ for $1 \leqslant j \leqslant n$.

(ii) Suppose that $\emptyset \neq D \subseteq [0, +\infty)^n$ is a finite set, $c_\beta \in \mathbb{C}$ for all $\beta = (\beta_1, \dots, \beta_n) \in D$ and

$$u(v_1, \dots, v_n) := \sum_{\beta \in D} c_\beta k^{\beta_1}(v_1) \cdot \dots \cdot k_{\beta_n}(v_n), \quad v_1 \in \mathbb{N}_0, \dots, v_n \in \mathbb{N}_0.$$

Suppose further that $a_j \geqslant 0$, $a_j(v) = k^{\gamma_j}(v)$ and $b_j(v) = k^{\delta_j}(v)$ for some nonnegative numbers $\gamma_j \geqslant 0$ and $\delta_j \geqslant 0$ such that $\gamma_j + \beta_j \geqslant m_j$ $(1 \leqslant j \leqslant n)$. Set

$$f_j(v) := k^{\delta_j+\gamma_j+\beta_j-m_j}(v + m_j) - \sum_{l=v+1}^{v+m_j} k^{\gamma_j+\beta_j-m_j}(v + m_j - l)k^{\delta_j}(l), \quad v \in \mathbb{N}_0.$$

We know that (see [472, Example 3]):

$$\Delta^\alpha k^\beta(\cdot) = k^{\beta-\alpha}(\cdot + \lceil \alpha \rceil), \quad \beta \geqslant \alpha > 0.$$

This simply implies

$$\mathbb{D}^\alpha_{\mathbf{a},\mathbf{b}} u(v_1, \dots, v_n) = \sum_{\beta \in D} c_\beta f_1(v_1) \cdot \dots \cdot f_n(v_n), \quad v_1 \in \mathbb{N}_0, \dots, v_n \in \mathbb{N}_0.$$

Remark 8.1.2. (i) Instead of the generalized Hilfer fractional derivatives and differences, we can consider here any other type of fractional derivatives of functions defined on the segment of the nonnegative real axis [236]. In such a way, we can extend the notion considered in this subsection and obtain much more general forms of the partial fractional derivatives.

(ii) It is well known that the composition of the Riemann–Liouville (Caputo) fractional derivatives of orders $\alpha > 0$ and $\beta > 0$ is not the Riemann–Liouville (Caputo) fractional derivative of order $\alpha + \beta$; see [645, Subsections 2.3.5–2.3.6] for more details. We can further extend the notion of fractional derivative $\mathbb{D}^\alpha_{\mathbf{a},\mathbf{b}} u$ by replacing some terms $D^{\alpha_j}_{a_j,b_j}$ in its definition by the finite compositions $D^{\alpha_{j_1}}_{a_{j_1},b_{j_1}} D^{\alpha_{j_2}}_{a_{j_2},b_{j_2}} \cdot \dots \cdot D^{\alpha_{j_s}}_{a_{j_s},b_{j_s}}$ of terms with respect to the variable x_j $(1 \leqslant j \leqslant n)$.

Let us recall that Clairaut's theorem on equality of mixed partial derivatives states that, if a function $u : \Omega \to \mathbb{R}$ defined on a nonempty set $\Omega \subseteq \mathbb{R}^n$ is given, as well as $x \in \mathbb{R}^n$ is a point such that some neighborhood $O(x)$ of it belongs to Ω, and $u(\cdot, \cdot)$ has continuous second partial derivatives on $O(x)$, then we have

$$\frac{\partial^2 u}{\partial x_i \partial x_j}(x) = \frac{\partial^2 u}{\partial x_j \partial x_i}(x).$$

This equality cannot be so easily interpreted for the generalized Hilfer partial fractional derivatives, because the equality

$$[D^{\alpha_1}_{a_1,b_1}(D^{\alpha_2}_{a_2,b_2} u)](x_1, x_2) = [D^{\alpha_2}_{a_2,b_2}(D^{\alpha_1}_{a_1,b_1} u)](x_1, x_2), \tag{479}$$

is not true, in general (of course, it is true in the case that $b = d$, $a = c$ and $m_1 = m_2$, at least almost everywhere). The formula (479) does not hold even for the Riemann–Liouville fractional derivatives and the Caputo fractional derivatives, as the following simple counterexample shows.

Example 8.1.3. Suppose that $0 < \alpha_1 < 1$, $0 < \alpha_2 < 1$ and $\alpha_1 \neq \alpha_2$. Let us consider the Caputo approach, in which $a_1(t) = a_2(t) = \delta(t)$, $b_1(t) = g_{1-\alpha_1}(t)$, $b_2(t) = g_{1-\alpha_2}(t)$ and $m_1 = m_2 = 1$. Then a simple computation shows that the equality (479) is equivalent with

$$\int_0^{x_1} g_{1-\alpha_2}(x_1 - r)\frac{d}{dr}\int_0^{x_2} g_{1-\alpha_1}(x_2 - l)\frac{\partial u}{\partial l}u(r,l)\,dl\,dr$$
$$= \int_0^{x_1} g_{1-\alpha_1}(x_1 - r)\frac{d}{dr}\int_0^{x_2} g_{1-\alpha_2}(x_2 - l)\frac{\partial u}{\partial l}u(r,l)\,dl\,dr. \tag{480}$$

Take now $u(x_1, x_2) := x_1x_2$ for $x_1 \geqslant 0$ and $x_2 \geqslant 0$. Then (480) is equivalent with

$$g_{2-\alpha_2}(x_1)\cdot g_{2-\alpha_1}(x_2) = g_{2-\alpha_1}(x_1)\cdot g_{2-\alpha_2}(x_2),$$

which is wrong. In the discrete setting, we cannot expect the validity of nontrivial fractional analogues of the equation (476).

We continue with the observation that the formulae [97, (1.13), (1.21)] can be straightforwardly extended to the multidimensional setting. For example, if $u \in L^1_{\mathrm{loc}}([0,\infty)^n)$ and $\alpha_j \geqslant 0$ for all $j \in \mathbb{N}_n$, then we have

$$D_R^{\alpha_1}D_R^{\alpha_2}\cdot\cdots\cdot D_R^{\alpha_n}J_{t_n}^{\alpha_n}\cdot\cdots\cdot J_{t_2}^{\alpha_2}J_{t_1}^{\alpha_1}u = u$$

and

$$\mathbf{D}_C^{\alpha_1}\mathbf{D}_C^{\alpha_2}\cdot\cdots\cdot \mathbf{D}_C^{\alpha_n}J_{t_n}^{\alpha_n}\cdot\cdots\cdot J_{t_2}^{\alpha_2}J_{t_1}^{\alpha_1}u = u,$$

with the meaning clear. The situation is a little bit complicated if we consider the second formulae in the equations [97, (1.13), (1.21)]; for example, in the two-dimensional setting, we have

$$J_{t_2}^{\alpha_2}J_{t_1}^{\alpha_1}D_R^{\alpha_1}D_R^{\alpha_2}u(x_1,x_2)$$
$$= u(x_1,x_2) - \sum_{k=0}^{m_2-1}\frac{\partial^k}{\partial x_2^k}[J_{t_2}^{m_2-\alpha_2} *_0 u](x_1,0)\cdot g_{\alpha_2+k+1-m_2}(x_2)$$
$$- \sum_{k=0}^{m_1-1}\left\{\int_0^{x_2} g_{\alpha_2}(x_2 - s)[D_R^{\alpha_2}u(x_1,x_2)]_{x_1=0,x_2=s}\,ds\right\}\cdot g_{\alpha_1+k+1-m_1}(x_1), \tag{481}$$

for any $(x_1,x_2) \in [0,\infty)^2$, provided that $u \in L^1_{\mathrm{loc}}([0,\infty)^2)$, $m_1 = \lceil\alpha_1\rceil$, $m_2 = \lceil\alpha_2\rceil$, for each $x_2 \geqslant 0$ the function $x_1 \mapsto D_R^{\alpha_2}u(x_1,x_2)$, $x_1 > 0$ is locally integrable and satisfies

$J_{t_1}^{m_1-\alpha_1} *_0 D_R^{\alpha_2} u \in W_{\mathrm{loc}}^{m_1,1}([0,\infty) : X)$, and for each $x_1 \geqslant 0$ we have $J_{t_2}^{m_2-\alpha_2} *_0 (\partial^{m_2-1}/\partial x_2^{m_2-1})u \in W_{\mathrm{loc}}^{m_2,1}([0,\infty) : X)$, as well as

$$J_{t_2}^{\alpha_2} J_{t_1}^{\alpha_1} \mathbf{D}_C^{\alpha_1} \mathbf{D}_C^{\alpha_2} u(x_1,x_2) = u(x_1,x_2) - \sum_{k=0}^{m_2-1} \left[\frac{\partial^k}{\partial x_2^k} u(x_1,0) \right] \cdot g_{k+1}(x_2)$$
$$- \sum_{k=0}^{m_1-1} \left[\int_0^{x_2} g_{\alpha_2}(x_2-s) \left[\frac{\partial^k}{\partial x_1^k} \mathbf{D}_C^{\alpha_2} u(x_1,x_2) \right]_{x_1=0,x_2=s} ds \right] \cdot g_{k+1}(x_1), \tag{482}$$

for any $(x_1,x_2) \in [0,\infty)^2$, provided that $m_1 = \lceil \alpha_1 \rceil$, $m_2 = \lceil \alpha_2 \rceil$, $u \in L_{\mathrm{loc}}^1([0,\infty)^2)$, for each $x_2 \geqslant 0$ the function $x_1 \mapsto f(x_1) := (\partial^{m_1-1}/\partial x_1^{m_1-1})\mathbf{D}_C^{\alpha_2} u(x_1,x_2)$, $x_1 \geqslant 0$ is continuous, $g_{m_1-\alpha_1} *_0 f \in W_{\mathrm{loc}}^{m_1,1}([0,\infty) : X)$, for each $x_1 \geqslant 0$ the function $x_2 \mapsto g(x_2) := (\partial^{m_2-1}/\partial x_2^{m_2-1})u(x_1,x_2)$, $x_2 \geqslant 0$ is continuous and $g_{m_2-\alpha_2} *_0 g \in W_{\mathrm{loc}}^{m_2,1}([0,\infty) : X)$.

8.1.2 Generalized Weyl fractional derivatives and differences

If $u : \mathbb{R} \to X$ is a locally integrable function, $\alpha \geqslant 0$ and $m = \lceil \alpha \rceil$, then the Weyl fractional derivative $D_W^\alpha u$ of function $u(\cdot)$ of order α is well-defined if the mapping $x \mapsto \int_{-\infty}^x g_{m-\alpha}(x-s)u(s)\,ds$, $x \in \mathbb{R}$ is well-defined and m-times continuously differentiable, by

$$[D_W^\alpha u](x) := \frac{d^m}{dx^m} \int_{-\infty}^{x} g_{m-\alpha}(x-s)u(s)\,ds, \quad x \in \mathbb{R};$$

cf. [592] for more details. Now we would like to propose the following notion.

Definition 8.1.4. Suppose that $a \in L_{\mathrm{loc}}^1([0,\infty))$, $u : \mathbb{R} \to X$ is a locally integrable function, $\alpha \geqslant 0$ and $m = \lceil \alpha \rceil$. The generalized Weyl fractional derivative $D_W^{\alpha,a} u$ of function $u(\cdot)$ is well-defined if the mapping $x \mapsto \int_{-\infty}^x a(x-s)u(s)\,ds$, $x \in \mathbb{R}$ is well-defined and m-times continuously differentiable, by

$$[D_W^{\alpha,a} u](x) := \frac{d^m}{dx^m} \int_{-\infty}^{x} a(x-s)u(s)\,ds, \quad x \in \mathbb{R}.$$

We call the function $x \mapsto I_{W,a}(x) := \int_{-\infty}^x a(x-s)u(s)\,ds$, $x \in \mathbb{R}$, if it is well-defined, the generalized Weyl a-integral of function $u(\cdot)$. If $a(t) = g_\zeta(t)$ for some $\zeta \in (0,1)$, then the class of functions for which the above integral absolutely converges and behaves nicely has first considered by M. J. Lighthill in [525], where it was called the class of "good functions." In general case, we have

$$\int_{-\infty}^{x} a(x-s)u(s)\,ds = \int_{0}^{+\infty} a(s)u(x-s)\,ds, \quad x \in \mathbb{R}$$

and the dominated convergence theorem implies

$$\frac{d}{dx^n}\int_{-\infty}^{x} a(x-s)u(s)\,ds = \int_{0}^{+\infty} a(s)u^{(n)}(x-s)\,ds, \quad x \in \mathbb{R},\ n \in \mathbb{N},$$

provided that there exists $m \in \mathbb{N}$ such that $\int_0^{+\infty} |a(s)|(1+s)^{-m}\,ds < +\infty$ and the function $u(\cdot)$ and all its derivatives are differentiable almost everywhere and for each $n \in \mathbb{N}$ and $\alpha \in \mathbb{N}_0$ there exists a finite real number $M_{n,\alpha} \geqslant 1$ such that $\|u^{(\alpha)}(x)\| \leqslant M_{n,\alpha}(1+|x|)^{-n}$, $x \in \mathbb{R}$; we call such functions "vector-valued good functions" and denote the corresponding class by $\mathbf{S}(X)$. If (G) holds, where:

(G) There exists an integer $m \in \mathbb{N}$ such that $\int_0^{+\infty} |a(s)|(1+s)^{-m}\,ds < +\infty$ and $\int_0^{+\infty} |b(s)|(1+s)^{-m}\,ds < +\infty$,

then we can repeat verbatim the argumentation from [592, Section 3, pp. 239–240] in order to see that the law of exponents for generalized Weyl integrals holds true:

$$I_{W,a}I_{W,b}u = I_{W,a*_0 b}u, \quad u \in \mathbf{S}(X); \tag{483}$$

here, we will only note that the Dirichlet integral formula given on [592, p. 239, l. -7–l. -4] in our new framework takes the form

$$\int_{t}^{w} a(x-t)\left[\int_{x}^{w} b(s-x)f(s)\,ds\right]dx = \int_{t}^{w} (a *_0 b)(s-t)f(s)\,ds,$$

which follows from an elementary change of variables in the double integral. Furthermore, if (G) holds, then we can repeat verbatim the argumentation from [592, Section 4, pp. 240–244] in order to see that the law of exponents for generalized Weyl derivatives holds true:

$$D_W^{\alpha,a}D_W^{\beta,b}u = D_W^{\lceil\alpha\rceil+\lceil\beta\rceil,a*_0 b}u, \quad u \in \mathbf{S}(X). \tag{484}$$

In connection with the above issue, we would like to note that the approach of R. Hilfer is insignificant for the definitions of Weyl fractional derivatives introduced above. We will only note here that the following formula holds true:

$$\int_{-\infty}^{x} b(x-s)\frac{d^m}{ds^m}\int_{-\infty}^{s} a(s-r)u(r)\,dr\,ds = \frac{d^m}{dx^m}\int_{-\infty}^{x} (a *_0 b)(x-s)u(s)\,ds, \quad x \in \mathbb{R}, \tag{485}$$

provided that $u \in \mathbf{S}(X)$ and (G) holds; furthermore, the assumption $u \in \mathbf{S}(X)$ can be slightly relaxed and all above mentioned statements can be slightly generalized keeping in mind the concrete value of integer $m \in \mathbb{N}$ satisfying (G).

Suppose now that $a : \mathbb{N}_0 \to \mathbb{C}$ and $f : \mathbb{Z} \to X$ are given sequences. If the series $\sum_{s=0}^{+\infty} a(s)f(v-s)$ is absolutely convergent for all $v \in \mathbb{Z}$, then we define

$$(\Delta_{W,a}f)(v) := \sum_{s=-\infty}^{v} a(v-s)f(s) = \sum_{s=0}^{+\infty} a(s)f(v-s), \quad v \in \mathbb{Z}. \tag{486}$$

Assume that the sequence $\Delta_{W,a}f : \mathbb{Z} \to X$ is well-defined and $m \in \mathbb{N}$. Then we put

$$(\Delta_{W,a,m}f)(v) := (\Delta^m \Delta_{W,a}f)(v), \quad v \in \mathbb{Z}.$$

It is worth noting that, if $m = \lceil \alpha \rceil$ and $a \equiv k^{m-\alpha}$ for some $\alpha > 0$, then the operator $\Delta_{a,m}$ reduces to the Weyl fractional derivative $D_W^\alpha f$ of sequence $f(\cdot)$ of order α; cf. [3, Definition 2.3]. Because of that, we will call the sequence $\Delta_{W,a,m}f$ the generalized Weyl (a, m)-fractional derivative of sequence $f(\cdot)$.

Concerning the discrete counterpart of formula (485), let us first define ($0 \leqslant \beta \leqslant 1$; $b : \mathbb{N}_0 \to \mathbb{C}$)

$$\Delta_W^{\alpha,\beta} f := \Delta^{\beta(m-\alpha)}\Delta^m \Delta^{(1-\beta)(m-\alpha)} f \quad \text{and} \quad \Delta_{a,b}f := \Delta_a \Delta^m \Delta_b f.$$

Then, under certain logical assumptions, we have (the multidimensional analogues of these formulae can be also achieved)

$$\Delta_W^{\alpha,\beta} f = D_W^\alpha f \quad \text{and} \quad \Delta_{a,b}f = \Delta_{a *_0 b}f. \tag{487}$$

Then both formulae can be proved in the same manner, with the help of the discrete Fubini theorem and the result established in [429, Theorem 3.12(ii)–(iii)]. For the sake of brevity, we will prove here the first formula in (487), only, extending thus the result established in [3, Remark 2.4]:

$$\begin{aligned}
[\Delta_W^{\alpha,\beta} f](v) &= \sum_{s=-\infty}^{v} k^{\beta(m-\alpha)}(v-s)[\Delta^m \Delta^{(1-\beta)(m-\alpha)} f](s) \\
&= \sum_{s=-\infty}^{v} k^{\beta(m-\alpha)}(v-s) \sum_{i=0}^{m} (-1)^{m-i}\binom{m}{i} [\Delta^{(1-\beta)(m-\alpha)} f](s+i) \\
&= \sum_{i=0}^{m} (-1)^{m-i}\binom{m}{i} \sum_{s=-\infty}^{v+i} k^{\beta(m-\alpha)}(v+i-s)[\Delta^{(1-\beta)(m-\alpha)} f](s) \\
&= \sum_{i=0}^{m} (-1)^{m-i}\binom{m}{i} \sum_{s=-\infty}^{v+i} k^{\beta(m-\alpha)}(v+i-s) \sum_{l=-\infty}^{s} k^{(1-\beta)(m-\alpha)}(s-l)f(l) \\
&= \sum_{i=0}^{m} (-1)^{m-i}\binom{m}{i} \sum_{s=-\infty}^{v+i} k^{m-\alpha}(v+i-s)f(s) = [D_W^\alpha f](v), \quad v \in \mathbb{Z}.
\end{aligned}$$

8.1.3 Continuation: multidimensional generalized Weyl fractional calculus

Suppose now that $a_j \in L^1_{loc}([0,\infty))$ for all $j \in \mathbb{N}_n$, $u : \mathbb{R}^n \to X$ is a locally integrable function and $\alpha_j \geqslant 0$ for all $j \in \mathbb{N}_n$. Define $\alpha := (\alpha_1, \ldots, \alpha_n)$ and

$$\mathbb{D}_W^{\alpha,\mathbf{a}} u(x_1,\ldots,x_n) := [D_W^{\alpha_1,a_1}(D_W^{\alpha_2,a_2}(\ldots(D_W^{\alpha_n,a_n} u(\cdot,\ldots,\cdot))\ldots))](x_1,\ldots,x_n), \tag{488}$$

for a. e. $(x_1,\ldots,x_n) \in \mathbb{R}^n$, provided that the right-hand side of (488) is well-defined. Here, we assume that the variables $x_1, x_2, \ldots, x_{n-1}$ are fixed in the computation of the term $D_W^{\alpha_n,a_n} u(x_1,\ldots,x_n), \ldots$, as well as that the variables $x_2, x_3, \ldots, x_n$ are fixed in the computation of the final term on the right-hand side of (488). We call $\mathbb{D}_W^{\alpha,\mathbf{a}} u$ the multidimensional generalized Weyl $(\alpha,\mathbf{a})$-fractional derivative of the function $u(\cdot)$. If $a_j \equiv g_{m_j-\alpha_j}$, where $m_j = \lceil \alpha_j \rceil$ for all $j \in \mathbb{N}_n$, then we call $\mathbb{D}_W^{\alpha} u := \mathbb{D}_W^{\alpha,\mathbf{a}} u$ the multidimensional generalized Weyl α-fractional derivative of function $u(\cdot)$; cf. also [682, p. 343] for the scalar-valued version of this notion. We call the function

$$\mathbf{x} \mapsto I_{W,\mathbf{a}}(\mathbf{x}) := \int_{-\infty}^{x_1}\int_{-\infty}^{x_2}\cdots\int_{-\infty}^{x_n} a_1(x_1-s_1)a_2(x_2-s_2)\cdot\cdots\cdot a_n(x_n-s_n)$$
$$\times\, u(s_1,s_2,\ldots,s_n)\, ds_1\, ds_2\cdots ds_n, \quad \mathbf{x} = (x_1,x_2,\ldots,x_n) \in \mathbb{R}^n,$$

if it is well-defined, the generalized Weyl $\mathbf{a}$-integral of function $u(\cdot)$.

Suppose now that $u : \mathbb{Z}^n \to X$, $a_j : \mathbb{N}_0 \to \mathbb{C}$ are given sequences and $m_j \in \mathbb{N}$ are given integers $(1 \leqslant j \leqslant n)$. Then we introduce the following multidimensional fractional difference operator:

$$\mathbb{D}_{W,\mathbf{a},\mathbf{m}} u(v_1,\ldots,v_n) := [\Delta_{W,a_1,m_1}(\Delta_{W,a_2,m_2}(\ldots(\Delta_{W,a_n,m_n} u(\cdot,\ldots,\cdot))\ldots))](v_1,\ldots,v_n), \tag{489}$$

for any $(v_1,\ldots,v_n) \in \mathbb{Z}^n$, provided that the right-hand side of (489) is well-defined. We call $\mathbb{D}_{W,\mathbf{a},\mathbf{m}} u$ the generalized multidimensional Weyl $(\mathbf{a},\mathbf{m})$-fractional derivative of $u(\cdot)$. If $m_j = \lceil \alpha_j \rceil$ and $a_j \equiv k^{m_j-\alpha_j}$ for $1 \leqslant j \leqslant n$, then we call $\mathbb{D}_{W,\mathbf{a},\mathbf{m}} u$ the generalized multidimensional Weyl α-fractional derivative of $u(\cdot)$, where $\alpha = (\alpha_1,\ldots,\alpha_n)$; in this case, we also write $\mathbb{D}_{W,\mathbf{a},\mathbf{m}} u = \mathbb{D}_W^{\alpha} u$.

Remark 8.1.5. It is clear that, in place of the generalized Weyl fractional derivatives and differences, we can consider here any other type of fractional derivatives of functions defined on the whole real axis (see, e. g., [236, 255, 600] and [682, Chapter 5]). We also recommend for reading research monograph [671] by A. Rougirel.

The formulae [592, (7.4), (7.6), (7.10), (7.12)–(7.13)] can be simply formulated in the multidimensional setting. For example, we have

$$D_W^{\alpha_1} D_W^{\alpha_2} \ldots D_W^{\alpha_n} e^{a_1x_1+a_2x_2+\cdots+a_nx_n} = a_1^{\alpha_1} a_2^{\alpha_2} \cdot\cdots\cdot a_n^{\alpha_n} e^{a_1x_1+a_2x_2+\cdots+a_nx_n}, \tag{490}$$

provided that $\alpha_j > 0$ and $a_j > 0$ for $1 \leqslant j \leqslant n$, with the meaning clear.

If all partial derivatives of a function $u : \mathbb{R}^n \to X$ are continuous almost everywhere and for each $m \in \mathbb{N}$ and $\alpha \in \mathbb{N}_0^n$, there exists a finite real number $M_{m,\alpha} \geqslant 1$ such that $\|u^{(\alpha)}(x)\| \leqslant M_{m,\alpha}(1 + |x|)^{-n}$, $x \in \mathbb{R}^n$, then we say that $u(\cdot)$ is a vector-valued good function of several variables; the corresponding class of vector-valued good functions will be denoted by $\mathbf{S}_n(X)$ henceforth. If $u \in \mathbf{S}_n(X)$, then the function $I_{W,\mathbf{a}}(\cdot)$ is infinitely differentiable and for each $\alpha \in \mathbb{N}_0^n$ and $\mathbf{x} \in \mathbb{R}^n$ we have

$$I_{W,\mathbf{a}}^{(\alpha)}(\mathbf{x}) = \int\limits_{[0,+\infty)^n} a_1(s_1)a_2(s_2)\cdot\dots\cdot a_n(s_n)\frac{\partial^\alpha u}{\partial x_1^{\alpha_1}\cdots\partial x_n^{\alpha_n}}(\mathbf{x}-\mathbf{s})\,d\mathbf{s}.$$

Furthermore, if the following condition holds:

(G1) There exists an integer $m \in \mathbb{N}$ such that $\int_0^{+\infty} |a_j(s)|(1+s)^{-m}\,ds < +\infty$ and $\int_0^{+\infty} |b_j(s)|(1+s)^{-m}\,ds < +\infty$ for all $j \in \mathbb{N}_n$,

then we can apply the Fubini theorem and (483) in order to see that the law of exponents for generalized multidimensional Weyl integrals holds true:

$$I_{W,\mathbf{a}}I_{W,\mathbf{b}}u = I_{W,\mathbf{a}*_0\mathbf{b}}u, \quad u \in \mathbf{S}_n(X), \tag{491}$$

where $\mathbf{a}*_0\mathbf{b} := (a_1 *_0 b_1, \dots, a_n *_0 b_n)$. If (G1) is valid, then the following multi-dimensional analogue of (484) holds:

$$\mathbb{D}_W^{\alpha,\mathbf{a}}\mathbb{D}_W^{\beta,\mathbf{b}}u = \mathbb{D}_W^{\lceil\alpha\rceil+\lceil\beta\rceil,\mathbf{a}*_0\mathbf{b}}u, \quad u \in \mathbf{S}_n(X), \tag{492}$$

where $\lceil\alpha\rceil + \lceil\beta\rceil := (\lceil\alpha_1\rceil + \lceil\beta_1\rceil, \dots, \lceil\alpha_n\rceil + \lceil\beta_n\rceil)$; in particular, we can clarify Clairaut's theorem on equality of mixed partial Weyl fractional derivatives of type (479).

The generation of C-regularized solution operator families in $L^p(\mathbb{R}^n)$ by the Weyl fractional differential operators of the form

$$A = \sum_{\alpha\in D} c_\alpha \mathbb{D}_W^\alpha u,$$

where D is a nonempty subset of $\mathbb{N}_0^n$ and $c_\alpha \in \mathbb{C}$ for all $\alpha \in D$, is a rather nontrivial problem. We will consider this issue somewhere else.

8.1.4 Multidimensional fractional calculus on some special regions of $\mathbb{R}^n$

Suppose that $f : I \to X$ and I has the above form. Suppose further that $\alpha_j \geqslant 0$ for all $j \in \mathbb{N}_n$ and $\alpha = (\alpha_1, \dots, \alpha_n)$. We define

$$\mathbb{D}^\alpha u(x_1, \dots, x_n) := [D^{\alpha_1}(D^{\alpha_2}(\dots(D^{\alpha_n}u(\cdot,\dots,\cdot))\dots))](x_1,\dots,x_n), \tag{493}$$

for a. e. $(x_1,\dots,x_n) \in I$, provided that the right-hand side of (493) is well-defined, where $D^{a_j} = D^{a_j}_{a_j,b_j}$ for some $a_j \in L^1_{\mathrm{loc}}(I_j)$ or $a_j(t) = \delta(t)$, and $b_j \in L^1_{\mathrm{loc}}(I_j)$ or $b_j(t) = \delta(t)$, provided that $I_j = [0,T_j)$, $I_j = [0,T_j]$ or $I_j = [0,+\infty)$, and $D^{a_j} = D^{a,a_j}_W$ with some $a_j \in L^1_{\mathrm{loc}}([0,\infty))$, if $I_j = \mathbb{R}$. We will not consider here the partial fractional derivatives of functions defined on some other regions of $\mathbb{R}^n$; for example, it could be interesting to consider the partial fractional derivatives of functions defined on convex polyhedrals in $\mathbb{R}^n$.

In the discrete setting, we will only consider the sets $I \subseteq \mathbb{Z}^n$, which have the form $I = I_1 \times I_2 \times \cdots \times I_n$, where $I_j = \mathbb{N}_0$ or $I_j = \mathbb{Z}$ for $1 \leqslant j \leqslant n$. If I has such a form and $u : I \to X$, then we define the partial fractional derivative $\mathbb{D}^a u(v_1,\dots,v_n)$ similarly as in the continuous setting; for example, in the two-dimensional setting, we can consider sequences defined on the set $I = \mathbb{N}_0 \times \mathbb{Z}$ or $I = \mathbb{Z} \times \mathbb{N}_0$.

We continue by providing the following illustrative example.

Example 8.1.6. Suppose that $n \geqslant 2$, $\emptyset \neq D \subseteq [0,+\infty)^n$ is a finite set, $c_\beta \in \mathbb{C}$ for all $\beta = (\beta_1,\dots,\beta_n) \in D$, $\beta_n > 0$ and

$$u(x_1,\dots,x_n) := \sum_{\beta\in D} c_\beta g_{\beta_1}(x_1)\cdot\dots\cdot g_{\beta_{n-1}}(x_{n-1})e^{\beta_n x_n}, \quad x_1 \geqslant 0,\dots,x_{n-1} \geqslant 0,\ x_n \in \mathbb{R}.$$

Suppose further that $a_j \geqslant 0$, $a_j(t) = g_{\gamma_j}(t)$ and $b_j(t) = g_{\delta_j}(t)$ for some non-negative numbers $\gamma_j \geqslant 0$ and $\delta_j \geqslant 0$ such that $\gamma_j + \beta_j \geqslant m_j$ $(1 \leqslant j \leqslant n-1)$. Let $D^{a_j}u = D^{a_j}_{a_j,b_j}u$ for $1 \leqslant j \leqslant n-1$, and let $D^{a_n}u = D^{a_n}_W$. If we define the functions $f_j(\cdot)$, for $1 \leqslant j \leqslant n-1$, as in Example 8.1.1(i), then we have

$$\mathbb{D}^a u(x_1,\dots,x_n) = \sum_{\beta\in D} c_\beta \beta_n^{a_n} f_1(x_1)\cdot\dots\cdot f_{n-1}(x_{n-1})e^{\beta_n x_n},$$

for any $x_1 \geqslant 0,\dots,x_{n-1} \geqslant 0$ and $x_n \in \mathbb{R}$; cf. also (490).

As in Example 8.1.1(i), we can construct a great number of various partial fractional differential equations having the function $u(x_1,\dots,x_n)$ as its solution; for example, we have

$$\begin{aligned}
&\mathbb{D}^a_{\mathbf{a},\mathbf{b}} u(x_1,\dots,x_n)\\
&= \left[\sum_{\beta\in D} c_\beta \beta_n^{a_n} \frac{x_1^{\delta_1+\gamma_1-m_1}}{\Gamma(\delta_1+\beta_1+\gamma_1-m_1)}\cdot\dots\cdot\frac{x_{n-1}^{\delta_{n-1}+\gamma_{n-1}-m_{n-1}}}{\Gamma(\delta_{n-1}+\beta_{n-1}+\gamma_{n-1}-m_{n-1})}\right]\cdot u(x_1,\dots,x_n),
\end{aligned}$$

for any $x_1 \geqslant 0,\dots,x_{n-1} \geqslant 0$ and $x_n \in \mathbb{R}$, provided that $\delta_j + \gamma_j > m_j$ for $1 \leqslant j \leqslant n-1$.

8.1.5 Multidimensional vector-valued Laplace transform

The multidimensional vector-valued Laplace transform has not attracted so much attention of the authors by now. Suppose that $f : [0,+\infty)^n \to X$ is a locally integrable function. Then the multidimensional vector-valued Laplace transform of $f(\cdot)$, denoted by $F(\cdot) = \tilde{f} = \mathcal{L}f$, is defined through

$$F(\lambda_1,\ldots,\lambda_n) := \lim_{T\to+\infty} \int\limits_{[0,T]^n} e^{-\lambda_1 t_1-\cdots-\lambda_n t_n} f(t_1,\ldots,t_n)\,dt_1\cdots dt_n$$

$$:= \int\limits_0^{+\infty}\cdots\int\limits_0^{+\infty} e^{-\lambda_1 t_1-\cdots-\lambda_n t_n} f(t_1,\ldots,t_n)\,dt_1\cdots dt_n, \tag{494}$$

if it is well-defined. We say that $f(\cdot)$ is Laplace transformable if and only if there exist real constants $\omega_1 \in \mathbb{R},\ldots,\omega_n \in \mathbb{R}$ such that $F(\lambda_1,\ldots,\lambda_n)$ is well-defined for $\operatorname{Re}\lambda_1 > \omega_1,\ldots,\operatorname{Re}\lambda_n > \omega_n$. This is always the case if there exist finite real constants $M \geqslant 1$ and $\omega_1 \in \mathbb{R},\ldots,\omega_n \in \mathbb{R}$ such that $\|f(t_1,\ldots,t_n)\| \leqslant M\exp(\omega_1 t_1 + \cdots + \omega_n t_n)$ for a. e. $t_1 \geqslant 0,\ldots,t_n \geqslant 0$; then $F(\lambda_1,\ldots,\lambda_n)$ is well-defined for $\operatorname{Re}\lambda_1 > \omega_1,\ldots,\operatorname{Re}\lambda_n > \omega_n$ and $F(\cdot)$ is analytic in this region of $\mathbb{C}^n$. The uniqueness theorem for Laplace transform holds in the multidimensional framework.

The numerical inversion of multidimensional vector-valued Laplace transform has been considered in many research articles by now (these papers can be easily located online and we will not quote them here). On the other hand, it seems that the complex inversion theorem for the multidimensional Laplace transform in both, the scalar-valued setting and the vector-valued setting, has not been properly formulated and proved by now. Concerning this issue, we will state and prove the following extension of [69, Theorem 2.5.1].

Theorem 8.1.7. *Suppose that $M > 0$, $\omega_1 \geqslant 0,\ldots,\omega_n \geqslant 0$, $\varepsilon_1 > 0,\ldots,\varepsilon_n > 0$ and $F : \{\lambda \in \mathbb{C} : \operatorname{Re}\lambda > \omega_1\} \times \cdots \times \{\lambda \in \mathbb{C} : \operatorname{Re}\lambda > \omega_n\} \to X$ is an analytic function such that*

$$\|F(\lambda_1,\ldots,\lambda_n)\| \leqslant M|\lambda_1|^{-1-\varepsilon_1}\cdot\ldots\cdot|\lambda_n|^{-1-\varepsilon_n}, \quad \operatorname{Re}\lambda_j > \omega_j \quad (1 \leqslant j \leqslant n). \tag{495}$$

Then there exist a real number $M_1 > 0$ and a continuous function $f : [0,+\infty)^n \to X$ such that

$$\|f(t_1,\ldots,t_n)\| \leqslant M_1\big[t_1^{\varepsilon_1}e^{\omega_1 t_1}\cdot\ldots\cdot t_n^{\varepsilon_n}e^{\omega_n t_n}\big] \quad \text{for all } t_1 \geqslant 0,\ldots,t_n \geqslant 0 \tag{496}$$

and $F(\lambda_1,\ldots,\lambda_n) = (\mathcal{L}f)(\lambda_1,\ldots,\lambda_n)$ for $\operatorname{Re}\lambda_j > \omega_j$ $(1 \leqslant j \leqslant n)$.

Proof. We will present the main details of proof, only. Let $a_j > \omega_j$ be pairwisely distinct numbers $(1 \leqslant j \leqslant n)$, and let

$$f(t_1,\ldots,t_n) := \frac{1}{(2\pi i)^n}\int\limits_{a_1-i\infty}^{a_1+i\infty}\cdots\int\limits_{a_n-i\infty}^{a_n+i\infty} e^{\lambda_1 t_1+\cdots+\lambda_n t_n}F(\lambda_1,\ldots,\lambda_n)\,d\lambda_1\cdots d\lambda_n, \tag{497}$$

for any $t_1 \geqslant 0,\ldots,t_n \geqslant 0$; it can be easily shown that, due to the estimate (495), the integral appearing in (497) is absolutely convergent so that $f(\cdot)$ is well-defined. The dominated convergence theorem implies that $f(\cdot)$ is continuous; moreover, we can use the Fubini theorem, the growth rate of $F(\cdot)$ and the computation carried out in the proof of the last mentioned theorem in order to see that there exists a constant $M_1 > 0$, independent of $a_1, \ldots, a_n$, such that

$$\|f(t_1,\ldots,t_n)\| \leqslant M_1\big[t_1^{\varepsilon_1}e^{a_1t_1}\cdot\ldots\cdot t_n^{\varepsilon_n}e^{a_nt_n}\big] \quad \text{for all } t_1 \geqslant 0,\ldots,t_n \geqslant 0.$$

On the other hand, an elementary contour argument shows that the definition of function $f(\cdot)$ does not depend on the choice of numbers $a_1 > \omega_1, \ldots, a_n > \omega_n$. In actual fact, we can fix the numbers $a_1 > \omega_1, \ldots, a_{n-1} > \omega_{n-1}$ and prove first that the definition of function $f(\cdot)$ does not depend on the choice of number $a_n > \omega_n$; after that, we can repeat this procedure $(n-1)$-times. Invoking this fact and letting $a_j \to \omega_j+$ for $1 \leqslant j \leqslant n$, we get (496). It remains to be proved that $F(\lambda_1,\ldots,\lambda_n) = (\mathcal{L}f)(\lambda_1,\ldots,\lambda_n)$ for $\operatorname{Re}\lambda_j > \omega_j$ $(1 \leqslant j \leqslant n)$. Let the numbers $\lambda_1,\ldots,\lambda_n$ enjoy the above properties and let $\omega_j < a_j < \operatorname{Re}\lambda_j$ for $1 \leqslant j \leqslant n$. Then the Fubini theorem and an elementary argumentation shows that

$$(\mathcal{L}f)(\lambda_1,\ldots,\lambda_n) = \frac{1}{(2\pi i)^n}\int_{a_1-i\infty}^{a_1+i\infty}\cdots\int_{a_n-i\infty}^{a_n+i\infty}\frac{F(z_1,\ldots,z_n)}{(\lambda_1-z_1)\cdot\ldots\cdot(\lambda_n-z_n)}\,dz_1\cdots dz_n.$$

Using the residue theorem and deforming the line $[a_n - i\infty, a_n + i\infty]$ into the union of the segment $[a_n - iR, a_n + iR]$ and the semicircle $a_n + \{Re^{i\theta} : -\pi/2 \leqslant \theta \leqslant \pi/2\}$, we get

$$(\mathcal{L}f)(\lambda_1,\ldots,\lambda_n) = \frac{1}{(2\pi i)^{n-1}}\int_{a_1-i\infty}^{a_1+i\infty}\cdots\int_{a_{n-1}-i\infty}^{a_{n-1}+i\infty}\frac{F(z_1,\ldots,z_{n-1},\lambda_n)}{(\lambda_1-z_1)\cdot\ldots\cdot(\lambda_{n-1}-z_{n-1})}\,dz_1\cdots dz_{n-1}.$$

Repeating this argument, we simply obtained the required equality. □

8.1.6 Fractional partial differential inclusions with Riemann–Liouville and Caputo derivatives

Suppose that $\alpha_1 \in [0,2)$, $\alpha_2 \in [0,2)$, $m_1 = \lceil\alpha_1\rceil$, $m_2 = \lceil\alpha_2\rceil$ and $\mathcal{A}$ is a closed MLO in X. In this subsection, we will provide certain results about the well-posedness of the following abstract two-dimensional Cauchy inclusions:

$$D_R^{\alpha_1}D_R^{\alpha_2}u(x_1,x_2) \in \mathcal{A}u(x_1,x_2) + f(x_1,x_2), \quad x_1 \geqslant 0,\ x_2 \geqslant 0, \tag{498}$$

subjected with the initial conditions of the form

$$\frac{\partial^k}{\partial x_2^k}\big[J_{t_2}^{m_2-\alpha_2} *_0 u\big](x_1,0) = f_k(x_1), \quad 0 \leqslant k \leqslant m_2-1; \tag{499}$$

$$\int_0^{x_2} g_{\alpha_2}(x_2-s)\Big[\frac{\partial^k}{\partial x_1^k}\big[J_{t_1}^{m_1-\alpha_1} *_0 D_R^{\alpha_2}u\big](x_1,x_2)\Big]_{x_1=0,x_2=s}\,ds = h_k(x_2), \quad 0 \leqslant k \leqslant m_1-1, \tag{500}$$

and

$$\mathbf{D}_C^{\alpha_1}\mathbf{D}_C^{\alpha_2}u(x_1,x_2) \in \mathcal{A}u(x_1,x_2)+f(x_1,x_2), \quad x_1 \geqslant 0,\ x_2 \geqslant 0, \tag{501}$$

subjected with the initial conditions of the form

$$\frac{\partial^k}{\partial x_2^k}u(x_1,0) = f_k(x_1), \quad 0 \leqslant k \leqslant m_2-1; \tag{502}$$

$$\int_0^{x_2} g_{\alpha_2}(x_2-s)\left[\frac{\partial^k}{\partial x_1^k}\mathbf{D}_C^{\alpha_2}u(x_1,x_2)\right]_{x_1=0,x_2=s} ds = h_k(x_2), \quad 0 \leqslant k \leqslant m_1-1. \tag{503}$$

Let us consider first the problem (501) equipped with the initial conditions (502)–(503). Assuming that $f \in L^1_{\mathrm{loc}}([0,\infty)^2 : X)$, all conditions for applying the formula (482) are satisfied and using the fact that, for every locally integrable function $u \in L^1_{\mathrm{loc}}([0,\infty)^2 : X)$, the assumption

$$J_{t_2}^{\alpha_2}J_{t_1}^{\alpha_1}u(x_1,x_2) = 0, \quad x_1 \geqslant 0,\ x_2 \geqslant 0$$

implies $u \equiv 0$, we get that the problem [(501); (502)–(503)] is equivalent with

$$\begin{aligned} &u(x_1,x_2) - \sum_{k=0}^{m_2-1} g_{k+1}(x_2)\cdot f_k(x_1) - \sum_{k=0}^{m_1-1} g_{k+1}(x_1)\cdot h_k(x_2) \\ &\in \mathcal{A}\int_0^{x_2} g_{\alpha_2}(x_2-r)\int_0^{x_1} g_{\alpha_1}(x_1-s)u(s,r)\,ds\,dr \\ &+ \int_0^{x_2} g_{\alpha_2}(x_2-r)\int_0^{x_1} g_{\alpha_1}(x_1-s)f(s,r)\,ds\,dr, \quad x_1 \geqslant 0,\ x_2 \geqslant 0, \end{aligned} \tag{504}$$

since $\mathcal{A}$ is closed. Similarly, if $f \in L^1_{\mathrm{loc}}([0,\infty)^2 : X)$ and all conditions for applying the formula (481) are satisfied, the problem [(498); (499)–(500)] is equivalent with

$$\begin{aligned} &u(x_1,x_2) - \sum_{k=0}^{m_2-1} g_{\alpha_2+k+1-m_2}(x_2)\cdot f_k(x_1) - \sum_{k=0}^{m_1-1} g_{\alpha_1+k+1-m_1}(x_1)\cdot h_k(x_2) \\ &\in \mathcal{A}\int_0^{x_2} g_{\alpha_2}(x_2-r)\int_0^{x_1} g_{\alpha_1}(x_1-s)u(s,r)\,ds\,dr \\ &+ \int_0^{x_2} g_{\alpha_2}(x_2-r)\int_0^{x_1} g_{\alpha_1}(x_1-s)f(s,r)\,ds\,dr, \quad x_1 \geqslant 0,\ x_2 \geqslant 0. \end{aligned} \tag{505}$$

We will use the following notion (cf. also [445, Definition 3.1.1(i)]).

Definition 8.1.8. It is said that a locally integrable function $u : [0,\infty)^2 \to X$ is:

(i) A solution of [(501); (502)–(503)] if

$$\int_0^{x_2} g_{a_2}(x_2 - r) \int_0^{x_1} g_{a_1}(x_1 - s) u(s,r)\, ds\, dr \in D(\mathcal{A})$$

and (505) holds for a. e. $x_1 \geqslant 0$ and $x_2 \geqslant 0$.

(ii) A strong solution of [(501); (502)–(503)] if there exists a locally integrable function $u_{\mathcal{A},a_1,a_2} : [0,\infty)^2 \to X$ such that

$$\int_0^{x_2} g_{a_2}(x_2 - r) \int_0^{x_1} g_{a_1}(x_1 - s) u_{\mathcal{A},a_1,a_2}(s,r)\, ds\, dr$$
$$\in \mathcal{A} \int_0^{x_2} g_{a_2}(x_2 - r) \int_0^{x_1} g_{a_1}(x_1 - s) u(s,r)\, ds\, dr \quad \text{for a. e. } x_1 \geqslant 0 \text{ and } x_2 \geqslant 0,$$

and

$$u(x_1,x_2) - \sum_{k=0}^{m_2-1} g_{k+1}(x_2) \cdot f_k(x_1) - \sum_{k=0}^{m_1-1} g_{k+1}(x_1) \cdot h_k(x_2)$$
$$= \int_0^{x_2} g_{a_2}(x_2 - r) \int_0^{x_1} g_{a_1}(x_1 - s) u_{\mathcal{A},a_1,a_2}(s,r)\, ds\, dr$$
$$+ \int_0^{x_2} g_{a_2}(x_2 - r) \int_0^{x_1} g_{a_1}(x_1 - s) f(s,r)\, ds\, dr \quad \text{for a. e. } x_1 \geqslant 0 \text{ and } x_2 \geqslant 0.$$

We similarly define the notion of a (strong) solution of problem [(498); (499)–(500)].

It is clear that any strong solution of [(501); (502)–(503)] ([(498); (499)–(500)]) is likewise a solution of the same problem and that the converse statement is not true, in general.

Let us now take a closer look at the abstract Cauchy inclusions (504) and (505). Applying the two-dimensional Laplace transform and the Fubini theorem, we get that the problem (504) is equivalent with

$$\int_0^{+\infty}\int_0^{+\infty} e^{-zx_1-\lambda x_2} u(x_1,x_2)\, dx_1\, dx_2 - \sum_{k=0}^{m_2-1} \lambda^{-1-k} \int_0^{+\infty} e^{-zx_1} f_k(x_1)\, dx_1$$
$$- \sum_{k=0}^{m_1-1} z^{-1-k} \int_0^{+\infty} e^{-\lambda x_2} h_k(x_2)\, dx_2$$
$$\in \mathcal{A}\left[z^{-a_1}\lambda^{-a_2} \int_0^{+\infty}\int_0^{+\infty} e^{-zx_1-\lambda x_2} u(x_1,x_2)\, dx_1\, dx_2 \right]$$

$$+ z^{-a_1}\lambda^{-a_2}\int_0^{+\infty}\int_0^{+\infty} e^{-zx_1-\lambda x_2} f(x_1,x_2)\,dx_1\,dx_2, \tag{506}$$

for all $z \in \mathbb{C}$ with $\operatorname{Re} z > \omega_1$ for some $\omega_1 > 0$ and $\lambda \in \mathbb{C}$ with $\operatorname{Re}\lambda > \omega_2$ for some $\omega_2 > 0$, under certain logical assumptions, as well as that the problem (505) is equivalent with

$$\begin{aligned}
&\int_0^{+\infty}\int_0^{+\infty} e^{-zx_1-\lambda x_2} u(x_1,x_2)\,dx_1\,dx_2 - \sum_{k=0}^{m_2-1} \lambda^{m_2-1-k-a_2}\int_0^{+\infty} e^{-zx_1} f_k(x_1)\,dx_1\\
&\quad - \sum_{k=0}^{m_1-1} z^{m_1-1-k-a_1}\int_0^{+\infty} e^{-\lambda x_2} h_k(x_2)\,dx_2\\
&\quad \in \mathcal{A}\left[z^{-a_1}\lambda^{-a_2}\int_0^{+\infty}\int_0^{+\infty} e^{-zx_1-\lambda x_2} u(x_1,x_2)\,dx_1\,dx_2\right]\\
&\quad + z^{-a_1}\lambda^{-a_2}\int_0^{+\infty}\int_0^{+\infty} e^{-zx_1-\lambda x_2} f(x_1,x_2)\,dx_1\,dx_2,
\end{aligned} \tag{507}$$

for all $z \in \mathbb{C}$ with $\operatorname{Re} z > \omega_1$ for some $\omega_1 > 0$ and $\lambda \in \mathbb{C}$ with $\operatorname{Re}\lambda > \omega_2$ for some $\omega_2 > 0$, under certain logical assumptions. After setting

$$\tilde{u}(z,\lambda) := \int_0^{+\infty}\int_0^{+\infty} e^{-zx_1-\lambda x_2} u(x_1,x_2)\,dx_1\,dx_2,$$

we get that the problem (506) is equivalent with

$$\begin{aligned}
(z^{a_1}\lambda^{a_2} - \mathcal{A})\tilde{u}(z,\lambda) \ni{}& \sum_{k=0}^{m_2-1} z^{a_1}\lambda^{a_2-1-k}\int_0^{+\infty} e^{-zx_1} f_k(x_1)\,dx_1\\
& - \sum_{k=0}^{m_1-1} z^{a_1-1-k}\lambda^{a_2}\int_0^{+\infty} e^{-\lambda x_2} h_k(x_2)\,dx_2 + \tilde{f}(z,\lambda),
\end{aligned} \tag{508}$$

for all $z \in \mathbb{C}$ with $\operatorname{Re} z > \omega_1$ and $\lambda \in \mathbb{C}$ with $\operatorname{Re}\lambda > \omega_2$, while the problem (507) is equivalent with

$$\begin{aligned}
(z^{a_1}\lambda^{a_2} - \mathcal{A})\tilde{u}(z,\lambda) \ni{}& \sum_{k=0}^{m_2-1} z^{a_1}\lambda^{m_2-1-k}\int_0^{+\infty} e^{-zx_1} f_k(x_1)\,dx_1\\
& - \sum_{k=0}^{m_1-1} z^{m_1-1-k}\lambda^{a_2}\int_0^{+\infty} e^{-\lambda x_2} h_k(x_2)\,dx_2 + \tilde{f}(z,\lambda),
\end{aligned} \tag{509}$$

for all $z \in \mathbb{C}$ with $\operatorname{Re} z > \omega_1$ and $\lambda \in \mathbb{C}$ with $\operatorname{Re}\lambda > \omega_2$. In the case that there exists an injective operator $C \in L(X)$, which commutes with $\mathcal{A}$ and condition (C01) clarified below holds, then the inclusion (508), respectively, (509), is equivalent with

$$\tilde{u}(z,\lambda) = (z^{\alpha_1}\lambda^{\alpha_2} - \mathcal{A})^{-1} C \sum_{k=0}^{m_2-1} z^{\alpha_1}\lambda^{\alpha_2-1-k} \int_0^{+\infty} e^{-zx_1} f_k(x_1)\,dx_1$$
$$- (z^{\alpha_1}\lambda^{\alpha_2} - \mathcal{A})^{-1} C \sum_{k=0}^{m_1-1} z^{\alpha_1-1-k}\lambda^{\alpha_2} \int_0^{+\infty} e^{-\lambda x_2} h_k(x_2)\,dx_2 + (z^{\alpha_1}\lambda^{\alpha_2} - \mathcal{A})^{-1} C\tilde{f}(z,\lambda), \tag{510}$$

for all $z \in \mathbb{C}$ with $\operatorname{Re} z > \omega_1$ and $\lambda \in \mathbb{C}$ with $\operatorname{Re}\lambda > \omega_2$, respectively,

$$\tilde{u}(z,\lambda) = (z^{\alpha_1}\lambda^{\alpha_2} - \mathcal{A})^{-1} C \sum_{k=0}^{m_2-1} z^{\alpha_1}\lambda^{m_2-1-k} \int_0^{+\infty} e^{-zx_1} f_k(x_1)\,dx_1$$
$$- (z^{\alpha_1}\lambda^{\alpha_2} - \mathcal{A})^{-1} C \sum_{k=0}^{m_1-1} z^{m_1-1-k}\lambda^{\alpha_2} \int_0^{+\infty} e^{-\lambda x_2} h_k(x_2)\,dx_2 + (z^{\alpha_1}\lambda^{\alpha_2} - \mathcal{A})^{-1} C\tilde{f}(z,\lambda), \tag{511}$$

for all $z \in \mathbb{C}$ with $\operatorname{Re} z > \omega_1$ and $\lambda \in \mathbb{C}$ with $\operatorname{Re}\lambda > \omega_2$.

Now we will formalize all this and state the following result by assuming some special conditions on the multivalued linear operator $\mathcal{A}$.

Theorem 8.1.9. *Suppose that $C \in L(X)$ is injective and commutes with $\mathcal{A}$, $f(\cdot;\cdot)$ is Laplace transformable and the following condition holds:*

(C01) *There exist real numbers $\omega_1 > 0$ and $\omega_2 > 0$ such that $z^{\alpha_1}\lambda^{\alpha_2} \in \rho_C(\mathcal{A})$ for all $z \in \mathbb{C}$ with $\operatorname{Re} z > \omega_1$ and $\lambda \in \mathbb{C}$ with $\operatorname{Re}\lambda > \omega_2$.*

Denote by D_1 the set of all indexes $k \in \mathbb{N}^0_{m_2-1}$ such that $f_k(\cdot)$ is not identically equal to the zero function and by D_2 the set of all indexes $k \in \mathbb{N}^0_{m_1-1}$ such that $h_k(\cdot)$ is not identically equal to the zero function. If the following condition holds:

(i) *For every $k \in D_1$, there exists a Laplace transformable function $u_k^1(\cdot;\cdot)$ such that*

$$\widetilde{u_k^1}(z,\lambda) = z^{\alpha_1}\lambda^{\alpha_2-1-k}(z^{\alpha_1}\lambda^{\alpha_2} - \mathcal{A})^{-1} C \int_0^{+\infty} e^{-zx_1} f_k(x_1)\,dx_1,$$

respectively,

$$\widetilde{u_k^1}(z,\lambda) = z^{\alpha_1}\lambda^{m_2-1-k}(z^{\alpha_1}\lambda^{\alpha_2} - \mathcal{A})^{-1} C \int_0^{+\infty} e^{-zx_1} f_k(x_1)\,dx_1,$$

for $\operatorname{Re} z > \omega_1$ and $\operatorname{Re}\lambda > \omega_2$.

(ii) *For every* $k \in D_2$, *there exists a Laplace transformable function* $u_k^2(\cdot;\cdot)$ *such that*

$$\widetilde{u_k^2}(z,\lambda) = z^{\alpha_1-1-k}\lambda^{\alpha_2}(z^{\alpha_1}\lambda^{\alpha_2} - \mathcal{A})^{-1}C\int_0^{+\infty} e^{-\lambda x_2}h_k(x_2)\,dx_2,$$

respectively,

$$\widetilde{u_k^2}(z,\lambda) = z^{m_1-1-k}\lambda^{\alpha_2}(z^{\alpha_1}\lambda^{\alpha_2} - \mathcal{A})^{-1}C\int_0^{+\infty} e^{-\lambda x_2}h_k(x_2)\,dx_2,$$

for $\operatorname{Re} z > \omega_1$ *and* $\operatorname{Re}\lambda > \omega_2$.

(iii) *There exists a Laplace transformable function* $u_k^2(\cdot;\cdot)$ *such that*

$$\widetilde{u_f}(z,\lambda) = (z^{\alpha_1}\lambda^{\alpha_2} - \mathcal{A})^{-1}C\tilde{f}(z,\lambda),$$

for $\operatorname{Re} z > \omega_1$ *and* $\operatorname{Re}\lambda > \omega_2$.

Then there exists a unique solution of problem $u(x_1,x_2)$ *of* [(501); (502)–(503)], *respectively,* [(498); (499)–(500)], *which is given by*

$$u(x_1,x_2) = \sum_{k\in D_1} u_k^1(x_1,x_2) + \sum_{k\in D_2} u_k^2(x_1,x_2) + u_f(x_1,x_2) \quad \text{for a. e. } x_1 \geqslant 0,\ x_2 \geqslant 0.$$

Furthermore, suppose that (i)–(iii) *and the following conditions hold:*

(i-s) *For every* $k \in D_1$, *there exists a Laplace transformable function* $u_k^1(\cdot;\cdot)$ *such that*

$$\widetilde{u_k^1}(z,\lambda) = z^{2\alpha_1}\lambda^{2\alpha_2-1-k}(z^{\alpha_1}\lambda^{\alpha_2} - \mathcal{A})^{-1}C\int_0^{+\infty} e^{-zx_1}f_k(x_1)\,dx_1 - z^{\alpha_1}\lambda^{\alpha_2-1-k}\int_0^{+\infty} e^{-zx_1}f_k(x_1)\,dx_1,$$

respectively,

$$\widetilde{u_k^1}(z,\lambda) = z^{2\alpha_1}\lambda^{\alpha_2+m_2-1-k}(z^{\alpha_1}\lambda^{\alpha_2} - \mathcal{A})^{-1}C\int_0^{+\infty} e^{-zx_1}f_k(x_1)\,dx_1 - z^{\alpha_1}\lambda^{m_2-1-k}\int_0^{+\infty} e^{-zx_1}f_k(x_1)\,dx_1,$$

for $\operatorname{Re} z > \omega_1$ *and* $\operatorname{Re}\lambda > \omega_2$.

(ii-s) *For every $k \in D_2$, there exists a Laplace transformable function $u_k^2(\cdot;\cdot)$ such that*

$$\widetilde{u_k^2}(z,\lambda) = z^{2a_1-1-k}\lambda^{2a_2}(z^{a_1}\lambda^{a_2} - \mathcal{A})^{-1}C\int_0^{+\infty} e^{-\lambda x_2}h_k(x_2)\,dx_2 - z^{a_1-1-k}\lambda^{a_2}C\int_0^{+\infty} e^{-\lambda x_2}h_k(x_2)\,dx_2,$$

respectively,

$$\widetilde{u_k^2}(z,\lambda) = z^{a_1+m_1-1-k}\lambda^{2a_2}(z^{a_1}\lambda^{a_2} - \mathcal{A})^{-1}C\int_0^{+\infty} e^{-\lambda x_2}h_k(x_2)\,dx_2 - z^{m_1-1-k}\lambda^{a_2}C\int_0^{+\infty} e^{-\lambda x_2}h_k(x_2)\,dx_2,$$

for $\operatorname{Re} z > \omega_1$ *and* $\operatorname{Re}\lambda > \omega_2$.

(iii-s) *There exists a Laplace transformable function $u_k^2(\cdot;\cdot)$ such that*

$$\widetilde{u_f}(z,\lambda) = z^{a_1}\lambda^{a_2}(z^{a_1}\lambda^{a_2} - \mathcal{A})^{-1}C\tilde{f}(z,\lambda) - C\tilde{f}(z,\lambda),$$

for $\operatorname{Re} z > \omega_1$ *and* $\operatorname{Re}\lambda > \omega_2$.

Then there exists a unique solution of problem $u(x_1,x_2)$ of [(501); (502)–(503)], *respectively,* [(498); (499)–(500)], *which is given by*

$$u(x_1,x_2) = \sum_{k\in D_1} u_k^1(x_1,x_2) + \sum_{k\in D_2} u_k^2(x_1,x_2) + u_f(x_1,x_2) \quad \textit{for a. e. } x_1 \geqslant 0,\ x_2 \geqslant 0. \tag{512}$$

Then the function $u(x_1,x_2)$, given by (512), *is a strong solution of problem $u(x_1,x_2)$ of* [(501); (502)–(503)], *respectively,* [(498); (499)–(500)].

Proof. Since we have assumed the conditions (i)–(iii), we simply infer that the function $u(x_1,x_2)$, given by (512), satisfies (510), respectively, (511). Arguing reversely, we get that (508), respectively, (509), holds true. Applying the inverse double Laplace transform, we get that (506), respectively, (507), holds true, which simply completes the proof of the first part of theorem. The second part of theorem follows similarly since, in this case, there exists a locally integrable function $u_{\mathcal{A},a_1,a_2}(\cdot;\cdot)$ such that $u_{\mathcal{A},a_1,a_2}(\cdot;\cdot) \in \mathcal{A}u(\cdot;\cdot)$ a. e. on $[0,+\infty)^2$, which can be proved by performing the double Laplace transform and (i-s)–(iii-s); see also [445, Theorem 1.2.4(i)]. □

The subsequent result follows immediately from Theorem 8.1.7 and Theorem 8.1.9 (we can similarly clarify the corresponding conditions ensuring the existence of a unique strong solution of problems under our consideration; we use the symbol $\tilde{\cdot}$ to denote both, the one-dimensional and the two-dimensional Laplace transform here, which will not cause any confusion).

Theorem 8.1.10. *Suppose that $f(\cdot;\cdot)$ is Laplace transformable and the following condition holds:*
(C1-s): (C1) *holds and there exist real numbers $M > 0$ and $\beta \in (0,1]$ such that*

$$\left\|(z^{a_1}\lambda^{a_2} - \mathcal{A})^{-1}C\right\| \leqslant \frac{M}{(1+|z|^{a_1}|\lambda|^{a_2})^{\beta}}, \quad \operatorname{Re} z > \omega_1,\ \operatorname{Re}\lambda > \omega_2. \tag{513}$$

Suppose further that the following conditions hold:
(i) *For every $k \in D_1$, there exist real numbers $M_{k,1} > 0$, $\varepsilon_{1,1}^k > 0$ and $\varepsilon_{1,2}^k > 0$ such that*

$$\left\||z|^{a_1}|\lambda|^{a_2-1-k}\frac{\|\tilde{f_k}(z)\|}{(1+|z|^{a_1}|\lambda|^{a_2})^{\beta}}\right\| \leqslant M_{k,1}|z|^{-1-\varepsilon_{1,1}^k}|\lambda|^{-1-\varepsilon_{1,2}^k}, \quad \operatorname{Re} z > \omega_1,\ \operatorname{Re}\lambda > \omega_2.$$

(ii) *For every $k \in D_2$, there exist real numbers $M_{k,2} > 0$, $\varepsilon_{2,1}^k > 0$ and $\varepsilon_{2,2}^k > 0$ such that*

$$\left\||z|^{a_1-1-k}|\lambda|^{a_2}\frac{\|\tilde{h_k}(\lambda)\|}{(1+|z|^{a_1}|\lambda|^{a_2})^{\beta}}\right\| \leqslant M_{k,1}|z|^{-1-\varepsilon_{2,1}^k}|\lambda|^{-1-\varepsilon_{2,2}^k}, \quad \operatorname{Re} z > \omega_1,\ \operatorname{Re}\lambda > \omega_2.$$

(iii) *There exist real numbers $M' > 0$, $\varepsilon_1 > 0$ and $\varepsilon_2 > 0$ such that*

$$\left\|\frac{\|\tilde{f}(z,\lambda)\|}{(1+|z|^{a_1}|\lambda|^{a_2})^{\beta}}\right\| \leqslant M'|z|^{-1-\varepsilon_1}|\lambda|^{-1-\varepsilon_2}, \quad \operatorname{Re} z > \omega_1,\ \operatorname{Re}\lambda > \omega_2.$$

Then there exists a unique continuous solution $u(x_1,x_2)$ of problem [(501); (502)–(503)], *respectively,* [(498); (499)–(500)], *and we have*

$$\|u(x_1,x_2)\| \leqslant M''\left[\sum_{k\in D_1} x_1^{\varepsilon_{1,1}}x_2^{\varepsilon_{1,2}}e^{\omega_1x_1+\omega_2x_2} + \sum_{k\in D_2} x_1^{\varepsilon_{2,1}}x_2^{\varepsilon_{2,2}}e^{\omega_1x_1+\omega_2x_2} + x_1^{\varepsilon_1}x_2^{\varepsilon_2}e^{\omega_1x_1+\omega_2x_2}\right],$$
$$x_1 \geqslant 0,\ x_2 \geqslant 0.$$

If $0 \notin D_1 \cup D_2$, then the requirements of Theorem 8.1.10 are satisfied in many important real situations, even for the degenerate Poisson heat operator $\Delta \cdot m(x)^{-1}$; cf. [445] and references cited therein for further information in this direction.

Remark 8.1.11. Suppose that $a_1 + a_2 < 2$. Then it is clear that the estimate (513) holds if $\Sigma_{(a_1+a_2)\pi/2} \subseteq \rho_C(\mathcal{A})$ and there exists $\beta \in (0,1]$ such that

$$\left\|(\lambda - \mathcal{A})^{-1}C\right\| \leqslant \frac{M}{(1+|\lambda|)^{\beta}}, \quad \lambda \in \Sigma_{(a_1+a_2)\pi/2}.$$

Disappointingly, we cannot prove that (513) holds if there exists a positive real number $a > 0$ such that $a + \Sigma_{(a_1+a_2)\pi/2} \subseteq \rho_C(\mathcal{A})$ and

$$\|(\lambda - \mathcal{A})^{-1}C\| \leqslant \frac{M}{(1+|\lambda|)^{\beta}}, \quad \lambda \in a + \Sigma_{(\alpha_1+\alpha_2)\pi/2}.$$

The main problem lies in the fact that, for every real number $\omega_1 > 0$, we have

$$\lim_{x \to \pm\infty} \operatorname{dist}(\{re^{i}\alpha_1 : r \geqslant 0\}, (\omega_1 + ix)^{\alpha_1}) = 0.$$

Remark 8.1.12. Suppose that $\alpha_1 + \alpha_2 \geqslant 2$. Then we can apply Theorem 8.1.10, with $C \neq I$, to a class of two-dimensional partial fractional differential equations involving the single-valued linear operators $\mathcal{A} = A$ whose C-resolvent is bounded by $(1 + |\cdot|)^{-1}$ on the set of form $\mathbb{C} \smallsetminus K$, where K is compact; see [445] for the corresponding examples. In particular, if $\alpha_1 = \alpha_2 = 1$, then we can analyze the well-posedness of problem

$$\frac{\partial^2}{\partial x_1 \partial x_2} u(x_1, x_2) = Au(x_1, x_2) + f(x_1, x_2), \quad x_1 \geqslant 0,\ x_2 \geqslant 0,$$

subjected with the initial conditions $u(x_1, 0) = f_0(x_1), x_1 \geqslant 0$ and $u(0, x_2) = u(0,0) + h_0(x_2)$, $x_2 \geqslant 0$.

Using the multidimensional generalizations of the formulae (481) and (482), we can similarly analyze the well-posedness of the abstract fractional Cauchy inclusions

$$D_R^{\alpha_1} D_R^{\alpha_2} \cdot \dots \cdot D_R^{\alpha_n} u(\mathbf{x}) \in \mathcal{A}u(\mathbf{x}) + f(\mathbf{x}), \quad \mathbf{x} = (x_1, x_2, \dots, x_n) \in [0, +\infty)^n$$

and

$$\mathbf{D}_C^{\alpha_1} \mathbf{D}_C^{\alpha_2} \cdot \dots \cdot \mathbf{D}_C^{\alpha_n} u(\mathbf{x}) \in \mathcal{A}u(\mathbf{x}) + f(\mathbf{x}), \quad \mathbf{x} = (x_1, x_2, \dots, x_n) \in [0, +\infty)^n,$$

subjected with certain initial conditions (for the scalar-valued case, see also [432, Section 3]). We leave all details to interested readers.

8.1.7 The abstract multiterm fractional partial differential equations with Riemann–Liouville and Caputo derivatives

In this subsection, we investigate the following operator extensions of the partial fractional differential equation (475):

$$\sum_{k=1}^{n} A_k D_R^{(0,\dots,\alpha_k,\dots,0)} u(x_1, \dots, x_k, \dots, x_n) = f(x_1, \dots, x_n), \quad x_1 \geqslant 0, \dots, x_n \geqslant 0, \tag{514}$$

subjected with the initial conditions

$$\left[\frac{\partial^j}{\partial x_k^j} J_{t_k}^{m_k - \alpha_k} u(x_1, \dots, x_n) \right]_{x_k = 0} = f_{k,j}(x_1, \dots, x_{k-1}, x_{k+1}, \dots, x_n), \tag{515}$$

for $1 \leqslant k \leqslant n$, $0 \leqslant j \leqslant m_k - 1$, and

$$\sum_{k=1}^{n} A_k D_C^{(0,\ldots,a_k,\ldots,0)} u(x_1,\ldots,x_k,\ldots,x_n) = f(x_1,\ldots,x_n), \quad x_1 \geqslant 0,\ldots,x_n \geqslant 0, \tag{516}$$

subjected with the initial conditions

$$\left[\frac{\partial^j}{\partial x_k^j} u(x_1,\ldots,x_n)\right]_{x_k=0} = f_{k,j}(x_1,\ldots,x_{k-1},x_{k+1},\ldots,x_n), \tag{517}$$

for $1 \leqslant k \leqslant n$, $0 \leqslant j \leqslant m_k - 1$, where A_k is a closed linear operator and $a_k \geqslant 0$ for $1 \leqslant k \leqslant n$. In order to do that, we essentially apply the multidimensional vector-valued Laplace transform.

We will use the following notion.

Definition 8.1.13. (i) By a mild LT-solution $u(x_1,\ldots,x_n)$ of [(514)–(515)], respectively, [(516)–(517)], we mean any Laplace transformable function $u(x_1,\ldots,x_n)$ such that the terms $D_R^{(0,\ldots,a_k,\ldots,0)} u(x_1,\ldots,x_k,\ldots,x_n)$, respectively, $D_C^{(0,\ldots,a_k,\ldots,0)} u(x_1,\ldots,x_k,\ldots,x_n)$, are well-defined and Laplace transformable for $1 \leqslant k \leqslant n$ as well as that the terms $(\partial^j/\partial x_k^j) J_{t_k}^{m_k-a_k} u(x_1,\ldots,x_n)$, respectively, $(\partial^j/\partial x_k^j) u(x_1,\ldots,x_n)$, are well-defined and continuous with respect to the variable x_j for $1 \leqslant k \leqslant n$, $0 \leqslant j \leqslant m_k - 1$,

$$\sum_{k=1}^{n} A_k \big(\mathcal{L} D_R^{(0,\ldots,a_k,\ldots,0)} u(x_1,\ldots,x_k,\ldots,x_n)\big)(\lambda_1,\ldots,\lambda_n) = \tilde{f}(\lambda_1,\ldots,\lambda_n), \tag{518}$$

for $\operatorname{Re}\lambda_j > \omega_j$ $(1 \leqslant j \leqslant n)$ and some nonnegative real numbers $\omega_1 \geqslant 0,\ldots,\omega_n \geqslant 0$, respectively, (518) holds with $D_R^{(0,\ldots,a_k,\ldots,0)} u(x_1,\ldots,x_k,\ldots,x_n)$ replaced with the term $D_C^{(0,\ldots,a_k,\ldots,0)} u(x_1,\ldots,x_k,\ldots,x_n)$ therein, and (515), respectively, (517), holds.

(ii) By a strong LT-solution $u(x_1,\ldots,x_n)$ of [(514)–(515)], respectively, [(516)–(517)], we mean any mild LT-solution $u(x_1,\ldots,x_n)$ of this problem, which additionally satisfies that the terms $A_k D_R^{(0,\ldots,a_k,\ldots,0)} u(x_1,\ldots,x_k,\ldots,x_n)$, resp. $A_k D_C^{(0,\ldots,a_k,\ldots,0)} u(x_1,\ldots,x_k,\ldots,x_n)$, are well-defined and Laplace transformable for $1 \leqslant k \leqslant n$.

The uniqueness theorem for Laplace transform and the closedness of operators A_k for $1 \leqslant k \leqslant n$ show that any strong LT-solution of [(514)–(515)], respectively, [(516)–(517)], satisfies that (514), respectively, (516), holds for a. e. $x_1 \geqslant 0,\ldots,x_n \geqslant 0$.

Our main result concerning the well-posedness of equations [(514)–(515)] and [(516)–(517)] reads as follows.

Theorem 8.1.14. *Suppose that $C \in L(X)$ is injective, A_k is a closed linear operator commuting with C and $a_k \geqslant 0$ for $1 \leqslant k \leqslant n$. Suppose further that there exist nonnegative real numbers $\omega_1 \geqslant 0,\ldots,\omega_n \geqslant 0$ such that the operator $\sum_{k=1}^{n} \lambda_k^{a_k} A_k$ is injective and $(\sum_{k=1}^{n} \lambda_k^{a_k} A_k)^{-1} C \in L(X)$ for $\operatorname{Re}\lambda_1 > \omega_1,\ldots,\operatorname{Re}\lambda_n > \omega_n$. Let the following conditions also hold:*

(i) *There exists a locally integrable, exponentially bounded function* $h(x_1,\dots,x_n)$ *for* $x_1 \geqslant 0,\dots,x_n \geqslant 0$ *satisfying that* $D_R^{(0,\dots,\alpha_k,\dots,0)}h(x_1,\dots,x_k,\dots,x_n)$, *respectively,* $D_C^{(0,\dots,\alpha_k,\dots,0)}h(x_1,\dots,x_k,\dots,x_n)$, *is well-defined, locally integrable and exponentially bounded* ($1 \leqslant k \leqslant n$), *the terms* $(\partial^j/\partial x_k^j)J_{t_k}^{m_k-\alpha_k}h(x_1,\dots,x_n)$, *respectively,* $(\partial^j/\partial x_k^j)h(x_1,\dots,x_n)$, *are well-defined and continuous with respect to the variable* x_j *for* $1 \leqslant k \leqslant n$, $0 \leqslant j \leqslant m_k - 1$ *and*

$$\tilde{h}(\lambda_1,\dots,\lambda_n) = \left(\sum_{k=1}^{n} \lambda_k^{\alpha_k} A_k\right)^{-1} C\tilde{f_0}(\lambda_1,\dots,\lambda_n), \quad \operatorname{Re}\lambda_1 > \omega_1,\dots,\operatorname{Re}\lambda_n > \omega_n,$$

where $f = Cf_0$.

(ii) *If* $1 \leqslant k \leqslant n$ *and* $0 \leqslant j \leqslant m_k - 1$, *then there exists a locally integrable, exponentially bounded function* $h_{k,j}(x_1,\dots,x_{k-1},x_{k+1},\dots,x_n)$ *for* $x_1 \geqslant 0,\dots,x_{k-1} \geqslant 0$, $x_{k+1} \geqslant 0,\dots,x_n \geqslant 0$ *satisfying that the terms* $D_R^{(0,\dots,\alpha_v,\dots,0)}h_{k,j}(x_1,\dots,x_{k-1},x_{k+1},\dots,x_n)$, *respectively,* $D_C^{(0,\dots,\alpha_v,\dots,0)}h_{k,j}(x_1,\dots,x_{k-1},x_{k+1},\dots,x_n)$, *are well-defined, locally integrable and exponentially bounded for* $1 \leqslant v \leqslant n$, *the terms* $(\partial^j/\partial x_v^j)J_{t_k}^{m_k-\alpha_k}h_{k,j}(x_1,\dots,x_{k-1},x_{k+1},\dots,x_n)$, *respectively,* $(\partial^j/\partial x_v^j)h_{k,j}(x_1,\dots,x_{k-1},x_{k+1},\dots,x_n)$ *are well-defined and continuous with respect to the variable* x_v *for* $1 \leqslant v \leqslant n$ *and*

$$\widetilde{h_{k,j}}(\lambda_1,\dots,\lambda_{k-1},\lambda_{k+1},\dots,\lambda_n) = \left(\sum_{k=1}^{n} \lambda_k^{\alpha_k} A_k\right)^{-1} CA_k\widetilde{f_{k,j,0}}(\lambda_1,\dots,\lambda_{k-1},\lambda_{k+1},\dots,\lambda_n),$$

provided that $\operatorname{Re}\lambda_1 > \omega_1,\dots,\operatorname{Re}\lambda_{k-1} > \omega_{k-1}$, $\operatorname{Re}\lambda_{k+1} > \omega_{k+1},\dots,\operatorname{Re}\lambda_n > \omega_n$, *where* $f_{k,j} = Cf_{k,j,0}$.

Then there exists a unique mild LT-*solution* $u(x_1,\dots,x_n)$ *of* [(514)–(515)], *respectively,* [(516)–(517)], *and we have*

$$u(x_1,\dots,x_n) = \sum_{k=1}^{n}\sum_{j=0}^{m_k-1} h_{k,j}(x_1,\dots,x_n) + h(x_1,\dots,x_n), \quad x_1 \geqslant 0,\dots,x_n \geqslant 0. \tag{519}$$

Furthermore, if the following conditions hold:

(i-s) *If* $1 \leqslant v \leqslant n$, *then the terms* $A_v h(x_1,\dots,x_v,\dots,x_n)$ *and* $D_R^{(0,\dots,\alpha_v,\dots,0)}A_v h(x_1,\dots,x_v,\dots,x_n)$, *respectively,* $D_C^{(0,\dots,\alpha_v,\dots,0)}A_v h(x_1,\dots,x_v,\dots,x_n)$, *are well-defined, locally integrable and exponentially bounded.*

(ii-s) *If* $1 \leqslant v \leqslant n$, $1 \leqslant k \leqslant n$ *and* $0 \leqslant j \leqslant m_k - 1$, *then the terms* $A_v h_{k,j}(x_1,\dots,x_{k-1},x_{k+1},\dots,x_n)$ *and* $D_R^{(0,\dots,\alpha_v,\dots,0)}A_v h_{k,j}(x_1,\dots,x_{k-1},x_{k+1},\dots,x_n)$, *respectively,* $D_C^{(0,\dots,\alpha_v,\dots,0)}A_v \times h_{k,j}(x_1,\dots,x_{k-1},x_{k+1},\dots,x_n)$, *are well-defined, locally integrable and exponentially bounded,*

then the function $u(x_1,\dots,x_n)$, *given by* (519), *is a strong* LT-*solution of* [(514)–(515)], *respectively,* [(516)–(517)].

Proof. Let $u(x_1,\dots,x_n)$ be given by (519), and let $\operatorname{Re}\lambda_1 > \omega_1,\dots,\operatorname{Re}\lambda_n > \omega_n$. Our assumptions imply that the term $D_R^{(0,\dots,\alpha_k,\dots,0)}u(x_1,\dots,x_k,\dots,x_n)$, respectively, $D_C^{(0,\dots,\alpha_k,\dots,0)}u(x_1,\dots,x_k,\dots,x_n)$, is well-defined as well as that we have (see also the equations [97, (1.22)–(1.23)] and the equation [443, (16)]):

$$D_R^{(0,\dots,\alpha_k,\dots,0)}u(x_1,\dots,x_k,\dots,x_n)$$
$$= \lambda_k^{\alpha_k}\tilde{u}(\lambda_1,\dots,\lambda_n) - \sum_{j=0}^{m_k-1}[\mathcal{L}_{t_1,\dots,t_{k-1},t_{k+1},\dots,t_n}f_{k,j}](\lambda_1,\dots,\lambda_{k-1},\lambda_{k+1},\dots,\lambda_n)\lambda_k^{m_k-1-j}, \quad (520)$$

respectively,

$$D_C^{(0,\dots,\alpha_k,\dots,0)}u(x_1,\dots,x_k,\dots,x_n)$$
$$= \lambda_k^{\alpha_k}\tilde{u}(\lambda_1,\dots,\lambda_n) - \sum_{j=0}^{m_k-1}[\mathcal{L}_{t_1,\dots,t_{k-1},t_{k+1},\dots,t_n}f_{k,j}](\lambda_1,\dots,\lambda_{k-1},\lambda_{k+1},\dots,\lambda_n)\lambda_k^{\alpha_k-1-j}, \quad (521)$$

where $\mathcal{L}_{t_1,\dots,t_{k-1},t_{k+1},\dots,t_n}$ denotes the multidimensional Laplace transform with respect to the variables $t_1,\dots,t_{k-1},t_{k+1},\dots,t_n$. Furthermore, our assumptions simply imply that

$$\tilde{u}(\lambda_1,\dots,\lambda_n) = \sum_{k=1}^{n}\left(\sum_{k=1}^{n}\lambda_k^{\alpha_k}A_k\right)^{-1} CA_k\sum_{j=0}^{m_k-1}[\mathcal{L}_{t_1,\dots,t_{k-1},t_{k+1},\dots,t_n}f_{k,j,0}]$$
$$\times(\lambda_1,\dots,\lambda_{k-1},\lambda_{k+1},\dots,\lambda_n) + \left(\sum_{k=1}^{n}\lambda_k^{\alpha_k}A_k\right)^{-1} C\widetilde{f_0}(\lambda_1,\dots,\lambda_n).$$

This yields

$$\left[\sum_{k=1}^{n}\lambda_k^{\alpha_k}A_k\right]\tilde{u}(\lambda_1,\dots,\lambda_n) - \sum_{k=1}^{n}A_k\sum_{j=0}^{m_k-1}[\mathcal{L}_{t_1,\dots,t_{k-1},t_{k+1},\dots,t_n}f_{k,j}](\lambda_1,\dots,\lambda_{k-1},\lambda_{k+1},\dots,\lambda_n)\lambda_k^{m_k-1-j}$$
$$= \tilde{f}(\lambda_1,\dots,\lambda_n),$$

respectively,

$$\left[\sum_{k=1}^{n}\lambda_k^{\alpha_k}A_k\right]\tilde{u}(\lambda_1,\dots,\lambda_n) - \sum_{k=1}^{n}A_k\sum_{j=0}^{m_k-1}[\mathcal{L}_{t_1,\dots,t_{k-1},t_{k+1},\dots,t_n}f_{k,j}](\lambda_1,\dots,\lambda_{k-1},\lambda_{k+1},\dots,\lambda_n)\lambda_k^{\alpha_k-1-j}$$
$$= \tilde{f}(\lambda_1,\dots,\lambda_n).$$

Keeping in mind the equations (520)–(521), it readily follows that the equations (518) and its analogue with Caputo fractional derivatives hold good. Therefore, the function $u(x_1,\dots,x_n)$ is a mild LT-solution of problem [(514)–(515)], respectively, [(516)–(517)]. The uniqueness of mild LT-solutions of this problem follows from a simple argumentation involving the injectiveness of the operator $\sum_{k=1}^{n}\lambda_k^{\alpha_k}A_k$ for $\operatorname{Re}\lambda_1 > \omega_1,\dots,\operatorname{Re}\lambda_n > \omega_n$ and the uniqueness theorem for Laplace transform. Finally, if the conditions (i-s) and (ii-s) hold, then we can simply prove that the function $A_v D_R^{(0,\ldots,a_v,\ldots,0)} A_v u(x_1,\ldots,x_n)$ is Laplace transformable and

$$\mathcal{L}[A_v D_R^{(0,\ldots,a_v,\ldots,0)} u(x_1,\ldots,x_n)] = A_v[\mathcal{L}u(x_1,\ldots,x_n)],$$

which simply completes the proof. □

Keeping in mind Theorem 8.1.7, we can apply Theorem 8.1.14 in many concrete situations, even if $a_k > 2$ for some indexes $k \in \mathbb{N}_n$; cf. [443] and [445] for more details. To wrap up this subsection, let us observe that we can similarly analyze some generalizations of the problems [(514)–(515)] and [(516)–(517)] with various types of generalized Laplace fractional derivatives, especially with the generalized Hilfer (a, b, α)-fractional derivatives.

8.1.8 Fractional partial difference equations with generalized Weyl derivatives

In this subsection, we will only explain how Theorem 3.5.5(i) can useful in the study of well-posedness of some classes of the abstract fractional partial difference equations with generalized Weyl derivatives. For some concrete applications of this result to the fractional partial difference equations with generalized Weyl derivatives, we will particularly consider the situation in which the sequences $a_i(\cdot)$ have the following form:

$$a_i(v_1,\ldots,v_n) = a_1^i(v_1)\cdot\cdots\cdot a_n^i(v_n), \quad (v_1,\ldots,v_n) \in \mathbb{N}_0^n \quad (1 \leqslant i \leqslant m). \tag{522}$$

Suppose now that $\mathrm{v}_1 \in \mathbb{N}_0^n,\ldots,\mathrm{v}_m \in \mathbb{N}_0^n$, $(S(\mathrm{v}))_{\mathrm{v}\in\mathbb{N}_0^n} \subseteq L(X)$ is a discrete $(k, C, B, (A_i)_{1\leqslant i\leqslant m}, (\mathrm{v}_i)_{1\leqslant i\leqslant m})$-existence family, $\sum_{\mathrm{v}\in\mathbb{N}_0^n} \|S(\mathrm{v})\| < +\infty$, (522) and the following conditions hold:

(a1) $f : \mathbb{Z}^n \to X$ is a bounded sequence, $k \in l^1(\mathbb{N}_0^n : \mathbb{C})$ and $\sum_{v=0}^{+\infty} |a_j^i(v)| < +\infty$ for $1 \leqslant j \leqslant n$ and $1 \leqslant i \leqslant m$, or

(b1) $f \in l^1(\mathbb{Z}^n : X)$, $k : \mathbb{N}_0^n \to \mathbb{C}$ is a bounded sequence and $a_j^i : \mathbb{Z} \to \mathbb{C}$ is a bounded sequence for $1 \leqslant j \leqslant n$ and $1 \leqslant i \leqslant m$.

Let $\mathbf{m} = (m_1,\ldots,m_n) \in \mathbb{N}^n$ be fixed, and let the sequences $u(\cdot)$ and $g(\cdot)$ be defined by (209) and (199), respectively. Then $u(\cdot)$ is bounded if (a1) holds, $u \in l^1(\mathbb{Z}^n : X)$ if (b1) holds, and a simple computation shows that we have

$$(\Delta_{v_1^{m_1},\ldots,v_n^{m_n}}^{m_1+\cdots+m_n} Bu)(\mathrm{v}) = A_1(\Delta_{W,a_1,\mathbf{m}}u)(\mathrm{v}+\mathrm{v}_i) + \cdots + A_m(\Delta_{W,a_m,\mathbf{m}}u)(\mathrm{v}+\mathrm{v}_m), \quad \mathrm{v} \in \mathbb{Z}^n. \tag{523}$$

Further on, if $a_1 = (a_1^1,\ldots,a_n^1) \in [0,+\infty)^n,\ldots,a_m = (a_1^m,\ldots,a_n^m) \in [0,+\infty)^n$, $m_j^i = \lceil a_j^i \rceil$ for $1 \leqslant j \leqslant n$ and $1 \leqslant i \leqslant m$, $a_j^i(v_j) = k^{m_j^i - a_j^i}(v_j)$ for $1 \leqslant j \leqslant n$ and $1 \leqslant i \leqslant m$,

$$m_j = m_j^1 = \cdots = m_j^n, \quad 1 \leqslant j \leqslant n,$$

then we have

$$(\Delta^{m_1+\cdots+m_n}_{v_1^{m_1},\ldots,v_n^{m_n}} Bu)(v) = A_1(\Delta^{\alpha_1}_W u)(v + v_i) + \cdots + A_m(\Delta^{\alpha_m}_W u)(v + v_m), \quad v \in \mathbb{Z}^n. \tag{524}$$

We can also analyze some other relatives of (523)–(524) as well as the existence and uniqueness of almost periodic type solutions to (523)–(524).

Let us finally note that R. Ponce has investigated, in [646], the well-posedness of the following abstract Volterra integrodifferential equation with Weyl fractional derivatives:

$$D^{\alpha}_W u(t) = Au(t) + \int_{-\infty}^{t} a(t-s)Au(s)\,ds + f(t, u(t)), \quad t \in \mathbb{R}, \tag{525}$$

where A is a closed linear operator, $\alpha > 0$ and $f(\cdot,\cdot)$ satisfies certain extra assumptions (cf. also the recent research article [297] by V. E. Fedorov and N. M. Skripka). We will only emphasize here that we can similarly consider the well-posedness of the following multivalued linear analogue of (525):

$$D^{\alpha}_W u(t) \in \mathcal{A}\left[u(t) + \int_{-\infty}^{t} a(t-s)u(s)\,ds\right] + f(t, u(t)), \quad t \in \mathbb{R},$$

provided that $\mathcal{A}$ is a closed subgenerator of an exponentially bounded $(g_\alpha, g_\alpha + (g_\alpha *_0 a))$-regularized C-resolvent family; cf. [445] for the notion.

Some final observations

In this section, we have introduced and analyzed several new types of partial fractional derivatives in the continuous setting and the discrete setting. We have investigated the well-posedness of some classes of the abstract fractional differential equations and the abstract fractional difference equations depending on several variables, providing also many illustrative examples and useful remarks.

Let us finally note that we can also consider several new types of partial fractional derivatives using the multidimensional convolution products

$$(\mathbf{a} *_0 \mathbf{b})(\mathbf{x}) := \int_0^{x_1} \cdots \int_0^{x_n} \mathbf{a}(x_1 - s_1, \ldots, x_n - s_n)\mathbf{b}(s_1, \ldots, s_n)\,ds_1 \cdots ds_n,$$

for $\mathbf{x} = (x_1, \ldots, x_n) \in [0, +\infty)^n$, where $\mathbf{a}, \mathbf{b} \in L^1_{loc}([0, +\infty)^n)$ and

$$(\mathbf{a} \circ \mathbf{b})(\mathbf{x}) := \int_{-\infty}^{x_1} \cdots \int_{-\infty}^{x_n} \mathbf{a}(x_1 - s_1, \ldots, x_n - s_n)\mathbf{b}(s_1, \ldots, s_n)\,ds_1 \cdots ds_n, \tag{526}$$

for $\mathbf{x} = (x_1, \ldots, x_n) \in \mathbb{R}^n$, where $\mathbf{a} \in L^1_{loc}([0, +\infty)^n)$ and $\mathbf{b} \in L^1_{loc}(\mathbb{R}^n)$.

If $\mathrm{j} = (j_1, \ldots, j_n) \in \mathbb{N}_0^n$ and $\mathrm{k} = (k_1, \ldots, k_n) \in \mathbb{N}_0^n$, then we write $\mathrm{j} \leqslant \mathrm{k}$ if and only if $j_m \leqslant k_m$ for all $1 \leqslant m \leqslant n$. In the discrete setting, we can consider several new types of partial fractional differences using the already considered multidimensional convolution products $*_0$ and $\circ$. It is clear that the equation (526) presents an extension of the generalized Weyl **a**-integral; if $\mathbf{a} \in L^1_{\mathrm{loc}}([0, +\infty)^n)$, $\mathbf{u} \in L^1_{\mathrm{loc}}(\mathbb{R}^n)$, $a_j \geqslant 0$ for $1 \leqslant j \leqslant n$ and $\alpha = (\alpha_1, \ldots, \alpha_n)$, then we also define

$$\mathbb{D}_W^{\alpha,\mathbf{a},1}\mathbf{u} := \frac{\partial^m}{\partial x_1^{m_1} \cdots \partial x_n^{m_n}}(\mathbf{a} \circ \mathbf{u})(\mathbf{x}), \quad \mathbf{x} = (x_1, \ldots, x_n) \in \mathbb{R}^n,$$

where $m_j = \lceil \alpha_j \rceil$ for $1 \leqslant j \leqslant n$ and $m = m_1 + \cdots + m_n$. It is worth noting that the formulae (491) and (492) continue to hold in this framework.

In the discrete framework, several new types of fractional partial difference operators can be introduced and analyzed using the multidimensional convolution products $*_0$, $\circ$ and the sequences $a : \mathbb{N}_0^n \to \mathbb{C}$, which do not have the form (522). We will consider such operators somewhere else.

8.2 Multidimensional Poisson transform and applications

As already mentioned several times, the vector-valued Poisson transform has been first considered in the pioneering paper [535] by C. Lizama, where the author has also presented some applications of Poisson transform to the abstract fractional difference equations. The main aim of this section is to extend some structural results from [535] to the higher-dimensional setting. We also further analyze here the multidimensional vector-valued Laplace transform and clarify certain relations between the solutions of the abstract (fractional) partial differential equations and the solutions of the abstract (fractional) difference equations of several variables. Our main results are Theorem 8.2.5, Theorem 8.2.10 and Theorem 8.2.11.

The organization and main ideas of this section can be simply explained as follows [471]. We first prove some new results about the multidimensional vector-valued Laplace transform; cf. Proposition 8.2.1. The applications of multidimensional vector-valued Poisson transforms to the abstract partial differential-difference equations are given in Section 8.2.1 and the applications of the multidimensional vector-valued Poisson transforms to the abstract fractional partial differential-difference equations are given in Section 8.2.2.

Let us introduce the following condition:

(GR) $f(\cdot)$ is Lebesgue measurable and there exist real constants $\omega_1 \in \mathbb{R}, \ldots, \omega_n \in \mathbb{R}$, $\eta_1 \in (-1, +\infty), \ldots, \eta_n \in (-1, +\infty)$ and $\zeta_1 \in (-1, +\infty), \ldots, \zeta_n \in (-1, +\infty)$ such that

$$\|f(t_1, \ldots, t_n)\| \leqslant M\big(t_1^{\eta_1} + t_1^{\zeta_1}\big) \cdot \cdots \cdot \big(t_n^{\eta_n} + t_n^{\zeta_n}\big) \exp(\omega_1 t_1 + \cdots + \omega_n t_n),$$
$$\text{for a. e. } t_1 \geqslant 0, \ldots, t_n \geqslant 0. \tag{527}$$

In this case, the Fubini theorem implies that the function $F(\lambda_1, \ldots, \lambda_n)$ is well-defined for $\operatorname{Re}\lambda_1 > \omega_1, \ldots, \operatorname{Re}\lambda_n > \omega_n$ and the Lebesgue dominated convergence theorem implies that $F(\cdot)$ is analytic in this region of $\mathbb{C}^n$.

The collection of all Lebesgue measurable functions $f(\cdot)$, which satisfies condition (GR), forms a vector space with the usual operations. Furthermore, if $f(\cdot)$ satisfies (GR) with $X = \mathbb{C}$, $g : [0, +\infty)^n \to X$ is Lebesgue measurable and there exist real constants $\omega_{1,g} \in \mathbb{R}, \ldots, \omega_{n,g} \in \mathbb{R}$, $\eta_{1,g} \in (-1, +\infty), \ldots, \eta_{n,g} \in (-1, +\infty)$ and $\zeta_{1,g} \in (-1, +\infty), \ldots, \zeta_{n,g} \in (-1, +\infty)$ such that

$$\|g(t_1, \ldots, t_n)\| \leqslant M(t_1^{\eta_{1,g}} + t_1^{\zeta_{1,g}}) \cdot \cdots \cdot (t_n^{\eta_{n,g}} + t_n^{\zeta_{n,g}}) \exp(\omega_{1,g}t_1 + \cdots + \omega_{n,g}t_n),$$
$$\text{for a. e. } t_1 \geqslant 0, \ldots, \ t_n \geqslant 0,$$

then the pointwise product $[fg](\cdot)$ also satisfies (GR), provided that

$$\min\{\eta_{j,g} + \eta_j,\ \eta_{j,g} + \zeta_j,\ \zeta_{j,g} + \eta_j,\ \zeta_{j,g} + \zeta_j : 1 \leqslant j \leqslant n\} > -1.$$

Now we will state and prove the following statements concerning the multidimensional Laplace transform, which will be sufficiently enough for our later purposes (cf. also the statements of [69, Theorem 1.5.1, Proposition 1.6.4], which will not be fully generalized to the multidimensional setting here).

Proposition 8.2.1. (i) *Suppose that $f : [0, +\infty)^n \to X$ satisfies* (GR). *Then we have*

$$F^{(v_1,\ldots,v_n)}(\lambda_1, \ldots, \lambda_n) = (-1)^{v_1+\cdots+v_n}\left(\mathcal{L}\left[\frac{\cdot_1^{v_1}}{v_1!} \cdot \cdots \cdot \frac{\cdot_n^{v_n}}{v_n!} f(\cdot_1, \ldots, \cdot_n)\right]\right)(\lambda_1, \ldots, \lambda_n),$$

for $\operatorname{Re}\lambda_1 > \omega_1, \ldots, \operatorname{Re}\lambda_n > \omega_n$ *and* $(v_1, \ldots, v_n) \in \mathbb{N}_0^n$.

(ii) *Suppose that $a \in L^1_{\mathrm{loc}}([0, +\infty)^n)$ satisfies* (GR) *with $X = \mathbb{C}$, the constants $\omega_1 \in \mathbb{R}, \ldots, \omega_n \in \mathbb{R}$ and the constants $\eta_1 \in (-1, +\infty), \ldots, \eta_n \in (-1, +\infty)$ and $\zeta_1 \in (-1, +\infty), \ldots, \zeta_n \in (-1, +\infty)$ replaced therein with the constants $\eta_{1,a} \in (-1, +\infty), \ldots, \eta_{n,a} \in (-1, +\infty)$ and $\zeta_{1,a} \in (-1, +\infty), \ldots, \zeta_{n,a} \in (-1, +\infty)$. Suppose further that $f \in L^1_{\mathrm{loc}}([0, +\infty)^n : X)$ satisfies* (GR) *with the same constants $\omega_1 \in \mathbb{R}, \ldots, \omega_n \in \mathbb{R}$; then $(a *_0 f)(\cdot) \in L^1_{\mathrm{loc}}([0, +\infty)^n)$ satisfies* (GR) *and we have*

$$\mathcal{L}(a *_0 f)(\lambda_1, \ldots, \lambda_n) = \mathcal{L}a(\lambda_1, \ldots, \lambda_n) \cdot \mathcal{L}f(\lambda_1, \ldots, \lambda_n), \tag{528}$$

for $\operatorname{Re}\lambda_1 > \omega_1, \ldots, \operatorname{Re}\lambda_n > \omega_n$.

Proof. Keeping in mind the estimate (527), the part (i) follows from a simple application of the Lebesgue dominated convergence theorem. It can be simply shown that $(a *_0 f)(\cdot)$ is Lebesgue measurable; furthermore, a simple computation involving the Fubini theorem and the identity $g_c *_0 g_d = g_{c+d}$ for $c,\ d > 0$ shows that there exist positive real constants $M,\ M_1 > 0$ such that

$$\begin{aligned}\|(a *_0 f)(x_1,\ldots,x_n)\| &\leqslant M^n \exp(\omega_1 x_1 + \cdots + \omega_n x_n)\\ &\quad \cdot \int_0^{x_1} \cdots \int_0^{x_n} \bigl((x_1-t_1)^{\eta_1} + (x_1-t_1)^{\zeta_1}\bigr) \cdot \cdots \cdot \bigl((x_n-t_n)^{\eta_n} + (x_n-t_n)^{\zeta_n}\bigr)\\ &\quad \times \bigl(t_1^{\eta_1} + t_1^{\zeta_1}\bigr) \cdot \cdots \cdot \bigl(t_n^{\eta_n} + t_n^{\zeta_n}\bigr)\, dt_1 \cdots dt_n\\ &\leqslant M_1^n \bigl(x_1^{\eta_{1;1}} + x_1^{\zeta_{1;1}}\bigr) \cdot \cdots \cdot \bigl(x_n^{\eta_{n;1}} + x_n^{\zeta_{n;1}}\bigr) \exp(\omega_1 x_1 + \cdots + \omega_n x_n),\end{aligned}$$

where $\eta_{j;1} = \min(\eta_{j,a}+\eta_j+1, \eta_{j,a}+\zeta_j+1, \zeta_{j,a}+\eta_j+1, \zeta_{j,a}+\zeta_j+1)$ and $\zeta_{j;1} = \max(\eta_{j,a}+\eta_j+1, \eta_{j,a}+\zeta_j+1, \zeta_{j,a}+\eta_j+1, \zeta_{j,a}+\zeta_j+1)$ for $1 \leqslant j \leqslant n$. The formula (528) can be deduced following the lines of the proofs of [230, Theorems 3.1, 4.1], where the author has considered the double Laplace transform in the scalar-valued setting. □

Now we will continue our recent investigation of Poisson-like transforms and explain how the already established results and ideas can be simply transferred to the multidimensional setting. Suppose that $u : [0,\infty)^n \to X$ is a given locally integrable function and the value of

$$\begin{aligned}[P(u)](\mathrm{v}) &:= [P(u)](v_1,\ldots,v_n)\\ &:= \int_{[0,\infty)^n} e^{-x_1-\cdots-x_n} \frac{x_1^{v_1}}{v_1!} \cdot \cdots \cdot \frac{x_n^{v_n}}{v_n!} u(x_1,\ldots,x_n)\, dx_1\, dx_2 \cdots dx_n\end{aligned}$$

is well-defined for all $v_1 \in \mathbb{N}_0, \ldots, v_n \in \mathbb{N}_0$. We call the mapping $u \mapsto P(u)$ the multidimensional Poisson transform; in terms of the multidimensional vector-valued Laplace transform, we have

$$[P(u)](v_1,\ldots,v_n) = \left(\mathcal{L}\left[\frac{\cdot_1^{v_1}}{v_1!} \cdot \cdots \cdot \frac{\cdot_n^{v_n}}{v_n!} u(\cdot_1,\ldots,\cdot_n)\right]\right)(1,\ldots,1), \quad (v_1,\ldots,v_n) \in \mathbb{N}_0^n.$$

We can simply prove that

$$\int_{[0,\infty)^n} \|u(x_1,\ldots,x_n)\|\, dx_1\, dx_2 \cdots dx_n < +\infty \quad \text{implies} \quad \sum_{\mathrm{v}\in\mathbb{N}_0^n} \|[P(u)](\mathrm{v})\| < +\infty.$$

We will analyze the multidimensional vector-valued Z-transform of sequences and its applications somewhere else, where we will also extend the statement of [535, Theorem 3.1] to the multidimensional setting. Concerning some applications of multidimensional Z-transform to the difference equations, we refer the reader to [413, Section 2.7] and the doctoral dissertation of P. Alper [47].

Now we will reconsider [535, Theorem 3.4] in the multidimensional setting.

Theorem 8.2.2. *Suppose that $a \in L^1_{\mathrm{loc}}([0,+\infty)^n)$ satisfies* (GR) *with $X = \mathbb{C}$, the constants $\omega_1 \in (-\infty,1),\ldots,\omega_n \in (-\infty,1)$ and the constants $\eta_1 \in (-1,+\infty),\ldots,\eta_n \in (-1,+\infty)$ and $\zeta_1 \in (-1,+\infty),\ldots,\zeta_n \in (-1,+\infty)$ replaced therein with the constants*

$\eta_{1,a} \in (-1,+\infty),\dots,\eta_{n,a} \in (-1,+\infty)$ *and* $\zeta_{1,a} \in (-1,+\infty),\dots,\zeta_{n,a} \in (-1,+\infty)$, *respectively. Suppose further that* $f \in L^1_{loc}([0,+\infty)^n : X)$ *satisfies* (GR) *with the same constants* $\omega_1 \in (-\infty,1),\dots,\omega_n \in (-\infty,1)$. *Then* $[P(a *_0 f)](v_1,\dots,v_n)$, $[P(a)](v_1,\dots,v_n)$ *and* $[P(f)](v_1,\dots,v_n)$ *exist for any* $(v_1,\dots,v_n) \in \mathbb{N}_0^n$; *furthermore, we have*

$$[P(a *_0 f)](v_1,\dots,v_n) = [P(a)](v_1,\dots,v_n)\cdot[P(f)](v_1,\dots,v_n),\quad (v_1,\dots,v_n) \in \mathbb{N}_0^n. \tag{529}$$

Proof. By Proposition 8.2.1(ii), we have that $(a *_0 f)(\cdot) \in L^1_{loc}([0,+\infty)^n)$ satisfies (GR) and (528) holds. It is clear that $[P(a*_0 f)](v_1,\dots,v_n)$, $[P(a)](v_1,\dots,v_n)$ and $[P(f)](v_1,\dots,v_n)$ exist for any $(v_1,\dots,v_n) \in \mathbb{N}_0^n$. Set now $G := \mathcal{L}(a *_0 f)$. Then we have

$$\begin{aligned}
&[P(a *_0 f)](v_1,\dots,v_n)\\
&= (-1)^{v_1+\dots+v_n}\frac{1}{v_1!}\cdot\dots\cdot\frac{1}{v_n!}G^{(v_1,\dots,v_n)}(1,\dots,1)\\
&= \frac{(-1)^{v_1+\dots+v_n}}{v_1!\cdot\dots\cdot v_n!}\big[[\mathcal{L}a]^{(v_1,\dots,v_n)}(\lambda_1,\dots,\lambda_n)\cdot[\mathcal{L}f]^{(v_1,\dots,v_n)}(\lambda_1,\dots,\lambda_n)\big]_{(\lambda_1,\dots,\lambda_n)=(1,\dots,1)}\\
&= \frac{(-1)^{v_1+\dots+v_n}}{v_1!\cdot\dots\cdot v_n!}\sum_{j\in\mathbb{N}_0^n;j\leq v}\binom{v_1}{j_1}\cdot\dots\cdot\binom{v_n}{j_n}\\
&\quad\times[\mathcal{L}a]^{(v_1-j_1,\dots,v_n-j_n)}(1,\dots,1)\cdot[\mathcal{L}f]^{(j_1,\dots,j_n)}(1,\dots,1)\\
&= \frac{1}{v_1!\cdot\dots\cdot v_n!}\sum_{j\in\mathbb{N}_0^n;j\leq v}\binom{v_1}{j_1}\cdot\dots\cdot\binom{v_n}{j_n}[P(a)](v-j)\cdot[P(f)](j)\\
&\quad\times(v_1-j_1)!\cdot\dots\cdot(v_n-j_n)!\cdot j_1!\cdot\dots\cdot j_n!\,[P(a)](v-j)\cdot[P(f)](j)\\
&= \sum_{j\in\mathbb{N}_0^n;j\leq v}[P(a)](v-j)\cdot[P(f)](j),\quad (v_1,\dots,v_n)\in\mathbb{N}_0^n,
\end{aligned}$$

where we have used Proposition 8.2.1(i)–(ii) and the Leibniz rule. This proves (529) and completes the proof. □

Remark 8.2.3. In the formulation of [535, Theorem 3.4], we must additionally assume that the Laplace transform of function $|a|(\cdot)$ exists at the point 1. Only in this way, we can apply [69, Proposition 1.6.4] as an essential ingredient in the proof of [535, Theorem 3.4], which does not work if $abs(a) < 1 < abs(|a|)$; here, we use the same notion and notation as in [69].

8.2.1 Applications to the abstract partial differential equations

In this subsection, we will present some applications of multidimensional vector-valued Poisson transform to the abstract partial differential equations.

Let us formally set $(x^v/v!) := 0$, if $-v \in \mathbb{N}$. We start with the following illustrative example.

Example 8.2.4. Let us consider the partial differential operator $u_{x_1x_1x_2}(\cdot,\cdot)$, in the dimension $n = 2$, and let us assume that the partial derivatives $u_{x_1x_1x_2}(\cdot,\cdot)$, $u_{x_1x_1}(\cdot,\cdot)$, $u_{x_1}(\cdot,\cdot)$ and the function $u(\cdot,\cdot)$ are continuous on $[0,\infty)^2$, as well as that

$$u_{x_1x_1x_2}(x_1,x_2) = g(x_1,x_2), \quad (x_1,x_2) \in [0,\infty)^2.$$

Let $v_1 \in \mathbb{N}_0$ and $v_2 \in \mathbb{N}_0$ be fixed. Applying the Fubini theorem and the partial integration with respect to the variable x_2, we get

$$\begin{aligned}&\int\limits_{[0,\infty)^2} e^{-x_1-x_2}\frac{x_1^{v_1+2}}{(v_1+2)!}\frac{x_2^{v_2+1}}{(v_2+1)!}u_{x_1x_1x_2}(x_1,x_2)\,dx_1\,dx_2\\ &\quad= \int\limits_{[0,\infty)^2} e^{-x_1-x_2}\frac{x_1^{v_1+2}}{(v_1+2)!}\left[\frac{x_2^{v_2+1}}{(v_2+1)!}-\frac{x_2^{v_2}}{v_2!}\right]u_{x_1x_1}(x_1,x_2)\,dx_1\,dx_2,\end{aligned} \tag{530}$$

provided that

$$\lim_{x_2\to+\infty} e^{-x_2}\frac{x_2^{v_2+1}}{(v_2+1)!}u_{x_1x_1}(x_1,x_2) = 0, \quad x_1 \geqslant 0 \tag{531}$$

and the both double integrals in (530) converges absolutely. Applying the Fubini theorem and the partial integration two times more, with respect to the variable x_1, we get

$$\begin{aligned}&\int\limits_{[0,\infty)^2} e^{-x_1-x_2}\frac{x_1^{v_1+2}}{(v_1+2)!}\frac{x_2^{v_2+1}}{(v_2+1)!}u_{x_1x_1x_2}(x_1,x_2)\,dx_1\,dx_2\\ &\quad= \int\limits_{[0,\infty)^2} e^{-x_1-x_2}\left[\frac{x_1^{v_1+2}}{(v_1+2)!}-2\frac{x_1^{v_1+1}}{(v_1+1)!}+\frac{x_1^{v_1}}{v_1!}\right]\left[\frac{x_2^{v_2+1}}{(v_2+1)!}-\frac{x_2^{v_2}}{v_2!}\right]u(x_1,x_2)\,dx_1\,dx_2,\end{aligned} \tag{532}$$

i. e.,

$$[\Delta^3_{x_1^2x_2}P(u)](v_1,v_2) = \int\limits_{[0,\infty)^2} e^{-x_1-x_2}\frac{x_1^{v_1+2}}{(v_1+2)!}\frac{x_2^{v_2+1}}{(v_2+1)!}g(x_1,x_2)\,dx_1\,dx_2,$$

provided that (531) holds, as well as

$$\lim_{x_1\to+\infty} e^{-x_1}\frac{x_1^{v_1+2}}{(v_1+2)!}u_{x_1}(x_1,x_2) = 0, \quad x_2 \geqslant 0,$$

$$\lim_{x_1\to+\infty} e^{-x_1}\left[\frac{x_1^{v_1+2}}{(v_1+2)!}-\frac{x_1^{v_1+1}}{(v_1+1)!}\right]u(x_1,x_2) = 0, \quad x_2 \geqslant 0,$$

the second integral in (532) converges absolutely and the integral

$$\int_{[0,\infty)^2} e^{-x_1-x_2}\left[\frac{x_1^{v_1+2}}{(v_1+2)!}-\frac{x_1^{v_1+1}}{(v_1+1)!}\right]\left[\frac{x_2^{v_2+1}}{(v_2+1)!}-\frac{x_2^{v_2}}{v_2!}\right]u_{x_1}(x_1,x_2)\,dx_1\,dx_2$$

converges absolutely.

In the general case, one can use the Fubini theorem and the partial integration in order to see that the following result holds true.

Theorem 8.2.5. *If D is a nonempty subset of $\mathbb{N}_0^n$, A_a is a closed linear operator on X for all $a \in D$, and*

$$\sum_{a\in D} A_a \frac{\partial^a u}{\partial x_1^{a_1}\cdots\partial x_n^{a_n}}(x_1,\dots,x_n)=\sum_{a\in D} A_a u^{(a)}(x_1,\dots,x_n)=g(x_1,\dots,x_n), \tag{533}$$

for any $(x_1,\dots,x_n)\in[0,\infty)^n$, then we have

$$\sum_{a\in D} A_a[\Delta^{|a|}_{v_1^{a_1}\cdot\cdots\cdot v_n^{a_n}}P(u)](v_1,\dots,v_n)$$

$$=\int_{[0,\infty)^n} e^{-x_1-\cdots-x_n}\frac{x_1^{v_1+a_1}}{(v_1+a_1)!}\cdot\cdots\cdot\frac{x_n^{v_n+a_n}}{(v_n+a_n)!}g(x_1,\dots,x_n)\,dx_1\,dx_2\cdots dx_n,$$

for any $(v_1,\dots,v_n)\in\mathbb{N}_0^n$, provided that $u:[0,\infty)^n\to X$ is continuous, the integral on the right-hand side of the above equality absolutely converges and, for every $v\in\mathbb{N}$, $(v_1,\dots,v_n)\in\mathbb{N}_0^n$ and for every multiindex $a=(a_1,\dots,a_n)\in D$, we have

0. *The integral*

$$\int_{[0,\infty)^n} e^{-x_1-\cdots-x_n}\frac{x_1^{v_1}}{v_1!}\cdot\cdots\cdot\frac{x_n^{v_n}}{v_n!}A_a u^{(a)}(x_1,\dots,x_n)\,dx_1\,dx_2\cdots dx_n$$

is convergent and the integral

$$\int_{[0,\infty)^n} e^{-x_1-\cdots-x_n}\frac{x_1^{v_1}}{v_1!}\cdot\cdots\cdot\frac{x_n^{v_n}}{v_n!}u^{(a)}(x_1,\dots,x_n),dx_1\,dx_2\cdots dx_n$$

is convergent.

1. *For every multiindex $(a_1,\dots,a_{n-1},\gamma_n)$, where $0\leqslant\gamma_n\leqslant a_n$, the mapping $\frac{\partial^{a_1+\cdots+a_{n-1}+\gamma_n}u}{\partial x_1^{a_1}\cdots\partial x_{n-1}^{a_{n-1}}\partial x_n^{\gamma_n}}$ is continuous on $[0,\infty)^n$,*

$$\int_{[0,\infty)^n} e^{-x_1-\cdots-x_n}\frac{x_1^{v_1}}{v_1!}\cdot\cdots\cdot\frac{x_n^{v_n}}{v_n!}\|u^{(a_1,\dots,a_{n-1},\gamma_n)}(x_1,\dots,x_n)\|\,dx_1\,dx_2\cdots dx_n<+\infty$$

and, for every multiindex $(a_1,\dots,a_{n-1},\beta_n)$, where $0\leqslant\beta_n<a_n$, we have

$$\lim_{x_n\to+\infty} e^{-x_n}\frac{x_n^v}{v!}\frac{\partial^{a_1+\cdots+a_{n-1}+\beta_n}u}{\partial x_1^{a_1}\cdots\partial x_{n-1}^{a_{n-1}}\partial x_n^{\beta_n}}(x_1,\dots,x_n)=0,\quad x_1\geqslant 0,\ x_2\geqslant 0,\dots,\ x_{n-1}\geqslant 0.$$

2. *For every multiindex $(\alpha_1,\ldots,\gamma_{n-1})$, where $0 \leqslant \gamma_{n-1} \leqslant \alpha_{n-1}$, the mapping $\frac{\partial^{\alpha_1+\cdots+\gamma_{n-1}}u}{\partial x_1^{\alpha_1}\cdots\partial x_{n-1}^{\gamma_{n-1}}}$ is continuous on $[0,\infty)^n$,*

$$\int\limits_{[0,\infty)^n} e^{-x_1-\cdots-x_n}\frac{x_1^{v_1}}{v_1!}\cdot\cdots\cdot\frac{x_n^{v_n}}{v_n!}\|u^{(\alpha_1,\ldots,\gamma_{n-1})}(x_1,\ldots,x_n)\|\,dx_1\,dx_2\cdots dx_n < +\infty$$

and, for every multiindex $(\alpha_1,\ldots,\beta_{n-1})$, where $0 \leqslant \beta_{n-1} < \alpha_{n-1}$, we have

$$\lim_{x_{n-1}\to+\infty} e^{-x_{n-1}}\frac{x_{n-1}^{v}}{v!}\frac{\partial^{\alpha_1+\cdots+\beta_{n-1}}u}{\partial x_1^{\alpha_1}\cdots\partial x_{n-1}^{\beta_{n-1}}}(x_1,\ldots,x_n) = 0, \quad x_1 \geqslant 0,\ x_2 \geqslant 0,\ldots,\ x_{n-2} \geqslant 0,\ x_n \geqslant 0;$$

$\ldots$;

n. *For every integer $\gamma_1 \in [0,\alpha_1]$, the mapping $(\partial^{\gamma_1}u/\partial x_1^{\gamma_1})$ is continuous on $[0,\infty)^n$,*

$$\int\limits_{[0,\infty)^n} e^{-x_1-\cdots-x_n}\frac{x_1^{v_1}}{v_1!}\cdot\cdots\cdot\frac{x_n^{v_n}}{v_n!}\left\|\frac{\partial^{\gamma_1}u}{\partial x_1^{\gamma_1}}(x_1,\ldots,x_n)\right\|\,dx_1\,dx_2\cdots dx_n < +\infty$$

and, for every integer $\beta_1 \in [0,\alpha_1)$, we have

$$\lim_{x_1\to+\infty} e^{-x_1}\frac{x_1^{v}}{v!}\frac{\partial^{\beta_1}u}{\partial x_1^{\beta_1}}(x_1,\ldots,x_n) = 0, \quad x_2 \geqslant 0,\ x_3 \geqslant 0,\ldots,\ x_n \geqslant 0.$$

Observe only that the prescribed assumptions imply, due to the assumption [0.] and the closedness of the operators A_α, that

$$\sum_{\alpha\in D} A_\alpha \int\limits_{[0,\infty)^n} e^{-x_1-\cdots-x_n}\frac{x_1^{v_1+\alpha_1}}{(v_1+\alpha_1)!}\cdot\cdots\cdot\frac{x_n^{v_n+\alpha_n}}{(v_n+\alpha_n)!}u^{(\alpha)}(x_1,\ldots,x_n)\,dx_1\,dx_2\cdots dx_n$$
$$= \int\limits_{[0,\infty)^n} e^{-x_1-\cdots-x_n}\frac{x_1^{v_1+\alpha_1}}{(v_1+\alpha_1)!}\cdot\cdots\cdot\frac{x_n^{v_n+\alpha_n}}{(v_n+\alpha_n)!}g(x_1,\ldots,x_n)\,dx_1\,dx_2\cdots dx_n$$

for any $(x_1,\ldots,x_n) \in [0,\infty)^n$, so that the required conclusion follows similarly as in Example 8.2.4, by means of the assumptions [1.]–[n.].

Remark 8.2.6. It is worth noting that we consider the equation (533) without initial conditions. If, for every multiindex $\alpha = (\alpha_1,\ldots,\alpha_n) \in D$, we impose the initial values

$$u_n(x_1,\ldots,x_{n-1}) = \frac{\partial^{\alpha_1+\cdots+\alpha_{n-1}+\beta_n}u}{\partial x_1^{\alpha_1}\cdots\partial x_{n-1}^{\alpha_{n-1}}\partial x_n^{\beta_n}}(x_1,\ldots,x_{n-1},0), \quad x_1 \geqslant 0,\ x_2 \geqslant 0,\ldots,\ x_{n-1} \geqslant 0,$$

in [1.], $\ldots$,

$$u_1(x_2,\ldots,x_n) = \frac{\partial^{\beta_1}u}{\partial x_1^{\beta_1}}(0,x_2,\ldots,x_n), \quad x_2 \geqslant 0,\ x_3 \geqslant 0,\ldots,\ x_n \geqslant 0,$$

in [n.], then the value of

$$\int\limits_{[0,\infty)^n} e^{-x_1-\cdots-x_n} \frac{x_1^{v_1}}{v_1!} \cdot \ldots \cdot \frac{x_n^{v_n}}{v_n!} g(x_1,\ldots,x_n)\, dx_1\, dx_2 \cdots dx_n, \quad (v_1,\ldots,v_n) \in \mathbb{N}_0^n,$$

can be computed in a similar manner. Details can be left to the attentive readers.

We will illustrate Theorem 8.2.5 with two well-known examples.

Example 8.2.7. If $t = x_1$, $x = x_2$, $v_1 = i$, $v_2 = j$, $a_{i,j} = [Pu](i,j)$ and

$$u_t = u_{xx}, \quad \text{respectively,} \quad u_{tt} = u_{xx},$$

then

$$a_{i,j+2} = 2a_{i+1,j+1} - a_{i+1,j}, \quad \text{respectively,} \quad -2a_{i+1,j+2} + a_{i,j+2} = -2a_{i+2,j+1} + a_{i+2,j}, \tag{534}$$

for any $i, j \in \mathbb{N}_0$, provided that the requirements of Theorem 8.2.5 hold. Concerning the uniqueness of solutions of differences equations in (534), we will only note here that the first of these equations is uniquely solvable for $(i,j) \in \mathbb{N}_0^2$, provided that the initial values $a_{i,0}$ and $a_{0,j}$ are given for all $i, j \in \mathbb{N}_0$, as well as that the second of these equations is uniquely solvable for $(i,j) \in \mathbb{N}_0^2$, provided that the initial values $a_{i,0}$, $a_{i,1}$, $a_{0,j}$ and $a_{1,j}$ are given for all $i, j \in \mathbb{N}_0$.

Example 8.2.8. In many published research articles by now, the authors have analyzed the well-posedness and qualitative properties of solutions to the following abstract (degenerate) higher-order differential equation:

$$A_n u^{(n)}(t) + A_{n-1} u^{(n-1)}(t) + \cdots + A_0 u(t) = f(t), \quad t \geqslant 0,$$

where A_j are differential operators with constant coefficients on the space $X = L^p(\mathbb{R}^n)$, where $1 \leqslant p \leqslant +\infty$; cf. [445, 792] and references cited therein for more details in this direction. If we set $t = x_1$ and denote the variables in $L^p(\mathbb{R}^n)$ by $x_2, \ldots, x_{n+1}$, we can provide a great number of applications of Theorem 8.2.5 to the abstract partial differential equations with constant coefficients, by applying also certain changes of variables with respect to the variables $x_2, \ldots, x_{n+1}$.

In order to avoid any form of repeating and plagiarism, we will only emphasize here the following important issues about the multidimensional Poisson-like transforms:

(i) It is worth noting that Theorem 8.2.5, Example 8.2.4 and Example 8.2.7 can be simply reformulated for the Poisson-like transform

$$[P_{a_1,\ldots,a_n}(u)](\mathrm{v}) := [P_{a_1,\ldots,a_n}(u)](v_1,\ldots,v_n) := \int\limits_{[0,\infty)^n} e^{-a_1x_1-\cdots-a_nx_n} \frac{x_1^{v_1}}{v_1!} \cdot \ldots \cdot \frac{x_n^{v_n}}{v_n!} u(x_1,\ldots,x_n)\, dx_1\, dx_2 \cdots dx_n,$$

where $a_j > 0$ and $v_j \in \mathbb{N}_0$ for $1 \leqslant j \leqslant n$. In such a way, we can extend [472, Theorem 4] to the higher-dimensional setting. The statement of [472, Theorem 5] can be also simply transferred to the higher-dimensional setting by the use of the Weyl convolution product $(a \circ b)(\cdot)$; cf. [472] for the notion.

(ii) Following our consideration from [472], where we have considered the Poisson-like transforms for not exponentially bounded functions, we can also put forward to consideration the following multidimensional transform:

$$\begin{aligned}
&[P_{\mathbf{a},\mathbf{b},\omega,\mathbf{j}}(u)](\mathrm{v}) \\
&\quad := [P_{\mathbf{a},\mathbf{b},\omega,\mathbf{j}}(u)](v_1,\ldots,v_n) \\
&\quad := \int\limits_{[0,\infty)^n} e^{-b_1(a_1x_1)^{j_1}-\cdots-b_n(a_nx_n)^{j_n}} \frac{(\omega_1x_1)^{v_1}}{v_1!}\cdot\ldots\cdot\frac{(\omega_nx_n)^{v_n}}{v_n!} u(x_1,\ldots,x_n)\,dx_1\,dx_2\cdots dx_n,
\end{aligned}$$

where $a_s \in \mathbb{R}$, $b_s \in \mathbb{R}\setminus\{0\}$, $\omega_s \in \mathbb{R}\setminus\{0\}$, $j_s \in \mathbb{N}$ and $v_s \in \mathbb{N}_0$ for $1 \leqslant s \leqslant n$. Applying the partial integration, we can simply show that for each multiindex $(\alpha_1,\ldots,\alpha_n) \in \mathbb{N}^n$, we have

$$\begin{aligned}
&[P_{\mathbf{a},\mathbf{b},\omega,\mathbf{j}}(u^{(\alpha_1,\ldots,\alpha_{n-1},\alpha_n)})](v_1+\alpha_1,\ldots,v_{n-1}+\alpha_{n-1},v_n+\alpha_n) \\
&\quad = -\omega_n[P_{\mathbf{a},\mathbf{b},\omega,\mathbf{j}}(u^{(\alpha_1,\ldots,\alpha_{n-1},\alpha_n-1)})](v_1+\alpha_1,\ldots,v_{n-1}+\alpha_{n-1},v_n+\alpha_n-1) \\
&\qquad + j_n a_n^{j_n} b_n \frac{(v_n+\alpha_n+j_n-1)!}{(v_n+\alpha_n)!}[P_{\mathbf{a},\mathbf{b},\omega,\mathbf{j}}(u^{(\alpha_1,\ldots,\alpha_{n-1},\alpha_n-1)})] \\
&\qquad \times (v_1+\alpha_1,\ldots,v_{n-1}+\alpha_{n-1},v_n+\alpha_n+j_n-1), \quad (v_1,\ldots,v_n) \in \mathbb{N}_0^n,
\end{aligned}$$

under certain logical assumptions. Proceeding in this way, we can find a form of the abstract nonautonomous difference equations of several variables, which corresponds to the abstract partial differential equation (533).

(iii) We have already examined the following Poisson-like transform:

$$v \mapsto y_{a,b,c,j,\omega}(v) := \int\limits_0^{+\infty} e^{-b(ct^{-1}+at)^j} \frac{(\omega t)^{v-\frac{1}{2}}}{\Gamma(v+\frac{1}{2})} u(t)\,dt, \quad v \in \mathbb{Z},$$

where $a \in \mathbb{R}$, $b,\ c,\ \omega \in \mathbb{R}\setminus\{0\}$ and $j \in \mathbb{N}$. The interested readers may try to introduce some multidimensional analogues of this transform as well as to reconsider Theorem 8.2.5 in this framework.

(iv) It is well known that a class of boundary value problems for the partial differential equations depending of variables x_1 and x_2, where $0 \leqslant x_1 \leqslant T$ and $x_2 \geqslant 0$, can be solved using the Fourier series and the method of separation of variables. The Poisson transforms can be also defined and analyzed for the functions defined on the closed rectangles; we can also prove an analogue of Theorem 8.2.5 in this framework.

8.2.2 Applications to the abstract fractional partial differential equations

In this subsection, we will present some applications of the multidimensional vector-valued Poisson transform to the abstract fractional partial differential equations.

We start with the following illustrative example.

Example 8.2.9. Suppose that $n \geqslant 2$, $u_j : [0,+\infty) \to \mathbb{C}$ is a locally integrable function $(1 \leqslant j \leqslant n-1)$ and $u_n : [0,+\infty) \to X$ is a locally integrable function. Set $u(x_1,\dots,x_n) := u_1(x_1)\cdot\dots\cdot u_n(x_n)$, $x_1 \geqslant 0,\dots,x_n \geqslant 0$. Then an elementary application of the Fubini theorem shows that

$$[P(u)](v_1,\dots,v_n) = [P(u_1)](v_1)\cdot\dots\cdot[P(u_n)](v_n), \quad (v_1,\dots,v_n) \in \mathbb{N}_0^n, \tag{535}$$

provided that the integrals, which define all terms in (535) converges absolutely; here, we do not make any terminologically difference between the multidimensional Poisson transform with respect to the variables $v_1,\dots,v_n$ and the one-dimensional Poisson transform with respect to the variable v_j $(1 \leqslant j \leqslant n)$.

Suppose that $f : [0,+\infty) \to X$ is a locally integrable function, $a > 0$ and $m = \lceil a \rceil$. Then a careful inspection of the proof of [4, Theorem 5.5] shows that the following equality holds true:

$$[P(D_t^a f)](v+m) = \Delta^a[P(f)](v), \quad v \in \mathbb{N}_0,$$

provided that $h = g_{m-a} *_0 f \in C^{m-1}([0,+\infty) : X)$, $h^{(m-1)}(\cdot)$ is locally absolutely continuous on $[0,+\infty)$ and there exist real numbers $M > 0$ and $\omega \in (0,1)$ such that $\|h^{(m)}(t)\| \leqslant Me^{\omega t}$ for a. e. $t \geqslant 0$. We have already proposed the following formula for the Caputo derivatives:

$$\begin{aligned}
&[P(\mathbf{D}_C^a f)](v+m) \\
&\quad = \Delta^a[P(f)](v) + \frac{(-1)^{v+m+1}}{(v+m)!}\sum_{k=0}^{m-1}(a-1-k)\cdot\dots\cdot(a-k-v-m)f^{(k)}(0), \quad v \in \mathbb{N}_0,
\end{aligned}$$

which holds under certain mild assumptions. Using the last two formulae and (535), we can simply compute the multidimensional Poisson transform of the fractional partial derivatives $\mathbb{D}^a u$ formed from the compositions of the Riemann–Liouville fractional derivatives of functions $u_j(\cdot)$ for $j \in J_1$ and the Caputo fractional derivatives of functions $u_j(\cdot)$ for $j \in J_2$, where $\mathbb{N}_n = J_1 \cup J_2$. For example, if $J_2 = \emptyset$, $a_1 \geqslant 0,\dots,a_n \geqslant 0$, $a = (a_1,\dots,a_n)$ and $m_j = \lceil a_j \rceil$ for $1 \leqslant j \leqslant n$, then we have

$$\begin{aligned}
&[P(D_{R,x_1}^{a_1}\cdot\dots\cdot D_{R,x_n}^{a_n} u)](v_1+m_1,\dots,v_n+m_n) \\
&\quad = [P(D_R^{a_1}u_1)](v_1+m_1)\cdot\dots\cdot[P(D_R^{a_n}u_n)](v_n+m_n) \\
&\quad = \Delta^{a_1}[P(u_1)](v_1)\cdot\dots\cdot\Delta^{a_n}[P(u_n)](v_n) \\
&\quad = [\Delta^a P(u)](v_1,\dots,v_n), \quad (v_1,\dots,v_n) \in \mathbb{N}_0^n,
\end{aligned}$$

under the following conditions:

(i) The integrals which define the terms $P(D_{R,x_1}^{\alpha_1}\cdot\cdots\cdot D_{R,x_n}^{\alpha_n}u), P(D_R^{\alpha_1}u_1),\ldots,P(D_R^{\alpha_n}u_n), P(u)$, $P(u_1),\ldots$ and $P(u_n)$ converge absolutely.

(ii) The functions $h_1 = g_{m_1-\alpha_1} *_0 u_1(\cdot),\ldots,h_n = g_{m_n-\alpha_n} *_0 u_n(\cdot)$ are $(m-1)$-times continuously differentiable on $[0,+\infty)$, the functions $h_1^{(m_1-1)}(\cdot),\ldots,h_n^{(m_n-1)}(\cdot)$ are locally absolutely continuous on $[0,+\infty)$ and there exist real numbers $M>0$ and $\omega\in(0,1)$ such that $\|h_j^{(m_j)}(t)\|\leqslant Me^{\omega t}$ for a. e. $t>0$ $(1\leqslant j\leqslant n)$.

We continue by stating the following general result (for simplicity, we denote henceforth $D_{R,x_1}^{\alpha_1}\cdot\cdots\cdot D_{R,x_n}^{\alpha_n}u = u^{(\alpha)}$).

Theorem 8.2.10. *Suppose that $u:[0,\infty)^n\to X$ is a locally integrable function, the term $u^{(\alpha)}$ is well-defined and the following conditions hold true:*

(i) *The integral, which defines the term $[P(u^{(\alpha)})](v_1+m_1,\ldots,v_n+m_n)$, converges absolutely for all $(v_1,\ldots,v_n)\in\mathbb{N}_0^n$.*

(ii) *The function $h_1 = J_{t_1}^{m_1-\alpha_1}D_{R,x_2}^{\alpha_2}\cdot\cdots\cdot D_{R,x_n}^{\alpha_n}u(x_1,x_2,\ldots,x_n)\in C^{m_1-1}([0,+\infty):X)$, $h_1^{(m_1-1)}(\cdot)$ is locally absolutely continuous on $[0,+\infty)$ and there exist real numbers $M>0$ and $\omega\in(0,1)$ such that $\|h_1^{(m_1)}(t)\|\leqslant Me^{\omega t}$ for a. e. $t\geqslant 0$.*

(iii) *The integral*

$$\int_{[0,+\infty)^n} e^{-x_1-x_2-\cdots-x_n}\frac{x_1^{v_1}}{v_1!}\frac{x_2^{v_2+m_2}}{(v_2+m_2)!}\cdot\cdots\cdot\frac{x_n^{v_n+m_n}}{(v_n+m_n)!}$$
$$\times D_{R,x_2}^{\alpha_2}\cdot\cdots\cdot D_{R,x_n}^{\alpha_n}u(x_1,x_2,\ldots,x_n)\,dx_1\,dx_2\cdots dx_n$$

converges absolutely for all $(v_1,\ldots,v_n)\in\mathbb{N}_0^n$.

(iv) *For every $j\in\{2,\ldots,n-1\}$, the integral*

$$\int_{[0,+\infty)^n} e^{-x_1-x_2-\cdots-x_n}\frac{x_1^{v_1}}{v_1!}\cdot\cdots\cdot\frac{x_j^{v_j}}{v_j!}\frac{x_{j+1}^{v_{j+1}+m_{j+1}}}{(v_{j+1}+m_{j+1})!}\cdot\cdots\cdot\frac{x_n^{v_n+m_n}}{(v_n+m_n)!}$$
$$\times D_{R,x_j}^{\alpha_j}\cdot\cdots\cdot D_{R,x_n}^{\alpha_n}u(x_1,x_2,\ldots,x_n)\,dx_1\,dx_2\cdots dx_n$$

converges absolutely for all $(v_1,\ldots,v_n)\in\mathbb{N}_0^n$ and the term which defines the term $[P(u)](v_1,\ldots,v_n)$ converges absolutely for all $(v_1,\ldots,v_n)\in\mathbb{N}_0^n$.

(v) *For every $j\in\{3,\ldots,n\}$, we have*

$$h_j = J_{t_{j-1}}^{m_{j-1}-\alpha_{j-1}}D_{R,x_j}^{\alpha_j}\cdot\cdots\cdot D_{R,x_n}^{\alpha_n}u(x_1,x_2,\ldots,x_n)\in C^{m_{j-1}-1}([0,+\infty):X),$$

$h_j^{(m_{j-1}-1)}(\cdot)$ is locally absolutely continuous on $[0,+\infty)$ and there exist real numbers $M>0$ and $\omega\in(0,1)$ such that $\|h_j^{(m_{j-1})}(t)\|\leqslant Me^{\omega t}$ for a. e. $t\geqslant 0$.

Then we have

$$[P(u^{(\alpha)})](v_1+m_1,\ldots,v_n+m_n)=[\Delta^{\alpha}P(u)](v_1,\ldots,v_n),\quad (v_1,\ldots,v_n)\in\mathbb{N}_0^n. \tag{536}$$

Proof. Keeping in mind condition (i), we can apply the Fubini theorem in order to see that

$$\begin{aligned}
&[P(u^{(\alpha)})](v_1+m_1,\ldots,v_n+m_n)\\
&=\int\limits_{[0,+\infty)^{n-1}} e^{-x_2-\cdots-x_n}\frac{x_2^{v_2+m_2}}{(v_2+m_2)!}\cdot\cdots\cdot\frac{x_n^{v_n+m_n}}{(v_n+m_n)!}\\
&\quad\times\left[\int\limits_0^{+\infty} e^{-x_1}\frac{x_1^{v_1+m_1}}{(v_1+m_1)!}D_{R,x_1}^{\alpha_1}\cdot\cdots\cdot D_{R,x_n}^{\alpha_n}u(x_1,x_2,\ldots,x_n)\,dx_1\right]dx_2\cdots dx_n.
\end{aligned}$$

Due to (ii)–(iii), we can apply [4, Theorem 5.5] and the Fubini theorem to obtain that

$$\begin{aligned}
&[P(u^{(\alpha)})](v_1+m_1,\ldots,v_n+m_n)\\
&=\int\limits_{[0,+\infty)^{n-1}} e^{-x_2-\cdots-x_n}\frac{x_2^{v_2+m_2}}{(v_2+m_2)!}\cdot\cdots\cdot\frac{x_n^{v_n+m_n}}{(v_n+m_n)!}\\
&\quad\times\left[\Delta_{x_1}^{\alpha_1}\int\limits_0^{+\infty} e^{-x_1}\frac{x_1^{v_1}}{v_1!}D_{R,x_2}^{\alpha_2}\cdot\cdots\cdot D_{R,x_n}^{\alpha_n}u(x_1,x_2,\ldots,x_n)\,dx_1\right]dx_2\cdots dx_n\\
&=\Delta_{x_1}^{\alpha_1}\int\limits_{[0,+\infty)^{n-1}} e^{-x_2-\cdots-x_n}\frac{x_2^{v_2+m_2}}{(v_2+m_2)!}\cdot\cdots\cdot\frac{x_n^{v_n+m_n}}{(v_n+m_n)!}\\
&\quad\times\left[\int\limits_0^{+\infty} e^{-x_1}\frac{x_1^{v_1}}{v_1!}D_{R,x_2}^{\alpha_2}\cdot\cdots\cdot D_{R,x_n}^{\alpha_n}u(x_1,x_2,\ldots,x_n)\,dx_1\right]dx_2\cdots dx_n\\
&=\Delta_{x_1}^{\alpha_1}\int\limits_{[0,+\infty)^{n}} e^{-x_1-x_2-\cdots-x_n}\frac{x_1^{v_1}}{v_1!}\frac{x_2^{v_2+m_2}}{(v_2+m_2)!}\cdot\cdots\cdot\frac{x_n^{v_n+m_n}}{(v_n+m_n)!}\\
&\quad\times D_{R,x_2}^{\alpha_2}\cdot\cdots\cdot D_{R,x_n}^{\alpha_n}u(x_1,x_2,\ldots,x_n)\,dx_1\,dx_2\cdots dx_n.
\end{aligned}$$

Keeping in mind the remaining assumptions and repeating the above procedure, we get

$$\begin{aligned}
&[P(u^{(\alpha)})](v_1+m_1,\ldots,v_n+m_n)\\
&=\Delta_{x_1}^{\alpha_1}\int\limits_{[0,+\infty)^{n}} e^{-x_1-x_2-\cdots-x_n}\frac{x_1^{v_1}}{v_1!}\frac{x_2^{v_2+m_2}}{(v_2+m_2)!}\cdot\cdots\cdot\frac{x_n^{v_n+m_n}}{(v_n+m_n)!}\\
&\quad\times D_{R,x_2}^{\alpha_2}\cdot\cdots\cdot D_{R,x_n}^{\alpha_n}u(x_1,x_2,\ldots,x_n)\,dx_1\,dx_2\cdots dx_n
\end{aligned}$$

$$= \Delta_{x_1}^{\alpha_1} \Delta_{x_2}^{\alpha_2} \int\limits_{[0,+\infty)^n} e^{-x_1-x_2-\cdots-x_n} \frac{x_1^{\nu_1}}{\nu_1!} \frac{x_2^{\nu_2}}{\nu_2!} \frac{x_3^{\nu_3+m_3}}{(\nu_3+m_3)!} \cdots \frac{x_n^{\nu_n+m_n}}{(\nu_n+m_n)!}$$

$$\times D_{R,x_2}^{\alpha_2} \cdots D_{R,x_n}^{\alpha_n} u(x_1,x_2,\ldots,x_n)\, dx_1\, dx_2 \cdots dx_n$$

$$= \Delta_{x_1}^{\alpha_1} \Delta_{x_2}^{\alpha_2} \cdots \Delta_{x_n}^{\alpha_n} \int\limits_{[0,+\infty)^n} e^{-x_1-x_2-\cdots-x_n} \frac{x_1^{\nu_1}}{\nu_1!} \cdots \frac{x_n^{\nu_n}}{\nu_n!} u(x_1,x_2,\ldots,x_n)\, dx_1\, dx_2 \cdots dx_n,$$

which completes the proof. $\qquad\square$

Now we will state and prove the following analogue of Theorem 8.2.5 for the fractional partial derivatives of the Riemann–Liouville type.

Theorem 8.2.11. *Suppose that D is a nonempty subset of $[0,+\infty)^n$ and A_α is a closed linear operator on X for all $\alpha \in D$; if $\alpha = (\alpha_1,\ldots,\alpha_n) \in D$, then we set $m_j = \lceil \alpha_j \rceil$ for $1 \leq j \leq n$. Suppose further that (533) holds for a. e. $(x_1,\ldots,x_n) \in [0,\infty)^n$. Then we have*

$$\sum_{\alpha \in D} A_\alpha [\Delta^\alpha P(u)](\nu_1,\ldots,\nu_n)$$

$$= \int\limits_{[0,\infty)^n} e^{-x_1-\cdots-x_n} \frac{x_1^{\nu_1+m_1}}{(\nu_1+m_1)!} \cdots \frac{x_n^{\nu_n+m_n}}{(\nu_n+m_n)!} g(x_1,\ldots,x_n)\, dx_1\, dx_2 \cdots dx_n,$$

for any $(\nu_1,\ldots,\nu_n) \in \mathbb{N}_0^n$, provided that the following conditions hold true:
(i) *Condition [0.] from the formulation of Theorem 8.2.5 holds.*
(ii) *For every multiindex $\alpha = (\alpha_1,\ldots,\alpha_n) \in D$, (536) holds true.*

Proof. Using conditions (i)–(ii) and [445, Theorem 1.2.3], the required statement simply follows from the next computation:

$$= \int\limits_{[0,\infty)^n} e^{-x_1-\cdots-x_n} \frac{x_1^{\nu_1+m_1}}{(\nu_1+m_1)!} \cdots \frac{x_n^{\nu_n+m_n}}{(\nu_n+m_n)!} g(x_1,\ldots,x_n)\, dx_1\, dx_2 \cdots dx_n$$

$$= \int\limits_{[0,\infty)^n} e^{-x_1-\cdots-x_n} \frac{x_1^{\nu_1+m_1}}{(\nu_1+m_1)!} \cdots \frac{x_n^{\nu_n+m_n}}{(\nu_n+m_n)!} \sum_{\alpha \in D} A_\alpha u^{(\alpha)}(x_1,\ldots,x_n)\, dx_1\, dx_2 \cdots dx_n$$

$$= \sum_{\alpha \in D} A_\alpha \int\limits_{[0,\infty)^n} e^{-x_1-\cdots-x_n} \frac{x_1^{\nu_1+m_1}}{(\nu_1+m_1)!} \cdots \frac{x_n^{\nu_n+m_n}}{(\nu_n+m_n)!} u^{(\alpha)}(x_1,\ldots,x_n)\, dx_1\, dx_2 \cdots dx_n$$

$$= \sum_{\alpha \in D} A_\alpha [P(u^{(\alpha)})](\nu_1+m_1,\ldots,\nu_n+m_n)$$

$$= \sum_{\alpha \in D} A_\alpha [\Delta^\alpha P(u)](\nu_1,\ldots,\nu_n), \quad (\nu_1,\ldots,\nu_n) \in \mathbb{N}_0^n. \qquad\square$$

It is worth noting that Theorem 8.2.11 can be successfully applied to the fractional partial equations considered in Section 8.1.6 and Section 8.1.7; cf. also [682, Chapter 5].

8.3 Notes and Appendices to Part III

Almost periodic solutions for a class of neutral integrodifferential equations

This part presents a novel contribution by delineating inherently natural conditions under which the specified class of neutral integrodifferential equations, defined in a Banach space, exhibits (unique) almost periodic mild solutions [495]. The versatile nature of the considered class, coupled with tailored accommodations and constraints, renders it applicable to modeling scenarios across diverse scientific domains. From applications in neuroscience and dynamical systems to addressing classical engineering problems, this class of equations emerges as a valuable and broadly applicable tool, amplifying its significance in various scientific contexts.

A continuous function $f : \mathbb{R} \times X \to X$ is said to be almost periodic in t uniformly for $u \in X$ if for each $\varepsilon > 0$ and for each compact subset K of X, the set of all real numbers τ such that

$$\|f(t+\tau,u) - f(t,u)\| \leqslant \varepsilon, \quad t \in \mathbb{R},\ u \in K$$

is relatively dense in $[0,\infty)$. If a continuous function $f : \mathbb{R} \times X \times Y \to X$ is given, then the number τ is said to be ε-period for $f(\cdot,\cdot,\cdot)$ if

$$\|f(t+\tau,u,v) - f(t,u,v)\| \leqslant \varepsilon, \quad t \in \mathbb{R},\ u \in X,\ v \in Y.$$

The set consisting of all ε-periods for $f(\cdot,\cdot,\cdot)$ will be denoted by $\vartheta_{X,Y}(f,\varepsilon)$. The continuous function $f : \mathbb{R} \times X \times Y \to X$ is said to be almost periodic in t uniformly for $(u,v) \in X \times Y$ if for each $\varepsilon > 0$ and for each compact subset E of $X \times Y$ such that the set $\vartheta_{X,Y}(f,\varepsilon)$ is relatively dense in $[0,\infty)$.

Here, we investigate the existence and uniqueness of almost periodic solutions of a neutral integrodifferential equation

$$u'(t) = Au(t) + f(t, u_t, Fu(t)), \quad t \in \mathbb{R}, \tag{537}$$

and the integral equation

$$Fu(t) = \int_{-\infty}^{t} k(t-s) g(s, u_s)\, ds,$$

where A is a linear operator on the Banach space X, $f : \mathbb{R} \times C \to X$ and $g : \mathbb{R} \times C \times X \to X$ are bounded functions on bounded sets, $k \in L^1(\mathbb{R} : \mathbb{C})$ is continuous, nonincreasing function and $u_r(t) = u(t+r)$, for $r \in [-m,0]$, $m \geqslant 0$ is a fixed constant, where C is the space of continuous functions from $[-m,0]$ to X equipped with the supremum norm.

For simplicity, we will consider the case when the operator A is the infinitesimal generator of a strongly continuous semigroup (C_0-semigroup) on X:

(A) $A : D(A) \subseteq X \to X$ is the infinitesimal generator of a strongly continuous semigroup $(T(t))_{t\geqslant 0}$, such that there exist constant C, $\sigma > 0$ such that $\|T(t)\| \leqslant Ce^{\sigma t}$, for $t \geqslant 0$.

Additionally, in certain statements, we will consider the following assumptions:

(C1)' The function $k(t) \in L^1(\mathbb{R} : \mathbb{C})$ is continuous and nonincreasing.

(C2)' $g \in AP(\mathbb{R} \times \mathcal{C} : X)$.

(C3)' $f \in AP(\mathbb{R} \times \mathcal{C} \times X : X)$.

(C4)' There exists a positive constant L_g such that $\|g(t,\phi) - g(t,\psi)\| \leqslant L_g\|\phi - \psi\|_{\mathcal{C}}$.

(C5)' There exists a positive constant L_f such that $\|f(t,\phi_1,\psi_1) - f(t,\phi_2,\psi_2)\| \leqslant L_f(\|\phi_1 - \phi_2\|_{\mathcal{C}} + \|\psi_1 - \psi_2\|)$.

(C6)' There exists a continuous and nondecreasing function $L_g : \mathbb{R} \to \mathbb{R}$ such that for each $v > 0$, and ϕ, $\psi \in \mathcal{C}$ such that $\|\phi\|_{\mathcal{C}} \leqslant v$ and $\|\psi\|_{\mathcal{C}} \leqslant v$, we have

$$\|g(t,\phi) - g(t,\psi)\| \leqslant L_g(v)\|\phi - \psi\|_{\mathcal{C}}, \quad t \in \mathbb{R},$$

where $L_g(0) = 0$.

(C7)' There exists a continuous and nondecreasing function $L_f : \mathbb{R} \to \mathbb{R}$ such that for each $v > 0$, and $\phi_i \in \mathcal{C}$, $\psi_i \in X$ such that $\|\phi_i\|_{\mathcal{C}} \leqslant v$ and $\|\psi_i\| \leqslant v$, $i = 1, 2$ we have

$$\|f(t,\phi_1,\psi_1) - f(t,\phi_2,\psi_2)\| \leqslant L_f(v)(\|\phi_1 - \phi_2\|_{\mathcal{C}} + \|\psi_1 - \psi_2\|), \quad t \in \mathbb{R},$$

where $L_f(0) = 0$.

(C8)' Let $f \in C(\mathbb{R}\times\mathcal{C}\times X : X)$. There exist a bounded measurable function $a_f : \mathbb{R} \to X$ and a constant b_f such that $\|f(t,u,v)\| \leqslant \|a_f(t)\| + b_f(\|u\|_{\mathcal{C}} + \|v\|)$, and $\sup_{t\in\mathbb{R}} \|a_f(t)\| = a_f^*$.

(C9)' Let $g \in C(\mathbb{R} \times \mathcal{C} : X)$. There exist a bounded measurable function $a_g : \mathbb{R} \to X$ and a constant b_g such that $\|g(t,u)\| \leqslant \|a_g(t)\| + b_g\|u\|_{\mathcal{C}}$, and $\sup_{t\in\mathbb{R}} \|a_g(t)\| = a_g^*$.

A mild solution of (537) is given by

$$u(t) = \int_{-\infty}^{t} T(t-s)f(s,u_s,Fu(s))\,ds,$$

for $t \in \mathbb{R}$.

Lemma 8.3.1. *Let* (A), (C1)'–(C3)' *and* (C5)' *hold. If* $u \in AP(\mathbb{R} : X)$, *then*

$$(\mathcal{S}u)(t) = \int_{-\infty}^{t} T(t-s)f(s,u_s,Fu(s))\,ds \in AP(\mathbb{R} : X).$$

Proof. If $u \in AP(\mathbb{R} : X)$, then it can be easily shown that $Fu \in AP(\mathbb{R} : X)$, $\phi(\cdot) = f(\cdot,u,Fu(\cdot)) \in AP(\mathbb{R} : X)$. It is also clear that for each $\varepsilon_1 \in \vartheta_{X,Y}(\phi,\cdot)$ and each compact subset E of $X \times Y$ we have

$$\|\phi(t-s+\omega) - \phi(t-s)\| \leqslant \varepsilon_1.$$

Also,

$$\begin{aligned}\|\mathcal{S}u(t+\omega)-\mathcal{S}u(t)\|_X &= \left\|\int_0^{+\infty} T(s)(\phi(t-s+\omega)-\phi(t-s))\,ds\right\| \\ &\leqslant \int_0^{+\infty} Ce^{-\sigma s}\|\phi(t-s+\omega)-\phi(t-s)\|_X\,ds \\ &\leqslant \varepsilon_1\int_0^{+\infty} Ce^{-\sigma s}\,ds = \frac{C\varepsilon_1}{\sigma}\leqslant \varepsilon,\end{aligned}$$

which yields that $\mathcal{S}u \in AP(\mathbb{R}:X)$. □

Theorem 8.3.2. *Let* (A) *and* (C1)'–(C5)' *hold. If* $\rho<1$, *where* $\rho=\frac{CL_f}{\sigma}(1+\|k\|_{L^1}L_g)$, *then* (537) *has a unique almost periodic mild solution.*

Proof. We define the operator $\mathcal{S}: AP(\mathbb{R}:X)\to AP(\mathbb{R}:X)$ by

$$(\mathcal{S}u)(t) := \int_{-\infty}^{t} T(t-s)f(s,u_s,Fu(s))\,ds, \quad t\in\mathbb{R}.$$

By Lemma 8.3.1, the operator $\mathcal{S}$ is well-defined. Now, let $u,\ v\in AP(\mathbb{R}:X)$. Then we obtain

$$\begin{aligned}\|\mathcal{S}u-\mathcal{S}v\|_X &\leqslant \int_{-\infty}^{t}\|T(t-s)(f(s,u_s,Fu(s))-f(s,v_s,Fv(s)))\|\,ds \\ &\leqslant C\int_{-\infty}^{t} e^{-\sigma(t-s)}\|f(s,u_s,Fu(s))-f(s,v_s,Fv(s))\|\,ds \\ &\leqslant CL_f\int_{-\infty}^{t} e^{-\sigma(t-s)}(\|u_s-v_s\|_C+\|Fu(s)-Fv(s)\|)\,ds \\ &\leqslant CL_f\int_{-\infty}^{t} e^{-\sigma(t-s)}\left(\|u_s-v_s\|_C+\int_{-\infty}^{t} k(t-s)\|g(s,u_s)-g(s,v_s)\|\right)ds \\ &\leqslant CL_f\int_{-\infty}^{t} e^{-\sigma(t-s)}(\|u-v\|_\infty(1+\|k\|_{L^1}L_g))\,ds \\ &\leqslant \frac{CL_f}{\sigma}(1+\|k\|_{L^1}L_g)\|u-v\|_\infty.\end{aligned}$$

Hence, by the Banach contraction mapping principle, $\mathcal{S}$ has a unique fixed point in $AP(\mathbb{R}:X)$, so (537) has a unique mild solution in $AP(\mathbb{R}:X)$. □

Theorem 8.3.3. *Let* (A), (C1)'–(C3)' *and* (C6)'–(C7)' *hold. If there is* $\nu > 0$ *such that* $\rho < 1$, *where*

$$\rho = \frac{C}{\sigma}\Big(L_f(\nu)(1 + L_g(\nu)\|k\|_{L^1}) + \frac{1}{\nu}L_f(\nu)\|k\|_{L^1} \cdot \sup_{t\in\mathbb{R}}\|g(t,0)\| + \frac{1}{\nu}\sup_{t\in\mathbb{R}}\|f(t,0,0)\|\Big),$$

then (537) *has a unique almost periodic mild solution, with* $\|u\|_\infty \leqslant \lambda$.

Proof. Note that f is bounded, so $f(\cdot,0)$ is also a bounded function in $\mathbb{R}$. We define the operator $\mathcal{R} : AP(\mathbb{R} : X) \to AP(\mathbb{R} : X)$ by

$$(\mathcal{R}u)(t) := \int_{-\infty}^{t} T(t-s)f(s,u_s,Fu(s))\,ds, \quad t \in \mathbb{R}.$$

Put $B_\nu := \{u \in AP(\mathbb{R} : X) : \|u\|_\infty \leqslant \nu\}$. For $u \in B_\nu$, we have

$$\begin{aligned}\|\mathcal{R}u(t)\| &\leqslant C\int_{-\infty}^{t} e^{-\sigma(t-s)}(\|f(s,u_s,Fu(s)) - f(s,0,0) + f(s,0,0)\|)\,ds\\ &\leqslant C\int_{-\infty}^{t} e^{-\sigma(t-s)}L_f(\nu)(\|u_s\|_C + \|Fu(s)\|)\,ds + C\int_{-\infty}^{t} e^{-\sigma(t-s)}\|f(s,0,0)\|\,ds\\ &\leqslant \frac{C\nu L_f(\nu)}{\sigma}(1 + L_g(\nu)\|k\|_{L^1}) + \frac{CL_f(\nu)}{\sigma}\|k\|_{L^1} \cdot \sup_{t\in\mathbb{R}}\|g(t,0)\| + \frac{C}{\sigma}\sup_{t\in\mathbb{R}}\|f(t,0,0)\| \leqslant \nu,\end{aligned}$$

so $\mathcal{R}u \in B_\nu$.

For $u,\ v \in B_\nu$, we obtain

$$\begin{aligned}\|\mathcal{R}u - \mathcal{R}v\| &\leqslant CL_f(\nu)\int_{-\infty}^{t} e^{-\sigma(t-s)}(\|u_s - v_s\|_C + \|Fu(s) - Fv(s)\|)\,ds\\ &\leqslant \frac{CL_f(\nu)}{\sigma}(1 + L_g(\nu)\|k\|_{L^1})\|u - v\|_\infty.\end{aligned}$$

Hence,

$$\|\mathcal{R}u - \mathcal{R}v\|_\infty \leqslant \frac{CL_f(\nu)}{\sigma}(1 + L_g(\nu)\|k\|_{L^1})\|u - v\|_\infty.$$

By the condition $\rho < 1$, using Banach contraction mapping principle, the equation (537) has a unique mild almost periodic solution. □

Theorem 8.3.4. *Let* (A), (C1)'–(C3)' *and* (C8)'–(C9)' *hold. Then the equation* (537) *has at least one solution.*

Proof. We define the closed ball B_r as

$$B_r = \{u \in AP(\mathbb{R} : X) \, : \, \|u\|_\infty < r\},$$

where $r \geqslant \frac{C(a_f^* + b_f \|k\|_{L^1} a_g^*)}{1 - C(b_f + \|k\|_{L^1} b_g)}$.

Let the operator $\mathcal{G} : AP(\mathbb{R} : X) \to AP(\mathbb{R} : X)$ be defined by

$$(\mathcal{G}u)(t) := \int_{-\infty}^{t} T(t-s) f(s, u_s, Fu(s))\, ds, \quad t \in \mathbb{R}.$$

Now, by applying (C8)–(C9), we obtain

$$\begin{aligned}
\|(\mathcal{G}u)(t)\| &= \left\| \int_{-\infty}^{t} T(t-s) f(s, u_s, Fu(s))\, ds \right\| \\
&\leqslant \int_{-\infty}^{t} \|T(t-s)\| \cdot \|f(s, u_s, Fu(s))\|\, ds \\
&\leqslant C \int_{-\infty}^{t} e^{-\sigma(t-s)} (\|a_f(t)\| + b_f(\|u\|_\infty + \|Fu(s)\|))\, ds \\
&= C \int_{-\infty}^{t} e^{-\sigma(t-s)} \left(a_f^* + b_f \left(\|u\|_\infty + \int_{-\infty}^{t} k(t-\theta) \|g(\theta, u_\theta)\|\, d\theta \right) \right) ds \\
&\leqslant C \int_{-\infty}^{t} e^{-\sigma(t-s)} (a_f^* + b_f(\|u\|_\infty + \|k\|_{L^1}(\|a_g(t)\| + b_g \|u\|_\infty)))\, ds \\
&\leqslant C(a_f^* + b_f \|k\|_{L^1} a_g^* + (b_f + \|k\|_{L^1} b_g) \|u\|_\infty) \leqslant r,
\end{aligned}$$

so $\mathcal{G} : B_r \to B_r$ and $\{\mathcal{G}u\}$ is uniformly continuous.

Now, we are going to prove that $\mathcal{G}$ is continuous. Let (u_n) be a sequence in B_r, such that $u_n \to u$, when $n \to \infty$. Then

$$(\mathcal{G}u_n)(t) = \int_{-\infty}^{t} T(t-s) f(s, (u_n)_s, Fu_n(s))\, ds,$$

where $(u_n)_s(t) = u_n(t+s)$.

Now, using Lebesgue dominated convergence theorem, and having on mind the continuity of the function f and g, we obtain

$$\lim_{n\to\infty} (\mathcal{G}u_n)(t) = \lim_{n\to\infty} \int_{-\infty}^{t} T(t-s) f(s, (u_n)_s, Fu_n(s))\, ds$$

$$= \int_{-\infty}^{t} T(t-s)f\Big(s, \lim_{n\to\infty}(u_n)_s, \lim_{n\to\infty} Fu_n(s)\Big)\,ds$$
$$= \int_{-\infty}^{t} T(t-s)f(s,u_s,Fu(s))\,ds = (\mathcal{G}u)(t).$$

Note that, in the upper equality, we used that

$$\begin{aligned}\lim_{n\to\infty} Fu_n(s) &= \lim_{n\to\infty}\int_{-\infty}^{t} k(t-s)g(s,(u_n)_s)\,ds\\ &= \int_{-\infty}^{t} k(t-s)g\Big(s,\lim_{n\to\infty}(u_n)_s\Big)\,ds = \int_{-\infty}^{t} k(t-s)g(s,u_s)\,ds = Fu(s).\end{aligned}$$

Next, we prove that $\{\mathcal{G}u\}$ is equicontinuous and the operator $\mathcal{G}$ is relatively compact. Let $u \in B_r$, and $t_1,\ t_2 \in \mathbb{R}$, $t_1 < t_2$ and $|t_1 - t_2| < \delta$ for some δ. We have

$$\begin{aligned}&\|(\mathcal{G}u)(t_2) - (\mathcal{G}u)(t_1)\|\\ &= \left\| \int_{-\infty}^{t_2} T(t_2-s)f(s,u_s,Fu(s))\,ds - \int_{-\infty}^{t_1} T(t_1-s)f(s,u_s,Fu(s))\,ds \right\|\\ &= \left\| \int_{-\infty}^{t_1} T(t_2-s)f(s,u_s,Fu(s))\,ds \right.\\ &\quad \left. + \int_{t_1}^{t_2} T(t_2-s)f(s,u_s,Fu(s))\,ds - \int_{-\infty}^{t_1} T(t_1-s)f(s,u_s,Fu(s))\,ds \right\|\\ &= \left\| \int_{-\infty}^{t_1} (T(t_2-s) - T(t_1-s))f(s,u_s,Fu(s))\,ds \right\| + \left\| \int_{t_1}^{t_2} T(t_2-s)f(s,u_s,Fu(s))\,ds \right\|\\ &\leqslant \int_{-\infty}^{t_1} \|T(t_1-s)(T(t_2-t_1)-I)f(s,u_s,Fu(s))\|\,ds + \int_{t_1}^{t_2} \|T(t_2-s)f(s,u_s,Fu(s))\|\,ds.\end{aligned}$$

Hence, $\|(\mathcal{G}u)(t_2) - (\mathcal{G}u)(t_1)\| \to 0$, when $t_1 \to t_2$. Now, by using Schauder's fixed-point theorem, we obtain existence of at least one solution of (537). □

The following example is a generalization of certain results in [522] (see also [77], [711] and [815]):

Example 8.3.5. Let us consider the Cohen–Grossberg neural network with delays given by the system

$$u_i'(t) = -a_i u_i(t) + \sum_{j=1}^{n} b_{ij}(t) f_j(u_j(t)) + \sum_{j=1}^{n} c_{ij} g_j(u_j(t - \tau_{ij}))$$
$$+ \sum_{j=1}^{n} d_{ij} \int_{-\infty}^{t} k_{ij}(t-s) v_j(u_j(s))\, ds + I_i(t), \tag{538}$$

for $i = 1, 2, \ldots, n$. In the context of this neural network model, n represents the number of units, $u_i(t)$ signifies the state of the ith unit at time t, $a_i > 0$ denotes the rate at which the ith unit resets its potential to the resting state in isolation, when detached from both the network and external inputs. The parameter $b_{ij}(t)$ indicates the strength of influence from the jth unit on the ith unit at time t, while $c_{ij}(t)$ represents the strength of the jth unit in the ith unit at time $t - \tau_{ij}$, where τ_{ij} corresponds to the transmission delay along the axon from the jth unit to the ith unit at time t. The terms f_j, g_j and v_j refer to the measured response or activation in response to incoming potentials for the jth unit, and $I_i(t)$ characterizes the varying external input signals directed to the ith unit at time t.

Let $b_{ij}(t), c_{ij}(t), d_{ij}(t), I_i(t) \in AP(\mathbb{R} : \mathbb{R})$, $k_{ij}(t) = e^{-t}$, $i, j = 1, 2, \ldots, n$. Additionally, let the following hold: There exist positive constants $L_{f_j}, L_{g_j}, L_{v_j}$ such that

$$|f_j(u) - f_j(v)| \leqslant L_{f_j}|u - v|, \quad |g_j(u) - g_j(v)| \leqslant L_{g_j}|u - v|, \quad |v_j(u) - v_j(v)| \leqslant L_{v_j}|u - v|,$$

for all $u, v \in \mathbb{R}$.

Let $A = \operatorname{diag}(-a_1, -a_2, \ldots, -a_n)$ and X be the Banach space of bounded continuous functions from $\mathbb{R}$ to $\mathbb{R}^n$. The semigroup generated by A is given by $T(t) = e^{tA} = \operatorname{diag}(e^{-ta_1}, e^{-ta_2}, \ldots, e^{-ta_n})$ and $\|T(t)\| \leqslant Ce^{-\sigma t}$, where $C = 1$ and $\sigma = \min_{1\leqslant i\leqslant n} a_i$. Hence, (A) holds. Let

$$\begin{aligned} &f(t, u(t-\tau), Fu(t)) \\ &= \Bigg(\sum_{j=1}^{n} b_{1j}(t) f_j(u_j(t)) + \sum_{j=1}^{n} c_{1j}(t) g_j(u_j(t - \tau_{1j})) \\ &\quad + \sum_{j=1}^{n} d_{1j}(t) \int_{-\infty}^{t} k_{1j}(y-s) v_j(u_j(s))\, ds + I_1(t), \ldots, \sum_{j=1}^{n} b_{nj}(t) f_j(u_j(t)) \\ &\quad + \sum_{j=1}^{n} c_{nj}(t) g_j(u_j(t-\tau_{nj})) + \sum_{j=1}^{n} d_{nj}(t) \int_{-\infty}^{t} k_{nj}(y-s) v_j(u_j(s))\, ds + I_n(t) \Bigg)^T. \end{aligned}$$

Note that (C4)'–(C5)' are fulfilled. Hence, by using Theorem 8.3.2, for $\rho = L_f(1 + L_g) < \sigma$, where

$$L_f = \max_{t\in\mathbb{R}} \Bigg(\sum_{i=1}^{n} \sum_{j=1}^{n} L_{f_j} b_{ij}(t) + \sum_{i=1}^{n} \sum_{j=1}^{n} L_{g_j} c_{ij}(t) \Bigg) + 1, \quad L_g = \max_{t\in\mathbb{R}} \Bigg(\sum_{i=1}^{n} \sum_{j=1}^{n} L_{v_j} d_{ij}(t) \Bigg),$$

the Cohen–Grossberg neural network (538) has a unique almost periodic solution.

Example 8.3.6. Let us consider the following neutral integrodifferential equation:

$$u'(t) = -au(t) + \frac{1}{2}\sin u(t-\tau) + \cos\frac{1}{3}\int_{-\infty}^{t} e^{-(t-s)}\cos(s-\tau)\,ds, \tag{539}$$

for $a \in \mathbb{R}$. We put $f(t,u,v) = \frac{1}{2}\sin u + \cos v$, $Fu = \int_{-\infty}^{t} k(t-s)g(s,u_s)\,ds$, $k(t) = e^{-t}$ and $g(t,u) = \frac{1}{3}\cos u$. The semigroup generated by A is given by $T(t) = e^{-at}$, so $\|T(t)\| \leqslant Ce^{-\sigma t}$, where $C = 1$ and $\sigma = -a$, so (A) is satisfied. Note that the functions $f(t,u,v)$ and $g(t,u)$ are almost periodic functions. Hence, the conditions (C1)'–(C3)' are fulfilled. Moreover, we have

$$|f(t,u,v)| \leqslant \frac{1}{2} + (|u| + |v|) \quad \text{and} \quad |g(t,u)| \leqslant \frac{1}{3} + |u|,$$

so $a_f^* = \frac{1}{2}$, $b_f = 1$, $a_g^* = \frac{1}{3}$ and $b_g = 1$. We conclude that (C8)'–(C9)' hold.

Now, by using Theorem 8.3.4, the equation (539) has at least one almost periodic solution.

In a similar fashion, we can consider the existence and uniqueness of almost periodic type solutions for the fractional neutral differential equation

$$D_{t,+}^{\alpha}u(t) = Au(t) + f(t,u_t,Fu(t)), \quad t \in \mathbb{R},$$

where $0 < \alpha < 1$, $D_{t,+}^{\alpha}\cdot$ is the Weyl–Liouville fractional derivative of order α and A is a subgenerator of a uniformly integrable (g_α, C)-regularized resolvent family. Let us finally note that the existence of positive periodic solutions to some classes of the first-order neutral differential equations with distributed deviating arguments has recently been analyzed by T. Candan in [162].

$\mathcal{S}$-asymptotically Bloch type periodic solutions for a class of abstract neutral fractional equations with ψ-Hilfer derivatives

In this part, we will briefly explain the main results established in a recent joint paper [191] with N. Chegloufa, B. Chaouchi and F. Boutaous. We will use the following notion.

Definition 8.3.7 ([717]). Let (a,b) $(-\infty \leqslant a < b \leqslant \infty)$ be a finite or infinite interval of the real line $\mathbb{R}$ and $\alpha > 0$. Let $\psi(x)$ be an increasing and positive monotone function on $(a,b]$, having a continuous derivative $\psi'(x)$ on (a,b). The left-sided fractional integral of a function $f(\cdot)$ with respect to $\psi(\cdot)$ on $[a,b]$ is defined by

$$I_{a+}^{\alpha,\psi}f(x) := \frac{1}{\Gamma(\alpha)}\int_{a}^{x} \psi'(x)(\psi(x)-\psi(t))^{\alpha-1}f(t)\,dt.$$

Definition 8.3.8. Let $n-1<\alpha<n$ with $n\in\mathbb{N}$, let $I=[a,b]$ be the interval such that $-\infty\leqslant a<b\leqslant\infty$ and let $f,\ \psi\in C^n([a,b])$ be two functions such that $\psi(\cdot)$ is increasing and $\psi'(x)\neq 0$ for all $x\in I$. The left-sided ψ-Hilfer fractional derivative ${}^H D_{a+}^{\alpha,\beta,\psi}$ of function $f(\cdot)$ of order α and type $0\leqslant\beta\leqslant 1$ is defined by

$$ {}^H D_{a+}^{\alpha,\beta,\psi} f(x) := I_{a+}^{\beta(n-\alpha),\psi}\left(\frac{1}{\psi'(x)}\frac{d}{dx}\right)^n I_{a+}^{(1-\beta)(n-\alpha),\psi} f(x). $$

Let $0<\alpha\leqslant 1$ and $0\leqslant\beta\leqslant 1$. Of concern is the following nonlinear fractional neutral functional differential equation with infinite delay:

$$ \begin{cases} {}^H D_{0+}^{\alpha,\beta,\psi}(u(t)-H(t,u_t)) = Au(t)+F(t,u(t),u_t), & t\geqslant 0,\\ u(t)=\varphi(t), & t\leqslant 0, \end{cases} \tag{540} $$

where A is the infinitesimal generator of an exponentially stable analytic semigroup $(T(t))_{t\geqslant 0}$ in Banach space X. Here, the fractional derivative ${}^H D_{0+}^{\alpha,\beta,\psi}(.)$ stands for the ψ-Hilfer fractional derivative of order α and type β, with respect to a positive real function $\psi(\cdot)$ fulfilling certain conditions (cf. Definition 8.3.8 with $a=0$). It is also worth mentioning that $u_t(\cdot)$ denotes here the classical history function defined by $u_t(s):=u(t+s),\ -\infty\leqslant s\leqslant 0$. The data $\varphi(\cdot)$ belongs to a suitable admissible phase space $\mathcal{B}$. In order to furnish a complete study of (540), we assume that $H(\cdot;\cdot)$ is a continuous function of the form $H:[0,+\infty)\times\mathcal{B}\to X$, while $F(\cdot;\cdot;\cdot)$ is a C^1-function of the form $F:[0,+\infty)\times X\times\mathcal{B}\to X$. In [191], we have provided sufficient conditions for the existence of an $\mathcal{S}$-asymptotically Bloch type periodic mild solution for problem (540). Moreover, we have analyzed the existence and uniqueness of the $\mathcal{S}$-asymptotically ω-antiperiodic (1/2)-mild solutions for a class of delayed partial neutral functional differential equations (cf. also the research article [387], where the authors have investigated the existence and uniqueness of asymptotic 1-periodic solutions to a class of abstract differential equations with infinite delay, and [826], where the authors have investigated the existence and asymptotic periodicity of solutions for a class of neutral integrodifferential evolution equations with infinite delay; the existence of almost automorphic and almost periodic solutions for the partial functional differential equations with delay has been also analyzed in [71], where the authors analyzed the reduction principle for partial functional differential equation without compactness conditions). More precisely, with the notion introduced in [191], we have proved the following results.

Theorem 8.3.9. (i) *Let A generate an exponentially stable analytic semigroup $(T(t))_{t\geqslant 0}$ in X, with the growth exponent $\nu_0<0$. For $\theta\in[0,1)$, we assume that $\mathcal{B}_\theta$ is a fading memory space, $\varphi\in\mathcal{B}_\theta$, $F:[0,+\infty)\times X_\theta\times\mathcal{B}_\theta\to X$ and $H:[0,+\infty)\times\mathcal{B}_\theta\to X_1$ are two continuous functions satisfying* (H1)–(H3) *with $\varphi(0)=H(0,\varphi)=0$. If*

$$ \left(\mu_1 L C_{\theta-1}+\frac{M_\theta\Gamma(1-\theta)(\mu_1 L+\max(L_1,\mu_1 L_2))}{|\nu_0|^{1-\theta}}\right)<1, $$

where

$$C_{\theta-1} := \left\|(-A)^{\theta-1}\right\|,$$

then the problem (540) *has a unique $\mathcal{S}$-asymptotically Bloch type periodic θ-mild solution.*

(ii) *Let A generate a compact and exponentially stable analytic semigroup $(T(t))_{t\geqslant 0}$ in X, with the growth exponent $\nu_0 < 0$. For $\theta \in [0,1)$, we assume that $\mathcal{B}_\theta$ is a fading memory space, $\varphi \in \mathcal{B}_\theta$, $F : [0,+\infty) \times X_\theta \times \mathcal{B}_\theta \to X$ and $H : [0,+\infty) \times \mathcal{B}_\theta \to X_1$ are continuous functions satisfying the conditions* (H3)–(H4) *with $\varphi(0) = H(0,\varphi) = 0$. If*

$$\left(L'C_{\theta-1}\mu_1 + (L'\mu_1 + (L_2' + L_1'\mu_1))M_\theta \frac{\Gamma(1-\theta)}{|\nu_0|^{1-\theta}}\right) < 1,$$

then the problem (540) *has an $\mathcal{S}$-asymptotically Bloch type periodic θ-mild solution.*

(iii) *Suppose that the functions $h(\cdot)$, $h_1(\cdot)$ and $h_2(\cdot)$ belong to the space $SAP_w([0,+\infty))$. If*

$$(1+\pi)l \sup_{t\geqslant 0}|h(t)| + \pi \max\Big(l_1 \sup_{t\geqslant 0}|h_1(t)|, l_2 \sup_{t\geqslant 0}|h_2(t)|\Big) < 1,$$

then the problem

$$\begin{cases} {}^H D_{0+}^{\frac{1}{2},\beta,\psi}(u(t,\xi) - h(t) \int\limits_{-\infty}^{t} (\int_0^\xi b(s-t)u(s,\eta)\,d\eta)ds) - \frac{\partial^2}{\partial \xi^2}u(t,\xi) & \\ \quad = h_2(t)\int_{-\infty}^t a(s-t)u(s,\xi)\,ds + h_1(t)f(\xi,u(t,\xi)), & \xi \in [0,\pi],\ t \geqslant 0, \\ u(t,0) = u(t,\pi) = 0, & t \geqslant 0, \\ u(\tau,\xi) = \varphi(\tau)(\xi), & \tau \leqslant 0 \end{cases}$$

has a unique $\mathcal{S}$-asymptotically ω-anti-periodic (1/2)*-mild solution.*

Specifically, we have discussed the existence and uniqueness of an $\mathcal{S}$-asymptotically ω-anti-periodic $\frac{1}{2}$-mild solution for the following problem:

$$\begin{cases} {}^H D_{0+}^{\frac{1}{2},\beta,\psi}(u(t,\xi) - h(t) \int\limits_{-\infty}^{t} (\int_0^\xi b(s-t)u(s,\eta)\,d\eta)ds) - \frac{\partial^2}{\partial \xi^2}u(t,\xi) & \\ \quad = h_2(t)\int_{-\infty}^t a(s-t)u(s,\xi)\,ds + h_1(t)f(\xi,u(t,\xi)), & \xi \in [0,\pi],\ t \geqslant 0, \\ u(t,0) = u(t,\pi) = 0, & t \geqslant 0, \\ u(\tau,\xi) = \varphi(\tau)(\xi), & \tau \leqslant 0; \end{cases} \tag{541}$$

here, $0 \leqslant \beta \leqslant 1$ and ${}^H D_{0+}^{\alpha,\beta,\psi}$ is the ψ-Hilfer fractional derivative of order $\frac{1}{2}$ and type β, with respect to the function ψ.

Suppose that $X := L^2([0,\pi])$ and $A : D(A) \subseteq X \to X$ is defined by

$$\begin{cases} Au := u'', \\ D(A) := \{u^{(j)} \in X : u(0) = u(\pi) = 0\}, \quad j \in \{0,1,2\}. \end{cases}$$

We know that A generates an exponentially stable analytic semigroup $(T(t))_{t\geqslant 0}$ on X. Moreover, A has the discrete spectrum $\sigma(A)$ with eigenvalues n^2, $n \in \mathbb{N}$, associated to a normalized eigenvectors $e_n(\xi) = \sqrt{2/\pi}\sin(n\xi)$. We note also that $\{e_n \mid n \in \mathbb{N}\}$ is an orthonormal basis of X. Hence, the associated semigroup $(T(t))_{t\geqslant 0}$ is explicitly given by

$$T(t)u = \sum_{n=1}^{\infty} e^{-n^2 t}\langle u, e_n\rangle e_n, \quad t \geqslant 0,\ u \in X.$$

Furthermore, $\|T(t)\| \leqslant e^{-t}$. On other side, the closed linear operator $(-A)^{-\frac{1}{2}}$ is well-defined and one has

$$\begin{cases} (-A)^{-\frac{1}{2}}u = \sum_{n=1}^{\infty} n\langle u, e_n\rangle e_n, \\ D((-A)^{\frac{1}{2}}) = \{u \in X : \sum_{n=1}^{\infty} n\langle u, e_n\rangle e_n \in X\}. \end{cases}$$

Note here that $D((-A)^{\frac{1}{2}})$ is the Banach space when equipped with the norm $\|u\|_{\frac{1}{2}} = \|u'\|$ for all $u \in X_{\frac{1}{2}}$.

Furthermore, we know that if $g(s) = 1 + |s|^n$ for some $n > 0$, then the space $C_g^0((-\infty,0], X_{\frac{1}{2}})$ is a fading memory space. Keeping this in mind, it readily follows that

$$\mu_1(t) = \sup_{t\leqslant 0} \frac{1}{1+|s|^n} = 1 \quad \text{and} \quad \mu_2(t) = \sup_{t\leqslant 0} \frac{1+|s+t|^n}{1+|s|^n} \leqslant 1.$$

Consider now the Banach space $\mathcal{B}_{\frac{1}{2}} := C_g^0((-\infty,0], X_{\frac{1}{2}})$, equipped with the norm

$$\|\phi\|_{\mathcal{B}_{\frac{1}{2}}} := \sup_{s\leqslant 0} \frac{\|\phi(s)\|_{\frac{1}{2}}}{1+|s|^n},$$

which is also equivalent to

$$\|\phi\|_{\mathcal{B}_{\frac{1}{2}}} = \sup_{s\leqslant 0} \frac{\|\phi'(s)\|}{1+|s|^n}.$$

To study the problem (541), we have imposed the following conditions:
- $f(\cdot,\cdot)$ is a continuous function satisfying the following conditions:
  - For $x, y \in X_{\frac{1}{2}}$, there exists $l_1 > 0$ such that

$$\|f(\cdot, x(\cdot)) - f(\cdot, y(\cdot))\| \leqslant l_1\|x - y\|.$$

- If λ be a complex number with $|\lambda| = 1$, then

$$f(\xi, \lambda x(\xi)) = \lambda f(\xi, x(\xi)), \quad \text{for every } \xi \in [0, \pi].$$

- The data function $\varphi(\cdot)$ belongs to the space $\in \mathcal{B}_{\frac{1}{2}}$.
- The functions $h(\cdot)$, $h_1(\cdot)$ and $h_2(\cdot)$ belong to the space $C^1([0, +\infty))$ with $h(0) = 0$.
- The functions $s \mapsto (1 + |s|^n)a(s)$ and $s \mapsto (1 + |s|^n)b(s)$ are integrable functions on $(-\infty, 0]$, and

$$l_2 = \int_{-\infty}^{0} (1 + |s|^n)|a(s)|ds, \quad l = \int_{-\infty}^{0} (1 + |s|^n)|b(s)|ds.$$

Now we can define the functions $F : [0, +\infty) \times X_{\frac{1}{2}} \times \mathcal{B}_{\frac{1}{2}} \to X$ and $H : [0, +\infty) \times \mathcal{B}_{\frac{1}{2}} \to X_1$ as follows:

$$F(t, x, \phi)(\xi) := \left(\int_{-\infty}^{0} a(s)\phi(s, \xi)\, ds + h_1(t) f(\xi, x(\xi)) \right) h_2(t),$$

and

$$H(t, \phi)(\xi) := \left(\int_{-\infty}^{0} \int_{0}^{\xi} b(s)\phi(s, \eta)\, d\eta\, ds \right) h(t).$$

Then we have the following result.

Theorem 8.3.10. *Suppose that the functions $h(\cdot)$, $h_1(\cdot)$ and $h_2(\cdot)$ belong to the space $SAP_w([0, +\infty))$. If*

$$(1 + \pi) l \sup_{t \geqslant 0} |h(t)| + \pi \max\left(l_1 \sup_{t \geqslant 0} |h_1(t)|, l_2 \sup_{t \geqslant 0} |h_2(t)| \right) < 1,$$

then the problem (541) *has a unique $\mathcal{S}$-asymptotically ω-anti-periodic $\frac{1}{2}$-mild solution.*

On the existence and uniqueness of pseudo $\mathcal{S}$-asymptotically periodic mild solutions for a class of neutral fractional evolution equations with nonlocal conditions

In our recent joint research article [192] with N. Chegloufa, B. Chaouchi and W.-S. Du, we have studied the existence and uniqueness of pseudo-$\mathcal{S}$-asymptotically periodic mild solutions for the following neutral fractional delayed evolution equation:

$$\begin{cases} {}^C D_t^q (u(t) - G(t, u_t)) + Au(t) = F(t, u_t), & t \geqslant 0, \\ u(t) = \varphi(t), & -r \leqslant t \leqslant 0, \end{cases} \tag{542}$$

where the fractional derivative ${}^C D_t^q$, $q \in (0, 1)$, is taken in the sense of Caputo approach and $(A, D(A))$ is a closed linear operator in a complex Banach space $(X, \|\cdot\|)$. We assume here that $-A$ generates a compact, exponentially stable analytic semigroup $(T(t))_{t \geqslant 0}$.

Let us recall that a bounded continuous function $f : [0,\infty) \to X$ is said to be pseudo-$\mathcal{S}$-asymptotically periodic if there exists $\omega > 0$ such that

$$\lim_{h\to+\infty} \frac{1}{h} \int_0^h \|f(t+\omega) - f(t)\|\, dt = 0.$$

The class of pseudo $\mathcal{S}$-asymptotically periodic functions was introduced by M. Pierri and V. Rolnik in [638] (2013); in that paper, the authors have considered the classical version of (542) with $q = 1$ and established several interesting results concerning the existence and uniqueness of pseudo-$\mathcal{S}$-asymptotically periodic mild solutions for such a problem. The class of pseudo-$\mathcal{S}$-asymptotically periodic functions is a natural generalization of the class of $\mathcal{S}$-asymptotically periodic functions. Consequently, the study [192] can be viewed as an extension and continuation of the research study [777] by M. Wei and Q. Li, where the solvability of problem (542) was discussed and optimal results about the existence and uniqueness of $\mathcal{S}$-asymptotically periodic mild solutions were successfully established. The abstract fractional integrodifferential neutral equation

$$\begin{cases} \frac{d}{dt}D(t,u_t) = \int_0^t g_{a-1}(t-s)AD(s,u_s)\, ds + f(t,u_t), & t \geqslant 0, \\ u_0 = \varphi, \end{cases}$$

where $1 < a < 2$, $(A, D(A))$ is a densely defined sectorial operator and $D = \varphi(0) + g(t,\varphi)$ with f, g and φ being suitable vector-valued functions, was analyzed by M. Yang and Q. Wang in [799], where the authors considered the existence and uniqueness of the pseudo-$\mathcal{S}$-asymptotically periodic mild solutions for the above problem.

With the notion introduced in [192], we have proved the following results.

Theorem 8.3.11. (i) *Assume that* (A1)–(A4) *hold and* $-A$ *generates a compact, exponentially stable analytic semigroup* $(T(t))_{t\geqslant 0}$ *on* X. *For* $\alpha \in [0,1)$, *we assume that* $\varphi \in B_\alpha$, $F : \mathbb{R}^+ \times B_\alpha \to X$ *and* $G : \mathbb{R}^+ \times B_\alpha \to X_1$ *are continuous functions and* $G(t,0) = 0$ *for* $t \geqslant 0$. *Then the problem* (542) *has at least one pseudo-*$\mathcal{S}$*-asymptotic* ω*-periodic* α*-mild solution.*

(ii) *Let* (A5)–(A8) *hold and let* $-A$ *be the generator of a compact, exponentially stable analytic semigroup* $T(t)_{(t\geqslant 0)}$. *For* $\alpha \in [0,1)$, *we assume that* $\varphi \in B_\alpha$, $F : \mathbb{R}^+ \times B_\alpha \to X$ *and* $G : \mathbb{R}^+ \times B_\alpha \to X_1$ *are continuous functions. Then* (542) *has a unique pseudo-*$\mathcal{S}$*-asymptotic* ω*-periodic* α*-mild solution.*

On the scientific work of A. B. Muravnik and N. V. Zaitseva

In this part, we will inscribe several recent results about the partial differential-difference equations obtained by the Russian mathematicians A. B. Muravnik and N. V. Zaitseva.

(i) Differential-difference elliptic equation with nonlocal potential

$$\Delta_x u(t,x) - \sum_{j=1}^{m} b_j u(t, x+h_j) + u_{tt}(t,x), \quad t>0,\ x \in \mathbb{R}^n, \tag{543}$$

where $b_j \geqslant 0$ and $h_j \in \mathbb{R}^n$ for all $j \in \mathbb{N}_n$, has been examined in [613] by applying the Fourier transform. The author has accompanied the initial-value condition $u_{|t=0} = u_0(x)$, $x \in \mathbb{R}^n$ to the above equation, where $u_0 \in L^1(\mathbb{R}^n)$. The integral representation of solution of convolution type to the above problem is given in [613, Theorem 1]; see also [611] and [612].

Here, we will only note that the convolution formula established in the proof of [613, Theorem 1] enables one to analyze the existence and uniqueness of generalized weighted Weyl almost periodic solutions of (543), provided that the boundary value $u_0(\cdot)$ has a certain generalized weighted Weyl almost periodic behavior (cf. [447, Example 6.3.4] for more details).

(ii) Suppose that $a_j,\ h_j \in \mathbb{R}$ for $j = 1, 2$. The existence of global classical solutions of hyperbolic differential-difference equation

$$u_{tt}(t,x) = a_1 u_{xx}(t, x-h_1) + a_2 u_{xx}(t, x-h_2), \quad t>0,\ x \in \mathbb{R},$$

has been sought in [806] by applying the Fourier transform and the Gelfand–Shilov classical operation scheme; see also [809].

(iii) In [807], the author has investigated the existence of smooth solutions of two hyperbolic differential-difference equations in the half-space $\{(t,x) : t>0,\ x \in \mathbb{R}^n\}$ by applying the Fourier transform and some elementary transformations. The first of these equations contains compositions of differential operators and shift operators with respect to each of the spatial variables:

$$u_{tt}(t,x) = a^2 \Delta_x u(t,x) + \sum_{j=1}^{n} b_j u_{x_j x_j}(x_1, \dots, x_{j-1}, x_j - h_j, x_{j+1}, \dots, x_n, t), \quad t>0,\ x \in \mathbb{R}^n,$$

where $a > 0$ and $b_j,\ h_j \in \mathbb{R}$ for all $j \in \mathbb{N}_n$. The second equation is very similar to the former one:

$$u_{tt}(t,x) = c^2 \Delta_x u(t,x) - \sum_{j=1}^{n} d_j u(x_1, \dots, x_{j-1}, x_j - l_j, x_{j+1}, \dots, x_n, t), \quad t>0,\ x \in \mathbb{R}^n,$$

where $c > 0$ and $d_j,\ l_j \in \mathbb{R}$ for all $j \in \mathbb{N}_n$. The main established results are [807, Theorems 1, 2]; see also [808] and [810].

We also refer the reader to the important research monographs [274] by L. E. El'sgol'ts, S. B. Norkin, [416] by G. A. Kamenskii, A. L. Skubachevskii, [614] by A. D. Myshkis, [695] by A. L. Skubachevskii and the survey articles [610] by A. B. Muravnik and [761] by V. V. Vlasov and D. A. Medvedev.

Discrete fractional operators based on the use of Z-transform of sequences
Recall that, if a sequence $(f_k)_{k\in\mathbb{N}_0}$ in X satisfies $\limsup_{k\to+\infty}\|f_k\|^{1/k} < r < +\infty$, then the Z-transform of $(f_k)_{k\in\mathbb{N}_0}$ is defined by

$$F(z) := F\{f_k\}(z) := \sum_{k=0}^{\infty}\frac{f(k)}{z^k}, \quad |z| > r$$

and $F(z)$ is analytic for $|z| > r$. If this is the case, then we have

$$\begin{aligned} F\{f_{k+j}\}(z) &= \sum_{k=0}^{\infty} x_{k+j}z^{-k} = z^j\sum_{k=0}^{\infty} x_{k+j}z^{-k-j} \\ &= z^j\left[F\{f_k\}(z) - \sum_{k=0}^{j-1} x_k z^{-k}\right], \quad |z| > r \quad (j\in\mathbb{N}). \end{aligned}$$

Since the Z-transform is linear, it follows that

$$\begin{aligned} F\{\Delta^m f_k\}(z) &= \sum_{j=0}^{m}(-1)^{m-j}\binom{m}{j}F\{f_{k+j}\}(z) \\ &= \sum_{j=0}^{m}(-1)^{m-j}\binom{m}{j}z^j\left[F\{f_k\}(z) - \sum_{k=0}^{j-1} x_k z^{-k}\right], \quad |z| > r \quad (m\in\mathbb{N}). \end{aligned} \tag{544}$$

Suppose now that $\alpha > 0$, $m = \lceil\alpha\rceil$, the sequences $(a_k)_{k\in\mathbb{N}_0}$ and $(b_k)_{k\in\mathbb{N}_0}$ are complex valued and $(u_k)_{k\in\mathbb{N}_0}$ is a given sequence in X so that $\limsup_{k\to+\infty}|a_k|^{1/k} < r < +\infty$, $\limsup_{k\to+\infty}|b_k|^{1/k} < r$ and $\limsup_{k\to+\infty}\|u_k\|^{1/k} < r$. Then we have $\limsup_{k\to+\infty}\|(a *_0 u)(k)\|^{1/k} < r$, $\limsup_{k\to+\infty}\|\Delta^m(a *_0 u)(k)\|^{1/k} < r$ and $\limsup_{k\to+\infty}\|b *_0 \Delta^m(a *_0 u)(k)\|^{1/k} < r$. Since the Z-transform is compatible with the finite convolution product $*_0$ and (544) holds, we can compute the Z-transform of the generalized Hilfer (a, b, α)-fractional derivative $\Delta^{\alpha}_{a,b}u$ as follows:

$$\begin{aligned} Z\{\Delta^{\alpha}_{a,b}u\}(z) &= F\{b_k\}(z)\cdot[\Delta^m(a *_0 u)(k)](z) \\ &= F\{b_k\}(z)\cdot\sum_{j=0}^{m}(-1)^{m-j}\binom{m}{j}z^j\left[F\{a_k\}(z)\cdot F\{u_k\}(z) - \sum_{k=0}^{j-1}(a *_0 u)(k)z^{-k}\right], \quad |z| > r. \end{aligned}$$

For the usual Hilfer fractional derivative, we have

$$F\{a_k\}(z)\cdot F\{b_k\}(z) = F\{(a *_0 b)(k)\}(z) = F\{1\}(z) = z/(z-1), \quad |z| > 1.$$

Motivated by this simple result, we would like to introduce the following discrete analogue of Definition 7.3.2.

Definition 8.3.12. Suppose that the following condition holds true:

(LD): $s \in \mathbb{N}_0$, $\omega > 0$, $w : \{z \in \mathbb{C} : |z| > \omega\} \to X$ and $w_k : \{z \in \mathbb{C} : |z| > \omega\} \to X$ are given functions, D_k is a nonempty subset of Z-valued sequences $(u_k)_{k\in\mathbb{N}_0}$ such that $\limsup_{k\to+\infty} \|u_k\|^{1/k} < +\infty$ and $x_k : D_k \times \{z \in \mathbb{C} : |z| > \omega\} \to X$ are given mappings $(0 \leqslant k \leqslant s)$. Set

$$W_d := \big(s, w(\cdot), (w_k(\cdot))_{0\leqslant k\leqslant s}, (D_k)_{0\leqslant k\leqslant s}, (x_k(\cdot;\cdot))_{0\leqslant k\leqslant s}\big).$$

The generalized discrete fractional derivative $D_Z^{W_d} u$ is defined for any function $u \in D_0 \cap \cdots \cap D_s$ such that there exist a sufficiently large number $r > \omega$ and an exponentially bounded sequence $h : \mathbb{N}_0 \to X$ such that

$$F\{h_k\}(z) = w(z)F\{u_k\}(z) - \sum_{k=0}^{s} w_k(z)x_k(u,z), \quad |z| > r.$$

In this case, we set $D_Z^{W_d} u := h$.

Besides the generalized Hilfer (a, b, α)-fractional derivatives, the notion introduced in Definition 8.3.12 also generalizes the notion of fractional derivatives considered in the research article [606], provided that $a = 0$ and $h = 1$ therein. It would be interesting to reconsider the results established in Proposition 7.3.4, Theorem 7.3.6, Theorem 7.3.7 and Sections 7.3.7–7.3.9 in the discrete setting. Also, it would be interesting to introduce and analyze some new classes of generalized discrete fractional derivatives of sequences defined on the whole integer line using the bilateral Z-transform of sequences.

Bibliography

[1] L. Abadias, E. Alvarez, *Asymptotic behavior for the discrete in time heat equation*, Mathematics, **10** (2022), 3128. https://doi.org/10.3390/math10173128

[2] L. Abadias, J. González-Camus, S. Rueda, *Time-step heat problem on the mesh: asymptotic behavior and decay rates*, Forum Math., **35** (2023), 1563–1582.

[3] L. Abadias, C. Lizama, *Almost automorphic mild solutions to fractional partial difference-differential equations*, Appl. Anal. (2015). https://doi.org/10.1080/00036811.2015.1064521

[4] L. Abadias, C. Lizama, P. Miana, M. Pillar Velasco, *On well-posedness of vector-valued fractional differential-difference equations*, Discrete Contin. Dyn. Syst., **39** (2019), 2679–2708.

[5] S. Abbas, *Weighted pseudo almost automorphic sequences and their applications*, Electron. J. Differ. Equ., **121** (2010), 1–14.

[6] S. Abbas, *A note on Weyl pseudo almost automorphic functions and their properties*, Math. Sci., **6** (2012), 29, 5 pp. https://doi.org/10.1186/2251-7456-6-29

[7] S. Abbas, B. Ahmad, M. Benchohra, A. Salim, *Fractional Difference, Differential Equations, and Inclusions: Analysis and Stability*, Morgan Kaufmann, Burlington, Massachusetts, 2024.

[8] S. Abbas, D. Bahuguna, *Almost periodic solutions of neutral functional differential equations*, Comput. Math. Appl., **55** (2008), 2593–2601.

[9] S. Abbas, Y.-K. Chang, M. Hafayed, *Stepanov pseudo almost automorphic sequences and their applications to difference equations*, Nonlinear Stud., **21** (2014), 99–111.

[10] S. Abbas, M. Kostić, *Metrical Weyl almost automorphy and applications*, Adv. Pure Appl. Math., in press. https://www.openscience.fr/Metrical-Weyl-almost-automorphy-and-applications

[11] M. S. Abdo, S. T. Thabet, B. Ahmad, *The existence and Ulam–Hyers stability results for Ψ-Hilfer fractional integro differential equations*, J. Pseudo-Differ. Oper. Appl., **11** (2020), 1757–1780.

[12] A. Abilassan, J. E. Restrepo, D. Suragan, *On a variant of multivariate Mittag-Leffler's function arising in the Laplace transform method*, Integral Transforms Spec. Funct., **34**(3) (2022), 244–260.

[13] N. Acar, *Development of nabla fractional calculus and a new approach to data fitting in time dependent cancer therapeutic study*, MSc Thesis, Western Kentucky University, 2012.

[14] M. Adivar, H. C. Koyuncuoğlu, *Almost automorphic solutions of discrete delayed neutral system*, J. Math. Anal. Appl., **435** (2016), 532–550.

[15] M. Adivar, H. C. Koyuncuoğlu, *Floquet theory based on new periodicity concept for hybrid systems involving q-difference equations*, Appl. Math. Comput., **273** (2016), 1208–1233.

[16] S. Afonso, E. Bonotto, M. da Silva, *Periodic solutions of neutral functional differential equations*, J. Differ. Equ., **350** (2023), 89–123.

[17] S. M. Afonso, A. L. Furtado, *Antiperiodic solutions for nth-order functional differential equations with infinite delay*, Electron. J. Differ. Equ., **44** (2016), 1–8.

[18] D. Agarwal, S. Dhama, M. Kostić, S. Abbas, *Periodicity, stability and synchronization of solution of hybrid coupled dynamic equations with multiple delays*, Math. Methods Appl. Sci., **47** (2024), 7616–7636.

[19] R. P. Agarwal, *Difference Equations and Inequalities*, Monographs and Textbooks in Pure and Applied Mathematics, vol. 155, Marcel Dekker, New York, 1992.

[20] R. P. Agarwal, M. Bohner, A. Özbekler, *Lyapunov Inequalities and Applications*, Springer, Switzerland AG, 2021. https://doi.org/10.1007/978-3-030-69029-8

[21] R. P. Agarwal, C. Cuevas, C. Lizama, *Regularity of Difference Equations on Banach Spaces*, Springer, Switzerland, 2014. https://doi.org/10.1007/978-3-319-06447-5

[22] R. P. Agarwal, C. Cuevas, M. V. S. Frasson, *Semilinear functional difference equations with infinite delay*, Math. Comput. Model., **55** (2012), 1083–1105.

[23] R. P. Agarwal, E. M. Elsayed, *Periodicity and stability of solutions of higher order rational difference equation*, Adv. Stud. Contemp. Math., **17**(2) (2008), 181–201.

[24] R. P. Agarwal, D. O'Regan, P. J. Y. Wong, *Constant-sign periodic and almost periodic solutions of a system of difference equations*, Comput. Math. Appl., **50** (2005), 1725–1754.

https://doi.org/10.1515/9783111689746-013

[25] R. P. Agarwal, P. J. Y. Wang, *Advanced Topics in Difference Equations*, Kluwer Academic Publisshers, Dordrecht/Boston/London, 1997.

[26] R. P. Agarwal, Y. Zhou, Y. He, *Existence of fractional neutral functional differential equations*, Comput. Math. Appl., **59** (2010), 1095–1100.

[27] C. H. Ahlbrandt, J. Ridenhour, *Floquet theory for time scales and Putzer representations of matrix logarithms*, J. Differ. Equ. Appl., **9** (2003), 77–92.

[28] B. Ahmad, K. Ntouyas, *Boundary value problems for q-difference inclusions*, Abstr. Appl. Anal., **2011** (2011), 292860, 15 pp. https://doi.org/10.1155/2011/292860

[29] W. A. Ahmood, *Extension of Laplace transform to multi-dimensional fractional integro-differential equations*, PhD Thesis, Universiti Putra Malaysia, 2017.

[30] E. Ait Dads, P. Cieutat, L. Lhachimi, *Positive pseudo almost periodic solutions for some nonlinear infnite delay integral equations*, Math. Comput. Model., **49** (2009), 721–739.

[31] E. Ait Dads, K. Ezzinbi, *Almost periodic solution for some neutral nonlinear integral equation*, Nonlinear Anal., **28** (1997), 1479–1489.

[32] E. Ait Dads, K. Ezzinbi, L. Lhachimi, *Discrete pseudo almost periodic solutions for some difference equations*, Adv. Pure Math., **1**(4) (2011), 118–127. https://doi.org/10.4236/apm.2011.14024

[33] E. Ait Dads, K. Ezzinbi, L. Lhachimi, *Pseudo almost periodic solutions for continuous algrebraic difference equations*, J. Nonlinear Evol. Equ. Appl., **5** (2019), 57–74.

[34] E. Ait Dads, S. Fatajou, Z. Zizi, *Stepanov Eberlein almost periodic functions and applications*, Math. Methods Appl. Sci., **46** (2023), 16761–16781.

[35] E. Ait Dads, L. Lhachimi, *On the quantitative and qualitative studies of the solutions for some difference equations*, J. Abstr. Differ. Equ. Appl., **7** (2016), 1–11.

[36] G. Akagi, U. Stefanelli, *Periodic solutions for doubly nonlinear evolution equations*, J. Differ. Equ., **251** (2011), 1790–1812.

[37] S. S. Akhtamova, T. Cuchta, A. P. Lyapin, *An approach to multidimensional discrete generating series*, Mathematics, **12**(1) (2024), 143. https://doi.org/10.3390/math12010143

[38] R. A. Ahmad, *Laplace transform of the product of two functions*, Ital. J. Pure Appl. Math., **44** (2020), 800–804.

[39] V. S. Alekseev, S. S. Akhtamova, A. P. Lyapin, *Discrete generating functions*, Math. Notes, **114**(5–6) (2023), 1087–1093.

[40] M. Alia, J. El Matloub, K. Ezzinbi, *Asymptotically almost periodic solutions for some partial differential inclusions in α-norm*, J. Evol. Equ., **24** (2024), 75. https://doi.org/10.1007/s00028-024-01007-z

[41] R. Almeida, A. B. Malinowska, M. T. T. Monteiro, *Fractional differential equations with a Caputo derivative with respect to a kernel function and their applications*, Math. Methods Appl. Sci., **41** (2018), 336–352.

[42] L. Alsedà i Soler, et al. (eds.), *Difference Equations, Discrete Dynamical Systems and Applications*, Springer Proceedings in Mathematics and Statistics, vol. 180, ICDEA, Barcelona, Spain, 2012.

[43] B. Alqahtani, S. Abbas, M. Benchohra, S. S. Alzaid, *Fractional q-difference inclusions in Banach spaces*, Mathematics, **8** (2020), 91. https://doi.org/10.3390/math8010091

[44] M. Alquran, K. Al-Khaled, J. Chattopadhyay, *Analytical solutions of fractional population diffusion model: residual power series*, Nonlinear Stud., **22** (2015), 31–39.

[45] E. Alvarez, S. Díaz, C. Lizama, *On the existence and uniqueness of (N,λ)-periodic solutions to a class of Volterra difference equations*, Adv. Differ. Equ., **2019** (2019), 105. https://doi.org/10.1186/s13662-019-2053-0

[46] E. Alvarez, S. Díaz, C. Lizama, *Existence of (N,λ)-periodic solutions for abstract fractional difference equations*, Mediterr. J. Math., **19** (2022), 47. https://doi.org/10.1007/s00009-021-01964-6

[47] P. Alper, *Higher-dimensional Z-transforms and nonlinear discrete systems*, Technological University, Electronics Laboratory, Delft-Netherlands, 1963.

[48] J. Alzabut, Y. Bolat, T. Abdeljawad, *Almost periodic dynamics of a discrete Nicholson's blowflies model involving a linear harvesting term*, Adv. Differ. Equ., **2012** (2012), 158. https://doi.org/10.1186/1687-1847-2012-158

[49] J. ALZABUT ET AL., *A survey on the oscillation of solutions for fractional difference equations*, Mathematics, **10** (2022), 894. https://doi.org/10.3390/math10060894

[50] J. ALZABUT ET AL., *Existence, uniqueness and synchronization of a fractional tumor growth model in discrete time with numerical results*, Results Phys., **54** (2023), 107030.

[51] J. ALZABUT ET AL., *Higher-order nabla difference equations of arbitrary order with forcing, positive and negative terms: non-oscillatory solutions*, Axioms, **12** (2023), 325. https://doi.org/10.3390/axioms12040325

[52] A. M. AMLEH, E. CAMOUZIS, G. LADAS, *On the dynamics of a rational difference equation, Part 1*, Int. J. Differ. Equ., **3** (2008), 1–35.

[53] A. M. AMLEH, E. CAMOUZIS, G. LADAS, *On the dynamics of a rational difference equation, Part 2*, Int. J. Differ. Equ., **3** (2008), 195–225.

[54] G. A. ANASTASSIOU, *Nabla discrete fractional calculus and nabla inequalities*, Math. Comput. Model., **51** (2010), 562–571.

[55] J. ANDRES, A. M. BERSANI, R. F. GRANDE, *Hierarchy of almost-periodic function spaces*, Rend. Mat. Appl. (7), **26** (2006), 121–188.

[56] J. ANDRES, D. PENNEQUIN, *On Stepanov almost-periodic oscillations and their discretizations*, J. Differ. Equ. Appl., **18** (2012), 1665–1682.

[57] J. ANDRES, D. PENNEQUIN, *On the nonexistence of purely Stepanov almost-periodic solutions of ordinary differential equations*, Proc. Am. Math. Soc., **140** (2012), 2825–2834.

[58] P. K. ANH, N. H. DU, L. C. LOI, *Singular difference equations: an overview*, Vietnam J. Math., **35**(4) (2007), 339–372.

[59] M. H. ANNABY, Z. S. MANSOUR, *q-Fractional Calculus and Equations*, Lecture Notes in Mathematics, Springer, Berlin, Heidelberg, 2012.

[60] M. S. APANOVICH, E. K. LEINARTAS, *On correctness of Cauchy problem for a polynomial difference operator with constant coefficients*, Bull. Irkutsk State Univ. Ser. Math., **26** (2018), 3–15.

[61] G. APREUTESEI, N. APREUTESEI, *Continuous dependence on data for bilocal difference equations*, J. Differ. Equ. Appl., **15** (2009), 511–527.

[62] G. APREUTESEI, N. APREUTESEI, *Second order difference inclusions of monotone type*, Math. Bohem., **137** (2012), 123–130.

[63] N. APREUTESEI, *Nonlinear Second Order Evolution Equations of Monotone Type and Applications*, Pushpa Publishing House, India, 2007.

[64] N. APREUTESEI, *On a class of difference equations of monotone type*, J. Math. Anal. Appl., **288** (2003), 833–851.

[65] D. ARAYA, R. CASTRO, C. LIZAMA, *Almost automorphic solutions of difference equations*, Adv. Differ. Equ., **2009** (2009), 591380, 15 pp. https://doi.org/10.1155/2009/591380

[66] A. ARBI, J. CAO, M. ES-SAIYDY, M. ZARHOUNI, M. ZITANE, *Dynamics of delayed cellular neutral networks in the Stepanov pseudo almost automorphic space*, Discrete Contin. Dyn. Syst., Ser. S, **15** (2022), 3097–3109.

[67] A. ARBI, N. TAHRI, *Stability analysis of inertial neural networks: a case of almost anti-periodic environment*, Math. Methods Appl. Sci., **45** (2022), 10476–10490.

[68] A. ARBI, N. TAHRI, *New results on time scales of pseudo Weyl almost periodic solution of delayed QVSICNNs*, Comput. Appl. Math., **41** (2022), 293. https://doi.org/10.1007/s40314-022-02003-0

[69] W. ARENDT, C. J. K. BATTY, M. HIEBER, F. NEUBRANDER, *Vector-Valued Laplace Transforms and Cauchy Problems*, Monographs in Mathematics, vol. 96, Birkhäuser, Basel, 2001.

[70] F. V. ATKINSON, J. R. HADDOCK, *Criteria for asymptotic constancy of solutions of functional differential equations*, J. Math. Anal. Appl., **91** (1983), 410–423.

[71] M. EL ATTAOUY, K. EZZINBI, G. M. N'GUÉRÉKATA, *Reduction principle for partial functional differential equation without compactness*, Electron. J. Differ. Equ., **39** (2023), 1–17.

[72] E. R. ATTIA, B. M. EL-MATARY, *New aspects for the oscillation of first-order difference equations with deviating arguments*, Opusc. Math., **42** (2022), 393–413.

[73] C. E. Avellar, J. K. Hale, *On the zeros of exponential polynomials*, J. Math. Anal. Appl., **73** (1980), 434–452.
[74] F. M. Atici, P. W. Eloe, *Initial value problems in discrete fractional calculus*, Proc. Am. Math. Soc., **137** (2009), 981–989.
[75] F. M. Atici, P. W. Eloe, *Discrete fractional calculus with the nabla operator*, Electron. J. Qual. Theory Differ. Equ., Spec. Ed. I, **3** (2009), 1–12.
[76] F. M. Atici, P. W. Eloe, *Two-point boundary value problems for finite fractional difference equations*, J. Differ. Equ. Appl., **17** (2011), 445–456.
[77] M. Ayachi, *Measure-pseudo almost periodic dynamical behaviors for BAM neural networks with D operator and hybrid time-varying delays*, Neurocomputing, **486** (2022), 160–173.
[78] N. V. Azbelev, V. P. Maksimov, L. F. Rakhmatullina, *Introduction to the Theory of Functional-Differential Equations*, Nauka, Moscow, 1991 (in Russian).
[79] A. Babakhani, *Theory of multidimensional Laplace transforms and boundary value problems*, PhD Thesis, Iowa State University, 1989.
[80] H. Bahouri, J.-Y. Chemin, R. Danchin, *Fourier Analysis and Nonlinear Partial Differential Equations*, Springer, Berlin-Heidelberg, 2011.
[81] D. Bainov, P. Simeonov, *Impulsive Differential Equations: Periodic Solutions and Applications*, Wiley, New York, 1993.
[82] C. T. H. Baker, Y. Song, *Periodic solutions of discrete Volterra equations*, Math. Comput. Simul., **64** (2004), 521–542.
[83] P. Baliarsingh, *On certain dynamic properties of difference sequences and the fractional derivatives*, Math. Methods Appl. Sci., **44** (2020), 3023–3035.
[84] P. Baliarsingh, M. Mursaleen, V. Rakočević, *A survey on the spectra of the difference operators over the Banach space c*, J. Phys., **115** (2021), 57. https://doi.org/10.1007/s13398-020-00997-y
[85] V. D. P. Balth, H. Bremmer, *Operational Calculus Based on Two-Sided Laplace Integral*, Cambridge University Press, London, 1950.
[86] J. Banasiak, *Mathematical Modelling in One Dimension. An Introduction via Difference and Differential Equations*, Cambridge University Press, 2013.
[87] N. R. Bantsur, E. P. Trofimchuk, *On the existence of T-periodic solutions of essentially nonlinear scalar differential equations with maxima*, Nelīnīīnī Koliv., **1** (1998), 1–5.
[88] N. R. Bantsur, O. P. Trofimchuk, *Existence and stability of periodic and almost periodic solutions of quasilinear equations with maxima*, Ukr. Math. J., **50** (1998), 847–856.
[89] V. Barbu, *Nonlinear Volterra equations in a Hilbert space*, SIAM J. Math. Anal., **6** (1975), 728–741.
[90] B. Basit, H. Güenzler, *Harmonic analysis for generalized vector-valued almost periodic and ergodic distributions*, Rend. Accad. Naz. Sci. XL Mem. Mat. Appl., **5** (2005), 35–54.
[91] B. Basit, H. Güenzler, *Generalized vector valued almost periodic and ergodic distributions*, J. Math. Anal. Appl., **314** (2006), 363–381.
[92] B. Basit, H. Güenzler, *Spectral criteria for solutions of evolution equations and comments on reduced spectra*, Far East J. Math. Sci.: FJMS, **65** (2012), 273–288.
[93] A. G. Baskakov, *Spectral analysis of differential operators with unbounded operator-valued coefficients, difference relations and semigroups of difference relations*, Izv. Math., **73** (2009), 215–278.
[94] A. G. Baskakov, A. Yu. Duplishcheva, *Difference operators and operator-valued matrices of the second order*, Izv. Math., **79** (2015), 217–232.
[95] A. Bašić, L. Smajlović, Z. Šabanac, *Discrete Bessel functions and discrete wave equation*, Result. Math., **79** (2024), 216. https://doi.org/10.1007/s00025-024-02235-y
[96] F. Bayart, É. Matheron, *Dynamics of Linear Operators*, Cambridge Tracts in Mathematics, vol. 179(1), Cambridge University Press, 2009.
[97] E. Bazhlekova, *Fractional evolution equations in Banach spaces*, PhD Thesis, Eindhoven University of Technology, Eindhoven, 2001.
[98] R. Bellman, K. L. Cooke, *Differential-difference equations*, USA Force Project Rand, R-374-PR, 1963.

[99] A. Bellour, E. Ait Dads, *Periodic solutions for nonlinear neutral delay integro-differential equations*, Electron. J. Differ. Equ., **100** (2015), 1–9.

[100] A. Bellow, V. Losert, *The weighted pointwise ergodic theorem and the individual ergodic theorem along subsequences*, Trans. Am. Math. Soc., **288** (1985), 307–345.

[101] P. R. Bender, *Some conditions for the existence of recurrent solutions to systems of ordinary differential equations*, PhD Thesis, Iowa State University, 1966. Retrospective Theses and Dissertations. 5304. https://lib.dr.iastate.edu/rtd/5304.

[102] M. Bensalah, M. Miraoui, M. Zorgui, *Pseudo asymptotically Bloch periodic functions: applications for some models with piecewise constant argument*, J. Elliptic Parabolic Equ., **10** (2024), 147–168.

[103] H. Bereketoglu, F. Karakoc, *Asymptotic constancy for a system of impulsive pantograph equations*, Acta Math. Hung., **145** (2015), 68–79.

[104] K. S. Berenhaut et al., *Periodic solutions of the rational difference equation*, J. Differ. Equ. Appl., **12** (2006), 183–189. https://doi.org/10.1080/10236190500539295

[105] L. Berezansky, E. Braverman, *On existence of positive solutions for linear difference equations with several delays*, Adv. Dyn. Syst. Appl., **1** (2006), 29–47.

[106] V. Bergelson et al., *Rationally almost periodic sequences, polynomial multiple recurrence and symbolic dynamics*, Ergod. Theory Dyn. Syst., **39** (2018), 2332–2383.

[107] A. Berger, S. Siegmund, Y. Yi, *On almost automorphic dynamics in symbolic lattices*, Ergod. Theory Dyn. Syst., **24** (2024), 677–696.

[108] J. Bergh, J. Löfström, *Interpolation Spaces*, Springer, Berlin, 1976.

[109] D. L. Bernstein, *The double Laplace integral*, PhD thesis, Brown University, 1939.

[110] D. L. Bernstein, *The double Laplace integral*, Duke Math. J., **8** (1941), 460–496.

[111] D. L. Bernstein, G. A. Coon, *Some properties of the double Laplace transformation*, Trans. Am. Math. Soc., **74** (1953), 135–176.

[112] A. S. Besicovitch, *Almost Periodic Functions*, Dover Publ., New York, 1954.

[113] P. Bezandry, *Existence of almost periodic solutions for semilinear stochastic evolution equations driven by fractional Brownian motion*, Electron. J. Differ. Equ., **2012**(156) (2012), 1–21.

[114] P. Bezandry, T. Diagana, *Existence of S^2-almost periodic solutions to a class of nonautonomous stochastic evolution equation*, Electron. J. Qual. Theory Differ. Equ., **35** (2008), 1–19.

[115] P. H. Bezandry, T. Diagana, S. Elaydi, *On the stochastic Beverton–Holt equation with survival rates*, J. Differ. Equ. Appl., **14** (2008), 175–190.

[116] M. S. Bichegkuev, *Conditions for solubility of difference inclusions*, Izv. Math., **72** (2008), 647–658.

[117] H. Bin, J. Yu, Z. Guo, *Nontrival periodic solutions for asymptotically linear resonant difference problem*, J. Math. Anal. Appl., **322** (2006), 477–488.

[118] G. D. Birkhoff, *Dynamical Systems*, revised ed., Amer. Math. Soc. Colloq. Publ., vol. IX, Amer. Math. Soc., Providence, RI, 1966.

[119] L. S. Block, W. A. Coppel, *Dynamics in One Dimension*, Springer, New York, 1992.

[120] J. Blot, D. Pennequin, *Existence and structure results on almost periodic solutions of difference equations*, J. Differ. Equ. Appl., **7** (2001), 383–402.

[121] S. Bochner, *Curvature and Betti numbers in real and complex vector bundles*, Rend. Semin. Mat. Univ. Politec. Torino, **15** (1955–1956).

[122] S. Bochner, *A new approach to almost periodicity*, Proc. Natl. Acad. Sci. USA, **48**(12) (1962), 2039–2043.

[123] M. Bohner, T. Cuchta, S. Streipert, *Delay dynamic equations on isolated time scales and the relevance of one-periodic coefficients*, Methods Appl. Anal., **45** (2022), 5821–5838.

[124] M. Bohner, V. F. Hatipoğlu, *Cobweb model with conformable fractional derivatives*, Math. Methods Appl. Sci., **41**(18) (2018), 9010–9017.

[125] M. Bohner, V. F. Hatipoğlu, *Dynamic cobweb models with conformable fractional derivatives*, Nonlinear Anal. Hybrid Syst., **32** (2019), 157–167.

[126] M. Bohner, J. M. Jonnalagadda, *Discrete fractional cobweb models*, Chaos Solitons Fractals, **162** (2022), 112451.

[127] M. Bohner, T. Kuchta, *The Bessel difference equation*, Proc. Am. Math. Soc., **145** (2017), 1567–1580.

[128] M. Bohner, T. Kuchta, *The generalized hypergeometric difference equation*, Demonstr. Math., **51** (2018), 62–75.

[129] M. Bohner, J. G. Mesquita, *Massera's theorem in quantum calculus*, Proc. Am. Math. Soc., **146** (2018), 4755–4766.

[130] M. Bohner, A. Peterson, *Laplace transform and Z-transform: unification and extension*, Methods Appl. Anal., **9** (2002), 151–157.

[131] H. Bohr, E. Fölner, *On some types of functional spaces: a contribution to the theory of almost periodic functions*, Acta Math., **76** (1944), 31–155.

[132] E. M. Bonotto, M. Federson, *Topological conjugation and asymptotic stability in impulsive semidynamical systems*, J. Math. Anal. Appl., **326** (2007), 869–881.

[133] E. M. Bonotto, L. P. Gimenes, G. M. Souto, *Asymptotically almost periodic motions in impulsive semidynamical systems*, Topol. Methods Nonlinear Anal., **49** (2017), 133–163.

[134] E. M. Bonotto, M. Z. Jimenez, *Weak almost periodic motions, minimality and stability in impulsive semidynamical systems*, J. Differ. Equ., **256** (2014), 1683–1701.

[135] E. M. Bonotto, M. Z. Jimenez, *On impulsive semidynamical systems: minimal, recurrent and almost periodic motions*, Topol. Methods Nonlinear Anal., **44** (2014), 121–141.

[136] A. Bostan, M. Bousquet-Mélou, S. Melczer, *Counting walks with large steps in an orthant*, J. Eur. Math. Soc., **23** (2021), 2221–2297.

[137] A. Bouakkaz, R. Khemis, *Positive periodic solutions for revisited Nicholson's blowflies equation with iterative harvesting term*, J. Math. Anal. Appl., **94**(2) (2021), 124663.

[138] M. Bousquet-Mélou, M. Petkovšek, *Linear recurrences with constant coefficients: the multivariate case*, Discrete Math., **225** (2000), 51–75.

[139] R. H. Boyer, *Discrete Bessel functions*, J. Math. Anal. Appl., **2** (1961), 509–524.

[140] C. Bouzar, F. Z. Tchouar, *Almost automorphic distributions*, Mediterr. J. Math., **14** (2017), 151. https://doi.org/10.1007/s00009-017-0953-3

[141] C. Bouzar, F. Z. Tchouar, *Asymptotic almost automorphy of functions and distributions*, Ural Math. J., **6** (2020), 54–70.

[142] R. N. Bracewell, *The Fourier Transform and its Applications*, 3rd ed., McGraw Hill, Boston, 2000.

[143] F. Brandt, M. Hieber, *Time periodic solutions to Hibler's sea ice model*, Nonlinearity, **36** (2023), 3109–3124.

[144] F. Brandt, M. Hieber, *Strong periodic solutions to quasilinear parabolic equations: an approach by the Da Prato--Grisvard theorem*, Bull. Lond. Math. Soc., **55** (2023), 1971–1993.

[145] T. Brikshavana, T. Sitthiwirattham, *On fractional Hahn calculus*, Adv. Differ. Equ., **2017** (2017), 354. https://doi.org/10.1186/s13662-017-1412-y

[146] G. Bruno, A. Pankov, *On convolution operators in the spaces of almost periodic functions and L^p spaces*, Z. Anal. Anwend., **19** (2000), 359–367.

[147] H. Brunner, *Theory and numerical analysis of Volterra functional equations*, Report. https://www-user.tu-chemnitz.de/~potts/cms/cms08/inhalt/brunnerscript.pdf.

[148] C. Bu, *Existence and uniqueness of almost periodic solution for a mathematical model of tumor growth*, J. Appl. Math. Phys., **10** (2022), 1013–1018.

[149] C. Budde, J. Kreulich, *On splittings and integration of almost periodic functions with and without geometry*, Semigroup Forum, **108** (2024), 335–364.

[150] D. Bugajewski, K. Kasprzak, A. Nawrocki, *Asymptotic properties and convolutions of some almost periodic functions with applications*, Ann. Mat. Pura Appl. (4), **202** (2023), 1033–1050. https://doi.org/10.1007/s10231-022-01270-2

[151] D. Bugajewski, A. Nawrocki, *Some remarks on almost periodic functions in view of the Lebesgue measure with applications to linear differential equations*, Ann. Acad. Sci. Fenn., Math., **42** (2017), 809–836.

[152] T. A. Burton, *Fixed points and differential equations with asymptotically constant or periodic solutions*, Electron. J. Qual. Theory Differ. Equ., **11** (2004), 1–31.

[153] T. A. Burton, T. Furumochi, *Asymptotic behavior of solutions of functional differential equations by fixed point theorems*, Dyn. Syst. Appl., **11** (2002), 499–519.
[154] H. Bustos, P. Figueroa, M. Pinto, *Poincaré–Perron problem for high order differential equations in the class of almost periodic type functions*, preprint, arXiv:2407.14444.
[155] A. Burqan et al., *ARA-residual power series method for solving partial fractional differential equations*, Alex. Eng. J., **62** (2023), 47–62.
[156] A. Cabada, N. Dimitrov, *Multiplicity results for nonlinear periodic fourth-order difference equations with parameter dependence and singularities*, J. Math. Anal. Appl., **371** (2010), 518–533.
[157] C. A. Cadavid, P. Hoyos, J. Jorgenson, L. Smajlović, J. D. Vélez, *Discrete diffusion-type equation on regular graphs and its applications*, J. Differ. Equ. Appl., **29** (2023), 455–488.
[158] X. Cai, J. Yu, Z. Guo, *Existence of periodic solutions for fourth-order difference equations*, Comput. Math. Appl., **50** (2005), 49–55.
[159] E. Camouzis, G. Ladas, *Dynamics of Third-Order Rational Difference Equations with Open Problems and Conjectures*, Advances in Discrete Mathematics and Applications, vol. 5, CRC Press, Boca Raton, 2008.
[160] E. Camouzis, G. Ladas, *When does periodicity destroy boundedness in rational equations?* J. Differ. Equ. Appl., **12** (2006), 961–979.
[161] S. L. Campbell, *Optimal control of discrete linear processes with quadratic cost*, Int. J. Syst. Sci., **9** (1978), 841–847.
[162] T. Candan, *Existence results for positive periodic solutions to first order neutral differential equations with distributed deviating arguments*, Hacet. J. Math. Stat., **53** (2024), 1326–1332.
[163] J. S. Cánovas, A. Linero Bas, G. Soler López, *On global periodicity of difference equations*, Taiwan. J. Math., **13** (2009), 1963–1983.
[164] J. Cao, B. Samet, Y. Zhou, *Asymptotically almost periodic mild solutions to a class of Weyl-like fractional difference equations*, Adv. Differ. Equ., **2019** (2019), 371. https://doi.org/10.1186/s13662-019-2316-9
[165] S. Castillo, M. Pinto, *Dichotomy and almost automorphic solution of difference system*, Electron. J. Qual. Theory Differ. Equ., **32** (2013), 1–17.
[166] J. Čermak, M. Urbánek, *On the asymptotics of solutions of delay dynamic equations on time scales*, Math. Comput. Model., **46** (2007), 445–458.
[167] A. V. Chaikovs'kyi, O. A. Lagoda, *Bounded solutions of difference equations in a Banach Space with asymptotically constant operator coefficient*, J. Math. Sci. (N.Y.), **272** (2023), 307–315, 2023.
[168] M. Cheng, Z. Liu, *The second Bogolyubov theorem and global averaging principle for SPDEs with monotone coefficients*, SIAM J. Math. Anal., **55** (2023), 1100–1144.
[169] F. Chérif, *Existence and global exponential stability of pseudo almost periodic solution for SICNNs with mixed delays*, J. Appl. Math. Comput., **39** (2012), 235–251.
[170] R. C. Singh Chandel, P. K. Vishwakarma, *Multidimensional fractional derivatives of the multiple hypergeometric functions of several variables*, Jñānābha, **24** (1994), 19–27.
[171] S. Chandragiri, *Difference equations and generating functions for some lattice path problems*, J. Sib. Fed. Univ. Math. Phys., **12**(5) (2019), 551–559.
[172] S. Chandragiri, *Counting lattice paths by using difference equations with non-constant coefficients*, Bull. Irkutsk State Univ. Ser. Math., **44** (2023), 55–70.
[173] Y.-K. Chang, P. Lü, *Weighted pseudo asymptotically antiperiodic sequential solutions to semilinear difference equations*, J. Differ. Equ. Appl., **27** (2021), 1482–1506.
[174] Y.-K. Chang, Y. Wei, *Pseudo S-asymptotically Bloch type periodic solutions to fractional integro-differential equations with Stepanov-like force terms*, Z. Angew. Math. Phys., **73** (2022), 17. https://doi.org/10.1007/s00033-022-01722-y
[175] B. Chaouchi, M. Kostić, *On the solvability of a fourth-order differential evolution equation on singular cylindrical domain in $\mathbb{R}^4$*, Math. Slovaca, **72** (2022), 911–924.
[176] B. Chaouchi, M. Kostić, H. C. Koyuncuoğlu, *Metrical Stepanov almost automorphy and applications*, Bull. Iranian Math. Soc., **50** (2024), 6. https://doi.org/10.1007/s41980-023-00840-1

[177] B. CHAOUCHI, M. KOSTIĆ, D. VELINOV, *Metrical almost periodicity, metrical approximations of functions and applications*, Turk. J. Math., **47** (2023), 769–793.

[178] S. CHAPMAN, *On non-integral orders of summability of series and integrals*, Proc. Lond. Math. Soc., **9** (1911), 369–409.

[179] A. CHAPOUTO, R. KILLIP, M. VISAN, *Bounded solutions of KdV: uniqueness and the loss of almost periodicity*, preprint, arXiv:2209.07501.

[180] G. E. CHATZARAKIS, L. HORVAT-DMITROVIĆ, M. PAŠIĆ, *Oscillation tests for difference equations with several non-monotone deviating arguments*, Math. Slovaca, **68** (2018), 1083–1096.

[181] G. E. CHATZARAKIS, J. MANOJLOVIĆ, S. PINELAS, I. P. STAVROULAKIS, *Oscillation criteria of difference equations with several deviating arguments*, Math. Bohem., in press. https://www.researchgate.net/publication/261709703

[182] V. B. L. CHAURASIA, R. S. DUBEY, *The n-dimensional generalized Weyl fractional calculus containing to n-dimensional H-transforms*, Gen. Math. Notes, **6** (2011), 61–72.

[183] A. CHÁVEZ, L. Q. HUATANGARI, *Vector valued piecewise continuous almost automorphic functions and some consequences*, J. Math. Anal. Appl., in press. https://doi.org/10.1016/j.jmaa.2024.128768

[184] A. CHÁVEZ, K. KHALIL, M. KOSTIĆ, M. PINTO, *Multi-dimensional almost automorphic type functions and applications*, Bull. Braz. Math. Soc. N.S., **53** (2022), 801–851. https://doi.org/10.1007/s00574-022-00284-x

[185] A. CHÁVEZ, K. KHALIL, M. KOSTIĆ, M. PINTO, *Stepanov multi-dimensional almost periodic functions and applications*, Filomat, **37** (2023), 3681–3713.

[186] D. N. CHEBAN, *Bohr/Levitan almost periodic and almost automorphic solutions of monotone difference equations with a strict monotone first integral*, J. Differ. Equ. Appl., **28** (2008), 510–546.

[187] D. N. CHEBAN, C. MAMMANA, *Invariant manifolds, global attractors and almost periodic solutions of nonautonomous difference equations*, Nonlinear Anal., **56** (2004), 465–484.

[188] D. N. CHEBAN, *Massera's theorem for asymptotically periodic scalar differential equations*, preprint, 2023. arXiv:2311.03005.

[189] D. N. CHEBAN, *Remotely almost periodic motions of dynamical systems*, hal-04698591, 2024.

[190] D. N. CHEBAN, Bohr-Levitan almost periodic and almost automorphic solutions of equation $x'(t) = f(t-1, x(t-1)) - f(t, x(t))$, in: U. Kähler, M. Reissig, I. Sabadini, J. Vindas (eds.), *Analysis, Applications, and Computations (ISAAC 2021)*, pp. 73–88, Trends in Mathematics, Birkhäuser, Cham, 2021. https://doi.org/10.1007/978-3-031-36375-7_3

[191] N. CHEGLOUFA, B. CHAOUCHI, F. BOUTAOUS, M. KOSTIĆ, *$\mathcal{S}$-Asymptotically Bloch type periodic solutions for abstract neutral fractional evolution equations with ψ-Hilfer derivatives*, J. Appl. Nonlinear Dyn., in press. https://www.researchgate.net/publication/372166107

[192] N. CHEGLOUFA, B. CHAOUCHI, M. KOSTIĆ, W.-S. DU, *On the study of pseudo-$\mathcal{S}$-asymptotically periodic mild solutions for a class of neutral fractional delayed evolution equations*, Axioms, **12** (2023), 800. https://doi.org/10.3390/axioms12080800

[193] C. CHEN, *Discrete Caputo delta fractional economic cobweb models*, Qual. Theory Dyn. Syst., **22** (2023), 8. https://doi.org/10.1007/s12346-022-00708-5

[194] C. CHEN, M. BOHNER, B. JIA, *Caputo fractional continuous cobweb models*, J. Comput. Appl. Math., **374** (2020), 112734.

[195] X. CHEN, F. LIN, *Almost periodic solutions of neutral functional differential equations*, Nonlinear Anal., **11**(2) (2010), 1182–1189.

[196] Y. CHEN, J. J. NIETO, D. O'REGAN, *Anti-periodic solutions for evolution equations associated with maximal monotone mappings*, Appl. Math. Lett., **24** (2011), 302–307.

[197] S. S. CHENG, *Partial Difference Equations*, CRC Press, London–New York, 2003.

[198] Y. CHENG, F. CONG, H. HUA, *Anti-periodic solutions for nonlinear evolution equations*, Adv. Differ. Equ., **1** (2012), 1–15.

[199] Z. CHENG, X. CUI, *Positive periodic solution to an indefinite singular equation*, Appl. Math. Lett., **112** (2021), 106740. https://doi.org/10.1016/j.aml.2020.106740

[200] K.-S. CHIU, *Numerical-analytic successive approximation method for the investigation of periodic solutions of nonlinear integro-differential systems with piecewise constant argument of generalized type*, Hacet. J. Math. Stat., **53** (2024), 1272–1290.

[201] M.-J. CHOI, *A condition for blow-up solutions to discrete semilinear wave equations on networks*, Appl. Anal., **101** (2022), 2008–2018.

[202] M. F. CHOQUEHUANCA, J. G. MESQUITA, A. PEREIRA, *Almost automorphic solutions of second-order equations involving time scales with boundary conditions*, Proc. Am. Math. Soc., **151** (2023), 1055–1070.

[203] J. CHU, Z. ZHANG, *Periodic solutions of singular differential equations with sign-changing potential*, Bull. Aust. Math. Soc., **82** (2010), 437–445.

[204] K. M. CHUDINOV, *Sharp explicit oscillation conditions for difference equations with several delays*, Georgian Math. J., **28** (2021), 207–218.

[205] P. CIEUTAT, *On Bochner's almost periodicity criterion*, Evol. Equ. Control Theory, **12** (2023), 1233–1246.

[206] A. CIMA, A. GASULL, F. MANOSAS, *On periodic rational difference equations of order k*, J. Differ. Equ. Appl., **10** (2004), 549–559. https://doi.org/10.1080/10236190410001667977

[207] I. CIORANESCU, *On the abstract Cauchy problem in spaces of almost periodic distributions*, J. Math. Anal. Appl., **148** (1990), 440–462.

[208] I. CIORANESCU, *The characterization of the almost periodic ultradistributions of Beurling type*, Proc. Am. Math. Soc., **116** (1992), 127–134.

[209] K. CONRAD, L^p-Spaces for $0 < p < 1$, https://kconrad.math.uconn.edu/blurbs/analysis/lpspace.pdf.

[210] K. L. COOKE, K. R. MEYER, *The condition of regular degeneration for singularly perturbed systems of linear differential-difference equations*, J. Math. Anal. Appl., **14** (1996), 83–106.

[211] C. CORDUNEANU, *Neutral functional equations of Volterra type*, Funct. Differ. Equ., **4** (1997), 265–270.

[212] C. CORDUNEANU, *Almost periodic discrete processes*, Libertas Math. (N.S.), **2** (1982), 159–169.

[213] C. CORDUNEANU, *Almost periodic solutions to nonlinear elliptic and parabolic equations*, Nonlinear Anal., **7** (1983), 357–363.

[214] M. G. CRANDALL, S.-O. LONDEN, J. A. NOHEL, *An abstract nonlinear Volterra integrodifferential equation*, J. Math. Anal. Appl., **64** (1978), 701–735.

[215] T. CUCHTA, *Discrete analogues of some classical special functions*, PhD thesis, Missouri University of Science and Technology, 2015.

[216] T. CUCHTA, *The heat equation on time scales*, Opusc. Math., **43** (2023), 47–491.

[217] T. CUCHTA, D. GROW, N. WINTZ, *Discrete matrix hypergeometric functions*, J. Math. Anal. Appl., **518**(2) (2023), 126716. https://doi.org/10.1016/j.jmaa.2022.126716

[218] T. CUCHTA, S. STREIPERT, *A discrete SIS model of fractional order*, Int. J. Dyn. Syst. Differ. Equ., **11** (2021), 275–286.

[219] P. CULL, M. FLAHIVE, R. ROBSON, *Difference Equations: From Rabbits to Chaos*, Springer, New York, 2005.

[220] J. M. CUSHING, *Periodic Kolmogorov systems*, SIAM J. Math. Anal., **13**(5) (1982), 811–827.

[221] J. M. CUSHING, *Periodic cycles of nonlinear discrete renewal equations*, J. Differ. Equ. Appl., **2** (1996), 117–137.

[222] J. J. DACUNHA, J. M. DAVIS, *A unified Floquet theory for discrete, continuous, and hybrid periodic linear systems*, J. Differ. Equ., **251** (2011), 2987–3027.

[223] J. V. DA C. SOUSA, E. C. DE OLIVEIRA, *Leibniz type rule: ψ-Hilfer fractional operator*, Commun. Nonlinear Sci. Numer. Simul., **77** (2019), 305–311.

[224] D. DAMANIK, M. LUKIĆ, A. VOBERG, P. YUDITSKII, *The Deift conjecture: a program to construct a counterexample*, preprint, arXiv:2111.09345.

[225] S. DANIELIAN, *Ecological synchrony and metapopulation persistence*, PhD Thesis, University of California, Riverside, 2022.

[226] L. I. DANILOV, *Measure-valued almost periodic functions and almost periodic selections of multivalued maps*, Sb. Math., **188** (1997), 3–24.

[227] L. I. DANILOV, *On a class of Besicovitch almost periodic type selections of multivalued maps*, Izv. IMI UdGU, **61** (2023), 57–75. https://doi.org/10.35634/2226-3594-2023-61-04

[228] F. M. Dannan, *Rational difference equations with parameter, boundedness and periodic solutions*, J. Differ. Equ. Appl. (2021). https://doi.org/10.1080/10236198.2021.1967945

[229] F. Dannan, S. Elyadi, P. Liu, *Periodic solutions of difference equations*, J. Differ. Equ. Appl., **6** (2000), 203–232.

[230] L. Debnath, *The double Laplace transforms and their properties with applications to functional, integral and partial differential equations*, Int. J. Appl. Comput. Math., **2**(2) (2016), 223–241.

[231] J. Debnath, R. S. Dahiya, *Theorems on multidimensional Laplace transform for solution of boundary value problems*, Comput. Math. Appl., **18** (1989), 1033–1056.

[232] P. M. de Carvalho-Neto, R. F. Júnior, *On the fractional version of Leibniz rule*, Math. Nachr., **293** (2020), 670–700.

[233] R. deLaubenfels, *Existence Families, Functional Calculi and Evolution Equations*, Lecture Notes in Mathematics, vol. 1570, Springer, New York, 1994.

[234] A. Demir, M. A. Bayrak, *A new approach for the solution of space-time fractional order heat-like partial differential equations by residual power series method*, Commun. Math. Appl., **10** (2019), 585–597.

[235] E. C. de Oliveira, S. Jarosz, J. Vaz Jr., *Fractional calculus via Laplace transform and its application in relaxation processes*, Commun. Nonlinear Sci. Numer. Simul., **69** (2019), 58–72.

[236] E. C. de Oliveira, J. A. T. Machado, *A review of definitions for fractional derivatives and integral*, Math. Probl. Eng., **2014** (2014), 238459, 6 pp. https://doi.org/10.1155/2014/238459

[237] J. de Vries, *Elements of Topological Dynamics*, Mathematics and Its Applications, vol. 257, Springer, 1993.

[238] M. Dehghan, N. Rastegar, *Stability and periodic character of a third order difference equation*, Math. Comput. Model., **54** (2011), 2560–2564.

[239] P. Deift, Some open problems in random matrix theory and the theory of integrable systems, in: *Integrable Systems and Random Matrices*, Contemp. Math., vol. 458, pp. 419–430, Amer. Math. Soc., Providence, RI, 2008.

[240] P. Deift, *Some open problems in random matrix theory and the theory of integrable systems. II*, SIGMA, **13** (2017), 016.

[241] L. Del Campo, M. Pinto, C. Vidal, *Bounded and periodic solutions in retarded difference equations using summable dichotomies*, Dyn. Syst. Appl., **21** (2012), 1–16.

[242] W. Desch, J. Prüss, *Counterexamples for linear abstract Volterra equations*, J. Integral Equ. Appl., **5** (1993), 29–45.

[243] C. Derbazi, Z. Baitiche, M. Benchohra, A. Cabada, *Initial value problem for nonlinear fractional differential equations with Ψ-Caputo derivative via monotone iterative technique*, Axioms, **9**(2) (2020), 57, 13 pp. https://doi.org/10.3390/axioms9020057

[244] M. M. Deza, M. Laurent, *Geometry of Cuts and Metrics*, Algorithms and Combinatorics, vol. 15, p. 27, Springer, Berlin, 1997. https://doi.org/10.1007/978-3-642-04295-9

[245] S. Dhama, S. Abbas, R. Sakthivel, *Stability and approximation of almost automorphic solutions on time scales for the stochastic Nicholson's blowflies model*, J. Integral Equ. Appl., **33** (2021), 31–51.

[246] T. Diagana, *Almost Automorphic Type and Almost Periodic Type Functions in Abstract Spaces*, Springer, New York, 2013.

[247] T. Diagana, S. Elaydi, A-A. Yakubu, *Population models in almost periodic environments*, J. Differ. Equ. Appl., **13** (2007), 239–260.

[248] T. Diagana, E. Hernández, M. Rabello, *Pseudo almost periodic solutions to some non-autonomous neutral functional differential equations with unbounded delay*, Math. Comput. Model., **45**(9–10) (2007), 1241–1252.

[249] T. Diagana, M. Kostić, Generalized almost automorphic and generalized asymptotically almost automorphic type functions in Lebesgue spaces with variable exponents $L^{p(x)}$, in: H. Forster (ed.), *Recent Studies in Differential Equations*, pp. 1–28, Nova Science Publishers, New York, 2020, Chapter 1.

[250] T. Diagana, M. M. Mbaye, *Square-mean almost periodic solutions to some singular stochastic differential equations*, Appl. Math. Lett., **54** (2016), 48–53.

[251] T. Diagana, D. Pennequin, *Almost periodic solutions for some semilinear singular difference equations*, J. Differ. Equ. Appl., **24**(1) (2017), 138–147. https://doi.org/10.1080/10236198.2017.1397142

[252] L. Díaz, R. Naulin, *A set of almost periodic discontinuous functions*, Pro Math., **20** (2006), 107–118.

[253] L. Díaz, T. J. Osler, *Differences of fractional order*, Math. Comput., **28** (1974), 185–202.

[254] J. Diblík, M. Fečkan, M. Pospíšil, *Nonexistence of periodic solutions and S-asymptotically periodic solutions in fractional difference equations*, Appl. Math. Comput., **257** (2015), 230–240.

[255] E. Diedrich, *The Fourier continuous derivative: A new approach to fractional differentiation*, preprint, 2023. https://doi.org/10.20944/preprints202310.0913.v4

[256] L. Diening, P. Harjulehto, P. Hästüso, M. Ruzicka, *Lebesgue and Sobolev Spaces with Variable Exponents*, Lecture Notes in Mathematics, vol. 2011, Springer, Heidelberg, 2011.

[257] K. Diethelm, *The Analysis of Fractional Differential Equations. An Application-Oriented Exposition Using Differential Operators of Caputo Type*, Springer, Berlin, 2010.

[258] B. X. Dieu, S. Siegmund, N. V. Minh, *A Katznelson--Tzafriri type theorem for almost periodic linear evolution equations*, Vietnam J. Math., **43** (2015), 403–415.

[259] D. S. Dilip, S. C. Babu, *Rectangular p-periodicity and p_q-periodicity of a rational difference system*, Ex. Counterex., **1** (2021), 100004.

[260] W. Dimbour, V. Valmorin, $\mathbb{S}$-Almost automorphic functions and applications, 2020. hal-03014691.

[261] H.-S. Ding, J.-D. Fu, G. M. N'Guérékata, *Positive almost periodic type solutions to a class of nonlinear difference equations*, Electron. J. Qual. Theory Differ. Equ., **25** (2011), 1–16.

[262] H.-S. Ding, W.-G. Jian, N. V. Minh, G. M. N'Guérékata, *Kadets type and Loomis type theorems for asymptotically almost periodic functions*, J. Differ. Equ., **373** (2023), 389–410.

[263] R. Doss, *On Riemann integrability and almost periodic functions*, Compos. Math., **12** (1954–1956), 271–283.

[264] T. Downarowicz, A. Iwanik, *Quasi-uniform convergence in compact dynamical systems*, Stud. Math., **89** (1988), 11–25.

[265] E. Drymonis, Y. Kostrov, Z. Kudlak, *On rational difference equations with nonnegative periodic coefficients*, Int. J. Differ. Equ., **1** (2012), 19–34.

[266] W.-S. Du, M. Kostić, D. Velinov, *Abstract impulsive Volterra integro-differential inclusions*, Fractal Fract., **7** (2023), 73. https://doi.org/10.3390/fractalfract7010073

[267] W.-S. Du, M. Kostić, D. Velinov, *Almost periodic type solutions of abstract impulsive Volterra integro-differential inclusions*, Fractal Fract., **7** (2023), 147. https://doi.org/10.3390/fractalfract7020147

[268] A. V. Dvornyk, V. I. Tkachenko, *Almost periodic solutions for systems with delay and nonfixed times of impulsive actions*, Ukr. Math. J., **68** (2017), 1673–1693.

[269] A. V. Dvornyk, O. O. Struk, V. I. Tkachenko, *Almost periodic solutions of Lotka–Volterra systems with diffusion and pulsed action*, Ukr. Math. J., **70** (2018), 197–216.

[270] A. I. Dvirnyj, V. I. Slyn'ko, *Stability in terms of two measures for a class of semilinear impulsive parabolic equations*, Sb. Math., **204** (2013), 485–507.

[271] T. Eisner, B. Farkas, M. Haase, R. Nagel, *Operator Theoretic Aspects of Ergodic Theory*, Graduate Text in Mathematics, vol. 272, Springer, Berlin, 2015.

[272] S. Elaydi, *An Introduction to Difference Equations*, 3rd ed., Undergraduate Texts in Mathematics, Springer, New York, NY, USA, 2005.

[273] S. Elaydi, *Stability and asymptoticity of Volterra difference equations: a progress report*, J. Comput. Appl. Math., **228** (2009), 504–513.

[274] L. E. El'sgol'ts, S. B. Norkin, *Introduction to the Theory of Differential Equations with Deviating Argument*, Nauka, Moscow, 1971.

[275] H. Eltayeb, A. Kilicman, I. Bachar, *On the application of multi-dimensional Laplace decomposition method for solving singular fractional pseudo-hyperbolic equations*, Fractal Fract., **6**(11) (2022), 690. https://doi.org/10.3390/fractalfract6110690

[276] H. Eltayeb, A. Kilicman, S. Mesloub, *Exact evaluation of infinite series using double Laplace transform technique*, Abstr. Appl. Anal. (2014). https://doi.org/10.1155/2014/327429

[277] K. J. Engel, *Operator Matrices and Systems of Evolution Equations*, Book Manuscript, 1996, https://www.researchgate.net/publication/266357142.

[278] L. Erbe, Q. Kong, B. G. Zhang, *Oscillation Theory for Functional Differential Equations*, Routledge, New York, 1994.

[279] B. Es-sebbar, K. Ezzinbi, F. Samir, M. T. Ziat, *Compact almost automorphic weak solutions for some monotone differential inclusions: applications to parabolic and hyperbolic equations*, J. Math. Anal. Appl., **486**(1) (2019), 123805. https://doi.org/10.1016/j.jmaa.2019.123805

[280] B. Es-sebbar, K. Ezzinbi, F. Samir, M. T. Ziat, *Almost periodicity and almost automorphy for some evolution equations using Favard's theory in uniformly convex Banach spaces*, Semigroup Forum, **94** (2017), 229–259. https://doi.org/10.1007/s00233-016-9810-0

[281] S. Etemad et al., *Quantum Laplace transforms for the Ulam–Hyers stability of certain q-difference equations of the Caputo-like type*, Fractal Fract., **8** (2024), 443. https://doi.org/10.3390/fractalfract8080443

[282] M. Ezekiel, *The Cobweb theory*, Q. J. Econ., **52** (1938), 255–280.

[283] K. Ezzinbi, M. A. Hachimi, *Existence of positive almost periodic solutions of functional equations via Hilbert's projective metric*, Topol. Methods Nonlinear Anal., **26** (1996), 1169–1176.

[284] M. Faheem, M. R. M. Rao, *Functional differential equations of delay type and nonlinear evolution in L_p-spaces*, J. Math. Anal. Appl., **123** (1987), 73–103.

[285] I. A. Falahah et al., *Synchronization of fractional partial difference equations via linear methods*, Axioms, **12** (2023), 728. https://doi.org/10.3390/axioms12080728

[286] L. Fang, N. N'gbo, Y. Xia, *Almost periodic solutions of a discrete Lotka–Volterra model via exponential dichotomy theory*, AIMS Math., **7**(3) (2022), 3788–3801. https://doi.org/10.3934/math.2022210

[287] A. Favini, A. Yagi, *Degenerate Differential Equations in Banach Spaces*, Pure and Applied Mathematics, Chapman and Hall/CRC, New York, 1998.

[288] M. Fečkan et al., *Caputo delta weakly fractional difference equations*, Fract. Calc. Appl. Anal., **25** (2022), 2222–2240.

[289] M. Fečkan, M. T. Khalladi, M. Kostić, A. Rahmani, *Multi-dimensional p-almost periodic type functions and applications*, Appl. Anal., **104** (2025), 142–168.

[290] V. E. Fedorov, Y. Apakov, A. Skorynin, *Analytic resolving families of operators for linear equations with Hilfer derivative*, J. Math. Sci. (N.Y.), **277** (2023), 385–402.

[291] V. E. Fedorov, A. D. Godova, *Integro-differential equations of Gerasimov type with sectorial operators*, Proc. Steklov Inst. Math., **325**(Suppl. 1) (2024), S99–S113.

[292] V. E. Fedorov, M. Kostić, *(F, G, C)-resolvent operator families and applications*, Mathematics, **11** (2023), 3505. https://doi.org/10.3390/math11163505

[293] V. E. Fedorov, M. Kostić, D. Velinov, *Abstract fractional differential inclusions with Hilfer derivatives*, J. Math. Sci. (N.Y.) (2024). https://doi.org/10.1007/s10958-024-07402-8

[294] V. E. Fedorov, M. V. Plekhanova, E. M. Izhberdeeva, *Initial value problems of linear equations with the Dzhrbashyan–Nersesyan derivative in Banach spaces*, Symmetry, **13** (2021), 1058. https://doi.org/10.3390/sym13061058

[295] V. E. Fedorov, M. V. Plekhanova, E. M. Izhberdeeva, *Analytic resolving families for equations with the Dzhrbashyan–Nersesyan fractional derivative*, Fractal Fract., **6** (2022), 541. https://doi.org/10.3390/fractalfract6100541

[296] V. E. Fedorov, M. V. Plekhanova, D. V. Melekhina, *Nonlinear inverse problems for equations with Dzhrbashyan–Nersesyan derivatives*, Fractal Fract., **7** (2023), 464. https://doi.org/10.3390/fractalfract7060464

[297] V. E. Fedorov, N. M. Skripka, *Evolution equations with Liouville derivative on $\mathbb{R}$ without initial conditions*, Mathematics, **12** (2024), 572. https://doi.org/10.3390/math12040572

[298] Z. Feng, Y. Wang, X. Ma, *Asymptotically almost periodic solutions for certain differential equations with piecewise constant arguments*, Adv. Differ. Equ., **2020** (2020), 242. https://doi.org/10.1186/s13662-020-02699-6

[299] B. Ferguson, G. Lim, *Discrete Time Dynamic Economic Models*, Taylor and Francis Group, London, 2003.

[300] F. Ferrari, *Weyl and Marchaud derivatives: a forgotten history*, Mathematics, **6** (2018), 6. https://doi.org/10.3390/math6010006

[301] R. A. C. Ferreira, *Discrete Fractional Calculus and Fractional Difference Equations*, Springer Briefs in Mathematics, Springer Cham, London–Berlin–New York, 2022.

[302] R. A. C. Ferreira, *Discrete fractional calculus and the Saalschutz theorem*, Bull. Math. Sci., **174**, 103086 (2022).

[303] P. Figueroa, M. Pinto, *Poincaré problem in the class of almost periodic type functions*, Bull. Belg. Math. Soc. Simon Stevin, **22** (2015), 177–198.

[304] A. M. Fink, *Almost Periodic Differential Equations*, Springer, Berlin, 1974.

[305] A. M. Fink, *Extensions of almost automorphic sequences*, J. Math. Anal. Appl., **27**(3) (1969), 519–523.

[306] A. M. Fink, J. A. Gatica, *Positive almost periodic solutions of some delay integral equations*, J. Differ. Equ., **83** (1990), 166–178.

[307] M. Folly-Gbetoula, *Dynamics and solutions of higher-order difference equations*, Mathematics, **11** (2023), 3693. https://doi.org/10.3390/math11173693

[308] T. Fort, *Linear difference equations and the Dirichlet series transform*, Am. Math. Mon., **62** (1955), 241.

[309] M. Friesl, A. Slavík, P. Stehlík, *Discrete-space partial dynamic equations on time scales and applications to stochastic processes*, Appl. Math. Lett., **37** (2014), 86–90.

[310] J. Fugére, S. Gaboury, R. Tremblay, *Leibniz rules and integral analogues for fractional derivatives via a new transformation formula*, Bull. Math. Anal. Appl., **4** (2012), 72–82.

[311] G. Fuhrmann, D. Kwietniak, *On tameness of almost automorphic dynamical systems for general groups*, Bull. Lond. Math. Soc., **52** (2020), 24–42.

[312] K. M. Furati, M. D. Kassim, N.-E. Tatar, *Existence and uniqueness for a problem involving Hilfer fractional derivative*, Comput. Math. Appl., **64** (2012), 1616–1626.

[313] T. Furumochi, *Periodic solutions of Volterra difference equations and attractivity*, Nonlinear Anal., **47** (2001), 4013–4024.

[314] T. Furumochi, S. Murakami, Y. Nagabuchi, *Volterra difference equations on a Banach space and abstract differential equations with piecewise continuous delays*, Jpn. J. Math., **30** (2004), 387–412.

[315] G. Gandolfo, *Economic Dynamics: Methods and Models*, 2nd ed., Advanced Textbooks in Economics, North-Holland Publishing Co., Amsterdam–New York, 1980.

[316] L. Gao, X. Sun, *Almost periodic solutions to impulsive stochastic delay differential equations driven by fractional Brownian motion with* $1/2 < H < 1$, Front. Phys., **9** (2021), 783125. https://doi.org/10.3389/fphy.2021.783125 Sec. Interdisciplinary Physics.

[317] F. Garciá-Ramos, T. Jäger, X. Ye, *Mean equicontinuity, almost automorphy and regularity*, Isr. J. Math., **243** (2021), 155–183.

[318] F. Garciá-Ramos, B. Marcus, *Mean sensitive, mean equicontinuous and almost periodic functions for dynamical systems*, Discrete Contin. Dyn. Syst., **39** (2019), 729–746.

[319] S. L. Gefter, A. L. Piven, *Implicit linear nonhomogeneous difference equation in Banach and locally convex spaces*, Zh. Mat. Fiz. Anal. Geom., **15** (2019), 336–353.

[320] M. I. Gil, *Difference Equations in Normed Spaces: Stability and Oscillation*, North Holand Math. Studies, vol. 206, North Holand, Amsterdam, 2007.

[321] A. Giusti et al., *A practical guide to Prabhakar fractional calculus*, Fract. Calc. Appl. Anal., **23** (2020), 9–54. https://doi.org/10.1515/fca-2020-0002

[322] A. Giusti et al., *On variable-order fractional linear viscoelasticity*, Fract. Calc. Appl. Anal., **27** (2024), 1564–1578. https://doi.org/10.1007/s13540-024-00288-y

[323] E. Glasner, *Enveloping semigroups in topological dynamics*, Topol. Appl., **154** (2007), 2344–2363.

[324] M. C. Gómez-Collado, *Almost periodic ultradistributions of Beurling and of Roumieu type*, Proc. Am. Math. Soc., **129** (2000), 2319–2329.

[325] C. González, A. Melado-Jiménez, *An application of Krasnoselskii fixed point theorem to the asymptotic behavior of solutions of difference equations in Banach spaces*, J. Math. Anal. Appl., **247** (2000), 290–299.

[326] C. González, A. Melado-Jiménez, *Asymptotic behavior of solutions to difference equations in Banach spaces*, Proc. Am. Math. Soc., **128** (2000), 1743–1749.

[327] J. González-Camus, *Well-posedness for fractional Cauchy problems involving discrete convolution operators*, Mediterr. J. Math., **20** (2023), 243. https://doi.org/10.1007/s00009-023-02443-w

[328] J. González-Camus, V. Keyantuo, C. Lizama, M. Warma, *Fundamental solutions for discrete dynamical systems involving the fractional Laplacian*, Math. Methods Appl. Sci., **42** (2019), 4688–4711.

[329] C. Goodrich, C. Lizama, *An unexpected property of fractional difference operators: Finite and eventual monotonicity*, Math. Methods Appl. Sci., **47** (2024), 5484–5508.

[330] C. Goodrich, A. C. Peterson, *Discrete Fractional Calculus*, Springer, Heidelberg, 2015.

[331] N. S. Gopal, J. M. Jonnalagadda, J. Alzabut, *Data dependence and existence and uniqueness for Hilfer nabla fractional difference equations*, Contemp. Math., **5** (2024), 780–796.

[332] K. Gopalsamy, *Stability and Oscillations in Delay Differential Equations of Population Dynamics*, Springer, Berlin, 1992.

[333] K. Gopalsamy, S. Mohamad, *Canonical solutions and almost periodicity in a discrete logistic equation*, Appl. Math. Comput., **113** (2000), 305–323. https://doi.org/10.1016/S0096-3003(99)00093-4

[334] K. Gopalsamy, S. Sariyasa, *Time delays and stimulus-dependent pattern formation in periodic environments in isolated neurons*, IEEE Trans. Neural Netw., **13** (2002), 551–563. https://doi.org/10.1109/TNN.2002.1000124

[335] W. H. Gottschalk, G. A. Hedlund, *Topological Dynamics*, American Mathematical Society Colloquium Publications, vol. 36, AMS, Providence, 1955.

[336] H. Gou, *On the S-asymptotically ω-periodic mild solutions for multi-term time fractional measure differential equations*, Topol. Methods Nonlinear Anal., **62** (2023), 569–590.

[337] H. Gou, B. Li, *Study on the mild solution of Sobolev type Hilfer fractional evolution equations with boundary conditions*, Chaos Solitons Fractals, **112** (2018), 168–179.

[338] H. W. Gould, *Combinatorial Identities: A Standardized Set of Tables Listing 500 Binomial Coefficient Summations*, Morgantown, recised edition, W. Va., 1972.

[339] S. P. Goyal, T. Mathur, *A theorem relating multidimensional generalized Weyl fractional integral, Laplace and Varma transforms with applications*, Tamsui Oxford Univ. J. Math. Sci., **19**(1) (2003), 41–54.

[340] R. Grau, A. Pereira, *Representations of abstract resolvent families on time scales via Laplace transform*, Fract. Calc. Appl. Anal., **27** (2023), 218–246.

[341] H. L. Gray, N. Fan Zhang, *On a new definition of the fractional difference*, Math. Compet., **50** (1988), 513–529.

[342] G. Gripenberg, *On the convergence of solutions of Volterra equations to almost-periodic functions*, Q. Appl. Math., **39** (1981), 363–373.

[343] G. Gripenberg, *Volterra integro-differential equations with accretive nonlinearity*, J. Differ. Equ., **60** (1985), 57–79.

[344] K.-G. Grosse-Erdmann, A. Peris, *Linear Chaos*, Springer, London, 2011.

[345] A. C. Grove, *An Introduction to the Laplace Transform and the z Transform*, Prentice Hall, New York, London, 1991.

[346] E. A. Grove, G. Ladas, *Periodicities in Nonlinear Difference Equations*, Chapman and Hall, CRC Press, Boca Raton, FL, 2005.

[347] H. Gu, J. J. Trujillo, *Existence of mild solution for evolution equation with Hilfer fractional derivative*, Appl. Math. Comput., **257** (2015), 344–354.

[348] G. M. N'Guérékata, *Almost Automorphic and Almost Periodic Functions in Abstract Spaces*, Kluwer Acad. Publ, Dordrecht, 2001.

[349] Z. Guo, J. Yu, *Existence of periodic and subharmonic solutions to subquadratic second-order equations*, J. Lond. Math. Soc., **68** (2003), 419–430.

[350] Z. Guo, J. Yu, *Periodic and subharmonic solutions for superquadratic discrete Hamiltonian systems*, Nonlinear Anal., **55** (2003), 969–983.

[351] Z. Guo, J. Yu, *Multiplicity results for periodic solutions to second order difference equations*, J. Dyn. Differ. Equ., **18** (2006), 943–960.

[352] Z. M. Guo, J. S. Yu, *Existence of periodic and subharmonic solutions for second-order superlinear difference equations*, Sci. China Ser. A, **46** (2003), 506–515.

[353] C. P. Gupta, *Existence and uniqueness results for the bending of an elastic beam equation at resonance*, J. Math. Anal. Appl., **135** (1988), 209–225.

[354] R. K. Gupta, P. Yadav, *Extended Lie method for mixed fractional derivatives, unconventional invariants and reduction, conservation laws and acoustic waves propagated via nonlinear dispersive equation*, Qual. Theory Dyn. Syst., **23** (2024), 203. https://doi.org/10.1007/s12346-024-01064-2

[355] J. R. Haddock, M. N. Nkashama, J. H. Wu, *Asymptotic constancy for linear neutral Volterra integrodifferential equations*, Tohoku Math. J., **41** (1989), 689–710.

[356] S. S. Haider, M. U. Rehman, T. Abdeljawad, *On Hilfer fractional difference operator*, Adv. Differ. Equ., **2020** (2020), 122. https://doi.org/10.1186/s13662-020-02576-2

[357] G. Haiyin, W. Ke, W. Fengying, D. Xiaohua, *Massera-type theorem and asymptotically periodic logistic equations*, Nonlinear Anal., **7** (2006), 1268–1283.

[358] S. Hajjaji, *Stepanov almost periodic solutions of some differential and integral equations with delays*, PhD Thesis, University of Sousse, 2023.

[359] R. Hakl, *On periodic-type boundary value problems for functional differential equations with positively homogeneous operator*, Miskolc Math. Notes, **5** (2004), 33–55.

[360] R. Hakl, *On a periodic type BVP for FDE with a positively homogeneous operator*, Mem. Differ. Equ. Math. Phys., **40** (2007), 17–54.

[361] R. Hakl, M. Pinto, V. Tkachenko, S. Trofimchuk, *Almost periodic evolution systems with impulse action at state-dependent moments*, J. Math. Anal. Appl., **446** (2017), 1030–1045.

[362] R. Hakl, E. Trofimchuk, S. Trofimchuk, *Periodic-type solutions for differential equations with positively homogeneous functionals*, J. Math. Sci. (N.Y.), **274** (2023), 126–141.

[363] A. Halanay, *Solutions périodiques et presque-périodiques des systèmes d'équations aux différences finies*, Arch. Ration. Mech. Anal., **12** (1964), 139–149.

[364] A. Halanay, V. Rasvan, *Periodic and almost periodic solutions for a class of systems described by coupled delay differential and difference equations*, Nonlinear Anal., **1** (1977), 197–206.

[365] E. J. Hale, *Fractional Leibniz rules in quasi-Banach function spaces and weighted bi-parameter settings*, PhD Thesis, Kansas Stae University, Manhattan, Kansas, 2024.

[366] J. K. Hale, S. M. Verduyn Lunel, *Introduction to Functional Diffrential Equations*, Springer, New York, 1993.

[367] T. Hamadneh et al., *The FitzHugh–Nagumo model described by fractional difference equations: stability and numerical simulation*, Axioms, **12** (2023), 806. https://doi.org/10.3390/axioms12090806

[368] Y. Hamaya, *Existence of almost periodic solutions of discrete Ricker delay models*, Int. J. Differ. Equ., **9** (2014), 187–205.

[369] Y. Hamaya, T. Itokazu, K. Saito, *Almost periodic solutions of nonlinear Volterra difference equations with unbounded delay*, Axioms, **4** (2015), 345–364. https://doi.org/10.3390/axioms4030345

[370] Y. Hamaya, K. Saito, *Almost periodic solutions in gross-substitute discrete dynamical systems*, Libertas Math. (N. S.), **38** (2018), 1–14.

[371] A. Haraux, P. Souplet, *An example of uniformly recurrent function which is not almost periodic*, J. Fourier Anal. Appl., **10** (2004), 217–220.

[372] P. Hasil, M. Veselý, *Critical oscillation constant for difference equations with almost periodic coefficients*, Abstr. Appl. Anal., **2012** (2012), 471435, 19 pp. https://doi.org/10.1155/2012/471435

[373] L. Haupt, T. Jäger, *Construction of smooth isomorphic and finite-to-one extensions of irrational rotations which are not almost automorphic*, preprint, 2023. arXiv:2312.04244.

[374] J. He, F. Kong, J. J. Nieto, H. Qiu, *Globally exponential stability of piecewise pseudo almost periodic solutions for neutral differential equations with impulses and delays*, Qual. Theory Dyn. Syst., **21** (2022), 48. https://doi.org/10.1007/s12346-022-00578-x

[375] J. W. He, C. Lizama, Y. Zhou, *The Cauchy problem for discrete time fractional evolution equations*, J. Comput. Appl. Math., **370**, 112683 (2020). https://doi.org/10.1016/j.cam.2019.112683

[376] W. H. He, J. D. Yin, Z. L. Zhou, *On quasi-weakly almost periodic points*, Sci. China Math., **56** (2013), 597–606.

[377] J. Hein, S. McCarthy, N. Gaswick, B. McKain, K. Speer, *Laplace transforms for the nabla difference operator*, Panam. Math. J., **21** (2011), 79–96.

[378] M. Hieber, A. Mahalov, R. Takada, *Time periodic and almost time periodic solutions to rotating stratified fluids subject to large forces*, J. Differ. Equ., **266** (2019), 977–1002.

[379] M. Hieber, T. H. Nguyen, *Stability and periodicity of solutions to the Oldroyd-B model on exterior domains*, Rend. Ist. Mat. Univ. Trieste, **52** (2020), 1–17.

[380] M. Hieber, C. Stinner, *Strong time periodic solutions to Keller–Segel systems: an approach by the quasilinear Arendt–Bu theorem*, J. Differ. Equ., **269** (2020), 1636–1655.

[381] R. Hilfer, *Applications of Fractional Calculus in Physics*, World Scientific, Singapore, 2000.

[382] R. Hilfer, Y. Luchko, Ž. Tomovski, *Operational method for solution of the fractional differential equations with the generalized Riemann–Liouville fractional derivatives*, Fract. Calc. Appl. Anal., **12** (2009), 299–318.

[383] S. Hilger, *Analysis on measure chains -- a unified approach to continuous and discrete calculus*, Result. Math., **18** (1990), 18–56.

[384] M. T. Holm, *The theory of discrete fractional calculus: development and application*, PhD Thesis, University of Nebraska-Lincoln, 2011.

[385] L. Hörmander, *An Introduction to Complex Analysis in Several Variables*, 3rd ed., North Holland, Amsterdam, 1990.

[386] Z. Hu, A. B. Mingarelli, *Bochner's theorem and Stepanov almost periodic functions*, Ann. Mat. Pura Appl., **187** (2008), 719–736.

[387] N. D. Huy, A. M. Le, V. T. Luong, N. N. Vien, *Asymptotic periodic solutions of differential equations with infinite delay*, preprint, 2023. arXiv:2309.02679.

[388] P. E. Hydon, *Difference Equations by Differential Equation Methods*, Cambridge University Press, 2014.

[389] A. O. Ignatyev, O. A. Ignatyev, *On the stability in periodic and almost periodic difference systems*, J. Math. Anal. Appl., **313** (2006), 678–688.

[390] M. N. Islam, Y. N. Raffoul, *Periodic solutions of neutral nonlinear system of differential equations with functional delay*, J. Math. Anal. Appl., **331** (2007), 1175–1186.

[391] G. M. Ismail et al., *Fractional residual power series method for the analytical and approximate studies of fractional physical phenomena*, Open Phys., **18** (2020), 799–805.

[392] T. Iwamiya, *Global existence of mild solutions to semilinear differential equations in Banach spaces*, Hiroshima Math. J., **16** (1986), 499–530.

[393] A. Iwanik, *Weyl almost periodic points in topological dynamics*, Colloq. Math., **56** (1988), 107–119.

[394] K. Jacobs, M. Keane, $0-1$-*Sequences of Toeplitz type*, Z. Wahrscheinlichkeitstheor. Verw. Geb., **13** (1969), 123–131.

[395] J. C. Jaeger, *The solution of boundary value problems by a double Laplace transformation*, Bull. Am. Math. Soc., **46** (1940), 687–693.

[396] D. L. Jagerman, *Difference Equations with Applications to Queues*, Marcel Dekker, Inc., New York-Basel, 2000.

[397] B. B. Jaimini, H. Nagar, *On multidimensional Weyl type fractional integral operator involving a general class of polynomials and multidimensional integral transforms*, J. Rajasthan Acad. Phys. Sci., **3** (2004), 237–245.

[398] A. Jaiswal, D. Bahuguna, *Hilfer fractional differential equations with almost sectorial operators*, Differ. Equ. Dyn. Syst., **31** (2023), 301–317.

[399] D. Ji, Y. Lu, *Stepanov-like pseudo almost automorphic solution to a parabolic evolution equation*, Adv. Differ. Equ., **2015** (2015), 341. https://doi.org/10.1186/s13662-015-0667-4

[400] B. Jia, L. Erbe, A. Peterson, *Comparison theorems and asymptotic behavior of solutions of discrete fractional equations*, Electron. J. Qual. Theory Differ. Equ., **89** (2015), 1–18.

[401] B. Jia, L. Erbe, A. Peterson, *Comparison theorems and asymptotic behavior of solutions of Caputo fractional equations*, Int. J. Differ. Equ., **11**(2) (2016), 163–178.

[402] H. Jiang et al., *Analytical solutions for the multi-term time-space Caputo–Riesz fractional advection–diffusion equations on a finite domain*, J. Math. Anal. Appl., **389** (2012), 1117–1127.

[403] R. A. Johnson, *On a Floquet theory for almost-periodic, two-dimensional linear systems*, J. Differ. Equ., **37** (1980), 184–205.

[404] R. A. Johnson, *Two-dimensional, almost periodic linear systems with proximal and recurrent behavior*, Proc. Am. Math. Soc., **82** (1981), 417–422.

[405] R. A. Johnson, *On almost-periodic linear differential systems of Millionščikov and Vinograd*, J. Math. Anal. Appl., **85** (1982), 452–460.

[406] R. A. Johnson, *Hopf bifurcation from nonperiodic solutions of differential equations. I. Linear theory*, J. Dyn. Differ. Equ., **1** (1989), 179–198.

[407] R. A. Johnson, Y. F. Yi, *Hopf bifurcation from non-periodic solutions of differential equations, II*, J. Differ. Equ., **107** (1994), 310–340.

[408] J. M. Jonnalagadda, N. S. Gopal, *On Hilfer-type nabla fractional differences*, Int. J. Differ. Equ., **15**(1) (2020), 91–107.

[409] J. M. Jonnalagadda, N. S. Gopal, *Linear Hilfer nabla fractional difference equations*, Int. J. Dyn. Syst. Differ. Equ., **11**(3–4) (2021), 322–340.

[410] C. Jordan, *Calculus of Finite Differences*, Chelsea Publishing Company, New York, 1950.

[411] G. Jumarie, *Modified Riemann–Liouville derivative and fractional Taylor series of nondifferentiable functions further results*, Comput. Math. Appl., **51** (2006), 1367–1376.

[412] E. I. Jury, *Sampled-Data Control Systems*, John Wiley and Sons, New York, 1958.

[413] E. I. Jury, *Theory and Appliction of the z-Transfrom Method*, R. E. Krieger Publishing Co., New York, 1964.

[414] V. Kac, P. Cheung, *Quantum Calculus*, Springer, New York, 2002.

[415] N. Kaldor, *A classificatory note on the determinateness of equilibrium*, Rev. Econ. Stud., **1** (1934), 122–136.

[416] G. A. Kamenskii, A. L. Skubachevskii, *Theory of Functional Differential Equations*, Mosk. Aviats. Inst, Moscow, 1992.

[417] R. Kamocki, *A new representation formula for the Hilfer fractional derivative and its application*, J. Comput. Appl. Math., **308** (2016), 39–45.

[418] K. Karthikeyan et al., *Almost sectorial operators on Ψ-Hilfer derivative fractional impulsive integro-differential equations*, Math. Methods Appl. Sci., **45** (2022), 8045–8059.

[419] P. Kasprzaka, A. Nawrocki, J. Signerska-Rynkowska, *Integrate-and-fire models with an almost periodic input function*, J. Differ. Equ., **264** (2018), 2495–2537.

[420] A. Katok, B. Hasselblatt, *Introduction to the Modern Theory of Dynamical Systems*, Cambridge University Press, 1997.

[421] E. R. Kaufmann, *A Kolmogorov predator–prey system on a time scale*, Dyn. Syst. Appl., **23**(4) (2014), 561–573.

[422] E. R. Kaufmann, N. Kosmatov, Y. N. Raffoul, *The connection between boundedness and periodicity in nonlinear functional neutral dynamic equations on a time scale*, Nonlinear Dyn. Syst. Theory, **9** (2009), 89–98.

[423] M. Keane, *Generalized Morse sequences*, Z. Wahrscheinlichkeitstheor. Verw. Geb., **10** (1968), 335–353.

[424] W. G. Kelley, A. C. Peterson, *Difference Equations: An Introduction with Applications*, 2nd. ed., Academic Press, San Diego, 2001.

[425] T. Kemmochi, *Discrete maximal regularity for abstract Cauchy problems*, Stud. Math., **234**(3) (2016), 241–263.
[426] C. M. Kent, W. Kosmala, M. A. Radin, S. Stević, *Solutions of the difference equation* $x_{n+1} = x_n x_{n-1} - 1$, Abstr. Appl. Anal., **2010** (2012), 469683. 13 pp. https://doi.org/10.1155/2010/469683
[427] L. Kexue, P. Jigen, *Laplace transform and fractional differential equations*, Appl. Math. Lett., **24** (2011), 2019–2023.
[428] L. Kexue, P. Jigen, J. Junxiong, *Cauchy problems for fractional differential equations with Riemann–Liouville fractional derivatives*, J. Funct. Anal., **263** (2012), 476–510.
[429] V. Keyantuo, C. Lizama, S. Rueda, M. Warma, *Asymptotic behavior of mild solutions for a class of abstract nonlinear difference equations of convolution type*, Adv. Differ. Equ., **2019** (2019), 251. https://doi.org/10.1186/s13662-019-2189-y
[430] M. T. Khalladi et al., *c-Almost periodic type functions and applications*, Nonauton. Dyn. Syst., **7** (2020), 176–193.
[431] A. A. Kilbas, H. M. Srivastava, J. J. Trujillo, *Theory and Applications of Fractional Differential Equations*, Elsevier Science B. V., Amsterdam, 2006.
[432] A. Kilicman, W. A. Ahmood, *Solving multi-dimensional fractional integro-differential equations with the initial and boundary conditions by using multi-dimensional Laplace Transform method*, Tbil. Math. J., **10** (2017), 105–115.
[433] A. Kisiolek, *Asymptotic behaviour of solutions of difference equations in Banach spaces*, Discuss. Math., Differ. Incl. Control Optim., **28** (2008), 5–13.
[434] V. L. Kocić, G. Ladas, *Global Behavior of Nonlinear Difference Equations of Higher Order with Applications*, Mathematics and its Applications, vol. 256, Kluwer Academic Publishers Group, Dordrecht, 1993.
[435] J. J. Koliha, *Metrics, Norms And Integrals: An Introduction To Contemporary Analysis*, World Scientific, Singapore, 2008.
[436] V. B. Kolmanovskii, *Limiting periodicity of the solutions for some Volterra difference equations*, Nonlinear Anal., **53** (2003), 669–681.
[437] V. B. Kolmanovskii, A. Myshkis, *Introduction to the Theory and Applications of Functional-Differential Equations*, Kluwer Academic Publishers, Dordrecht, 1999.
[438] H. Komatsu, *Ultradistributions, I. Structure theorems and a characterization*, J. Fac. Sci., Univ. Tokyo, Sect. 1A, Math., **20** (1973), 25–105.
[439] H. Komatsu, *Ultradistributions, II. The kernel theorem and ultradistributions with support in a manifold*, J. Fac. Sci., Univ. Tokyo, Sect. 1A, Math., **24** (1977), 607–628.
[440] H. Komatsu, *Ultradistributions, III. Vector valued ultradistributions. The theory of kernels*, J. Fac. Sci., Univ. Tokyo, Sect. 1A, Math., **29** (1982), 653–718.
[441] K. Konstaninidis, G. Papaschinopoulos, C. J. Schinas, *Asymptotic behaviour of the solutions of systems of partial linear homogeneous and nonhomogeneous difference equations*, Math. Methods Appl. Sci., **43** (2020), 3925–3935.
[442] M. Kostić, *Generalized Semigroups and Cosine Functions*, Mathematical Institute SANU, Belgrade, 2011.
[443] M. Kostić, *Abstract Volterra Integro-Differential Equations*, Taylor and Francis Group/CRC Press/Science Publishers, Boca Raton, Fl, 2015.
[444] M. Kostić, *Almost Periodic and Almost Automorphic Solutions to Integro-Differential Equations*, W. de Gruyter, Berlin, 2019.
[445] M. Kostić, *Abstract Degenerate Volterra Integro-Differential Equations*, Mathematical Institute SANU, Belgrade, 2020.
[446] M. Kostić, *Chaos for Linear Operators and Abstract Differential Equations*, Nova Science Publishers, New York, 2020.
[447] M. Kostić, *Selected Topics in Almost Periodicity*, W. de Gruyter, Berlin, 2022.
[448] M. Kostić, *Metrical Almost Periodicity and Applications to Integro-Differential Equations*, W. de Gruyter, Berlin, 2023.

[449] M. Kostić, *Vector-valued almost periodic ultradistributions and their generalizations*, Mat. Bilt., **42** (2018), 5–20.

[450] M. Kostić, *Composition principles for generalized almost periodic functions*, Bull. Cl. Sci. Math. Nat. Sci. Math., **43** (2018), 65–80.

[451] M. Kostić, *p-Almost periodic type functions in* $\mathbb{R}^n$, Chelyab. Fiz.-Mat. Zh., **7** (2022), 80–96.

[452] M. Kostić, *Metrical almost periodicity and applications*, Ann. Pol. Math., **129** (2022), 219–254. https://doi.org/10.4064/ap220510-15-11

[453] M. Kostić, *Multi-dimensional weighted ergodic components in general metric*, Discuss. Math., Differ. Incl. Control Optim., **42** (2022), 101–125.

[454] M. Kostić, *Asymptotically p-almost periodic type functions in general metric*, Ann. Univ. Craiova, Ser. Mat. Comput., **49** (2022), 358–370.

[455] M. Kostić, *Multi-dimensional Besicovitch almost periodic type functions and applications*, Commun. Pure Appl. Anal., **21** (2022), 4215–4250.

[456] M. Kostić, *Stepanov and Weyl classes of multi-dimensional p-almost periodic type functions*, Electron. J. Math. Anal. Appl., **10** (2022), 11–35.

[457] M. Kostić, *Hölder p-almost periodic type functions in* $\mathbb{R}^n$, Adv. Math. Sci. Appl., **31** (2022), 293–325.

[458] M. Kostić, *Stepanov p-almost periodic functions in general metric*, Facta Univ., Ser. Math. Inform., **37** (2022), 345–358.

[459] M. Kostić, *A note on c-almost periodic ultradistributions and c-almost periodic hyperfunctions*, Funct. Anal. Approx. Comput., **14** (2022), 15–21.

[460] M. Kostić, *Abstract degenerate Volterra inclusions in locally convex spaces*, Electron. J. Differ. Equ., **63** (2023), 1–55.

[461] M. Kostić, *Weyl almost automorphic functions and applications*, Adv. Pure Appl. Math., **14**(4) (2023), 1–36. https://doi.org/10.21494/ISTE.OP.2023.0998

[462] M. Kostić, *Weyl p-almost periodic functions in general metric*, Math. Slovaca, **73** (2023), 465–484.

[463] M. Kostić, *Generalized almost periodic functions with values in ordered Banach spaces*, submitted, https://www.researchgate.net/publication/364347447.

[464] M. Kostić, *Besicovitch multi-dimensional almost automorphic type functions and applications*, J. Nonlinear Evol. Equ. Appl., **3** (2023), 35–52.

[465] M. Kostić, *Multi-dimensional p-almost periodic type distributions*, An. Ştiinţ. Univ. “Al.I. Cuza” Iaşi. Inform. (N. S.), **69** (2023), 163–181.

[466] M. Kostić, *Metrically generalized p-almost periodic sequences and applications*, Funct. Anal. Approx. Comput., **16** (2024), 49–56.

[467] M. Kostić, *Asymptotically almost periodic solutions of abstract nonlinear Volterra equations*, J. Nonlinear Evol. Equ. Appl., **7** (2024), 99–105.

[468] M. Kostić, *Leibniz rules for fractional derivatives of non-differentiable functions*, Integral Transforms Spec. Funct., in press, 1–17 pp. https://doi.org/10.1080/10652469.2024.2414806

[469] M. Kostić, *Almost periodic functions and almost automorphic functions in general measure*, Discuss. Math., Differ. Incl. Control Optim., **44** (2024), 5–18. https://doi.org/10.7151/dmdico.1244.

[470] M. Kostić, *Abstract Volterra difference inclusions*, Bol. Soc. Mat. Mex., in press. https://www.researchgate.net/publication/377397057

[471] M. Kostić, *Multi-dimensional Poisson transform and applications*, Thang Long J. Sci., Math. Sci., **3** (2024), 29–49. https://science.thanglong.edu.vn/index.php/volc/article/view/159

[472] M. Kostić, *Abstract multi-term fractional difference equations*, submitted, 2024. arXiv:2403.19697v1

[473] M. Kostić, *Abstract degenerate difference equations*, submitted, https://www.researchgate.net/publication/377926222.

[474] M. Kostić, *Generalized vectorial almost periodicity*, Kragujevac J. Math., in press. https://www.researchgate.net/publication/384286187.

[475] M. Kostić, *p-Almost periodic ultradistributions in* $\mathbb{R}^n$, Funct. Anal. Approx. Comput., **16**(3) (2024), 55–63.

[476] M. Kostić, B. Chaouchi, W.-S. Du, D. Velinov, *Generalized p-almost periodic sequences and applications*, Fractal Fract., **7** (2023), 410. https://doi.org/10.3390/fractalfract7050410

[477] M. Kostić, W.-S. Du, *Stepanov almost periodic type functions and applications to abstract impulsive Volterra integro-differential inclusions*, Fractal Fract., **7** (2023), 736. https://doi.org/10.3390/fractalfract7100736

[478] M. Kostić, V. Fedorov, S. Pilipović, D. Velinov, *c-Almost periodic type distributions*, Chelyab. Fiz.-Mat. Zh., **6** (2020), 190–207.

[479] M. Kostić, V. E. Fedorov, *Abstract fractional differential inclusions with generalized Laplace derivatives*, J. Math. Sci. (N.Y.) (2024). https://doi.org/10.1007/s10958-024-07406-4

[480] M. Kostić, V. E. Fedorov, *Multi-dimensional Weyl almost periodic type functions and applications*, Appl. Anal. Discrete Math., **17** (2023), 446–473.

[481] M. Kostić, V. E. Fedorov, H. C. Koyuncuoğlu, *Metrical Bochner criterion and metrical Stepanov almost periodicity*, Chelyab. Fiz.-Mat. Zh., **24** (2024), 90–100.

[482] M. Kostić, H. C. Koyuncuoğlu, *Generalized almost periodic solutions of Volterra difference equations*, Malaya J. Mat., **11**(S) (2023), 149–165.

[483] M. Kostić, H. C. Koyuncuoğlu, *Multi-dimensional almost automorphic type sequences and applications*, Georgian Math. J., **31** (2024), 453–471.

[484] M. Kostić, H. C. Koyuncuoğlu, V. E. Fedorov, *Almost automorphic solutions to nonlinear difference equations*, Mathematics, **11**(23) (2023), 4824. https://doi.org/10.3390/math11234824

[485] M. Kostić, H. C. Koyuncuoğlu, J. M. Jonnalagadda, *On the asymptotic behaviour of the solutions of neutral Hilfer fractional difference equations*, submitted, https://www.researchgate.net/publication/381047206.

[486] M. Kostić, H. C. Koyuncuoğlu, T. Katican, *Asymptotic constancy for solutions of abstract non-linear fractional equations with delay and generalized Hilfer (a, b, α)-derivatives*, Chaos Solitons Fractals, **191**, 115934 (2025). https://doi.org/10.1016/j.chaos.2024.115934

[487] M. Kostić, H. C. Koyuncuoğlu, Y. N. Raffoul, *Positive periodic solutions for certain kinds of delayed q-difference equations with biological background*, Ann. Funct. Anal., **15** (2024), 5. https://doi.org/10.1007/s43034-023-00306-9

[488] M. Kostić, H. C. Koyuncuoğlu, Y. N. Raffoul, *More on the affine-periodic solutions of discrete dynamical systems: Massera's criterion, affine-periodic Floquet decomposition, and existence results*, submitted, https://www.researchgate.net/publication/373171539.

[489] M. Kostić, H. C. Koyuncuoğlu, Y. N. Raffoul, *(h, k)-Dichotomy on time scales and its application to Volterra integro-dynamic systems*, submitted, www.researchgate.net/publication/384079849.

[490] M. Kostić, H. C. Koyuncuoğlu, D. Velinov, *Measure theorethical appoach to almost periodicity*, submitted, https://www.researchgate.net/publication/375645038.

[491] M. Kostić, H. C. Koyuncuoğlu, D. Velinov, *Abstract fractional difference inclusions*, Izv. Math., in press. https://www.researchgate.net/publication/377160469.

[492] M. Kostić, H. C. Koyuncuoğlu, D. Velinov, *Stepanov-like pseudo S-asymptotically (ω, c)-periodic solutions fof a class of stochastic Volterra integro-differential equations*, Axioms, **13**(12) (2024), 871. https://doi.org/10.3390/axioms13120871

[493] M. Kostić, V. Kumar, *Remotely c-almost periodic type functions in $\mathbb{R}^n$*, Arch. Math. (Brno), **58** (2022), 85–104.

[494] M. Kostić, V. Kumar, M. Pinto, *Stepanov multi-dimensional almost automorphic type functions and applications*, J. Nonlinear Evol. Equ. Appl., **1** (2022), 1–24.

[495] M. Kostić, D. Velinov, *Almost periodic solutions for a class of neutral integro-differential equations*, Bol. Soc. Parana. Mat., in press. https://www.researchgate.net/publication/375891059

[496] H. C. Koyuncuoğlu, M. Adivar, *Almost periodic solutions of Volterra difference systems*, Demonstr. Math., **50** (2017), 320–329. https://doi.org/10.1515/dema-2017-0030

[497] H. C. Koyuncuoğlu, M. Adivar, *On the affine-periodic solutions of discrete dynamical systems*, Turk. J. Math., **42** (2018), 2260–2269.

[498] H. C. Koyuncuoğlu, Ö. Ö. Kaymak, J. M. Jonnalagadda, *Cobweb models with Hilfer nabla fractional differences*, preprint, https://www.researchgate.net/publication/379604827.

[499] H. C. Koyuncuoğlu, M. Kostić, Ö. Ö. Kaymak, T. Katican, *Periodic solutions of Kolmogorov systems on quantum time scales*, submitted, https://www.researchgate.net/publication/376835319.

[500] H. C. Koyuncuoğlu, Y. N. Raffoul, N. Turhan, *Asymptotic constancy for the solutions of Caputo fractional differential equations with delay*, Symmetry, **15** (2023), 88. https://doi.org/10.3390/sym15010088

[501] H. C. Koyuncuoğlu, N. Turhan, M. Adivar, *An asymptotic result for a certain type of delay dynamic equation with biological background*, Math. Methods Appl. Sci., **43** (2020), 7303–7310.

[502] K. D. Kucche, A. D. Mali, J. V. da, C. Sousa, *On the nonlinear Ψ-Hilfer fractional differential equations*, Comput. Appl. Math., **38** (2019), 1–25.

[503] M. R. S. Kulenović, G. Ladas, *Dynamics of Second Order Rational Difference Equations, with Open Problems and Conjectures*, CRC Press, Boca Raton, 2002.

[504] Dj. Kurepa, *Tableaux ramifiés d'ensembles, espaces pseudodistaciés*, C. R. Acad. Sci. Paris, **198** (1934), 1563–1565.

[505] A. C. Lazer, S. Solimini, *On periodic solutions of nonlinear differential equations with singularities*, Proc. Am. Math. Soc., **99** (1987), 10–114.

[506] E. K. Leinartas, *Multiple Laurent series and difference equations*, Sib. Math. J., **45** (2004), 321–326.

[507] E. K. Leinartas, M. Passare, A. K. Tsikh, *Multidimensional versions of Poincaré's theorem for difference equations*, Sb. Math., **199** (2008), 1505–1521.

[508] M. Levitan, *Almost Periodic Functions*, G.I.T.T.L, Moscow, 1953 (in Russian).

[509] B. M. Levitan, V. V. Zhikov, *Almost Periodic Functions and Differential Equations*, Cambridge Univ. Press, London, 1982.

[510] H. Levy, F. Lessman, *Finite Difference Equations*, Macmillan Co., New York, 1961.

[511] M. Li, C. Chen, F.-B. Li, *On fractional powers of generators of fractional resolvent families*, J. Funct. Anal., **259** (2010), 2702–2726.

[512] Q. Li, X. Wu, *Existence and asymptotic behavior of square-mean S-asymptotically periodic solutions for fractional stochastic evolution equation with delay*, Fract. Calc. Appl. Anal., **26** (2023), 718–750.

[513] Q. Li, X. Wu, *Existence and asymptotic behavior of square-mean S-asymptotically periodic solutions of fractional stochastic evolution equations*, Discrete Contin. Dyn. Syst., Ser. S, **17** (2024), 664–689.

[514] Y. Li, X. Huang, *Weyl almost periodic functions on time scales and their Fourier series*, preprint.

[515] Y. Li, S. Shen, *Compact almost automorphic function on time scales and its application*, Qual. Theory Dyn. Syst., **20** (2021), 86. https://doi.org/10.1007/s12346-021-00522-5

[516] Y. Li, L. Yang, W. Wu, *Almost periodic solutions for a class of discrete systems with Allee-effect*, Appl. Math., **59** (2014), 191–203.

[517] Z. Li, M. Haan, F. Chen, *Almost periodic solutions of a discrete almost periodic logistic equation with delay*, Appl. Math. Comput., **232** (2014), 743–751.

[518] Z. Li, X. Zhu, *Existence of chaos for a simple delay difference equation*, Adv. Differ. Equ., **2015** (2015), 39. https://doi.org/10.1186/s13662-015-0374-1

[519] J. Liang, G. M. N'Guérékata, T.-J. Xiao, *$\alpha \sim I$ relatively dense sets and weighted pseudo almost periodic phenomenon*, Proc. Am. Math. Soc., **148** (2020), 687–696.

[520] J. Liang, Y. Liu, C. Gao, *Almost periodic solution to singular systems*, Chin. Sci. Bull., **43** (1998), 698–700.

[521] S. Liang, R. Wu, L. Chen, *Laplace transform of fractional order differential equations*, Electron. J. Differ. Equ., **139** (2015), 1–15.

[522] T. Liang, Y. Q. Yang, Y. Liu, L. Li, *Existence and global exponential stability of almost periodic solutions to Cohen–Grossberg neural networks with distributed delays on time scales*, Neurocomputing, **123** (2014), 207–215.

[523] Z. Liang, X. Zhang, S. Li, *Periodic solutions of an indefinite singular planar differential system*, Mediterr. J. Math., **20** (2023), 41. https://doi.org/10.1007/s00009-022-02237-6

[524] Z. Liang, X. Zhang, S. Li, Z. Zhou, *Periodic solutions of a class of indefinite singular differential equations*, Electron. Res. Arch., **31** (2023), 2139–2148. https://doi.org/10.3934/era.2023110

[525] M. J. LIGHTHILL, *Introduction to Fourier Transform and Generalised Functions*, Cambridge University Press, Cambridge, 1959.

[526] D.-S. LIN, Y.-K. CHANG, *Pseudo S-asymptotically (ω, c)-periodic sequential solutions to some semilinear difference equations in Banach spaces*, J. Integral Equ. Appl., **36**(4) (2024), 447–469.

[527] D.-S. LIN, Y.-K. CHANG, *Pseudo (ω, c)-periodic solutions to Volterra difference equations in Banach spaces*, Comput. Appl. Math., in press.

[528] A. LINERO-BAS, V. MANOSA, D. NIEVES-ROLDÁN, *On the accumulation points of non-periodic orbits of a difference equation of fourth order*, preprint, arXiv:2306.12061v1.

[529] A. LINERO-BAS, D. NIEVES-ROLDÁN, *Periods of a max-type equation*, J. Differ. Equ. Appl., **27** (2021), 1608–1645.

[530] A. LINERO-BAS, D. NIEVES-ROLDÁN, A survey on max-type difference equations, in: S. Elaydi, M. R. S. Kulenović, S. Kalabušić (eds.), *Advances in Discrete Dynamical Systems, Difference Equations and Applications (ICDEA 2021)*, Springer Proceedings in Mathematics and Statistics, vol. 416, pp. 123–154, Springer, Cham, 2023. https://doi.org/10.1007/978-3-031-25225-9_6

[531] J. LIOUVILLE, *Mémoire sur le calcul des différentielles à indices quelconques*, J. Éc. Polytech., **13** (1832), 71–162.

[532] Y. LIU ET AL., *Multiplicity of positive almost periodic solutions and local asymptotical stability for a kind of time-varying fishing model with harvesting term*, Eng. Lett., **28** (2020), 4, EL/28/4/31.

[533] Z. LIU, L. CHEN, *Positive periodic solution of a general discrete non-autonomous difference system of plankton allelopathy with delays*, J. Comput. Appl. Math., **197** (2006), 446–456.

[534] C. LIZAMA, *l_p-maximal regularity for fractional difference equations on UMD spaces*, Math. Nachr., **288**(17/18) (2015), 2079–2092.

[535] C. LIZAMA, *The Poisson distribution, abstract fractional difference equations, and stability*, Proc. Am. Math. Soc., **145** (2017), 3809–3827.

[536] C. LIZAMA, J. G. MESQUITA, *Almost automorphic solutions of non-autonomous difference equations*, J. Math. Anal. Appl., **407** (2013), 339–349.

[537] C. LIZAMA, M. MURILLO-ARCILA, *L^p-maximal regularity for a class of fractional difference equations on UMD spaces: the case $1 < \alpha < 2$*, Banach J. Math. Anal., **11** (2017), 188–206.

[538] C. LIZAMA, M. MURILLO-ARCILA, *On a connection between the N-dimensional fractional Laplacian and $1-D$ operators on lattices*, J. Math. Anal. Appl., **511** (2022), 126051.

[539] C. LIZAMA, M. MURILLO-ARCILA, *On proportional hybrid operators in the discrete setting*, Math. Methods Appl. Sci. (2024), 1–21. https://doi.org/10.1002/mma.10551

[540] C. LIZAMA, M. MURILLO-ARCILA, C. LEAL, *Lebesgue regularity for differential difference equations with fractional damping*, Math. Methods Appl. Sci., **41** (2018), 2535–2545.

[541] C. LIZAMA, R. PONCE, *Solutions of abstract integro-differential equations via Poisson transformation*, Math. Methods Appl. Sci., **44** (2021), 2495–2505.

[542] C. LIZAMA, L. RONCAL, *Hölder–Lebesgue regularity and almost periodicity for semidiscrete equations with a fractional laplacian*, Discrete Contin. Dyn. Syst., **38** (2018), 1365–1403.

[543] N. T. LOAN, P. T. XUAN, *Almost periodic solutions of the parabolic-elliptic Keller–Segel system on the whole space*, Arch. Math. (Basel), **123** (2024), 431–446. https://doi.org/10.1007/s00013-024-02023-8

[544] N. T. LOAN ET AL., *Periodic solutions of the parabolic–elliptic Keller--Segel system on whole spaces*, Math. Nachr., **297** (2024), 3003–3023.

[545] L. C. LOI, *Subadjoint equations of index-1 linear singular difference equations*, Vietnam J. Math., **41** (2013), 81–96.

[546] L. C. LOI, N. H. DU, P. K. ANH, *On linear implicit non-autonomous systems of difference equations*, J. Differ. Equ. Appl., **8**(12) (2002), 1085–1105.

[547] S.-O. LONDEN, *On an integral equations in a Hilbert space*, SIAM J. Math. Anal., **8** (1977), 950–970.

[548] W. LONG, H.-S. DING, *Composition theorems of Stepanov almost periodic functions and Stepanov-like pseudo-almost periodic functions*, Adv. Differ. Equ., **2011** (2011), 654695, 12 pp. https://doi.org/10.1155/2011/654695

[549] M. E. Lord, A. R. Mitchell, *A new approach to the method of nonlinear variation of parameters*, Appl. Math. Comput., **4** (1978), 95–105.

[550] P. Lü, Y.-K. Chang, *Pseudo antiperiodic solutions to Volterra difference equations*, Mediterr. J. Math., **20** (2023), 36. https://doi.org/10.1007/s00009-022-02238-5

[551] P. Lü, Y.-K. Chang, *Pseudo S-asymptotically ω-antiperiodic solutions for SICNNs with mixed delays*, Neural Process. Lett., **55** (2023), 5401–5423. https://doi.org/10.1007/s11063-022-11091-2

[552] S. Lu, *A new result on the existence of periodic solutions for Liénard equations with a singularity of repulsive type*, J. Inequal. Appl., **2017** (2017), 37. https://doi.org/10.1186/s13660-016-1285-8

[553] Y. Luchko, *On the 1st-level general fractional derivatives of arbitrary order*, Fractal Fract., **7** (2023), 183. https://doi.org/10.3390/fractalfract7020183

[554] Y. Luchko, *On a generic fractional derivative associated with the Riemann–Liouville fractional integral*, Axioms, **13** (2024), 604. https://doi.org/10.3390/ axioms13090604

[555] Y. Luchko, J. J. Trujillo, *Caputo-type modification of the Erdélyi-Kober fractional derivative*, Fract. Calc. Appl. Anal., **10** (2007), 249–267.

[556] C. Ludwin, *Blood alcohol content*, Undergrad. J. Math. Model. One+Two, **3**(2) (2011), 1.

[557] A. Lunardi, *Interpolation Theory*, 3rd ed., Appunti. Scuola Normale Superiore di Pisa (Nuova Serie) [Lecture Notes. Scuola Normale Superiore di Pisa (New Series)], vol. 16, Edizioni della Normale, Pisa, 2018.

[558] Y. Luo, W. Wang, *Existence for impulsive semilinear functional differential inclusions*, Qual. Theory Dyn. Syst., **20** (2021), 22. https://doi.org/10.1007/s12346-021-00457-x

[559] V. T. Luong, N. D. Huy, N. V. Minh, N. N. Vien, *On asymptotic periodic solutions of fractional differential equations and applications*, Proc. Am. Math. Soc., **151** (2023), 5299–5312.

[560] W. Lv, *Existence of solutions for discrete fractional boundary value problems with a p-Laplacian operator*, Adv. Differ. Equ., **2012** (2012), 163.

[561] W. Lv, *Solvability for discrete fractional boundary value problems with a p-Laplacian operator*, Discrete Dyn. Nat. Soc., **2013** (2013), 679290.

[562] W. Lv, J. Feng, *Nonlinear discrete fractional mixed type sum-difference equation boundary value problems in Banach spaces*, Adv. Differ. Equ., **2014** (2014), 184. https://doi.org/10.1186/1687-1847-2014-184

[563] Z. Lv et al., *Solvability of a boundary value problem involving fractional difference equations*, Axioms, **12** (2023), 650. https://doi.org/10.3390/axioms12070650

[564] A. P. Lyapin, S. Chandragiri, *Generating functions for vector partition functions and a basic recurrence relation*, J. Differ. Equ. Appl., **25** (2019), 1052–1061.

[565] A. P. Lyapin, T. Cuchta, *Sections of the generating series of a solution to a difference equation in a simplicial cone*, Bull. Irkutsk State Univ. Ser. Math., **42** (2022), 75–89.

[566] P. Magal, S. Ruan, *Theory and Applications of Abstract Semilinear Cauchy Problems*, Springer, Berlin, 2018.

[567] J.-H. Mai, W.-H. Sun, *Almost periodic points and minimal sets in ω-regular spaces*, Topol. Appl., **154** (2007), 2873–2879.

[568] M. O. Mamchuev, *A boundary balue problem for a system of multidimensional fractional partial differential equations*, Vestn. Samar. Gos. Univ. Estestvennonauchn. Ser., **8** (2008), 164–175.

[569] J. C. Martin, *Substitution minimal flows*, Bull. Am. Math. Soc., **77** (1971), 610–612.

[570] P. Martinez-Amores, *Periodic solutions for coupled systems of differential-difference and difference equations*, Ann. Mat., **121** (1979), 171–186.

[571] J. Marcinkiewicz, *Une remarque sur les espaces de M. Besicovitch*, C. R. Acad. Sci. Paris, **208** (1939), 57–159.

[572] N. G. Markley, M. E. Paul, *Almost automorphic symbolic minimal sets without unique ergodicity*, Isr. J. Math., **34** (1979), 259–272.

[573] I. I. Marmershtein, *Necessary and sufficient criteria for the boundedness of the solutions of certain systems of linear difference equations in Banach space*, Izv. Akad. Nauk SSSR Math. USSR Izv. Ser. Mat., **35** (1971), 719–729.

[574] D. Maroncelli, *Periodic solutions to nonlinear second-order difference equations with two-dimensional kernel*, Mathematics, **12** (2024), 849. https://doi.org/10.3390/math12060849

[575] D. Maroncelli, J. Rodríguez, *Periodic behaviour of nonlinear, second-order, discrete dynamical systems*, J. Differ. Equ. Appl., **22** (2015), 280–294.

[576] R. März, *On linear differential–algebraic equations and linearizations*, Appl. Numer. Math., **18** (1995), 267–292.

[577] H. Matsunaga, S. Murakami, Y. Nagabuchi, Y. Nakano, *Formal adjoint equations and asymptotic formula for solutions of Volterra difference equations with infinite delay*, J. Differ. Equ. Appl., **18** (2012), 57–88.

[578] M. Maqbul, D. Bahuguna, *Almost periodic solutions for Stepanov-almost periodic differential equations*, Differ. Equ. Dyn. Syst., **22** (2014), 251–264.

[579] C. Maulén, S. Castillo, M. Kostić, M. Pinto, *Remotely almost periodic solutions of ordinary differential equations*, J. Math., **2021** (2021), 9985454. https://doi.org/10.1155/2021/9985454

[580] M. G. Mazhgikhova, *The Cauchy problem for the delay differential equation with Dzhrbashyan–Nersesyan fractional derivative*, Vestn. KRAUNC. Fiz.-Mat. Nauk., **42** (2023), 98–107.

[581] M. M. Mbaye, Almost periodic solution of some stochastic difference equations, in: M. A. McKibben, M. Webster (eds.), *Brownian Motion: Elements, Dynamics, and Applications*, pp. 1–17, Nova Science Publishers, Inc., New York, 2007.

[582] M. McKibben, *Discovering Evolution Equations with Applications: Volume 1 – Deterministic Equations*, Chapman and Hall/CRC, Boca Raton, 2010.

[583] D. C. mcMahon, T. S. Wu, *On weak mixing and local almost periodicity*, Duke Math. J., **39** (1972), 333–343.

[584] R. Meise, D. Vogt, *Introduction to Functional Analysis*, Oxf. Grad. Texts Math., Clarendon Press, New York, 1997. Translated from the German by M. S. Ramanujan and revised by the authors.

[585] B. Meknani, J. Zhang, T. Abdelhamid, *Pseudo-almost periodic C^0 solutions to the evolution equations with nonlocal initial conditions*, Appl. Anal., **3** (2021), 1–11.

[586] I. V. Melnikova, A. I. Filinkov, *Abstract Cauchy Problems: Three Approaches*, Chapman and Hall/CRC, Boca Raton, 2001.

[587] F. Meng, Q. Wiu, *Reducibility for a class of two-dimensional almost periodic system with small perturbation*, Adv. Pure Math., **11** (2021), 950–962.

[588] X. Meng, J. Jiao, L. Chen, *Global dynamics behaviors for a nonautonomous Lotka–Volterra almost periodic dispersal system with delays*, Nonlinear Anal., **68** (2008), 3633–3645.

[589] A. Michalowicz, S. Stoiński, *On the almost periodic functions in the sense of Levitan*, Comment. Math. Prace Mat., **67** (2007), 149–159.

[590] J. Migda, *Asymptotic behavior of solutions to difference equations in Banach spaces*, Electron. J. Qual. Theory Dyn. Syst., **88** (2021), 1–17. https://doi.org/10.14232/ejqtde.2021.1.88

[591] M. Migda, E. Schmeidel, M. Zdanowicz, *Periodic solutions of a 2-dimensional system of neutral difference equations*, Discrete Contin. Dyn. Syst., Ser. B, **23** (2018), 359–367. https://doi.org/10.3934/dcdsb.2018024

[592] K. S. Miller, B. Ross, *An Introduction to the Fractional Calculus and Fractional Differential Equations*, Wiley, New York, 1993.

[593] R. K. Miller, R. L. Wheeler, *Asymptotic behavior for a linear Volterra integral equation in Hilbert space*, J. Differ. Equ., **23** (1977), 270–284.

[594] P. Milnes, *Almost automorphic functions and totally bounded groups*, Rocky Mt. J. Math., **7**(2) (1977), 231–250.

[595] N. V. Minh, *Asymptotic behavior of individual orbits of discrete systems*, Proc. Am. Math. Soc., **137** (2009), 3025–3035.

[596] N. V. Minh, *On the asymptotic behaviour of Volterra difference equations*, J. Differ. Equ. Appl., **19** (2013), 1317–1330.

[597] N. V. Minh, H. Matsunaga, N. D. Huy, V. T. Luong, *A Katznelson–Tzafriri type theorem for difference equations and applications*, Proc. Am. Math. Soc., **150** (2022), 1105–1114.

[598] N. V. Minh, H. Matsunaga, N. D. Huy, V. T. Luong, *A spectral theory of polynomially bounded sequences and applications to the asymptotic behavior of discrete systems*, Funkc. Ekvacioj, **65** (2022), 261–285.

[599] M. Miraoui, M. Zorgui, *Measure pseudo Bloch periodic solutions for some difference and differential equations with piecewise constant argument*, Rocky Mt. J. Math., in press.

[600] D. Mitrović, *On a Leibniz type formula for fractional derivatives*, Filomat, **27** (2013), 1141–1146.

[601] P. O. Mohammeda, T. Abdeljawadb, F. K. Hamasalh, *Discrete Prabhakar fractional difference and sum operators*, Chaos Solitons Fractals, **150** (2021), 111182, 11 pp. https://doi.org/10.1016/j.chaos.2021.111182

[602] G. P. Mophou, G. M. N'Guérékata, *On some classes of almost automorphic functions and applications to fractional differential equations*, Comput. Math. Appl., **59** (2010), 1310–1317.

[603] C. A. Morales, *Equicontinuity on semi-locally connected spaces*, Topol. Appl., **198** (2016), 101–106.

[604] C. J. Moreno, *The zeros of exponential polynomials (I)*, Compos. Math., **26** (1973), 69–78.

[605] K. Mourad, B. Ichrak, *Sobolev type neutral integro-differenrial systems involving (k, ψ)-Hilfer fractional derivative and nonlocal conditions: trajectory controllability*, J. Math. Ser. (N.S.), 2024. https://doi.org/10.1007/s10958-024-07203-z

[606] D. Mozyrska, M. Wyrwas, *The $\mathcal{Z}$-transform method and delta type fractional difference operators*, Discrete Dyn. Nat. Soc., **2015** (2015), 852734, 12 pp. https://doi.org/10.1155/2015/852734

[607] T. A. Mughrabi, *Multi-dimensional Laplace transforms and applications*, PhD Thesis, Iowa State University, 1988.

[608] S. Murakami, Y. Nagabuchi, *Invariant manifolds for abstract functional differential equations and related Volterra difference equations in a Banach space*, Funkc. Ekvacioj, **50** (2007), 133–170.

[609] S. Murakami, Y. Nagabuchi, *Stability properties and asymptotic almost periodicity for linear Volterra difference equations in a Banach space*, Jpn. J. Math., **31** (2005), 193–223.

[610] A. B. Muravnik, *Functional differential parabolic equations: integral transformations and qualitative properties of solutions of the Cauchy problem*, J. Math. Sci. (N.Y.), **216** (2016), 345–496.

[611] A. B. Muravnik, *Elliptic differential-difference equations in the half-space*, Math. Notes, **67** (2020), 727–732.

[612] A. B. Muravnik, *Half-plane differential-difference elliptic problems with general-kind nonlocal potentials*, Complex Var. Elliptic Equ., **67** (2022), 1101–1120.

[613] A. B. Muravnik, *Differential-difference elliptic equations with nonlocal potentials in half-spaces*, Mathematics, **2023** (2698), 11. https://doi.org/10.3390/math11122698

[614] A. D. Myshkis, *Linear Differential Equations with Delayed Argument*, Nauka, Moscow, 1972 (in Russian).

[615] Y. Nagabuchi, *Decomposition of phase space for linear Volterra difference equations in a Banach space*, Funkc. Ekvacioj, **49** (2006), 269–290.

[616] A. M. Nagy, S. Assidi, B. Makhlouf, *Convergence of solutions for perturbed and unperturbed cobweb models with generalized Caputo derivative*, Bound. Value Probl., **2022** (2022), 89. https://doi.org/10.1186/s13661-022-01671-5

[617] A. M. Nakhushev, *Fractional Calculus and Its Applications*, Fizmatlit, Moscow, 2003 (in Russian).

[618] A. Nawrocki, *On some applications of convolution to linear differential equations with Levitan almost periodic coefficients*, Topol. Methods Nonlinear Anal., **50** (2017), 489–512.

[619] A. Nawrocki, *Diophantine approximations and almost periodic functions*, Demonstr. Math., **50** (2017), 100–104.

[620] A. Nawrocki, *On some generalizations of almost periodic functions and their applications*, PhD Thesis, Adam Mickiewicz University, Poznán, 2018 (in Polish).

[621] F. Neubrander, *Wellposedness of higher order abstract Cauchy problems*, Trans. Am. Math. Soc., **295** (1986), 257–290.

[622] T. T. Ngoc, P. T. Xuan, *Existence and stability of almost periodic solutions of Boussinesq systems in Morrey-type spaces*, preprint, https://www.researchgate.net/publication/381614173.

[623] C. Niu, X. Chen, *Almost periodic sequence solutions of a discrete Lotka–Volterra competitive system with feedback control*, Nonlinear Anal., **10** (2009), 3152–3161.

[624] R. Nussbanum, *A periodicity threshold theorem for some nonlinear integral equations*, SIAM J. Math. Anal., **9** (1978), 356–376.

[625] R. Oeckl, *Ra-Notes-19/10/2010*, https://matmor.unam.mx/~robert/cur/2011-1.

[626] S. Oharu, T. Takahashi, *Characterization of nonlinear semigroups associated with semilinear evolution equations*, Trans. Am. Math. Soc., **311** (1989), 593–619.

[627] M. D. Ortigueira, *Discrete-time fractional difference calculus: origins, evolutions, and new formalisms*, Fractal Fract., **7** (2023), 502. https://doi.org/10.3390/fractalfract7070502

[628] M. D. Ortigueira, J. A. Tenreiro Machado, *New discrete-time fractional derivatives based on the bilinear transformation: definitions and properties*, J. Adv. Res., **25** (2020), 1–10.

[629] T. J. Osler, *Leibniz rule for the fractional derivatives and an application to infinite series*, SIAM J. Appl. Math., **18** (1970), 658–674.

[630] T. J. Osler, *A further extension of the Leibniz rule to fractional derivatives and its relation to Parseval's formula*, SIAM J. Math. Anal., **3** (1972), 1–16.

[631] T. J. Osler, *An integral analog of the Leibniz rule*, Math. Comput., **26** (1972), 903–915.

[632] S. K. Panchal, P. V. Dole, A. D. Khandagale, *k-Hilfer–Prabhakar fractional derivatives and applications*, Indian J. Math., **59** (2017), 367–383.

[633] A. A. Pankov, *Bounded and Almost Periodic Solutions of Nonlinear Operator Differential Equations*, Kluwer Acad. Publ., Dordrecht, 1990.

[634] N. S. Papageorgiou, V. D. Radulescu, D. D. Repovš, *Periodic solutions for a class of evolution inclusions*, Comput. Math. Appl., **75** (2018), 3047–3065. https://doi.org/10.1016/j.camwa.2018.01.031

[635] J. Y. Park, Y. C. Kwumn, J. M. Jeong, *Existence of periodic solutions for delay evolution integrodifferential equations*, Math. Comput. Model., **65** (2004), 597–603.

[636] M. E. Paul, *Construction of almost automorphic symbolic minimal flows*, Gen. Topol. Appl., **6** (1976), 45–56.

[637] D. Pennequin, *Existence of almost periodic solutions of discrete time equations*, Discrete Contin. Dyn. Syst., **7** (2001), 51–60.

[638] M. Pierri, V. Rolnik, *On pseudo $\mathcal{S}$-asymptotically periodic functions*, Bull. Aust. Math. Soc., **87** (2013), 238–254.

[639] V. Pilipauskaitė, D. Surgailis, *Fractional operators and fractionally integrated random fields on $\mathbb{Z}^{\nu}$*, Fractal Fract., **8** (2024), 353. https://doi.org/10.3390/fractalfract8060353

[640] S. Pilipović, *Characterizations of bounded sets in spaces of ultradistributions*, Proc. Am. Math. Soc., **120** (1994), 1191–1206.

[641] S. Pilipović, *Tempered ultradistributions*, Boll. Unione Mat. Ital., **7**(2-B) (1998), 235–251.

[642] E. Pinney, *Ordinary Difference-Differential Equations*, University of California Press, 1959.

[643] M. Pinto, G. Robledo, *Existence and stability of almost periodic solutions in impulsive neural network models*, Appl. Math. Comput., **217** (2010), 4167–4177.

[644] M. Pinto, S. Trofimchuk, *Stability and existence of multiple periodic solutions for a quasilinear differential equation with maxima*, Proc. R. Soc. Edinb. A, **130** (2000), 1103–1118.

[645] I. Podlubny, *Fractional Differential Equations*, Academic Press, New York, 1999.

[646] R. Ponce, *Bounded mild solutions to fractional integrodifferential equations in Banach spaces*, Semigroup Forum, **87** (2013), 377–392.

[647] R. Ponce, *Identification of the order in fractional discrete systems*, Math. Methods Appl. Sci., **47** (2024), 9758–9768.

[648] T. R. Prabhakar, *A singular integral equation with a generalized Mittag–Leffler function in the kernel*, Yokohama Math. J., **19** (1971), 7–15.

[649] D. B. Price, *On the Laplace transform for distributions*, SIAM J. Math. Anal., **6**(1) (1975), 49–80. https://doi.org/10.1137/0506006

[650] J. G. Proakis, D. G. Manolakis, *Digital Signal Processing: Principles, Algorithms, and Applications*, 4th ed., Prentice Hall, London, 2007.

[651] C. Promsakon et al., *Existence and uniqueness of solutions for fractional-differential equation with boundary condition using nonlinear multi-fractional derivatives*, Math. Probl. Eng., **2024** (2024), 6844686, 7 pp. https://doi.org/10.1155/2024/6844686
[652] J. Prüss, *Evolutionary Integral Equations and Applications*, Birkhäuser-Verlag, Basel, 1993.
[653] A. V. Pskhu, *Equations with Fractional-Order Partial Derivatives*, Nauka, Moscow, 2005 (in Russian).
[654] A. V. Pskhu, *Boundary value problem for a multidimensional fractional partial differential equation*, Differ. Equ., **47** (2011), 385–395.
[655] P. Pych-Taberska, M. Topolewska, *Approximation of almost periodic functions by convolution type operators*, Rev. Mat. Complut., **9** (1996), 131–147.
[656] S. Radchenko, V. Samoilenko, P. Samusenko, Asymptotic solutions of singularly perturbed linear differential-algebraic equations with periodic coefficients, Mat. Stud., **59** (2023), 187–200.
[657] Y. Raffoul, *Positive periodic solutions in neutral nonlinear differential equations*, Electron. J. Qual. Theory Differ. Equ., **16** (2007), 1–10.
[658] Y. Raffoul, *Discrete population models with asymptotically constant or periodic solutions*, Int. J. Differ. Equ., **6** (2011), 143–152.
[659] Y. N. Raffoul, E. Yankson, *Positive periodic solutions in neutral delay difference equations*, Adv. Dyn. Syst. Appl., **5** (2010), 123–130.
[660] Md. Mahfujur Rahman et al., *The travelling wave solutions of space-time fractional partial differential equations by modified Kudryashov method*, J. Appl. Math. Phys., **8** (2020), 2683–2690.
[661] A. Rahmani et al., *Proportional Caputo fractional differential inclusions in Banach spaces*, Symmetry, **14** (2022), 1941. https://doi.org/10.3390/sym14091941
[662] R. K. Raina, *A note on the multidimensional Weyl fractional operator*, Proc. Indian Acad. Sci. Math. Sci., **101** (1991), 179–181.
[663] N. Rani, A. Fernandez, *Mikusiński's operational calculus for Prabhakar fractional calculus*, Integral Transforms Spec. Funct., **33** (2022), 945–965.
[664] A. Reich, *Präkompakte Gruppen und Fastperiodizität*, Math. Z., **116** (1970), 218–234.
[665] S. Reich, I. Shafrir, *An existence theorem for a difference inclusion in general Banach spaces*, J. Math. Anal. Appl., **160** (1991), 406–412.
[666] S. Rogosin, M. Dubatovskaya, *Letnikov vs. Marchaud: a survey on two prominent constructions of fractional derivatives*, Mathematics, **6**(1) (2018), 3. https://doi.org/10.3390/math6010003
[667] D. Rojas, *Resonance of bounded isochronous oscillators*, Nonlinear Anal., **192** (2020), 111680. https://doi.org/10.1016/j.na.2019.111680
[668] D. Rojas, P. J. Torres, *Periodic bouncing solutions of the Lazer–-Solimini equation with weak repulsive singularity*, Nonlinear Anal., **64** (2022), 103441. https://doi.org/10.1016/j.nonrwa.2021.103441
[669] A. N. Ronto, *On periodic solutions of systems with "maxima"*, Dopov. Nats. Akad. Nauk Ukr. Mat. Prirodozn. Tekh. Nauki, **12** (1999), 27–31.
[670] M. Rosenzweig, *L^p spaces for* $0 < p < 1$, https://matthewhr.files.wordpress.com/2012/09/lp-spaces-for-p-in-01.pdf.
[671] A. Rougirel, *Unified Theory for Fractional and Entire Differential Operators: An Approach via Differential Quadruplets and Boundary Restriction Operators*, Birkhäuser-Verlag, Basel, 2024.
[672] V. Rozko, *A class of almost periodic motions in pulsed system*, Differ. Uravn., **8** (1972), 2012–2022.
[673] V. P. Rubanik, *Oscillations of Quasilinear Systems with Retardation*, Nauka, Moscow, 1969.
[674] A. A. Ryzhkova, I. A. Trishina, *On periodic at infinity functions*, Nauch. Ved. Belgorod. Univ. Ser. Mat. Fiz., **36** (2014), 71–75.
[675] A. A. Ryzhkova, I. A. Trishina, *Almost periodic at infinity solutions of difference equations*, Izv. Sarat. Univ. Mat. Meh. Inform., **15** (2015), 45–49.
[676] R. Saadeh, A. Burqan, *Adapting a new formula to generalize multidimensional transforms*, Math. Methods Appl. Sci., **46** (2023), 15285–15304.
[677] L. T. Sac, P. T. Xuan, *Well-posedness and stability for a class of solutions of semi-linear diffusion equations with rough coefficients*, Filomat, in press. arXiv:2109.05185v5

[678] K. Saito, *Periodic solutions in gross-substitute discrete dynamical systems*, Libertas Math. (N. S.), **39** (2019), 1–12.

[679] S. Salahshour, A. Ahmadian, T. Allahviranloo, *A new fractional dynamic cobweb model based on nonsingular kernel derivatives*, Chaos Solitons Fractals, **145** (2021), 110755.

[680] A. Salim, S. Abbas, M. Benchohra, J. E. Lazreg, *Caputo fractional q-difference equations in Banach spaces*, J. Innov. Appl. Math. Comput. Sci., **3**(1) (2023), 1–14. https://doi.org/10.58205/jiamcs.v3i1.67

[681] S. Salsa, *Partial Differential Equations in Action: From Modelling to Theory*, Springer, Milano, 2008.

[682] S. G. Samko, A. A. Kilbas, O. I. Marichev, *Fractional Derivatives and Integrals: Theory and Applications*, Gordon and Breach, New York, 1993.

[683] A. M. Samoilenko, N. A. Perestyuk, *Impulsive Differential Equations*, World Scientific, Singapore, 1995.

[684] A. M. Samoilenko, S. I. Trofimchuk, *Unbounded functions with almost periodic differences*, Ukr. Math. J., **43** (1992), 1306–1309.

[685] P. Sattayatham, S. Tangmanee, W. Wei, *On periodic solutions of nonlinear evolutions equations in Banach spaces*, J. Math. Anal. Appl., **276**(1) (2002), 98–108.

[686] L. Schwartz, *Theorie des Distributions*, 2 vols., Hermann, Paris, 1950–1951.

[687] G. Sedaghat, *Global attractivity in nonlinear higher order difference equations in Banach algebras*, preprint, arXiv:1203.0227.

[688] G. R. Sell, W. Shen, Y. Yi, *Topological dynamics and differential equations*, Contemp. Math., **00** (1997), 1–19.

[689] S. Sengul, *Discrete fractional calculus and its applications to tumor growth*, Masters Theses Specialist Project, Paper 161. http://digitalcommons.wku.edu/theses/161

[690] D. Sforza, *Existence in the large for a semilinear integrodifferential equation with infinite delay*, J. Differ. Equ., **120** (1995), 289–303.

[691] Y. Shi, G. Chen, *Discrete chaos in Banach spaces*, Sci. China Ser. A, **34** (2004), 595–609, 2004.

[692] Y. Shi, P. Yu, G. Chen, *Chaotification of dynamical systems in Banach spaces*, Int. J. Bifurc. Chaos, **16** (2006), 2615–2636.

[693] J. A. Siddiqi, *Infinite matrices summing every almost periodic sequence*, Pac. J. Math., **39** (1971), 235–251.

[694] B. Simon, *A Comprehensive Course in Analysis*, American Mathematical Society, Providence, Rhode Island, 2015.

[695] A. L. Skubachevskii, *Elliptic Functional–Differential Equations and Applications*, Birkhäuser-Verlag, Basel–Boston–Berlin, 1997.

[696] A. Slavik, *Asymptotic behavior of solutions to the multidimensional semidiscrete diffusion equation*, Electron. J. Qual. Theory Differ. Equ., **9** (2022), 1–9. https://doi.org/10.14232/ejqtde.2022.1.9

[697] A. Slavik, *Discrete Bessel functions and partial difference equations*, J. Differ. Equ. Appl., **24** (2017), 425–437.

[698] A. Slavik, *Discrete-space systems of partial dynamic equations and discrete-space wave equation*, Qual. Theory Dyn. Syst., **16** (2017), 299–315.

[699] A. Slavík, P. Stehlík, *Dynamic diffusion-type equations on discrete-space domains*, J. Math. Anal. Appl., **427** (2015), 525–545.

[700] V. E. Slyusarchuk, *Bounded and periodic solutions of nonlinear functional differential equations*, Sb. Math., **203** (2012), 135–160.

[701] V. E. Slyusarchuk, *Conditions for the existence of almost periodic solutions of nonlinear difference equations with discrete argument*, J. Math. Sci. (N.Y.), **201** (2014), 391–399.

[702] V. E. Slyusarchuk, *Conditions for almost periodicity of bounded solutions of nonlinear differential equations unsolved with respect to the derivative*, Ukr. Math. J., **66** (2014), 432–442.

[703] V. E. Slyusarchuk, *Conditions for almost periodicity of bounded solutions of non-linear differential-difference equations*, Izv. Math., **78** (2014), 1232–1243.

[704] V. E. Slyusarchuk, *Conditions for the existence of almost periodic solutions of nonlinear difference equations in Banach space*, Mat. Zametki, **97** (2015), 277–285.

[705] V. E. Slyusarchuk, *Almost-periodic solutions of discrete equations*, Izv. Math., **80** (2016), 403–416.

[706] V. E. Slyusarchuk, *Necessary and sufficient conditions for the existence and uniqueness of a bounded solution of the equation $dx(t)/dt = f(x(t) + h_1(t)) + h_2(t)$*, Sb. Math., **208** (2017), 88–103.

[707] G. D. Smith, G. D. Smith, G. D. S. Smith, *Numerical Solution of Partial Differential Equations: Finite Difference Methods*, Oxford University Press, Oxford, UK, 1985.

[708] H. L. Smith, *Periodic competitive differential equations and the discrete dynamics of competitive maps*, J. Differ. Equ., **64** (1986), 165–194.

[709] H. L. Smith, *Periodic solutions of periodic competitive and cooperative systems*, SIAM J. Math. Anal., **17** (1986), 1289–1318.

[710] H. L. Smith, H. R. Thieme, *Dynamical Systems and Population Persistence*, Graduate Studies in Mathematics, vol. 118, American Mathematical Society, Providence, 2011.

[711] Q. K. Song, J. D. Cao, *Stability analysis of Cohen–Grossberg neural networks with both time-varying and continuously distributed delays*, J. Comput. Appl. Math., **197**(1) (2006), 188–203.

[712] Y. Song, *Almost periodic solutions of discrete Volterra equations*, J. Math. Anal. Appl., **314** (2006), 174–194.

[713] Y. Song, *Periodic and almost periodic solutions of functional difference equations with finite delay*, Adv. Differ. Equ., **2007** (2007), 68023, 15 pp. https://doi.org/10.1155/2007/68023

[714] Y. Song, C. T. H. Baker, *Perturbation theory for discrete Volterra equations*, J. Differ. Equ. Appl., **9** (2003), 969–987.

[715] Y. Song, C. T. H. Baker, *Perturbations of Volterra difference equations*, J. Differ. Equ. Appl., **10** (2004), 379–397.

[716] Y. Song, H. Tian, *Periodic and almost periodic solutions of nonlinear discrete Volterra equations with unbounded delay*, J. Comput. Appl. Math., **205** (2007), 859–870.

[717] J. V. C. Sousa, E. C. de Oliveira, *On the ψ-Hilfer fractional derivative*, Commun. Nonlinear Sci. Numer. Simul., **60** (2018), 72–91.

[718] J. V. da, C. Sousa, F. Jarad,
T. Abdeljawad, *Existence of mild solutions to Hilfer fractional evolution equations in Banach space*, Ann. Funct. Anal., **12** (2021), 12. https://doi.org/10.1007/s43034-020-00095-5

[719] M. R. Spiegel, *Schaum's outline of: Theory and Problems of Calculus of Finite Differences and Difference Equations*, McGraw-Hill Inc., New York, 1971.

[720] H. M. Srivastava, R. C. Singh Chandel, P. K. Vishwakarma, *Fractional derivatives of certain generalized hypergeometric functions of several variables*, J. Math. Anal. Appl., **184** (1994), 560–572.

[721] H. M. Srivastava, D. Raghavan, S. Nagarajan, *A comparative study of the stability of some fractional-order cobweb economic models*, Rev. R. Acad. Cienc. Exactas Fís. Nat., Ser. A Mat., **116** (2022), 98. https://doi.org/10.1007/s13398-022-01239-z

[722] H. M. Srivastava, S. B. Yakubovich, Yu. F. Luchko, *The convolution method for the development of new Leibniz rules involving fractional derivatives and of their integral analogues*, Integral Transforms Spec. Funct., **1** (1993), 119–134.

[723] G. T. Stamov, *Almost Periodic Solutions of Impulsive Differential Equations*, Springer, Berlin/Heidelberg, Germany, 2012.

[724] W. Stepanoff, *Über einige Verallgemeinerungen der fast periodischen Funktionen*, Math. Ann., **95** (1926), 473–498.

[725] S. Stoíński, *Real-valued functions almost periodic in variation*, Funct. Approx. Comment. Math., **22** (1993), 141–148.

[726] S. Stoíński, *Almost periodic function in the Lebesgue measure*, Comment. Math. Prace Mat., **34** (1994), 189–198.

[727] S. Stoíński, *L_α-almost periodic functions*, Demonstr. Math., **28** (1995), 689–696.

[728] S. Stoíński, *On compactness of almost periodic functions in the Lebesgue measure*, Fasc. Math., **30** (1999), 171–175.

[729] S. Stoíński, *Almost Periodic Functions*, Scientific Publisher AMU, blocationPoznań 2008 (in Polish).

[730] S. H. Streipert, G. S. K. Wolkowicz, *Derivation and dynamics of discrete population models with distributed delay in reproduction*, Math. Biosci., **376** (2024), 109279.

[731] R. I. Strichartz, *Magnified curves on a clat torus, determination of almost periodic functions, and the Riemann–Lebesgue lemma*, Proc. Am. Math. Soc., **107** (1989), 755–759.

[732] J. Stryja, *Analysis of almost-periodic functions*, Mgr. Thesis, Palacký University, Olomouc, 2001 (in Czech).

[733] Q. Sun, *The existence of periodic solutions to fourth-order nonlinear difference equations*, Master Thesis, Guangzhou University, 2008.

[734] G. A. Sviridyuk, V. E. Fedorov, *Linear Sobolev Type Equations and Degenerate Semigroups of Operators*, Inverse and Ill-Posed Problems (Book 42), VSP, Utrecht, Boston, 2003.

[735] R. J. Swift, *Almost periodic Harmonizable processes*, Georgian Math. J., **3** (1996), 275–292.

[736] L. Székelyhidi, Difference equations in several variables, in: *Discrete Spectral Synthesis and Its Applications*, Springer Monographs in Mathematics, Springer, Dordrecht, 2006. https://doi.org/10.1007/978-1-4020-4637-7_6

[737] V. E. Tarasov, *On chain rule for fractional derivatives*, Commun. Nonlinear Sci. Numer. Simul., **30** (2016), 1–4.

[738] V. E. Tarasov, *General fractional calculus: multi-kernel approach*, Mathematics, **9** (2021), 1501. https://doi.org/10.3390/math9131501

[739] V. E. Tarasov, *General fractional dynamics*, Mathematics, **9** (2021), 1464. https://doi.org/10.3390/math9131464

[740] V. E. Tarasov, *Fractional dynamics with depreciation and obsolescence: equations with Prabhakar fractional derivatives*, Mathematics, **10** (2022), 1540. https://doi.org/10.3390/math10091540

[741] V. E. Tarasov, *Fractional economic dynamics with memory*, Mathematics, **12** (2024), 2411. https://doi.org/10.3390/math12152411

[742] V. E. Tarasov, *Exact finite-difference calculus: beyond set of entire functions*, Mathematics, **12** (2024), 972. https://doi.org/10.3390/math12070972

[743] J. Tariboon, S. Ntouyas, P. Agarwal, *New concepts of fractional quantum calculus and applications to impulsive fractional q-difference equations*, Adv. Differ. Equ., **2015** (2015), 18. https://doi.org/10.1186/s13662-014-0348-8

[744] R. Terras, *Almost automorphic functions on topological groups*, Indiana Univ. Math. J., **21**(8) (1972), 759–773.

[745] T. V. Thuy, P. T. Xuan, *Well-posedness and stability for a class of solutions of semi-linear diffusion equations with rough coefficients*, preprint, 2021, arXiv:2109.05185v5.

[746] N. Touafek, E. M. Elsayed, *On the solutions of systems of rational difference equations*, Math. Comput. Model., **55** (2012), 1987–1997.

[747] R. Tremblay, S. Gaboury, *Applications of the integral analogue of a new Leibniz-type rule for fractional derivatives to functions of several variables*, Int. J. Contemp. Math. Sci., **7** (2012), 1877–1887.

[748] H. Triebel, *Interpolation Theory, Function Spaces, Differential Operators*, North-Holland, Amsterdam, 1978.

[749] I. A. Trishina, *Almost periodic at infinity functions relative to the subspace of functions integrally decreasing at infinity*, Izv. Sarat. Univ. Mat. Meh. Inform., **17** (2017), 402–418.

[750] I. A. Trishina, *Functions slowly varying at infinity*, Vestn. Voronezh. Univ. Ser. Fiz. Mat., **4** (2017), 134–144.

[751] O. Trofymchuk, E. Liz, S. Trofimchuk, *The peak-end rule and its dynamic realization through differential equations with maxima*, Nonlinearity, **36** (2023), 507–536.

[752] Y. Z. Tsypkin, *Theory of Pulse Systems*, State Press for Physics and Mathematical Literature, Moscow, 19S8 (in Russian).

[753] S.-T. Tu, T.-C. Wu, H. M. Srivastava, *Commutativity of the Leibniz rules in fractional calculus*, Comput. Math. Appl., **40** (2000), 303–312.

[754] V. V. Uchaikin, *The Method of Fractional Derivatives*, Artishok, Ulyanovsk, 2008 (in Russian).

[755] H. D. URSELL, *Parseval's theorem for almost-periodic functions*, Proc. Lond. Math. Soc., **s2-32** (1931), 402–440.
[756] L. QI, R. YUAN, *Piecewise continuous almost automorphic functions and Favard's theorems for impulsive differential equations in honor of Russell Johnson*, J. Dyn. Differ. Equ. (2022). https://doi.org/10.1007/s10884-020-09879-8
[757] W. QI, Y. LI, *Weyl almost anti-periodic solution to a neutral functional semilinear differential equation*, Electron. Res. Arch., **31**(3) (2023), 1662–1672. https://doi.org/10.3934/era.2023086
[758] W. A. VEECH, *Almost automorphic functions on groups*, Am. J. Math., **87**(3) (1965), 719–751.
[759] W. A. VEECH, *On a theorem of Bochner*, Ann. Math., **86**(1) (1967), 117–137.
[760] M. VESELÝ, *Constructions of almost periodic sequences and functions and homogeneous linear difference and differential equations*, PhD Thesis, Masaryk University, Brno, 2011.
[761] V. V. VLASOV, D. A. MEDVEDEV, *Functional–differential equations in Sobolev spaces and related problems of spectral theory*, J. Math. Sci. (N.Y.), **164** (2010), 659–841.
[762] A. R. VOLKOVA, V. E. FEDOROV, D. M. GORDIEVSKIKH, *On solvability of some classes of equations with Hilfer derivative in Banach spaces*, Chelyab. Fiz.-Mat. Zh., **7** (2022), 11–19 [in Russian].
[763] I. I. VRABIE, *Compactness Methods for Nonlinear Evolutions*, CRC Press, Boca Raton, 1995.
[764] I. I. VRABIE, *Almost periodic solutions for nonlinear delay evolutions with nonlocal initial conditions*, J. Evol. Equ., **13** (2013), 693–714.
[765] I. I. VRABIE, *Delay evolution equations with mixed nonlocal plus local initial conditions*, Commun. Contemp. Math., **17**(02) (2015), 1350035.
[766] H. A. WAHASH, M. S. ABDO, S. K. PANCHAL, *Fractional integrodifferential equations with nonlocal conditions and generalized Hilfer fractional derivative*, Ufa Math. J., **11** (2019), 151–171.
[767] P. WALTERS, *An Introduction to Ergodic Theory*, Graduate Texts in Mathematics, vol. 79, Springer, New York–Berlin, 1982.
[768] C. WANG, Y. TAN, R. P. AGARWAL, *Almost periodic fractional fuzzy dynamic equations on timescales: a survey*, Math. Methods Appl. Sci., **47** (2024), 2345–2401.
[769] J. R. WANG, X. XIANG, W. WEI, *Periodic solutions of a class of integrodifferential impulsive periodic systems with time-varying generating operators on Banach space*, Electron. J. Qual. Theory Differ. Equ., **4** (2009), 1–17.
[770] Q. WANG, R. XU, *On Hilfer generalized proportional nabla fractional difference operators*, Mathematics, **10** (2022), 2654. https://doi.org/10.3390/math10152654
[771] Q. WANG, R. XU, *A review of definitions of fractional differences and sums*, Math. Found. Comput., **6** (2023), 136–160.
[772] S. WANG, *The existence of almost periodic solution: via coincidence degree theory*, Bound. Value Probl., **2016** (2016), 71. https://doi.org/10.1186/s13661-016-0576-9
[773] W. WANG, X. YANG, *Positive periodic solutions for neutral functional difference equations*, Int. J. Difference Equ., **7** (2012), 99–109.
[774] X. WANG, X. FU, *p-th Besicovitch almost periodic solutions in distribution for semi-linear non-autonomous stochastic evolution equations*, Bull. Malays. Math. Sci. Soc., **47** (2024), 23. https://doi.org/10.1007/s40840-023-01613-z
[775] X. WANG, X. FU, *Existence and stability of p-th Weyl almost automorphic solutions in distribution for neutral stochastic functional differential equations*, Comput. Appl. Math., in press.
[776] G. WEBB, *Continuous nonlinear perturbations of linear accretive operators in Banach spaces*, J. Funct. Anal., **10** (1972), 191–203.
[777] M. WEI, Q. LI, *Existence and uniqueness of $\mathcal{S}$-asymptotically periodic α-mild solutions for neutral fractional delayed evolution equation*, Appl. Math. J. Chin. Univ., **37** (2022), 228–245.
[778] Y. WEI ET AL., *Discussion on the Leibniz rule and Laplace transform of fractional derivatives using series representation*, Integral Transforms Spec. Funct., **31** (2020), 304–322.
[779] Y. WEI, S. LIU, Y.-K. CHANG, *Stepanov type pseudo Bloch periodic functions and applications to some evolution equations in Banach spaces*, Rocky Mt. J. Math., in press. https://projecteuclid.org/journals/rmjm/rocky-mountain-journal-of-mathematics/DownloadAcceptedPapers/230521

[780] S. Willard, *General Topology*, Addison-Wesley Publishing Company, 1970.
[781] H. Wu, Z. Tan, X. Hu, H. Xiao, *The existence of multiple periodic solutions to a class of fourth-order difference equations*, Math. Probl. Eng., **2022** (2022), 7694885, 9 pp. https://doi.org/10.1155/2022/7694885
[782] J. Wu, Y. Liu, *Two periodic solutions of neutral difference systems depending on two parameters*, J. Comput. Appl. Math., **206** (2007), 713–725.
[783] Y. H. Xia, *Almost periodic solution of a population model: via spectral radius of matrix*, Bull. Malays. Math. Sci. Soc., **37** (2014), 249–259.
[784] Y. H. Xia, S. S. Chen, *Quasi-uniformly asymptotic stability and existence of almost periodic solutions of difference equations with applications in population dynamic systems*, J. Differ. Equ. Appl., **14** (2008), 59–81. https://doi.org/10.1080/10236190701470407
[785] Z. Xia, *Weighted pseudo periodic solutions of neutral functional differential equations*, Electron. J. Differ. Equ., **191** (2014), 1–17.
[786] Z. Xia, *Discrete weighted peudo asymptotic periodicity of second order difference equations*, Discrete Dyn. Nat. Soc., **2014** (2014), 949487, 8 pp. https://doi.org/10.1155/2014/949487
[787] Z. Xia, *Dynamics of pseudo almost periodic solution for impulsive neoclassical growth model*, ANZIAM J., **58** (2017), 359–367.
[788] Z. Xia, D. Wang, *Pseudo almost automorphic mild solution of nonautonomous stochastic functional integro-differential equations*, Filomat, **32** (2018), 1233–1250.
[789] Z. Xia, D. Wang, *Asymptotic behavior of mild solutions for nonlinear fractional difference equations*, Fract. Calc. Appl. Anal., **21**(2) (2018), 527–551.
[790] Z. Xia, D. Wang, C.-F. Wen, J.-C. Yao, *Pseudo asymptotically periodic mild solutions of semilinear functional integro-differential equations in Banach spaces*, Math. Methods Appl. Sci., **2017** (2017), 1–23.
[791] H. Xiao, B. Zheng, *The existence of multiple periodic solutions of nonautonomous delay differential equations*, J. Appl. Math., **2011** (2011), 829107.
[792] T.-J. Xiao, J. Liang, *The Cauchy Problem for Higher-Order Abstract Differential Equations*, Springer, Berlin, 1998.
[793] X. Xiang, N. U. Ahmed, *Existence of periodic solutions of semilinear evolution equations with time lags*, Nonlinear Anal., **18** (1992), 1063–1070.
[794] X. Xing, *Positive periodic solution for second-order singular semipositone differential equations*, Abstr. Appl. Anal., **2013** (2013), 310469, 7 pp. https://doi.org/10.1155/2013/310469
[795] J. C. Xiong, *Set of almost periodic points of a continuous self-map of an interval*, Acta Math. Sin. New Ser., **2** (1986), 73–77.
[796] R. Xu, M. A. J. Chaplain, F. A. Davidson, *Periodic solutions of a discrete nonautonomous Lotka–Volterra predator–prey model with time delays*, Discrete Contin. Dyn. Syst., Ser. B, **4** (2004), 823–831.
[797] X. Xue, Y. Cheng, *Existence of periodic solutions of nonlinear evolution inclusions in Banach spaces*, Nonlinear Anal., **11** (2010), 459–471.
[798] Y. Xue, X. Xie, F. Chen, R. Han, *Almost periodic solution of a discrete commensalism system*, Discrete Dyn. Nat. Soc., **2015** (2015), 295483. https://doi.org/10.1155/2015/295483
[799] M. Yang, Q. Wang, *Pseudo asymptotically periodic solutions for fractional integro-differential neutral equations*, Sci. China Math., **62** (2019), 1705–1718.
[800] L. Ye, Y. Liu, *Pseudo-almost periodic C^0-solution for evolution inclusion with mixed nonlocal plus local initial conditions*, Hacet. J. Math. Stat., **52** (2023), 1480–1491.
[801] T. Yoshizawa, *Asymptotically almost periodic solutions of an almost periodic system*, Funkc. Ekvacioj, **12** (1969), 23–40.
[802] R. Yuan, *Almost periodic solutions of a class of singularly perturbed differential equations with piecewise constant argument*, Nonlinear Anal., **37** (1999), 841–859.
[803] R. Yuan, *On the almost periodic solution of a class of singularly perturbed differential equations with piecewise constant argument*, Int. J. Quant. Theory Differ. Equ. Appl., **2** (2008), 69–100.
[804] R. Yuan, *On Favard's theorems*, J. Differ. Equ., **249** (2010), 1884–1916.

[805] S. Zaidman, *Almost-Periodic Functions in Abstract Spaces*, Pitman Research Notes in Math., vol. 126, Pitman, Boston, 1985.
[806] N. V. Zaitseva, *On global classical solutions of hyperbolic differential-difference equations*, Dokl. Math., **101** (2020), 115–116.
[807] N. V. Zaitseva, *Classical solutions of hyperbolic differential–difference equations in a half-space*, Differ. Equ., **57** (2021), 1629–1639.
[808] N. V. Zaitseva, *Classical solutions of hyperbolic differential-difference equations with several nonlocal terms*, Lobachevskii J. Math., **42** (2021), 231–236.
[809] N. V. Zaitseva, *Classical solutions of hyperbolic equations with nonlocal potentials*, Dokl. Math., **103** (2021), 127–129.
[810] N. V. Zaitseva, *Classical solutions of a multidimensional hyperbolic differential-difference equation with shifts of various directions in the potentials*, Math. Notes, **112** (2022), 872–880.
[811] C. Zhang, L. Jiang, *Remotely almost periodic solutions to systems of differential equations with piecewise constant argument*, Appl. Math. Lett., **21** (2008), 761–768. https://doi.org/10.1016/j.aml.2007.08.007
[812] G. Zhang, *Variational methods for fourth-order difference boundary value problem and second order difference system*, Master Thesis, Lanzhou University, 2011.
[813] H. Zhang, *New results on the positive pseudo almost periodic solutions for a generalized model of hematopoiesis*, Electron. J. Qual. Theory Differ. Equ., **24** (2014), 1–10.
[814] J. Zhang, S. Bu, *Maximal regularity for fractional difference equations of order $2 < \alpha < 3$ on UMD spaces*, Electron. J. Differ. Equ., **20** (2024), 1–17. https://doi.org/10.58997/ejde.2024.20
[815] J. Y. Zhang, Y. Suda, T. Iwasa, *Absolutely exponential stability of a class of neural networks with unbounded delay*, Neural Netw., **17**(3) (2004), 391–397.
[816] L.-L. Zhang, H.-X. Li, *Weighted pseudo almost periodic solutions of second order neutral differential equations with piecewise constant argument*, Nonlinear Anal., **74** (2011), 6770–6780.
[817] L.-L. Zhang, H.-X. Li, *Almost automorphic solutions for differential equations with piecewise argument*, Bull. Aust. Math. Soc., **90** (2014), 99–112.
[818] R. Y. Zhang, Z. C. Wang, Y. Chen, J. Wu, *Periodic solutions of a single species discrete population model with periodic harvest/stock*, Comput. Math. Appl., **39** (2000), 77–90.
[819] S. Zhang, *Almost periodic solutions of difference systems*, Chin. Sci. Bull., **43** (1998), 2041–2046.
[820] S. Zhang, *Existence of almost periodic solution for difference systems*, Ann. Differ. Equ., **16** (2000), 184–206.
[821] S. Zhang, P. Liu, K. Gopalsamy, *Almost periodic solutions of nonautonomous linear difference equations*, Appl. Anal., **81** (2022), 281–301.
[822] S. Zhang, G. Zheng, *Almost periodic solutions of delay difference systems*, Appl. Math. Comput., **131** (2002), 497–516. https://doi.org/10.1016/S0096-3003(01)00165-5
[823] T. Zhang, Y. Li, *Global exponential stability of discrete-time almost automorphic Caputo–Fabrizio BAM fuzzy neural networks via exponential Euler technique*, Knowl.-Based Syst., **246**, 108675 (2022).
[824] D. Zhao, *Periodic solutions and homoclinic orbits to fourth-order difference equations*, Master Thesis, vol. 1, 2007.
[825] J. Zhao, R. Ma, *Ambrosetti–Prodi-type results for a class of difference equations with nonlinearities indefinite in sign*, Open Math., **20** (2022), 783–790.
[826] J. Zhu, X. Fu, *Existence and asymptotic periodicity of solutions for neutral integro-differential evolution equations with infinite delay*, Math. Slovaca, **72** (2022), 121–140.
[827] Q.-F. Zou, H.-S. Ding, *Almost periodic solutions for a nonlinear integro-differential equation with neutral delay*, J. Nonlinear Sci. Appl., **9** (2016), 4500–4508.

Index

https://doi.org/10.1515/9783111689746-014

De Gruyter Studies in Mathematics

Volume 64
Dorina Mitrea, Irina Mitrea, Marius Mitrea, Michael Taylor
The Hodge–Laplacian. Boundary Value Problems on Riemannian Manifolds, 2nd Edition, 2025
ISBN 978-3-11-148098-5, e-ISBN 978-3-11-148140-1, e-ISBN (ePUB) 978-3-11-148389-4

Volume 100
Changxing Miao, Ruipeng Shen
Regularity and Scattering of Dispersive Wave Equation. Multiplier Method and Morawetz Estimate, 2025
ISBN 978-3-11-148754-0, e-ISBN 978-3-11-148835-6, e-ISBN (ePUB) 978-3-11-148940-7

Volume 99
Marcus Laurel, Marius Mitrea
Weighted Morrey Spaces. Calderón-Zygmund Theory and Boundary Problems, 2024
ISBN 978-3-11-145816-8, e-ISBN 978-3-11-145827-4, e-ISBN (ePUB) 978-3-11-146145-8

Volume 98
Peter J. Brockwell, Alexander M. Lindner
Continuous-Parameter Time Series, 2024
ISBN 978-3-11-132499-9, e-ISBN 978-3-11-132503-3, e-ISBN (ePUB) 978-3-11-132520-0

Volume 97
Ştefan Ovidiu I. Tohăneanu
Commutative Algebra Methods for Coding Theory, 2024
ISBN 978-3-11-121292-0, e-ISBN 978-3-11-121479-5, e-ISBN (ePUB) 978-3-11-121538-9

Volume 25
Karl H. Hofmann, Sidney A. Morris
The Structure of Compact Groups. A Primer for the Student – A Handbook for the Expert, 5th Edition, 2023
ISBN 978-3-11-117163-0, e-ISBN 978-3-11-117260-6, e-ISBN (ePUB) 978-3-11-117405-1

Volume 96
Francesco Aldo Costabile, Maria Italia Gualtieri, Anna Napoli
Polynomial Sequences. Basic Methods, Special Classes, and Computational Applications, 2023
ISBN 978-3-11-075723-1, e-ISBN 978-3-11-075724-8, e-ISBN (ePUB) 978-3-11-075732-3

www.ingramcontent.com/pod-product-compliance
Lightning Source LLC
LaVergne TN
LVHW081928010825
817679LV00006B/313